Study Guide and Solutions Manual

for

McMurry's

Organic Chemistry
A Biological Approach

Susan McMurry
Cornell University

THOMSON
BROOKS/COLE

Australia • Brazil • Canada • Mexico • Singapore • Spain • United Kingdom • United States

Printed in the United States of America
1 2 3 4 5 6 7 10 09 08 07 06

Printer: Thomson/West

0-495-01530-X
Cover Image: Radiograph by Albert Koetsier, www.beyondlight.com

Thomson Higher Education
10 Davis Drive
Belmont, CA 94002-3098
USA

Preface

What enters your mind when you hear the words "organic chemistry?" Some of you may think, "the chemistry of life," or "the chemistry of carbon." Other responses might include "pre-med, "pressure," "difficult," or "memorization." Although formally the study of the compounds of carbon, the discipline of organic chemistry encompasses many skills that are common to other areas of study. Organic chemistry is as much a liberal art as a science, and mastery of the concepts and techniques of organic chemistry can lead to improved competence in other fields.

As you work on the problems that accompany the text, you will bring to the task many problem-solving techniques. For example, planning an organic synthesis requires the skills of a chess player; you must plan your moves while looking several steps ahead, and you must keep your plan flexible. Structure-determination problems are like detective problems, in which many clues must be assembled to yield the most likely solution. Naming organic compounds is similar to the systematic naming of biological specimens; in both cases, a set of rules must be learned and then applied to the specimen or compound under study.

The problems in the text fall into two categories: drill and complex. Drill problems, which appear throughout the text and at the end of each chapter, test your knowledge of one fact or technique at a time. You may need to rely on memorization to solve these problems, which you should work on first. More complicated problems require you to recall facts from several parts of the text and then use one or more of the problem-solving techniques mentioned above. As each major type of problem—synthesis, nomenclature, or structure determination—is introduced in the text, a solution is extensively worked out in this *Solutions Manual.*

Here are several suggestions that may help you with problem solving:

1. The text is organized into chapters that describe individual functional groups. As you study each functional group, *make sure that you understand the structure and reactivity of that group*. In case your memory of a specific reaction fails you, you can rely on your general knowledge of functional groups for help.

2. *Use molecular models.* It is difficult to visualize the three-dimensional structure of an organic molecule when looking at a two-dimensional drawing. Models will help you to appreciate the structural aspects of organic chemistry and are indispensable tools for understanding stereochemistry.

3. Every effort has been made to make this *Solutions Manual* as clear, attractive, and error-free as possible. Nevertheless, you should *use the Solutions Manual in moderation*. The principal use of this book should be to check answers to problems you have already worked out. The *Solutions Manual* should not be used as a substitute for effort; at times, struggling with a problem is the only way to teach yourself.

4. *Look through the appendices at the end of the Solutions Manual.* Some of these appendices contain tables that may help you in working problems; others present information related to the history of organic chemistry.

Although the *Solutions Manual* is written to accompany *Organic Chemistry,* it contains several unique features. Each chapter of the *Solutions Manual* begins with an outline of the text that can be used for a concise review of the text material and can also serve as a reference. After every few chapters a Review Unit has been inserted. In most cases, the chapters covered in the Review Units are related to each other, and the units are planned to appear at approximately the place in the textbook where a test might be given. Each unit lists the vocabulary for the chapters covered, the skills needed to solve problems, and several important points that might need reinforcing or that restate material in the text from a slightly different point of view. Finally, the small self-test that has been included allows you to test yourself on the material from more than one chapter.

I have tried to include many types of study aids in this *Solutions Manual*. Nevertheless, this book can only serve as an adjunct to the larger and more complete textbook. If *Organic Chemistry* is the guidebook to your study of organic chemistry, then the *Solutions Manual* is the roadmap that shows you how to find what you need.

Acknowledgments I would like to thank my husband, John McMurry, for offering me the opportunity to write this book many years ago and for supporting my efforts while this edition was being prepared. Although many people at Brooks/Cole Publishing company have given me encouragement during this project, special thanks are due to Ellen Bitter. I also would like to acknowledge the contributions of Debra Boehmier, John Scheffer and Rodney Boyer, whose comments, suggestions and accuracy checks were indispensable.

Contents

Solutions to Problems

Appendices

Chapter 1 – Structure and Bonding

Chapter Outline

I. Atomic Structure (Sections 1.1 – 1.3).
   A. Introduction to atomic structure (Section 1.1).
      1. Atoms consist of a dense, positively charged nucleus surrounded by negatively charged electrons.
         a. The nucleus is made up of positively charged protons and uncharged neutrons.
         b. The nucleus contains most of the mass of the atom.
         c. Electrons move about the nucleus at a distance of about 10^{-10} m.
      2. The atomic number (Z) gives the number of protons in the nucleus.
      3. The mass number (A) gives the total number of protons and neutrons.
      4. All atoms of a given element have the same value of Z.
         a. Atoms of a given element can have different values of A.
         b. Atoms of the same element with different values of A are called isotopes.
   B. Orbitals (Section 1.2).
      1. The distribution of electrons in an atom can be described by a wave equation.
         a. The solution to a wave equation is an orbital, represented by Ψ.
         b. Ψ^2 predicts the volume of space in which an electron is likely to be found.
      2. There are four different kinds of orbitals (*s*, *p*, *d*, *f*).
         a. The *s* orbitals are spherical.
         b. The *p* orbitals are dumbbell-shaped.
         c. Four of the five *d* orbitals are cloverleaf-shaped.
      3. An atom's electrons are organized into shells.
         a. The shells differ in the numbers and kinds of orbitals they have.
         b. Electrons in different orbitals have different energies.
         c. Each orbital can hold two electrons.
      4. The two lowest-energy electrons are in the 1*s* orbital.
         a. The 2*s* orbital is the next in energy.
            Each *p* orbital has a region of zero density, called a node.
         b. The next three orbitals are $2p_x$, $2p_y$ and $2p_z$, which have the same energy .
   C. Electron Configuration (Section 1.3).
      1. The ground-state electron configuration of an atom is a listing of the orbitals occupied by the electrons of the atom.
      2. Rules for predicting the ground-state electron configuration of an atom:
         a. Orbitals with the lowest energy levels are filled first.
            The order of filling is 1*s*, 2*s*, 2*p*, 3*s*, 3*p*, 4*s*, 3*d*.
         b. Only two electrons can occupy each orbital, and they must be of opposite spin.
         c. If two or more orbitals have the same energy, one electron occupies each until all are half-full (Hund's rule). Only then does a second electron occupy one of the orbitals.
            All of the electrons in a half-filled shell have the same spin.

II. Chemical Bonding Theory (Sections 1.4 – 1.5).
   A. Development of chemical b0nding theory (Section 1.4).
      1. Kekulé and Couper proposed that carbon has four "affinity units" – carbon is tetravalent.
      2. Other scientists suggested that carbon can form double bonds, triple bonds and rings.

3. Van't Hoff and Le Bel proposed that the 4 atoms to which carbon forms bonds sit at the corners of a regular tetrahedron.
4. In a drawing of a tetrahedral carbon, a wedged line represents a bond pointing toward the viewer, and a dashed line points behind the plane of the page.

B. Covalent bonds.
1. Atoms bond together because the resulting compound is more stable than the individual atoms.
   a. Atoms tend to achieve the electron configuration of the nearest noble gas.
   b. Atoms in groups 1A, 2A and 7A either lose electrons or gain electrons to form ionic compounds.
   c. Atoms in the middle of the periodic table share electrons by forming covalent bonds.
2. The number of covalent bonds formed by an atom depends on the number of electrons it has and on the number it needs to achieve an octet.
3. Covalent bonds can be represented two ways.
   a. In Lewis structures, bonds are represented as pairs of dots.
   b. In line-bond structures, bonds are represented as lines drawn between two atoms.
4. Valence electrons not used for bonding are called lone-pair electrons. Lone-pair electrons are represented as dots.

C. Valence bond theory (Section 1.5).
1. Covalent bonds are formed by the overlap of two atomic orbitals, each of which contains one electron. The two electrons have opposite spins.
2. Each of the bonded atoms retains its atomic orbitals, but the electron pair of the overlapping orbitals is shared by both atoms.
3. The greater the orbital overlap, the stronger the bond.
4. Bonds formed by the head-on overlap of two atomic orbitals are cylindrically symmetrical and are called σ bonds.
5. Bond strength is the measure of the amount of energy needed to break a bond.
6. Bond length is the optimum distance between nuclei.
7. Every bond has a characteristic bond length and bond strength.

III. Hybridization (Sections 1.6 – 1.10).

A. sp^3 Orbitals (Sections 1.6, 1.7).
1. Structure of methane (Section 1.6).
   a. When carbon forms 4 bonds with hydrogen, one $2s$ orbital and three $2p$ orbitals combine to form four equivalent atomic orbitals (sp^3 hybrid orbitals).
   b. These orbitals are tetrahedrally oriented.
   c. Because these orbitals are unsymmetrical, they can form stronger bonds than unhybridized orbitals can.
   d. These bonds have a specific geometry and a bond angle of 109.5°.
2. Structure of ethane (Section 1.7).
   a. Ethane has the same type of hybridization as occurs in methane.
   b. The C–C bond is formed by overlap of two sp^3 orbitals.
   c. Bond lengths, strengths and angles are very close to those of methane.

B. sp^2 Orbitals (Section 1.8).
1. If one carbon $2s$ orbital combines with two carbon $2p$ orbitals, three hybrid sp^2 orbitals are formed, and one p orbital remains unchanged.
2. The three sp^2 orbitals lie in a plane at angles of 120°, and the p orbital is perpendicular to them.
3. Two different types of bonds form between two carbons.
   a. A σ bond forms from the overlap of two sp^2 orbitals.
   b. A π bond forms by sideways overlap of two p orbitals.
   c. This combination is known as a carbon-carbon double bond.

4. Ethylene is composed of a carbon-carbon double bond and four σ bonds formed between the remaining four sp^2 orbitals of carbon and the 1*s* orbitals of hydrogen.
   The double bond of ethylene is both shorter and stronger than the C–C bond of ethane.

C. *sp* Orbitals (Section 1.9).
1. If one carbon 2*s* orbital combines with one carbon 2*p* orbital, two hybrid *sp* orbitals are formed, and two *p* orbitals are unchanged.
2. The two *sp* orbitals are 180° apart, and the two *p* orbitals are perpendicular to them and to each other.
3. Two different types of bonds form.
   a. A σ bond forms from the overlap of two *sp* orbitals.
   b. Two π bonds form by sideways overlap of four *p* orbitals.
   c. This combination is known as a carbon-carbon triple bond.
4. Acetylene is composed of a carbon-carbon triple bond and two σ bonds formed between the remaining two *sp* orbitals of carbon and the 1*s* orbitals of hydrogen.
   The triple bond of acetylene is the strongest carbon-carbon bond.

D. Hybridization of nitrogen, oxygen, phosphorus and sulfur (Section 1.10).
1. Covalent bonds between other elements can be described by using hybrid orbitals.
2. Both the nitrogen atom in ammonia and the oxygen atom in water form sp^3 hybrid orbitals.
   The lone-pair electrons in these compounds occupy sp^3 orbitals.
3. The bond angles between hydrogen and the central atom is often less than 109° because the lone-pair electrons take up more room than the σ bond.
4. Because of their positions in the third row, phosphorus and sulfur can form more than the typical number of covalent bonds.

IV. Molecular orbital theory (Section 1.11).

A. Molecular orbitals arise from a mathematical combination of atomic orbitals and belong to the entire molecule.
1. Two 1*s* orbitals can combine in two different ways.
   a. The additive combination is a bonding MO and is lower in energy than the two hydrogen 1*s* atomic orbitals.
   b. The subtractive combination is an antibonding MO and is higher in energy than the two hydrogen 1*s* atomic orbitals.
2. A node is a region between nuclei where electrons aren't found.
   If a node occurs between two nuclei, the nuclei repel each other.
3. The number of MOs in a molecule is the same as the number of atomic orbitals combined.

V. Chemical structures (Section 1.12).

A. Drawing chemical structures.
1. Condensed structures don't show C–H bonds and don't show the bonds between CH_3, CH_2 and CH units.
2. Skeletal structures are simpler still.
   a. Carbon atoms aren't usually shown.
   b. Hydrogen atoms bonded to carbon aren't usually shown.
   c. Other atoms are shown.

Solutions to Problems

1.1 (a) To find the ground-state electron configuration of an element, first locate its atomic number. For oxygen, the atomic number is 8; oxygen thus has 8 protons and 8 electrons. Next, assign the electrons to the proper energy levels, starting with the lowest level. Fill each level *completely* before assigning electrons to a higher energy level.

Notice that the 2*p* electrons are in different orbitals. According to *Hund's rule*, we must place one electron into each orbital of the same energy level until all orbitals are half-filled.

Oxygen

2*p* ↑↓ ↑ ↑

2*s* ↑↓

1*s* ↑↓

Remember that only two electrons can occupy the same orbital, and that they must be of opposite spin.

A different way to represent the ground-state electron configuration is to simply write down the occupied orbitals and to indicate the number of electrons in each orbital. For example, the electron configuration for oxygen is $1s^2\ 2s^2\ 2p^4$.

(b) Phosphorus, with an atomic number of 15, has 15 electrons. Assigning these to energy levels:

Phosphorus

3*p* ↑ ↑ ↑

3*s* ↑↓

2*p* ↑↓ ↑↓ ↑↓

2*s* ↑↓

1*s* ↑↓

The more concise way to represent ground-state electron configuration for phosphorus: $1s^2\ 2s^2\ 2p^6\ 3s^2\ 3p^3$

(c) $1s^2\ 2s^2\ 2p^6\ 3s^2\ 3p^4$

Sulfur

3*p* ↑↓ ↑ ↑

3*s* ↑↓

2*p* ↑↓ ↑↓ ↑↓

2*s* ↑↓

1*s* ↑↓

1.2 **Strategy:** The elements of the periodic table are organized into groups that are based on the number of outer-shell electrons each element has. For example, an element in group 1A has one outer-shell electron, and an element in group 5A has five outer-shell electrons. To find the number of outer-shell electrons for a given element, use the periodic table to locate its group.

Solution:
(a) Magnesium (group 2A) has two electrons in its outermost shell.
(b) Cobalt is a transition metal, which has two electrons in the 4*s* subshell, plus seven electrons in its 3*d* subshell.
(c) Selenium (group 6A) has six electrons in its outermost shell.

1.3 **Strategy:** A solid line represents a bond lying in the plane of the page, a wedged bond represents a bond pointing out of the plane of the page toward the viewer, and a dashed bond represents a bond pointing behind the plane of the page.

Solution

H
Cl—C---Cl
Cl

Chloroform

1.4

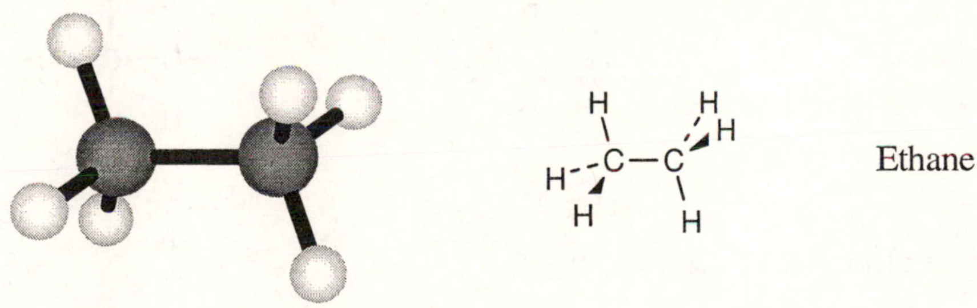

1.5 **Strategy:** Identify the group of the central element to predict the number of covalent bonds the element can form.

Solution: (a) Carbon (Group 4A) has four electrons in its valence shell and forms four bonds to achieve the noble-gas configuration of neon. A likely formula is CH_2Cl_2.

(b) Sulfur belongs to Group 6A and needs to share two electrons in order to make an octet. One electron is shared with carbon and the other is shared with a hydrogen. A likely formula is CH_3SH.

(c) CH_3NH_2 (see Problem 1.6 c for the structure).

1.6 **Strategy:** Start by drawing the electron-dot structure of the molecule.

(1) Determine the number of valence, or outer-shell electrons for each atom in the molecule. For ethanol, we know that carbon has four valence electrons, hydrogen has one valence electron, and oxygen has six valence electrons.

```
 .
.C.      4 x 2 = 8
 .
H.       1 x 6 = 6
 ..
:O.      6 x 1 = 6
 .       ---------
           20    total valence electrons
```

(2) Next, use two electrons for each single bond.

```
   H H
   .. ..
 H:C:C:O:H
   .. ..
   H H
```

(3) Finally, use the remaining electrons to achieve an noble gas configuration for all atoms. To convert to a line-bond structure, replace the bonding electron pairs with a line.

Solution:

Molecule — *Lewis structure* — *Line-bond structure*

(a) CH_3CH_2OH

```
   H H                  H  H
   .. ..                |  |    ..
 H:C:C:O:H          H—C—C—O—H
   .. ..                |  |    ..
   H H                  H  H
```

(b) H_2S

8 valence electrons

```
   ..                ..
 H:S:             H—S:
   ..                |
   H                 H
```

(c) CH_3NH_2

14 valence electrons

```
   H H              H  H
   .. ..            |  |
 H:C:N:H         H—C—N—H
   ..  ..           |  ..
   H                H
```

(d) $N(CH_3)_3$

26 valence electrons

```
   H          H            H        H
   ..   ..    ..           |   ..   |
 H:C  : N  :  C:H       H—C—N—C—H
   ..   ..    ..           |   |    |
   H   H:C:H  H            H   |    H
         ..                 H—C—H
         H                     |
                               H
```

1.7 Each of the two carbons has 4 valence electrons. Two electrons are used to form the carbon-carbon bond, and the 6 electrons that remain can form bonds with a maximum of 6 hydrogens. Thus, the formula C_2H_7 is not possible.

1.8 **Strategy:** Connect the carbons and add hydrogens so that all carbons are bonded to four different atoms.

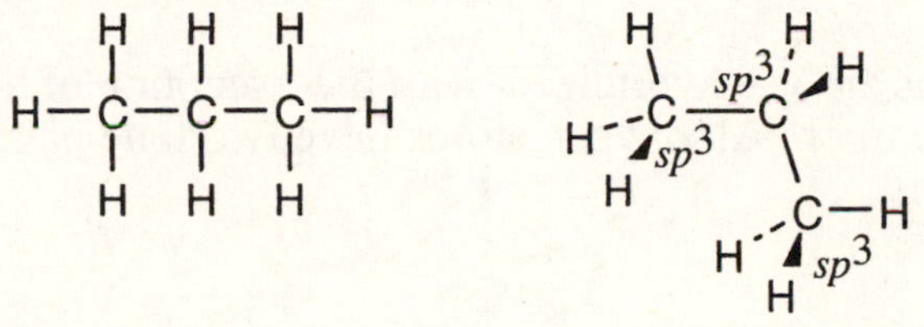

H H H
| | |
H—C—C—C—H
| | |
H H H

Propane

The geometry around all carbon atoms is tetrahedral, and all bond angles are approximately 109°.

1.9

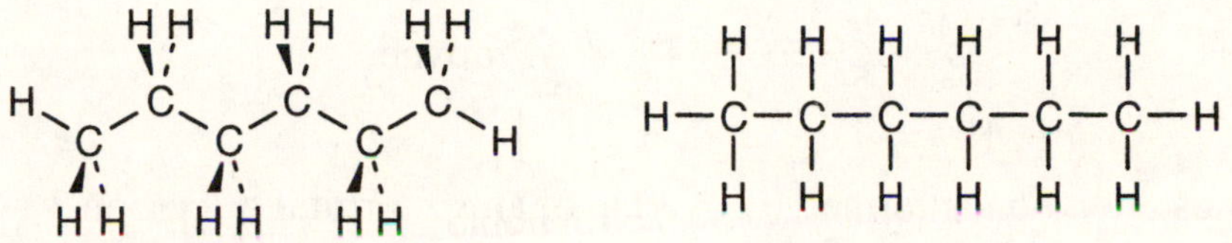

Hexane

1.10

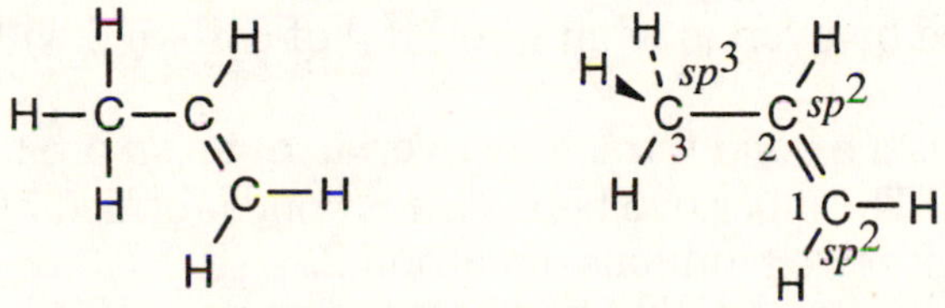

Propene

The C3–H bonds are σ bonds formed by overlap of an sp^3 orbital of carbon 3 with an s orbital of hydrogen.

The C2–H and C1–H bonds are σ bonds formed by overlap of an sp^2 orbital of carbon with an s orbital of hydrogen.

The C2–C3 bond is a σ bond formed by overlap of an sp^3 orbital of carbon 3 with an sp^2 orbital of carbon 2.

There are two C1–C2 bonds. One is a σ bond formed by overlap of an sp^2 orbital of carbon 1 with an sp^2 orbital of carbon 2. The other is a π bond formed by overlap of a p orbital of carbon 1 with a p orbital of carbon 2. All four atoms connected to the carbon-carbon double bond lie in the same plane, and all bond angles between these atoms are 120°. The bond angle between hydrogen and the sp^3-hybridized carbon is 109 °.

1.11

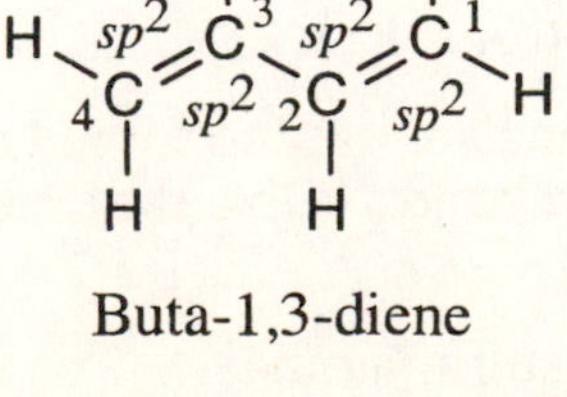

All atoms lie in the same plane, and all bond angles are approximately 120 °.

Buta-1,3-diene

1.12

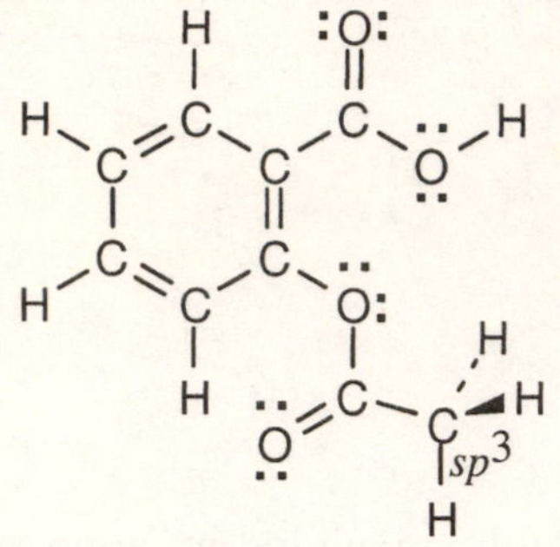

Aspirin.

All carbons are sp^2 hybridized, with the exception of the indicated carbon. All oxygen atoms have two lone pairs of electrons.

1.13

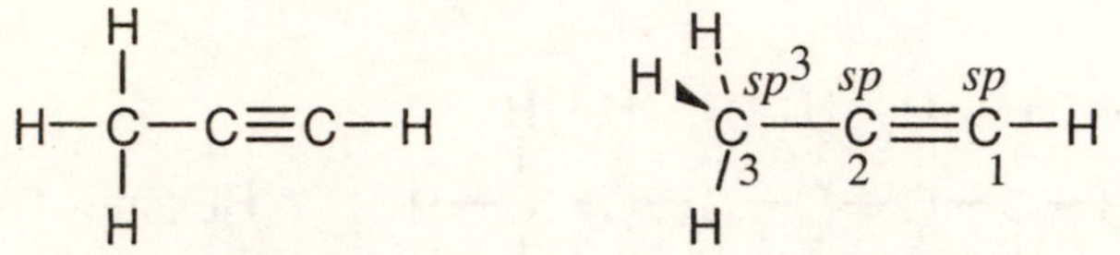

H—C—C≡C—H (with H above and below the first C)

H, H, H on C3 (sp^3) — C2 (sp) ≡ C1 (sp) — H

Propyne

The C3-H bonds are σ bonds formed by overlap of an sp^3 orbital of carbon 3 with an *s* orbital of hydrogen.

The C1-H bond is a σ bond formed by overlap of an *sp* orbital of carbon 1 with an *s* orbital of hydrogen.

The C2-C3 bond is a σ bond formed by overlap of an *sp* orbital of carbon 2 with an sp^3 orbital of carbon 3.

There are three C1-C2 bonds. One is a σ bond formed by overlap of an *sp* orbital of carbon 1 with an *sp* orbital of carbon 2. The other two bonds are π bonds formed by overlap of two *p* orbitals of carbon 1 with two *p* orbitals of carbon 2.

The three carbon atoms of propyne lie in a straight line; the bond angle is 180 °. The $H–C_1≡C_2$ bond angle is also 180 °. The bond angle between hydrogen and the sp^3-hybridized carbon is 109 °.

1.14

(a)

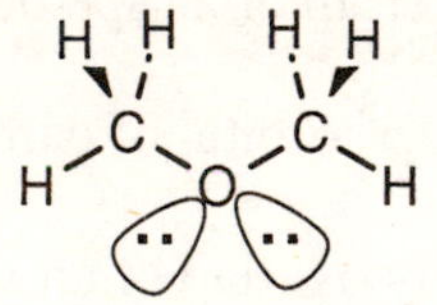

The sp^3-hybridized oxygen atom has tetrahedral geometry.

(b)

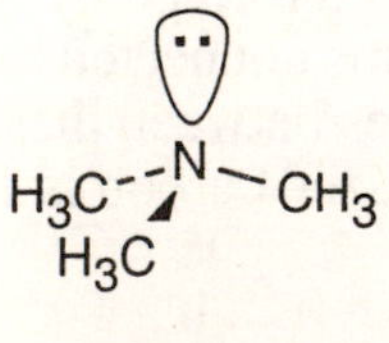

H_3C–N(CH_3)–CH_3

Tetrahedral geometry at nitrogen and carbon.

(c)

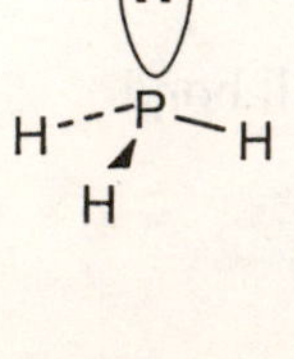

H–P(H)–H

Like nitrogen, phosphorus has five outer-shell electrons. PH_3 has tetrahedral geometry.

(d)

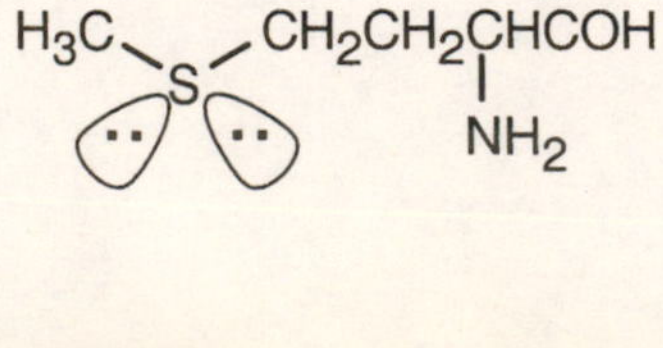

H_3C–S–$CH_2CH_2CH(NH_2)COOH$

The sp^3-hybridized sulfur atom has tetrahedral geometry.

1.15 **Strategy:** Remember that the end of a line represents a carbon atom with 3 hydrogens, a two-way intersection represents a carbon atom with 2 hydrogens, a three-way intersection represents a carbon with 1 hydrogen and a four-way intersection represents a carbon with no hydrogens.

Solution:

(a)

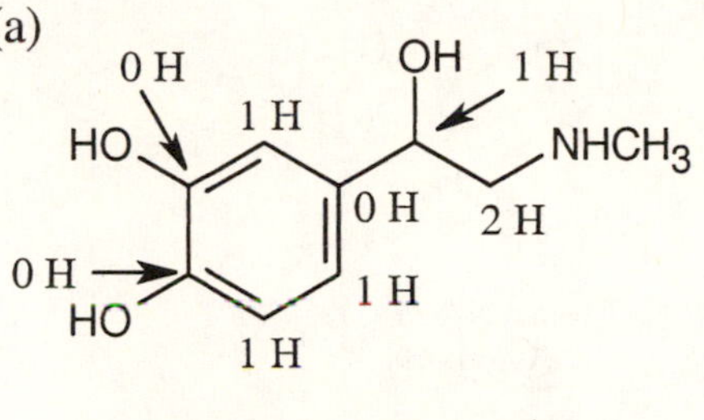

Adrenaline – $C_9H_{13}NO_3$

(b)

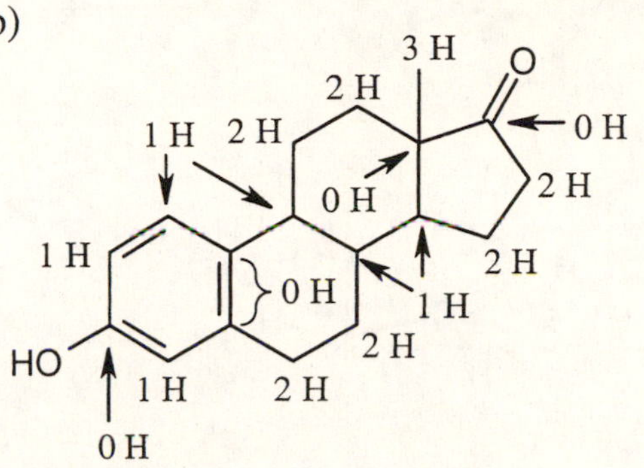

Estrone – $C_{18}H_{22}O_2$

1.16 Several possible skeletal structures can satisfy each molecular formula.

(a) C_5H_{12}

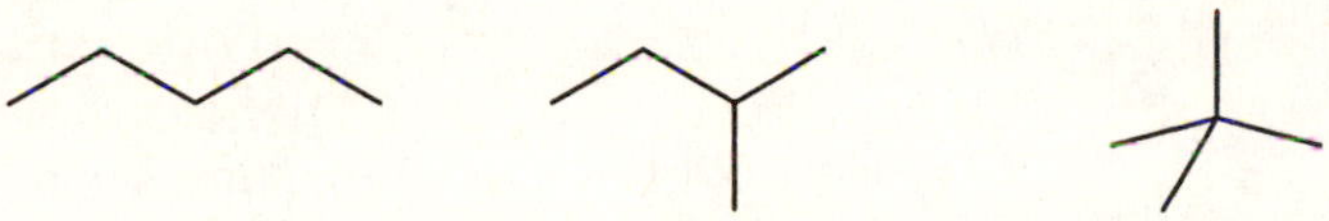

(b) C_2H_7N

(c) C_3H_6O

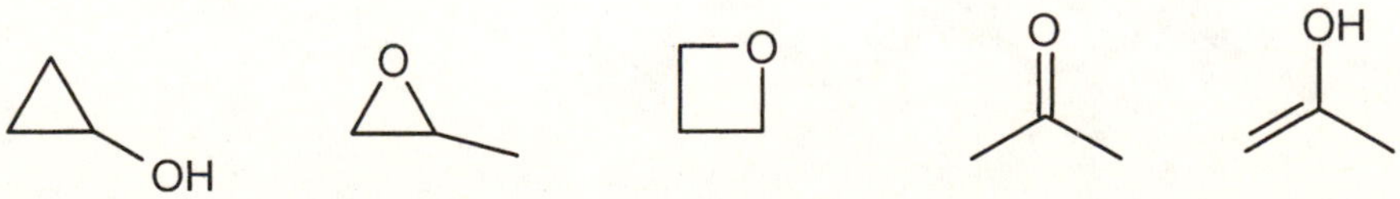

(d) C_4H_9Cl

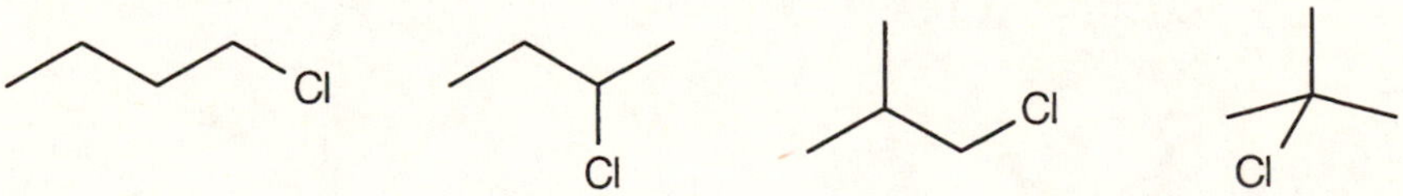

1.17

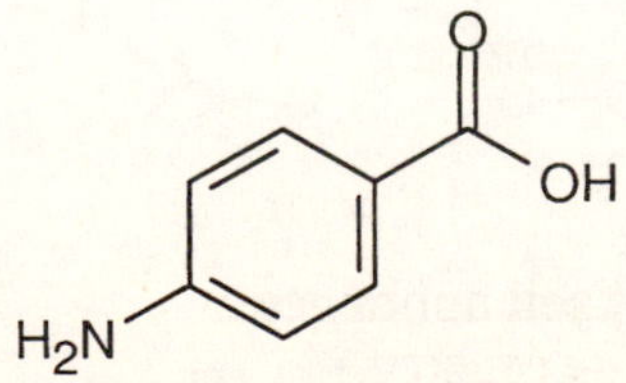

PABA

Visualizing Chemistry

1.18

(a)

(b)

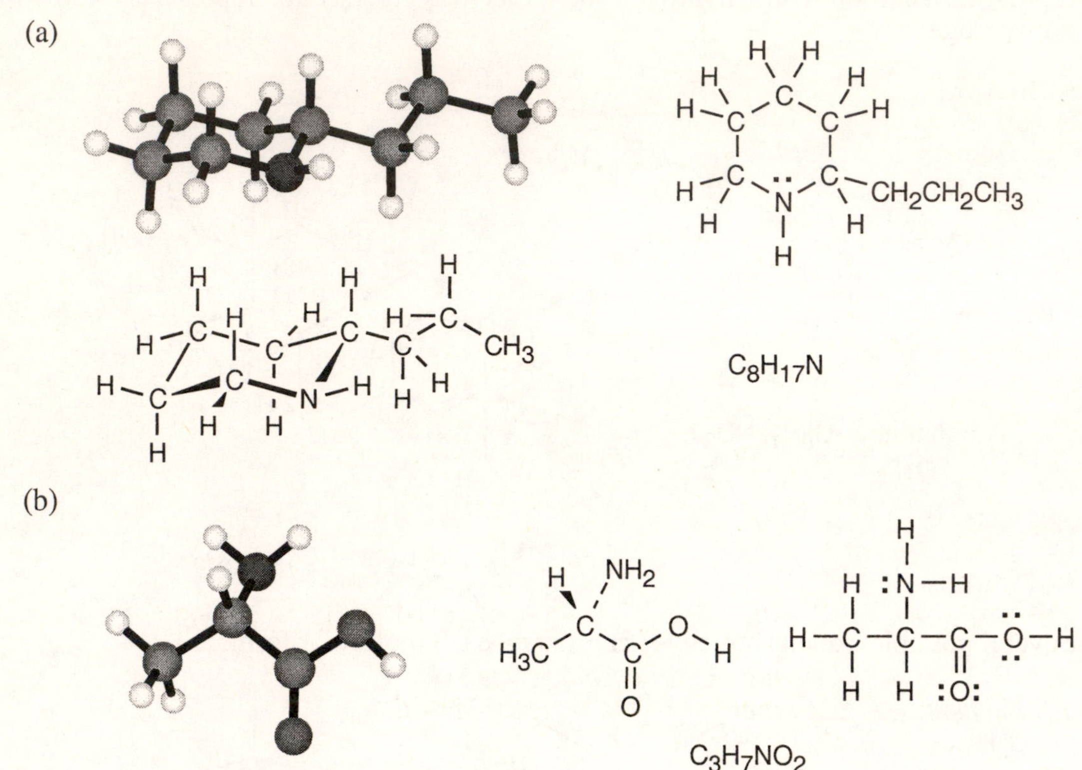

1.19 Citric acid ($C_6H_8O_7$) contains seven oxygen atoms, each of which has two electron lone pairs. Three of the oxygens form double bonds with carbon.

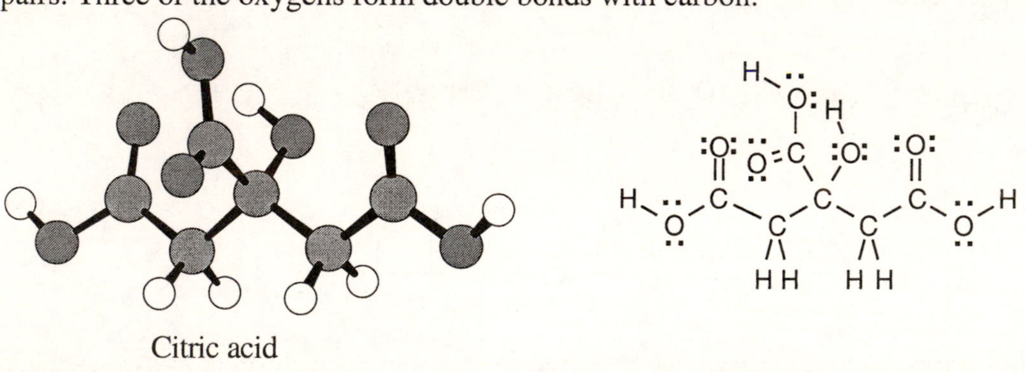

Citric acid

1.20

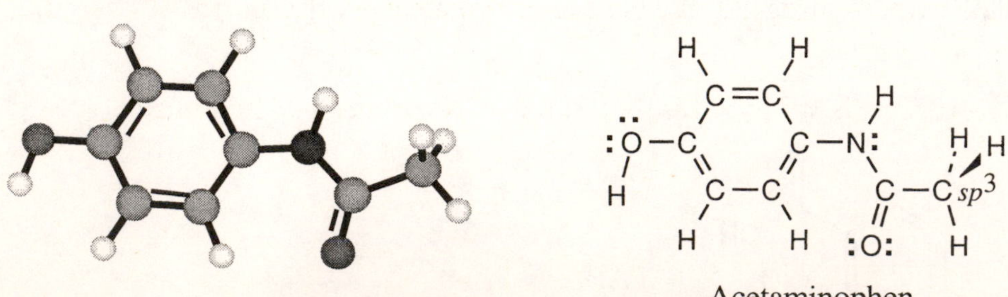

Acetaminophen

All carbons are sp^2 hybridized, except for the carbon indicated as sp^3. The two oxygen atoms and the nitrogen atom have lone pair electrons, as shown.

1.21

Aspartame

Additional Problems

1.22

	Element	*Atomic Number*	*Number of valence electrons*
(a)	Zinc	30	2
(b)	Iodine	53	7
(c)	Silicon	14	4
(d)	Iron	26	2 (in 4*s* subshell), 6 (in 3*d* subshell)

1.23

	Element	*Atomic Number*	*Ground-state Electron configuration*
(a)	Potassium	19	$1s^2\,2s^2\,2p^6\,3s^2\,3p^6\,4s^1$
(b)	Arsenic	33	$1s^2\,2s^2\,2p^6\,3s^2\,3p^4\,4s^2\,3d^{10}\,4p^3$
(c)	Aluminum	13	$1s^2\,2s^2\,2p^6\,3s^2\,3p^1$
(d)	Germanium	32	$1s^2\,2s^2\,2p^6\,3s^2\,3p^6\,4s^2\,3d^{10}\,4p^2$

1.24 (a) NH_2OH (b) $AlCl_3$ (c) CF_2Cl_2 (d) CH_2O

1.25

```
   H
   ..
H: C : C : : : N:
   ..
   H
```

Acetonitrile

In the compound acetonitrile, nitrogen has eight electrons in its outer electron shell. Six are used in the carbon-nitrogen triple bond, and two are a nonbonding electron pair.

1.26 The H_3C– carbon is sp^3 hybridized, and the –CN carbon is sp hybridized.

1.27

```
         ..
H       :Cl:
 \      /
  C = C
 /      \
H        H
```

Vinyl chloride

Vinyl chloride has 18 valence electrons. Eight electrons are used for 4 single bonds, 4 electrons are used in the carbon-carbon double bond, and 6 electrons are in the 3 lone pairs that surround chlorine.

1.28

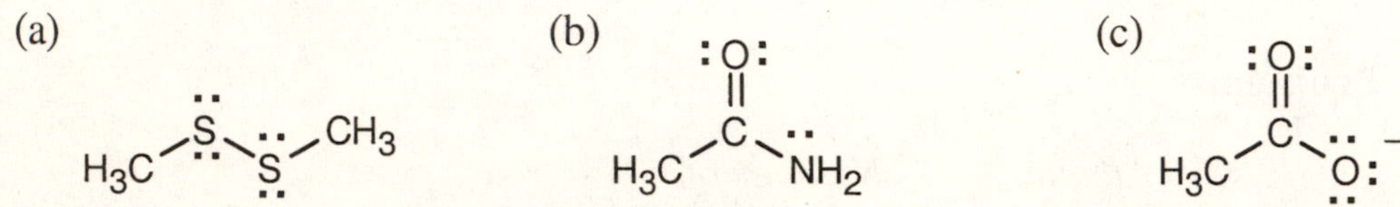

1.29 In molecular formulas of organic molecules, carbon is listed first, followed by hydrogen. All other elements are listed in alphabetical order.

Compound	*Molecular Formula*
(a) Aspirin	$C_9H_8O_4$
(b) Vitamin C	$C_6H_8O_6$
(c) Nicotine	$C_{10}H_{14}N_2$
(d) Glucose	$C_6H_{12}O_6$

1.30 To work a problem of this sort, you must examine all possible structures consistent with the rules of valence. You must systematically consider all possible attachments, including those that have branches, rings and multiple bonds.

```
(a)   H  H  H        (b)  H  H          (c)  H  H                 H       H
      |  |  |             |  |               |  |                 |       |
    H-C--C--C-H         H-C--N-H           H-C--C-O-H           H-C-O-C-H
      |  |  |             |                  |  |                 |       |
      H  H  H             H                  H  H                 H       H

(d)   H  H  H              H  Br H
      |  |  |              |  |  |
    H-C--C--C-Br         H-C--C--C-H
      |  |  |              |  |  |
      H  H  H              H  H  H

(e)   H  O               H                    O
      |  ||              |             H    /   \    H
    H-C--C-H         H-C=C-O-H           C ---- C
      |                  |             H          H
      H                  H
```

(f) H H H H H H H H H H H
H—C—C—C—H H—C—C—C—H H—C—C—N—C—H H—C—N—C—H
H H N H H N H H H H H H C H
H H H H H H H

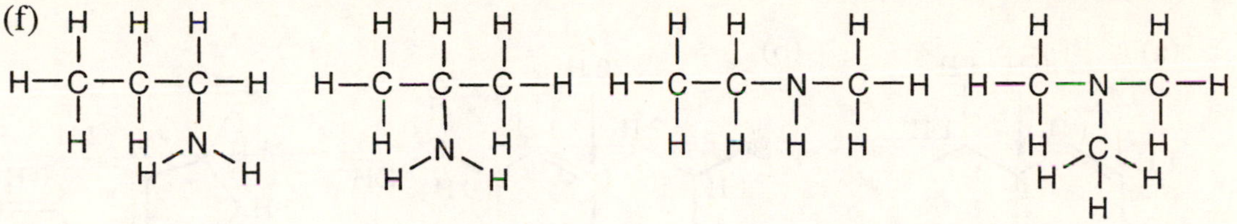

1.31

(a) *sp*^{3} *sp*^{3} *sp*^{3} $CH_3CH_2CH_3$

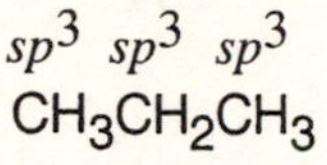

(b) *sp*^{3} H_3C *sp*^{2} *sp*^{2} $C=CH_2$ H_3C *sp*^{3}

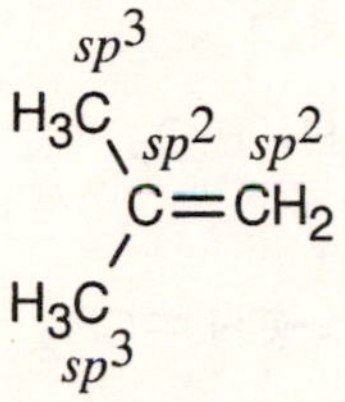

(c) *sp*^{2} *sp*^{2} *sp* *sp* $H_2C=CH-C\equiv CH$

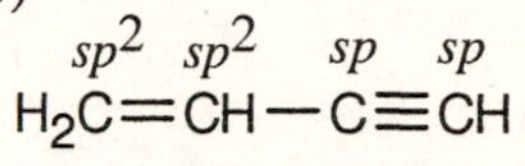

(d) O *sp*^{3} C *sp*^{2} CH_3 OH

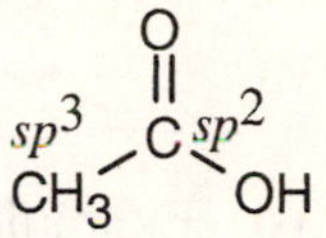

1.32

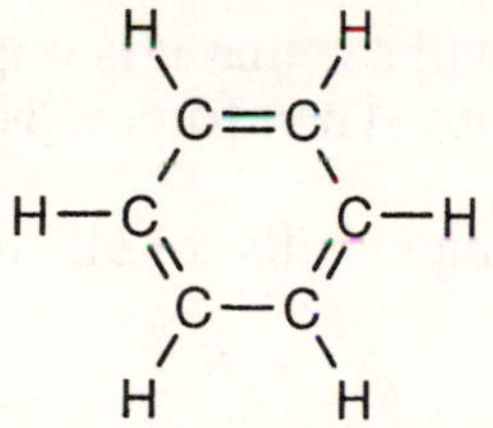

H H C=C H—C C—H C—C H H Benzene

All carbon atoms of benzene are *sp*^{2} hybridized, and all bond angles of benzene are 120°. Benzene is a planar molecule.

1.33

(a) O C *sp*^{2} H_2C 120° OH NH_2

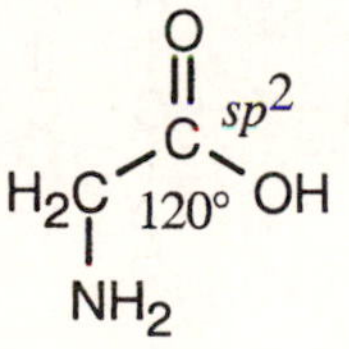

Glycine

(b) N *sp*^{2} H H C 120° C C C H C H H

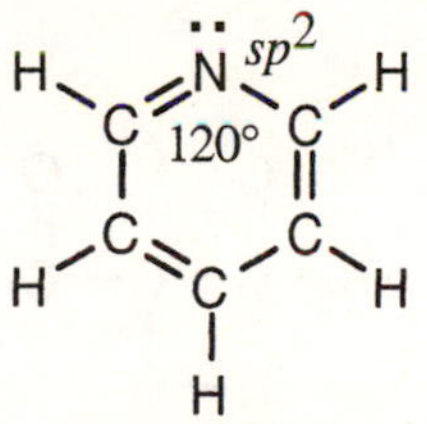

Pyridine

(c) O H_3C 109° C OH *sp*^{3} C H OH

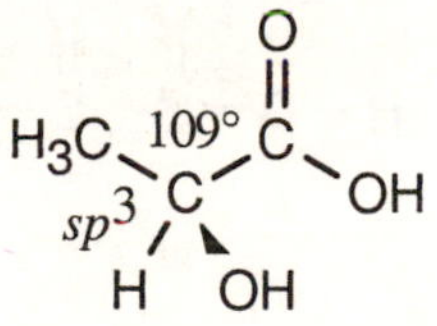

Lactic acid

1.34

(a) NH

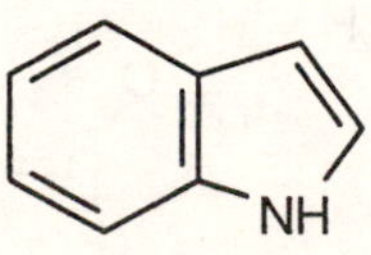

(b)

(c) Cl Cl

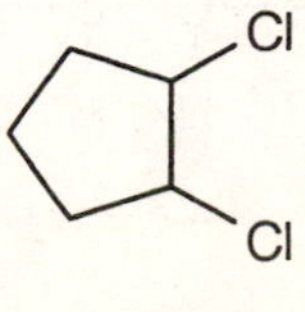

(d) O O

1.35

(a) $C_{10}H_{11}N$ (b) $C_{11}H_{11}BrO_2$ (c) $C_9H_{12}O$

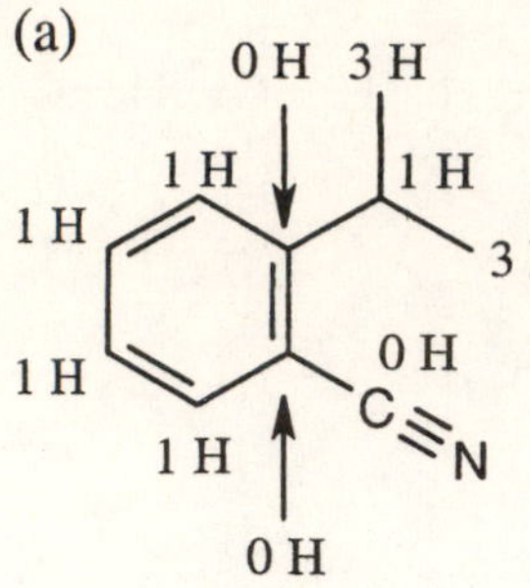

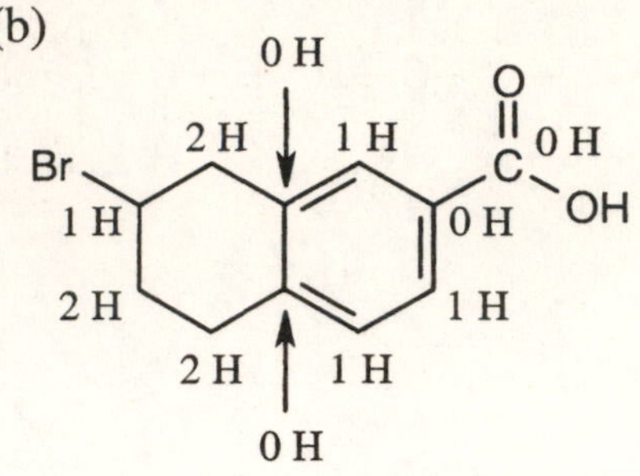

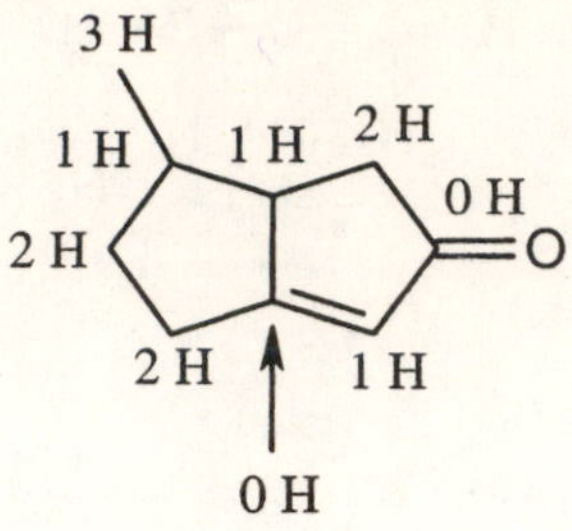

1.36 Examples:

(a) $CH_3CH_2CH{=}CH_2$ (b) $H_2C{=}CH{-}CH{=}CH_2$ (c) $H_2C{=}CH{-}C{\equiv}CH$

1.37 (a) The 4 valence electrons of carbon can form bonds with a maximum of 4 hydrogens. Thus, it is not possible for the compound CH_5 to exist.
(b) If you try to draw a molecule with the formula C_2H_6N, you will see that it is impossible for both carbons and nitrogen to have a complete octet of electrons. Therefore, C_2H_6N is unlikely to exist.
(c) A compound with the formula $C_3H_5Br_2$ doesn't have filled outer shells for all atoms and is thus unlikely to exist.

1.38

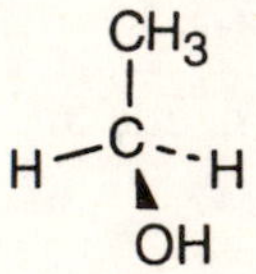

Ethanol

1.39

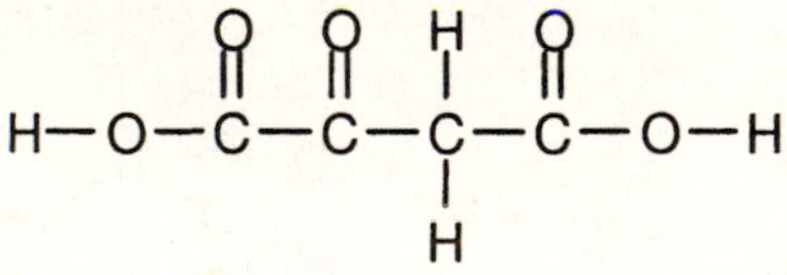

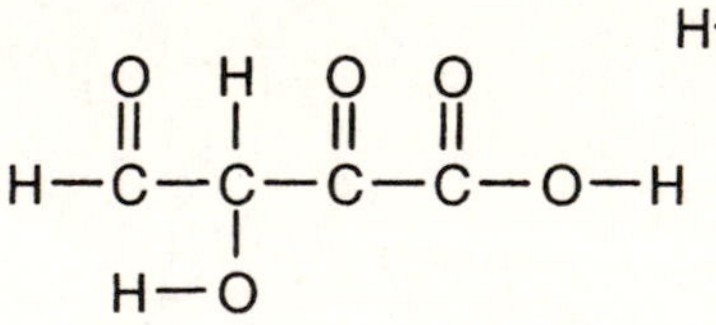

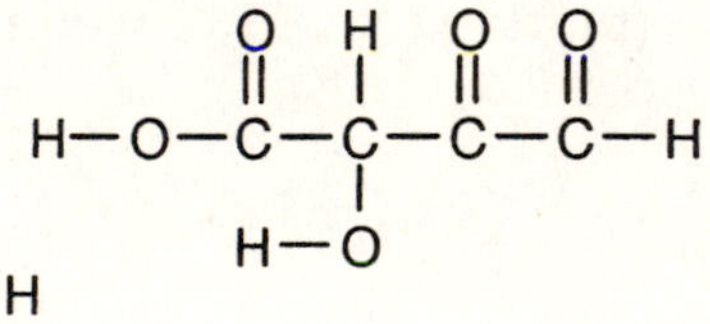

1.40

(a) (b) 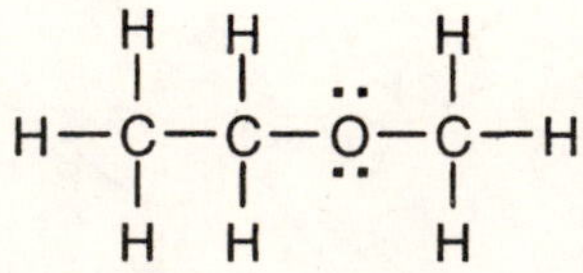 (c) 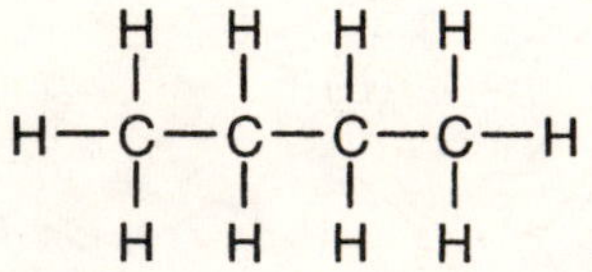 (d)

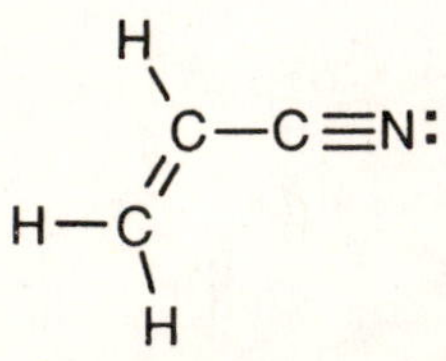

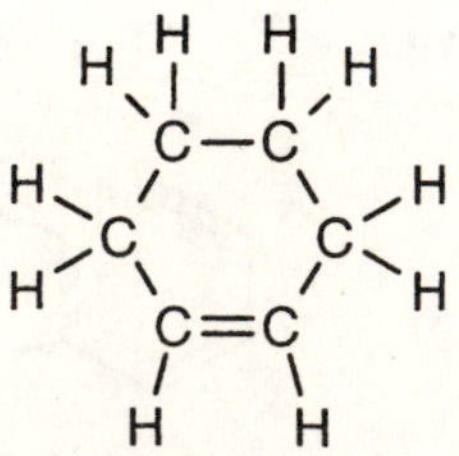

1.41

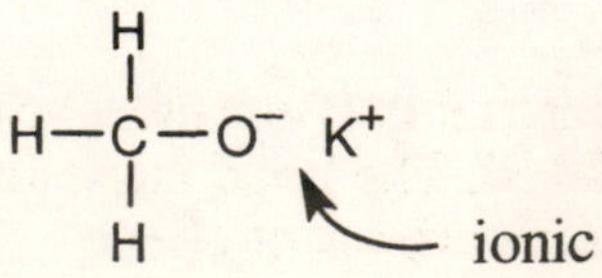

ionic

All other bonds are covalent.

1.42

(a)

Procaine

(b)

Vitamin C

1.43

Pyridoxal phosphate

The bond angles formed by atoms having sp^3 hybridization are approximately 109 °.
The bond angles formed by atoms having sp^2 hybridization are approximately 120 °.

1.44 In a compound containing a carbon–carbon triple bond, atoms bonded to the *sp*–hybridized carbons must lie in a straight line. It is not possible to form a five-membered ring if four carbons must have a linear relationship.

1.45 The π bonding molecular orbital in ethylene results from the combination of two *p* atomic orbitals *with the same algebraic sign.* (A second molecular orbital can form by the combination of two *p* orbitals with opposite algebraic signs, but it is an antibonding orbital.)

1.46

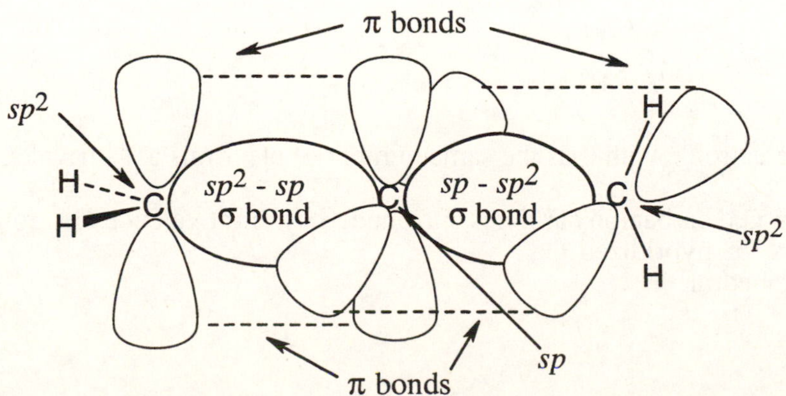

The central carbon of allene forms two σ bonds and two π bonds. The central carbon is *sp*–hybridized, and the two terminal carbons are sp^2–hybridized. The bond angle formed by the three carbons is 180°, indicating linear geometry for the carbons of allene.

1.47

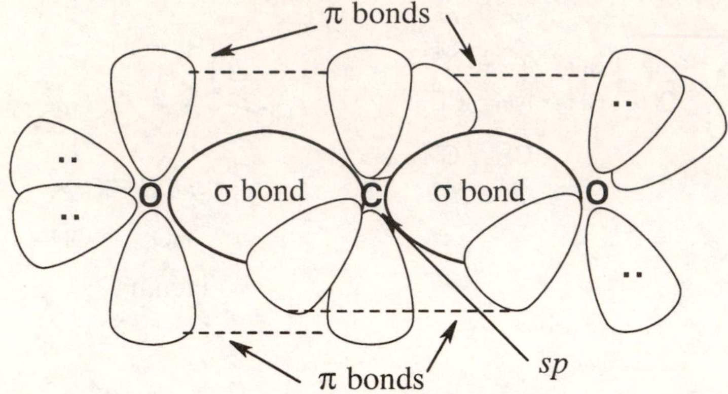

Carbon dioxide is a linear molecule.

1.48

Caffeine

All of the indicated atoms are sp^2-hybridized.

1.49 (a) The positively charged carbon atom is surrounded by six valence shell electrons; carbon has three valence electrons, and each hydrogen brings three valence electrons.
(b) The positively charged carbon is sp^2–hybridized.
(c) A carbocation is planar about the positively charged carbon.

1.50

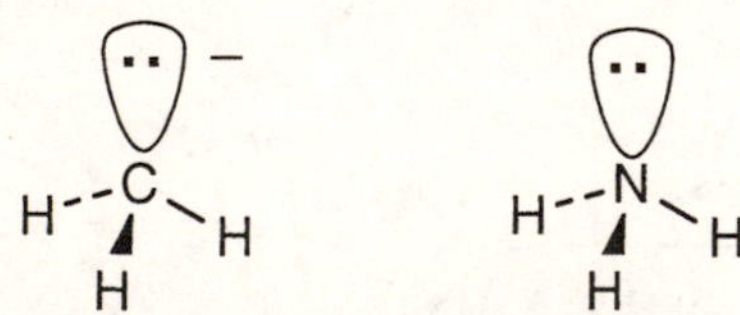

(a) A carbanion is isoelectronic with (has the same number of electrons as) a trivalent nitrogen compound.
(b) The negatively charged carbanion carbon is surrounded by eight valence electrons.
(c) The carbon atom is sp^3-hybridized.
(d) A carbanion is tetrahedral.

1.51 According to the Pauli Exclusion Principle, two electrons in the same orbital must have opposite spins. Thus, the two electrons of triplet (spin-unpaired) methylene must occupy different orbitals. In triplet methylene, *sp*-hybridized carbon forms one bond to each of two hydrogens. Each of the two unpaired electrons occupies a *p* orbital. In singlet (spin-paired) methylene the two electrons can occupy the same orbital because they have opposite spins. Including the two C–H bonds, there are a total of three occupied orbitals. We predict sp^2 hybridization and planar geometry for singlet methylene.

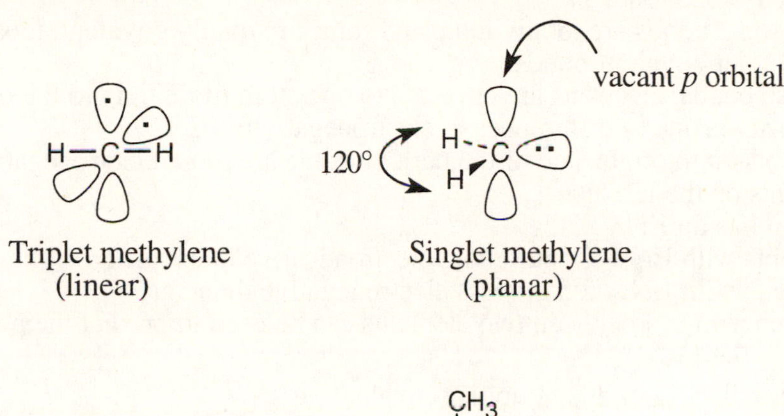

Triplet methylene (linear) Singlet methylene (planar)

1.52

$CH_3CH_2CH_2CH_3$ $CH_3CH(CH_3)CH_3$

The two compounds differ in the way that the carbon atoms are connected.

1.53

$H_2C{=}CH{-}CH_3$ (cyclopropane, C_3H_6 ring with H atoms)

One compound has a double bond, and the other has a ring.

1.54

CH_3CH_2OH CH_3OCH_3

The two compounds differ in the location of the oxygen atom.

1.55

$CH_3CH_2CH{=}CH_2$ $CH_3CH{=}CHCH_3$ $H_2C{=}C(CH_3)_2$

The compounds differ in the way that the carbon atoms are connected and in the location of the double bond.

Chapter 2 – Polar Covalent Bonds; Acids and Bases

Chapter Outline

I. Polar covalent bonds (Sections 2.1 – 2.3).
  A. Electronegativity (Section 2.1).
    1. Although some bonds are totally ionic and some are totally covalent, most chemical bonds are polar covalent bonds.
       In these bonds, electrons are attracted to one atom more than to the other atom.
    2. Bond polarity is due to differences in electronegativity (EN).
       a. Elements on the right side of the periodic table are more electronegative than elements on the left side.
       b. Carbon has an EN of 2.5.
       c. Elements with EN > 2.5 are more electronegative than carbon.
       d. Elements with EN < 2.5 are less electronegative than carbon.
    3. The difference in EN between two elements can be used to predict the polarity of a bond.
       a. If ΔEN < 0.4, a bond is nonpolar covalent.
       b. If ΔEN is between 0.4 and 2.0, a bond is polar covalent.
       c. If ΔEN > 2.0, a bond is ionic.
       d. The symbols $\delta+$ and $\delta-$ are used to indicate partial charges.
       e. A crossed arrow is used to indicate bond polarity.
          The tail of the arrow is electron-poor, and the head of the arrow is electron-rich.
    4. Electrostatic potential maps are also used to show electron-rich (red) and electron-poor (blue) regions of molecules.
    5. An inductive effect is an atom's ability to polarize a bond.
  B. Dipole moment (Section 2.2).
    1. Dipole moment is the measure of a molecule's overall polarity.
    2. Dipole moment (μ) = Q x r, where Q = charge and r = distance between charges.
       Dipole moment is measured in debyes (D).
    3. Dipole moment can be used to measure charge separation.
    4. Water and ammonia have large values of D; methane and ethane have D = 0.
  C. Formal charge (Section 2.3).
    1. Formal charge is a type of electronic "bookkeeping".
    2.
$$(\mathrm{FC}) = \left[\begin{matrix}\text{\# of valence}\\ \text{electrons}\end{matrix}\right] - \left[\frac{\text{\# of bonding electrons}}{2}\right] - \left[\begin{matrix}\text{\# nonbonding}\\ \text{electrons}\end{matrix}\right]$$
    3. Formal charge is helpful in predicting reactivity.

II. Resonance (Sections 2.4 – 2.6).
  A. Chemical structures and resonance (Section 2.4).
    1. Some molecules (acetate ion, for example) can be drawn as two (or more) different electron-dot structures.
       a. These structures are called resonance structures.
       b. The true structure of the molecule is intermediate between the structures.
       c. The true structure is called a resonance hybrid.
    2. Resonance structures differ only in the placement of π and nonbonding electrons.
       All atoms occupy the same positions.
    3. Resonance is an important concept in organic chemistry.

B. Rules for resonance forms (Section 2.5).
  1. Individual resonance forms are imaginary, not real.
  2. Resonance forms differ only in the placement of their π or nonbonding electrons.
     A curved arrow is used to indicate the movement of electrons, not atoms.
  3. Different resonance forms of a molecule don't have to be equivalent.
     If resonance forms are nonequivalent, the structure of the actual molecule resembles the more stable resonance form(s).
  4. Resonance forms must be valid electron-dot structures and obey normal rules of valency.
  5. The resonance hybrid is more stable than any individual resonance form.

C. A useful technique for drawing resonance forms (Section 2.6).
  1. Any three-atom grouping with a multiple bond adjacent to a nonbonding *p* orbital has two resonance forms.
  2. One atom in the grouping has a lone electron pair, a vacant orbital or a single electron.
  3. By recognizing these three-atom pieces, resonance forms can be generated.

III. Acids and bases (Sections 2.7 – 2.11).

A. Brønsted-Lowry definition (Section 2.7).
  1. A Brønsted-Lowry acid donates an H^+ ion; a Brønsted-Lowry base accepts H^+.
  2. The product that results when a base gains H^+ is the conjugate acid of the base; the product that results when an acid loses H^+ is the conjugate base of the acid.

B. Acid and base strength (Section 2.8 – 2.10).
  1. A strong acid reacts almost completely with water (Section 2.8).
  2. The strength of an acid in water is indicated by K_a, the acidity constant.
  3. Strong acids have large acidity constants, and weaker acids have smaller acidity constants.
  4. The pK_a is normally used to express acid strength.
     a. $pK_a = -\log K_a$
     b. A strong acid has a small pK_a, and a weak acid has a large pK_a.
     c. The conjugate base of a strong acid is a weak base, and the conjugate base of a weak acid is a strong base.
  5. Predicting acid–base reactions from pK_a (Section 2.9).
     a. An acid with a low pK_a reacts with the conjugate base of an acid with a high pK_a
     b. In other words, the products of an acid–base reaction are more stable than the reactants.
  6. Organic acids and organic bases (Section 2.10).
     a. There are two main types of organic acids:
        i. Acids that contain hydrogen bonded to oxygen.
        ii. Acids that have hydrogen bonded to the carbon next to a C=O group.
     b. The main type of organic base contains a nitrogen atom with a lone electron pair.

C. Lewis acids and bases (Section 2.11).
  1. A Lewis acid accepts an electron pair.
     a. A Lewis acid may have either a vacant low-energy orbital or a polar bond to hydrogen.
     b. Examples include metal cations, halogen acids, Group 3 compounds and transition-metal compounds.
  2. A Lewis base has a pair of nonbonding electrons.
     a. Most oxygen- and nitrogen-containing organic compounds are Lewis bases.
     b. Many organic Lewis bases have more than one basic site.
  3. A curved arrow shows the movement of electrons from a Lewis base to a Lewis acid.

IV. Noncovalent interactions in molecules (Section 2.12).
   A. Dipole-dipole interactions occur between polar molecules as a result of electrostatic interactions among dipoles.
   B. Dispersion forces result from the constantly changing electron distribution within molecules.
      These forces are small but their cumulative effect may be important.
   C. Hydrogen bonding.
      1. Hydrogen bonds form between a hydrogen bonded to an electronegative atom and an unshared electron pair on another electronegative atom.
      2. Hydrogen bonds are extremely important in living organisms.
      3. Hydrophilic substances dissolve in water because they are capable of forming hydrogen bonds.
      4. Hydrophobic substances don't form hydrogen bonds.

Answers to Problems

2.1 After solving this problem, use Figure 2.2 to check your answers. The larger the number, the more electronegative the element.

More electronegative	*Less electronegative*
(a) H (2.1)	Li (1.0)
(b) Br (2.8)	B (2.0)
(c) Cl (3.0)	I (2.5)
(d) C (2.5)	H (2.1)

Carbon is slightly more electronegative than hydrogen.

2.2 As in Problem 2.1, use Figure 2.2.

(a) $\overset{\delta+}{H_3C}—\overset{\delta-}{Cl}$

(b) $\overset{\delta+}{H_3C}—\overset{\delta-}{NH_2}$

(c) $\overset{\delta-}{H_2N}—\overset{\delta+}{H}$

(d) $H_3C—SH$
Carbon and sulfur have identical electronegativies.

(e) $\overset{\delta-}{H_3C}—\overset{\delta+}{MgBr}$

(f) $\overset{\delta+}{H_3C}—\overset{\delta-}{F}$

2.3 **Strategy:** Use Figure 2.2 to find the electronegativities of each element. Calculate ΔEN and rank the answers in order of increasing ΔEN.

Solution:

Carbon:	EN = 2.5
Lithium:	EN = 1.0
	ΔEN = 1.5

Carbon:	EN = 2.5
Potassium:	EN = 0.8
	ΔEN = 1.7

Fluorine:	EN = 4.0
Carbon:	EN = 2.5
	ΔEN = 1.5

Carbon:	EN = 2.5
Magnesium:	EN = 1.2
	ΔEN = 1.3

Oxygen	EN = 3.5
Carbon:	EN = 2.5
	ΔEN = 1.0

The most polar bond has the largest ΔEN. Thus, in order of increasing bond polarity:

$$H_3C\text{—}OH < H_3C\text{—}MgBr < H_3C\text{—}Li,\ H_3C\text{—}F < H_3C\text{—}K$$

2.4 In an electrostatic potential map, the color red indicates regions of a molecule that are electron-rich. The map shows that nitrogen is the most electronegative atom in methylamine, and the direction of polarity of the C–N bond is:

δ–
$:NH_2$
H–C(H)(H) δ+

Methylamine

2.5

H H
C
HO C OH
H H

The dipole moment of ethylene glycol is zero because the bond polarities of the two carbon-oxygen bonds cancel.

2.6 **Strategy:** For each bond, identify the more electronegative element, and draw an arrow that points from the less electronegative element to the more electronegative element. Estimate the sum of the individual dipole moments to arrive at the dipole moment for the entire molecule.

Solution:

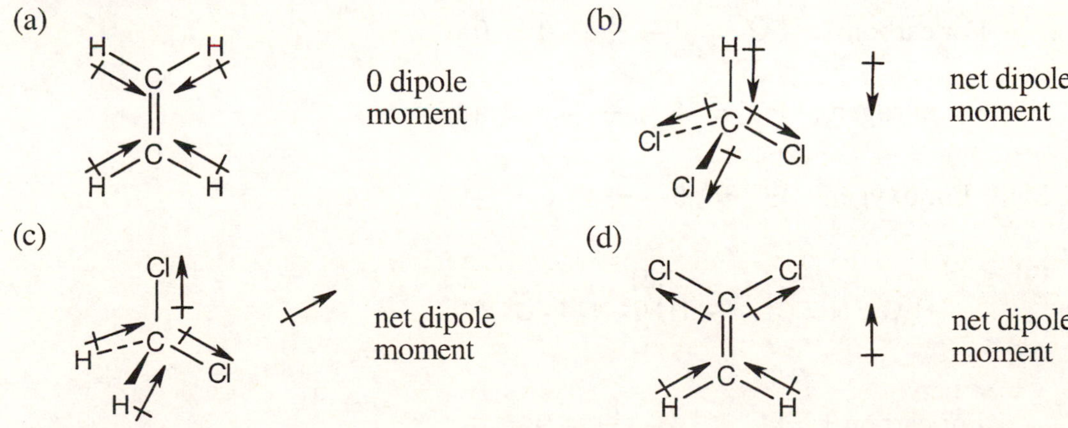

2.7 **Strategy:** To find the formal charge of an atom in a molecule, follow these two steps:
(1) Draw a electron-dot structure of the molecule.
(2) Use the formula in Section 2.3 to determine formal charge for each atom. The periodic table shows the number of valence electrons of the element, and the electron-dot structure shows the number of bonding and nonbonding electrons.

Solution:

$$\text{Formal charge (FC)} = \left[\begin{matrix}\text{\# of valence}\\ \text{electrons}\end{matrix}\right] - \left[\frac{\text{\# of bonding electrons}}{2}\right] - \left[\begin{matrix}\text{\# nonbonding}\\ \text{electrons}\end{matrix}\right]$$

(a)

$$H_2C{=}N{=}\ddot{N}\colon \quad = \quad H\colon\underset{}{\overset{H}{\ddot{C}}}\colon\colon \overset{1}{N}\colon\colon \overset{2}{\ddot{N}}\colon$$

For carbon: $FC = 4 - \frac{8}{2} - 0 = 0$

For nitrogen 1: $FC = 5 - \frac{8}{2} - 0 = +1$

For nitrogen 2: $FC = 5 - \frac{4}{2} - 4 = -1$

Remember: *Valence electrons* are the electrons characteristic of a specific element. *Bonding electrons* are those electrons involved in bonding to other atoms. *Nonbonding electrons* are those electrons in lone pairs.

(b)

$$H_3C{-}C{\equiv}N{-}\ddot{\underset{..}{O}}\colon = H\colon\overset{H}{\underset{H}{\overset{1}{\ddot{\underset{..}{C}}}}}\colon\overset{2}{C}\colon\colon\colon N\colon\ddot{\underset{..}{O}}\colon$$

For carbon 1: $FC = 4 - \frac{8}{2} - 0 = 0$

For carbon 2: $FC = 4 - \frac{8}{2} - 0 = 0$

For nitrogen : $FC = 5 - \frac{8}{2} - 0 = +1$

For oxygen: $FC = 6 - \frac{2}{2} - 6 = -1$

(c)

$$H_3C{-}N{\equiv}C\colon \quad = H\colon\overset{H}{\underset{H}{\overset{1}{\ddot{\underset{..}{C}}}}}\colon N\colon\colon\colon \overset{2}{C}\colon$$

For carbon 1: $FC = 4 - \frac{8}{2} - 0 = 0$

For carbon 2: $FC = 4 - \frac{6}{2} - 2 = -1$

For nitrogen : $FC = 5 - \frac{8}{2} - 0 = +1$

2.8

$$\text{Formal charge (FC)} = \left[\begin{matrix}\text{\# of valence}\\ \text{electrons}\end{matrix}\right] - \left[\frac{\text{\# of bonding electrons}}{2}\right] - \left[\begin{matrix}\text{\# nonbonding}\\ \text{electrons}\end{matrix}\right]$$

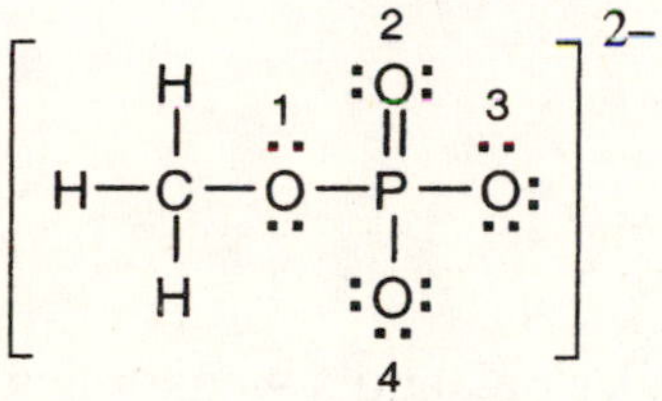

Methyl phosphate

For oxygen 1: $FC = 6 - \frac{4}{2} - 4 = 0$

For oxygen 2: $FC = 6 - \frac{4}{2} - 4 = 0$

For oxygen 3 : $FC = 6 - \frac{2}{2} - 6 = -1$

For oxygen 4: $FC = 6 - \frac{2}{2} - 6 = -1$

Oxygen atoms 3 and 4 each have a formal charge of –1.

2.9 **Strategy:** Look for three-atom groupings that contain a multiple bond next to an atom with a *p* orbital. Exchange the positions of the bond and the electrons in the *p* orbital to draw the resonance form of each grouping.

Solution:

(a) Methyl phosphate anion has 3 three-atom groupings and thus has 3 resonance forms.

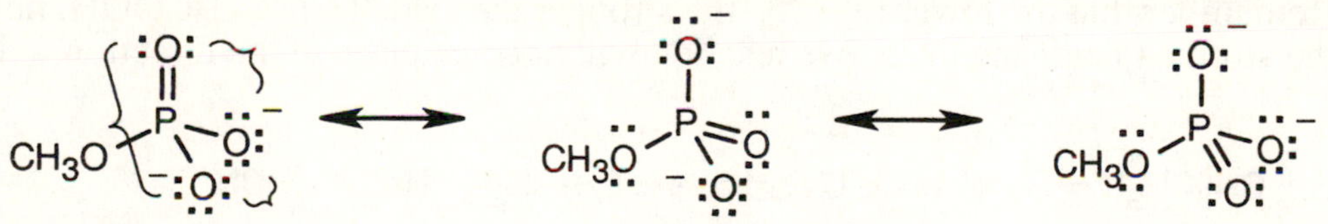

Recall from Chapter 1 that phosphorus, a third-row element, can form more than four covalent bonds

(b)

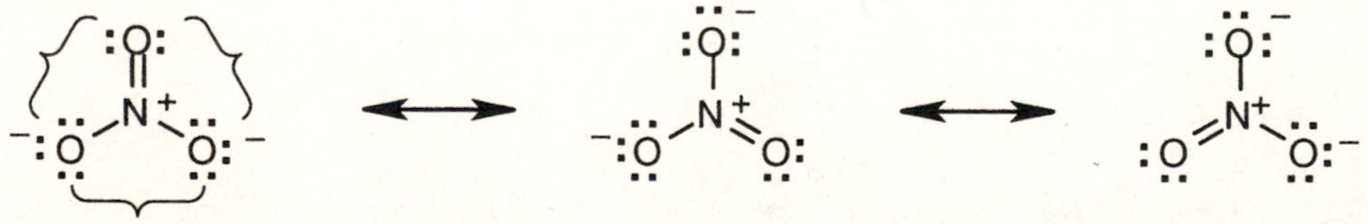

(c)

$H_2C{=}CH{-}\overset{+}{C}H_2 \longleftrightarrow H_2\overset{+}{C}{-}CH{=}CH_2$

(d)

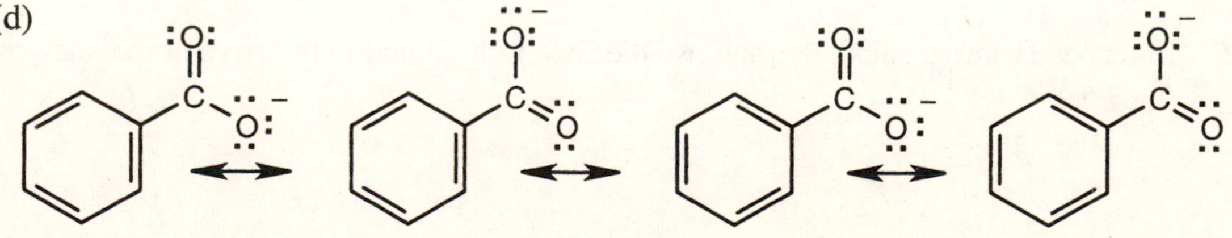

2.10 **Strategy:** When an acid loses a proton, the product is the conjugate base of the acid. When a base gains a proton, the product is the conjugate acid of the base.

Solution:

$$\underset{\text{Acid}}{H{-}NO_3} + \underset{\text{Base}}{:NH_3} \rightleftharpoons \underset{\text{Conjugate base}}{NO_3^-} + \underset{\text{Conjugate acid}}{NH_4^+}$$

2.11 Recall from Section 2.8 that a stronger acid has a smaller pK_a and a weaker acid has a larger pK_a. Accordingly, phenylalanine ($pK_a = 1.83$) is a stronger acid than tryptophan ($pK_a = 2.83$).

2.12 HO–H is a stronger acid than H_2N–H. Since H_2N^- is a stronger base than HO^-, the conjugate acid of H_2N^- (H_2N–H) is a weaker acid than the conjugate acid of HO^- (HO–H).

2.13 **Strategy:** Use Table 2.3 to find the strength of each acid. A reaction takes place as written if the stronger acid is the reactant.

Solution:

(a)

$$\underset{\substack{pK_a = 9.3 \\ \text{Weaker acid}}}{H{-}CN} + CH_3CO_2^-\ Na^+ \xrightarrow{?} Na^+\ {}^-CN + \underset{\substack{pK_a = 4.7 \\ \text{Stronger acid}}}{CH_3CO_2H}$$

Remember that the lower the pK_a, the stronger the acid. Thus CH_3COOH, not HCN, is the stronger acid, and the above reaction will not take place in the direction written.

(b)

$$\underset{\substack{pK_a = 16 \\ \text{Weaker acid}}}{CH_3CH_2O{-}H} + Na^+\ {}^-CN \xrightarrow{?} CH_3CH_2O^-\ Na^+ + \underset{\substack{pK_a = 9.3 \\ \text{Stronger acid}}}{HCN}$$

Using the same reasoning as in part (a), we can see that the above reaction will not occur.

2.14

$$\underset{\substack{pK_a = 19 \\ \text{Stronger acid}}}{H_3C{-}C({=}O){-}CH_3} + Na^+\ {}^-{:}\ddot{N}H_2 \xrightarrow{?} H_3C{-}C({=}O){-}CH_2{:}^-\ Na^+ + \underset{\substack{pK_a \approx 36 \\ \text{Weaker acid}}}{:NH_3}$$

The above reaction will take place as written.

2.15 Enter –9.31 into a calculator and use the INV LOG function to arrive at the answer $K_a = 4.9 \times 10^{-10}$.

2.16 **Strategy:** Locate the electron pair(s) of the Lewis base and draw a curved arrow from the electron pair to the Lewis acid. The electron pair moves from the atom at the tail of the arrow (Lewis base) to the atom at the point of the arrow (Lewis acid).

Solution: (Note: electron dots have been omitted from Cl^- to reduce clutter.)

(a)

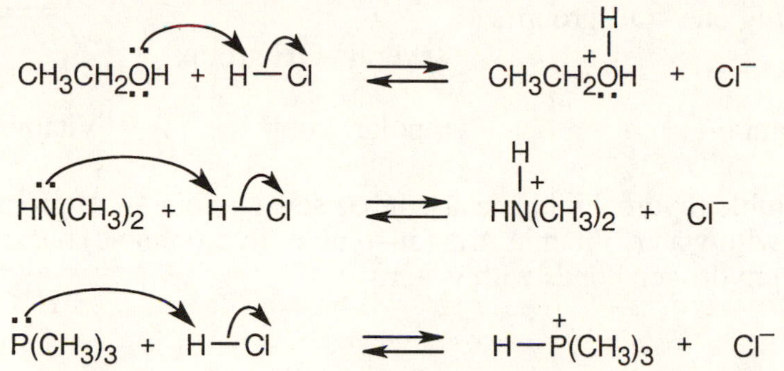

$$CH_3CH_2\ddot{O}H + H{-}Cl \rightleftharpoons CH_3CH_2\overset{+}{\ddot{O}}H(H) + Cl^-$$

$$H\ddot{N}(CH_3)_2 + H{-}Cl \rightleftharpoons H\overset{+}{N}(H)(CH_3)_2 + Cl^-$$

$$\ddot{P}(CH_3)_3 + H{-}Cl \rightleftharpoons H{-}\overset{+}{P}(CH_3)_3 + Cl^-$$

(b)

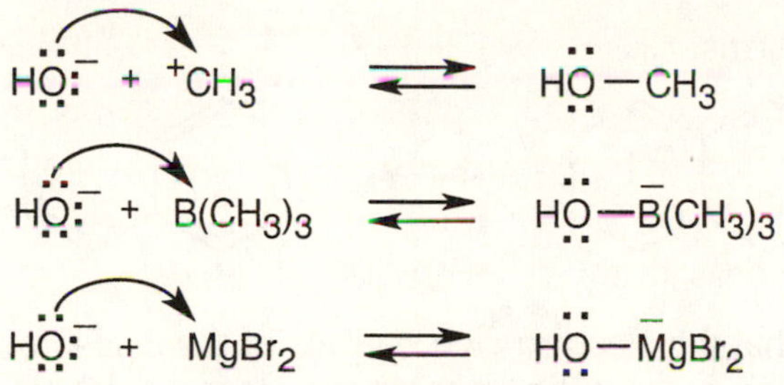

$$H\ddot{O}{:}^- + {}^+CH_3 \rightleftharpoons H\ddot{O}{-}CH_3$$

$$H\ddot{O}{:}^- + B(CH_3)_3 \rightleftharpoons H\ddot{O}{-}\overset{-}{B}(CH_3)_3$$

$$H\ddot{O}{:}^- + MgBr_2 \rightleftharpoons H\ddot{O}{-}\overset{-}{Mg}Br_2$$

2.17 (a) The nitrogen on the left is more electron-rich and more basic. The indicated hydrogen is most electron-poor (bluest) and is most acidic.

more basic (red) → N; most acidic (blue) → N–H

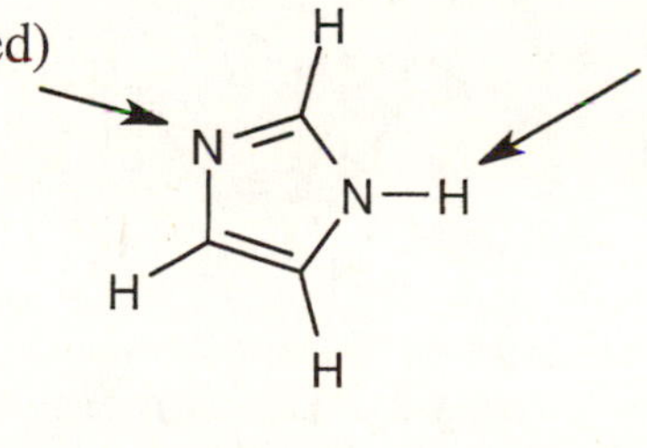

Imidazole

(b)

A–H + :N (imidazole) ⇌ H–N⁺ (imidazolium) ⟷ resonance form with + on the other N

imidazole N–H + :B ⇌ imidazolide anion (N:⁻) ⟷ resonance form with − on the other N

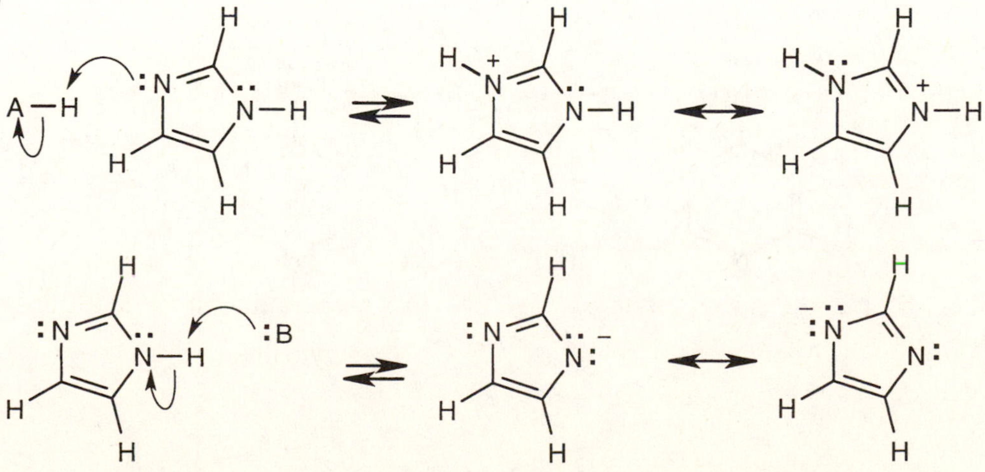

2.18

only one –OH group

several –OH groups

Vitamin A

* = polar group

Vitamin C

Vitamin C is water-soluble (hydrophilic) because it has several polar –OH groups that can form hydrogen bonds with water. Vitamin A is fat-soluble (hydrophobic) because most of its structure can't form hydrogen bonds with water.

Visualizing Chemistry

2.19 Naphthalene has three resonance forms.

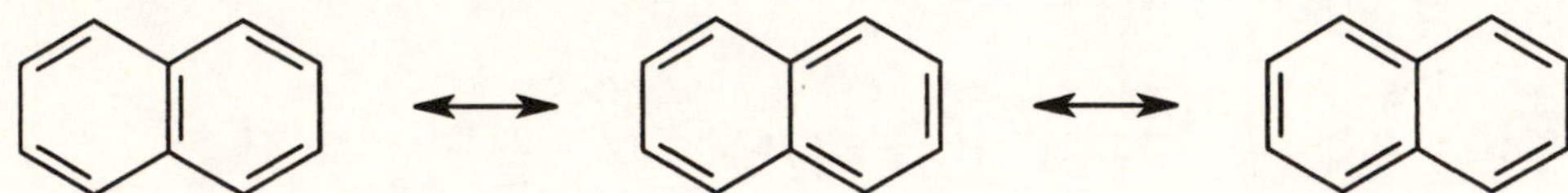

2.20 Electrostatic potential maps show that the electron-rich regions of the cis isomer lie on the same side of the double bond, leading to a net dipole moment. Because the electron-rich regions of the trans isomer are symmetrical about the double bond, the individual bond dipole moments cancel, and the isomer has no overall dipole moment.

net dipole moment

zero dipole moment

cis-1,2-Dichloroethylene

trans-1,2-Dichloroethylene

2.21

Adenine

Cytosine

2.22 Electrostatic potential maps show that the most electron-rich atom in (a) is oxygen, not nitrogen. In (b), nitrogen is electron-rich and is thus more basic than the nitrogen in (a). In (a), the hydrogen atoms bonded to nitrogen are more acidic (electron-poor) than all other hydrogens in both molecules.

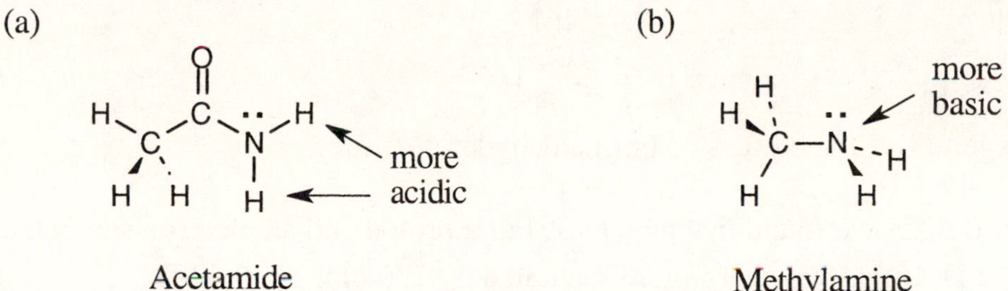

Additional Problems

2.23 Use Figure 2.2 if you need help. The most electronegative element is starred.

(a) $CH_2\overset{*}{F}Cl$ (b) $\overset{*}{F}CH_2CH_2CH_2Br$ (c) $H\overset{*}{O}CH_2CH_2NH_2$ (d) $CH_3\overset{*}{O}CH_2Li$

2.24

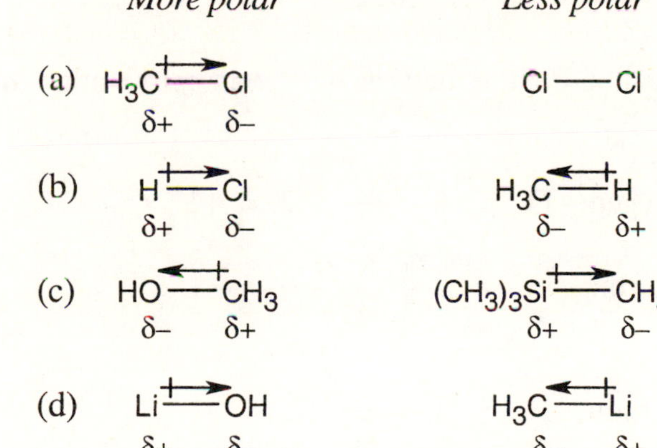

2.25

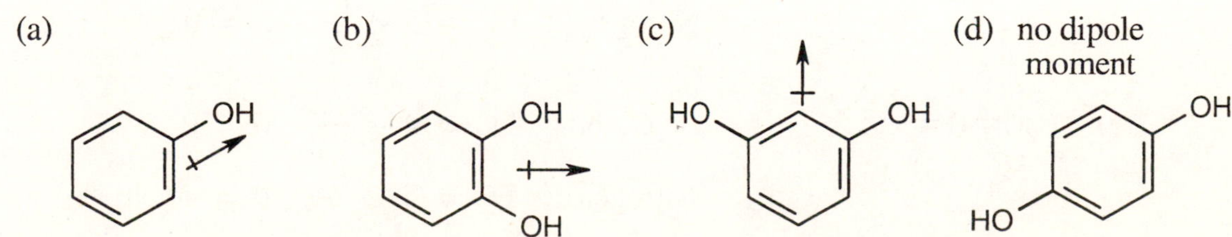

2.26 In phosgene, the individual bond polarities tend to cancel, but in formaldehyde, the bond polarities add to each other. Thus, phosgene has a smaller dipole moment than formaldehyde.

Phosgene Formaldehyde

2.27 (a) In Section 2.2, we found that $\mu = Q \times r$. For a proton and an electron separated by 100 pm, $\mu = 4.8$ D. If the two charges are separated by 136 pm, $\mu = 6.53$ D.

(b) Since the observed dipole moment is 1.08 D, the H–Cl bond has (1.08 D / 6.53 D) x 100 % = 16.5 % ionic character.

2.28 The magnitude of a dipole moment depends on both electronegativity and distance between atoms. Fluorine is more electronegative than chlorine, but a C–F bond is shorter than a C–Cl bond. Thus, the dipole moment of CH_3F is smaller than that of CH_3Cl.

2.29 The observed dipole moment is due to the lone pair electrons on sulfur.

$$H_3C{-}\ddot{\underset{\cdot\cdot}{S}}H$$

2.30 To save space, molecules are shown as line-bond structures with lone pairs, rather than as Lewis structures.

(a) $(CH_3)_2\ddot{O}BF_3$

$$\text{Oxygen: FC} = 6 - \frac{6}{2} - 2 = +1$$

$$\text{Boron: FC} = 3 - \frac{8}{2} - 0 = -1$$

(b) $H_2\ddot{C}{-}\overset{1}{N}{\equiv}\overset{2}{N}{:}$

$$\text{Carbon: FC} = 4 - \frac{6}{2} - 2 = -1$$

$$\text{Nitrogen 1: FC} = 5 - \frac{8}{2} - 0 = +1$$

$$\text{Nitrogen 2: FC} = 5 - \frac{6}{2} - 2 = 0$$

(c) $H_2C{=}\overset{1}{N}{=}\overset{2}{\underset{\cdot\cdot}{N}}{:}$

$$\text{Carbon: FC} = 4 - \frac{8}{2} - 0 = 0$$

$$\text{Nitrogen 1: FC} = 5 - \frac{8}{2} - 0 = +1$$

$$\text{Nitrogen 2: FC} = 5 - \frac{4}{2} - 4 = -1$$

(d) $\ddot{O}=\ddot{O}-\ddot{O}$: (atoms numbered 1, 2, 3)

Oxygen 1: $FC = 6 - \frac{4}{2} - 4 = 0$

Oxygen 2: $FC = 6 - \frac{6}{2} - 2 = +1$

Oxygen 3: $FC = 6 - \frac{2}{2} - 6 = -1$

(e) $H_2\ddot{C}-P(CH_3)_3$

Carbon: $FC = 4 - \frac{6}{2} - 2 = -1$

Phosphorus: $FC = 5 - \frac{8}{2} - 0 = +1$

(f) pyridine N–$\ddot{O}$:

Nitrogen: $FC = 5 - \frac{8}{2} - 0 = +1$

Oxygen: $FC = 6 - \frac{2}{2} - 6 = -1$

2.31

(a) $(CH_3)_3N-\ddot{O}$:

Oxygen: $FC = 6 - \frac{2}{2} - 6 = -1$

Nitrogen: $FC = 5 - \frac{8}{2} - 0 = +1$

(b) $H_3C-\ddot{N}-N\equiv N$: (nitrogens numbered 1, 2, 3)

Nitrogen 1: $FC = 5 - \frac{4}{2} - 4 = -1$

Nitrogen 2: $FC = 5 - \frac{8}{2} - 0 = +1$

Nitrogen 3: $FC = 5 - \frac{6}{2} - 2 = 0$

(c) $H_3C-\ddot{N}=N=\ddot{N}$: (nitrogens numbered 1, 2, 3)

Nitrogen 1: $FC = 5 - \frac{6}{2} - 2 = 0$

Nitrogen 2: $FC = 5 - \frac{8}{2} - 0 = +1$

Nitrogen 3: $FC = 5 - \frac{4}{2} - 4 = -1$

2.32 Resonance forms do not differ in the position of nuclei. The two structures in (a) are not resonance forms because the carbon and hydrogen atoms outside the ring occupy different positions in the two forms.

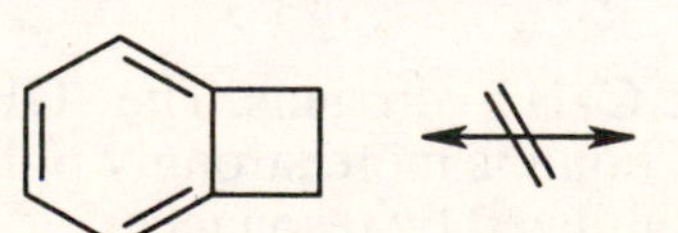 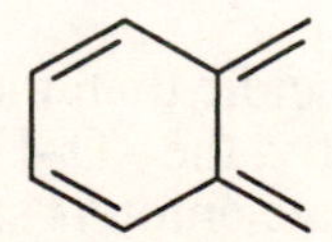

not resonance structures

The pairs of structures in parts (b), (c), and (d) represent resonance forms.

2.33

(a)

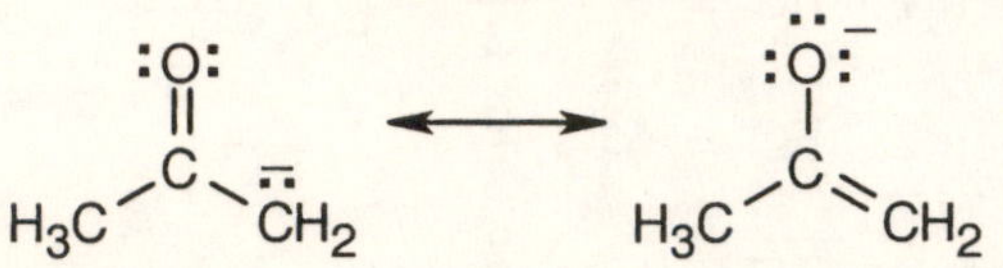

$H_3C{-}C({=}O){-}\ddot{C}H_2^{-} \longleftrightarrow H_3C{-}C(O^-){=}CH_2$

(b)

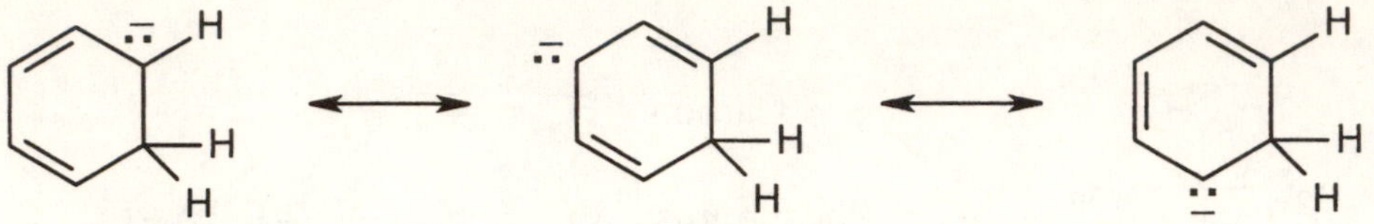

(c)

$H_2N{-}C(NH_2){=}NH_2^+ \longleftrightarrow H_2N{-}C({=}NH_2^+){-}NH_2 \longleftrightarrow {}^+H_2N{=}C(NH_2){-}NH_2 \longleftrightarrow H_2N{-}C^+(NH_2){-}NH_2$

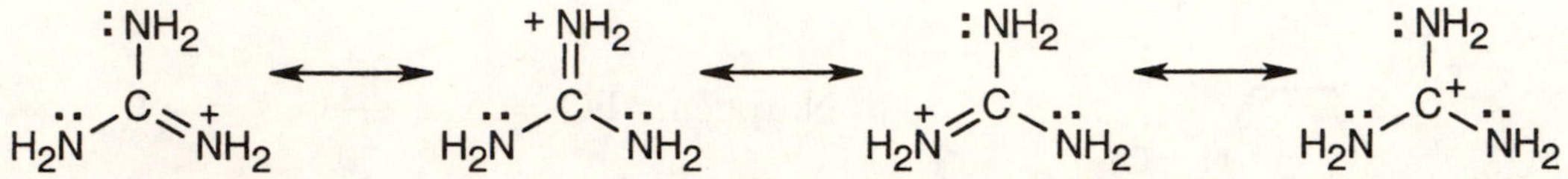

The last resonance structure is a minor contributor because its carbon lacks a complete electron octet.

(d)

$H_3C{-}\ddot{S}{-}\overset{+}{C}H_2 \longleftrightarrow H_3C{-}\overset{+}{S}{=}CH_2$

(e)

$H_2C{=}CH{-}CH{=}CH{-}\overset{+}{C}H{-}CH_3 \longleftrightarrow H_2C{=}CH{-}\overset{+}{C}H{-}CH{=}CH{-}CH_3$

$\longleftrightarrow \overset{+}{H_2C}{-}CH{=}CH{-}CH{=}CH{-}CH_3$

2.34 The two structures are not resonance forms because the positions of the carbon atoms are different in the two forms.

2.35

$CH_3\ddot{O}H + HCl \rightleftharpoons CH_3\ddot{O}H_2^+ + Cl^-$

$CH_3\ddot{O}H + Na^+\ {}^-\ddot{N}H_2 \rightleftharpoons CH_3\ddot{O}{:}^-\ Na^+ + {:}NH_3$

2.36

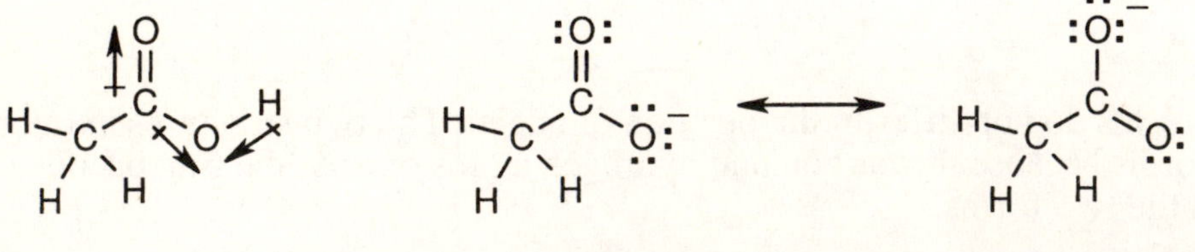

H
H–C–C(=O)–O–H
H

$H_3C{-}C({=}O){-}O^- \longleftrightarrow H_3C{-}C(O^-){=}O$

The O–H hydrogen of acetic acid is more acidic than the C–H hydrogens. The –OH oxygen is electronegative, and, consequently, the –O–H bond is more strongly polarized than the –C–H bonds. In addition, the acetate anion is stabilized by resonance.

2.37 The Lewis acids shown below can accept an electron pair either because they have a vacant orbital or because they can donate H^+. The Lewis bases have nonbonding electron pairs.

Lewis acids: $AlBr_3$, BH_3, HF, $TiCl_4$ (vacant *d* orbitals)

Lewis bases: $CH_3CH_2\ddot{N}H_2$, $H_3C{-}\ddot{\underset{..}{S}}{-}CH_3$

2.38 In maleic acid, the individual dipole moments add to produce a net dipole moment for the whole molecule. The individual dipole moments in fumaric acid cancel, resulting in a zero dipole moment.

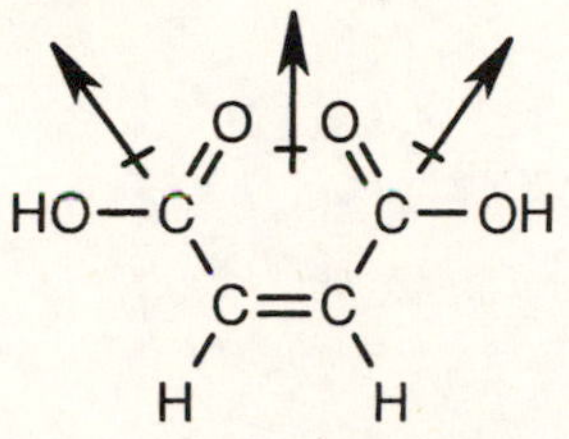

Maleic acid: HO—C(=O)—CH=CH—C(=O)—OH (cis)

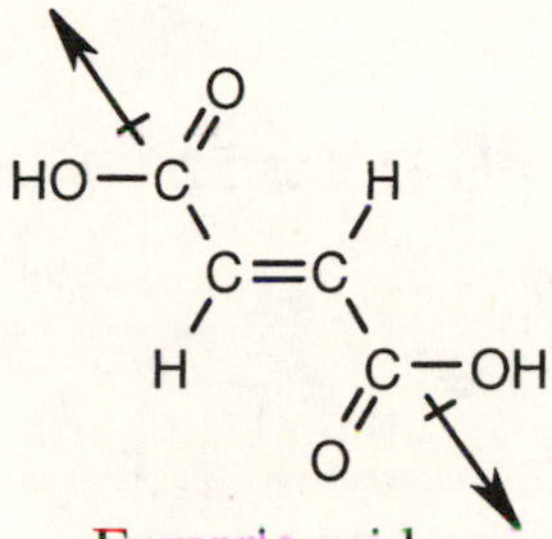

Fumaric acid: HO—C(=O)—CH=CH—C(=O)—OH (trans)

Maleic acid Fumaric acid

2.39 The substances with the largest values of pK_a are the least acidic.

Least acidic ⟶ *Most acidic*

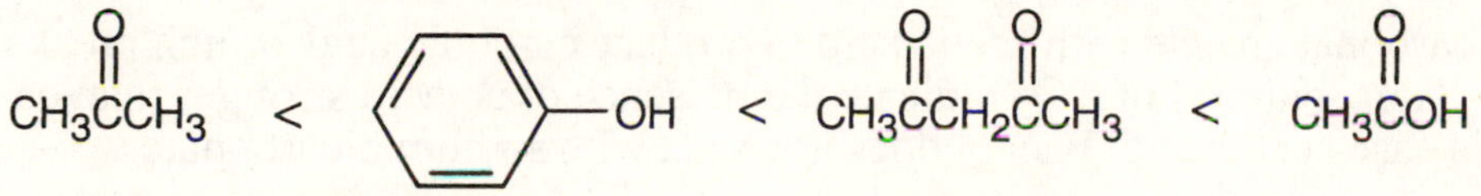

CH_3CCH_3 (O) < C_6H_5—OH < $CH_3CCH_2CCH_3$ (O, O) < CH_3COH (O)

pK_a = 19 pK_a = 9.9 pK_a = 9 pK_a = 4.8

2.40 To react completely (> 99.9%) with NaOH, an acid must have a pK_a at least 3 units smaller than the pK_a of H_2O. Thus, all substances in the previous problem except acetone react completely with NaOH.

2.41 The stronger the acid (smaller pK_a), the weaker its conjugate base. Since NH_4^+ is a stronger acid than $CH_3NH_3^+$, CH_3NH_2 is a stronger base than NH_3.

2.42

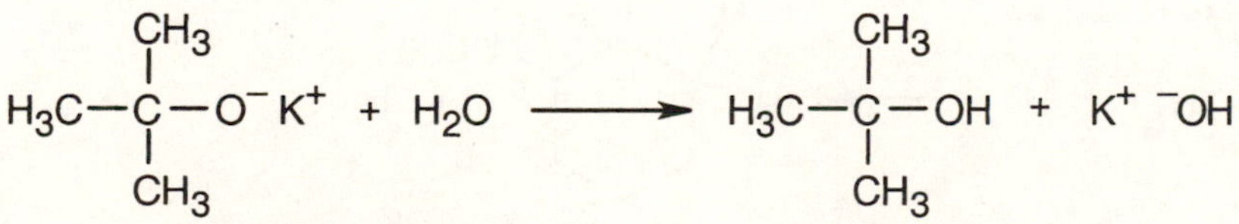

$$(CH_3)_3C{-}O^-K^+ + H_2O \longrightarrow (CH_3)_3C{-}OH + K^+\ {}^-OH$$

pK_a = 15.7 pK_a ≈ 18
stronger acid weaker acid

The reaction takes place as written because water is a stronger acid than *tert*-butyl alcohol. Thus, a solution of potassium *tert*-butoxide in water can't be prepared.

2.43

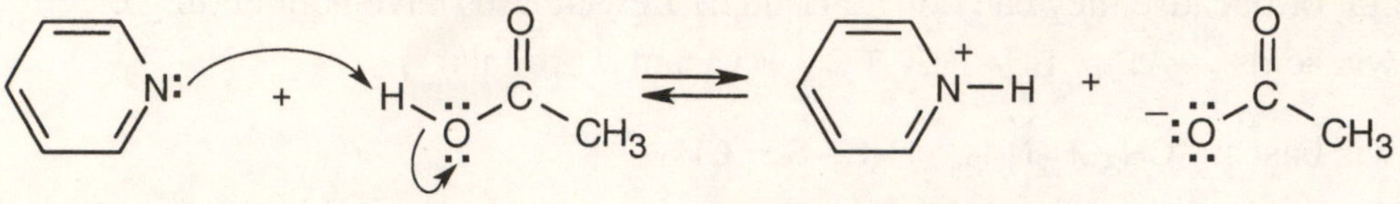

Pyridine (N:) + $H-\ddot{O}-C(=O)-CH_3$ ⇌ pyridinium ($\overset{+}{N}-H$) + $^{-}:\ddot{O}-C(=O)-CH_3$

2.44 (a) Acetone: $K_a = 5 \times 10^{-20}$ (b) Formic acid: $K_a = 1.8 \times 10^{-4}$

2.45 (a) Nitromethane: $pK_a = 10.30$ (b) Acrylic acid: $pK_a = 4.25$

2.46

$$\underset{[0.050\ M]}{\text{Formic acid}} + H_2O \overset{K_a}{\rightleftharpoons} \underset{[x]}{\text{Formate}^-} + \underset{[x]}{H_3O^+}$$

$$K_a = 1.8 \times 10^{-4} = \frac{x^2}{0.050 - x}$$

If you let $0.050 - x = 0.050$, then $x = 3.0 \times 10^{-3}$ and pH = 2.52. If you calculate x exactly, then $x = 2.9 \times 10^{-3}$ and pH = 2.54.

2.47 Only acetic acid will react with sodium bicarbonate. Acetic acid is the only substance in Problem 2.39 that is a stronger acid than carbonic acid.

2.48 Sodium bicarbonate reacts with acetic acid to produce carbonic acid, which breaks down to form CO_2. Thus, bubbles of CO_2 indicate the presence of an acid stronger than carbonic acid, in this case acetic acid. Phenol does not react with sodium bicarbonate.

2.49 Reactions (a) and (c) are reactions between Brønsted-Lowry acids and bases; the stronger acid and stronger base are identified. Reactions (b) and (d) occur between Lewis acids and bases.

(a)

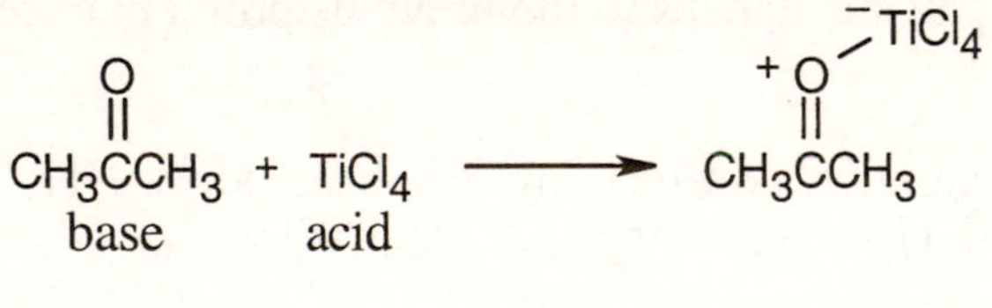

$$\underset{\text{base}}{CH_3C(=O)CH_3} + \underset{\text{acid}}{TiCl_4} \longrightarrow CH_3C(=\overset{+}{O}-\overset{-}{TiCl_4})CH_3$$

(b)

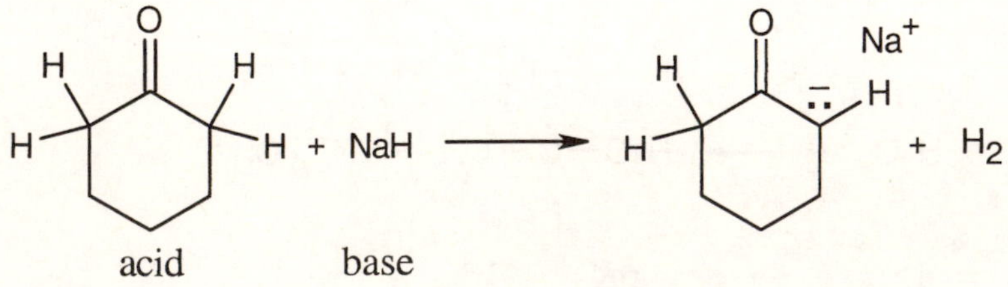

cyclohexanone (acid) + NaH (base) ⟶ enolate carbanion Na^+ + H_2

(c)

morpholine (base) + BH_3 (acid) ⟶ morpholine–$\overset{+}{N}-\overset{-}{BH_3}$ adduct

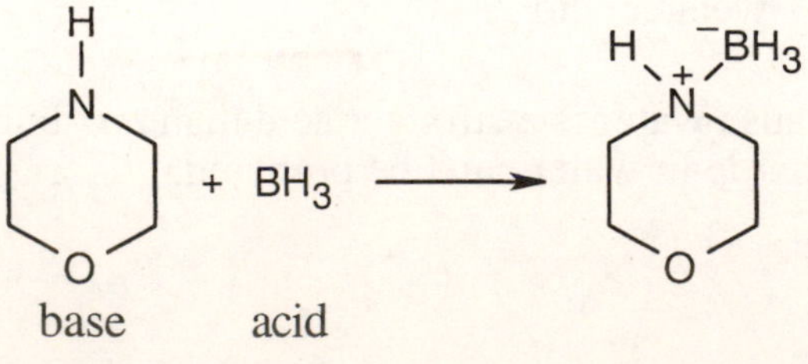

2.50 Pairs (a) and (d) represent resonance structures; pairs (b) and (c) do not. For two structures to be resonance forms, all atoms must be in the same positions.

2.51

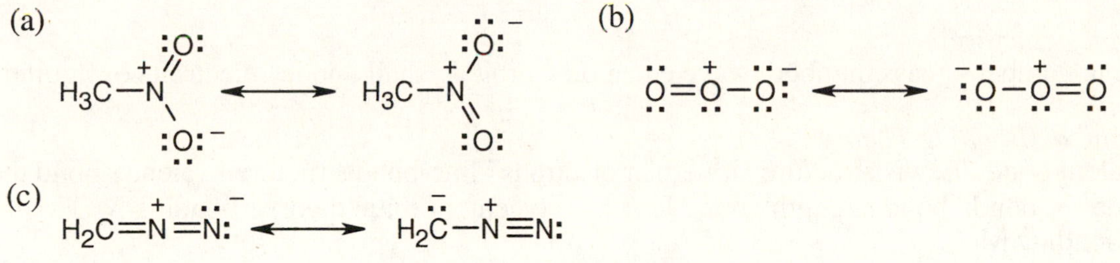

(a) (b)

(c)

2.52

(a) (b) (c) (d)

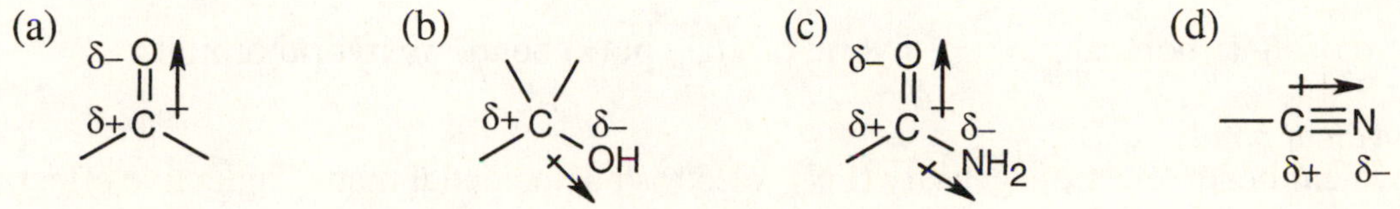

2.53

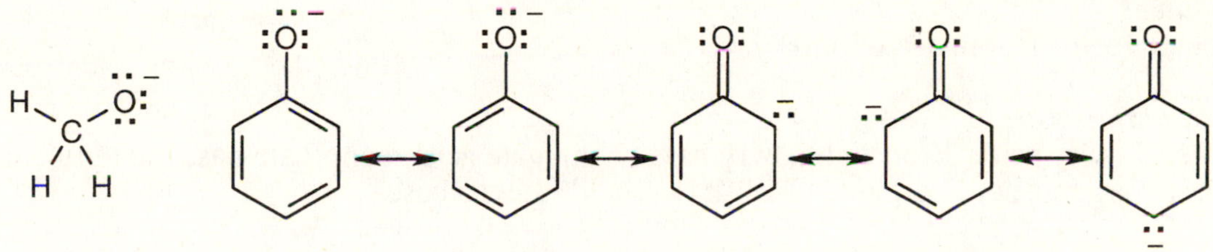

When phenol loses a proton, the resulting anion is stabilized by resonance.

2.54 The cation pictured can be represented by two resonance forms. Reaction with water can occur at either positively charged carbon, resulting in two products.

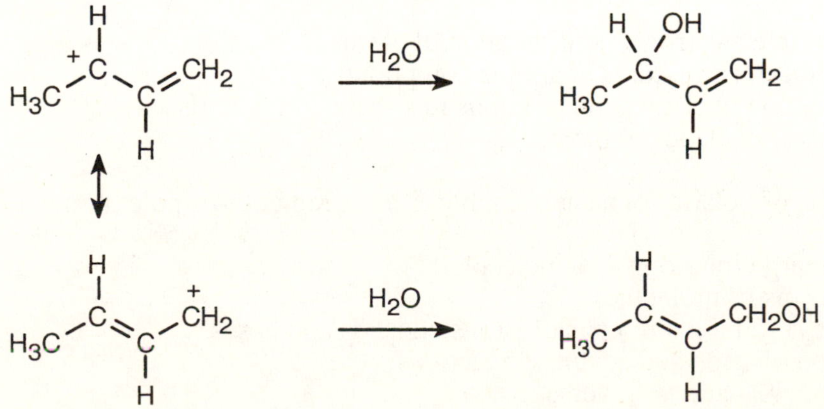

Review Unit 1: Bonds and Bond Polarity

Major Topics Covered (with vocabulary:)

Atomic Structure:
atomic number mass number wave equation orbital shell node electron configuration

Chemical Bonding Theory:
covalent bond Lewis structure lone-pair electrons line-bond structure valence-bond theory sigma (σ) bond bond strength bond length molecular orbital theory bonding MO antibonding MO

Hybridization:
sp^3 hybrid orbital bond angle sp^2 hybrid orbital pi (π) bond *sp* hybrid orbital

Polar covalent bonds:
polar covalent bond electronegativity (EN) electrostatic potential maps inductive effect dipole moment formal charge dipolar molecule

Resonance:
resonance form resonance hybrid

Acids and Bases:
Brønsted-Lowry acid Brønsted-Lowry base conjugate acid conjugate base acidity constant K_a pK_a organic acid organic base Lewis acid Lewis base

Chemical Structures:
condensed structure skeletal structure space-filling models ball-and-stick models

Types of Problems:

After studying these chapters you should be able to:

- Predict the ground state electronic configuration of atoms.
- Draw Lewis electron-dot structures of simple compounds.
- Predict and describe the hybridization of bonds in simple compounds.
- Predict bond angles and shapes of molecules.

- Predict the direction of polarity of a chemical bond, and predict the dipole moment of a simple compound.
- Calculate formal charge for atoms in a molecule.
- Draw resonance forms of molecules.
- Predict the relative acid/base strengths of Brønsted acids and bases.
- Predict the direction of Brønsted acid/base reactions.
- Calculate: pK_a from K_a, and vice versa.
  pH of a solution of a weak acid.
- Identify Lewis acids and bases.
- Draw chemical structures from molecular formulas, and vice versa.

Points to Remember:

* In order for carbon, with valence shell electron configuration of $2s^22p^2$, to form four sp^3 hybrid orbitals, it is necessary that one electron be promoted from the $2s$ subshell to the $2p$ subshell. Although this promotion requires energy, the resulting hybrid orbitals are able to form stronger bonds, and compounds containing these bonds are more stable.

* Assigning formal charge to atoms in a molecule is helpful in showing where the electrons in a bond are located. Even if a bond is polar covalent, in some molecules the electrons "belong" more to one of the atoms than the other. This "ownership" is useful for predicting the outcomes of chemical reactions, as we will see in later chapters.

* Resonance structures are representations of the distribution of π and nonbonding electrons in a molecule. Electrons don't move around in the molecule, and the molecule doesn't change back and forth, from structure to structure. Rather, resonance structures are an attempt to show, by conventional line-bond drawings, the electron distribution of a molecule that can't be represented by any one structure.

* As in general chemistry, acid-base reactions are of fundamental importance in organic chemistry. Organic acids and bases, as well as inorganic acids and bases, occur frequently in reactions, and large numbers of reactions are catalyzed by Brønsted acids and bases and Lewis acids and bases.

Self-Test:

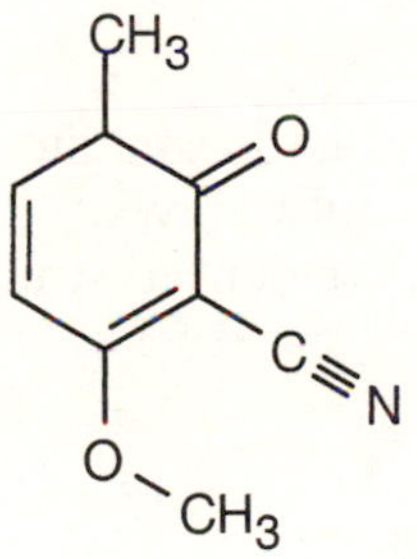

CH_3 O C≡N O CH_3

O N C H_3C CH_3 H C F F F

O N—N

A
Ricinine
(a toxic component of castor beans)

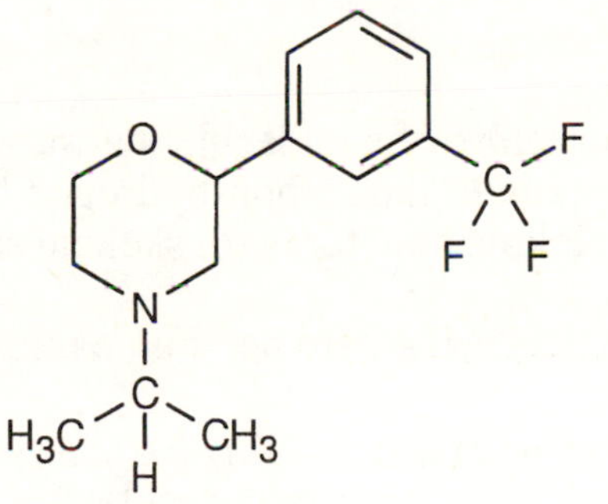

B
Oxaflozane
(an antidepressant)

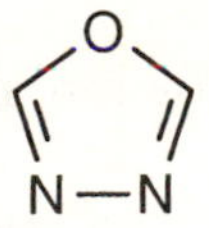

C
1,3,4-Oxadiazole

For **A** (ricinine) and **B** (oxaflozane): Add all missing electron lone pairs. Identify the hybridization of all carbons. Indicate the direction of bond polarity for all bonds with Δ EN ≥ 0.5. In each compound, which bond is the most polar? Convert **A** and **B** to molecular formulas.

Draw a resonance structure for **B**. Which atom (or atoms) of **B** can act as a Lewis base?

Add missing electron lone pairs to **C**. Is it possible to draw resonance forms for **C**? If so, draw at least one resonance form, and describe it.

Multiple Choice:

1. Which element has $4s^24p^2$ as its valence shell electronic configuration?
(a) Ca (b) C (c) Al (d) Ge

2. Which compound (or group of atoms) has an oxygen with a +1 formal charge?
(a) NO_3^- (b) O_3 (c) acetone anion (d) acetate anion

The following questions involve these acids: (i) HW ($pK_a = 2$); (ii) HX ($pK_a = 6$); (iii) HY ($pK_a = 10$); (iv) HZ ($pK_a = 20$).

3. Which of the above acids react almost completely with water to form hydroxide ion?
(a) none of them (b) all of them (c) HY and HZ (d) HZ

4. The conjugate bases of which of the above acids react almost completely with water to form hydroxide ion?
(a) none of them (b) all of them (c) HZ (d) HY and HZ

5. If you want to convert HX to X^-, which bases can you use?
(a) W^- (b) Y^- (c) Z^- (d) Y^- or Z^-

6. If you add equimolar amounts of HW, X^- and HY to a solution, what are the principal species in the resulting solution?
(a) HW, HX, HY (b) W^-, HX, HY (c) HW, X^-, HY (d) HW, HX, Y^-

7. What is the approximate pH difference between a solution of 1 M HX and a solution of 1 M HY?
(a) 2 (b) 3 (c) 4 (d) 6

8. If you wanted to write the structure of a molecule that shows carbon and hydrogen atoms as groups, without indicating many of the carbon-hydrogen bonds, you would draw a:
(a) molecular formula (b) Kekulé structure (c) skeletal structure (d) condensed structure

9. Which of the following molecules has zero net dipole moment?

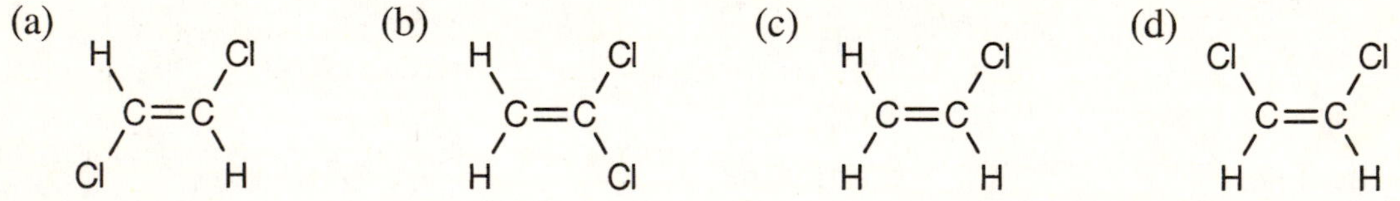

```
(a)                  (b)                  (c)                  (d)
  H       Cl           H       Cl           H       Cl           Cl      Cl
   \     /              \     /              \     /              \     /
    C=C                  C=C                  C=C                  C=C
   /     \              /     \              /     \              /     \
 Cl       H            H       Cl           H       H            H       H
```

10. In which of the following bonds is carbon the more electronegative element?
(a) C—Br (b) C—I (c) C—P (d) C—S

Chapter 3 – Organic Compounds: Alkanes and Their Stereochemistry

Chapter Outline

I. Functional Groups (Section 3.1).
   A. Functional groups are groups of atoms within a molecule that have a characteristic chemical behavior.
   B. The chemistry of every organic molecule is determined by its functional groups.
   C. Functional groups described in this text can be grouped into three categories:
      1. Functional groups with carbon-carbon multiple bonds.
      2. Groups in which carbon forms a single bond to an electronegative atom.
      3. Groups with a carbon-oxygen double bond.

II. Alkanes (Sections 3.2 – 3.5).
   A. Alkanes and alkane isomers (Section 3.2).
      1. Alkanes are formed by overlap of carbon sp^3 orbitals.
      2. Alkanes are described as saturated hydrocarbons.
         a. They are hydrocarbons because they contain only carbon and hydrogen.
         b. They are saturated because all bonds are single bonds.
         c. The general formula for alkanes is C_nH_{2n+2}.
      3. For alkanes with four or more carbons, the carbons can be connected in more than one way.
         a. If the carbons are in a row, the alkane is a straight-chain alkane.
         b. If the carbon chain has a branch, the alkane is a branched-chain alkane.
      4. Alkanes with the same molecular formula are isomers.
         a. Isomers whose atoms are connected differently are constitutional isomers.
            Constitutional isomers are always different compounds with different properties but with the same molecular formula.
         b. A given alkane can be drawn in many ways.
      5. Straight-chain alkanes are named according to the number of carbons in their chain.
   B. Alkyl groups (Section 3.3).
      1. An alkyl group is the partial structure that results from the removal of a hydrogen atom from an alkane.
         a. Alkyl groups are named by replacing the -ane of an alkane by -yl.
         b. *n*-Alkyl groups are formed by removal of an end carbon of a straight-chain alkane.
         c. Branched-chain alkyl groups are formed by removal of a hydrogen atom from an internal carbon.
            The prefixes *sec*- and *tert*- refer to the degree of substitution at the branching carbon atom.
      2. There are four possible degrees of alkyl substitution for carbon.
         a. A primary carbon is bonded to one other carbon.
         b. A secondary carbon is bonded to two other carbons.
         c. A tertiary carbon is bonded to three other carbons.
         d. A quaternary carbon is bonded to four other carbons.
         e. The symbol **R** refers to the rest of the molecule.
      3. Hydrogens are also described as primary, secondary and tertiary.
         a. Primary hydrogens are bonded to primary carbons (RCH_3).
         b. Secondary hydrogens are bonded to secondary carbons (R_2CH_2).
         c. Tertiary hydrogens are bonded to tertiary carbons (R_3CH).

C. Naming alkanes (Section 3.4).
  1. The system of nomenclature used in this book is the IUPAC system.
     In this system, a chemical name has a prefix, a parent and a suffix.
     i. The parent shows the number of carbons in the principal chain.
     ii. The suffix identifies the functional group family.
     iii. The prefix shows the location of functional groups.
  2. Naming an alkane:
     a. Find the parent hydrocarbon.
        i. Find the longest continuous chain of carbons, and use its name as the parent name.
        ii. If two chains have the same number of carbons, choose the one with more branch points.
     b. Number the atoms in the parent chain.
        i. Start numbering at the end nearer the first branch point.
        ii. If branching occurs an equal distance from both ends, begin numbering at the end nearer the second branch point.
     c. Identify and number the substituents.
        i. Give each substituent a number that corresponds to its position on the parent chain.
        ii. Two substituents on the same carbon receive the same number.
     d. Write the name as a single word.
        i. Use hyphens to separate prefixes and commas to separate numbers.
        ii. Use the prefixes, *di-*, *tri-*, *tetra-* if necessary, but don't use them for alphabetizing.
     e. Name a complex substituent as if it were a compound, and set it off in parentheses.
        i. Some simple branched-chain alkyl groups have common names.
        ii. The prefix *iso* is used for alphabetizing, but *sec-* and *tert-* are not.

D. Properties of alkanes (Section 3.5).
  1. Alkanes are chemically inert to most laboratory reagents.
  2. Alkanes react with O_2 and Cl_2.
  3. The boiling points and melting points of alkanes increase with increasing molecular weight.
     a. This effect is due to weak dispersion forces between molecules.
     b. The strength of these forces increases with increasing molecular weight.
  4. Increased branching lowers an alkane's boiling point.

III. Conformations of straight-chain alkanes (Sections 3.6 – 3.7).

A. Conformations of ethane (Section 3.6).
  1. Rotation about a single bond produces isomers that differ in conformation.
     These isomers have the same connections of atoms and can't be isolated.
  2. These isomers can be represented in two ways:
     a. Sawhorse representations view the C–C bond from an oblique angle.
     b. Newman projections represent the two carbons as a circle.
  3. There is a barrier to rotation that makes some conformers of lower energy than others.
     a. The lowest energy conformer (staggered conformation) occurs when all C–H bonds are as far from each other as possible.
     b. The highest energy conformer (eclipsed conformation) occurs when all C–H bonds are as close to each other as possible.
     c. Between these two conformations lie an infinite number of other conformations.

4. The staggered conformation is 12 kJ/mol lower in energy than the eclipsed conformation.
    a. This energy difference is due to torsional strain from interactions between C–H bonding orbitals on one carbon and C–H antibonding orbitals on an adjacent carbon, which stabilize the staggered conformer.
    b. The torsional strain resulting from a single C–H interaction is 4.0 kJ/mol.
    c. The barrier to rotation can be represented on a graph of potential energy vs. angle of rotation.

B. Conformations of other alkanes (Section 3.7).
1. Conformations of propane.
    a. Propane also shows a barrier to rotation that is 14 kJ/mol.
    b. The eclipsing interaction between a C–C bond and a C–H bond is 6.0 kJ/mol.
2. Conformations of butane.
    a. Not all staggered conformations of butane have the same energy; not all eclipsed conformations have the same energy.
        i. In the lowest energy conformation (anti) the two large methyl groups are as far from each other as possible.
        ii. The eclipsed conformation that has two methyl–hydrogen interactions and a H–H interaction is 16 kJ/mol higher in energy than the anti conformation.
        iii. The conformation with two methyl groups 60° apart (gauche conformation) is 3.8 kJ/mol higher in energy than the anti conformation.
            This energy difference is due to steric strain – the repulsive interaction that results from forcing atoms to be closer together than their atomic radii allow.
        iv. The highest energy conformations occur when the two methyl groups are eclipsed.
            This conformation is 19 kJ/mol less stable than the anti conformation.
            The value of a methyl-methyl eclipsing interaction is 11 kJ/mol.
    b. The most favored conformation for any straight-chain alkane has carbon-carbon bonds in staggered arrangements and large substituents anti to each other.
    c. At room temperature, bond rotation occurs rapidly, but a majority of molecules adopt the most stable conformation.

Solutions to Problems

3.1 Notice that certain functional groups have different designations if other functional groups are present in a molecule. For example, a hydrocarbon containing a carbon–carbon double bond and no other functional group is an alkene; if other groups are present, the group is referred to as a carbon–carbon double bond. Similarly, a hydrocarbon containing a benzene ring and only carbon- and hydrogen-containing substituents is an arene; if other groups are present, the ring is labeled an aromatic ring.

(a)

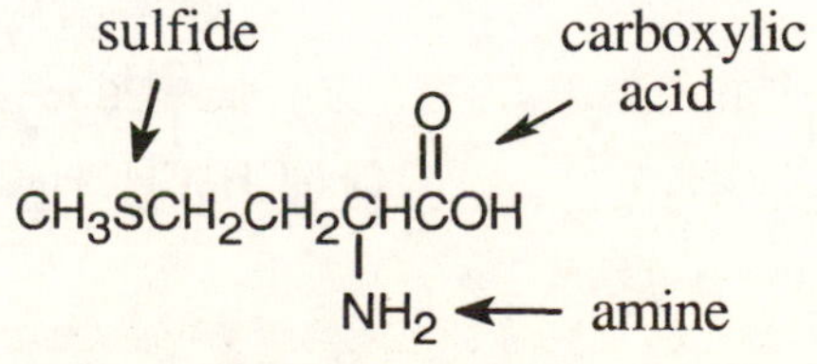

sulfide → $CH_3SCH_2CH_2CH(NH_2)COOH$ ($CH_3SCH_2CH_2CHCOH$ with C=O, carboxylic acid; NH_2 ← amine)

Methionine

(b)

CO_2H ← carboxylic acid; CH_3; aromatic ring

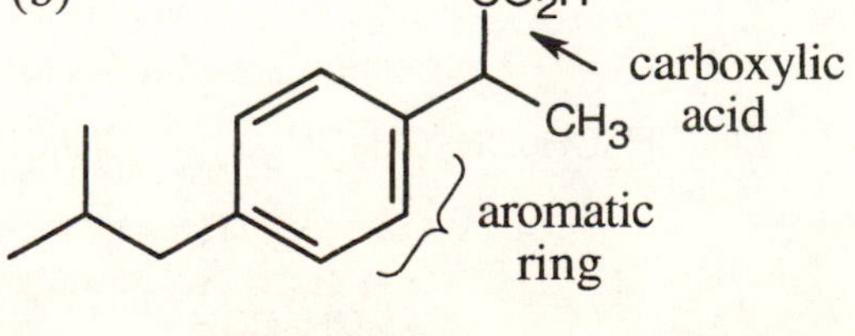

Ibuprofen

(c)

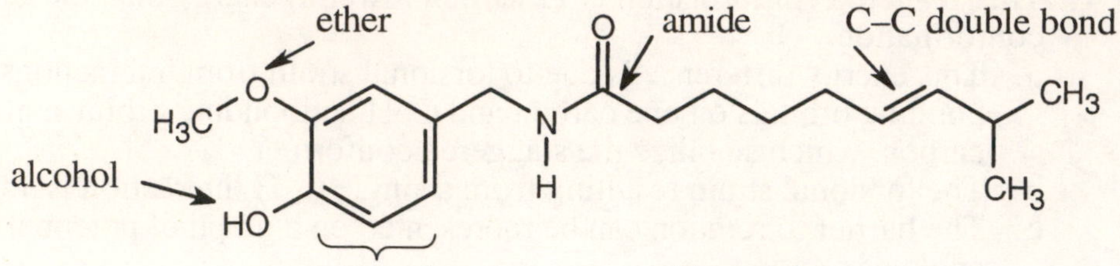

ether, amide, C–C double bond, alcohol, aromatic ring

Capsaicin

3.2

(a) CH_3OH Methanol

(b) Toluene

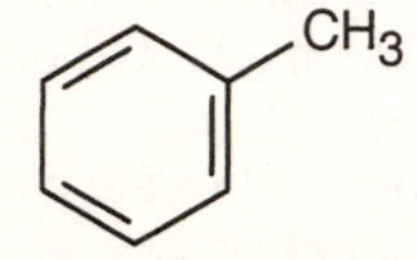

(c) CH_3COH (with C=O) Acetic acid

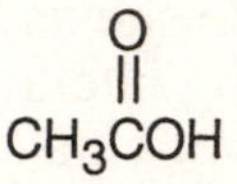

(d) CH_3NH_2 Methylamine

(e) $CH_3CCH_2NH_2$ (with C=O) Aminoacetone

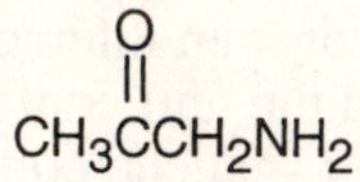

(f) Buta-1,3-diene

3.3

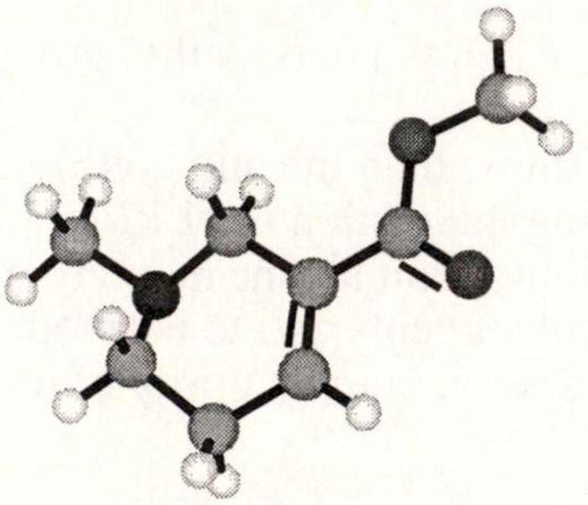

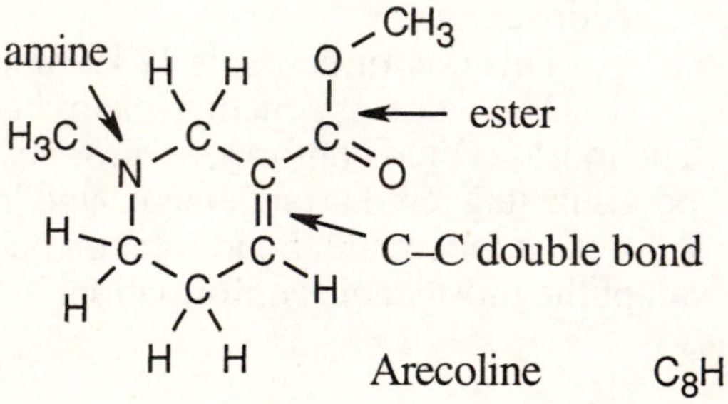

amine, ester, C–C double bond

Arecoline $C_8H_{13}NO_2$

3.4 **Strategy:** We know that carbon forms four bonds and hydrogen forms one bond. Thus, draw all possible six-carbon skeletons and add hydrogens so that all carbons have four bonds. To draw all possible skeletons in this problem: (1) Draw the six-carbon straight-chain skeleton. (2) Draw a five-carbon chain, identify the different types of carbon atoms on the chain, and add a –CH_3 group to each of the different types of carbons, generating two skeletons. (3) Repeat the process with the four-carbon chain to give rise to the last two skeletons. Add hydrogens to the remaining carbons to complete the structures.

Solution:

$CH_3CH_2CH_2CH_2CH_2CH_3$

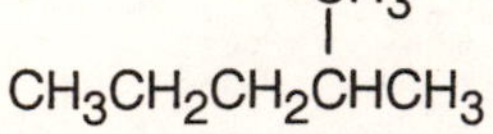

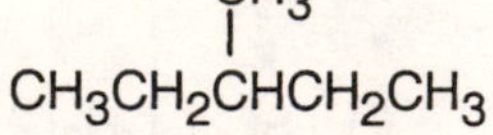

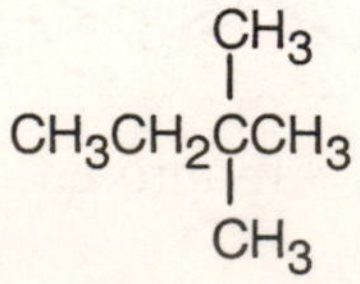

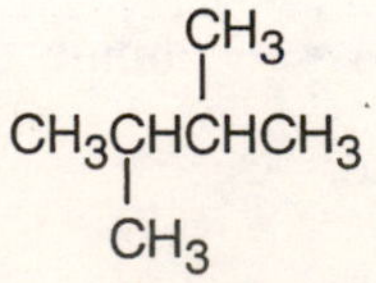

$CH_3CH_2CH_2CH(CH_3)CH_3$

$CH_3CH_2CH(CH_3)CH_2CH_3$

$CH_3CH_2C(CH_3)_2CH_3$

$CH_3CH(CH_3)CH(CH_3)CH_3$

3.5 (a) Nine isomeric esters of formula $C_5H_{10}O_2$ can be drawn.

$$\overset{O}{\overset{\|}{}}\quad CH_3CH_2CH_2\overset{O}{\overset{\|}{C}}OCH_3 \qquad CH_3\underset{CH_3}{\underset{|}{C}}H\overset{O}{\overset{\|}{C}}OCH_3 \qquad CH_3CH_2\overset{O}{\overset{\|}{C}}OCH_2CH_3$$

$$CH_3\overset{O}{\overset{\|}{C}}OCH_2CH_2CH_3 \qquad CH_3\overset{O}{\overset{\|}{C}}O\overset{CH_3}{\overset{|}{C}}HCH_3 \qquad H\overset{O}{\overset{\|}{C}}OCH_2CH_2CH_2CH_3$$

$$H\overset{O}{\overset{\|}{C}}O\overset{CH_3}{\overset{|}{C}}HCH_2CH_3 \qquad H\overset{O}{\overset{\|}{C}}OCH_2\overset{CH_3}{\overset{|}{C}}HCH_3 \qquad H\overset{O}{\overset{\|}{C}}O\underset{CH_3}{\underset{|}{\overset{CH_3}{\overset{|}{C}}}}CH_3$$

(b)

$$CH_3CH_2SSCH_2CH_3 \qquad CH_3SSCH_2CH_2CH_3 \qquad CH_3SS\underset{CH_3}{\underset{|}{C}}HCH_3$$

3.6 (a) Two alcohols have the formula C_3H_8O.

$$CH_3CH_2CH_2OH \quad \text{and} \quad CH_3\overset{OH}{\overset{|}{C}}HCH_3$$

(b) Four bromoalkanes have the formula C_4H_9Br.

$$CH_3CH_2CH_2CH_2Br \qquad CH_3CH_2\overset{Br}{\overset{|}{C}}HCH_3 \qquad CH_3\overset{CH_3}{\overset{|}{C}}HCH_2Br \qquad CH_3\underset{CH_3}{\underset{|}{\overset{Br}{\overset{|}{C}}}}CH_3$$

(c) Four thioesters have the formula C_4H_8OS.

$$CH_3CH_2\overset{O}{\overset{\|}{C}}SCH_3 \qquad CH_3CH_2S\overset{O}{\overset{\|}{C}}CH_3 \qquad H\overset{O}{\overset{\|}{C}}S\overset{CH_3}{\overset{|}{C}}HCH_3 \qquad H\overset{O}{\overset{\|}{C}}SCH_2CH_2CH_3$$

3.7

$$CH_3CH_2CH_2CH_2CH_2\text{–}\!\wr \qquad CH_3CH_2CH_2\underset{CH_3}{\underset{|}{C}}H\text{–}\!\wr \qquad CH_3CH_2\underset{CH_2CH_3}{\underset{|}{C}}H\text{–}\!\wr \qquad CH_3CH_2\underset{CH_3}{\underset{|}{C}}HCH_2\text{–}\!\wr$$

$$CH_3\underset{CH_3}{\underset{|}{C}}HCH_2CH_2\text{–}\!\wr \qquad CH_3CH_2\underset{CH_3}{\underset{|}{\overset{CH_3}{\overset{|}{C}}}}\text{–}\!\wr \qquad CH_3\overset{CH_3}{\overset{|}{C}}H\underset{CH_3}{\underset{|}{C}}H\text{–}\!\wr \qquad CH_3\underset{CH_3}{\underset{|}{\overset{CH_3}{\overset{|}{C}}}}CH_2\text{–}\!\wr$$

3.8

(a)

$$\begin{array}{c} p \\ CH_3 \\ | \\ CH_3CHCH_2CH_2CH_3 \\ p\ \ t\ \ s\ \ s\ \ p \end{array}$$

(b)

$$\begin{array}{c} p\ \ t\ \ p \\ CH_3CHCH_3 \\ | \\ CH_3CH_2CHCH_2CH_3 \\ p\ \ s\ \ t\ \ s\ \ p \end{array}$$

(c)

$$\begin{array}{c} p \qquad p \\ CH_3 \quad CH_3 \\ | \qquad | \\ CH_3CHCH_2-C-CH_3 \\ p\ \ t\ \ s \quad |q\ \ p \\ CH_3 \\ p \end{array}$$

p = primary; s = secondary; t = tertiary; q = quaternary

3.9

(a)

$$\begin{array}{c} p \\ CH_3 \\ | \\ CH_3CHCH_2CH_2CH_3 \\ p\ \ t\ \ s\ \ s\ \ p \end{array}$$

(b)

$$\begin{array}{c} p\ \ t\ \ p \\ CH_3CHCH_3 \\ | \\ CH_3CH_2CHCH_2CH_3 \\ p\ \ s\ \ t\ \ s\ \ p \end{array}$$

(c)

$$\begin{array}{c} p \qquad p \\ CH_3 \quad CH_3 \\ | \qquad | \\ CH_3CHCH_2-C-CH_3 \\ p\ \ t\ \ s \quad |\ \ p \\ CH_3 \\ p \end{array}$$

3.10

(a)

$$\begin{array}{c} CH_3 \\ t\ | \\ CH_3CHCHCH_3 \\ |\ t \\ CH_3 \end{array}$$

(b)

$$\begin{array}{c} \boxed{CH_3CHCH_3} \\ | \\ CH_3CH_2CHCH_2CH_3 \end{array}$$

(c)

$$\begin{array}{c} q \searrow CH_3 \\ | \\ CH_3CH_2CCH_3 \\ s \quad | \\ CH_3 \end{array}$$

3.11

(a)

$$CH_3CH_2CH_2CH_2CH_3$$

Pentane

$$\begin{array}{c} CH_3 \\ | \\ CH_3CH_2CHCH_3 \end{array}$$

2-Methylbutane

$$\begin{array}{c} CH_3 \\ | \\ CH_3CCH_3 \\ | \\ CH_3 \end{array}$$

2,2-Dimethylpropane

(b) **Strategy:**

Step 1: Find the longest continuous carbon chain and use it as the parent name. In (b), the longest chain is a hexane (boxed).
Step 2: Identify the substituents. In (b), both substituents are methyl groups.
Step 3: Number the substituents. In (b), numbering can start from either end of the carbon chain, and the methyl groups are in the 3- and 4- positions.
Step 4: Name the compound. Remember that the prefix *di-* must be used when two substituents are the same. The IUPAC name is 3,4-dimethylhexane.

Solution:

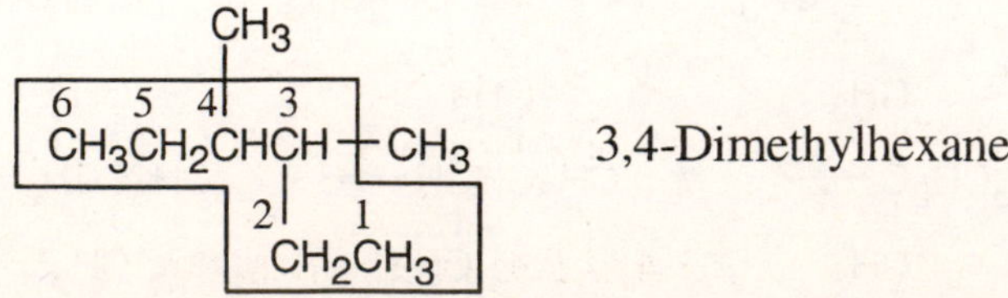

3,4-Dimethylhexane

(c)

$$\mathrm{CH_3CH(CH_3)CH_2CH(CH_3)CH_3}$$

2,4-Dimethylpentane

(d)

$$\mathrm{CH_3C(CH_3)_2CH_2CH_2CH(CH_3)CH_2CH_3}$$

2,2,5-Trimethylheptane

3.12 **Strategy:** When you are asked to draw the structure corresponding to a given name, draw the parent carbon chain, attach the specified groups to the proper carbons, and fill in the necessary hydrogens.

Solution:

(a)

$$\mathrm{CH_3CH_2CH_2CH_2CH_2CH(CH_3)CH(CH_3)CH_2CH_3}$$

3,4-Dimethylnonane

(b)

$$\mathrm{CH_3CH_2CH_2C(CH_3)_2—CH(CH_2CH_3)CH_2CH_3}$$

3-Ethyl-4,4-dimethylheptane

(c)

$$\mathrm{CH_3CH_2CH_2CH_2CH(CH_2CH_2CH_3)CH_2C(CH_3)_3}$$

2,2-Dimethyl-4-propyloctane

(d)

$$\mathrm{CH_3CH(CH_3)CH_2C(CH_3)_2CH_3}$$

2,2,4-Trimethylpentane

3.13

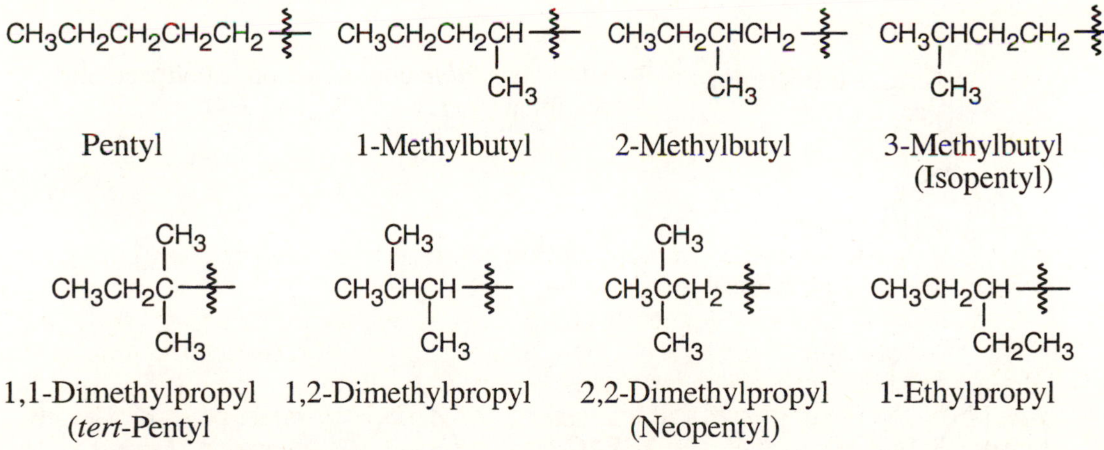

3.14

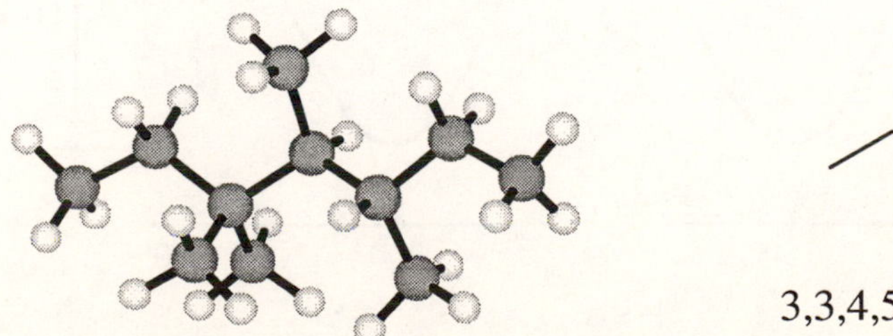

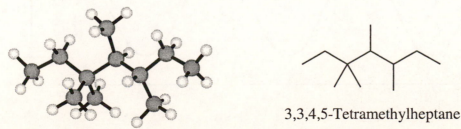

3,3,4,5-Tetramethylheptane

3.15

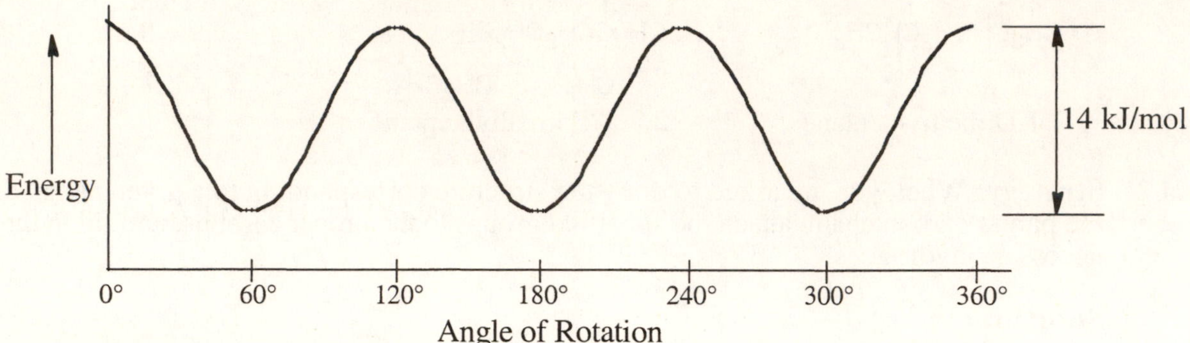

3.16

(a)

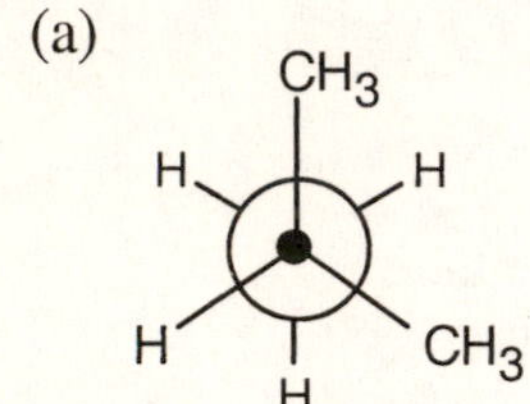

The *most stable* conformation is staggered and occurs at 60°, 180°, and 300°.

(b)

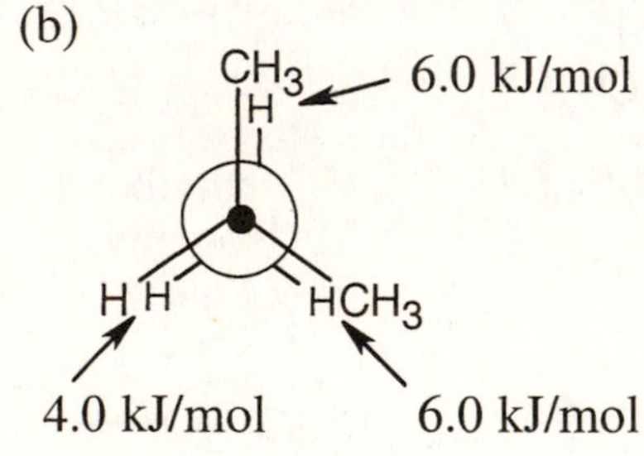

The *least stable* conformation is eclipsed and occurs at 0°, 120°, 240°, and 360°.

(c),(d)

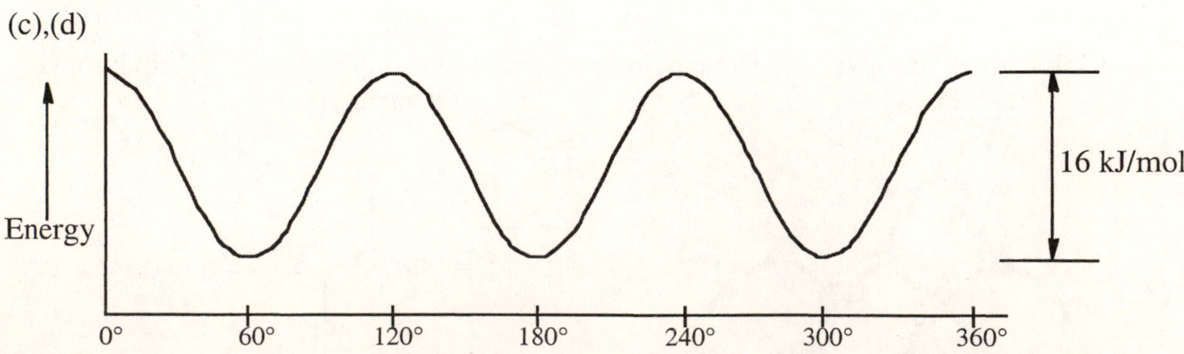

3.17

CH_3, H_3C, H, H, CH_3, CH_3

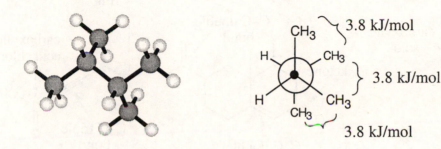

This conformation of 2,3-dimethylbutane is the most stable because it is staggered and has the fewest CH_3–CH_3 gauche interactions.

3.18 The conformation is a staggered conformation in which the hydrogens on carbons 2 and 3 are 60° apart. Draw the Newman projection.

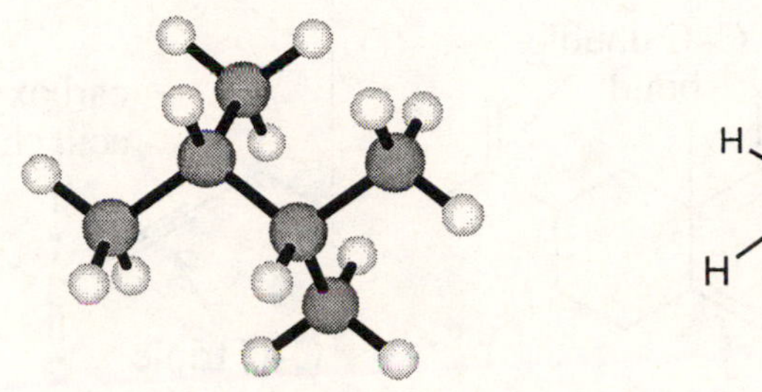

CH_3 3.8 kJ/mol
H CH_3 3.8 kJ/mol
H CH_3
CH_3 3.8 kJ/mol

The Newman projection shows three gauche interactions, each of which has an energy cost of 3.8 kJ/mol. The total strain energy is 11.4 kJ/mol.

Visualizing Chemistry

3.19

(a)

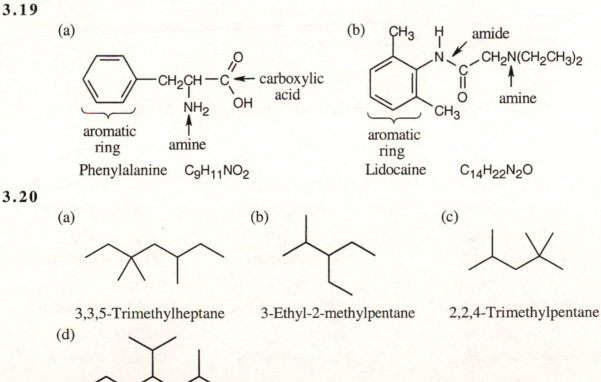

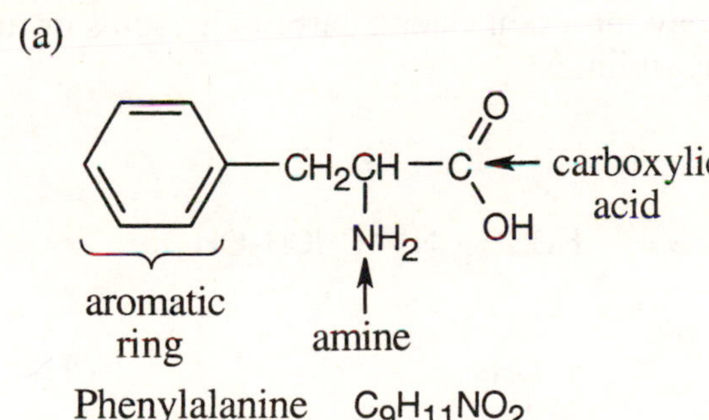

Phenylalanine $C_9H_{11}NO_2$

(b)

CH_3 H amide
$CH_2N(CH_2CH_3)_2$
N–C=O
amine
CH_3
aromatic ring

Lidocaine $C_{14}H_{22}N_2O$

3.20

(a) (b) (c)

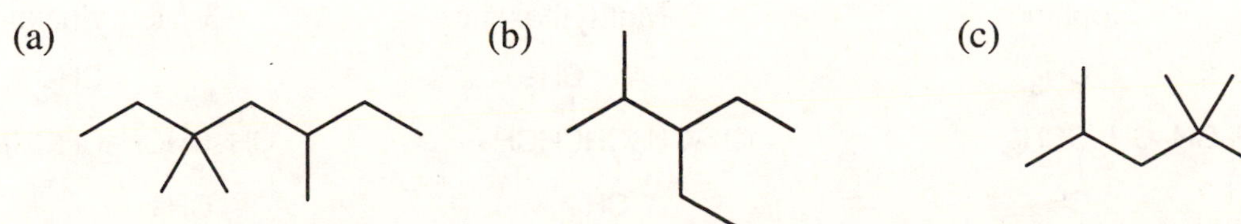

3,3,5-Trimethylheptane 3-Ethyl-2-methylpentane 2,2,4-Trimethylpentane

(d)

4-Isopropyl-2-methylheptane

Additional Problems

3.21

(a) aromatic ring; CH_2OH — alcohol; $NHCH_3$ — amine

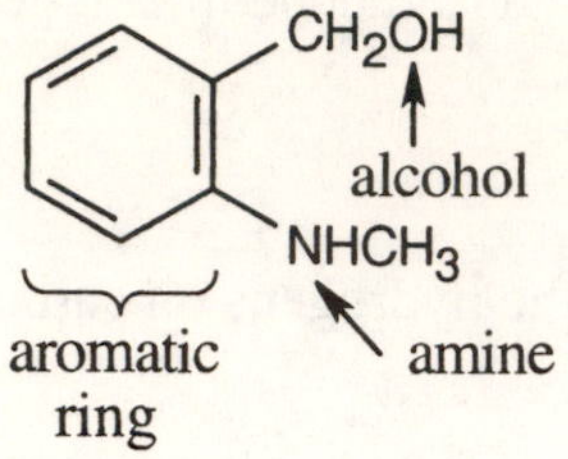

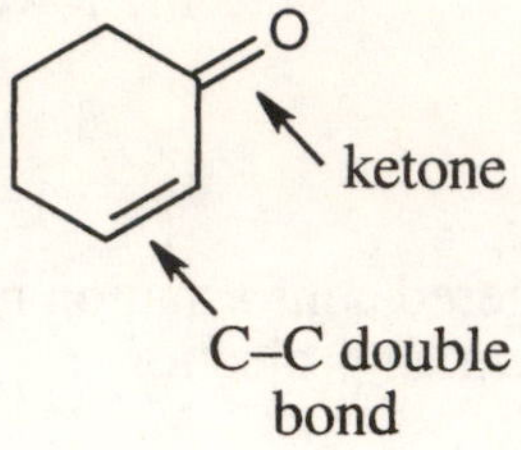

(b) ketone; C–C double bond

(c) aromatic ring; amide

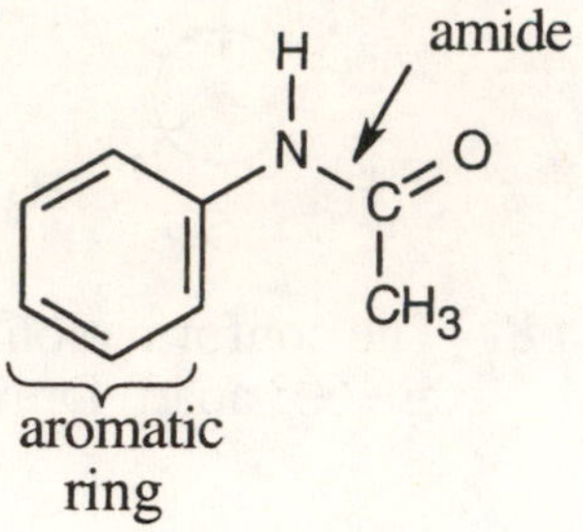

(d) $CH_3CH(NH_2)COOH$ — carboxylic acid; amine

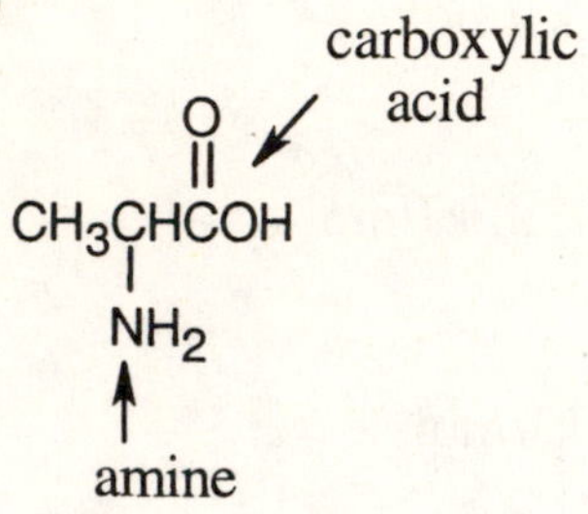

(e) ketone; C–C double bond; C–C double bond

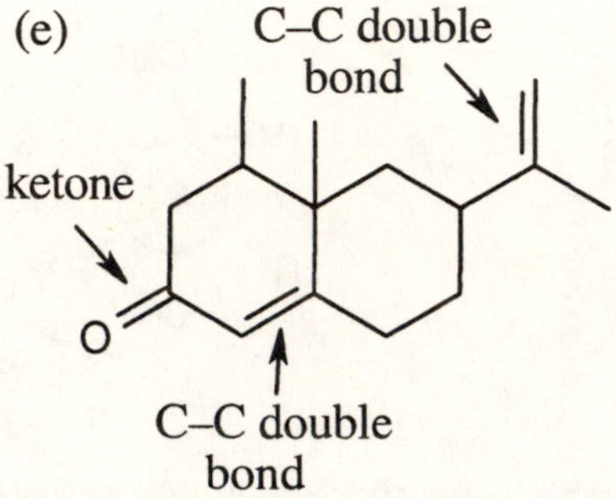

(f) carboxylic acid chloride; C–C triple bond

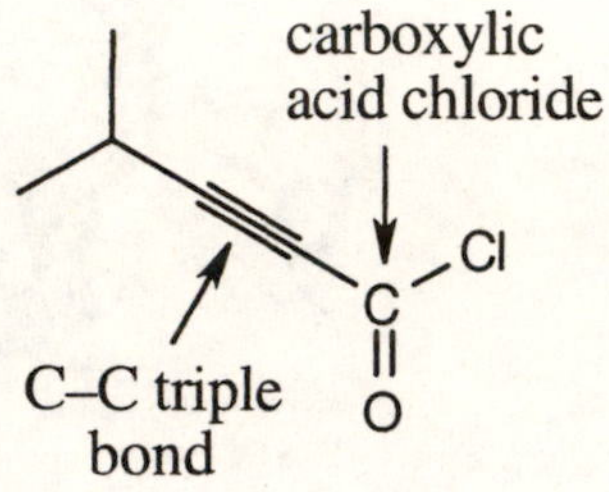

3.22 (a) Eighteen isomers have the formula C_8H_{18}. Three are pictured.

$CH_3CH_2CH_2CH_2CH_2CH_2CH_2CH_3$ $\quad$ $CH_3CH_2CH_2CH(CH_2CH_3)CH_2CH_3$ $\quad$ $CH_3C(CH_3)_2—C(CH_3)_2CH_3$

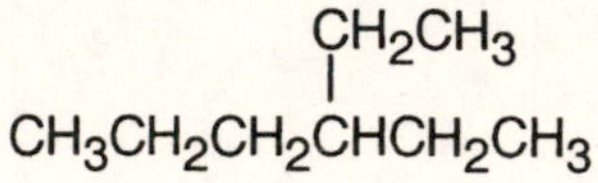

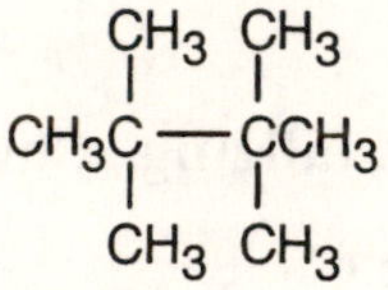

(b) Structures with the formula $C_4H_8O_2$ may represent esters, carboxylic acids or many other complicated molecules. Three possibilities:

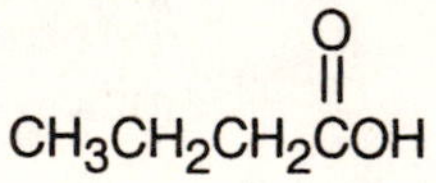

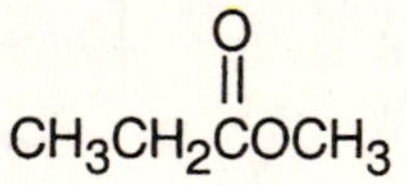

$CH_3CH_2CH_2COH$ (C=O) $\quad$ $CH_3CH_2COCH_3$ (C=O) $\quad$ $HOCH_2CH{=}CHCH_2OH$

3.23

$CH_3CH_2CH_2CH_2CH_2CH_2CH_3$ — Heptane

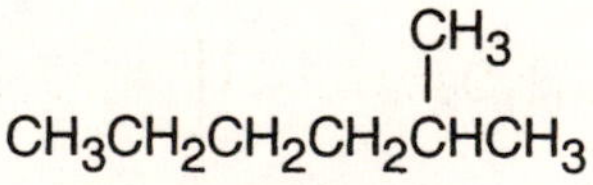

$CH_3CH_2CH_2CH_2CH(CH_3)CH_3$ — 2-Methylhexane

$CH_3CH_2CH_2CH(CH_3)CH_2CH_3$ — 3-Methylhexane

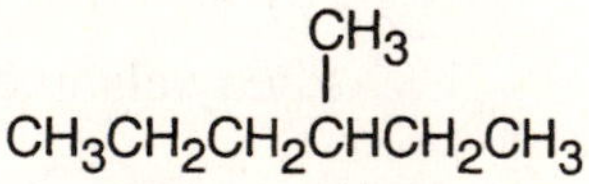

$CH_3CH_2CH_2C(CH_3)_2CH_3$ — 2,2-Dimethylpentane

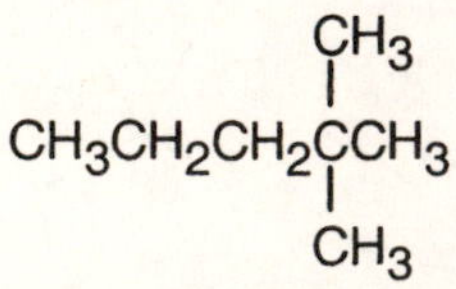

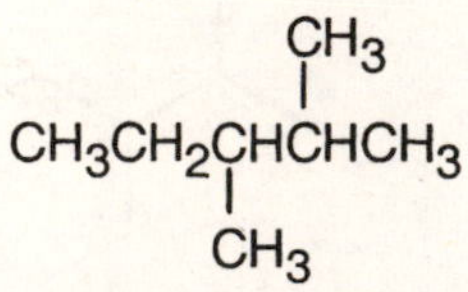

$CH_3CH_2CH(CH_3)CH(CH_3)CH_3$ — 2,3-Dimethylpentane

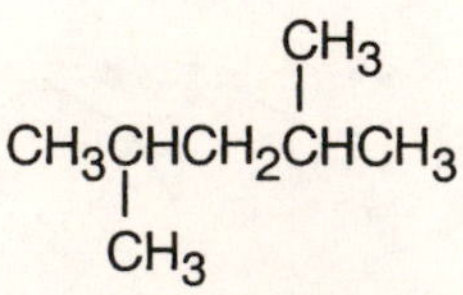

$CH_3CH(CH_3)CH_2CH(CH_3)CH_3$ — 2,4-Dimethylpentane

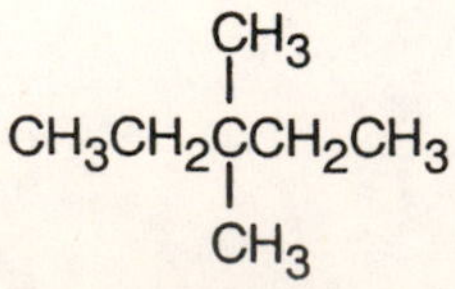

$CH_3CH_2C(CH_3)_2CH_2CH_3$ — 3,3-Dimethylpentane

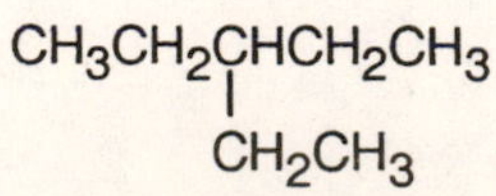

$CH_3CH_2CH(CH_2CH_3)CH_2CH_3$ — 3-Ethylpentane

$CH_3CH(CH_3)—C(CH_3)_2CH_3$ — 2,2,3-Trimethylbutane

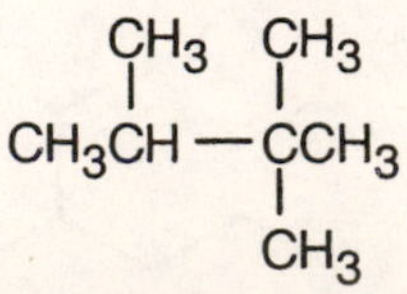

3.24

(a)

$CH_3CH(Br)CH(CH_3)CH_3$ — same

$CH_3C(CH_3)HCH(Br)CH_3$ — same

$CH_3CH(Br)CH(CH_3)CH_3$ — same

(b)

HO, OH (1,2-benzenediol) — same

HO, HO (1,2-benzenediol) — same

HO, OH (1,3-benzenediol) — different

(c)

$CH_3CH_2CH(\overset{1}{C}H_2OH)CH_2\overset{5}{C}H(CH_3)CH_3$ — different

$HO\overset{1}{C}H_2CH(CH_3)CH_2CH(CH_2\overset{6}{C}H_3)CH_3$ — same

$\overset{6}{C}H_3CH_2CH(CH_3)CH_2CH(CH_3)\overset{1}{C}H_2OH$ — same

Give the number "1" to the carbon bonded to –OH, and count to find the longest chain containing the –OH group.

3.25 The isomers may be either alcohols or ethers.

$CH_3CH_2CH_2CH_2—OH$ $CH_3CH_2CH(OH)CH_3$ $CH_3CH(CH_3)CH_2—OH$ $CH_3C(CH_3)(OH)CH_3$

$CH_3CH_2CH_2—OCH_3$ $CH_3CH(OCH_3)CH_3$ $CH_3CH_2—OCH_2CH_3$

3.26 Different answers to this problem and to Problem 3.27 are acceptable.

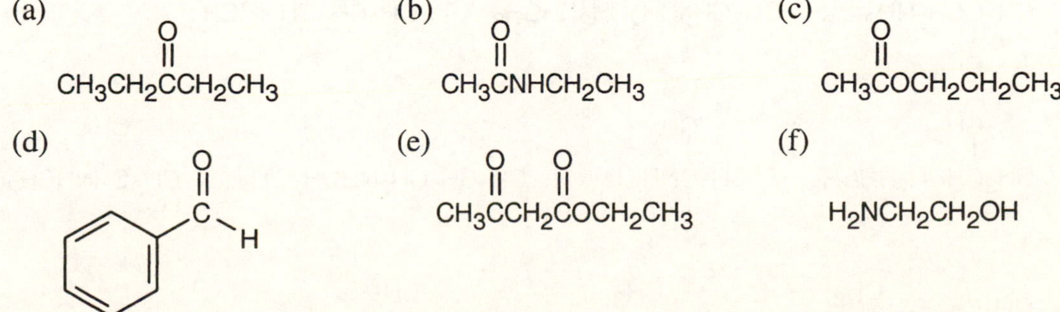

3.27

(a)

$$\overset{O}{\parallel}$$

C_4H_8O: $CH_3CCH_2CH_3$

(b)

C_5H_9N: $CH_3CH_2CH_2CH_2C{\equiv}N$

(c)

$C_4H_6O_2$: $HCCH_2CH_2CH$ (O, ‖ above each terminal C)

(d)

$C_6H_{11}Br$: $CH_3CH_2CH{=}CHCH_2CH_2Br$

(e)

C_6H_{14}: $CH_3CH_2CH_2CH_2CH_2CH_3$

(f)

C_6H_{12}: [hexagon]

(g)

C_5H_8: $CH_3CH{=}CHCH{=}CH_2$

(h)

C_5H_8O: $H_2C{=}CHCCH_2CH_3$ (O, ‖ above the third C)

3.28 First, draw all straight-chain isomers. Then proceed to the simplest branched structure.

(a) There are four alcohol isomers with the formula $C_4H_{10}O$.

$CH_3CH_2CH_2CH_2OH$

OH
$CH_3CH_2CHCH_3$

CH_3
CH_3CHCH_2OH

OH
CH_3CCH_3
CH_3

(b) There are 17 isomers of $C_5H_{13}N$. Nitrogen can be bonded to one, two or three alkyl groups.

$CH_3CH_2CH_2CH_2CH_2NH_2$

NH_2
$CH_3CH_2CH_2CHCH_3$

NH_2
$CH_3CH_2CHCH_2CH_3$

$CH_3CH_2CHCH_2NH_2$
CH_3

NH_2
$CH_3CH_2CCH_3$
CH_3

NH_2
$CH_3CHCHCH_3$
CH_3

$H_2NCH_2CH_2CHCH_3$
CH_3

CH_3
$CH_3CCH_2NH_2$
CH_3

$CH_3CH_2CH_2CH_2NHCH_3$

CH_3
$CH_3CH_2CHNHCH_3$

CH_3
$CH_3CHCH_2NHCH_3$

CH_3
CH_3CNHCH_3
CH_3

$CH_3CH_2CH_2NHCH_2CH_3$

CH_3
$CH_3CHNHCH_2CH_3$

CH_3
$CH_3CH_2CH_2NCH_3$

CH_3
CH_3CHNCH_3
CH_3

CH_3
$CH_3CH_2NCH_2CH_3$

(c) There are 3 ketone isomers with the formula $C_5H_{10}O$.

$$CH_3CH_2CH_2\overset{O}{\overset{\|}{C}}CH_3 \qquad CH_3CH_2\overset{O}{\overset{\|}{C}}CH_2CH_3 \qquad CH_3\underset{CH_3}{\underset{|}{C}}H\overset{O}{\overset{\|}{C}}CH_3$$

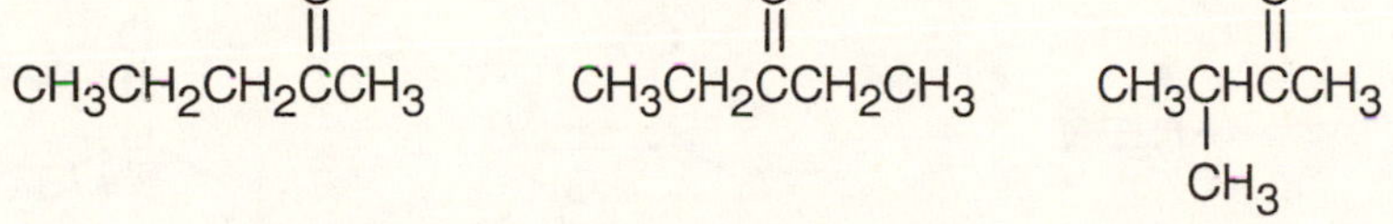

(d) There are 4 isomeric aldehydes with the formula $C_5H_{10}O$. Remember that the aldehyde functional group can occur only at the end of a chain.

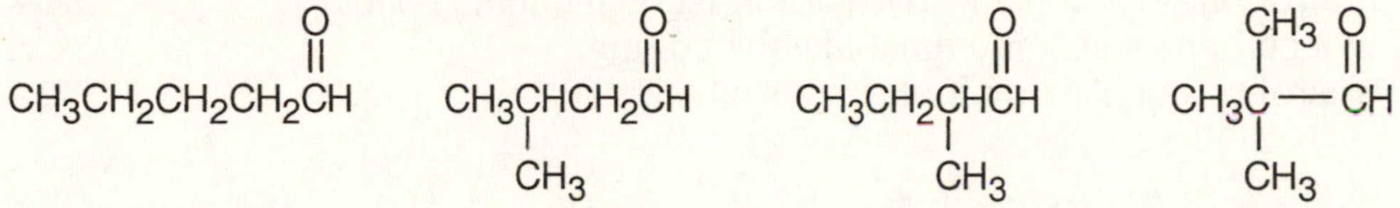

$$CH_3CH_2CH_2CH_2\overset{O}{\overset{\|}{C}}H \qquad CH_3\underset{CH_3}{\underset{|}{C}}HCH_2\overset{O}{\overset{\|}{C}}H \qquad CH_3CH_2\underset{CH_3}{\underset{|}{C}}H\overset{O}{\overset{\|}{C}}H \qquad CH_3\overset{CH_3}{\overset{|}{\underset{CH_3}{\underset{|}{C}}}}-\overset{O}{\overset{\|}{C}}H$$

(e) There are 4 esters with the formula $C_4H_8O_2$.

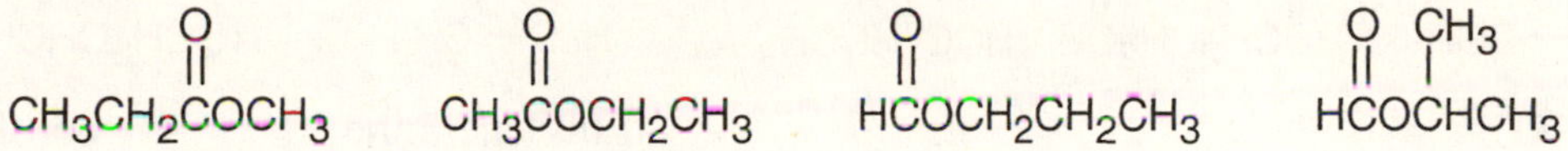

$$CH_3CH_2\overset{O}{\overset{\|}{C}}OCH_3 \qquad CH_3\overset{O}{\overset{\|}{C}}OCH_2CH_3 \qquad H\overset{O}{\overset{\|}{C}}OCH_2CH_2CH_3 \qquad H\overset{O}{\overset{\|}{C}}O\overset{CH_3}{\overset{|}{C}}HCH_3$$

(f) There are 3 ethers with the formula $C_4H_{10}O$.

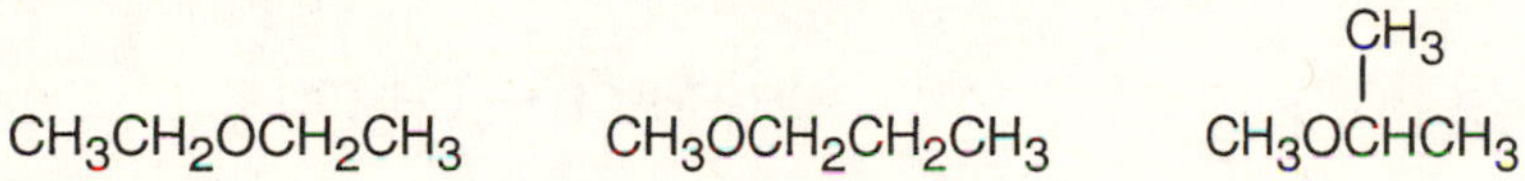

$$CH_3CH_2OCH_2CH_3 \qquad CH_3OCH_2CH_2CH_3 \qquad CH_3O\overset{CH_3}{\overset{|}{C}}HCH_3$$

3.29

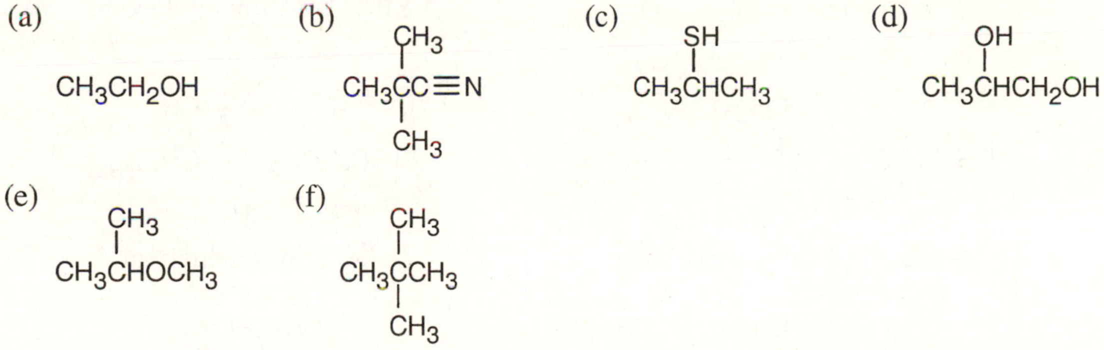

(a) CH_3CH_2OH

(b) $CH_3\overset{CH_3}{\overset{|}{\underset{CH_3}{\underset{|}{C}}}}C\equiv N$

(c) $CH_3\overset{SH}{\overset{|}{C}}HCH_3$

(d) $CH_3\overset{OH}{\overset{|}{C}}HCH_2OH$

(e) $CH_3\overset{CH_3}{\overset{|}{C}}HOCH_3$

(f) $CH_3\overset{CH_3}{\overset{|}{\underset{CH_3}{\underset{|}{C}}}}CH_3$

3.30

$$CH_3CH_2CH_2CH_2CH_2Br \qquad CH_3CH_2CH_2\overset{Br}{\overset{|}{C}}HCH_3 \qquad CH_3CH_2\overset{Br}{\overset{|}{C}}HCH_2CH_3$$

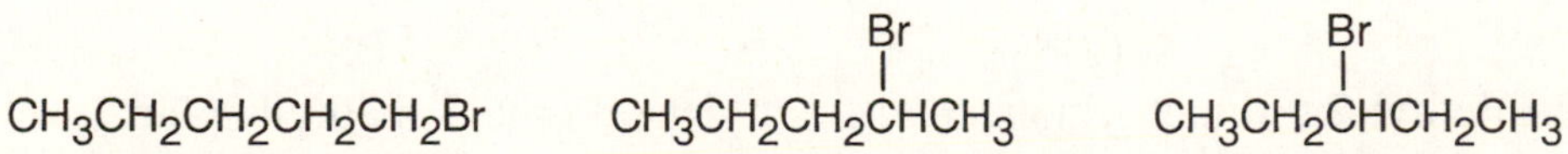

1-Bromopentane 2-Bromopentane 3-Bromopentane

3.31

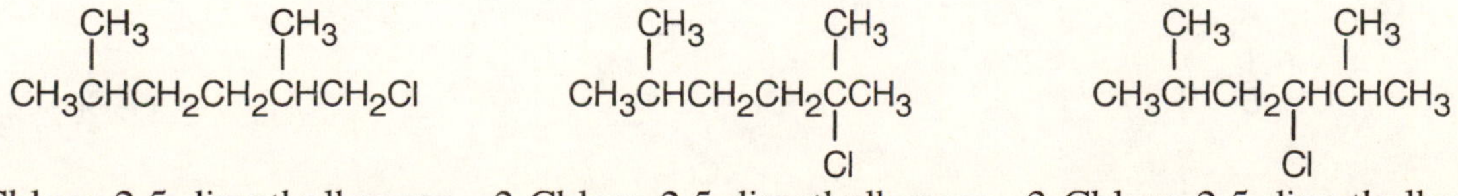

$$CH_3\overset{CH_3}{\overset{|}{C}}HCH_2CH_2\overset{CH_3}{\overset{|}{C}}HCH_2Cl \qquad CH_3\overset{CH_3}{\overset{|}{C}}HCH_2CH_2\overset{CH_3}{\overset{|}{\underset{Cl}{\underset{|}{C}}}}CH_3 \qquad CH_3\overset{CH_3}{\overset{|}{C}}HCH_2\underset{Cl}{\underset{|}{C}}H\overset{CH_3}{\overset{|}{C}}HCH_3$$

1-Chloro-2,5-dimethylhexane 2-Chloro-2,5-dimethylhexane 3-Chloro-2,5-dimethylhexane

3.32

(a) $\overset{O}{\overset{\|}{C}}$ ← sp^2 (ketone carbonyl carbon)

(b) sp → $—C{\equiv}N$

(c) $\overset{O}{\overset{\|}{C}}$ ← sp^2, with OH

(d) $\overset{O}{\overset{\|}{C}}$ ← sp^2, with S

3.33 (a) Although it is stated that biacetyl contains no rings or carbon–carbon double bonds, it is obvious from the formula for biacetyl that some sort of multiple bond must be present. The structure for biacetyl contains two carbon-oxygen double bonds.
(b) Ethyleneimine contains a three-membered ring.
(c) Glycerol contains no multiple bonds or rings.

(a) $CH_3C(=O)—C(=O)CH_3$
Biacetyl

also possible: ($HC(=O)CH_2CH_2C(=O)H$ $HC(=O)CH_2C(=O)CH_3$)

(b) $H_2C—CH_2$ ring with N–H
Ethyleneimine

(c) $HOCH_2CH(OH)CH_2OH$
Glycerol

3.34

(a) $CH_3CH_2CH_2CH_2CH_2CH(CH_3)CH_3$
2-Methylheptane

(b) $CH_3CH_2CH(CH_2CH_3)—CH_2C(CH_3)_2CH_3$
4-Ethyl-2,2-dimethylhexane

(c) $CH_3CH_2CH_2CH_2C(CH_3)(CH_2CH_3)—CH(CH_3)CH_2CH_3$
4-Ethyl-3,4-dimethyloctane

(d) $CH_3CH_2CH_2C(CH_3)_2CH_2CH(CH_3)CH_3$
2,4,4-Trimethylheptane

(e) $CH_3CH_2CH_2CH_2CH(CH_3)CH_2C(CH_2CH_3)_2—CH(CH_3)CH_3$
3,3-Diethyl-2,5-dimethylnonane

(f) $CH_3CH_2CH_2CH(CH(CH_3)CH_3)CH(CH_3)CH_2CH_3$
4-Isopropyl-3-methylheptane

3.35

(a) $CH_3CH(CH_3)CH_3$
2-Methylpropane

(b) $CH_3C(CH_3)_2CH_3$
2,2-Dimethylpropane

(c) $CH_3CH_2CH_2CH_2CH_2CH_3$
Hexane

3.36

(a)

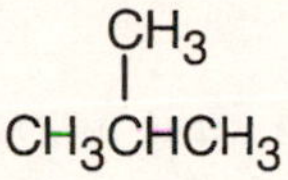

$$\begin{array}{c} CH_3 \\ | \\ CH_3CHCH_3 \end{array}$$

2-Methylpropane

(b)

CH_3CH_3

Ethane

3.37

(a) $CH_3CH_2CH_2CH_2CH_2Br$

(b) Cyclobutyl with CH_2OCH_3 substituent

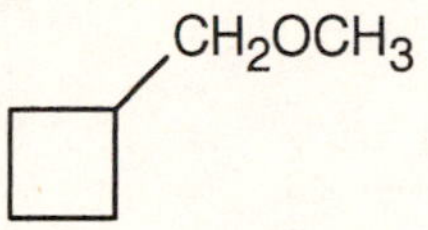

(c)

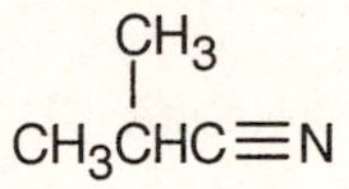

$$\begin{array}{c} CH_3 \\ | \\ CH_3CHC{\equiv}N \end{array}$$

(d) Cyclopentyl with CH_2OH substituent

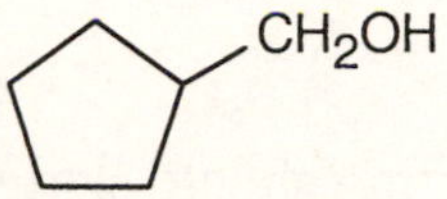

(e) There are no aldehyde isomers. However, the structure below is a ketone isomer.

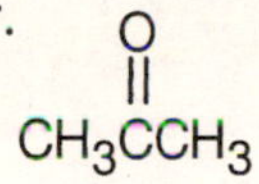

$$\begin{array}{c} O \\ \| \\ CH_3CCH_3 \end{array}$$

(f) Benzene ring with COOH and CH_3 (ortho)

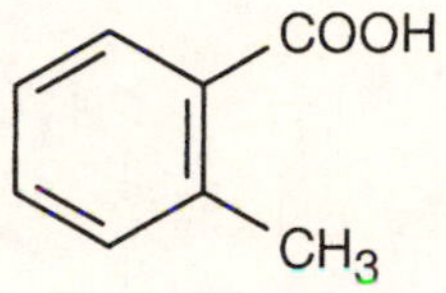

3.38

(a)

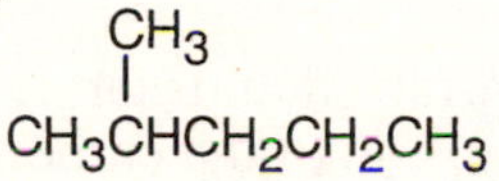

$$\begin{array}{c} CH_3 \\ | \\ CH_3CHCH_2CH_2CH_3 \end{array}$$

2-Methylpentane

(b)

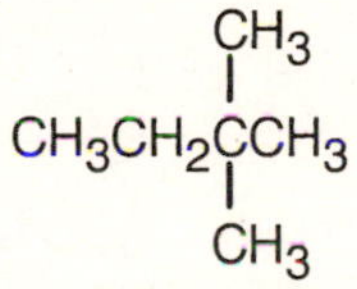

$$\begin{array}{c} CH_3 \\ | \\ CH_3CH_2CCH_3 \\ | \\ CH_3 \end{array}$$

2,2-Dimethylbutane

(c)

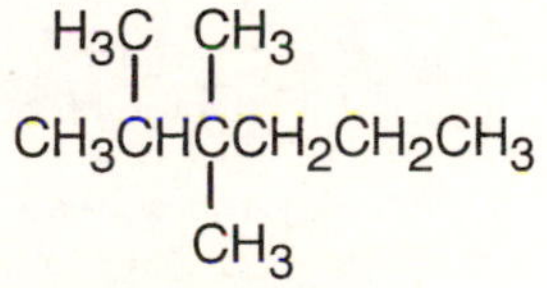

$$\begin{array}{c} H_3C\ \ CH_3 \\ |\ \ \ | \\ CH_3CHCCH_2CH_2CH_3 \\ | \\ CH_3 \end{array}$$

2,3,3-Trimethylhexane

(d)

$$\begin{array}{c} CH_2CH_3 \quad CH_3 \\ | \qquad\qquad | \\ CH_3CH_2CHCH_2CH_2CHCH_3 \end{array}$$

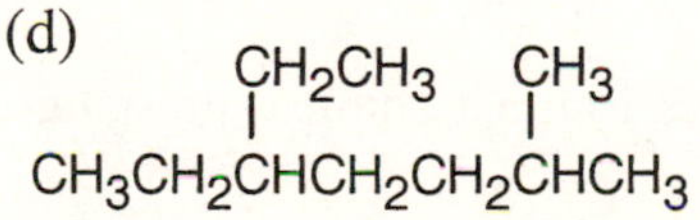

5-Ethyl-2-methylheptane

(e)

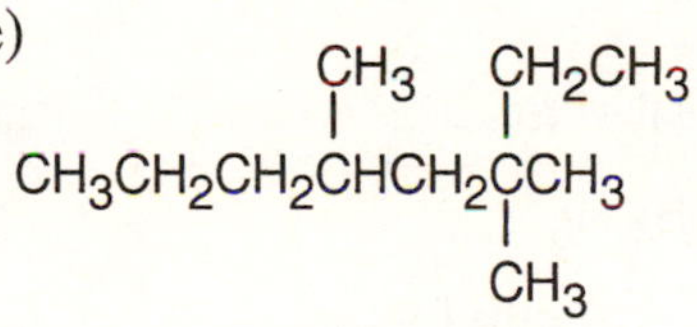

$$\begin{array}{c} CH_3 \quad CH_2CH_3 \\ | \qquad | \\ CH_3CH_2CH_2CHCH_2CCH_3 \\ | \\ CH_3 \end{array}$$

3,3,5-Trimethyloctane

(f)

$$\begin{array}{c} H_3C \quad CH_3 \\ | \qquad | \\ CH_3C{-}CCH_2CH_2CH_3 \\ | \qquad | \\ H_3C \quad CH_3 \end{array}$$

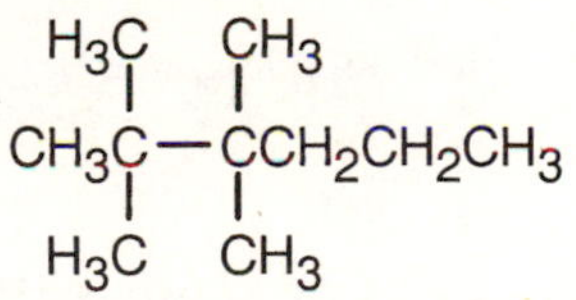

2,2,3,3-Tetramethylhexane

3.39

$CH_3CH_2CH_2CH_2CH_2CH_3$

Hexane

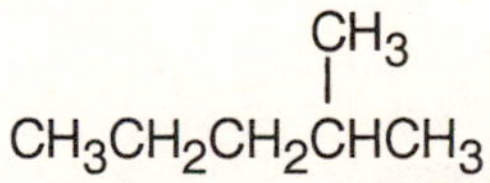

$$\begin{array}{c} CH_3 \\ | \\ CH_3CH_2CH_2CHCH_3 \end{array}$$

2-Methylpentane

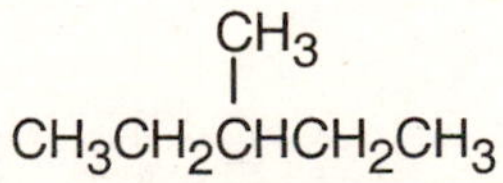

$$\begin{array}{c} CH_3 \\ | \\ CH_3CH_2CHCH_2CH_3 \end{array}$$

3-Methylpentane

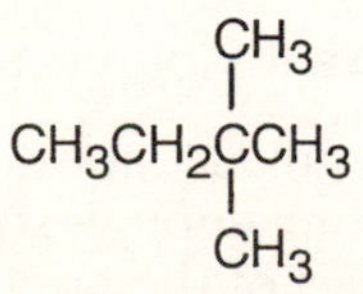

$$\begin{array}{c} CH_3 \\ | \\ CH_3CH_2CCH_3 \\ | \\ CH_3 \end{array}$$

2,2-Dimethylbutane

$$\begin{array}{c} CH_3 \\ | \\ CH_3CHCHCH_3 \\ | \\ CH_3 \end{array}$$

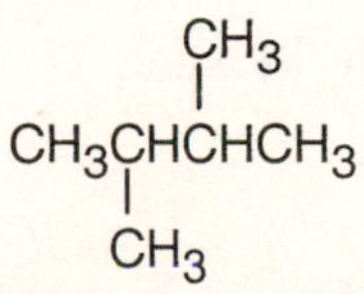

2,3-Dimethylbutane

3.40 *Structure and Correct Name* — *Error*

(a)

$$\begin{array}{l} \quad CH_2CH_3 \quad\quad CH_3 \\ \quad | \;\; 8 \quad\quad\quad\;\; | \\ CH_3CHCH_2CH_2CH_2CCH_3 \\ \quad\quad\quad\quad\quad\quad\quad | \; 1 \\ \quad\quad\quad\quad\quad\quad\quad CH_3 \end{array}$$

Correct name: 2,2,6-Trimethyloctane

The longest chain is an octane.

(b)

$$\begin{array}{l} \quad CH_3 \\ \quad | \\ CH_3CHCHCH_2CH_2CH_3 \\ 1 \quad\quad | \quad\quad\quad\quad 6 \\ \quad\quad CH_2CH_3 \end{array}$$

Correct name: 3-Ethyl-2-methylhexane

The longest chain is a hexane. Numbering should start from the opposite end of the carbon chain, nearer the first branch.

(c)

$$\begin{array}{l} \quad\quad\quad CH_3 \\ \quad\quad\quad | \\ CH_3CH_2C{-}CHCH_2CH_3 \\ 1 \quad\quad\;\; | \quad\; | \quad\quad\;\; 6 \\ \quad\quad CH_3 \; CH_2CH_3 \end{array}$$

Correct name: 4-Ethyl-3,3-dimethylhexane

Numbering should start from the opposite end of the carbon chain. See step 2(b) in Section 3.4.

(d)

$$\begin{array}{l} \quad\quad\quad CH_3 \; CH_3 \\ \quad\quad\quad | \quad\quad | \\ CH_3CH_2CH{-}CCH_2CH_2CH_2CH_3 \\ 1 \quad\quad\quad\quad\quad | \quad\quad\quad\quad\quad 8 \\ \quad\quad\quad\quad\quad CH_3 \end{array}$$

Correct name: 3,4,4-Trimethyloctane

Numbering should start from the opposite end of the carbon chain.

(e)

$$\begin{array}{l} \quad\quad\quad\quad\quad CH_3 \\ \quad\quad\quad\quad\quad | \\ CH_3CH_2CH_2CHCH_2CHCH_3 \\ 8 \quad\quad\quad\quad\quad\quad\quad\quad | \\ \quad\quad\quad\quad\quad\quad CH_3CHCH_3 \\ \quad\quad\quad\quad\quad\quad\quad\quad\quad 1 \end{array}$$

Correct name: 2,3,5-Trimethyloctane

The longest chain is an octane.

3.41

(a)

$$\begin{array}{l} \quad CH_3 \; CH_2CH_3 \\ \quad | \quad\quad | \\ CH_3CCH_2CCH_2CH_3 \\ \quad | \quad\quad | \\ \quad CH_3 \; CH_2CH_3 \end{array}$$

4,4-Diethyl-2,2-dimethylhexane

(b)

$$\begin{array}{l} \quad\quad\quad\quad\quad\quad\quad\quad\quad\quad\quad CH_3 \\ \quad\quad\quad\quad\quad\quad\quad\quad\quad\quad\quad | \\ \quad\quad\quad\quad\quad\quad\quad CH_2CH_2CHCH_3 \\ \quad\quad\quad\quad\quad\quad\quad | \\ CH_3CH_2CH_2CH_2CH_2CHCH_2CH_2CH_2CH_2CH_3 \end{array}$$

6-(3-Methylbutyl)-undecane

Remember that you must choose an alkane whose principal chain is long enough so that the substituent does not become part of the principal chain.

3.42

(a), (b)

CH_3
$CH_3CH_2-CHCH_3$
↑
C2–C3 bond

2-Methylbutane

3.8 kJ/mol CH_3 H_3C H H H CH_3

Most stable conformation

CH_3 ← 6.0 kJ/mol H
H H H_3C CH_3
4.0 kJ/mol 11.0 kJ/mol

Least stable comformation

The energy difference between the two conformations is (11.0 + 6.0 + 4.0) kJ/mol – 3.8 kJ/mol = 17.0 kJ/mol.

(c) Consider the least stable conformation to be at zero degrees. Keeping the front of the projection unchanged, rotate the back by 60° to obtain each conformation.

CH_3 H H H CH_3 CH_3 3.8 kJ/mol

CH_3 H ← 6.0 kJ/mol H CH_3 H CH_3 6.0 kJ/mol 6.0 kJ/mol

3.8 kJ/mol CH_3 H_3C H H H CH_3

<u>at 60°</u>: energy = 3.8 kJ/mol <u>at 120°</u>: energy = 18.0 kJ/mol <u>at 180°</u>: energy = 3.8 kJ/mol

CH_3 CH_3 ← 11.0 kJ/mol H H H CH_3 4.0 kJ/mol 6.0 kJ/mol

CH_3 3.8 kJ/mol H CH_3 3.8 kJ/mol H CH_3 H

<u>at 240°</u>: energy = 21.0 kJ/mol <u>at 300°</u>: energy = 7.6 kJ/mol

Use the lowest energy conformation as the energy minimum. The highest energy conformation is 17.2 kJ/mol higher in energy than the lowest energy conformation.

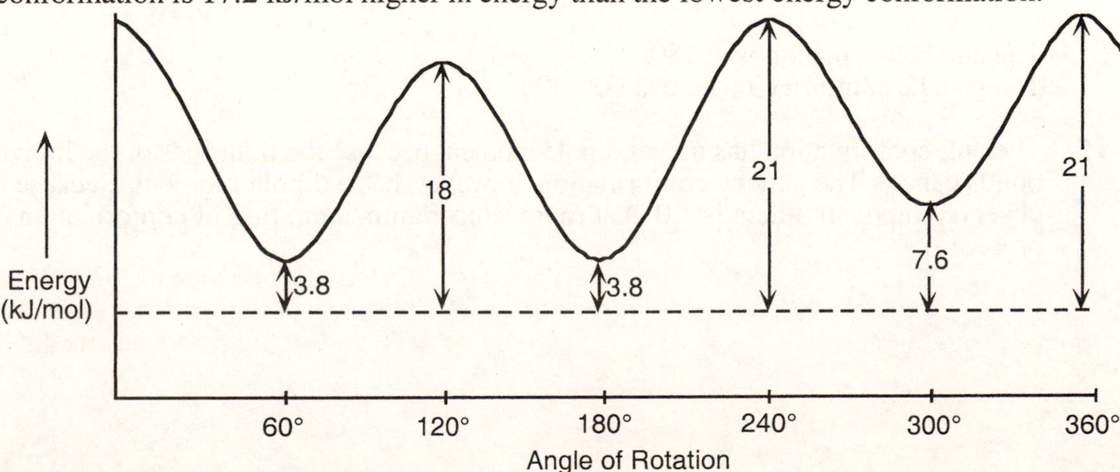

3.43

2 CH_3–CH_3 gauche
= 2 x 3.8 kJ/mol
= 7.6 kJ/mol

3 CH_3–CH_3 gauche
= 3 x 3.8 kJ/mol
= 11.4 kJ/mol

3 CH_3–CH_3 gauche
= 3 x 3.8 kJ/mol
= 11.4 kJ/mol

3.44 Since we are not told the values of the interactions for 1,2–dibromoethane, the diagram can only be qualitative.

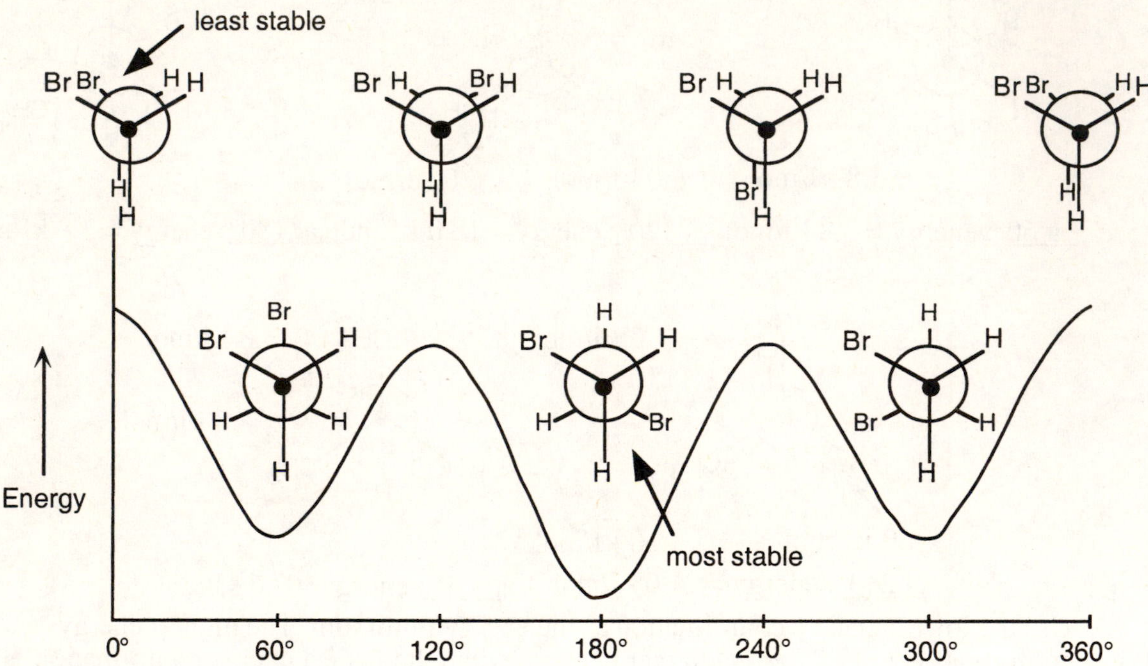

The anti conformation is at 180°.
The gauche conformations are at 60°, 300°.

3.45 The anti conformation has no net dipole moment because the polarities of the individual bonds cancel. The gauche conformation, however, has a dipole moment. Because the observed dipole moment is 1.0 D at room temperature, a mixture of conformations must be present.

3.46 The highest energy conformation of bromoethane has a strain energy of 15 kJ/mol. Because this includes two H–H eclipsing interactions of 4.0 kJ/mol each, the value of an H–Br eclipsing interaction is 15 kJ/mol – 2(4.0 kJ/mol) = 7.0 kJ/mol.

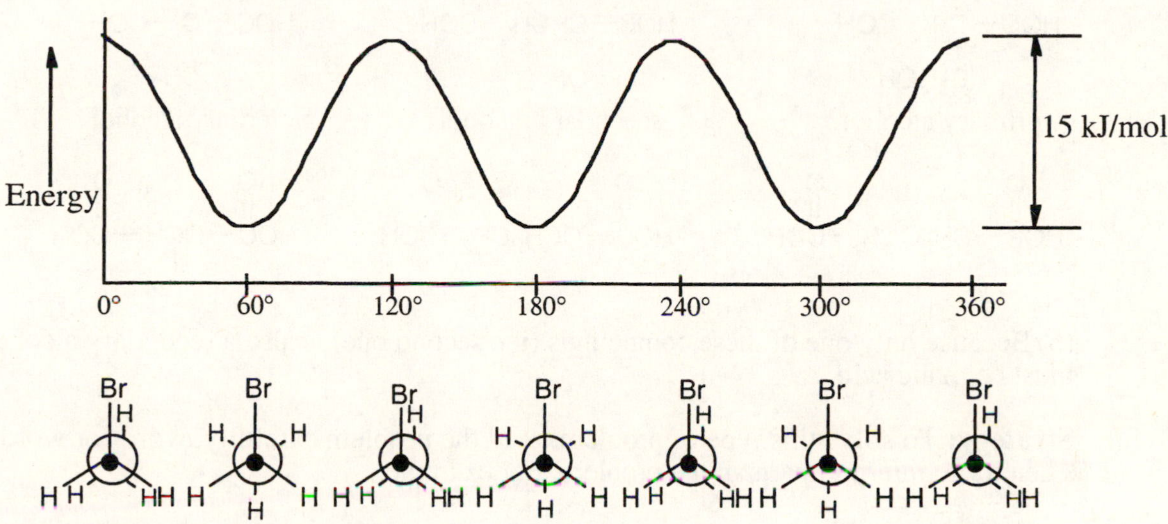

3.47 The best way to draw pentane is to make a model and to copy it onto the page. A model shows the relationship among atoms, and its drawing shows how these relationships appear in two dimensions. From your model, you should be able to see that all atoms are staggered in the drawing.

3.48

3.49 **Strategy:** (a) Because malic acid has two $-CO_2H$ groups, the formula for the rest of the molecule is C_2H_4O. Possible structures for malic acid are:

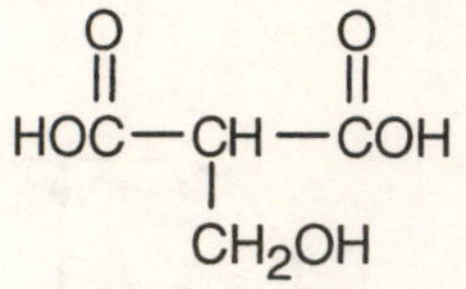

$HOOC-CH(CH_2OH)-COOH$ primary alcohol

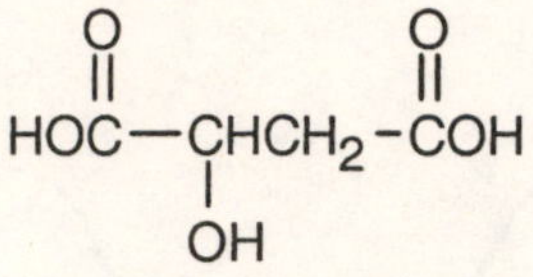

$HOOC-CH(OH)CH_2-COOH$ secondary alcohol

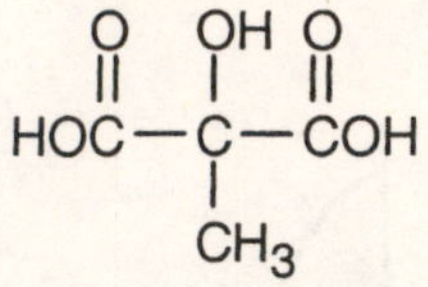

$HOOC-C(OH)(CH_3)-COOH$ tertiary alcohol

$HOOC-CH_2OCH_2-COOH$

$HOOC-OCH_2CH_2-COOH$

$HOOC-OCH(CH_3)-COOH$

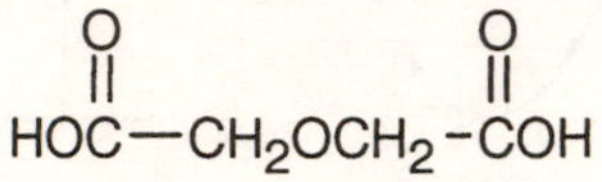

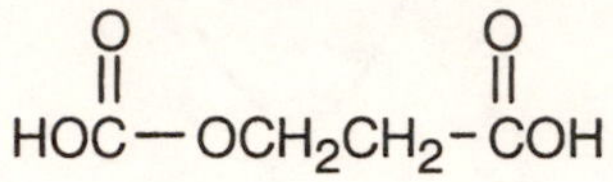

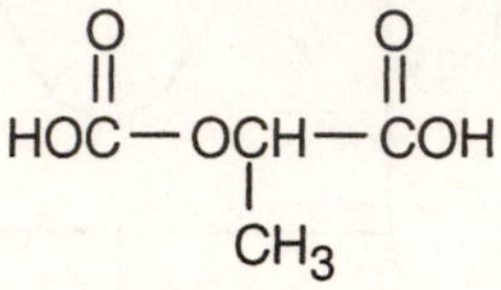

(b) Because only one of these compounds (the second one) is also a secondary alcohol, it must be malic acid.

3.50 **Strategy:** To solve this type of problem, read the problem carefully, word for word. Then try to interpret parts of the problem. For example:

1) Formaldehyde is an aldehyde, $H-C(=O)-H$
2) It trimerizes — that is, 3 formaldehydes come together to form a compound $C_3H_6O_3$. Because no atoms are eliminated, all of the original atoms are still present.
3) There are no carbonyl groups. This means that trioxane cannot contain any $-C{=}O$ functional groups. If you look back to Table 3.1, you can see that the only oxygen-containing functional groups that can be present are either ethers or alcohols.
4) A monobromo derivative is a compound in which one of the –H's has been replaced by a –Br. Because only one monobromo derivative is possible, we know that there can only be one type of hydrogen in trioxane. The only possibility for trioxane is:

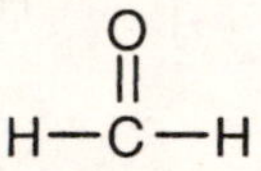

Solution:

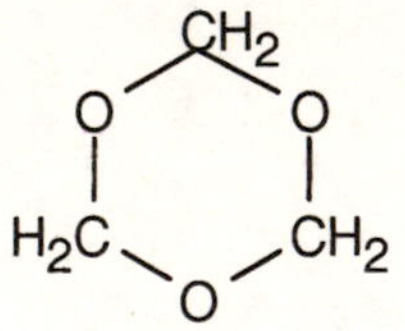

(cyclic $-CH_2-O-CH_2-O-CH_2-O-$ ring) Trioxane

3.51 A puckered ring allows all the bonds in the ring to have a nearly tetrahedral bond angle. (If the ring were flat, C-C-C bond angles would be 120°.)

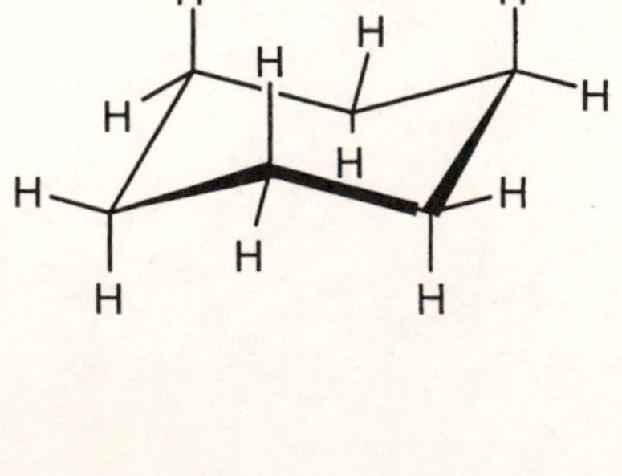

3.52

H H H CH_3

H_3C CH_3 H_3C H

In one of the 1,2-dimethylcyclohexanes the two methyl groups are on the same side of the ring, and in the other isomer the methyl groups are on opposite sides.

Chapter 4 – Organic Compounds: Cycloalkanes and Their Stereochemistry

Chapter Outline

I. Cycloalkanes (Sections 4.1 – 4.2).
  A. Cycloalkanes have the general formula C_nH_{2n}, if they have one ring.
  B. Naming cycloalkanes (Section 4.1).
    1. Find the parent.
      a. If the number of carbon atoms in the ring is larger than the number in the largest substituent, the compound is named as an alkyl-substituted cycloalkane.
      b. If the number of carbon atoms in the ring is smaller than the number in the largest substituent, the compound is named as an cycloalkyl-substituted alkane.
    2. Number the substituents.
      a. Start at a point of attachment and number the substituents so that the second substituent has the lowest possible number.
      b. If necessary, proceed to the next substituent until a point of difference is found.
      c. If two or more substituents might potentially receive the same number, number them by alphabetical priority.
      d. Halogens are treated in the same way as alkyl groups.
  C. Cis–trans isomerism in cycloalkanes (Section 4.2).
    1. Unlike open-chain alkanes, cycloalkanes have much less rotational freedom.
      a. Very small rings are rigid.
      b. Large rings have more rotational freedom.
    2. Cycloalkanes have a "top" side and a "bottom" side.
      a. If two substituents are on the same side of a ring, the ring is cis-disubstituted.
      b. If two substituents are on opposite sides of a ring,the ring is trans-disubstituted.
    3. Substituents in the two types of disubstituted cycloalkanes are connected in the same order but differ in spatial orientation.
      a. These cycloalkenes are stereoisomers that are known as cis–trans isomers.
      b. Cis–trans isomers are stable compounds that can't be interconverted.

II. Conformations of cycloalkanes (Sections 4.3 – 4.8).
  A. General principles (Section 4.3).
    1. Baeyer strain theory.
      a. A. von Baeyer suggested that rings other than those of 5 or 6 carbons were too strained to exist.
      b. This concept of angle strain is true for smaller rings, but larger rings can be easily prepared.
    2. The nature of ring strain.
      a. Rings tend to adopt puckered conformations.
      b. Several factors account for ring strain.
        i. Angle strain occurs when bond angles are distorted from their normal values.
        ii. Torsional strain is due to eclipsing of bonds.
        iii. Steric strain results when atoms approach too closely.
  B. Conformations of small rings (Section 4.4).
    1. Cyclopropane.
      a. Cyclopropane has bent bonds.
      b. Because of bent bonds, cyclopropane is more reactive than other cycloalkanes.

2. Cyclobutane.
   a. Cyclobutane has less angle strain than cyclopropane but has more torsional strain.
   b. Cyclobutane has almost the same total strain as cyclopropane.
   c. Cyclobutane is slightly bent to relieve torsional strain, but this increases angle strain.
3. Cyclopentane
   a. Cyclopentane has little angle strain but considerable torsional strain.
   b. To relieve torsional strain, cyclopentane adopts a puckered conformation.
      In this conformation, one carbon is bent out of plane; hydrogens are nearly staggered.

C. Conformations of cyclohexane (Sections 4.5 – 4.8).
1. Chair cyclohexane (Section 4.5).
   a. The chair conformation of cyclohexane is strain-free.
   b. In a standard drawing of cyclohexane, the lower bond is in front.
2. Axial and equatorial bonds in cyclohexane (Section 4.6).
   a. There are two kinds of positions on a cyclohexane ring.
      i. Six axial hydrogens are perpendicular to the plane of the ring.
      ii. Six equatorial hydrogens are roughly in the plane of the ring.
   b. Each carbon has one axial hydrogen and one equatorial hydrogen.
   c. Each side of the ring has alternating axial and equatorial hydrogens.
   d. All hydrogens on the same side of the ring are cis.
3. Conformational mobility of cyclohexanes (Section 4.6).
   a. Different chair conformations of cyclohexanes interconvert by a ring-flip.
   b. After a ring-flip, an axial bond becomes an equatorial bond, and vice versa.
   c. The energy barrier to interconversion is 45 kJ/mol, making interconversion rapid at room temperature.
4. Conformations of monosubstituted cyclohexanes (Section 4.7).
   a. Both conformations aren't equally stable at room temperature.
      In methylcyclohexane, 95% of molecules have the methyl group in the equatorial position.
   b. The energy difference is due to 1,3-diaxial interactions.
      i. These interactions are due to steric strain.
      ii. They are the same interactions as occur in gauche butane.
   c. Axial methylcyclohexane has two gauche interactions that cause it to be 7.6 kJ/mol less stable than equatorial methylcyclohexane.
   d. All substituents are more stable in the equatorial position.
      The size of the strain depends on the size of the group.
5. Conformations of disubstituted cyclohexanes (Section 4.8).
   a. In *cis*-1,2-dimethylcyclohexane, one methyl group is axial and one is equatorial in both chair conformations, which are of equal energy.
   b. In *trans*-1,2-dimethylcyclohexane, both methyl groups are either both axial or both equatorial.
      i. The conformation with both methyl groups axial is 11.4 kJ/mol less stable than the conformation with both groups equatorial.
      ii. The trans isomer exists almost exclusively in the diequatorial conformation.
   c. This type of conformational analysis can be carried out for most substituted cyclohexanes.

D. Conformations of polycyclic molecules (Section 4.9).
1. Decalin has two rings that can be either cis-fused or trans-fused.
   The two decalins are nonconvertible.
2. Steroids have four fused rings.

Solutions to Problems

4.1 **Strategy:** The steps for naming a cycloalkane are very similar to the steps used for naming an open-chain alkane.
Step 1: Name the parent cycloalkane. In (a), the parent is cyclohexane. If the compound has an alkyl substituent with more carbons than the ring size, the compound is named as a cycloalkyl-substituted alkane.
Step 2: Identify the substituents. In (a), both substituents are methyl groups.
Step 3: Number the substituents so that the second substituent receives the lowest possible number. In (a), the substituents are in the 1- and 4- positions.
Step 4: Name the compound. If two different alkyl groups are present, cite them alphabetically. Halogen substituents follow the same rules as alkyl substituents.

Solution:

(a) CH_3 CH_3 1,4-Dimethylcyclohexane

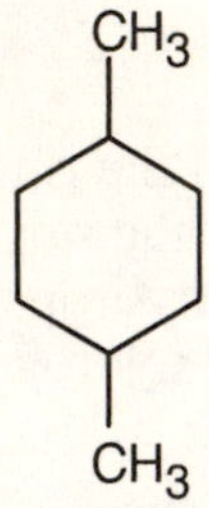

(b) $CH_2CH_2CH_3$ CH_3 1-Methyl-3-propylcyclopentane

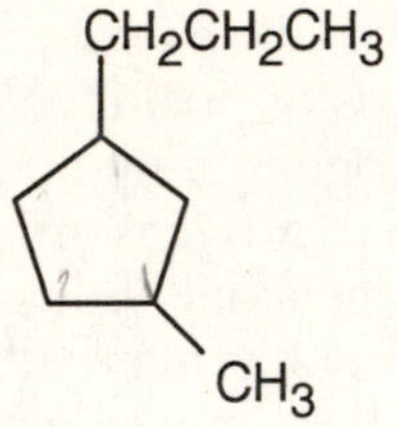

(c) 3-Cyclobutylpentane

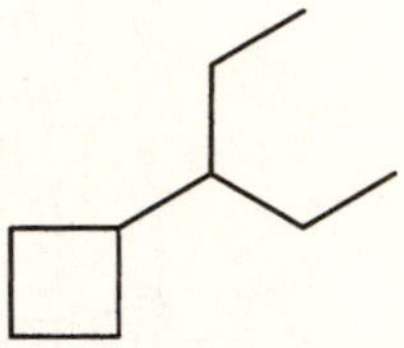

(d) CH_2CH_3 Br 1-Bromo-4-ethylcyclodecane

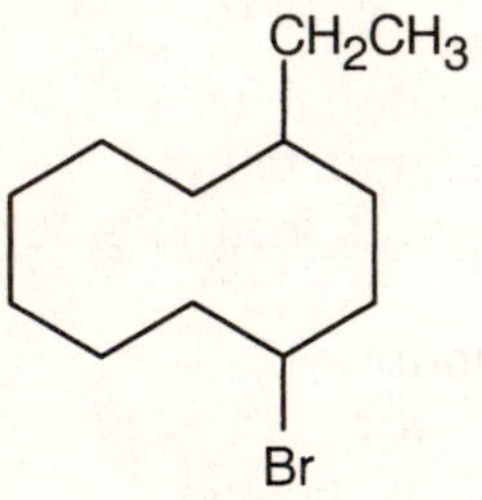

(e) CH_3 $CH(CH_3)_2$ 1-Isopropyl-2-methylcyclohexane

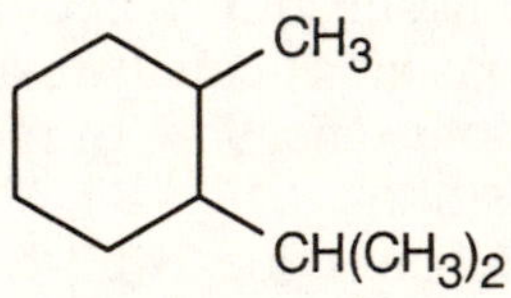

(f) Br CH_3 $C(CH_3)_3$ 4-Bromo-1-*tert*-butyl-2-methylcycloheptane

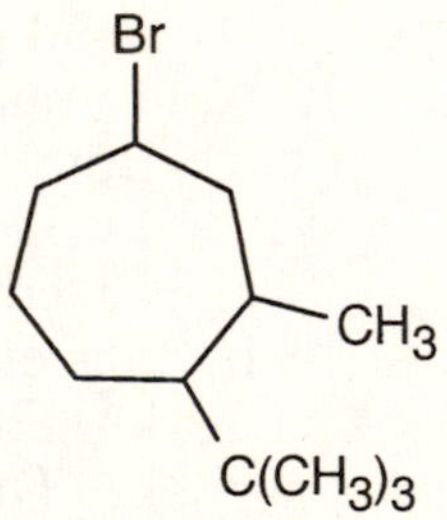

4.2 To draw a substituted cycloalkane, simply draw the ring and attach substituents in the specified positions.

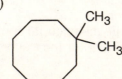

(a) CH_3 CH_3 1,1-Dimethylcyclooctane

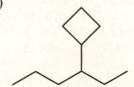

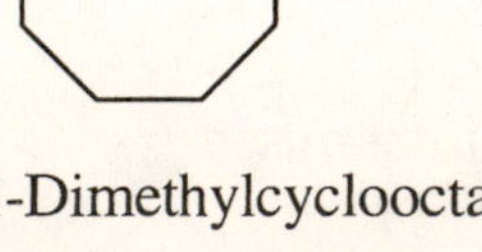

(b) 3-Cyclobutylhexane

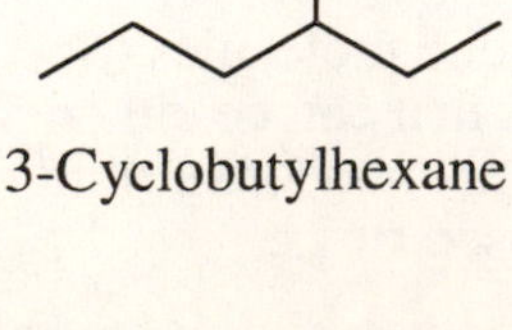

(c) Cl Cl 1,2-Dimethylcyclopentane

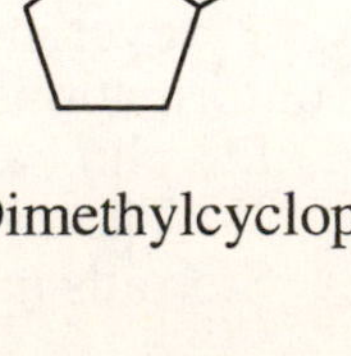

(d)

1,3-Dibromo-5-methylcyclohexane

4.3

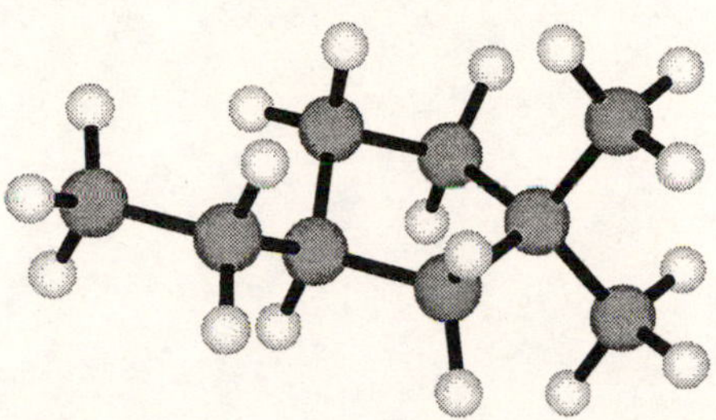

3-Ethyl-1,1-dimethylcyclopentane

4.4 **Strategy:** Two substituents are cis if they both have either dashed or wedged bonds. The substituents are trans if one has a wedged bond and the other has a dashed bond.

Solution:

(a)

trans-1-Chloro-4-methylcyclohexane

(b)

cis-1-Ethyl-3-methylcycloheptane

4.5

(a)

trans-1-Bromo-3-methylcyclohexane

(b)

cis-1,2-Dimethylcyclobutane

(c)

trans-1-*tert*-Butyl-2-ethylcyclohexane

4.6

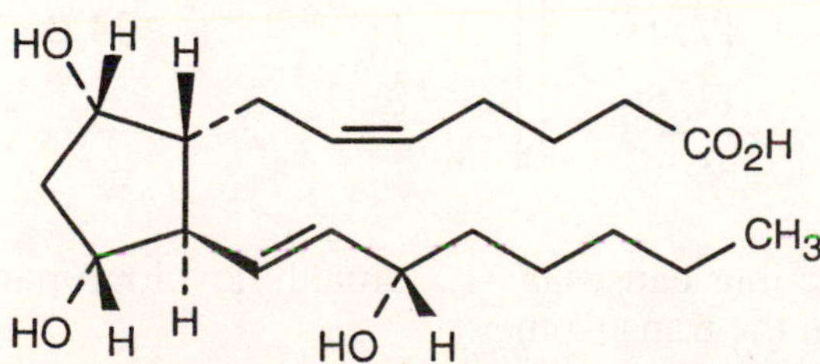

Prostaglandin $F_{2\alpha}$

The two hydroxyl groups are cis because they both point behind the plane of the page. The carbon chains have a trans relationship (one is dashed and the other is wedged).

4.7

(a)

cis-1,2-Dimethylcyclopentane

(b)

cis-1-Bromo-3-methylcyclobutane

4.8

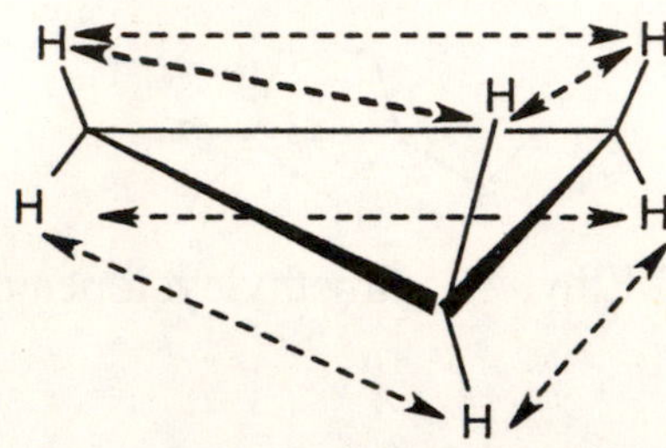

All hydrogen atoms on the same side of the cyclopropane ring are eclipsed by neighboring hydrogens. If we draw each hydrogen-hydrogen interaction, we count six eclipsing interactions. Since each of these interactions costs 4.0 kJ/mol, all six cost 24.0 kJ/mol. 24 kJ/mol ÷115 kJ/mol = 0.21; thus, 21% of the total strain energy of cyclopropane is due to torsional strain.

4.9

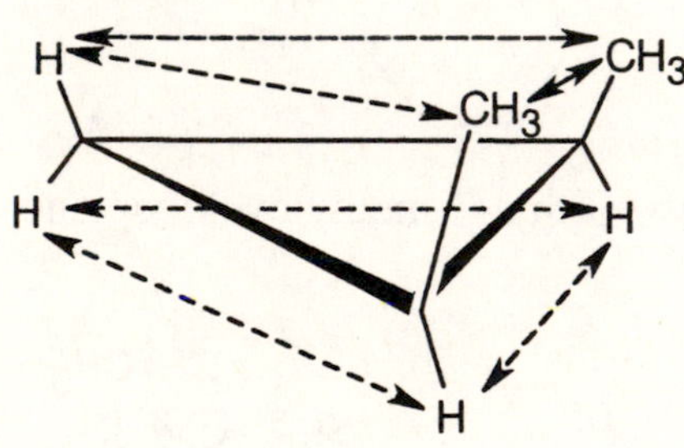

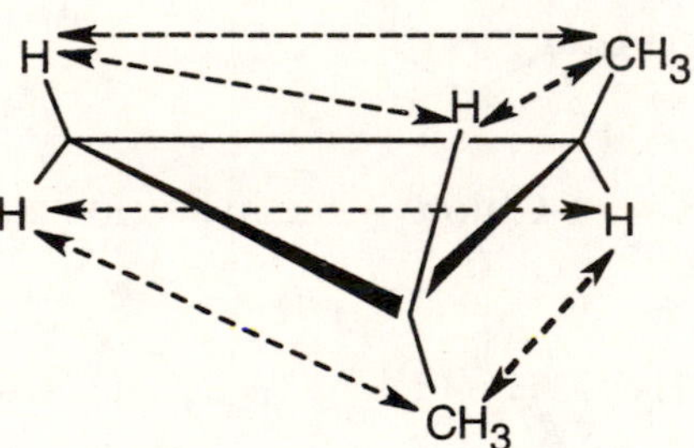

		cis isomer		*trans isomer*	
Eclipsing interaction	Energy cost (kJ/mol)	# of interactions	Total energy cost (kJ/mol)	# of interactions	Total energy cost (kJ/mol)
H–H	4.0	3	12.0	2	8.0
H–CH_3	6.0	2	12.0	4	24.0
CH_3–CH_3	11	1	11	0	0
			35		32

The added energy cost of eclipsing interactions causes cis-1,2-dimethylcyclopropane to be of higher energy and to be less stable than the trans isomer.

4.10

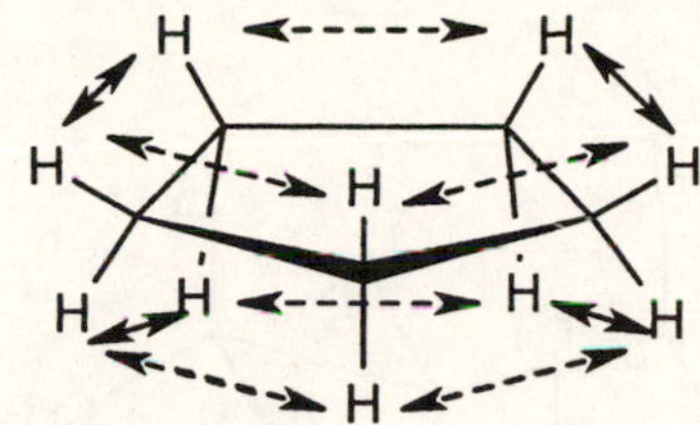

If cyclopentane were planar, it would have ten hydrogen–hydrogen interactions with a total energy cost of 40 kJ/mol. The measured total strain energy of 26 kJ/mol indicates that 14 kJ/mol of eclipsing strain in cyclopentane (35%) has been relieved by puckering.

4.11

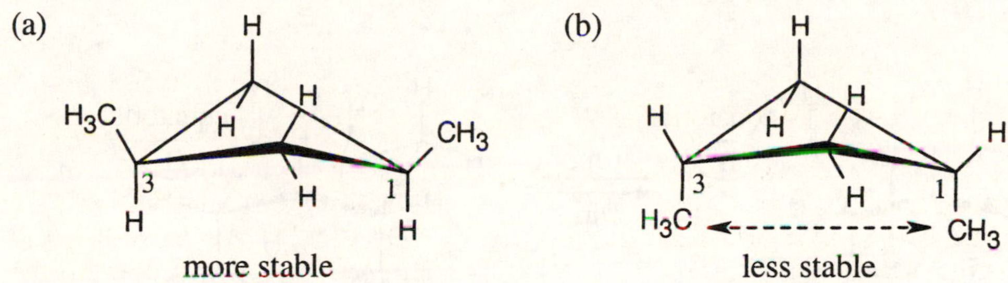

The methyl groups are farther apart in the more stable conformation of *cis*-1,3-dimethylcyclobutane.

4.12 Use the technique in Section 4.5 to draw the cyclohexane ring. Figure 4.10 shows how to attach axial and equatorial bonds.

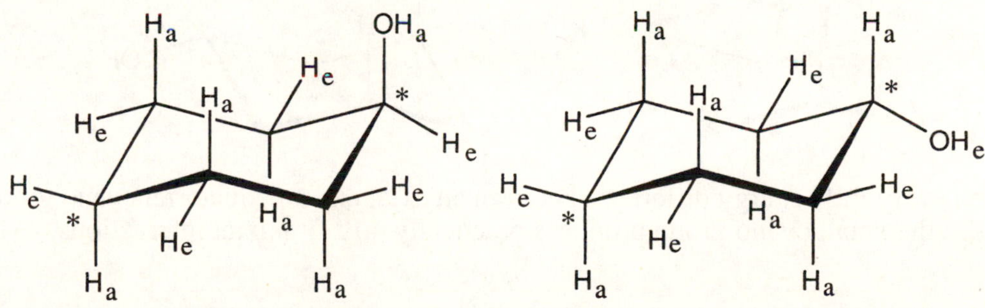

The conformation with –OH in the equatorial position is more stable.
Note: The starred ring carbons lie in the plane of the paper.

4.13 In trans-1,4-disubstituted cyclohexanes, the methyl substituents are either both axial or both equatorial.

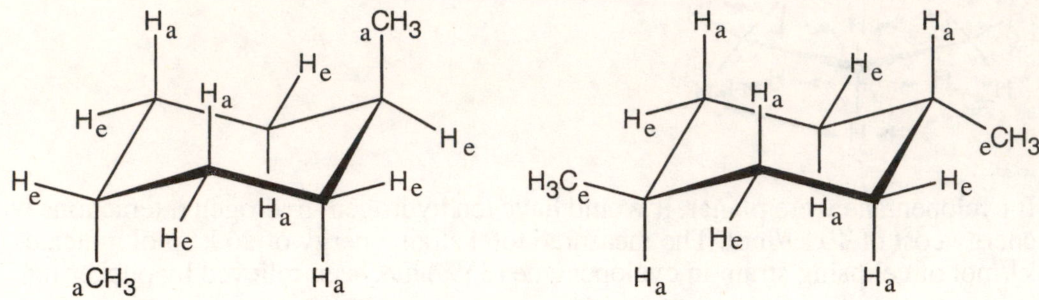

trans-1,4-Dimethylcyclohexane

4.14

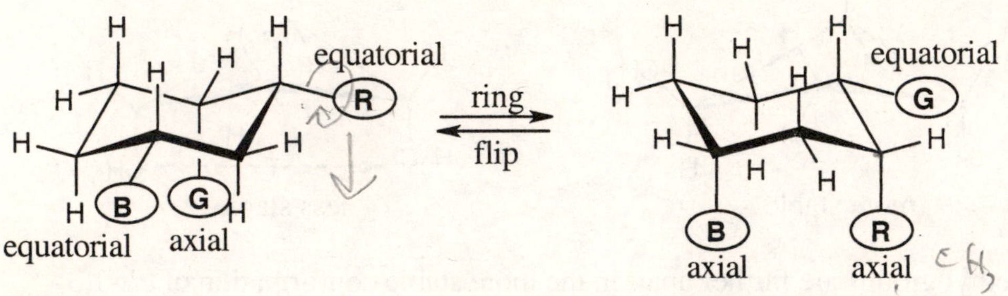

4.15 Table 4.1 shows that an axial hydroxyl group causes 2 x 2.1 kJ/mol of steric strain. Thus, the energy difference between axial and equatorial cyclohexanol is 4.2 kJ/mol.

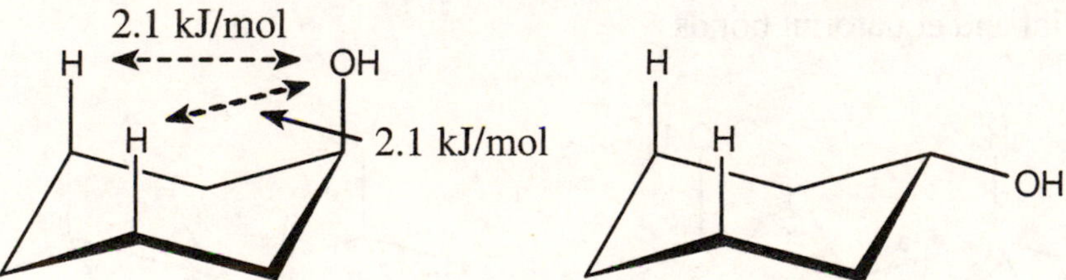

4.16 There is very little energy difference between an axial and an equatorial cyano group because the small cyano group produces practically no 1,3 diaxial interactions.

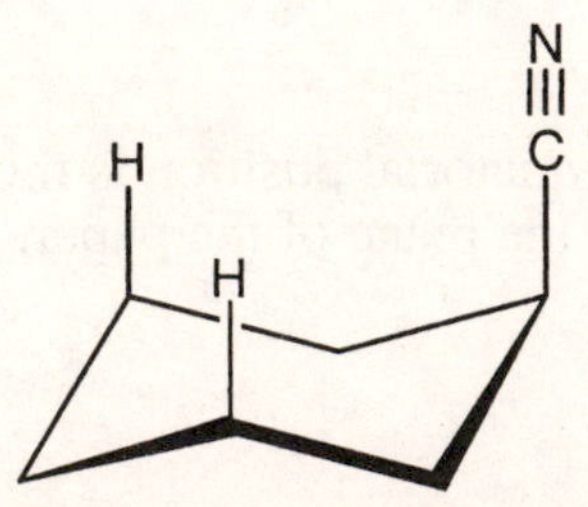

Cyclohexanecarbonitrile

4.17 **Strategy:** Draw the two chair conformations of each molecule, and look for gauche and 1,3-diaxial interactions. Use Table 4.2 to estimate the values of the interactions. Calculate the total strain; the conformation with the smaller value for strain energy is more stable.

Solution:

(a)

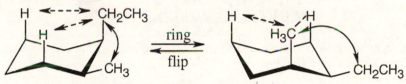

trans-1-Chloro-3-methylcyclohexane

2 ($H–CH_3$) = 7.6 kJ/mol 2 (H–Cl) = 2.0 kJ/mol

The second conformation is more stable than the first.

(b)

cis-1-Ethyl-2-methylcyclohexane

one $CH_3–CH_2CH_3$ gauche interaction = 3.8 kJ/mol	one $CH_3–CH_2CH_3$ gauche interaction = 3.8 kJ/mol
2 ($H–CH_2CH_3$) = 8.0 kJ/mol	2 ($H–CH_3$) = 7.6 kJ/mol
Total = 11.8 kJ/mol	Total = 11.4 kJ/mol

The second conformation is more stable than the first.

(c)

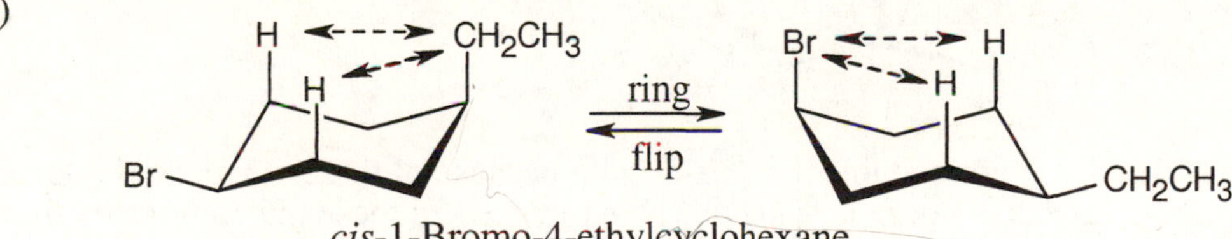

cis-1-Bromo-4-ethylcyclohexane

2 ($H–CH_2CH_3$) = 8.0 kJ/mol 2 (H–Br) = 2.0 kJ/mol

The second conformation is more stable than the first.

(d)

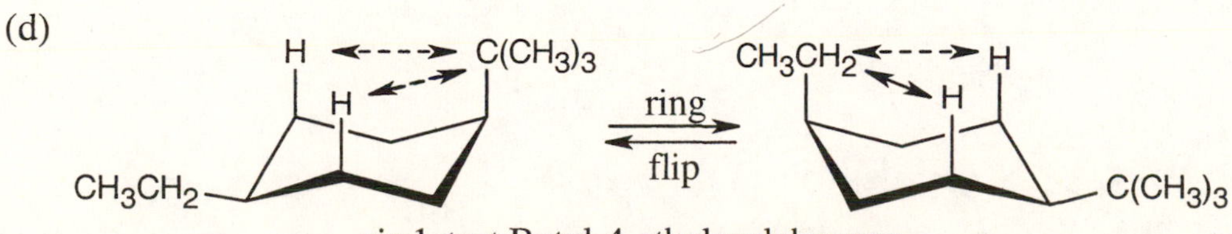

cis-1-*tert*-Butyl-4-ethylcyclohexane

2 [$H–C(CH_3)_3$] = 22.8 kJ/mol 2 ($H–CH_2CH_3$) = 8.0 kJ/mol

The second conformation is more stable than the first.

4.18 **Strategy:** The three substituents have the orientations shown in the first structure. To decide if the conformation shown is the more stable conformation or the less stable conformation, perform a ring-flip on the illustrated conformation and do a calculation like those in the previous problem. Notice that each conformation has a Cl–CH_3 gauche interaction, but we don't need to know its energy cost because it is present in both conformations.

Solution:

ring flip

$$2\ [\text{H–CH}_3] = 7.6\ \text{kJ/mol} \qquad 2\ [\text{H–CH}_3] = 7.6\ \text{kJ/mol}$$
$$\underline{2\ [\text{H–Cl}] \quad = 2.0\ \text{kJ/mol}}$$
$$= 9.6\ \text{kJ/mol}$$

The conformation shown in the model is the less stable chair form.

4.19

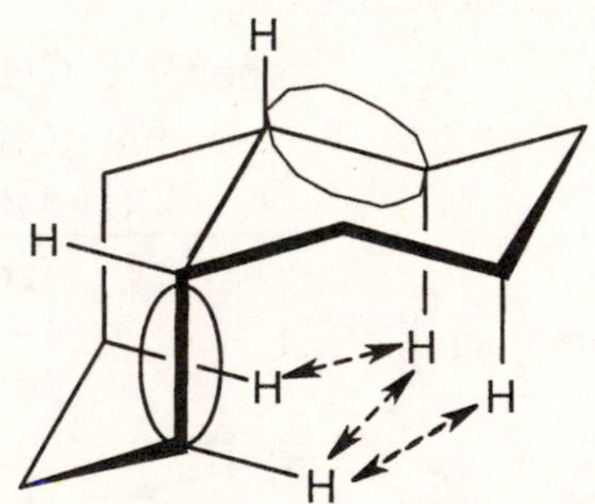

Trans-decalin is more stable than *cis*-decalin because of three 1,3–diaxial interactions present in the cis isomer. You may be able to visualize these interactions by thinking of the circled parts of *cis*-decalin as similar to axial methyl groups. The gauche interactions that occur with axial methyl groups also occur in *cis*-decalin.

Visualizing Chemistry

4.20

(a)

cis-1-Ethyl-3-methyl-cyclopentane

(b)

1,1,4-Trimethylcyclohexane

4.21

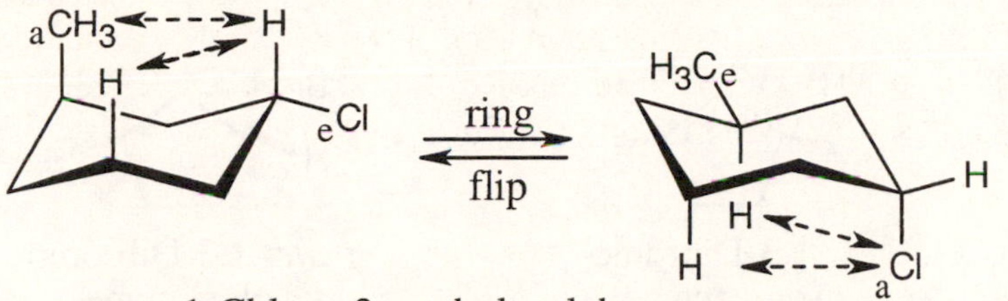

aCH3 H H eCl ring flip H3Ce H H H Cl a

trans-1-Chloro-3-methylcyclohexane

2 ($H–CH_3$) = 7.6 kJ/mol 2 (H–Cl) = 2.0 kJ/mol

The conformation shown (the left structure) is the less stable conformation.

4.22

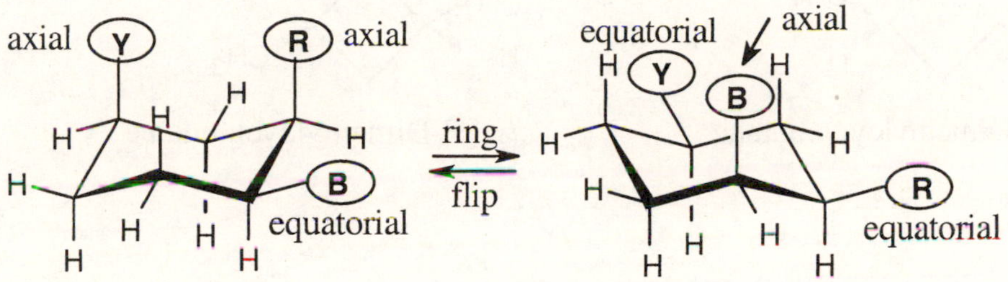

axial Y R axial B equatorial ring flip equatorial axial Y B R equatorial

4.23

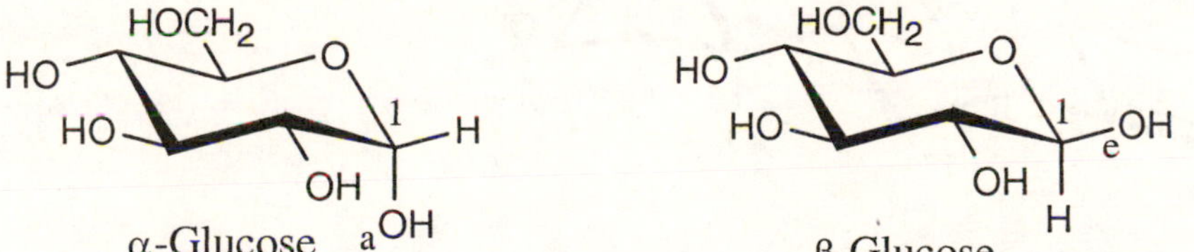

$HOCH_2$ HO O 1 H HO OH aOH

α-Glucose

$HOCH_2$ HO O 1 eOH HO OH H

β-Glucose

The only difference between α-glucose and β-glucose is in the orientation of the –OH group at carbon 1; the –OH group is axial in α-glucose, and it is equatorial in β-glucose. You would expect β-glucose to be more stable because it has all substituents in the equatorial position.

Additional Problems

4.24

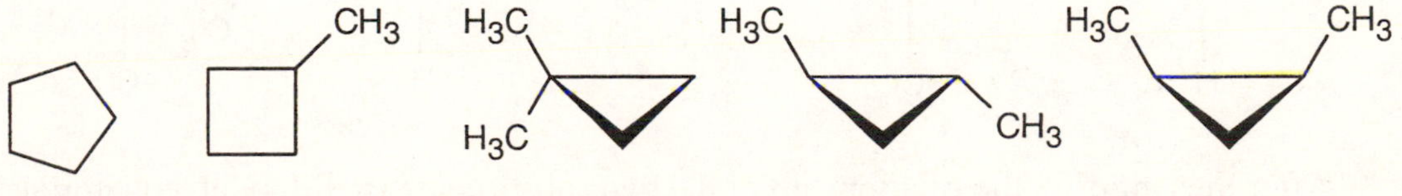

CH_3 H_3C H_3C H_3C CH_3 H_3C CH_3

The last two structures are cis-trans isomers.

4.25

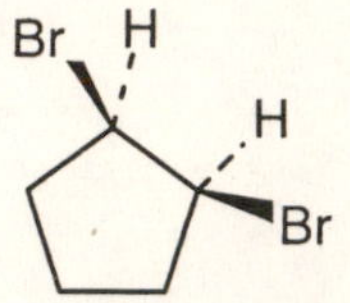

cis-1,2-Dibromocyclopentane

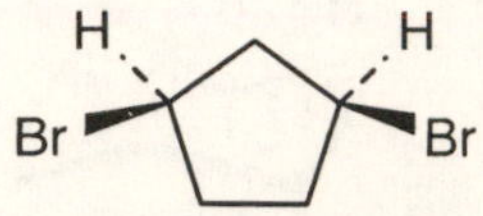

cis-1,3-Dibromocyclopentane

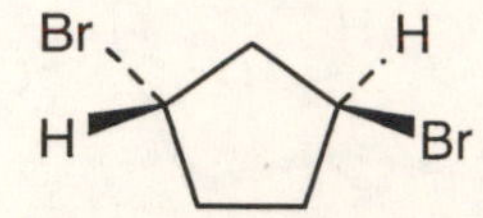

trans-1,3-Dibromocyclopentane

constitutional isomers of *cis*-1,2-dibromocyclopentane

4.26

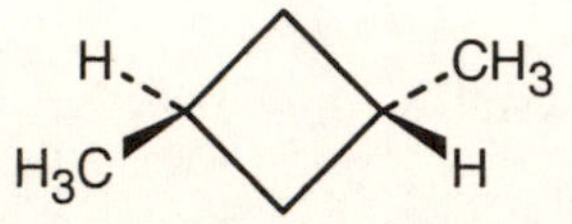

trans-1,3-Dimethylcyclobutane

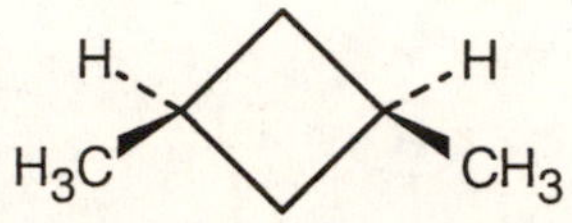

cis-1,3-Dimethylcyclobutane

4.27

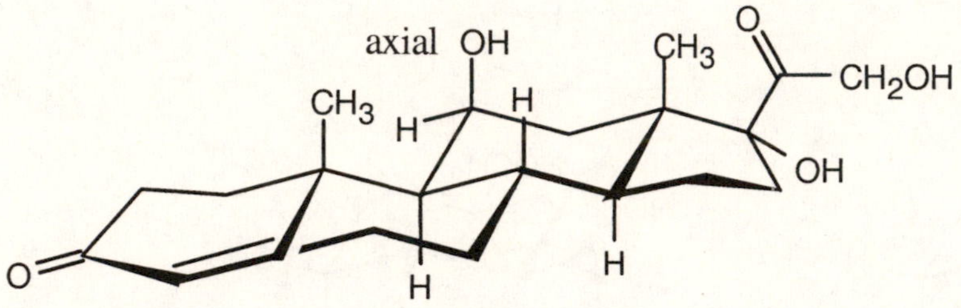

axial OH

4.28 Make a model of *cis*-1,2-dichlorocyclohexane. Notice that all cis substituents are on the same side of the ring and that two adjacent cis substituents have an axial–equatorial relationship. Now, perform a ring-flip on the cyclohexane.

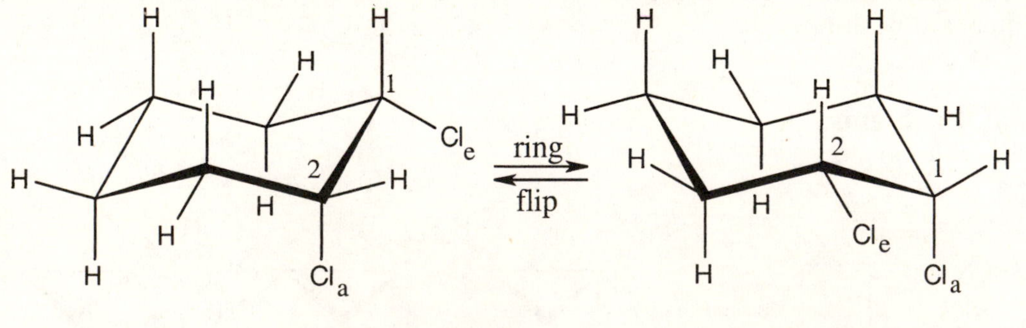

ring flip

After the ring-flip, the relationship of the two substituents is still axial–equatorial. No two adjacent cis substituents can be converted to being both axial or both equatorial. Don't forget that there are only two chair conformations of any given cyclohexane.

4.29 For a *trans*-1,2-disubstituted cyclohexane, two adjacent substituents must be either both axial or both equatorial.

A ring flip converts two adjacent axial substituents to equatorial substituents, and vice versa. As in Problem 4.28, no two adjacent trans substituents can have an axial-equatorial relationship.

4.30

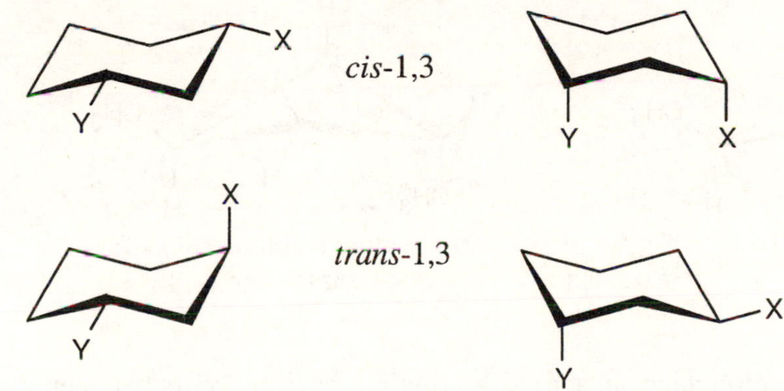

A cis-1,3-disubstituted isomer exists almost exclusively in the diequatorial conformation, which has no 1,3-diaxial interactions. The trans isomer must have one group axial, leading to 1,3-diaxial interactions. Thus, the trans isomer is less stable than the cis isomer.

4.31

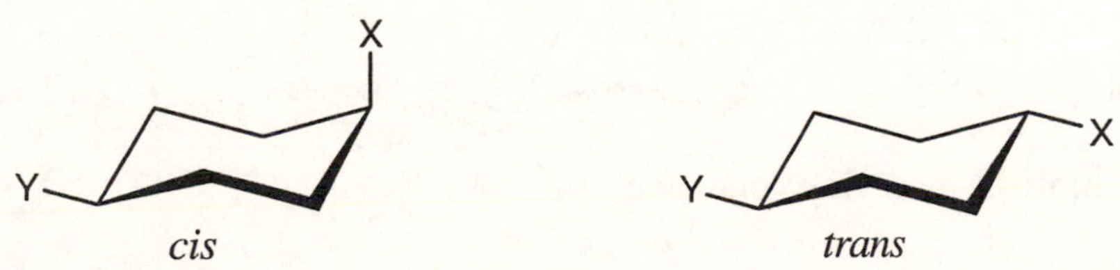

The *trans*-1,4-isomer is more stable.

4.32

less stable
cis

more stable
trans

Two types of interaction are present in *cis*-1,2-dimethylcyclobutane. One interaction occurs between the two methyl groups, which are almost eclipsed. The other is an across-the-ring interaction between methyl group at position 1 of the ring and a hydrogen at position 3. Because neither of these interactions are present in trans isomer, it is more stable than the cis isomer.

more stable
cis

less stable
trans

In *trans*-1,3-dimethylcyclobutane, an across-the-ring interaction occurs between the methyl group at position 3 of the ring and a hydrogen at position 1. Because no interactions are present in the cis isomer, it is more stable than the trans isomer.

4.33

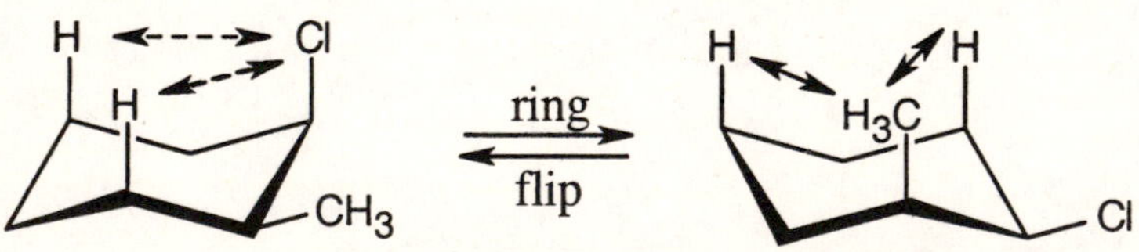

cis-1-Chloro-2-methylcyclohexane

Use Table 4.2 to find the values of 1,3-diaxial interactions. For the first conformation, the steric strain is 2 x 1.0 kJ/mol = 2.0 kJ/mol. The steric strain in the second conformation is 2 x 3.8 kJ/mol, or 7.6 kJ/mol. The first conformation is more stable than the second conformation by 5.6 kJ/mol.

4.34

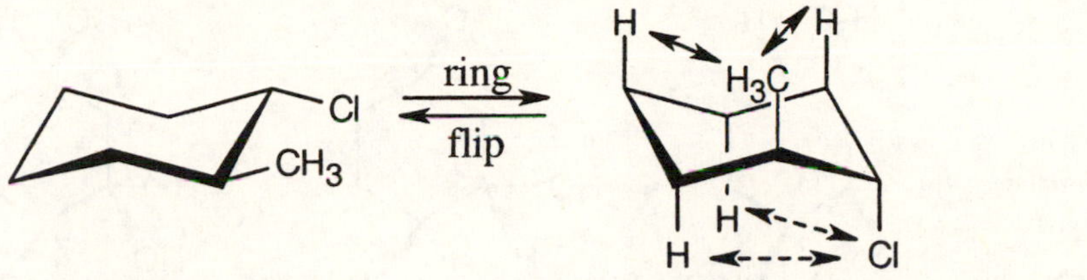

ring flip

H H H_3C Cl CH_3 H H Cl

no 1,3-diaxial interactions

2 x 3.8 kJ/mol = 7.6 kJ/mol
2 x 1.0 kJ/mol = 2.0 kJ/mol
= 9.6 kJ/mol

trans-1-Chloro-2-methylcyclohexane

The first conformation is more stable than the second conformation by a maximum of 9.6 kJ/mol. (A gauche interaction between the two substituents in the diequatorial conformation reduces the value of the energy difference.)

4.35

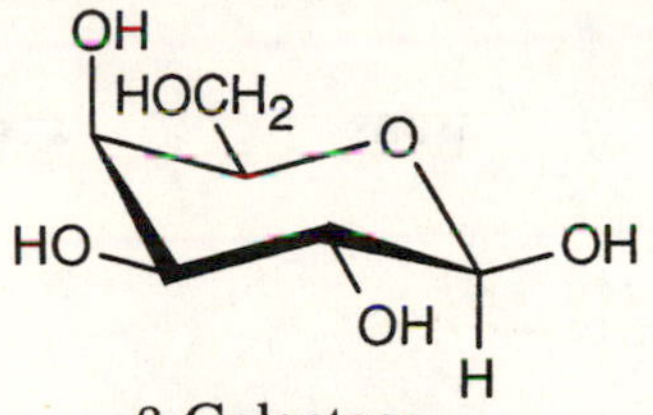

OH $HOCH_2$ O HO OH OH H

β-Galactose

In this conformation, all substituents, except for one hydroxyl group, are equatorial.

4.36 From the flat-ring drawing you can see that the methyl group and the –OH group have a cis relationship, and the isopropyl group has a trans relationship to both of these groups. Draw a chair cyclohexane ring and attach the groups with the correct relationship.

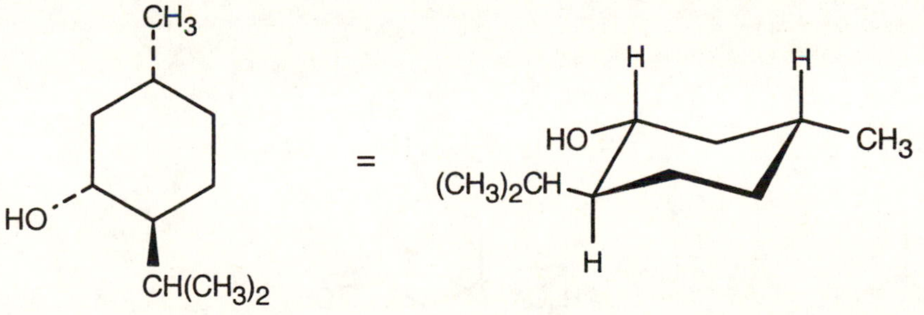

CH_3 HO $CH(CH_3)_2$ = H H HO CH_3 $(CH_3)_2CH$ H

In this conformation, all substituents are equatorial. Now, perform a ring flip.

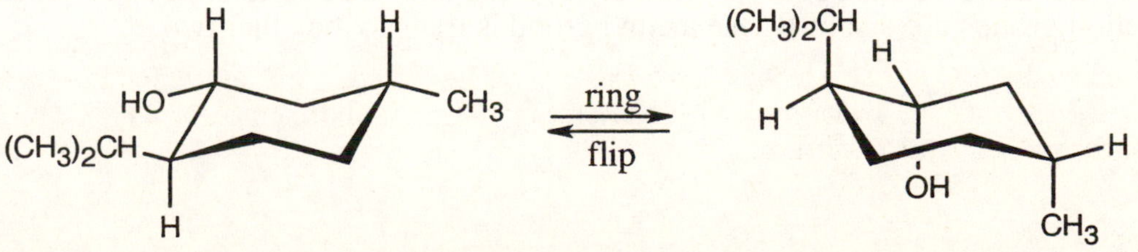

H H HO CH_3 $(CH_3)_2CH$ H ring flip $(CH_3)_2CH$ H H H OH CH_3

The second conformation is less stable because all substituents are axial.

4.37

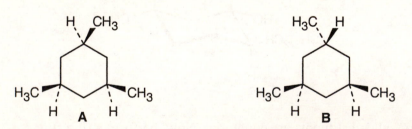

Menthol

Isomers of menthol

The substituents on the ring have the following relationships:

	Menthol	Isomer A	Isomer B	Isomer C
$—CH(CH_3)_2$, $—CH_3$	trans	trans	cis	cis
$—CH(CH_3)_2$, $—OH$	trans	cis	trans	cis
$—CH_3$, $—OH$	cis	trans	trans	cis

4.38

CH_2OH O OH red OH green OH black OH blue

cis relationship: red–green, blue–black

trans relationship: red–blue, green–blue
red–black, green–black

4.39

Two cis–trans isomers of 1,3,5–trimethylcyclohexane are possible. In one isomer (**A**), all methyl groups are cis; in **B**, one methyl group is trans to the other two.

4.40 Make sure you know the difference between axial-equatorial and cis-trans. Axial substituents are parallel to the axis of the ring; equatorial substituents lie around the equator of the ring. Cis substituents are on the same side of the ring; trans substituents are on opposite side of the ring.

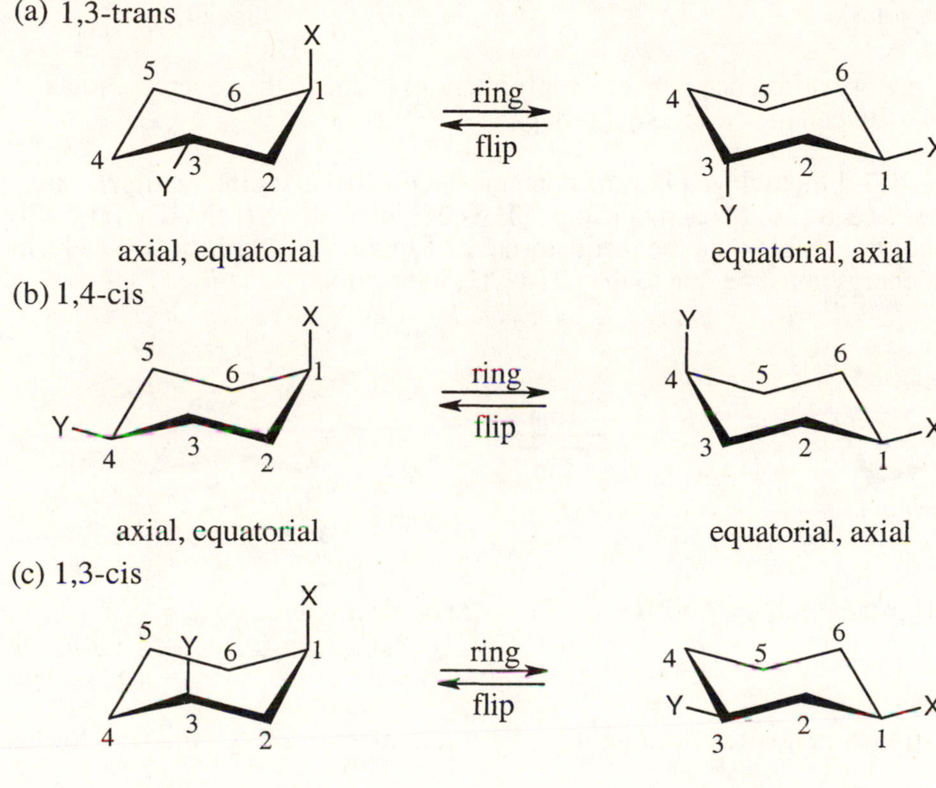

(d) 1,5-trans is the same as 1,3-trans

(e) 1,5-cis is the same as 1,3-cis

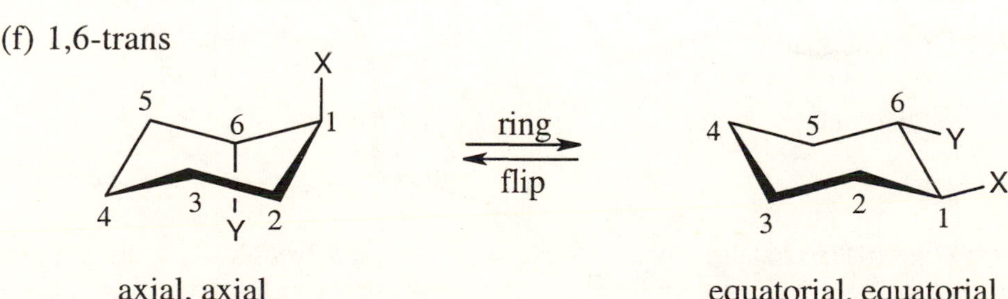

1,6-trans is the same as 1,2-trans.

4.41

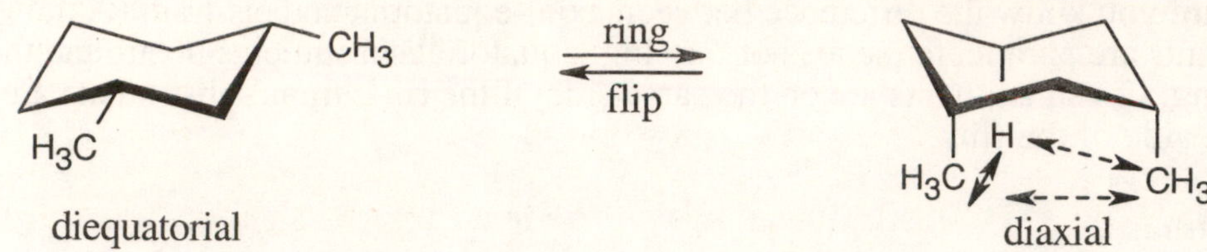

The large energy difference between conformations is due to the severe 1,3 diaxial interaction between the two methyl groups.

4.42 Diaxial cis-1,3-dimethylcyclohexane contains three 1,3–diaxial interactions -- two H–CH_3 interactions of 3.8 kJ/mol each, and one CH_3–CH_3 interaction. If the diaxial conformation is 23 kJ/mol less stable than the diequatorial, 23 kJ/mol – 2(3.8 kJ/mol) ≈ 15 kJ/mol of this strain energy must be due to the CH_3–CH_3 interaction.

4.43

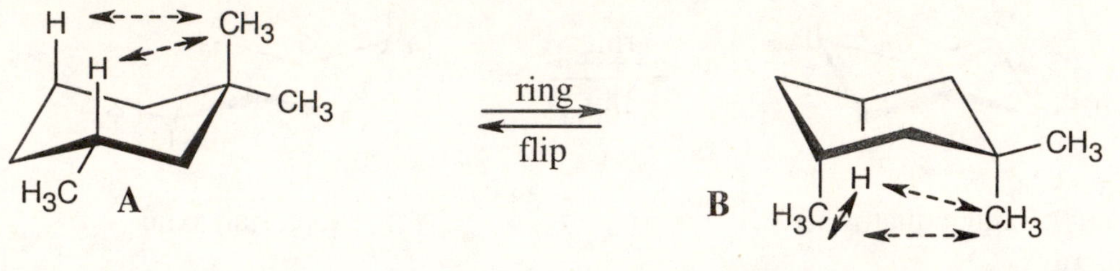

2 H – CH_3 interactions = 7.6 kJ/mol

2 H – CH_3 interactions = 7.6 kJ/mol
1 CH_3 – CH_3 interaction = 15 kJ/mol
≈ 23 kJ/mol

Conformation **A** is favored because it is 15 kJ/mol lower in energy than conformation **B**.

4.44 Note: In working with decalins, it is essential to use models. Many structural features of decalins that are obvious with models are not easily visualized with drawings.

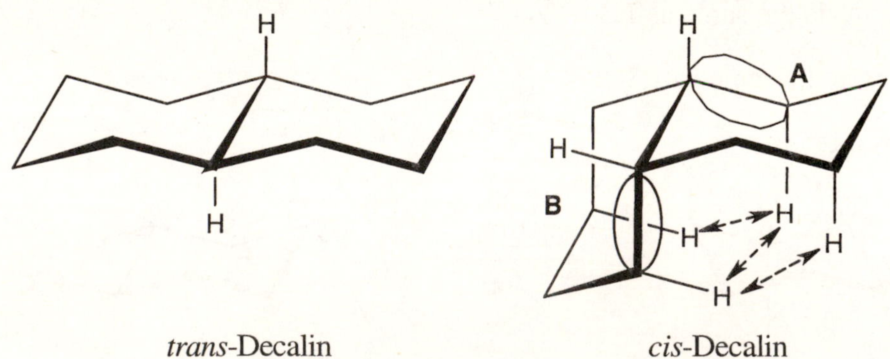

trans-Decalin *cis*-Decalin

No 1,3-diaxial interactions are present in *trans*-decalin.

At the ring junction of *cis*-decalin, one ring acts as an axial substituent of the other (see circled bonds). The circled part of ring **B** has two 1,3-diaxial interactions with ring **A** (indicated by arrows). Similarly, the circled part of ring **A** has two 1,3-diaxial interactions with ring **B**; one of these interactions is the same as an interaction of part of the **B** ring with ring **A**. These three 1,3-diaxial interactions have a total energy cost of 3 x 3.8 kJ/mol = 11.4 kJ/mol. *Cis*-decalin is therefore less stable than *trans*-decalin by 11.4 kJ/mol.

4.45 A ring-flip converts an axial substituent into an equatorial substituent and vice versa. At the ring junction of *trans*-decalin, each ring is a trans–trans diequatorial substituent of the other. If a ring-flip were to occur, the two rings would become axial substituents of each other. You can see with models that a diaxial ring junction is impossibly strained. Consequently, *trans*-decalin does not ring-flip.

The rings of *cis*-decalin are joined by an axial bond and an equatorial bond. After a ring-flip, the rings are still linked by an equatorial and an axial bond. No additional strain or interaction is introduced by a ring-flip of *cis*-decalin.

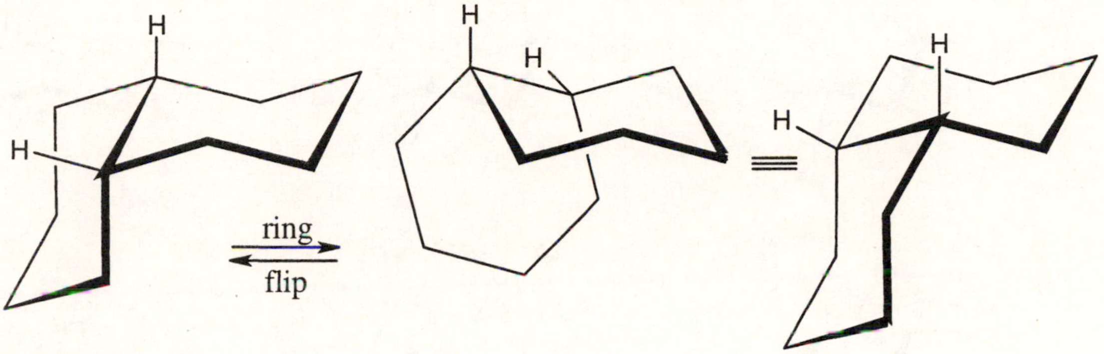

H H ring flip H H H H

4.46 **Strategy:** In the flat-ring structure shown, all –OH groups have a trans relationship except for the starred group. If all of the groups had a trans relationship, the most stable conformation would have all –OH groups in the equatorial position. We expect that the most stable conformation of this structure has one group in the axial position.

Draw both rings and add –OH groups having the indicated relationships. Perform a ring-flip on the structure you have drawn to arrive at the other conformation.

Solution:

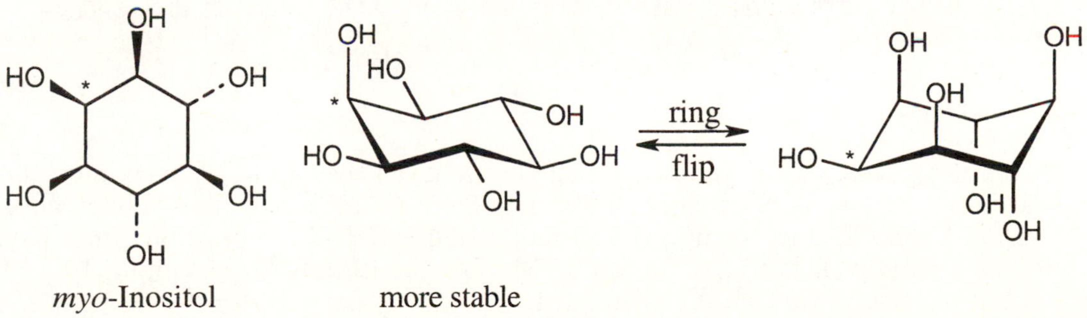

myo-Inositol

more stable

ring flip

4.47

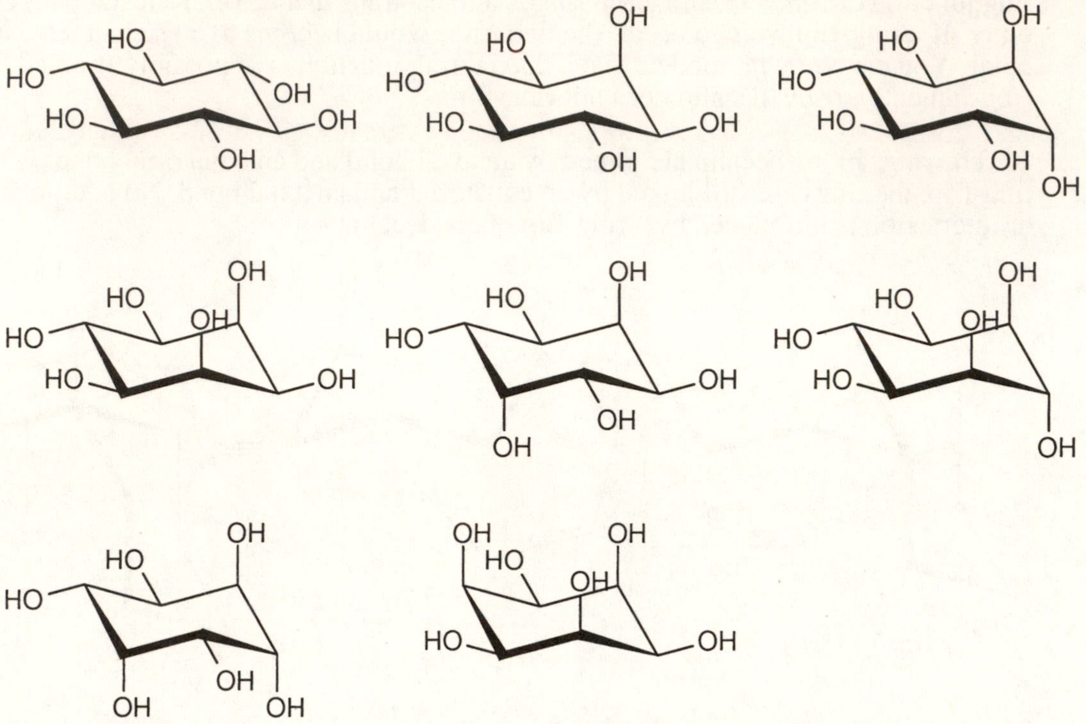

There are eight cis-trans stereoisomers of *myo*-inositol. The first isomer is the most stable because all hydroxyl groups can assume an equatorial conformation.

4.48

Conformation **A** of *cis*-1-chloro-3-methylcyclohexane has no 1,3–diaxial interactions and is the more stable conformation. Steric strain in **B** is due to one CH_3–H interaction (3.8 kJ/mol), one Cl–H interaction (1.0 kJ/mol) and one CH_3–Cl interaction. Since the total-strain energy of B is 15.5 kJ/mol, 15.5 kJ/mol - 3.8 kJ/mol - 1.0 kJ/mol = 10.7 kJ/mol of strain is caused by a CH_3–Cl interaction.

4.49 A steroid ring system is rigid, and ring-flips don't occur. Thus, substituents such as the methyl groups shown remain axial. Substituents on the same side of the ring system as the methyl groups are in alternating axial and equatorial positions. Thus, an up substituent at C3 (a) is equatorial.

Substituents on the bottom side of the ring system also alternate axial and equatorial positions. A substituent at C7 (b) is axial, and one at C11 (c) is equatorial

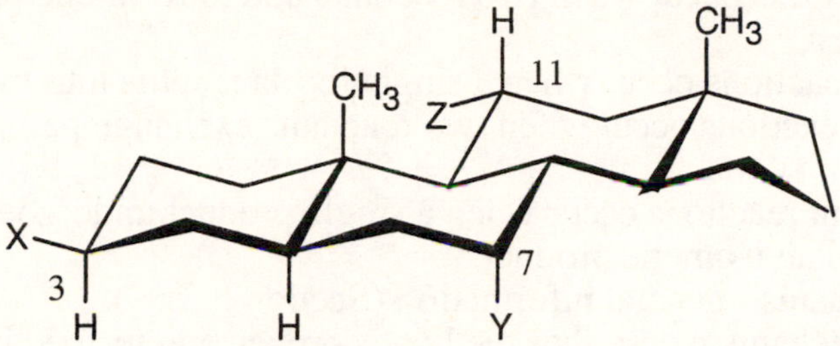

H CH3
11
CH3 Z
X
7
3
H H Y

4.50

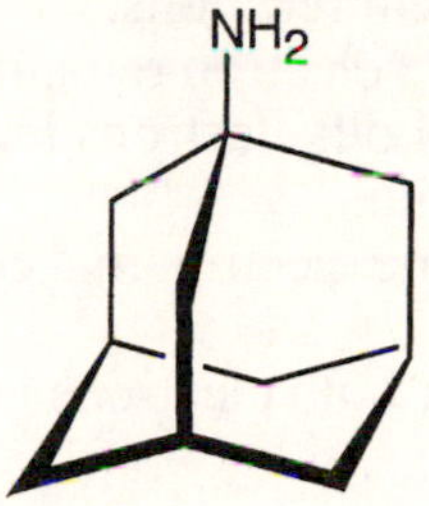

NH2

Amantadine

4.51

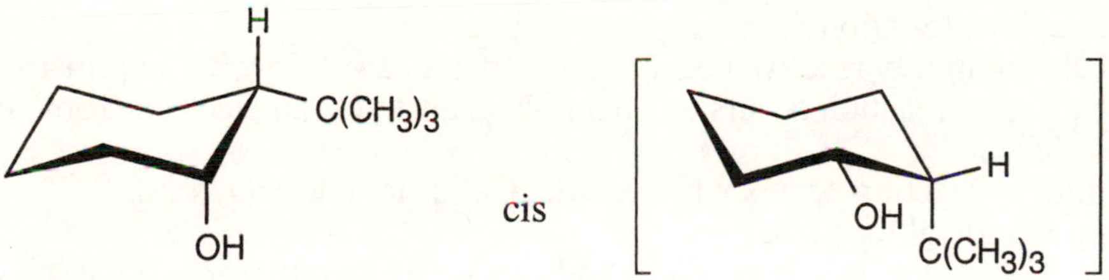

H
C(CH3)3
OH
cis
H
OH
C(CH3)3

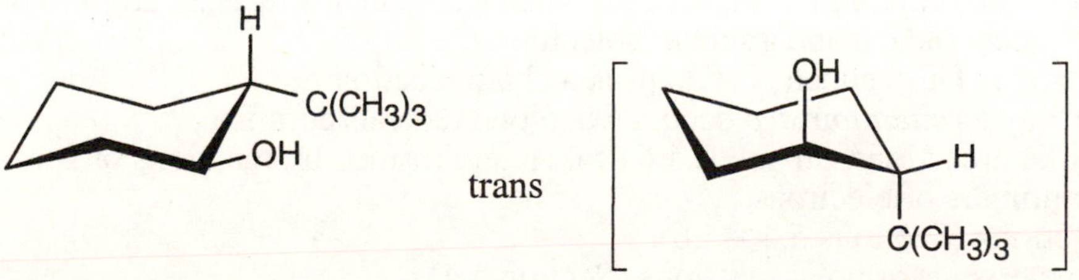

H
C(CH3)3
OH
trans
OH
H
C(CH3)3

All four conformations of the two isomers are illustrated. The second conformation of each pair has a high degree of steric strain, and thus each isomer adopts the first conformation. Since only the cis isomer has the hydroxyl group in the necessary axial position, it oxidizes faster than the trans isomer.

Chapter 5 – An Overview of Organic Reactions

Chapter Outline

I. Organic Reactions (Sections 5.1 – 5.6).
   A. Kinds of organic reactions (Section 5.1).
      1. Addition reactions occur when two reactants add to form one product, with no atoms left over.
      2. Elimination reactions occur when a single reactant splits into two products.
      3. Substitution reactions occur when two reactants exchange parts to yield two new products.
      4. Rearrangement reactions occur when a single product undergoes a rearrangement of bonds to yield an isomeric product.
   B. Reaction mechanisms - general information (Section 5.2).
      1. A reaction mechanism describes the bonds broken and formed in a chemical reaction, and accounts for all reactants and products.
      2. Bond breaking and formation in chemical reactions.
         a. Bond breaking is symmetrical if one electron remains with each fragment.
         b. Bond breaking is unsymmetrical if both electrons remain with one fragment and the other fragment has a vacant orbital.
         c. Bond formation is symmetrical if one electron in a covalent bond comes from each reactant.
         d. Bond formation is unsymmetrical if both electrons in a covalent bond come from one reactant.
      3. Types of reactions.
         a. Radical reactions involve symmetrical bond breaking and bond formation.
         b. Polar reactions involve unsymmetrical bond breaking and bond formation.
   C. Radical reactions (Section 5.3).
      1. Radicals are highly reactive because they contain an atom with an unpaired electron.
      2. A substitution reaction occurs when a radical abstracts an electron from another molecule.
      3. An addition reaction occurs when a radical adds to a double bond.
      4. Steps in a radical reaction.
         a. The *initiation step* produces radicals by the symmetrical breaking of a bond.
         b. The *propagation steps* occur when a radical abstracts an atom to produce a new radical and a stable molecule.
            This sequence of steps is a chain reaction.
         c. A *termination step* occurs when two radicals combine.
      5. In radical reactions, all bonds are broken and formed by reactions of species with odd numbers of electrons.
   D. Polar reactions (Sections 5.4 – 5.6).
      1. Characteristics of polar reactions (Section 5.4).
         a. Polar reactions occur as a result of positive and negative centers within molecules.
         b. These charge differences are usually due to electronegativity differences between atoms.
            i. Differences may also be due to interactions of functional groups with solvents, as well as with Lewis acids or bases.
            ii. Some bonds in which one atom is polarizable may also behave as polar bonds.

c. In polar reactions, electron-rich sites in one molecule react with electron-poor sites in another molecule.
d. The movement of an electron pair in a polar reaction is shown by a curved arrow.
e. The reacting species:
   i. A nucleophile is a compound with an electron-rich atom.
   ii. An electrophile is a compound with an electron-poor atom.
   iii. Some compounds can behave as both nucleophiles and as electrophiles
f. Many polar reactions can be explained in terms of acid-base reactions.

2. An example of a polar reaction: addition of H_2O to ethylene (Section 5.5).
   a. This reaction is known as an electrophilic addition.
   b. The π electrons in ethylene behave as a nucleophile.
   c. The reaction begins by the attack of the π electrons on the electrophile H_3O^+, forming a new C–H bond plus H_2O.
   d. The resulting intermediate carbocation reacts with water to form a C–O bond.
   e. A second water molecule acts as a base to remove H^+.
3. Rules for using curved arrows in polar reaction mechanisms (Section 5.6).
   a. Electrons must move from a nucleophilic source to an electrophilic sink.
   b. The nucleophile can be either negatively charged or neutral.
   c. The electrophile can be either positively charged or neutral.
   d. The octet rule must be followed.

II. Describing a reaction (Sections 5.7 – 5.10).
   A. Equilibria, rates, and energy changes (Section 5.7).
      1. All chemical reactions are equilibria that can be expressed by an equilibrium constant K_{eq} that shows the ratio of products to reactants.
         a. If $K_{eq} > 1$, [products] > [reactants].
         b. If $K_{eq} < 1$, [reactants] > [products].
      2. For a reaction to proceed as written, the energy of the products must be lower than the energy of the reactants.
         a. The energy change that occurs during a reaction is described by $\Delta G°$, the Gibbs free-energy change.
         b. Favorable reactions have negative $\Delta G°$ and are exergonic.
         c. Unfavorable reactions have positive $\Delta G°$ and are endergonic.
      3. $\Delta G°$ is composed of two terms – $\Delta H°$, and $\Delta S°$, which is temperature-dependent.
         a. $\Delta H°$ is a measure of the change in total bonding energy during a reaction.
            i. If $\Delta H°$ is negative, a reaction is exothermic.
            ii. If $\Delta H°$ is positive, a reaction is endothermic.
         b. $\Delta S°$ (entropy) is a measure of the freedom of motion of a reaction.
            i. A reaction that produces two product molecules from one reactant molecule has positive entropy.
            ii. A reaction that produces one product molecule from two reactant molecules has negative entropy.
         c. $\Delta G° = \Delta H° - T\,\Delta S°$.
      4. None of these expressions predict the rate of a reaction.
   B. Bond dissociation energies (Section 5.8).
      1. The bond dissociation energy (D) measures the heat needed to break a bond.
      2. Each bond has a characteristic strength.
      3. It is possible to calculate $\Delta H°$ for a reaction by using values of D.

4. Values of D are of limited usefulness.
   a. Calculations of ΔH° can't provide values of ΔS° or ΔG°.
   b. Calculations of ΔH° can't give information about reaction rates.
   c. Values of D are for gas-phase reactions and don't account for the effects of solvent.

C. Energy diagrams and transition states (Section 5.9).
   1. Reaction energy diagrams show the energy changes that occur during a reaction.
      The vertical axis represents energy changes, and the horizontal axis represents the progress of a reaction.
   2. The transition state is the highest-energy species in this reaction.
      It is possible for a reaction to have more than one transition state.
   3. The difference in energy between the reactants and the transition state is the energy of activation $\Delta G^\ddagger$.
      Values of $\Delta G^\ddagger$ range from 40 – 150 kJ/mol.
   4. After reaching the transition state, a molecule can go on to form products or can revert to starting material.
   5. Every reaction has its own energy profile.

D. Intermediates (Section 5.10).
   1. In a reaction of at least two steps, an intermediate is the species that lies at the energy minimum between two transition states.
   2. Even though an intermediate lies at an energy minimum between two transition states, it is a high-energy species and usually can't be isolated.
   3. Each step of a reaction has its own $\Delta G^\ddagger$ and ΔG°, but the total reaction has an overall ΔG°.
   4. Biological reactions take place in several small steps, each of which has a small value of $\Delta G^\ddagger$.

III. A Comparison of biological and laboratory reactions (Section 5.11).
   A. Laboratory reactions are carried out in organic solvents; biological reactions occur in aqueous medium.
   B. Laboratory reactions take place over a wide variety of temperatures; biological reactions take place at the temperature of the organism.
   C. Laboratory reactions are uncatalyzed, or use simple catalysts; biological reactions are enzyme-catalyzed.
   D. Laboratory reagents are usually small and simple; biological reactions involve large, complex coenzymes.
   E. Biological reactions have high specificity for substrate, whereas laboratory reactions are relatively nonspecific.

Solutions to Problems

5.1

(a) $CH_3Br + KOH \longrightarrow CH_3OH + KBr$ substitution

(b) $CH_3CH_2OH \longrightarrow H_2C{=}CH_2 + H_2O$ elimination

(c) $H_2C{=}CH_2 + H_2 \longrightarrow CH_3CH_3$ addition

5.2

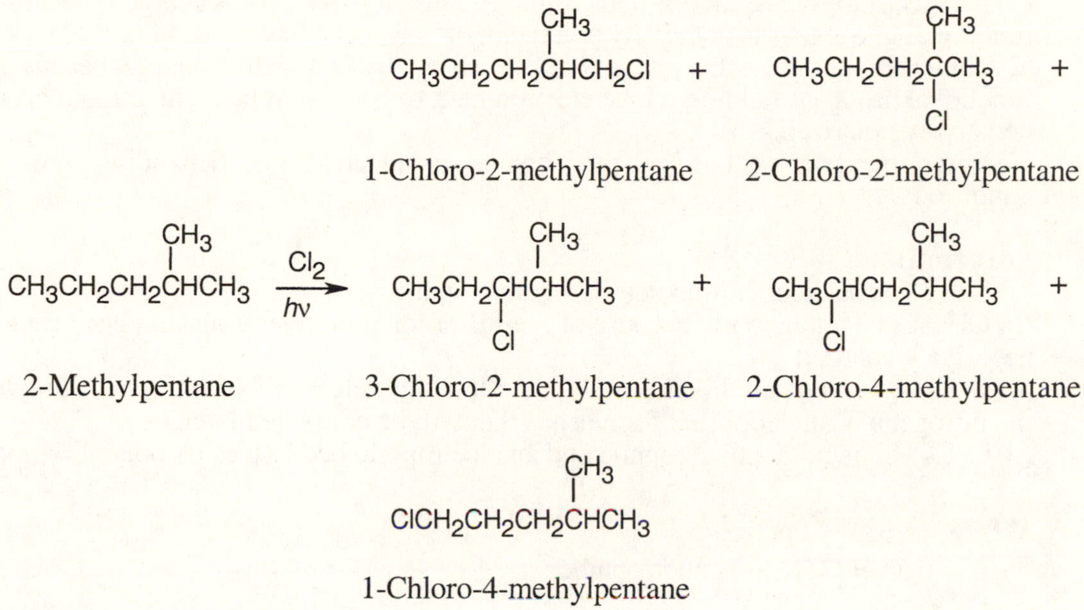

$$CH_3CH_2CH_2CH(CH_3)CH_3 \xrightarrow[h\nu]{Cl_2} CH_3CH_2CH_2CH(CH_3)CH_2Cl + CH_3CH_2CH_2C(CH_3)(Cl)CH_3 + CH_3CH_2CH(Cl)CH(CH_3)CH_3 + CH_3CH(Cl)CH_2CH(CH_3)CH_3 + ClCH_2CH_2CH_2CH(CH_3)CH_3$$

2-Methylpentane → 1-Chloro-2-methylpentane + 2-Chloro-2-methylpentane + 3-Chloro-2-methylpentane + 2-Chloro-4-methylpentane + 1-Chloro-4-methylpentane

5.3 **Strategy:** The tails of the arrows show the location of the bond to be broken, and the heads show where the electrons are moving. In radical reactions, the arrow is a fishhook (half-headed).

Solution:

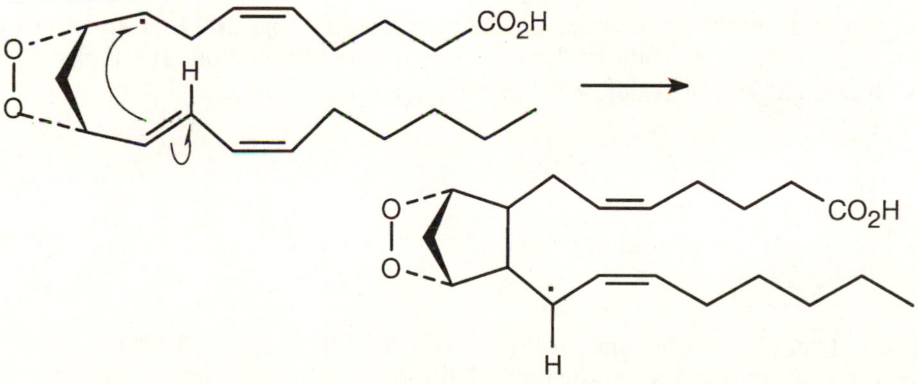

O, O, H, CO_2H → O, O, CO_2H, H

The reaction is a radical addition to a double bond and a rearrangement.

5.4 **Strategy:** Keep in mind:
(1) An electrophile is electron-poor, either because it is positively charged, because it has a functional group that is positively polarized, or because it has a vacant orbital.
(2) A nucleophile is electron-rich, either because it has a negative charge, because it has a functional group containing a lone electron pair, or because it has a functional group that is negatively polarized.
(3) Some species can act as both nucleophiles and electrophiles, depending on the reaction conditions.

Solution:
(a) The electron-poor carbon acts as an electrophile.
(b) CH_3S^- is a nucleophile because of the sulfur lone-pair electrons and because it is negatively charged.
(c) $C_4H_6N_2$ is a nucleophile because of the lone-pair electrons of nitrogen. (Only one of the nitrogens is nucleophilic, for reasons that will be explained later.)
(d) CH_3CHO is both a nucleophile and an electrophile because of its polar C=O bond.

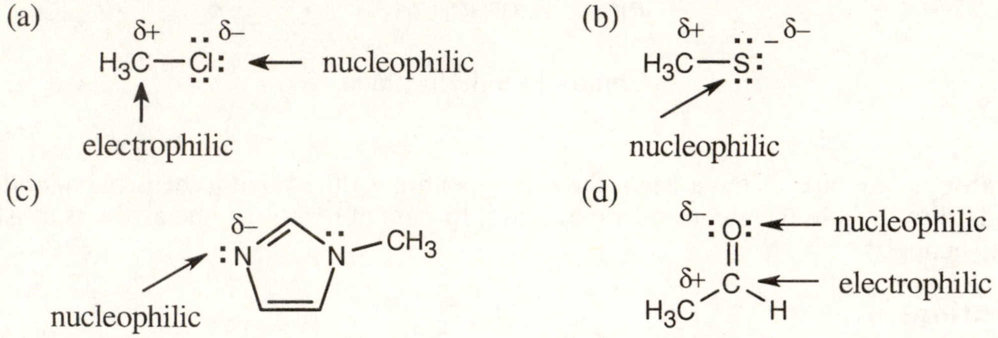

(a) δ+ δ– $H_3C—Cl:$ ← nucleophilic; ↑ electrophilic

(b) δ+ δ– $H_3C—S:^-$; nucleophilic

(c) δ– :N ... N—CH_3 ; nucleophilic

(d) δ– :O: ← nucleophilic; δ+ C ← electrophilic; H_3C, H

5.5 BF_3 is likely to be an electrophile because the electrostatic potential map indicates that the boron atom is electron-poor (blue). The Lewis structure shows that BF_3 lacks a complete electron octet and can accept an electron pair from a nucleophile.

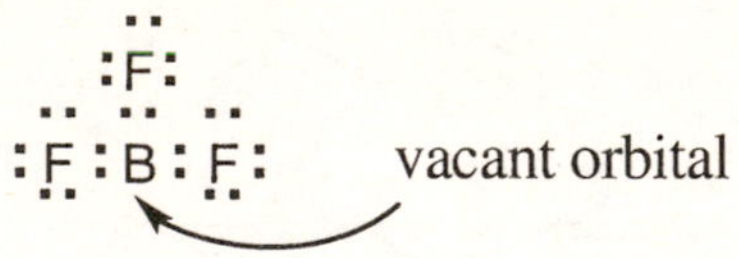

:F:B:F: (with :F: above) — vacant orbital

5.6 **Strategy:** Reaction of cyclohexene with H_2O is an electrophilic addition reaction in which water adds to a double bond to produce an alcohol.

Solution:

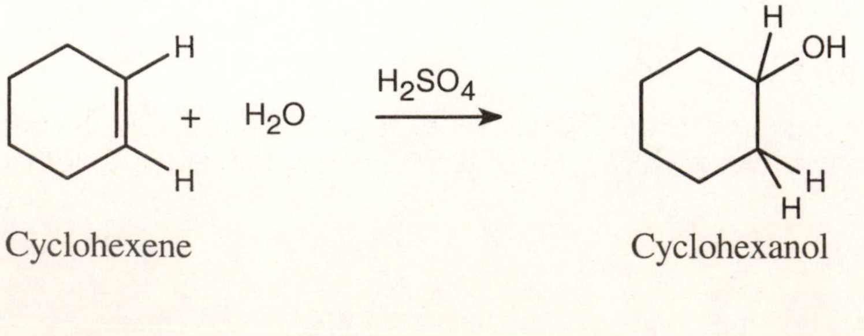

Cyclohexene + H_2O $\xrightarrow{H_2SO_4}$ Cyclohexanol

5.7 The mechanism is pictured in Figure 5.3. The steps are: (1)Reaction of the double bond with H_3O^+, forming a carbocation; (2) formation of a C–O bond by electron pair donation from H_2O to form the protonated addition product; (3) removal of H^+, regenerating H_3O^+ and forming 2-methylpropan-2-ol.

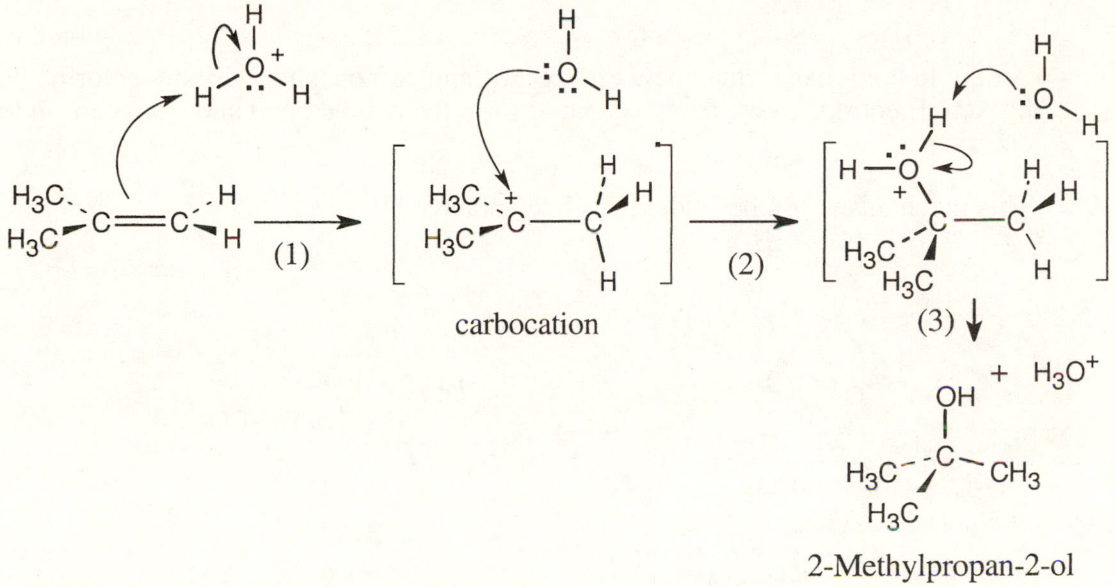

carbocation

2-Methylpropan-2-ol

5.8 **Strategy:** For curved arrow problems, follow these steps:
(1) Locate the bonding changes. In (a), a bond from nitrogen to chlorine has formed, and a Cl–Cl bond has broken.
(2) Identify the nucleophile and electrophile (in (a), the nucleophile is ammonia and the electrophile is one Cl in the Cl_2 molecule), and then draw a curved arrow whose tail is near the nucleophile and whose head is near the electrophile.
(3) Check to see that all bonding changes are accounted for. In (a), we must draw a second arrow to show the polar bond-breaking of Cl_2 to form Cl^-.

Solution:

(a)

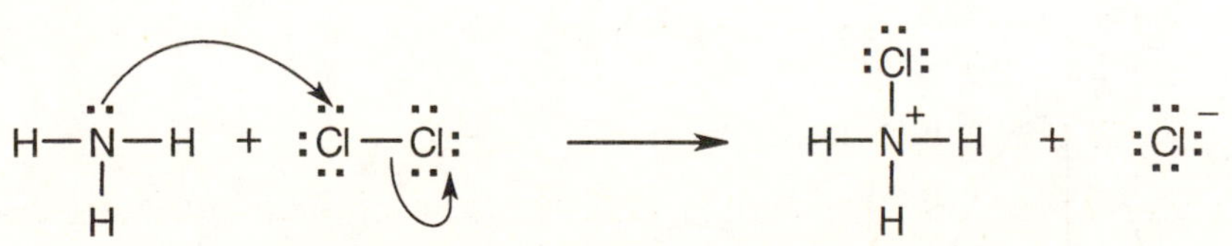

(b)

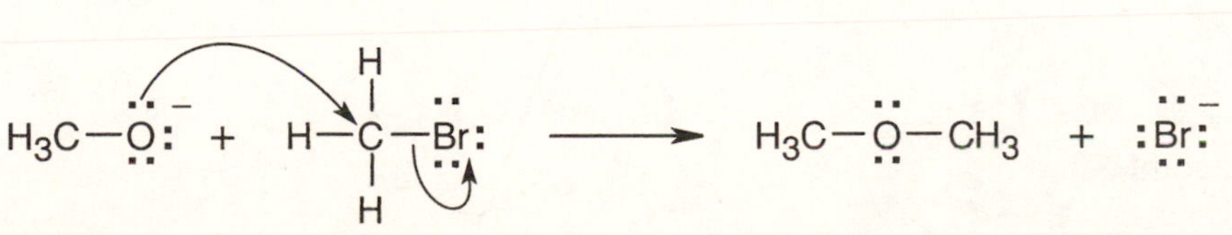

A bond has formed between oxygen and the carbon of bromomethane, and the bond between carbon and bromine has broken. CH_3O^- is the nucleophile and bromomethane is the electrophile.

(c)

A double bond has formed between oxygen and carbon, and a carbon–chlorine bond has broken. Electrons move from oxygen to form the double bond and from carbon to chlorine.

5.9 This mechanism will be studied in a later chapter.

5.10 A negative value of $\Delta G°$ indicates that a reaction is favorable. Thus, a reaction with $\Delta G° = -44$ kJ/mol is more favorable than a reaction with $\Delta G° = +44$ kJ/mol.

5.11 A large K_{eq} is related to a large negative $\Delta G°$. Consequently, a reaction with $K_{eq} = 1000$ is more exergonic than a reaction with $K_{eq} = 0.001$.

5.12 A reaction with $\Delta G^{\ddagger} = 45$ kJ/mol is faster than a reaction with $\Delta G^{\ddagger} = 70$ kJ/mol because a larger value for $\Delta G^{\ddagger}$ indicates a slower reaction.

5.13

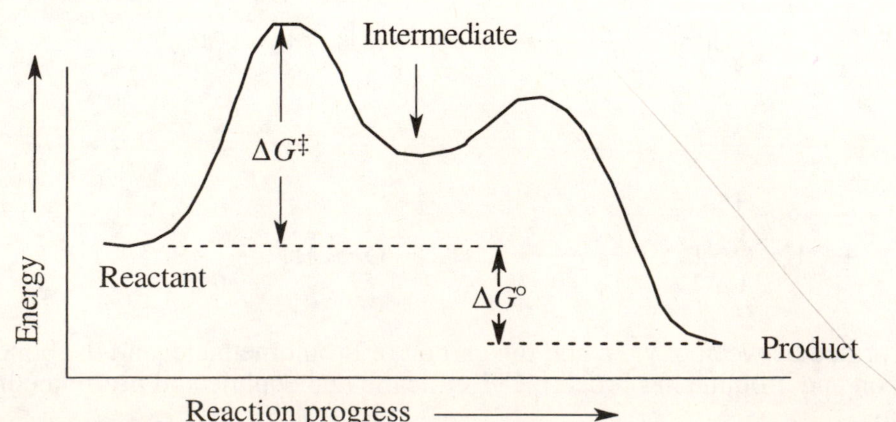

Visualizing Chemistry

5.14

$$CH_3CH_2CH_2CH{=}CH_2 + H_2O \xrightarrow{H_2SO_4} CH_3CH_2CH_2\overset{OH}{\overset{|}{C}}HCH_3$$

or

$$CH_3CH_2CH{=}CHCH_3 + H_2O \xrightarrow{H_2SO_4} CH_3CH_2CH_2\overset{OH}{\overset{|}{C}}HCH_3$$

5.15

CH_2

or

CH_3

$\xrightarrow{H_3O^+}$ [$+$ CH_3] $\xrightarrow{H_3O^+}$ OH CH_3

5.16 (a) The electrostatic potential map shows that the formaldehyde oxygen is electron-rich, and the carbon–oxygen bond is polarized. The carbon atom is thus relatively electron-poor and is likely to be electrophilic.
(b) The sulfur atom is more electron-rich than the other atoms of methanethiol and is likely to be nucleophilic.

5.17

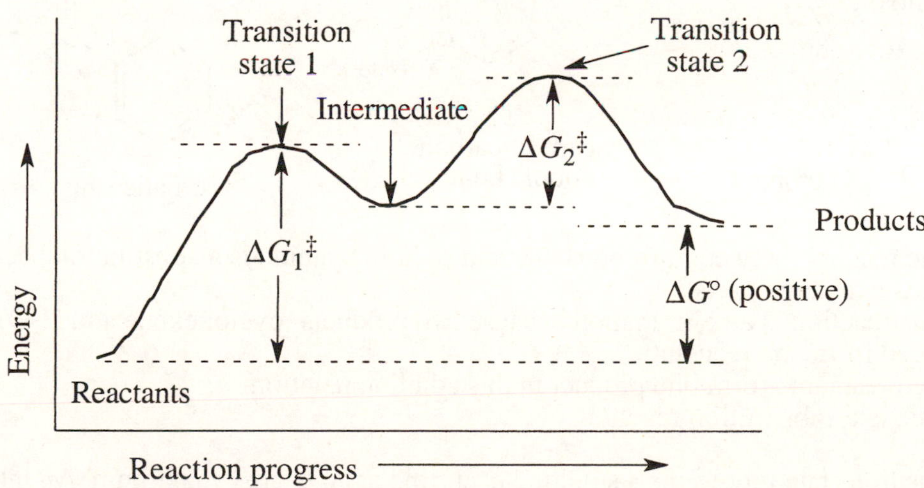

(a) $\Delta G°$ is positive.
(b) There are two steps in the reaction.
(c) There are two transition states, as indicated on the diagram.

5.18

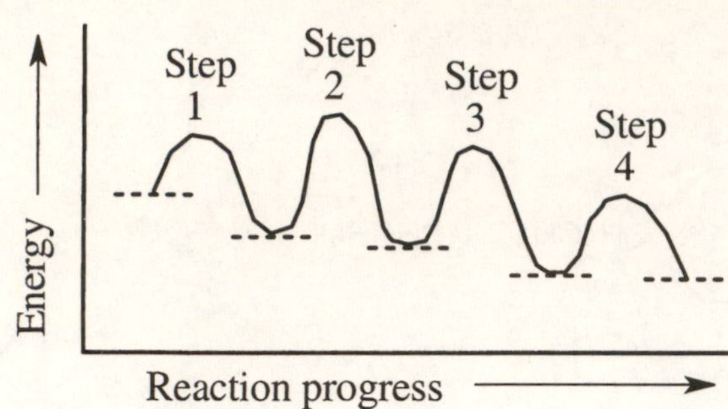

(a) The reaction involves four steps, noted above.
(b) Step 1 is the most exergonic because the energy difference between reactant and product ($\Delta G°$) is greatest.
(c) Step 2 is slowest because it has the largest value of $\Delta G^{\ddagger}$.

Additional Problems

5.19

(a) δ+ δ– $CH_3CH_2C{\equiv}N$ — nitrile

(b) δ+ O δ+ CH_3, δ– — ether

(c) ketone, ester — $CH_3CCH_2C{-}OCH_3$ (δ– O, δ– O; δ+ δ+ δ–)

(d) carbon-carbon double bonds; δ– O, δ+ ← ketone; δ+, δ– O ← ketone

(e) δ– O, δ+ C, δ– NH_2 — amide; carbon-carbon double bond

(f) δ– O, δ+ C ← aldehyde, H; aromatic ring

5.20 (a) The reaction between bromoethane and sodium cyanide is a substitution because two reagents exchange parts.
(b) This reaction is an elimination because two products (cyclohexene and H_2O) are produced from one reactant
(c) Two reactants form one product in this addition reaction.
(d) This is a substitution reaction.

5.21 A transition state represents a structure occurring at an energy maximum. An intermediate occurs at an energy minimum between two transition states. Even though an intermediate may be of such high energy that it cannot be isolated, it is still of lower energy than the transition states surrounding it.

5.22

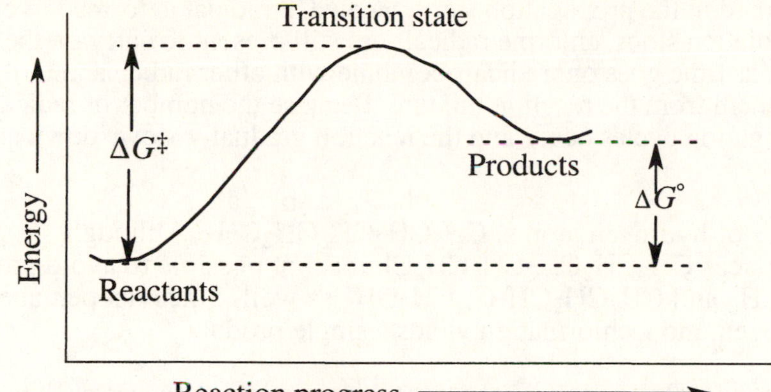

$\Delta G°$ is positive because $K_{eq} < 1$.

5.23

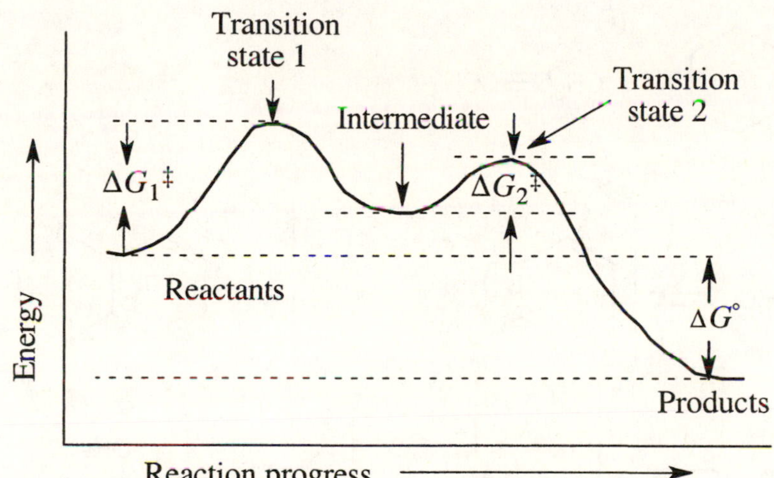

$\Delta G°$ is negative because $K_{eq} > 1$.

5.24 Problem 5.23 shows a reaction energy diagram of a two-step exergonic reaction. Step 2 is faster than step 1 because $\Delta G^{\ddagger}_2 < \Delta G^{\ddagger}_1$.

5.25

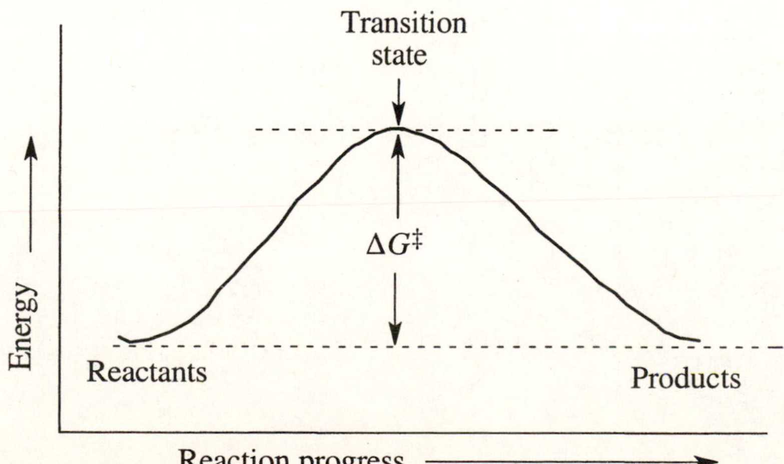

A reaction with $K_{eq} = 1$ has $\Delta G° = 0$.

5.26 Irradiation initiates the chlorination reaction by producing chlorine radicals. For every chlorine radical consumed in the propagation steps, a new Cl· radical is formed to carry on the reaction. After irradiation stops, chlorine radicals are still present to carry on the propagation steps, but, as time goes on, radicals combine with other radicals in termination reactions that remove them from the reaction mixture. Because the number of radicals decreases, fewer propagation cycles occur and the reaction gradually slows down and stops.

5.27 Pentane has three types of hydrogen atoms, $\overset{a}{CH_3}\overset{b}{CH_2}\overset{c}{CH_2}\overset{b}{CH_2}\overset{a}{CH_3}$. Although monochlorination produces $CH_3CH_2CH_2CH_2CH_2Cl$, it is not possible to avoid producing $CH_3CH_2CH_2CH(Cl)CH_3$ and $CH_3CH_2CH(Cl)CH_2CH_3$ as well. Since neopentane has only one type of hydrogen, monochlorination yields a single product.

5.28 The following compounds yield single monohalogenation products because each has only one kind of hydrogen atom.

(a) CH_3CH_3 (c) [cyclohexane] (f) $CH_3C{\equiv}CCH_3$

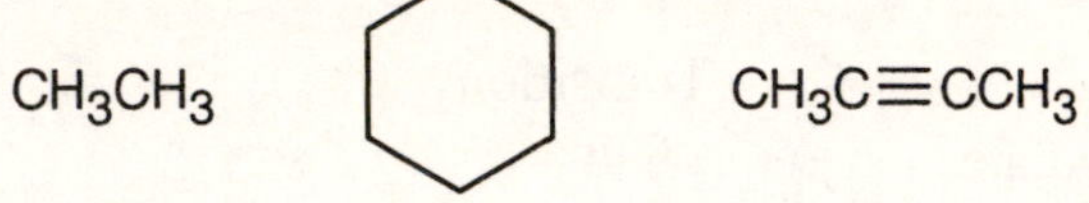

5.29

(a) [Mechanism: benzene (H) + D–Cl ⇌ [arenium ion with D, H, + ; :Cl:⁻] ⇌ deuterated benzene (D, H) + H–Cl]

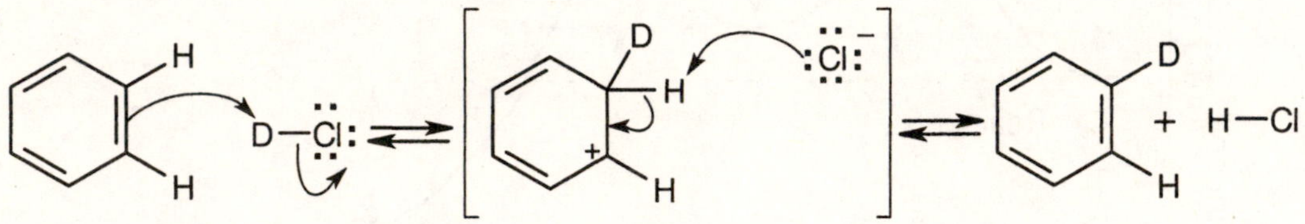

(b) [Mechanism: methyloxirane (CH₃) + H–Cl ⇌ [protonated epoxide (H, +O), :Cl:⁻] ⇌ Cl–CH₂–CH(OH)–CH₃]

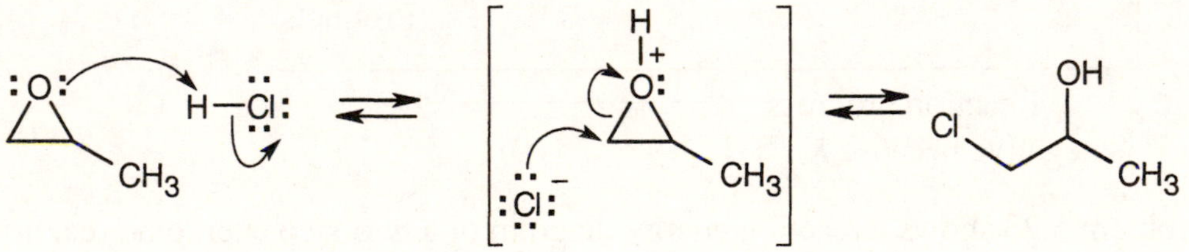

5.30

(a) [Mechanism: $(H_3C)_2C(OH)(OCH_3)$ + :OH⁻ ⇌ H_3C–C(=O)–CH_3 + H_2O + :OCH₃⁻]

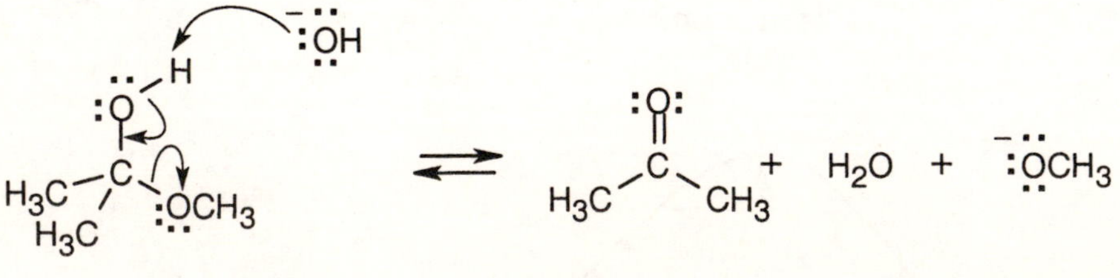

(b) [Mechanism: HO:⁻ + H_3C–C(=O)–CH_3 (H atoms on α-carbon) ⇌ H_2O + H_2C=C(O⁻)–CH_3]

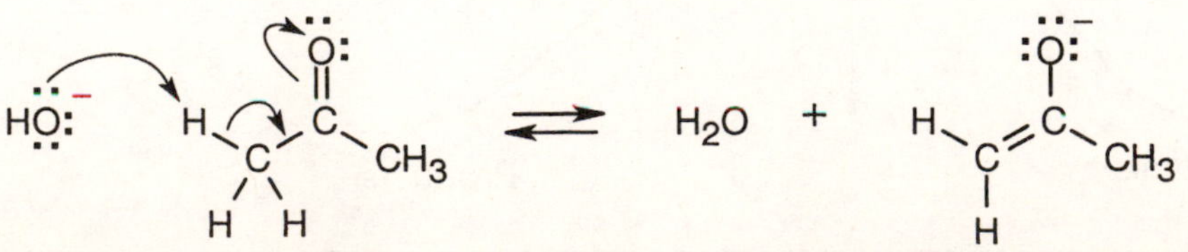

5.31

(a)

$$K_{eq} = \frac{[\text{Products}]}{[\text{Reactants}]} = \frac{0.70}{0.30} = 2.3$$

(b) Section 5.9 states that reactions that occur spontaneously have $\Delta G^{\ddagger}$ of less than 80 kJ/mol at room temperature. Since this reaction proceeds slowly at room temperature, $\Delta G^{\ddagger}$ is probably close to 80 kJ/mol.

(c)

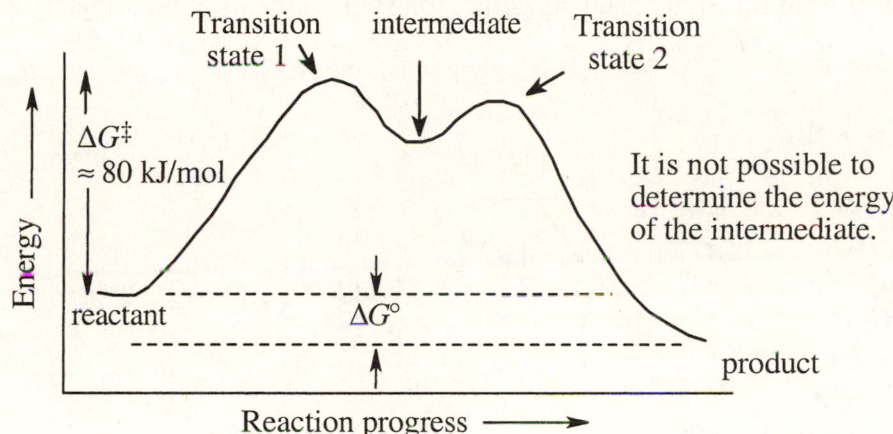

5.32

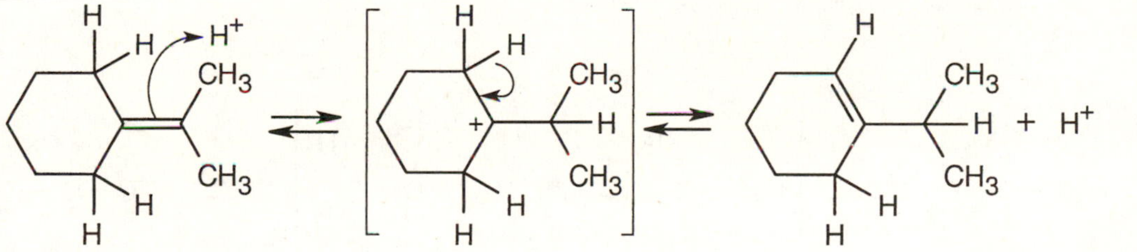

5.33

$$(CH_3)_3C{-}Cl \underset{\text{T.S. 1}}{\rightleftharpoons} \left[(CH_3)_3C^{+} \underset{\text{T.S. 2}}{\overset{H_2O}{\rightleftharpoons}} (CH_3)_3C{-}O^{+}H_2 \right] \underset{\text{T.S. 3}}{\overset{H_2O}{\rightleftharpoons}} (CH_3)_3C{-}O{-}H + H_3O^{+} + Cl^{-}$$

Intermediate 1 Intermediate 2

(a) $\Delta G^{\ddagger}$ for the first step is approximately 80 kJ/mol because the reaction takes place slowly at room temperature. $\Delta G^{\ddagger}$ values for the second and third steps are smaller - perhaps 60 kJ/mol for Step 2, and 40 kJ/mol for Step 3. ΔG° is approximately zero.

(b)

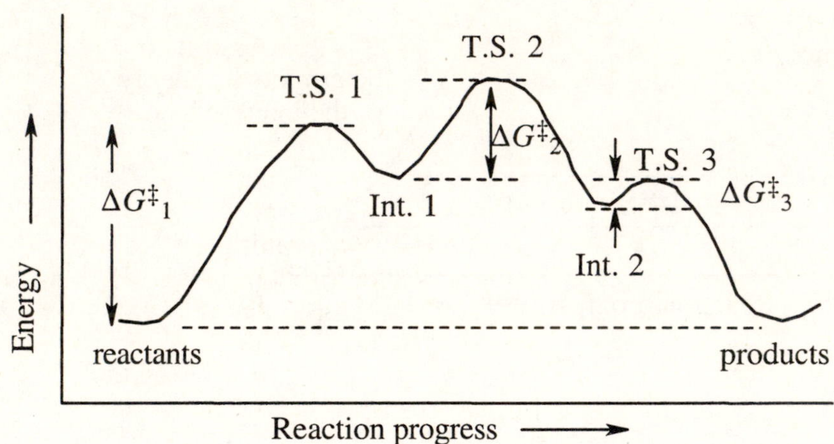

5.34

$$(CH_3)_3C{-}\ddot{Cl}: \rightleftharpoons \left[(CH_3)_3C^{+} + :\ddot{Cl}:^{-} \xrightarrow{H_2\ddot{O}} (CH_3)_3C{-}\ddot{O}^{+}H_2 \right] \xrightarrow{H_2\ddot{O}} (CH_3)_3C{-}O{-}H + H_3O^{+}$$

5.35 $\Delta G^{\circ} = \Delta H^{\circ} - T\Delta S^{\circ}$

$= -75$ kJ/mol – (298 K) (0.054 kJ/K·mol)

$= -75$ kJ/mol – 16 kJ/mol

$= -91$ kJ/mol

The reaction is exothermic because ΔH° is negative, and it is exergonic because ΔG° is negative.

5.36 Each arrow represents either the formation of a bond or the breaking of a bond. The numbers over the arrows identify the bonds broken and formed.

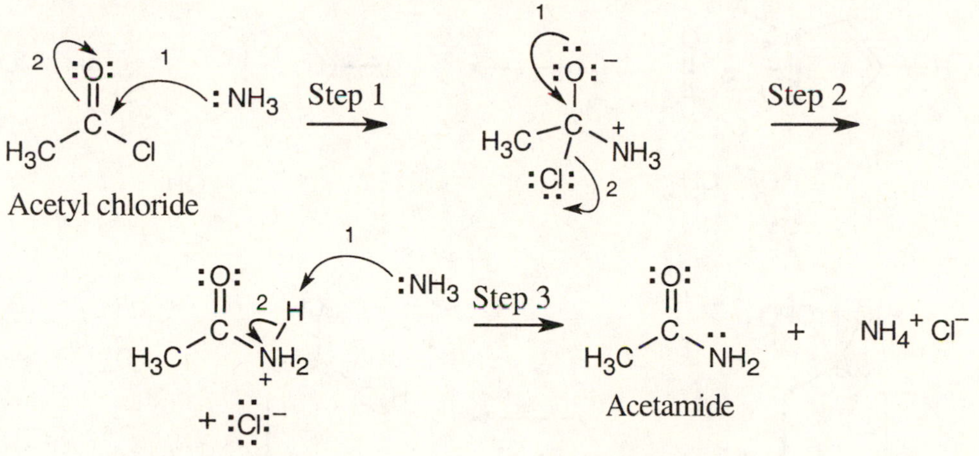

Acetyl chloride

Step 1

Step 2

Step 3

Acetamide

$NH_4^+\ Cl^-$

	Bonds formed	*Bonds broken*
Step 1:	C–N (1)	C–O (2)
Step 2:	C–O (1)	C–Cl (2)
Step 3:	N–H (1)	N–H (2)

5.37

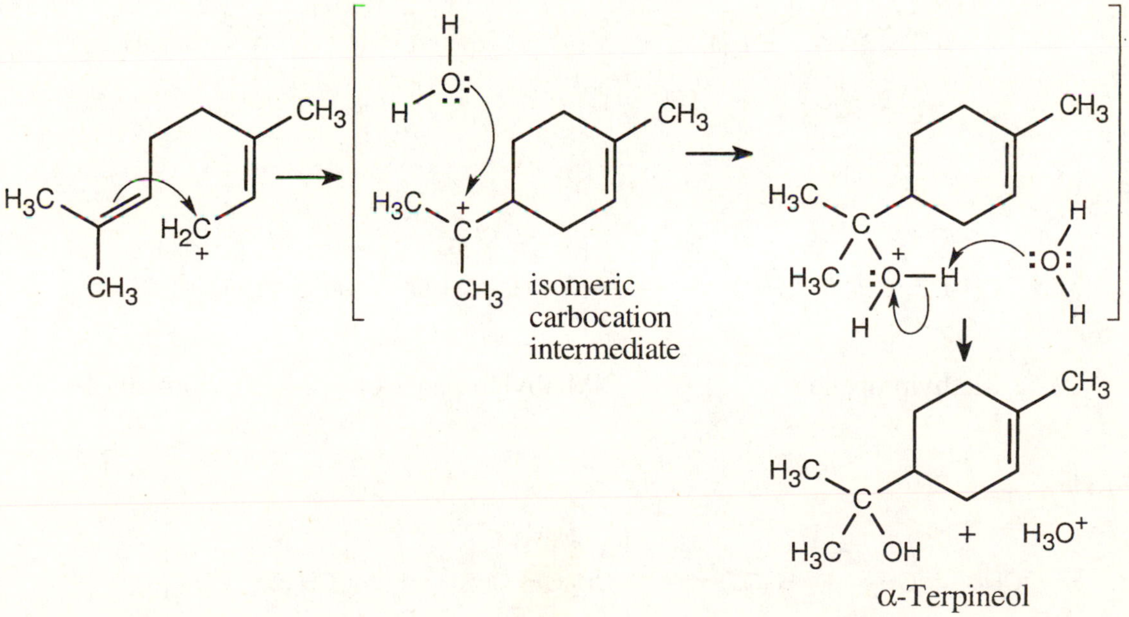

isomeric carbocation intermediate

$+\ H_3O^+$

α-Terpineol

5.38

(a)

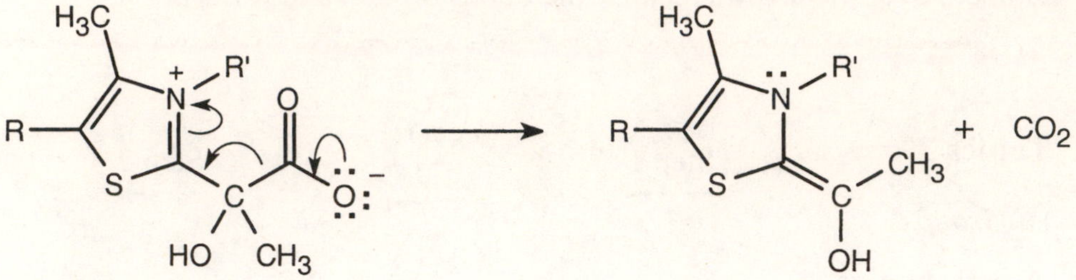

(b)

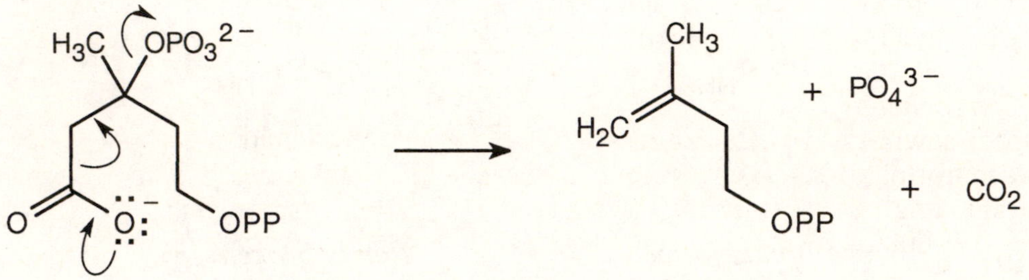

(c)

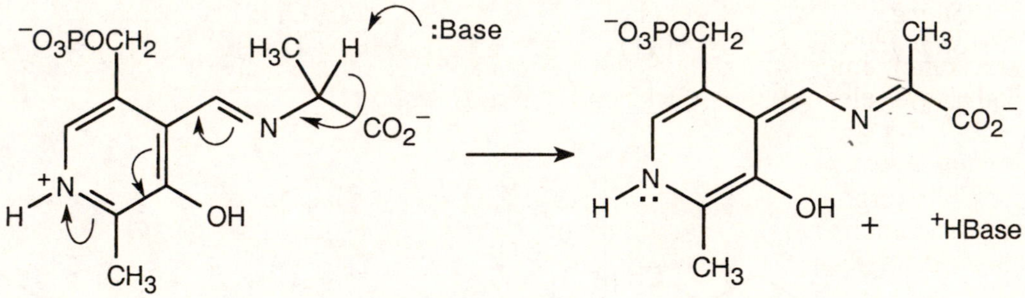

5.39

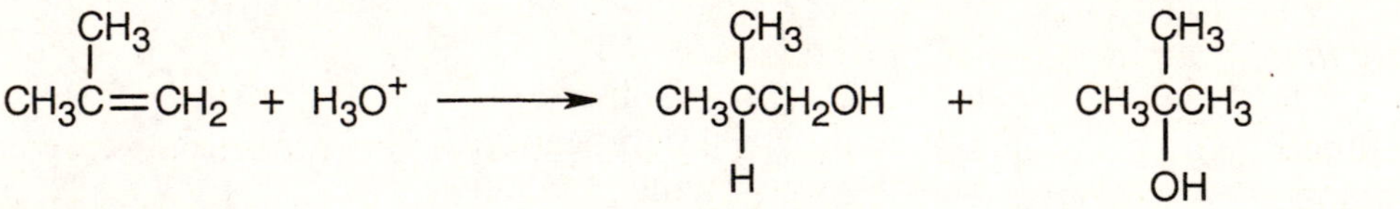

2-Methylpropene 2-Methylpropan-1-ol 2-Methylpropan-2-ol

5.40

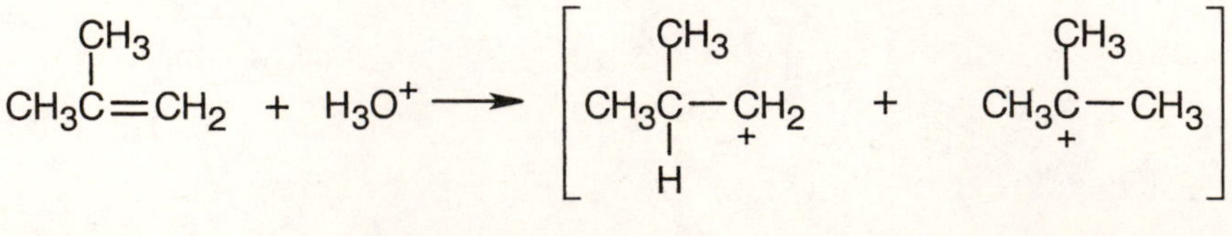

The second carbocation is more stable because more alkyl substituents are bonded to the positively charged carbon.

Review Unit 2: Alkanes and Their Stereochemistry: Organic Reactions

Major Topics Covered (with Vocabulary):

Functional Groups.

Alkanes:
saturated aliphatic straight-chain alkane branched-chain alkane isomer constitutional isomer alkyl group primary, secondary, tertiary, quaternary carbon IUPAC system of nomenclature primary, secondary, tertiary hydrogen paraffin cycloalkane cis-trans isomer stereoisomer

Alkane Stereochemistry:
conformer sawhorse representation Newman projection staggered conformation eclipsed conformation torsional strain dihedral angle anti conformation gauche conformation steric strain angle strain heat of combustion chair conformation axial group equatorial group ring-flip 1,3-diaxial interaction conformational analysis polycyclic molecules

Organic Reactions:
addition reaction elimination reaction substitution reaction rearrangement reaction reaction mechanism homolytic heterolytic homogenic heterogenic radical reaction polar reaction initiation propagation termination electronegativity polarizability curved arrow electrophile nucleophile carbocation

Describing a Reaction:
K_{eq} ΔG° exergonic endergonic enthalpy entropy heat of reaction exothermic endothermic bond dissociation energy reaction energy diagram transition state activation energy reaction intermediate

Types of Problems:

After studying these chapters, you should be able to:

- Identify functional groups, and draw molecules containing a given functional group.
- Draw all isomers of a given molecular formula.
- Name and draw alkanes, alkyl groups, and cycloalkanes, including cis-trans isomers.
- Identify carbons and hydrogens as being primary, secondary or tertiary.
- Draw energy vs. angle of rotation graphs for single bond conformations.
- Draw Newman projections of bond conformations and predict their relative stability.

- Understand the geometry of, and predict the stability of, cycloalkanes having fewer than 6 carbons.
- Draw and name substituted cyclohexanes.
- Predict the stability of substituted cyclohexanes by estimating steric interactions.

- Identify reactions as polar, radical, substitution, elimination, addition, or rearrangement reactions.
- Understand the mechanism of radical reactions.

- Identify reagents as electrophiles or nucleophiles.
- Use curved arrows to draw reaction mechanisms.
- Understand the concepts of equilibrium and rate.
- Calculate K_{eq} and $\Delta G°$ of reactions.
- Draw reaction energy diagrams and label them properly.

Points to Remember:

* In identifying the functional groups in a compound, some groups have different designations that depend on the number and importance of other groups in the molecule. For example, a compound containing an –OH group and few other groups is probably named as an alcohol, but when several other groups are present , the –OH group is referred to as a hydroxyl group. There is a priority list of functional groups in the Appendix of this book, and this priority order will become more apparent as you progress through the text.

* It is surprising how many errors can be made in naming compounds as simple as alkanes. Why is this? Often the problem is a result of just not paying attention. It is very easy to undercount or overcount the $-CH_2-$ groups in a chain and to misnumber substituents. Let's work through a problem, using the rules in Section 3.4.

```
   CH3 CH2CH3
    |   |
CH3CCH2CHCH2CH3
    |
   CH3
```

<u>Find</u> the longest chain. In the above compound, the longest chain is a hexane (Try all possibilities; there are two different six-carbon chains in the compound.) <u>Identify</u> the substituents. The compound has two methyl groups and an ethyl group. It's a good idea to list these groups to keep track of them. <u>Number</u> the chain and the groups. Try both possible sets of numbers, and see which results in the lower combination of numbers. The compound might be named either as a 2,2,4-trisubstituted hexane or a 3,5,5-trisubstituted hexane, but the first name has a lower combination of numbers. <u>Name</u> the compound, remembering the prefix *di-* and remembering to list substituents in alphabetical order. The correct name for the above compound is 4-ethyl-2,2-dimethylhexane.
The acronym FINN (from the first letters of each step listed above) may be helpful.

* When performing a ring-flip on a cyclohexane ring, keep track of the positions on the ring. Compound A undergoes a ring-flip to compound B, not to the compound C. I

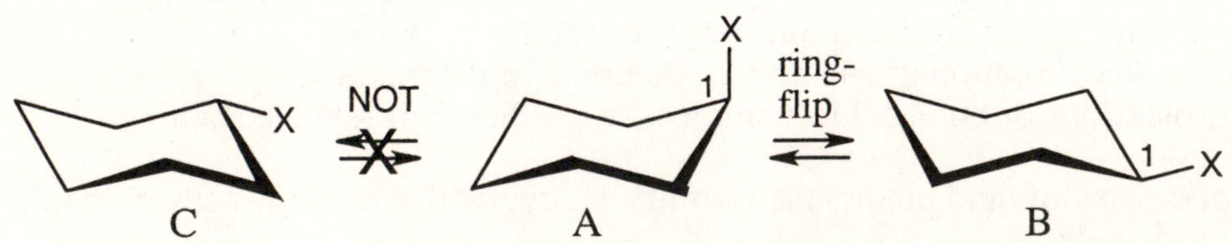

* In virtually all cases, a compound is of lower energy than the free elements of which it is composed. Thus, energy is released when a compound is formed from its component elements, and energy is required when bonds are broken. Entropy decreases when a compound is formed from its component elements (because disorder decreases). For two compounds of similar structure, less energy is required to break all bonds of the higher energy compound than is required to break all bonds of the lower energy compound.

Self-test

CH_3 $CH(CH_3)_2$
$CH_3CHCH_2CH_2CCH_2CHCH_2CH_3$
A CH_3 CH_2CH_3

CH_3 CH_3
$CH_3CHCH_2CH_2CH_2CHNHCH(CH_3)_2$
B
Metron S (an antihistamine)

Name **A**, and identify carbons as primary, secondary, tertiary or quaternary.
B is an amine with two alkyl groups. Name these groups and identify alkyl hydrogens as primary, secondary or tertiary.

CH_3 CH_3 O
$CHCOCH_3$
N
CCH_2OCH_3
C CH_3 O Metalaxyl (a fungicide)

CH_3
H_3C Br
D

Identify all functional groups of **C** (metalaxyl).

Indicate the cis/trans relationship of the substituents in **D**. Draw both possible chair conformations, and calculate the energy difference between them.

O
H_3C C CH_3 + H_2O $\rightleftharpoons$ HO OH H_3C C CH_3
E

What type of reaction is occurring in **E**? Would you expect that the reaction occurs by a polar or a radical mechanism? If K_{eq} for the reaction at 298 K is 10^{-3}, what sign do you expect for $\Delta G°$? Would you expect $\Delta S°$ to be negative or positive? What about $\Delta H°$?

Multiple Choice

1. Which of the following functional groups doesn't contain a carbonyl group?
(a) aldehyde (b) ester (c) ether (d) ketone

2. Which of the following compounds contains primary, secondary, tertiary and quaternary carbons?
(a) 2,2,4-Trimethylhexane (b) Ethylcyclohexane (c) 2-Methyl-4-ethylcyclohexane
(d) 1,1-Dimethylcyclohexane

3. How many isomers of the formula $C_4H_8Br_2$ are there?
(a) 4 (b) 6 (c) 8 (d) 9

4. The lowest energy conformation of 2-methylbutane occurs:
(a) when all methyl groups are anti (b) when all methyl groups are gauche
(c) when two methyl groups are anti (d) when two methyl groups are eclipsed

5. The strain in a cyclopentane ring is due to:
(a) angle strain (b) torsional strain (c) steric stain (d) angle strain and torsional strain

6. In which molecule do the substituents in the more stable conformation have a diequatorial relationship?
(a) cis-1,2 disubstituted (b) cis-1,3 disubstituted (c) trans-1,3-disubstituted
(d) cis-1,4 disubstituted

7. Which of the following molecules is not a nucleophile?
(a) BH_3 (b) NH_3 (c) HO^- (d) $H_2C{=}CH_2$

8. Which of the following reactions probably has the greatest entropy increase?
(a) addition reaction (b) elimination reaction (c) substitution reaction (d) rearrangement

9. At a specific temperature T, a reaction has negative $\Delta S°$ and $K_{eq} > 1$. What can you say about $\Delta G°$ and $\Delta H°$?
(a) $\Delta G°$ is negative and $\Delta H°$ is positive (b) $\Delta G°$ and $\Delta H°$ are both positive (c) $\Delta G°$ and $\Delta H°$ are both negative (d) $\Delta G°$ is negative but you can't predict the sign of $\Delta H°$.

10. In which of the following situations is $\Delta G^‡$ likely to be smallest?
(a) a slow exergonic reaction (b) a fast exergonic reaction (c) a fast endergonic reaction
(d) a slow endergonic reaction

Chapter 6 – Alkenes and Alkynes

Chapter Outline

I. Introduction to alkene and alkyne chemistry (Sections 6.1 – 6.5).
   A. Calculating a molecule's degree of unsaturation (Section 6.1).
      1. The degree of unsaturation of a molecule describes the number of multiple bonds and/or rings a molecule has.
      2. To calculate degree of unsaturation of a compound, first determine the equivalent hydrocarbon formula of the compound.
         a. Add the number of halogens to the number of hydrogens.
         b. Subtract one hydrogen for every nitrogen.
         c. Ignore the number of oxygens.
      3. Calculate the number of pairs of hydrogens that would be present in an alkane C_nH_{2n+2} that has the same number of carbons as the equivalent hydrocarbon of the compound of interest. The difference is the degree of unsaturation.
   B. Naming alkenes and alkynes (Section 6.2).
      1. Find the longest chain containing the double or triple bond, and name it.
      2. Number the carbon atoms in the chain, beginning at the end nearer the double bond.
      3. Number the substituents and write the name.
         a. Name the substituents alphabetically.
         b. Indicate the position of the double or triple bond.
         c. The suffix -ene is used for alkenes; the suffix -yne is used for alkynes.
         d. Use the suffixes -diene, -diyne, -triene , etc. if more than one double or triple bond is present.
   C. Double bond geometry (Sections 6.3 – 6.4).
      1. Electronic structure of alkenes (Section 6.3).
         a. Carbon atoms in a double bond are sp^2-hybridized.
         b. The two carbons in a double bond form one σ bond and one π bond.
         c. Free rotation doesn't occur around double bonds.
         d. 268 kJ/mol of energy is required to break a π bond.
      2. Cis–trans isomerism .
         a. A disubstituted alkene can have substituents either on the same side of the double bond (cis) or on opposite sides (trans).
         b. These isomers don't interconvert because free rotation about a double bond isn't possible.
         c. Cis–trans isomerism doesn't occur if one carbon in the double bond is bonded to identical substituents.
      3. *E,Z* isomerism (Section 6.4).
         a. The *E,Z* system is used to describe the arrangement of substituents around a double bond that can't be described by the cis–trans system.
         b. Sequence rules for *E,Z* isomers:
            i. For each double bond carbon, rank its substituents by atomic number. An atom with a high atomic number receives a higher priority than an atom with a lower atomic number.
            ii. If a decision can't be reached, look at the second or third atom until a difference is found.
            iii. Multiple-bonded atoms are equivalent to the same number of single-bonded atoms.

D. Stability of alkenes (Section 6.5).
  1. Cis alkenes are less stable than trans alkenes because of steric strain between double bond substituents.
  2. Stabilities of alkenes can be determined experimentally by measuring:
     a. Cis–trans equilibrium constants.
     b. Heats of hydrogenation – the most useful method.
  3. The heat of hydrogenation of a cis isomer is a larger negative number than the heat of hydrogenation of a trans isomer.
     This indicates that a cis isomer is of higher energy and is less stable than a trans isomer.
  4. Alkene double bonds become more stable with increasing substitution for two reasons:
     a. Hyperconjugation – a stabilizing interaction between the antibonding π orbital and a filled C–H σ orbital on an adjacent substituent.
     b. More substituted double bonds have more of the stronger sp^2–sp^3 bonds.

II. Electrophilic addition reactions (Sections 6.6 - 6.10).

A. Addition of H_2O to alkenes (Sections 6.6 - 6.7).
  1. Mechanism of addition (Section 6.6).
     a. The electrons of the nucleophilic π bond attack the electrophile H_3O^+.
     b. Two electrons from the π bond form a new σ bond between –H and an alkene carbon.
     c. The carbocation intermediate reacts with H_2O to form a C–OH_2^+ bond.
     d. Water acts as a base to remove H^+, yielding an alcohol product.
     e. Reactions also occur with HBr, HCl and HI.
     f. Alkynes also undergo similar reactions.
  2. Organic reactions are often written in different ways to emphasize different points.
  3. Orientation of addition: Markovnikov's Rule (Section 6.7).
     a. In the addition of HX to a double bond, H attaches to the carbon with fewer substituents, and X attaches to the carbon with more substituents.
     b. If the carbons have the same number of substituents, a mixture of products results.

B. Carbocation structure and stability (Section 6.8).
  1. Carbocations are planar: the unoccupied *p* orbital extends above and below the plane.
  2. The stability of carbocations increases with increasing substitution.
     a. Carbocations can be stabilized by inductive effects of neighboring alkyl groups.
     b. Carbocation can be stabilized by hyperconjugation: The more alkyl groups on the carbocation, the more opportunities there are for hyperconjugation.

C. The Hammond Postulate (Section 6.9).
  1. The transition state for an endergonic reaction step resembles the product of that step.
  2. The transition state for an exergonic reaction step resembles the reactant for that step.
  3. In an electrophilic addition reaction, the transition state for alkene protonation resembles the carbocation intermediate.
  4. More stable carbocations form faster because their transition states are also stabilized.

D. Carbocation rearrangements (Section 6.10).
  1. In some electrophilic addition reactions, products from carbocation rearrangements are formed.
  2. The appearance of these products supports the two-step electrophilic addition mechanism, in which an intermediate carbocation is formed.
  3. Intermediate carbocations can rearrange to more stable carbocations by either a hydride shift or by an alkyl shift.
  4. In both cases a group moves to an adjacent positively charged carbon, taking its bonding electron pair with it.

Solutions to Problems

6.1 **Strategy:** Because two hydrogens must be removed from a saturated compound to introduce an unsaturation, a compound's degree of unsaturation refers to the number of pairs of hydrogens by which it's formula differs from that of the corresponding saturated compound. For example, a saturated alkane with eight carbons has the formula C_8H_{18}. The compound in (a), C_8H_{14}, has four fewer (or two pairs fewer) hydrogens and thus has a degree of unsaturation of 2. C_8H_{14} may have two double bonds, or two rings, or one of each, or a triple bond.

Solution:

Compound	*Degree of unsaturation*	*Compound*	*Degree of unsaturation*
(a) C_8H_{14}	2	(b) C_5H_6	3
(c) $C_{12}H_{20}$	3		

Strategy: Unlike the hydrocarbons in parts (a)–(c), the compounds in parts (d)–(f) contain additional elements. Review the rules for these elements.

Solution:
(d) For C_6H_5N, subtract one hydrogen for each nitrogen present to find the formula of the equivalent hydrocarbon – C_6H_4. Compared to the alkane C_6H_{14}, the compound of formula C_6H_4 has 10 fewer hydrogens, or 5 fewer hydrogen pairs, and has a degree of unsaturation of 5.
(e) $C_6H_5NO_2$ also has 5 degrees of unsaturation because oxygen doesn't affect the equivalent hydrocarbon formula of a compound.
(f) A halogen atom is equivalent to a hydrogen atom in calculating the equivalent hydrocarbon formula. For $C_8H_9Cl_3$, the equivalent hydrocarbon formula is C_8H_{12}, and the degree of unsaturation is 3.

6.2

Compound	*Degree of Unsaturation*	*Structures*
(a) C_4H_8	1	

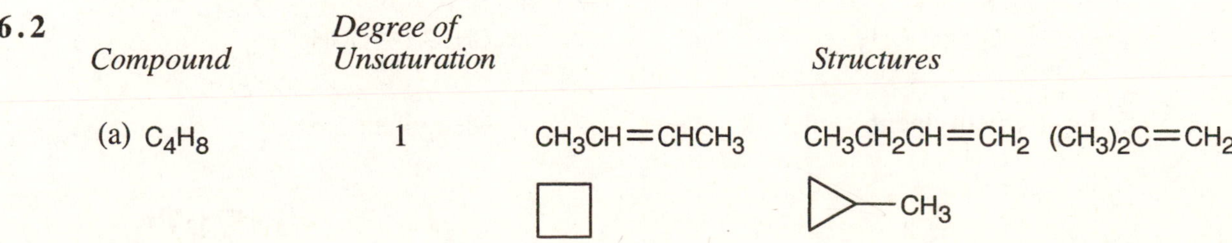

(b) C_4H_6 2 $CH_2{=}CHCH{=}CH_2$ $CH_3CH{=}C{=}CH_2$ $CH_3C{\equiv}CCH_3$

$CH_3CH_2C{\equiv}CH$

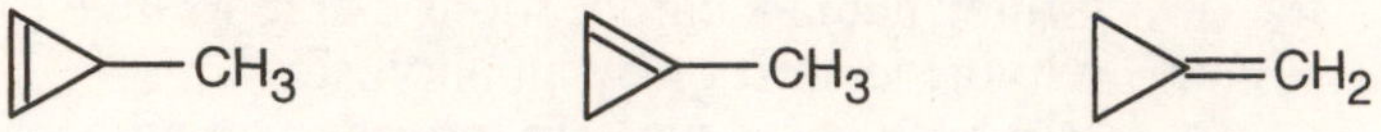

$-CH_3$ $-CH_3$ $=CH_2$

(c) C_3H_4 2 $H_2C{=}C{=}CH_2$ $CH_3C{\equiv}CH$

6.3 A C_{16} hydrocarbon with 11 degrees of unsaturation (three rings and eight double bonds) has a formula $C_{16}H_{34} - H_{22} = C_{16}H_{12}$. Adding two hydrogens (because of the two nitrogens) and subtracting one hydrogen (because of the chlorine), gives the formula $C_{16}H_{13}ClN_2O$ for Diazepam.

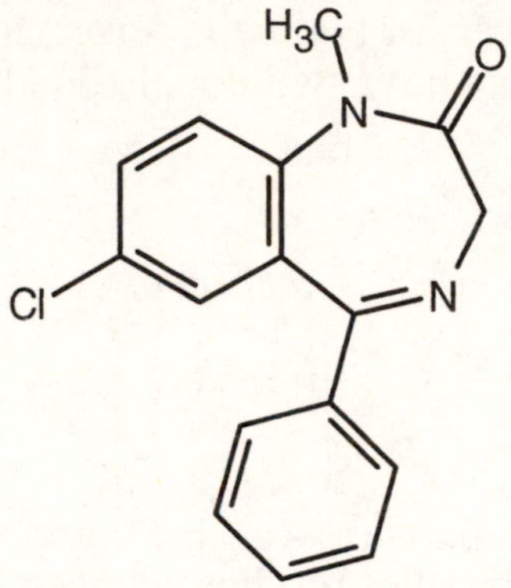

Diazepam

6.4 (1) Find the longest chain containing the double bond and name it. In (a), the longest chain is a pentene.
(2) Identify the substituents. There are three methyl groups in (a).
(3) Number the substituents, remembering that the double bond receives the lowest possible number. The methyl groups are attached to C3 and C4 (two methyl groups).
(4) Name the compound, remembering to use the prefix "tri-" before "methyl" and remembering to use a number to signify the location of the double bond. The name of the compound in (a) is 3,4,4-trimethylpent-1-ene.

(a) $H_2C{=}CHCH(CH_3){-}C(CH_3)_2CH_3$ (numbered 1 2 3 4 5)

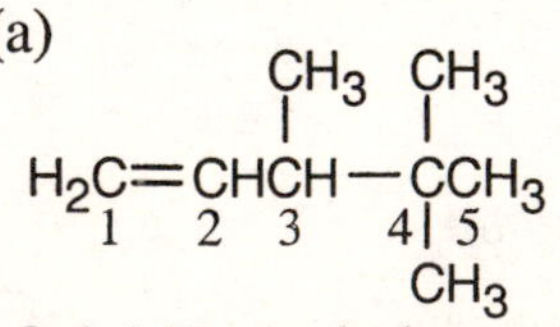

3,4,4-Trimethylpent-1-ene

(b) $CH_3CH_2CH{=}C(CH_3)CH_2CH_3$

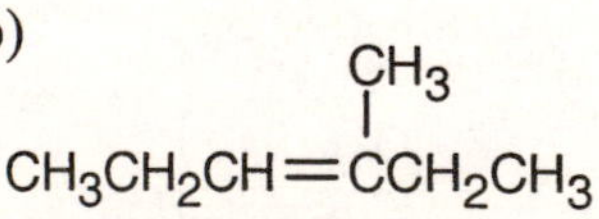

3-Methylhex-3-ene

(c) $CH_3CH{=}CHCH(CH_3)CH{=}CHCH(CH_3)CH_3$

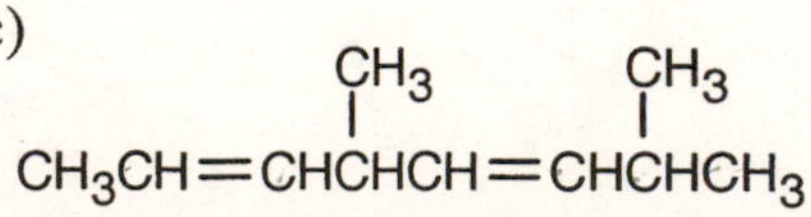

4,7-Dimethylocta-2,5-diene

(d) $CH_3CH_2CH_2CH{=}CHCH(CH(CH_3)CH_2CH_3)CH_2CH_3$

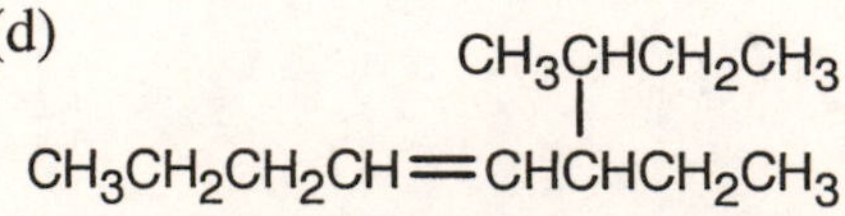

6-Ethyl-7-methylnon-4-ene

(e)

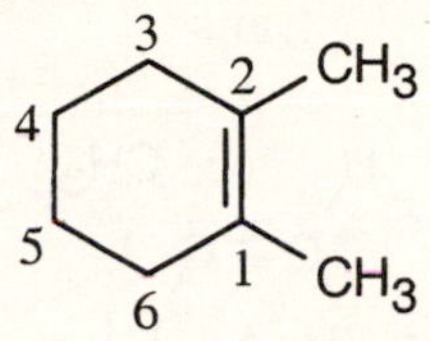

1,2-Dimethylcyclohexene

(f)

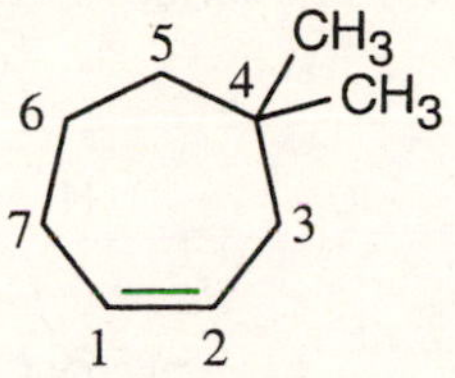

4,4-Dimethylcycloheptene

(g)

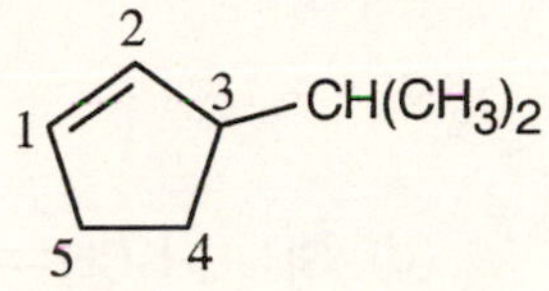

3-Isopropylcyclopentene

6.5

(a)

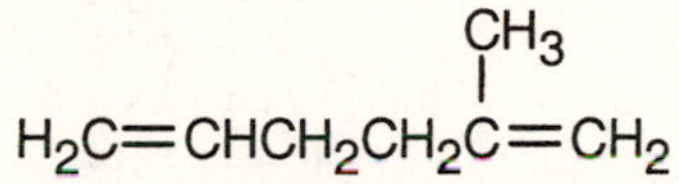

$H_2C{=}CHCH_2CH_2C(CH_3){=}CH_2$

2-Methylhexa-1,5-diene

(b)

$CH_3CH_2CH_2CH{=}C(CH_2CH_3)C(CH_3)_3$

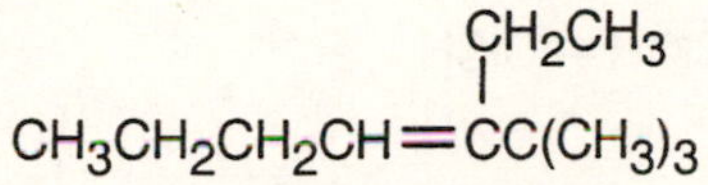

3-Ethyl-2,2-dimethylhept-3-ene

(c)

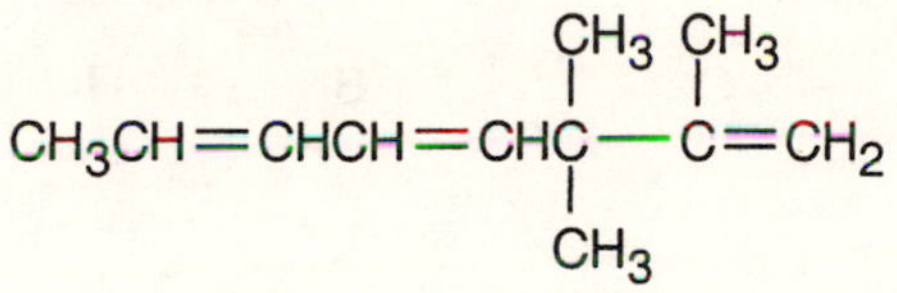

$CH_3CH{=}CHCH{=}CHC(CH_3)_2{-}C(CH_3){=}CH_2$

2,3,3-Trimethylocta-1,4,6-triene

(d)

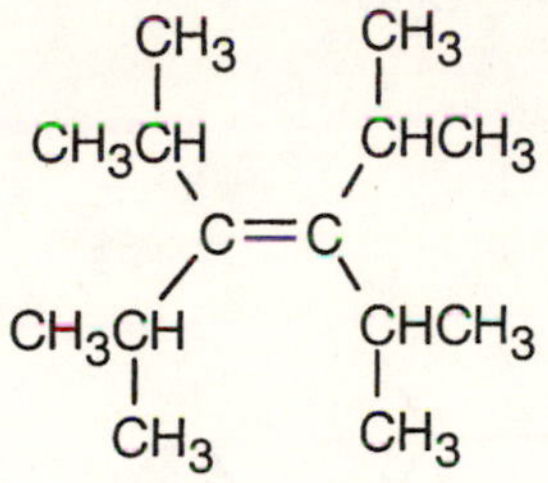

$((CH_3)_2CH)_2C{=}C(CH(CH_3)_2)_2$

3,4-Diisopropyl-2,5-dimethylhex-3-ene

6.6

(a)

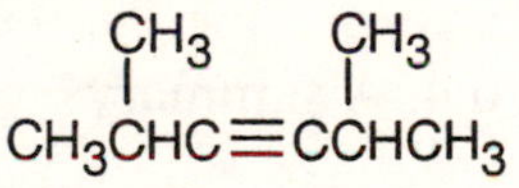

$CH_3CH(CH_3)C{\equiv}CCH(CH_3)CH_3$

2,5-Dimethylhex-3-yne

(b)

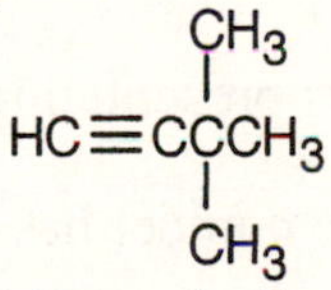

$HC{\equiv}CC(CH_3)_2CH_3$

3,3-Dimethylbut-1-yne

(c)

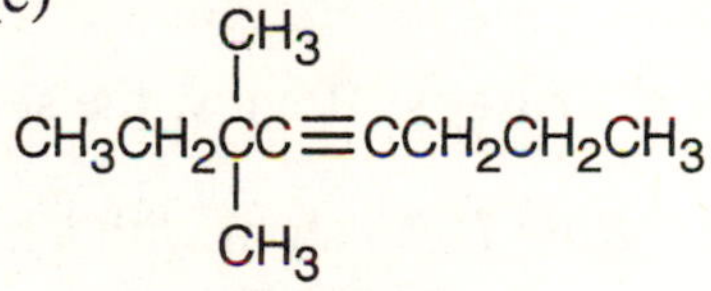

$CH_3CH_2C(CH_3)_2C{\equiv}CCH_2CH_2CH_3$

3,3-Dimethyloct-4-yne

(d)

$CH_3CH_2C(CH_3)_2C{\equiv}CCH(CH_3)CH_3$

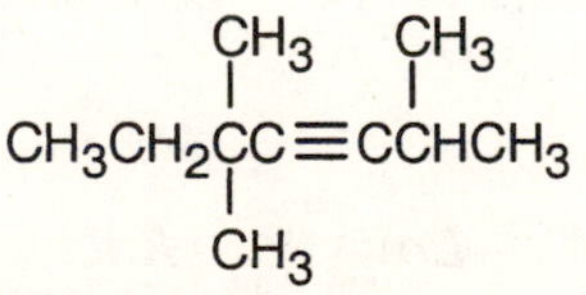

2,5,5-Trimethylhept-3-yne

(e)

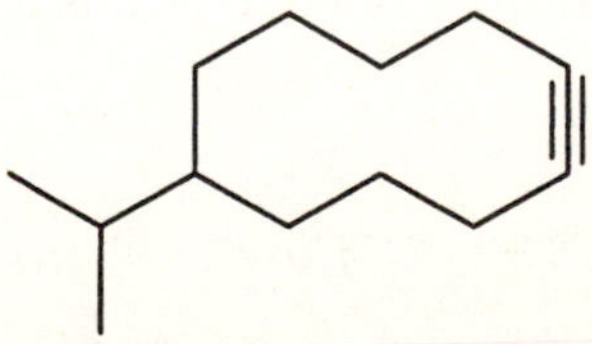

6-Isopropylcyclodecyne

6.7

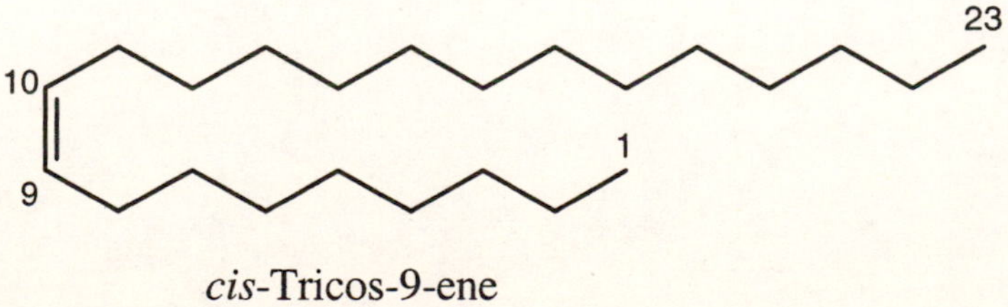

cis-Tricos-9-ene

6.8 Compounds in (c), (e), and (f) can exist as cis-trans isomers.

```
                                          cis                    trans

                                     H       H              H        CH3
                                      \     /                \      /
(c) CH3CH2CH=CHCH3                     C=C                    C=C
                                      /     \                /      \
                                CH3CH2       CH3       CH3CH2        H

                                     H       H              H        Cl
                                      \     /                \      /
(e) ClCH=ClCH                          C=C                    C=C
                                      /     \                /      \
                                    Cl       Cl            Cl        H

                                     H       H              H        Cl
                                      \     /                \      /
(f) BrCH=CHCl                          C=C                    C=C
                                      /     \                /      \
                                    Br       Cl            Br        H
```

6.9

(a)

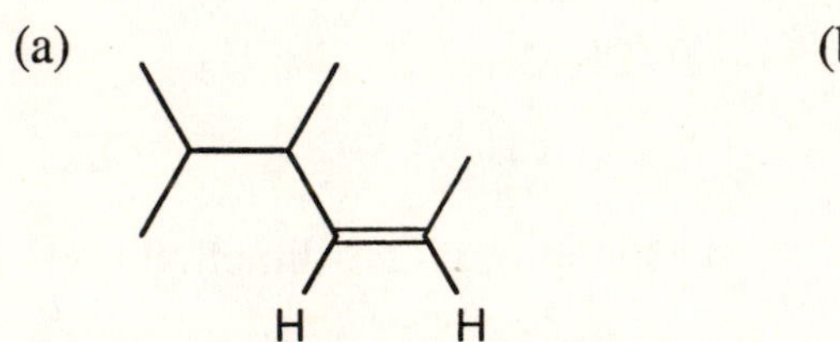

cis-4,5-Dimethylhex-2-ene

(b)

H

H

trans-6-Methylhept-3-ene

6.10 **Strategy:** Review the sequence rules presented in Section 6.4. A summary:

Rule 1: An atom with a higher atomic number has priority over an atom with a lower atomic number.
Rule 2: If a decision can't be reached by using Rule 1, look at the second, third, or fourth atom away from the double-bond carbon until a decision can be made.
Rule 3: Multiple-bonded atoms are equivalent to the same number of single-bonded atoms.

Solution:

High	*Low*	*Rule*	*High*	*Low*	*Rule*
(a) –Br	–H	1	(b) –Br	–Cl	1
(c) $–CH_2CH_3$	$–CH_3$	2	(d) –OH	$–NH_2$	1
(e) $–CH_2OH$	$–CH_3$	2	(f) –CH=O	$–CH_2OH$	3

6.11 *Highest priority* ——————> *Lowest Priority*

(a) –Cl, –OH, $–CH_3$, –H

(b) $–CH_2OH$, $–CH{=}CH_2$, $–CH_2CH_3$, $–CH_3$

(c) $–CO_2H$, $–CH_2OH$, $–C{\equiv}N$, $–CH_2NH_2$

(d) $–CH_2OCH_3$, $–C{\equiv}N$, $–C{\equiv}CH$, $–CH_2CH_3$

6.12

(a)

Low	H_3C		CH_2OH	Low
		C=C		*Z*
High	CH_3CH_2		Cl	High

First, consider the substituents on the right side of the double bond. –Cl ranks higher than $–CH_2OH$ by Rule 1 of the Cahn–Ingold-Prelog rules. On the left side of the double bond, $–CH_2CH_3$ ranks higher than $–CH_3$

(b)

High	Cl		CH_2CH_3	Low
		C=C		*E*
Low	CH_3O		$CH_2CH_2CH_3$	High

(c)

High	H_3C (on cyclopentane ring)		CO_2H	High
		C=C		*Z*
Low			CH_2OH	Low

Notice that the upper substituent on the left side of the double bond is of higher priority because of the methyl group attached to the ring.

(d)

Low	H		CN	High
		C=C		*E*
High	H_3C		CH_2NH_2	Low

6.13

High	vinyl		CO_2CH_3	High
		C=C		*Z*
Low	isopropyl		CH_2OH	Low

6.14

	More stable	*Less stable*
(a)	$(H_3C)_2C=CH_2$	$CH_3CH_2CH=CH_2$
	2-Methylpropene disubstituted double bond	But-1-ene monosubstituted double bond
(b)	$CH_3CH_2CH_2(H)C=C(H)CH_3$ *E*	$CH_3CH_2CH_2(H)C=C(H)CH_3$ *Z*
	trans-Hex-2-ene no steric strain	*cis*-Hex-2-ene steric strain of groups on the same side of the double bond
(c)	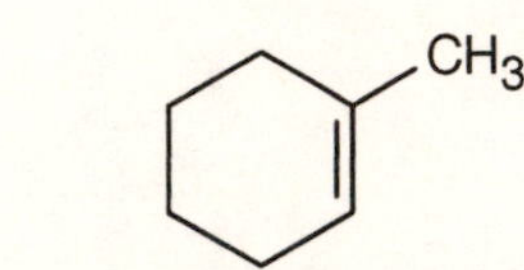	3-Methylcyclohexene structure (CH_3)
	1-Methylcyclohexene trisubstituted double bond	3-Methylcyclohexene disubstituted double bond

6.15 **Strategy:** All these reactions are electrophilic additions of HX to an alkene. Use Markovnikov's rule to predict orientation.

Solution:

(a)

$$\text{Cyclohexene} + HCl \longrightarrow \text{Chlorocyclohexane}$$

Chlorocyclohexane

(b)

$$CH_3C(CH_3){=}CHCH_2CH_3 + HBr \longrightarrow CH_3C(CH_3)(Br)CH_2CH_2CH_3$$

2-Bromo-2-methylpentane

In accordance with Markovnikov's Rule, H forms a bond to the carbon with fewer substituents, and Br forms a bond to the carbon with more substituents.

(c)

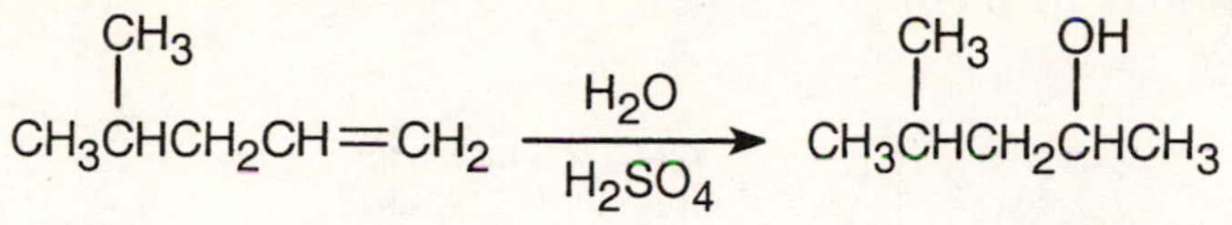

$$\underset{\text{}}{CH_3\overset{CH_3}{\overset{|}{C}}HCH_2CH{=}CH_2} \xrightarrow[H_2SO_4]{H_2O} CH_3\overset{CH_3}{\overset{|}{C}}HCH_2\overset{OH}{\overset{|}{C}}HCH_3$$

(d)

Methylenecyclohexane ($=CH_2$) + HBr ⟶ (cyclohexane with Br and CH_3 on the same carbon)

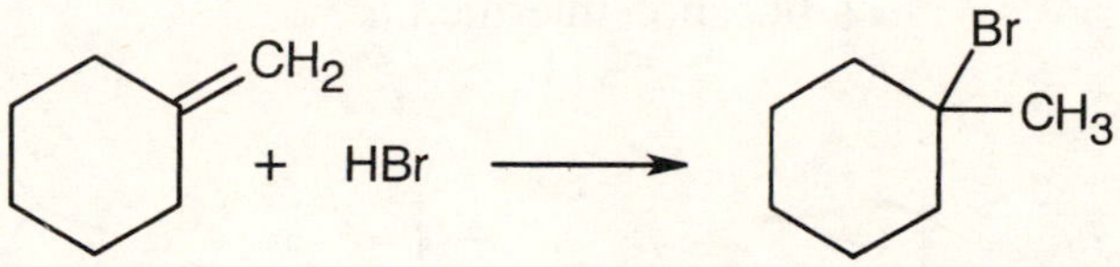

1-Bromo-1-methylcyclohexane

6.16 **Strategy:** Think backwards in choosing the alkene starting material for synthesis of the desired haloalkanes. Remember that halogen is bonded to one end of the double bond and that more than one starting material can give rise to the desired product.

Solution:

(a)

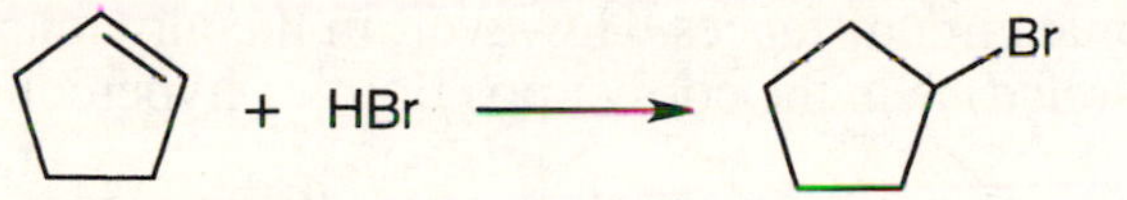

Cyclopentene + HBr ⟶ (cyclopentane with Br)

Cyclopentene

(b)

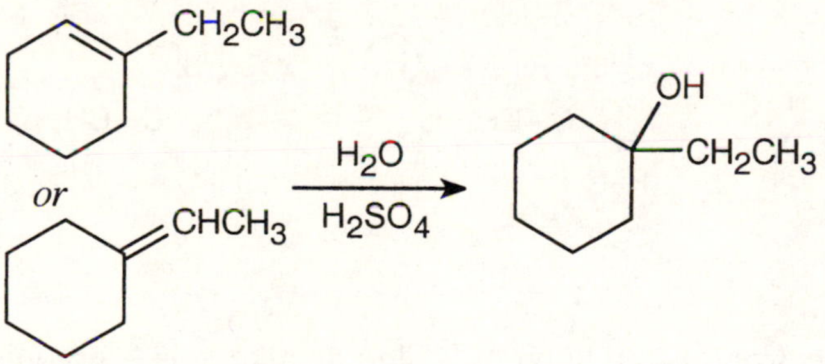

(1-ethylcyclohexene, with CH_2CH_3) *or* (ethylidenecyclohexane, $=CHCH_3$) $\xrightarrow[H_2SO_4]{H_2O}$ (cyclohexane with OH and CH_2CH_3 on the same carbon)

(c)

$$CH_3CH_2CH{=}CHCH_2CH_3 + HBr \longrightarrow CH_3CH_2\overset{Br}{\overset{|}{C}}HCH_2CH_2CH_3$$

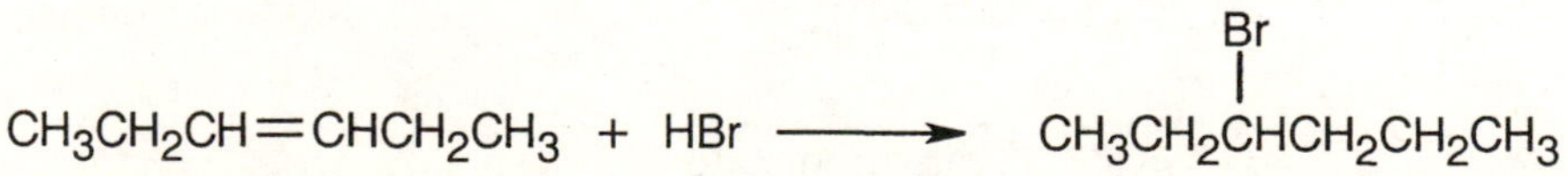

Hex-3-ene

(d)

(vinylcyclohexane) + HCl ⟶ (cyclohexyl–CH(Cl)–CH_3)

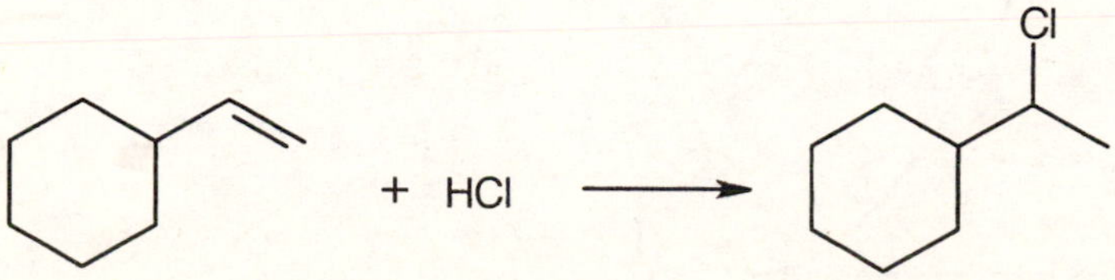

6.17 The more stable carbocation is formed.

(a)

$CH_3CH_2C(CH_3){=}CHCH(CH_3)CH_3 + HBr \longrightarrow [CH_3CH_2\overset{+}{C}(CH_3)CH_2CH(CH_3)CH_3] \longrightarrow CH_3CH_2CBr(CH_3)CH_2CH(CH_3)CH_3$

carbocation intermediate

(b)

(cyclopentylidene)$=CHCH_3 \xrightarrow[H_2SO_4]{H_2O}$ [(cyclopentyl cation)$^+$$-CH_2CH_3$] $\longrightarrow$ 1-ethylcyclopentanol (OH, CH_2CH_3)

carbocation intermediate

6.18 Two representations of this secondary carbocation are shown on the left below. This secondary carbocation can experience hyperconjugation of the vacant *p* orbital with two adjacent C–H bonds under normal circumstances. However, in the alignment shown in the drawing, only one hydrogen (circled) is in the correct position for hyperconjugation with the carbocation carbon.
Because there is rotation about the carbon-carbon bonds, all of the C–H bonds starred in the representation on the far right can be involved in hyperconjugation at some time.

$(H_3C)_2CH\overset{+}{C}HCH_3$

6.19 The second step in the electrophilic addition of HCl to an alkene is exergonic. According to the Hammond Postulate, the transition state should resemble the carbocation intermediate.

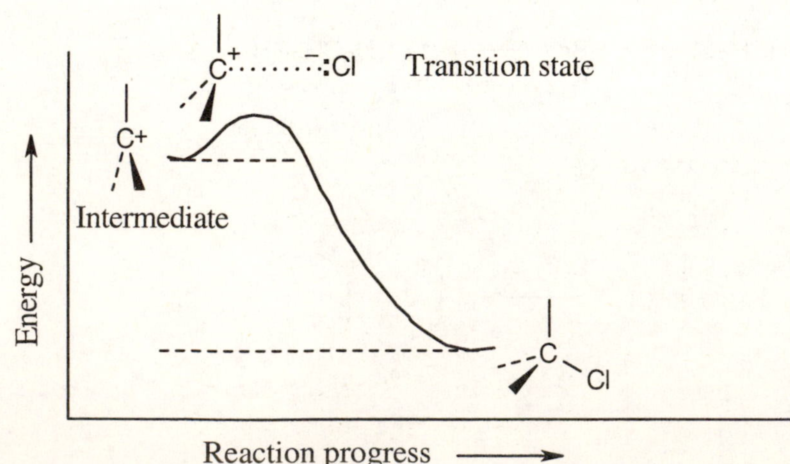

6.20

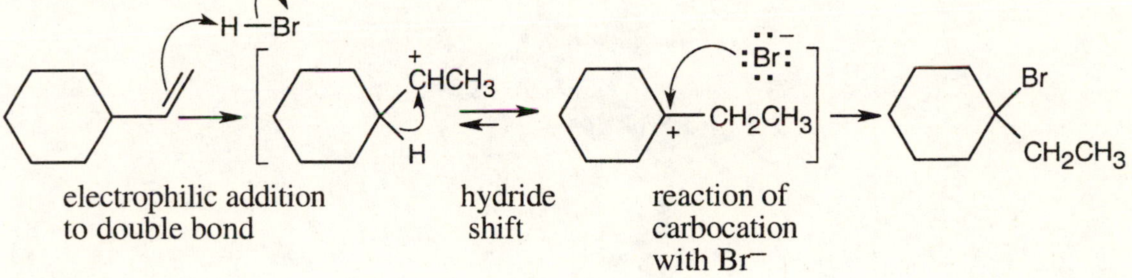

H—Br

$\overset{+}{C}HCH_3$ H   :Br:⁻ CH_2CH_3 +   Br CH_2CH_3

electrophilic addition to double bond   hydride shift   reaction of carbocation with Br^-

Visualizing Chemistry

6.21

(a) 2,4,5-Trimethylhex-2-ene

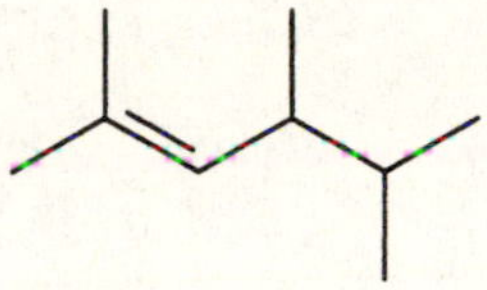

(b) 1-Ethyl-3,3-dimethylcyclohexene

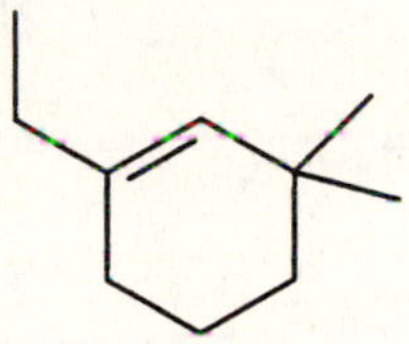

6.22

(a) High Cl   Low   Low   C=O High   H   *E*

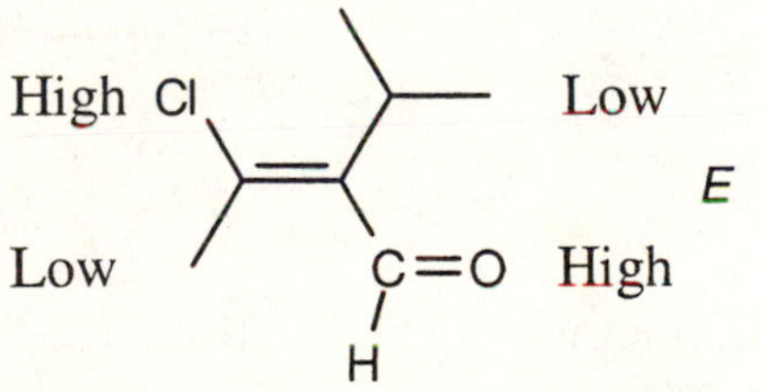

(b) High OCH_3 High   C=O   OH   Low   Low   *Z*

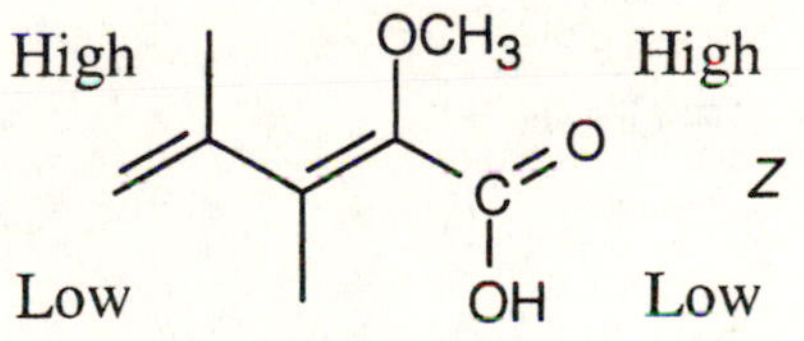

6.23

$=CHCH_3$ *or* $-CH_2CH_3$ HCl $\overset{+}{}-CH_2CH_3$ Cl CH_2CH_3

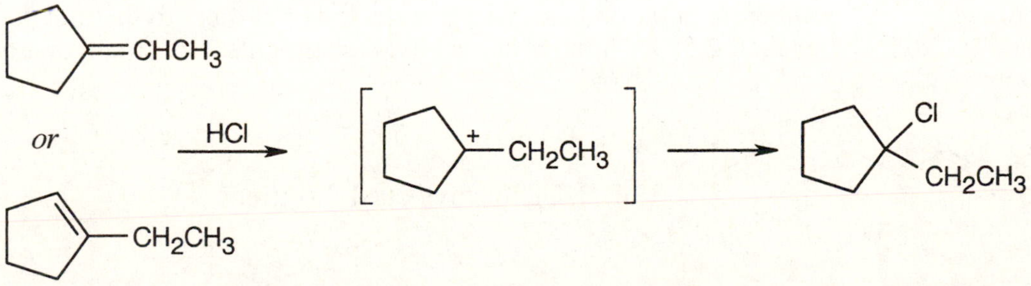

Either of the two compounds shown can react with HCl to form the illustrated tertiary carbocation. In the conformation shown, the three circled hydrogens are aligned for hyperconjugation with the vacant *p* orbital. Because of conformational mobility, the three starred hydrogens are also able to be involved in hyperconjugation.

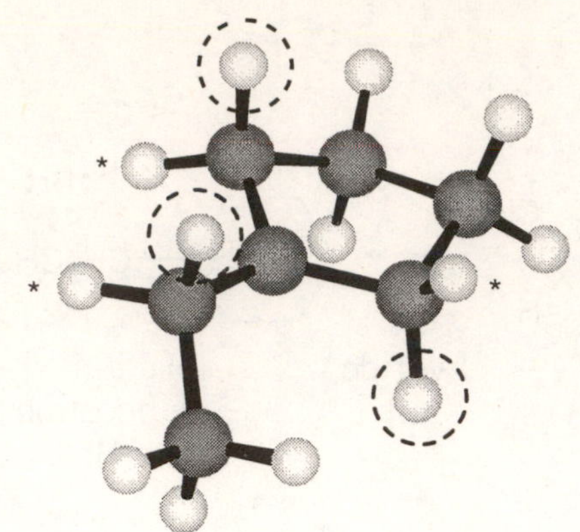

Additional Problems

6.24

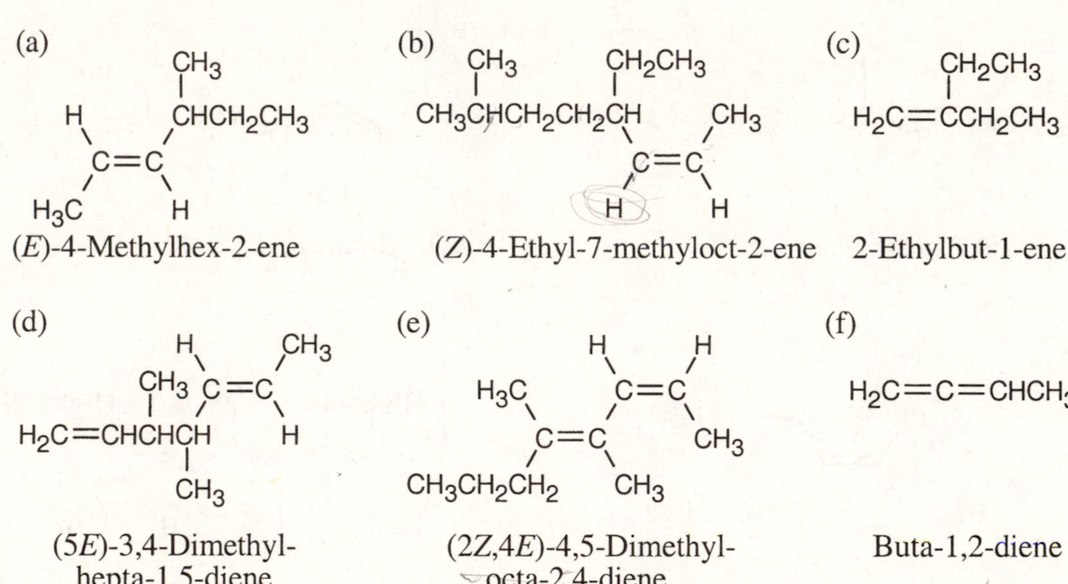

(a) (*E*)-4-Methylhex-2-ene

(b) (*Z*)-4-Ethyl-7-methyloct-2-ene

(c) 2-Ethylbut-1-ene

(d) (5*E*)-3,4-Dimethylhepta-1,5-diene

(e) (2*Z*,4*E*)-4,5-Dimethylocta-2,4-diene

(f) Buta-1,2-diene

6.25 Because the longest carbon chain contains 8 carbons and 3 double bonds, ocimene is an *octatriene*. Start numbering at the end that will give the lower number to the first double bond (1,3,6 is lower than 2,5,7). Number the methyl substituents and, finally, name the compound.

(3*E*)-3,7-Dimethylocta-1,3,6-triene

6.26

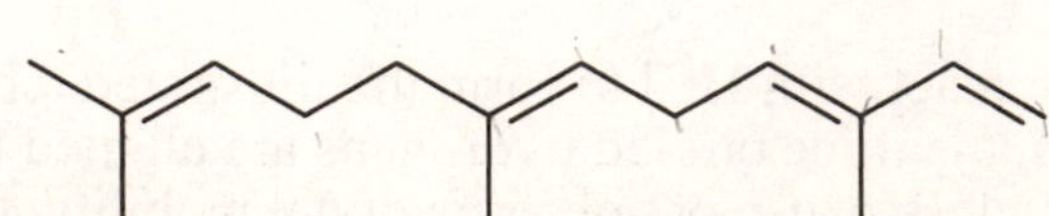

(3*E*,6*E*)-3,7,11-Trimethyldodeca-1,3,6,10-tetraene

6.27

(a)

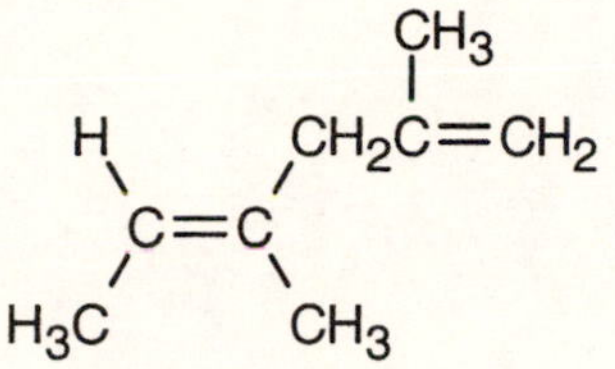

$H(CH_3)C{=}C(CH_3)CH_2C(CH_3){=}CH_2$

(4*E*)-2,4-Dimethylhexa-1,4-diene

(b)

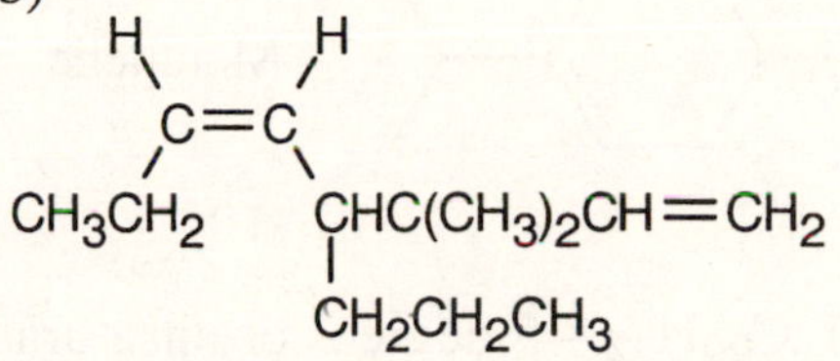

$CH_3CH_2CH{=}CHCH(CH_2CH_2CH_3)C(CH_3)_2CH{=}CH_2$

cis-3,3-Dimethyl-4-propylocta-1,5-diene

(c)

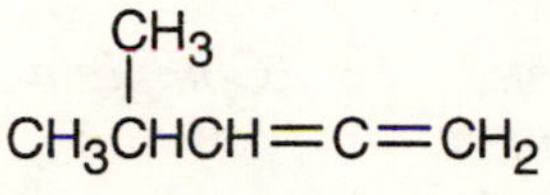

$CH_3CH(CH_3)CH{=}C{=}CH_2$

4-Methylpenta-1,2-diene

(d)

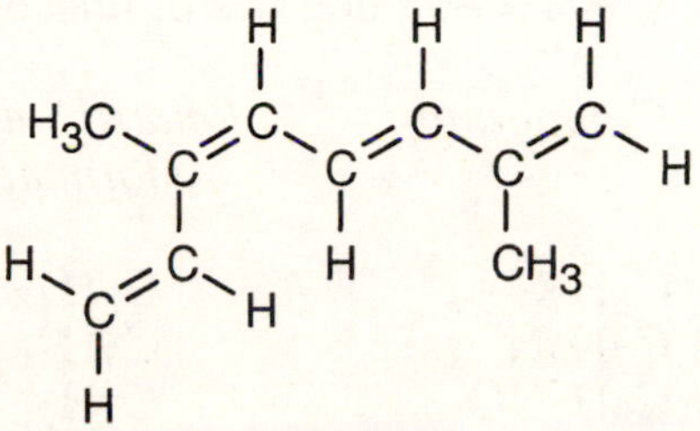

$H_2C{=}CH{-}C(CH_3){=}CH{-}CH{=}C(CH_3){-}CH{=}CH_2$

(3*E*,5*Z*)-2,6-Dimethylocta-1,3,5,7-tetraene

(e)

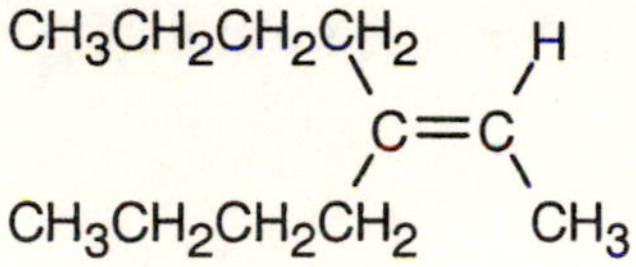

$(CH_3CH_2CH_2CH_2)_2C{=}CHCH_3$

3-Butylhept-2-ene

(f)

$(CH_3)_3CCH{=}CHC(CH_3)_3$

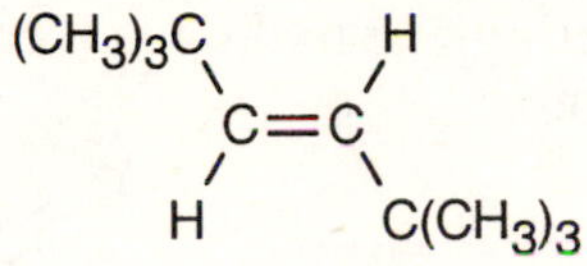

trans-2,2,5,5-Tetramethylhex-3-ene

6.28

$CH_3CH_2CH_2CH_2C{\equiv}CH$ Hex-1-yne

$CH_3CH_2CH_2C{\equiv}CCH_3$ Hex-2-yne

$CH_3CH_2C{\equiv}CCH_2CH_3$ Hex-3-yne

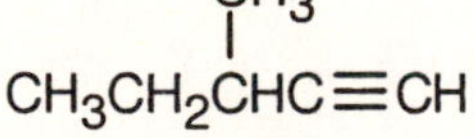

$CH_3CH_2CH(CH_3)C{\equiv}CH$ 3-Methylpent-1-yne

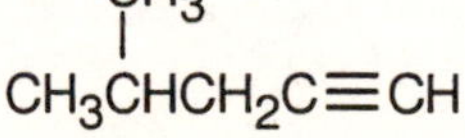

$CH_3CH(CH_3)CH_2C{\equiv}CH$ 4-Methylpent-1-yne

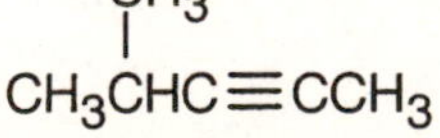

$CH_3CH(CH_3)C{\equiv}CCH_3$ 4-Methylpent-2-yne

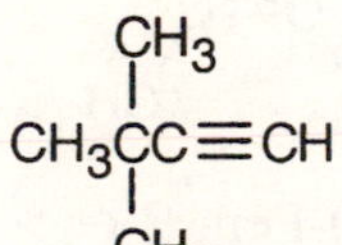

$CH_3C(CH_3)_2C{\equiv}CH$ 3,3-Dimethylbut-1-yne

6.29 $CH_3C{\equiv}C{-}C{\equiv}C{-}C{\equiv}C{-}C{\equiv}C{-}C{\equiv}C{-}CH{=}CH_2$ Tridec-1-ene-3,5,7,9,11-pentayne

6.30

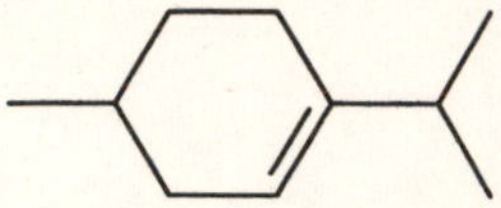

Menthene

6.31

(a) $C_{20}H_{32}$ – 5 degrees of unsaturation
(b) $C_9H_{16}Br_2$ – one degree of unsaturation.
(c) $C_{10}H_{12}N_2O_3$ – 6 degrees of unsaturation.
(d) $C_{20}H_{32}ClN$ – 5 degrees of unsaturation.
(e) $C_{40}H_{56}$ –13 degrees of unsaturation.

6.32

Compound	*Related Saturated formula*	*Degree of unsaturation*	*Complete formula*
(a) $C_8H_?O_2$	C_8H_{18}	3	$C_8H_{12}O_2$
(b) $C_7H_?N$	C_7H_{16}	2	$C_7H_{13}N$
(c) $C_9H_?NO$	C_9H_{20}	4	$C_9H_{13}NO$

6.33 Solve this problem in the same way as we solved problems 6.3 and 6.32. A C_{22} hydrocarbon with 12 degrees of unsaturation (four rings and eight double bonds) has a formula $C_{22}H_{46} - H_{24} = C_{22}H_{22}$. Adding two hydrogens (because of the two nitrogens) and subtracting one hydrogen (because of the chlorine), gives the formula $C_{22}H_{23}ClN_2O_2$ for loratadine.

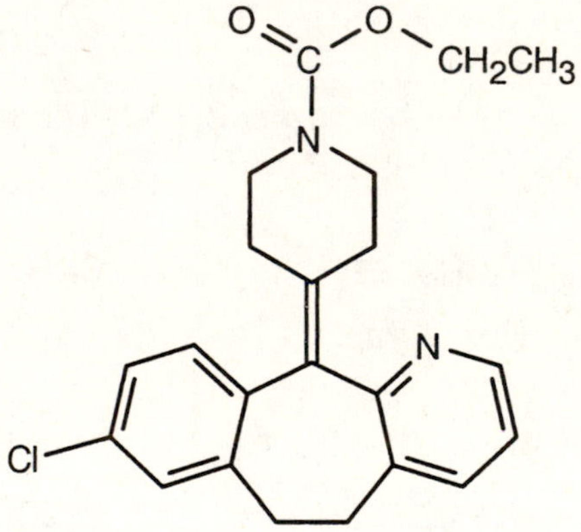

O C O CH_2CH_3 N N Cl

Loratadine

6.34

$CH_3CH_2CH_2CH{=}CH_2$

Pent-1-ene

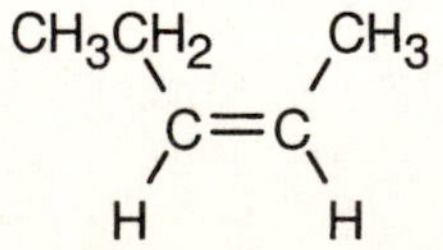

CH_3CH_2 CH_3 C=C H H

(*Z*)-Pent-2-ene

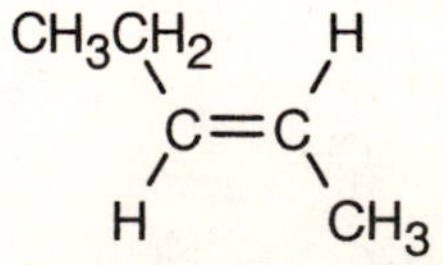

CH_3CH_2 H C=C H CH_3

(*E*)-Pent-2-ene

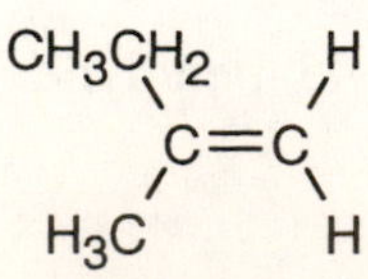

CH_3CH_2 H C=C H_3C H

2-Methylbut-1-ene

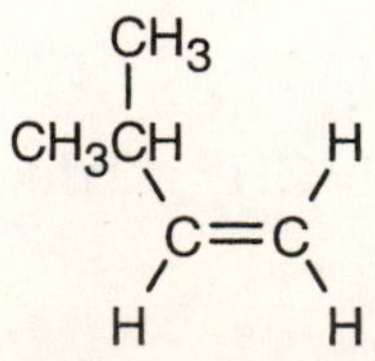

CH_3 CH_3CH H C=C H H

3-Methylbut-1-ene

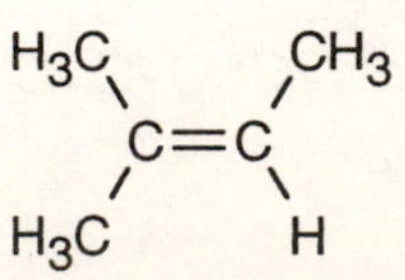

H_3C CH_3 C=C H_3C H

2-Methylbut-2-ene

6.35

$CH_3CH_2CH_2CH_2CH{=}CH_2$

Hex-1-ene

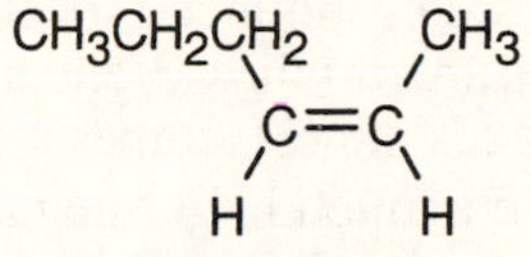

$$\begin{array}{ccc} CH_3CH_2CH_2 & & CH_3 \\ & C{=}C & \\ H & & H \end{array}$$

(*Z*)-Hex-2-ene

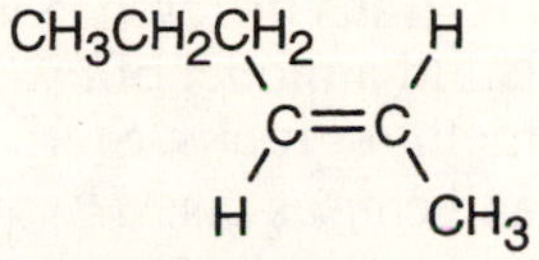

$$\begin{array}{ccc} CH_3CH_2CH_2 & & H \\ & C{=}C & \\ H & & CH_3 \end{array}$$

(*E*)-Hex-2-ene

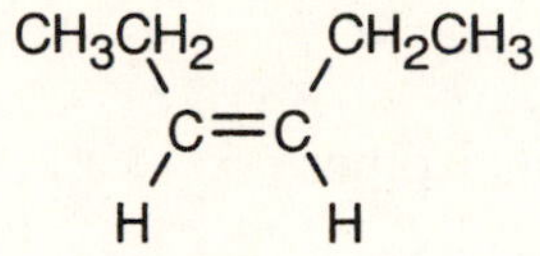

$$\begin{array}{ccc} CH_3CH_2 & & CH_2CH_3 \\ & C{=}C & \\ H & & H \end{array}$$

(*Z*)-Hex-3-ene

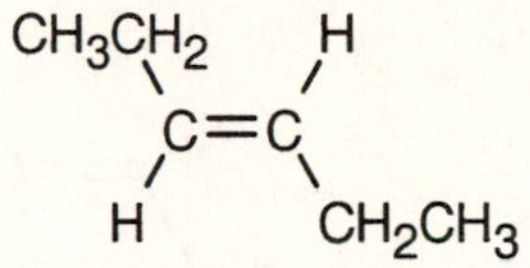

$$\begin{array}{ccc} CH_3CH_2 & & H \\ & C{=}C & \\ H & & CH_2CH_3 \end{array}$$

(*E*)-Hex-3-ene

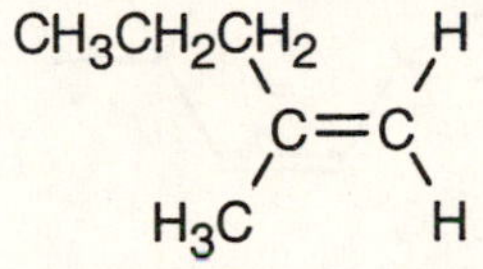

$$\begin{array}{ccc} CH_3CH_2CH_2 & & H \\ & C{=}C & \\ H_3C & & H \end{array}$$

2-Methylpent-1-ene

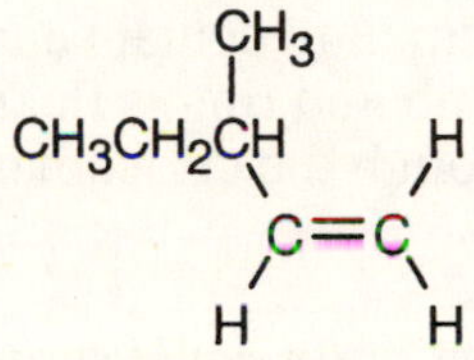

$$\begin{array}{ccc} CH_3 & & \\ | & & \\ CH_3CH_2CH & & H \\ & C{=}C & \\ H & & H \end{array}$$

3-Methylpent-1-ene

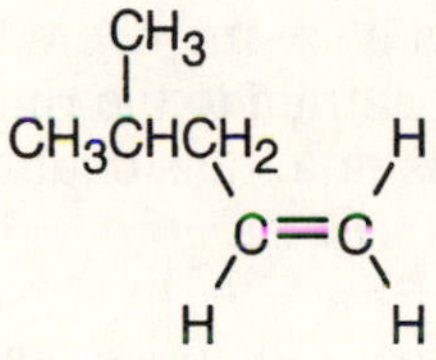

$$\begin{array}{ccc} CH_3 & & \\ | & & \\ CH_3CHCH_2 & & H \\ & C{=}C & \\ H & & H \end{array}$$

4-Methylpent-1-ene

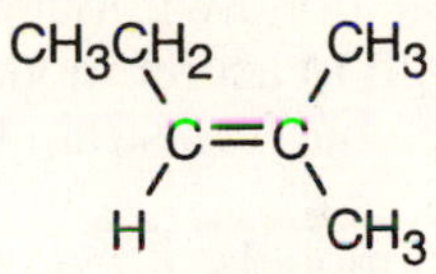

$$\begin{array}{ccc} CH_3CH_2 & & CH_3 \\ & C{=}C & \\ H & & CH_3 \end{array}$$

2-Methylpent-2-ene

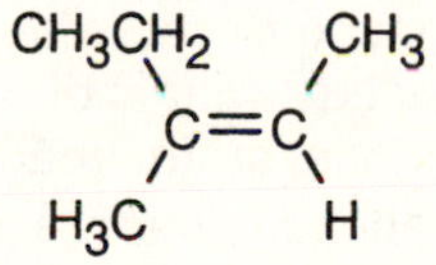

$$\begin{array}{ccc} CH_3CH_2 & & CH_3 \\ & C{=}C & \\ H_3C & & H \end{array}$$

(*Z*)-3-Methylpent-2-ene

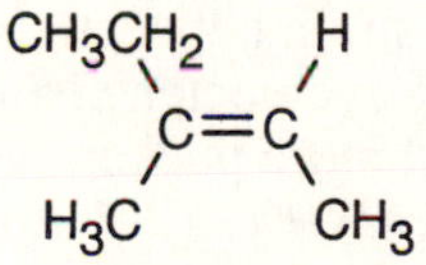

$$\begin{array}{ccc} CH_3CH_2 & & H \\ & C{=}C & \\ H_3C & & CH_3 \end{array}$$

(*E*)-3-Methylpent-2-ene

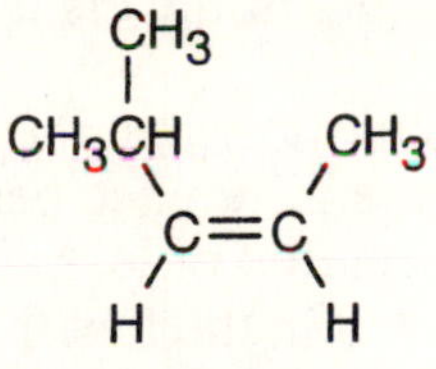

$$\begin{array}{ccc} CH_3 & & \\ | & & \\ CH_3CH & & CH_3 \\ & C{=}C & \\ H & & H \end{array}$$

(*Z*)-4-Methylpent-2-ene

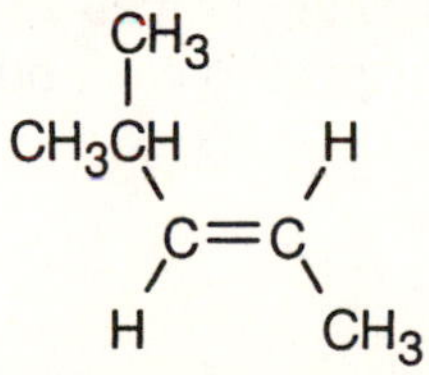

$$\begin{array}{ccc} CH_3 & & \\ | & & \\ CH_3CH & & H \\ & C{=}C & \\ H & & CH_3 \end{array}$$

(*E*)-4-Methylpent-2-ene

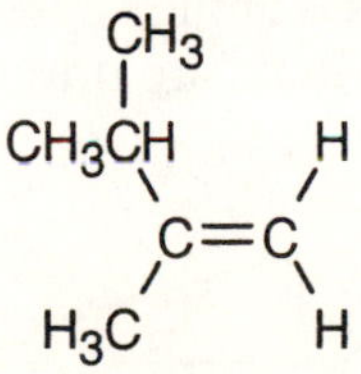

$$\begin{array}{ccc} CH_3 & & \\ | & & \\ CH_3CH & & H \\ & C{=}C & \\ H_3C & & H \end{array}$$

2,3-Dimethylbut-1-ene

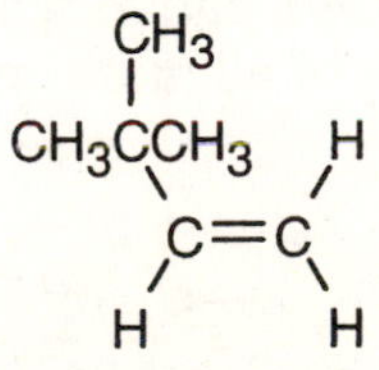

$$\begin{array}{ccc} CH_3 & & \\ | & & \\ CH_3CCH_3 & & H \\ & C{=}C & \\ H & & H \end{array}$$

3,3-Dimethylbut-1-ene

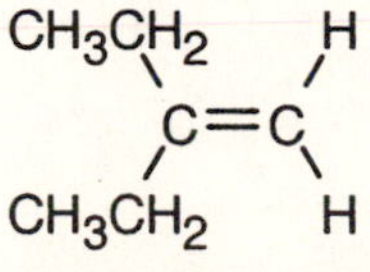

$$\begin{array}{ccc} CH_3CH_2 & & H \\ & C{=}C & \\ CH_3CH_2 & & H \end{array}$$

2-Ethylbut-1-ene

$$\begin{array}{ccc} H_3C & & CH_3 \\ & C{=}C & \\ H_3C & & CH_3 \end{array}$$

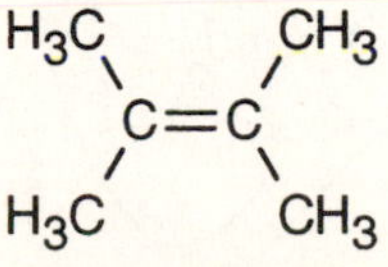

2,3-Dimethylbut-2-ene

6.36 As expected, the two trans compounds are more stable than their cis counterparts. The cis–trans difference is much more pronounced for the tetramethyl compound, however. Build a model of *cis*-2,2,5,5-tetramethylhex-3-ene and notice the extreme crowding of the methyl groups. Steric strain makes the cis isomer much less stable than the trans isomer and causes cis $\Delta H°_{hydrog}$ to have a much larger negative value than trans $\Delta H°_{hydrog}$ for the hexene isomers.

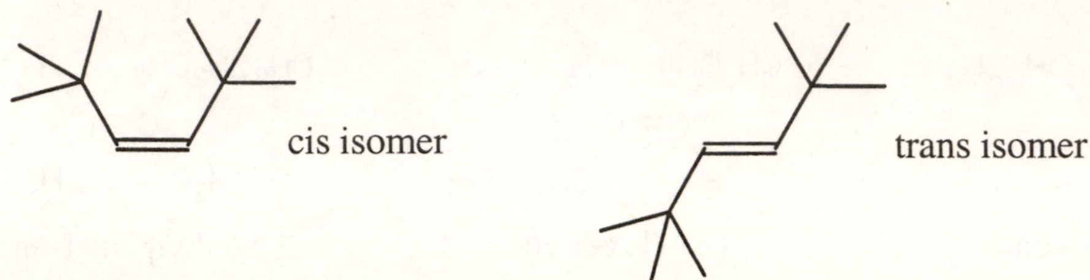

6.37 A model of cyclohexene shows that a six-membered ring is too small to contain a trans double bond without causing severe strain to the ring. A ten-membered ring is flexible enough to accommodate either a cis or a trans double bond, although the cis isomer has somewhat less strain than the trans isomer.

6.38 Build models of the two cyclooctenes and notice the large amount of torsional strain in *trans*-cyclooctene relative to *cis*-cyclooctene. This torsional strain, in addition to angle strain, causes the trans isomer to be of higher energy and to have a $\Delta H°_{hydrog}$ larger than the $\Delta H°_{hydrog}$ of the cis isomer.

6.39 Models show that the difference in strain between the two cyclononene isomers is smaller than the difference between the two cyclooctene isomers.This reduced strain is due to a combination of less angle strain and more puckering to relieve torsional strain and is reflected in the fact that the values of $\Delta H°_{hydrog}$ for the two cyclononene isomers are relatively close. Nevertheless, the trans isomer is still more strained than the cis isomer.

6.40 The central carbon of allene forms two σ bonds and two π bonds. The central carbon is *sp*-hybridized, and the carbon–carbon bond angle is 180°, indicating linear geometry for the carbons of allene. The terminal $=CH_2$ units are oriented 90° with respect to each other.

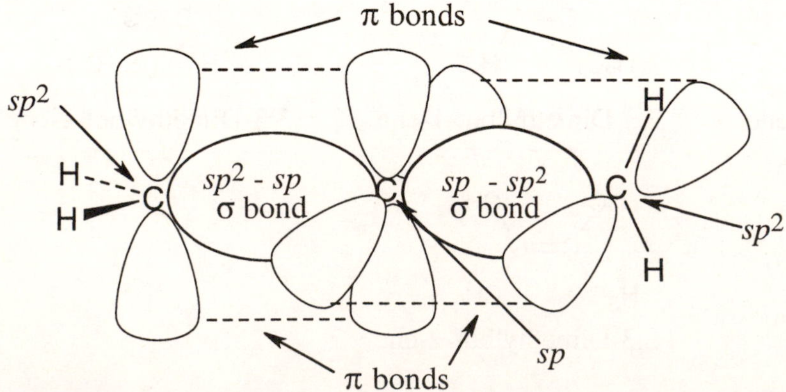

6.41 The heat of hydrogenation for a typical diene is 2 x ($\Delta H°_{hydrog}$ of an alkene) = –252 kJ/mol. Thus, allene, with $\Delta H°_{hydrog}$ = –295 kJ/mol is 43 kJ/mol higher in energy than a typical diene and is less stable.

6.42

(a)

$$CH_3CH_2CH{=}C(CH_3)CH_2CH_3 \xrightarrow[H_2SO_4]{H_2O} CH_3CH_2CH_2C(CH_3)(OH)CH_2CH_3$$

(b)

1-ethylcyclopentene + HBr ⟶ 1-bromo-1-ethylcyclopentane (CH_2CH_3, Br)

(c)

3-methylcyclohexene $\xrightarrow{HBr}$ 1-bromo-3-methylcyclohexane + 1-bromo-2-methylcyclohexane

Two products are formed because the two possible carbocations are of similar stability.

(d)

$$H_2C{=}CHCH_2CH_2CH_2CH{=}CH_2 + 2\ HCl \longrightarrow CH_3CH(Cl)CH_2CH_2CH_2CH(Cl)CH_3$$

6.43

(a)

methylenecyclohexane (CH_2) + HBr ⟶ 1-bromo-1-methylcyclohexane (Br, CH_3)

(b)

octahydronaphthalene (bicyclic alkene) + HBr ⟶ bromodecalin (Br at ring fusion)

(c)

$$CH_3CH{=}CHCH(CH_3)CH_3 + HBr \longrightarrow CH_3CH(Br)CH_2CH(CH_3)CH_3 + CH_3CH_2CH(Br)CH(CH_3)CH_3$$

6.44 *Highest priority ——————> Lowest Priority*

(a) –I, –Br, $–CH_3$, –H

(b) $–OCH_3$, –OH, $–CO_2H$, –H

(c) $–CO_2CH_3$, $–CO_2H$, $–CH_2OH$, $–CH_3$

(d) $–COCH_3$, $–CH_2CH_2OH$, $–CH_2CH_3$, $–CH_3$

(e) $–CH_2Br$, –C≡N, $–CH_2NH_2$, $–CH{=}CH_2$

(f) $–CH_2OCH_3$, $–CH_2OH$, $–CH{=}CH_2$, CH_2CH_3

6.45

(a)

High $HOCH_2$ CH_3 High

C=C Z

Low H_3C H Low

(b)

Low HO_2C H Low

C=C Z

High Cl OCH_3 High

(c)

High NC CH_3 Low

C=C E

Low CH_3CH_2 CH_2OH High

(d)

High CH_3O_2C $CH{=}CH_2$ High

C=C Z

Low HO_2C CH_2CH_3 Low

6.46

(a)

CH_3

3-Methylcyclohexene

(b)

2,3-Dimethylcyclopentene

(c)

Ethylcyclobutadiene

(d)

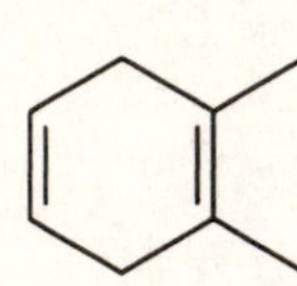

1,2-Dimethylcyclohexa-1,4-diene

(e)

5-Methylcyclohexa-1,3-diene

(f)

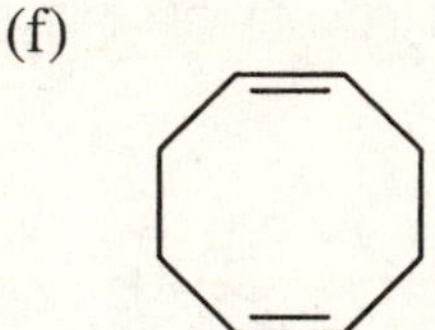

Cycloocta-1,5-diene

6.47

(a)

1 2 3 4 5 6 7 8

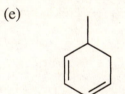

(3*E*,5*Z*)-Octa-1,3,5-triene

(b)

1 2 3 4 5 6 7

6-[(*Z*)-But-1-enyl]-cyclohepta-1,4-diene

(c)

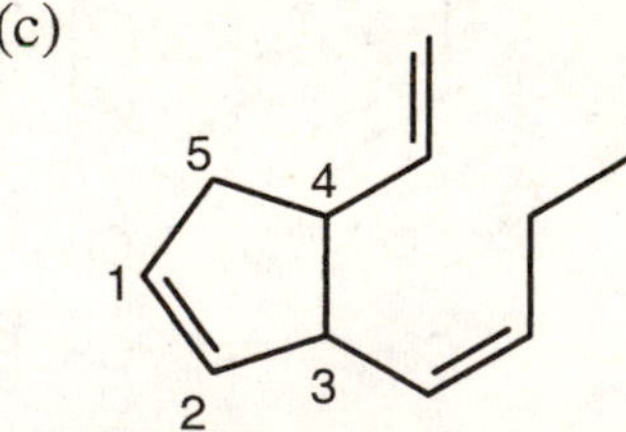

3-[(*Z*)-But-1-enyl]-4-vinylcyclopentene

6.48

(a)

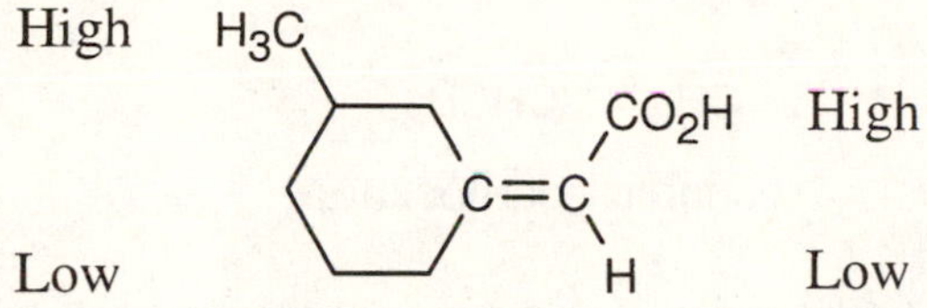

High H_3C (on cyclohexane ring) ; CO_2H High

C=C

Low (ring) ; H Low

Z (correct)

(b)

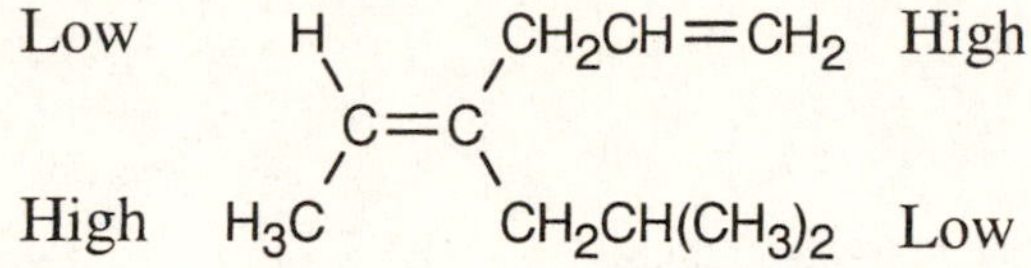

Low H ; $CH_2CH{=}CH_2$ High

C=C

High H_3C ; $CH_2CH(CH_3)_2$ Low

E (correct)

(c)

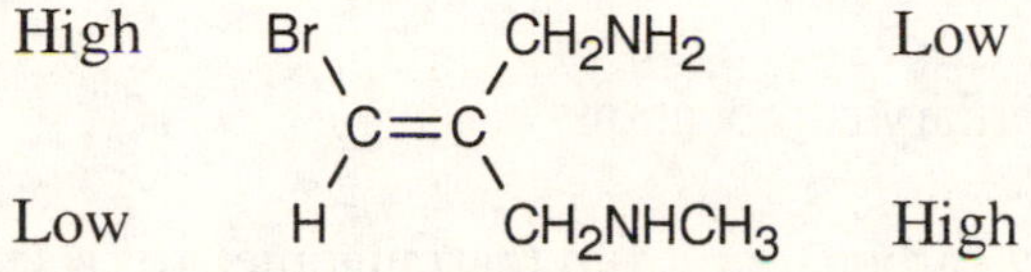

High Br ; CH_2NH_2 Low

C=C

Low H ; CH_2NHCH_3 High

E (incorrect)

(d)

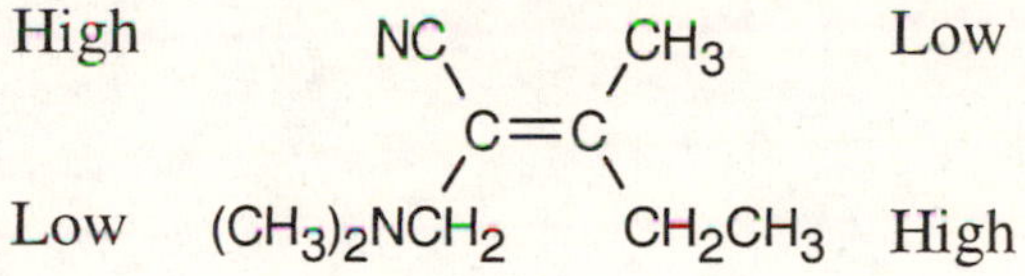

High NC ; CH_3 Low

C=C

Low $(CH_3)_2NCH_2$; CH_2CH_3 High

E (correct)

(e)

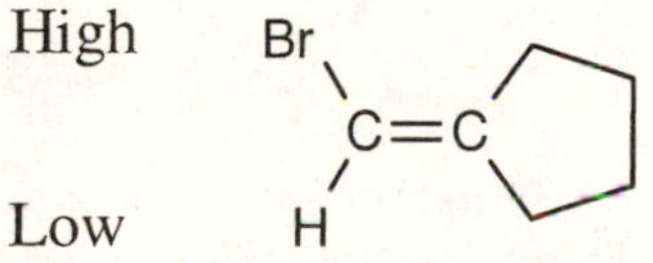

High Br

C=C (cyclopentylidene)

Low H

This compound doesn't show *E-Z* isomerism.

(f)

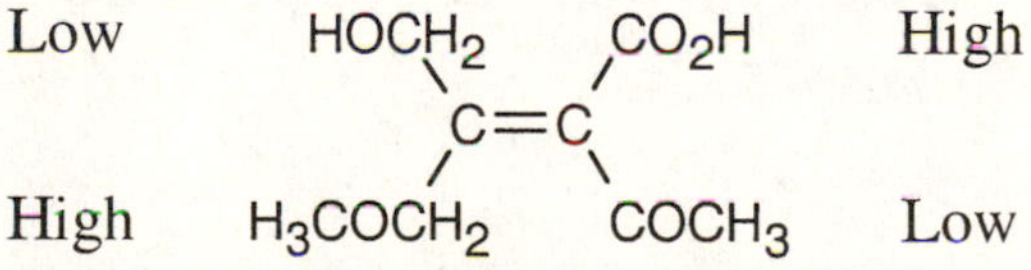

Low $HOCH_2$; CO_2H High

C=C

High H_3COCH_2 ; $COCH_3$ Low

E (correct)

6.49 Treatment of the *tert*-butyl ester with trifluoroacetic acid cleaves the –$OC(CH_3)_3$ group and replaces it with an –OH group, which has a lower priority than the –OCH_3 group on the upper carbon and the –$OC(CH_3)_3$ group that was removed. The result is a change in the *E,Z* designation around the double bond without breaking any of the bonds attached to the double-bond carbons.

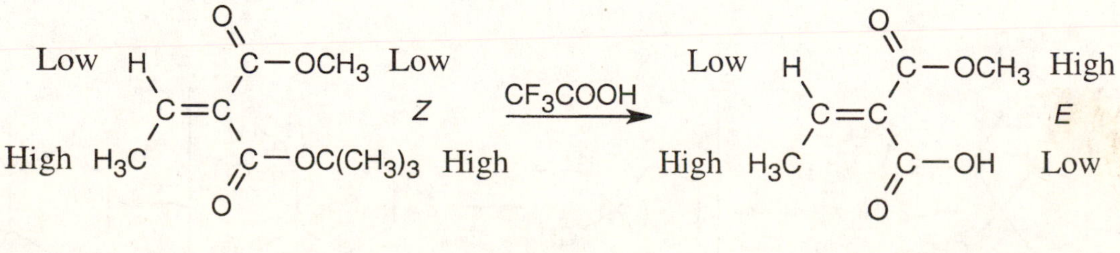

Low H ; $C({=}O){-}OCH_3$ Low

C=C *Z*

High H_3C ; $C({=}O){-}OC(CH_3)_3$ High

$\xrightarrow{CF_3COOH}$

Low H ; $C({=}O){-}OCH_3$ High

C=C *E*

High H_3C ; $C({=}O){-}OH$ Low

6.50

(a)

$CH_3CH_2\overset{H}{C}HCH_2^+ \longrightarrow CH_3CH_2\overset{+}{C}HCH_3$

primary carbocation → secondary carbocation

The primary carbocation rearranges to the more stable secondary carbocation by a hydride shift.

(b)

$CH_3\overset{H}{\underset{CH_3}{C}}-\overset{+}{C}HCH_3 \longrightarrow CH_3\overset{+}{\underset{CH_3}{C}}-CH_2CH_3$

secondary carbocation → tertiary carbocation

This hydride shift produces a tertiary carbocation from rearrangement of a secondary carbocation.

(c)

CH_3, $\overset{+}{C}H_2$ → CH_3, +

primary carbocation → tertiary carbocation

An alkyl shift forms a tertiary carbocation from a primary carbocation. In this example, rearrangement involves migration of the electrons from one of the cyclobutane ring bonds.

6.51

H—Cl: → [+, H, H; 3° carbocation ⇌ hydride shift ⇌ :Cl:⁻, +, H; 3° carbocation] → Cl

6.52

H—Cl:, CH_3 → [+, CH_3 ⇌ bond shift (alkyl shift) ⇌ CH_3, CH_3, +, CH_3, :Cl:⁻] → CH_3, CH_3, CH_3, Cl

6.53

Reaction of the π electrons of the double bond with H^+ yields the carbocation pictured on the far right. A bond shift (alkyl shift) produces the bracketed intermediate, which reacts with Br^- to yield 1-bromo-2-methylcyclobutane.

6.54 The reaction is exergonic because it is spontaneous. According to the Hammond Postulate, the transition state should resemble the isobutyl cation.

6.55

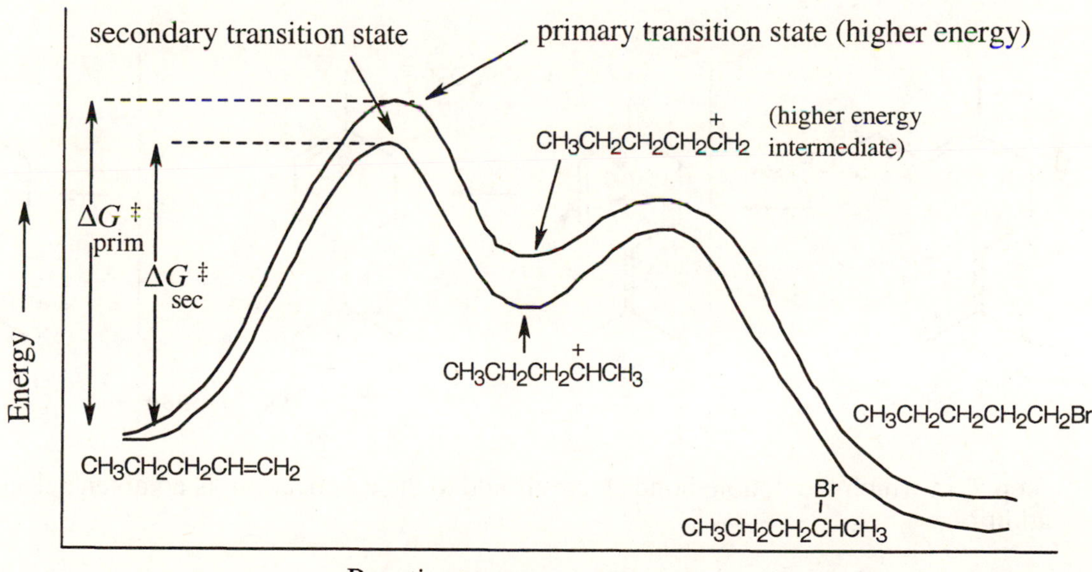

6.56

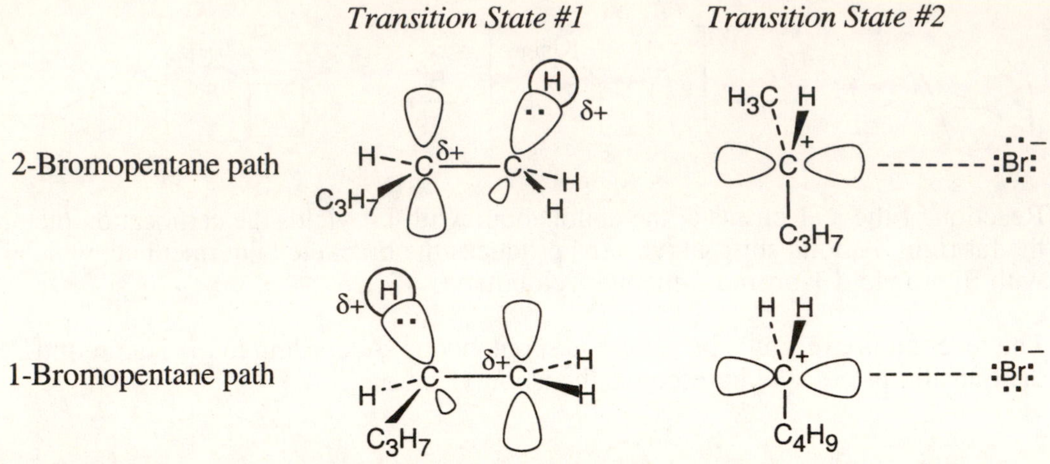

Transition State #1 *Transition State #2*

2-Bromopentane path

1-Bromopentane path

The first step (carbocation formation) is endergonic for both reaction paths, and both transition states resemble the carbocation intermediates. Transition states for the exergonic second step also resemble the carbocation intermediate. Transition state #1 for 1-bromopentane is more like the carbocation intermediate than is transition state #1 for 2-bromopentane.

6.57

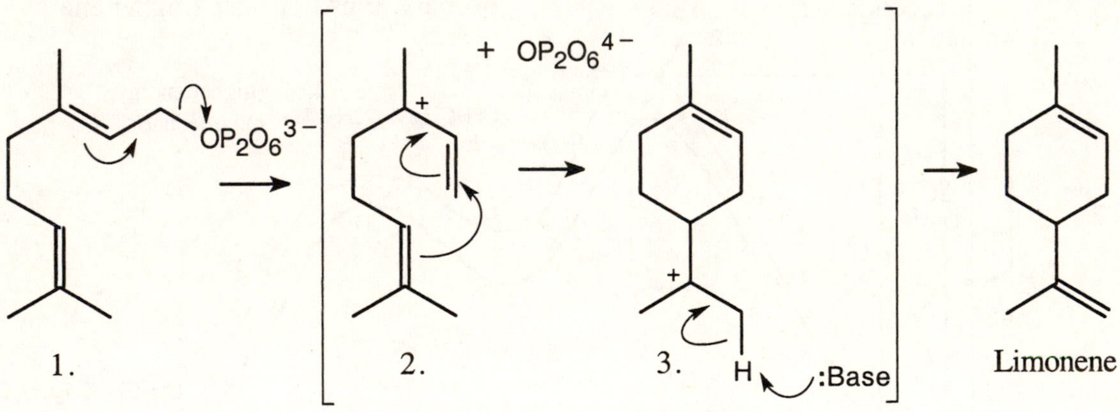

Limonene

Step 2, in which the double bond electrons add to the carbocation, is an alkene electrophilic addition.

6.58

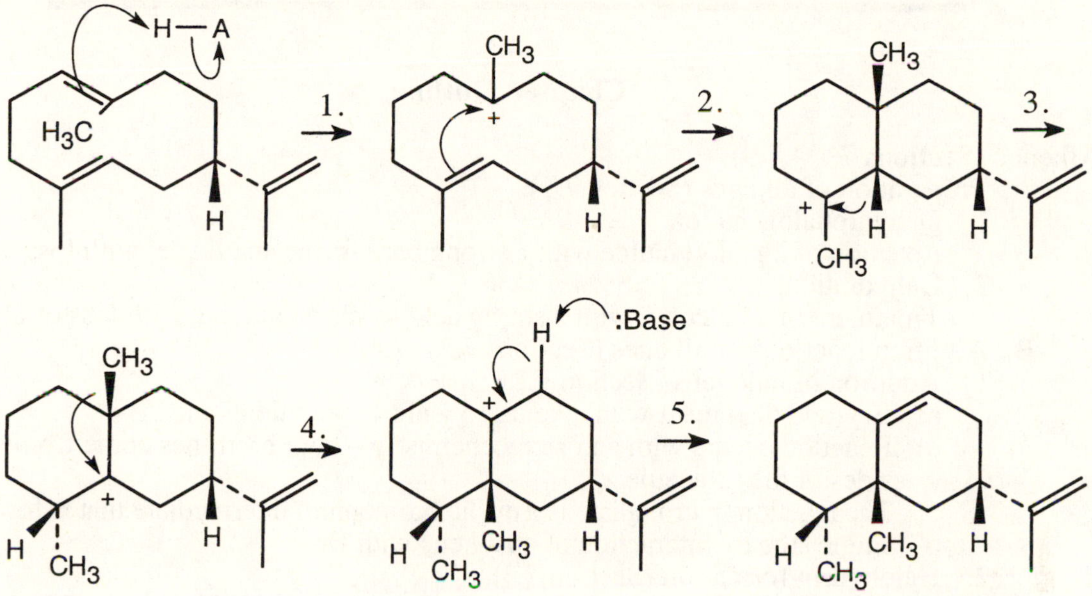

H—A

1. → CH3 + 2. → CH3 + H CH3 3. →

CH3 + H CH3 H 4. → H :Base + CH3 H CH3 H 5. → H CH3 CH3 H

epi-Aristolochene

Step 2 is an alkene electrophilic addition, and steps 3 and 4 involve carbocation rearrangements.

6.59 Reaction of 1-chloropropane with the Lewis acid $AlCl_3$ forms a carbocation. The less stable propyl carbocation undergoes a hydride shift to produce the more stable isopropyl carbocation, which reacts with benzene to give isopropylbenzene.

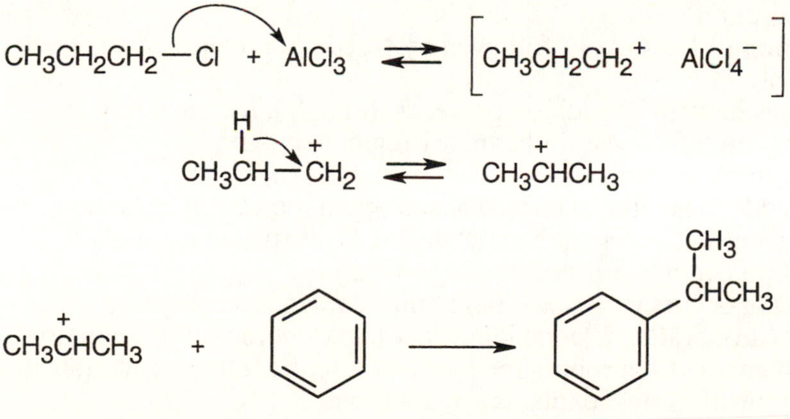

$$CH_3CH_2CH_2{-}Cl + AlCl_3 \rightleftharpoons [CH_3CH_2CH_2^+ \quad AlCl_4^-]$$

$$CH_3CH(H){-}\overset{+}{C}H_2 \rightleftharpoons CH_3\overset{+}{C}HCH_3$$

$$CH_3\overset{+}{C}HCH_3 + C_6H_6 \longrightarrow C_6H_5CH(CH_3)CH_3$$

6.60

$$CH_3CH(CH_3){-}C(CH_3){=}CH_2 \xrightarrow{HBr} CH_3CH(CH_3){-}C(CH_3)(Br){-}CH_3 \xrightarrow[CH_3OH]{KOH} CH_3C(CH_3){=}C(CH_3){-}CH_3$$

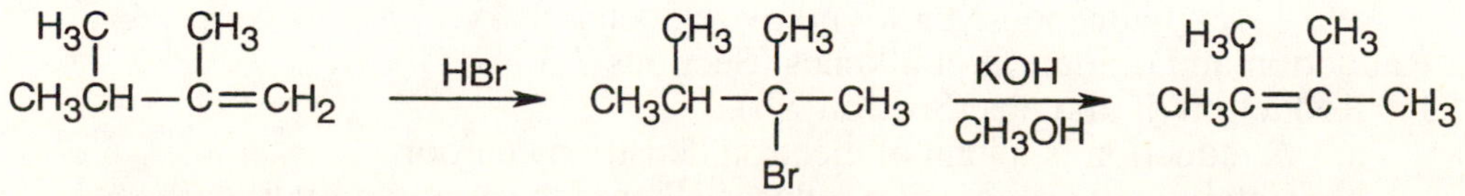

2,3-Dimethylbut-1-ene — 2-Bromo-2,3-dimethylbutane — 2,3-Dimethylbut-2-ene

The product, 2,3-dimethylbut-2-ene, is formed by elimination of HBr from 2-bromo-2,3-dimethylbutane. The product has the more substituted double bond.

Chapter 7 – Reactions of Alkenes and Alkynes

Chapter Outline

I. Alkenes (Sections 7.1 – 7.9).
  A. Preparation of alkenes (Section 7.1).
    1. Dehydrohalogenation.
       Reaction of an alkyl halide with a strong base forms an alkene, with loss of HX.
    2. Dehydration.
       Treatment of an alcohol with a strong acid forms an alkene, with loss of H_2O.
  B. Addition reactions of alkenes (Sections 7.2 – 7.4).
    1. Addition of halogens (Section 7.2).
      a. Br_2 (and Cl_2) react with alkenes to yield 1,2-dihaloalkanes.
      b. Reaction occurs with anti stereochemistry – both bromines come from opposite sides of the molecule.
      c. The reaction intermediate is a cyclic bromonium intermediate that is formed in a single step by interaction of an alkene with Br^+.
    2. Halohydrin formation (Section 7.3).
      a. Alkenes add HO–X (X = Br or Cl) when they react with halogens in the presence of H_2O.
      b. The added nucleophile (H_2O) intercepts the bromonium ion to yield a bromohydrin.
    3. Addition of water to alkenes (Section 7.4).
      a. Hydration.
        i. Water adds to alkenes to yield alcohols in the presence of a strong acid catalyst.
        ii. Although this reaction is important industrially, reaction conditions are too severe for most molecules.
      b. Oxymercuration.
        i. Addition of $Hg(OAc)_2$, followed by $NaBH_4$, converts an alkene to an alcohol.
        ii. The mechanism of addition proceeds through a mercurinium ion.
        iii. The reaction follows Markovnikov regiochemistry.
      c. Hydroboration/oxidation.
        i. BH_3 adds to an alkene to produce an organoborane intermediate.
        ii. Treatment of the organoborane with H_2O_2 forms an alcohol.
        iii. Addition occurs with syn stereochemistry.
        iv. Addition occurs with non-Markovnikov regiochemistry.
            Hydroboration is complementary to oxymercuration/reduction.
      d. The mechanism of hydroboration involves a four-center, cyclic transition state.
        i. This transition state explains syn addition.
        ii. Stabilization of the transition state by a substituted double-bond carbon also explains non-Markovnikov regiochemistry.
  C. Reduction and oxidation of alkenes (Sections 7.5 – 7.7).
    1. Reduction of alkenes (Section 7.5).
      a. A reduction is a gain of electron density by carbon.
      b. Catalytic hydrogenation reduces alkenes to saturated hydrocarbons.
      c. The catalysts used are Pt and Pd.
         Catalytic hydrogenation is a heterogeneous process that takes place on the surface of the catalyst.

d. Hydrogenation occurs with syn stereochemistry.
The reaction is sensitive to the steric environment around the double bond.
e. Alkenes are much more reactive than other functional groups.

2. Oxidation of alkenes (Sections 7.6 – 7.7).
a. An oxidation is a loss of electron density by carbon
b. Epoxidation (Section 7.6).
i. An epoxide is a cyclic ether in a three-membered ring.
ii. Epoxides are prepared by treating alkenes with a peroxyacid.
The reaction occurs by a one-step mechanism.
iii. Epoxides can also be formed by treatment of a halohydrin with base.
c. Hydroxylation (Section 7.7).
i. Acid-catalyzed ring-opening of an epoxide yields a diol.
ii. OsO_4 causes the addition of two –OH groups to an alkene to form a diol.
iii. Hydroxylation occurs through a cyclic osmate with syn stereochemistry.

D. Alkene polymers (Sections 7.8 – 7.9).
a. Many types of polymers can be formed by radical polymerization of alkene monomers (Section 7.8).
b. There are 3 steps in a polymerization reaction.
i. Initiation involves homolytic cleavage of a weak bond to form a radical
The radical adds to an alkene to generate an alkyl radical.
ii. The alkyl radical adds to another alkene molecule to yield a second radical.
This step is repeated many, many times.
iii. Termination occurs when two radical fragments combine.
c. Mechanisms of radical reactions are shown by using fishhook arrows.
d. As in electrophilic addition reactions, the more stable radical (more substituted) is formed in preference to the less stable radical.
e. Biological additions of radicals (Section 7.9).
Biochemical radical reactions are more controlled than laboratory radical reactions.

II. Conjugated dienes (Sections 7.10 – 7.11).
A. Stability of conjugated dienes (Section 7.10)
1. Heats of hydrogenation show that conjugated dienes are somewhat more stable than nonconjugated dienes.
2. Because conjugated dienes are more stable and contain less energy, they release less heat on hydrogenation.
3. The stability of 1,3-butadiene may be due to the greater amount of *s* character of the C–C single bond between the double bonds.
4. Molecular orbital theory offers another explanation.
a. If we combine 4 adjacent *p* orbitals, we generate a set of 4 molecular orbitals.
b. Bonding electrons go into the lower two MOs.
c. The lowest MO has a bonding interaction between C2 and C3 that gives that bond partial double-bond character.
d. The π electrons of butadiene are delocalized over the entire π framework.

B. Reactions of conjugated dienes (Section 7.11).
Conjugated dienes react in electrophilic addition reactions to give products of both 1,2 addition and 1,4 addition.
a. Addition of an electrophile gives an allylic carbocation intermediate that is resonance-stabilized.
b. Addition of the nucleophile in the second step of the reaction can occur at either end of the allylic carbocation to yield two products.

III. Reactions of alkynes (Section 7.12).
   A. Alkynes, although less reactive, undergo addition reactions similar to those of alkenes.
      1. Additions show Markovnikov regiochemistry.
      2. Hydrogenation with Lindlar catalyst gives an alkene with cis stereochemistry.
   B. Terminal alkynes are weakly acidic.
      1. Treatment with a strong base produces an acetylide anion.
      2. An acetylide anion is more stable because the charge resides in an *s* orbital that is lower in energy.

Solutions to Problems

7.1

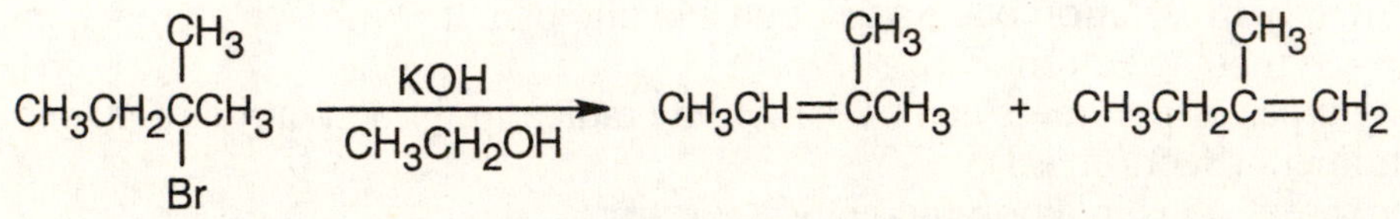

$$CH_3CH_2C(CH_3)(Br)CH_3 \xrightarrow[CH_3CH_2OH]{KOH} CH_3CH{=}C(CH_3)CH_3 + CH_3CH_2C(CH_3){=}CH_2$$

Dehydrobromination can occur in two directions to yield a mixture of products.

7.2

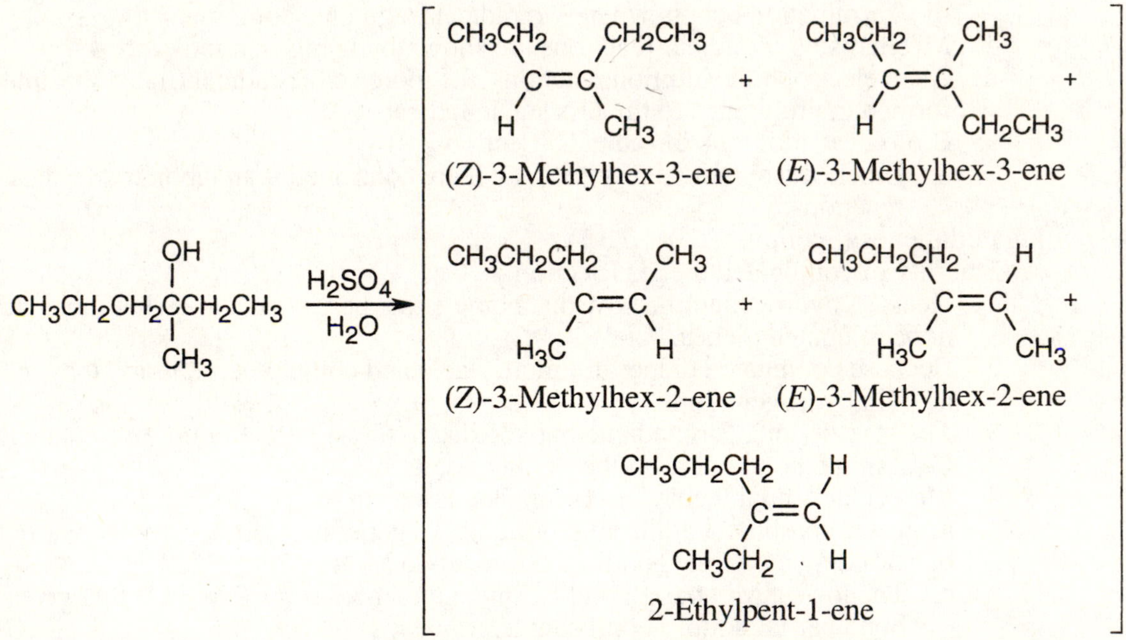

$$CH_3CH_2CH_2C(OH)(CH_3)CH_2CH_3 \xrightarrow[H_2O]{H_2SO_4}$$

(*Z*)-3-Methylhex-3-ene: $(CH_3CH_2)(H)C{=}C(CH_2CH_3)(CH_3)$ +
(*E*)-3-Methylhex-3-ene: $(CH_3CH_2)(H)C{=}C(CH_3)(CH_2CH_3)$ +
(*Z*)-3-Methylhex-2-ene: $(CH_3CH_2CH_2)(H_3C)C{=}C(CH_3)(H)$ +
(*E*)-3-Methylhex-2-ene: $(CH_3CH_2CH_2)(H_3C)C{=}C(H)(CH_3)$ +
2-Ethylpent-1-ene: $(CH_3CH_2CH_2)(CH_3CH_2)C{=}C(H)(H)$

Five alkene products, including *E*, *Z* isomers, might be obtained by dehydration of 3-methylhexan-3-ol.

7.3

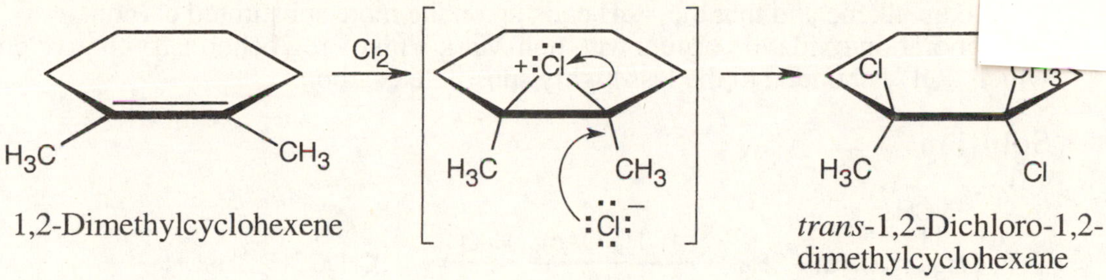

H3C CH3
1,2-Dimethylcyclohexene
Cl2
+:Cl:
H3C CH3
:Cl:−
Cl CH3
H3C Cl
trans-1,2-Dichloro-1,2-dimethylcyclohexane

The chlorines are trans to one another in the product, as are the methyl groups.

7.4

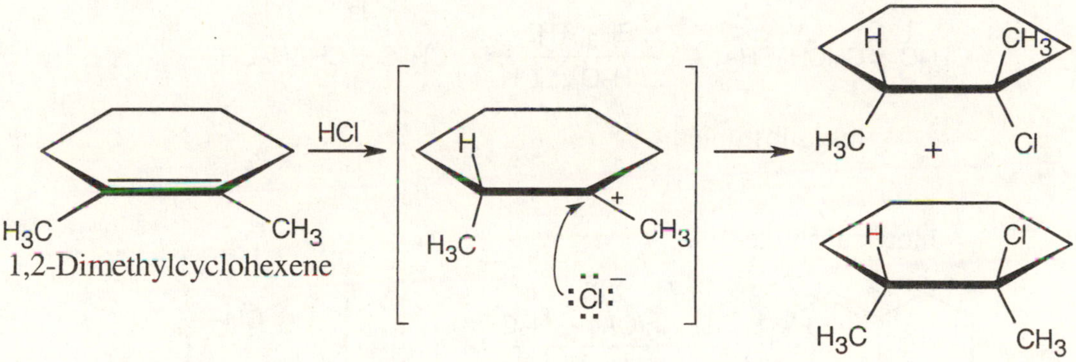

H3C CH3
1,2-Dimethylcyclohexene
HCl
H
H3C CH3
+
:Cl:−
H CH3
H3C Cl
+
H Cl
H3C CH3

Addition of hydrogen halides involves formation of an open carbocation, not a cyclic halonium ion intermediate. The carbocation, which is sp^2-hybridized and planar, can react with chloride from either top or bottom, yielding products in which the two methyl groups can be either cis or trans to each other.

7.5 Reaction of the alkene with Br_2 forms a cyclic bromonium ion. When this bromonium ion is opened by water, a partial positive charge develops at the carbon whose bond to bromine is being cleaved.

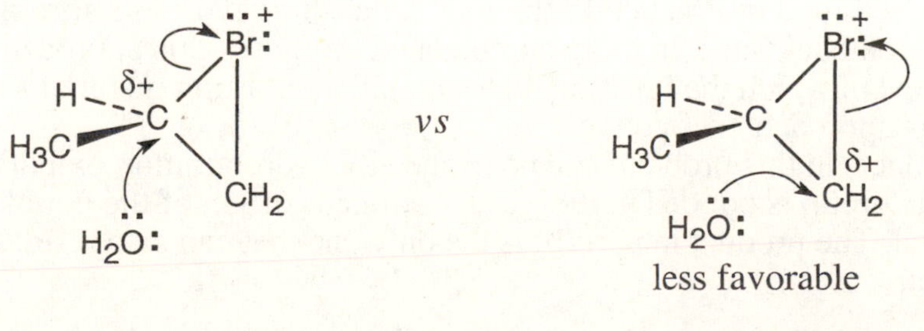

Br:+
H δ+ C
H3C
CH2
H2O:
vs
Br:+
H C
H3C
δ+ CH2
H2O:
less favorable

Since a secondary carbon can stabilize this charge better than a primary carbon, opening of the bromonium ion occurs at the secondary carbon to yield the Markovnikov product.

7.6 **Strategy:** Keep in mind that oxymercuration is equivalent to Markovnikov addition of H_2O to an alkene and that the –OH ends up on the more substituted carbon. Hydroboration/oxidation occurs with non-Markovnikov regiochemistry to give products in which –OH is bonded to the less highly substituted carbon.

Solution:

(a)

$$CH_3C(CH_3){=}CHCH_2CH_3 \xrightarrow[2.\ NaBH_4]{1.\ Hg(OAc)_2,\ H_2O} CH_3C(CH_3)(OH)CH_2CH_2CH_3$$

(arrow at $C(CH_3)$: more substituted)

$$CH_3C(CH_3){=}CHCH_2CH_3 \xrightarrow[2.\ H_2O_2,\ {}^{-}OH]{1.\ BH_3,\ THF} CH_3C(CH_3)(H){-}CH(OH)CH_2CH_3$$

(arrow at $=CH$: less substituted)

(b)

more substituted

1-Ethylidenecyclohexane (arrow at ring carbon) $\xrightarrow[2.\ NaBH_4]{1.\ Hg(OAc)_2,\ H_2O}$ 1-ethylcyclohexanol (OH on ring carbon, CH_3)

less substituted

1-Ethylidenecyclohexane (arrow at $=CH$) $\xrightarrow[2.\ H_2O_2,\ {}^{-}OH]{1.\ BH_3,\ THF}$ 1-cyclohexylethanol (OH, CH_3)

7.7 **Strategy:** As described in Practice Problem 7.2, the strategy in this sort of problem begins with a look backward. In more complicated syntheses this approach is essential, but even in problems in which the functional group(s) in the starting material and the reagents are known, this approach is effective.

All the products in this problem can be formed by hydroboration/oxidation of a double bond. The –OH group is bonded to the less substituted carbon of the double bond in the starting material. The product in part (b) is the only one that can also result from oxymercuration.

Solution:

(a)

$$CH_3CH(CH_3)CH{=}CH_2 \xrightarrow[2.\ H_2O_2,\ {}^{-}OH]{1.\ BH_3,\ THF} CH_3CH(CH_3)CH_2CH_2OH$$

(b)

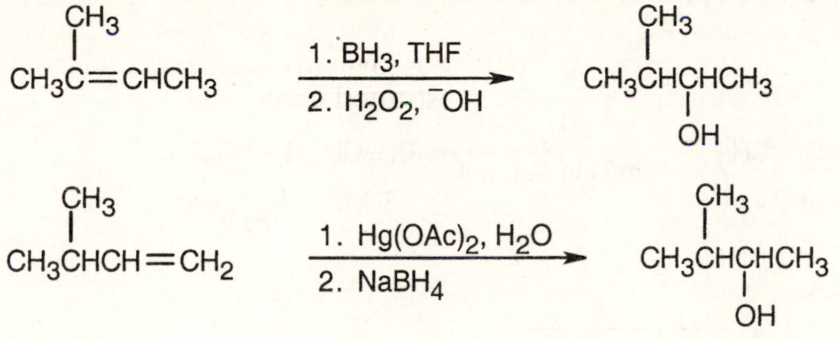

(c)

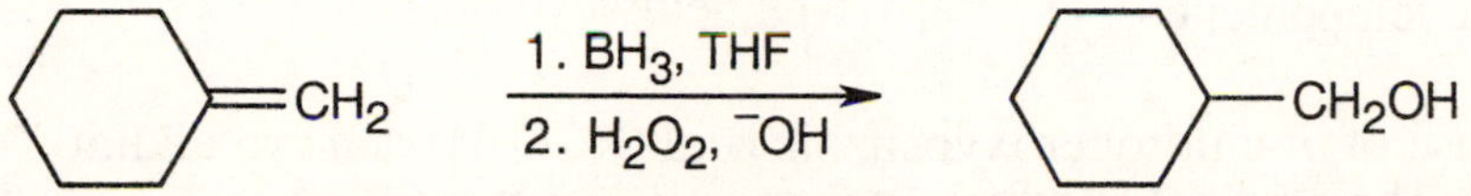

7.8 The drawings below show the transition states resulting from addition of BH_3 to the double bond of the cycloalkene. Addition can occur on either side of the double bond.

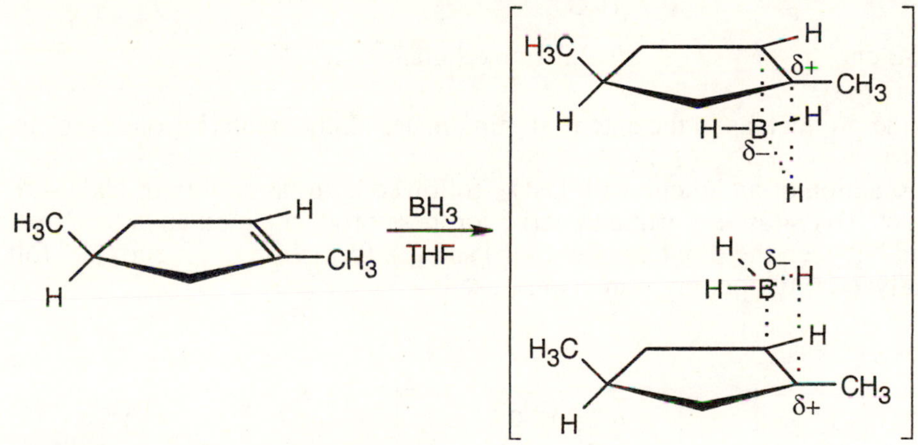

Reaction of the two neutral alkylborane adducts with hydrogen peroxide gives two alcohol isomers. In one isomer, the two methyl groups have a cis relationship, and in the other isomer, they have a trans relationship.

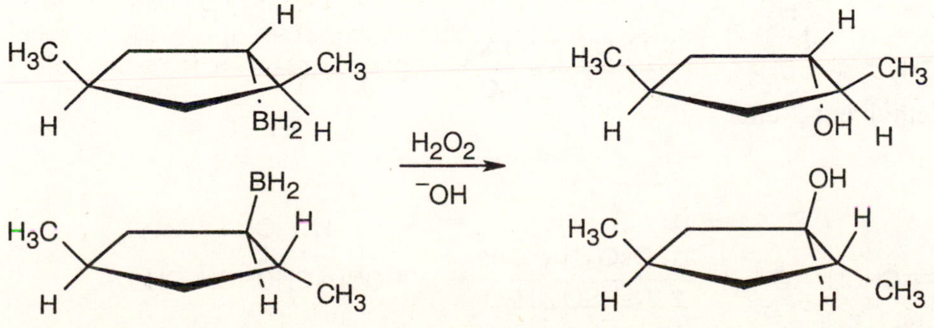

7.9 Catalytic hydrogenation produces alkanes from alkenes.

(a)

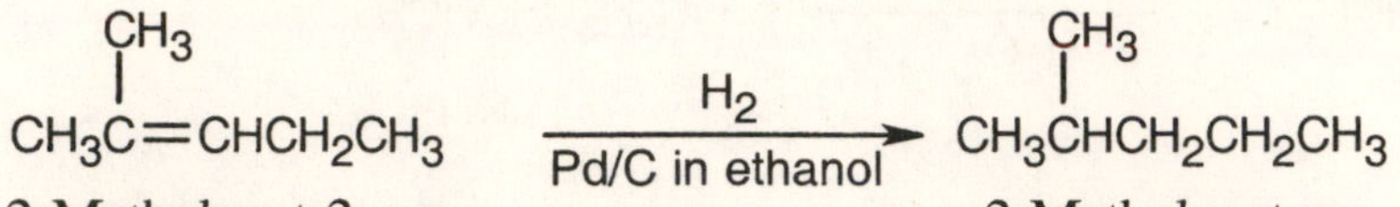

$$CH_3\overset{CH_3}{\overset{|}{C}}{=}CHCH_2CH_3 \xrightarrow[\text{Pd/C in ethanol}]{H_2} CH_3\overset{CH_3}{\overset{|}{C}}HCH_2CH_2CH_3$$

2-Methylpent-2-ene → 2-Methylpentane

(b)

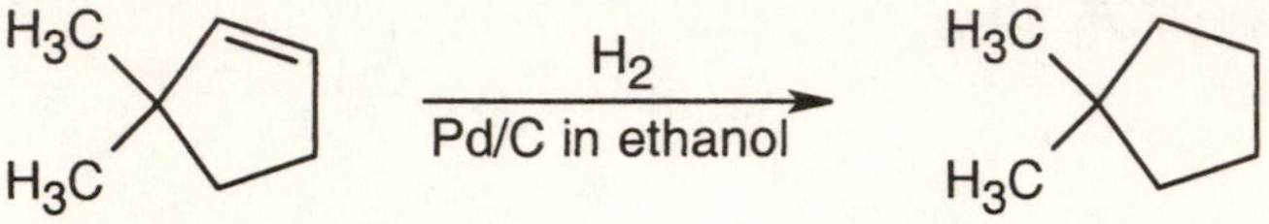

H_3C, H_3C → $\xrightarrow[\text{Pd/C in ethanol}]{H_2}$ H_3C, H_3C

3,3-Dimethylcyclopentene → 1,1-Dimethylcyclopentane

7.10 Epoxidation by use of *m*-chloroperoxybenzoic acid (RCO_3H) is a syn addition of oxygen to a double bond. The original bond stereochemistry is retained.

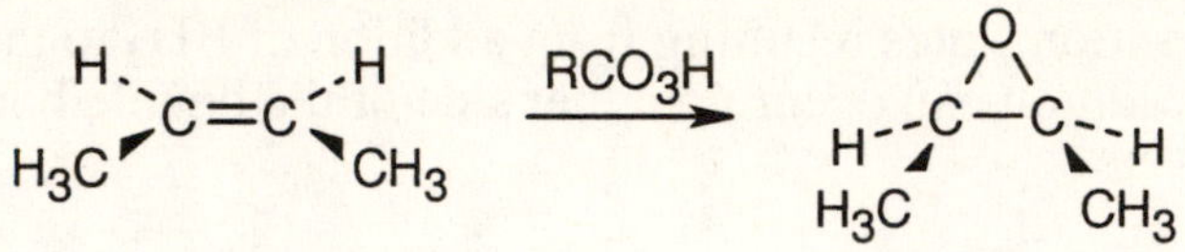

H, H, H_3C, C=C, CH_3 $\xrightarrow{RCO_3H}$ O, H, C—C, H, H_3C, CH_3

cis-But-2-ene → *cis*-2,3-Epoxybutane

In the epoxide product, as in the alkene starting material, the methyl groups are cis.

7.11 **Strategy:** Reaction of an alkene with OsO_4, followed by treatment with $NaHSO_3$, yields a diol product. To choose a starting material for these products, pick an alkene that has a double bond between the diol carbons. In (b) and (c), formation of an epoxide, followed by acid-catalyzed ring opening, can also be used.

Solution:

(a)

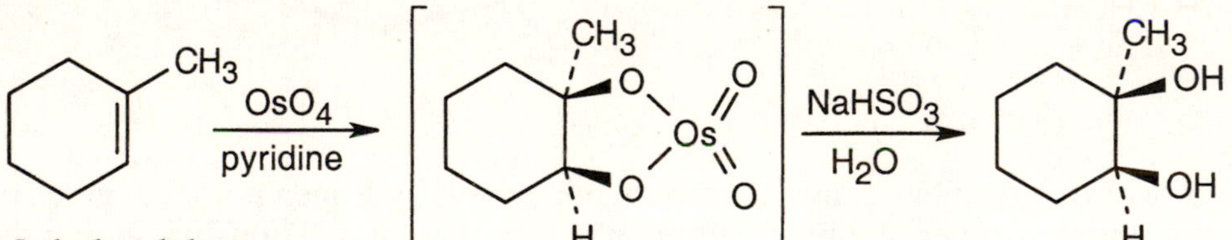

CH_3 $\xrightarrow[\text{pyridine}]{OsO_4}$ [CH_3, O, O, Os, O, O, H] $\xrightarrow[H_2O]{NaHSO_3}$ CH_3, OH, OH, H

1-Methylcyclohexene

(b)

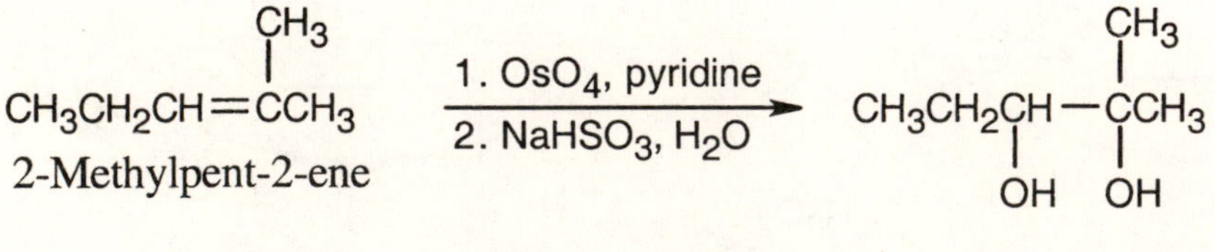

$$CH_3CH_2CH{=}\overset{CH_3}{\overset{|}{C}}CH_3 \xrightarrow[\text{2. NaHSO}_3\text{, H}_2\text{O}]{\text{1. OsO}_4\text{, pyridine}} CH_3CH_2\underset{OH}{\underset{|}{C}}H{-}\overset{CH_3}{\overset{|}{\underset{OH}{\underset{|}{C}}}}CH_3$$

2-Methylpent-2-ene

(c)

$$CH_2{=}CHCH{=}CH_2 \xrightarrow[\text{2. NaHSO}_3\text{, H}_2\text{O}]{\text{1. OsO}_4\text{, pyridine}} HOCH_2\overset{HO}{\overset{|}{C}}H\overset{OH}{\overset{|}{C}}HCH_2OH$$

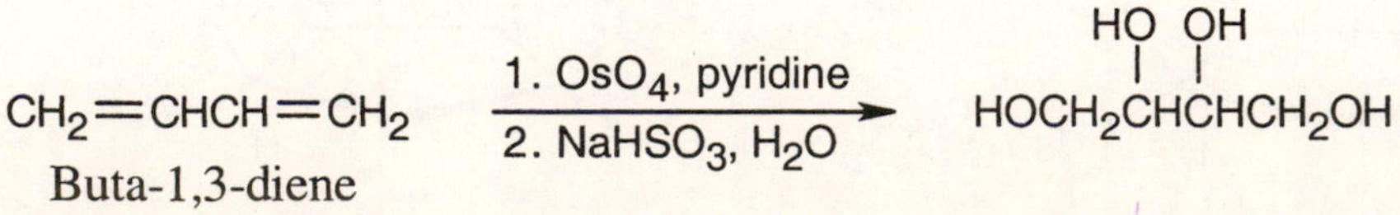

Buta-1,3-diene

7.12 **Strategy:** Find the smallest repeating unit in each polymer. This is the monomer unit.

	Monomer	*Polymer*
(a)	$H_2C{=}CHOCH_3$	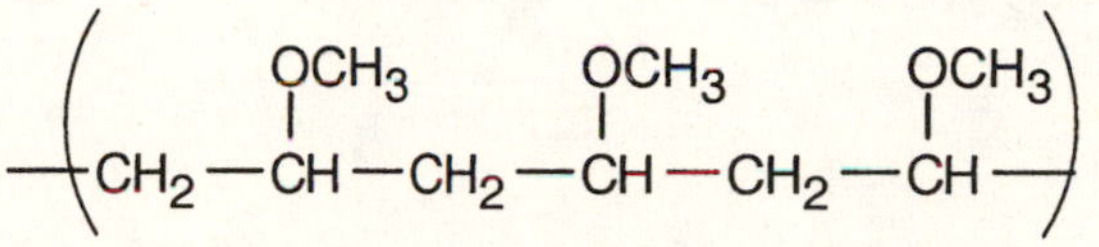 $\left(CH_2{-}CH(OCH_3){-}CH_2{-}CH(OCH_3){-}CH_2{-}CH(OCH_3)\right)$
(b)	$ClHC{=}CHCl$	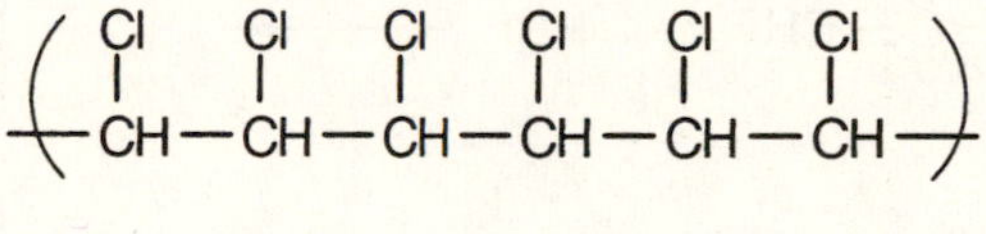 $\left(CH(Cl){-}CH(Cl){-}CH(Cl){-}CH(Cl){-}CH(Cl){-}CH(Cl)\right)$

7.13 One radical abstracts a hydrogen atom from a second radical, and the remaining two electrons create a double bond.

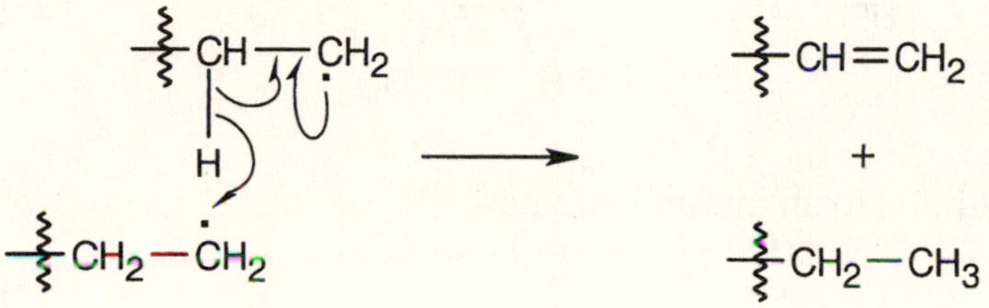

$$\sim CH(H){-}\dot{C}H_2 \;+\; \sim CH_2{-}\dot{C}H_2 \longrightarrow \sim CH{=}CH_2 \;+\; \sim CH_2{-}CH_3$$

7.14

$CH_3CH{=}CHCH{=}CH_2$ Penta-1,3-diene

Product	*Name*	*Results from:*
$CH_3CH{=}CHCH(Cl)CH_3$	4-Chloropent-2-ene	1,2 addition 1,4 addition
$CH_3CH_2CH(Cl)CH{=}CH_2$	3-Chloropent-1-ene	1,2 addition
$CH_3CH_2CH{=}CHCH_2Cl$	1-Chloropent-2-ene	1,4 addition

7.15

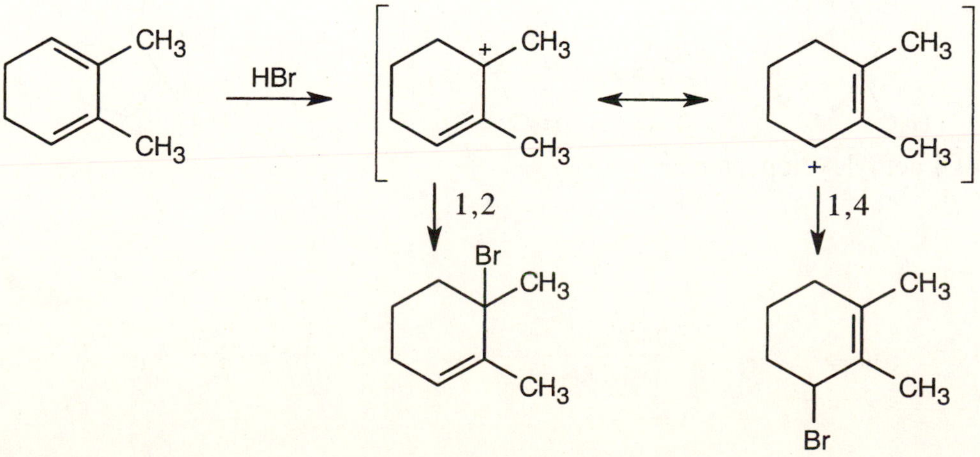

1,2-Dimethylcyclohexa-1,3-diene (ring with CH_3, CH_3) $\xrightarrow{HBr}$ [allylic carbocation with + on the CH_3-bearing carbon $\longleftrightarrow$ resonance form with + at the ring carbon adjacent to the double bond bearing CH_3, CH_3]

1,2 → Br on the carbon bearing CH_3 (ring with CH_3, CH_3)

1,4 → Br on the ring carbon adjacent to the $C{=}C$ bearing CH_3, CH_3

7.16 Markovnikov addition is observed with alkynes as well as with alkenes.

(a)

$$CH_3CH_2CH_2C{\equiv}CH \;+\; 2\,Cl_2 \longrightarrow CH_3CH_2CH_2CCl_2CHCl_2$$

(b)

Cyclopentyl–C≡CH + 1 HBr ⟶ Cyclopentyl–C(Br)=CH₂

(c)

$$CH_3CH_2CH_2CH_2C{\equiv}CCH_3 \;+\; 1\,HBr \longrightarrow$$

(CH₃CH₂CH₂CH₂)(H)C=C(Br)(CH₃) + (CH₃CH₂CH₂CH₂)(Br)C=C(H)(CH₃)

Two products result from addition to an internal alkyne.

Visualizing Chemistry

7.17

(a)

$$CH_3\overset{CH_3}{\overset{|}{C}}{=}CH\,CH_2\overset{CH_3}{\overset{|}{C}}HCH_2CH_3 \xrightarrow{RCO_3H}$$ epoxide ⟶ (H_3O^+) ⟶ $CH_3C(CH_3)(OH)\text{–}CH(OH)CH_2CH(CH_3)CH_2CH_3$

2,5-Dimethylhept-2-ene

(b)

3,3-Dimethylcyclopentene $\xrightarrow{RCO_3H}$ epoxide $\xrightarrow{H_3O^+}$ trans-diol (HO, H; OH, H)

3,3-Dimethylcyclopentene

7.18

(a)

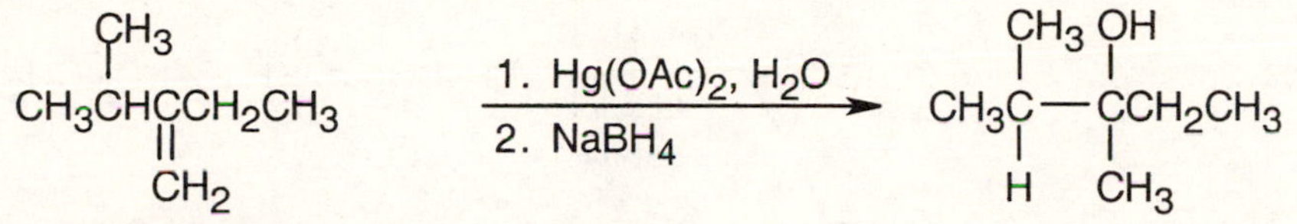

$$CH_3CH(CH_3)C(=CH_2)CH_2CH_3 \xrightarrow[\text{2. } NaBH_4]{\text{1. } Hg(OAc)_2,\ H_2O} CH_3C(CH_3)(H)-C(OH)(CH_3)CH_2CH_3$$

2-Ethyl-3-methylbut-1-ene

or

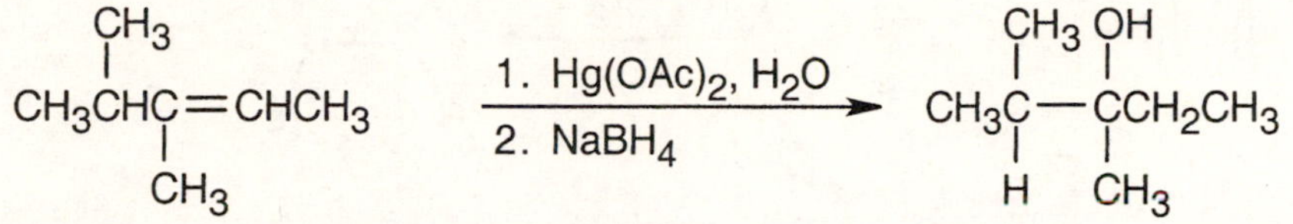

$$CH_3CH(CH_3)C(CH_3)=CHCH_3 \xrightarrow[\text{2. } NaBH_4]{\text{1. } Hg(OAc)_2,\ H_2O} CH_3C(CH_3)(H)-C(OH)(CH_3)CH_2CH_3$$

3,4-Dimethylpent-2-ene

Only oxymercuration/reduction can be used to produce an alcohol that has –OH bonded to the more substituted carbon. A third alkene, 2,3-dimethylpent-2-ene, gives a mixture of tertiary alcohols when treated with either BH_3 or $Hg(OAc)_2$.

(b)

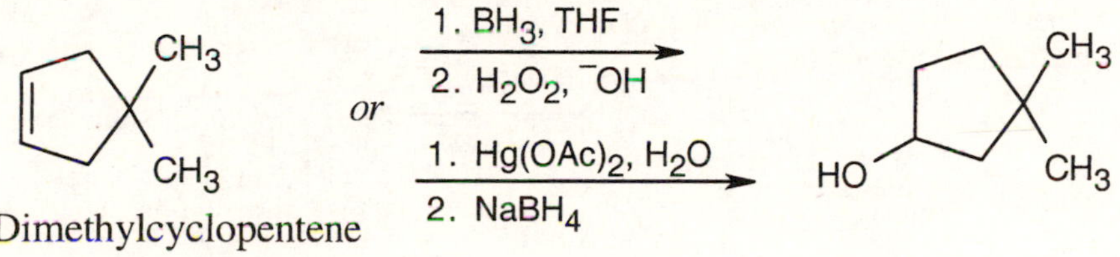

4,4-Dimethylcyclopentene

1. BH_3, THF
2. H_2O_2, ^-OH

or

1. $Hg(OAc)_2$, H_2O
2. $NaBH_4$

→ 3,3-dimethylcyclopentanol (HO, CH_3, CH_3)

Both hydroboration/oxidation and oxymercuration yield the same alcohol product from the symmetrical alkene starting material.

7.19

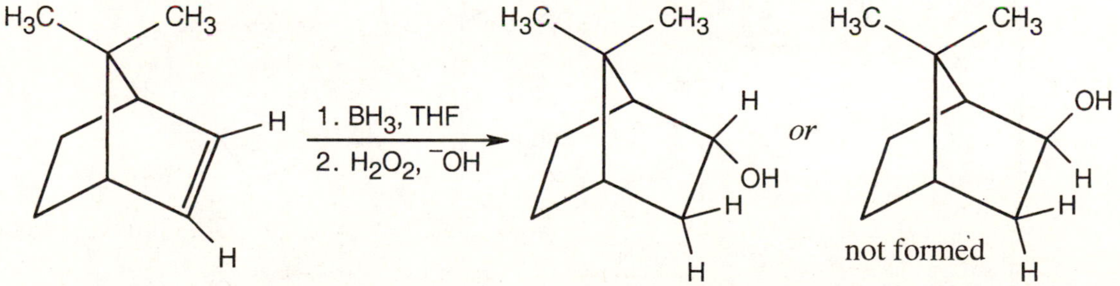

1. BH_3, THF
2. H_2O_2, ^-OH

(bicyclic alkene with H_3C, CH_3 on the bridge; H, H on the double bond) → (product with OH on the bottom face, H on top) *or* (product with OH on the top face) not formed

Two possible alcohols might be formed by hydroboration/oxidation of the alkene shown. One product results from addition of BH_3 to the top face of the double bond (not formed), and the other product results from addition to the bottom face of the double bond (formed). Addition from the top face does not occur because a methyl group on the bridge of the bicyclic ring system blocks approach of the borane.

7.20

(a)

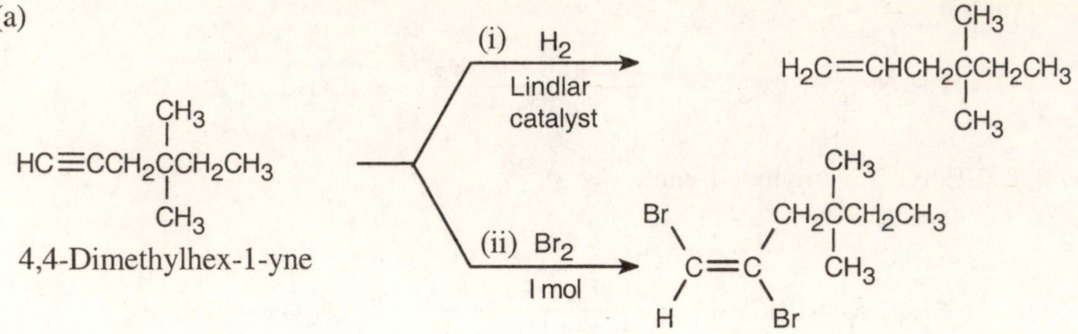
HC≡CCH2CCH2CH3 (CH3, CH3)
4,4-Dimethylhex-1-yne
(i) H2
Lindlar catalyst
H2C=CHCH2CCH2CH3 (CH3, CH3)
(ii) Br2
1 mol
Br, H, Br, CH2CCH2CH3, CH3, CH3
C=C

(b)

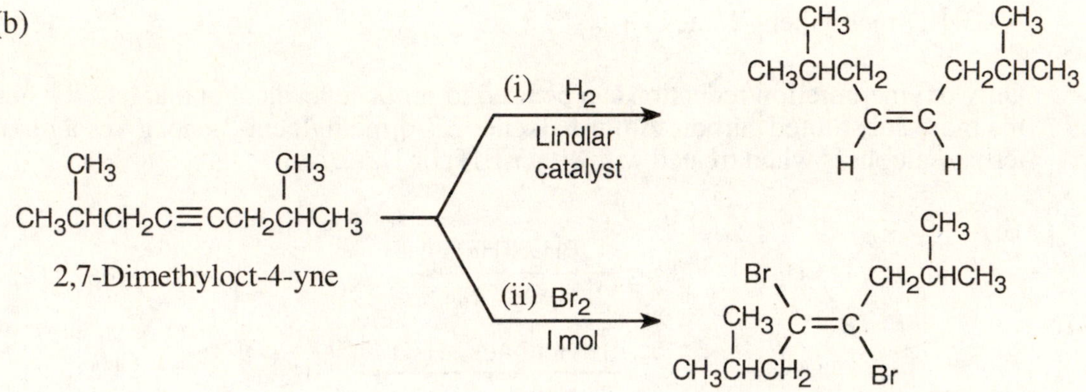
CH3CHCH2C≡CCH2CHCH3 (CH3, CH3)
2,7-Dimethyloct-4-yne
(i) H2
Lindlar catalyst
CH3CHCH2, CH2CHCH3, CH3, CH3, H, H
C=C
(ii) Br2
1 mol
Br, CH2CHCH3, CH3, CH3CHCH2, Br
C=C

7.21

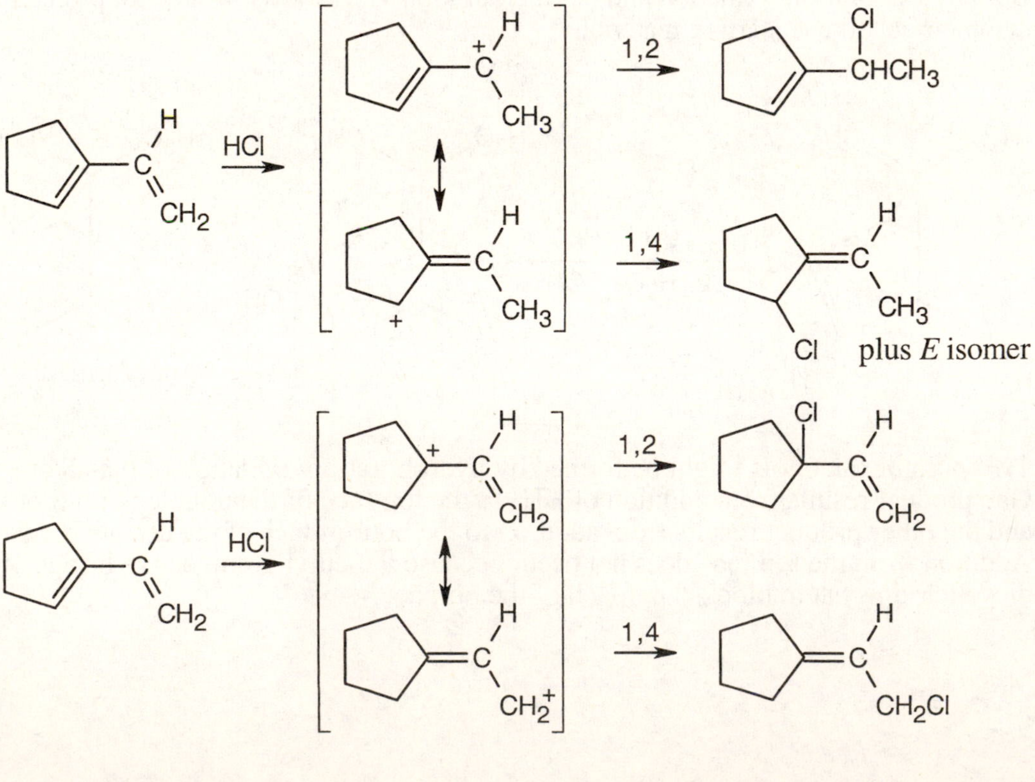
HCl
1,2
1,4
CHCH3
CH3
CH2
CH2Cl
Cl
plus E isomer

Additional Problems

7.22 Excluding double-bond isomers:

Conjugated dienes:

$CH_3CH{=}CHCH{=}CH_2$
Penta-1,3-diene

$H_2C{=}CHC(CH_3){=}CH_2$
2-Methylbuta-1,3-diene

Cumulated dienes:

$CH_3CH_2CH{=}C{=}CH_2$
Penta-1,2-diene

$CH_3CH{=}C{=}CHCH_3$
Penta-2,3-diene

$H_2C{=}C{=}C(CH_3)_2$
3-Methylbuta-1,2-diene

Nonconjugated diene:

$H_2C{=}CHCH_2CH{=}CH_2$
Penta-1,4-diene

7.23

(a) $C_6H_5CH{=}CH_2 \xrightarrow{H_2/Pd} C_6H_5CH_2CH_3$

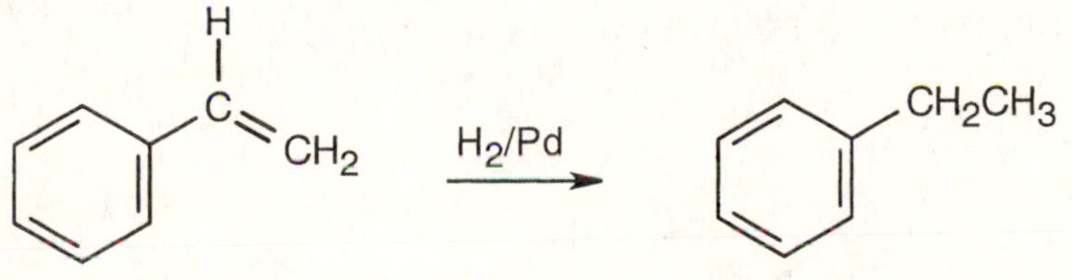

(b) $C_6H_5CH{=}CH_2 \xrightarrow{Br_2} C_6H_5CH(Br)CH_2Br$

(c) $C_6H_5CH{=}CH_2 \xrightarrow[H_2O]{Cl_2} C_6H_5CH(OH)CH_2Cl$

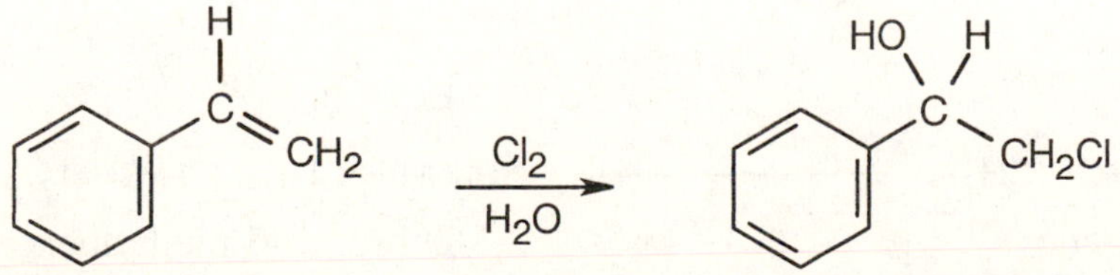

Halogen is bonded to the less substituted carbon (see problem 7.5)

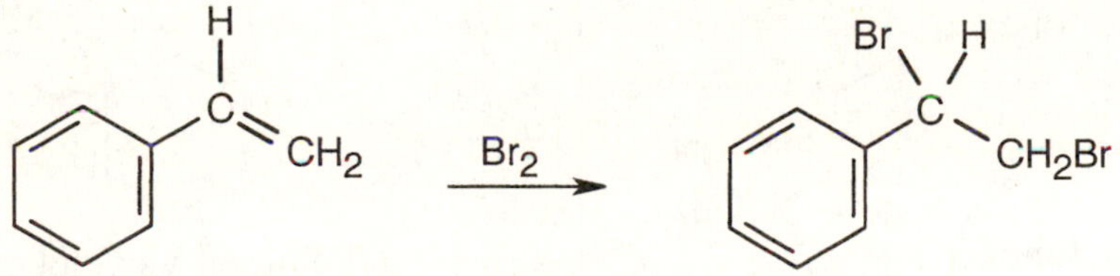

(d) $C_6H_5CH{=}CH_2 \xrightarrow[2.\ NaHSO_3,\ H_2O]{1.\ OsO_4,\ pyridine} C_6H_5CH(OH)CH_2OH$

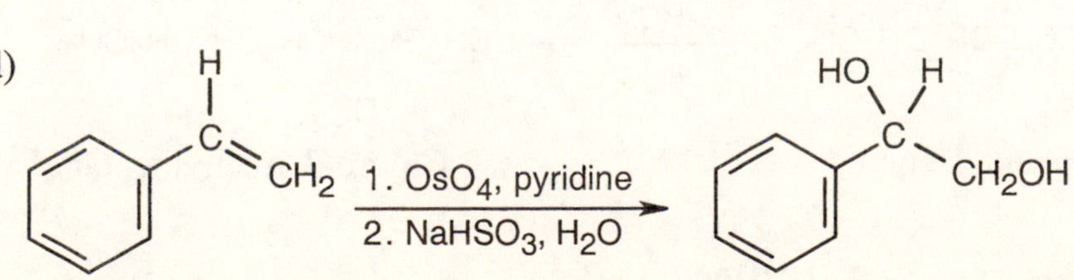

(e)

$$C_6H_5CH=CH_2 \xrightarrow[\text{2. } H_2O_2,\ ^-OH]{\text{1. } BH_3,\ THF} C_6H_5CH_2CH_2OH$$

(f)

$$C_6H_5CH=CH_2 \xrightarrow{RCO_3H} C_6H_5\text{-epoxide}$$

RCO_3H = *meta*-Chloroperoxybenzoic acid

7.24

(a)

$CH_3CH_2CH_2CH_2C(CH_3)=CH_2$	2-Methylhex-1-ene
$CH_3CH_2CH_2CH=C(CH_3)_2$	2-Methylhex-2-ene
$CH_3CH_2CH=CHCH(CH_3)_2$	2-Methylhex-3-ene
$CH_3CH=CHCH_2CH(CH_3)_2$	5-Methylhex-2-ene
$H_2C=CHCH_2CH_2CH(CH_3)_2$	5-Methylhex-1-ene

$\xrightarrow{H_2/Pd}$ $CH_3CH_2CH_2CH_2CH(CH_3)CH_3$
2-Methylhexane

(b)

3,3-Dimethylcyclohexene

4,4-Dimethylcyclohexene

$\xrightarrow{H_2/Pd}$ 1,1-Dimethylcyclohexane

(c)

$$CH_3CH{=}CHCH_2CH(CH_3)_2 \xrightarrow{Br_2} CH_3CHBr{-}CHBrCH_2CH(CH_3)_2$$

5-Methylhex-2-ene → 2,3-Dibromo-5-methylhexane

(d)

$$H_2C{=}CHCH(CH_3)CH_2CH_2CH_2CH_3 \xrightarrow{HCl} CH_3CHClCH(CH_3)CH_2CH_2CH_2CH_3$$

3-Methylhept-1-ene → 2-Chloro-3-methylheptane

(e)

$$CH_3CH_2CH_2CH{=}CH_2 \xrightarrow[\text{2. } NaBH_4]{\text{1. } Hg(OAc)_2,\ H_2O} CH_3CH_2CH_2CH(OH)CH_3$$

Pent-1-ene → Pentan-2-ol

7.25

(a)

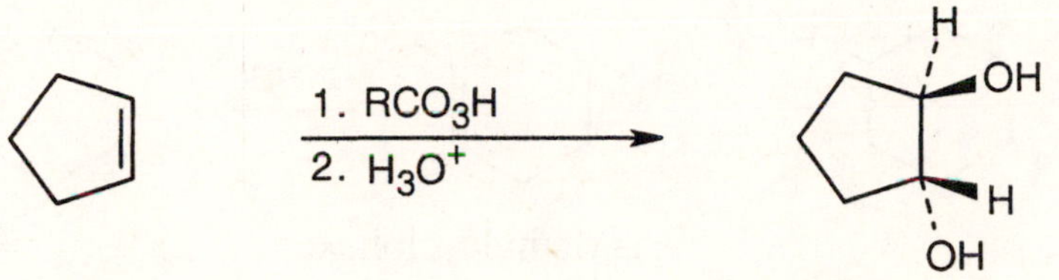
1. RCO_3H
2. H_3O^+
H
OH
H
OH

(b)

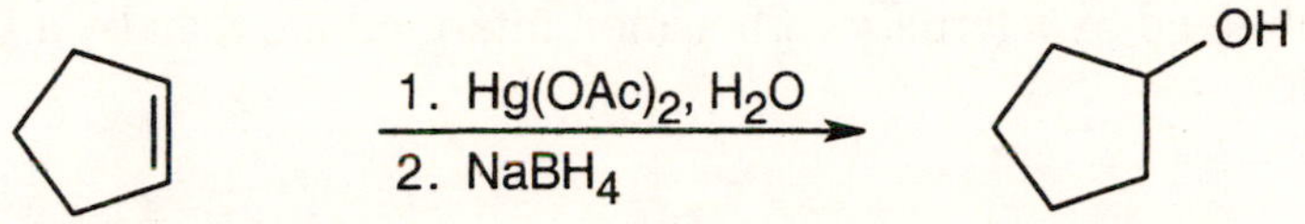
1. $Hg(OAc)_2$, H_2O
2. $NaBH_4$
OH

Hydroboration/oxidation is another route to this product.

(c)

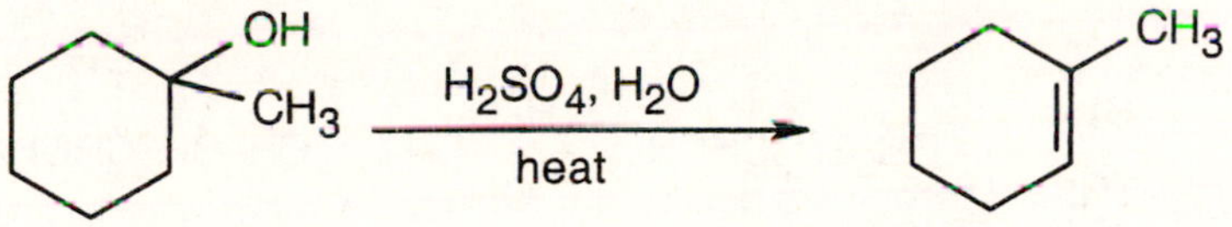
OH
CH_3
H_2SO_4, H_2O
heat
CH_3

(d)

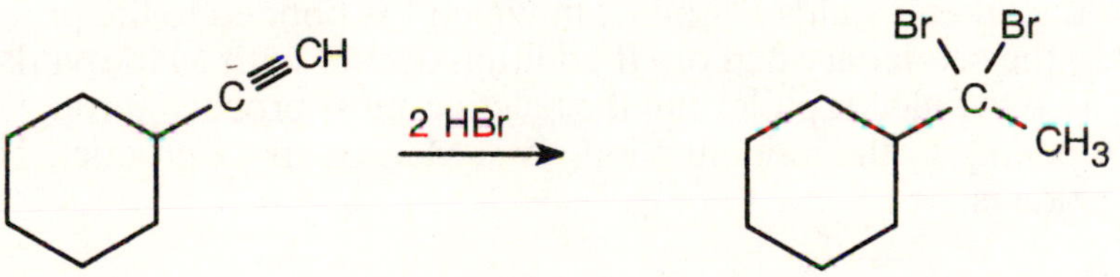
C≡CH
2 HBr
Br
Br
C
CH_3

(e)

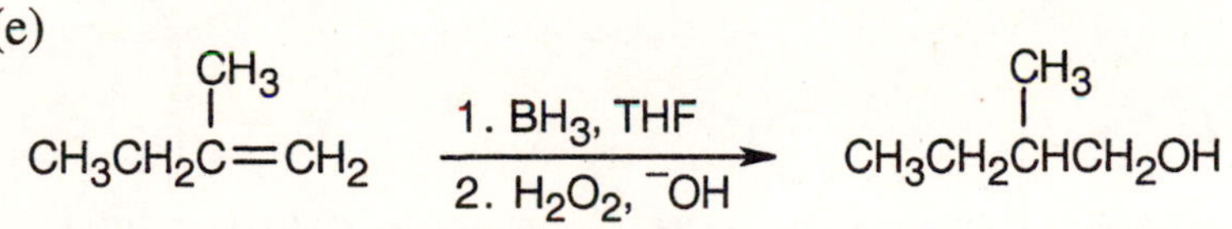
CH_3
$CH_3CH_2C{=}CH_2$
1. BH_3, THF
2. H_2O_2, ^-OH
CH_3
$CH_3CH_2CHCH_2OH$

7.26 Recall the mechanism of hydroboration and note that the hydrogen added to the double bond comes from borane. The product of hydroboration with BD_3 has deuterium bonded to the more substituted carbon; –D and –OH are cis to one another.

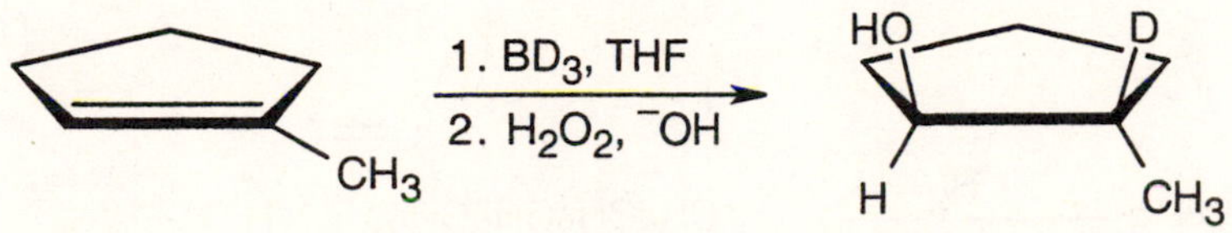
CH_3
1. BD_3, THF
2. H_2O_2, ^-OH
HO
D
H
CH_3

7.27

Cyclohexene $\xrightarrow{HBr}$ Bromocyclohexane

1-Methylcyclohexene $\xrightarrow{HBr}$ 1-Bromo-1-methylcyclohexane

Addition to 1-methylcyclohexene occurs at a faster rate. Positive charge generated during the reaction is better stabilized by a tertiary carbocation intermediate than by a secondary carbocation intermediate.

7.28

(a)

$$CH_3CH{=}CHCH_3 \xrightarrow{HBr} CH_3CH_2CH(Br)CH_3$$

But-2-ene

(b)

$$CH_3CH{=}CHCH_3 \xrightarrow[THF]{BH_3} CH_3CH_2CH(BH_2)CH_3 \xrightarrow[^-OH]{H_2O_2} CH_3CH_2CH(OH)CH_3$$

But-2-ene

7.29

Addition of IN_3 to the alkene yields a product in which I is bonded to the primary carbon and N_3 is bonded to the secondary carbon. If addition occurs with Markovnikov orientation, I^+ must be the electrophile, and the reaction must proceed through an iodonium ion intermediate. Opening of the iodonium ion gives Markovnikov product. The bond polarity of iodine azide is:

$$\overset{+\!\!\longrightarrow}{I{-}N_3}$$

$$CH_3CH_2CH{=}CH_2 \xrightarrow{I-N_3} \left[CH_3CH_2\overset{I^+}{CH{-}CH_2} \quad :N_3^- \right] \longrightarrow CH_3CH_2CH(N_3)CH_2I$$

7.30

$$CH_3CH_2CH_2CH_2C{\equiv}CH \xrightarrow[\text{Lindlar catalyst}]{H_2} \text{cis } CH_3CH_2CH_2CH_2(H)C{=}CH_2 \xrightarrow{RCO_3H} \text{epoxide}$$

7.31 Any unsubstituted cyclic 1,3-diene gives the same product from 1,2- and 1,4 addition. For example:

Cyclopenta-1,3-diene $\xrightarrow{HCl}$ 3-chlorocyclopentene (Cl)

7.32

$$H_3C-C(=O)-CH_3 \xrightarrow[2.\ H_3O^+]{1.\ ^-:C\equiv CH} (H_3C)_2C(OH)-C\equiv CH \xrightarrow[\text{Lindlar catalyst}]{H_2} (H_3C)_2C(OH)-CH=CH_2 \xrightarrow[H_2O,\ \Delta]{H_2SO_4} H_2C=C(CH_3)-CH=CH_2$$

2-Methylbuta-1,3-diene

7.33

CH_3O-steroid 17-ketone $\xrightarrow[2.\ H_3O^+]{1.\ ^-:C\equiv CH}$ CH_3O-steroid with 17-OH and 17-C≡CH

Mestranol

7.34 (a) Addition of HBr occurs with Markovnikov regiochemistry – bromine adds to the more substituted carbon.
(b) Hydroxylation of double bonds produces cis, not trans, diols.
(c) Because hydroboration is a syn addition, the –H and the –OH added to the double bond must be cis to each other.

7.35 (a) This alcohol can't be synthesized selectively by hydroboration/oxidation. Consider the two possible starting materials.

1.

$$CH_3CH_2CH_2CH=CH_2 \xrightarrow[2.\ H_2O_2,\ ^-OH]{1.\ BH_3,\ THF} CH_3CH_2CH_2CH_2CH_2OH$$

Pent-1-ene yields only the primary alcohol.

2.

$$CH_3CH_2CH=CHCH_3 \xrightarrow[2.\ H_2O_2,\ ^-OH]{1.\ BH_3,\ THF} CH_3CH_2CH(OH)CH_2CH_3 + CH_3CH_2CH_2CH(OH)CH_3$$

Pent-2-ene yields a mixture of alcohols.

(b)

$$(CH_3)_2C{=}C(CH_3)_2 \xrightarrow[\text{2. } H_2O_2,\ ^-OH]{\text{1. } BH_3,\ THF} (CH_3)_2CH\overset{OH}{\overset{|}{C}}(CH_3)_2$$

2,3-Dimethylbut-2-ene yields the desired alcohol exclusively.

(c) This alcohol can't be formed cleanly by a hydroboration reaction. The –H and –OH added to a double bond must be cis to each other.

(d) The product shown is not a hydroboration product; hydroboration yields an alcohol in which ^-OH is bonded to the less substituted carbon.

7.36

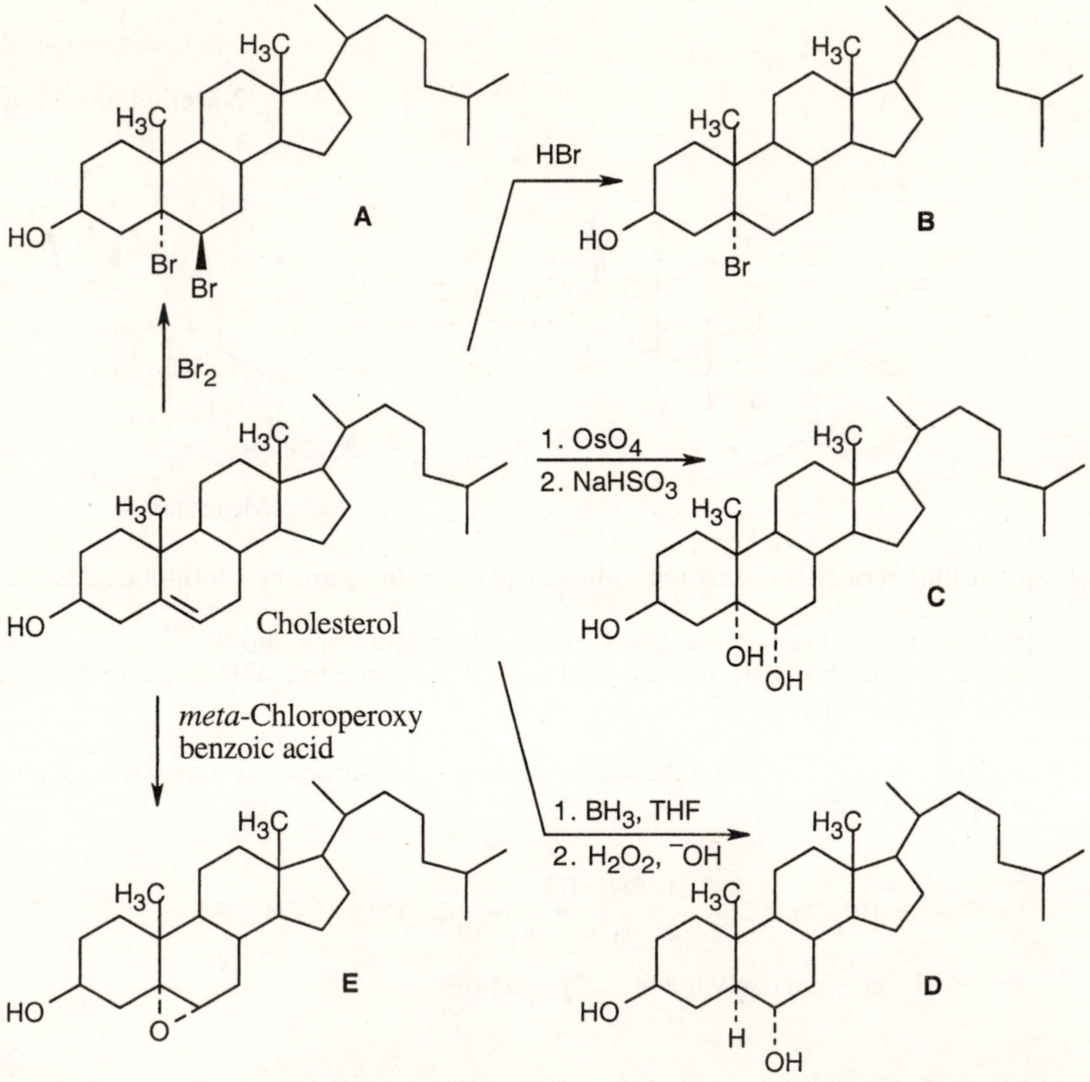

H_3C H_3C HO Br Br **A**

HBr

H_3C H_3C HO Br **B**

Br_2

H_3C H_3C HO Cholesterol

1. OsO_4
2. $NaHSO_3$

H_3C H_3C HO OH OH **C**

meta-Chloroperoxy benzoic acid

H_3C H_3C HO O **E**

1. BH_3, THF
2. H_2O_2, ^-OH

H_3C H_3C HO H OH **D**

7.37

$$H_2C{=}C(CH_3){-}COCH_3\ (\text{C=O}) \longrightarrow \left(CH_2{-}C(CO_2CH_3)(CH_3){-}CH_2{-}C(CO_2CH_3)(CH_3){-}CH_2{-}C(CO_2CH_3)(CH_3)\right)$$

7.38

many $H_2C{=}CH$–N(pyrrolidone) $\longrightarrow$ $\left(CH_2{-}CH{-}CH_2{-}CH{-}CH_2{-}CH\right)$ (each CH bearing an N-pyrrolidone ring)

N-Vinylpyrrolidone | Poly(vinyl pyrrolidone)

7.39

$$(H_3C)_2C{=}CH_2 + H{-}A \rightleftarrows \left[H_3C{-}\overset{+}{C}(CH_3){-}CH_3 + CH_3\ddot{O}{-}H \rightleftarrows CH_3\overset{+}{O}(H){-}C(CH_3)_2CH_3 + :A^- \right] \rightleftarrows CH_3OC(CH_3)_3 + HA$$

protonation of double bond | nucleophilic attack of methanol on carbocation | loss of proton

The above mechanism is the same as the mechanism shown in Section 7.4 with one exception: In this problem, methanol, rather than water, is the nucleophile, and an ether, rather than an alcohol, is the observed product.

7.40

$$HC{=}CH_2\ (H_2C{-}H_2C{-}CH_2OH) + Br{-}Br \longrightarrow \left[\text{cyclic } Br^+ \text{ ion} + Br^- \longrightarrow \text{ring with } {+}\ddot{O}{-}H + :OH_2 \right] \longrightarrow H_3O^+ + \text{HC–CH}_2\text{Br (tetrahydrofuran ring)}$$

formation of cyclic bromonium ion | nucleophilic attack of –OH on bromonium ion | loss of H^+

The above mechanism is the same as that for halohydrin formation, shown in Section 7.3. In this case, the nucleophile is the hydroxyl group of pent-4-en-1-ol.

7.41

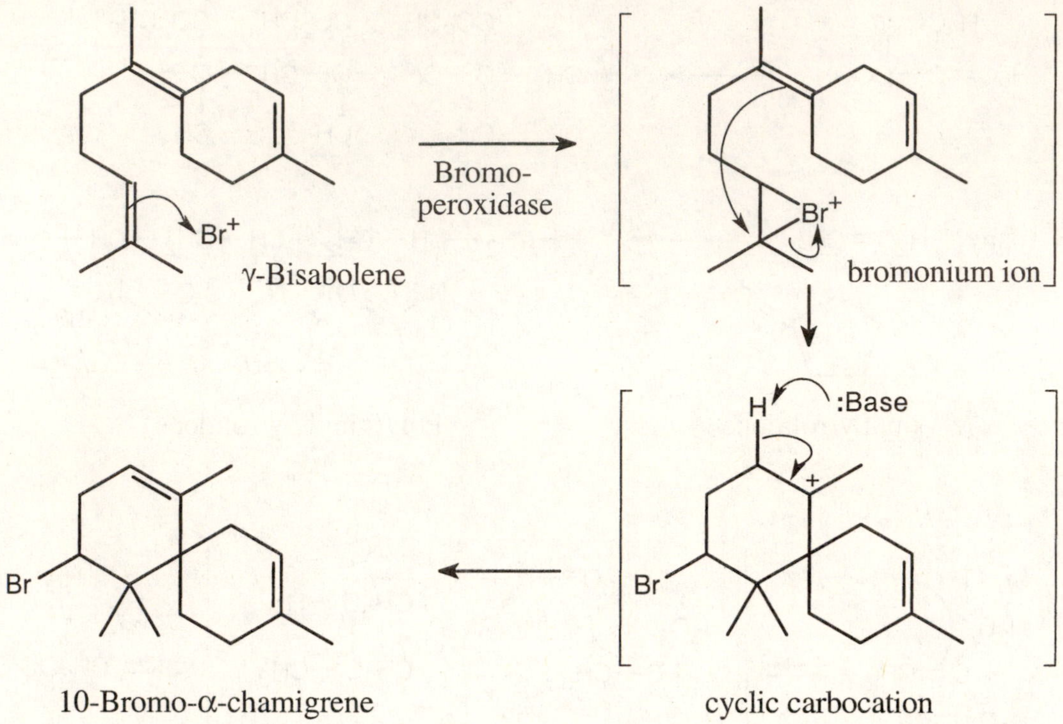

Bromo-peroxidase

Br+

γ-Bisabolene

Br+

bromonium ion

H :Base

Br

Br

10-Bromo-α-chamigrene

cyclic carbocation

7.42

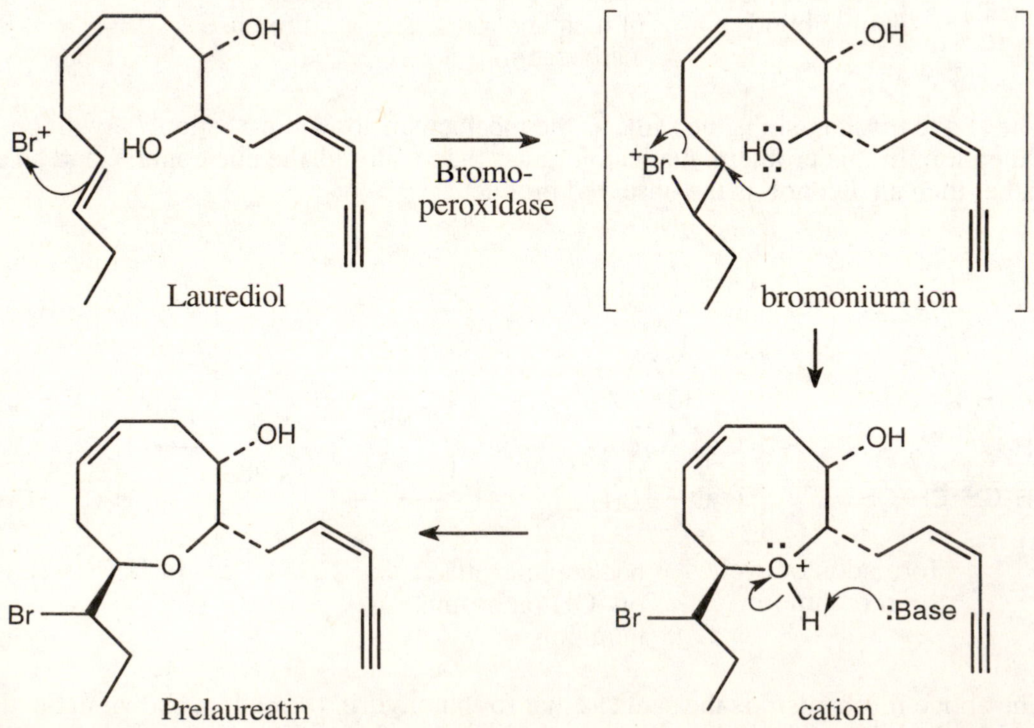

OH

Br+ HO

Laurediol

Bromo-peroxidase

OH

+Br HO

bromonium ion

OH

Br O

Prelaureatin

OH

O+ H :Base

Br

cation

7.43 (a) Bromine dissolved in CH_2Cl_2 has a reddish-brown color. When an alkene such as cyclopentene is added to the bromine solution, the double bond reacts with bromine, and the color disappears. This test distinguishes cyclopentene from cyclopentane, which does not react with Br_2. Alternatively, each compound can be treated with H_2/Pd. The alkene takes up H_2, and the alkane is unreactive.
(b) An aromatic compound such as benzene is unreactive to the Br_2/CH_2Cl_2 reagent and can be distinguished from hex-2-ene, which decolorizes Br_2/CH_2Cl_2. Also, an aromatic compound doesn't take up H_2 under reaction conditions used for hydrogenation of alkenes.

7.44 Make models of the cis and trans diols. Notice that it is much easier to form a five-membered cyclic periodate from the cis diol A than from the trans diol B.Because any factor that lowers the energy of a transition state or intermediate also lowers $\Delta G^‡$ and increases the rate of reaction, we predict that diol cleavage should proceed more slowly for trans diols than for cis diols.

7.45

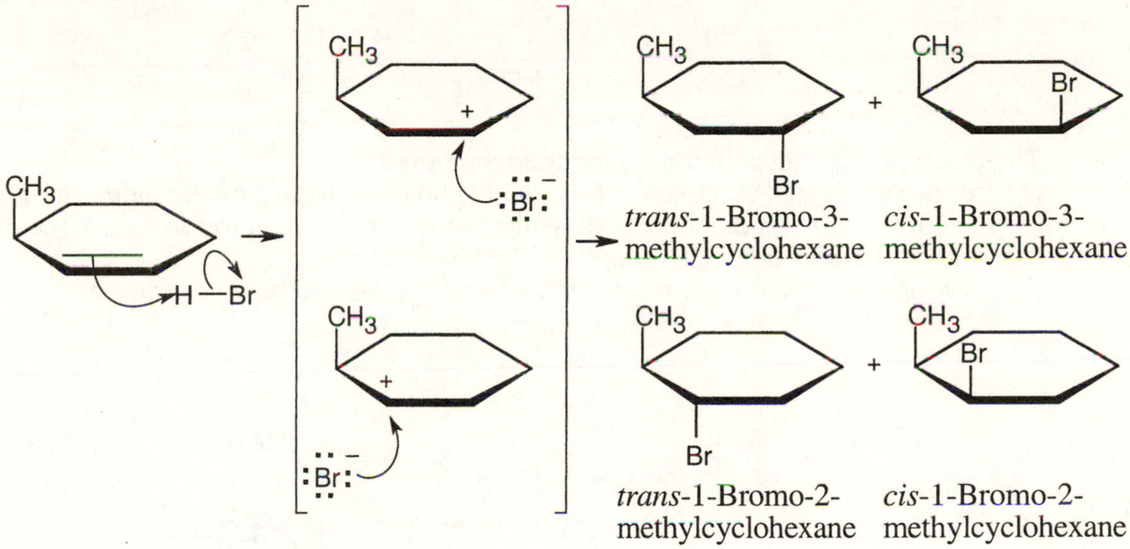

CH3 H—Br → [CH3 + :Br:⁻ ; CH3 + :Br:⁻] →

CH3 Br *trans*-1-Bromo-3-methylcyclohexane + CH3 Br *cis*-1-Bromo-3-methylcyclohexane

CH3 Br *trans*-1-Bromo-2-methylcyclohexane + CH3 Br *cis*-1-Bromo-2-methylcyclohexane

In the reaction of 3-methylcyclohexene with HBr, two intermediate carbocations of approximately equal stability are formed. Both react with bromide ion from top and bottom faces to give four different products.

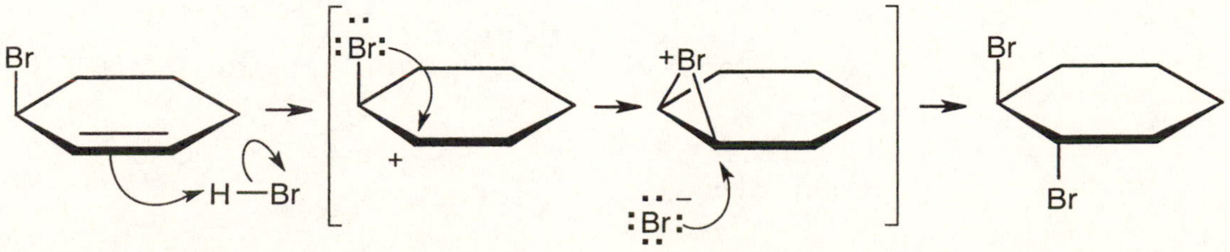

Br H—Br → [:Br: + → +Br :Br:⁻] → Br Br

The most stable cation intermediate from protonation of 3-bromocyclohexene is a cyclic bromonium ion, which reacts with Br^- from the opposite side to yield anti product.

7.46

CO$_2$CH$_3$ 1. → [CO$_2$CH$_3$, Hg$^+$, AcO + $^-$OAc 2. → CO$_2$CH$_3$, +, Hg, AcO] 3. ↓ [CO$_2$CH$_3$, H, :Base, +, Hg, AcO] 4. → CO$_2$CH$_3$, Hg, AcO + HOAc

OAc–Hg–OAc

The reaction mechanism involves the following steps:
(1) Addition of Hg(OAc)$_2$ to one of the double bonds to form a cyclic mercurinium ion;
(2) Reaction of a second double bond with the mercurinium ion to form a six-membered ring and a different carbocation;
(3) A second cyclization forms the other ring and yields another carbocation;
(4) Removal of –H gives a double bond.

7.47

OAc–Hg–OAc → [$\overset{+}{\text{HgOAc}}$, HÖCH$_3$ → HgOAc, $\overset{+}{\text{O}}$CH$_3$, H, $^-$:OAc] ↓ HgOAc, OCH$_3$ + HOAc —NaBH$_4$→ OCH$_3$

Cyclohexyl methyl ether

The above mechanism is the same as the mechanism shown in Section 7.4 with one exception: In this problem, methanol, rather than water, is the nucleophile, and an ether, rather than an alcohol, is the observed product.

7.48 Hydroboration of 2-methylpent-2-ene at 160°C is reversible. The initial organoborane intermediate can eliminate BH_3 in either of two ways, yielding either 2-methylpent-2-ene or 4-methylpent-2-ene, which in turn can undergo reversible hydroboration to yield either 4-methylpent-2-ene or 4-methylpent-1-ene. The effect of these reversible reactions is to migrate the double bond along the carbon chain. A final hydroboration then yields the most stable (primary) organoborane, which is oxidized to form 4-methylpentan-1-ol.

2-Methylpent-2-ene ⇌ (BH_3 / $-BH_3$) [organoborane, BH_2] ⇌ ($-BH_3$ / BH_3) 4-Methylpent-2-ene ⇌ (BH_3 / $-BH_3$) [organoborane, BH_2]

⇌ ($-BH_3$ / BH_3) 4-Methylpent-1-ene ⇌ (BH_3 / $-BH_3$) [organoborane, BH_2] → (H_2O_2 / NaOH, H_2O) 4-Methylpentan-1-ol (OH)

7.49

cis-But-2-ene → (OsO_4, pyridine) cyclic osmate → ($NaHSO_3$, H_2O) diol (HO, OH; H, H_3C; H, CH_3)

trans-But-2-ene → (OsO_4, pyridine) cyclic osmate → ($NaHSO_3$, H_2O) diol (HO, OH; H, H_3C; CH_3, H)

Formation of the cyclic osmate, which occurs with syn stereochemistry, retains the cis-trans stereochemistry of the double bond because osmate formation is a single-step reaction. Treatment of the osmate with $NaHSO_3$ does not affect the stereochemistry of the carbon–oxygen bond. The diol produced from *cis*-but-2-ene is a stereoisomer of the diol produced from *trans*-but-2-ene. We'll study this type of isomerism in Chapter 9.

Chapter 8 – Aromatic Compounds

Chapter Outline

I. Introduction to aromatic compounds (Sections 8.1 – 8.2).
   A. Naming aromatic compounds (Section 8.1).
      1. Many aromatic compounds have nonsystematic names.
      2. Monosubstituted benzenes are named in the same way as other hydrocarbons, with -benzene as the parent name.
         a. Alkyl-substituted benzenes are named in two ways:
            i. If the alkyl substituent has six or fewer carbons, the hydrocarbon is named as an alkyl-substituted benzene.
            ii. If the alkyl substituent has more than six carbons, the compound is named as a phenyl-substituted alkane.
         b. The $C_6H_5CH_2$– group is a benzyl group.
      3. Disubstituted benzenes are named by the ortho(*o*), meta(*m*), para(*p*) system.
         a. A benzene ring with two substituents in a 1,2 relationship is *o*-disubstituted.
         b. A benzene ring with two substituents in a 1,3 relationship is *m*-disubstituted.
         c. A benzene ring with two substituents in a 1,4 relationship is *p*-disubstituted.
         d. The *o, m, p*–system of nomenclature is also used in describing reactions.
      4. Benzenes with more than two substituents are named by numbering the position of each substituent.
         a. Number so that the lowest possible combination of numbers is used.
         b. Substituents are listed alphabetically.
      5. Any of the nonsystematic names can be used as a parent name.
   C. Structure and stability of benzene (Section 8.2).
      1. Stability of benzene.
         a. Benzene doesn't undergo typical alkene reactions.
         b. Benzene reacts slowly with Br_2 to give substitution, not addition, product.
         c. $\Delta H°_{hydrog}$ of benzene is 150 kJ/mol less than that predicted for 3 x $\Delta H°_{hydrog}$ of cyclohexene, indicating that benzene has extra stability.
      2. Structure of benzene.
         a. All carbon–carbon bonds of benzene have the same length.
         b. The electron density in all bonds is identical.
         c. Benzene is planar, with all bond angles 120°.
         d. All carbons are sp^2-hybridized and identical, and each carbon has an electron in a *p* orbital perpendicular to the plane of the ring.
         e. Resonance theory explains that benzene is a resonance hybrid of two forms.
         f. Benzene is represented in this book as one line-bond structure, rather than as a hexagon with a circle to represent the double bonds.
   D. Molecular orbital picture of benzene.
      1. It is impossible to define 3 localized π bonds; the electrons are delocalized over the ring.
      2. Six molecular orbitals (MOs) can be constructed for benzene.
         a. The 3 lower-energy MOs are bonding MOs.
         b. The 3 higher energy MOs are antibonding.
         c. One pair of bonding orbitals is degenerate, as is one pair of antibonding orbitals.
         d. The 6 bonding electrons of benzene occupy the 3 bonding orbitals and are delocalized over the ring.

II. Aromaticity (Sections 8.3 – 8.5).
  A. The Hückel $4n + 2$ rule (Section 8.3).
    1. For a compound to be aromatic, it must possess the qualities we have already mentioned and, in addition, must fulfill Hückel's Rule.
    2. Hückel's Rule: A molecule is aromatic only if it has a planar, monocyclic system of conjugation with a total of $4n + 2$ π electrons (where n is an integer).
    3. Molecules with (4, 8, 12 ...) π electrons are not aromatic.
    4. Examples:
      a. Cyclobutadiene ($n = 4$) is not aromatic.
      b. Benzene ($n = 6$) is aromatic.
      c. Planar cyclooctatetraene ($n = 8$) is not aromatic.
        i. Cyclooctatetraene is stable, but its chemical behavior is like an alkene, rather than an aromatic compound.
        ii. Cyclooctatetraene is tub-shaped, and its bonds have two different lengths.
    5. Why $4n + 2$?.
      a. For aromatic compounds, there is a single lowest-energy MO that can accept two π electrons.
      b. The next highest levels occur in degenerate pairs that can accept 4 π electrons.
      c. For all aromatic compounds and ions, a stable species occurs only when ($4n + 2$) π electrons are available to completely fill the bonding MOs.
  B. Aromatic heterocycles (Section 8.4).
    1. A heterocycle (a cyclic compound containing one or more elements in addition to carbon) can also be aromatic.
    2. Pyridine.
      a. The nitrogen atom of pyridine contributes one π electron to the π system of the ring, making pyridine aromatic.
      b. The nitrogen lone pair is not involved with the ring π system.
    3. Pyrrole.
      a. The nitrogen of pyrrole contributes both lone-pair electrons to the ring π system making pyrrole aromatic.
      b. The nitrogen atom makes a different contribution to the π ring system in pyrrole and in pyridine.
  C. Polycyclic aromatic compounds (Section 8.5).
    1. Although Hückel's Rule strictly applies only to monocyclic compounds, some polycyclic compounds show aromatic behavior.
    2. Naphthalene and anthracene are two common aromatic polycyclic compounds.
      a. Both can be represented by several resonance forms.
      b. Both show chemical and physical properties common to aromatic compounds.
      c. Both have a Hückel number of π electrons.
    3. There are heterocyclic analogues of polycyclic aromatic compounds, many of which have biological importance.

III. Reactions of aromatic compounds (Sections 8.6 – 8.9).
  A. Electrophilic substitutions (Sections 8.6 – 8.8).
    1. Characteristics of electrophilic aromatic substitutions (Section 8.6).
      a. Many substituents (–Cl, –Br, –I, –R, –COR, $–NO_2$, $–SO_3H$, –OH) can be introduced onto an aromatic ring.
      b. Aromatic rings are less reactive than alkene double bonds.
      c. Reaction takes place by substitution, rather than addition, in order to preserve the aromaticity of the aromatic ring.
    2. Halogenation.
      a. Chlorine and bromine react in the presence of $FeCl_3$ or $FeBr_3$ to yield halogenated rings.
      b. Iodination occurs only in the presence of an oxidizing agent.

2. Nitration.
   a. A mixture of HNO_3 and H_2SO_4 is used for nitration.
   b. The reactive electrophile is NO_2^+.
      iii. Products of nitration can be reduced with Fe or Sn to yield an arylamine.
3. Sulfonation
   a. Rings can be sulfonated by a mixture of SO_3 and H_2SO_4 to yield sulfonic acids.
   b. The reactive electrophile is either SO_3 or HSO^{3+}.
   c. Sulfonation is reversible.
4. Hydroxylation occurs in biological systems.

B. Alkylation of aromatic rings (Section 8.7).
1. The Friedel-Crafts reaction introduces an alkyl group onto an aromatic ring.
2. An alkyl chloride, plus an $AlCl_3$ catalyst, produces an electrophilic carbocation.
3. There are several limitations to using the Friedel-Crafts reaction.
   a. Only alkyl halides – not aryl or vinylic halides – can be used.
   b. Friedel-Crafts reactions don't succeed on rings that have amino substituents or deactivating groups.
   c. Polyalkylation is often seen.
   d. Rearrangements of the alkyl carbocation often occur.
      Rearrangements may occur by hydride shifts or by alkyl shifts.

C. Acylation of aromatic rings.
1. Friedel-Crafts acylation occurs when an aromatic ring reacts with a carboxylic acid chloride.
2. The reactive electrophile is an acyl cation, which doesn't rearrange.
3. Polyacylation never occurs in acylation reactions.

D. Substituent effects in substituted aromatic rings (Section 8.8).
1. Types of substituent effects.
   a. Substituents affect the reactivity of an aromatic ring.
   b. Substituents affect the orientation of further substitution.
2. Substituents can be grouped into three groups:
   a. Ortho/para directing activators.
   b. Ortho/para-directing deactivators.
   c. Meta-directing deactivators.
3. Two kinds of effects are responsible for reactivity and orientation.
   a. Inductive effects are due to differences in bond polarity.
   b. Resonance effects are due to overlap of a *p* orbital of a substituent with a *p* orbital on an aromatic ring.
      i. Carbonyl, cyano and nitro substituents withdraw electrons.
         These substituents have the structure –Y=Z.
      ii. Halogen, hydroxyl, alkoxyl and amino substituents donate electrons.
         These substituents have the structure –Y:.
      iii. Resonance effects are greatest at the ortho and para positions.
   c. Resonance and inductive effects don't always act in the same direction.

E. Explanation of substituent effects.
1. All activating groups donate electrons to an aromatic ring.
2. All deactivating groups withdraw electrons from a ring.
3. Ortho-and para-directing activators.
   a. Alkyl groups inductively donate electrons to a ring.
   b. OH, NH_2 donate electrons by resonance involving the ring and the group.
   c. The intermediates of *o,p*-reaction are more stabilized by resonance than are intermediates of *m*-reaction.
4. Meta-directing deactivators.
   a. Meta-directing deactivators act through both inductive and resonance effects.
   b. Because resonance effects destabilize ortho and para positions the most, substitution ion occurs at the meta position.

F. Oxidation and reduction of organic compounds (Section 8.9).
   1. Oxidation of alkylbenzene side chains.
      a. Strong oxidizing agents cause the oxidation of alkyl side chains with benzylic hydrogens.
      b. The products of side-chain oxidation are benzoic acids.
      c. Reaction proceeds by a complex radical mechanism.
   2. Reduction of aromatic compounds; catalytic hydrogenation.
      a. It is possible to selectively reduce alkene bonds in the presence of aromatic rings because rings are relatively inert to catalytic hydrogenation.
      b. With a stronger catalyst, aromatic rings can be reduced.

G. Synthesis of substituted benzenes (Section 8.10).
   1. To synthesize substituted benzenes, it is important to introduce groups so that they have the proper orienting effects.
   2. It is best to use retrosynthetic analysis to plan a synthesis.

Solutions to Problems

8.1 An ortho disubstituted benzene has two substituents in a 1,2 relationship. A meta disubstituted benzene has two substituents in a 1,3 relationship. A para disubstituted benzene has two substituents in a 1,4 relationship.

(a) Cl, CH_3 — meta disubstituted

(b) NO_2, Br — para disubstituted

(c) SO_3H, OH — ortho disubstituted

8.2 Remember to give the lowest possible numbers to substituents on trisubstituted rings.

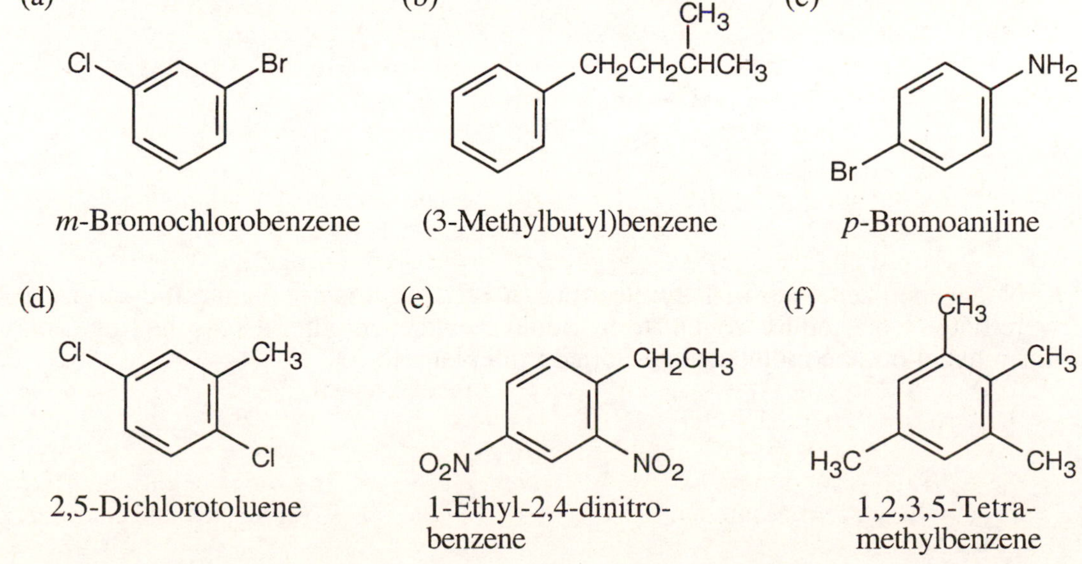

(a) *m*-Bromochlorobenzene (b) (3-Methylbutyl)benzene (c) *p*-Bromoaniline

(d) 2,5-Dichlorotoluene (e) 1-Ethyl-2,4-dinitrobenzene (f) 1,2,3,5-Tetramethylbenzene

8.3

(a) *p*-Bromochlorobenzene

(b) *p*-Bromotoluene

(c) *m*-Chloroaniline

(d) 1-Chloro-3,5-dimethylbenzene

8.4

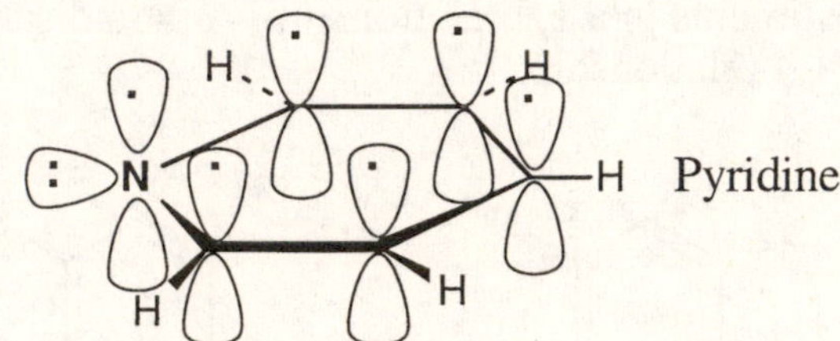

The electronic descriptions of pyridine and benzene are very similar. The pyridine ring is formed by the σ overlap of carbon and nitrogen sp^2 orbitals. In addition, six *p* orbitals, perpendicular to the plane of the ring, hold six electrons. These six *p* orbitals form six π molecular orbitals that allow electrons to be delocalized over the π system of the pyridine ring. The lone pair of nitrogen electrons occupies an sp^2 orbital that lies in the plane of the ring.

8.5

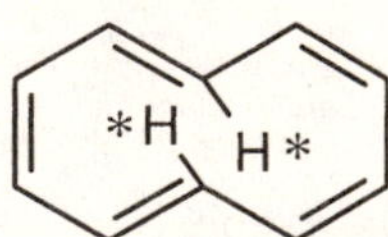

Cyclodecapentaene has $4n + 2$ π electrons (n = 2), but it is not planar. If cyclodecapentaene were flat, the starred hydrogen atoms would crowd each other across the ring. To avoid this interaction, the molecule is distorted from planarity.

8.6

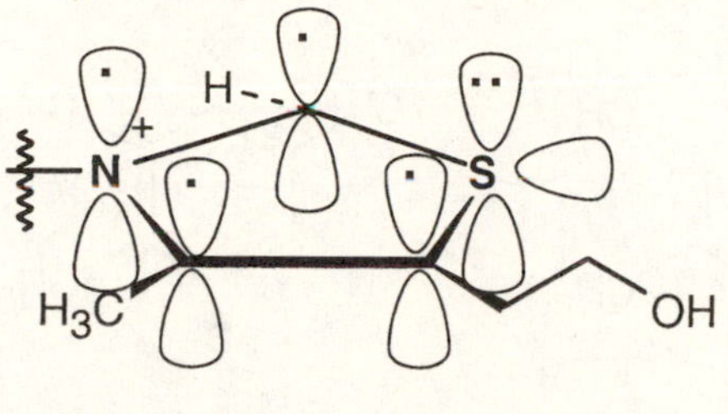

H–
N+
S
H$_3$C
OH

The heterocyclic thiazolium ion contains six π electrons. Each carbon contributes one electron, the nitrogen contributes one electron, and the sulfur contributes two electrons to the ring π system. The thiazolium ion is aromatic because it has 6 π electrons in a cyclic, conjugated system.

8.7

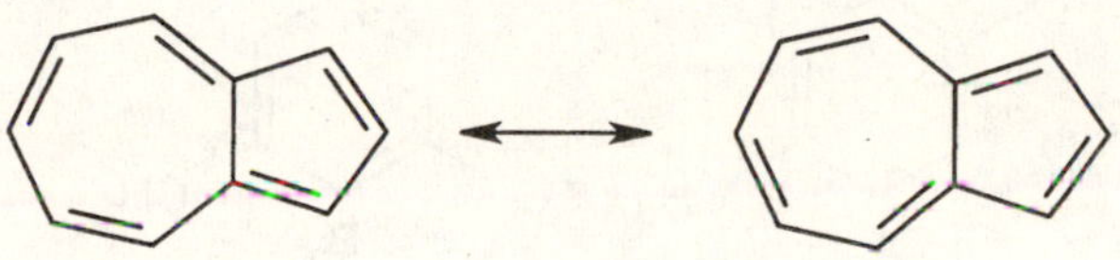

Azulene

Azulene is aromatic because it has a conjugated cyclic π electron system containing ten π electrons (a Hückel number).

8.8

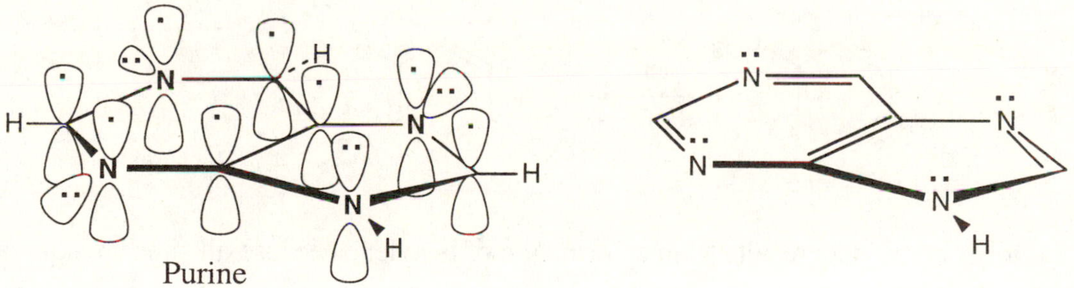

Purine

Purine is a ten-π-electron aromatic molecule. The N–H nitrogen atom in the five-membered ring donates both electrons of its lone pair to the π electron system, and each of the other three nitrogens donates one electron to the π electron system.

8.9

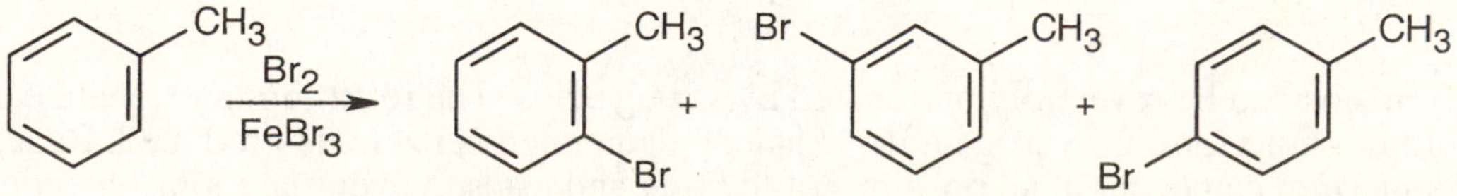

CH_3 $\xrightarrow[FeBr_3]{Br_2}$ o-Bromotoluene + m-Bromotoluene + p-Bromotoluene

o-Bromotoluene *m*-Bromotoluene *p*-Bromotoluene

8.10

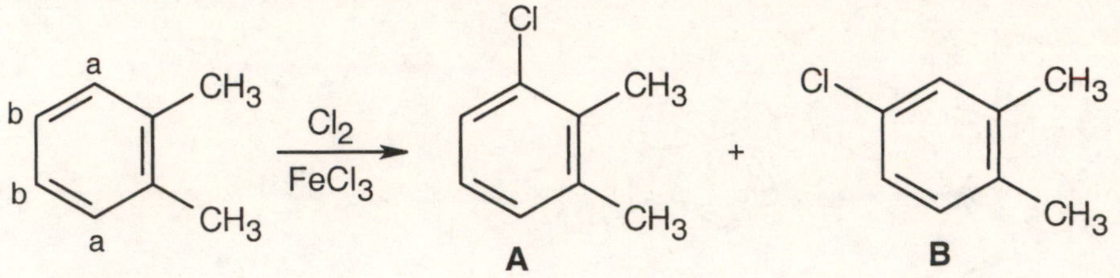

o-Xylene

Chlorination at position "a" of *o*-xylene yields product **A**, and chlorination at position "b" yields product **B**.

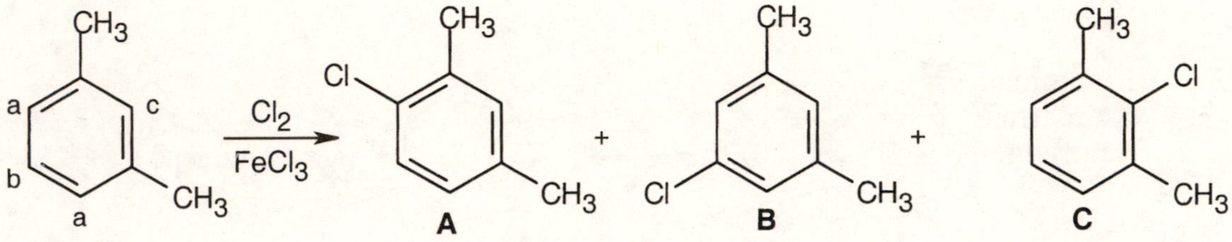

m-Xylene

Three products might be formed on chlorination of *m*–xylene. Product **C** is unlikely to form because substitution rarely occurs between two meta substituents. Product **B** is also unlikely to form for reasons to be explained later in the chapter.

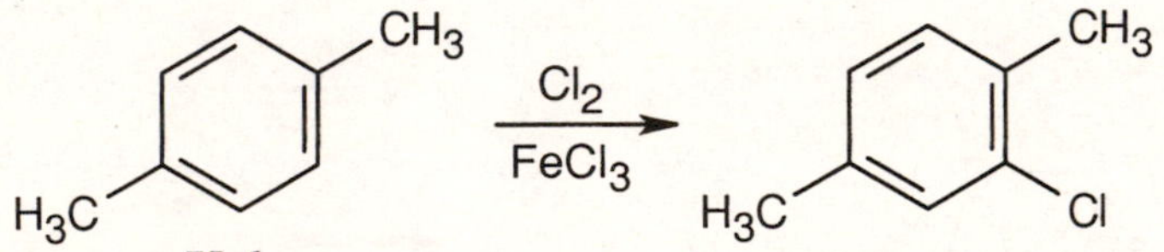

p-Xylene

Only one product results from chlorination of *p*–xylene because all sites are equivalent.

8.11

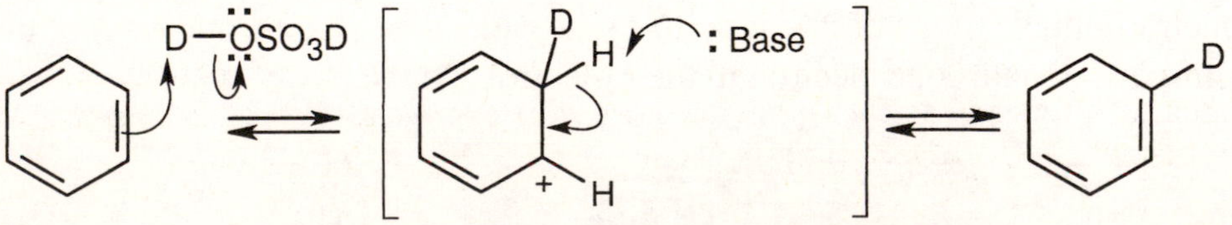

Benzene can be reversibly protonated by strong acids. The resulting intermediate can lose either deuterium or hydrogen. If –H is lost, deuterated benzene is produced. Reaction with deuterium can occur at all positions of the ring and leads to eventual replacement of all hydrogens by deuterium.

8.12 **Strategy:** Carbocation rearrangements of alkyl halides occur (1) if the initial carbocation is primary or secondary, and (2) if it is possible for the initial carbocation to rearrange to a more stable secondary or tertiary cation.

Solution:

(a) Although $CH_3\overset{+}{C}H_2$ is a primary carbocation, it can't rearrange to a more stable cation.

(b) $CH_3CH_2CH(Cl)CH_3$ forms a secondary carbocation that doesn't rearrange.

(c) $CH_3CH_2\overset{+}{C}H_2$ rearranges to the more stable $CH_3\overset{+}{C}HCH_3$.

(d) $(CH_3)_3C\overset{+}{C}H_2$ (primary) undergoes an alkyl shift to yield $(CH_3)_2\overset{+}{C}CH_2CH_3$ (tertiary).

(e) The cyclohexyl carbocation doesn't rearrange.

In summary:
No rearrangement: (a) CH_3CH_2Cl, (b) $CH_3CH_2CH(Cl)CH_3$, (e) chlorocyclohexane
Rearrangement: (c) $CH_3CH_2CH_2Cl$, (d) $(CH_3)_3CCH_2Cl$

8.13

$$CH_3CH(CH_3)CH_2{-}Cl \xrightarrow{AlCl_3} \left[CH_3C(CH_3)(H){-}\overset{+}{C}H_2 \;\; AlCl_4^- \right] \rightleftharpoons CH_3\overset{+}{C}(CH_3)CH_3 \;\; AlCl_4^-$$

Isobutyl carbocation (primary) — *tert*-Butyl carbocation (tertiary)

$$C_6H_6 + CH_3\overset{+}{C}(CH_3)CH_3 \;\; AlCl_4^- \longrightarrow \left[\text{arenium ion with } C(CH_3)_3 \text{ and } H \right] \;\; Cl{-}\overset{-}{A}lCl_3 \longrightarrow C_6H_5C(CH_3)_3 + HCl + AlCl_3$$

tert-Butylbenzene

The isobutyl carbocation, initially formed when 1-chloro-2-methylpropane and $AlCl_3$ react, rearranges by a hydride shift to give the more stable *tert*–butyl carbocation, which can then alkylate benzene to form *tert*–butylbenzene.

8.14 **Strategy:** To identify the carboxylic acid chloride used in the Friedel–Crafts acylation of benzene, break the bond between benzene and the ketone carbon and replace it with –Cl.

Solution:

(a)

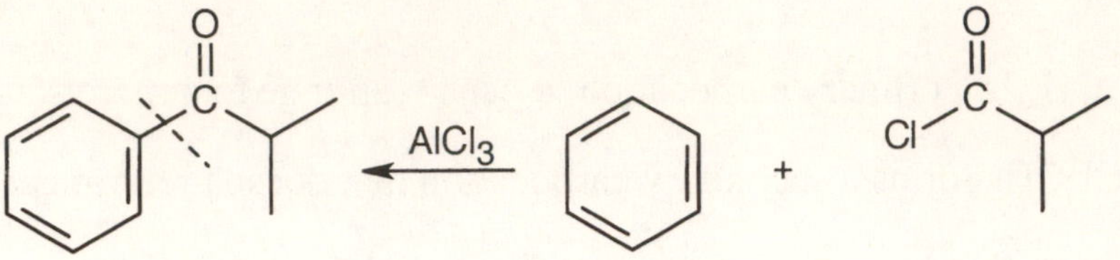

$AlCl_3$ ← +

(b)

$AlCl_3$ ← +

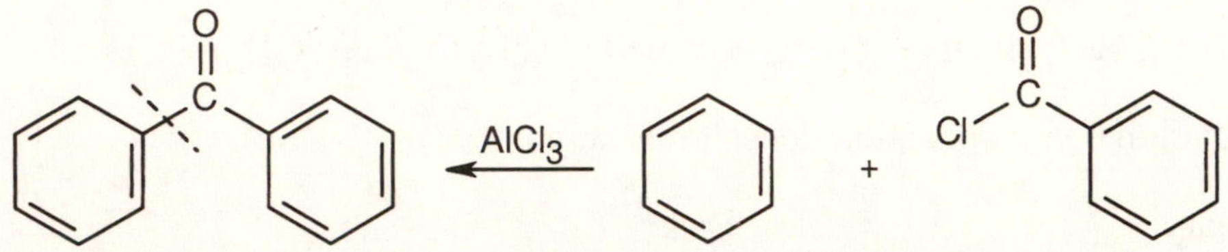

8.15 Use Figure 8.15 to find the activating and deactivating effects of groups.

Most Reactive ⟶ *Least Reactive*

(a) Phenol > toluene > benzene > nitrobenzene

(b) Phenol > benzene > chlorobenzene > benzoic acid

(c) Aniline > benzene > bromobenzene > benzaldehyde

8.16 **Strategy:** Refer to Figure 8.15 in the text for the directing effects of substituents. You should memorize the effects of the most important groups. As in Worked Example 8.3, identify the directing effect of the substituent, and draw the product.

Solution:

(a)

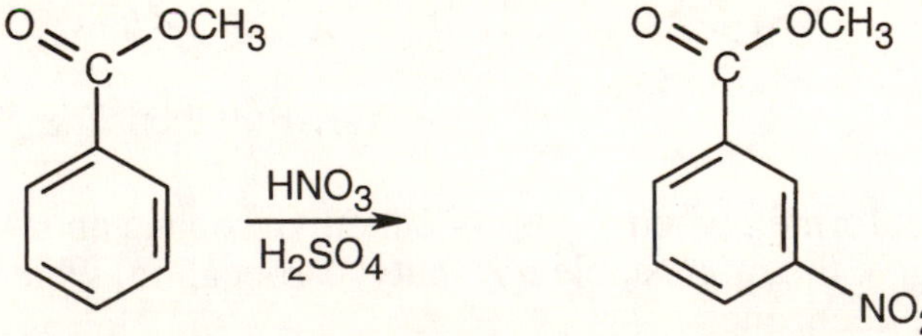

$\xrightarrow[H_2SO_4]{HNO_3}$

The $-CO_2CH_3$ is a deactivator, and it is a meta–director.

(b)

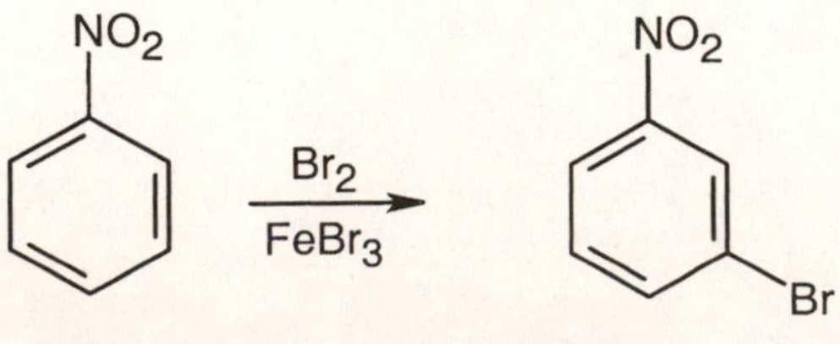

$\xrightarrow[FeBr_3]{Br_2}$

The $-NO_2$ group is a meta–director.

(c)

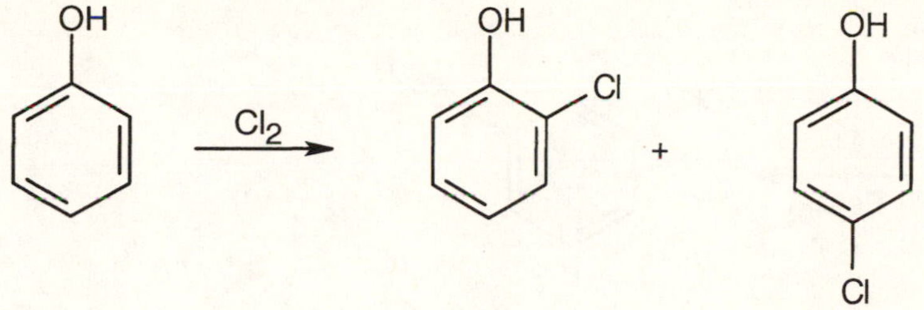

OH, Cl in ring; Cl_2 → ortho-chlorophenol + para-chlorophenol

(d)

NH_2; Br_2 → ortho-bromoaniline + para-bromoaniline

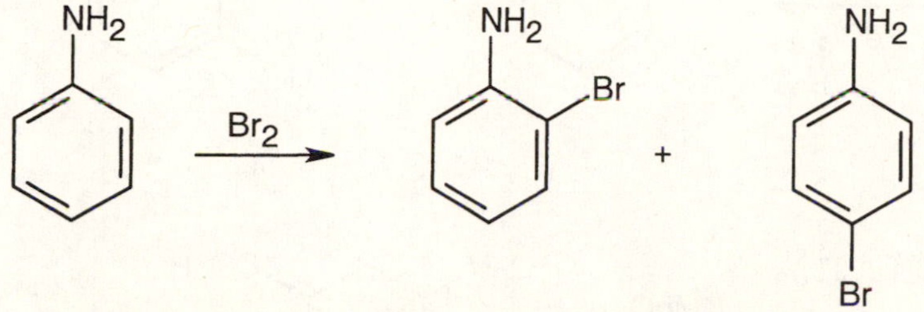

No catalyst is necessary because aniline is highly activated.

8.17

Ortho intermediate:

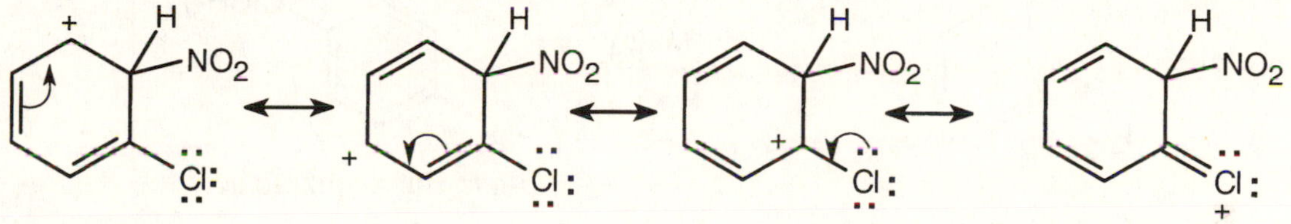

Para intermediate:

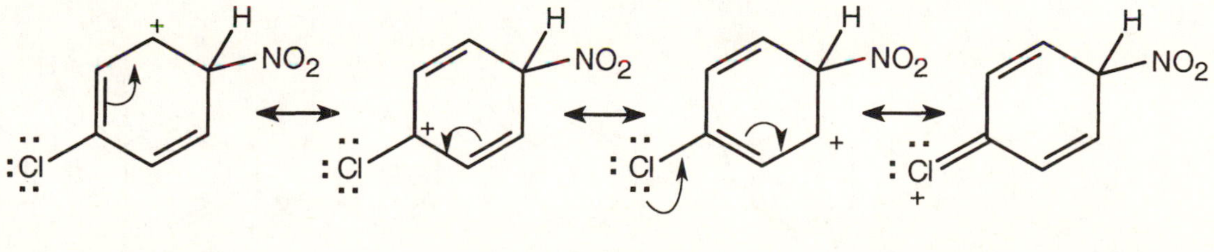

Meta intermediate:

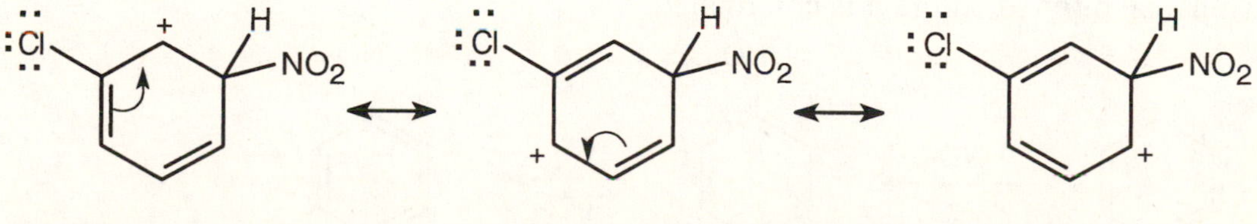

The ortho and para intermediates can be stabilized by resonance donation of electrons from chlorine. This stabilization doesn't occur for the meta intermediate.

8.18

(a)

$$\text{C}_6\text{H}_5\text{C}{\equiv}\text{N} \xrightarrow[\text{FeCl}_3]{\text{Cl}_2} m\text{-ClC}_6\text{H}_4\text{C}{\equiv}\text{N}$$

(b)

$$\text{C}_6\text{H}_5\text{Br} \xrightarrow[\text{FeCl}_3]{\text{Cl}_2} o\text{-BrC}_6\text{H}_4\text{Cl} + p\text{-BrC}_6\text{H}_4\text{Cl}$$

8.19

(a)

$$m\text{-O}_2\text{NC}_6\text{H}_4\text{CH(CH}_3)_2 \xrightarrow[\text{H}_2\text{O}]{\text{KMnO}_4} m\text{-O}_2\text{NC}_6\text{H}_4\text{CO}_2\text{H}$$

m-Nitrobenzoic acid

(b)

$$p\text{-H}_3\text{CC}_6\text{H}_4\text{C(CH}_3)_3 \xrightarrow[\text{H}_2\text{O}]{\text{KMnO}_4} p\text{-HO}_2\text{CC}_6\text{H}_4\text{C(CH}_3)_3$$

p-tert-Butylbenzoic acid

Treatment with $KMnO_4$ oxidizes the methyl group but leaves the *tert*-butyl group untouched.

8.20

$$\text{C}_6\text{H}_5\text{COCl} + \text{C}_6\text{H}_6 \xrightarrow{\text{AlCl}_3} \text{C}_6\text{H}_5\text{COC}_6\text{H}_5 \xrightarrow[\text{Pd}]{\text{H}_2} \text{C}_6\text{H}_5\text{CH}_2\text{C}_6\text{H}_5$$

8.21 (a) In order to synthesize the product with the correct orientation of substituents, benzene must be nitrated before it is chlorinated.

$$\text{C}_6\text{H}_6 \xrightarrow[\text{H}_2\text{SO}_4]{\text{HNO}_3} \text{C}_6\text{H}_5\text{NO}_2 \xrightarrow[\text{FeCl}_3]{\text{Cl}_2} m\text{-ClC}_6\text{H}_4\text{NO}_2$$

m-Chloronitrobenzene

(b) Chlorine can be introduced into the correct position if benzene is first acylated. The chlorination product can then be reduced.

CH_3COCl, $AlCl_3$ → ; Cl_2, $FeCl_3$ → ; H_2, Pd →

m-Chloroethylbenzene

(c) Two routes are possible. Note that Friedel–Crafts alkylation of benzene or chlorobenzene yields rearranged product.

CH_3CH_2COCl, $AlCl_3$ → ; H_2, Pd → ; Cl_2, $FeCl_3$ →

Cl_2, $FeCl_3$ → ; CH_3CH_2COCl, $AlCl_3$ → ; H_2, Pd →

p-Chloropropylbenzene

(d) In planning this pathway, remember that the ring must be sulfonated after Friedel–Crafts alkylation (because a sulfonated ring is too deactivated for alkylation to take place). Performing the reactions in this order allows the first two groups to direct bromine to the same position.

CH_3Cl, $AlCl_3$ → ; SO_3, H_2SO_4 → ; Br_2, $FeBr_3$ →

3-Bromo-2-methylbenzenesulfonic acid

8.22 (a) Friedel–Crafts acylation, like Friedel-Crafts alkylation, does not occur at an aromatic ring carrying a strongly electron-withdrawing group.

(b) There are two problems with this synthesis as it is written:

1. Rearrangement often occurs during Friedel–Crafts alkylations using primary halides.
2. Even if *p*-chloropropylbenzene could be synthesized, introduction of the second –Cl group would occur ortho to the alkyl group.

A possible route to this compound:

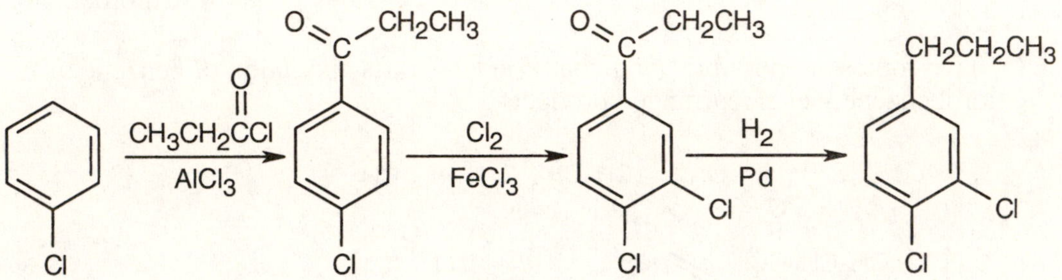

O CH2CH3 C | O || CH_3CH_2CCl $AlCl_3$ | Cl → Cl_2 $FeCl_3$ → O CH2CH3 C | Cl Cl → H_2 Pd → $CH_2CH_2CH_3$ Cl Cl

Visualizing Chemistry

8.23

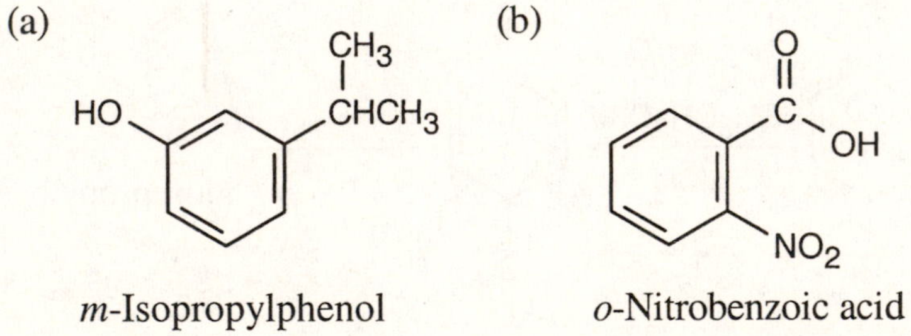

(a) CH_3 HO CHCH3 *m*-Isopropylphenol

(b) O || C OH NO_2 *o*-Nitrobenzoic acid

8.24 Three resonance forms for the carbocation of the formula $C_{13}H_9$ are shown below, and more can be drawn. These forms show that the positive charge of the carbocation can be stabilized in the same way as an allylic or benzylic carbocation is stabilized – by overlap with the neighboring π electrons of the ring system.

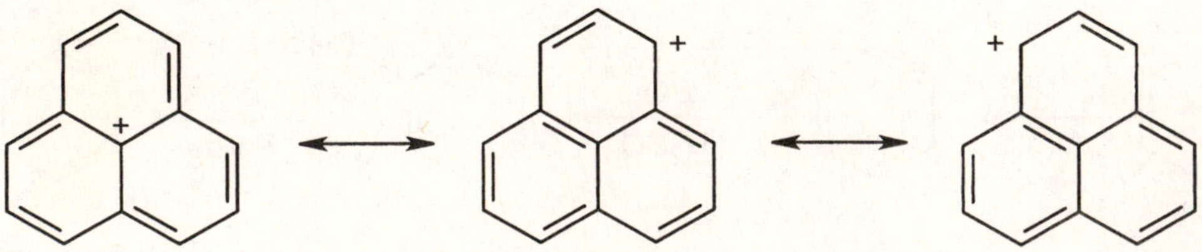

+ ⟷ + ⟷ +

8.25 (a) The methoxyl group is an *o,p*-director.

(i)

$$\text{C}_6\text{H}_5\text{OCH}_3 \xrightarrow[\text{FeBr}_3]{\text{Br}_2} p\text{-BrC}_6\text{H}_4\text{OCH}_3 + o\text{-BrC}_6\text{H}_4\text{OCH}_3$$

p-Bromomethoxybenzene *o*-Bromomethoxybenzene

(ii)

$$\text{C}_6\text{H}_5\text{OCH}_3 \xrightarrow[\text{AlCl}_3]{\text{CH}_3\text{COCl}} p\text{-H}_3\text{CC(=O)C}_6\text{H}_4\text{OCH}_3 + o\text{-CH}_3\text{C(=O)C}_6\text{H}_4\text{OCH}_3$$

p-Methoxyacetophenone *o*-Methoxyacetophenone

(b) Both functional groups direct substituents to the same position.

(i)

$$p\text{-HC(=O)C}_6\text{H}_4\text{CH}_3 \xrightarrow[\text{FeBr}_3]{\text{Br}_2} \text{3-Br-4-CH}_3\text{C}_6\text{H}_3\text{CHO}$$

3-Bromo-4-methylbenzaldehyde

(ii)

$$p\text{-HC(=O)C}_6\text{H}_4\text{CH}_3 \xrightarrow[\text{AlCl}_3]{\text{CH}_3\text{COCl}} \text{3-CH}_3\text{C(=O)-4-CH}_3\text{C}_6\text{H}_3\text{CHO}$$

3-Acetyl-4-methylbenzaldehyde

8.26 Toluene is more reactive toward electrophilic substitution than (trifluoromethyl)benzene. The electronegativity of the three fluorine atoms causes the trifluoromethyl group to be electron-withdrawing and deactivating toward electrophilic substitution. The electrostatic potential map shows that the aromatic ring of (trifluoromethyl)benzene is more electron-poor, and thus less reactive, than the ring of toluene.

8.27 Assume that you can separate ortho and para products of nitration. The reverse sequence of reactions won't work because nitrobenzene is unreactive to Friedel–Crafts alkylation.

$$\text{C}_6\text{H}_6 \xrightarrow[\text{AlCl}_3]{\text{CH}_3\text{Cl}} \text{C}_6\text{H}_5\text{CH}_3 \xrightarrow[\text{H}_2\text{SO}_4]{\text{HNO}_3} p\text{-O}_2\text{NC}_6\text{H}_4\text{CH}_3$$

Additional Problems

8.28

(a) 2-Methyl-5-phenylhexane

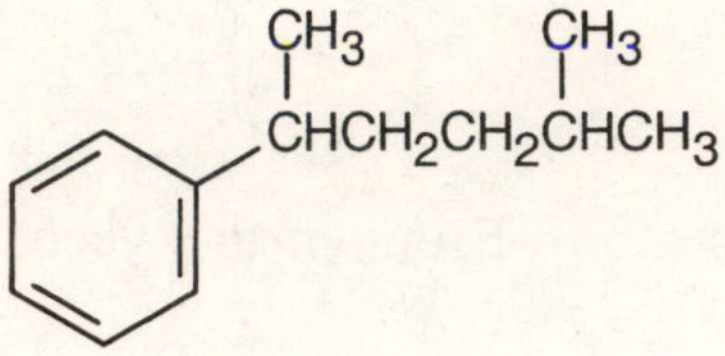

(b) *m*-Bromobenzoic acid

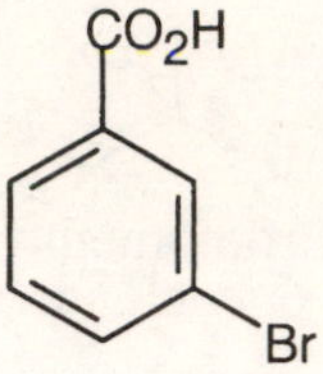

(c) 1-Bromo-3,5-dimethylbenzene

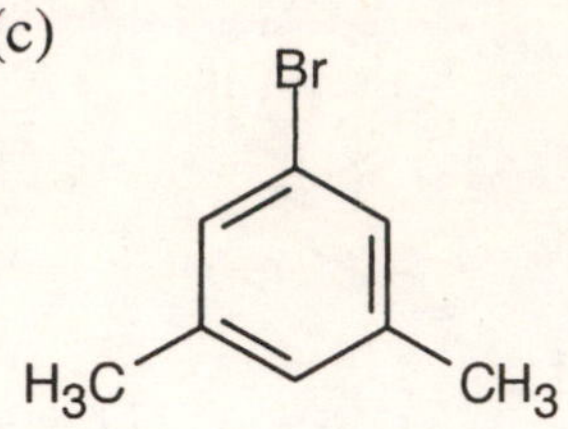

(d) *o*-Bromopropylbenzene

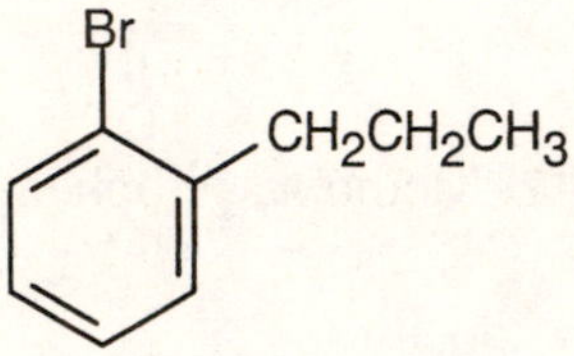

(e) 1-Fluoro-2,4-dinitrobenzene

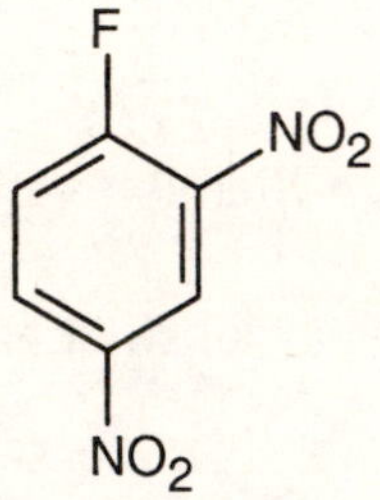

(f) *p*-Chloroaniline

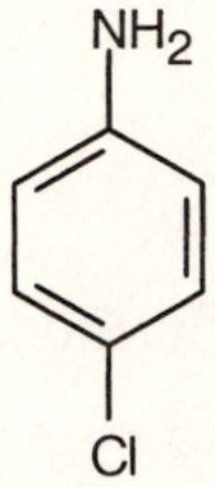

8.29

(a) 3-Methyl-2-nitrobenzoic acid

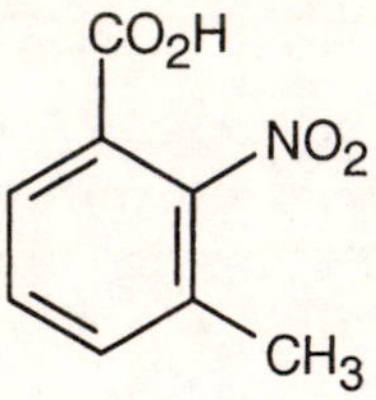

(b) Benzene-1,3,5-triol

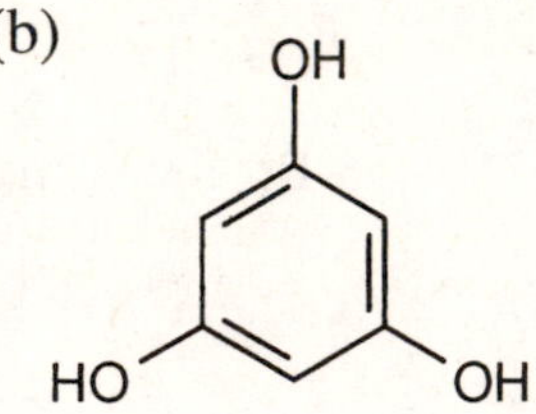

(c) 3-Methyl-2-phenylhexane

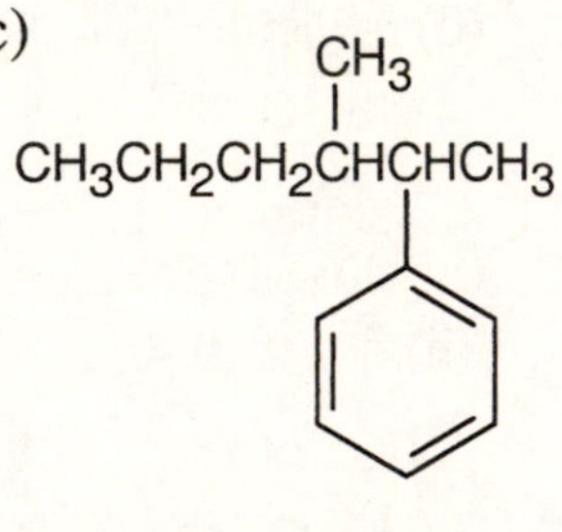

(d) *o*-Aminobenzoic acid

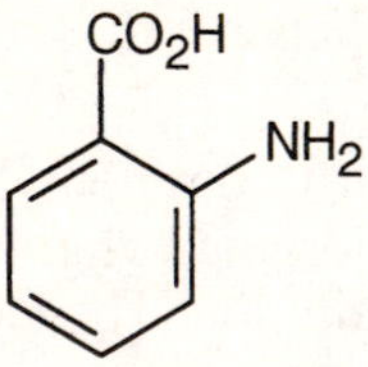

(e) *m*-Bromophenol

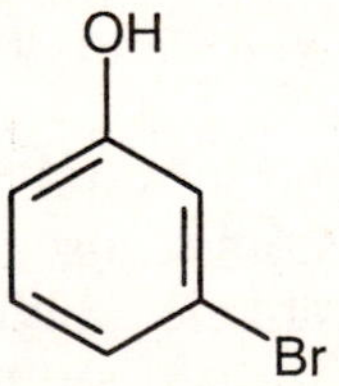

(f) 2,4,6-Trinitrophenol

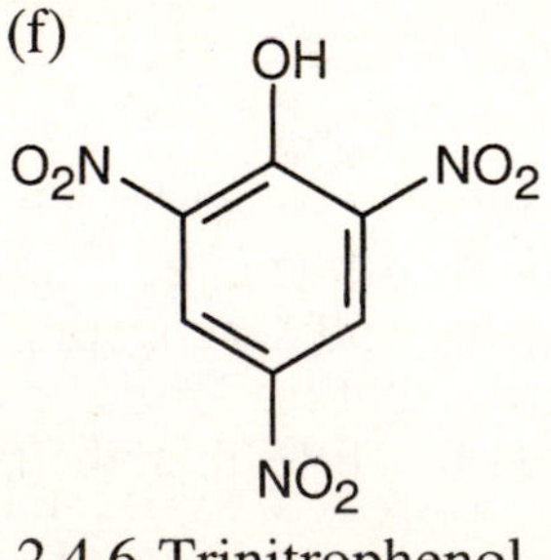

(g) *p*-Iodonitrobenzene

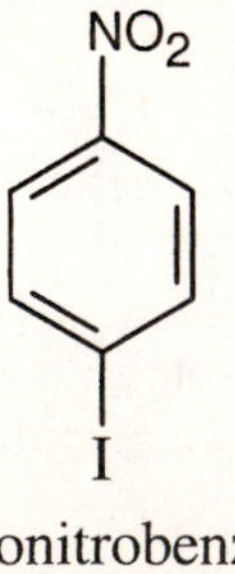

8.30 All aromatic compounds of formula C_7H_7Cl have one ring and three double bonds.

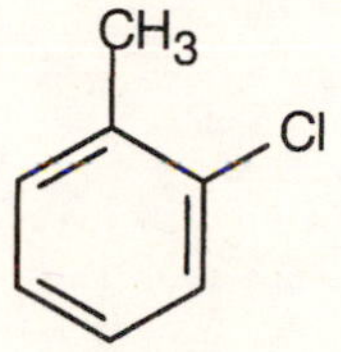

o-Chlorotoluene 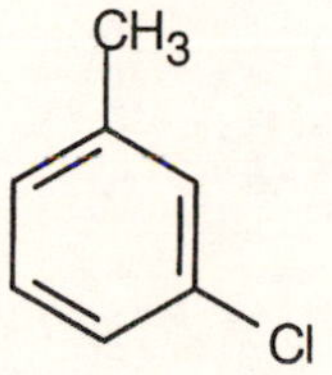 *m*-Chlorotoluene 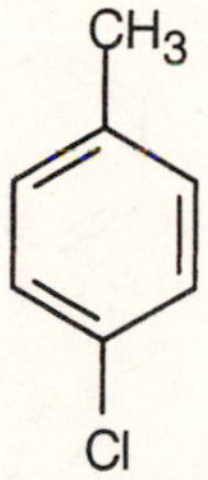 *p*-Chlorotoluene 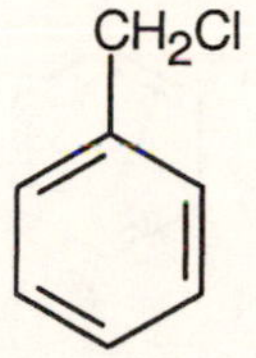 Benzyl chloride (Chloromethyl)-benzene

8.31 *Most reactive* ————————> *Least reactive*

Phenol > Toluene > *p*-Bromotoluene > Bromobenzene

Benzoic acid and nitrobenzene don't undergo Friedel–Crafts alkylations.

8.32 *Most reactive* ————————> *Least reactive*

(a) Benzene > Chlorobenzene > *o*-Dichlorobenzene
(b) Phenol > Nitrobenzene > *p*-Bromonitrobenzene
(c) *o*-Xylene > Fluorobenzene > Benzaldehyde
(d) *p*-Methoxybenzonitrile > *p*-Methylbenzonitrile > Benzonitrile

8.33 All compounds in this problem have four double bonds and/or rings and must be substituted benzenes, if they are to be aromatic. They may be substituted by methyl, ethyl, propyl, or butyl groups.

(a)

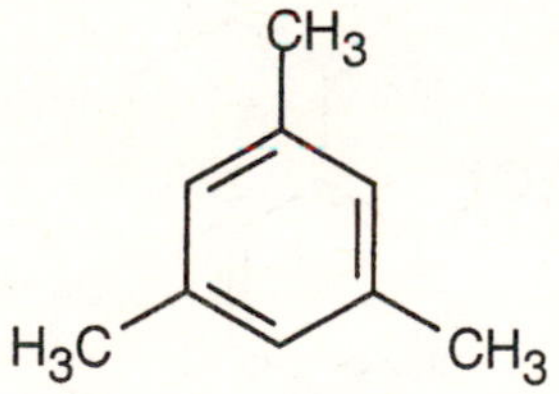

(b)

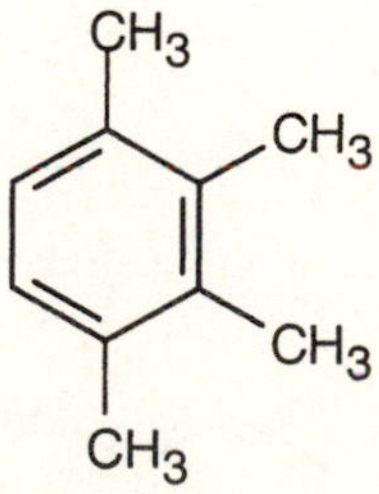

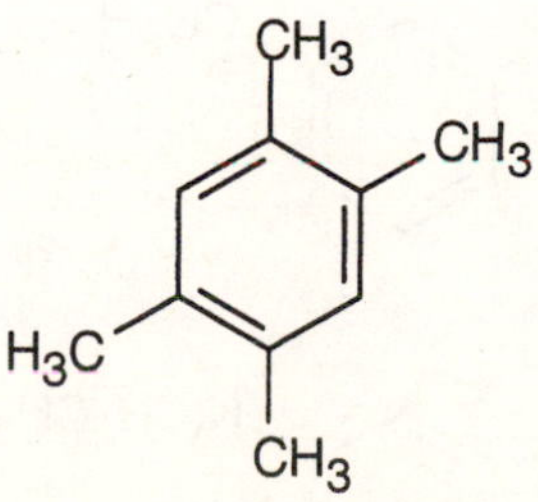

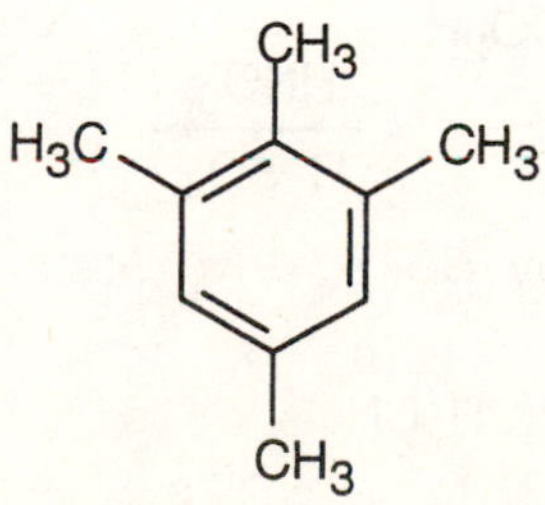

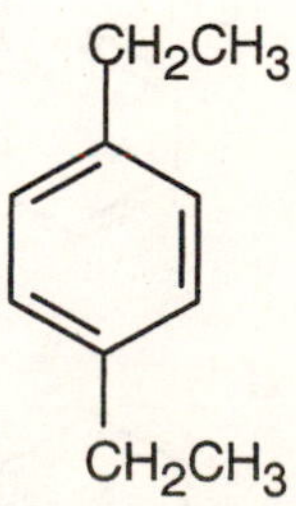

(c)

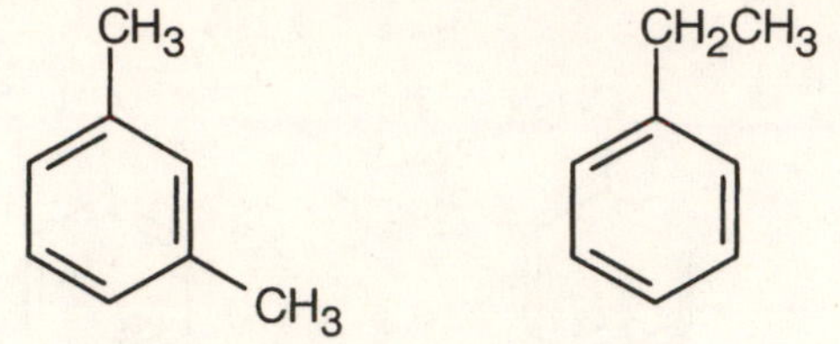

CH_3 CH_2CH_3

CH_3

(d)

CH_2CH_3 CH_2CH_3 $CH(CH_3)_2$ CH_3 $CH_2CH_2CH_3$

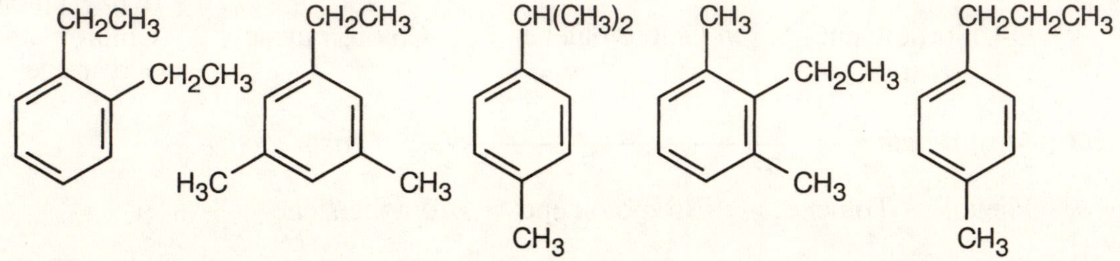

CH_2CH_3 CH_2CH_3

H_3C CH_3 CH_3

CH_3 CH_3

8.34

(a)

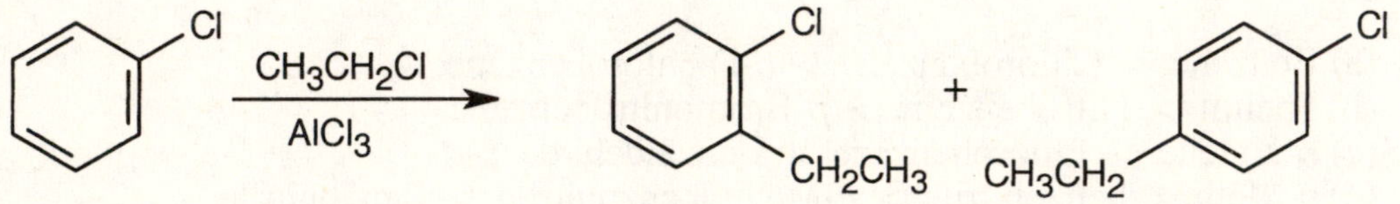

Cl $\xrightarrow[AlCl_3]{CH_3CH_2Cl}$ Cl, CH_2CH_3 + Cl, CH_3CH_2

(b)

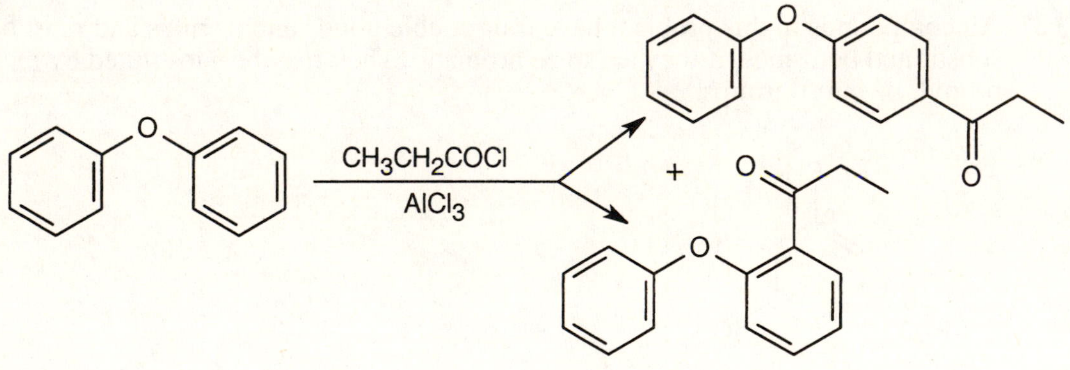

O $\xrightarrow[AlCl_3]{CH_3CH_2COCl}$ O, O + O, O

(c)

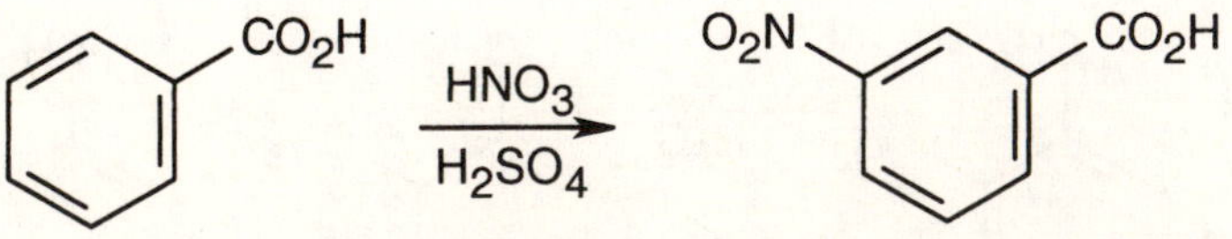

CO_2H $\xrightarrow[H_2SO_4]{HNO_3}$ O_2N CO_2H

(d)

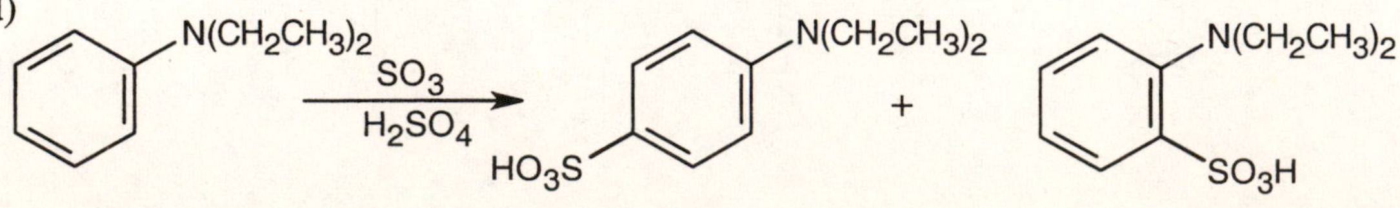

$N(CH_2CH_3)_2$ $\xrightarrow[H_2SO_4]{SO_3}$ $N(CH_2CH_3)_2$, HO_3S + $N(CH_2CH_3)_2$, SO_3H

8.35

	Group:	*Identification:*	*Reason:*
(a)	$-\ddot{N}(CH_3)_2$	*o,p*-activator	Reaction intermediates are stabilized by electron donation by the amine nitrogen.
(b)	cyclopentyl	*o,p*-activator	Reaction intermediates are stabilized by the electron donating inductive effect of the alkyl group.
(c)	$-\ddot{O}CH_2CH_3$	*o,p*-activator	Reaction intermediates are stabilized by electron donation by the ether oxygen.
(d)	$-C(=\ddot{O}:)-C_6H_{11}$	*m*-deactivator	Reaction intermediates are destabilized by electron withdrawal by the carbonyl oxygen.

8.36

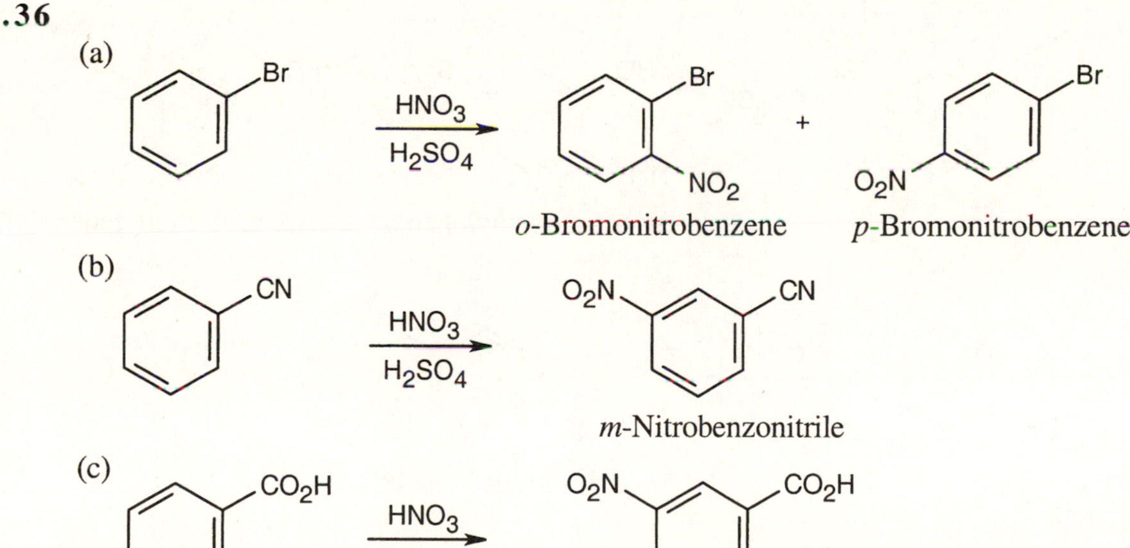

(d)

NO_2 $\xrightarrow[H_2SO_4]{HNO_3}$ O_2N NO_2

m-Dinitrobenzene

(e)

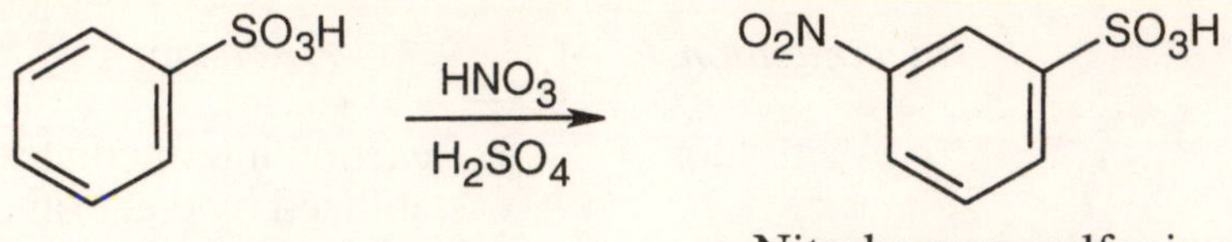

$C_6H_5SO_3H \xrightarrow[H_2SO_4]{HNO_3}$ O_2N–C_6H_4–SO_3H

m-Nitrobenzenesulfonic acid

(f)

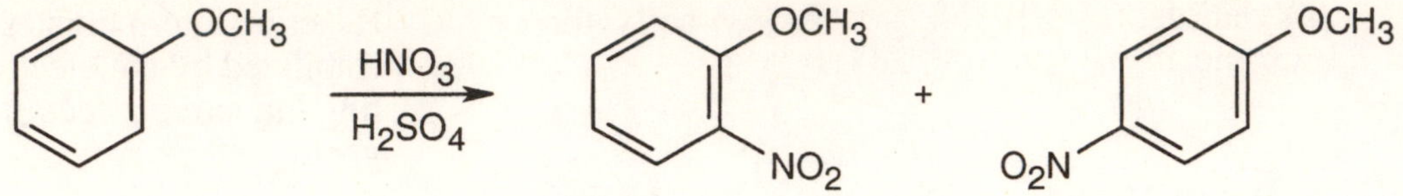

$C_6H_5OCH_3 \xrightarrow[H_2SO_4]{HNO_3}$ o-O_2N–C_6H_4–OCH_3 + p-O_2N–C_6H_4–OCH_3

o-Methoxynitrobenzene *p*-Methoxynitrobenzene

Only methoxybenzene reacts faster than benzene.

8.37

(a)

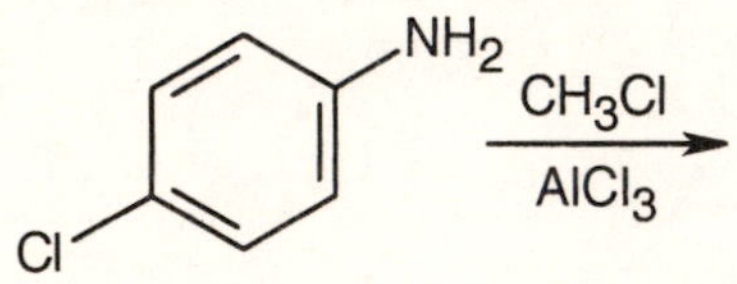

p-Cl–C_6H_4–$NH_2 \xrightarrow[AlCl_3]{CH_3Cl}$ No reaction. $AlCl_3$ combines with $-\ddot{N}H_2$ to form a complex that deactivates the ring toward Friedel-Crafts alkylation.

(b)

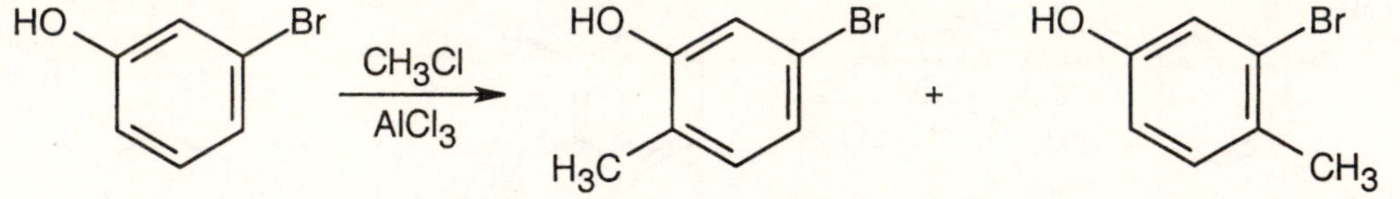

m-HO–C_6H_4–Br $\xrightarrow[AlCl_3]{CH_3Cl}$ 5-Bromo-2-methylphenol + 3-Bromo-4-methylphenol

5-Bromo-2-methylphenol 3-Bromo-4-methylphenol

(c)

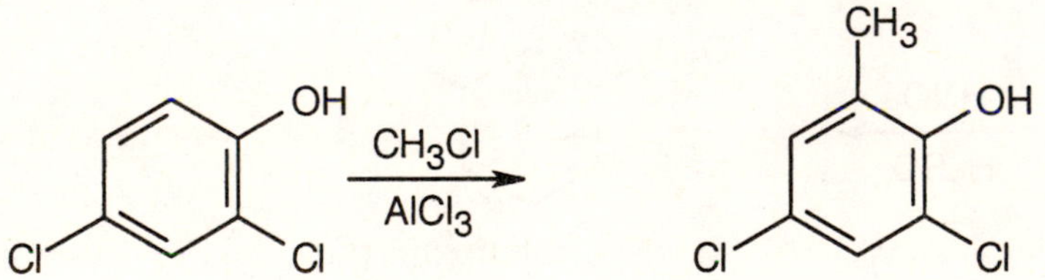

2,4-Dichlorophenol $\xrightarrow[AlCl_3]{CH_3Cl}$ 2,4-Dichloro-6-methylphenol

2,4-Dichloro-6-methylphenol

(d)

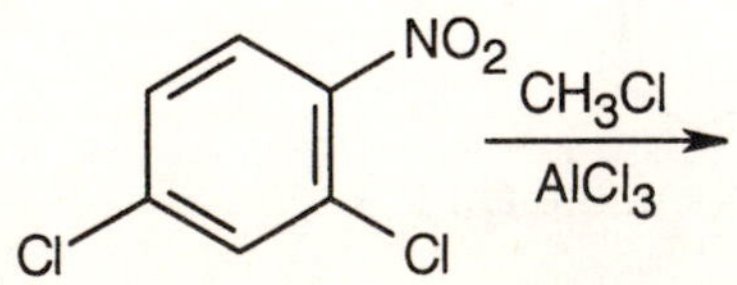

2,4-Dichloronitrobenzene $\xrightarrow[AlCl_3]{CH_3Cl}$ No reaction. The ring is deactivated.

(e)

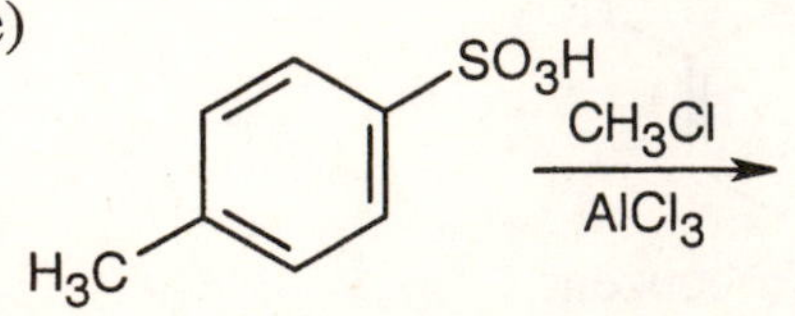

p-H_3C–C_6H_4–$SO_3H \xrightarrow[AlCl_3]{CH_3Cl}$ No reaction.The ring is deactivated.

(f)

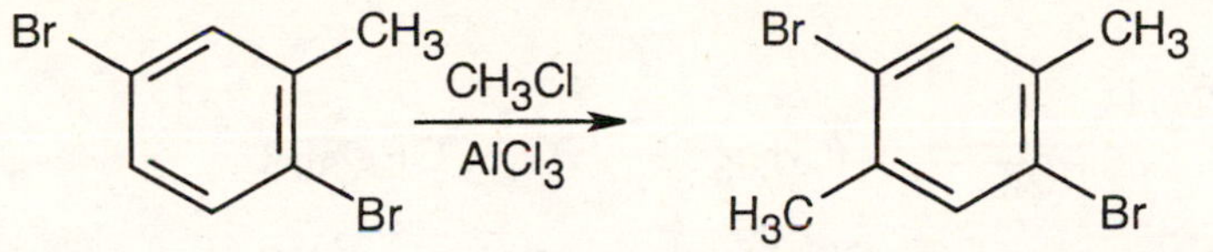
Br
CH3
CH3Cl
AlCl3
Br
Br
CH3
H3C
Br

1,4-Dibromo-2,5-dimethylbenzene

Alkylation occurs in the indicated position because the methyl group is more activating than bromine, and because substitution rarely takes place between two groups.

8.38

(a)

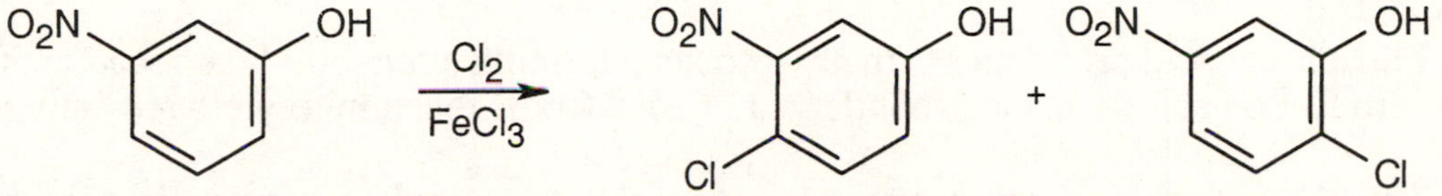
O2N
OH
Cl2
FeCl3
O2N
OH
+
O2N
OH
Cl
Cl

4-Chloro-3-nitrophenol 2-Chloro-5-nitrophenol

The –OH group directs the orientation of substitution.

(b)

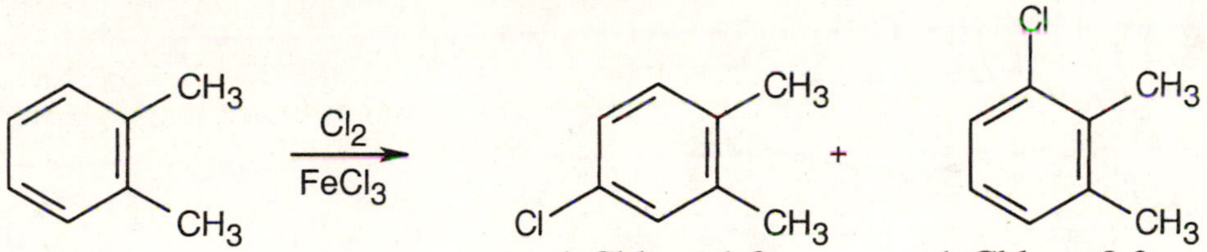
CH3
CH3
Cl2
FeCl3
CH3
Cl
CH3
+
Cl
CH3
CH3

4-Chloro-1,2-dimethylbenzene 1-Chloro-2,3-dimethylbenzene

(c)

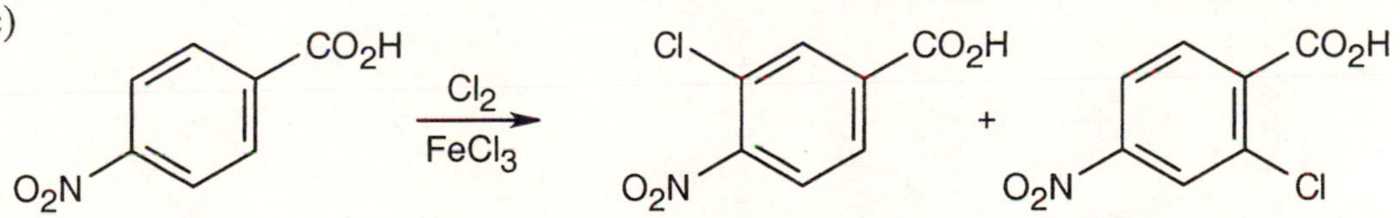
CO2H
O2N
Cl2
FeCl3
Cl
CO2H
O2N
+
CO2H
O2N
Cl

3-Chloro-4-nitrobenzoic acid 2-Chloro-4-nitrobenzoic acid

Both groups are deactivating to a similar extent, and both possible products form.

(d)

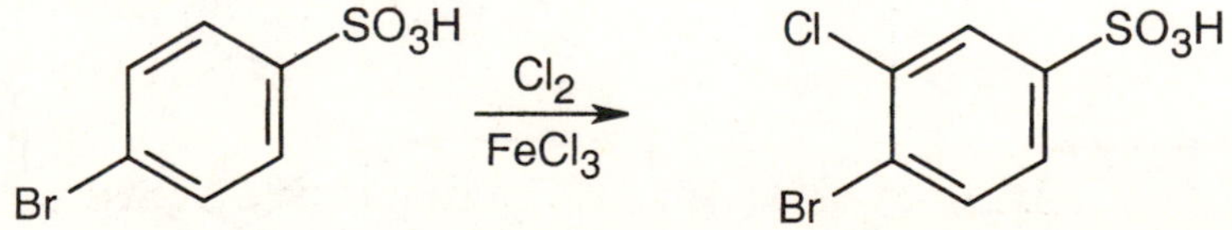
SO3H
Br
Cl2
FeCl3
Cl
SO3H
Br

4-Bromo-3-chlorobenzenesulfonic acid

8.39

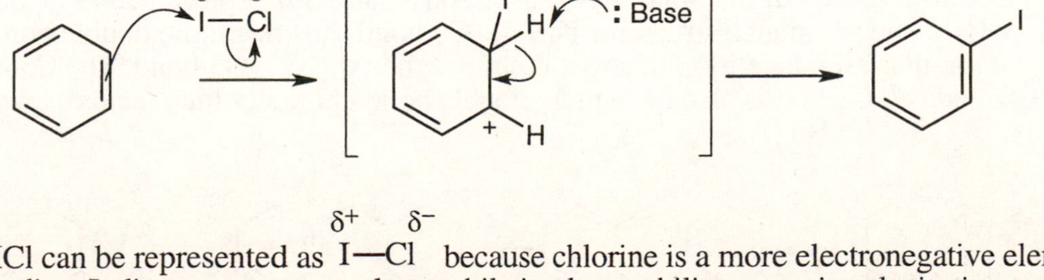
δ+ δ−
I—Cl
I
H
: Base
+
H
I

ICl can be represented as $\overset{\delta+}{\mathrm{I}}\text{—}\overset{\delta-}{\mathrm{Cl}}$ because chlorine is a more electronegative element than iodine. Iodine can act as an electrophile in electrophilic aromatic substitution reactions.

8.40

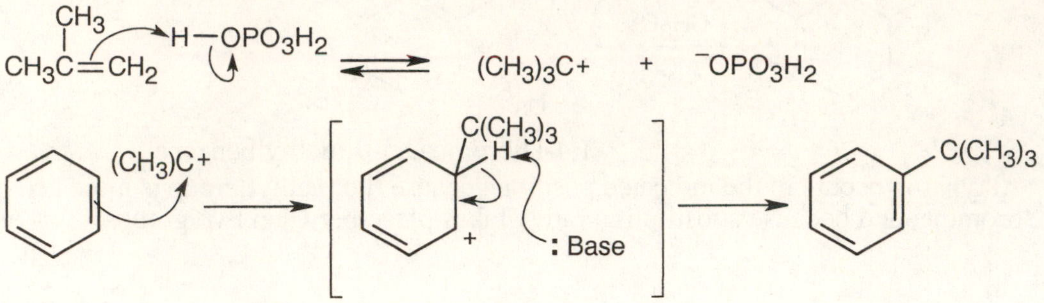

Phosphoric acid protonates 2-methylpropene, forming a *tert*-butyl carbocation. The carbocation acts as an electrophile in a Friedel–Crafts reaction to yield *tert*-butylbenzene.

8.41 The 1,2,4-triazole ring is aromatic because it has 6 π electrons in a cyclic, conjugated system.

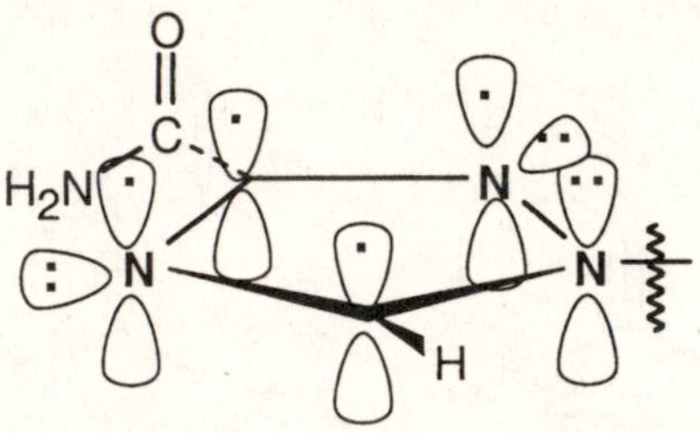

1,2,4-Triazole ring

8.42 The isoxazole ring is aromatic for the same reasons as the 1,2,4-triazole ring is aromatic.

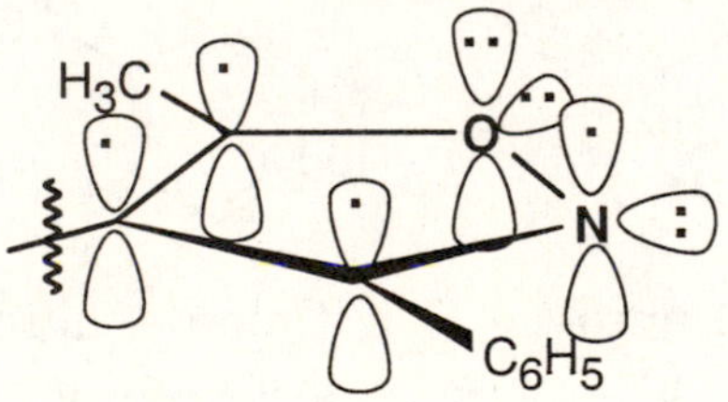

Isoxazole ring

8.43

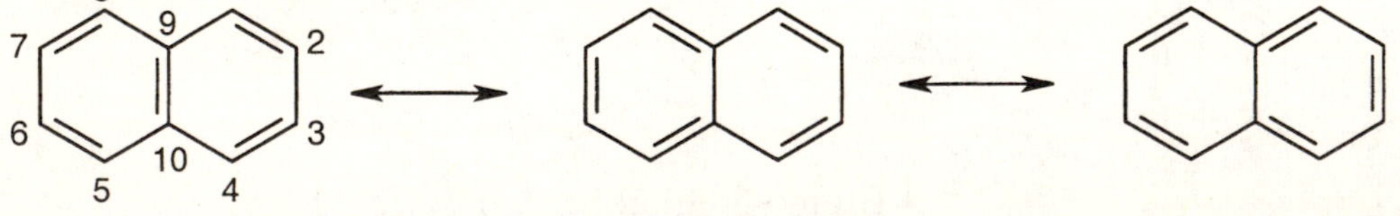

The bond between carbons 1 and 2 is represented as a double bond in two of the three resonance structures, but the bond between carbons 2 and 3 is represented as a double bond in only one resonance structure. The C1–C2 bond thus has more double bond character in the resonance hybrid, and it is shorter than the C2–C3 bond. The C3–C4, C5–C6, and C7–C8 bonds also have more double bond character than the remaining bonds.

8.44

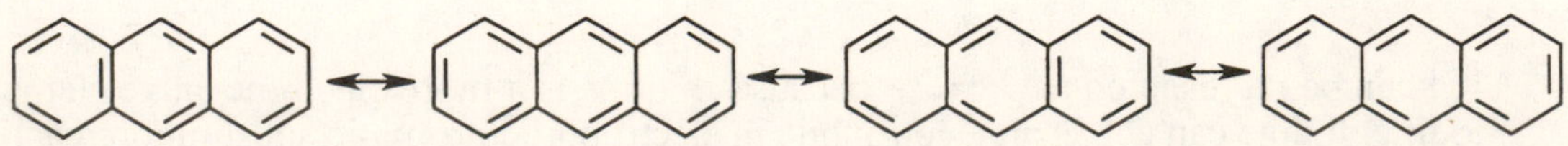

8.45, 8.46

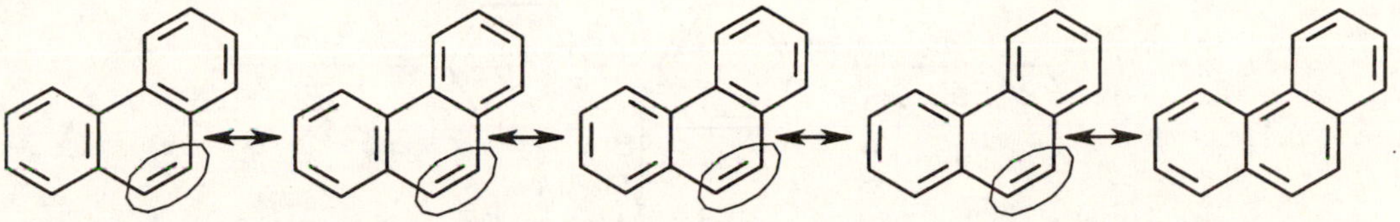

The circled bond is represented as a double bond in four of the five resonance forms of phenanthrene. This bond has more double-bond character and thus is shorter than the other carbon–carbon bonds of phenanthrene.

8.47 The aromatic ring is deactivated toward electrophilic aromatic substitution by the combined electron-withdrawing inductive effect of electronegative nitrogen and oxygen. The lone pair of electrons of nitrogen can, however, stabilize by resonance the ortho and para substituted intermediates but not the meta intermediate.

Ortho attack:

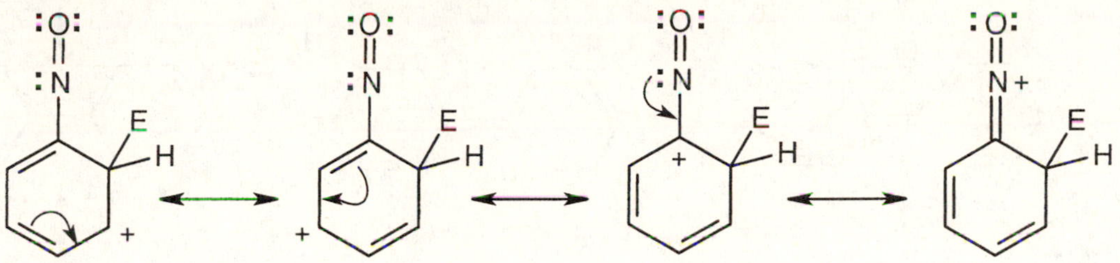

Meta attack:

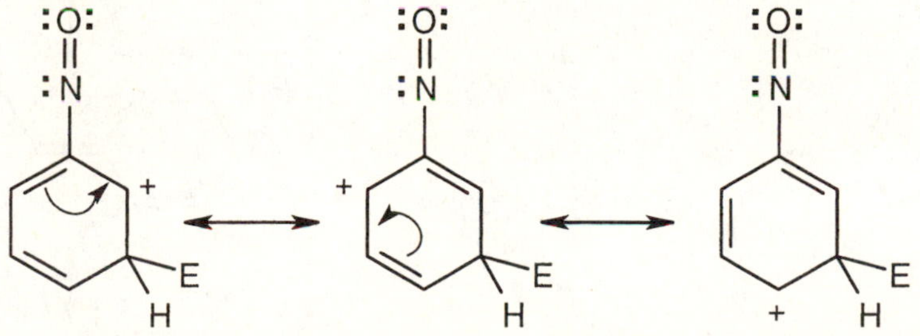

Para attack:

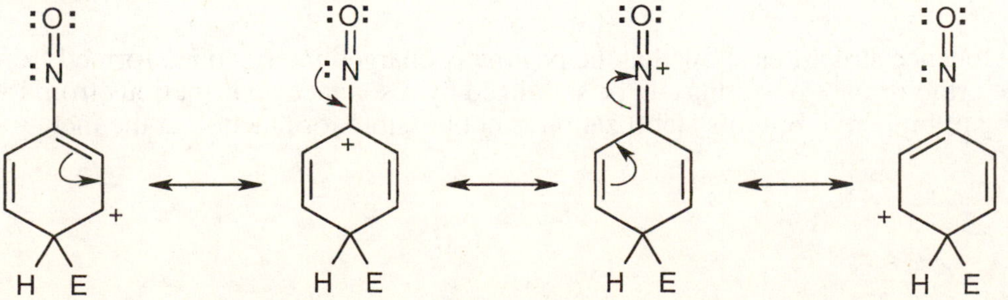

8.48

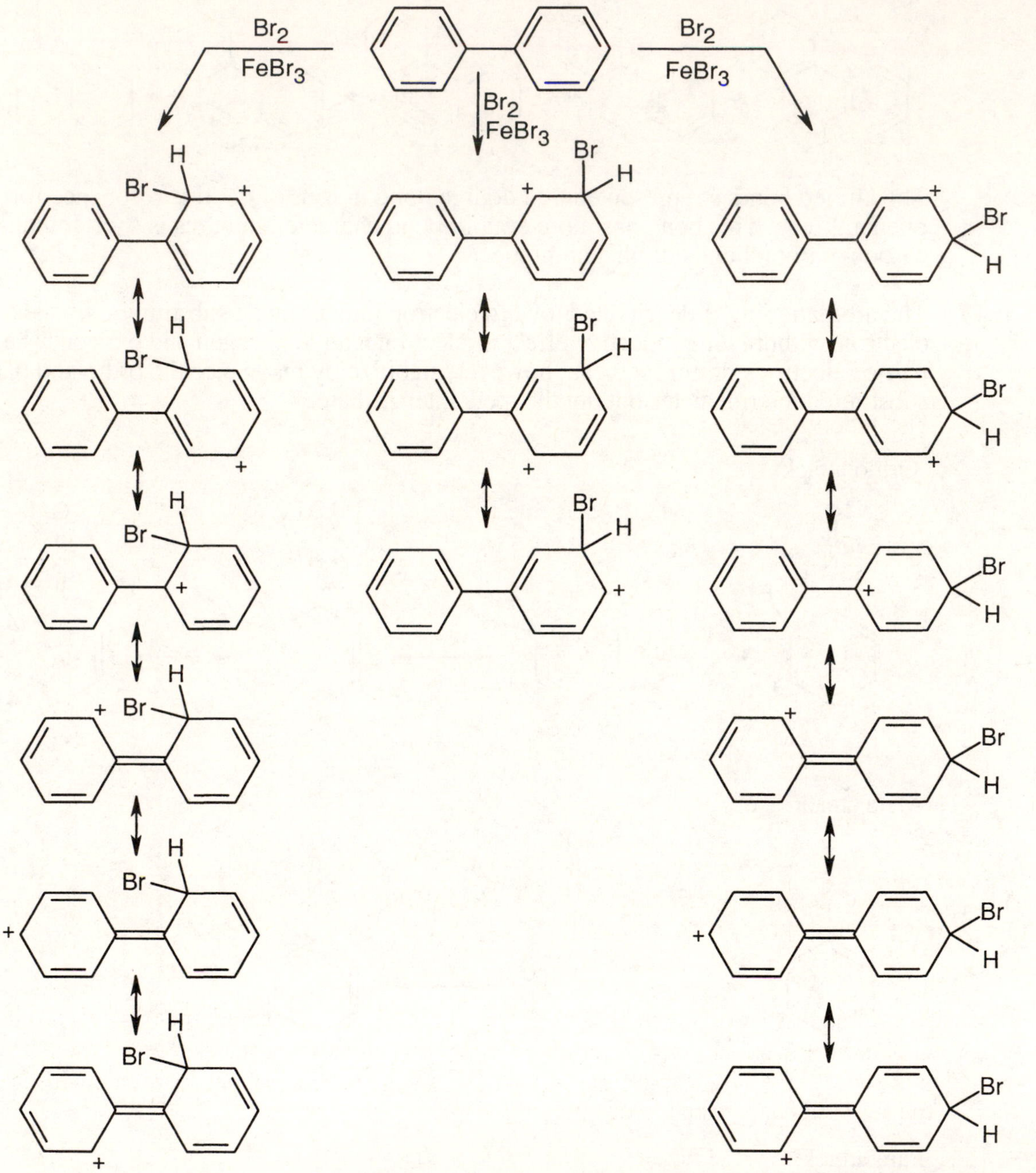

Br$_2$ FeBr$_3$

Br H +

Resonance structures show that the positively charged intermediate formed from reaction at the ortho or para positions can be stabilized by resonance contributions from the second ring of biphenyl, but this stabilization is not possible for reaction at the meta position.

8.49

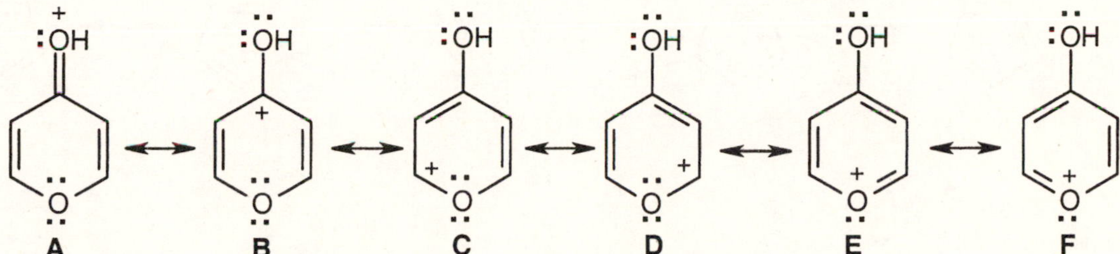

Protonation of 4-pyrone gives structure **A**, which has resonance forms **B**, **C**, **D**, **E** and **F**. In **E** and **F**, a lone pair of electrons of the ring oxygen is delocalized into the ring to produce a six π electron system, which should be aromatic according to Hückel's rule.

8.50 The second resonance form of *N*-phenylsydnone shows the aromaticity of the five-membered ring more clearly. In this form, the ring oxygen contributes two electrons to the ring π system, each nitrogen contributes one electron and each carbon contributes one electron (the carbonyl oxygen bears a formal negative charge). This 6 π electron, cyclic conjugated system obeys Hückel's rule and is expected to show aromatic behavior.

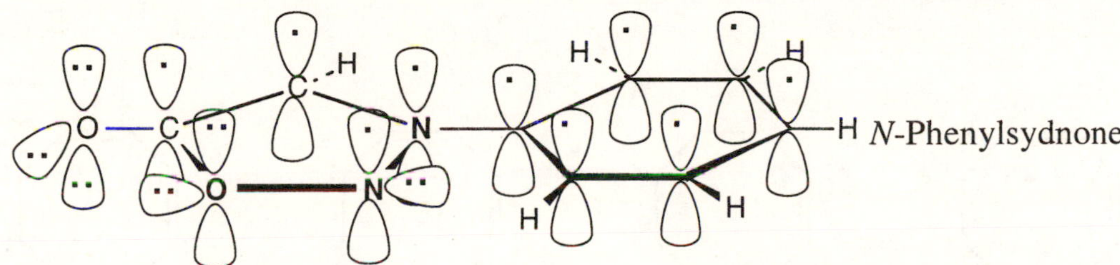

8.51 An acyl substituent is deactivating. Once an aromatic ring has been acylated, it is much less reactive to further substitution. An alkyl substituent is activating, however, so an alkyl-substituted ring is more reactive than an unsubstituted ring, and polysubstitution occurs readily.

8.52 Both ions have six π electrons in a cyclic conjugated system and are aromatic. In the cyclopentadienyl anion, four of the five carbons have one π electron, and the fifth carbon has two. In the cycloheptatrienyl cation, six of the seven carbons have one π electron, and the seventh has a vacant *p* orbital.

The cyclopentadienyl anion has five equivalent resonance forms that delocalize the negative charge. The cycloheptatrienyl has seven equivalent resonance forms.

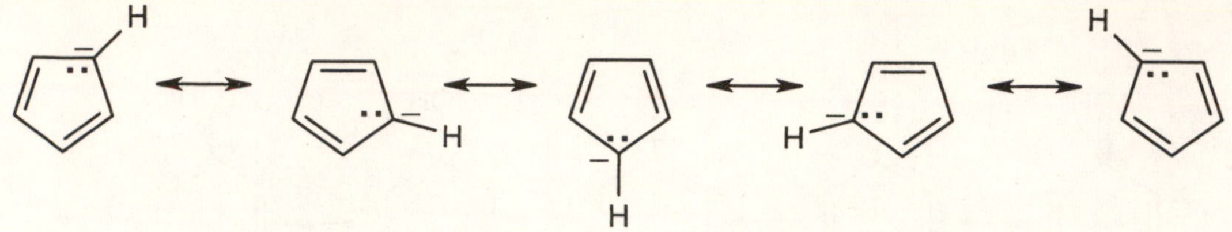

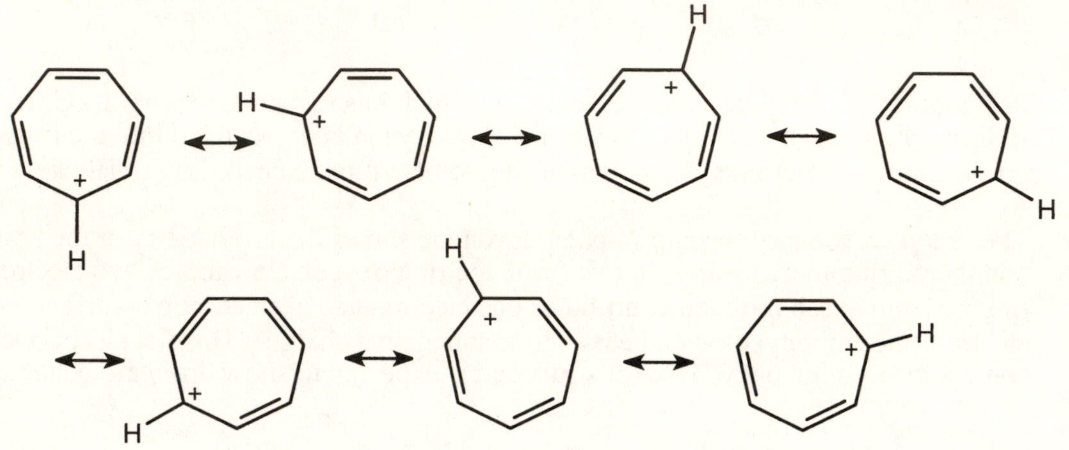

8.53

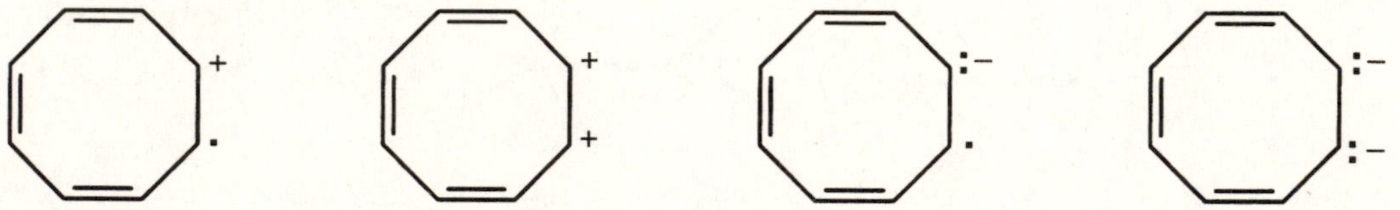

7 π electrons 6 π electrons 9 π electrons 10 π electrons

The dication (6 π electrons) and the dianion (10 π electrons) are the most stable because they both have a Hückel number of π electrons in a cyclic conjugated system.

8.54 When directly bonded to a ring, the –CN group is a meta-directing deactivator for both inductive and resonance reasons. In 3-phenylpropanenitrile, however, the saturated side chain does not allow resonance interactions of –CN with the aromatic ring, and the –CN group is too far from the ring for its inductive effect to be strongly felt. The side chain acts as an alkyl substituent, and ortho–para substitution is observed.

In 3-phenylpropenenitrile, the –CN group interacts with the ring through the π electrons of the side chain. Resonance forms show that –CN deactivates the ring toward electrophilic substitution, and substitution occurs at the meta position.

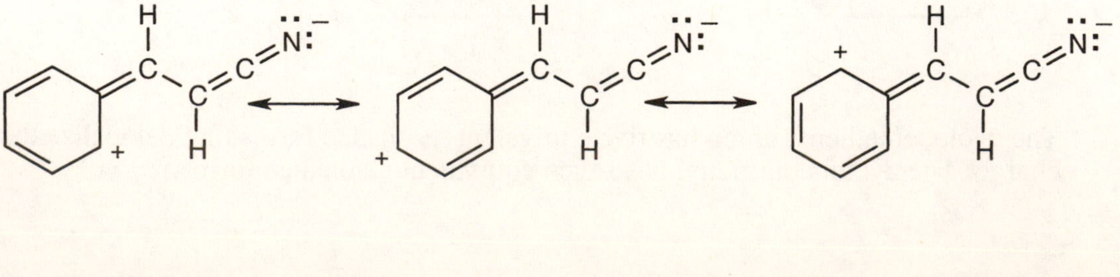

8.55

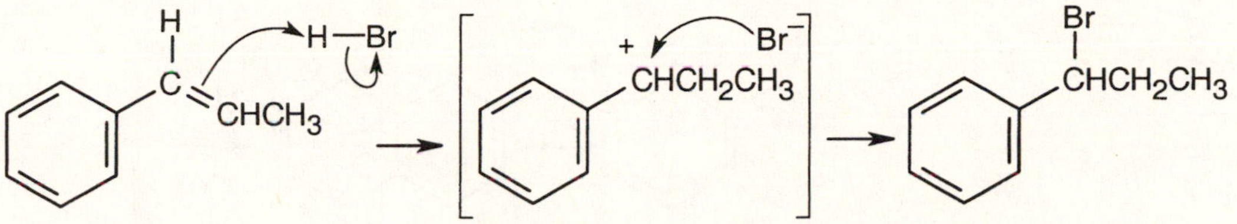

Protonation of the double bond at carbon 1 of 1-phenylpropene leads to an intermediate carbocation that can be stabilized by resonance involving the benzene ring.

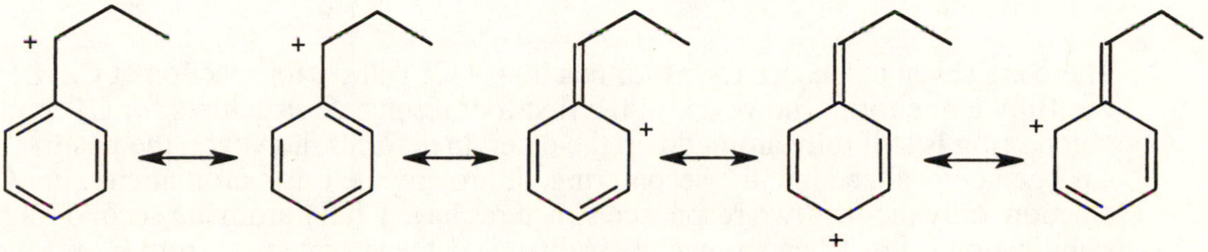

8.56

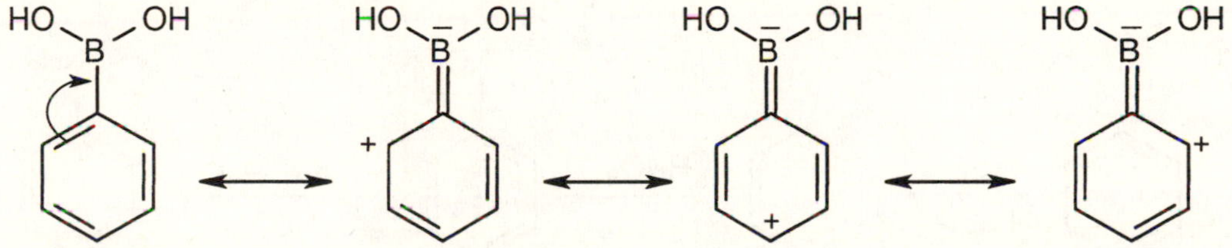

The trivalent boron atom in phenylboronic acid has only six outer shell electrons and is a Lewis acid. It is possible to write resonance forms for phenylboronic acid in which an electron pair from the phenyl ring is delocalized onto boron. In these resonance forms, the ortho and para positions of phenylboronic acid are the most electron-deficient, and substitutions occur primarily at the meta position.

8.57 Resonance forms for thc intermediate from reaction at C1:

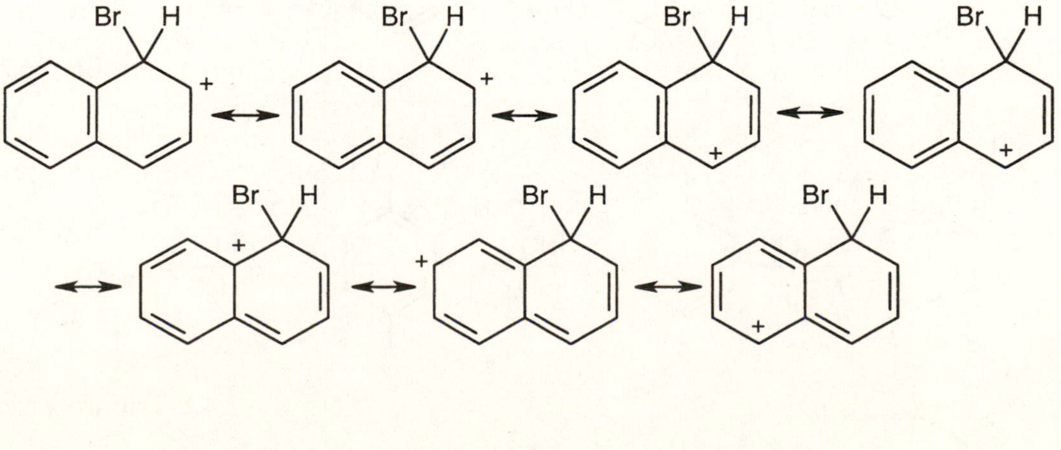

Resonance forms for the intermediate from reaction at C2:

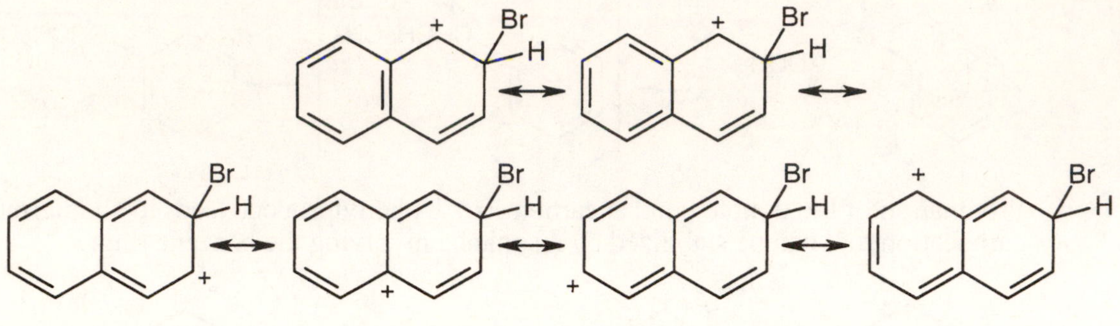
Br
H
+
Br
H
+
Br
H
+
Br
H
+
Br
H
+
Br
H
+

There are seven resonance forms for reaction at C1 and six for reaction at C2. Look carefully at the forms, however. In the first four resonance structures for C1 reaction, the second ring is still fully aromatic. In the other three forms, however, the positive charge has been delocalized into the second ring, destroying the ring's aromaticity. For C2 reaction, only the first two resonance structures have a fully aromatic second ring. Since stabilization is lost when aromaticity is disrupted, the intermediate from C2 reaction is less stable than the intermediate from C1 reaction, and C1 reaction is favored.

8.58

(a)

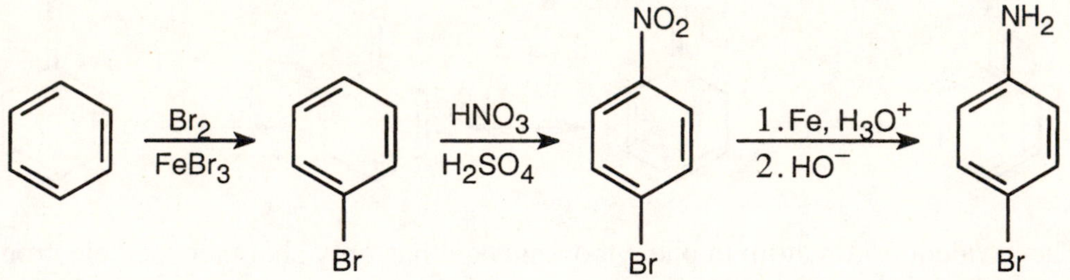
Br2
FeBr3
Br
HNO3
H2SO4
NO2
Br
1. Fe, H3O+
2. HO−
NH2
Br

p-Bromoaniline

(b)

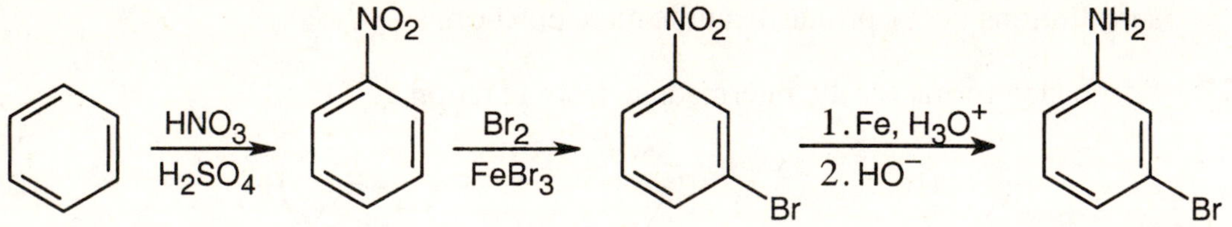
HNO3
H2SO4
NO2
Br2
FeBr3
NO2
Br
1. Fe, H3O+
2. HO−
NH2
Br

m-Bromoaniline

(c)

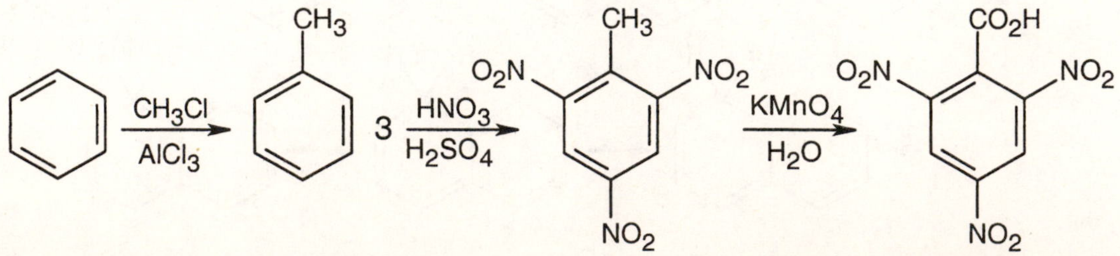
CH3Cl
AlCl3
CH3
3 HNO3
H2SO4
CH3
O2N
NO2
NO2
KMnO4
H2O
CO2H
O2N
NO2
NO2

2,4,6-Trinitrobenzoic acid

(d)

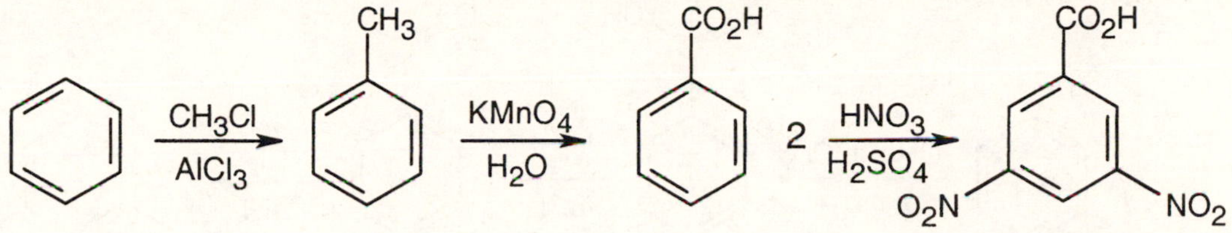

Benzene $\xrightarrow[AlCl_3]{CH_3Cl}$ Toluene (CH_3) $\xrightarrow[H_2O]{KMnO_4}$ Benzoic acid (CO_2H) $\xrightarrow[H_2SO_4]{2\ HNO_3}$ (CO_2H; O_2N, NO_2)

3,5-Dinitrobenzoic acid

8.59

(a)

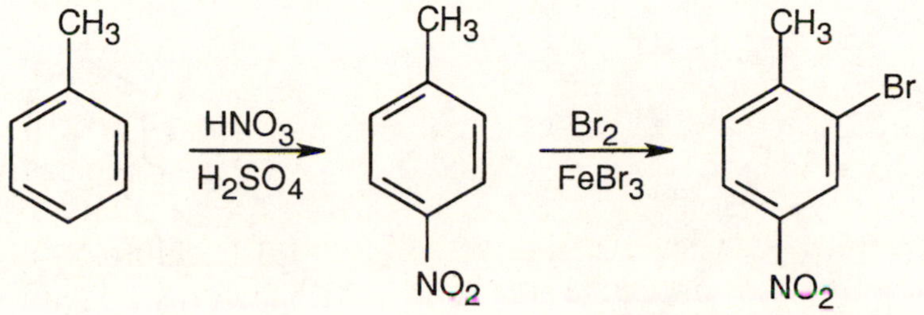

Toluene (CH_3) $\xrightarrow[H_2SO_4]{HNO_3}$ (CH_3; NO_2) $\xrightarrow[FeBr_3]{Br_2}$ (CH_3; Br; NO_2)

2-Bromo-4-nitrotoluene

(b)

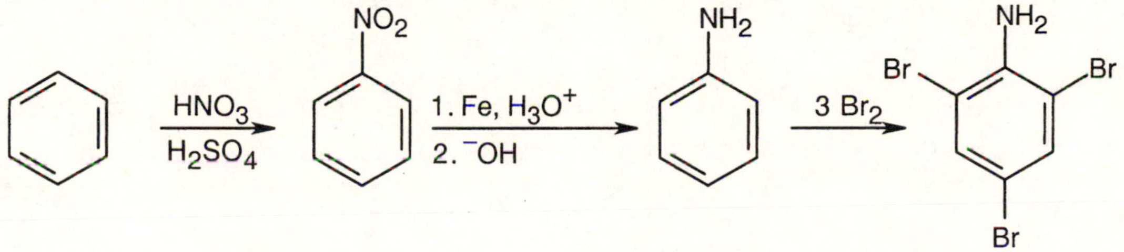

Benzene $\xrightarrow[H_2SO_4]{HNO_3}$ (NO_2) $\xrightarrow[2.\ ^{-}OH]{1.\ Fe,\ H_3O^+}$ (NH_2) $\xrightarrow{3\ Br_2}$ (NH_2; Br, Br; Br)

2,4,6-Tribromo-
aniline

(c)

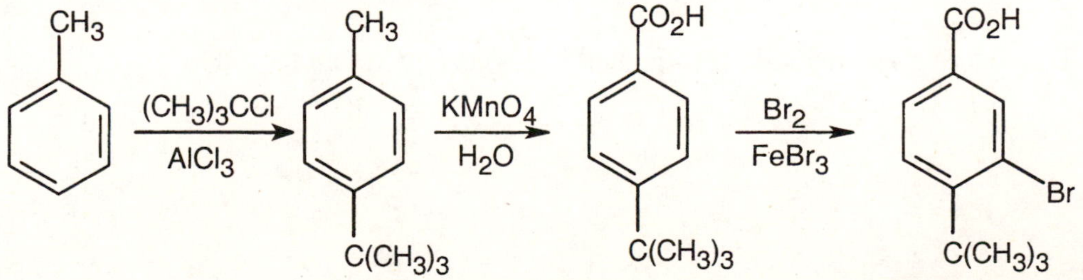

Toluene (CH_3) $\xrightarrow[AlCl_3]{(CH_3)_3CCl}$ (CH_3; $C(CH_3)_3$) $\xrightarrow[H_2O]{KMnO_4}$ (CO_2H; $C(CH_3)_3$) $\xrightarrow[FeBr_3]{Br_2}$ (CO_2H; Br; $C(CH_3)_3$)

3-Bromo-4-*tert*-
butylbenzoic acid

(d)

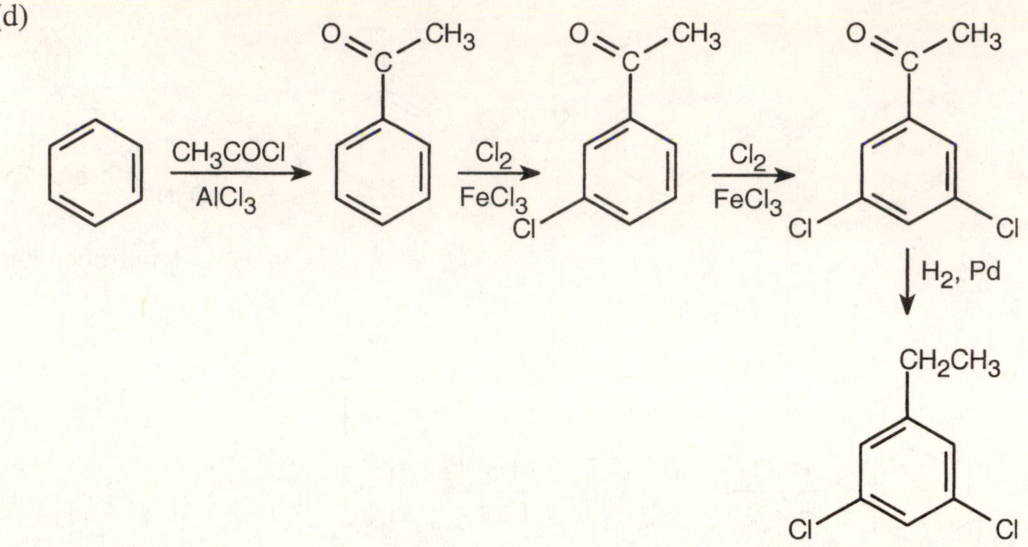

CH3COCl / AlCl3 → Cl2 / FeCl3 → Cl2 / FeCl3 → H2, Pd

1,3-Dichloro-5-ethylbenzene

8.60

$$HO{-}OH \xrightarrow[\text{catalyst}]{\text{acid}} HO{-}\overset{+}{O}H_2$$ reactive electrophile

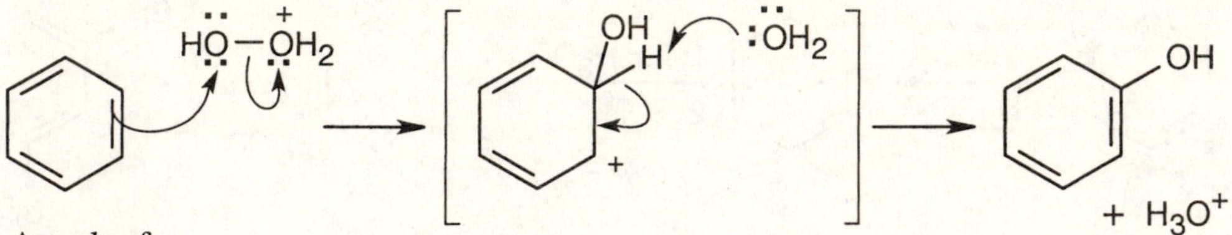

Attack of π electrons on reactive electrophile

Loss of proton

+ H_3O^+

The reactive electrophile (protonated H_2O_2) is equivalent to ^+OH.

8.61

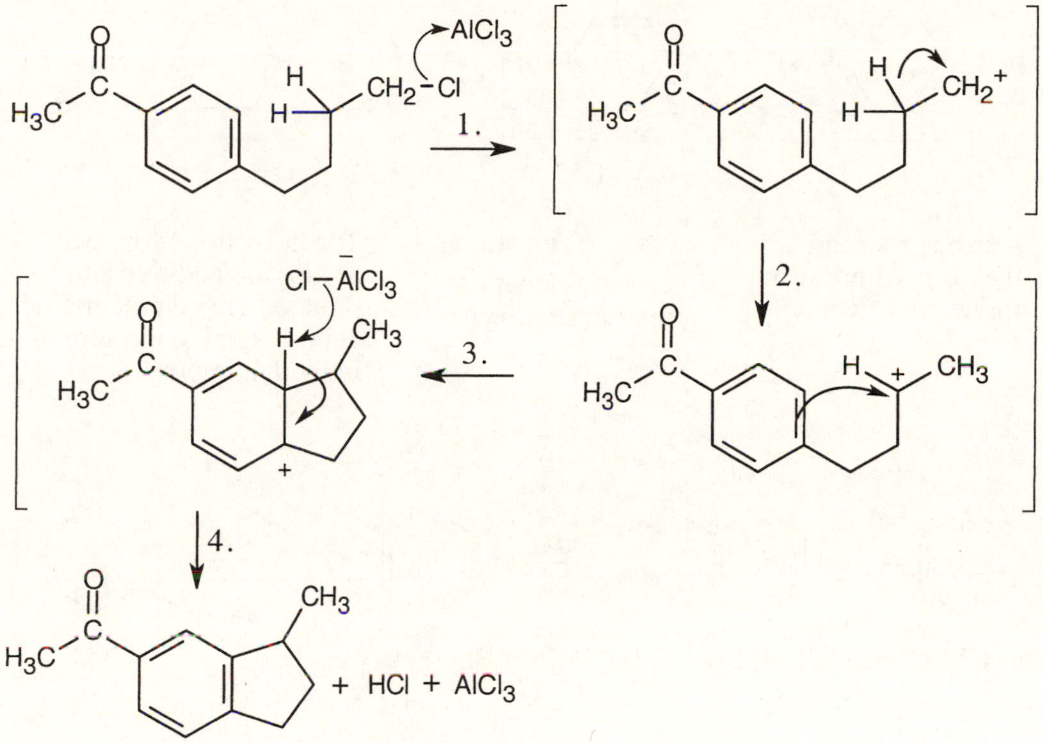

Step 1: Formation of primary carbocation.
Step 2: Rearrangement to secondary carbocation.
Step 3: Reaction of ring π electrons with carbocation,
Step 4: Loss of H^+.

This reaction takes place despite the fact that an electron-withdrawing group is attached to the ring. Apparently, the cyclization reaction is strongly favored.

8.62

(1)

$$^{-}\!:C\equiv\ddot{O}^{+} \xrightleftharpoons{HCl,\ AlCl_3} H-C\equiv\ddot{O}^{+}\ \ AlCl_4^{-}$$

Carbon monoxide is protonated to form an acyl cation.

(2)

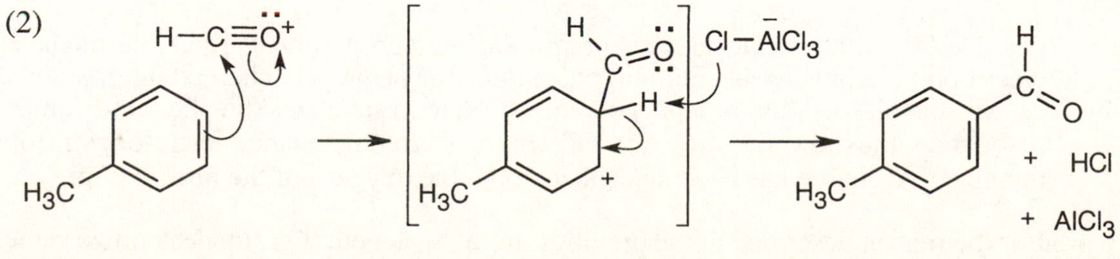

The acyl cation reacts with benzene by a Friedel–Crafts acylation mechanism.

8.63

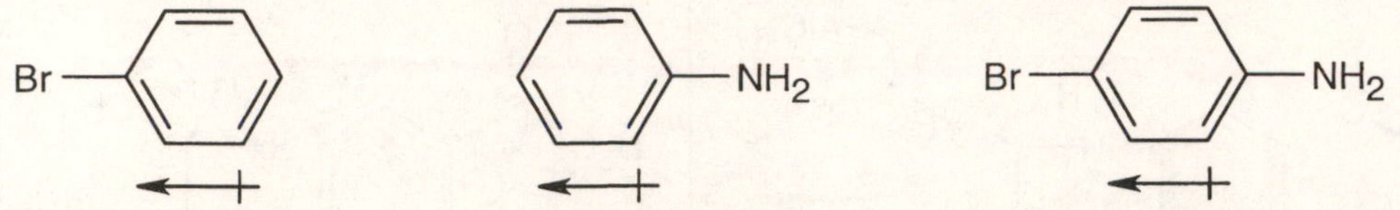

Br–C₆H₄ (bromobenzene) μ = 1.53 D; C₆H₅–NH₂ μ = 1.52 D; Br–C₆H₄–NH₂ μ = 2.91 D

–Br has a strong electron-withdrawing inductive effect.

$-NH_2$ has a strong electron-donating resonance effect.

The polarities of the two groups are additive and produce a net dipole moment almost equal to the sum of the individual moments.

8.64

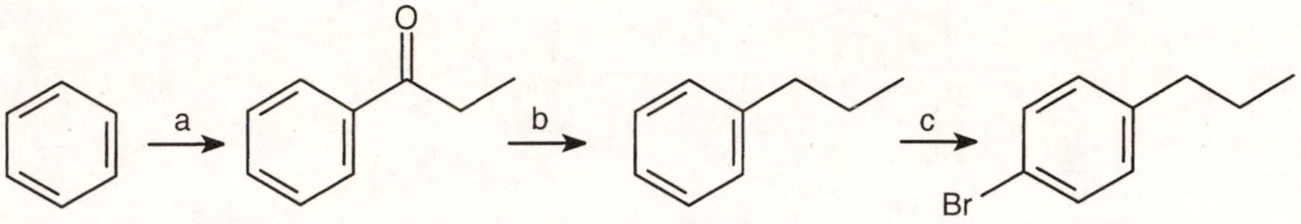

(a) CH_3CH_2COCl, $AlCl_3$; (b) H_2, Pd; (c) Br_2, $FeBr_3$.

8.65

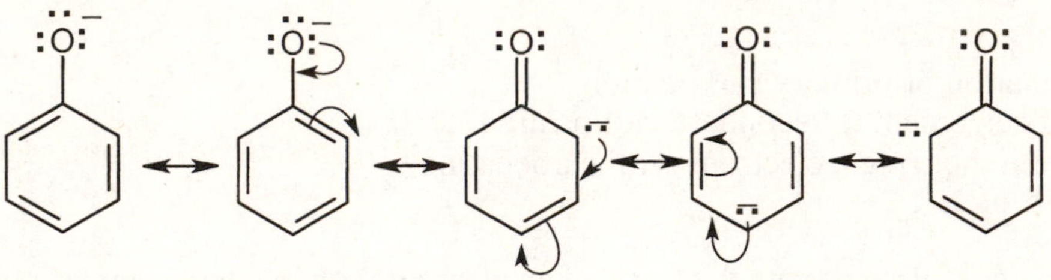

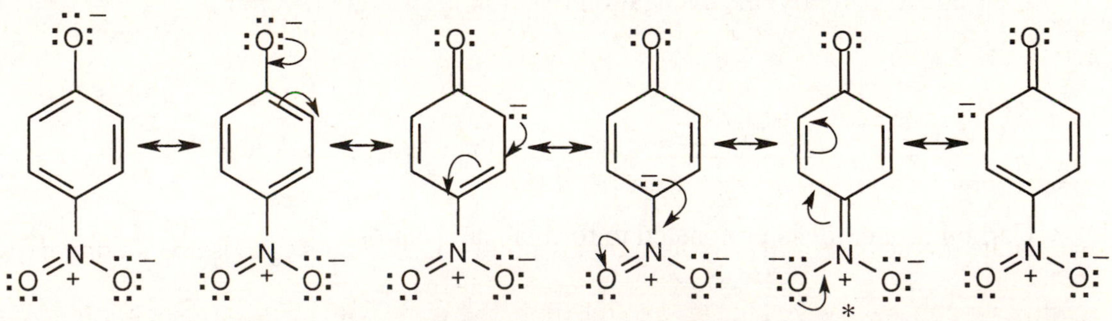

An electron-withdrawing substituent *destabilizes* a positively charged intermediate (as in electrophilic aromatic substitution) but *stabilizes* a negatively charged intermediate. In the case of the dissociation of a phenol, an $-NO_2$ group stabilizes the phenoxide anion by resonance, thus lowering ΔG^o and pK_a. In the starred resonance form for *p*-nitrophenol, the negative charge has been delocalized onto the oxygens of the nitro group.

8.66 For the reason described in the previous problem, a methyl group destabilizes a negatively charged intermediate, thus raising ΔG^o and pK_a, making *p*-methylphenol less acidic than phenol. A bromo substituent is electron-withdrawing and should lower the pK_a of a substituted phenol, but to a lesser degree than a nitro group.

Review Unit 3: Alkenes, Alkynes, and Aromatic Compounds

Major Topics Covered (with vocabulary):

Introduction to alkenes:
degree of unsaturation methylene group vinyl group allyl group cis–trans isomerism *E,Z* isomerism sequence rules heat of hydrogenation hyperconjugation

Electrophilic addition reactions:
electrophilic addition reaction regiospecific Markovnikov's rule Hammond Postulate carbocation rearrangement hydride shift

Other reactions of alkenes:
dehydrohalogenation dehydration anti stereochemistry bromonium ion halohydrin hydration oxymercuration hydroboration syn stereochemistry hydrogenation epoxidation hydroxylation diol conjugation allylic 1,4-addition

Polymerization reactions:
polymer monomer chain branching radical polymerization cationic polymerization

Alkynes:
alkyne enyne Lindlar catalyst acetylide anion

Aromaticity:
aromatic arene phenyl group benzyl group ortho, meta, para substitution degenerate Hückel $4n + 2$ rule heterocycle polycyclic aromatic compound

Chemistry of aromatic compounds:
electrophilic aromatic substitution sulfonation Friedel-Crafts alkylation polyalkylation Friedel-Crafts acylation ortho- and para-directing activator ortho- and para-directing deactivator meta-directing deactivator inductive effect resonance effect benzylic position

Types of Problems:

After studying these chapters you should be able to:

- Calculate the degree of unsaturation of any compound, including those containing N, O, and halogen.
- Name acyclic and cyclic alkenes and alkynes, and draw structures corresponding to names.
- Assign *E,Z* priorities to groups.
- Assign cis-trans and *E,Z* designations to double bonds.
- Predict the relative stability of alkene double bonds.

- Formulate mechanisms of electrophilic addition reactions.
- Predict the products of reactions involving alkenes and alkynes.
- Choose the correct alkene or alkyne starting material to yield a given product.
- Carry out syntheses involving alkenes and alkynes.

- Name and draw substituted benzenes.
- Draw resonance structures and molecular orbital diagrams for benzene and other cyclic conjugated molecules.
- Use Hückel's rule to predict aromaticity.
- Draw orbital pictures of cyclic conjugated molecules.

- Predict the products of electrophilic aromatic substitution reactions.
- Formulate the mechanisms of electrophilic aromatic substitution reactions.
- Understand the activating and directing effects of substituents on aromatic rings, and use inductive and resonance arguments to predict orientation and reactivity.
- Synthesize substituted benzenes.

Points to Remember:

* Calculating the degree of unsaturation is an absolutely essential technique in the structure determination of <u>all</u> organic compounds. It is the starting point for deciding which functional groups are or aren't present in a given compound, and eliminates many possibilities. When a structure determination problem is given, always calculate the degree of unsaturation first.

* All cis–trans isomers can also be described by the *E,Z* designation, but not all *E,Z* isomers can be described by the cis–trans designation.

* Not all hydrogens bonded to carbons adjacent to a carbocation can take part in hyperconjugation at the same time. At any given instant, some of the hydrogens have C-H bonds that lie in the plane of the carbocation and are not suitably oriented for hyperconjugative overlap

* Although it is very important to work backwards when planning an organic synthesis, don't forget to pay attention to the starting material, also. Planning a synthesis is like solving a maze from the middle outward: keeping your eye on the starting material can keep you from running into a dead end.

* To be aromatic, a molecule must be planar, cyclic, conjugated, and it must have $4n + 2$ electrons in its π system.

* The carbocation intermediate of electrophilic aromatic substitution loses a proton to yield the aromatic product. In all cases, a base is involved with proton removal, but the nature of the base varies with the type of substitution reaction. Although this book shows the loss of the proton, it often doesn't show the base responsible for proton removal. This doesn't imply that the proton flies off, unassisted; it just means that the base involved has not been identified in the problem.

* Activating groups achieve their effects by making an aromatic ring more electron-rich and reactive toward electrophiles. Ortho-para directing groups achieve their effects by stabilizing the positive charge that results from ortho or para addition of an electrophile to the aromatic ring. The intermediate resulting from addition to a ring with an ortho-para director usually has one resonance form that is especially stable. The intermediate resulting from addition to a ring with a meta-director usually has a resonance form that is especially unfavorable when addition occurs ortho or para to the functional group. Meta-substitution results because it is less unfavorable than ortho-para substitution.

Self-Test:

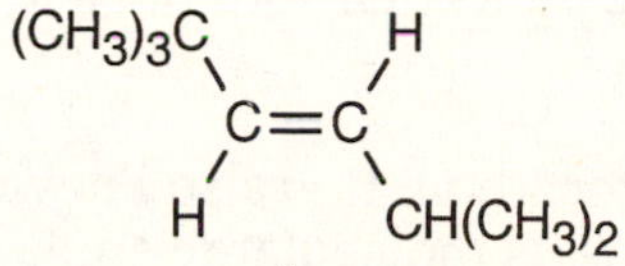
(CH3)3C
H
C=C
H
CH(CH3)2
A

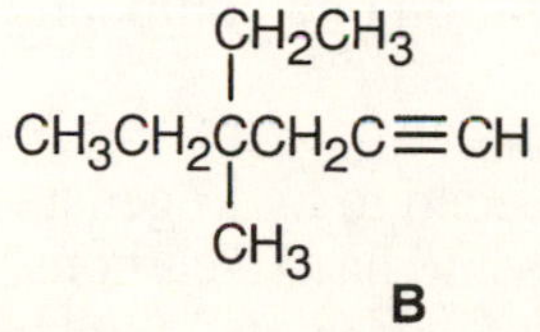
CH2CH3
CH3CH2CCH2C≡CH
CH3
B

Provide names for **A** and **B** (include bond stereochemistry). Predict the products of reaction of **A** and **B** with (a) 1 equiv HBr (b) H_2, Pd/C (c) BH_3, THF, then H_2O_2, HO^-. Give a reagent that reacts with **A** but not **B**. Give a reagent that reacts with **B**, but not **A**.

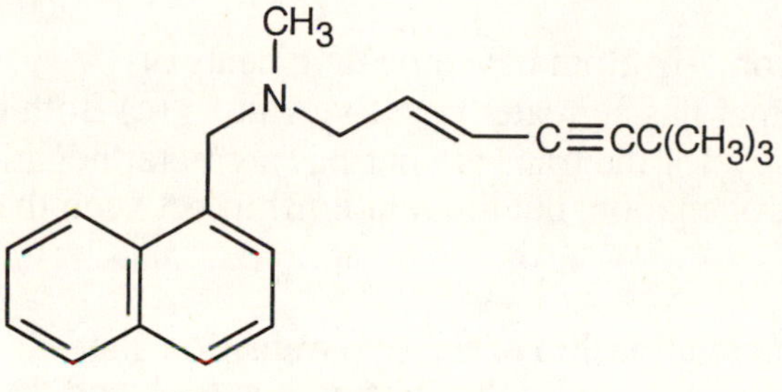
CH3
N
C≡CC(CH3)3
C Terbinafine (an antifungal)

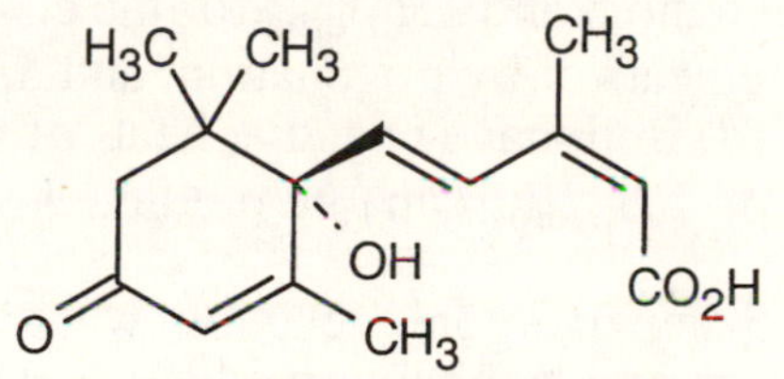
H3C
CH3
CH3
OH
CO2H
O
CH3
D Abscisic acid (a plant hormone)

What is the configuration of the double bond in the side chain of **C**? What is the configuration of the triple bond after hydrogenation with Lindlar catalyst? If **C** is treated with $Br/FeBr_3$, in which ring and at which position(s) is bromination likely to occur?

Give *E,Z* configurations for the double bonds in **D**.

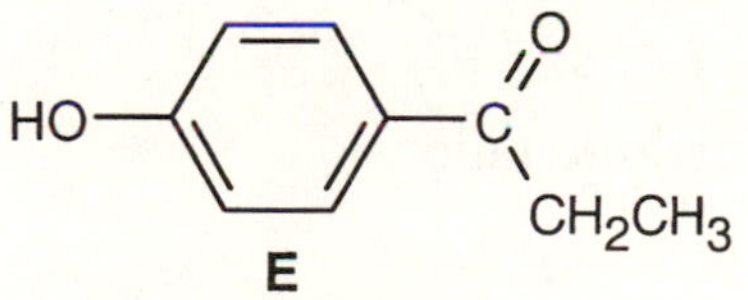
O
HO
C
CH2CH3
E
Paroxypropione

Paroxypropione (**E**) is a hormone inhibitor. Predict the products of reaction of **E** with: (a) Br_2, $FeBr_3$; (b) CH_3Cl, $AlCl_3$; (c) $KMnO_4$, H_3O^+; (d) H_2, Pd/C. If the product of (c) is treated with the reagents in (a) or (b), does the orientation of substitution change?

Multiple Choice:

1. What is the degree of unsaturation of a compound whose molecular formula is $C_{11}H_{13}N$?
(a) 4 (b) 5 (c) 6 (d) 7

2. Two equivalents of H_2 are needed to hydrogenate a hydrocarbon. It is also known that the compound contains two rings and has 15 carbons. What is its molecular formula?
(a) $C_{15}H_{22}$ (b) $C_{15}H_{24}$ (c) $C_{15}H_{28}$ (d) $C_{15}H_{32}$

3. Which group is of lower priority than $–CH{=}CH_2$?
(a) $–CH(CH_3)_2$ (b) $–CH{=}C(CH_3)_2$ (c) $–C{\equiv}CH$ (d) $–C(CH_3)_3$

4. What is the usual relationship between the heats of hydrogenation of a pair of cis/trans alkene isomers?
(a) Both have positive heats of hydrogenation (b) Both have negative heats of hydrogenation, and ΔH_{hydrog} for the cis isomer has a greater negative value (c) Both have negative heats of hydrogenation, and ΔH_{hydrog} for the trans isomer has a greater negative value (d) Both have negative heats of hydrogenation, but the relationship between the two values of ΔH_{hydrog} can't be predicted.

5. In a two-step exergonic reaction, what is the relationship of the two transition states?
(a) both resemble the intermediate (b) the first resembles the starting material, and the second resembles the product (c) the first resembles the intermediate and the second resembles the product (d) there is no predictable relationship between the two transition states

6. For synthesis of an alcohol, acid-catalyzed hydration of an alkene is useful in all of the following instances except:
(a) when an alkene has no acid-sensitive groups (b) when an alkene is symmetrical
(c) when a large amount of the alcohol is needed
(d) when two possible carbocation intermediates are of similar stability.

7. A reaction that produces a diol from an alcohol is a:
(a) hydration (b) hydrogenation (c) hydroboration (d) hydroxylation

8. Which of the following compounds is aromatic?

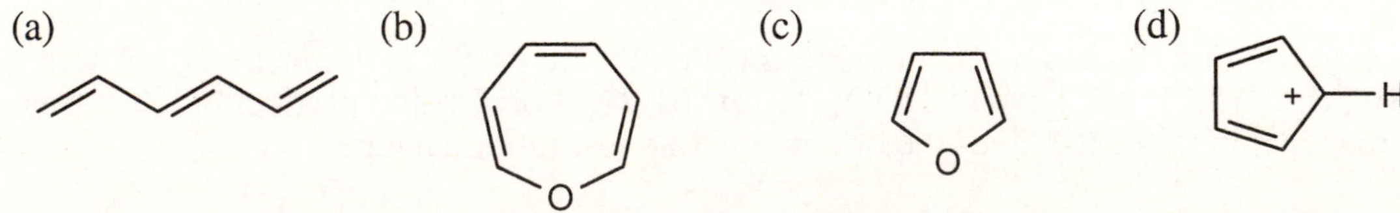

9. Which of the following functional groups isn't a meta-directing deactivator?
(a) $–NO_2$ (b) $–N{=}O$ (c) $–N(CH_3)_3^+$ (d) $–NHCOCH_3$

10. Which of the following compounds can't be synthesized by an electrophilic aromatic substitution reaction that we have studied?
(a) *m*-Cresol (b) *p*-Chloroaniline (c) 2,4-Toluenedisulfonic acid (d) *m*-Bromotoluene

Chapter 9 – Stereochemistry

Chapter Outline

I. Handedness (Sections 9.1 – 9.4).
  A. Enantiomers and tetrahedral carbon (Section 9.1).
    When four different groups are bonded to a carbon atom, two different arrangements are possible.
    a. These arrangements are mirror images.
    b. The two mirror-image molecules are enantiomers.
  B. The reason for handedness in molecules: Chirality (Section 9.2).
    1. Molecules that are not superimposable on their mirror-images are chiral.
      a. A molecule is not chiral if it contains a plane of symmetry.
      b. A molecule with no plane of symmetry is chiral.
    2. A carbon bonded to four different groups is a chirality center.
    3. Any $–CH_2–$ or $–CH_3$ carbon is achiral.
  C. Optical activity (Section 9.3).
    1. Solutions of certain substances rotate the plane of plane-polarized light. These substances are said to be optically active.
    2. The amount of rotation can be measured with a polarimeter.
    3. The degree of rotation can also be measured.
      a. A compound whose solution rotates plane-polarized light to the right is dextrorotatory.
      b. A compound whose solution rotates plane-polarized light to the left is levorotatory.
    4. Specific rotation.
      a. The amount of rotation depends on concentration, path length and wavelength.
      b. Specific rotation is the observed rotation of a sample with concentration = 1 g/mL, sample path length of 1 dm, and light of wavelength = 589 nm.
      c. Specific rotation is a physical constant characteristic of a given optically active compound.
  D. Pasteur's discovery of enantiomerism (Section 9.4).
    1. Pasteur discovered two different types of crystals in a solution that he was evaporating.
    2. The crystals were mirror images.
    3. Solutions of each of the two types of crystals were optically active, and their specific rotations were equal in magnitude but opposite in sign.
    4. Pasteur postulated that some molecules are handed and thus discovered the phenomenon of enantiomerism.

II. Stereoisomers and configurations (Sections 9.5 – 9.8).
  A. Specification of configurations of stereoisomers (Section 9.5).
    1. Rules for assigning configurations at a chirality center:
      a. Assign priorities to each group bonded to the carbon by using Cahn-Ingold-Prelog rules (Section 6.6).
      b. Orient the molecule so that the group of lowest priority is pointing to the rear.
      c. Draw a curved arrow from group 1 to group 2 to group 3.
      d. If the arrow rotates clockwise, the chirality center is *R*, and if the arrow rotates counterclockwise, the chirality center is *S*.
    2. The sign of rotation is not related to *R,S* designation.
    3. X-ray experiments have proven *R,S* conventions to be correct.

B. Diastereomers (Section 9.6).
1. A molecule with two chirality centers can have four possible stereoisomers.
a. The stereoisomers group into two pairs of enantiomers.
b. A stereoisomer from one pair is the diastereomer of a stereoisomer from the other pair.
2. Diastereomers are stereoisomers that are not mirror images.
3. Epimers are diastereomers whose configuration differs at only one chirality center.

C. Meso compounds (Section 9.7).
1. A meso compound occurs when a compound with two chirality centers possesses a plane of symmetry.
2. A meso compound is achiral despite having two chirality centers.

D. Racemic mixtures and their resolution (Section 9.8).
1. A racemic mixture is a 50:50 mixture of two enantiomers.
Racemic mixtures show zero optical rotation.
2. Some racemic mixtures can be resolved into their component enantiomers.
a. If a racemic mixture of a carboxylic acid reacts with a chiral amine, the product ammonium salts are diastereomers.
b. The diastereomeric salts differ in chemical and physical properties and can be separated.
c. The original enantiomers can be recovered by acidification.

III. A review of isomerism (Section 9.9).
A. Constitutional isomers differ in connections between atoms.
1. Skeletal isomers have different carbon skeletons.
2. Functional isomers contain different functional groups.
3. Positional isomers have functional groups in different positions.

B. Stereoisomers have the same connections between atoms, but different geometry.
1. Enantiomers have a mirror-image relationship.
2. Diastereomers are non-mirror-image stereoisomers.
a. Configurational diastereomers.
b. Cis–trans isomers differ in the arrangement of substituents in a double bond or ring.

IV. Stereochemistry of reactions (Sections 9.10 – 9.11).
A. Addition of H_2O to an achiral alkene (Section 9.10).
1. When H_2O adds to an achiral alkene, a racemic mixture of products is formed.
2. The achiral cationic intermediate can react from either side to produce a racemic mixture.
3. Alternatively, the transition states for topside reaction and bottom side reaction are enantiomers and have the same energy.
4. Enzyme-catalyzed reactions give a single enantiomer, even when the substrate is achiral.

B. Addition of H_2O to a chiral alkene (Section 9.11).
1. When H^+ adds to a chiral alkene, the intermediate carbocation is chiral.
2. The original chirality center is unaffected by the reaction.
3. Reaction of H_2O with the carbocation doesn't occur with equal probability from either side, and the resulting product is an optically active mixture of diastereomeric alcohols.
4. Reaction of a chiral reactant with an achiral reactant leads to unequal amounts of diastereomeric products.

V. Chirality at atoms other than carbon (Section 9.12).
  A. Other elements with tetrahedral atoms can be chirality centers.
  B. Trivalent nitrogen can, theoretically, be chiral, but rapid inversion of the nitrogen lone pair interconverts the enantiomers.
  C. Chiral phosphines and sulfonium salts can be isolated because their rate of inversion is slower.

VI. Prochirality (Section 9.13).
  A. A molecule is prochiral if it can be converted from achiral to chiral in a single chemical step.
  B. Identifying prochirality.
    1. For sp^2 carbon, draw the plane that includes the atoms bonded to the sp^2 carbon.
      a. Assign priorities to the groups bonded to the carbon.
      b. Draw an curved arrow from group 1 to group 2 to group 3.
      c. The face of the plane on which the curved arrow rotates clockwise is the *Re* face.
      d. The face on which the arrow rotates counterclockwise is the *Si* face.
    2. An atom that is sp^3-hybridized may have a prochirality center if, when one of its attached groups is replaced, it becomes a chirality center.
      a. For $–CH_2X$, imagine a replacement of one hydrogen with deuterium.
      b. Rank the groups, including the deuterium.
      c. If the replacement leads to *R* chirality, the atom is *pro-R.*
      d. If the replacement leads to *S* chirality, the atom is *pro-S.*
  C. Many biochemical reactions involve prochiral compounds.

VII. Chirality in nature (Section 9.14).
  A. Different enantiomers of a chiral molecule have different properties in nature.
    1. (+)-Limonene has the odor of oranges, and (–)-limonene has the odor of lemons.
    2. Racemic fluoxetine is an antidepressant, but the *S* enantiomer is effective against migraine.
  B. In nature, a molecule must fit into a chiral receptor, and only one enantiomer usually fits.

Solutions to Problems

9.1 Objects having a plane of symmetry are achiral.
Chiral: screw, beanstalk, shoe.
Achiral: screwdriver.

9.2 **Strategy:** Use the following rules to locate centers that are *not* chirality centers.
1. All $–CH_3$ and $–CX_3$ carbons are not chirality centers.
2. All $–CH_2$ and $–CX_2–$ carbons are not chirality centers.
3. All $—\overset{|}{C}=\overset{|}{C}—$ and $—C\equiv C—$ carbons are not chirality centers

By rule 3, all aromatic ring carbons are not chirality centers.

Solution:

(a) Coniine

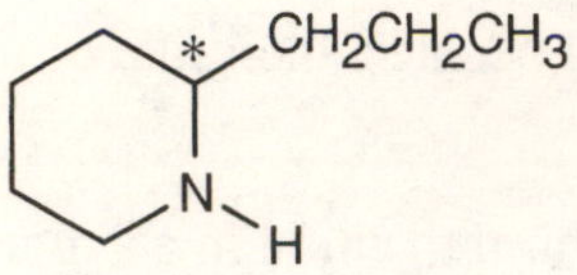

(b) Menthol

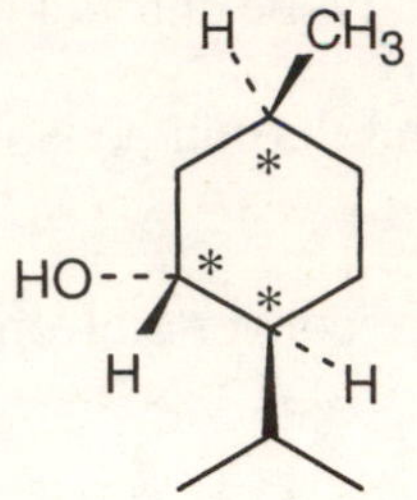

(c) Dextromethorphan

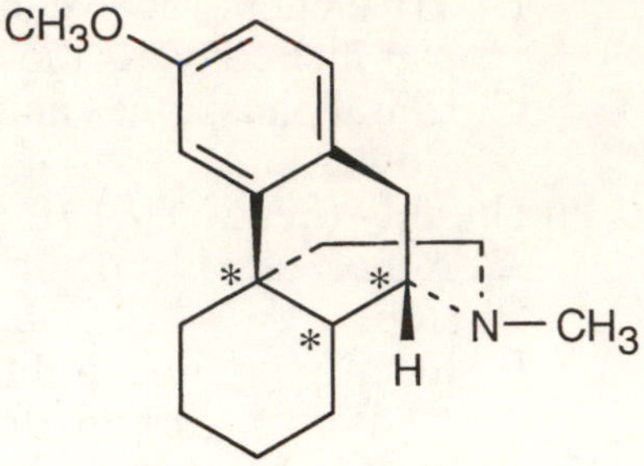

9.3

Alanine

9.4

(a) Threose

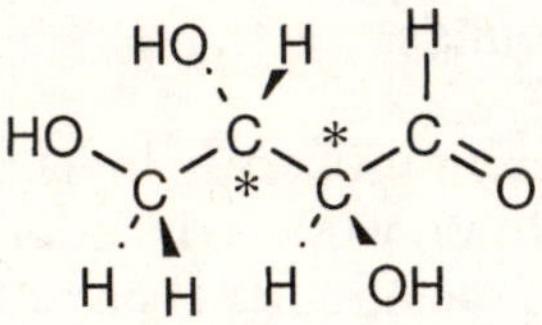

(b) Enflurane

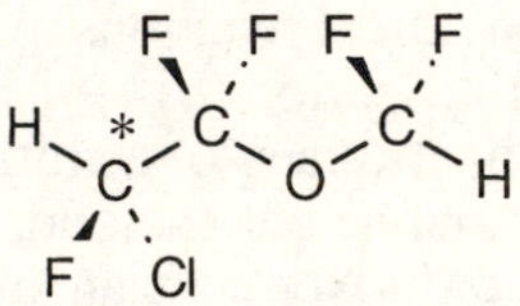

9.5 By convention, a (–) rotation indicates rotation to the left, and thus cocaine is levorotatory.

9.6

Strategy:

Use the formula $[\alpha]_D = \dfrac{\alpha}{l \times C}$, where

$[\alpha]_D$ = specific rotation

α = observed rotation

l = path length of cell (in dm)

C = concentration (in g/mL)

In this problem: $\alpha = 1.21°$

l = 5.00 cm = 0.500 dm

C = 1.50 g/10.0 mL = 0.150 g/mL

Solution: $$[\alpha]_D = \frac{+1.21°}{0.500 \text{ dm} \times 0.150 \text{ g/mL}} = +16.1°$$

9.7 Use the sequence rules in Section 9.5.

(a) By Rule 1, –H is of lowest priority, and –OH is of highest priority. By Rule 2, $-CH_2CH_2OH$ is of higher priority than $-CH_2CH_3$.

Highest ⟶ *Lowest*

$-OH$, $-CH_2CH_2OH$, $-CH_2CH_3$, $-H$

(b) By Rule 3, $-CO_2H$ is considered as $-\overset{\displaystyle -O\ \ O-}{\overset{\backslash\ /}{C}}-OH$. Because 3 oxygens are attached to a $-CO_2H$ carbon and only one oxygen is attached to $-CH_2OH$, $-CO_2H$ is of higher priority than $-CH_2OH$. $-CO_2CH_3$ is of higher priority than $-CO_2H$ by Rule 2, and –OH is of highest priority by Rule 1.

Highest ⟶ *Lowest*

(b) $-OH$, $-CO_2CH_3$. $-CO_2H$, $-CH_2OH$

(c) $-NH_2$, $-CN$, $-CH_2NHCH_3$, $-CH_2NH_2$

(d) $-SSCH_3$, $-SH$, $-CH_2SCH_3$, $-CH_3$

9.8 **Strategy:** All stereochemistry problems are easier if you use models. Part (a) will be solved by two methods – with models and without models.
(a) *With models:* Build a model of (a). Orient the model so that group 4 is pointing to the rear. Note the direction of rotation of arrows that go from group 1 to group 2 to group 3. The arrows point counterclockwise, and the configuration is *S*.

Without models: Imagine yourself looking at the molecule, with the group of lowest priority pointing to the back. Your viewpoint would be at the upper right of the molecule, and you would see group 1 on the left, group 3 on the right and group 2 at the bottom. The arrow of rotation travels counterclockwise, and the configuration is *S*.

Solution:

(a)

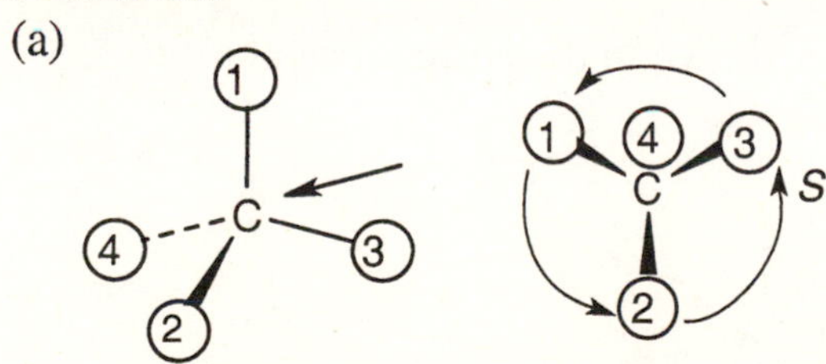

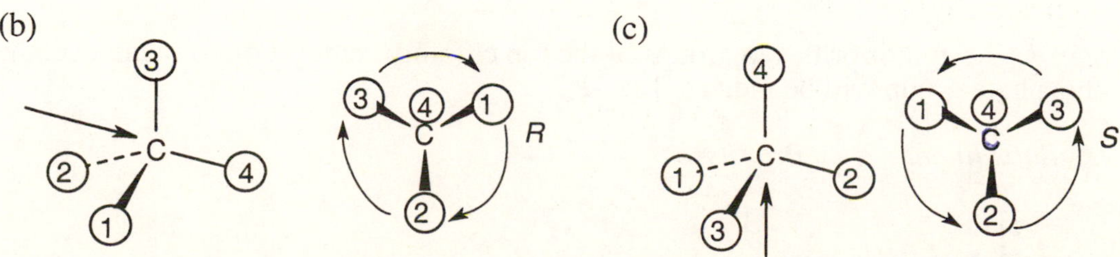

9.9

(a)

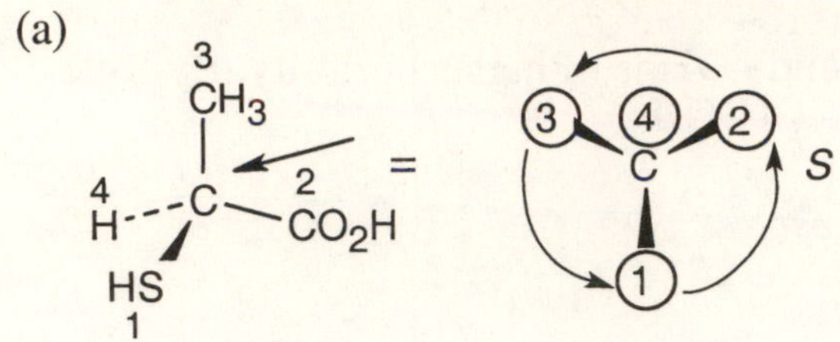

(b)

1 OH; 3 H_3C—C---CO_2H 2; H 4 = (2) (4) (1) C (3) S

(c)

H 2; C=O; 4 H; C; OH 1; CH_2OH 3 = (1) (4) (2) C (3) R

In (b), the observer is behind the page, looking out and down toward the right. In (c), the observer is behind the page looking out and up to the left

9.10

OH

*

$CH_3CH_2CH_2CCH_3$ (*S*)-Pentan-2-ol

H

Substituent	*Priority*
—OH	1
—$CH_2CH_2CH_3$	2
—CH_3	3
—H	4

HO H CH_3 C $CH_2CH_2CH_3$

H S HO---C—$CH_2CH_2CH_3$ CH_3

9.11 Fortunately, the illustration of methionine is shown in the correct orientation.

3 $CH_3SCH_2CH_2$ 4 H 2 CO_2H (*S*)-Methionine; C; NH_2 1

9.12 *Step 1.* Assign priorities to groups of the top chirality center. For convenience, the phosphate group will be indicated as –**P**

Substituent	*Priority*
–OH	1
–CHO	2
–CH(OH)CH_2O**P**	3
–H	4

Step 2. Orient the model so that the lowest priority group of the first chirality center points to the rear. If you aren't using a model, orient yourself so that you are 180° from the lowest priority group. You would be looking out of the page, upward to the left.

Step 3. Note the direction of rotation of the arrow that travels from to group 1 to group 2 to group 3. The rotation is clockwise, and the configuration is *R*.

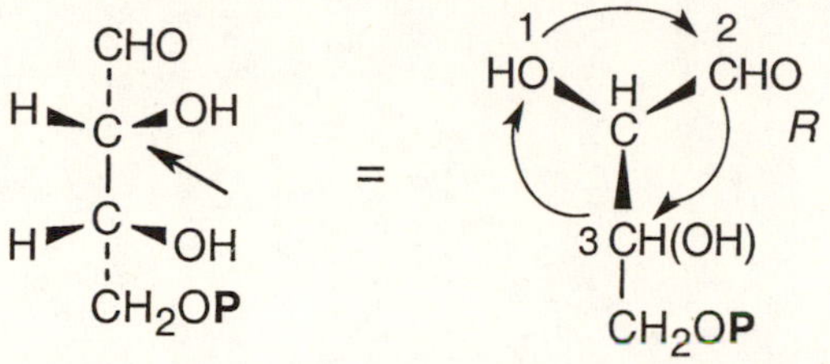

CHO, H, C, OH, C, H, OH, CH_2O**P** = 1 HO, H, C, 2 CHO, *R*, 3 CH(OH), CH_2O**P**

Step 4. Repeat steps 1-3 for the next chirality center.

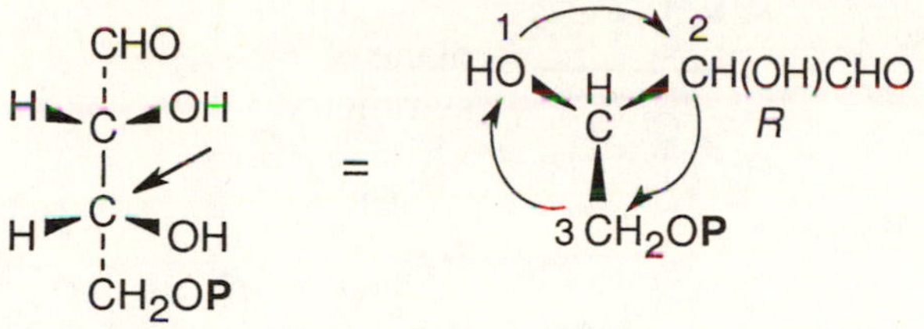

CHO, H, C, OH, C, H, OH, CH_2O**P** = 1 HO, H, C, 2 CH(OH)CHO, *R*, 3 CH_2O**P**

(a) *R,R* (b) *S,R* (c) *R,S* (d) *S,S*
a, d are enantiomers and are diastereomeric with b, c.
b, c are enantiomers and are diastereomeric with a, d.
Structure (a) is D-erythrose, structure (d) is its enantiomer, and structures (b) and (c) are its diastereomers.

9.13

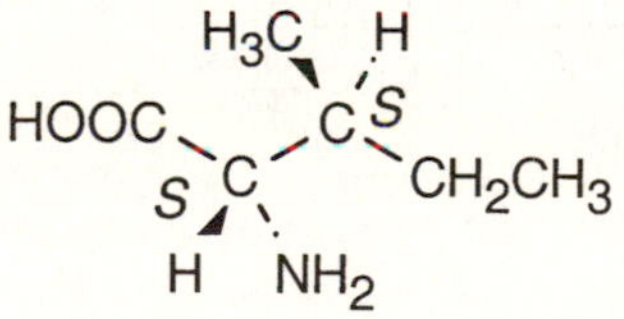

H_3C H, HOOC, C *S*, C, CH_2CH_3, *S*, H NH_2 Isoleucine

9.14

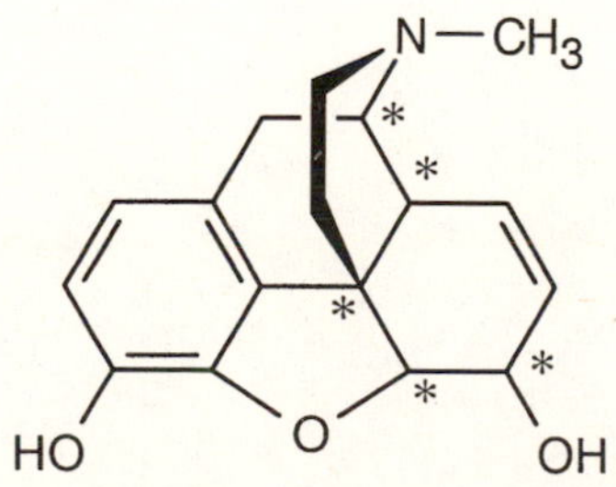

N—CH_3, HO, O, OH Morphine

Morphine has five chirality centers and, in principle, can have $2^5 = 32$ stereoisomers. Most of these stereoisomers are too strained to exist.

9.15 Strategy: To decide if a structure represents a meso compound, try to locate a plane of symmetry that divides the molecule into two halves that are mirror images. Molecular models are always helpful.

Solution.

(a)

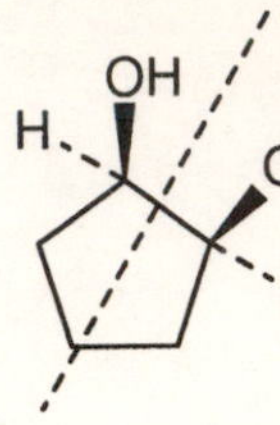

plane of symmetry

meso

(b) and (c) are not meso structures.

(d)

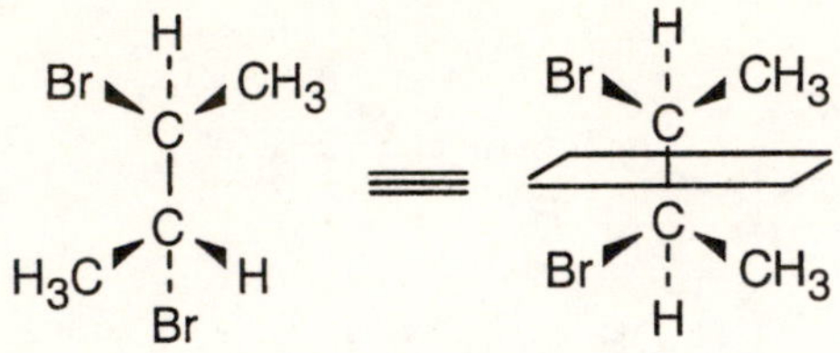

plane of symmetry

meso

9.16 Strategy: For a molecule to exist as a meso form, it must possess a plane of symmetry. Butane-2,3-diol can exist as a pair of enantiomers *or* as a meso compound, depending on the configurations at carbons 2 and 3.

Solution:

(a)

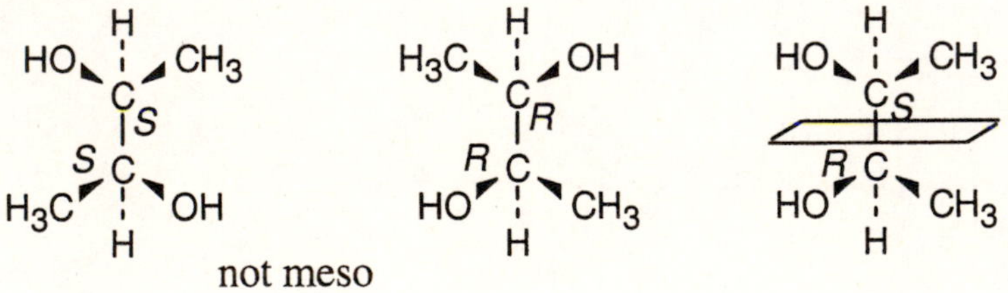

not meso

plane of symmetry

meso

(b) Pentane-2,3-diol has no symmetry plane and thus can't exist in a meso form.

(c) Pentane-2,4-diol can exist in a meso form.

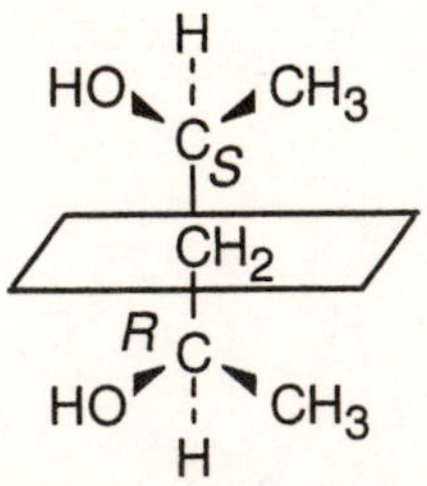

plane of symmetry

Pentane-2,4-diol can also exist as a pair of enantiomers (2*R*,4*R*) and (2*S*,4*S*) that are not meso compounds.

9.17 The molecule represents a meso compound. The symmetry plane passes through the carbon bearing the –OH group and between the two ring carbons that are bonded to methyl groups.

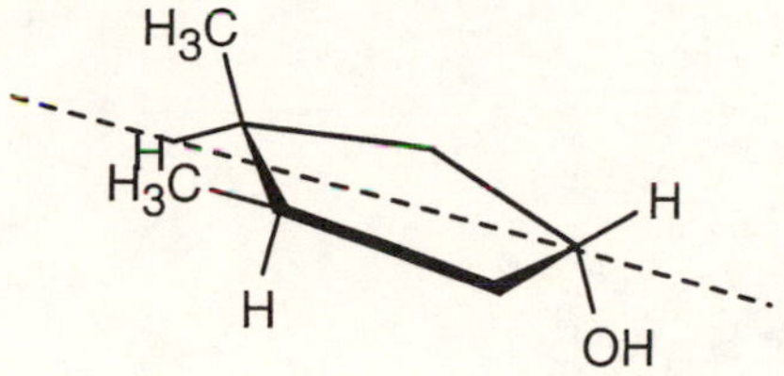

plane of symmetry

9.18

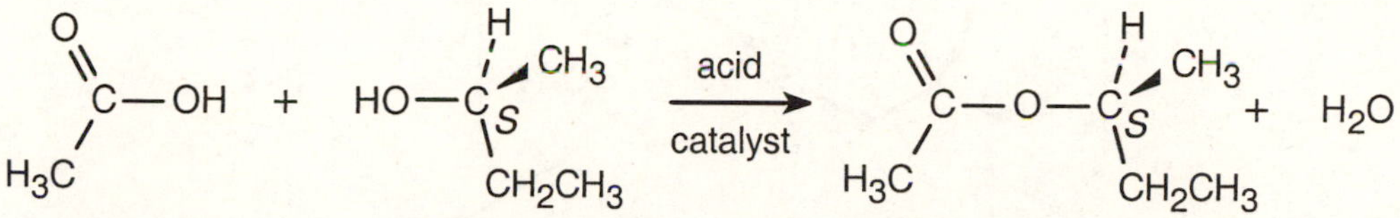

Acetic acid + (*S*)-Butan-2-ol → (acid catalyst) *sec*-Butyl acetate + H_2O

The product is the pure *S*-ester. No new chirality centers are formed during the reaction, and the configuration at the chirality center of (*S*)-butan-2-ol is unchanged.

9.19

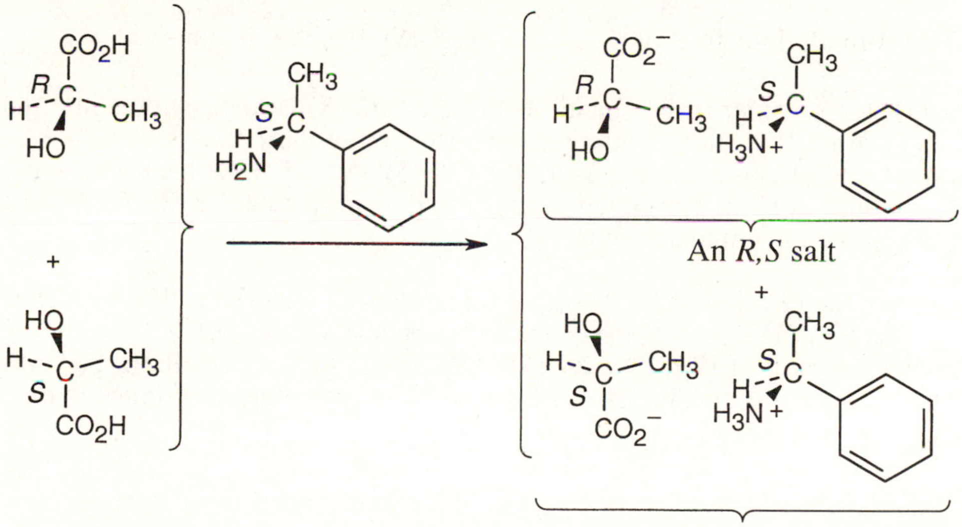

An *R,S* salt

+

An *S,S* salt

The two product salts have the configurations (*R,S*) and (*S,S*) and are diastereomers.

9.20 (a)

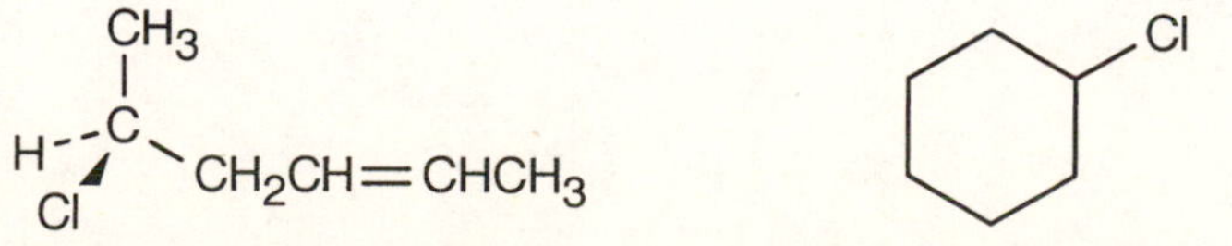

(*S*)-5-Chlorohex-2-ene   Chlorocyclohexane

These two compounds are constitutional isomers.

(b) The two dibromopentane stereoisomers are diastereomers.

9.21 Look back to Figure 9.16, which shows the reaction of (*R*)-4-methylhex-1-ene with H_3O^+. In a similar way, we can write a reaction mechanism for the reaction of H_3O^+ with (*S*)-4-methylhex-1-ene.

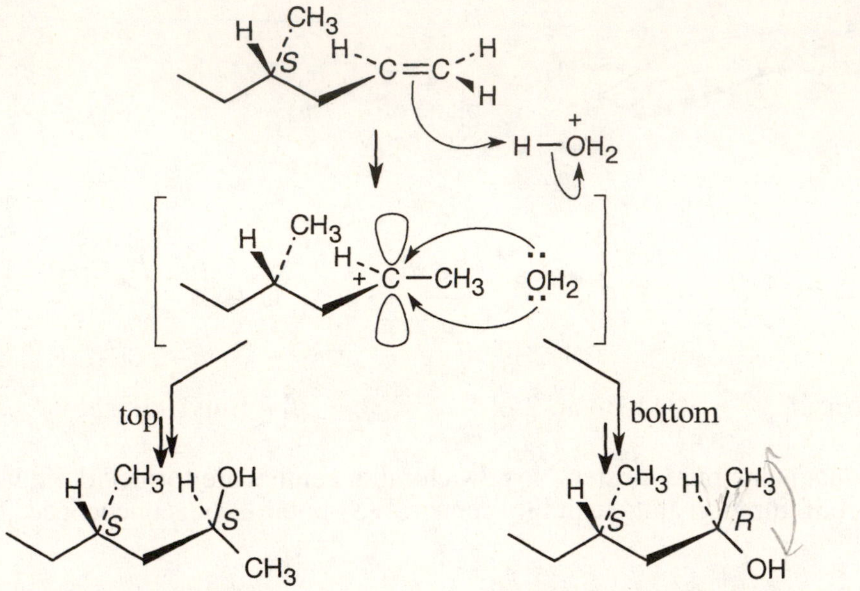

top

bottom

(2*S*,4*S*)-4-methylhexan-2-ol

(2*R*,4*S*)-4-methylhexan-2-ol

The (2*S*,4*S*) stereoisomer is the enantiomer of the (2*R*,4*R*) isomer (shown in Figure 9.16), and the transition states leading to the formation of these two isomers are enantiomeric and of equal energy. Thus, the (2*S*,4*S*) and (2*R*,4*R*) enantiomers are formed in equal amounts. A similar argument can be used to show that the (2*R*,4*S*)and (2*S*,4*R*) isomers are formed in equal amounts.

The product of the reaction of H_3O^+ with racemic starting material is thus a racemic mixture of the four possible stereoisomers and is optically inactive.

Note that the ratio of (2*R*,4*R* + 2*S*,4*S*):(2*R*,4*S* + 2*S*,4*R*) is not 50:50. Nevertheless, the product mixture is optically inactive because the enantiomers are formed in equal amounts.

9.22

Two enantiomeric carbocations are formed. Each carbocation can react with H_2O from either the top or the bottom to yield a total of four stereoisomers. The same argument used in Problem 9.21 can be used to show that the (1*S*,3*R*) and (1*R*,3*S*) enantiomers are formed in equal amounts, and the (1*S*,3*S*) and (1*R*,3*R*) isomers are formed in equal amounts. The result is a non-50:50 mixture of two racemic pairs.

9.23 **Strategy:** For each molecule, replace the left hydrogen with ^{2}H. Give priorities to the groups and assign *R,S* configuration to the chirality center. If the configuration is *R*, the replaced hydrogen is *pro-R*, and if the configuration is *S*, the replaced hydrogen is *Pro-S*.

Solution:

(a)

(*S*)-Glyceraldehyde

(b)

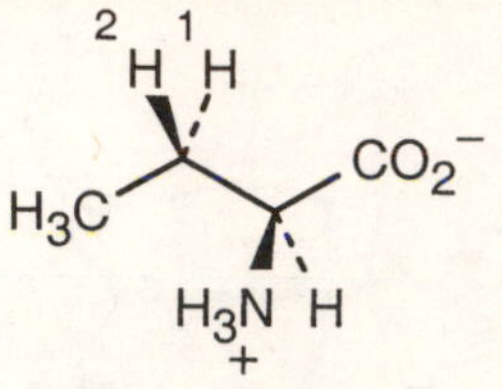

(S)-Alanine

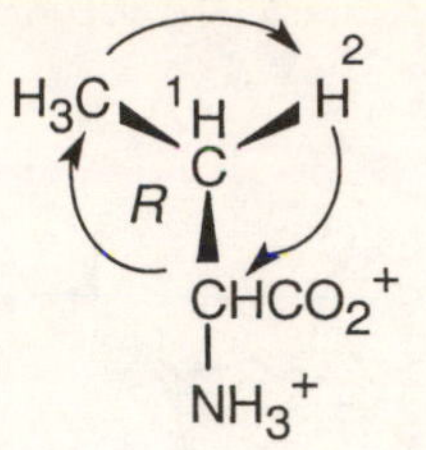

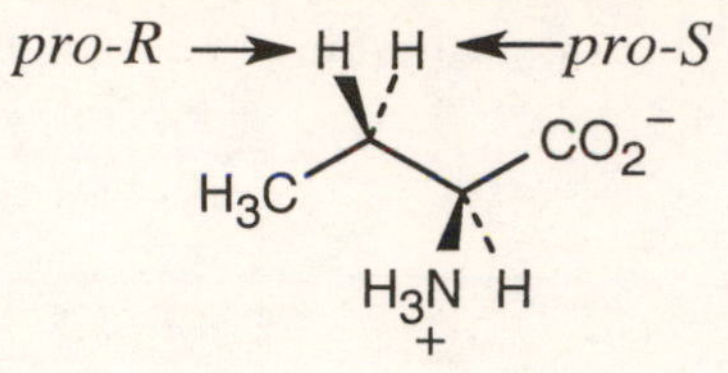

9.24 **Strategy:** Draw the plane that includes the sp^2 carbon and its substituents, and rank the substituents. For the upper face, draw the arrow that proceeds from group 1 to group 2 to group 3. If the direction of rotation is clockwise, the face is the *Re* face; if rotation is counterclockwise, the face is the *Si* face.

Solution:

(a) *Re* face

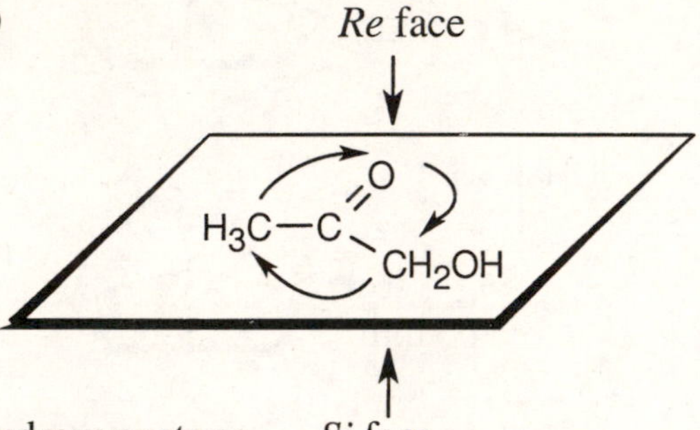

Hydroxyacetone *Si* face

(b) *Re* face

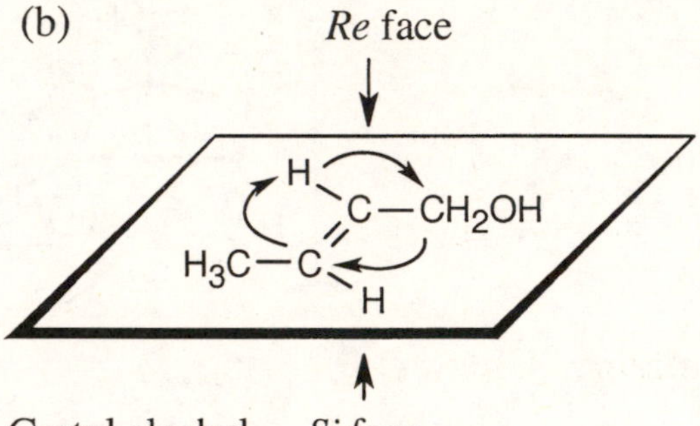

Crotyl alcohol *Si* face

9.25 **Strategy:** Use the strategy in the previous problem to identify the faces of the plane that contains the sp^2 carbon. Draw the product that results from reaction at the *Re* face, and assign configuration to the chirality center.

Solution:

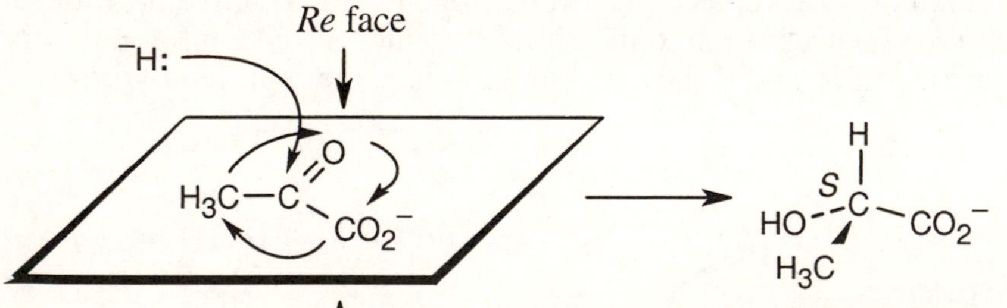

Re face

Si face

(S)-Lactate

9.26 Addition of –OH takes place on the *Re* face of aconitate. Addition of H also occurs on the *Re* face and yields (2*R*,3*S*)-isocitrate.

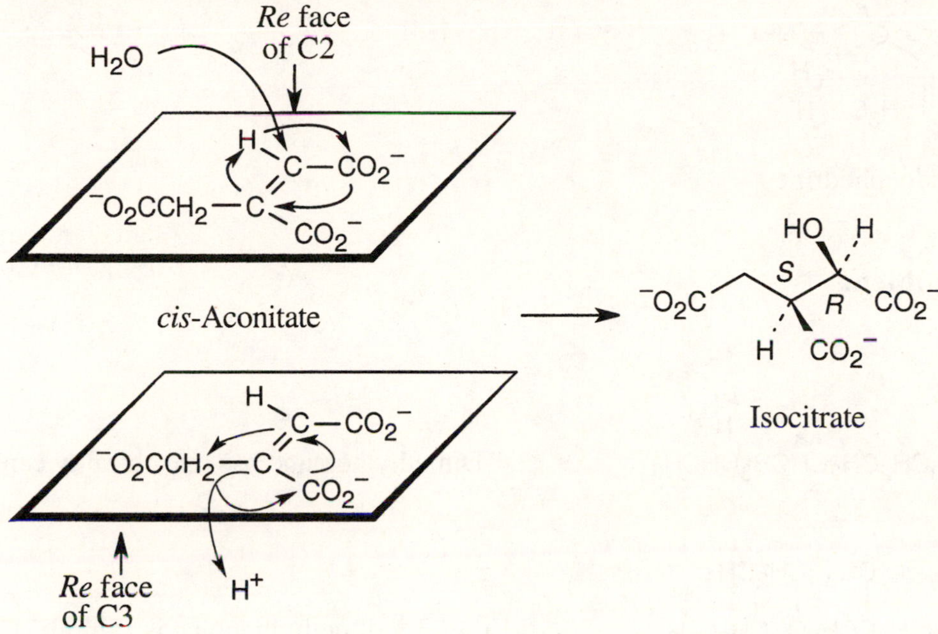

Visualizing Chemistry

9.27 Structures (a), (b), and (d) are identical (*R* enantiomer), and (c) represents the *S* enantiomer.

9.28

(a)

(*S*)-Serine

S

(b)

(*R*)-Adrenaline

R

R = ring

9.29

(a)

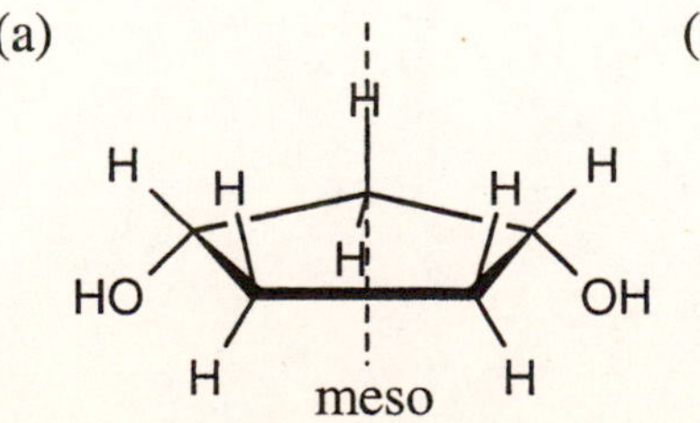

(b)

meso

(c)

9.30

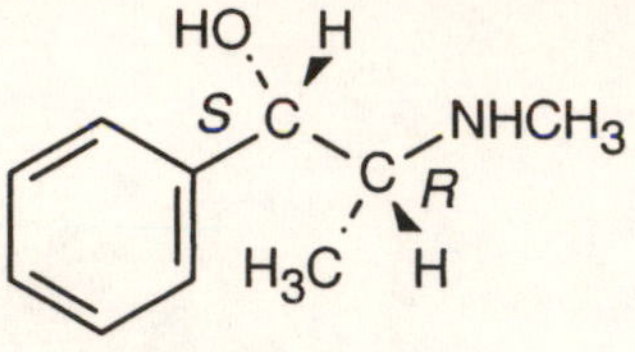

$$\text{Pseudoephedrine: Ph–C(}S\text{)(OH)(H)–C(}R\text{)(NHCH}_3\text{)(CH}_3\text{)(H)}$$

Pseudoephedrine

Additional Problems

9.31

(a)

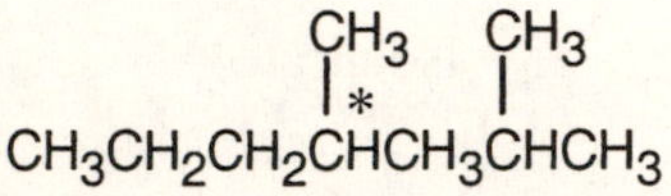

$$\underset{}{CH_3CH_2CH_2\overset{*}{\underset{|}{C}}H(CH_3)CH_2CH(CH_3)CH_3}$$

$CH_3CH_2CH_2\overset{*}{C}H(CH_3)CH_2CH(CH_3)CH_3$ 2,4-Dimethylheptane has one chirality center.

(b)

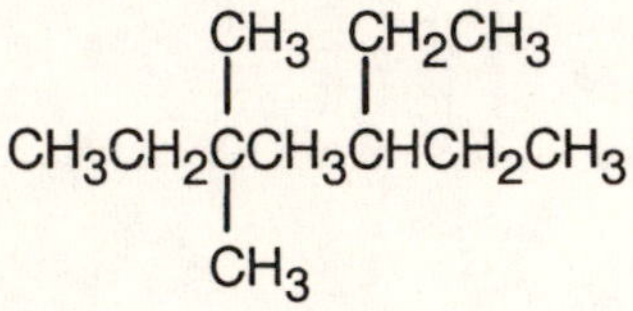

$CH_3CH_2C(CH_3)_2CH_2CH(CH_2CH_3)CH_2CH_3$ 5-Ethyl-3,3-dimethylheptane is achiral.

(c)

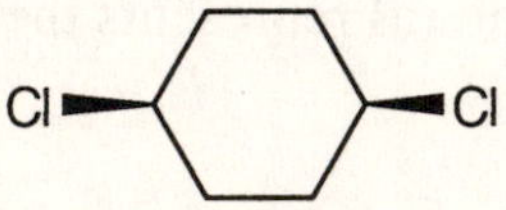

Cl–(cyclohexane)–Cl *cis*-1,4-Dichlorocyclohexane is achiral. Notice the plane of symmetry that passes through the –Cl groups.

(d) $CH_3C{\equiv}C\overset{*}{C}H(CH_3)\overset{*}{C}H(CH_3)C{\equiv}CCH_3$ 4,5-Dimethylocta-2,6-diyne has two chirality centers.

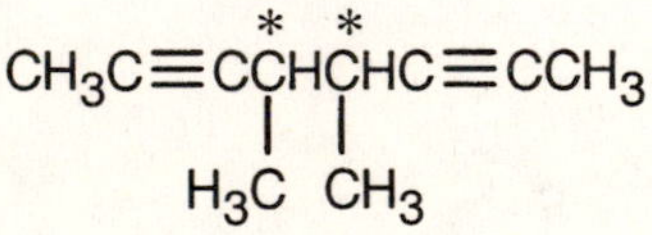

The chirality of (d) depends on the configuration at both of the chirality centers. The (*R,R*) and (*S,S*) isomers are chiral enantiomers; the (*R,S*) isomer is an achiral meso compound.

9.32

(a) $CH_3CH_2CH_2\overset{*}{C}H(Cl)CH_3$
2-Chloropentane

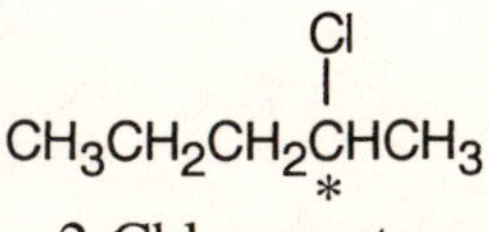

(b) $CH_3CH_2CH_2CH_2\overset{*}{C}H(OH)CH_3$
Hexan-2-ol

(c) $CH_3CH_2\overset{*}{C}H(CH_3)CH{=}CH_2$
3-Methylpent-1-ene

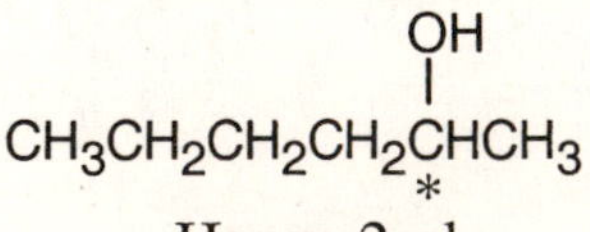

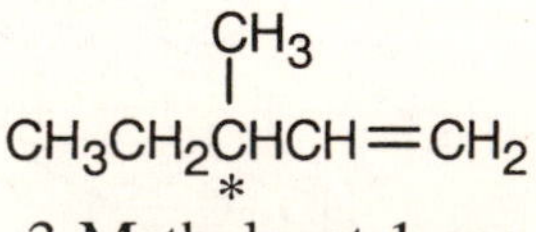

(d) $CH_3CH_2CH_2CH_2\overset{*}{C}H(CH_3)CH_2CH_3$
3-Methylheptane

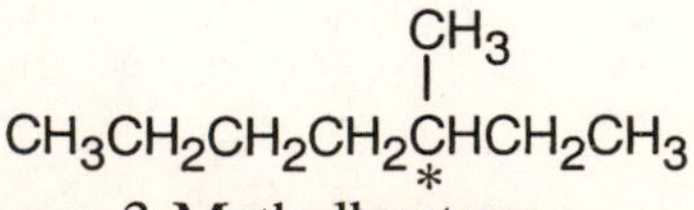

9.33

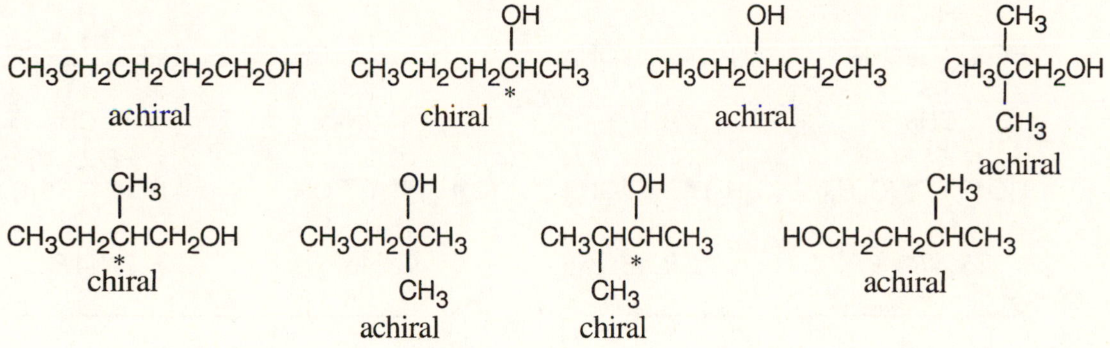

$CH_3CH_2CH_2CH_2CH_2OH$ achiral

$CH_3CH_2CH_2\overset{*}{C}H(OH)CH_3$ chiral

$CH_3CH_2CH(OH)CH_2CH_3$ achiral

$CH_3C(CH_3)_2CH_2OH$ achiral

$CH_3CH_2\overset{*}{C}H(CH_3)CH_2OH$ chiral

$CH_3CH_2C(OH)(CH_3)CH_3$ achiral

$CH_3CH(CH_3)\overset{*}{C}H(OH)CH_3$ chiral

$HOCH_2CH_2CH(CH_3)CH_3$ achiral

9.34

(a) $CH_3CH_2\overset{*}{C}H(OH)CH_3$ (b) $CH_3CH_2\overset{*}{C}H(CH_3)CO_2H$ (c) $CH_3\overset{*}{C}H(Br)\overset{*}{C}H(OH)CH_3$ (d) $CH_3\overset{*}{C}H(OH)CHO$

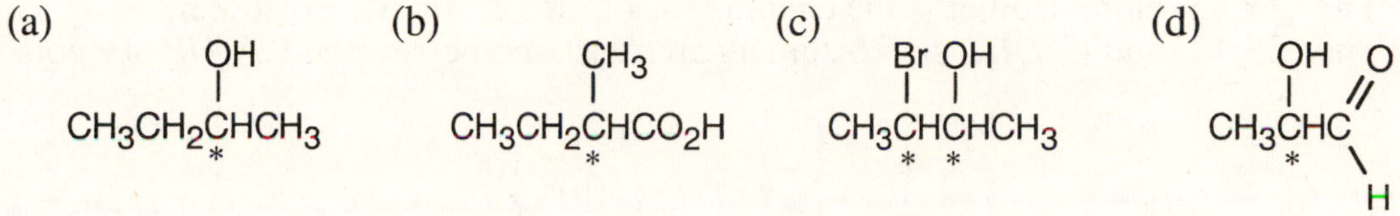

9.35

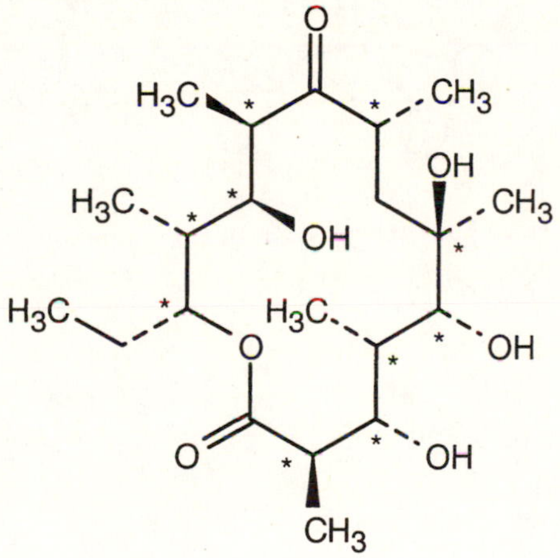

Erythronolide B

Erythronolide B has ten chirality centers.

9.36

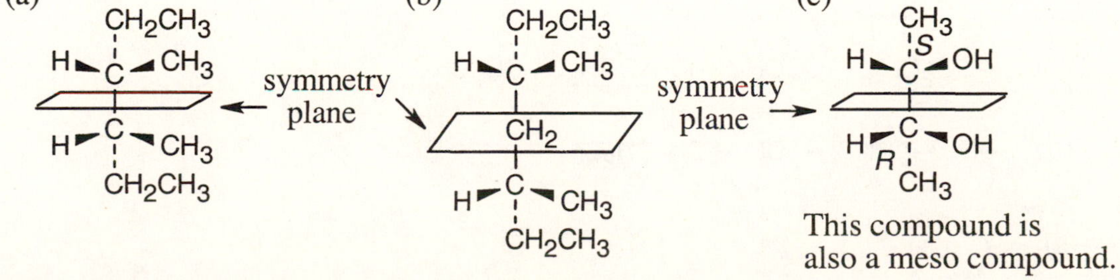

(a) $CH_3CH_2CH(CH_3)CH(CH_3)CH_2CH_3$ — symmetry plane

(b) $CH_3CH_2CH(CH_3)CH_2CH(CH_3)CH_2CH_3$ — symmetry plane

(c) $CH_3CH(OH)CH(OH)CH_3$ (S, R) — symmetry plane

This compound is also a meso compound.

9.37 The specific rotation of (2*R*,3*R*)-dichloropentane is equal in magnitude and opposite in sign to the specific rotation of (2*S*,3*S*)-dichloropentane because the compounds are enantiomers. There is no predictable relationship between the specific rotations of the (2*R*,3*S*) and (2*R*,3*R*) isomers because they are diastereomers.

9.38–9.39

H
HO C CH3
S
CH2
R C
HO (C4H9)
H

H
H3C C OH
R
CH2
S C
(C4H9) OH
H

H
HO C CH3
S
CH2
S C
(C4H9) OH
H

H
H3C C OH
R
CH2
R C
HO (C4H9)
H

2S,4R *2R,4S* *2S,4S* *2R,4R*

enantiomers enantiomers

The (2*R*,4*S*) stereoisomer is the enantiomer of the (2*S*,4*R*) stereoisomer.
The (2*S*,4*S*) and (2*R*,4*R*) stereoisomers are diastereomers of the (2*S*,4*R*) stereoisomer.

9.40

(a)

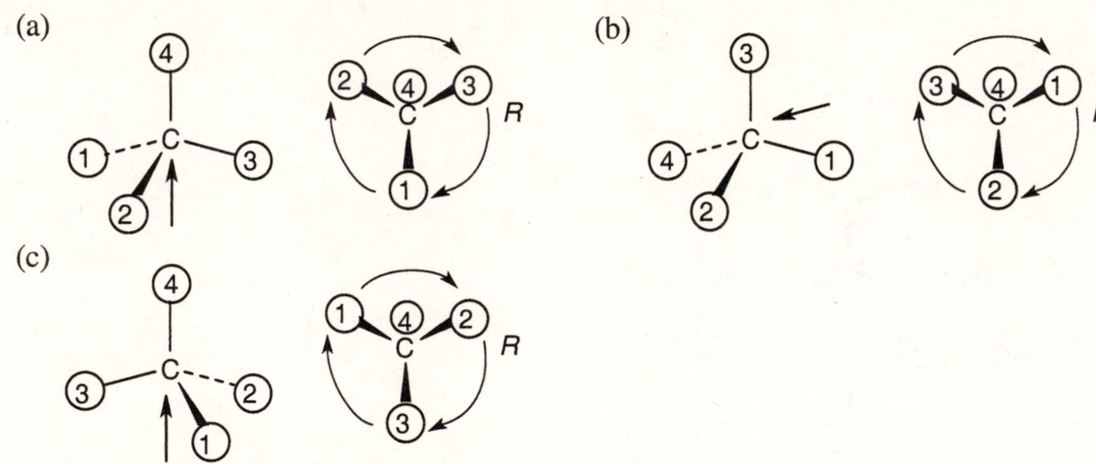

9.41

Highest ⟶ *Lowest*

(a) —$C(CH_3)_3$, —$CH{=}CH_2$, —$CH(CH_3)_2$, —CH_2CH_3

(b) , —C≡CH, —$C(CH_3)_3$, —$CH{=}CH_2$

(c) —CO_2CH_3, —$COCH_3$, —CH_2OCH_3, —CH_2CH_3

(d) —Br, —CH_2Br, —CN, —CH_2CH_2Br

9.42

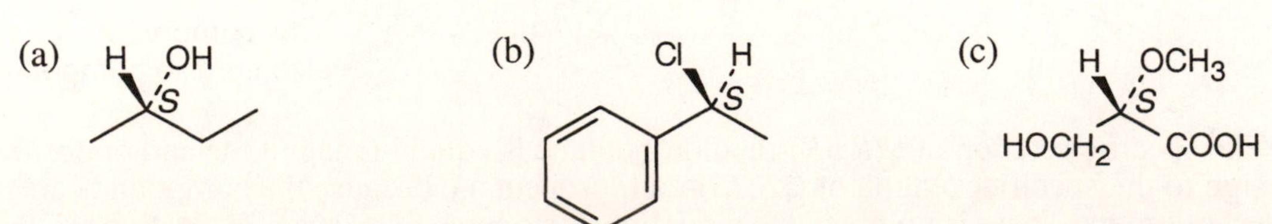

9.43

(a) (b) 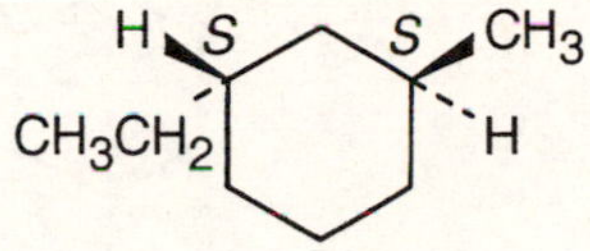 (c)

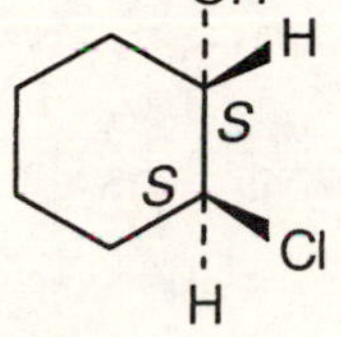

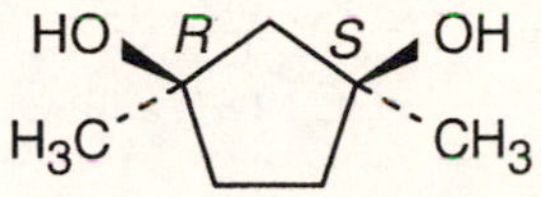

9.44

(a) (b)

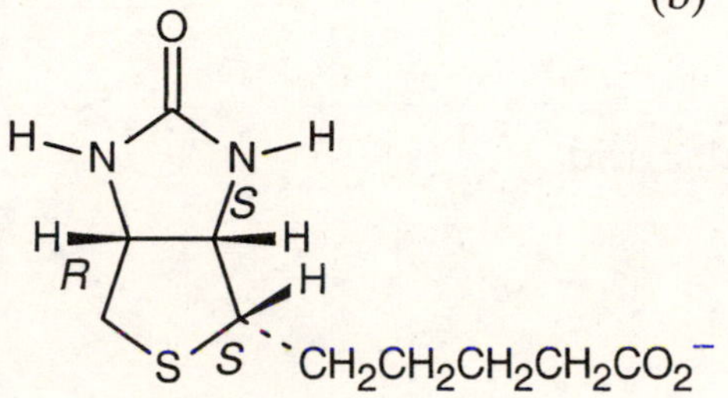

Biotin

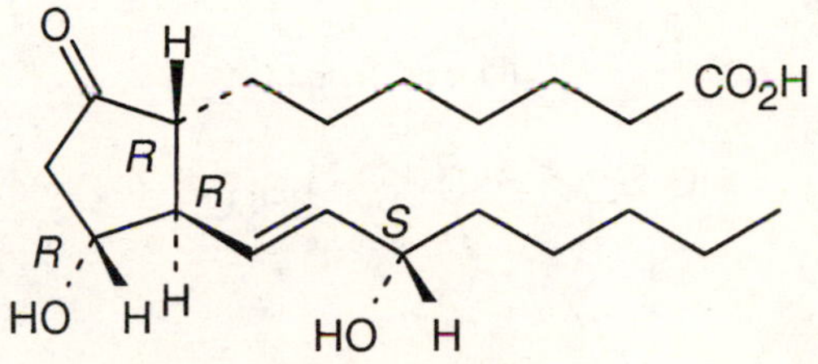

Prostaglandin E_1

9.45

9.46

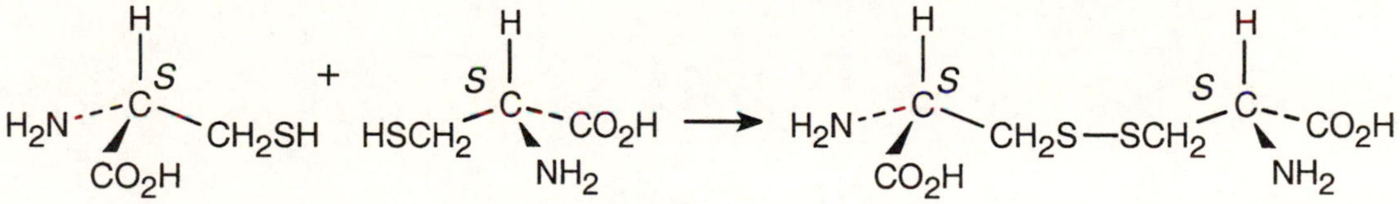

Cystine has the (*S*,*S*) configuration and is optically active.

9.47 Identical molecules: b (*S* enantiomer), c (*R* enantiomer), d (*S* enantiomer).
Pair of enantiomers: a

9.48

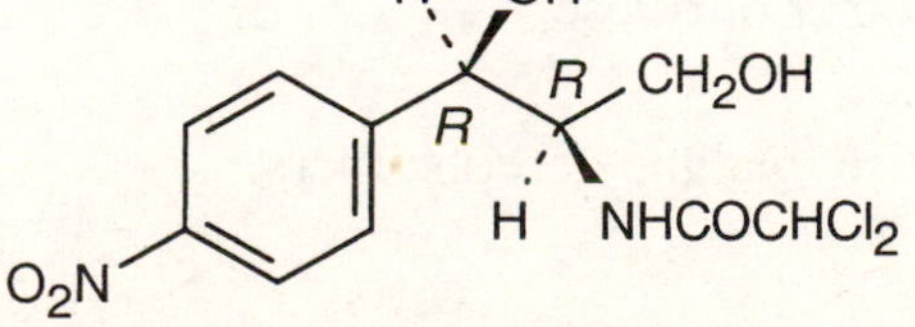

Chloramphenicol

9.49

(a) H OH H OH; H_3C *R* CH_2CH_2 *S* CH_3; plane of symmetry

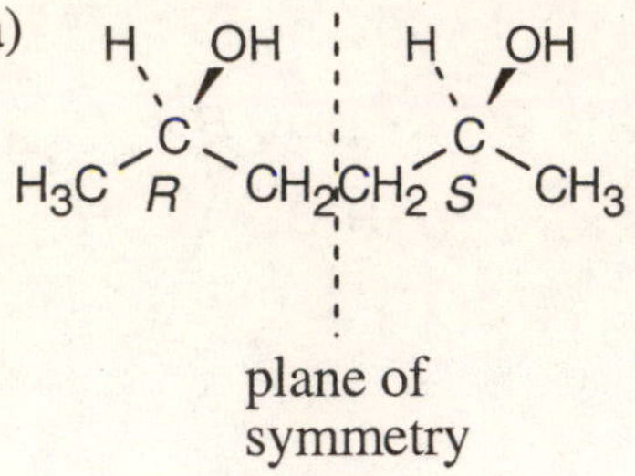

(b) H_3C CH_3; H H; *R* *S*; plane of symmetry

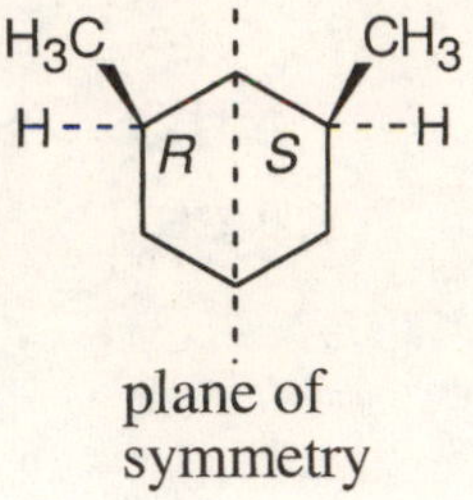

(c) H_3C H *S*; plane of symmetry; OH; H *R* H_3C

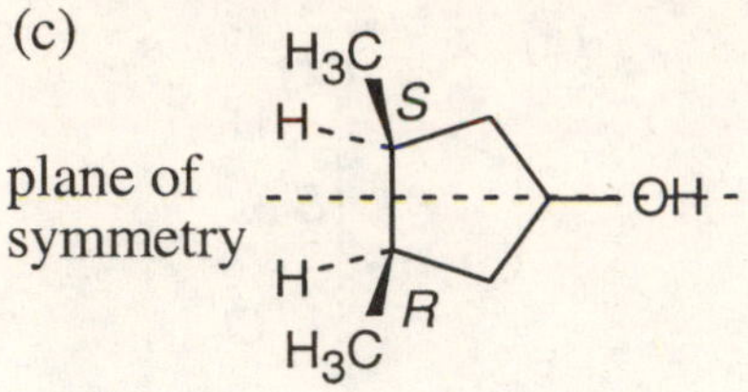

9.50

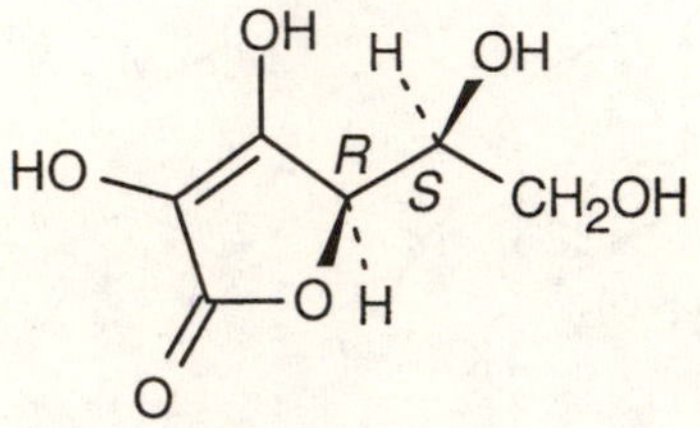

OH H OH; HO; *R* *S* CH_2OH; O H; O

Ascorbic acid

9.51

(a) Cl; H CH_3; *S*; H_3C H; H

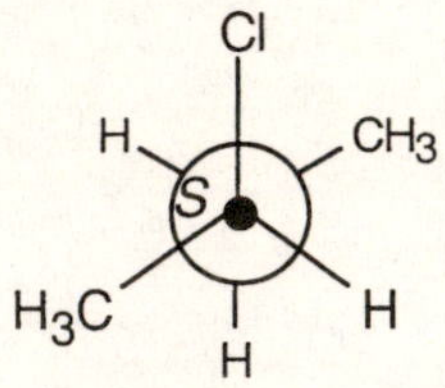

(b) H *R*; H_3C OH; H_3C CH_3; H

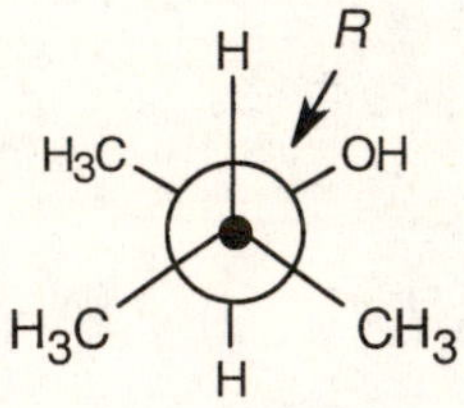

9.52

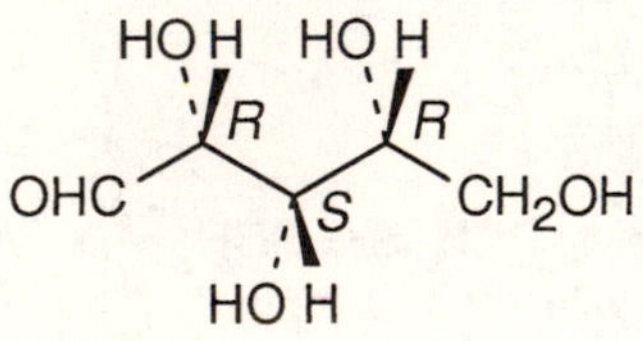

HO H HO H; *R* *R*; OHC *S* CH_2OH; HO H

(+)-Xylose

9.53 (a) - (c)

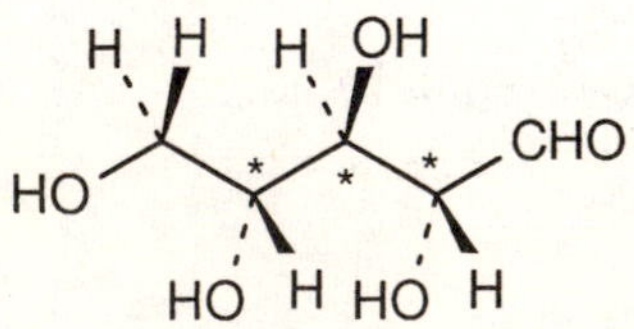

H H H OH; HO CHO; HO H HO H

Ribose

H H HO H; HO CHO; H OH H OH

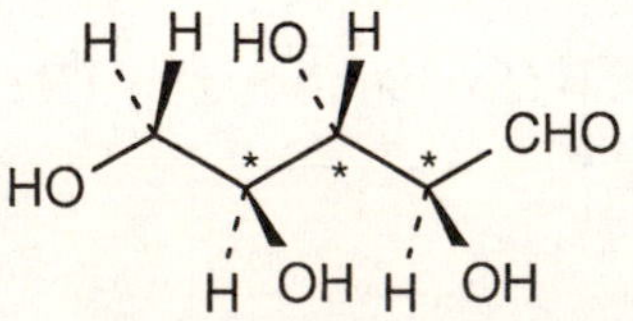

Enantiomer of ribose

Ribose has three chirality centers, which give rise to eight stereoisomers.

(d) Ribose has six diastereomers.

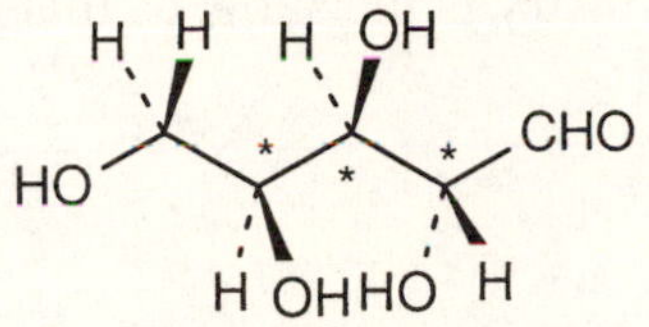

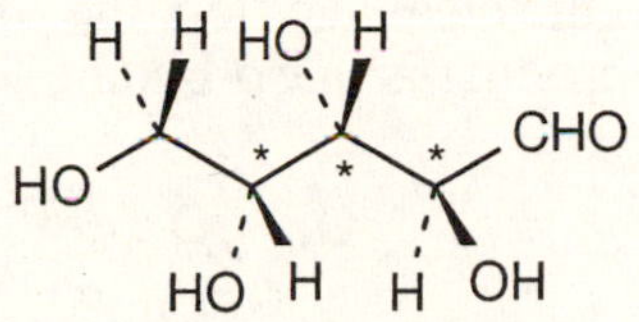

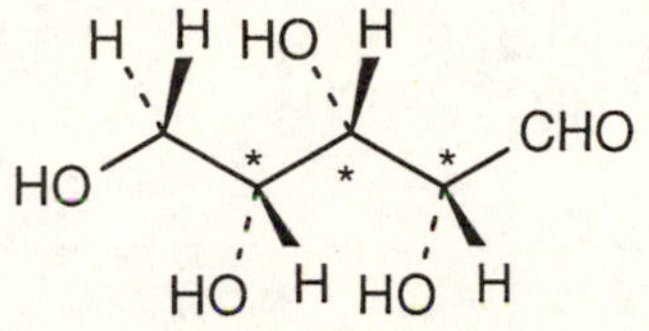

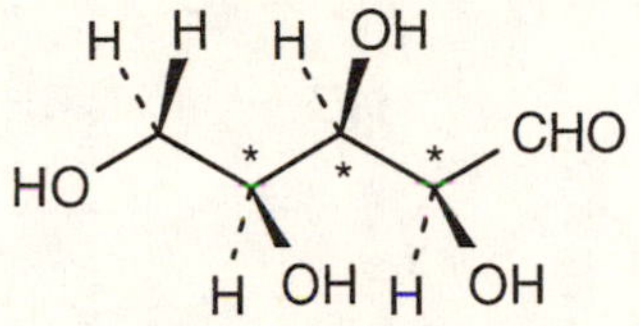

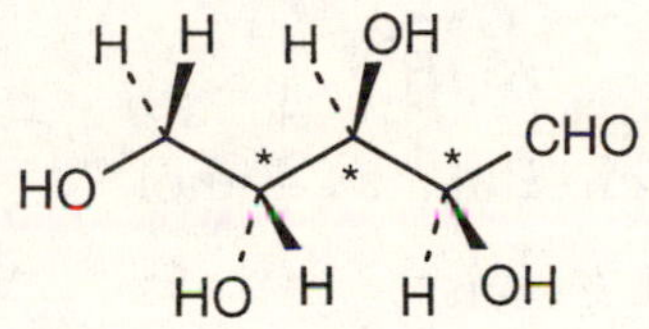

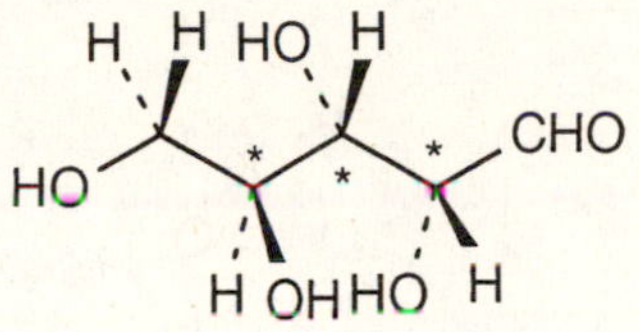

9.54 Ribitol is an optically inactive meso compound. Catalytic hydrogenation converts the aldehyde functional group into a hydroxyl group and makes the two halves of ribitol mirror images of each other.

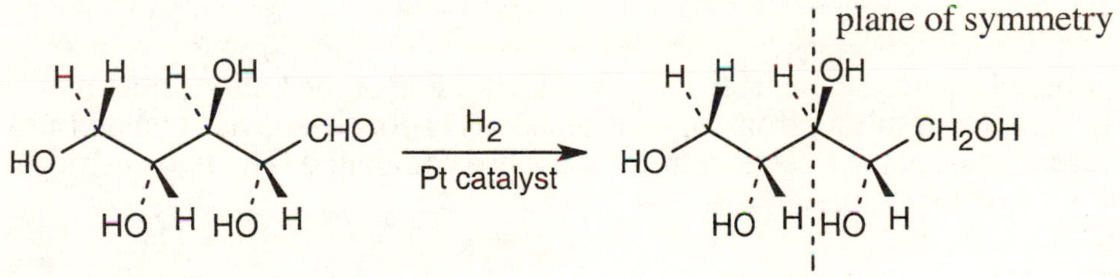

Ribose → (H_2, Pt catalyst) → Ribitol

plane of symmetry

9.55 The initial product of hydroxylation of a double bond is a cyclic *osmate*, which is reduced on treatment with $NaHSO_3$ to cleave the osmium-oxygen bonds. Since no carbon-oxygen bonds are broken in the cleavage step, the final stereochemistry is the same as that of the initial adduct. The product is *meso*-butane-2,3-diol.

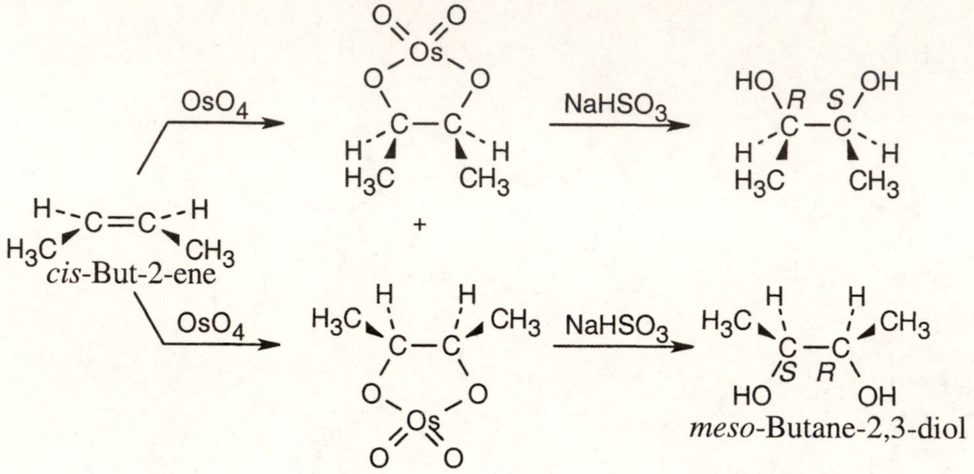

OsO_4 $NaHSO_3$

cis-But-2-ene

meso-Butane-2,3-diol

9.56

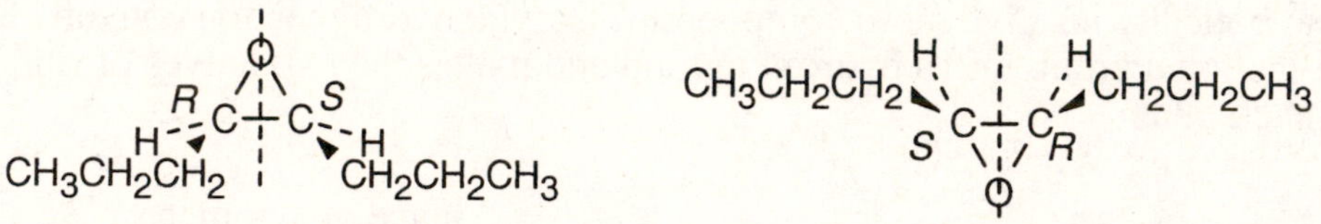

Peroxycarboxylic acids can react from either the top side or the bottom side of a double bond. The epoxide resulting from top-side attack on *cis*-oct-4-ene has two chirality centers, but because it has a plane of symmetry, it is a meso compound. The two epoxides are identical.

9.57

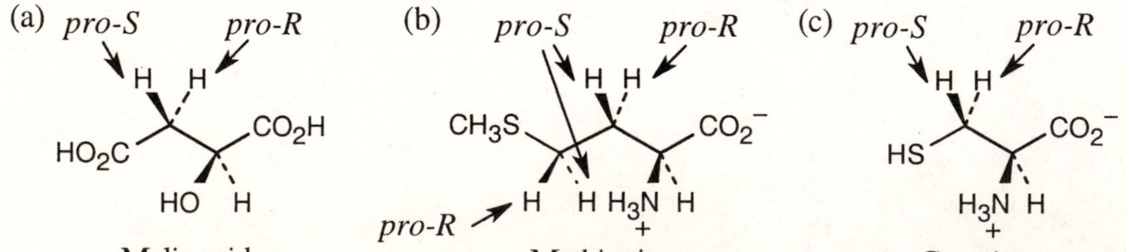

(a) *pro-S* *pro-R*

Malic acid

(b) *pro-S* *pro-R* *pro-R*

Methionine

(c) *pro-S* *pro-R*

Cysteine

9.58

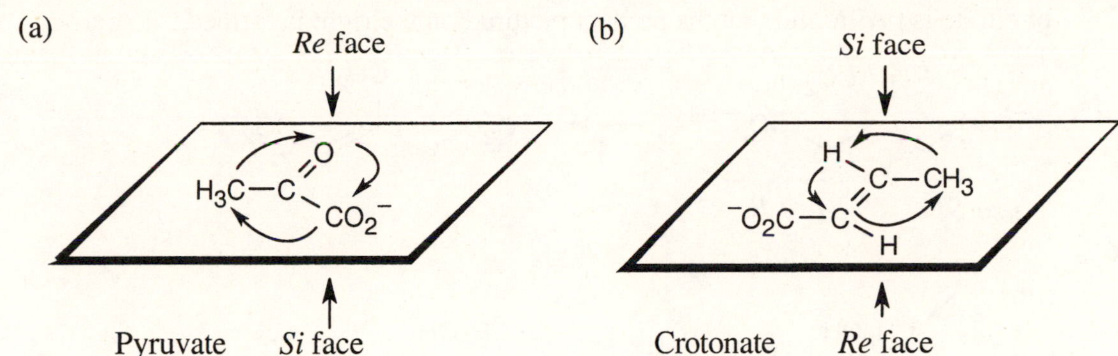

9.59

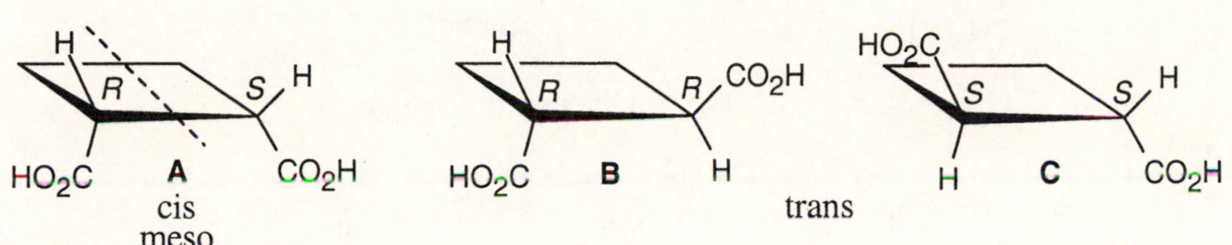

B and **C** are enantiomers and are optically active. Compound **A** is their diastereomer and is a *meso* compound, which is not optically active.

The two isomeric cyclobutane-1,3-dicarboxylic acids are achiral and are optically inactive.

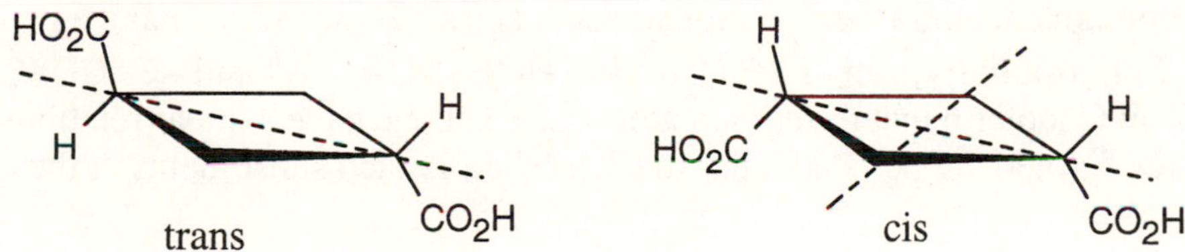

9.60 Remember that *each* sp^2 carbon has a *Re* face and a *Si* face.

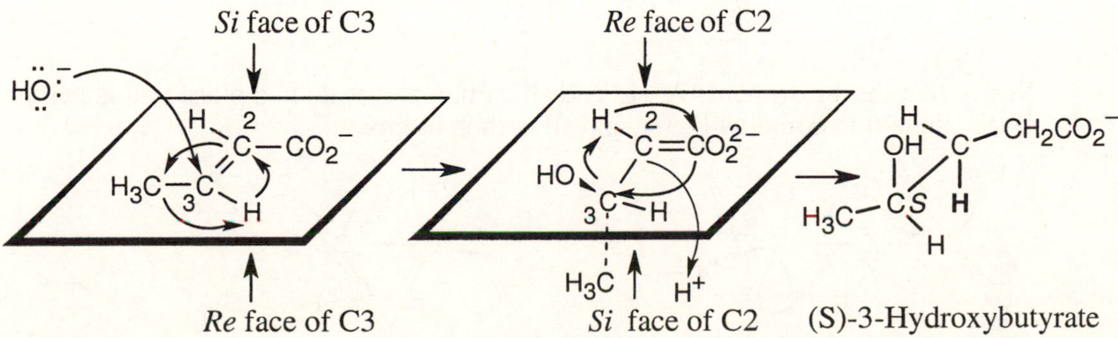

9.61 Performing the replacement test to assign *pro-R/pro-S* prochirality shows that the right arm of citrate is *pro-R* and that the product pictured on the right is formed.

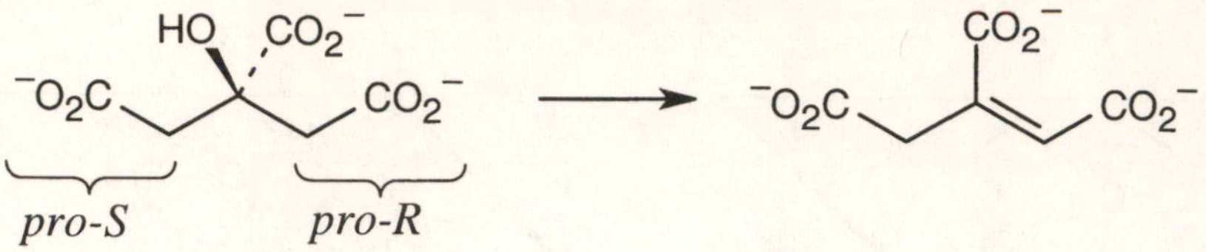

pro-S *pro-R*

9.62

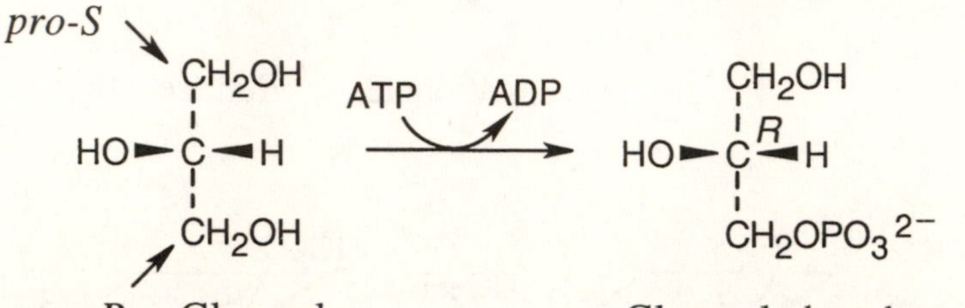

pro-S

pro-R Glycerol

Glycerol phosphate

9.63

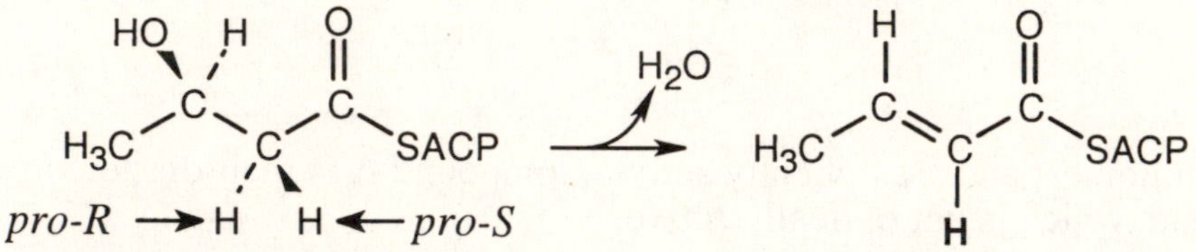

pro-R *pro-S*

(*R*)-3-Hydroxybutyryl ACP

trans-Crotonyl ACP

The reaction removes the *pro-R* hydrogen.

9.64 Mycomycin contains no chiral carbon atoms, yet is chiral. To see why, make a model of mycomycin. For simplicity, call –CH=CHCH=CHCH$_2$CO$_2$H "A" and –C≡CC≡CH "B". Remember from Chapter 6 that the carbon atoms of an allene have a linear relationship and that the π bonds formed are perpendicular to each other. Attach substituents at the sp^2 carbons.

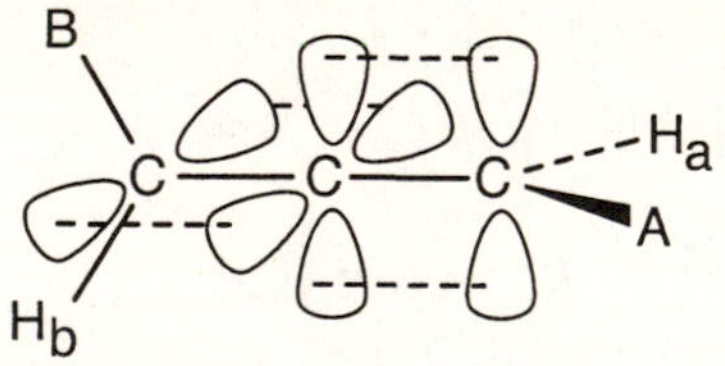

Notice that the substituents A, H$_a$, and all carbon atoms lie in a plane that is perpendicular to the plane that contains B, H$_b$, and all carbon atoms.

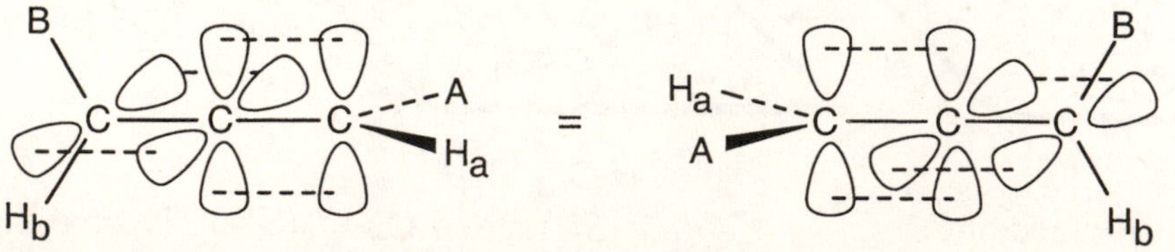

Now, make another model identical to the first, except for an exchange of A and H$_a$. This new allene is not superimposable on the original allene. The two allenes are enantiomers and are chiral because they possess no plane of symmetry.

9.65 4-Methylcyclohexylideneacetic acid is chiral for the same reason that mycomycin (Problem 9.64) is chiral: It possesses no plane of symmetry and is not superimposable on its mirror image. As in the case of allenes, the two groups at one end of the molecule lie in a plane perpendicular to the plane that contains the two groups at the other end.

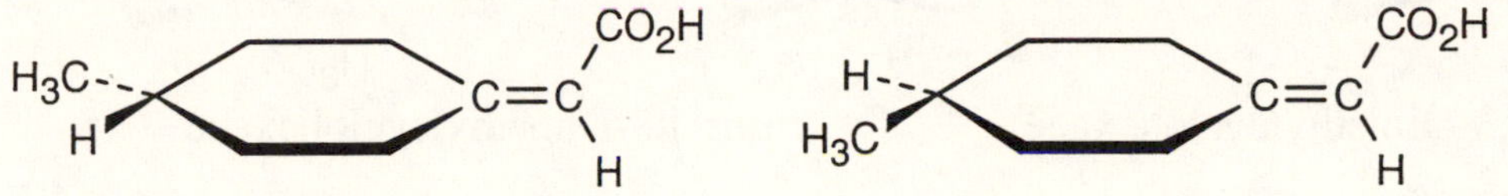

H_3C, H, C=C, CO_2H, H — H, H_3C, C=C, CO_2H, H

9.66

(a)

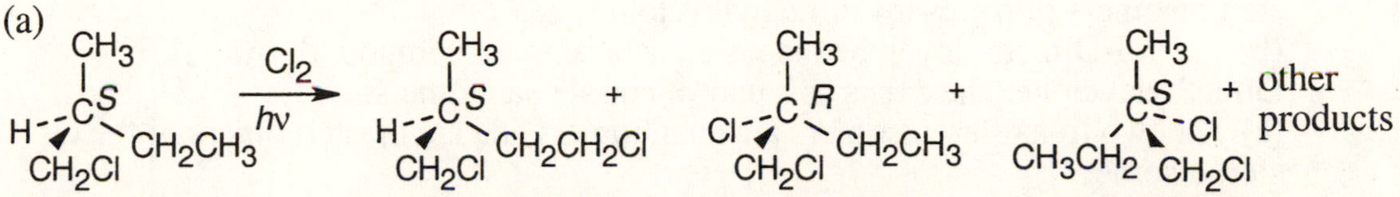

(S)-1-Chloro-2-methylbutane $\xrightarrow[h\nu]{Cl_2}$ (S)-1,4-Dichloro-2-methylbutane + (R)-1,2-Dichloro-2-methylbutane + (S)-1,2-Dichloro-2-methylbutane + other products

(R)-1,2-Dichloro-2-methylbutane and (S)-1,2-Dichloro-2-methylbutane: 1:1 mixture

(b) Chlorination at carbon 4 yields an optically active product because the chirality center at C2 is not affected. Chlorination at carbon 2 yields an optically inactive racemic product.

(c) Radical chlorination reactions taking place at a chirality center occur with racemization; radical chlorination reactions at a site other than the chirality center do not affect the stereochemistry of the chirality center.

9.67 Both of the diastereomers shown below are meso compounds with three chirality centers. In each structure the central carbon is bonded to four different groups (a group with *R* configuration, a group with *S* configuration, –OH, and –H).

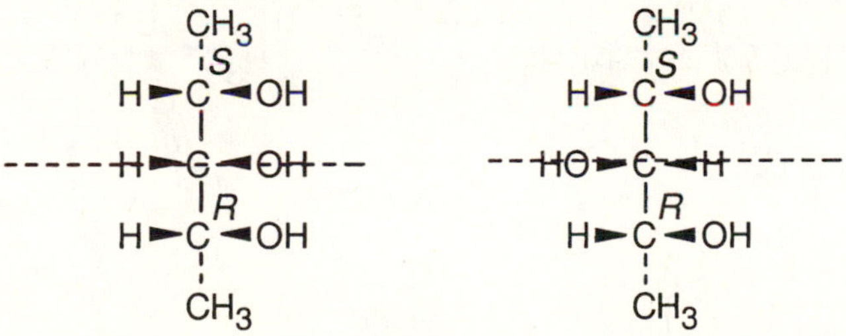

9.68

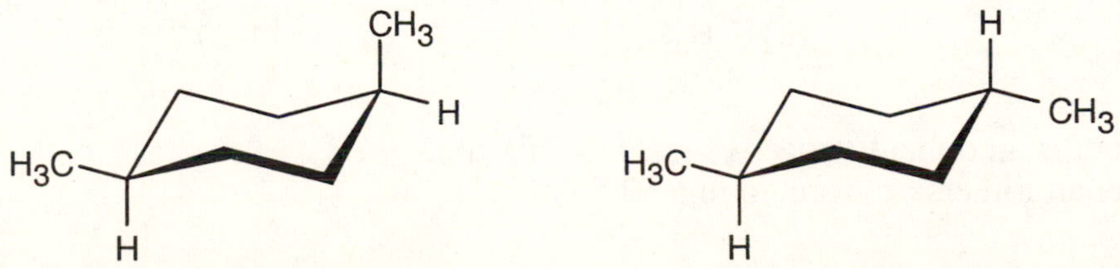

cis-1,4-Dimethylcyclohexane — *trans*-1,4-Dimethylcyclohexane

(a) There is only one stereoisomer of each of the 1,4-dimethylcyclohexanes.
(b) Neither 1,4-dimethylcyclohexane is chiral.
(c) The two 1,4-dimethylcyclohexanes are diastereomers.

9.69

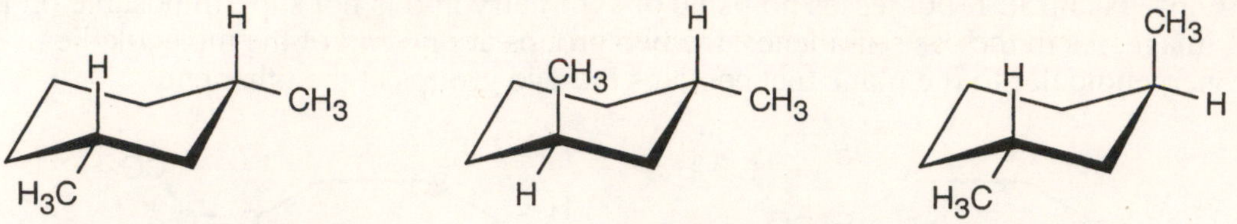

H H CH$_3$
H CH$_3$ CH$_3$ CH$_3$ H H
H$_3$C H H$_3$C

cis-1,3-Dimethylcyclohexane *trans*-1,3-Dimethylcyclohexane

(a) There is one stereoisomer of *cis*-1,3-dimethylcyclohexane, and there are two stereoisomers of *trans*-1,3-dimethylcyclohexane.
(b) *cis*-1,3-Dimethylcyclohexane is an achiral meso compound. *trans*-1,3-Dimethylcyclohexane exists as a pair of chiral enantiomers.
(c) The two trans stereoisomers are enantiomers and are diastereomers of the cis stereoisomer.

9.70

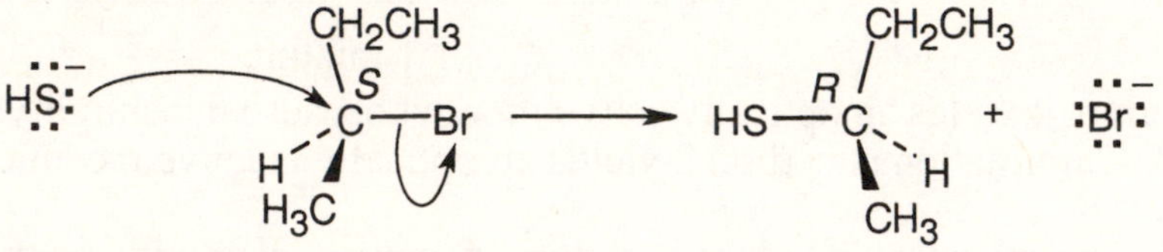

HS^- + (S)-$CH_3CH_2CH(CH_3)Br$ ⟶ (R)-$HS-CH(CH_3)CH_2CH_3$ + Br^-

The product is (*R*)-butane-2-thiol.

9.71 The reaction proceeds by addition of acetylide anion to the carbonyl group and occurs with equal probability from either side of the planar ketone carbon.

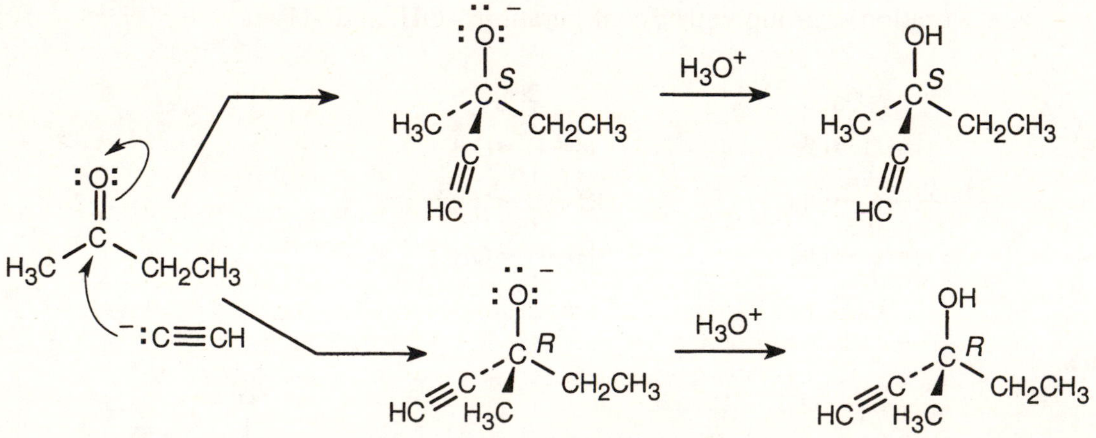

$H_3C-CO-CH_2CH_3$ + $^-$:C≡CH ⟶ (*S*) alkoxide $\xrightarrow{H_3O^+}$ (*S*) alcohol

⟶ (*R*) alkoxide $\xrightarrow{H_3O^+}$ (*R*) alcohol

(a) The product is an optically inactive racemic mixture.
(b) The two enantiomers are formed in a 50:50 ratio.

Chapter 10 – Alkyl Halides: Nucleophilic Substitutions and Eliminations

Chapter Outline

I. Alkyl halides (Sections 10.1 –10.3).
  A. Naming alkyl halides (Section 10.1).
    1. Rules for naming alkyl halides:
      a. Find the longest chain and name it as the parent.
         If a double or triple bond is present, the parent chain must contain it.
      b. Number the carbon atoms of the parent chain, beginning at the end nearer the first substituent, whether alkyl or halo.
      c. Number each substituent.
        i. If more than one of the same kind of substituent is present, number each, and use the prefixes *di-*, *tri-*, *tetra-* and so on.
        ii. If different halogens are present, number all and list them in alphabetical order.
      d. If the parent chain can be numbered from either end, start at the end nearer the substituent that has alphabetical priority.
    2. Some alkyl halides are named by first citing the name of the alkyl group and then citing the halogen.
  B. Preparation of alkyl halides (Section 10.2).
    1. Alkyl halides can be prepared by reacting alkenes with HX or X_2.
    2. Alkyl halides from alcohols.
      a. Tertiary alkyl chlorides, bromides or iodides can be prepared by the reaction of a tertiary alcohol with HCl, HBr or HI.
         Reaction of secondary or primary alcohols occurs under more drastic conditions, which may destroy other acid-sensitive functional groups.
      b. Primary and secondary alkyl halides can be formed by treatment of the corresponding alcohols with $SOCl_2$ or PBr_3.
         Reaction conditions are mild, and the reagents don't interfere with other functional groups.
  C. Reactions of alkyl halides: Grignard reagents (Section 10.3).
    1. Organohalides react with Mg to produce organomagnesium halides, RMgX.
       These compounds are known as Grignard reagents.
    2. Grignard reagents can be formed from alkyl, alkenyl and aryl halides.
       Steric hindrance is no barrier to formation of Grignard reagents.
    3. The carbon bonded to Mg is negatively polarized and is nucleophilic.
    4. Grignard reagents react with weak acids to form hydrocarbons..

II. S_N2 reactions (Sections 10.4 – 10.6).
  A. Background of S_N2 reactions (Section 10.4).
    1. Walden discovered that (+) malic acid and (–) malic acid could be interconverted.
       This discovery meant that one or more reactions must have occurred with inversion of configuration at the chirality center.
    2. Others discovered that the nucleophilic substitution of tosylate ion by acetate ion occurred with inversion of configuration.
    3. They concluded that the nucleophilic substitution reactions of primary and secondary alkyl halides always proceed with inversion of configuration.

B. Kinetics and mechanism of S_N2 reactions (Section 10.5).
  1. Kinetics
    a. Reaction rate is the rate at which reactants are converted to products.
    b. The kinetics of a reaction measure the relationship between reactant concentrations and product concentrations.
    c. In an S_N2 reaction, reaction rate depends on the concentration of both alkyl halide and nucleophile.
      This type of reaction is a second-order reaction.
    d. In a second-order reaction, rate = k x [RX] x [Nu]
      The constant k is the rate constant; its units are (L/mol·s).
  2. Mechanism of the S_N2 reaction.
    a. The mechanism of an S_N2 reaction must account for both the observed stereochemistry and the observed kinetics.
    b. The reaction takes place in a single step, without intermediates.
      i. The nucleophile attacks the substrate from a direction directly opposite to the leaving group.
      ii. This type of reaction accounts for inversion of configuration.
      iii. In the transition state, the new bond forms at the same time as the old bond breaks.
      iv. Negative charge is shared between the attacking nucleophile and the leaving group.
      v. The three remaining bonds to carbon are in a planar arrangement.
      vi. Both substrate and nucleophile are involved in the step whose rate is measured.

C. Characteristics of the S_N2 reaction (Section 10.6).
  1. Changes in the energy levels of reactants or of the transition state affect the reaction rate.
  2. Changes in the substrate.
    a. Reaction rate is decreased if the substrate is bulky.
    b. Substrates, in order of increasing reactivity: tertiary, neopentyl, secondary, primary, methyl.
    c. S_N2 reactions can occur only at relatively unhindered sites.
    d. Vinylic and aryl halides are unreactive to S_N2 substitutions.
  3. Changes in the nucleophile.
    a. Any species can act as a nucleophile if it has an unshared electron pair.
      If the nucleophile has a negative charge, the product is neutral.
      If the nucleophile is neutral, the product is positively charged.
    b. The reactivity of a nucleophile is dependent on reaction conditions.
    c. In general, nucleophilicity parallels basicity.
    d. Nucleophilicity increases going down a column of the periodic table.
    e. Negatively charged nucleophiles are usually more reactive than neutral nucleophiles.
  4. Changes in the leaving group.
    a. In general, the best leaving groups are best able to stabilize negative charge.
    b. Usually, the best leaving groups are the weakest bases.
    c. Good leaving groups lower the energy of the transition state.
    d. Poor leaving groups include F^-, HO^-. RO^-, and H_2N^-.
  5. Changes in the solvent.
    a. Polar, protic solvents slow S_N2 reactions by lowering the reactivity of the nucleophile .
    b. Polar, aprotic solvents raise the ground-state energy of the nucleophile and make it more reactive.

6. A summary:
   i. Steric hindrance in the substrate raises the energy of the transition state, increasing $\Delta G^{\ddagger}$, and deceasing the reaction rate.
   ii. More reactive nucleophiles have a higher ground-state energy, decreasing $\Delta G^{\ddagger}$, and increasing the reaction rate.
   iii. Good leaving groups decrease the energy of the transition state, decreasing $\Delta G^{\ddagger}$, and increasing the reaction rate.
   iv. Polar protic solvents solvate the nucleophile, lowering the ground-state energy, increasing $\Delta G^{\ddagger}$, and decreasing the reaction rate. Polar aprotic solvents don't solvate the nucleophile, raising the ground-state energy, decreasing $\Delta G^{\ddagger}$, and increasing the reaction rate.

III. S_N1 Reactions (Sections 10.7 –10.8).
- A. Background of S_N1 reactions (Section 10.7).
  1. Under certain reaction conditions, tertiary halides are much more reactive than primary and methyl halides.
  2. These reactions must be occurring by a different mechanism.
- B. Kinetics and mechanism of the S_N1 reaction.
  1. Kinetics.
     a. The rate of reaction of a tertiary alkyl halide with water depends only on the concentration of the alkyl halide.
     b. The reaction is a first order process, with reaction rate = k x [RX].
     c. The rate expression shows that only RX is involved in the slowest, or rate-limiting, step, and the nucleophile is involved in a different, faster step.
     d. The rate expression also shows that there must be at least two steps in the reaction.
     e. In an S_N1 reaction, slow dissociation of the substrate is followed by rapid reaction with the nucleophile.
  2. Stereochemistry of S_N1 reactions.
     a. An S_N1 reaction of an enantiomer produces racemic product because an S_N1 reaction proceeds through a planar, achiral intermediate.
     b. Few S_N1 reactions proceed with complete racemization.
        The ion pair formed by the leaving group and the carbocation sometimes shields one side of the carbocation from attack before the leaving group can diffuse away.
- C. Characteristics of the S_N1 reaction (Section 10.8).
  1. As in S_N2 reactions, factors that lower $\Delta G^{\ddagger}$ favor faster reactions.
  2. Changes in the substrate.
     a. The more stable the carbocation intermediate, the faster the S_N1 reaction.
     b. Substrates, in order of increasing reactivity: methyl, primary, secondary and allyl and benzyl, tertiary.
     c. Allylic and benzylic substrates are also reactive in S_N2 reactions.
  3. Changes in the leaving group.
     a. The best leaving groups are the conjugate bases of strong acids.
     b. In S_N1 reactions, water can act as a leaving group.
  4. Changes in the nucleophile have no effect on S_N1 reactions.
  5. Changes in the solvent.
     a. Polar solvents (high dielectric constant) increase the rates of S_N1 reactions.
     b. Polar solvents stabilize the carbocation intermediate more than the reactants and lower $\Delta G^{\ddagger}$.
     c. Polar solvents stabilize by orienting themselves around the carbocation, with electron-rich ends facing the positive charge.

6. A summary:
    a. The best substrates are those that form stable carbocations.
    b. Good leaving groups lower the energy of the transition state leading to carbocation formation and increase the reaction rate.
    c. The nucleophile doesn't affect the reaction rate, but it must be nonbasic.
    d. Polar solvents stabilize the carbocation intermediate and increase the reaction rate.

IV. Biological substitution reactions (Section 10.9).
  A. Both S_N1 and S_N2 reactions occur often in biochemical pathways.
  B. In S_N1 reactions, the leaving group is often an organodiphosphate.
  C. S_N2 reactions are involved in biological methylations.

V. Elimination reactions (Sections 10.10 – 10.13).
  A. Introduction (Section 10.10).
    1. In addition to bringing about substitution, a basic nucleophile can also cause elimination of HX from an alkyl halide to form a carbon–carbon double bond.
    2. A mixture of double-bond products is usually formed, but the product with the more substituted double bond is the major product.
       This observation is the basis of Zaitsev's rule.
    3. Double-bond formation can occur by several mechanistic routes, but at this point, we will study only three mechanisms.
  B. The E2 reaction (Section 10.11)
    1. General features.
      a. An E2 reaction occurs when an alkyl halide is treated with strong base.
      b. The reaction occurs in one step, without intermediates.
      c. E2 reactions follow second-order kinetics.
      d. The deuterium isotope effect.
        i. In a reaction in which a C–H bond is cleaved in the rate-limiting step, substitution of –D for –H results in a decrease in rate.
        ii. Because this effect is observed in E2 reactions, these reactions must involve C–H bond breaking in the rate-limiting step.
      e. E2 reactions always occur with periplanar geometry.
        i. Periplanar geometry is required because of the need for overlap of the sp^3 orbitals of the reactant as they become π orbitals in the product.
        ii. Anti periplanar geometry is preferred because it allows the substituents of the two carbons to assume a staggered relationship.
        iii. Syn periplanar geometry occurs only when anti periplanar geometry isn't possible.
      e. The preference for anti periplanar geometry results in the formation of double bonds with specific *E,Z* configurations.
      f. The chemistry of substituted cyclohexanes is controlled by their conformations.
         The preference for antiperiplanar geometry for E2 reactions can be met only if the atoms to be eliminated have a trans-diaxial relationship.
  C. The El reaction (Section 10.12).
    1. An El reaction occurs when the intermediate carbocation of an S_N1 loses H^+ to form a C=C bond.
    2. El reactions usually occur in competition with S_N1 reactions.
    3. El reactions show first-order kinetics.
    4. There is no geometric requirement for the groups to be eliminated, and the most stable (Zaitsev) product is formed.
    5. No deuterium isotope effect is observed: C–H bond-breaking occurs after the rate-limiting step.

D. The E1cB reaction.
   1. The E1cB reaction takes place through a carbanion.
   2. The rate-limiting step involves base-induced abstraction of a proton.
   3. Often the leaving group is poor.
   4. A carbonyl group stabilizes the anion.
   5. The E1cB is fairly common in biochemical pathways (Section 10.13).

III. Summary of reactivity (Section 10.14).
   A. Primary halides.
      1. S_N2 reaction is usually observed.
      2. E1 reaction occurs if a strong, bulky base is used.
   B. Secondary halides.
      1. S_N2 and E2 reactions occur in competition.
      2. Strong bases promote E2 elimination.
      3. Secondary halides (especially allylic and benzylic halides) can react by S_N1 and E1 routes if weakly basic nucleophiles and protic solvents are used.
   C. Tertiary halides.
      1. Under basic conditions, E2 elimination is favored.
      2. S_N1 and E1 products are formed under nonbasic conditions.
   D. E1cB occurs with all of the above halides if the leaving group is two carbons away from a carbonyl group.

Solutions to Problems

10.1 The rules that were given for naming alkanes in Section 3.4 are used for alkyl halides. A halogen is treated as an alkyl substituent.

(a) $CH_3CH_2CH_2CH_2I$

1-Iodobutane

(b) $CH_3CH(CH_3)CH_2CH_2Cl$

1-Chloro-3-methylbutane

(c) $BrCH_2CH_2CH_2C(CH_3)_2CH_2Br$

1,5-Dibromo-2,2-dimethylpentane

(d) $CH_3C(CH_3)(Cl)CH_2CH_2Cl$

1,3-Dichloro-3-methylbutane

(e) $CH_3CH(I)CH(CH_2CH_2Cl)CH_2CH_3$

1-Chloro-3-ethyl-4-iodopentane

(f) $CH_3CH(Br)CH_2CH_2CH(Cl)CH_3$

2-Bromo-5-chlorohexane

10.2

(a) $CH_3CH_2CH_2C(CH_3)_2-CHClCH_3$

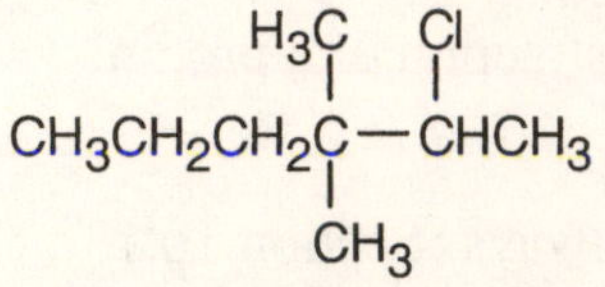

2-Chloro-3,3-dimethylhexane

(b) $CH_3CH_2CH_2CCl_2-CH(CH_3)CH_3$

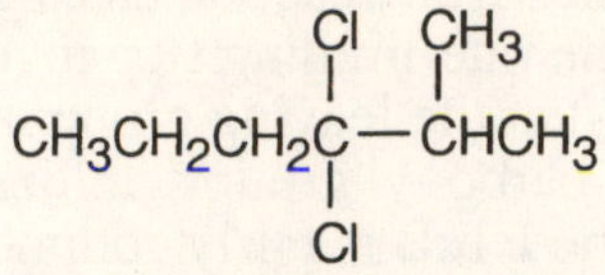

3,3-Dichloro-2-methylhexane

(c) $CH_3CH_2C(Br)(CH_2CH_3)CH_2CH_3$

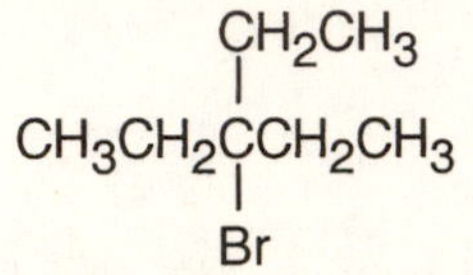

3-Bromo-3-ethylpentane

(d) $(H_3C)_2HC-C_6H_{10}Br_2$

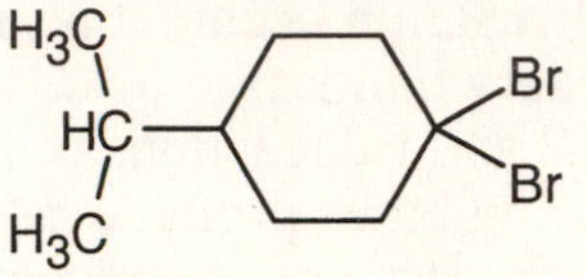

1,1-Dibromo-4-isopropylcyclohexane

(e) $CH_3CH_2CH_2CH_2CH_2CH(CH_3CH_2CHCH_3)CH_2CHClCH_3$

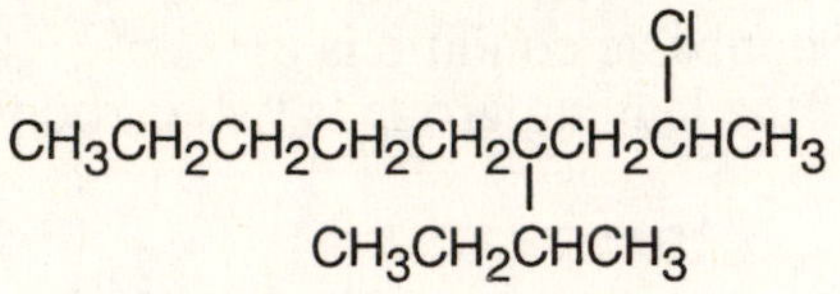

4-*sec*-Butyl-2-chlorononane

(f) $(CH_3)_3C-C_6H_{10}Br_2$

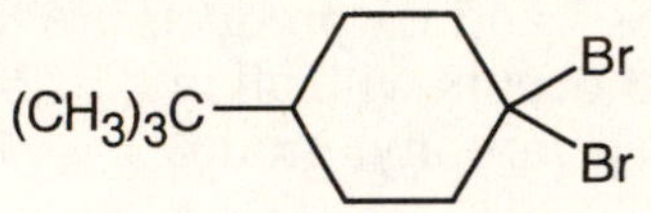

1,1-Dibromo-4-*tert*-butylcyclohexane

10.3 Remember that halogen acids are used for converting tertiary alcohols to alkyl halides. PBr_3 and $SOCl_2$ are used for converting secondary and primary alcohols to alkyl halides.

(a) $CH_3C(OH)(CH_3)CH_3 \xrightarrow[\text{ether}]{HCl} CH_3C(Cl)(CH_3)CH_3$

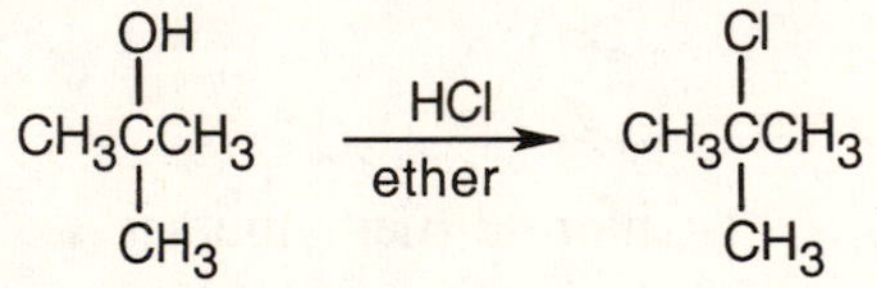

(b) $CH_3CH(OH)CH_2CH(CH_3)CH_3 \xrightarrow[\text{ether}]{PBr_3} CH_3CH(Br)CH_2CH(CH_3)CH_3$

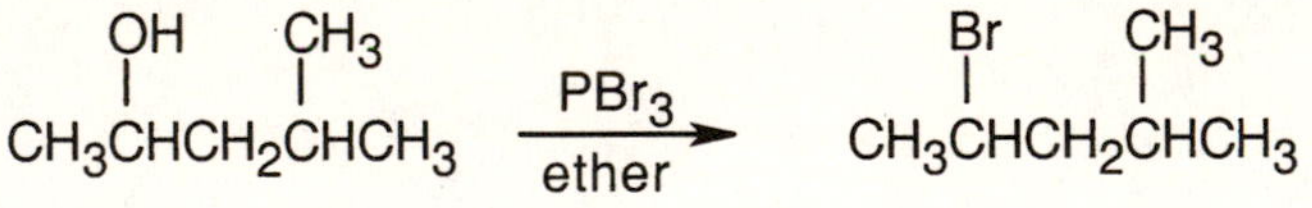

(c) $HOCH_2CH_2CH_2CH_2CH(CH_3)CH_3 \xrightarrow[\text{ether}]{PBr_3} BrCH_2CH_2CH_2CH_2CH(CH_3)CH_3$

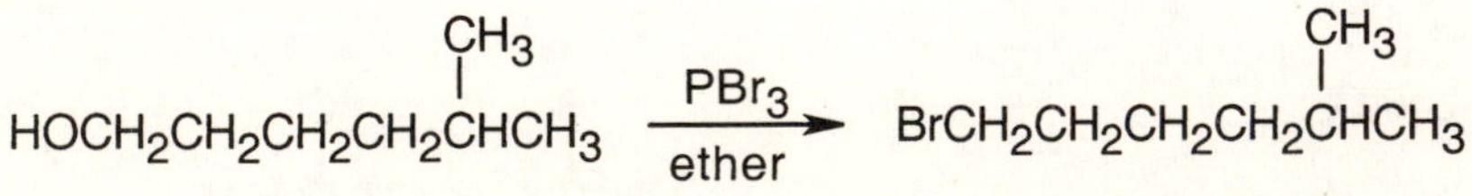

(d) $CH_3CH_2CH(CH_3)CH_2C(OH)(CH_3)CH_3 \xrightarrow[\text{ether}]{HCl} CH_3CH_2CH(CH_3)CH_2C(Cl)(CH_3)CH_3$

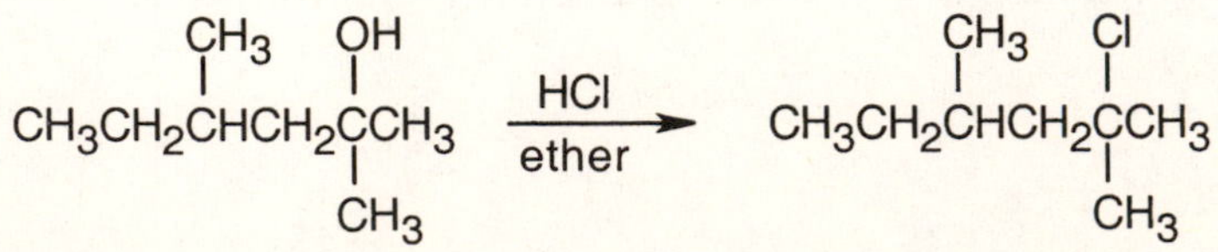

10.4 Table 7.3 shows that the pK_a of CH_3–H is 60. Since CH_4 is a very weak acid, $^-$:CH_3 is a very strong base. Alkyl Grignard reagents are similar in base strength to $^-$:CH_3, but alkynyl Grignard reagents are somewhat weaker bases. Both reactions (a) and (b) occur as written.

(a)

$$\underset{\text{stronger base}}{CH_3MgBr} + \underset{\text{stronger acid}}{H-C\equiv C-H} \longrightarrow \underset{\text{weaker acid}}{CH_4} + \underset{\text{weaker base}}{H-C\equiv C-MgBr}$$

(b)

$$\underset{\text{stronger base}}{CH_3MgBr} + \underset{\text{stronger acid}}{NH_3} \longrightarrow \underset{\text{weaker acid}}{CH_4} + \underset{\text{weaker base}}{H_2N-MgBr}$$

10.5 Just as Grignard reagents react with *proton* donors to convert R–MgX into R–H, they also react with *deuterium* donors to convert R–MgX into R–D. In this case:

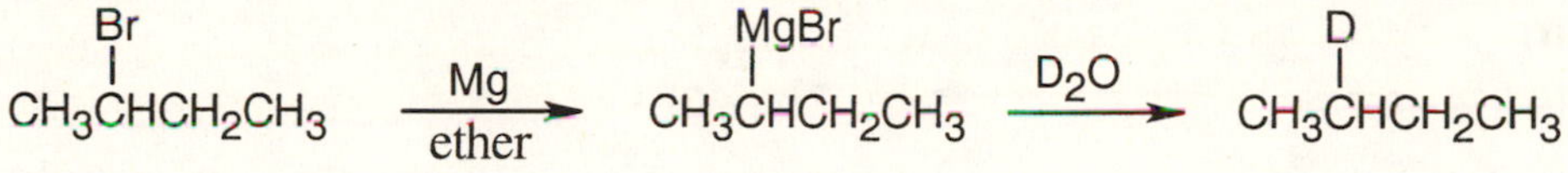

$$CH_3CH(Br)CH_2CH_3 \xrightarrow[\text{ether}]{Mg} CH_3CH(MgBr)CH_2CH_3 \xrightarrow{D_2O} CH_3CH(D)CH_2CH_3$$

10.6 **Strategy:** As described in Practice Problem 10.1, identify the leaving group and the chirality center. Draw the product carbon skeleton, inverting the configuration at the chirality center, and replace the leaving group with the nucleophilic reactant.

Solution:

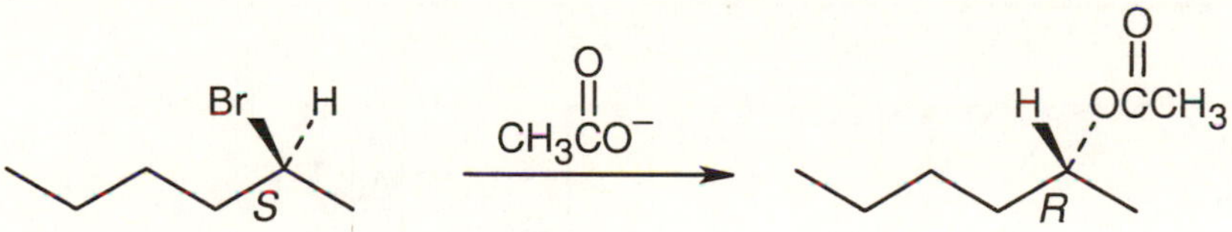

(S)-bromoalkane (Br, H) $\xrightarrow{CH_3CO_2^-}$ (R)-acetate (H, $OCCH_3$ with C=O)

10.7 Use the suggestions in the previous problem to draw the correct product.

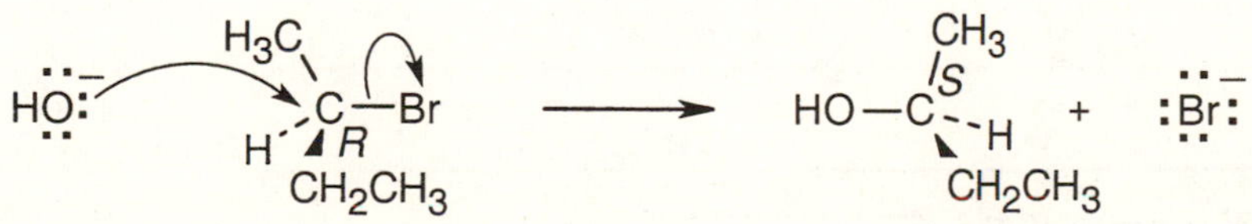

$$HO:^- + H_3C(H)(CH_2CH_3)C\text{–}Br\ (R) \longrightarrow HO\text{–}C(CH_3)(H)(CH_2CH_3)\ (S) + :Br:^-$$

10.8

$$HS^- + H_3C\text{–}CH(CH_3)\text{–}CH_2\text{–}CH(Br)\text{–}CH_3\ (S) \longrightarrow H_3C\text{–}CH(CH_3)\text{–}CH_2\text{–}CH(SH)\text{–}CH_3\ (R) + Br^-$$

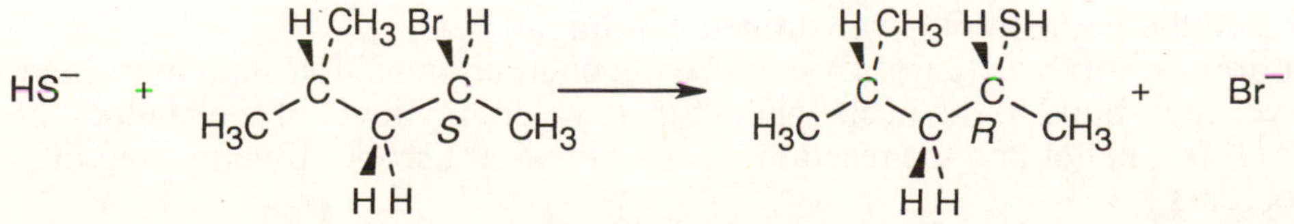

10.9 All of the nucleophiles in this problem are relatively reactive.

(a)

$$CH_3CH_2CH_2CH_2Br + NaI \longrightarrow CH_3CH_2CH_2CH_2I + NaBr$$

(b)

$$CH_3CH_2CH_2CH_2Br + KOH \longrightarrow CH_3CH_2CH_2CH_2OH + KBr$$

(c)

$$CH_3CH_2CH_2CH_2Br + NH_3 \longrightarrow CH_3CH_2CH_2CH_2NH_3^+ \ Br^-$$

10.10 (a) $(CH_3)_2N^-$ is more nucleophilic because it is more basic than $(CH_3)_2NH$.
(b) $(CH_3)_3N$ is more nucleophilic than $(CH_3)_3B$. $(CH_3)_3B$ is non-nucleophilic because it has no lone electron pair.
(c) H_2S is more nucleophilic than H_2O because nucleophilicity increases in going down a column of the periodic table.

10.11 **Strategy:** In this problem, we are comparing two effects – the effect of the substrate and the effect of the leaving group. Tertiary substrates are less reactive than secondary substrates, which are less reactive than primary substrates.

Solution:

Least reactive ⟶ *Most reactive*

CH_3NH_2	< $(CH_3)_2CHCl$	< CH_3Cl	< CH_3OTos
poor leaving group	secondary carbon	primary carbon	excellent leaving group

10.12

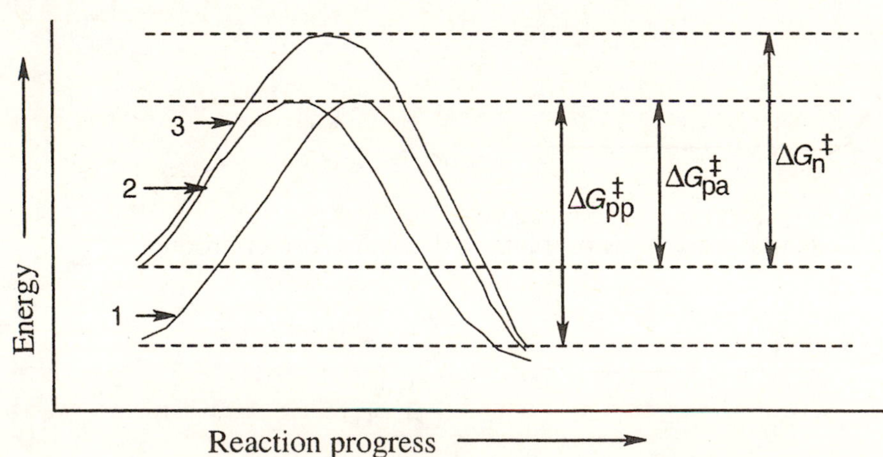

Polar protic solvents (curve 1) stabilize the charged transition state by solvation and also stabilize the nucleophile by hydrogen bonding.

Polar aprotic solvents (curve 2) stabilize the charged transition state by solvation, but do not hydrogen-bond to the nucleophile. Since the energy level of the nucleophile is higher, $\Delta G^‡$ is smaller and the reaction is faster in polar aprotic solvents than in polar protic solvents.

Nonpolar solvents (curve 3) stabilize neither the nucleophile nor the transition state. $\Delta G^‡$ is therefore higher in nonpolar solvents than in polar aprotic solvents, and the reaction rate is slower.

10.13

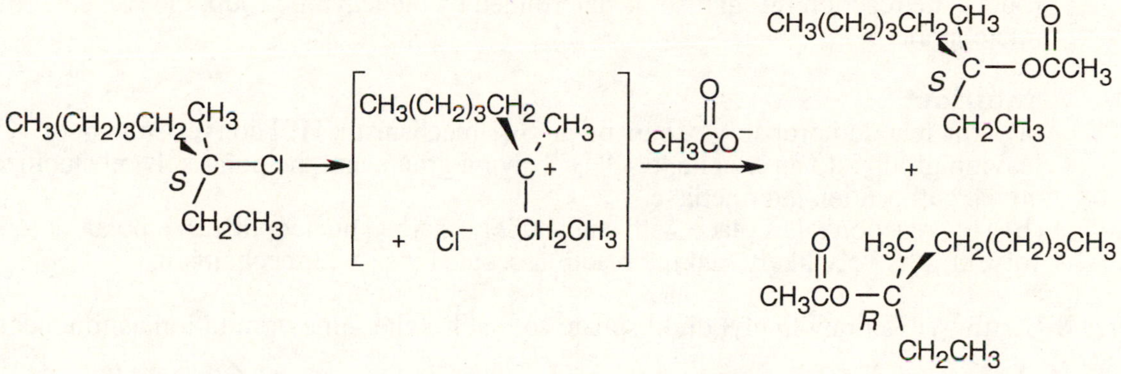

$CH_3(CH_2)_3CH_2(CH_3)C(S)(CH_2CH_3)$—Cl → [$CH_3(CH_2)_3CH_2(CH_3)C^+(CH_2CH_3)$ + Cl^-] $\xrightarrow{CH_3CO^-\ (O=)}$ $CH_3(CH_2)_3CH_2(CH_3)C(S)(CH_2CH_3)$—$OC(=O)CH_3$ + $CH_3C(=O)O$—$C(R)(CH_3)(CH_2(CH_2)_3CH_3)(CH_2CH_3)$

Reaction with acetate can occur on either side of the planar, achiral carbocation intermediate, resulting in a mixture of both the *R* and *S* enantiomeric acetates. The ratio of enantiomers is probably close to 50:50.

10.14 The *S* substrate reacts with water to form a mixture of *R* and *S* alcohols. The ratio of enantiomers is close to 50:50.

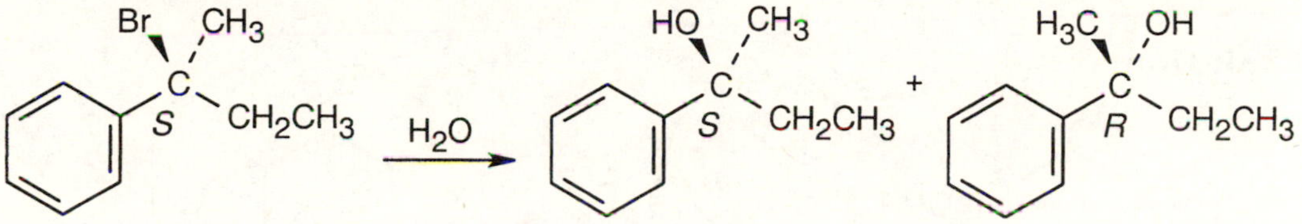

$C_6H_5C(S)(Br)(CH_3)CH_2CH_3 \xrightarrow{H_2O} C_6H_5C(S)(HO)(CH_3)CH_2CH_3 + C_6H_5C(R)(H_3C)(OH)CH_2CH_3$

10.15 S_N1 reactivity is related to carbocation stability. Thus, substrates that form the most stable carbocations are the most reactive in S_N1 reactions.

Least reactive ⟶ *Most reactive*

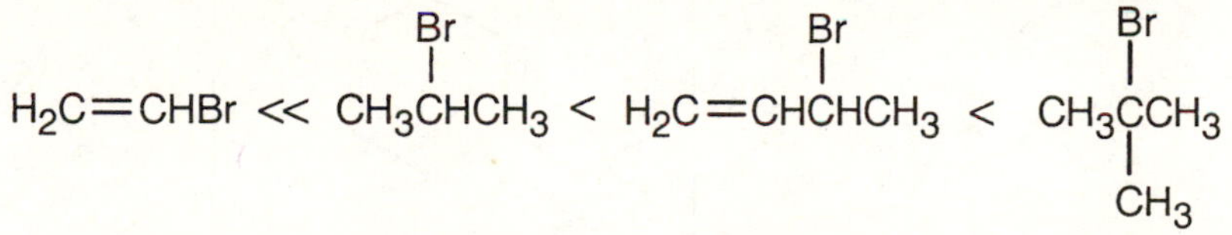

$H_2C{=}CHBr \ll CH_3CH(Br)CH_3 < H_2C{=}CHCH(Br)CH_3 < CH_3C(Br)(CH_3)CH_3$

vinylic — secondary — allylic — tertiary

10.16

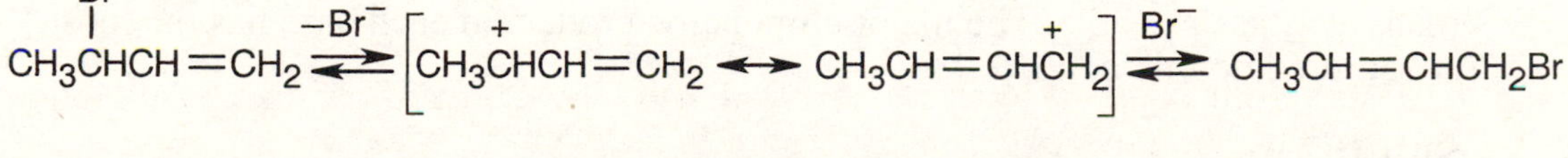

$CH_3CH(Br)CH{=}CH_2 \underset{}{\overset{-Br^-}{\rightleftharpoons}} [CH_3\overset{+}{C}HCH{=}CH_2 \longleftrightarrow CH_3CH{=}CH\overset{+}{C}H_2] \overset{Br^-}{\rightleftharpoons} CH_3CH{=}CHCH_2Br$

The two bromobutenes form the same allylic carbocation in the rate-limiting step.

10.17 Strategy: Both substrates have allylic groups and can react either by an S_N1 or an S_N2 route. The reaction mechanism is determined by the leaving group, the solvent, or the nucleophile.

Solution:
(a) This reaction probably occurs by an S_N1 mechanism. HCl converts the poor –OH leaving group into an excellent $–OH_2^+$ leaving group, and the polar solvent stabilizes the allylic carbocation intermediate.
(b) This reaction takes place with a negatively charged nucleophile in a polar, aprotic solvent. It is very likely that the reaction occurs by an S_N2 mechanism.

10.18 Strategy: Redraw linalyl diphosphate so that has the same orientation as limonene.

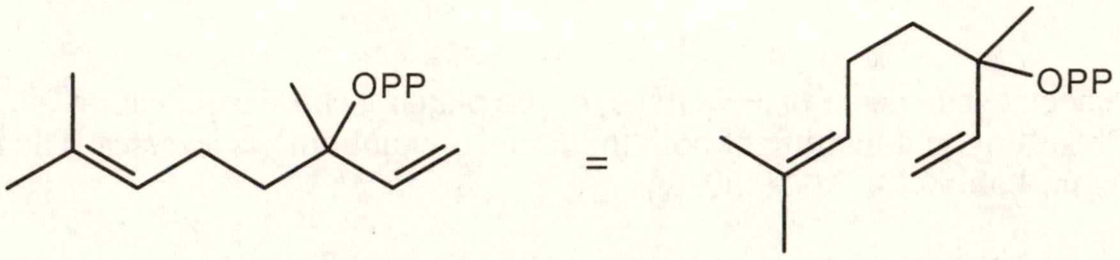

OPP = OPP

After dissociation of PP_i, the cation cyclizes by attack of the double bond π electrons. Removal of an –H yields limonene.

Solution:

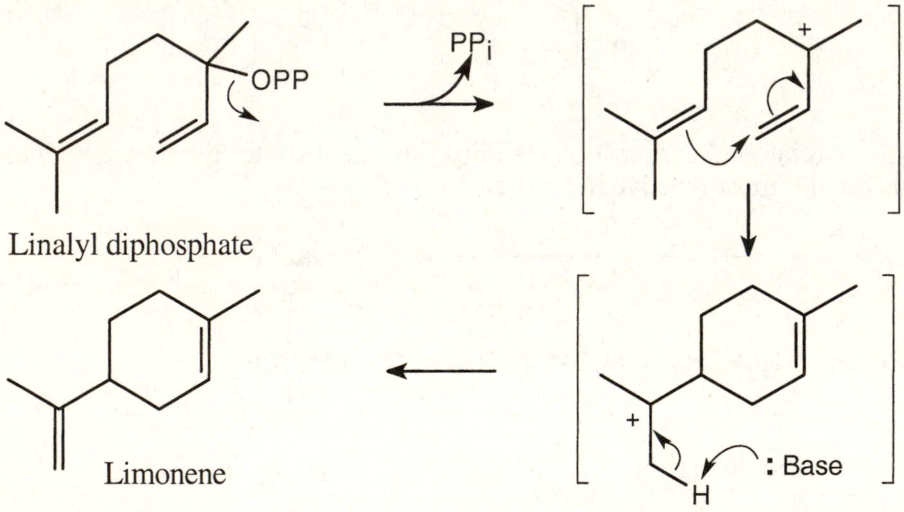

OPP → PP_i → [+] ↓ [+ H : Base] ← Limonene

Linalyl diphosphate

Limonene

10.19 Strategy: Form the double bond by removing HX from the alkyl halide reactant in as many ways as possible. The major elimination product in each case has the most substituted double bond.

Solution:
(a)

$$\underset{\text{}}{CH_3CH_2\overset{Br}{\overset{|}{C}}H\overset{CH_3}{\overset{|}{C}}HCH_3} \xrightarrow{\text{Base}} CH_3CH_2CH{=}\overset{CH_3}{\overset{|}{C}}CH_3 + CH_3CH{=}CH\overset{CH_3}{\overset{|}{C}}HCH_3$$

major (trisubstituted double bond) minor (disubstituted double bond)

(b)

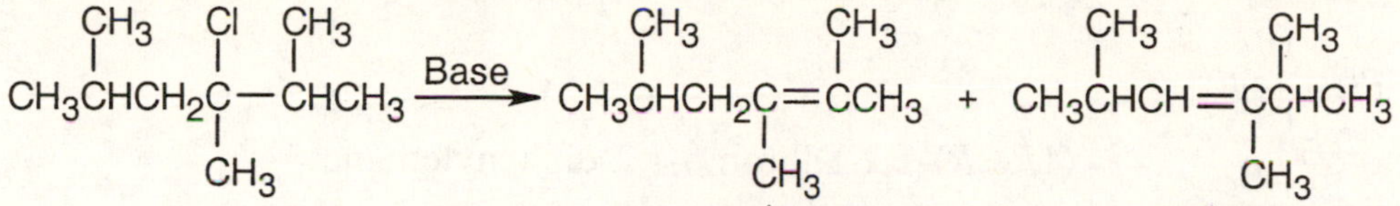

$$CH_3CH(CH_3)CH_2C(Cl)(CH_3)\text{—}CH(CH_3)CH_3 \xrightarrow{\text{Base}} CH_3CH(CH_3)CH_2C(CH_3)\text{=}C(CH_3)CH_3 + CH_3CH(CH_3)CH\text{=}C(CH_3)CH(CH_3)CH_3$$

major (tetrasubstituted double bond) — minor (trisubstituted double bond)

$$+\ CH_3CH(CH_3)CH_2C(\text{=}CH_2)CH(CH_3)CH_3$$

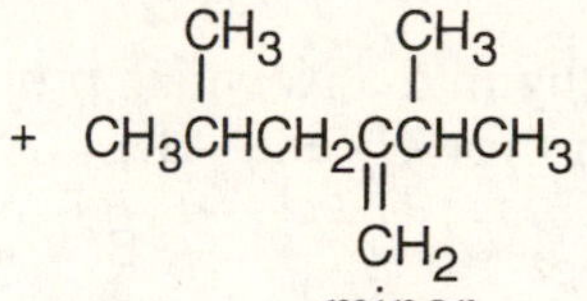

minor
(disubstituted double bond)

(c)

$$C_6H_{11}\text{—}CH(Br)CH_3 \xrightarrow{\text{Base}} C_6H_{10}\text{=}CHCH_3 + C_6H_{11}\text{—}CH\text{=}CH_2$$

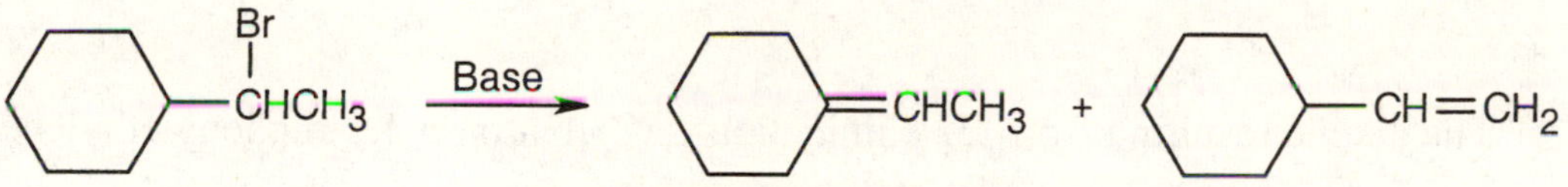

major (trisubstituted double bond) — minor (monosubstituted double bond)

10.20 Strategy: For maximum yield, the alkyl halide reactant should not give a mixture of products on elimination.

Solution:

(a)

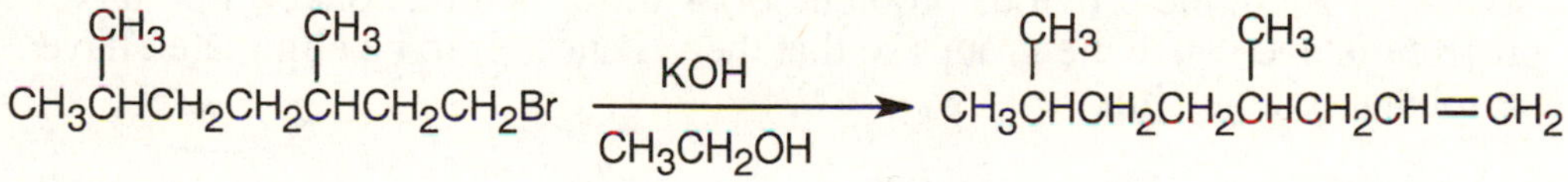

$$CH_3CH(CH_3)CH_2CH_2CH(CH_3)CH_2CH_2Br \xrightarrow[CH_3CH_2OH]{KOH} CH_3CH(CH_3)CH_2CH_2CH(CH_3)CH_2CH\text{=}CH_2$$

(b)

$$\text{4-bromo-1,2-dimethylcyclopentane (Br, } CH_3\text{, } CH_3\text{)} \xrightarrow[CH_3CH_2OH]{KOH} \text{4,5-dimethylcyclopentene (} CH_3\text{, } CH_3\text{)}$$

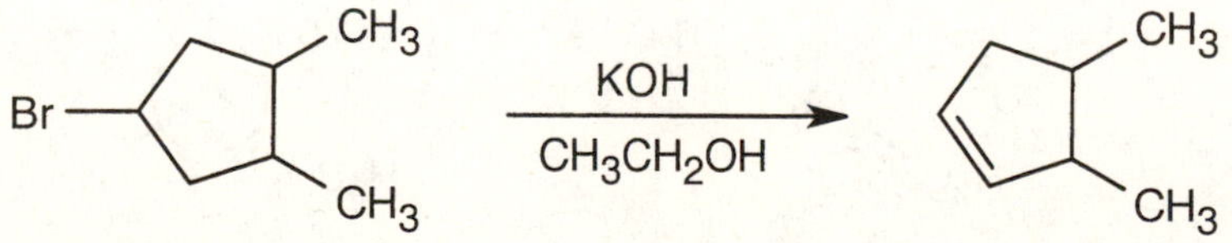

10.21 Strategy: Draw the reactant with correct stereochemistry.

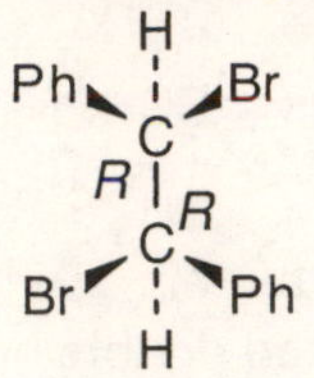

(1*R*,2*R*)-1,2-Dibromo-1,2-diphenylethane

Convert this drawing into a Newman projection, and draw the conformation having anti periplanar geometry for –H and –Br.

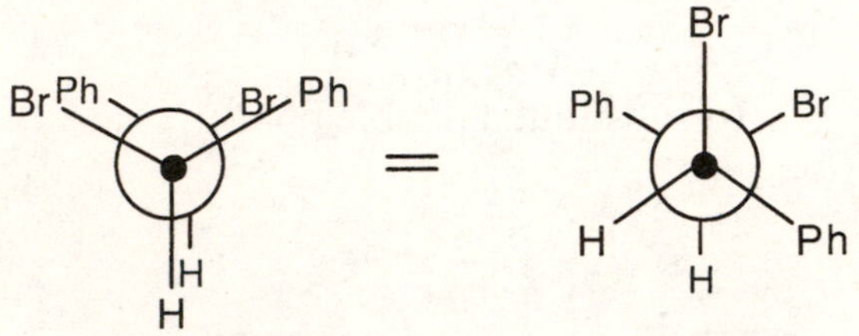

The alkene resulting from E2 elimination is (*Z*)-1-bromo-1,2-diphenylethylene.

Solution:

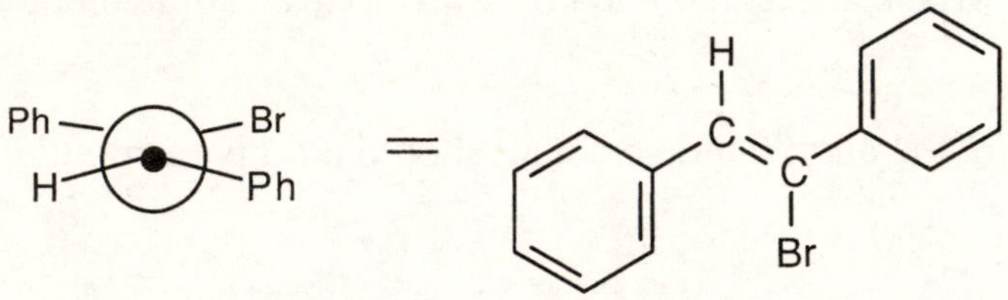

10.22 Strategy: As in the previous problem, draw the structure, convert it to a Newman projection, and rotate the groups so that the –H and –Br to be eliminated have an anti periplanar relationship.

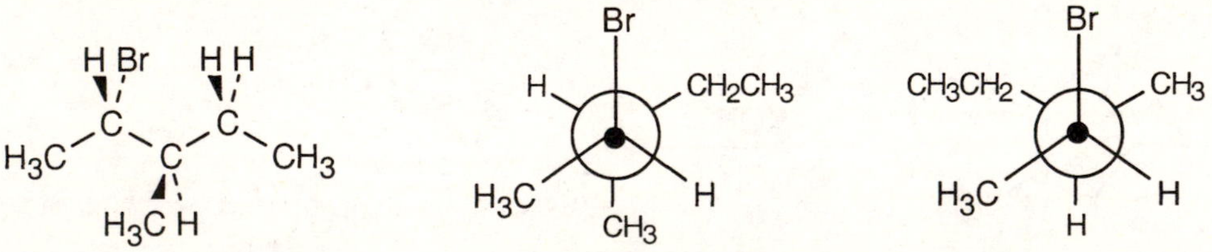

The major product is (*Z*)-3-methylpent-2-ene. A small amount of 3-methylpent-1-ene is also formed.

Solution:

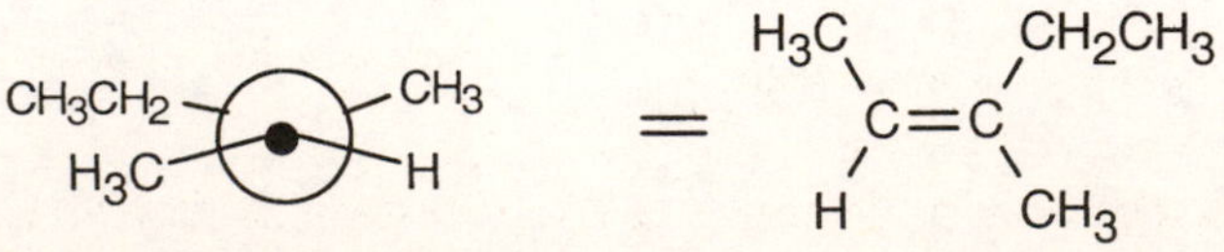

(*Z*)-3-Methylpent-2-ene

10.23

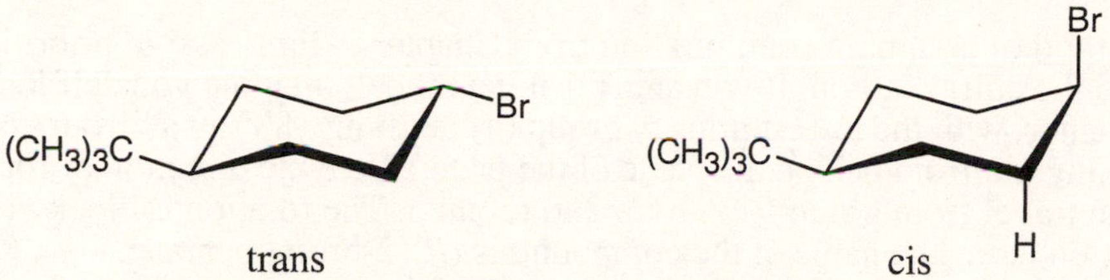

$(CH_3)_3C$– (chair, Br) trans

$(CH_3)_3C$– (chair, Br, H) cis

The more stable conformations of each of the two isomers are pictured above; the larger *tert*–butyl group is always equatorial in the more stable conformation. The cis isomer reacts faster under E2 conditions because –Br and –H are in the anti-periplanar arrangement that favors E2 elimination.

10.24 Use the summary in Section 10.14 if you need help.

(a)

$$CH_3CH_2CH_2CH_2Br + NaN_3 \longrightarrow CH_3CH_2CH_2CH_2N_3$$

primary — substitution product

The reaction occurs by an S_N2 mechanism because the substrate is primary, the nucleophile is nonbasic, and the product is a substitution product.

(b)

$$CH_3CH_2CHClCH_2CH_3 + KOH \longrightarrow CH_3CH_2CH{=}CHCH_3$$

secondary — strong base — elimination product

This is an E2 reaction since a secondary halide reacts with a strong base to yield an elimination product.

(c)

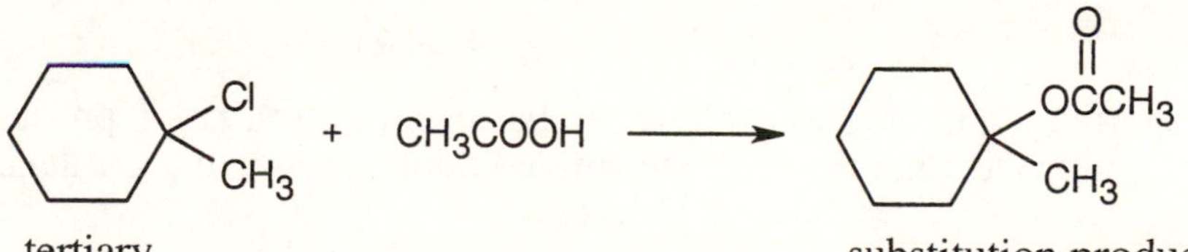

1-chloro-1-methylcyclohexane (Cl, CH_3) + CH_3COOH ⟶ 1-methylcyclohexyl acetate ($OCCH_3$ with C=O, CH_3)

tertiary — substitution product

This is an S_N1 reaction. Tertiary substrates form substitution products only by the S_N1 route.

(d)

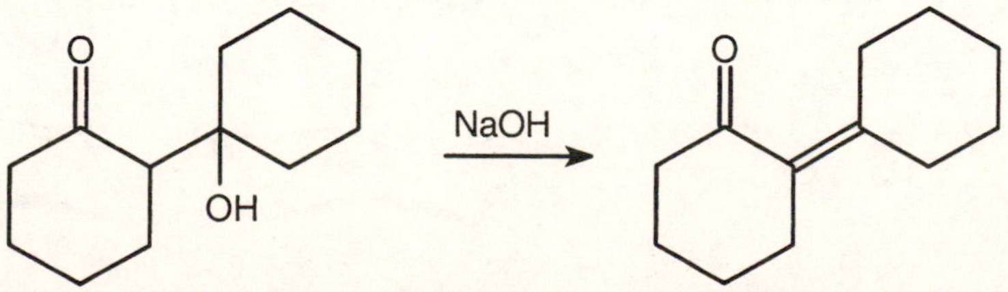

2-(1-hydroxycyclohexyl)cyclohexanone (O, OH) $\xrightarrow{NaOH}$ 2-cyclohexylidenecyclohexanone (O)

This is an E1cB reaction because the leaving group is two carbons away from a carbonyl group.

Visualizing Chemistry

10.25 Recall the method of assigning configuration from Chapter 9. First, assign priorities to the groups around the chiral carbon. If you aren't using a model, imagine yourself looking at the chirality center, with the lowest priority group (4) pointing 180° away from you. (You would be looking out from behind the plane of the page.) Note the direction of rotation of the arrows that travel from group 1 to group 2 to group 3. The rotation is clockwise, and the configuration is *R*. The name of the compound is (*R*)-2-bromopentane.

Reaction of (*S*)-pentan-2-ol with PBr_3 to form (*R*)-2-bromopentane occurs with a change in stereochemistry because the configuration at the chirality center changes from *S* to *R* (S_N2 reaction).

10.26 (a)

(i) $CH_3CH_2Cl + Na^+ \ {}^-SCH_3 \longrightarrow CH_3CH_2SCH_3 + NaCl$

(ii) $CH_3CH_2Cl + Na^+ \ {}^-OH \longrightarrow CH_3CH_2OH + NaCl$

Both reactions yield S_N2 substitution products because the substrate is primary and both nucleophiles are good. E2 elimination product is unlikely to form because the base ^-OH isn't bulky.

(b)

(i) $CH_3C(Cl)(CH_3)CH_2CH_3 + Na^+ \ {}^-SCH_3 \longrightarrow (H_3C)_2C{=}CHCH_3$ (major) $+ H_2C{=}C(CH_2CH_3)CH_3$ (minor) $+ HSCH_3 + NaCl$

(ii) $CH_3C(Cl)(CH_3)CH_2CH_3 + Na^+ \ {}^-OH \longrightarrow (H_3C)_2C{=}CHCH_3$ (major) $+ H_2C{=}C(CH_2CH_3)CH_3$ (minor) $+ H_2O + NaCl$

The substrate is tertiary, and the nucleophiles are basic. Two elimination products are expected; the major product has the more substituted double bond, in accordance with Zaitsev's rule.

(c)

(i) chlorocyclohexane (Cl) $+ Na^+ \ {}^-SCH_3 \longrightarrow$ cyclohexyl–SCH_3 $+ NaCl$

(ii) chlorocyclohexane (Cl) $+ Na^+ \ {}^-OH \longrightarrow$ cyclohexyl–OH $+$ cyclohexene $+ H_2O + NaCl$

In (i), the secondary substrate reacts with the good, but weakly basic, nucleophile to yield substitution product. In (ii), NaOH is a poorer nucleophile but a stronger base, and both substitution and elimination product are formed.

10.27

Reaction of the secondary bromide with the weakly basic acetate nucleophile occurs by an S_N2 route, with inversion of configuration, to produce the *R* acetate.

10.28

The *S* substrate has a secondary allylic chloride group and a primary hydroxyl group. S_N2 reaction occurs at the secondary carbon to give the *R* cyano product because hydroxide is a poor leaving group.

10.29

Rotate the left side of the molecule so that the groups to be eliminated have an anti periplanar relationship. The double bond in the product has the *E* configuration.

Additional Problems

10.30

(a) $CH_3CH(CH_3)CH(Br)CH(Br)CH_2CH(CH_3)CH_3$

3,4-Dibromo-2,6-dimethylheptane

(b) $CH_3CH{=}CHCH_2CH(I)CH_3$

5-Iodohex-2-ene

(c) $CH_3C(Br)(CH_3)CH_2CH(Cl)CH(CH_3)CH_3$

2-Bromo-4-chloro-2,5-dimethylhexane

(d) $CH_3CH_2CH(CH_2Br)CH_2CH_2CH_3$

3-(Bromomethyl)hexane

(e) $ClCH_2CH_2CH_2C{\equiv}CCH_2Br$

1-Bromo-6-chlorohex-2-yne

10.31

(a)

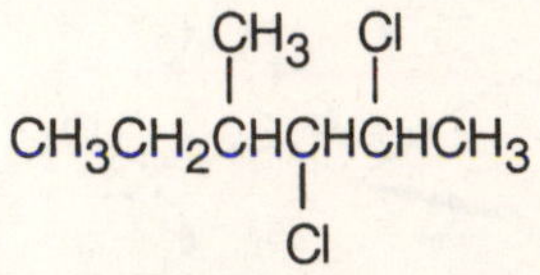

```
     CH3 Cl
     |   |
CH3CH2CHCHCHCH3
         |
         Cl
```

2,3-Dichloro-4-methylhexane

(b)

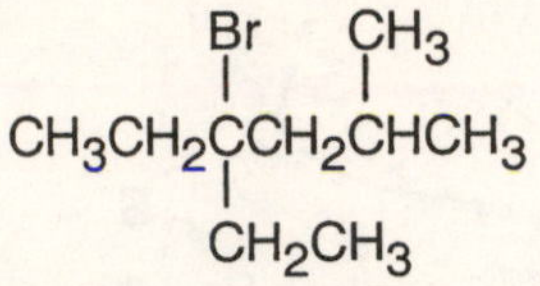

```
      Br   CH3
      |    |
CH3CH2CCH2CHCH3
      |
      CH2CH3
```

4-Bromo-4-ethyl-2-methylhexane

(c)

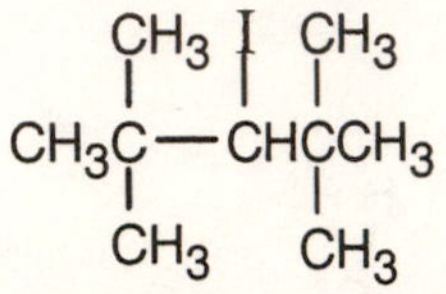

```
   CH3 I CH3
   |   | |
CH3C---CHCCH3
   |     |
   CH3   CH3
```

3-Iodo-2,2,4,4-tetramethylpentane

(d)

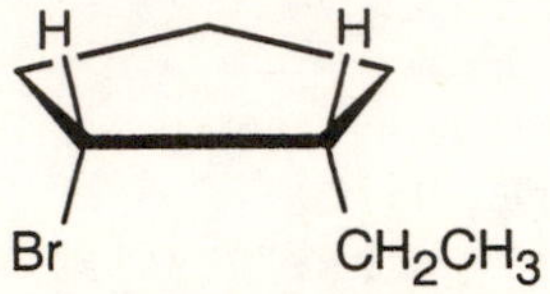

H H

Br CH_2CH_3

cis-1-Bromo-2-ethylcyclopentane

10.32 A Grignard reagent can't be prepared from a compound containing an acidic functional group because the Grignard reagent is immediately quenched by the proton source. For example, the $-CO_2H$, $-OH$, $-NH_2$, and $RC\equiv CH$ functional groups are not compatible with formation of Grignard reagents.

10.33 (a) CH_3I reacts faster than CH_3Br because I^- is a better leaving group than Br^-.

(b) CH_3CH_2I reacts faster with OH^- in dimethylsulfoxide (DMSO) than in ethanol. Ethanol, a protic solvent, hydrogen-bonds with hydroxide ion and decreases its reactivity.

(c) Under the S_N2 conditions of this reaction, CH_3Cl reacts faster than $(CH_3)_3CCl$ because approach of the nucleophile to the bulky $(CH_3)_3CCl$ molecule is hindered.

(d) $H_2C{=}CHCH_2Br$ reacts faster because vinylic halides such as $H_2C{=}CHBr$ are unreactive to substitution reactions.

10.34

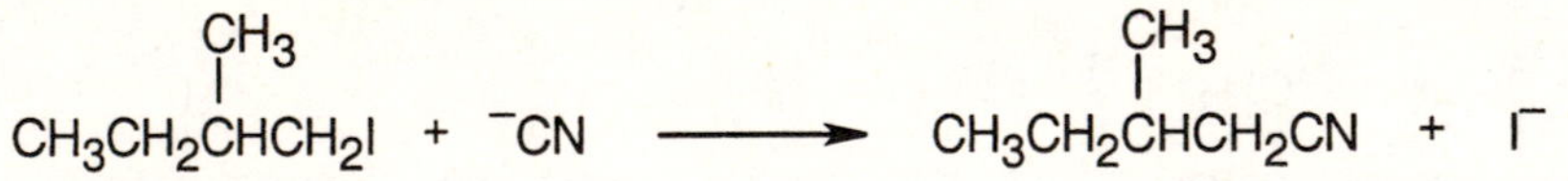

```
       CH3                            CH3
       |                              |
CH3CH2CHCH2I  +  -CN  ------>  CH3CH2CHCH2CN  +  I-
```

primary halide

This is an S_N2 reaction, whose rate depends on the concentration of both alkyl halide and nucleophile. Rate = k x [RX] x [Nu:$^-$]

(a) Halving the concentration of cyanide ion and doubling the concentration of alkyl halide doesn't change the reaction rate.

(b) Tripling the concentrations of both cyanide ion and alkyl halide causes a ninefold increase in reaction rate.

10.35

$$CH_3CH_2C(CH_3)_2I \text{ (tertiary halide)} + CH_3CH_2OH \longrightarrow CH_3CH_2C(CH_3)_2OCH_2CH_3 + HI$$

This is an S_N1 reaction, whose rate depends only on the concentration of 2-iodo-2-methylbutane. Rate = k x [RX].

(a) Tripling the concentration of alkyl halide triples the rate of reaction.

(b) Halving the concentration of ethanol by dilution with diethyl ether reduces the polarity of the solvent and decreases the rate.

10.36

(a)

$$CH_3CH_2CH_2CH_2Br + NaCN \longrightarrow CH_3CH_2CH_2CH_2CN + NaBr$$

(b)

$$H_3C{-}Br + {}^{-}O{-}C(CH_3)_3 \longrightarrow H_3C{-}O{-}C(CH_3)_3 + NaBr$$

or

$$(CH_3)_3C{-}Cl + CH_3OH \longrightarrow \underset{\text{(major)}}{(CH_3)_3C{-}O{-}CH_3} + HCl + \underset{\text{(minor)}}{(CH_3)_2C{=}CH_2}$$

(c)

$$CH_3CH_2CH_2Br + \text{excess } NH_3 \longrightarrow \underset{\text{major}}{CH_3CH_2CH_2NH_3^{+}\,Br^{-}}$$

10.37 (a) The difference in this pair of reactions is in the *leaving group*. Since $^{-}$OTos is a better leaving group than $^{-}$Cl (see Section 10.6), S_N2 displacement by iodide on CH_3–OTos proceeds faster.

(b) The *substrates* in these two reactions are different. Bromoethane is a primary bromoalkane, and bromocyclohexane is a secondary bromoalkane. Since S_N2 reactions proceed faster at primary than secondary carbon atoms, S_N2 displacement on bromoethane is a faster reaction.

(c) Ethoxide ion and cyanide ion are different *nucleophiles*. Since CN^- is more reactive than $CH_3CH_2O^-$ in S_N2 reactions, S_N2 displacement on 2-bromopropane by CN^- proceeds at a faster rate.

(d) The *solvent* in each reaction is different. The S_N2 reaction on bromoethane in polar, aprotic acetonitrile proceeds faster than in nonpolar toluene.

10.38 Because 1-bromopropane is a primary haloalkane, the reaction proceeds by either an S_N2 or E2 mechanism, depending on the basicity and the amount of steric hindrance in the nucleophile.

(a)

$$CH_3CH_2CH_2Br + NaNH_2 \longrightarrow CH_3CH_2CH_2NH_2 + NaBr$$

(b)

$$CH_3CH_2CH_2Br + K^+ \ {}^-OC(CH_3)_3 \longrightarrow CH_3CH{=}CH_2 + HOC(CH_3)_3 + KBr$$

$K^+ \ {}^-OC(CH_3)_3$ is a strong, bulky base that brings about elimination as well as substitution.

(c)

$$CH_3CH_2CH_2Br + NaI \longrightarrow CH_3CH_2CH_2I + NaBr$$

(d)

$$CH_3CH_2CH_2Br + NaCN \longrightarrow CH_3CH_2CH_2CN + NaBr$$

(e)

$$CH_3CH_2CH_2Br \xrightarrow{Mg} CH_3CH_2CH_2MgBr \xrightarrow{H_2O} CH_3CH_2CH_3$$

10.39 To predict nucleophilicity, remember these guidelines:

(1) In comparing nucleophiles that have the same attacking atom, nucleophilicity parallels basicity. In other words, a more basic nucleophile is a more effective nucleophile.

(2) Nucleophilicity increases in going down a column of the periodic table.

(3) A negatively charged nucleophile is usually more reactive than a neutral nucleophile.

	More Nucleophilic	*Less Nucleophilic*	*Reason*
(a)	${}^-NH_2$	NH_3	Rule 1 or 3
(b)	$CH_3CO_2^-$	H_2O	Rule 1 or 3
(c)	F^-	BF_3	BF_3 is not a nucleophile
(d)	$(CH_3)_3P$	$(CH_3)_3N$	Rule 2
(e)	I^-	Cl^-	Rule 2
(f)	${}^-C{\equiv}N$	${}^-OCH_3$	Reactivity chart, Section 10.6

10.40 An alcohol is converted to an ether by two different routes in this series of reactions. The two resulting ethers have identical structural formulas but differ in the sign of specific rotation. Therefore, at some step or steps in these reaction sequences, inversion of configuration at the chiral carbon must have occurred. Let's study each step of the series to find where inversion is occurring.

$CH_3CH(OH)CH_2C_6H_5$, $[\alpha]_D = +33.0°$ —③ TosCl→ $CH_3CH(OTos)CH_2C_6H_5$, $[\alpha]_D = +31.1°$

① K ↓ (from the alcohol): $[CH_3CH(O^-K^+)CH_2C_6H_5]$

② CH_3CH_2Br ↓: $CH_3CH(OCH_2CH_3)CH_2C_6H_5$, $[\alpha]_D = +23.5°$

④ CH_3CH_2OH, heat ↓ (from the tosylate): $CH_3CH(OCH_2CH_3)CH_2C_6H_5$, $[\alpha]_D = -19.9°$

In step 1, the alcohol reacts with potassium metal to produce a potassium alkoxide. Since the bond between carbon and oxygen has not been broken, no inversion occurs in this step.

The potassium alkoxide acts as a nucleophile in the S_N2 displacement on CH_3CH_2Br in step 2. It is the C–Br bond of bromoethane, however, not the C–O bond of the alkoxide, that is broken. No inversion at the carbon chirality center occurs in step 2.

The starting alcohol reacts with tosyl chloride in step 3. Again, because the O–H bond, rather than the C–O bond, of the alcohol is broken, no inversion occurs at this step.

Inversion must therefore occur at step 4 when the ^-OTos group is displaced by CH_3CH_2OH. The C–O bond of the tosylate (–OTos) is broken, and a new C–O bond is formed.

Notice the specific rotations of the two enantiomeric products. The product of steps 1 and 2 should be enantiomerically pure because neither reaction has affected the C–O bond. Reaction 4 proceeds with some racemization at the chirality center to give a smaller absolute value of $[\alpha]_D$.

10.41

(a)

$CH_3CH_2CHBrCH_3 \xrightarrow{^-OEt}$ $CH_3CH_2CH{=}CH_2$ + cis-$CH_3CH{=}CHCH_3$ (H₃C and CH₃ …) + trans-$CH_3CH{=}CHCH_3$

(b)

$H_2C{=}CHBr$, like other vinylic organohalides, does not undergo nucleophilic substitutions.

(c)

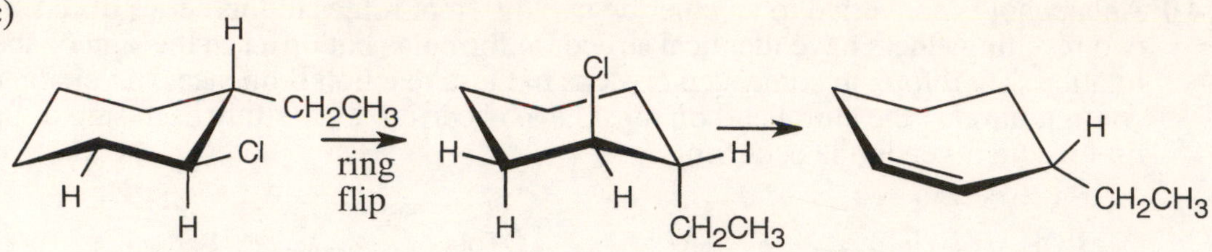

This alkyl halide gives the less substituted cycloalkene (non-Zaitsev product). Elimination to form the Zaitsev product does not occur because the –Cl and –H involved cannot assume the anti-periplanar geometry preferred for E2 elimination.

(d)

$$(CH_3)_3C{-}OH \; + \; HCl \xrightarrow{0°} (CH_3)_3C{-}Cl \; + \; H_2O$$

10.42 If reaction had proceeded with complete inversion, the product would have had a specific rotation of +53.6°. If complete racemization had occurred, $[\alpha]_D$ would have been zero. The observed rotation was +5.3°. Since $\frac{+5.3°}{+53.6°} = 0.099$, 9.9% of the original tosylate was inverted. The remaining 90.1% of the product must have been racemized.

10.43 (a) Substitution does not take place with secondary alkyl halides when a strong, bulky base is used. Elimination occurs instead and produces $H_2C{=}CHCH_2CH_3$ and $CH_3CH{=}CHCH_3$.
(b) Fluoroalkanes don't undergo S_N2 reactions because F^- is a poor leaving group.
(c) $SOCl_2$ in pyridine converts primary and secondary alcohols to chlorides by an S_N2 mechanism. 1-Methylcyclohexan-1-ol is a tertiary alcohol and does not undergo S_N2 substitution. Instead, E2 elimination occurs to give 1-methylcyclohexene.

10.44 S_N1 reactivity:

Least reactive ⟶ *Most reactive*

(a)

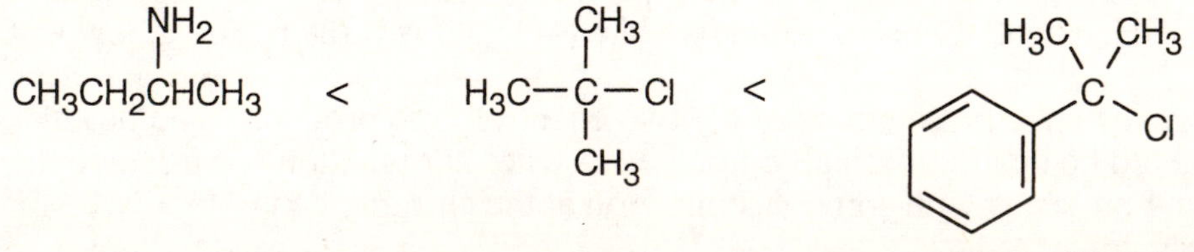

most stable carbocation

(b)

$(CH_3)_3COH$ < $(CH_3)_3CCl$ < $(CH_3)_3CBr$

best leaving group

(c)

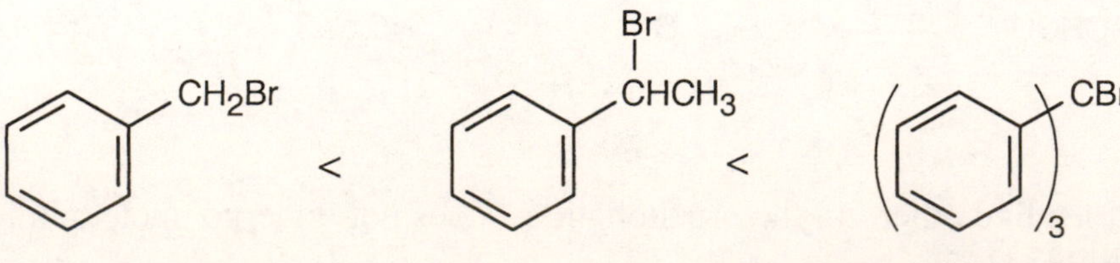

most stable carbocation

10.45 S_N2 reactivity:

Least reactive ⟶ *Most reactive*

(a)

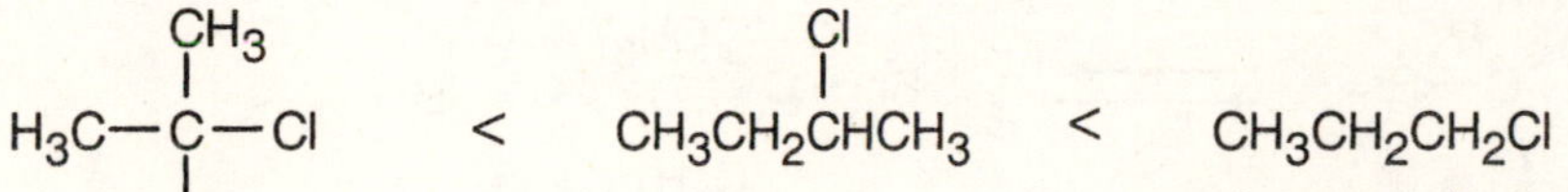

$H_3C-C(CH_3)_2-Cl$ < $CH_3CH_2CH(Cl)CH_3$ < $CH_3CH_2CH_2Cl$
primary substrate

(b)

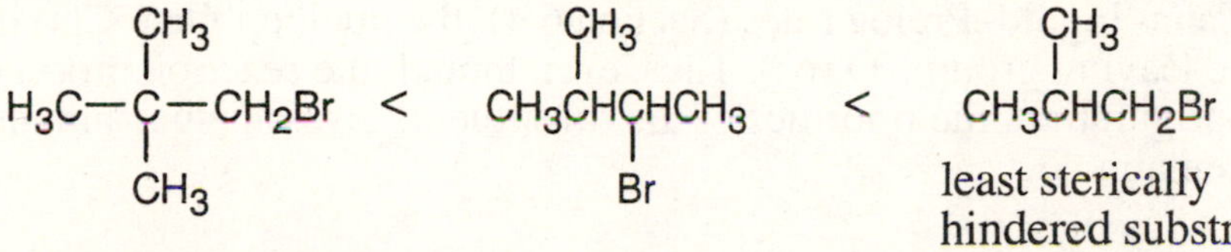

$H_3C-C(CH_3)_2-CH_2Br$ < $CH_3CH(CH_3)CH(Br)CH_3$ < $CH_3CH(CH_3)CH_2Br$
least sterically hindered substrate

(c)

$CH_3CH_2CH_2OCH_3$ < $CH_3CH_2CH_2Br$ < $CH_3CH_2CH_2OTos$
best leaving group

10.46 (*R*)-2-Bromooctane is a secondary bromoalkane, which undergoes S_N2 substitution. Since S_N2 reactions proceed with inversion of configuration, the configuration at the carbon chirality center is inverted. (This does not necessarily mean that all *R* isomers become *S* isomers after an S_N2 reaction. The *R,S* designation refers to the priorities of groups, which may change when the nucleophile is varied.)

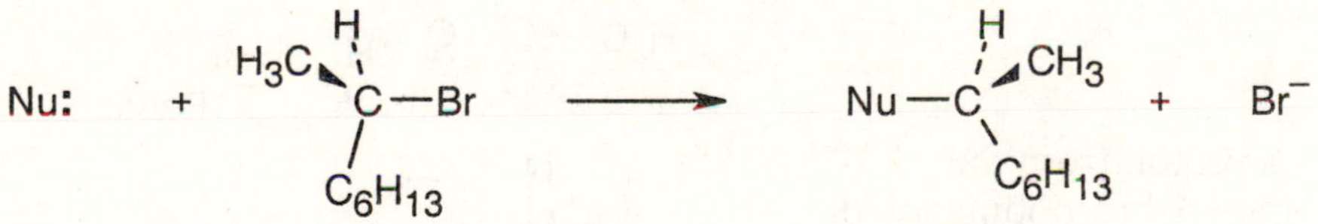

Nu: + H_3C, H, C_6H_{13} C—Br ⟶ Nu—C(H)(CH_3)(C_6H_{13}) + Br^-

(*R*)-2-Bromooctane

	Nucleophile	*Product*
(a)	^-CN	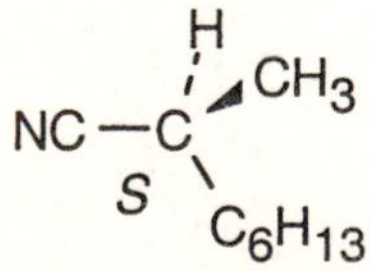 NC—C(H)(CH_3)(C_6H_{13}), *S*
(b)	$CH_3CO_2^-$	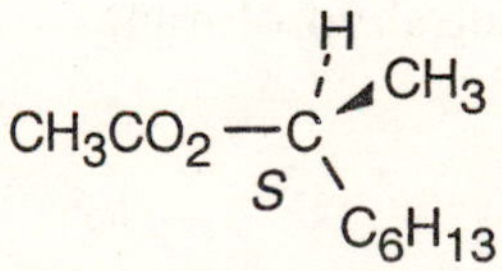 CH_3CO_2—C(H)(CH_3)(C_6H_{13}), *S*
(c)	CH_3S^-	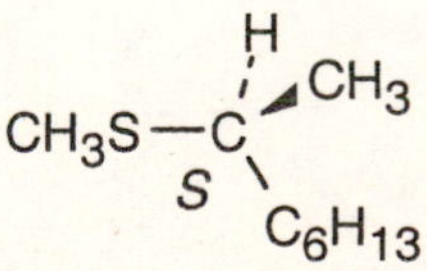 CH_3S—C(H)(CH_3)(C_6H_{13}), *S*

10.47 After 50% of the starting material has reacted, the reaction mixture consists of 50% (*R*)-2-bromooctane and 50% (*S*)-2-bromooctane. At this point, the *R* starting material is completely racemized.

Br^- + (R)-H_3C, H, C_6H_{13} C—Br ⇌ Br—C (S) H, CH_3, C_6H_{13} + Br^-

10.48 According to Cahn–Ingold–Prelog rules (Section 6.4), the nucleophile ($^-$CN) has a lower priority than the leaving group ($^-$OTos). Thus, even though the reaction proceeds with inversion of configuration, the priorities of the substituents also change, and the configuration remains *S*.

H (4), OTos (1), H_3C (3), CH_2OCH_3 (2), C *S* —NaCN→ NC (2), H (4), H_3C (3), CH_2OCH_3 (1), C *S*

10.49

Base removes a proton from the hydroxyl group of 4-bromobutan-1-ol.

$SN2$ displacement of Br^- by O^- yields the cyclic ether tetrahydrofuran.

$BrCH_2CH_2CH_2CH_2OH$ + :Base$^-$ ⇌ [$BrCH_2CH_2CH_2CH_2O^-$] + H:Base

→ Tetrahydrofuran + $:Br:^-$

10.50

Cl, H, Ph, H, Ph, H
A
=
Cl, Ph, H, H, Ph, H
B

Both Newman projections place –H and –Cl in the correct anti periplanar geometry for E2 elimination.

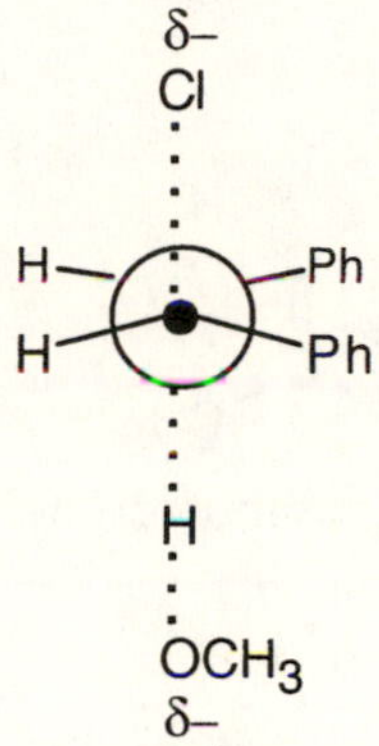

δ– Cl, H, Ph, H, Ph, H, OCH$_3$ δ–
A‡

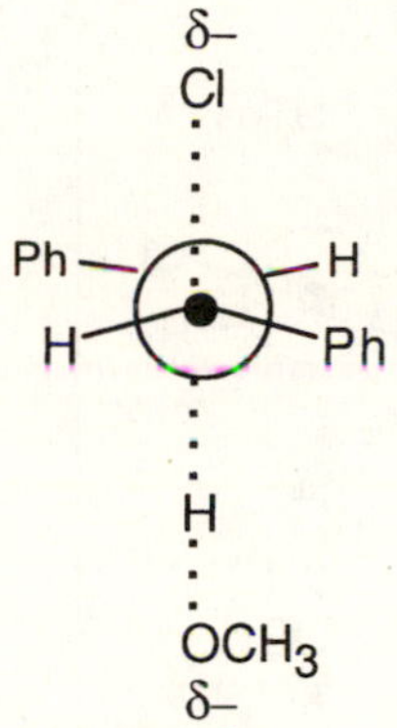

δ– Cl, Ph, H, H, Ph, H, OCH$_3$ δ–
B‡

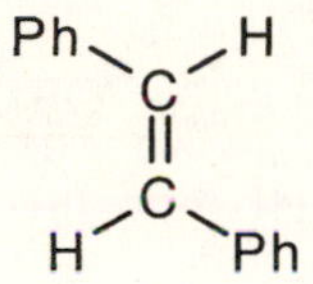

Ph(H)C=C(H)Ph

trans-1,2-Diphenylethylene

Either transition state **A**‡ or **B**‡ can form when 1-chloro-1,2-diphenylethane undergoes E2 elimination. Crowding of the two phenyl groups in **A**‡ makes this transition state (and the product resulting from it) of higher energy than transition state **B**‡. Formation of the product from **B**‡ is therefore favored, and *trans*-1,2-diphenylethylene is the major product.

10.51

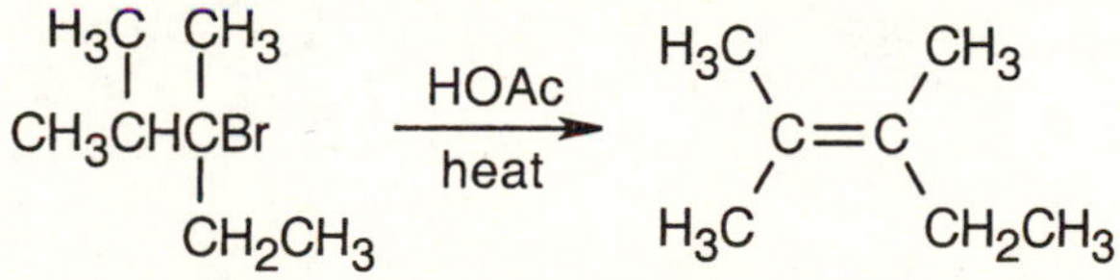

$CH_3CH(CH_3)C(CH_3)(Br)CH_2CH_3$ $\xrightarrow[\text{heat}]{\text{HOAc}}$ $(H_3C)_2C{=}C(CH_3)CH_2CH_3$

The alkene shown above has the most highly substituted double bond, and, according to Zaitsev's rule, is the major product. The following minor products may also form.

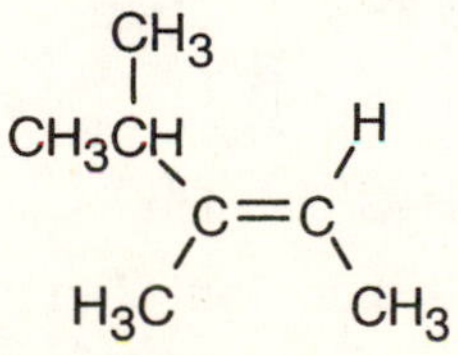

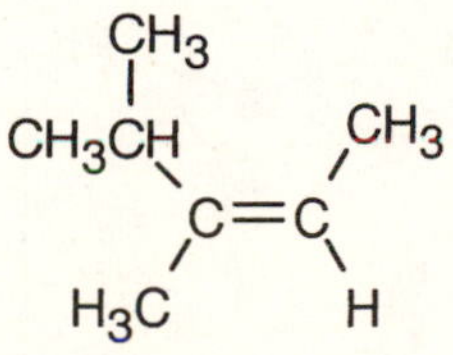

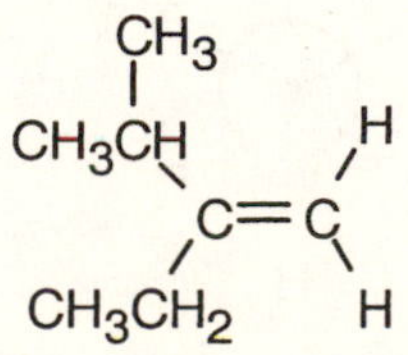

$(CH_3)_2CH(H_3C)C{=}C(H)CH_3$ $(CH_3)_2CH(H_3C)C{=}C(CH_3)H$ $(CH_3)_2CH(CH_3CH_2)C{=}CH_2$

10.52

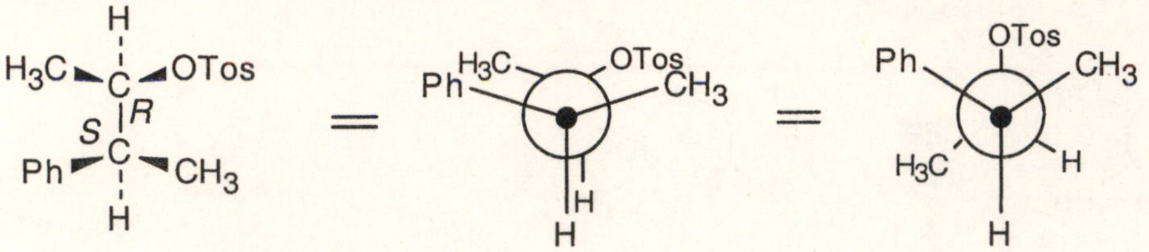

H, H₃C, OTos, R, S, Ph, CH₃, H = Ph, H₃C, OTos, CH₃, H, H = Ph, OTos, CH₃, H₃C, H, H

Draw a Newman projection of the tosylate of (2*R*,3*S*)-3-phenylbutan-2-ol, and rotate the projection until the –OTos and the –H on the adjoining carbon atom are anti periplanar. Even though this conformation has several gauche interactions, it is the only conformation in which –OTos and –H are 180° apart.

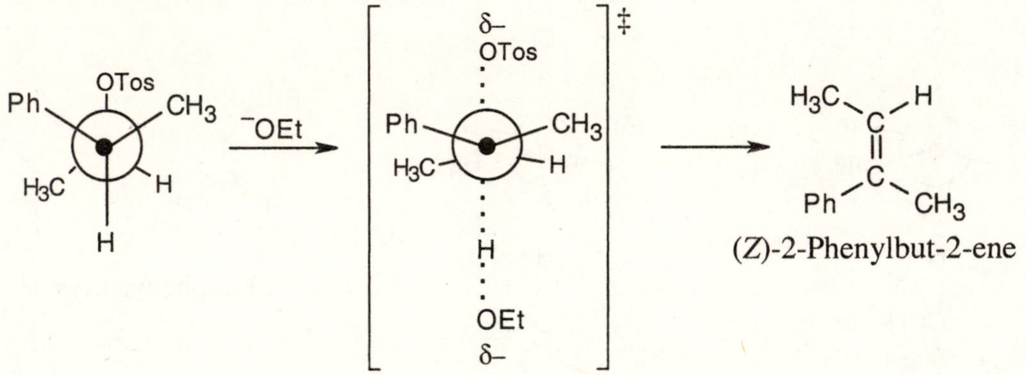

Ph, OTos, CH₃, H₃C, H, H → $^-$OEt → [δ– OTos, Ph, CH₃, H₃C, H, H, OEt δ–]‡ → H₃C, H, C=C, Ph, CH₃

(*Z*)-2-Phenylbut-2-ene

Elimination yields the *Z* isomer of 2-phenylbut-2-ene.

10.53 Using the same argument used in the previous problem, one can show that elimination from the tosylate of (2*R*,3*R*)-3-phenylbutan-2-ol gives the *E* alkene.

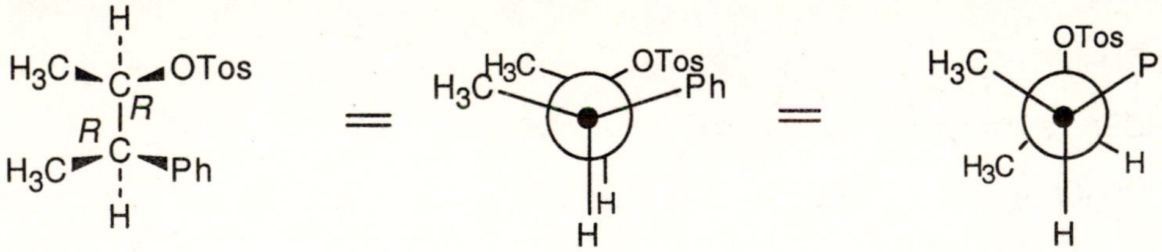

H, H₃C, OTos, R, R, H₃C, Ph, H = H₃C, H₃C, OTos, Ph, H, H = H₃C, OTos, Ph, H₃C, H, H

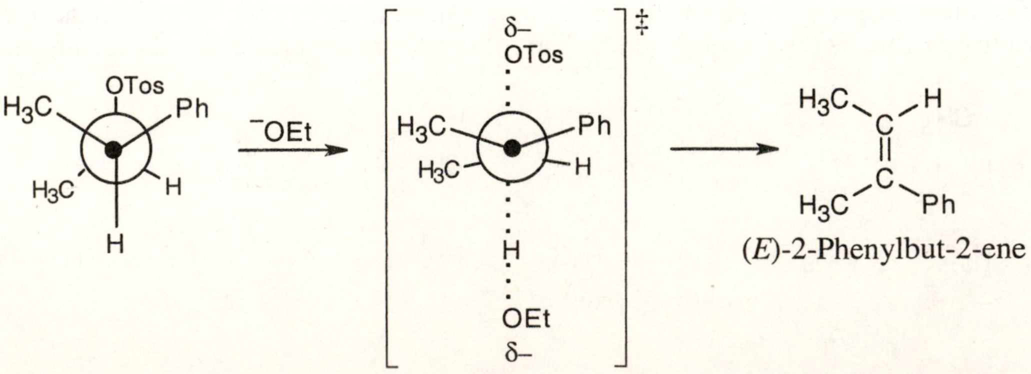

H₃C, OTos, Ph, H₃C, H, H → $^-$OEt → [δ– OTos, H₃C, Ph, H₃C, H, H, OEt δ–]‡ → H₃C, H, C=C, H₃C, Ph

(*E*)-2-Phenylbut-2-ene

The (2*S*,3*S*) isomer also forms the *E* alkene; the (2*S*,3*R*) isomer yields the *Z* alkene.

10.54

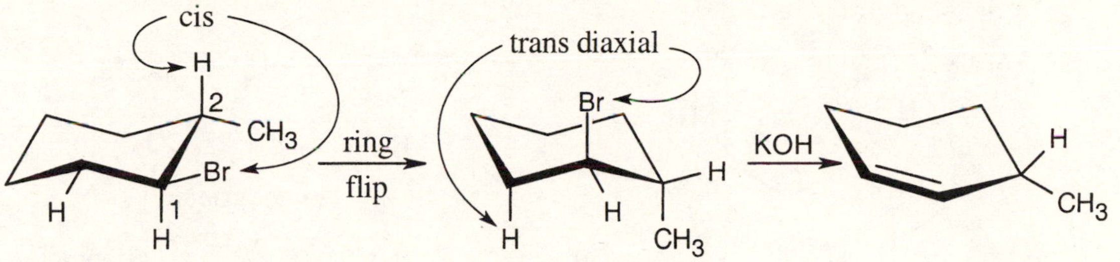

E2 reactions require that the two atoms to be eliminated have a periplanar relationship. Since it's impossible for bromine and the hydrogen at C2 to be periplanar, elimination occurs in the non-Zaitsev direction to yield 3-methylcyclohexene.

10.55

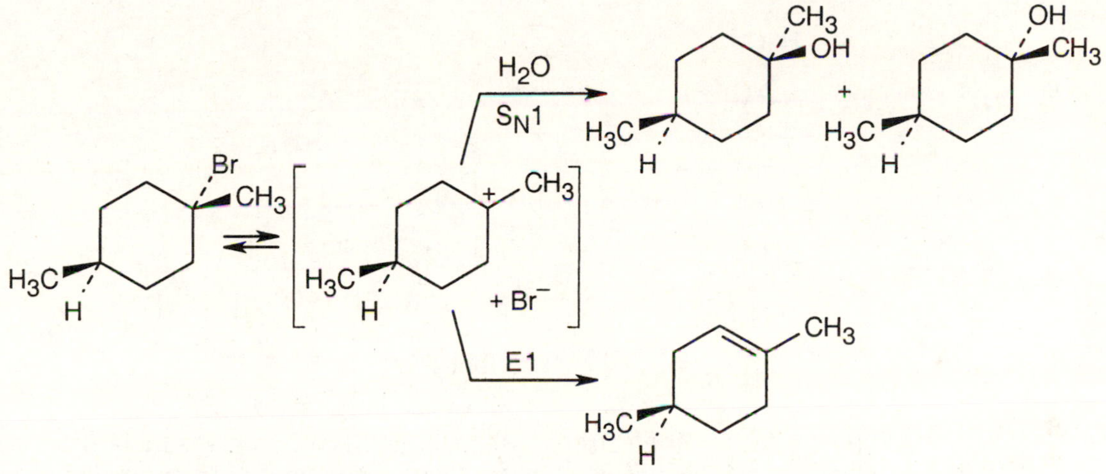

This tertiary bromoalkane reacts by S_N1 and E1 routes to yield alcohol and alkene products. The alcohol products are diastereomers.

10.56 Step 1: NAD^+ oxidizes an alcohol to a ketone.

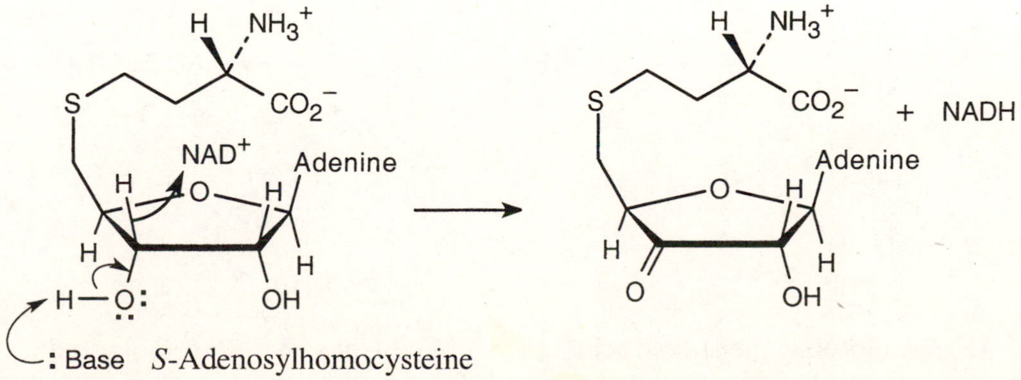

Step 2: Base brings about an E1cB elimination reaction that has homocysteine as the leaving group.

H—A   H NH$_3^+$ S   CO$_2^-$ Adenine   O   H   H   H   : Base O   OH   →   HS   H NH$_3^+$ CO$_2^-$

Homocysteine

+

H$_2$C=   O   H   Adenine   H   O   OH

10.57

$:N{\equiv}C:^- + H_3C(H)(H)C{-}I \longrightarrow :N{\equiv}C{-}C(CH_3)(H)(H) + I^-$

Formal charge = −1

$^-:C{\equiv}N: + H_3C(H)(H)C{-}I \longrightarrow {}^-:C{\equiv}\overset{+}{N}{-}C(CH_3)(H)(H) + I^-$

Formal charge = +1

S_N2 attack by the lone pair electrons associated with carbon gives the nitrile product. Attack by the lone pair electrons associated with nitrogen yields isonitrile product.

10.58 NH_2^- acts as a base to form the acetylide anion, which undergoes S_N2 reaction with a primary bromoalkane to yield the internal alkyne.

$RC{\equiv}C{-}H + :NH_2^- \longrightarrow [\,:NH_3 \quad RC{\equiv}C:^- \quad Br{-}CH_2R'\,] \longrightarrow RC{\equiv}CH_2R' + Br^-$

10.59

HO$_2$C   H   C=C   Cl   CO$_2$H

(*Z*)-2-Chlorobut-2-enedioic acid

HO$_2$C   H   C=C   HO$_2$C   Cl

(*E*)-2-Chlorobut-2-enedioic acid

Hydrogen and chlorine are anti to each other in the *Z* isomer and are syn in the *E* isomer. Since the *Z* isomer reacts fifty times faster than the *E* isomer, elimination must proceed more favorably when the substituents to be eliminated are anti to one another. This is the same stereochemical result that occurs in E2 eliminations of alkyl halides.

10.60 Since butan-2-ol is a secondary alcohol, substitution can occur by either an S_N1 or S_N2 route, depending on reaction conditions. Two factors favor an S_N1 mechanism in this case. (1) The reaction is run under acidic conditions in a polar, protic solvent. (2) Dilute acid converts a poor leaving group (^-OH) into a good leaving group (OH_2), which dissociates easily.

Protonation of the hydroxyl oxygen..

is followed by loss of water to form a planar carbocation.

Reaction with water from either side of the planar carbocation yields racemic product.

10.61 The chiral tertiary alcohol (*R*)-3-methylhexan-3-ol reacts with HBr by an S_N1 pathway. HBr protonates the hydroxyl group, which dissociates to yield a planar, achiral carbocation. Reaction with bromide ion can occur from either side of the carbocation to produce (±)3-bromo-3-methylhexane. The mechanism is similar to that shown in the previous problem.

10.62 Since carbon–deuterium bonds are slightly stronger than carbon–hydrogen bonds, more energy is required to break a C–D bond than to break a C–H bond. In a reaction where either a carbon–deuterium or a carbon–hydrogen bond can be broken in the rate-limiting step, C–H bond-breaking occurs faster because the energy of activation for C–H breakage is lower.

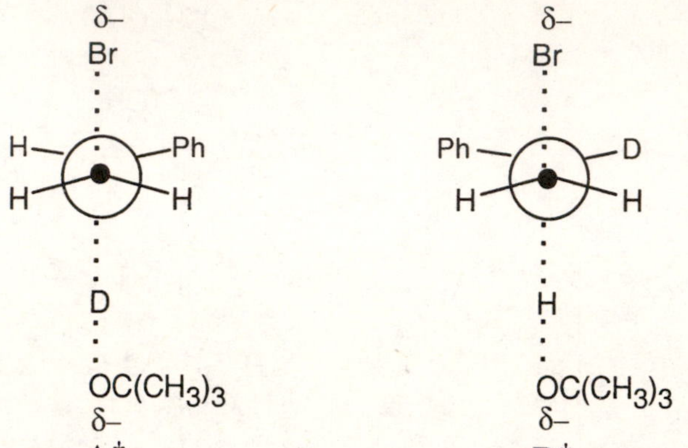

δ– Br H Ph H H D $OC(CH_3)_3$ δ– **A‡**

δ– Br Ph D H H H $OC(CH_3)_3$ δ– **B‡**

Transition state **A**‡ is of higher energy than transition state **B**‡ because more energy is required to break the C–D bond. The product that results from transition state **B**‡ is thus formed in greater abundance.

10.63

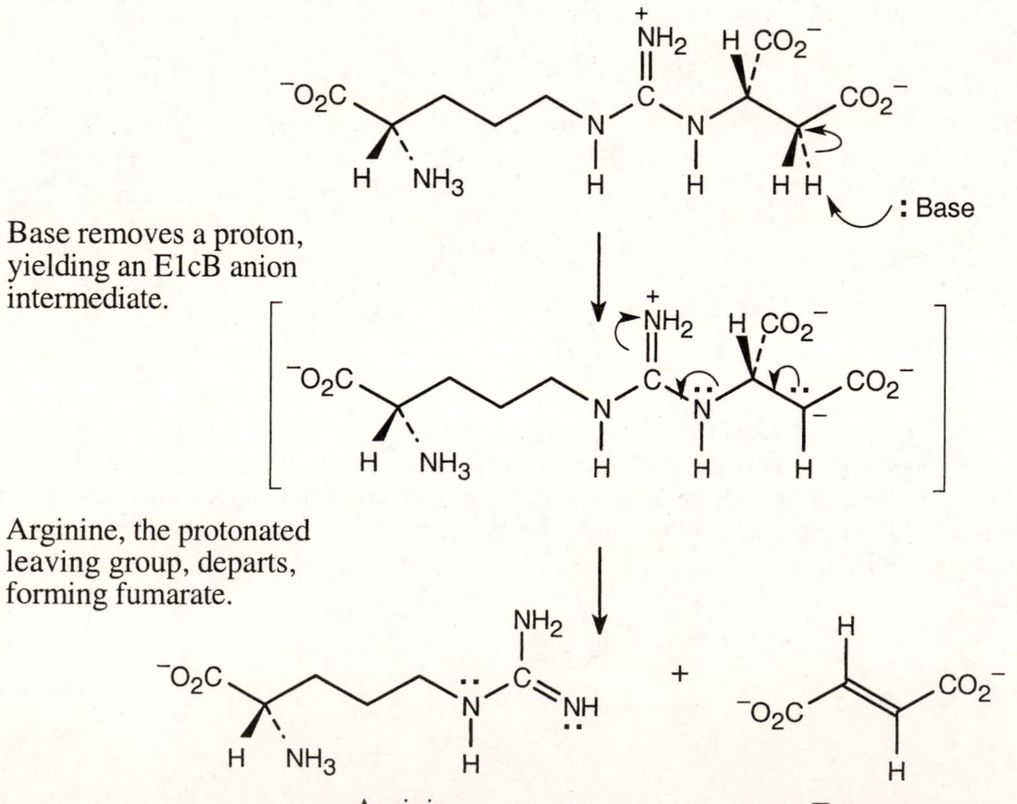

Base removes a proton, yielding an E1cB anion intermediate.

Arginine, the protonated leaving group, departs, forming fumarate.

Arginine

Fumarate

10.64

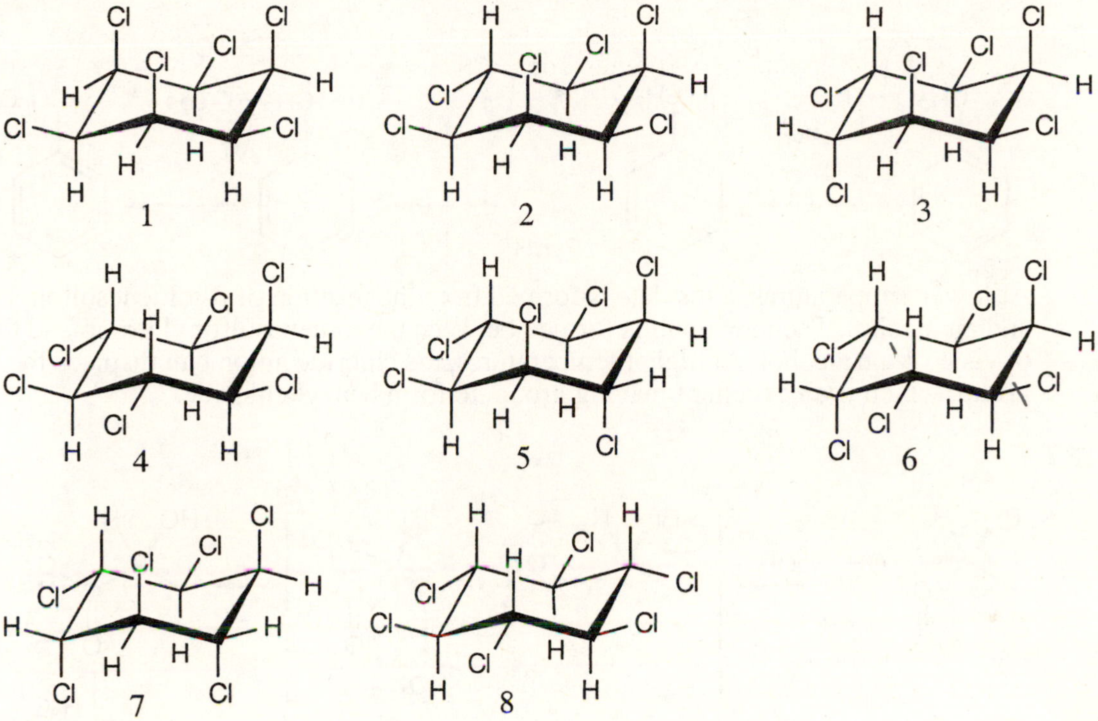

Diastereomer *8* reacts much more slowly than other isomers in an E2 reaction because no pair of hydrogen and chlorine atoms can adopt the anti-periplanar orientation preferred for E2 elimination.

10.65

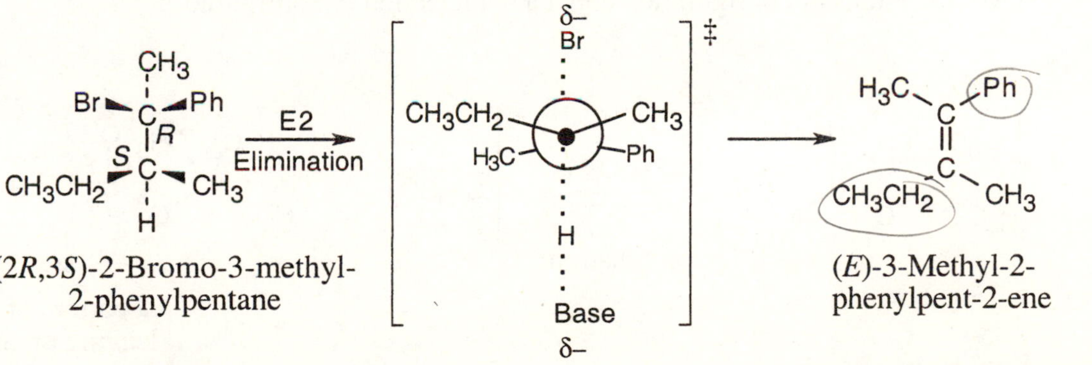

(2*R*,3*S*)-2-Bromo-3-methyl-2-phenylpentane

E2 Elimination

(*E*)-3-Methyl-2-phenylpent-2-ene

The (2*S*,3*R*) isomer also yields *E* product.

10.66

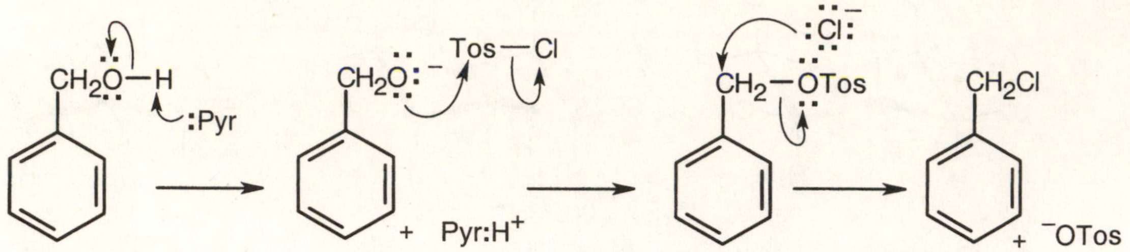

At lower temperatures, a tosylate is formed from the reaction of *p*-toluenesulfonyl chloride and an alcohol. The new bond is formed between the toluenesulfonyl group and the oxygen of the alcohol. At higher temperatures, the chloride anion can displace the –OTos group, which is an excellent leaving group, to form benzyl chloride.

10.67

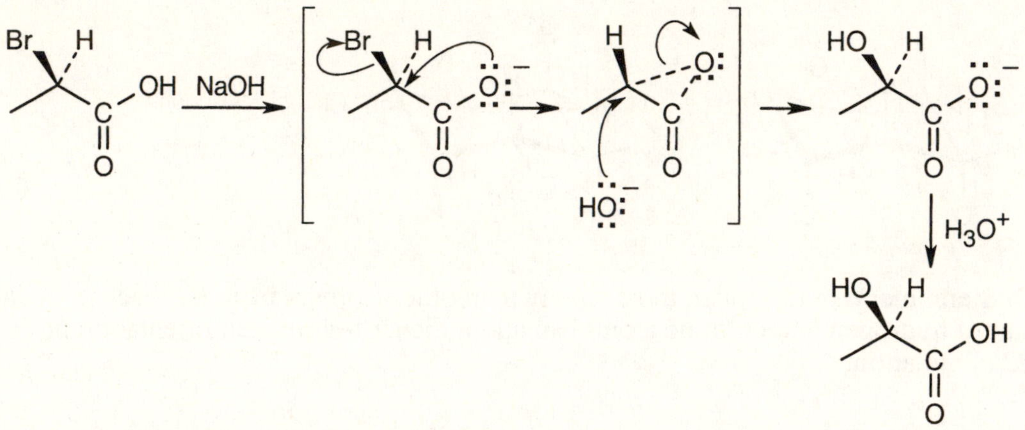

Two inversions of configuration equal a net retention of configuration.

10.68

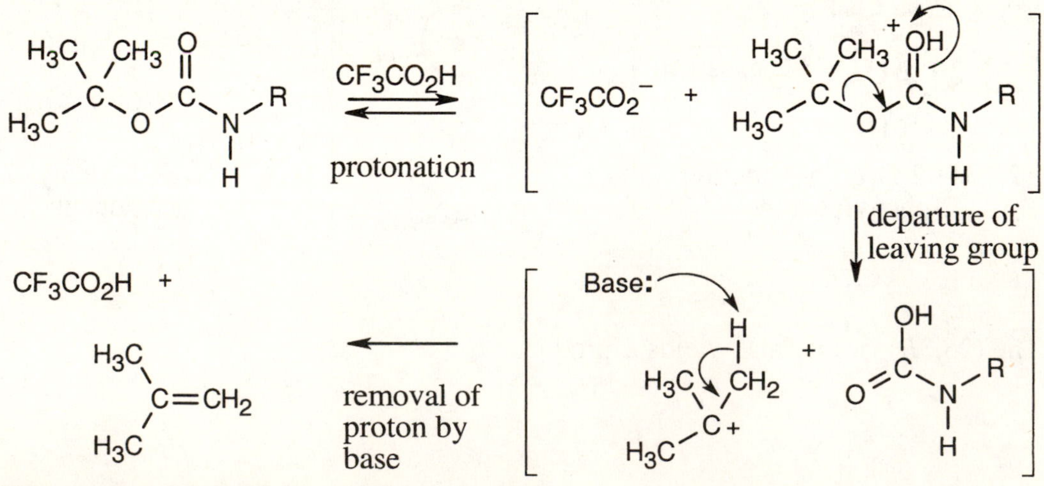

This reaction proceeds by an E1 mechanism.

10.69 Notice that the chiral methyl group has the (*R*) configuration in both *N*-methyltetrahydrofolate and in methionine. This fact suggests that methylation proceeds with two inversions of configuration which, in fact, has been shown to be the case.

10.70

$$CH_3CH_2CH_2CH_2CH_2NH_2 \xrightarrow{\text{excess } CH_3I} CH_3CH_2CH_2CH_2CH_2\overset{+}{N}(CH_3)_3\ I^-$$

$$CH_3CH_2CH_2CH_2CH_2\overset{+}{N}(CH_3)_3\ I^- \xrightarrow{Ag_2O,\ H_2O} CH_3CH_2CH_2CH{=}CH_2 + :N(CH_3)_3$$

The intermediate is a charged quaternary ammonium compound that results from S_N2 substitutions on three CH_3I molecules by the amine nitrogen. E2 elimination occurs because the neutral $N(CH_3)_3$ molecule is a good leaving group.

10.71

$$\text{cyclohexyl-}\ddot{\underset{..}{O}}:^- + CH_3I \longrightarrow \text{cyclohexyl-}OCH_3 + I^-$$

This is an excellent method of ether preparation because iodomethane is very reactive in S_N2 displacements.

$$\text{cyclohexyl-}I + CH_3\ddot{\underset{..}{O}}:^- \longrightarrow \text{cyclohexyl-}OCH_3 + \text{cyclohexene} + CH_3OH + I^-$$

Reaction of a secondary alkyl halide with a basic nucleophile yields both substitution and elimination products. This is a less satisfactory method of ether preparation.

Review Unit 4: Stereochemistry; Alkyl Halides; Substitutions and Eliminations

Major Topics Covered (with vocabulary):

Handedness:
stereoisomer enantiomer chiral plane of symmetry achiral chirality center plane-polarized light optical activity levorotatory dextrorotatory specific rotation

Stereoisomers and configuration:
configuration absolute configuration diastereomer meso compound racemate resolution prochirality *re* face *si* face prochirality center *pro-R* *pro-S*

Stereochemistry of reactions.

Alkyl halides:
allylic position delocalization Grignard reagent

Substitution reactions:
nucleophilic substitution reaction Walden inversion reaction rate kinetics second-order reaction rate constant S_N2 reaction bimolecular nucleophilicity leaving group solvation S_N1 reaction first-order reaction rate-limiting step ion pair dielectric polarization

Elimination reactions:
Zaitsev's rule E2 reaction syn periplanar geometry anti periplanar geometry deuterium isotope effect E1 reaction

Types of Problems:

After studying these chapters, you should be able to:

- Calculate the specific rotation of an optically active compound.
- Locate chirality centers, assign priorities to substituents, and assign *R,S* designations to chirality centers.
- Given a stereoisomer, draw its enantiomer and/or diastereomers.
- Locate the symmetry plane of a meso compound.
- Predict the stereochemistry of reaction products, using the concept of prochirality when appropriate.

- Draw, name, and synthesize alkyl halides.
- Formulate the mechanisms of S_N2, S_N1 and elimination reactions.
- Predict the effect of substrate, nucleophile, leaving group and solvent on substitution and elimination reactions.
- Predict the products of substitution and elimination reactions.
- Classify substitution and elimination reactions by type.

Points to Remember:

* A helpful strategy for assigning *R,S* designations: Using models, build two enantiomers by adding four groups to each of two tetrahedral carbons. Number the groups 1–4, to represent priorities of groups at a tetrahedral carbon, and assign a configuration to each carbon. Attach a label that indicates the configuration of each enantiomer. Keep these two enantiomers, and use them to check your answer every time that you need to assign *R,S* configurations to a chiral atom.

* When assigning *pro-R* or *pro-S* designations to a hydrogen, mentally replace the hydrogen that points out of the plane of the page. The other hydrogen is then positioned for prochirality assignment without manipulating the molecule. If the designation is *R*, the replaced hydrogen is *pro-R*; if the designation is *S*, the replaced hydrogen is *pro-S*.

* In naming alkyl halides by the IUPAC system, remember that a halogen is named as a substituent on an alkane. When numbering the alkyl halide, the halogens are numbered in the same way as alkyl groups and are cited alphabetically.

* Predicting the outcome of substitutions and eliminations is only straightforward in certain cases. For primary halides, S_N2 and E2 reactions are predicted. For tertiary halides, S_N1, E2 and E1(to a certain extent) are the choices. The possibilities for secondary halides are more complicated. In addition, many reactions yield both substitution and elimination products, and both inversion and retention of configuration may occur in the same reaction.

Self-Test:

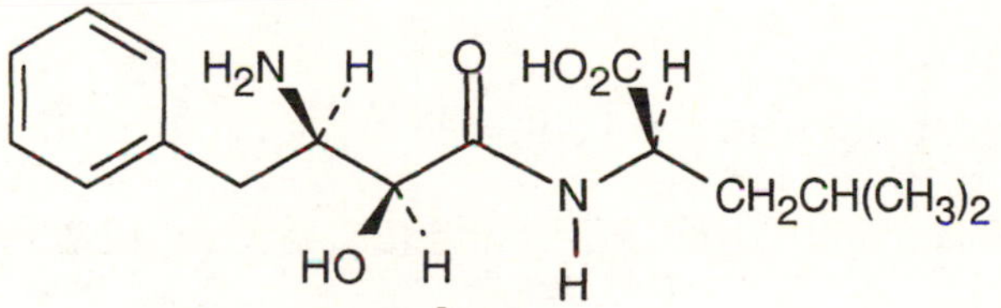

H_2N H O HO_2C H N $CH_2CH(CH_3)_2$ HO H H

A
Ubenimex
(an antitumor drug)

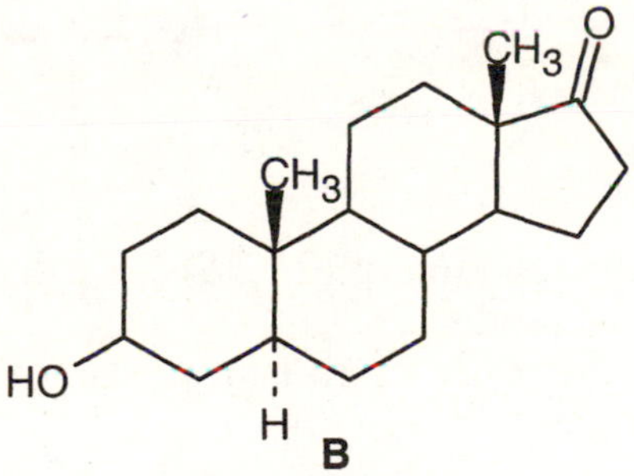

CH_3 O CH_3 HO H

B
Epiandosterone
(an androgen)

Assign *R,S* designations to the chiral carbons in **A**. Indicate the chirality centers in **B**. How many stereoisomers of **B** are possible?

$CH_3CH_2CHCH(CH_3)CH_3$ with Br on C-3 (structure: CH_3 above, Br below $CH_3CH_2CHCHCH_3$)

C

$HC{\equiv}CC(CH_2CH_3)(OH)CH{=}CHCl$ (structure: CH_2CH_3 above, OH below $HC{\equiv}CCCH{=}CHCl$)

D

Ethchlorvynol
(a sedative)

Name **C**. Draw all stereoisomers of **C**, label them, and describe their relationship. Predict the products of reaction of **C** with:(a) NaOH; (b) Mg, then H_2O. Draw the E2 elimination products formed from **C**, and give *E,Z* designations to the double bond configurations.

Draw the *R* enantiomer of **D**. Predict the products of reaction of **D** with: (a) HBr; (b) product of (a) + aqueous ethanol. Describe the reactivity of the –Cl atom in substitution and elimination reactions.

Multiple Choice:

1. A meso compound and a racemate are identical in all respects except:
(a) molecular formula (b) degree of rotation of plane-polarized light
(c) connectivity of atoms (d) physical properties

2. Which of the following projections represents an *R* enantiomer?

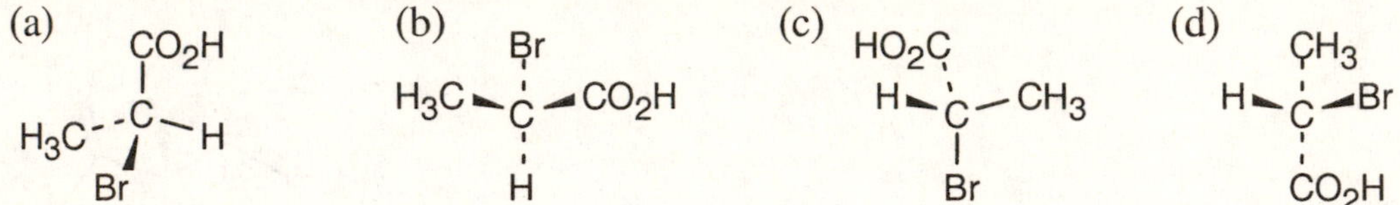

3. In the reaction of (2*R*,3*S*)-3-methyl-2-pentanol with tosyl chloride, what is the configuration of the product?
(a) a mixture of all four possible stereoisomers (b) (2*R*,3*S*) and (2*S*,3*S*) (c) (2*R*,3*S*)
(d) (2*S*,3*S*)

4. Which of the following statements is false?
(a) 2,3-Dibromopentane has four stereoisomers. (b) 1,5-Dibromopentane has a prochiral carbon. (c) (2*R*,4*R*)-2,4-Dibromopentane is a meso compound. (d) 3-Bromopentane has a *pro-R* arm and a *pro-S* arm.

5. A Grignard reagent:
(a) is a strong acid (b) can be used to introduce deuterium into a hydrocarbon (c) is useful biochemically (d) only reacts with alkyl halides

6. Which of the following reactions is an oxidation?
(a) hydroxylation (b) hydration (c) hydrogenation (d) addition of HBr

7. All of the following are true of S_N2 reactions except:
(a) The rate varies with the concentration of nucleophile (b) The rate varies with the type of nucleophile (c) The nucleophile is involved in the rate-determining step (d) The rate of the S_N2 reaction of a substrate and a nucleophile is the same as the rate of the E2 reaction of the same two compounds.

8. Which of the following is true of S_N1 reactions?
(a) The rate varies with the concentration of nucleophile (b) The rate varies with the type of nucleophile (c) The rate is increased by use of a polar solvent. (d) The nucleophile is involved in the rate-determining step.

9. Which base is best for converting 1-bromohexane to 1-hexene?
(a) $(CH_3)_3CO^-$ (b) ^-CN (c) ^-OH (d) $^-C{\equiv}CH$

10. Which of the following is both a good nucleophile and a good leaving group?
(a) ^-OH (b) ^-CN (c) ^-Cl (d) ^-I

Chapter 11 – Structure Determination: Mass Spectrometry, Infrared Spectroscopy, and Ultraviolet Spectroscopy

Chapter Outline

I. Mass Spectrometry (Sections 11.1 – 11.4).
  A. General features of mass spectrometry (Section 11.1).
    1. Purpose of mass spectrometry.
      a. Mass spectrometry is used to measure the molecular weight of a compound.
      b. Mass spectrometry can also provide information on the structure of an unknown compound.
    2. Technique of mass spectrometry for small molecules.
      a. A small amount of sample is vaporized into the *ionization source* and is bombarded by a stream of high-energy electrons.
      b. An electron is dislodged, producing a cation radical.
      c. Most of the cation radicals fragment; the fragments may be positively charged or neutral.
      d. In the *deflector*, a strong magnetic field deflects the positively charged fragments, which are separated by *m/z* ratio.
      e. A *detector* records the fragments as peaks.
    3. Important terms.
      a. The mass spectrum is presented as a bar graph, with masses *(m/z)* on the *x* axis and intensity on the *y* axis.
      b. The base peak is the tallest peak and is assigned an intensity of 100%.
      c. The parent peak, or molecular ion (M^+), corresponds to the unfragmented cation radical.
      d. In large molecules, the base peak is often not the molecular ion.
  B. Interpreting mass spectra (Sections 11.2 – 11.3).
    1. Molecular weight (Section 11.2).
      a. Mass spectra can frequently provide the molecular weight of a sample.
        i. Double-focusing mass spectrometers can provide mass measurements accurate to 0.0001 amu.
        ii. Some samples fragment so easily that M^+ is not seen.
      b. If you know the molecular weight of the sample, you can often deduce its molecular formula.
      c. There is often a peak at M+1 that is due to contributions from ^{13}C and ^{2}H.
    2. Fragmentation patterns of hydrocarbons.
      a. Fragmentation patterns can be used to identify a known compound, because a given compound has a unique fragmentation "fingerprint".
      b. Fragmentation patterns can also provide structural information.
        i. Most hydrocarbons fragment into carbocations and radicals.
        ii. The positive charge remains with the fragment most able to stabilize it.
        iii. It is often difficult to assign structures to fragments.
        iv. For hexane, major fragments correspond to the loss of methyl, ethyl, propyl, and butyl radicals.
    3. Fragmentation patterns of common functional groups (Section 11.3).
      a. Alcohols.
        i. Alcohols can fragment by alpha cleavage, in which a C–C bond next to the –OH group is broken.
          The products are a cation and a radical.
        ii. Alcohols can also dehydrate, leaving an alkene cation radical with a mass 18 units less than M^+.

b. Amines also undergo alpha cleavage, forming a cation and a radical.
c. Carbonyl compounds.
i. Aldehydes and ketones with a hydrogen 3 carbons from the carbonyl group can undergo the McLafferty rearrangement.
The products are a cation radical and a neutral alkene.
ii. Aldehydes and ketones also undergo alpha cleavage, which breaks a bond between the carbonyl group and a neighboring carbon.
The products are a cation and a radical.

C. Mass spectrometry in biological systems: TOF instruments (Section 11.4).
1. Time-of-flight instruments are used to produce charged molecules with little fragmentation.
2. The ionizer can be either ESI or MALDI.
a. In an ESI source, the sample is dissolved in a polar solvent and sprayed through a steel capillary tube.
i. As the sample exits, it is subjected to a high voltage, which protonates the sample.
ii. The solvent is evaporated, yielding protonated sample molecules.
b. In a MALDI source, the sample is absorbed onto a matrix compound.
i. The matrix compound is ionized by a burst of laser light.
ii. The matrix compound transfers energy to the sample, protonating it.
3. The samples are focused into a small packet and given a burst of energy.
Each molecule moves at a velocity that depends on the square of its mass.
4. The analyzer is an electrically grounded tube that detects the molecules by velocity.

II. Spectroscopy and the electromagnetic spectrum (Section 11.5).
A. The nature of radiant energy.
1. Different types of electromagnetic radiation make up the electromagnetic spectrum.
2. Electromagnetic radiation behaves both as a particle and as a wave.
3. Electromagnetic radiation can be characterized by three variables.
a. The wavelength (λ) measures the distance from one maximum to the next.
b. The frequency (ν) measures the number of wave maxima that pass a fixed point per unit time.
c. The amplitude is the height measured from the midpoint to the maximum.
4. Wavelength times frequency equals the speed of light.
5. Electromagnetic energy is transmitted in discrete energy bundles called quanta.
a. $\varepsilon = h \times \nu$
b. Energy varies directly with frequency but inversely with wavelength.
c. $E = 1.20 \times 10^{-2}$ kJ/mol $\div\ \lambda$ (cm) for a "mole" of photons.

B. Electromagnetic radiation and organic molecules.
1. When an organic compound is struck by a beam of electromagnetic radiation, it absorbs radiation of certain wavelengths, and transmits radiation of other wavelengths.
2. If we determine which wavelengths are absorbed and which are transmitted, we can obtain an absorption spectrum of the compound.
For an infrared spectrum:
i. The horizontal axis records wavelength.
ii. The vertical axis records percent transmittance.
iii. The baseline runs across the top of the spectrum.
iv. Energy absorption is a downward spike.
3. The energy a molecule absorbs is distributed over the molecule.
4. There are many types of spectroscopies that differ in the region of the electromagnetic spectrum that is being used.

III. Infrared Spectroscopy (Sections 11.6 – 11.8).
   A. Infrared radiation (Section 11.6).
      1. The infrared (IR) region of the electromagnetic spectrum extends from 7.8×10^{-7} m to 10^{-4} m.
         a. Organic chemists use the region from 2.5×10^{-6} m to 2.5×10^{-5} m.
         b. Wavelengths are usually given in μm, and frequencies are expressed in wavenumbers, which are the reciprocal of wavelength.
         c. The useful range of IR radiation is 4000 cm^{-1} – 400 cm^{-1}; this corresponds to energies of 48.0 kJ/mol – 4.80 kJ/mol.
      2. IR radiation causes bonds to stretch and bend and causes other molecular vibrations.
      3. Energy is absorbed at a specific frequency that corresponds to the frequency of the vibrational motion of a bond.
      4. If we measure the frequencies at which IR energy is absorbed, we can find out the kinds of bonds a compound contains and identify functional groups.
   B. Interpreting IR spectra (Sections 11.7 – 11.8).
      1. General principles (Section 11.7).
         a. Most molecules have very complex IR spectra.
            i. This complexity means that each molecule has a unique fingerprint that allows it to be identified by IR spectroscopy.
            ii. Complexity also means that not all absorptions can be identified.
         b. Most functional groups have characteristic IR absorption bands that don't change from one compound to another.
         c. The significant regions of IR absorptions :
            i. 4000 cm^{-1}–2500 cm^{-1} corresponds to absorptions by C–H, O–H, and N–H bonds.
            ii. 2500 cm^{-1}–2000 cm^{-1} corresponds to triple-bond stretches.
            iii. 2000 cm^{-1}–1500 cm^{-1} corresponds to double bond stretches.
            iv. The region below 1500 cm^{-1} is the fingerprint region, where many complex bond vibrations occur.
         d. The frequency of absorption of different bonds depends on two factors:
            i. The strength of the bond.
            ii. The difference in mass between the two atoms in the bond.
      2. Interpreting IR spectra (Section 11.8).
         a. Alkanes.
            i. C–C absorbs at 800–1300 cm^{-1}.
            ii C–H absorbs at 2850–2960 cm^{-1}.
         b. Alkenes.
            i. =C–H absorbs at 3020–3100 cm^{-1}.
            ii. C=C absorbs at 1650–1670 cm^{-1}.
            iii. $RCH{=}CH_2$ absorbs at 910 and 990 cm^{-1}.
            iv. $R_2C{=}CH_2$ absorbs at 890 cm^{-1}.
         c. Alkynes.
            i. –C≡C– absorbs at 2100–2260 cm^{-1}.
            ii. ≡C–H absorbs at 3300 cm^{-1}.
         d. Aromatic compounds.
            i. –C=H absorbs at 3030 cm^{-1}.
            ii. Two bands (1500 cm^{-1} and 1600 cm^{-1}) are due to ring motions.
            iii. Strong absorptions, due to ring substitution patterns, appear in the range 690 cm^{-1} to 900 cm^{-1}.

e. The alcohol O–H bond absorbs at 3400–3650 cm^{-1}.
f. The N–H bond of amines absorbs at 3300–3500 cm^{-1}.
g. Carbonyl compounds.
   i. Saturated aldehydes absorb at 1730 cm^{-1}; unsaturated aldehydes absorb at 1705 cm^{-1}.
   ii. Saturated ketones absorb at 1715 cm^{-1}; unsaturated ketones absorb at 1690 cm^{-1}.
   iii. Saturated esters absorb at 1735 cm^{-1}; unsaturated esters absorb at 1715 cm^{-1}.

III. Ultraviolet spectroscopy (Sections 11.9 – 11.11).
   A. Principles of ultraviolet spectroscopy (Section 11.9).
      1. The ultraviolet region of interest is between the wavelengths 200 nm and 400 nm.
      2. The energy absorbed is used to promote a π electron in a conjugated system from one orbital to another.
      3. For example, when buta-1,3-diene is irradiated with ultraviolet light, a π electron is promoted from the highest occupied molecular orbital (HOMO) to the lowest unoccupied molecular orbital.
      4. UV radiation of 217 nm is necessary to promote this transition.
      5. This transition is known as a $\pi \rightarrow \pi^*$ transition.
   B. The ultraviolet spectrum.
      1. A UV spectrum is a plot of absorbance vs. wavelength.
         a. The absorbance is $A = \log [I_0/I]$.
         b. I_0 = intensity of incident light.
         c. I = intensity of transmitted light.
         d. The baseline is zero absorbance.
      2. For a specific substance, A is related to the molar absorptivity (ε).
         a. Molar absorptivity is the absorbance of a sample whose concentration is 1 mol/L with a pathlength of 1 cm..
         b. $A = \varepsilon \times c \times l$
         c. The range of ε is 5,000 – 25,000 L/mol·cm.
      3. UV spectra usually consist of a single broad peak, whose maximum is λ_{max}.
   C. Interpreting UV spectra (Section 11.10).
      1. The wavelength necessary for a $\pi \rightarrow \pi^*$ transition depends on the energy difference between HOMO and LUMO.
      2. By measuring this difference, it is possible to learn about the extent of conjugation in a molecule.
      3. As the extent of conjugation increases, λ_{max} increases.
      4. Different types of conjugated systems have characteristic values of λ_{max}.
   D. Colored organic compounds (Section 11.11).
      Compounds with extensive systems of conjugated bonds absorb in the visible range of the electromagnetic spectrum (400 – 800 nm).

Solutions to Problems

11.1 **Strategy:** If the isotopic masses of the atoms C, H, and O had integral values of 12 amu, 1 amu and 16 amu, many molecular formulas would correspond to a molecular weight of 288 amu. Because isotopic masses are not integral, however, only one molecular formula is associated with a molecular ion at 288.2089 amu.

To reduce the number of possible formulas, assume that the difference in molecular weight between 288 and 288.2089 is due mainly to hydrogen. Divide 0.2089 by 0.00783, the amount by which the atomic weight of one ^{1}H atom differs from 1. The answer, 26.67, gives a "ballpark" estimate of the number of hydrogens in testosterone. Then, divide 288 by 12, to determine the maximum number of carbons. Since 288 ÷ 12 = 24, we know that testosterone can have no more than 22 carbons if it also includes hydrogen and oxygen. Make a list of *reasonable* molecular formulas containing C, H and O whose mass is 288 and that contain 20-30 hydrogens. Tabulate these, and calculate their exact masses using the values in the text.

	Isotopic mass			
Molecular formula	Mass of carbons	Mass of hydrogens	Mass of oxygens	Mass of molecular ion
$C_{20}H_{32}O$	240.0000 amu	32.2504 amu	15.9949 amu	288.2453 amu
$C_{19}H_{28}O_2$	228.0000	28.2191	31.9898	288.2089
$C_{18}H_{24}O_3$	216.0000	24.1879	47.9847	288.1726

Solution: The only possible formula for testosterone is $C_{19}H_{28}O_2$.

11.2

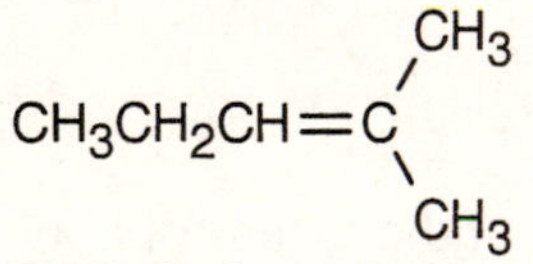

$CH_3CH_2CH{=}C(CH_3)_2$ — 2-Methylpent-2-ene

$CH_3CH_2CH_2CH{=}CHCH_3$ — Hex-2-ene

Fragmentations often occur to give a relatively stable carbocation.

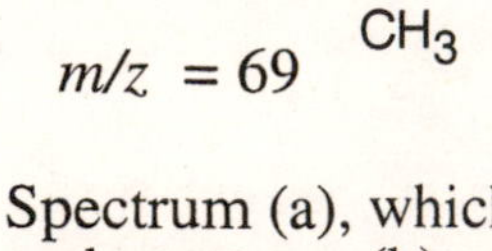

$^+CH_2CH{=}C(CH_3)_2$ — m/z = 69

$^+CH_2CH{=}CHCH_3$ — m/z = 55

Spectrum (a), which has a dominant peak at m/z = 69, corresponds to 2-methylpent-2-ene, and spectrum (b), which has m/z = 55 as its base peak, corresponds to hex-2-ene.

11.3 **Strategy:** In a mass spectrum, the molecular ion is both a cation and a radical. When it fragments, two kinds of cleavage can occur. (1) Cleavage can form a radical and a cation (the species observed in the mass spectrum). Alpha cleavage shows this type of pattern. (2) Cleavage can form a neutral molecule and a different radical cation (the species observed in the mass spectrum). Alcohol dehydration and the McLafferty rearrangement show this cleavage pattern.

For each compound, calculate the mass of the molecular ion and identify the functional groups present. Draw the fragmentation products and calculate their masses.

Solution:

(a)

$[H_3C{-}C({=}O){\vdots}CH_2CH_2CH_3]^{+\cdot}$ $\xrightarrow[\text{cleavage}]{\text{Alpha}}$ $[H_3C{-}C{=}O]^{+}$ + $\cdot CH_2CH_2CH_3$

$M^+ = 86$ $\quad m/z = 43$

$[H_3C{\vdots}C({=}O){-}CH_2CH_2CH_3]^{+\cdot}$ $\xrightarrow[\text{cleavage}]{\text{Alpha}}$ $H_3C\cdot$ + $[O{=}C{-}CH_2CH_2CH_3]^{+}$

$M^+ = 86$ $\quad m/z = 71$

In theory, alpha cleavage can take place on either side of the carbonyl group, to produce the cations with $m/z = 43$ and $m/z = 71$. In practice, cleavage occurs on the more substituted side of the carbonyl group, and the first cation, with $m/z = 43$, is observed.

(b)

[cyclohexanol (OH)]$^{+\cdot}$ $\xrightarrow{\text{Dehydration}}$ [cyclohexene]$^{+\cdot}$ + H_2O

$M^+ = 100$ $\quad m/z = 82$

Dehydration of cyclohexanol produces a cation radical with $m/z = 82$.

(c)

$[CH_3CH(CH_3)CH_2C(O)CH_3]^{+\cdot}$ $\xrightarrow{\text{McLafferty rearrangement}}$ [six-membered transition state: H_3C–C(H)(–CH_2–H)–CH_2–C(=O)–CH_3]$^{+\cdot}$ $\rightarrow$ $[H_2C{=}C(OH){-}CH_3]^{+\cdot}$ + $CH_2{=}CH{-}CH_3$

$M^+ = 100$ $\quad m/z = 58$

The cation radical fragment resulting from McLafferty rearrangement has $m/z = 58$.

(d)

$[H_3C{\vdots}CH_2{-}N(CH_2CH_3){-}CH_2CH_3]^{+\cdot}$ $\xrightarrow[\text{cleavage}]{\text{Alpha}}$ $H_3C\cdot$ + $[CH_2{-}N(CH_2CH_3){-}CH_2CH_3]^{+}$

$M^+ = 101$ $\quad m/z = 86$

Alpha cleavage of triethylamine yields a cation with $m/z = 86$.

11.4 Strategy: The molecule, 2-methylpentan-2-ol, produces fragments that result from both dehydration and from alpha cleavage. Two different alpha cleavage products can be detected.

Solution:

$$\left[CH_3CH_2CH_2C(CH_3)_2OH\right]^{+\cdot} \xrightarrow{\text{Dehydration}} \left[CH_3CH_2CH{=}C(CH_3)_2\right]^{+\cdot} + H_2O$$

$M^+ = 102$ $\qquad$ $m/z = 84$

Alpha cleavage (down):

$$\left[CH_3CH_2CH_2C(OH)CH_3\right]^{+} + H_3C\cdot$$

$m/z = 87$

Alpha cleavage (diagonal):

$$CH_3CH_2CH_2\cdot + \left[HO\text{–}C(CH_3)\text{–}CH_3\right]^{+}$$

$m/z = 59$

11.5 Strategy: At first glance, we know that: (1) energy increases as wavelength decreases, and (2) the wavelength of X-radiation is shorter than the wavelength of infrared radiation. Thus, we estimate that an X ray is of higher energy than an infrared ray.

Solution:

$\varepsilon = h\nu = hc/\lambda$; $h = 6.62 \times 10^{-34}$ J·s; $c = 3.00 \times 10^8$ m/s

for $\lambda = 10^{-6}$ m (infrared radiation):

$$\varepsilon = \frac{(6.62 \times 10^{-34}\ \text{J·s})(3.00 \times 10^8\ \text{m/s})}{1.0 \times 10^{-6}\ \text{m}} = 2.0 \times 10^{-19}\ \text{J}$$

for $\lambda = 3.0 \times 10^{-9}$ m (X radiation):

$$\varepsilon = \frac{(6.62 \times 10^{-34}\ \text{J·s})(3.00 \times 10^8\ \text{m/s})}{3.0 \times 10^{-9}\ \text{m}} = 6.6 \times 10^{-17}\ \text{J}$$

Confirming our estimate, the calculation shows that an X ray is of higher energy than infrared radiation.

When comparing radiation expressed by different units, convert radiation in m to radiation in Hz by the equation:

$$\nu = \frac{c}{\lambda} = \frac{3.00 \times 10^8\ \text{m/s}}{9.0 \times 10^{-6}\ \text{m}} = 3.3 \times 10^{13}\ \text{Hz}$$

The equation $\varepsilon = h\nu$ shows that the greater the value of ν, the greater the energy. Thus, radiation with $\nu = 3.3 \times 10^{13}$ Hz ($\lambda = 9.0 \times 10^{-6}$ m) is higher in energy than radiation with $\nu = 4.0 \times 10^9$ Hz.

11.6 Use the expression for energy shown in this section. Convert frequency to wavelength when you need to.

(a) $E = \dfrac{1.20 \times 10^{-4} \text{ kJ/mol}}{\lambda \text{ (in m)}} = \dfrac{1.20 \times 10^{-4} \text{ kJ/mol}}{5.0 \times 10^{-11}}$

$= 2.4 \times 10^{6}$ kJ/mol for a gamma ray.

(b) $E = 4.0 \times 10^{4}$ kJ/mol for an X ray.

(c) $\nu = \dfrac{c}{\lambda};\ \lambda = \dfrac{c}{\nu} = \dfrac{3.0 \times 10^{8} \text{ m/s}}{6.0 \times 10^{15} \text{ Hz}} = 5.0 \times 10^{-8}\text{m}$

$E = \dfrac{1.20 \times 10^{-4} \text{ kJ/mol}}{5.0 \times 10^{-8}} = 2.4 \times 10^{3}$ kJ/mol for ultraviolet light

(d) $E = 2.8 \times 10^{2}$ kJ/mol for visible light.

(e) $E = 6.0$ kJ/mol for infrared radiation

(f) $E = 4.0 \times 10^{-2}$ kJ/mol for microwave radiation.

11.7 (a) A compound with a strong absorption at 1710 cm^{-1} contains a carbonyl group and is either a ketone or aldehyde.

(b) A compound with a nitro group has a strong absorption at 1540 cm^{-1}.

(c) A compound showing both carbonyl (1720 cm^{-1}) and –OH (2500-3000 cm^{-1} broad) absorptions is a carboxylic acid.

11.8 **Strategy:** To use IR spectroscopy to distinguish between isomers, find a strong IR absorption that is present in one isomer but absent in the other.

Solution:

(a)

CH_3CH_2OH	CH_3OCH_3
Strong hydroxyl band at 3400 – 3640 cm^{-1}	No band in the region 3400 – 3640 cm^{-1}

(b)

$CH_3CH_2CH_2CH_2CH{=}CH_2$	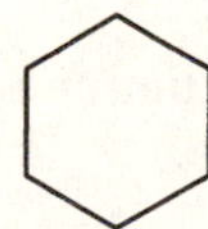
Alkene bands at 3020–3100 cm^{-1} and at 1640–1680^{-1}.	No bands in alkene region.

(c)

$CH_3CH_2CO_2H$	$HOCH_2CH_2CHO$
Strong, broad band at 2500–3100 cm^{-1}	Strong band at 3400–3640 cm^{-1}

11.9 (a) An ester next to a double bond absorbs at 1715 cm^{-1}. The alkene double bond absorbs at 1640–1680 cm^{-1}.

(b) The aldehyde carbonyl group absorbs at 1730 cm^{-1}. The alkyne C≡C bond absorbs at 2100–2260 cm^{-1}, and the alkyne H–C bond absorbs at 3300 cm^{-1}.

(c) The most important absorptions for this compound are due to the alcohol group (a broad, intense band at 3400–3650 cm^{-1}) and to the carboxylic acid group, which has a C=O absorption in the range 1710 –1760 cm^{-1} and a broad O–H absorption in the range 2500–3100 cm^{-1}. Absorptions due to the aromatic ring [3030 cm^{-1} (w) and 1450–1600 cm^{-1} (m)] may also be seen.

11.10

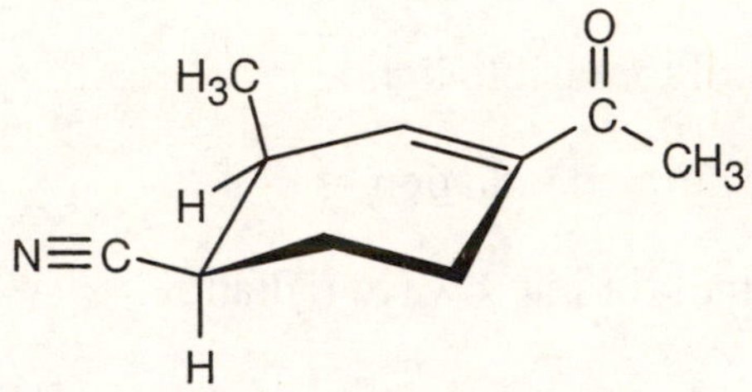

H_3C O C CH_3 H N≡C H

The compound contains nitrile and ketone groups, as well as a carbon–carbon double bond. The nitrile absorption occurs at 2210–2260 cm^{-1}. The ketone shows an absorption at 1690 cm^{-1}, a value lower than the usual value because the ketone is next to the double bond. The double bond absorption occurs at 1640–1680 cm^{-1}.

11.11

$$200 \text{ nm} = 200 \times 10^{-9} \text{ m} = 2 \times 10^{-7} \text{ m}$$

$$400 \text{ nm} = 400 \times 10^{-9} \text{ m} = 4 \times 10^{-7} \text{ m}$$

for $\lambda = 2 \times 10^{-7}$ m:

$$E = \frac{1.20 \times 10^{-4} \text{ kJ/mol}}{\lambda \text{ (in m)}} = \frac{1.20 \times 10^{-4} \text{ kJ/mol}}{2.0 \times 10^{-7}} = 6.0 \times 10^{2} \text{ kJ/mol}$$

for $\lambda = 4 \times 10^{-7}$ m:

$$E = \frac{1.20 \times 10^{-4} \text{ kJ/mol}}{\lambda \text{ (in m)}} = \frac{1.20 \times 10^{-4} \text{ kJ/mol}}{4.0 \times 10^{-7}} = 3.0 \times 10^{2} \text{ kJ/mol}$$

The energy of electromagnetic radiation in the region of the spectrum from 200 nm to 400 nm is 300 – 600 kJ/mol.

	UV	IR
Energy (in kJ/mol)	300 – 600	4.8 – 48

The energy required for UV transitions is greater than the energy required for IR transitions.

11.12

$$\varepsilon = \frac{A}{C \times l}$$

Where ε = molar absorptivity (in L/mol·cm)

A = absorbance

l = sample path length (in cm)

C = concentration (in mol/L)

In this problem:

$\varepsilon = 50{,}100 = 5.01 \times 10^4$ L/mol ·cm

$l = 1.00$ cm

$A = 0.735$

$$C = \frac{A}{\varepsilon \times l} = \frac{0.735}{5.01 \times 10^4 \text{ L/mol·cm} \times 1.00 \text{ cm}} = 1.47 \times 10^{-5} \text{ M}$$

11.13 All compounds having alternating single and multiple bonds should show ultraviolet absorption in the range 200–400 nm. Only compound (a) is not UV-active. All of the compounds pictured below are UV active.

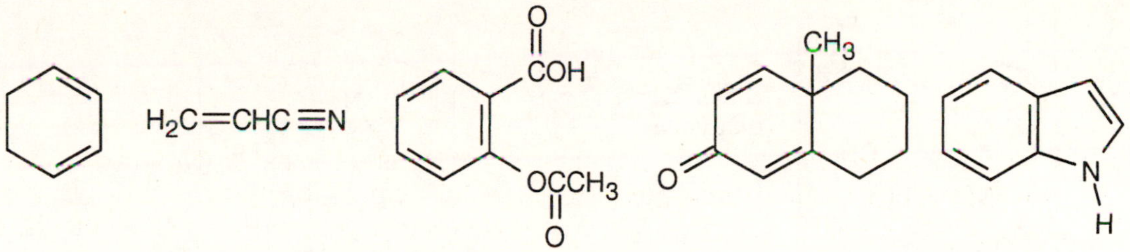

$H_2C{=}CHC{\equiv}N$ COH OCCH₃ CH₃ O N H

Visualizing Chemistry

11.14 (a) The mass spectrum of this ketone shows fragments resulting from both McLafferty rearrangement and alpha cleavage.
McLafferty rearrangement:

$[CH_3CH_2CH(CH_3)COCH_2CH_3]^{+\cdot}$ $M^+ = 114$ —McLafferty rearrangement→ H H H C O H C C C CH₂CH₃ H H CH₃

↓

$H_2C{=}CH_2$ + $[H_3C\text{–}CH{=}C(OH)\text{–}CH_2CH_3]^{+\cdot}$ $m/z = 86$

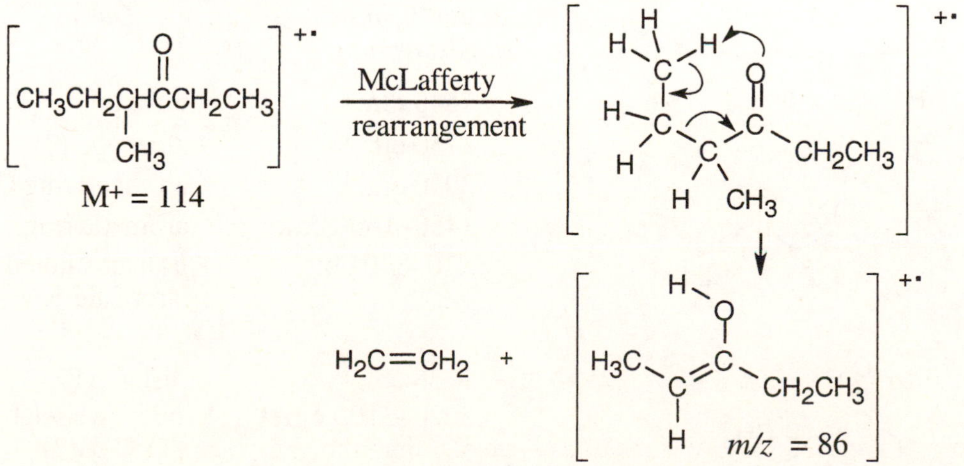

Alpha cleavage:

$[CH_3CH_2CH(CH_3)-C(=O) + CH_2CH_3]^{+\cdot}$ —Alpha cleavage→ $[CH_3CH_2CH(CH_3)-C{=}O]^{+}$ + $\cdot CH_2CH_3$

$M^+ = 114$; $m/z = 85$

$[CH_3CH_2CH(CH_3) + C(=O)-CH_2CH_3]^{+\cdot}$ —Alpha cleavage→ $CH_3CH_2CH^{\cdot}(CH_3)$ + $[O{=}C-CH_2CH_3]^{+}$

$M^+ = 114$; $m/z = 57$

(b) Two different fragments can arise from alpha cleavage of this amine:

[pyrrolidine N–CH(CH_3)CH_3]$^{+\cdot}$ —Alpha cleavage→ $CH_3^{\cdot}$ + [pyrrolidine N–CH(CH_3)]$^{+}$

$M^+ = 113$; $m/z = 98$

[pyrrolidine N–CH(CH_3)CH_3]$^{+\cdot}$ —Alpha cleavage→ [$CH_2CH_2CH_2$–N(H_2C)–CH(CH_3)CH_3]$^{+\cdot}$

$M^+ = 113$; $m/z = 113$

The second product results from cleavage of a bond in the five-membered ring.

11.15

Compound	*Significant IR Absorption*	*Due to:*
(a) p-nitrophenylacetaldehyde (O=C(H)–CH₂–C₆H₄–NO₂; labels 1 nitro, 2 aldehyde, 3 ring)	1540 cm^{-1}	nitro group (1)
	1730 cm^{-1}	aldehyde (2)
	3030 cm^{-1},	aromatic ring C–H(3)
	1450–1600 cm^{-1}	aromatic ring C=C(3)
	810–840 cm^{-1}	*p*-disubstituted aromatic ring
(b) $H_3C-O-C(=O)-CH(CH_3)-CH{=}CH_2$ (labels 1 ester, 2 vinylic C–H, 3 C=CH₂)	1735 cm^{-1}	ester (1)
	3020–3100 cm^{-1}	vinylic stretch C–H(2)
	910 cm^{-1}, 990 cm^{-1}	$C{=}CH_2$ bend(3)

(c)

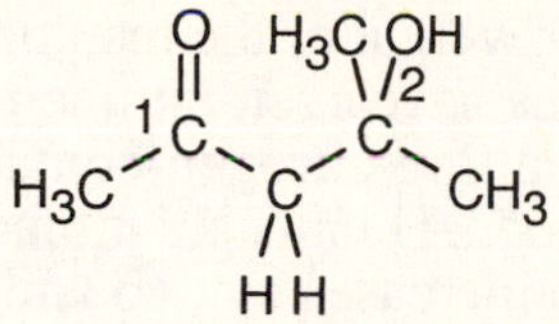

1715 cm^{-1}	ketone (1)
3400–3650 cm^{-1}	alcohol (2)

11.16 The compound in 15(a) shows a UV absorption because of its conjugated double bonds. No other compounds in Problems 11.14 and 11.15 show UV absorptions.

Additional Problems

11.17

M^+	Molecular Formula	Degree of Unsaturation	Possible Structure
(a) 132	$C_{10}H_{12}$	5	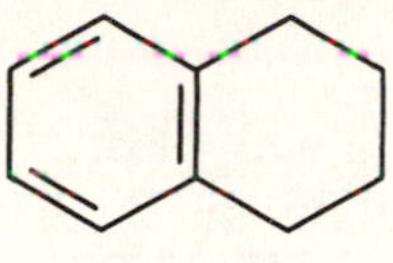
(b) 166	$C_{13}H_{10}$	9	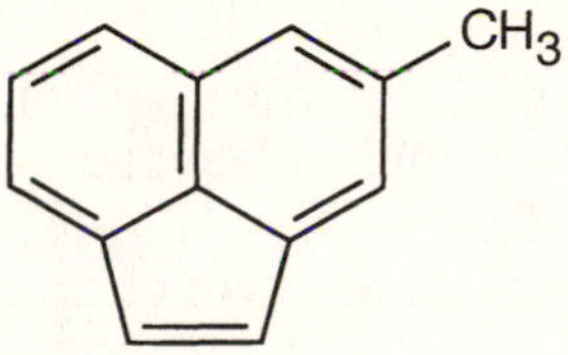
	$C_{12}H_{22}$	2	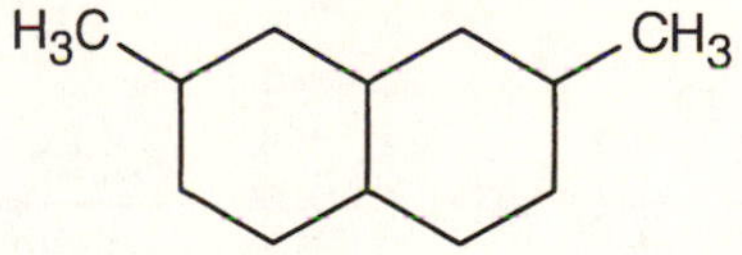
(c) 84	C_6H_{12}	1	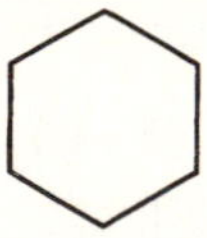

11.18 As a first approximation, reasonable molecular formulas for camphor are $C_{10}H_{16}O$, $C_9H_{12}O_2$, $C_8H_8O_3$. The actual formula is $C_{10}H_{16}O$ (M^+ = 152.1201).

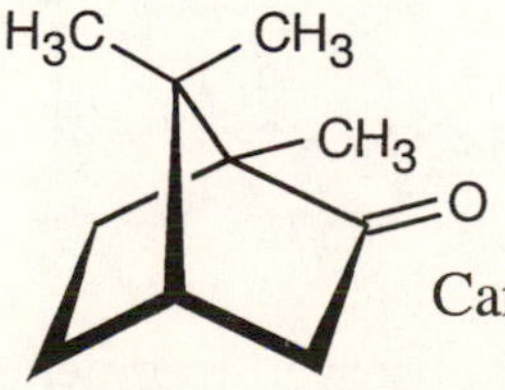

Camphor

11.19 Carbon is tetravalent, and nitrogen is trivalent. If a C–H unit (formula weight 13) is replaced by an N atom (formula weight 14), the molecular weight of the resulting compound increases by one. Since all neutral hydrocarbons have even-numbered molecular weights (C_nH_{2n+2}, C_nH_{2n}, and so forth) the resulting nitrogen-containing compounds have odd-numbered molecular weights. If two C–H units are replaced by two N atoms, the molecular weight of the resulting compound increases by two and remains an even number.

11.20 Because M^+ is an odd number, pyridine contains an odd number of nitrogen atoms. If pyridine contained one nitrogen atom (atomic weight 14) the remaining atoms would have a formula weight of 65, corresponding to $–C_5H_5$. C_5H_5N is, in fact, the molecular formula of pyridine.

11.21 In order to simplify this problem, neglect the ^{13}C and ^2H isotopes in determining the molecular ions of these compounds.

(a) The formula weight of $–CH_3$ is 15, and the atomic masses of the two bromine isotopes are 79 and 81. The two molecular ions of bromoethane occur at $M^+ = 94$ (50.7%) and $M^+ = 96$ (49.3%).

(b) The formula weight of $–C_6H_{13}$ is 85, and the atomic masses of the two chlorine isotopes are 35 and 37. The two molecular ions of 1-chlorohexane occur at $M^+ = 120$ (75.8%) and $M^+ = 122$ (24.2%).

11.22 (a) The molecular formula of the ketone is $C_5H_{10}O$, and the fragments correspond to the products of alpha cleavage (McLafferty rearrangement fragments have even-numbered values of *m/z*). Draw all possible ketone structures, show the charged products of alpha cleavage, and note which fragments correspond to those listed.

$$\left[H_3C\text{-}C(=O)\text{-}CH_2CH_2CH_3\right]^{+\bullet} \xrightarrow{\text{Alpha cleavage}} \left[H_3C\text{-}C{=}O\right]^+ + \left[O{=}C\text{-}CH_2CH_2CH_3\right]^+$$

$M^+ = 86$; $m/z = 43$; $m/z = 71$

$$\left[H_3C\text{-}C(=O)\text{-}CH(CH_3)CH_3\right]^{+\bullet} \xrightarrow{\text{Alpha cleavage}} \left[H_3C\text{-}C{=}O\right]^+ + \left[O{=}C\text{-}CH(CH_3)CH_3\right]^+$$

$M^+ = 86$; $m/z = 43$; $m/z = 71$

$$\left[CH_3CH_2\text{-}C(=O)\text{-}CH_2CH_3\right] \xrightarrow{\text{Alpha cleavage}} \left[CH_3CH_2\text{-}C{=}O\right] + \left[O{=}C\text{-}CH_2CH_3\right]$$

$M^+ = 86$; $m/z = 57$; $m/z = 57$

Either of the first two compounds shows the observed fragments in its mass spectrum.

(b) $C_5H_{12}O$ is the formula of an alcohol with $M^+ = 88$. The fragment at $m/z = 70$ is due to the product of dehydration of M^+. The other two fragments are a result of alpha cleavage. Draw the possible C_5 alcohol isomers, and draw their products of alpha cleavage. The tertiary alcohol shown fits the data.

$$\left[CH_3CH_2-\underset{CH_3}{\overset{OH}{C}}-CH_3\right]^{+\cdot} \xrightarrow{\text{Alpha cleavage}} \left[\underset{CH_3CH_2}{}\overset{OH}{C}\underset{CH_3}{}\right]^{+} + \left[\underset{H_3C}{}\overset{OH}{C}\underset{CH_3}{}\right]^{+}$$

$M^+ = 88$; $m/z = 73$; $m/z = 59$

$$\left[CH_3CH_2-\underset{CH_3}{\overset{OH}{C}}-CH_3\right]^{+\cdot} \xrightarrow{\text{Dehydration}} \left[(H_3C)(H)C{=}C(CH_3)_2\right]^{+\cdot}$$

$M^+ = 88$; $m/z = 70$

11.23

$$CH_3CH_2CH_2\overset{CH_3}{\overset{|}{C}}HCH_3$$

2-Methylpentane

$\downarrow e^-$

$$\left[CH_3CH_2CH_2\overset{CH_3}{\overset{|}{C}}HCH_3\right]^{+\cdot} + e^-$$

$CH_3CH_2CH_2\overset{+}{C}HCH_3$	$\overset{+}{C}H_2\overset{CH_3}{\overset{|}{C}}HCH_3$	$\overset{+}{C}HCH_3$ (with CH_3)	$CH_3\overset{+}{C}H_2$
$m/z = 71$	$m/z = 57$	$m/z = 43$	$m/z = 29$

The molecular ion, at $m/z = 86$, is present in very low abundance. The base peak, at $m/z = 43$, represents a stable secondary carbocation.

11.24 Before doing the hydrogenation, familiarize yourself with the mass spectra of cyclohexene and cyclohexane. Note that M^+ is different for each compound. After the reaction is underway, inject a sample from the reaction mixture into the mass spectrometer. If the reaction is finished, the mass spectrum of the reaction mixture should be superimposable with the mass spectrum of cyclohexane.

11.25 (a) This ketone shows mass spectrum fragments that are due to alpha cleavage and to the McLafferty rearrangement. The molecular ion occurs at $M^+ = 148$, and major fragments have $m/z = 120$, 105, and 71. (Note that only charged species are shown.)

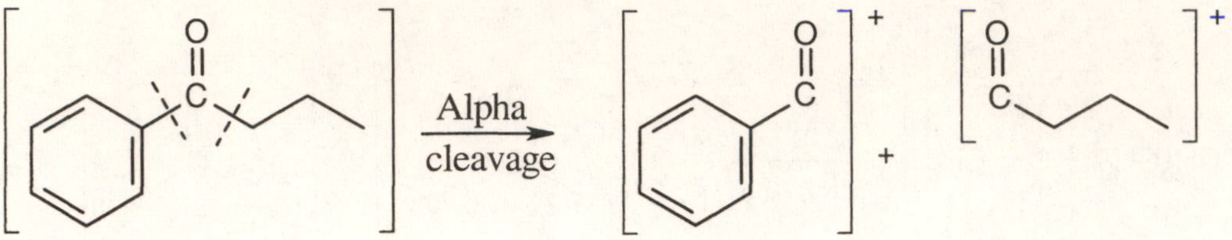

Alpha cleavage

$C_{10}H_{12}O$ $M^+ = 148$ — $m/z = 105$ + $m/z = 71$

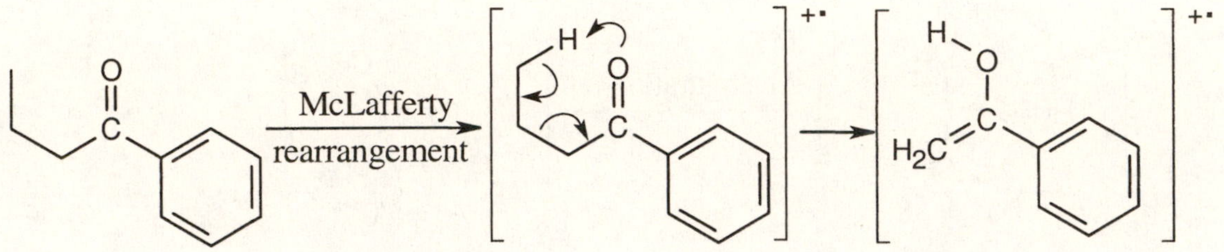

McLafferty rearrangement

$m/z = 120$

(b) The fragments in the mass spectrum of this alcohol ($C_8H_{16}O$) result from dehydration and alpha cleavage. Major fragments have m/z values of 128 (the same value as the molecular ion), 110 and 99.

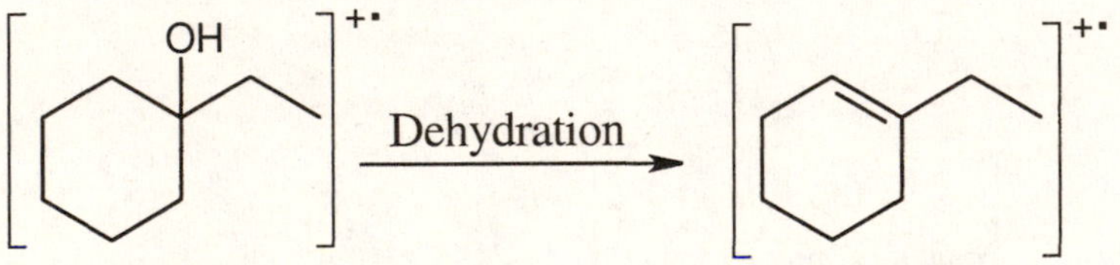

Dehydration

$M^+ = 128$ — $m/z = 110$

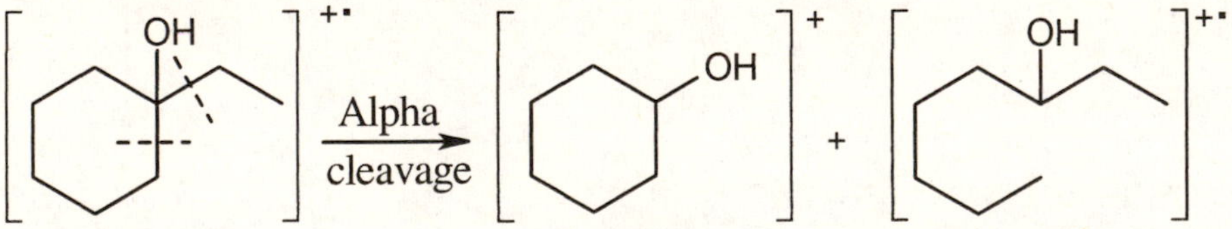

Alpha cleavage

$M^+ = 128$ — $m/z = 99$ + $m/z = 128$

(c) Amines fragment by alpha cleavage. In this problem, cleavage occurs in the ring, producing a fragment with the same value of m/z as the molecular ion (99).

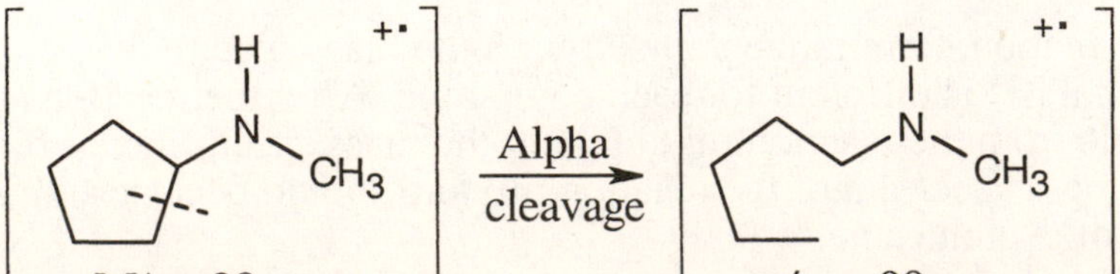

Alpha cleavage

$M^+ = 99$ — $m/z = 99$

11.26 $CH_3CH_2C{\equiv}CH$ shows absorptions at 2100-2260 cm^{-1} (C≡C) and at 3300 cm^{-1} (C≡C–H) that are due to the terminal alkyne bond.

$H_2C{=}CHCH{=}CH_2$ has absorptions in the regions 1640-1680 cm^{-1} and 3020–3100 that are due to the double bonds. It also shows absorptions at 910 cm^{-1} and 990 cm^{-1} that are due to monosubstituted alkene bonds. No absorptions occur in the alkyne region.

$CH_3C{\equiv}CCH_3$. For reasons we won't discuss, symmetrically substituted alkynes such as but-2-yne do not show a C≡C bond absorption in the IR. This alkyne is distinguished from the other isomers in that it shows no absorptions in either the alkyne or alkene regions.

11.27 Two enantiomers have identical physical properties (other than the sign of specific rotation). Thus, their IR spectra are also identical.

11.28 Diastereomers have different physical properties and chemical behavior, and their IR spectra are also different.

11.29 (a) Absorptions at 3300 cm^{-1} and 2150 cm^{-1} are due to a terminal triple bond. Possible structures:

$CH_3CH_2CH_2C{\equiv}CH$ $(CH_3)_2CHC{\equiv}CH$

(b) An IR absorption at 3400 cm^{-1} is due to a hydroxyl group. Since no double bond absorption is present, the compound must be a cyclic alcohol.

OH CH₂OH CH₃ OH CH₃ HO

(c) An absorption at 1715 cm^{-1} is due to a ketone. The only possible structure is $CH_3CH_2COCH_3$.

(d) Absorptions at 1600 cm^{-1} and 1500 cm^{-1} are due to an aromatic ring. Possible structures:

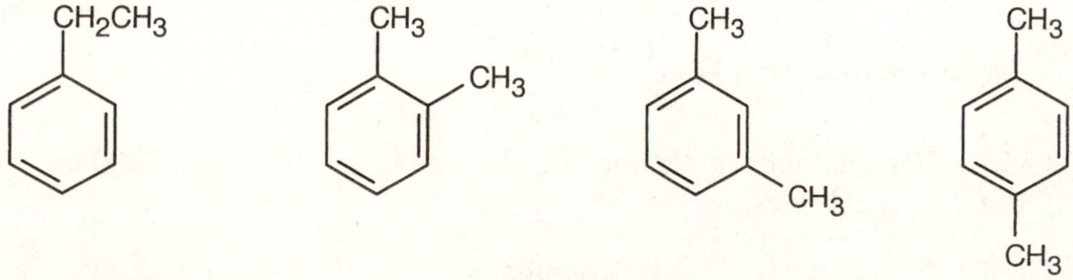

CH_2CH_3 CH_3 CH_3 CH_3 CH_3 CH_3 CH_3

11.30 (a) $HC{\equiv}CCH_2NH_2$
Alkyne absorptions at
3300 cm^{-1}, 2100–2260 cm^{-1}
Amine absorption at
3300-3500 cm^{-1}

$CH_3CH_2C{\equiv}N$
Nitrile absorption at
2210–2260 cm^{-1}

(b) CH_3COCH_3
Strong ketone absorption
at 1715 cm^{-1}

CH_3CH_2CHO
Strong aldehyde absorption
at 1730 cm^{-1}

11.31 Spectrum (b) differs from spectrum (a) in several respects. Note in particular the absorptions at 715 cm^{-1} (strong), 1140 cm^{-1} (strong), 1650 cm^{-1} (medium), and 3000 cm^{-1} (medium) in spectrum (b). The absorptions at 1650 cm^{-1} (C=C stretch) and 3000 cm^{-1} (=C–H stretch) can be found in Table 11.1. They allow us to assign spectrum (b) to cyclohexene and spectrum (a) to cyclohexane.

11.32 Only absorptions with medium to strong intensity are listed.

(a) [benzoic acid: C_6H_5COOH]

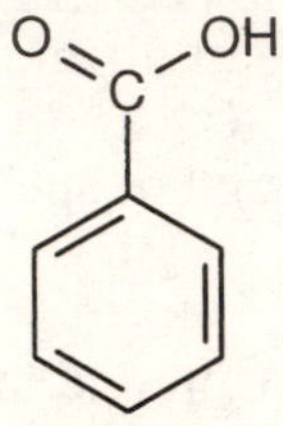

aromatic ring C=C
1450–1600 cm^{-1}
aromatic ring C–H
3030 cm^{-1}
carboxylic acid C=O
1710–1760 cm^{-1}
carboxylic acid O–H
2500–3100 cm^{-1}
monosubstituted aromatic ring
690–710 cm^{-1}
730–770 cm^{-1}

(b) [methyl benzoate: $C_6H_5COOCH_3$]

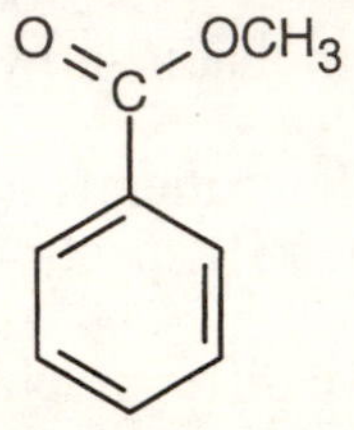

aromatic ring C=C
1450–1600 cm^{-1}
aromatic ring C–H
3030 cm^{-1}
aromatic ester
1715 cm^{-1}
monosubstituted aromatic ring
690–710 cm^{-1}
730–770 cm^{-1}

(c) [p-hydroxybenzonitrile: $N\equiv C-C_6H_4-OH$]

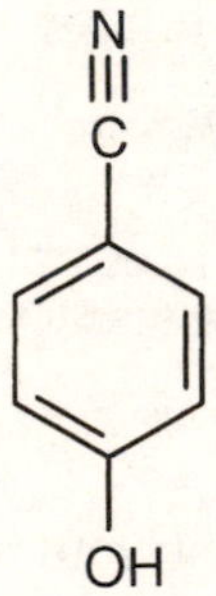

aromatic ring C=C
1450–1600 cm^{-1}
aromatic ring C–H
3030 cm^{-1}
alcohol O–H
3400–3650 cm^{-1}
nitrile C≡N
2210–2260 cm^{-1}
p-disubstituted aromatic ring
810–840 cm^{-1}

(d) [cyclohex-3-enone]

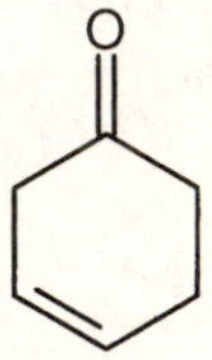

alkene C=C
1640–1680 cm^{-1}
alkene =C–H
3020–3100 cm^{-1}
ketone
1715 cm^{-1}

(e) $CH_3\overset{O}{\overset{\|}{C}}CH_2CH_2\overset{O}{\overset{\|}{C}}OCH_3$

ester
1735 cm^{-1}

ketone
1715 cm^{-1}

11.33 (a) $CH_3C{\equiv}CCH_3$ exhibits no terminal ≡C–H stretching vibration at 3300 cm^{-1}, as $CH_3CH_2C{\equiv}CH$ does.

(b) $CH_3COCH{=}CHCH_3$, a ketone conjugated with a double bond, shows a strong ketone absorption at 1690 cm^{-1}; $CH_3COCH_2CH{=}CH_2$ shows a ketone absorption at 1715 cm^{-1} and monosubstituted alkene absorptions at 910 cm^{-1} and 990 cm^{-1}.

(c) CH_3CH_2CHO exhibits an aldehyde band at 1730 cm^{-1}; $H_2C{=}CHOCH_3$ shows characteristic monosubstituted alkene absorptions at 910 cm^{-1} and 990 cm^{-1}.

11.34

	Compound	*Distinguishing Absorption*	*Due to:*
(a)	$CH_3CH_2\overset{O}{\overset{\Vert}{C}}CH_3$	1715 cm^{-1}	C=O (ketone)
(b)	$(CH_3)_2CHCH_2C{\equiv}CH$	2100–2260 cm^{-1}	C≡C
		3300 cm^{-1}	C≡C—H
(c)	$(CH_3)_2CHCH_2CH{=}CH_2$	910 cm^{-1}; 990 cm^{-1}	$RCH{=}CH_2$
		1640–1680 cm^{-1}	C=C
		3020–3100 cm^{-1}	=C—H
(d)	$CH_3CH_2CH_2\overset{O}{\overset{\Vert}{C}}OCH_3$	1735 cm^{-1}	C=O (ester)
(e)	$C_6H_5\overset{O}{\overset{\Vert}{C}}CH_3$	1690 cm^{-1}	ketone next to aromatic ring
		1450–1600 cm^{-1}	aromatic ring
		3030 cm^{-1}	aromatic ring
		690–710 cm^{-1} 730–770 cm^{-1}	monosubstituted aromatic ring
(f)	HO–C_6H_4–CHO (meta)	1710 cm^{-1}	aldehyde next to aromatic ring
		3400–3650 cm^{-1}	alcohol
		1450–1600 cm^{-1}	aromatic ring
		3030 cm^{-1}	aromatic ring
		690–710 cm^{-1} 810–850 cm^{-1}	*m*-disubstituted aromatic ring

11.35

1-Methylcyclohexanol $\xrightarrow{H_3O^+}$ 1-Methylcyclohexene

The infrared spectrum of the starting alcohol shows a broad absorption at 3400-3640 cm^{-1}, due to an O–H stretch. The alkene product exhibits medium intensity absorbances at 1645-1670 cm^{-1} and at 3000-3100 cm^{-1}. Monitoring the disappearance of the alcohol absorption makes it possible to decide when reaction is complete. It is also possible to monitor the *appearance* of the alkene absorptions.

11.36

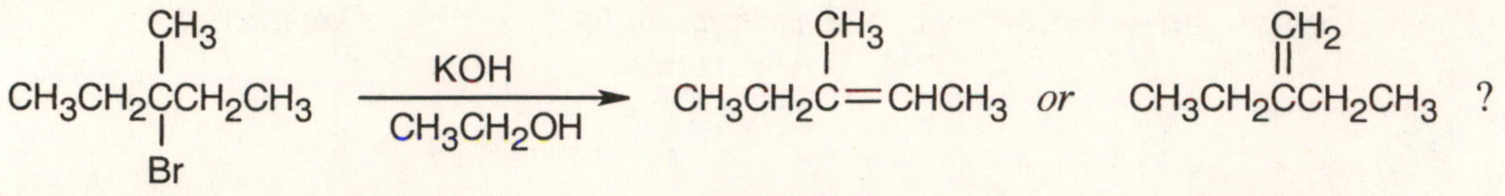

$$\underset{\text{Br}}{\overset{\text{CH}_3}{CH_3CH_2\overset{|}{\underset{|}{C}}CH_2CH_3}} \xrightarrow[\text{CH}_3\text{CH}_2\text{OH}]{\text{KOH}} \overset{\text{CH}_3}{CH_3CH_2\overset{|}{C}{=}CHCH_3} \text{ or } \overset{\text{CH}_2}{CH_3CH_2\overset{||}{C}CH_2CH_3} \;?$$

3-Bromo-3-methylpentane — 3-Methylpent-2-ene — 2-Ethylbut-1-ene

The IR spectra of both products show the characteristic absorptions of alkenes in the regions 3020-3100 cm^{-1} and 1650 cm^{-1}. However, in the region 700-1000 cm^{-1}, 2-ethylbut-1-ene shows a strong absorption at 890 cm^{-1} that is typical of 2,2-disubstituted $R_2C{=}CH_2$ alkenes. The presence or absence of this peak should help to identify the product. (3-Methylpent-2-ene is the major product of the dehydrobromination reaction.)

11.37 The following expressions are needed:

$\varepsilon = h\nu = hc/\lambda = hc\tilde{\nu}$, where $\tilde{\nu}$ is the wavenumber. The last expression shows that, as $\tilde{\nu}$ increases, the energy needed to cause IR absorption increases, indicating greater bond strength. Thus an ester C=O bond ($\tilde{\nu}$ = 1735 cm^{-1}) is stronger than a ketone C=O bond ($\tilde{\nu}$ = 1715 cm^{-1}).

11.38 Possible molecular formulas containing carbon, hydrogen, and oxygen and having M^+ = 150 are $C_{10}H_{14}O$, $C_9H_{10}O_2$, and $C_8H_6O_3$. The first formula has four degrees of unsaturation, the second has five degrees of unsaturation, and the third has six degrees of unsaturation. Since carvone has three double bonds (including the ketone) and one ring, $C_{10}H_{14}O$ is the correct molecular formula for carvone.

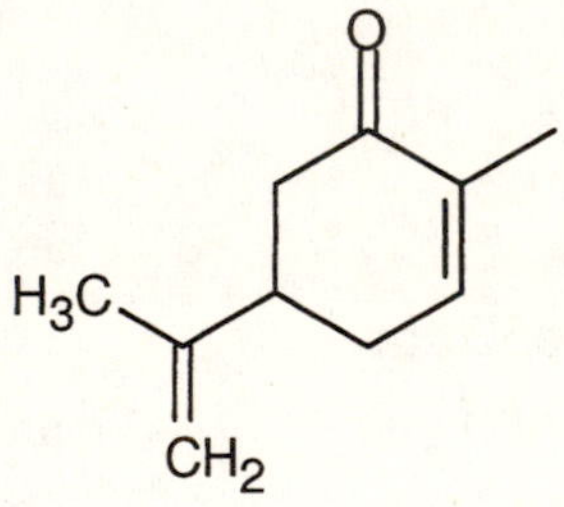

H_3C, CH_2

Carvone

11.39 The intense absorption at 1690 cm^{-1} is due to a ketone conjugated with a double bond.

11.40 To absorb in the 200–400 nm range, an alkene must be conjugated. Since the double bonds of allene aren't conjugated, allene doesn't absorb light in the UV region.

11.41 Only compounds having alternating multiple bonds show $\pi \rightarrow \pi^*$ ultraviolet absorptions in the 200–400 nm range. Of the compounds shown, only pyridine (b) absorbs in this range.

11.42 The value of λ_{max} in the ultraviolet spectrum of dienes becomes larger with increasing alkyl substitution. Since energy is inversely related to λ_{max}, the energy needed to produce ultraviolet absorption decreases with increasing substitution.

Diene	*# of –CH₃ groups*	λ_{max}(nm)	$\lambda_{max} - \lambda_{max}$ (butadiene)
$H_2C{=}CH{-}CH{=}CH_2$	0	217	0
$H_2C{=}CH{-}C(CH_3){=}CH_2$	1	220	3
$H_3C{-}CH{=}CH{-}CH{=}CH_2$	1	223	6
$H_2C{=}C(CH_3){-}C(CH_3){=}CH_2$	2	226	9
$H_3C{-}CH{=}CH{-}CH{=}CH{-}CH_3$	2	227	10
$H_3C{-}C(CH_3){=}CH{-}C(CH_3){=}CH_2$	3	232	15
$H_3C{-}C(CH_3){=}CH{-}CH{=}C(CH_3){-}CH_3$	4	240	23

Each alkyl substituent causes an increase in λ_{max} of approximately 5 nm.

11.43

$H_2C=CH-CH=CH-CH=CH_2$

Hexa-1,3,5-triene
λ_{max} = 258 nm

$H_2C=CH-CH=C(CH_3)-C(CH_3)=CH_2$

2,3-Dimethylhexa-1,3,5-triene
λ_{max} ≈ 268 nm

In Problem 11.42, we concluded that one alkyl group increases λ_{max} of a conjugated diene by 5 nm. Since 2,3-dimethylhexa-1,3,5-triene has two methyl substituents, its UV λ_{max} should be about 10 nm longer than the λ_{max} of hexa-1,3,5-triene.

11.44

$$C = \frac{A}{\varepsilon \times l} = \frac{0.065}{11{,}900 \text{ L/mol·cm} \times 1.00 \text{ cm}} = \frac{6.5 \times 10^{-2}}{1.19 \times 10^{4} \text{ L/mol}} = 5.5 \times 10^{-6}\text{M}$$

11.45 The peak of maximum intensity (base peak) in the mass spectrum occurs at $m/z = 67$. This peak does *not* represent the molecular ion, however, because M^+ of a hydrocarbon must be an even number. Careful inspection reveals the molecular ion peak at $m/z = 68$. $M^+ = 68$ corresponds to a hydrocarbon of molecular formula C_5H_8 with a degree of unsaturation of two.

Fairly intense peaks in the mass spectrum occur at $m/z = 67$, 53, 40, 39, and 27. The peak at $m/z = 67$ corresponds to loss of one hydrogen atom, and the peak at $m/z = 53$ represents loss of a methyl group. The unknown hydrocarbon thus contains a methyl group.

Significant IR absorptions occur at 2130 cm^{-1} (–C≡C– stretch) and at 3320 cm^{-1} (≡C–H stretch). These bands indicate that the unknown hydrocarbon is a terminal alkyne. Possible structures for C_5H_8 are $CH_3CH_2CH_2C{\equiv}CH$ and $(CH_3)_2CHC{\equiv}CH$. [Pent-1-yne is correct.]

11.46 The molecular ion, $M^+ = 70$, corresponds to the molecular formula C_5H_{10}. This compound has one double bond or ring.

The base peak in the mass spectrum occurs at $m/z = 55$. This peak represents loss of a methyl group from the molecular ion and indicates the presence of a methyl group in the unknown hydrocarbon. All other peaks occur with low intensity.

In the IR spectrum, it is possible to distinguish absorptions at 1660 cm^{-1} and at 3000 cm^{-1} due to a double bond. (The 2960 cm^{-1} absorption is rather hard to detect because it occurs as a shoulder on the alkane C–H stretch at 2850-2960 cm^{-1}.)

Since no absorptions occur in the region 890 cm^{-1} – 990 cm^{-1}, we can exclude terminal alkenes as possible structures. The remaining possibilities for C_5H_{10} are $CH_3CH_2CH{=}CHCH_3$ and $(CH_3)_2C{=}CHCH_3$. [2-Methylbut-2-ene is correct.]

11.47

(a) $CH_3CH_2CH(CH_3)CHO$

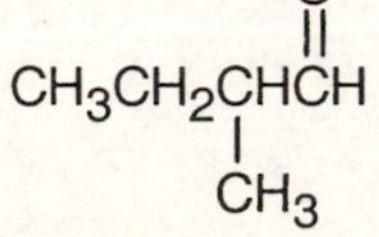

(b) $CH_3CH_2CH_2CH_2C{\equiv}N$ $\quad$ $CH_3C(CH_3)_2C{\equiv}N$ $\quad$ $CH_3CH(CH_3)CH_2C{\equiv}N$

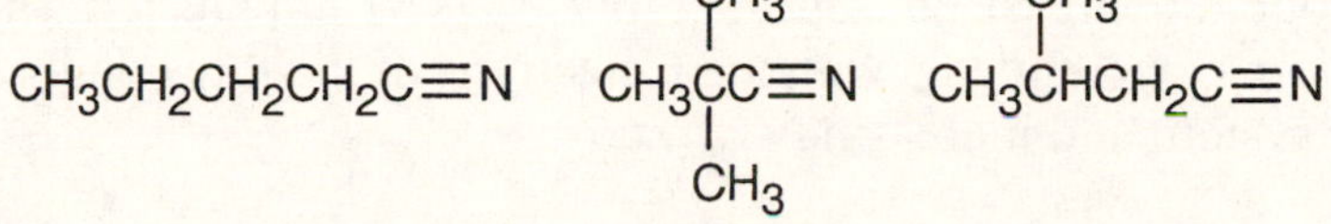

11.48 The simplest way to distinguish between the two isomers is by taking their IR spectra. The aldehyde carbonyl group absorbs at 1730 cm^{-1}, and the ketone carbonyl group absorbs at 1715 cm^{-1}.

The mass spectra of the two isomers also differ. Like ketones, aldehydes also undergo alpha cleavage and McLafferty rearrangements.

McLafferty rearrangement:

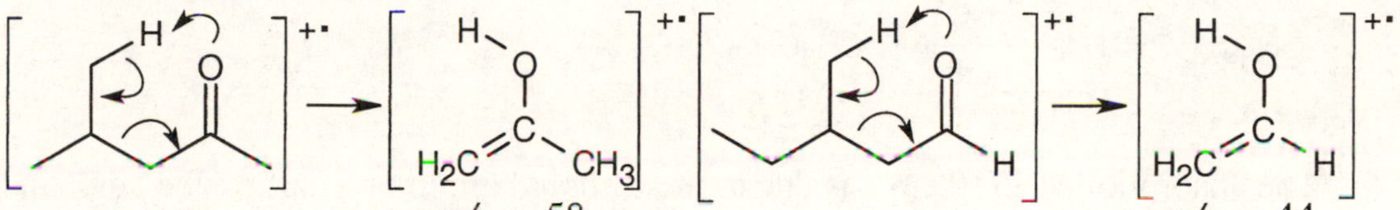

$H_2C{=}C(OH)CH_3$ $^{+\cdot}$ $\quad$ m/z = 58 $\qquad$ $H_2C{=}C(OH)H$ $^{+\cdot}$ $\quad$ m/z = 44

The fragments from the McLafferty rearrangements differ in values of m/z.

Alpha cleavage:

m/z = 85 $\qquad$ m/z = 43

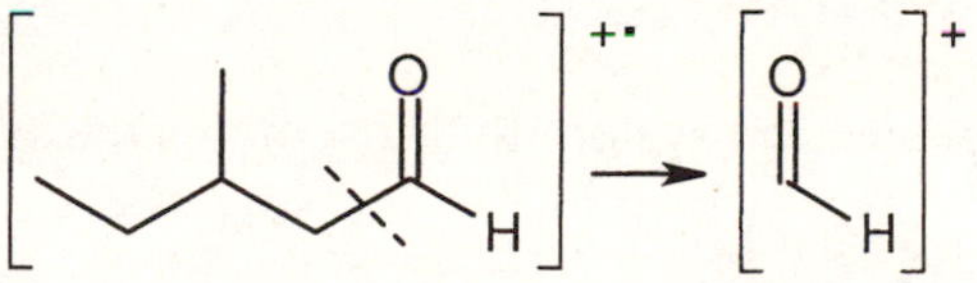

m/z = 29

The fragments resulting from alpha cleavage also differ in values of m/z.

11.49

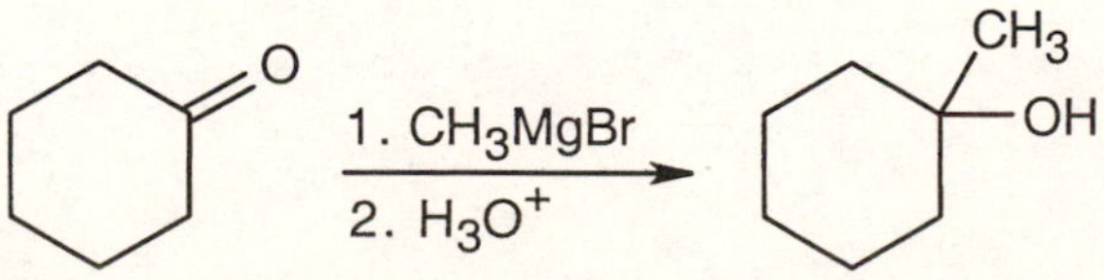

Cyclohexanone $\xrightarrow[\text{2. } H_3O^+]{\text{1. } CH_3MgBr}$ 1-methylcyclohexanol (CH_3, OH)

The absorption at 3400 cm^{-1} is due to a hydroxyl group.

11.50 Double bonds can be conjugated not only with other multiple bonds but also with the lone-pair electrons of atoms such as oxygen and nitrogen. *p*-Toluidine has the same number of double bonds as benzene, yet its λ_{max} is 31 nm greater. The electron pair of the nitrogen atom can conjugate with the π electrons of the three double bonds of the ring, extending the π system and increasing λ_{max}.

11.51

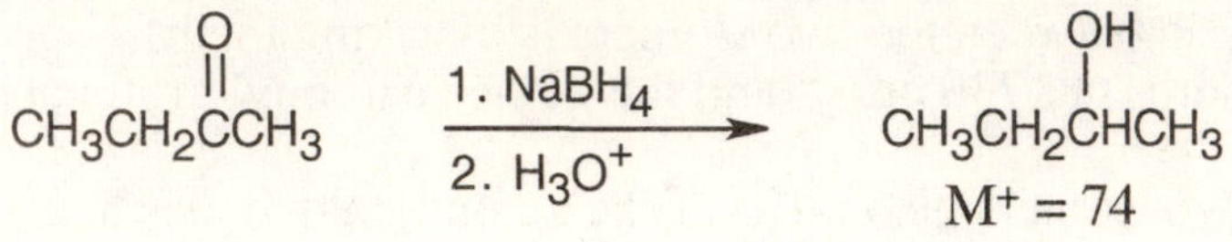

$$CH_3CH_2\overset{O}{\overset{\|}{C}}CH_3 \xrightarrow[2.\ H_3O^+]{1.\ NaBH_4} CH_3CH_2\overset{OH}{\overset{|}{C}}HCH_3$$

$M^+ = 74$

11.52

$$CH_3CH_2C{\equiv}N \xrightarrow[\text{heat}]{H_3O^+} CH_3CH_2\overset{O}{\overset{\|}{C}}OH$$

$M^+ = 74$

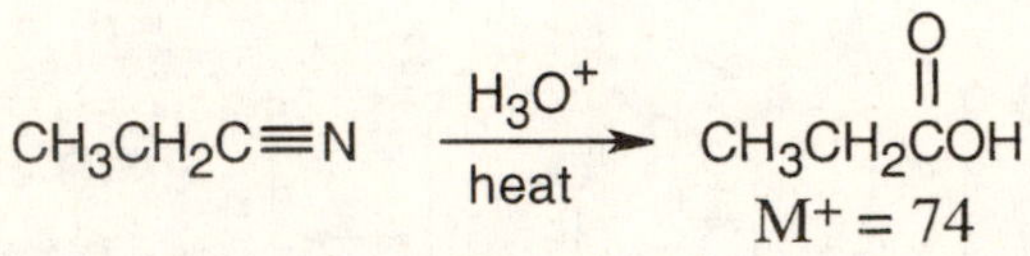

The absorption at 1710 cm^{-1} is due to the carbonyl group of a carboxylic acid, and the absorption at 2500–3100 cm^{-1} is due to the –OH group of the carboxylic acid.

11.53 The lone pair electrons from nitrogen can overlap with the double bond π electrons in a manner similar to the overlap of the π electrons of two conjugated double bonds. This electron contribution from nitrogen makes an enamine double bond electron-rich.

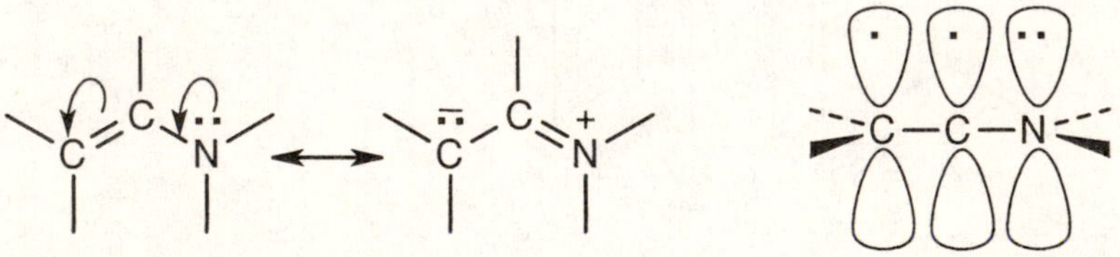

C=C–N: ⟷ :C–C=N⁺ C–C–N

The orbital picture of an enamine shows a 4 *p*-electron system that resembles the system of a conjugated diene.

Chapter 12 – Structure Determination: Nuclear Magnetic Resonance Spectroscopy

Chapter Outline

I. Principles of Nuclear Magnetic Resonance Spectroscopy (Sections 12.1 – 12.3)
   A. Theory of NMR Spectroscopy (Section 12.1)
      1. Many nuclei behave as if they were spinning about an axis.
         a. The positively charged nuclei produce a magnetic field that can interact with an externally applied magnetic field.
         b. The ^{13}C nucleus and the ^{1}H nucleus behave in this manner.
         c. In the absence of an external magnetic field the spins of magnetic nuclei are randomly oriented.
      2. When a sample containing these nuclei is placed between the poles of a strong magnet, the nuclei align themselves either with the applied field or against the applied field.
         The parallel orientation is slightly lower in energy and is slightly favored.
      3. If the sample is irradiated with radiofrequency energy of the correct frequency, the nuclei of lower energy absorb energy and "spin-flip" to the higher energy state.
         a. The magnetic nuclei are in resonance with the applied radiation.
         b. The frequency of the rf radiation needed for resonance depends on the magnetic field strength and on the identity of the magnetic nuclei.
            i. In a strong magnetic field, higher frequency rf energy is needed.
            ii. At a magnetic field strength of 4.7 T, rf energy of 200 MHz is needed to bring a ^{1}H nucleus into resonance, and energy of 50 MHz is needed for ^{13}C.
      4. Nuclei with an odd number of protons and nuclei with an odd number of neutrons show magnetic properties.
   B. The nature of NMR absorptions (Section 12.2).
      1. Not all ^{13}C nuclei and not all ^{1}H nuclei absorb at the same frequency.
         a. Each magnetic nucleus is surrounded by electrons that set up their own magnetic fields.
         b. These small fields oppose the applied field and shield the magnetic nuclei.
            i. $\boldsymbol{B}_{\text{effective}} = \boldsymbol{B}_{\text{applied}} - \boldsymbol{B}_{\text{local}}$
            ii. This expression shows that the magnetic field felt by a nucleus is less than the applied field.
         c. These shielded nuclei absorb at slightly different values of magnetic field strength.
         d. A sensitive NMR spectrometer can detect these small differences.
         e. Thus, NMR spectra can be used to map the carbon–hydrogen framework of a molecule.
      2. NMR spectra.
         a. The horizontal axis shows effective field strength, and the vertical axis shows intensity of absorption.
         b. Each peak corresponds to a chemically distinct nucleus.
         c. Zero absorption is at the bottom.
         d. Absorptions due to both ^{13}C and ^{1}H can't both be observed at the same time.
      3. Operation of an NMR spectrometer
         a. A solution of a sample is placed in a thin glass tube between the poles of a magnet.
         b. The strong magnetic field causes the nuclei to align in either of the two possible orientations.

c. The strength of the applied magnetic field is varied, holding the rf frequency constant.
d. Chemically distinct nuclei come into resonance at slightly different values of ***B***.
e. A detector monitors the absorption of rf energy
f. The signal is amplified and recorded.

4. Time scale of NMR absorptions.
a. The time scale (10^{-3} s) of NMR spectra is much slower than that of most other spectra.
b. If a process occurs faster than the time scale of NMR, absorptions are observed as "time-averaged processes.
NMR records only a single spectrum.
c. NMR can be used to measure rates and activation energies of fast processes.
i. Because cyclohexane ring-flips are very fast at room temperature, only a single peak is observed for equatorial and axial hydrogens at room temperature.
ii. At –90°, both axial and equatorial hydrogens can be identified.

C. Chemical Shifts (Section 12.3).
1. Field strength increases from left (downfield) to right (upfield).
a. Nuclei that absorb downfield require a lower field strength for resonance and are deshielded.
b. Nuclei that absorb upfield require a higher field strength and are shielded.
2. TMS(tetramethylsilane) is used as a reference point in both ^{13}C NMR and ^{1}H NMR.
The TMS absorption occurs upfield of other absorptions, and is set as the zero point.
3. The chemical shift is the position on the chart where a nucleus absorbs energy.
4. NMR charts are calibrated by using an arbitrary scale – the delta scale.
a. One δ equals 1 ppm of the spectrometer operating frequency.
b. By using this system, all chemical shifts occur at the same value of δ, regardless of the spectrometer operating frequency.
5. NMR absorptions occur over a narrow range.
a. ^{1}H absorptions occur 0–10 δ downfield from TMS.
b. ^{13}C absorptions occur 1–220 δ downfield from TMS.
c. The chances of accidental overlap can be reduced by using an instrument with a higher field strength.

II. ^{13}C NMR spectroscopy (Sections 12.4 – 12.7).
A. Signal averaging and FT-NMR (Section 12.4).
1. The low natural abundance of ^{13}C (1.1%) makes it difficult to observe ^{13}C peaks because of background noise.
2. If hundreds of individual runs are averaged, the background noise cancels.
This technique takes a long time.
3. In FT-NMR, all signals are recorded simultaneously.
a. The sample is irradiated with a pulse of rf energy that covers all useful frequencies.
b. The resulting complex signal must be mathematically manipulated before display.
c. FT-NMR takes only a few seconds per spectrum.
4. FT-NMR and signal averaging provide increased speed and sensitivity.
a. Only a few mg of sample are needed for ^{13}C NMR spectra.
b. Only a few μg of sample are needed for ^{1}H NMR spectra.

B. Characteristics of ^{13}C NMR spectroscopy (Section 12.5).
   1. Each distinct carbon shows a single line.
   2. The chemical shift depends on the electronic environment within a molecule.
      a. Carbons bonded to electronegative atoms absorb downfield.
      b. Carbons with sp^3 hybridization absorb in the range 0–90 δ.
      c. Carbons with sp^2 hybridization absorb in the range 110–220 δ.
         Carbonyl carbons absorb in the range 160–220 δ.
   3. Symmetry reduces the number of absorptions.
   4. Peaks aren't uniform in size.

C. DEPT ^{13}C NMR spectra (Section 12.6).
   1. With DEPT experiments, the number of hydrogens bonded to each carbon can be determined.
   2. DEPT experiments are run in three stages.
      a. A broadband decoupled spectrum gives the chemical shifts of all carbons.
      b. A DEPT–90 spectrum shows signals due only to CH carbons.
      c. A DEPT–135 spectrum shows CH_3 and CH resonances as positive signals, and CH_2 resonances as negative signals.
   3. Interpretation of DEPT spectra.
      a. Subtract all peaks in the DEPT–135 spectrum from the broadband-decoupled spectrum to find C.
      b. Use DEPT–90 spectrum to identify CH.
      c. Use negative DEPT–135 peaks to identify CH_2.
      d. Subtract DEPT–90 peaks from positive DEPT–135 peaks to identify CH_3.

D. Uses of ^{13}C NMR spectroscopy (Section 12.7).
   ^{13}C NMR spectroscopy can show the number of nonequivalent carbons in a molecule and can identify symmetry in a molecule.

III. ^{1}H NMR Spectroscopy (Sections 12.8 – 12.13).

A. Proton equivalence (Section 12.8).
   1. ^{1}H NMR can be used to determine the number of nonequivalent protons in a molecule.
   2. If it is not possible to decide quickly if two protons are equivalent, replace each proton by –X.
      a. If the protons are unrelated, the products formed by replacement are constitutional isomers.
      b. If the protons are chemically identical, the same product will form, regardless of which proton is replaced, and the protons are homotopic.
      c. If the replacement products are enantiomers, the protons are enantiotopic.
      d. If the molecule contains a chirality center, the replacement products are diastereomers, and the protons are diastereotopic.

B. Chemical shifts in ^{1}H NMR spectroscopy (Section 12.9).
   1. Chemical shifts are determined by the local magnetic fields surrounding magnetic nuclei.
      a. More strongly shielded nuclei absorb upfield.
      b. Less shielded nuclei absorb downfield.
   2. Most ^{1}H NMR chemical shifts are in the range 0–10 δ.
      a. Protons that are sp^3-hybridized absorb at higher field strength.
      b. Protons that are sp^2-hybridized absorb at lower field strength.
      c. Protons on carbons that are bonded to electronegative atoms absorb at lower field strength.

3. The ^{1}H NMR spectrum can be divided into 5 regions:
   a. Saturated (0–1.5 δ).
   b. Allylic (1.5–2.5 δ).
   c. H bonded to C next to an electronegative atom (2.5–4.5 δ).
   d. Vinylic (4.5–6.5 δ).
   e. Aromatic (6.5–8.0).
   f. Aldehyde and carboxylic acid protons absorb even farther downfield.

C. Integration of ^{1}H NMR signals: proton counting (Section 12.10).
   1. The area of a peak is proportional to the number of protons causing the peak.
   2. Integrated peak areas are superimposed over a spectrum as a stair-step line.
   3. To compare two peaks, measure their relative heights.

D. Spin-spin splitting (Section 12.11).
   1. The tiny magnetic field produced by one nucleus can affect the magnetic field felt by other neighboring nuclei.
   2. Protons that have n equivalent neighboring protons show a peak in their ^{1}H NMR spectrum that is split into $n + 1$ smaller peaks (a multiplet).
   3. This splitting is caused by the coupling of spins of neighboring nuclei.
   4. The distance between peaks in a multiplet is called the coupling constant (J).
      a. The value of J is usually 0–18 Hz.
      b. The value of J is determined by the geometry of the molecule and is independent of the spectrometer operating frequency.
      c. The value of J is shared between both groups of hydrogens whose spins are coupled.
      d. By comparing values of J, it is possible to know the atoms whose spins are coupled.
   5. Three rules for spin-spin splitting in ^{1}H NMR:
      a. Chemically identical protons don't show spin-spin splitting.
      b. The signal of a proton with n equivalent neighboring protons is split into a multiplet of $n + 1$ peaks with coupling constant J.
      c. Two groups of coupled protons have the same value of J.
   6. Spin-spin splitting isn't seen in ^{13}C NMR.
      a. Although spin-spin splitting can occur between carbon and other magnetic nuclei, the spectrometer operating conditions suppress it.
      b. Coupling between the spins of two ^{13}C nuclei isn't seen because of the low probability that two ^{13}C nuclei would be adjacent.

E. Complex spin-spin splitting (Section 12.12).
   1. At times the signals in a ^{1}H NMR absorption overlap accidentally.
   2. Also, signals may be split by two or more nonequivalent kinds of protons.
      To understand the effect of multiple coupling, it helps to draw a tree diagram.

F. Uses of ^{1}H NMR spectroscopy (Section 12.13).
   ^{1}H NMR can be used to identify the products of reactions.

Solutions to Problems

12.1

$$E = \frac{1.20 \times 10^{-4} \text{ kJ/mol}}{\lambda \text{ (in m)}}$$

$$\lambda = \frac{c}{\nu} = \frac{3.0 \times 10^{8} \text{ m/s}}{\nu}; \quad \nu = 187 \text{ MHz} = 1.87 \times 10^{8} \text{ Hz}$$

$$\lambda = \frac{3.0 \times 10^{8} \text{ m/s}}{1.87 \times 10^{8} \text{ Hz}} = 1.60 \text{ m}$$

$$E = \frac{1.20 \times 10^{-4} \text{ kJ/mol}}{1.60} = 7.5 \times 10^{-5} \text{ kJ/mol}$$

Compare this value with $E = 8.0 \times 10^{-5}$ kJ/mol for ^{1}H. It takes slightly less energy to spin-flip a ^{19}F nucleus than to spin-flip a ^{1}H nucleus.

12.2

```
b H      CH3 a
   \    /
    C=C
   /    \
c H      Cl
```

2-Chloropropene has three kinds of protons. Protons b and c differ because one is cis to the chlorine and the other is trans.

12.3

$$\delta = \frac{\text{Observed chemical shift (in Hz)}}{200 \text{ MHz}}$$

(a) $\delta = \frac{1454 \text{ Hz}}{200 \text{ MHz}} = 7.27\ \delta$ for $CHCl_3$ (b) $\delta = \frac{610 \text{ Hz}}{200 \text{ MHz}} = 3.05\ \delta$ for CH_3Cl

(c) $\delta = \frac{693 \text{ Hz}}{200 \text{ MHz}} = 3.46\ \delta$ for CH_3OH (d) $\delta = \frac{1060 \text{ Hz}}{200 \text{ MHz}} = 5.30\ \delta$ for CH_2Cl_2

12.4

(a)

$$\delta = \frac{\text{Observed chemical shift (\# Hz away from TMS)}}{\text{Spectrometer frequency in MHz}}$$

Units of δ are parts per million. In this problem, $\delta = 2.1$ ppm

$$2.1 \text{ ppm} = \frac{\text{Observed chemical shift}}{200 \text{ (MHz)}}$$

$$420 \text{ Hz} = \text{Observed chemical shift}$$

(b) If the ^{1}H NMR spectrum of acetone were recorded at 500 MHz, the position of absorption would still be 2.1 δ because measurements given in ppm or δ units are independent of the operating frequency of the NMR spectrometer.

(c) $2.1\ \delta = \frac{\text{Observed chemical shift}}{500 \text{ (MHz)}}$; Observed chemical shift $= 1050$ Hz

12.5

(a)

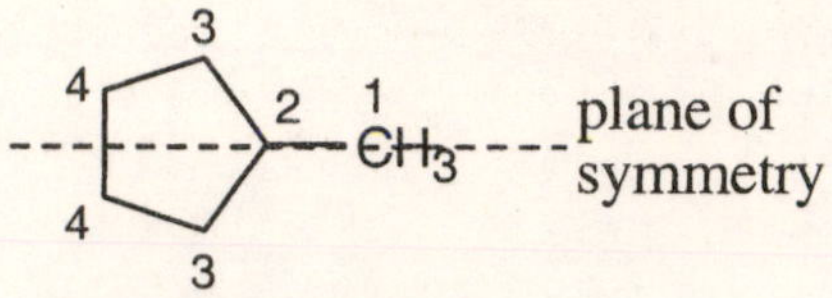

Methylcyclopentane

Four resonance lines are observed because of symmetry.

(b)

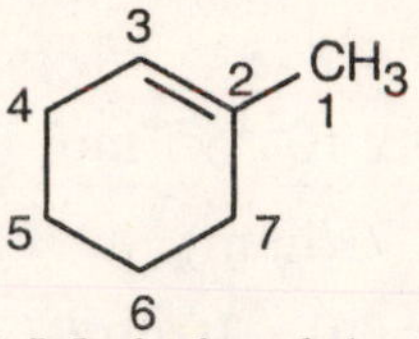

1-Methylcyclohexene

Seven lines are seen because no two carbons are equivalent

(c)

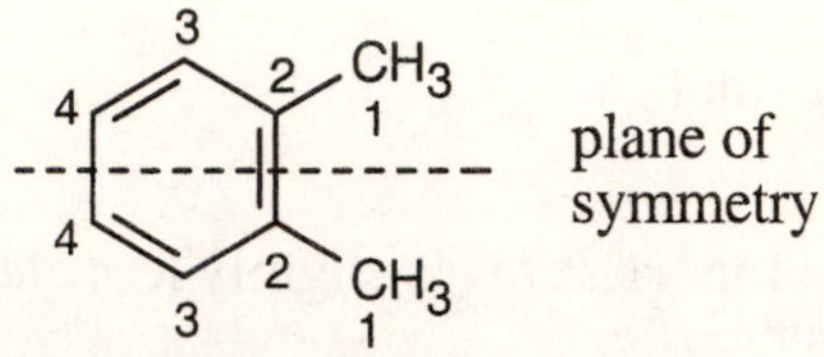

1,2-Dimethylbenzene

Four resonance lines are seen.

(d)

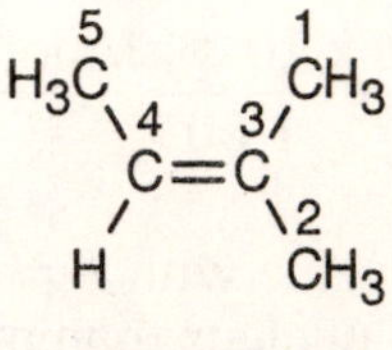

2-Methylbut-2-ene

Five resonance lines are observed. Carbons 1 and 2 are nonequivalent because of the double bond stereochemistry.

(e)

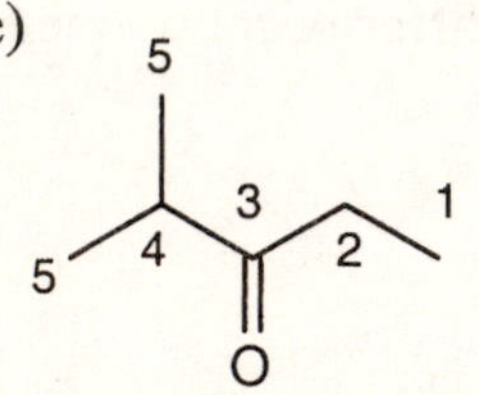

Five resonance lines are seen.

(f)

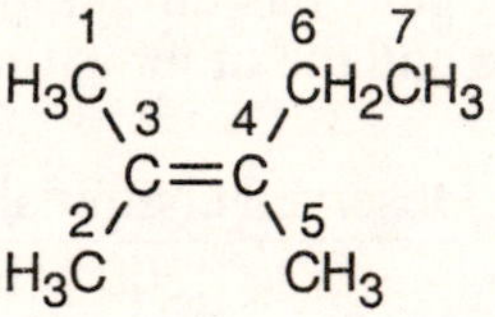

Seven resonance lines are seen.

12.6

(a)

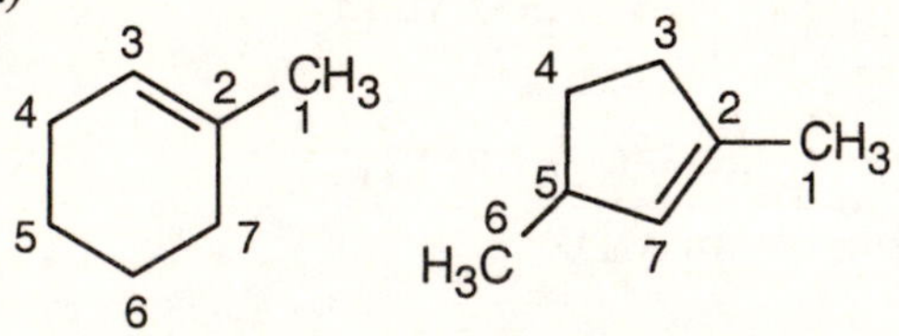

(b)

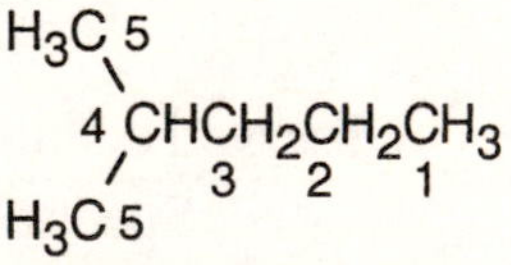

Two of the 6 carbons are equivalent.

(c)

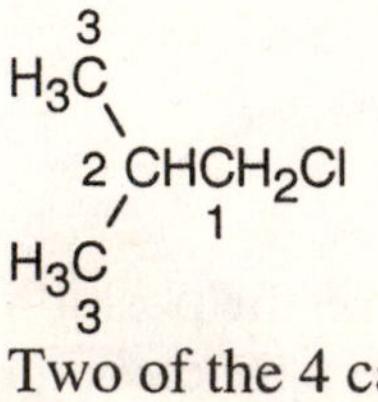

Two of the 4 carbons are equivalent.

12.7 Methyl propanoate has 4 unique carbons, and each one absorbs in a specific region of the ^{13}C spectrum. The absorption (4) has the lowest value of δ and occurs in the $-CH_3$ region of the ^{13}C spectrum. Absorption (3) occurs in the $-CH_2-$ region. The methyl group (1) is next to an electronegative atom and absorbs downfield from the other two absorptions. The carbonyl carbon (2) absorbs far downfield.

$CH_3CH_2COCH_3$ (C=O at carbon 2; carbons numbered 4 3 2 1)

δ (ppm)	*Assignment*
9.3	4
27.6	3
51.4	1
174.6	2

12.8 **Strategy:** The top spectrum shows all eight ^{13}C NMR peaks. The middle spectrum (DEPT-90) shows only peaks due to CH carbons. From the DEPT-90 spectrum, the absorption at 124 δ can be assigned to the vinyl carbon (5), and the absorption at 68 δ can be assigned to the –OH carbon (2).

The DEPT-135 spectrum shows all but the quaternary carbon (6), which appears in the top spectrum at 132 δ. The top half of the DEPT-135 spectrum shows absorptions due to CH_3 carbons and CH carbons (which we have already identified). The 3 remaining peaks on the top of the DEPT-135 spectrum are due to methyl groups. Although we haven't learned enough to identify these peaks, the peak at 23 δ is due to carbon (1). The other two peaks arise from carbons (7) (26 δ) and (8) (18 δ).

The bottom half of the DEPT-135 shows the two CH_2 carbons. Carbon (3) absorbs at 39 δ (negative), and carbon (4) absorbs at 24 δ (negative).

Solution:

6-Methylhept-5-en-2-ol

Carbon	*Chemical Shift (δ)*
1	23
2	68
3	39 (negative)
4	24 (negative)
5	124
6	132
7	26
8	18

12.9 **Strategy:** Identify the carbons as CH_3, CH_2, CH or quaternary, and use Figure 12.7 to find approximate values for chemical shifts. (When an actual spectrum is given, it is easier to assign the carbons to the chemical shifts.) Remember: DEPT-90 spectra identify CH carbons, and DEPT-135 spectra identify CH_3 carbons (positive peaks), CH carbons (positive peaks already identified), and CH_2 carbons (negative peaks). Quaternary carbons are identified in the broadband-decoupled spectrum, in which all peaks appear.

Solution:

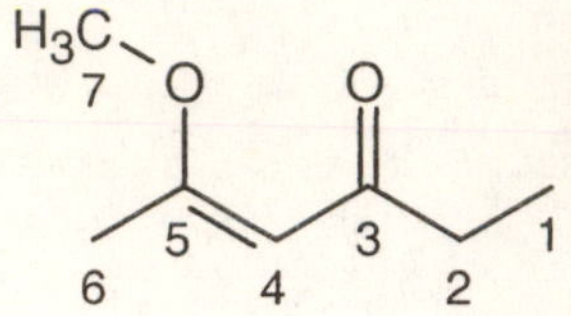

Carbon	Chemical Shift (δ)	DEPT-90?	DEPT-135?
1	10–30	no	yes (positive)
2	30–50	no	yes (negative)
3	160–220	no	no
4	110–150	yes	yes (positive)
5	110–150	no	no
6	10–30	no	yes (positive)
7	50–90	no	yes (positive)

12.10 **Strategy:** Always start this type of problem by calculating the degree of unsaturation of the unknown compound. $C_{11}H_{16}$ has 4 degrees of unsaturation. Since the unknown hydrocarbon is aromatic, a benzene ring accounts for all four degrees of unsaturation.

Next, look for elements of symmetry. Although the molecular formula indicates 11 carbons, only 7 peaks appear in the ^{13}C NMR spectrum, indicating a plane of symmetry. Four of the 7 peaks are due to aromatic carbons, indicating a benzene ring that is probably monosubstituted. (Prove to yourself that a monosubstituted benzene ring has 4 different kinds of carbons).

Solution: The DEPT-90 spectrum shows that 3 of the kinds of carbons in the aromatic ring are CH carbons. The positive peaks in the DEPT-135 spectrum include these three peaks, along with the peak at 29.5 δ, which is due to a CH_3 carbon. The negative peak in the DEPT-135 spectrum is due to a CH_2 carbon.

Two peaks remain unidentified and are thus quaternary carbons; one of them is aromatic.

At this point, the unknown structure is a monosubstituted benzene ring with a substituent that contains CH_2, C, and CH_3 carbons. A structure for the unknown compound that satisfies all data:

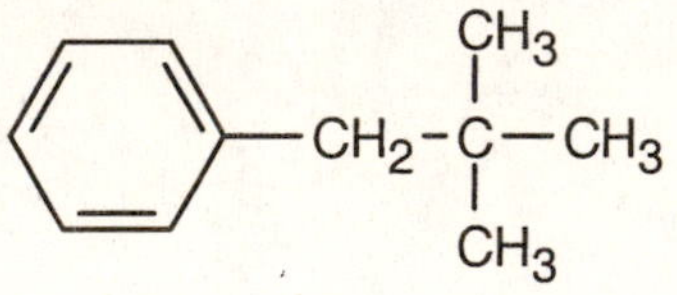

$C_6H_5-CH_2-C(CH_3)_2-CH_3$

12.11

$CH_3CH_2CH_2CH_2C{\equiv}CH$ + HBr ⟶ $CH_3CH_2CH_2CH_2C(Br){=}CH_2$ 2-Bromohex-1-ene *or* $CH_3CH_2CH_2CH_2CH{=}CHBr$? 1-Bromohex-1-ene

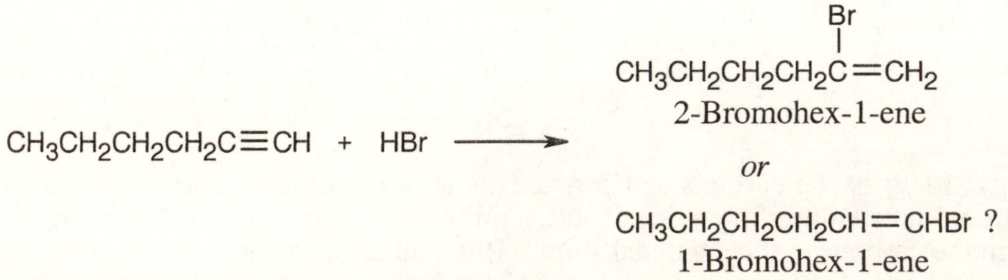

The two possible products are easy to distinguish by using ^{13}C NMR. 2-Bromohex-1-ene, the actual product formed, shows no peaks in its DEPT-90 ^{13}C NMR spectrum because it has no CH carbons. The other possible product, 1-bromohex-1-ene, shows 2 peaks in its DEPT-90 spectrum.

12.12 Strategy: First, check for protons that are unrelated (none appear in this problem). Next, look for molecules that already have chirality centers. Replacement of a $-CH_2-$ proton by X in (d) and replacement of a $-CH_3$ proton in (e) produces a second chirality center, and the two possible replacement products are diastereomers. Thus, the indicated protons in (d) and (e) are diastereotopic.

Finally, for the other molecules, mentally replace each of the two hydrogens in the indicated set with X, a different group. In (a), the resulting products are enantiomers, and the protons are enantiotopic. Replacement of the protons in (b) produces two chirality centers (the carbon bearing the hydroxyl group is now chiral) and the indicated protons are diastereotopic. Replacement of one of the methyl protons in each of the groups in (c) produces a pair of double-bond isomers that are diastereomers; these protons are diastereotopic. The protons in (f) are homotopic.

Solution:

(a) enantiotopic

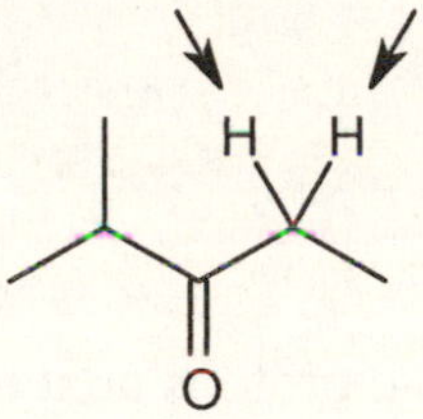

(b) diastereotopic

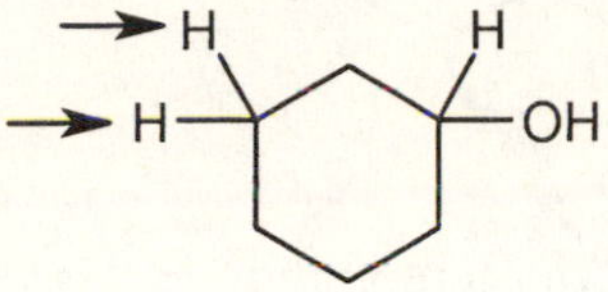

(c) diastereotopic

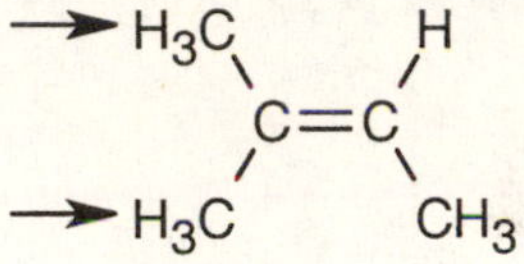

(d) diastereotopic

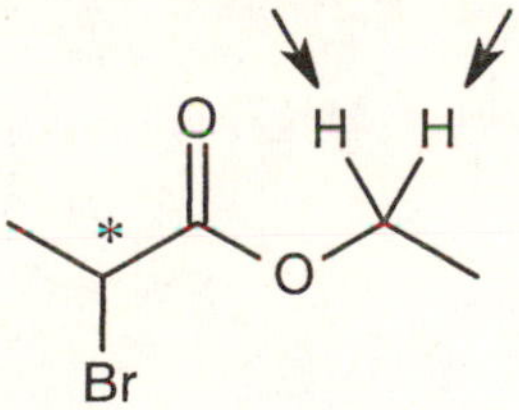

(e) diastereotopic

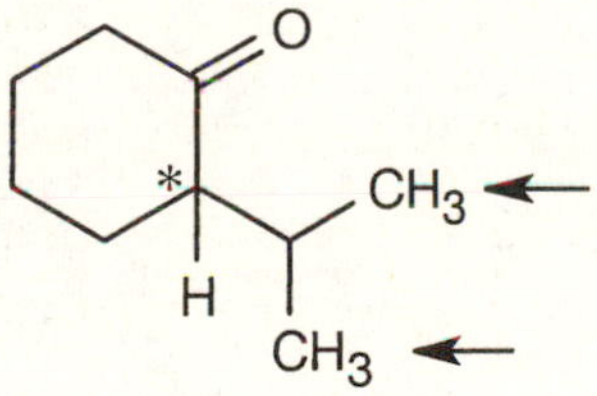

(f) homotopic

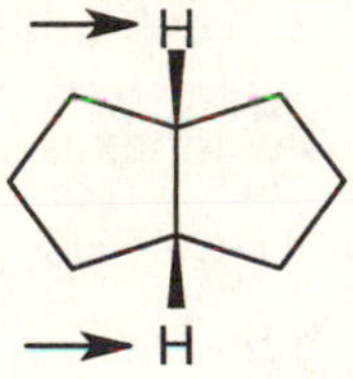

12.13

Compound	Kinds of non-equivalent protons	Compound	Kinds of non-equivalent protons
(a) $\overset{1}{C}H_3\overset{2}{C}H_2Br$	2	(b) 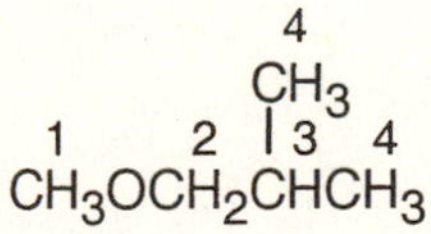 $\overset{1}{C}H_3O\overset{2}{C}H_2\overset{3}{C}H(\overset{4}{C}H_3)\overset{4}{C}H_3$	4
(c) $\overset{1}{C}H_3\overset{2}{C}H_2\overset{3}{C}H_2NO_2$	3	(d) p-$CH_3C_6H_4$–H (1: CH_3; 2, 3, 4: ring H) 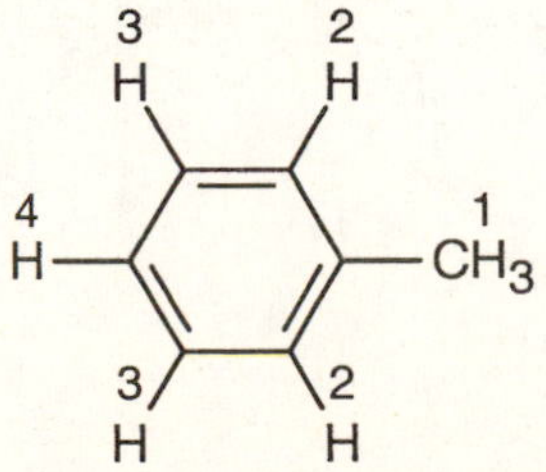	4

Compound	Kinds of non-equivalent protons	Compound	Kinds of non-equivalent protons
(e) 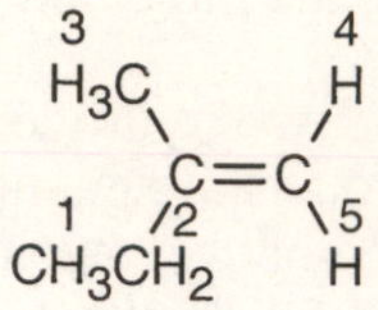 $H_3C(3)(CH_3CH_2(1,2))C{=}CH(4)H(5)$ The two vinylic protons are nonequivalent.	5	(f) 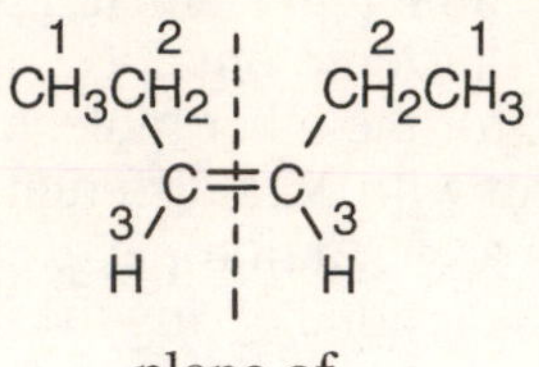 $CH_3CH_2(1,2)CH{=}CHCH_2CH_3$ plane of symmetry	3

12.14

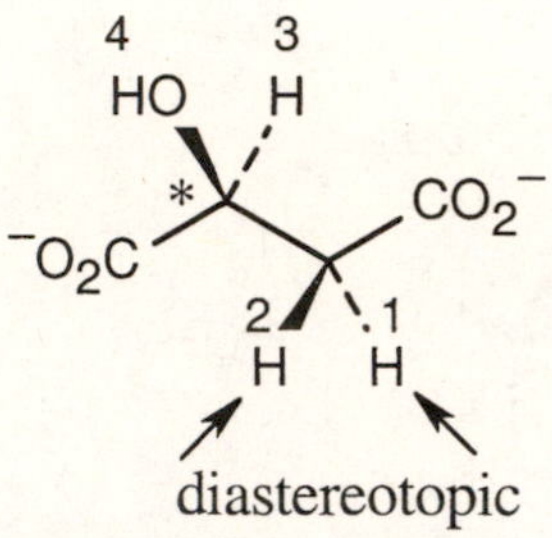

(S)-Malate

diastereotopic

Because (*S*)-malate already has a chirality center (starred), the two protons next to it are diastereotopic and absorb at different values. The ^{1}H NMR spectrum of(*S*)-malate has four absorptions.

12.15

	Compound	δ	Kind of proton
(a)	Cyclohexane	1.43	secondary alkyl
(b)	CH_3COCH_3	2.17	methyl ketone
(c)	Benzene	7.37	aromatic
(d)	CH_2Cl_2	5.30	protons adjacent to two halogens
(e)	OHCCHO	9.70	aldehyde
(f)	$(CH_3)_3N$	2.12	methyl protons adjacent to nitrogen

12.16

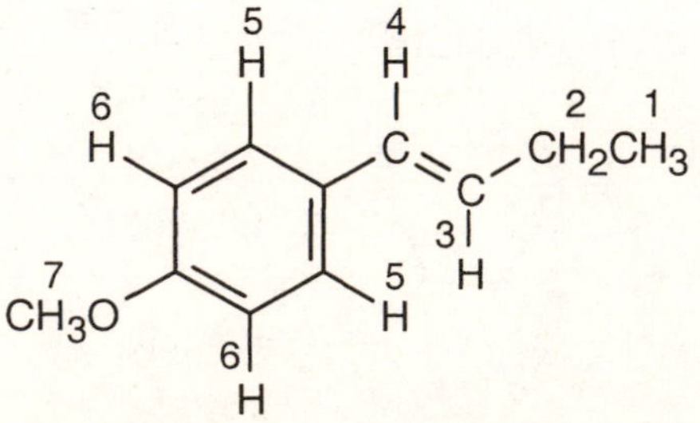

Proton	δ	Kind of proton
1	1.0	primary alkyl
2	1.8	allylic
3	6.1	vinylic
4	6.3	vinylic (different from proton 3)
5	7.2	aromatic
6	6.8	aromatic
7	3.8	ether

This compound has seven different kinds of protons. Notice that the two protons labeled 5 are equivalent to each other, as are the two protons labeled 6, because of rotation around the bond joining the aromatic ring and the side chain.

12.17

H_3C—C_6H_4—CH_3 *p*-Xylene

There are two absorptions in the ^{1}H NMR spectrum of *p*–xylene. The four ring protons absorb at 7.0 δ, and the six methyl-group protons absorb at 2.3 δ. The peak ratio of methyl protons:ring protons is 3:2.

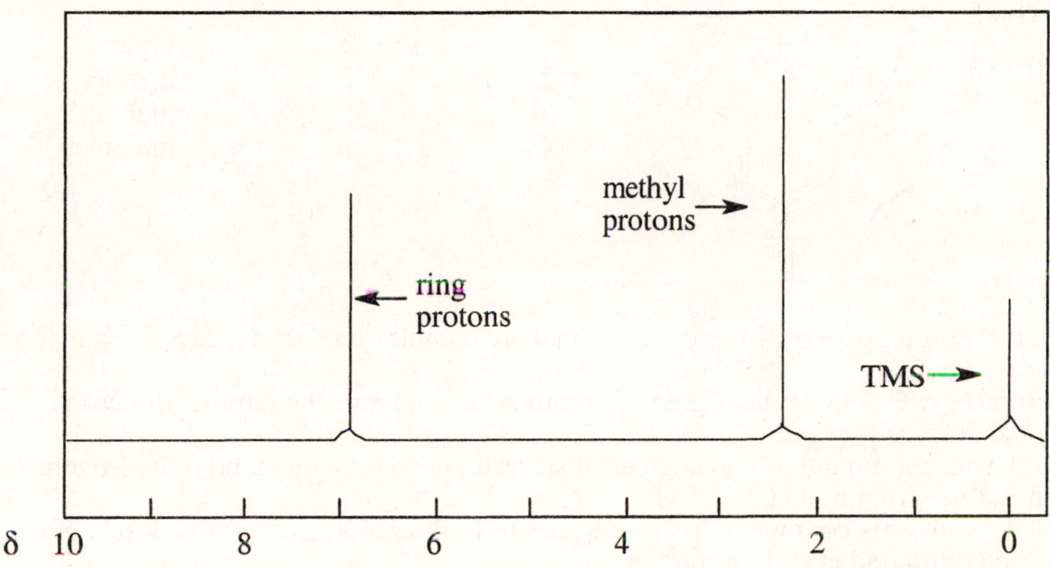

12.18

	Compound	Proton	Number of Adjacent Protons	Splitting
(a)	$\overset{1}{CHBr_2}\overset{2}{CH_3}$	1	3	quartet
		2	1	doublet
(b)	$\overset{1}{CH_3}O\overset{2}{CH_2}\overset{3}{CH_2}Br$	1	0	singlet
		2	2	triplet
		3	2	triplet
(c)	$Cl\overset{1}{CH_2}\overset{2}{CH_2}\overset{1}{CH_2}Cl$	1	2	triplet
		2	4	quintet
(d)	$\overset{1}{CH_3}\overset{2}{CH}(\overset{1}{CH_3})C(=O)O\overset{3}{CH_2}\overset{4}{CH_3}$	1	1	doublet
		2	6	septet
		3	3	quartet
		4	2	triplet

Compound	Proton	Number of Adjacent Protons	Splitting
(e) $CH_3CH_2COCH(CH_3)CH_3$ (C=O; protons numbered 1, 2, 3, 4; 4 = CH_3)	1	2	triplet
	2	3	quartet
	3	6	septet
	4	1	doublet
(f) (structure with protons labeled 1, 2, 3, 4)	1	2	triplet
	2	1	doublet
	3	1	multiplet
	4	1	multiplet

The splitting patterns for protons 3 and 4 are complex and are not explained in the text.

12.19 **Strategy:** Calculate the degree of unsaturation, and note the number of peaks to see if symmetry is present.
(a) This compound has no degrees of saturation and only one kind of hydrogen. The only possible structure is CH_3OCH_3.
(b) Again, this compound has no degrees of unsaturation and has two kinds of hydrogens. The compound is 2-chloropropane.
(c) This compound, with no degrees of unsaturation, has two different kinds of hydrogen, each of which has two neighboring hydrogens.
(d) $C_4H_8O_2$; one degree of unsaturation and 3 different kinds of hydrogen.

Solution:

(a) CH_3OCH_3

(b) $CH_3CH(Cl)CH_3$

(c) $ClCH_2CH_2OCH_2CH_2Cl$

(d) $CH_3CH_2C(=O)OCH_3$ *or* $CH_3C(=O)OCH_2CH_3$

12.20 The molecular formula ($C_4H_{10}O$) indicates that the compound has no multiple bonds or rings. The ^1H NMR spectrum shows two signals, corresponding to two types of hydrogens in the ratio 33:50, or 2:3. Since the unknown contains 10 hydrogens, four protons are of one type and six are of the other type.

The upfield signal at 1.2 δ is due to saturated primary protons. The downfield signal at 3.5 δ is due to protons on carbon adjacent to an electronegative atom — in this case, oxygen.

The signal at 1.2 δ is a triplet, indicating two neighboring protons. The signal at 3.5 δ is a quartet, indicating three neighboring protons. The compound is diethyl ether, $CH_3CH_2OCH_2CH_3$.

12.21

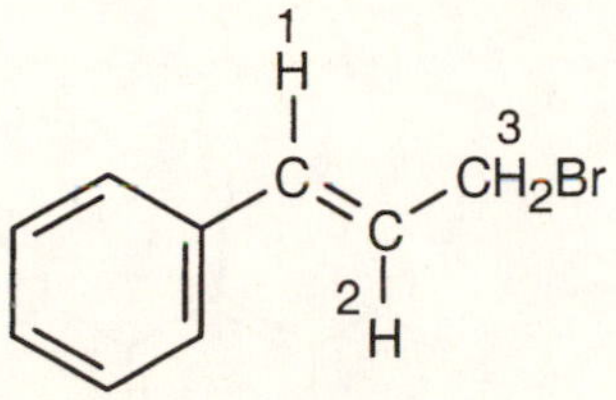

(*E*)-3-Bromo-1-phenylprop-1-ene

Coupling of the C2 proton to the Cl vinylic proton occurs with $J = 16$ Hz and causes the signal of the C2 proton to be split into a doublet. The C2 proton is also coupled to the two C3 protons with $J = 8$ Hz. This splitting causes each leg of the C2 proton doublet to be split into a triplet, producing six lines in all. Because of the size of the coupling constants, two of the lines coincide, and a quintet is observed.

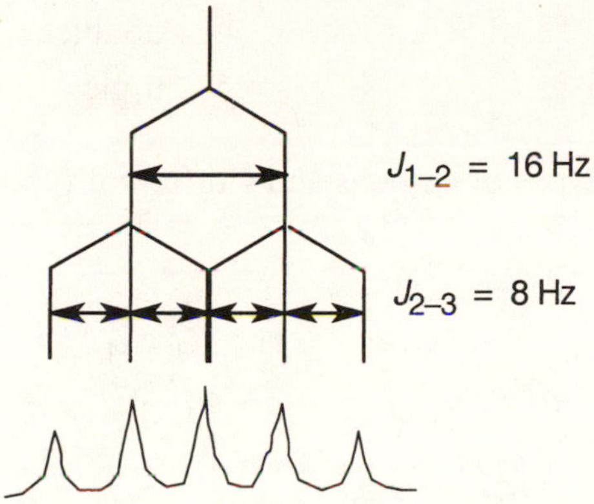

J_{1-2} = 16 Hz

J_{2-3} = 8 Hz

12.22

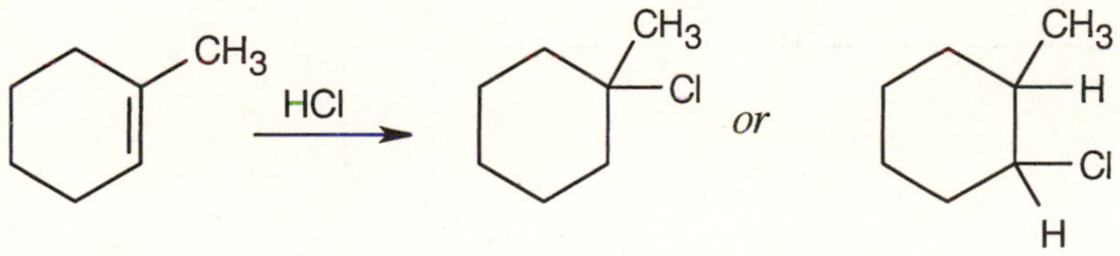

HCl

or

Focus on the ^{1}H NMR methyl group absorption. In the first product, the methyl group signal is unsplit; in the other product, it appears as a doublet. In addition, the second product shows a downfield absorption in the 2.5 δ – 4.0 δ region due to the proton bonded to a carbon that is also bonded to an electronegative atom. If you were to take the ^{1}H NMR spectrum of the reaction product, you would find an unsplit methyl group, and you could conclude that the product was 1-chloro-1-methylcyclohexane.

Visualizing Chemistry

12.23

(a)

1. doublet
2. septet
3. singlet

(b)

1. singlet
2. doublet
3. doublet
4. doublet
5. triplet

12.24 The compound has 5 different types of carbons and 4 different types of hydrogens.

^{13}C

^{1}H

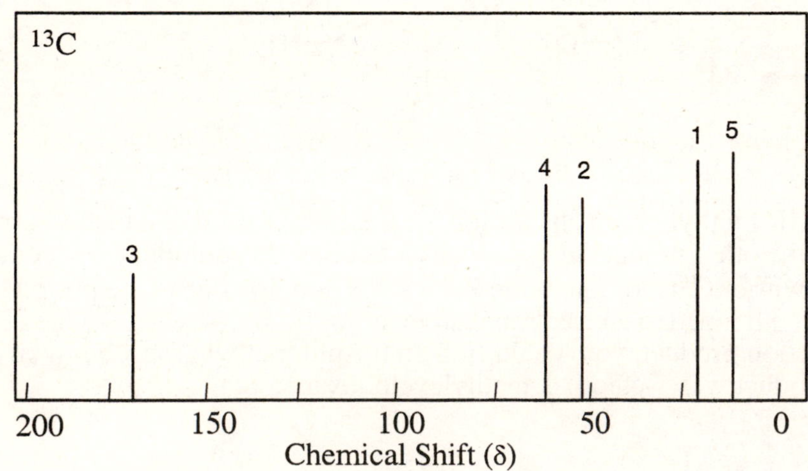

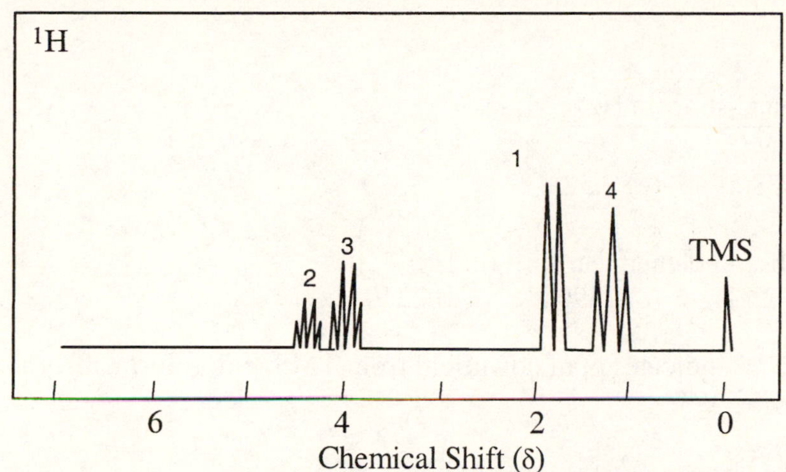

12.25 If you assign *R,S* configurations to the two carbons bonded to the methyl group, it is apparent that *cis*-1,2-dimethylcyclohexane is a meso compound. When the cyclohexane ring undergoes a ring-flip, the ring passes through an intermediate that has a plane of symmetry. Both the ^{13}C NMR spectrum and the ^{1}H NMR spectrum show 4 peaks.

12.26 (a) Because cysteine has a chirality center, the indicated protons are diastereotopic.
(b) Imagine replacing first one, then the other, of the indicated protons with a substituent X. The two resulting compounds would be enantiomers.

(a) diastereotopic

(b) enantiotopic

Additional Problems

12.27

$$\delta = \frac{\text{Observed chemical shift (in Hz)}}{\text{200 MHz}}$$

(a) 2.18 δ (b) 4.78 δ (c) 7.52 δ

12.28 δ x 300 MHz = Observed chemical shift (in Hz)
(a) 630 Hz (b) 1035 Hz (c) 1890 Hz (d) 2310 Hz

12.29 (a) Since the symbol "δ" indicates ppm downfield from TMS, chloroform absorbs at 7.3 ppm.

(b)

$$\delta = \frac{\text{Observed chemical shift (in Hz)}}{\text{Spectrometer frequency in MHz}}$$

$$7.3 \text{ ppm} = \frac{\text{chemical shift}}{\text{360 MHz}}; \ 7.3 \text{ ppm x 360 MHz} = \text{chemical shift}$$

$$2600 \text{ Hz} = \text{chemical shift}$$

(c) The value of δ is still 7.3 because the chemical shift measured in δ is independent of the operating frequency of the spectrometer.

12.30

(a)

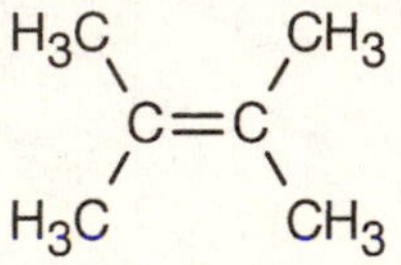

^{13}C: 2 absorptions
^{1}H: 1 absorption

(b)

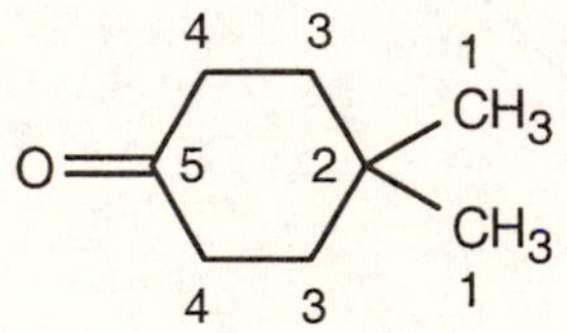

^{13}C: 5 absorptions
(at room temperature)

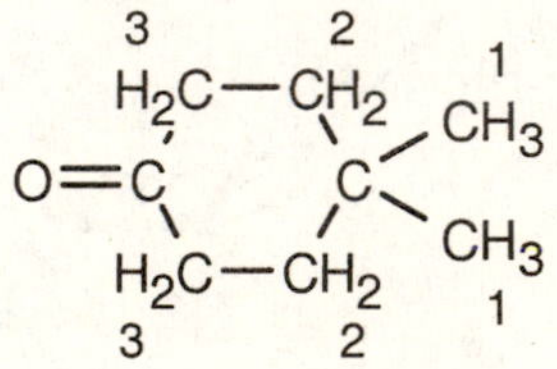

^{1}H: 3 absorptions
(at room temperature)

(c)

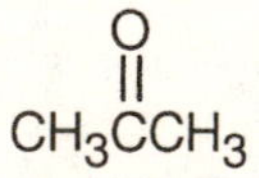

^{13}C: 2 absorptions
^{1}H: 1 absorption

(d)

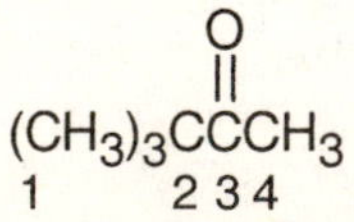

^{13}C: 4 absorptions

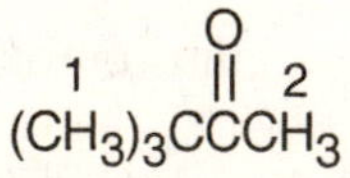

^{1}H: 2 absorptions

(e)

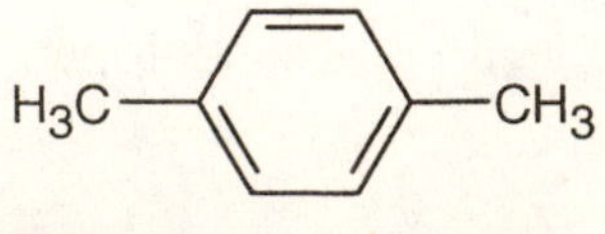

^{13}C: 3 absorptions
^{1}H: 2 absorptions

(f)

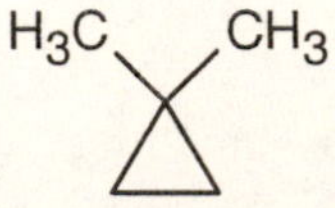

^{13}C: 3 absorptions
^{1}H: 2 absorptions

12.31–12.32

Compound	*Number of* ^{13}C *Absorptions*	*Carbons Showing Peaks in DEPT-135* ^{13}C *NMR Spectrum*: *Positive Peaks*	*Negative Peaks*	*No Peaks*
(a) 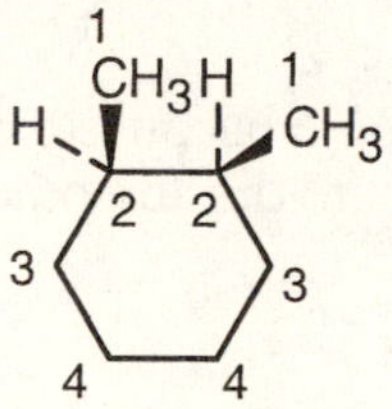	4	carbons 1,2	carbons 3,4	
(b) $\overset{1}{CH_3}\overset{2}{CH_2}O\overset{3}{CH_3}$	3	carbons 1,3	carbon 2	
(c) 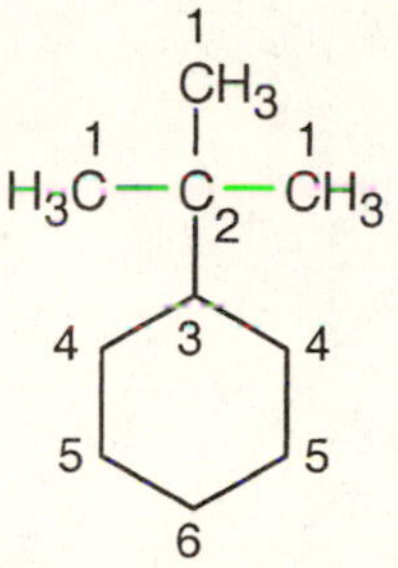	6	carbons 1,3	carbons 4,5,6	carbon 2
(d) 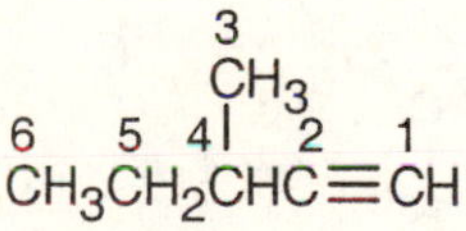 $\overset{6}{CH_3}\overset{5}{CH_2}\overset{4}{CH}(\overset{3}{CH_3})\overset{2}{C}{\equiv}\overset{1}{CH}$	6	carbons 1,3,4,6	carbon 5	carbon 2
(e) 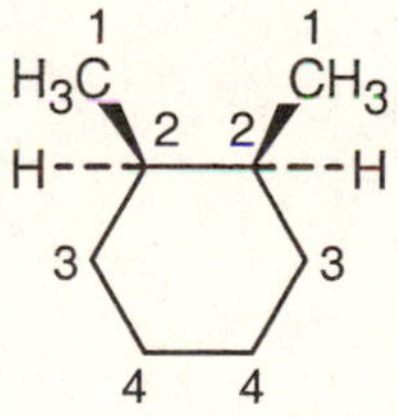	4	carbons 1,2	carbons 3,4	
(f) 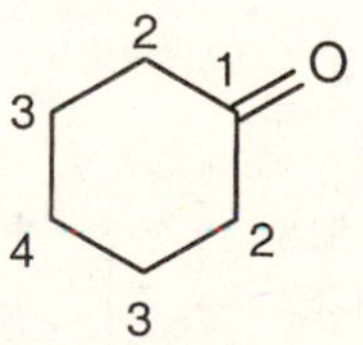	4		carbons 2,3,4	carbon 1

12.33 A nucleus that absorbs at 6.50 δ is less shielded than a nucleus that absorbs at 3.20 δ and thus requires a weaker applied field to come into resonance. A shielded nucleus feels a smaller effective field, and a stronger applied field is needed to bring it into resonance.

12.34 (a) enantiotopic (b) diastereotopic (c) diastereotopic

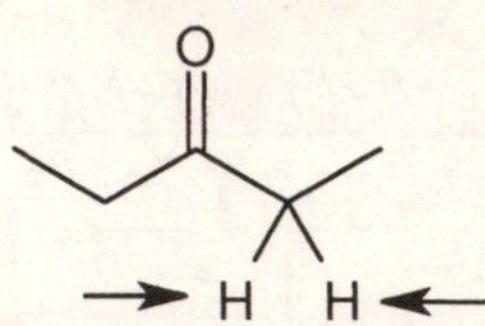

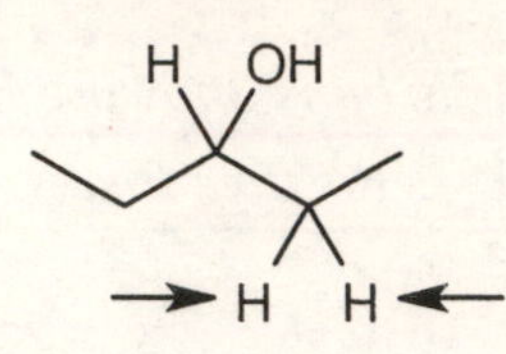

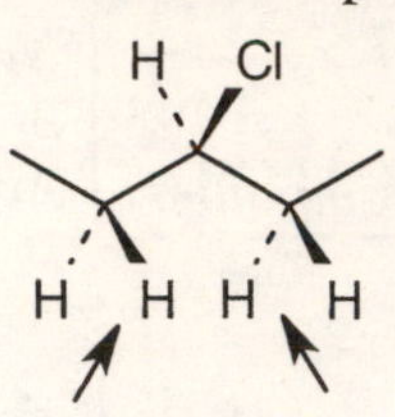

Refer to Problem 12.12 for help. The protons in (c) are diastereotopic because the molecules that result from replacement of the indicated hydrogens are diastereomers (prove it to yourself with models).

12.35

Compound	*Kinds of non-equivalent protons*	*Compound*	*Kinds of non-equivalent protons*
(a) 1 1 H_3C CH_3 2 2 3 3 4	4	(b) 1 2 3 4 $CH_3CH_2CH_2OCH_3$	4
(c) 1 1 2 2 2 2 1 1	2	(d) 3 H 1 H 4 C C 5 6 4 5 H 2	6
(e) 3 H 2 H C C 4 5 $COOCH_2CH_3$ 1 H	5		

12.36 (a) homotopic (b) enantiotopic (c) diastereotopic

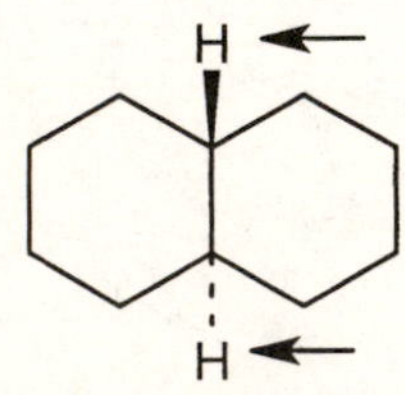

H H

H_3C CH_2 H_3C

12.37

Lowest Chemical Shift ⟶ *Highest Chemical Shift*

CH_4	<	Cyclohexane	<	CH_3COCH_3	<	CH_2Cl_2,	$H_2C{=}CH_2$	<	Benzene
0.23		1.43		2.17		5.30	5.33		7.37

12.38

Compound	Number of peaks	Peak Assignment	Splitting Pattern	
(a) $\overset{1}{(CH_3)_3}\overset{2}{CH}$	2	1	doublet	(9H)
		2	multiplet (dectet)	(1H)
(b) $\overset{1}{CH_3}\overset{2}{CH_2}\overset{O}{\overset{\parallel}{C}}O\overset{3}{CH_3}$	3	1	triplet	(3H)
		2	quartet	(2H)
		3	singlet	(3H)
(c) 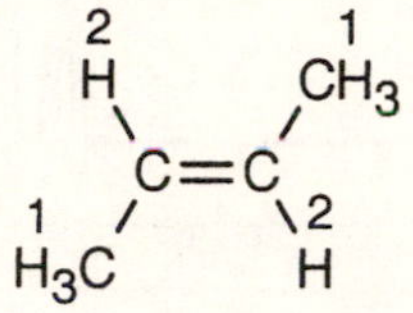 (2)H and (1)CH_3 on one carbon, (1)H_3C and (2)H on the other: $H(CH_3)C=C(H)CH_3$	2	1	doublet	(6H)
		2	quartet	(2H)

12.39

Compound	Peak Assignment	Splitting Pattern	
$\overset{1}{CH_3}\overset{2}{CH_2}\overset{O}{\overset{\parallel}{C}}O\overset{3}{CH}(\overset{4}{CH_3})_2$	1	triplet	(3H)
	2	quartet	(2H)
	3	septet	(1H)
	4	doublet	(6H)

12.40 First, check each isomer for structural differences that are obviously recognizable in the ^1H NMR spectrum. If it is not possible to pick out distinguishing features immediately, it may be necessary to sketch an approximate spectrum of each isomer for comparison.

(a) $CH_3CH{=}CHCH_2CH_3$ has two vinylic protons with chemical shifts at 5.4 – 5.5 δ. Because ethylcyclopropane shows no signal in this region, it should be easy to distinguish one isomer from the other.

(b) $CH_3CH_2OCH_2CH_3$ has two kinds of protons, and its ^1H NMR spectrum consists of two peaks — a triplet and a quartet. $CH_3OCH_2CH_2CH_3$ has four different types of protons, and its spectrum is more complex. In particular, the methyl group bonded to oxygen shows an unsplit singlet absorption.

(c) Each compound shows three peaks in its ^1H NMR spectrum. The ester, however, shows a downfield absorption due to the $–CH_2–$ hydrogens next to oxygen. No comparable peak shows in the spectrum of the ketone.

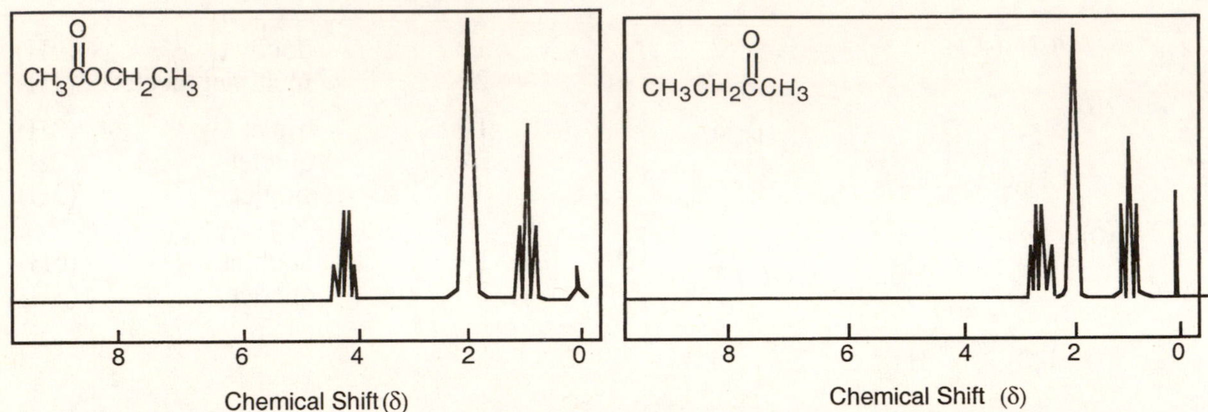

d) Each isomer contains four different kinds of protons — two kinds of methyl protons and two kinds of vinylic protons. For the first isomer, the methyl peaks are both singlets, whereas for the second isomer, one peak is a singlet and one is a doublet.

12.41

(a) $(CH_3)_4C$ (b) (c)

12.42

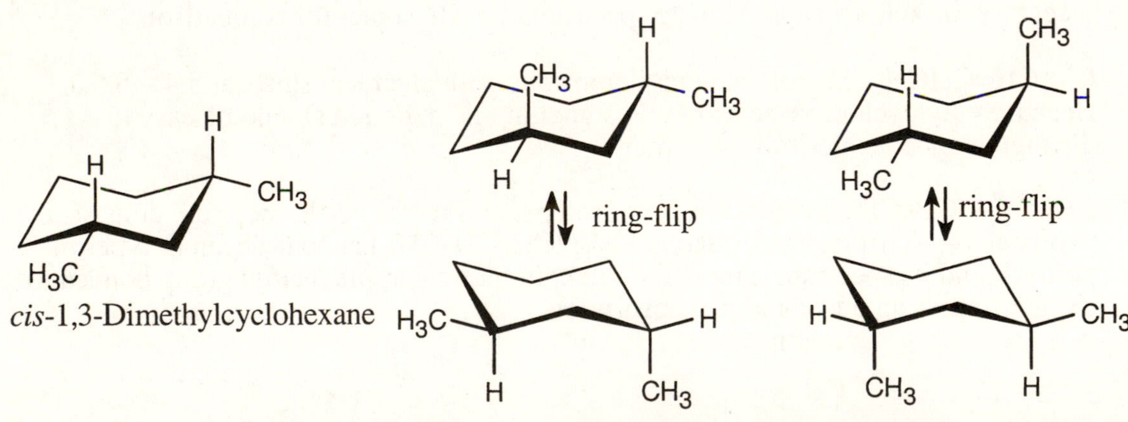

cis-1,3-Dimethylcyclohexane is a meso compound. Because of symmetry, it shows 5 absorptions in its ^{13}C NMR spectrum. *trans*-1,3-Dimethylcyclohexane exists as a pair of enantiomers. At room temperature, both enantiomers undergo ring-flips at a rate that is faster than the time frame of an NMR spectrum and that averages absorptions due to nonequivalent carbons. Like the cis isomer, the racemic mixture of trans enantiomers shows 5 absorptions in its ^{13}C spectrum.

12.43 (a),(b) C_3H_6O contains one double bond or ring. Possible structures for C_3H_6O include:

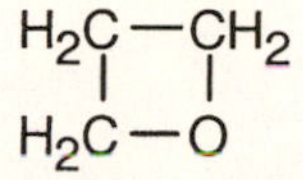

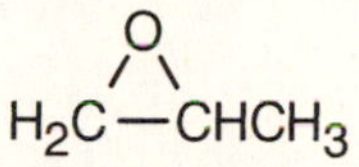

$H_2C{-}CH_2$ / $H_2C{-}O$ (four-membered ring)	$H_2C{-}CHCH_3$ bridged by O (three-membered ring)	$H_2C{=}CHOCH_3$
Cyclic ether	Cyclic ether	Ether, double bond

$H_2C{=}CHCH_2OH$	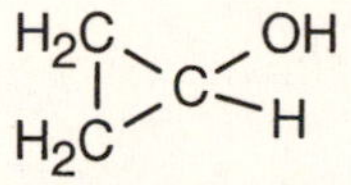 ring of H_2C, H_2C, C with OH and H	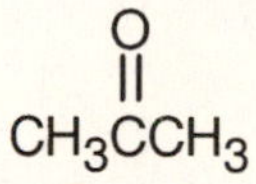 $CH_3C(=O)CH_3$	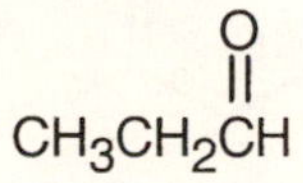 $CH_3CH_2C(=O)H$
Alcohol, double bond	Cyclic alcohol	Ketone (acetone)	Aldehyde

(c) Saturated ketones absorb at 1715 cm^{-1} in the infrared. Only the last two compounds above show an infrared absorption in this region.

(d) Since the unknown compound of this problem is a ketone and shows only one ^1H NMR absorption (in the methyl ketone region), it must be acetone.

12.44 Either ^1H NMR or ^{13}C NMR can be used to distinguish among these isomers. In either case, it is first necessary to find the number of different kinds of protons or carbon atoms.

Compound	*Kinds of Protons*	*Kinds of Carbon atoms*	*Number of ^1H NMR peaks*	*Number of ^{13}C NMR peaks*
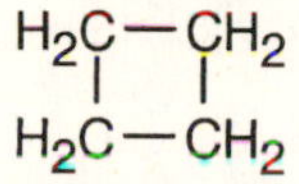 $H_2C{-}CH_2$ / $H_2C{-}CH_2$ (cyclobutane)	1	1	1	1
$H_2C{=}CHCH_2CH_3$	5	4	5	4
$CH_3CH{=}CHCH_3$	2	2	2	2
$(CH_3)_2C{=}CH_2$	2	3	2	3

^{13}C NMR is the simplest method for identifying these compounds because each isomer differs in the number of absorptions in its ^{13}C NMR spectrum. ^1H NMR can also be used to distinguish among the isomers because the two isomers that show two ^1H NMR peaks differ in their splitting patterns.

12.45

		Number of Peaks	Distinguishing Absorptions
3-methylcyclohex-2-enone (CH_3)	^{13}C	7	Two vinylic peaks
	^{1}H	5	Unsplit vinylic peak, relative area 1
3-cyclopentenyl methyl ketone ($C(=O)CH_3$)	^{13}C	5	One vinylic peak
	^{1}H	4	Split vinylic peak, relative area 2

The two isomers have different numbers of peaks in both ^{1}H NMR and ^{13}C NMR. In addition, the distinguishing absorptions in the vinylic region of both the ^{1}H and ^{13}C spectra make it possible to identify each isomer by its NMR spectrum.

The ketone IR absorption of 3-methylcyclohex-2-enone occurs near 1690 cm^{-1} because the double bond is next to the ketone group. The ketone IR absorption of 3-cyclopentenyl methyl ketone occurs near 1715 cm^{-1}, the usual position for ketone absorption.

3-Methylcyclohex-2-enone shows a UV absorption because its double bonds are conjugated.

12.46 The unknown compound has no degrees of unsaturation and has two different kinds of hydrogens. The unknown compound is $BrCH_2CH_2CH_2Br$.

12.47

(a) $(CH_3)_2CHC(=O)CH_3$ (1 = $(CH_3)_2$; 3 = CH; 2 = CH_3 of ketone)

1 = 0.95 δ (isopropyl group)
2 = 2.10 δ (methyl ketone)
3 = 2.43 δ (isopropyl group)

(b) $H_3C(Br)C=CH_2$ (1 = H_3C; 2, 3 = the two vinylic H)

1 = 2.32 δ (methyl group attached to double bond)
2,3 = 5.35 δ, 5.54 δ (vinylic H)

12.48 Possible structures for $C_4H_7ClO_2$ are $CH_3CH_2CO_2CH_2Cl$ and $ClCH_2CO_2CH_2CH_3$. Chemical shift data can distinguish between them.

$CH_3CH_2C(=O)OCH_2Cl$ — **A**

$ClCH_2C(=O)OCH_2CH_3$ — **B**

In **A**, the protons attached to the carbon bonded to both oxygen and chlorine ($-OCH_2Cl$) absorb far downfield (5.0 – 6.0 δ). Because no signal is present in this region of the ^{1}H NMR spectrum given, the unknown must be **B**. In addition, the quartet absorbing at 4.3 δ is typical of a CH_2 group next to an electronegative atom and coupled with a methyl group.

12.49

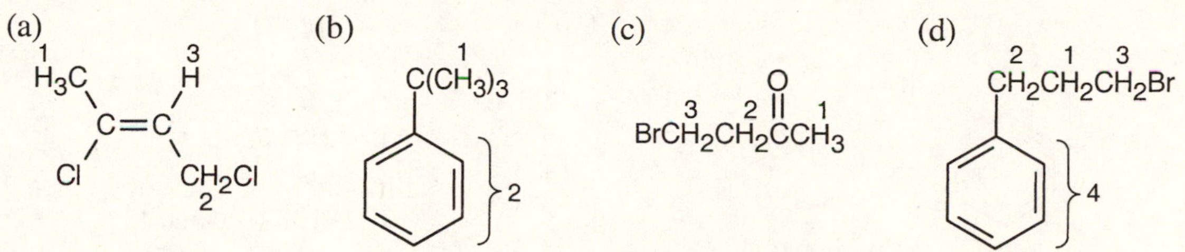

(a) 1 = 2.18 δ (allylic)
2 = 4.16 δ (H,Cl bonded to same C)
3 = 5.71 δ (vinylic)
The *E* isomer is also a satisfactory answer

(b) 1 = 1.30 δ (saturated)
2 = 7.30 δ (aromatic)

(c) 1 = 2.11 δ (next to C=O)
2 = 3.52 δ (next to C=O, Br)
3 = 4.40 δ (H,Br bonded to same C)

(d) 1 = 2.15 δ
2 = 2.75 δ (benzylic)
3 = 3.38 δ (H,Br bonded to same C)
4 = 7.22 δ (aromatic)

In (b) and (d), the aromatic ring hydrogens accidentally have the same chemical shift.

12.50

(a) $(CH_3)_2CHCH_2Br$

(b) $CH_3CHCH_2CH_2Cl$ (with Cl on C-2)

12.51

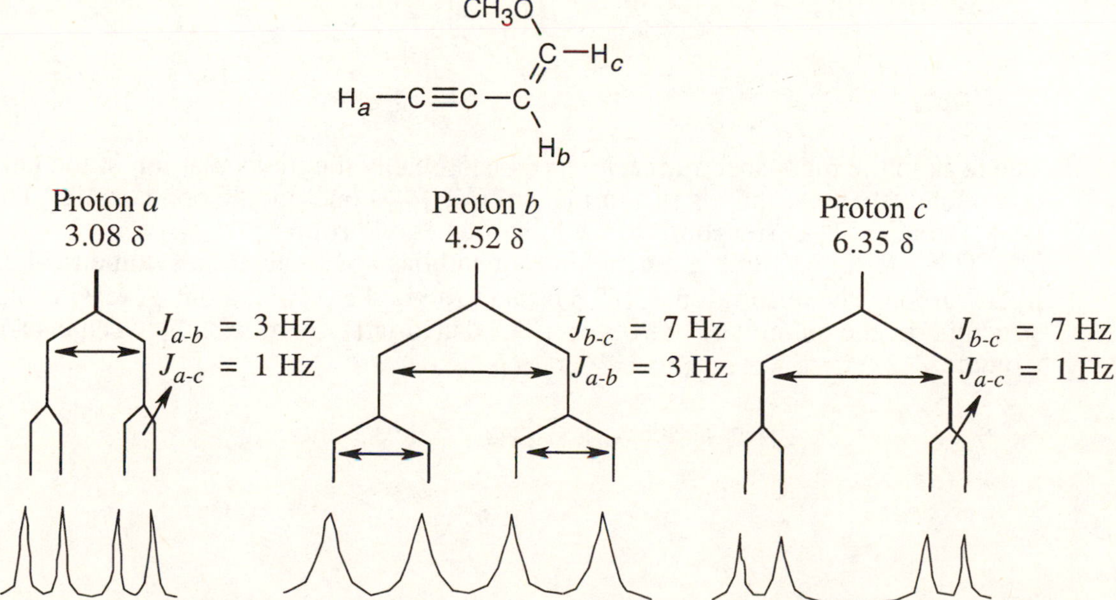

12.52

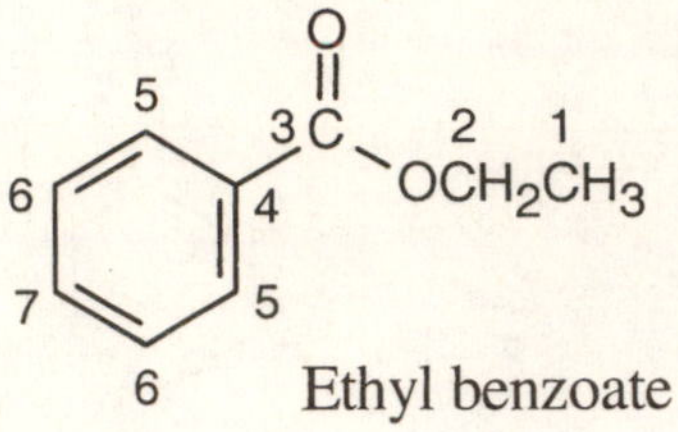

Ethyl benzoate

Carbon	δ (ppm)
1	14
2	61
3	166
4	127–133 (4 peaks)
5	
6	
7	

12.53 Compound **A** (4 multiple bonds and/or rings) must be symmetrical because it exhibits only six peaks in its ^{13}C NMR spectrum. Saturated carbons account for two of these peaks (δ = 15, 28 ppm), and unsaturated carbons account for the other four (δ = 119, 129, 131, 143 ppm).

^{1}H NMR shows a triplet (3 H at 1.1 δ), and a quartet (2 H at 2.5 δ), indicating the presence of an ethyl group. The other signals (4 H at 6.9 – 7.3 δ are due to aromatic protons.

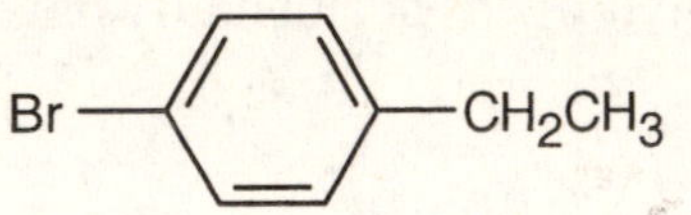

Br—C_6H_4—CH_2CH_3 **A**

12.54

(a) $CH_3CH_2CH_2COCH_3$

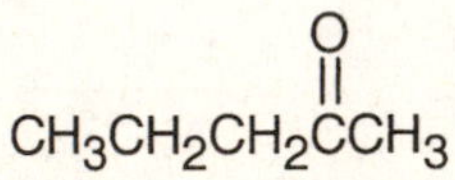

(b) p-bromotoluene (Br, CH_3)

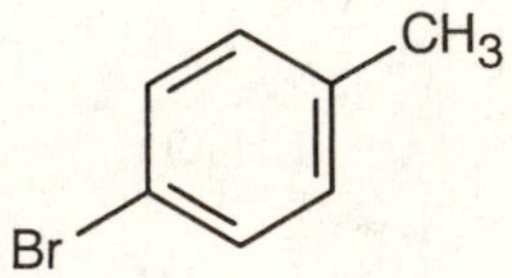

(c) $C_6H_5CH_2CH_2Br$

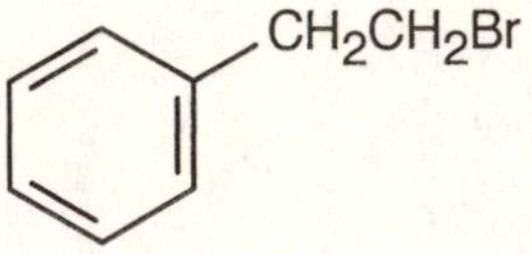

12.55 The peak in the mass spectrum at *m/z* = 84 is probably the molecular ion of the unknown compound and corresponds to a formula of C_6H_{12} — one double bond or ring. The base peak, at *m/z* = 55, corresponds to the loss of an ethyl group.

^{13}C NMR shows three different kinds of carbons and indicates a symmetrical hydrocarbon. The absorption at 132 δ is due to a vinylic carbon atom. A reasonable structure for the unknown is hex-3-ene. The data do not distinguish between cis and trans isomers.

$CH_3CH_2CH{=}CHCH_2CH_3$ Hex-3-ene

12.56 Compound **A**, a hydrocarbon having $M^+ = 96$, has the formula C_7H_{12}, indicating two degrees of unsaturation. Because it reacts with BH_3, Compound **A** contains a double bond. From the broadband decoupled ^{13}C NMR spectrum, we can see that C_7H_{12} is symmetrical, since it shows only five peaks.

The DEPT-135 spectrum of Compound **A** indicates three different CH_2 carbons, one $=CH_2$ carbon and one $-C=$ carbon; the last two carbons are shown to be sp^2-hybridized by their chemical shifts. In the DEPT-135 spectrum of Compound **B**, the absorptions due to the double bond carbons have been replaced by a CH carbon and a CH_2 carbon bonded to an electronegative group.

1. BH_3, THF
2. H_2O_2, ^-OH

Compound **A**

1	106.9 δ
2	149.7 δ
3	35.7 δ
4, 5	26.8 δ, 28.7 δ

Compound **B**

1	68.2 δ
2	40.5 δ
3,4,5	29.9 δ, 26.9 δ, 26.1 δ

12.57 The IR absorption indicates that **C** is an alcohol. From M^+, we can arrive at a molecular formula of $C_5H_{10}O$, which indicates one degree of unsaturation. The broadband-decoupled spectrum shows five peaks; two are due to a double bond, and one is due to a carbon bonded to an electronegative atom (O).

The DEPT spectra show that **C** contains 4 CH_2 carbons and one CH carbon, and that **C** has a monosubstituted double bond. $HOCH_2CH_2CH_2CH{=}CH_2$ is a likely structure for **C**.

12.58 Compound **D** is very similar to Compound **C**. The DEPT spectra make it possible to distinguish between the isomers. **D** has 2 CH carbons, one CH_3 carbon, and 2 CH_2 carbons, and, like **C**, has a monosubstituted double bond. The peak at 74.4 δ is due to a secondary alcohol.

$CH_3CH_2CH(OH)CH{=}CH_2$ Compound **D**

12.59 Compound **E**, $C_7H_{12}O_2$, has two degrees of unsaturation and has two equivalent carbons because its broadband-decoupled spectrum shows only 6 peaks. Two carbons absorb in the vinylic region of the spectrum; because one is a CH carbon and the other is a CH_2 carbon, **E** contains a monosubstituted double bond. The peak at 165.8 δ, not seen in the DEPT spectra, is due to a carbonyl group.

$H_2C{=}CHC(=O)OCH_2CH(CH_3)CH_3$ (carbons numbered 6, 5, 4, 3, 2, 1; CH_3 branch = 1)

Compound **E**

Carbon	*δ (ppm)*
1	19.1
2	28.0
3	70.5
4	165.8
5	129.8
6	129.0

12.60

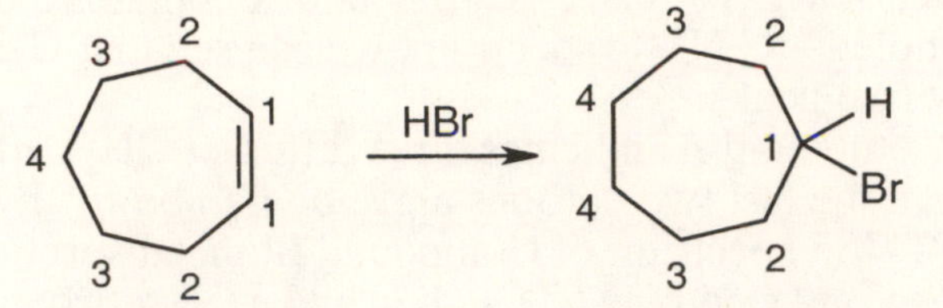

$$\text{Compound F} \xrightarrow{\text{HBr}} \text{Compound G}$$

Compound F		Compound G	
Carbon	*δ (ppm)*	*Carbon*	*δ (ppm)*
1	132.4	1	56.0
2	32.2	2	39.9
3,4	29.3	3,4	27.7
	27.6		25.1

12.61 Make a model of one enantiomer of 3-methylbutan-2-ol and orient it as a staggered Newman projection along the C2-C3 bond. The *S* enantiomer is pictured.

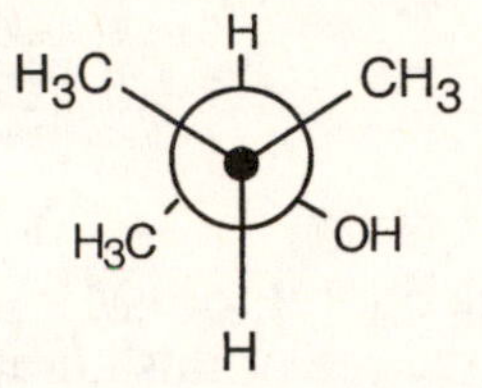

H, H_3C, CH_3, H_3C, OH, H

Because of the chirality center at C2, the two methyl groups at the front of the projection are diastereotopic. Since the methyl groups aren't equivalent, their carbons show slightly different signals in the ^{13}C NMR.

12.62 Commercial pentane-2,4-diol is a mixture of three stereoisomers: (*R,R*), (*S,S*), and (*R,S*). The meso isomer shows three signals in its ^{13}C NMR spectrum. Its diastereomers, the *R,R* and *S,S* enantiomeric pair, also show three signals, but two of these signals occur at different δ values from the meso isomer. This is expected, because diastereomers differ in physical and chemical properties. One resonance from the meso compound accidentally overlaps with one signal from the enantiomeric pair.

12.63 The product (M^+= 88) has the formula $C_4H_8O_2$. the IR absorption indicates that the product is an ester. The ^1H NMR shows an ethyl group and an $-OCH_3$ group.

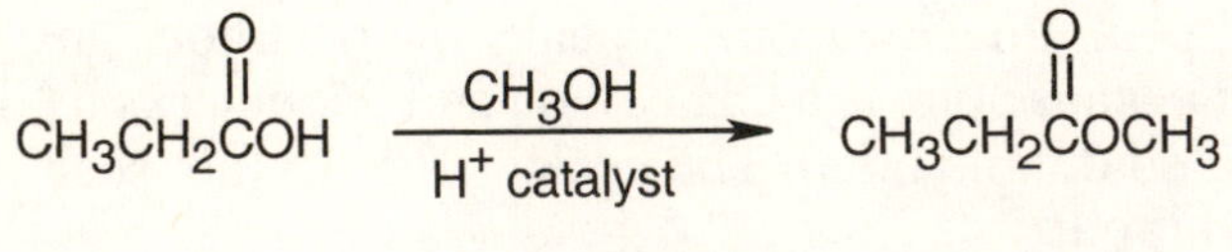

$$CH_3CH_2\overset{O}{\overset{\|}{C}}OH \xrightarrow[\text{H}^+ \text{ catalyst}]{CH_3OH} CH_3CH_2\overset{O}{\overset{\|}{C}}OCH_3$$

12.64 The product is a methyl ketone.

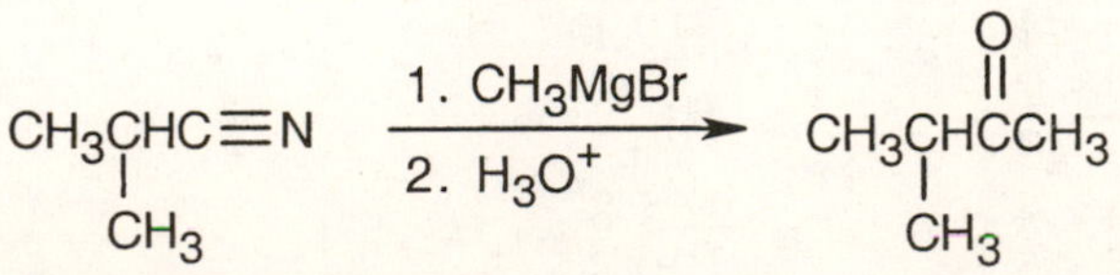

$$CH_3\underset{CH_3}{\underset{|}{C}}HC{\equiv}N \xrightarrow[2.\ H_3O^+]{1.\ CH_3MgBr} CH_3\underset{CH_3}{\underset{|}{C}}H\overset{O}{\overset{\|}{C}}CH_3$$

Review Unit 5: Spectroscopy

Major Topics Covered (with vocabulary):

Mass Spectrometry:
cation radical   mass spectrum   base peak   double-focusing mass spectrometer   molecular ion   alpha cleavage   McLafferty rearrangement   dehydration   time-of-flight mass analyzer

The Electromagnetic Spectrum:
electromagnetic radiation   wavelength   frequency   hertz   amplitude   quanta   absorption spectrum

Infrared Spectroscopy:
wavenumber   fingerprint region

Ultraviolet spectroscopy:
highest occupied molecular orbital (HOMO)   lowest unoccupied molecular orbital (LUMO)   molar absorptivity

Nuclear Magnetic Resonance Spectroscopy:
nuclear magnetic resonance   rf energy   effective magnetic field   shielding   downfield   upfield   chemical shift   delta scale   FT-NMR   DEPT ^{13}C NMR   homotopic   enantiotopic   diastereotopic   integration   multiplet   spin-spin splitting   coupling $n + 1$ rule   coupling constant   tree diagram

Types of Problems:

After studying these chapters, you should be able to:

- Write molecular formulas corresponding to a given molecular ion.
- Use mass spectra to determine molecular weights and base peaks, to distinguish between hydrocarbons, and to identify selected functional groups by their fragmentation patterns.
- Calculate the energy of electromagnetic radiation, and convert from wavelength to wavenumber and *vice versa*.
- Identify functional groups by their infrared absorptions.
- Use IR and MS to monitor reaction progress.
- Calculate the energy required for UV absorption, and use molar absorptivity to calculate concentration.
- Predict if and where a compound absorbs in the ultraviolet region.

- Calculate the relationship between delta value, chemical shift, and spectrometer operating frequency.
- Identify nonequivalent carbons and hydrogens, and predict the number of signals appearing in the ^{1}H NMR and ^{13}C NMR spectra of compounds.
- Assign resonances to specific carbons or hydrogens of a given structure.
- Propose structures for compounds, given their NMR spectra.
- Predict splitting patterns, using tree diagrams if necessary.
- Use NMR to distinguish between isomers and to identify reaction products.

Points to Remember:

* In mass spectrometry, the molecular ion is a cation radical. Further fragmentations of the molecular ion can be of two types – those that produce a cation plus a radical, and those that produce a different cation radical plus a neutral atom. In all cases, the fragment bearing the charge – whether cation or cation radical – is the one that is detected.

* Although mass spectrometry has many uses in research, we are interested in it for only a limited amount of data. The most important piece of information it provides for us is the molecular weight of an unknown. A mass spectrum can also show if an unknown is branched or straight-chain (branched hydrocarbons have more complex spectra than their straight-chain isomers). Finally, if we know if certain groups are present, we can obtain structural information about an unknown compound. For example if we know that a ketone is present, we can look for peaks that correspond to alpha cleavage and/or McLafferty rearrangement fragments.

* The position of an IR absorption is related to both the strength of the bond and to the nature of the two atoms that form the bond. For example, a carbon-carbon triple bond absorbs a higher frequency than a carbon-carbon double bond, which absorbs at a higher frequency than a carbon-carbon single bond. Bonds between two atoms of significantly different mass absorb at higher frequencies than bonds between two atoms of similar mass.

* Not all IR absorptions are due to bond stretches. Many of the absorptions in the fingerprint region of an IR spectrum are due to bending and out-of-plane motions.

* It is confusing, but true, that larger δ values in an NMR spectrum are associated with nuclei that are less shielded, and that these nuclei require a lower field strength for resonance. Nuclei with small values of δ are more shielded and require a higher field strength for resonance.

* Both ^{13}C NMR and ^{1}H NMR are indispensable for establishing the structure of an organic compound. ^{13}C NMR indicates if a molecule is symmetrical and shows the types of carbons in a molecule (by DEPT NMR). ^{1}H NMR shows how the carbons are connected (by spin-spin splitting) and how many protons are in the molecule (by integration). Both types of spectra show (by chemical shift) the electronic environment of the magnetic nuclei.

Self-Test:

Compound **A** is a hydrocarbon with $M^+ = 78$. What is its molecular formula? What is its degree of unsaturation? Draw three possible formulas for **A**. The ^{13}C NMR spectrum of **A** shows 3 peaks – at 18.5 δ, 69.4 δ and 82.4 δ, and the ^{1}H NMR spectrum shows two peaks. What is the structure of **A**? What significant absorptions would you see in the IR spectrum of **A**?

Compound **B** has the molecular formula C_8H_{14}, and shows 3 peaks in its ^{1}H NMR spectrum – at 1.7 δ (6H), 2.1 δ (4H) and 4.7 δ (4H). All 3 peaks are singlets. **B** also shows an IR absorption at 890 cm^{-1}. What is a possible structure for **B**? If you're still not sure, the following peaks were observed in the ^{13}C NMR spectrum of **B**: 22 δ, 36 δ, 110 δ, 146 δ. The peaks at 36 δ and 110 δ were negative signals in the DEPT-135 spectrum, and the peak at 22 δ was a positive signal.

Compound **C** is a hydrocarbon with $M^+ = 112$. What are possible molecular formulas for **C**? The five peaks in the ^{1}H NMR spectrum of **C** are all singlets and occur at the following δ values: 0.9 δ (9 H), 1.8 δ (3 H), 1.9 δ (2 H), 4.6 δ (1 H) and 4.8 δ (1 H). An IR absorption at 890 cm^{-1} is also present What is the structure of **C**?

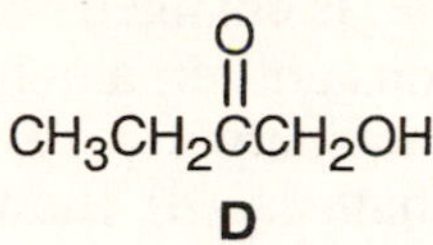

Describe the ^{13}C NMR and ^{1}H NMR spectra of **D**. For the ^{1}H NMR spectrum, include the spin-spin splitting patterns, peak areas, and positions of the chemical shifts. Give two significant absorptions that you might see in the IR spectrum. Would you expect to see products of McLafferty rearrangement in the mass spectrum of **D**? Of alpha cleavage?

Multiple choice:

1. Which of the following formulas could not arise from a compound with $M^+ = 142$ that contains C, H, and possibly O?
(a) $C_{11}H_{10}$ (b) $C_{10}H_8O$ (c) $C_9H_{18}O$ (d) $C_8H_{14}O_2$

2. Which of the following mass spectrum fragments is a cation, rather than a cation radical?
(a) molecular ion (b) product of alpha cleavage (c) product of McLafferty rearrangement (d) product of dehydration of an alcohol

3. Which element contributes significantly to $(M+1)^+$?
(a) N (b) H (c) C (d) O

4. In which type of spectroscopy is the wavelength of absorption the longest?
(a) NMR spectroscopy (b) infrared spectroscopy (c) ultraviolet spectroscopy (d) X-ray spectroscopy

5. Which functional group is hard to detect in an IR spectrum?
(a) aldehyde (b) –C≡C– (c) alcohol (d) ether

6. IR spectroscopy is especially useful for:
(a) determining if an alkyne triple bond is at the end of a carbon chain or is in the middle
(b) predicting the type of carbonyl group that is present in a compound
(c) deciding if a double bond is monosubstituted or disubstituted
(d) all of these situations

7. If a nucleus is strongly shielded:
(a) The effective field is smaller than the applied field, and the absorption is shifted downfield. (b) The effective field is larger than the applied field, and the absorption is shifted upfield. (c) The effective field is smaller than the applied field, and the absorption is shifted upfield. (d) The effective field is larger than the applied field, and the absorption is shifted downfield.

8. When the operating frequency of an ^1H NMR spectrometer is changed:
(a) The value of chemical shift in δ and of the coupling constant remain the same. (b) The values of chemical shift in Hz and of the coupling constant change. (c) The value of chemical shift in Hz remains the same, but the coupling constant changes. (d) The values of chemical shift in δ and of the coupling constant change.

9. ^{13}C NMR can provide all of the following data except:
(a) the presence or absence of symmetry in a molecule (b) the connectivity of the carbons in a molecule (c) the chemical environment of a carbon (d) the number of hydrogens bonded to a carbon

10. Which kind of carbon is detected in DEPT-90 ^{13}C NMR spectroscopy?
(a) primary carbon (b) secondary carbon (c) tertiary carbon (d) quaternary carbon

11. The protons on carbon 3 of (*R*)-2-bromobutane are:
(a) homotopic (b) enantiotopic (c) diastereotopic (d) unrelated

Chapter 13 – Alcohols, Phenols, and Thiols; Ethers and Sulfides

Chapter Outline

I. Alcohols, phenols and sulfides (Sections 13.1 – 13.6).
   A. Naming alcohols, phenols and sulfides (Section 13.1).
      1. Alcohols are classified as primary, secondary or tertiary, depending on the number of organic groups bonded to the –OH carbon.
      2. Rules for naming alcohols.
         a. The longest chain containing the –OH group is the parent chain, and the parent name replaces *-e* with *-ol*.
         b. Numbering begins at the end of the chain nearer the –OH group.
         c. The substituents are numbered according to their position on the chain and cited in alphabetical order.
      3. Phenols are named according to rules discussed in Section 8.1.
      4. Thiols (sulfur analogs of alcohols) are named by the same system as alcohols, with the suffix *-thiol* replacing *-ol*.
         The –SH group is a mercapto- group.
   B. Properties of alcohols, phenols and sulfides (Section 13.2).
      1. Alcohols have sp^3 hybridization and a nearly tetrahedral bond angle.
      2. Alcohols and phenols have elevated boiling points, relative to hydrocarbons, due to hydrogen bonding.
         a. In hydrogen bonding, an –OH hydrogen is attracted to a lone pair of electrons on another molecule, resulting in a weak electrostatic force that holds the molecules together.
         b. These weak forces must be overcome in boiling.
      3. Alcohols and phenols are weakly acidic as well as weakly basic.
      4. Alcohols and phenols dissociate to a slight extent to form alkoxide ions and phenoxide ions.
      5. Acidity of alcohols.
         a. Alcohols are similar in acidity to water.
         b. Alkyl substituents decrease acidity by preventing solvation of the alkoxide ion.
         c. Electron-withdrawing substituents increase acidity by delocalizing negative charge.
         d. Alcohols don't react with weak bases, but they do react with alkali metals and strong bases.
      6. Acidity of phenols and thiols.
         a. Phenols and thiols are a million times more acidic than alcohols and are soluble in dilute NaOH.
         b. Phenol acidity is due to resonance stabilization of the phenoxide anion.
         c. Electron-withdrawing substituents increase phenol acidity, and electron-donating substituents decrease phenol acidity.
   C. Preparation of alcohols (Section 13.3).
      1. Familiar methods.
         a. Hydration of alkenes.
            i. Hydroboration/oxidation yields non-Markovnikov products.
            ii. Oxymercuration/reduction yields Markovnikov products.
         b. 1,2-diols can be prepared by OsO_4 hydroxylation, followed by reduction.
            i. This reaction occurs with syn stereochemistry.
            ii. Ring-opening of epoxides produces 1,2-diols with anti stereochemistry.

2. Reduction of carbonyl compounds.
   a. Aldehydes are reduced to primary alcohols.
   b. Ketones are reduced to secondary alcohols.
      Either $NaBH_4$(milder) or $LiAlH_4$(more reactive) can be used to reduce aldehydes and ketones.
   c. Carboxylic acids and esters are reduced to primary alcohols with $LiAlH_4$.
      i. These reactions occur by addition of hydride to the positively polarized carbon of a carbonyl group.
      ii. Water adds to the alkoxide intermediate to yield alcohol product.
3. Reaction of carbonyl compounds with Grignard reagents.
   a. RMgX adds to carbonyl compounds to give alcohol products.
      i. Reaction of RMgX with formaldehyde yields primary alcohols.
      ii. Reaction of RMgX with aldehydes yields secondary alcohols.
      iii. Reaction of RMgX with ketones yields tertiary alcohols.
      iv. Reaction of RMgX with esters yields tertiary alcohols with at least two identical R groups bonded to the alcohol carbon.
      v. No reaction occurs with carboxylic acids because the acidic hydrogen quenches the Grignard reagent.
   b. Limitations of the Grignard reaction.
      i. Grignard reagents can't be prepared from reagents containing other reactive functional groups.
      ii. Grignard reagents can't be prepared from compounds having acidic hydrogens.
   c. Grignard reagents behave as carbon anions and add to the carbonyl carbon.
      A proton from water is added to the alkoxide intermediate during workup to produce the alcohol.

D. Reactions of alcohols (Sections 13.4 – 13.5).
1. Dehydration to yield alkenes (Section 13.4).
   a. Tertiary alcohols can undergo acid-catalyzed dehydration with warm aqueous H_2SO_4.
      i. Zaitsev products are usually formed.
      ii. The severe conditions needed for dehydration of secondary and primary alcohols restrict this method to tertiary alcohols.
      iii. Tertiary alcohols react fastest because the intermediate carbocation formed in this E1 reaction is most stable.
   b. Secondary and primary alcohols are dehydrated with $POCl_3$ in pyridine.
      i. This reaction occurs by an E2 mechanism.
      ii. Pyridine serves as a base and as a solvent.
2. Conversion into esters.
   a. Alcohols react with carboxylic acids to yield esters.
   b. In the lab, acid chlorides, rather than carboxylic acids, are used.
   c. Similar processes occur in living organisms.
3. Oxidation of alcohols and phenols (Section 13.5).
   a. Primary alcohols can be oxidized to aldehydes or carboxylic acids.
   b. Secondary alcohols can be oxidized to ketones.
   c. Tertiary alcohols aren't oxidized.
   d. Oxidation to ketones and carboxylic acids can be carried out with $KMnO_4$, CrO_3, or $Na_2Cr_2O_7$.
   e. Oxidation of a primary alcohol to an aldehyde is achieved with PCC.
      PCC is also used on sensitive alcohols.

f. Oxidation occurs by a mechanism closely related to an E2 mechanism.
The reaction involves a chromate intermediate.
g. Strong oxidizing agents convert phenols to quinones.
i. Reaction with Fremy's salt to form a quinone occurs by a radical mechanism.
ii. The redox reaction quinone $\rightarrow$ hydroquinone occurs readily.
iii. Ubiquinones are an important class of biochemical oxidizing agents that function as a quinone/hydroquinone redox system.

E. Preparation and reactions of thiols (Section 13.6).
1. Thiols stink!
2. Thiols may be prepared by S_N2 displacement with a sulfur nucleophile.
a. The reaction may proceed to form sulfides.
b. Better yields occur when thiourea is used.
3. Thiols can be oxidized by Br_2 or I_2 to yield disulfides, RSSR.
The reaction can be reversed by treatment with zinc and acid.

II. Ethers and sulfides (Sections 13.7 – 13.10).
A. Naming ethers and sulfides(Section 13.7).
1. Ethers with no other functional groups are named by citing the two organic substituents and adding the word "ether".
When other functional groups are present, the ether is an alkoxy substituent.
2. Sulfides (sulfur analogs of ethers) are named by the same system as ethers, with *sulfide* replacing *ether*.
The –SR group is an alkylthio- group.
B. Preparation of ethers (Section 13.8).
1. Symmetrical ethers can be synthesized by acid-catalyzed dehydration of alcohols.
2. Williamson ether synthesis.
a. Metal alkoxides react with primary alkyl halides and tosylates to form ethers.
b. The alkoxides are prepared by reacting an alcohol with a strong base, such as NaH.
Reaction of the free alcohol with the halide can be achieved with Ag_2O.
c. The reaction occurs via an S_N2 mechanism.
i. The halide component must be primary.
ii. In cases where one ether component is hindered, reaction should occur between the alkoxide of the more hindered reagent and the halide of the less hindered reagent.
d. A variation of the Williamson synthesis uses Ag_2O.
C. Reactions of ethers (Section 13.9).
1. In general, ethers are unreactive.
2. Acidic cleavage.
a. Strong acids can be used to cleave ethers.
b. Cleavage can occur by S_N2 or S_N1 routes.
i. Primary and secondary alcohols react by an S_N2 mechanism, in which the halide attacks the ether at the less hindered site.
This route selectively produces one halide and one alcohol.
ii. Tertiary, benzylic and allylic ethers react by either an S_N1 or an E1 route.
3. Base-catalyzed ring-opening of epoxides.
a. Base-catalyzed ring opening occurs because of the reactivity of the strained epoxide ring.
b. Ring-opening also occurs when epoxides react with Grignard reagents, forming a product with two more carbons than the starting alkyl halide.
c. Ring-opening takes place by an S_N2 mechanism, in which the nucleophile attacks the less hindered epoxide carbon.

D. Sulfides (Section 13.10).
1. Treatment of a thiol with base yields a thiolate anion, which can react with an alkyl halide to form a sulfide.
2. Thiolate anions are excellent nucleophiles.
3. Dialkyl sulfides can react with alkyl halides to form trialkylsulfonium salts, which are also good alkylating agents.
Many biochemical reactions use trialkylsulfonium groups as alkylating agents.
4. Sulfides are easily oxidized to sulfoxides (R_2SO) and sulfones (R_2SO_2).
Dimethyl sulfoxide is used as a polar aprotic solvent.

III. Spectroscopy of alcohols, phenols and ethers (Section 13.11).
A. IR spectroscopy.
1. Both alcohols and phenols show –OH stretches in the region 3300–3600 cm^{-1}.
a. Unassociated alcohols show a peak at 3600 cm^{-1}.
b. Hydrogen-bonded alcohols show a broader peak at 3300–3400 cm^{-1}.
2. Alcohols show a C–O stretch near 1050 cm^{-1}.
3. Phenols show aromatic bands at 1500–1600 cm^{-1}.
4. Phenol shows monosubstituted aromatic bands at 690 and 760 cm^{-1}.
5. Ethers are difficult to identify by IR spectroscopy because many other absorptions occur at 1050–1150 cm^{-1}, where ethers absorb.
B. NMR spectroscopy.
1. In ^{13}C NMR spectroscopy, carbons bonded to –OH groups absorb in the range 50–80 δ.
2. ^{1}H NMR.
a. Hydrogens on carbons bearing –OH groups absorb in the range 3.5–4.5 δ.
The hydroxyl hydrogen doesn't split these signals.
b. D_2O exchange can be used to locate the O–H signal.
c. Spin-spin splitting occurs between the oxygen-bearing carbon and neighboring –H.
d. Phenols show aromatic ring absorptions, as well as an O–H absorption in the range 3–8 δ.
e. Hydrogens on a carbon next to an ether oxygen absorb downfield (3.4–4.5 δ).
f. Hydrogens on a carbon next to an epoxide oxygen absorb at a slightly higher field (2.5–3.5 δ).
C. Mass Spectrometry.
1. Alcohols undergo alpha cleavage to give a neutral radical and an oxygen-containing cation.
2. Alcohols also undergo dehydration to give an alkene radical cation.

Solutions to Problems

13.1 The parent chain must contain the hydroxyl group, and the hydroxyl group(s) should receive the lowest possible number.

(a) OH, OH; $CH_3CHCH_2CHCHCH_3$; CH_3
5-Methylhexane-2,4-diol

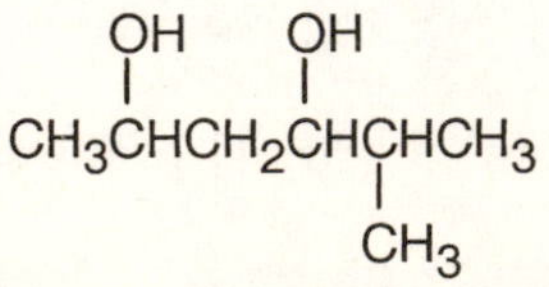

(b) OH; $C_6H_5CH_2CH_2CCH_3$; CH_3
2-Methyl-4-phenylbutan-2-ol

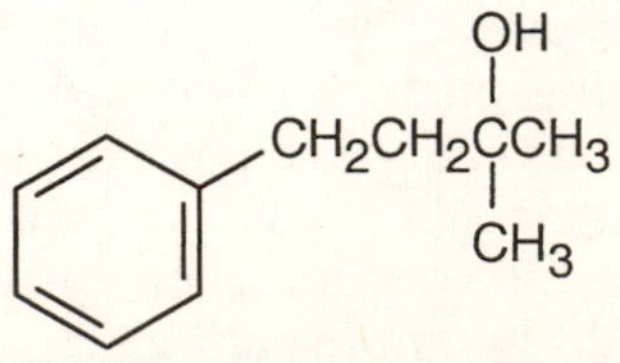

(c) HO; CH_3; CH_3
4,4-Dimethylcyclohexanol

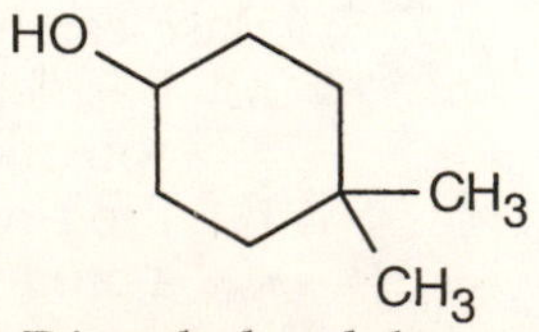

(d) *trans*-2-Bromocyclopentanol

(e) 2-Methylheptane-4-thiol

(f) Cyclopent-2-ene-1-thiol

13.2

(a) 2-Ethylbut-2-en-1-ol

(b)

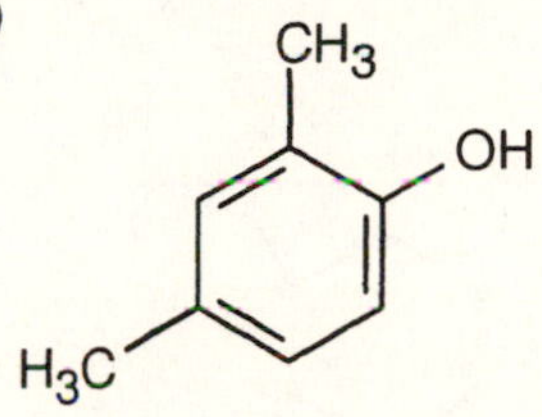

Cyclohex-3-en-1-ol

(c) *trans*-3-Chlorocyclohexanol

(d) Pentane-1,4-dithiol

(e) 2,4-Dimethylphenol

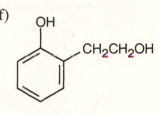

(f) *o*-(2-Hydroxyethyl)phenol

13.3

Least acidic ⟶ *Most acidic*

(a) *p*-Methylphenol < Phenol < *p*-(Trifluoromethyl)phenol
phenol with electron-donating groups; phenol with electron-withdrawing groups

(b) Benzyl alcohol < Phenol < *p*-Hydroxybenzoic acid
alcohol; phenol; carboxylic acid

13.4 We saw in Chapter 8 that a nitro group is electron-withdrawing. Since electron-withdrawing groups stabilize phenoxide anions, *p*-nitrobenzyl alcohol is more acidic than benzyl alcohol. The methoxy group, which is electron-donating, destabilizes an alkoxide ion, making *p*-methoxybenzyl alcohol less acidic than benzyl alcohol.

13.5 For the same reason that phenols are more acidic than alcohols, thiophenols are more acidic than thiols. An aromatic ring can stabilize the negative charge more effectively than an alkyl group.

13.6

(a)

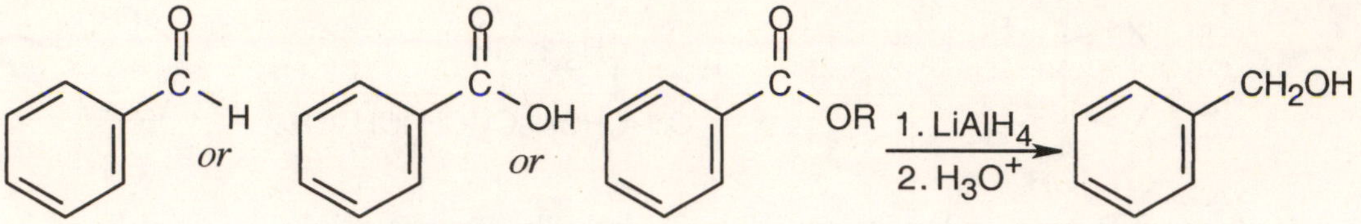

Benzaldehyde (C_6H_5CHO) *or* benzoic acid (C_6H_5COOH) *or* ester (C_6H_5COOR) $\xrightarrow[2.\ H_3O^+]{1.\ LiAlH_4}$ $C_6H_5CH_2OH$

Benzyl alcohol may be the reduction product of an aldehyde, a carboxylic acid, or an ester.

(b)

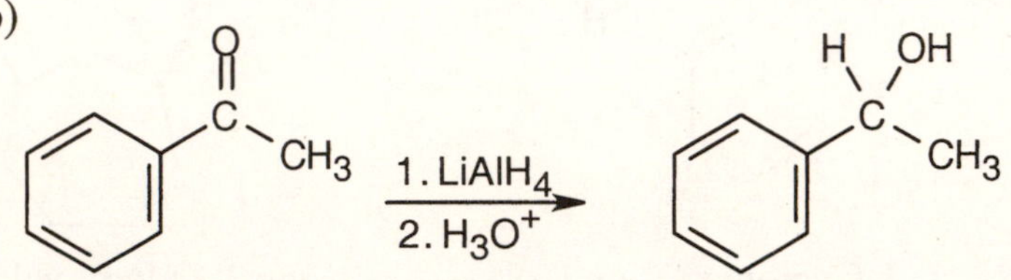

$C_6H_5COCH_3$ $\xrightarrow[2.\ H_3O^+]{1.\ LiAlH_4}$ $C_6H_5CH(OH)CH_3$

Reduction of a ketone yields this secondary alcohol. $NaBH_4$ can also be used as the reducing agent.

(c)

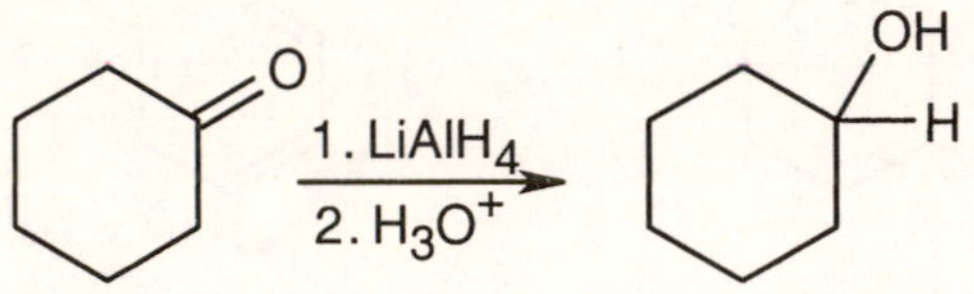

Cyclohexanone $\xrightarrow[2.\ H_3O^+]{1.\ LiAlH_4}$ cyclohexanol

$NaBH_4$ can also be used.

(d)

$(CH_3)_2CHCHO$ *or* $(CH_3)_2CHCO_2H$ *or* $(CH_3)_2CHCO_2R$ $\xrightarrow[2.\ H_3O^+]{1.\ LiAlH_4}$ $(CH_3)_2CHCH_2OH$

13.7 Both products have an –OH and a methyl group bonded to what was formerly a ketone carbon.

(a)

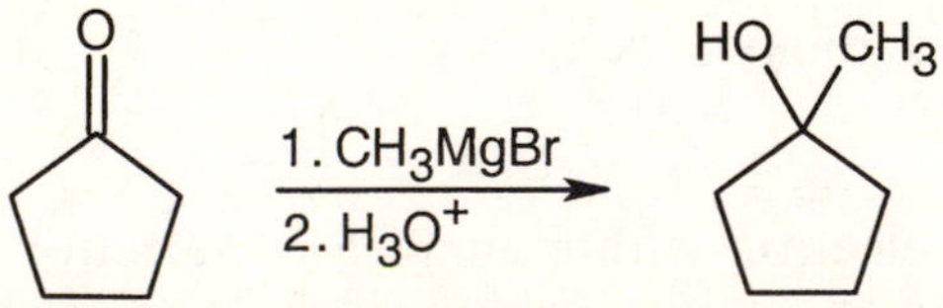

Cyclopentanone $\xrightarrow[2.\ H_3O^+]{1.\ CH_3MgBr}$ 1-methylcyclopentanol (HO, CH_3)

(b)

$CH_3CH_2CH_2\overset{O}{\overset{\|}{C}}CH_2CH_3$ $\xrightarrow[2.\ H_3O^+]{1.\ CH_3MgBr}$ $CH_3CH_2CH_2C(OH)(CH_3)CH_2CH_3$

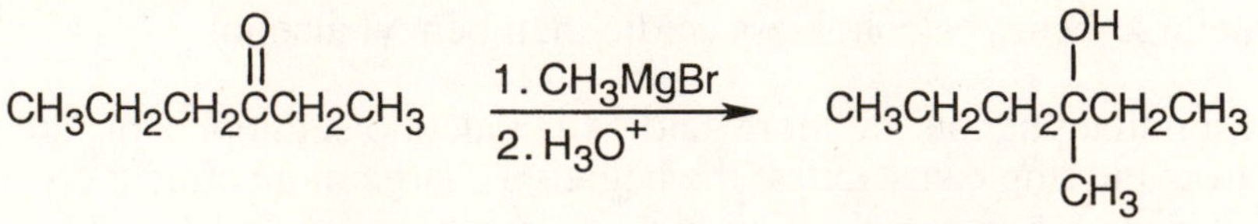

13.8 **Strategy:** First, identify the type of alcohol. If the alcohol is primary, it can only be synthesized from formaldehyde plus the appropriate Grignard reagent. If the alcohol is secondary, it is synthesized from an aldehyde and a Grignard reagent. (Usually, there are two combinations of aldehyde and Grignard reagent). A tertiary alcohol is synthesized from a ketone and a Grignard reagent. If all three groups on the tertiary alcohol are different, there are often three different combinations of ketone and Grignard reagent. If two of the groups on the alcohol carbon are the same, the alcohol may also be synthesized from an ester and two equivalents of Grignard reagent.

Solution:

(a) 2-Methylpropan-2-ol is a tertiary alcohol. To synthesize a tertiary alcohol, start with a ketone.

$$CH_3\overset{O}{\overset{\|}{C}}CH_3 \xrightarrow[2.\ H_3O^+]{1.\ CH_3MgBr} CH_3\overset{OH}{\overset{|}{\underset{CH_3}{\underset{|}{C}}}}CH_3$$

If two or more alkyl groups bonded to the carbon bearing the –OH group are the same, an alcohol can be synthesized from an ester and a Grignard reagent.

$$CH_3\overset{O}{\overset{\|}{C}}OR \xrightarrow[2.\ H_3O^+]{1.\ 2\ CH_3MgBr} CH_3\overset{OH}{\overset{|}{\underset{CH_3}{\underset{|}{C}}}}CH_3$$

2-Methylpropan-2-ol

(b) Since 1-methylcyclohexanol is a tertiary alcohol, start with a ketone.

$$\text{Cyclohexanone} \xrightarrow[2.\ H_3O^+]{1.\ CH_3MgBr} \text{(cyclohexane ring with OH and } CH_3 \text{ on C-1)}$$

1-Methylcyclohexanol

(c) Three possible combinations of ketone plus Grignard reagent can be used to synthesize this tertiary alcohol.

$C_6H_5COCH_2CH_3$ $\xrightarrow[2.\ H_3O^+]{1.\ CH_3MgBr}$

or

$C_6H_5COCH_3$ $\xrightarrow[2.\ H_3O^+]{1.\ CH_3CH_2MgBr}$ $C_6H_5C(OH)(CH_3)CH_2CH_3$

2-Phenylbutan-2-ol

or

$CH_3CH_2COCH_3$ $\xrightarrow[2.\ H_3O^+]{1.\ C_6H_5MgBr}$

(d) Formaldehyde must be used to synthesize this primary alcohol.

$HCHO$ $\xrightarrow[2.\ H_3O^+]{1.\ C_6H_5MgBr}$ $C_6H_5CH_2OH$

Benzyl alcohol

13.9 **Strategy:** First, interpret the structure of the alcohol. This alcohol, 1-ethylcyclohexanol, is a tertiary alcohol that can be synthesized from a ketone. Only one combination of ketone and Grignard reagent is possible.

Solution:

Cyclohexanone $\xrightarrow[2.\ H_3O^+]{1.\ CH_3CH_2MgBr}$ 1-ethylcyclohexanol (OH, CH_2CH_3)

13.10

(a)

$CH_3CH_2CH(OH)CH(CH_3)_2$ $\xrightarrow[pyridine]{POCl_3}$ $(CH_3CH_2)(H)C{=}C(CH_3)(CH_3)$ (major) + $(H)(H_3C)C{=}C(CH(CH_3)_2)(H)$ (minor) + $(H_3C)(H)C{=}C(CH(CH_3)_2)(H)$ (minor)

The major product has the more substituted double bond.

(b)

$$\text{trans-2-Methylcyclohexanol (H at C1 and H at C2 shown)} \xrightarrow[\text{pyridine}]{POCl_3} \text{3-Methylcyclohexene}$$

3-Methylcyclohexene

In E2 elimination, dehydration proceeds most readily when the two groups to be eliminated have a trans–diaxial relationship. In this compound, the only hydrogen with the proper stereochemical relationship to the –OH group is at C6. Thus, the non-Zaitsev product 3-methylcyclohexene is formed.

(c)

$$\text{cis-2-Methylcyclohexanol} \xrightarrow[\text{pyridine}]{POCl_3} \text{1-Methylcyclohexene}$$

1-Methylcyclohexene

Here, the hydrogen at C2 is trans to the hydroxyl, and dehydration yields the Zaitsev product, 1-methylcyclohexene.

13.11 **Strategy:** Aldehydes are synthesized from oxidation of primary alcohols, and ketones are synthesized from oxidation of secondary alcohols.

Solution:

(a)

$$C_6H_5CH(OH)CH_3 \xrightarrow[H_3O^+]{CrO_3} C_6H_5COCH_3$$

(b)

$$CH_3CH(CH_3)CH_2OH \xrightarrow[CH_2Cl_2]{PCC} CH_3CH(CH_3)CHO$$

(c)

$$\text{Cyclopentanol} \xrightarrow[H_3O^+]{CrO_3} \text{Cyclopentanone}$$

13.12

	Starting material	CrO_3, H_3O^+ Product	PCC Product
(a)	$CH_3CH_2CH_2CH_2CH_2CH_2OH$	$CH_3CH_2CH_2CH_2CH_2CO_2H$	$CH_3CH_2CH_2CH_2CH_2CHO$
(b)	$CH_3CH_2CH_2CH_2CH(OH)CH_3$	$CH_3CH_2CH_2CH_2C(=O)CH_3$	$CH_3CH_2CH_2CH_2C(=O)CH_3$
(c)	$CH_3CH_2CH_2CH_2CH_2CHO$	$CH_3CH_2CH_2CH_2CH_2CO_2H$	no reaction

13.13 Thiourea is used to prepare thiols from alkyl halides.

$$CH_3CH{=}CHC(=O)OCH_3 \xrightarrow[2.\ H_3O^+]{1.\ LiAlH_4} CH_3CH{=}CHCH_2OH \xrightarrow{PBr_3} CH_3CH{=}CHCH_2Br$$

Methyl but-2-enoate

$$CH_3CH{=}CHCH_2Br \xrightarrow[2.\ ^-OH,\ H_2O]{1.\ (H_2N)_2C{=}S} CH_3CH{=}CHCH_2SH$$

But-2-ene-1-thiol

13.14 Ethers can be named either as alkoxy-substituted compounds or by citing the two groups bonded to oxygen, followed by the word "ether". For sulfides, replace *ether* with the word *sulfide* , and replace *alkoxy* with *alkylthio*

(a) $(CH_3)_2CHOCH(CH_3)_2$

2-Isopropoxypropane *or* Diisopropyl ether

(b) Cyclopentyl–$OCH_2CH_2CH_3$

Propoxycyclopentane *or* Cyclopentyl propyl ether

(c) Br–C_6H_4–OCH_3 (para)

p-Bromoanisole *or* *p*-Bromomethoxybenzene

(d) Cyclohexene with OCH_3 on C1

1-Methoxycyclohexene

(e) $C_6H_5CH_2SCH_3$

Benzyl methyl sulfide

(f) $CH_3SCH_2CH{=}CH_2$

Allyl methyl sulfide

13.15 The first step of the dehydration mechanism is protonation of an alcohol. Water is then displaced by another molecule of alcohol to form an ether. If two different alcohols are present, either one can be protonated and either one can displace water, yielding a mixture of products.

If this procedure were used with ethanol and propan-1-ol, the products would be diethyl ether, ethyl propyl ether, and dipropyl ether. If there were equimolar amounts of the alcohols, and if they were of equal reactivity, the product ratio would be diethyl ether : ethyl propyl ether : dipropyl ether = 1:2:1.

13.16 Remember that the halide in the Williamson ether synthesis should be primary or methyl to avoid competing elimination reactions. The alkoxide anions shown are formed by treating the corresponding alcohols with NaH.

(a)

$$\left.\begin{array}{l} CH_3CH_2CH_2O^- + CH_3Br \\ \quad\textit{or} \\ CH_3CH_2CH_2Br + CH_3O^- \end{array}\right\} \longrightarrow CH_3CH_2CH_2OCH_3 + Br^-$$

Methyl propyl ether

(b)

$$C_6H_5O^- + CH_3Br \longrightarrow C_6H_5OCH_3 + Br^-$$

Methyl phenyl ether (Anisole)

(c)

$$(CH_3)_2CHO^- + C_6H_5CH_2Br \longrightarrow (CH_3)_2CHOCH_2C_6H_5 + Br^-$$

Benzyl isopropyl ether

(d)

$$(CH_3)_3CCH_2O^- + CH_3CH_2Br \longrightarrow (CH_3)_3CCH_2OCH_2CH_3 + Br^-$$

Ethyl 2,2-dimethylpropyl ether

13.17 The compounds most reactive in the Williamson ether synthesis are also most reactive in any S_N2 reaction (review Chapter 10 if necessary).

Most reactive ⟶ *Least reactive*

(a)

CH_3CH_2Br	>	$CH_3CHBrCH_3$	>>	C_6H_5Br
primary halide		secondary halide		aryl halide (not reactive)

(b)

CH_3CH_2Br	>	CH_3CH_2Cl	>>	$CH_3CH{=}CHI$
better leaving group		poorer leaving group		vinylic (not reactive)

13.18 (a) First, notice the substitution pattern of the ether. Bonded to the ether oxygen are a primary alkyl group and a tertiary benzyl group. When one group is tertiary, cleavage occurs by an S_N1 route to give a tertiary halide and a primary alcohol.

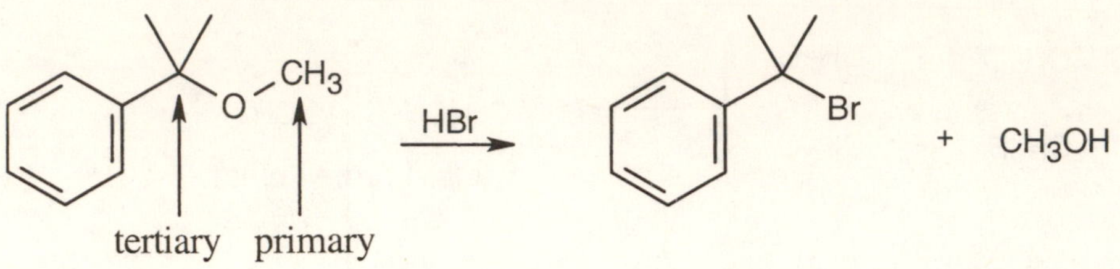

HBr → + CH_3OH

tertiary primary

(b) In this problem, the groups are primary and secondary alkyl groups. Br^- attacks at the less hindered primary group, and oxygen remains with the secondary group to give a secondary alcohol.

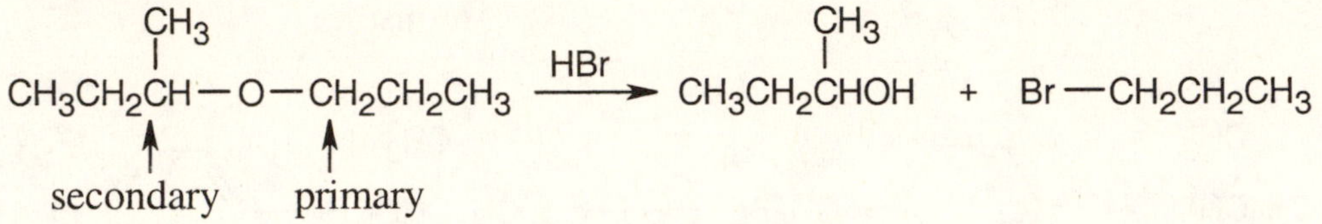

$$CH_3CH_2CH(CH_3)-O-CH_2CH_2CH_3 \xrightarrow{HBr} CH_3CH_2CH(CH_3)OH + Br-CH_2CH_2CH_3$$

secondary primary

13.19

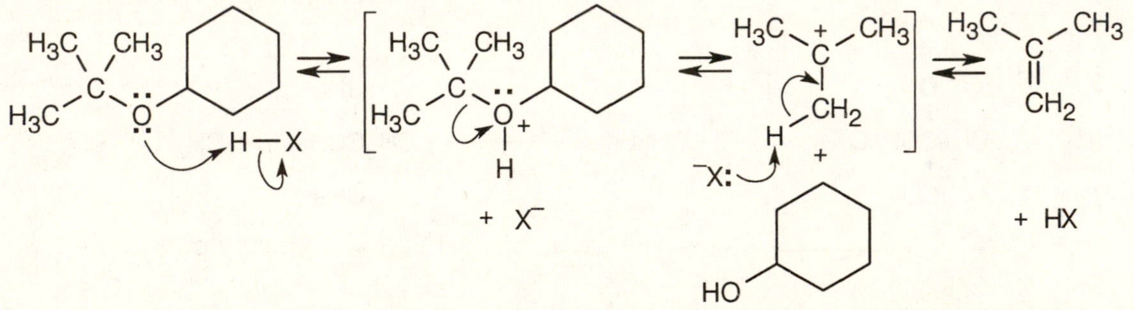

H–X + X^- $^-$X: + HX HO

The first step of this acid-catalyzed E1 ether cleavage is protonation of the ether oxygen. The protonated intermediate collapses to form cyclohexanol and a tertiary carbocation. The carbocation then loses a proton to form an alkene, 2-methylpropene.The acid used for cleavage is trifluoroacetic acid.

Visualizing Chemistry

13.20

(a) (*R*)-5-Methylhexan-3-ol

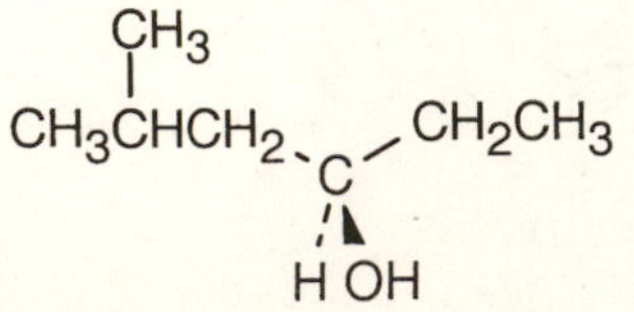

(b) *cis*-3-Methylcyclohexanol

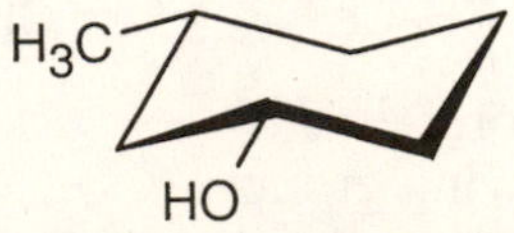

(c) (*S*)-1-Cyclopentylethanethiol

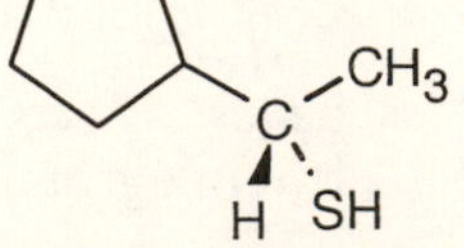

(d)

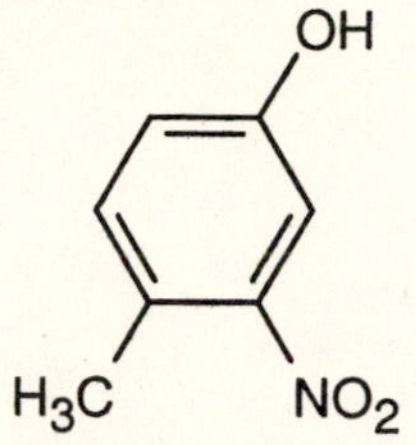

OH
H_3C NO_2
4-Methyl-3-nitrophenol

13.21 The reduction products are racemic mixtures.

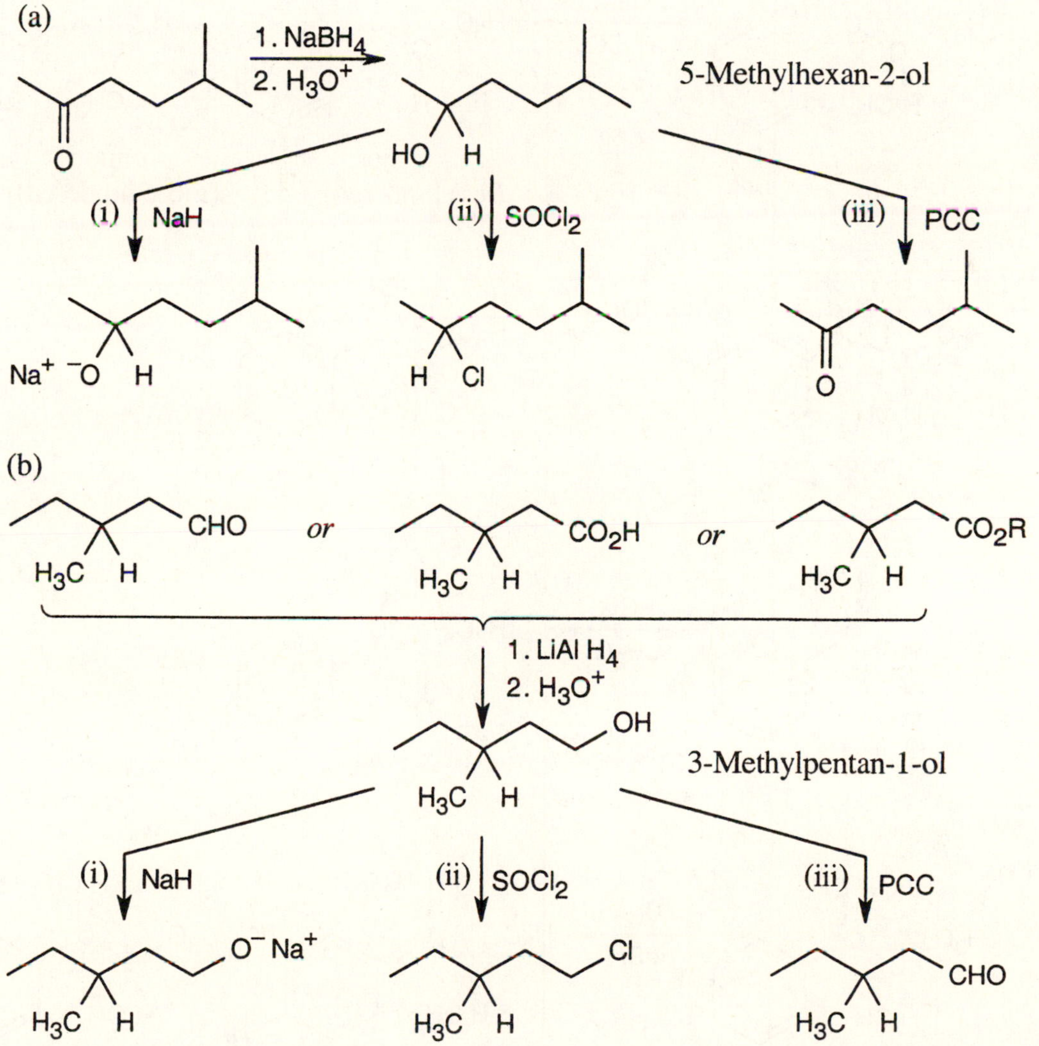

(a)

1. $NaBH_4$
2. H_3O^+

HO H

5-Methylhexan-2-ol

(i) NaH (ii) $SOCl_2$ (iii) PCC

Na^+ ^-O H

H Cl

O

(b)

CHO *or* CO_2H *or* CO_2R

H_3C H H_3C H H_3C H

1. $LiAlH_4$
2. H_3O^+

OH
H_3C H
3-Methylpentan-1-ol

(i) NaH (ii) $SOCl_2$ (iii) PCC

O^- Na^+ Cl CHO

H_3C H H_3C H H_3C H

13.22

(a)

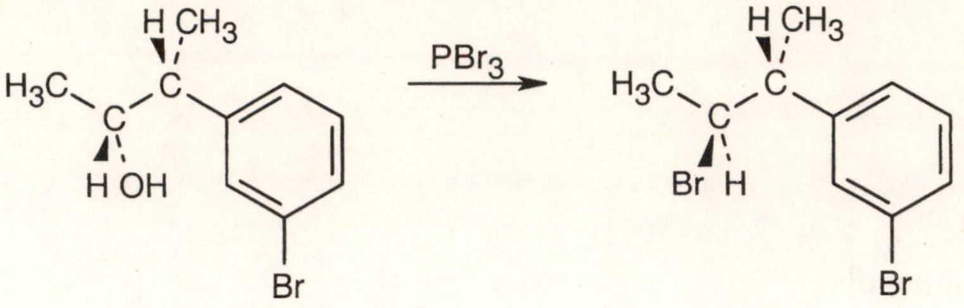

PBr$_3$

(b)

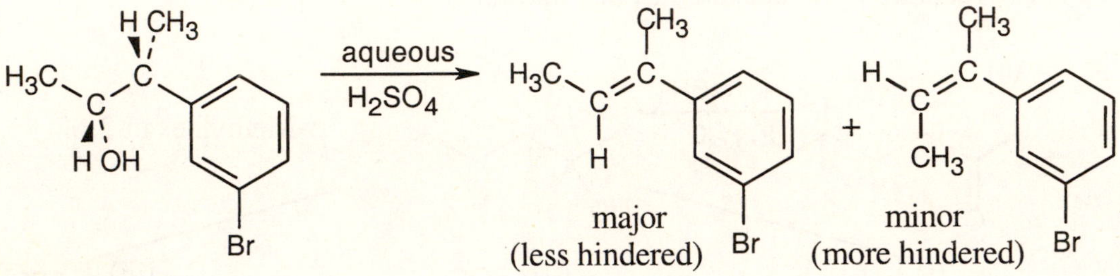

aqueous H_2SO_4

major (less hindered) + minor (more hindered)

(c)

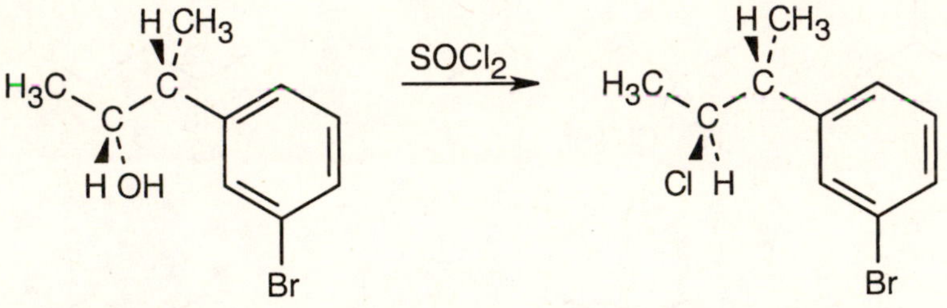

$SOCl_2$

(d)

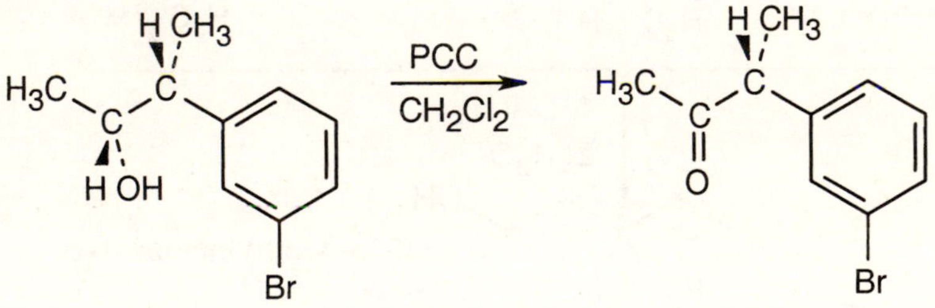

PCC CH_2Cl_2

(e)

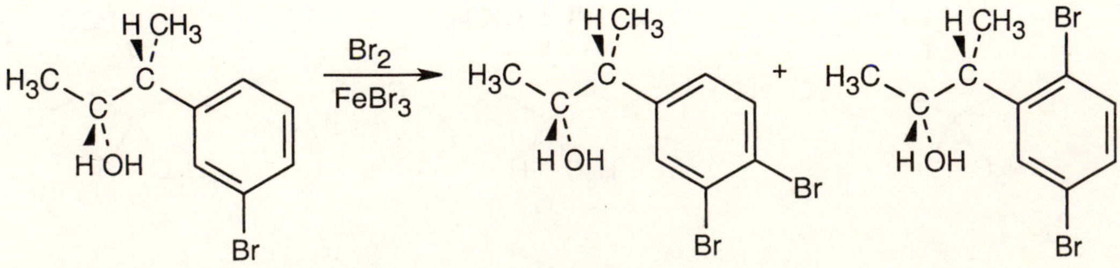

Br_2 $FeBr_3$

13.23

(a)

$$\mathrm{CH_3CH(CH_3)CH_2CH_2C(=O)OCH_3} \xrightarrow[\text{2. } H_3O^+]{\text{1. NaBH}_4} \text{no reaction}$$

(b)

$$\mathrm{CH_3CH(CH_3)CH_2CH_2C(=O)OCH_3} \xrightarrow[\text{2. } H_3O^+]{\text{1. LiAlH}_4} \mathrm{CH_3CH(CH_3)CH_2CH_2CH_2OH} + \mathrm{HOCH_3}$$

(c)

$$\mathrm{CH_3CH(CH_3)CH_2CH_2C(=O)OCH_3} \xrightarrow[\text{2. } H_3O^+]{\text{1. 2 } CH_3CH_2MgBr} \mathrm{CH_3CH(CH_3)CH_2CH_2C(OH)(CH_2CH_3)CH_2CH_3} + \mathrm{HOCH_3}$$

Additional Problems

13.24

(a) $\mathrm{HOCH_2CH_2CH(CH_3)CH_2OH}$

2-Methylbutane-1,4-diol

(b) $\mathrm{CH_3CH(OH)CH(CH_2CH_2CH_3)CH_2CH_3}$

3-Ethylhexan-2-ol

(c) *cis*-Cyclobutane-1,3-diol

(d) *cis*-2-Methylcyclo-hept-4-en-1-ol

(e) *cis*-3-Phenylcyclopentanol

(f) Cyclopentyl–$\mathrm{OCH(CH_3)CH_2CH_3}$

sec-Butyl cyclopentyl ether

(g) *o*-Ethylbenzenethiol (SH, CH_2CH_3)

(h) *o*-Dimethoxybenzene (OCH_3, OCH_3)

(i) $\mathrm{CH_3CH_2CH(CH_3)CH_2SCH(CH_3)CH_3}$

Isoropyl 2-methylbutyl sulfide

13.25 None of these alcohols has multiple bonds or rings.

$CH_3CH_2CH_2CH_2CH_2OH$ — Pentan-1-ol

$CH_3CH_2CH_2\overset{*}{C}H(OH)CH_3$ — Pentan-2-ol

$CH_3CH_2CH(OH)CH_2CH_3$ — Pentan-3-ol

$CH_3CH_2\overset{*}{C}H(CH_3)CH_2OH$ — 2-Methylbutan-1-ol

$CH_3CH_2C(OH)(CH_3)CH_3$ — 2-Methylbutan-2-ol

$CH_3\overset{*}{C}H(OH)CH(CH_3)CH_3$ — 3-Methylbutan-2-ol

$HOCH_2CH_2CH(CH_3)CH_3$ — 3-Methylbutan-1-ol

$CH_3C(CH_3)_2CH_2OH$ — 2,2-Dimethylpropan-1-ol

Pentan-2-ol, 2-methylbutan-1-ol and 3-methylbutan-2-ol have chiral carbons.

13.26 Primary alcohols react with CrO_3 in aqueous acid to form carboxylic acids, secondary alcohols yield ketones, and tertiary alcohols are unreactive to oxidation. Of the eight alcohols in the previous problem, only 2-methylbutan-2-ol is unreactive to CrO_3 oxidation.

$$CH_3CH_2CH_2CH_2CH_2OH \xrightarrow[H_3O^+]{CrO_3} CH_3CH_2CH_2CH_2CO_2H$$

$$CH_3CH_2CH_2CH(OH)CH_3 \xrightarrow[H_3O^+]{CrO_3} CH_3CH_2CH_2C(=O)CH_3$$

$$CH_3CH_2CH(OH)CH_2CH_3 \xrightarrow[H_3O^+]{CrO_3} CH_3CH_2C(=O)CH_2CH_3$$

$$CH_3CH_2CH(CH_3)CH_2OH \xrightarrow[H_3O^+]{CrO_3} CH_3CH_2CH(CH_3)CO_2H$$

$$CH_3CH(OH)CH(CH_3)CH_3 \xrightarrow[H_3O^+]{CrO_3} CH_3C(=O)CH(CH_3)CH_3$$

$$HOCH_2CH_2CH(CH_3)CH_3 \xrightarrow[H_3O^+]{CrO_3} HO_2CCH_2CH(CH_3)CH_3$$

$$CH_3C(CH_3)_2CH_2OH \xrightarrow[H_3O^+]{CrO_3} CH_3C(CH_3)_2CO_2H$$

13.27

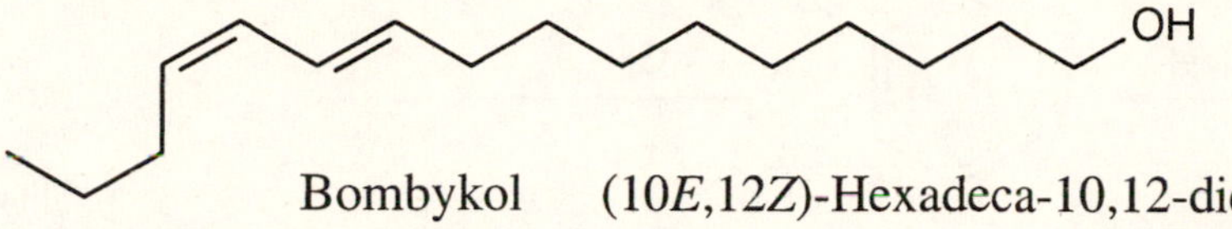

Bombykol (10*E*,12*Z*)-Hexadeca-10,12-dien-1-ol

13.28

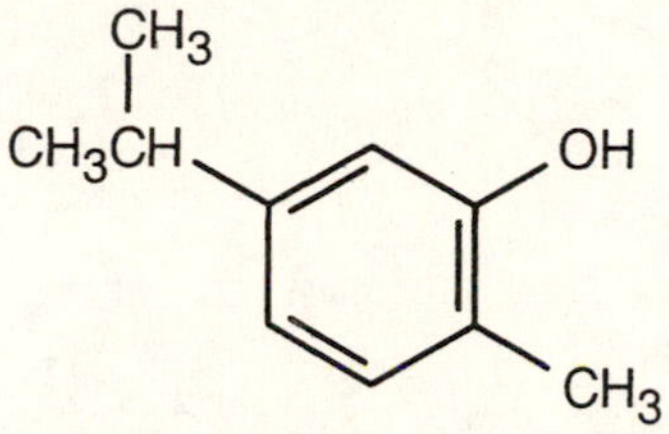

$CH_3CH(CH_3)$–, OH, CH_3

5-Isopropyl-2-methylphenol
(Carvacrol)

13.29

(a) Cyclohexyl–OCH_2CH_3 $\xrightarrow[H_2O]{HI}$ Cyclohexyl–OH + CH_3CH_2I

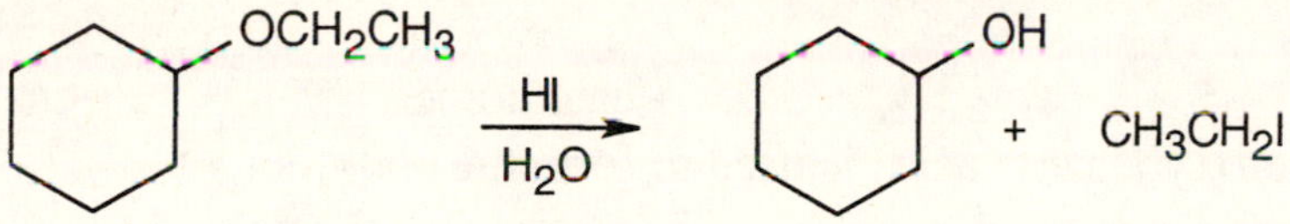

(b) Phenyl–$OC(CH_3)_3$ $\xrightarrow{CF_3CO_2H}$ Phenyl–OH + H_3C–$C(CH_3)$=CH_2

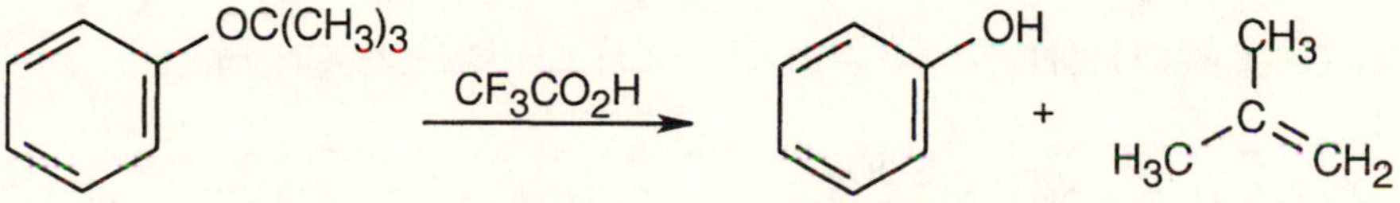

(c) $(CH_3)_3CCH_2OCH_2CH_3$ $\xrightarrow[H_2O]{HI}$ $(CH_3)_3CCH_2OH$ + CH_3CH_2I

13.30

(a) Phenyl–OH $\xrightarrow{NaH}$ Phenyl–$O^- Na^+$ $\xrightarrow{CH_3CH_2Br}$ Phenyl–O–CH_2CH_3

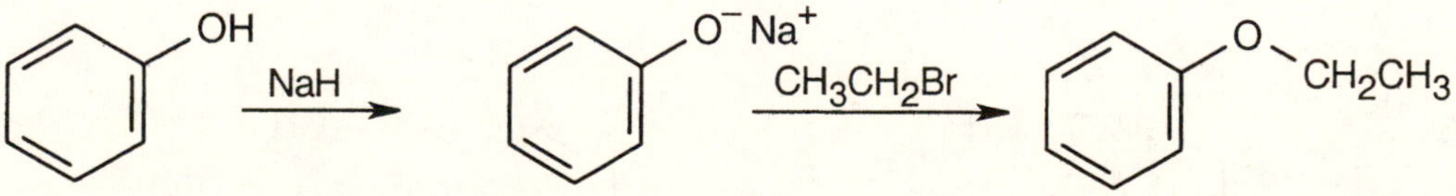

(b)

Benzyl alcohol ($C_6H_5CH_2OH$) $\xrightarrow{NaH}$ $C_6H_5CH_2O^-Na^+$ $\xrightarrow{CH_3Br}$ $C_6H_5CH_2OCH_3$

or

$CH_3OH \xrightarrow{NaH} CH_3O^-Na^+$ $\xrightarrow{C_6H_5CH_2Br}$ $C_6H_5CH_2OCH_3$

(c)

Cyclohexene $\xrightarrow{RCO_3H}$ cyclohexene oxide $\xrightarrow[HOCH_3]{^-OCH_3}$ trans-2-methoxycyclohexanol (OH, OCH_3) + enantiomer $\xrightarrow[2.\ CH_3I]{1.\ NaH}$ trans-1,2-dimethoxycyclohexane (OCH_3, OCH_3) + enantiomer

RCO_3H (*m*-chloroperoxybenzoic acid) is used to form the epoxide.

13.31

(a) $CH_3CH_2CH_2CH_2CH_2OH \xrightarrow{PBr_3} CH_3CH_2CH_2CH_2CH_2Br$

(b) $CH_3CH_2CH_2CH_2CH_2OH \xrightarrow{SOCl_2} CH_3CH_2CH_2CH_2CH_2Cl$

(c) $CH_3CH_2CH_2CH_2CH_2OH \xrightarrow[H_3O^+]{CrO_3} CH_3CH_2CH_2CH_2CO_2H$

(d) $CH_3CH_2CH_2CH_2CH_2OH \xrightarrow[CH_2Cl_2]{PCC} CH_3CH_2CH_2CH_2CHO$

13.32

(a) $C_6H_5CH_2CH_2OH \xrightarrow[pyridine]{POCl_3} C_6H_5CH{=}CH_2$

2-Phenylethanol — Styrene

Acid-catalyzed dehydration also forms styrene.

(b) $C_6H_5CH_2CH_2OH \xrightarrow[CH_2Cl_2]{PCC} C_6H_5CH_2CHO$

Phenylacetaldehyde

PCC = $C_5H_6N^+CrO_3Cl^-$ (pyridinium chlorochromate)

(c)

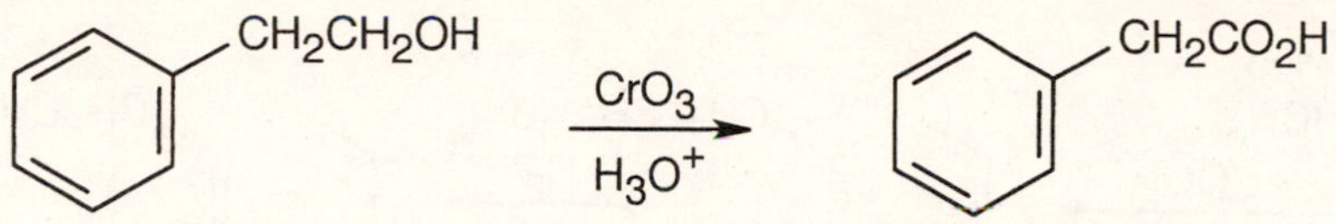

$C_6H_5CH_2CH_2OH \xrightarrow[H_3O^+]{CrO_3} C_6H_5CH_2CO_2H$

Phenylacetic acid

(d)

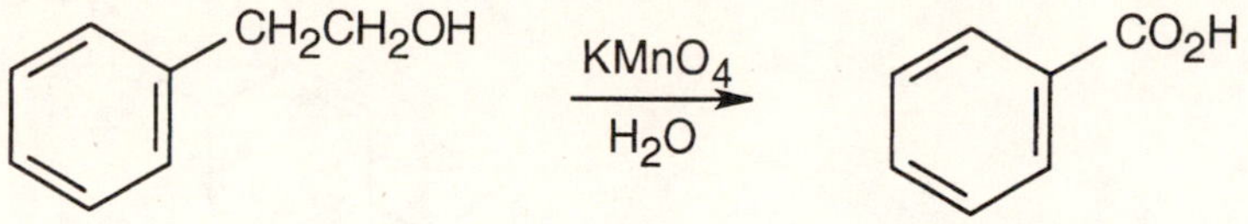

$C_6H_5CH_2CH_2OH \xrightarrow[H_2O]{KMnO_4} C_6H_5CO_2H$

Benzoic acid

(e)

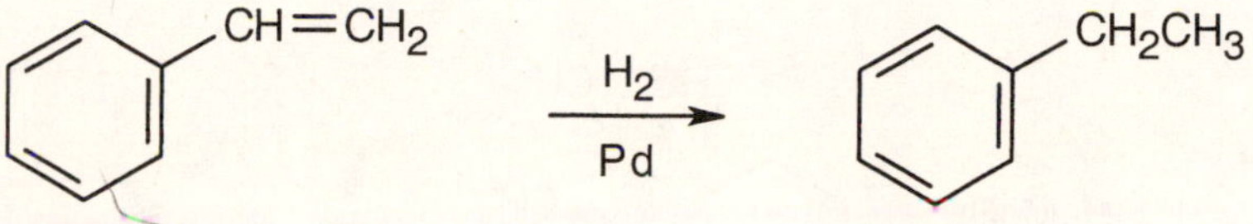

$C_6H_5CH{=}CH_2 \xrightarrow[Pd]{H_2} C_6H_5CH_2CH_3$

from (a) Ethylbenzene

(f)

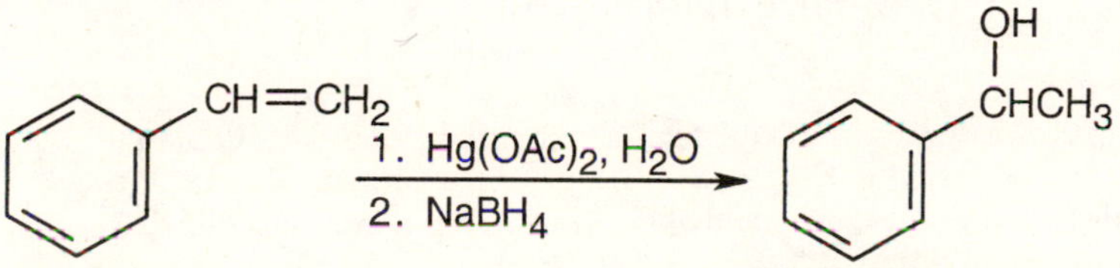

$C_6H_5CH{=}CH_2 \xrightarrow[2.\ NaBH_4]{1.\ Hg(OAc)_2,\ H_2O} C_6H_5CH(OH)CH_3$

from (a) 1-Phenylethanol

13.33

(a)

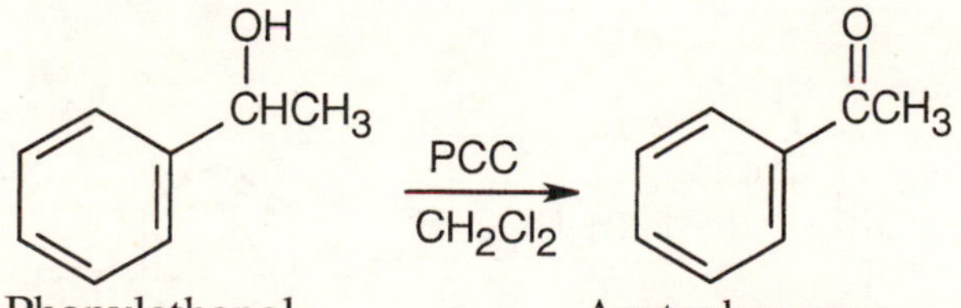

$C_6H_5CH(OH)CH_3 \xrightarrow[CH_2Cl_2]{PCC} C_6H_5COCH_3$

1-Phenylethanol Acetophenone

CrO_3 can also be used.

(b)

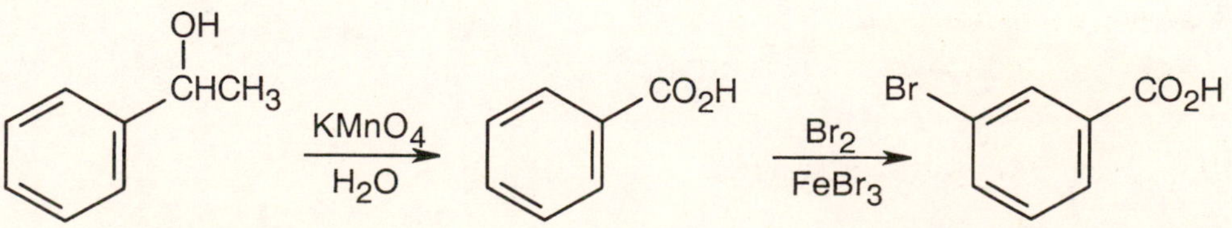

$C_6H_5CH(OH)CH_3 \xrightarrow[H_2O]{KMnO_4} C_6H_5CO_2H \xrightarrow[FeBr_3]{Br_2} m\text{-}BrC_6H_4CO_2H$

m-Bromobenzoic acid

(c)

$$C_6H_5CH(OH)CH_3 \xrightarrow[\text{pyridine}]{POCl_3} C_6H_5CH{=}CH_2 \xrightarrow[\text{Pd}]{H_2} C_6H_5CH_2CH_3 \xrightarrow{Cl_2,\ FeCl_3} p\text{-}ClC_6H_4CH_2CH_3$$

p-Chloroethylbenzene

(d)

$$C_6H_5COCH_3 \xrightarrow[\text{2. } H_3O^+]{\text{1. } CH_3MgBr} C_6H_5C(OH)(CH_3)_2$$

from (a)

2-Phenylpropan-2-ol

(e)

$$C_6H_5CH(OH)CH_3 \xrightarrow{NaH} C_6H_5CH(O^-\,Na^+)CH_3 + H_2 \xrightarrow{CH_3Br} C_6H_5CH(OCH_3)CH_3 + NaBr$$

Methyl 1-phenylethyl ether

(f)

$$C_6H_5CH(OH)CH_3 \xrightarrow{PBr_3} C_6H_5CH(Br)CH_3 \xrightarrow[\text{2. } ^-OH,\ H_2O]{\text{1. } (H_2N)_2C{=}S} C_6H_5CH(SH)CH_3$$

1-Phenylethanethiol

13.34 In some of these problems, different combinations of Grignard reagent and carbonyl compound are possible. Remember that aqueous acid is added to the initial Grignard adduct to yield the alcohol.

(a)

$CH_3CHO + CH_3CH_2MgBr$

or

$CH_3CH_2CHO + CH_3MgBr$

$\longrightarrow CH_3CH(OH)CH_2CH_3$

Butan-2-ol

(b)

$CH_3CH_2CHO + CH_3CH_2MgBr \longrightarrow CH_3CH_2CH(OH)CH_2CH_3$

Pentan-3-ol

(c)

$H_2C{=}C(CH_3){-}MgBr + CH_2O \longrightarrow H_2C{=}C(CH_3)CH_2OH$

2-Methylprop-2-en-1-ol

(d)

$C_6H_5COC_6H_5 + C_6H_5MgBr$

or

$C_6H_5COOR + 2\ C_6H_5MgBr$

$\longrightarrow (C_6H_5)_3C{-}OH$

Triphenylmethanol

(e)

$C_6H_5COCH_3 + CH_3MgBr$

or

$C_6H_5COOR + 2\ CH_3MgBr$

or

$C_6H_5MgBr + CH_3COCH_3$

$\longrightarrow C_6H_5C(OH)(CH_3)CH_3$

2-Phenylpropan-2-ol

(f)

1-Cyclohexenyl-MgBr $+ CH_2O \longrightarrow$ 1-Cyclohexenyl-CH_2OH

13.35

	Alcohol	*Carbonyl precursor(s)*	
(a)	$CH_3CH_2CH_2CH_2C(CH_3)_2CH_2OH$	$CH_3CH_2CH_2CH_2C(CH_3)_2CHO$	$CH_3CH_2CH_2CH_2C(CH_3)_2CO_2H$
		$CH_3CH_2CH_2CH_2C(CH_3)_2CO_2R$	
(b)	$(CH_3)CCH(OH)CH_3$	$(CH_3)CC(=O)CH_3$	
(c)	cyclohexyl–$CH(OH)CH_2CH_3$	cyclohexyl–$C(=O)CH_2CH_3$	

13.36 In these compounds you want to reduce some, but not all, of the functional groups present. To do this, you must choose the correct reducing agent.

(a) $C_6H_5CH{=}CHCO_2H \xrightarrow[\text{Pd/C}]{H_2} C_6H_5CH_2CH_2CO_2H$

H_2 with a palladium catalyst hydrogenates the side-chain double bond without affecting the carbonyl group or the aromatic ring.

(b) $C_6H_5CH{=}CHCO_2H \xrightarrow[2.\ H_3O^+]{1.\ LiAlH_4} C_6H_5CH{=}CHCH_2OH$

$LiAlH_4$ reduces carbonyl groups without affecting carbon–carbon double bonds.

(c) $C_6H_5CH{=}CHCH_2OH \xrightarrow{PBr_3} C_6H_5CH{=}CHCH_2Br$

(from b)

$\downarrow$ 1. $(H_2N)_2C{=}S$ 2. H_2O, NaOH

$C_6H_5CH{=}CHCH_2SH$

13.37

(a)

1. BH_3, THF
2. H_2O_2, ^{-}OH

OH

1. NaH
2. CH_3CH_2Br

OCH_2CH_3

+ H_2

(b)

OCH_3 H H_3C H

HI / H_2O

OH H H_3C H + CH_3I

PBr_3

H Br H_3C H

(c)

RCO_3H

$C(CH_3)_3$ → O, $C(CH_3)_3$ + O, $C(CH_3)_3$

H_3O^+

H OH H OH $C(CH_3)_3$

13.38

Grignard Reagent + *Carbonyl Compound* → *Product (after dilute acid workup)*

(a)

CH_3MgBr + $CH_3\overset{O}{\overset{\|}{C}}CH_3$

or

2 CH_3MgBr + $CH_3\overset{O}{\overset{\|}{C}}OR$

→ $(CH_3)COH$

(b)

CH_3CH_2MgBr + O (cyclohexanone) → OH, CH_2CH_3

(c)

CH_3CH_2MgBr + $CH_3CH_2\overset{O}{\overset{\|}{C}}$—$C_6H_5$

or

2 CH_3CH_2MgBr + $RO\overset{O}{\overset{\|}{C}}$—$C_6H_5$

or

C_6H_5—MgBr + $CH_3CH_2\overset{O}{\overset{\|}{C}}CH_2CH_3$

→ $CH_3CH_2\overset{OH}{\overset{|}{C}}CH_2CH_3$ (with C_6H_5 on the central carbon)

(d)

$CH_3CH_2CH_2MgBr + CH_3C(=O)C_6H_5$

or

$CH_3MgBr + CH_3CH_2CH_2C(=O)C_6H_5$

or

$C_6H_5MgBr + CH_3CH_2CH_2C(=O)CH_3$

$\longrightarrow CH_3CH_2CH_2C(OH)(C_6H_5)CH_3$

(e)

$p\text{-}H_3C\text{-}C_6H_4CH_2MgBr + H_2C{=}O \longrightarrow p\text{-}H_3C\text{-}C_6H_4CH_2CH_2OH$

(f)

$CH_3MgBr + C_5H_9CH_2C(=O)CH_3$

or

$2\ CH_3MgBr + C_5H_9CH_2C(=O)OR$

or

$C_5H_9CH_2MgBr + CH_3C(=O)CH_3$

$\longrightarrow C_5H_9CH_2C(OH)(CH_3)_2$

13.39

$ClCH_2CH_2CH_2CH_2OH \xrightarrow{^{-}H:} [ClCH_2CH_2CH_2CH_2O^{-}] + H_2 \longrightarrow$ tetrahydrofuran $+ Cl^{-}$

Strong base deprotonates the alcohol hydrogen

The alkoxide displaces Cl⁻ to form tetrahydrofuran

13.40

$\text{tetrahydrofuran} \xrightarrow[H_2O]{HI} ICH_2CH_2CH_2CH_2OH$

13.41

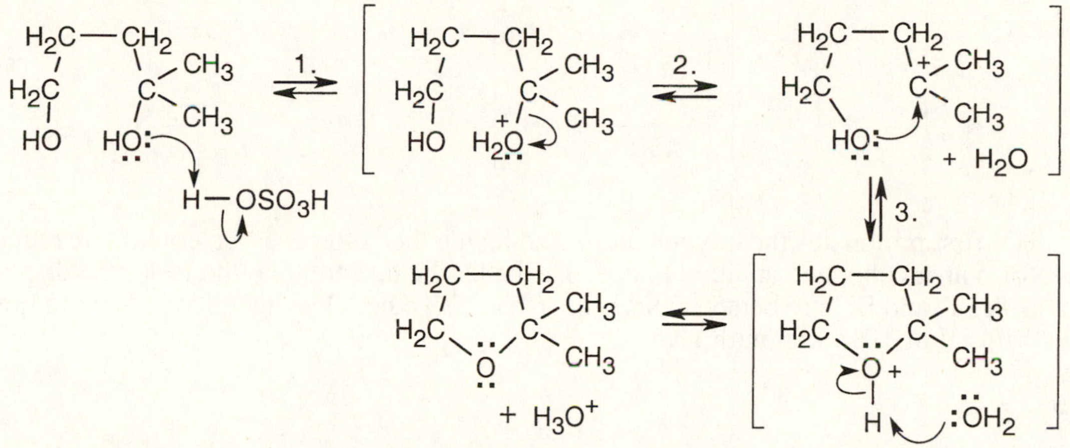

The reaction involves: (1) protonation of the tertiary hydroxyl group; (2) loss of water to form a tertiary carbocation; (3) nucleophilic attack on the carbocation by the second hydroxyl group. The tertiary hydroxyl group is more likely to be eliminated because the resulting carbocation is more stable.

13.42

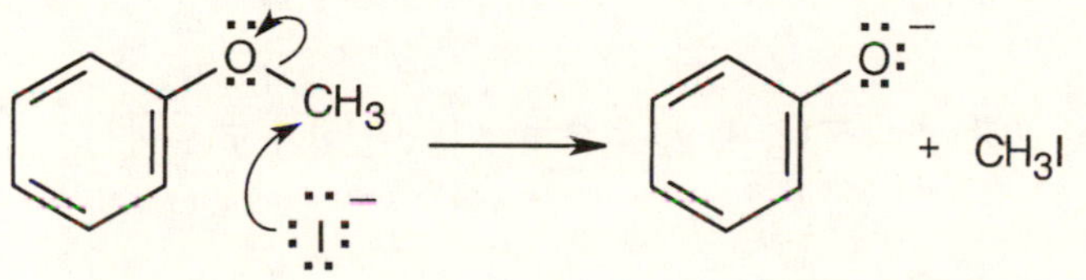

This reaction is an S_N2 displacement and can't occur at an aryl carbon. DMF is a polar aprotic solvent that increases the rate of an S_N2 reaction by making anions more nucleophilic.

13.43

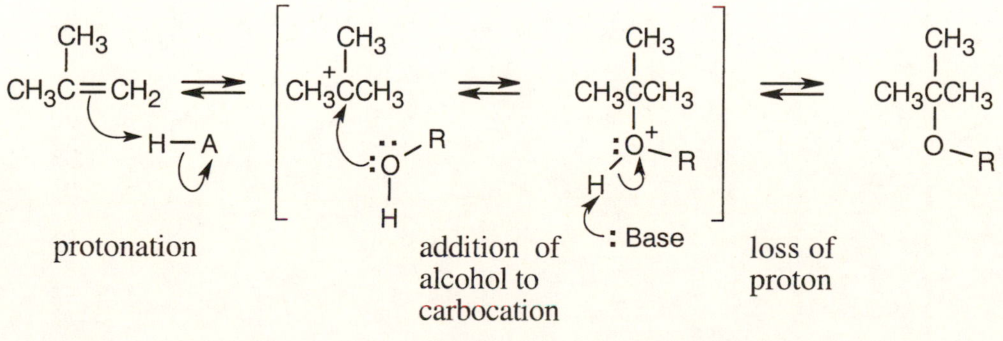

protonation

addition of alcohol to carbocation

loss of proton

Notice that this reaction is the reverse of acid-catalyzed cleavage of a tertiary ether (Problem 13.19).

13.44

$$R\ddot{O}R' + H{-}X \rightleftharpoons \left[R\overset{+}{\ddot{O}}(H)R' \quad :\ddot{X}:^- \right] \longrightarrow R{-}X + HO{-}R'$$

HX first protonates the oxygen atom, and halide then effects a nucleophilic reaction to form an alcohol and an alkyl halide. The better the nucleophile, the faster the displacement. Since I^- and Br^- are better nucleophiles than Cl^-, ether cleavage proceeds more rapidly with HI or HBr than with HCl.

13.45

$$(CH_3)_3C{-}CH(\ddot{O}H)CH_3 + H{-}\ddot{O}SO_3H \underset{\text{of –OH}}{\overset{\text{protonation}}{\rightleftharpoons}} \left[(CH_3)_3C{-}CH(\overset{+}{O}H_2)CH_3\right] \xrightarrow{\text{loss of } H_2O} \left[(CH_3)_2(CH_3)C{-}\overset{+}{C}HCH_3 + H_2O\right]$$

$$\downarrow \text{alkyl shift}$$

$$(CH_3)_2C{=}C(CH_3)_2 + H_3O^+ \xleftarrow{\text{loss of } H^+} \left[(CH_3)_2\overset{+}{C}{-}C(CH_3)(H)CH_3 \quad :\ddot{O}H_2\right]$$

13.46 This mechanism consists of the same steps as are seen in Problem 13.45. Two different alkyl shifts result in two different cycloalkenes.

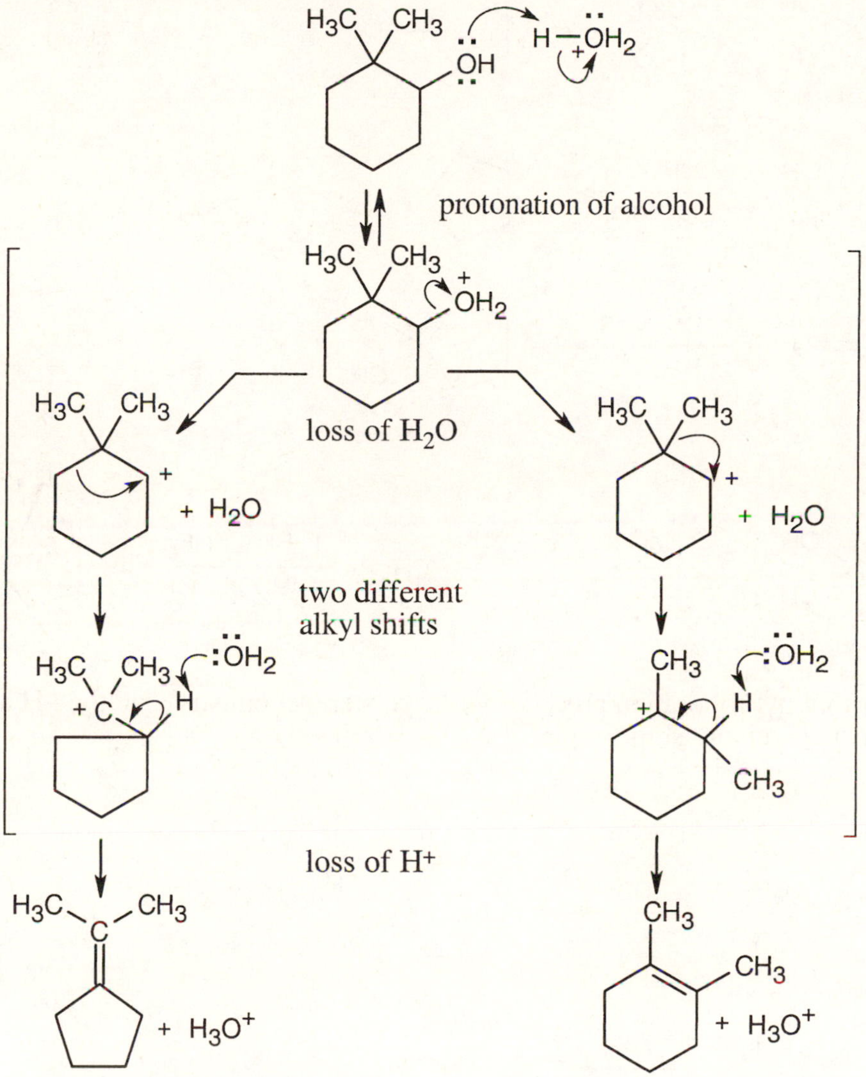

protonation of alcohol

loss of H_2O

$+ H_2O$

$+ H_2O$

two different alkyl shifts

loss of H^+

$+ H_3O^+$

$+ H_3O^+$

Isopropylidenecyclopentane

1,2-Dimethylcyclohexene

13.47

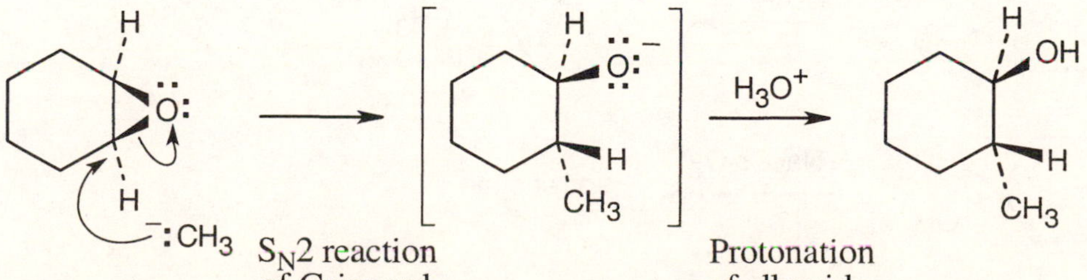

S_N2 reaction of Grignard reagent

H_3O^+

Protonation of alkoxide anion

The methyl group and the hydroxyl group have a trans relationship.

13.48

(a)

Cyclopentanol (OH, H) $\xrightarrow[H_3O^+]{\text{PCC or } CrO_3}$ cyclopentanone (=O)

(b)

Cyclopentanol (OH, H) $\xrightarrow[\text{pyridine}]{POCl_3}$ cyclopentene

(c)

Cyclopentanone (=O) from (a) $\xrightarrow[2.\ H_3O^+]{1.\ CH_3MgBr}$ 1-methylcyclopentanol (OH, CH_3)

(d)

1-methylcyclopentanol (OH, CH_3) from (c) $\xrightarrow{H_3O^+}$ 1-methylcyclopentene (CH_3) $\xrightarrow[2.\ H_2O_2,\ ^{-}OH]{1.\ BH_3,\ THF}$ product (H OH; H, CH_3)

Remember that hydroboration proceeds with syn stereochemistry, and the –H and –OH added have a cis relationship.

13.49

(a)

1-methylcyclohexanol (CH_3, OH) $\xrightarrow{HBr}$ 1-bromo-1-methylcyclohexane (CH_3, Br)

(b)

1-methylcyclohexanol (CH_3, OH) $\xrightarrow{NaH}$ (CH_3, $O^-\ Na^+$) + H_2

(c)

1-methylcyclohexanol (CH_3, OH) $\xrightarrow{H_2SO_4}$ 1-methylcyclohexene (CH_3)

(d)

1-methylcyclohexanol (CH_3, OH) $\xrightarrow{Na_2Cr_2O_7}$ no reaction

Tertiary alcohols aren't oxidized by sodium dichromate.

13.50 Remember that electron-withdrawing groups stabilize phenoxide anions and increase acidity. Electron-donating groups decrease phenol acidity.

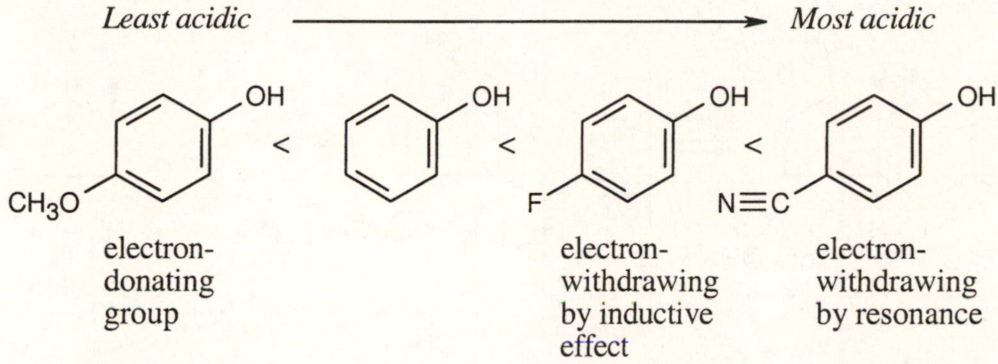

Least acidic ⟶ *Most acidic*

CH_3O–C₆H₄–OH < C₆H₅–OH < F–C₆H₄–OH < N≡C–C₆H₄–OH

electron-donating group (p-methoxyphenol); electron-withdrawing by inductive effect (p-fluorophenol); electron-withdrawing by resonance (p-cyanophenol)

13.51

H_3C–CH_2–C(=O)–CH_3 $\xrightarrow[2.\ H_3O^+]{1.\ NaBH_4}$ (S)-H_3C–CH_2–CH(OH)–CH_3 + (R)-H_3C–CH_2–CH(OH)–CH_3

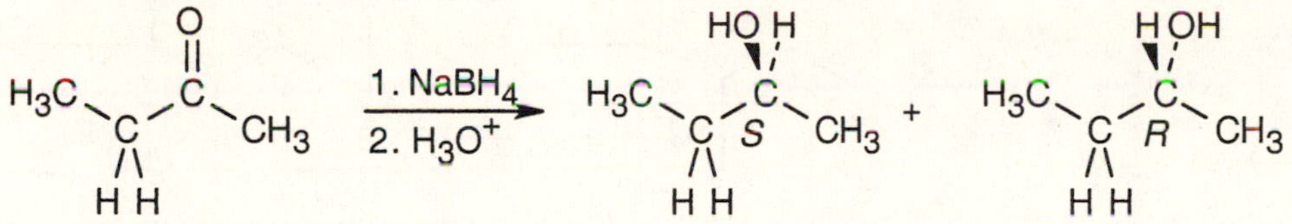

Reaction of butan-2-one with $NaBH_4$ produces a racemic mixture of (*R*)-butan-2-ol and (*S*)-butan-2-ol.

13.52

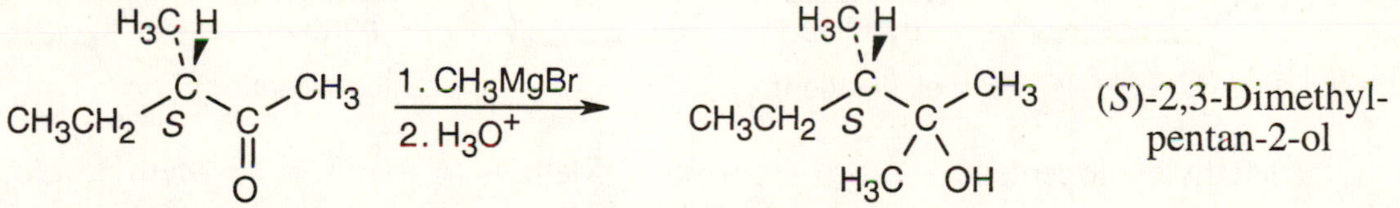

CH_3CH_2–CH(CH_3)(*S*)–C(=O)–CH_3 $\xrightarrow[2.\ H_3O^+]{1.\ CH_3MgBr}$ CH_3CH_2–CH(CH_3)(*S*)–C(CH_3)(H_3C)(OH) (*S*)-2,3-Dimethylpentan-2-ol

Addition of methylmagnesium bromide to the carbonyl group doesn't produce a new chirality center and doesn't affect the chirality center already present. The product is pure (*S*)-2,3-dimethylpentan-2-ol.

13.53

H–OH$_2^+$; :OH ; H$_3$C ; H ; protonation of –OH ; $^+$OH$_2$; loss of H_2O ; + H_2O ; alkyl shift ; H_2O: ; H ; CH_3 ; H_3O^+ + ; loss of H^+

This is a carbocation rearrangement involving the shift of an alkyl group. The sequence of steps is the same as those seen in Problems 13.45 and 13.46.

13.54

$POCl_3$ pyridine (syn elimination) ; $POCl_3$ pyridine (anti elimination) ; H ; OH ; H_3C

1-Methylcyclopentene ; *trans*-2-Methylcyclopentanol ; 3-Methylcyclopentene

The more stable dehydration product is 1-methylcyclopentene, which can be formed only via syn elimination. The product of anti elimination is 3-methylcyclopentene. Since this product predominates, the requirement of anti periplanar geometry must be more important than formation of the more stable product.

13.55 The pinacol rearrangement follows a sequence of steps similar to other rearrangements we have studied in this chapter. The second hydroxyl group assists in the alkyl shift.

$$\underset{CH_3}{CH_3\overset{HO\ :\ddot{O}H}{C-C}(CH_3)_2} + H-\overset{+}{\ddot{O}}H_2 \underset{\text{protonation of –OH}}{\rightleftharpoons} \left[\underset{CH_3}{CH_3\overset{HO\ \ \overset{+}{\ddot{O}}H_2}{C-C}(CH_3)_2}\right] \xrightarrow{\text{loss of } H_2O} \left[\underset{CH_3}{CH_3\overset{H\ddot{O}:}{C-\overset{+}{C}}(CH_3)_2} + H_2\ddot{O}\right]$$

alkyl shift

$$H_3O^+ + CH_3\overset{:O:}{\overset{\|}{C}}C(CH_3)_3 \underset{\text{loss of } H^+}{\rightleftharpoons} \left[H_2\ddot{O}: \quad CH_3\overset{H-\overset{+}{O}:}{\overset{\|}{C}}-\underset{CH_3}{C}(CH_3)_2\right]$$

13.56 The hydroxyl group is axial in the cis isomer, which is expected to oxidize faster than the trans isomer. (Remember that the bulky *tert*-butyl group is always equatorial in the more stable conformation.)

cis-4-*tert*-Butylcyclohexanol $\xrightarrow[\text{H}_3\text{O}^+]{\text{CrO}_3}$ (faster) 4-*tert*-butylcyclohexanone

trans-4-*tert*-Butylcyclohexanol $\xrightarrow[\text{H}_3\text{O}^+]{\text{CrO}_3}$ (slower)

13.57

Cyclohexanone $\xrightarrow[\text{2. } H_3O^+]{\text{1. LiAlH}_4}$ Cyclohexanol (–OH) $\xrightarrow{PBr_3}$ Bromocyclohexane (–Br) $\xrightarrow{\text{Mg, ether}}$ Cyclohexylmagnesium bromide (–MgBr) $\xrightarrow[\text{2. } H_3O^+]{\text{1. Cyclohexanone}}$ 1-Cyclohexylcyclohexanol (OH) $\xrightarrow{H_2SO_4}$ Bicyclohexylidene

13.58

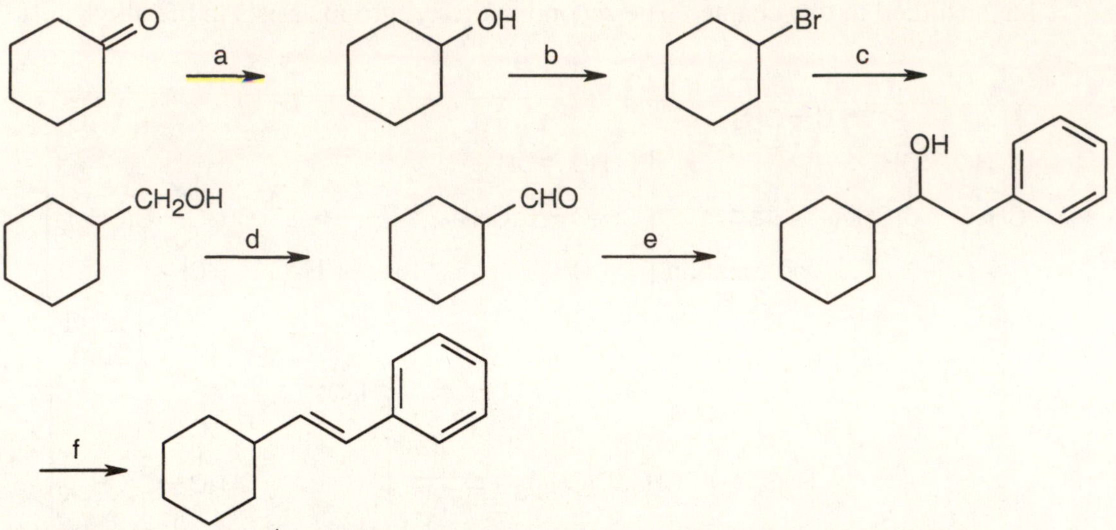

(a) $NaBH_4$, then H_3O^+ (b) PBr_3 (c) Mg, ether, then CH_2O (d) PCC, CH_2Cl_2 (e) $C_6H_5CH_2MgBr$, then H_3O^+ (f) $POCl_3$, pyridine.

13.59

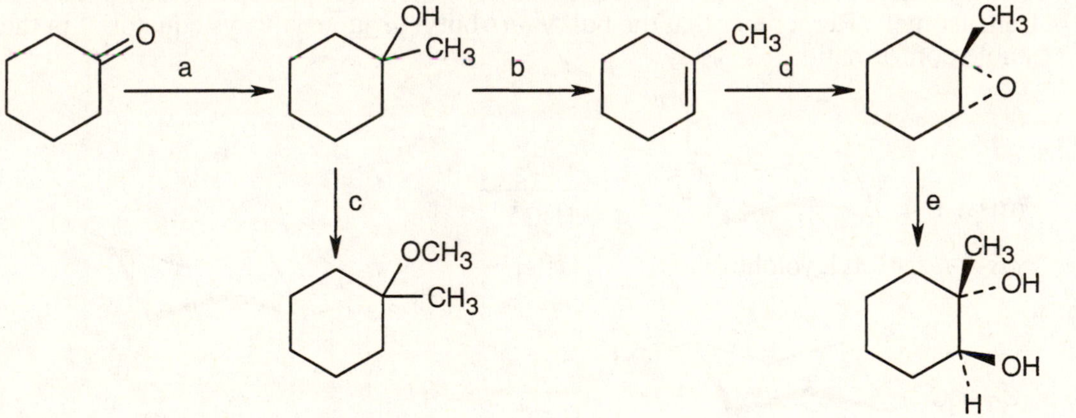

(a) CH_3MgBr, ether; (b) H_2SO_4, H_2O; (c) NaH, then CH_3I; (d) *m*-$ClC_6H_4CO_3H$; (e) H_3O^+.

13.60 Disparlure, $C_{19}H_{38}O$, contains one degree of unsaturation, which the ^1H NMR absorption at 2.8 δ identifies as an epoxide ring.

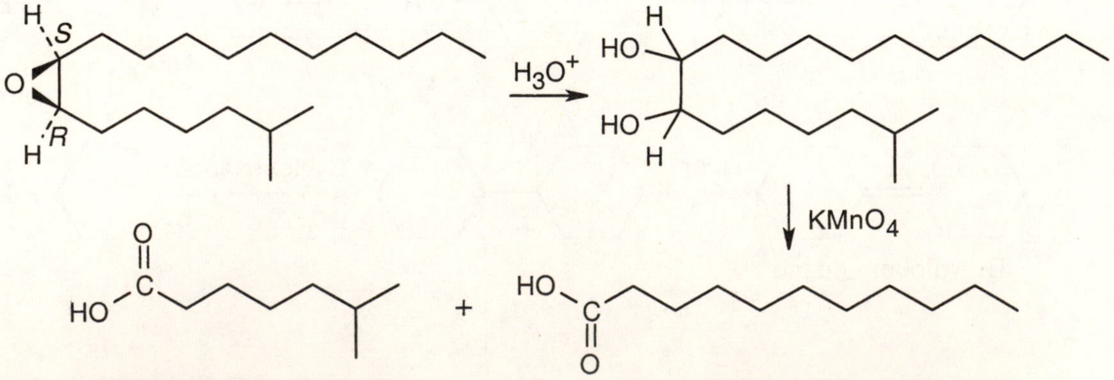

6-Methylheptanoic acid

Undecanoic acid

13.61

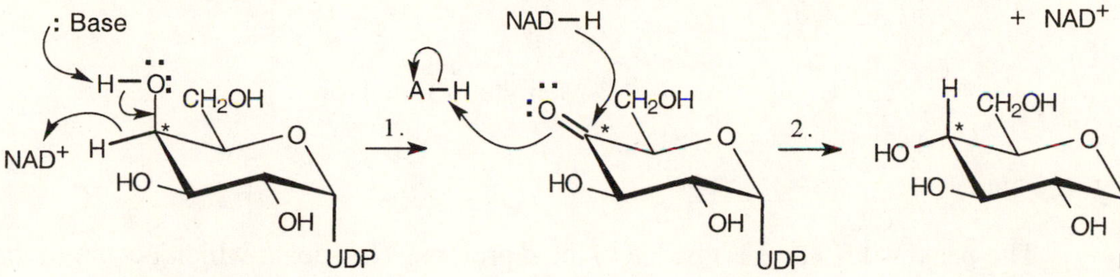

: Base

NAD—H

+ NAD^+

UDP-Galactose

UDP-Glucose

Step 1: Base deprotonates the C4 hydroxyl group while NAD^+ oxidizes the alcohol to a ketone by the mechanism shown in Figure 13.5.
Step 2: When the ketone is reduced by the NADH formed in Step 1, the configuration at the starred carbon is inverted, and UDP-glucose is formed.

13.62 $M^+ = 116$ corresponds to a sulfide of molecular formula $C_6H_{12}S$, indicating one degree of unsaturation. The IR absorption at 890 cm^{-1} is due to a $=CH_2$ group.

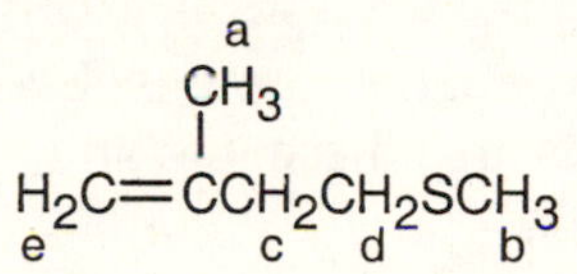

$H_2C{=}C(CH_3)CH_2CH_2SCH_3$ (e = H_2C, a = CH_3, c = CH_2, d = CH_2, b = SCH_3)

2-Methyl-4(methylthio)but-1-ene

a = 1.74 δ
b = 2.11 δ
c = 2.27 δ
d = 2.57 δ
e = 4.73 δ

13.63

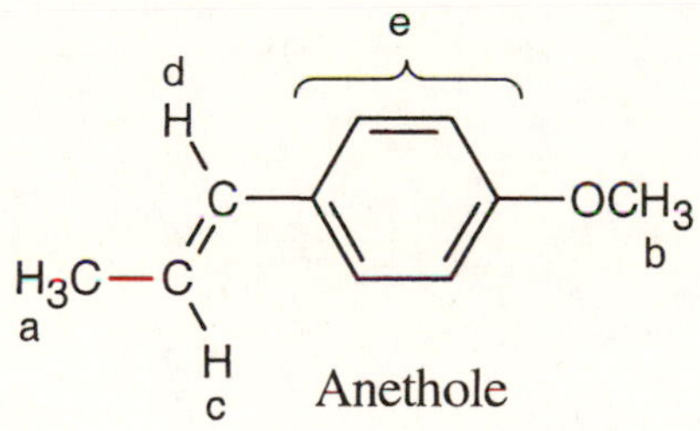

Anethole

Peak	*Chemical shift*	*Multiplicity*	*Split by:*
a	1.83 δ	doublet	c
b	3.75 δ	singlet	
c	6.08 δ	two quartets	a,d
d	6.28 δ	doublet	c
e	6.80 δ, 7.23 δ	multiplet	

13.64 Strategy:

1. $C_8H_{18}O_2$ has *no* double bonds or rings.
2. The IR band at 3350 cm^{-1} shows the presence of a hydroxyl group.
3. The compound is symmetrical (simple NMR).
4. There is no splitting.

Solution:

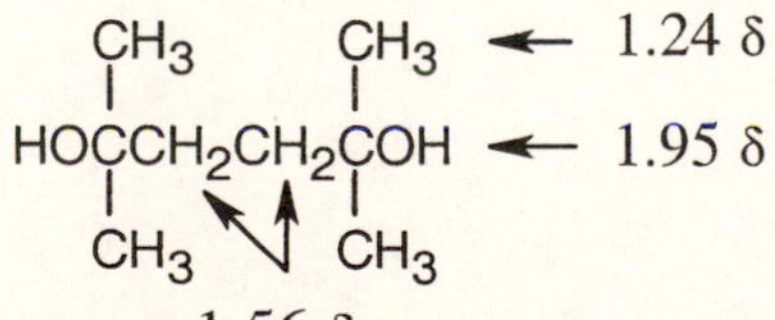

$HOC(CH_3)_2CH_2CH_2C(CH_3)_2OH$

CH_3 ← 1.24 δ
OH ← 1.95 δ
CH_2: 1.56 δ

2,5-Dimethylhexane-2,5-diol

13.65

```
e          d
H          CH3
  \       /
   C=C
  /       \
H          CH2CH2OH
f          c   b  a
```

3-Methylbut-3-en-1-ol

The peak at 1.75 δ (3 H) is due to the *d* protons. This peak, which occurs in the allylic region of the spectrum, is unsplit.

The peak at 2.13 δ (1 H) is due to the –OH proton *a*.

The peak at 2.30 δ (2 H) is due to protons *c*. The peak is a triplet because of splitting by the adjacent *b* protons.

The peak at 3.70 δ (2 H) is due to the *b* protons. The adjacent oxygen causes the peak to be downfield, and the adjacent $-CH_2-$ group splits the peak into a triplet.

The peaks at 4.78 δ and 4.85 δ (2 H) are due to protons *e* and *f*.

13.66 (a) The infrared absorption at 3400 cm^{-1} indicates the presence of an alcohol. The weak absorption at 1640 cm^{-1} is due to a C=C stretch.

(b) (1) The absorptions at 1.63 δ and 1.70 δ are due to unsplit methyl protons. Because the absorptions are shifted slightly downfield, the protons are adjacent to an unsaturated center.

(2) The broad singlet at 3.83 δ is due to an alcohol proton.

(3) The doublet at 4.15 δ is due to two protons bonded to a carbon bearing an electronegative atom, oxygen, in this case.

(4) The proton absorbing at 5.70 δ is a vinylic proton.

(c)

```
H3C          CH2OH
   \        /
    C=C
   /        \
H3C          H
      A
```

3-Methylbut-2-en-1-ol

13.67 (a) $C_5H_{12}O$, $C_4H_8O_2$, $C_3H_4O_3$

(b) The ^1H NMR data show that the compound has twelve protons.

(c) The IR absorption at 3600 cm^{-1} shows that the compound is an alcohol.

(d) The compound contains five carbons, two of which are identical.

(e) $C_5H_{12}O$ is the molecular formula of the compound.

(f), (g)

```
      b
     OH
     |
CH3CCH2CH3
 c  |  d   a
   CH3
    c
```

a = 0.9 δ
b = 1.0 δ
c = 1.2 δ
d = 1.4 δ

2-Methylbutan-2-ol

13.68

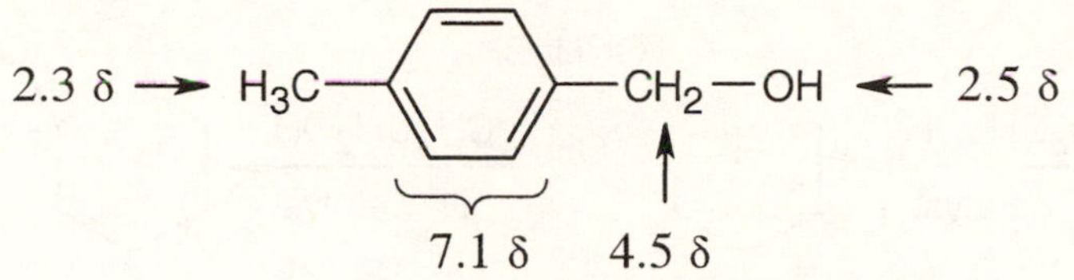

$2.3\ \delta \rightarrow H_3C-C_6H_4-CH_2-OH \leftarrow 2.5\ \delta$

7.1 δ 4.5 δ

p-Methylbenzyl alcohol

13.69

Structural formula: $C_8H_{10}O$ contains 4 multiple bonds and/or rings.

Infrared: The broad band at 3500 cm^{-1} indicates a hydroxyl group. The absorptions at 1500 cm^{-1} and 1600 cm^{-1} are due to an aromatic ring. The absorption at 830 cm^{-1} shows that the ring is *p*-disubstituted. Compound **A** is probably a phenol.

^1H NMR: The triplet at 1.18 δ (3 H) is coupled with the quartet at 2.56 δ (2 H). These two absorptions are due to an ethyl group.

The peaks at 6.75 δ-7.05 δ (4 H) are due to aromatic ring protons. The symmetrical splitting pattern of these peaks indicate that the aromatic ring is *p*-disubstituted.

The absorption at 5.50 δ (1 H) is due to an –OH proton.

Compound **A**

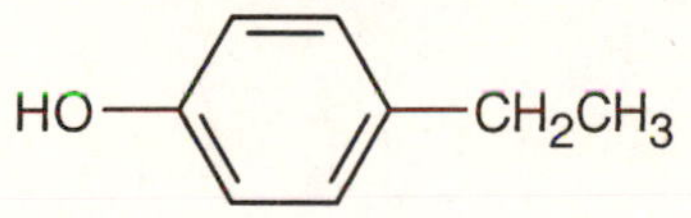

$HO-C_6H_4-CH_2CH_3$

p-Ethylphenol

13.70

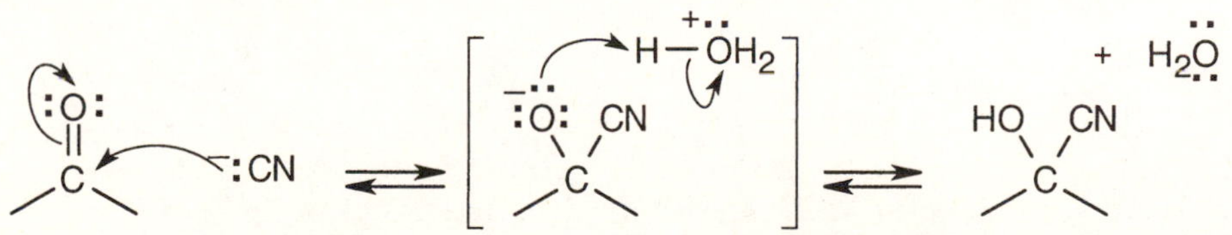

$R_2C{=}O + {}^-{:}CN \rightleftharpoons [R_2C(O^-)CN + H-OH_2^+] \rightleftharpoons R_2C(OH)CN + H_2O$

The nucleophile $^-$CN adds to the positively polarized carbonyl carbon.

The tetrahedral intermediate is protonated to give the addition product, a cyanohydrin,.

13.71

(a)

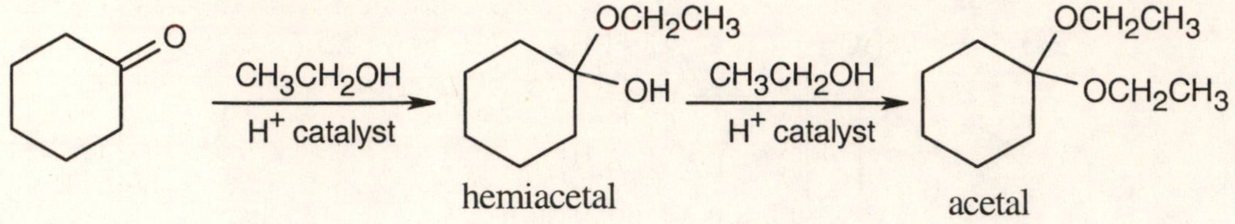

OCH_2CH_3 CH_3CH_2OH / H^+ catalyst → hemiacetal (OCH_2CH_3, OH) CH_3CH_2OH / H^+ catalyst → acetal (OCH_2CH_3, OCH_2CH_3)

(b)

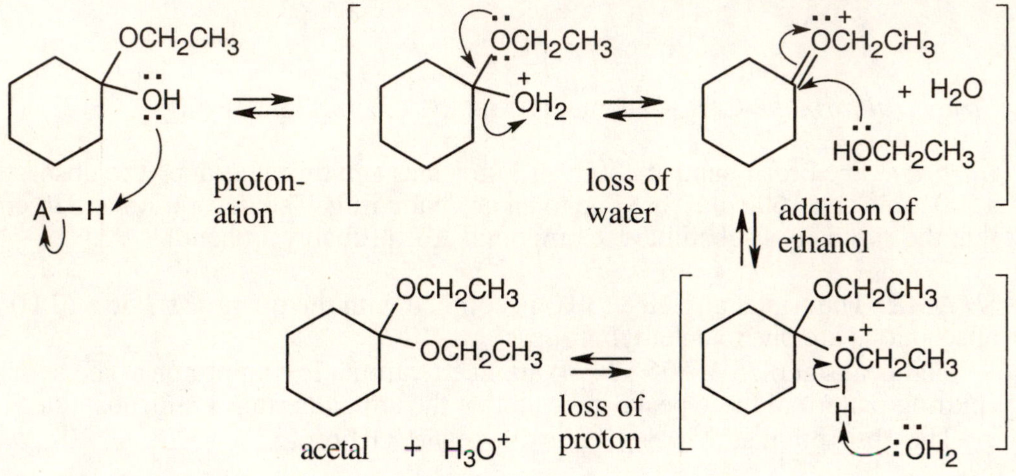

OCH_2CH_3, OH; A—H; proton- ation ⇌ [OCH_2CH_3, $\overset{+}{O}H_2$] ⇌ loss of water [$\overset{+}{O}CH_2CH_3$ + H_2O; $HOCH_2CH_3$]

⇅ addition of ethanol

[OCH_2CH_3, $\overset{+}{O}CH_2CH_3$, H, $:OH_2$] ⇌ loss of proton: acetal (OCH_2CH_3, OCH_2CH_3) + H_3O^+

A Preview of Carbonyl Chemistry

Chapter Outline

I. The carbonyl group.
  A. Kinds of carbonyl compounds.
    1. All carbonyl compounds contain an acyl group (R–C=O).
    2. The groups bonded to the acyl group can be of two types:
      a. Groups that can't act as leaving groups.
        Examples: aldehydes and ketones.
      b. Groups that can act as leaving groups.
        Examples: carboxylic acids, esters, amides, acid halides, lactones, acid anhydrides, lactams.
  B. Nature of the carbonyl group.
    1. The carbonyl carbon is *sp*^{2}-hybridized.
      a. A π bond is formed between carbon and oxygen.
      b. Carbonyl compounds are planar about the double bond.
    2. The carbon-oxygen bond is polar.
      a. The carbonyl carbon acts as an electrophile.
      b. The carbonyl oxygen acts as a nucleophile.

II. Reactions of carbonyl compounds.
  A. Nucleophilic addition reactions of aldehydes and ketones.
    1. A nucleophile adds to the carbonyl carbon.
    2. The resulting tetrahedral intermediate has two fates:
      a. The negatively charged oxygen can be protonated to form an alcohol.
      b. Loss of water leads to formation of a C=Nu double bond.
  B. Nucleophilic acyl substitution reactions.
    1. A nucleophile adds to the carbonyl carbon.
    2. The resulting tetrahedral intermediate expels a leaving group to form a new carbonyl compound.
    3. This type of reaction takes place with carbonyl compounds other than aldehydes and ketones.
  C. Alpha substitution reactions.
    1. Reaction can occur at the position next to the carbonyl carbon (α position).
      a. This type of reaction is possible because of the acidity of alpha hydrogens.
      b. Reaction with a strong base forms an enolate anion, which behaves as a nucleophile.
    2. All carbonyl compounds can undergo α substitution reactions.
  D. Carbonyl condensation reactions.
    1. Carbonyl condensation reactions occur when two carbonyl compounds react with each other.
    2. The enolate of one carbonyl compound adds to the carbonyl group of a second compound.

Solutions to Problems

1. According to the electrostatic potential maps, the carbonyl carbon of acetyl chloride is more electrophilic and the oxygen of acetone is more nucleophilic. This makes sense, because acetyl chloride has two electron-withdrawing groups that make its carbonyl carbon electron-poor and thus electrophilic. Because acetyl chloride has two electron-withdrawing groups, neither group is as nucleophilic as the carbonyl oxygen of acetone.

2. The reaction of cyanide ion with acetone is a nucleophilic addition reaction.

$$H_3C-\overset{O}{\overset{\|}{C}}-CH_3 \xrightarrow{^-CN} \left[(H_3C)_2\overset{O^-}{\overset{|}{C}}-CN \right] \xrightarrow{H_3O^+} (H_3C)_2\overset{OH}{\overset{|}{C}}-CN$$

Cyanide anion adds to the positively polarized carbonyl carbon to form a tetrahedral intermediate. This intermediate is protonated to yield acetone cyanohydrin.

3. (a) This reaction is a *nucleophilic acyl substitution*. Ammonia adds to acetyl chloride, and chloride is eliminated, resulting in formation of an amide.

   (b) In this *nucleophilic addition* reaction, addition of the nucleophile is followed by loss of water.

   (c) Two molecules of cyclopentanone react in this *carbonyl condensation*.

Chapter 14 – Aldehydes and Ketones: Nucleophilic Addition Reactions

Chapter Outline

I. General information about aldehydes and ketones (Sections 14.1 – 14.3).
  A. Naming aldehydes and ketones (Section 14.1).
    1. Naming aldehydes.
      a. Aldehydes are named by replacing the *-e* of the corresponding alkane with *-al*.
      b. The parent chain must contain the –CHO group.
      c. The aldehyde carbon is always carbon 1.
      d. When the –CHO group is attached to a ring, the suffix *-carbaldehyde* is used.
    2. Naming ketones.
      a. Ketones are named by replacing the *-e* of the corresponding alkane with *-one*.
      b. Numbering starts at the end of the carbon chain nearer to the carbonyl carbon.
      c. The word *acyl* is used when a RCO– group is a substituent.
  B. Preparation of aldehydes and ketones (Section 14.2).
    1. Preparation of aldehydes.
      a. Oxidation of primary alcohols with PCC.
      b. Reduction of carboxylic acid derivatives.
    2. Preparation of ketones.
      a. Oxidation of secondary alcohols.
      b. Friedel-Crafts acylation of aromatic compounds.
  C. Oxidation of aldehydes (Section 14.3).
    1. Aldehydes can be oxidized to carboxylic acids by many reagents.
      a. CrO_3 is used for normal aldehydes.
      b. Oxidation occurs through intermediate 1,1-diols.
    2. Ketones are generally inert to oxidation, but can be oxidized to carboxylic acids with strong oxidizing agents.

II. Nucleophilic addition reactions of aldehydes and ketones (Sections 14.4 – 14.11).
  A. Characteristics of nucleophilic addition reactions (Section 14.4).
    1. Mechanism of nucleophilic addition reactions.
      a. A nucleophile attacks the electrophilic carbonyl carbon from a direction 75° from the plane of the carbonyl group.
      b. The carbonyl group rehybridizes from sp^2 to sp^3, and a tetrahedral alkoxide intermediate is produced.
      c. The attacking nucleophile may be neutral or negatively charged.
        Neutral nucleophiles usually have a hydrogen atom that can be eliminated.
      d. The tetrahedral intermediate has two fates:
        i. The intermediate can be protonated to give an alcohol.
        ii. The carbonyl oxygen can be eliminated as –OH to give a product with a C=Nu double bond.
      e. Possible reversibility and acid/base catalysis are important features of nucleophilic addition reactions.
    2. Relative reactivity of aldehydes and ketones.
      a. Aldehydes are usually more reactive than ketones in nucleophilic addition reactions for two reasons:
        i A nucleophile can approach the carbonyl group of an aldehyde with more ease because only one alkyl group is in the way.
        ii. Aldehyde carbonyl groups are more strongly polarized and electrophilic because they are less stabilized by the inductive effect of alkyl groups.

b. Aromatic aldehydes are less reactive than aliphatic aldehydes because the electron-donating aromatic ring makes the carbonyl carbon less electrophilic.

B. Nucleophilic addition reactions (Section 14.5 – 14.10).
1. Hydration (Section 14.5).
a. Water adds to aldehydes and ketones to give 1,1-diols (often referred to as gem-diols or hydrates).
b. The reaction is reversible, but generally, the equilibrium favors the carbonyl compound.
c. Reaction is slow in pure water, but is catalyzed by both aqueous acid and base.
i. The base-catalyzed reaction is an addition of –OH, followed by protonation of the tetrahedral intermediate by water.
ii. In the acid-catalyzed reaction, the carbonyl oxygen is protonated, and neutral water adds to the carbonyl carbon.
d. The catalysts have different effects.
i. Base catalysis makes water a better nucleophile.
ii. Acid catalysis makes the carbonyl carbon a better electrophile.
e. Reactions of carbonyl groups with H-X, where X is electronegative, are reversible; the equilibrium favors the aldehyde or ketone.
2. Addition of Grignard and hydride reagents (Section 14.6).
a. Addition of Grignard reagents.
i. Mg^{2+} complexes with oxygen, making the carbonyl group more electrophilic.
ii. $R:^-$ adds to the carbonyl carbon to form a tetrahedral intermediate.
iii. Water is added in a separate step to protonate the intermediate, yielding an alcohol.
iv. Grignard reactions are irreversible because $R:^-$ is not a leaving group.
b. Hydride addition.
i. $LiAlH_4$ and $NaBH_4$ act as if they are $H:^-$ donors and add to carbonyl compounds to form tetrahedral alkoxide intermediates.
ii. In a separate step, water is added to protonate the intermediate, yielding an alcohol.
3. Addition of amines (Section 14.7).
a. Amines add to aldehydes and ketones to form imines and enamines.
b. Imines are formed when a primary amine adds to an aldehyde or ketone.
i. The process is acid-catalyzed.
ii. A proton transfer converts the initial adduct to a carbinolamine.
iii. Acid-catalyzed elimination of water yields an imine.
iv. The reaction rate maximum occurs at pH = 4.5. At this pH, [H] is high enough to catalyze elimination of water, but low enough so that the amine is nucleophilic.
v. Some imine derivatives are useful for characterizing aldehydes and ketones.
c. Enamines are produced when aldehydes and ketones react with secondary amines.
The mechanism is similar to that of imine formation, except a proton from the α carbon is lost in the dehydration step.
4. Addition of alcohols: acetal formation (Section 14.8).
a. In the presence of an acid catalyst, two equivalents of an alcohol can add to an aldehyde or ketone to produce an acetal.
i. The initial intermediate is a hemiacetal (a hydroxy ether).
ii. Protonation of –OH, loss of water, with formation of an oxonium ion, and addition of a second molecule of ROH yields the acetal.

b. The reaction is reversible, so changing the reaction conditions can drive the reaction in either direction.
c. Diols are often used, forming cyclic acetals.

5. The Wittig Reaction (Section 14.9).
   a. The Wittig reaction converts an aldehyde or ketone to an alkene.
   b. Steps in the Wittig reaction:
      i. An alkyl halide reacts with triphenylphosphine to form an alkyltriphenylphosphonium salt.
      ii. Butyllithium converts the salt to an ylide.
      iii. The ylide adds to an aldehyde or ketone to from a dipolar intermediate.
      iv. The intermediate forms a four-membered ring, which decomposes to form the alkene and triphenylphosphine oxide.
   c. Uses of the Wittig reaction.
      i. The Wittig reaction can be used to produce mono-, di-, and trisubstituted alkenes, but steric hindrance keeps tetrasubstituted alkenes from forming.
      ii. The Wittig reaction produces pure alkenes of known stereochemistry (excluding *E,Z* isomers).
6. Biological reductions (Section 14.10).
   a. The Cannizzaro reaction is unique in that the tetrahedral intermediate of addition of a nucleophile to an aldehyde can expel a leaving group.
   b. Steps in the Cannizzaro reaction.
      i. HO^- adds to an aldehyde with no α hydrogens to form a tetrahedral intermediate.
      ii. H^- is expelled and adds to another molecule of aldehyde.
      iii. The result is a disproportionation reaction, in which one molecule of aldehyde is oxidized and a second molecule is reduced.
   c. The Cannizzaro reaction isn't synthetically useful, but it resembles the mode of action of the enzyme cofactor NADH.
7. Conjugate addition to α,β-unsaturated aldehydes and ketones (Section 14.11).
   a. Steps in conjugate addition.
      i. Because the double bond of an α,β-unsaturated aldehyde/ketone is conjugated with the carbonyl group, addition can occur at the β position, which is an electrophilic site.
      ii. Protonation of the α carbon of the enolate intermediate results in a product having a carbonyl group and a nucleophile with a 1,3 relationship.
   b. Conjugate addition of amines.
      i. Primary and secondary amines add to α,β-unsaturated aldehydes and ketones.
      ii. The conjugate addition product is often formed exclusively.
   c. Conjugate addition of water
      i. Water can add to yield β-hydroxy aldehydes and ketones.
      ii. Conjugate addition of water also occurs in living systems.

III. Spectroscopy of aldehydes and ketones (Section 14.12).
   A. IR spectroscopy.
      1. The C=O absorption of aldehydes and ketones occurs in the range 1660–1770 cm^{-1}.
         a. The exact position of absorption can be used to distinguish between an aldehyde and a ketone.
         b. The position of absorption also gives information about other structural features, such as unsaturation and angle strain.
         c. The absorption values are constant from one compound to another.
      2. Aldehydes also show absorptions in the range 2720–2820 cm^{-1}.

B. NMR spectroscopy.
  1. ^1H NMR spectroscopy.
    a. Aldehyde protons absorb near 10 δ, and show spin-spin coupling with protons on the adjacent carbon.
    b. Hydrogens on the carbon next to a carbonyl group absorb near 2.0–2.3 δ. Methyl ketone protons absorb at 2.1 δ.
  2. ^{13}C NMR spectroscopy.
    a. The carbonyl-group carbons absorb in the range 190–215 δ.
    b. These absorptions characterize aldehydes and ketones.
    c. Unsaturation lowers the value of δ.

C. Mass spectrometry.
  1. Some aliphatic aldehydes and ketones undergo McLafferty rearrangement.
    a. A hydrogen on the γ carbon is transferred to the carbonyl oxygen, the bond between the α carbon and the β carbon is broken, and a neutral alkene fragment is produced.
    b. The remaining cation radical is detected.
  2. Alpha cleavage.
    a. The bond between the carbonyl group and the α carbon is cleaved.
    b. The products are a neutral radical and an acyl cation, which is detected.

Solutions to Problems

14.1 Remember that the principal chain must contain the aldehyde or ketone group, the aldehyde carbon is carbon 1 in an acyclic compound, and the suffix *-carbaldehyde* is used when the aldehyde group is attached to a ring.

(a) $CH_3CH_2C(=O)CH(CH_3)_2$

2-Methylpentan-3-one

(b) $C_6H_5CH_2CH_2CHO$

3-Phenylpropanal

(c) $CH_3C(=O)CH_2CH_2CH_2C(=O)CH_2CH_3$

Octane-2,6-dione

(d) CH_3, H, H, CHO

trans-2-Methylcyclohexane-carbaldehyde

(e) $CH_3CH{=}CHCH_2CH_2CHO$

Hex-4-enal

(f) O, H_3C, H, H, CH_3

cis-2,5-Dimethyl-cyclohexanone

14.2

(a) $CH_3CH(CH_3)CH_2CHO$

3-Methylbutanal

(b) $CH_3CH(Cl)CH_2C(=O)CH_3$

4-Chloropentan-2-one

(c) $C_6H_5CH_2CHO$

Phenylacetaldehyde

(d) *cis*-3-*tert*-Butylcyclohexanecarbaldehyde

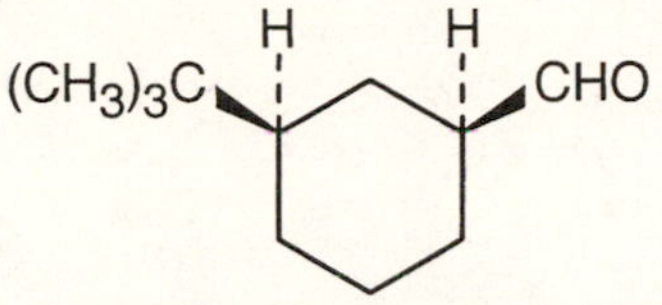

(e) $H_2C{=}C(CH_3)CH_2CHO$

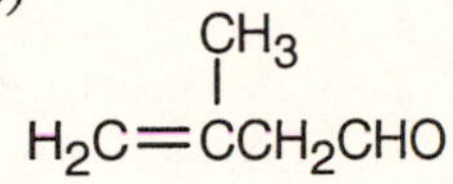

3-Methylbut-3-enal

(f) $CH_3CH_2CH(CH_3)CH_2CH_2CH(CH_3CHCl)CHO$

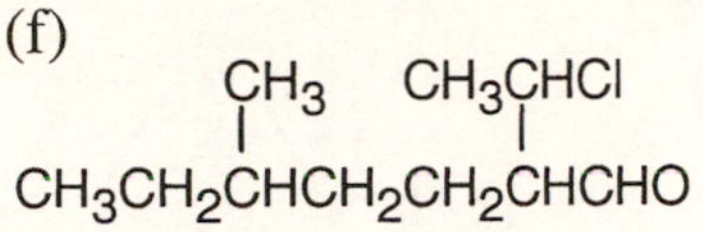

2-(1-Chloroethyl)-5-methylheptanal

14.3 All of these methods are familiar.

(a) Benzene $\xrightarrow[AlCl_3]{CH_3COCl}$ acetophenone ($C_6H_5COCH_3$) $\xrightarrow[FeBr_3]{Br_2}$ m-bromoacetophenone

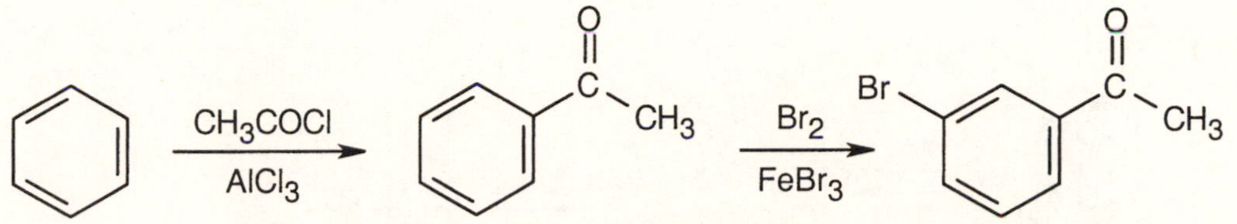

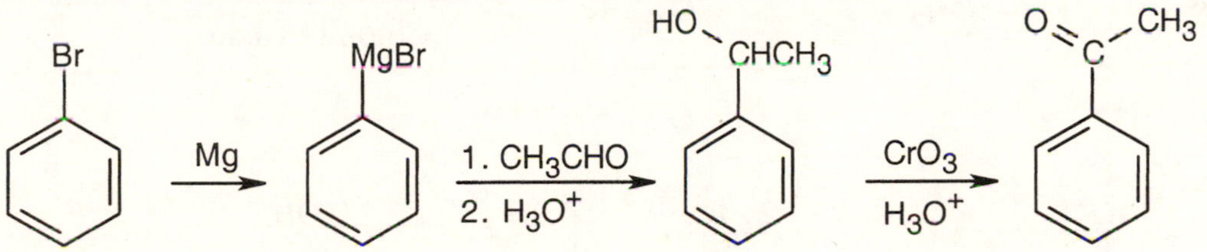

(b) Bromobenzene $\xrightarrow{Mg}$ C_6H_5MgBr $\xrightarrow[2.\ H_3O^+]{1.\ CH_3CHO}$ $C_6H_5CH(OH)CH_3$ $\xrightarrow[H_3O^+]{CrO_3}$ $C_6H_5COCH_3$

(c) 1-Methylcyclohexene $\xrightarrow[2.\ H_2O_2,\ ^-OH]{1.\ BH_3,\ THF}$ trans-2-methylcyclohexanol $\xrightarrow[H_3O^+]{CrO_3}$ 2-methylcyclohexanone

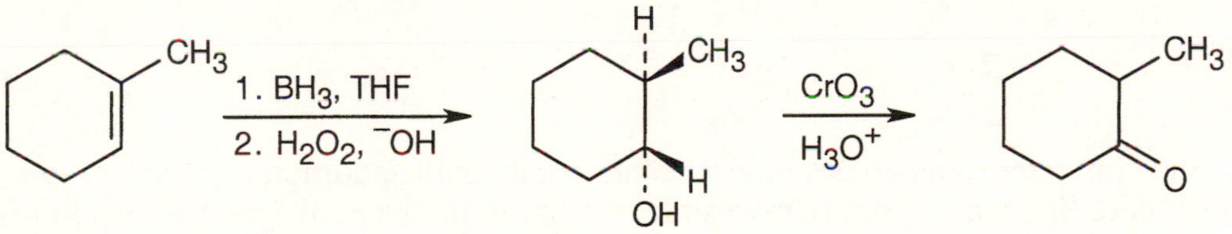

14.4

Cyclohexanone (δ– on O, δ+ on C) + $^-$:CN $\rightleftarrows$ [alkoxide tetrahedral intermediate, $:\ddot{O}:^-$, CN] $\underset{}{\overset{H_2O}{\rightleftarrows}}$ HO, CN (cyanohydrin) + HO^-

nucleophilic addition

protonation of tetrahedral intermediate

cyanohydrin

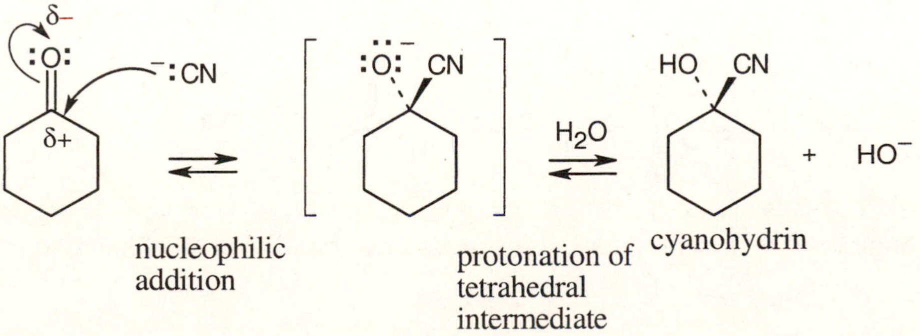

Cyanide anion adds to the positively polarized carbonyl carbon to form a tetrahedral intermediate. This intermediate is protonated to yield the cyanohydrin.

14.5

O_2N–C₆H₄–CHO CH_3O–C₆H₄–CHO

electron-withdrawing   electron-donating

The electron-withdrawing nitro group makes the aldehyde carbon of *p*-nitrobenzaldehyde more electron-poor (more electrophilic) and more reactive toward nucleophiles than the aldehyde carbon of *p*–methoxybenzaldehyde.

14.6

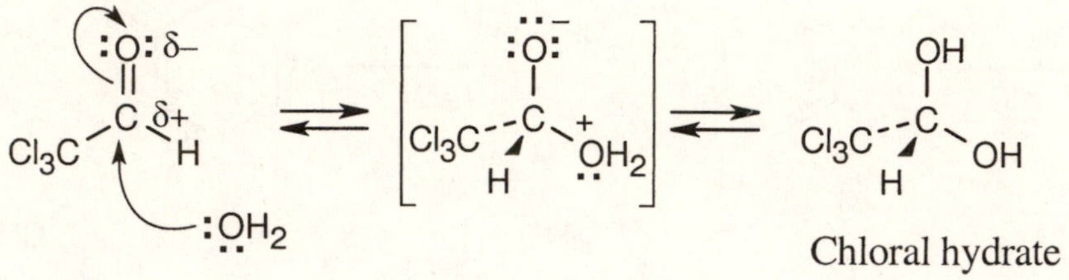

Cl_3C, δ–, δ+, H, $:OH_2$, Cl_3C, H, OH_2, OH, Cl_3C, H, OH

Chloral hydrate

14.7

R, R, $:OH_2$, R, R, OH_2, OH, R, R, OH

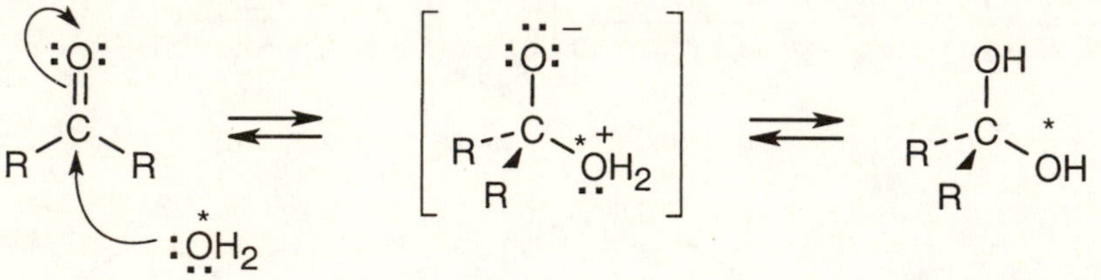

The above mechanism is similar to other nucleophilic addition mechanisms we have studied. Since all steps are reversible, we can write the above mechanism in reverse to show how labeled oxygen is incorporated into an aldehyde or ketone.

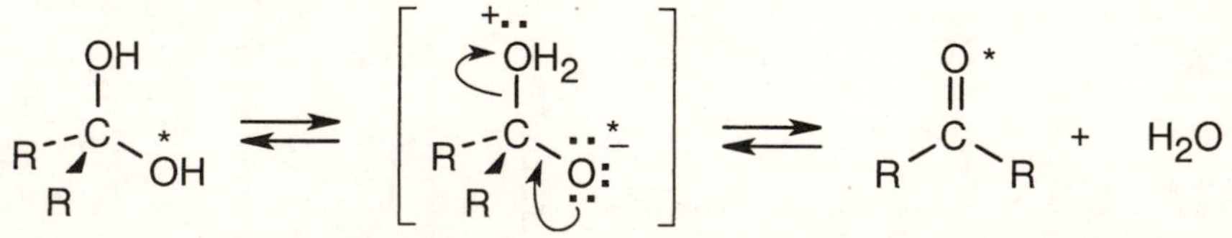

OH, R, R, OH, OH_2, R, R, O, R, R, + H_2O

This exchange is very slow in water but proceeds more rapidly when either acid or base is present.

14.8 **Strategy:** Reaction of a ketone or aldehyde with a primary amine yields an imine, in which C=O has been replaced by C=NR. Reaction of a ketone or aldehyde with a secondary amine yields an enamine, in which C=O has been replaced by C–NR_2, and the double bond has moved.

Solution:

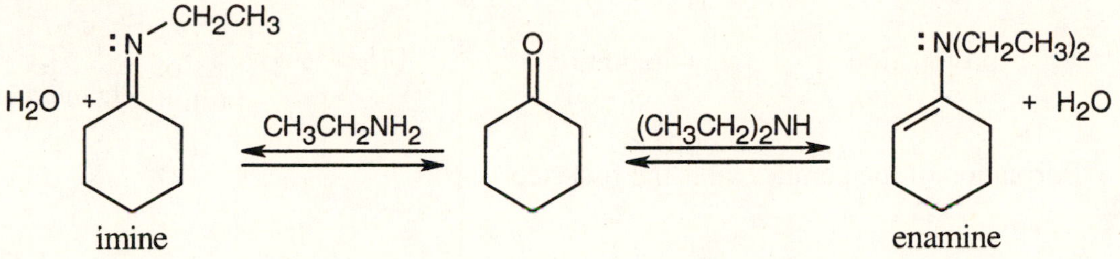

H_2O + imine ⇌ ($CH_3CH_2NH_2$) cyclohexanone ⇌ ($(CH_3CH_2)_2NH$) enamine + H_2O

imine (N–CH_2CH_3)   enamine ($N(CH_2CH_3)_2$)

14.9

protonation of nitrogen ⇌ addition of water ⇌ loss of proton ⇌ proton transfer ⇌ loss of amine

H_2O^+–H, :NR, HNR+, :OH_2, HNR, H, OH, HNR + H_3O^+, H_2NR^+, O:⁻, RNH_2 + :O:

carbinolamine

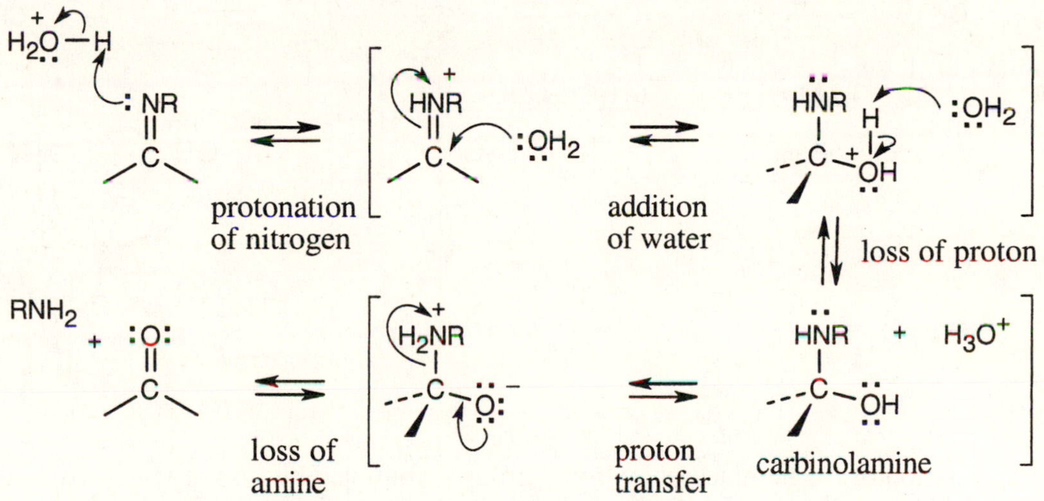

14.10 The structure is an enamine, which is prepared from a ketone and a secondary amine. Find the components and draw the reaction.

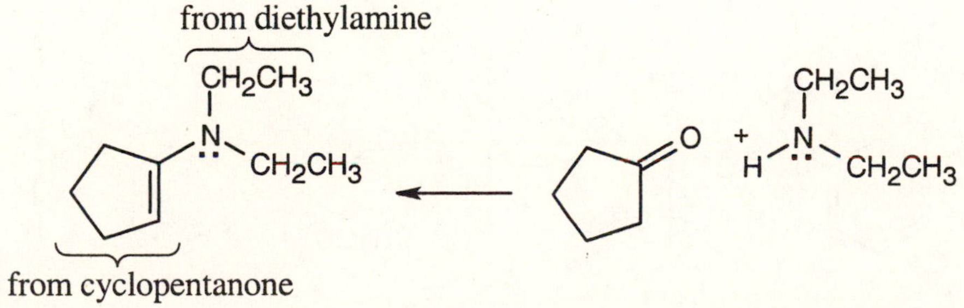

from diethylamine

CH_2CH_3, N, CH_2CH_3 ⟵ cyclopentanone + H–N(CH_2CH_3)–CH_2CH_3

from cyclopentanone

14.11

(a)

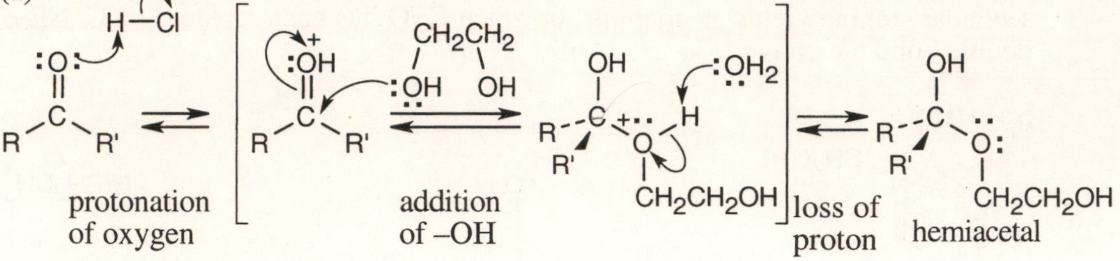

:O: H—Cl → [:OH⁺ ... CH2CH2 :OH OH ...] → hemiacetal

protonation of oxygen — addition of –OH — loss of proton — hemiacetal

Formation of the hemiacetal is the first step.

(b)

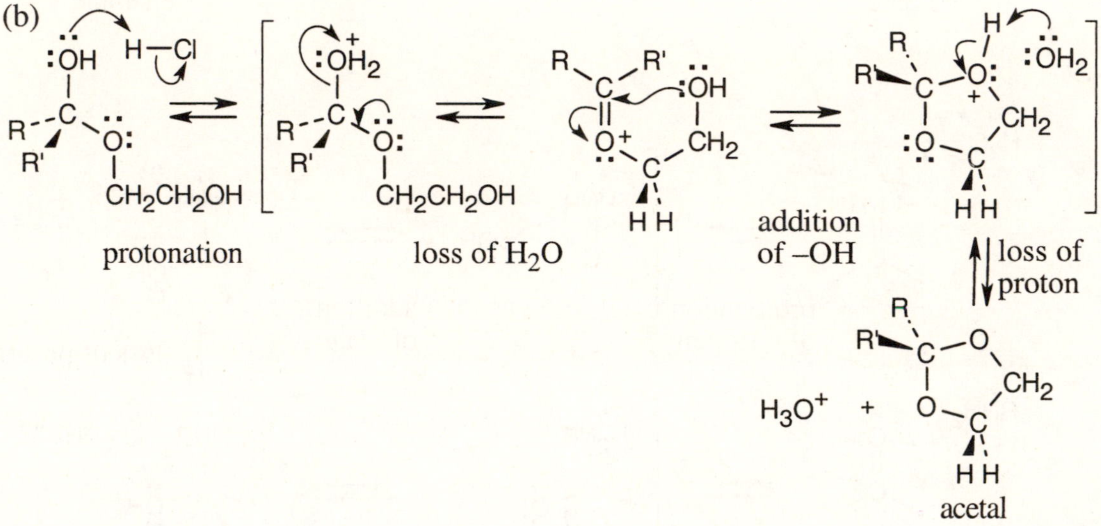

protonation — loss of H_2O — addition of –OH — loss of proton

H_3O^+ + acetal

Protonation of the hemiacetal hydroxyl group is followed by loss of water. Attack by the second hydroxyl group of ethylene glycol forms the cyclic acetal ring.

14.12 **Strategy:** Locate the two identical –OR groups to identify the alcohol that was used to form the acetal. (The illustrated acetal was formed from methanol.) Replace these two –OR groups by =O to find the carbonyl compound.

Solution:

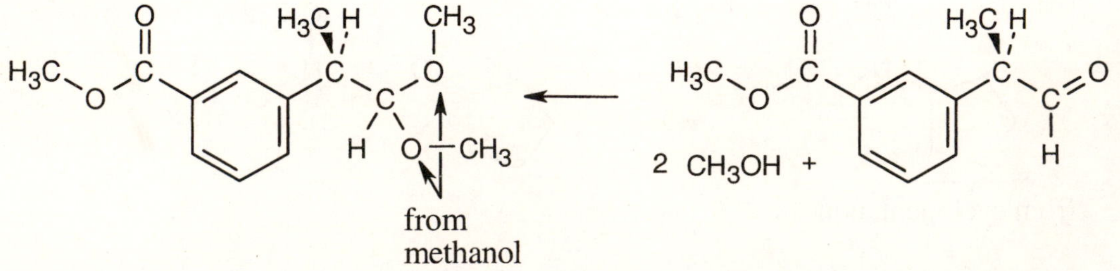

from methanol

2 CH_3OH +

14.13 Locate the double bond that is formed by the Wittig reaction. The simpler or less substituted component comes from the ylide, and the more substituted component comes from the aldehyde or ketone. Triphenylphosphine oxide is a byproduct of all these reactions.

(a)

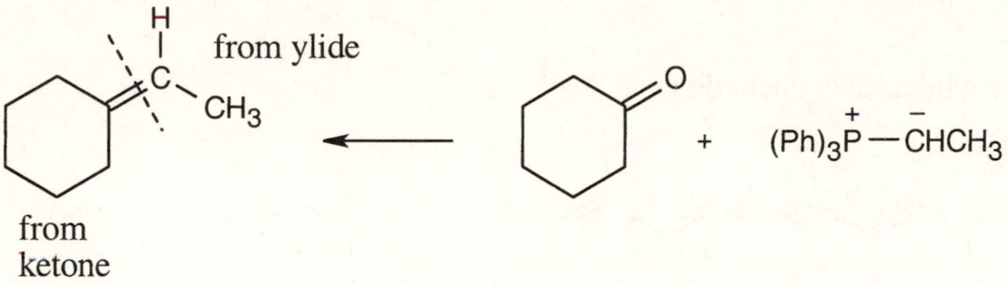

H, C, CH_3 from ylide; from ketone ⟵ cyclohexanone + $(Ph)_3\overset{+}{P}-\overset{-}{C}HCH_3$

(b)

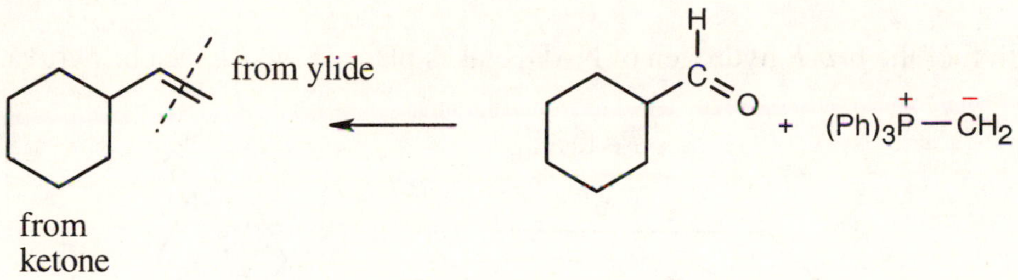

from ylide; from ketone ⟵ cyclohexanecarbaldehyde (H–C=O) + $(Ph)_3\overset{+}{P}-\overset{-}{C}H_2$

(c)

$CH_3CH_2CH_2CH{=}C(CH_3)CH_3$ (from ylide | from ketone) ⟵ $(Ph)_3\overset{+}{P}-\overset{-}{C}HCH_2CH_2CH_3$ + $CH_3\overset{O}{\overset{\|}{C}}CH_3$

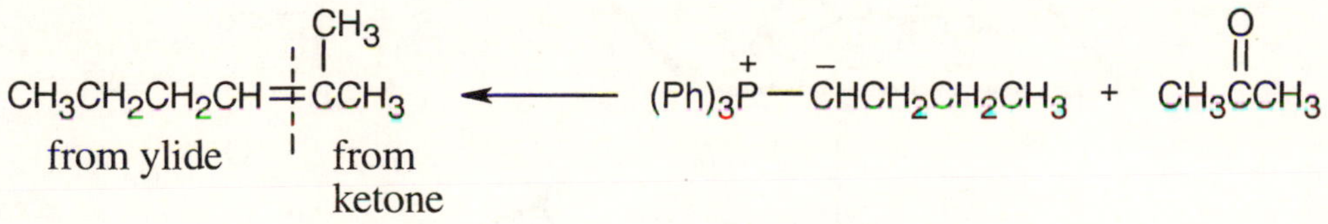

(d)

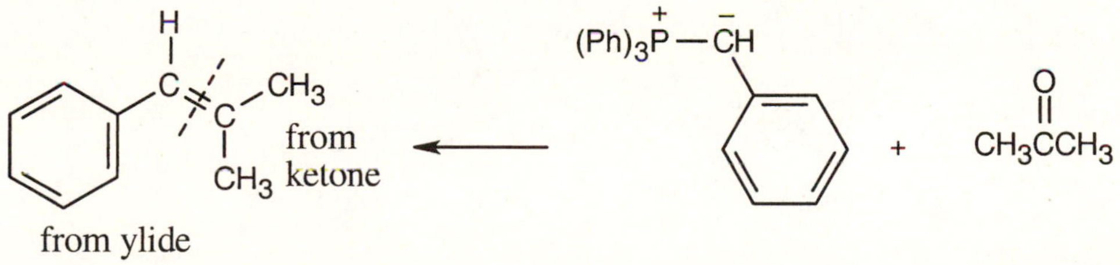

H, C, C, CH_3, CH_3; from ketone; from ylide ⟵ $(Ph)_3\overset{+}{P}-\overset{-}{C}H$–Ph + $CH_3\overset{O}{\overset{\|}{C}}CH_3$

(e)

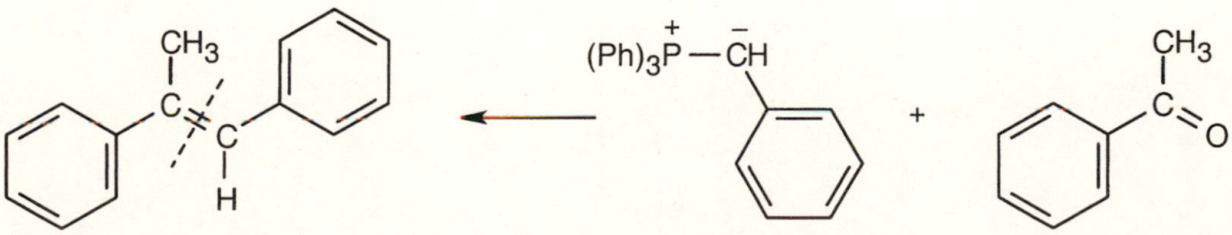

CH_3, C, C, H ⟵ $(Ph)_3\overset{+}{P}-\overset{-}{C}H$–Ph + Ph–C($CH_3$)=O

(f)

CH_2 from ylide; from ketone ⟵ cyclohex-2-enone + $(Ph)_3\overset{+}{P}-\overset{-}{C}H_2$

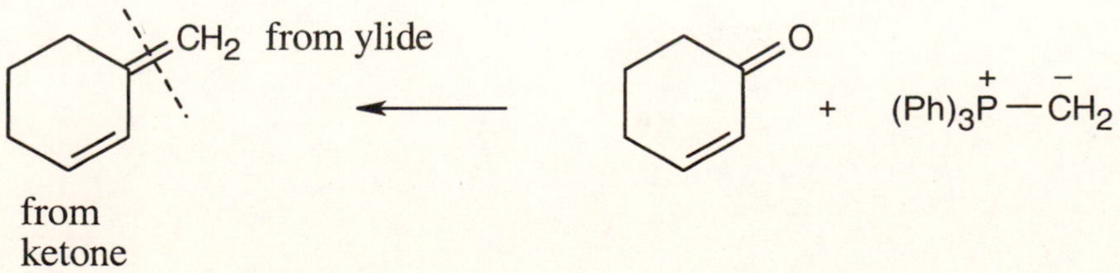

14.14

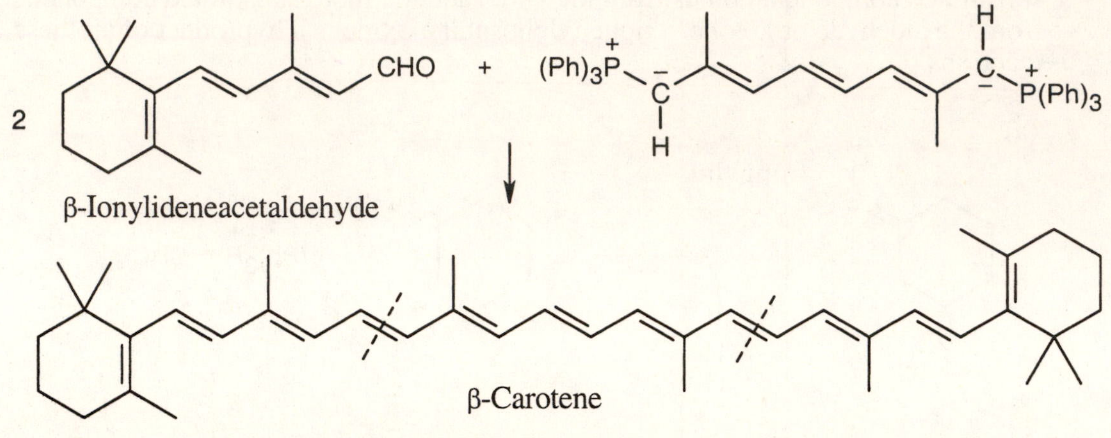

2 β-Ionylideneacetaldehyde CHO + $(Ph)_3\overset{+}{P}$–$\overset{-}{C}$H ... $\overset{-}{C}$H–$\overset{+}{P}(Ph)_3$

↓

β-Carotene

14.15 Addition of the *pro-R* hydrogen of NADH takes place at the *Re* face of pyruvate.

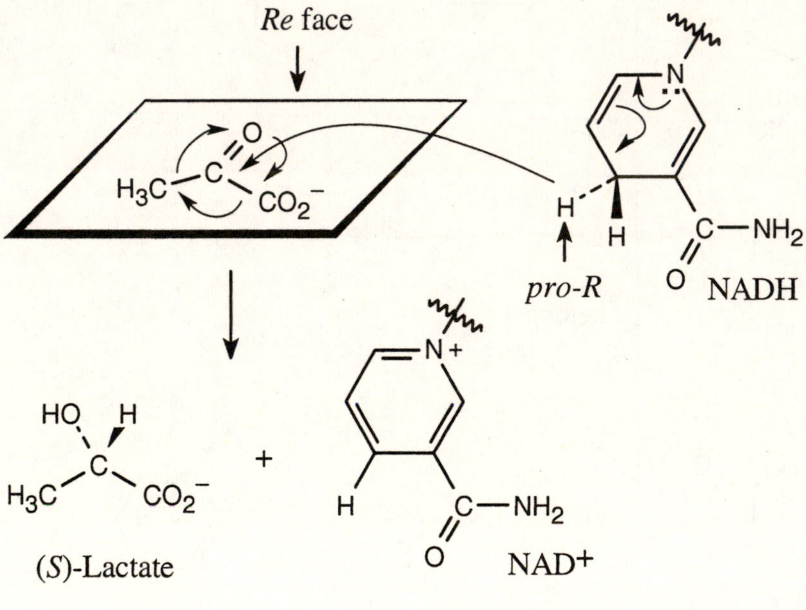

Re face

H_3C–C(=O)–CO_2^-

H H *pro-R* C–NH_2 O NADH

↓

HO H H_3C–C–CO_2^- + H C–NH_2 O NAD^+

(*S*)-Lactate

14.16 The –OH group adds to the *Re* face at carbon 2, and $-H^+$ adds to the *Re* face at carbon 3, to yield (2*R*,3*S*)-isocitrate.

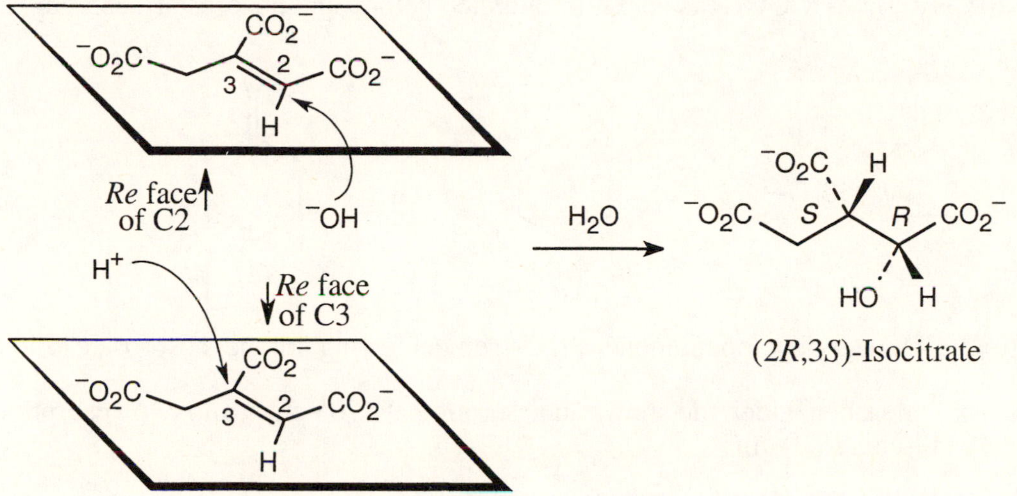

Re face of C2

^-OH

H_2O

H^+

Re face of C3

(2*R*,3*S*)-Isocitrate

14.17

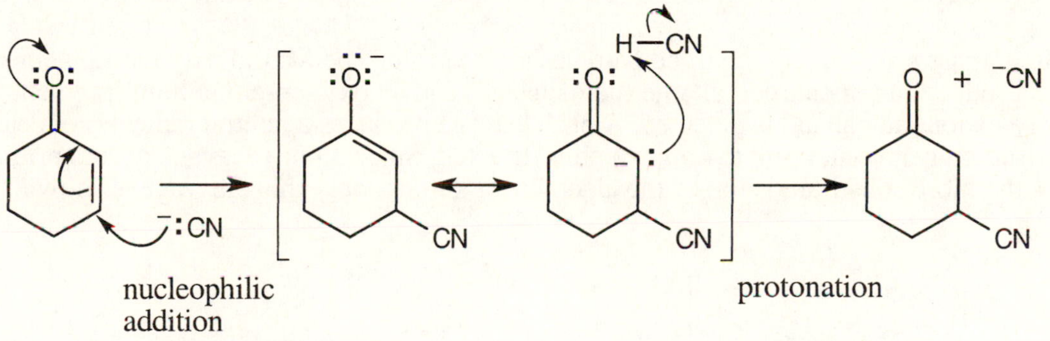

nucleophilic addition

protonation

14.18

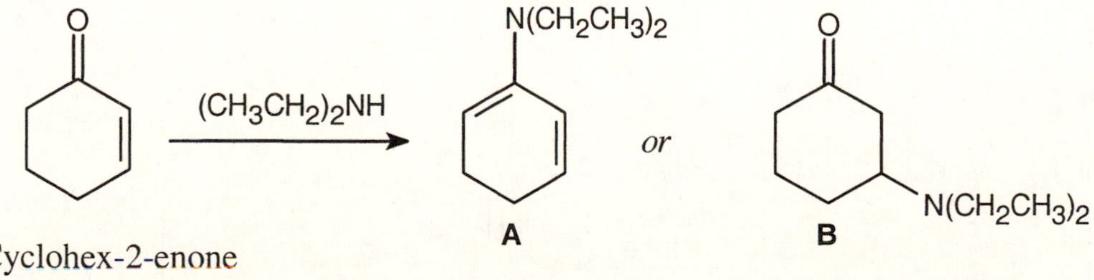

$(CH_3CH_2)_2NH$

$N(CH_2CH_3)_2$

or

Cyclohex-2-enone

A

B

Cyclohex-2-enone is a cyclic α,ß-unsaturated ketone whose carbonyl IR absorption occurs at 1685 cm^{-1}. If direct addition product **A** is formed, the carbonyl absorption will vanish. If conjugate addition produces **B**, the carbonyl absorption will shift to 1715 cm^{-1}, where 6-membered-ring saturated ketones absorb.

14.19 (a) $H_2C{=}CHCH_2COCH_3$ absorbs at 1715 cm^{-1}. (Pent-4-en-2-one is not an α,ß-unsaturated ketone.)

(b) $CH_3CH{=}CHCOCH_3$ is an α,ß-unsaturated ketone and absorbs at 1685 cm^{-1}.

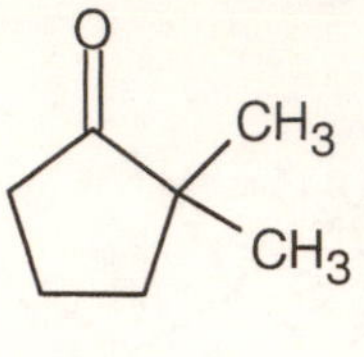

(c) 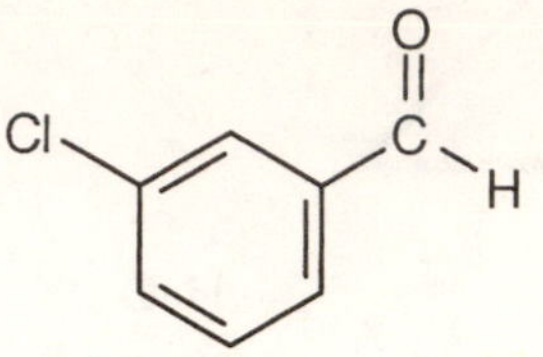 (d) 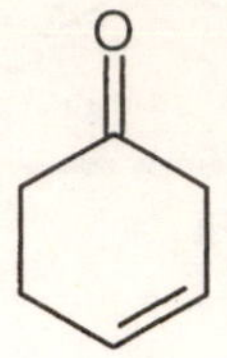 (e)

(c) 2,2-Dimethylcyclopentanone, a five-membered-ring ketone, absorbs at 1750 cm^{-1}.

(d) *m*-Chlorobenzaldehyde shows an absorption at 1705 cm^{-1} and two absorptions at 2720 cm^{-1} and 2820 cm^{-1}.

(e) Cyclohex-3-enone absorbs at 1715 cm^{-1}.

(f) $CH_3CH_2CH_2CH{=}CHCHO$ is an α,ß-unsaturated aldehyde and absorbs at 1705 cm^{-1}.

14.20 In mass spectra, only charged particles are detected. The McLafferty rearrangement produces an uncharged alkene (not detected) and an oxygen-containing fragment, which is a cation radical and is detected. Alpha cleavage produces a neutral radical (not detected) and an oxygen-containing cation, which is detected. Since alpha cleavage occurs primarily on the more substituted side of the aldehyde or ketone, only this cleavage is shown.

(a)

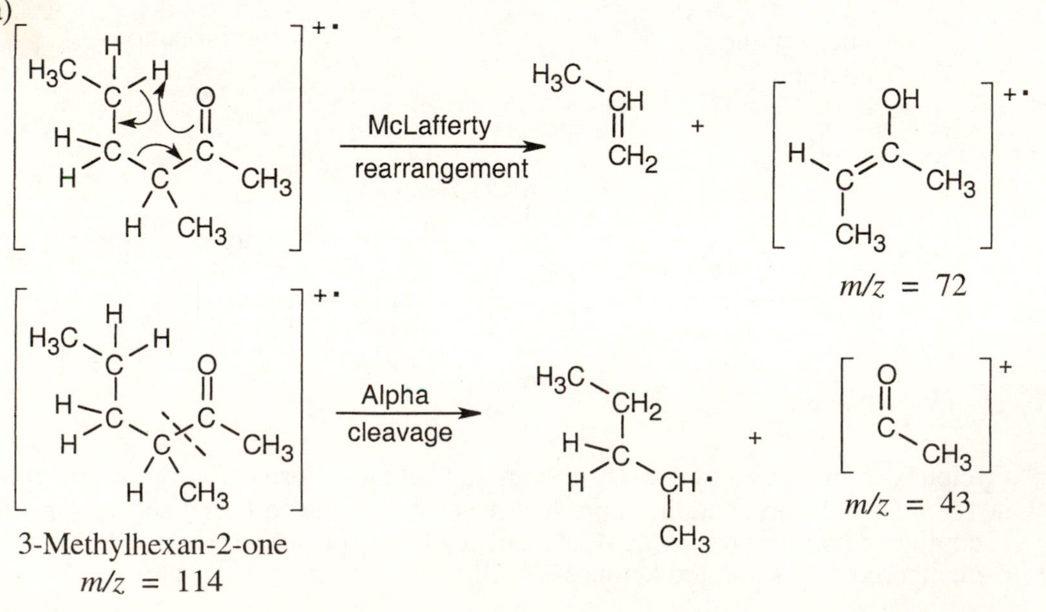

McLafferty rearrangement

m/z = 72

Alpha cleavage

m/z = 43

3-Methylhexan-2-one
m/z = 114

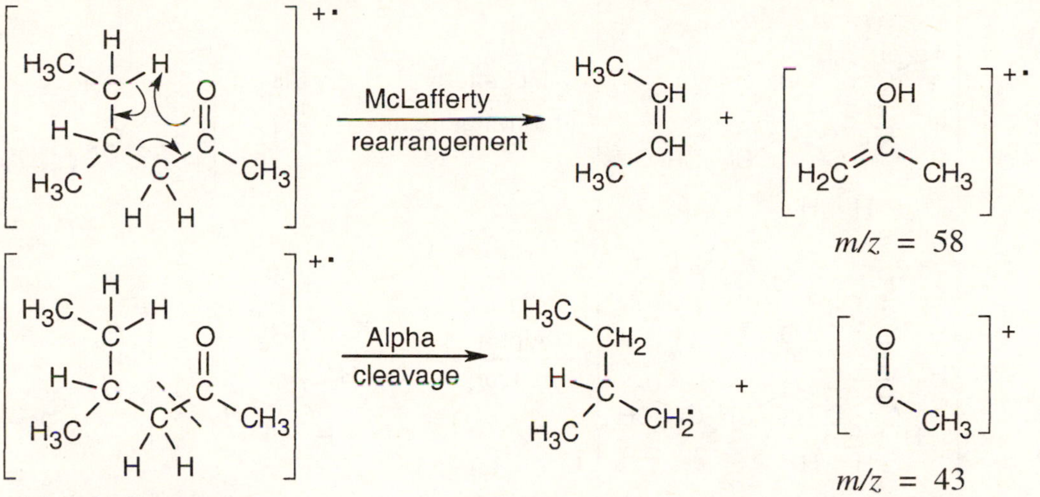

McLafferty rearrangement

$m/z = 58$

Alpha cleavage

$m/z = 43$

4-Methylhexan-2-one
$m/z = 114$

Both isomers exhibit peaks at $m/z = 43$ due to α-cleavage. The products of McLafferty rearrangement, however, occur at different values of m/z and can be used to identify each isomer.

(b)

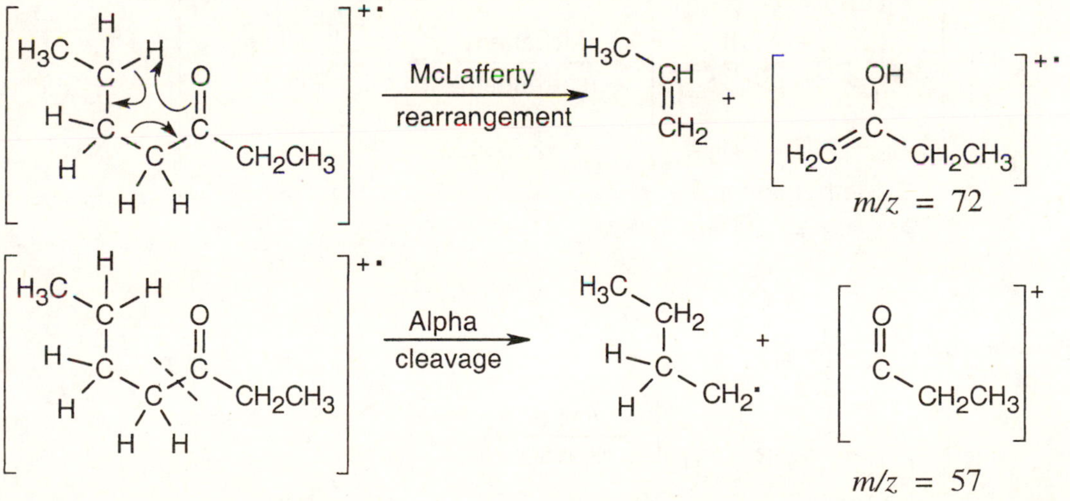

McLafferty rearrangement

$m/z = 72$

Alpha cleavage

$m/z = 57$

Heptan-3-one $m/z = 114$

McLafferty rearrangement

$CH_2{=}CH_2$ + $[H_2C{=}C(OH)CH_2CH_2CH_3]^{+\cdot}$

m/z = 86

Alpha cleavage

$CH_3CH_2CH_2^{\cdot}$ + $[O{=}C{-}CH_2CH_2CH_3]^{+}$

m/z = 71

Heptan-4-one m/z = 114

The isomers can be distinguished on the basis of both α-cleavage products (m/z = 57 *vs* m/z = 71) and McLafferty rearrangement products (m/z = 72 *vs* m/z = 86).

(c)

McLafferty rearrangement

$H_3C{-}CH{=}CH_2$ + $[H{-}C(CH_3){=}C(OH){-}H]^{+\cdot}$

m/z = 58

2-Methylpentanal
m/z = 100

Alpha cleavage

$H_3C{-}CH_2{-}CH_2{-}\dot{C}H{-}CH_3$ + $[O{=}C{-}H]^{+}$

m/z = 29

2-Methylpentanal
m/z = 100

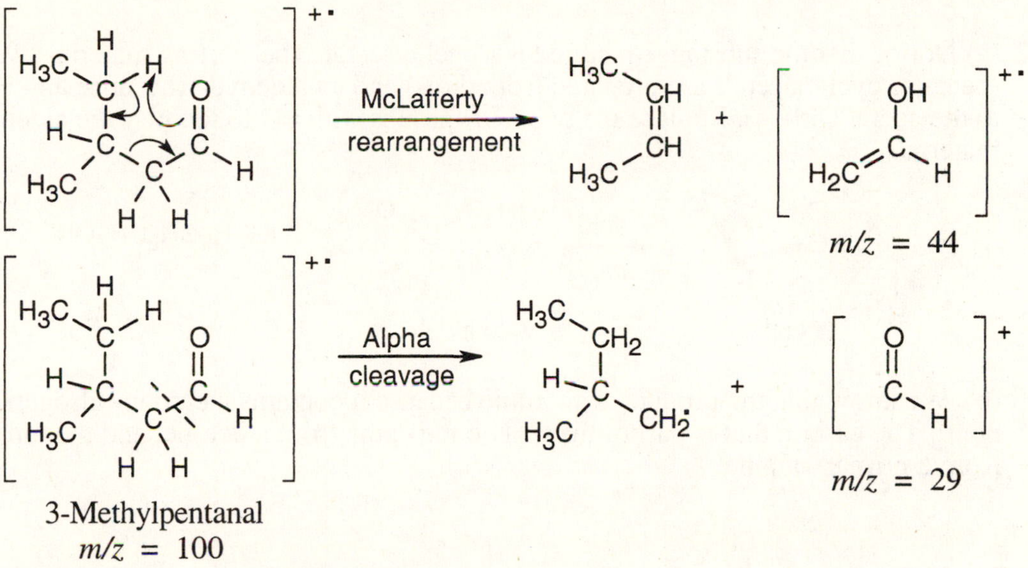

McLafferty rearrangement

m/z = 44

Alpha cleavage

m/z = 29

3-Methylpentanal
m/z = 100

The fragments from McLafferty rearrangement, which occur at different values of m/z, serve to distinguish the two isomers.

14.21 **IR:** The only important IR absorption for the compound is seen at 1750 cm^{-1}, where 5-membered ring ketones absorb.

Mass spectrum: The products of alpha cleavage, which occurs in the ring, have the same mass as the molecular ion.

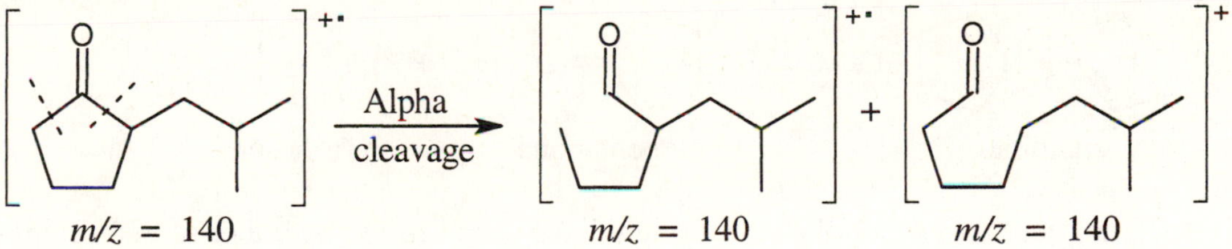

Alpha cleavage

m/z = 140 m/z = 140 m/z = 140

The charged fragment resulting from the McLafferty rearrangement appears at m/z = 84.

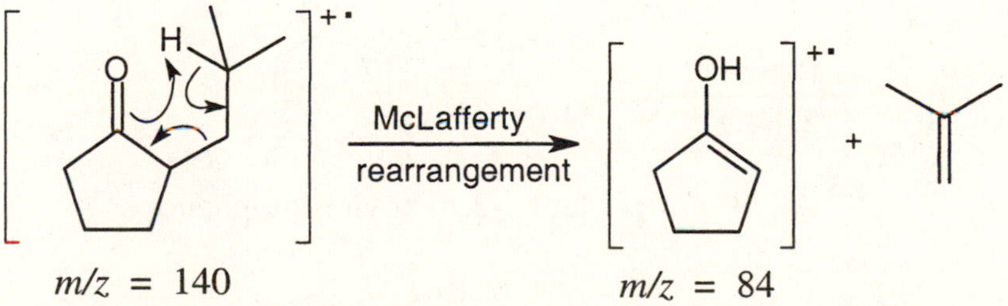

McLafferty rearrangement

m/z = 140 m/z = 84

Visualizing Chemistry

14.22 (a) Notice that the substance pictured is a cyclic acetal. The starting materials were a diol, (because cyclic acetals are prepared from diols) and an aldehyde (because an –H is bonded to the acetal carbon). Replace the two –OR groups with =O to identify the aldehyde starting material.

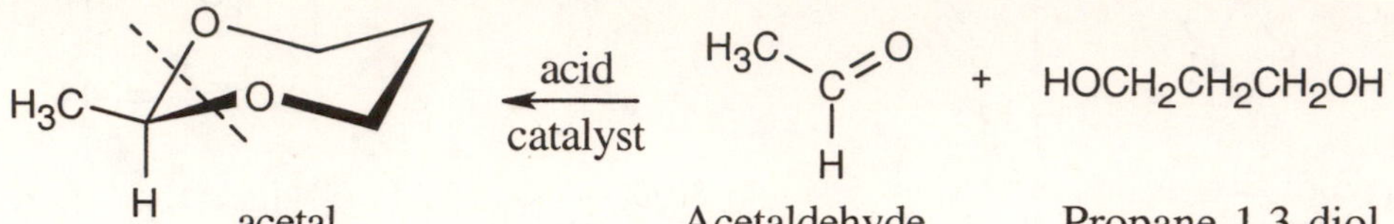

acetal ← (acid catalyst) — Acetaldehyde (H_3C–CH=O) + Propane-1,3-diol ($HOCH_2CH_2CH_2OH$)

(b) We know that the product is an imine because it contains a carbon-nitrogen double bond. The carbon that is part of the C=N bond came from a ketone, and the nitrogen came from a primary amine.

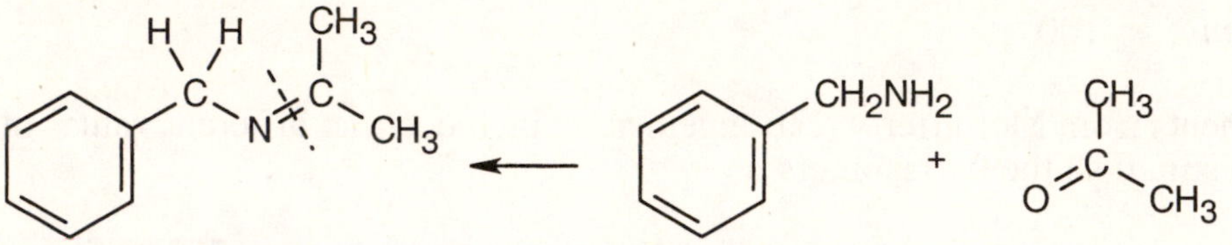

imine ← Benzylamine (CH_2NH_2) + Acetone

(c) The product is an enamine, formed from a ketone and a secondary amine. Nitrogen is bonded to the carbon that once bore the carbonyl oxygen.

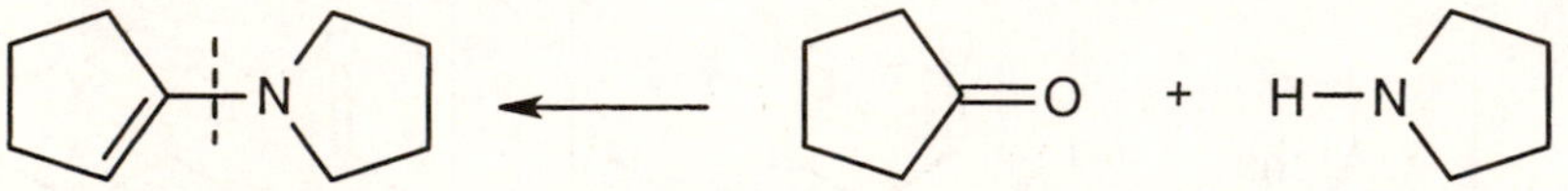

enamine ← Cyclopentanone + Pyrrolidine (H–N)

(d) The secondary alcohol product might have been formed by either of two routes – by reduction of a ketone or by Grignard addition to an aldehyde.

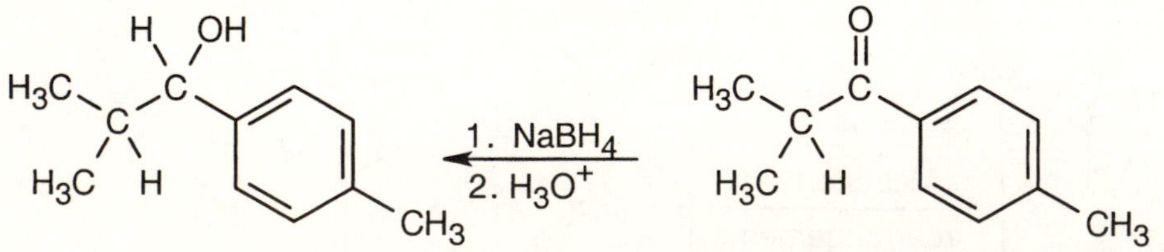

Product ← 1. $NaBH_4$ 2. H_3O^+ — *p*-Methyl-2-methylpropiophenone

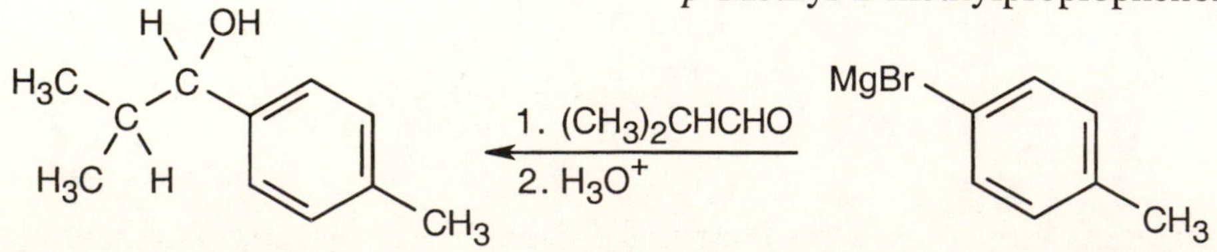

or

Product ← 1. $(CH_3)_2CHCHO$ 2. H_3O^+ — p-methylphenyl-MgBr

14.23 The intermediate results from the addition of an amine to a ketone. The product is an enamine because the amine nitrogen in the carbinolamine intermediate comes from a secondary amine.

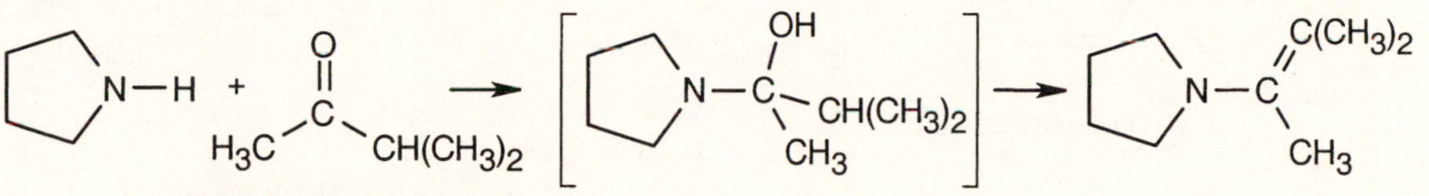

3-Methylbutan-2-one

14.24 (a) The nitrogen atom is sp^2-hybridized, and nitrogen and the carbons bonded to it lie in a plane.
(b) A p orbital holds the lone-pair electrons of nitrogen.
(c) The p orbital holding the lone-pair electrons of nitrogen is aligned for overlap with the π electrons of the enamine double bond. With this geometry, the nitrogen lone-pair electrons can be conjugated with the double bond.

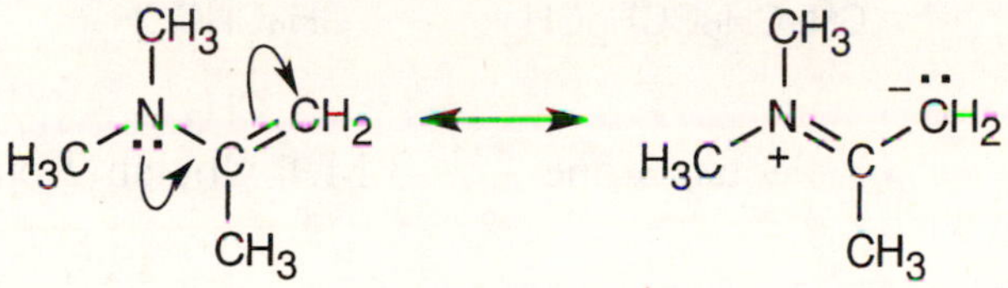

Additional Problems

14.25

(a) CH_3CCH_2Br (C=O) — Bromoacetone

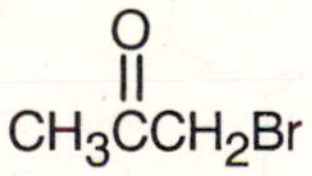

(b) (*S*)-2-Hydroxypropanal

(c) $CH_3CH_2CH_2CH_2CCH(CH_3)_2$ (C=O) — 2-Methylheptan-3-one

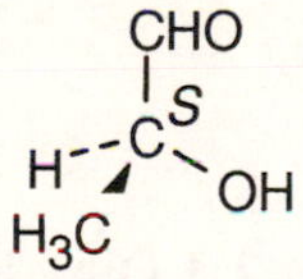

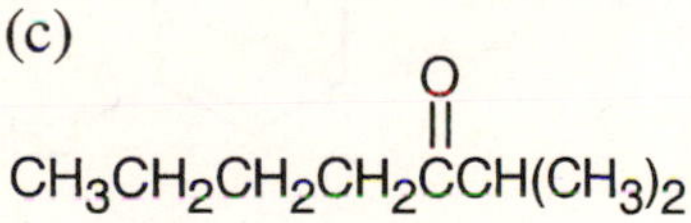

(d) (2*S*,3*R*)-2,3,4-Trihydroxybutanal

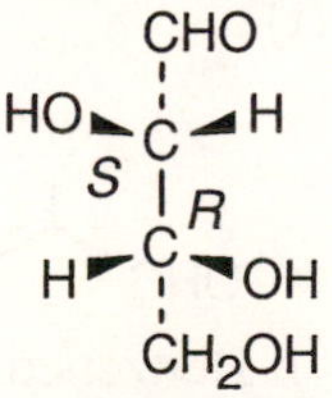

(e) $(CH_3)_3CCC(CH_3)_3$ (C=O) — 2,2,4,4-Tetramethylpentan-3-one

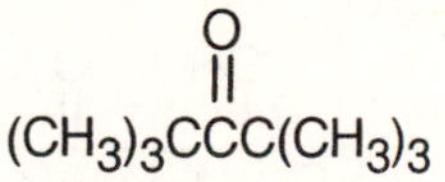

(f) $CH_3C{=}CHCCH_3$ (CH_3, C=O) — 4-Methylpent-3-en-2-one

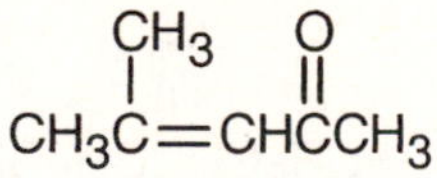

(g) OHCCH$_2$CH$_2$CHO — Butanedial

(h) C_6H_5–CH=CHCHO — 3-Phenylprop-2-enal

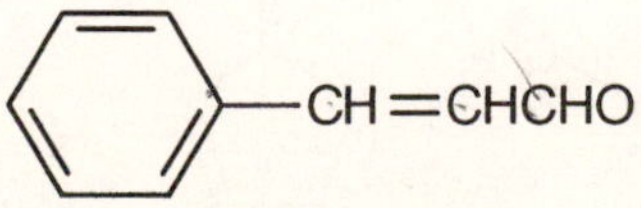

(i) 6,6-Dimethylcyclohexa-2,4-dienone

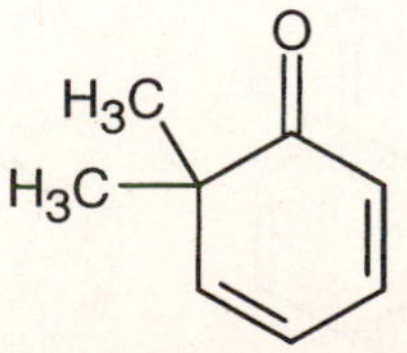

(j)

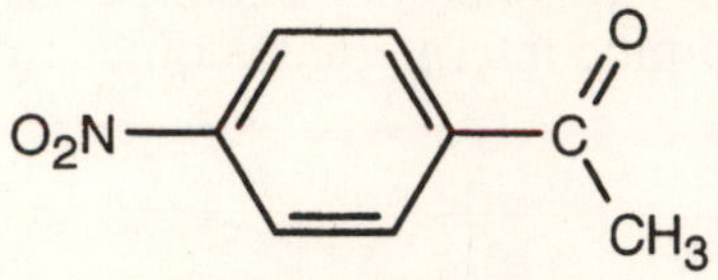

O_2N–C_6H_4–$C(=O)CH_3$

p-Nitroacetophenone

14.26 Only 2-methylbutanal is chiral.

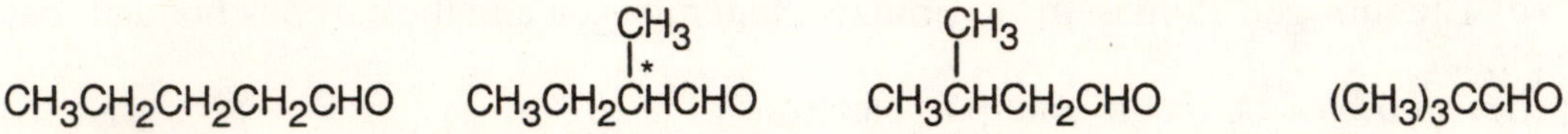

$CH_3CH_2CH_2CH_2CHO$	$CH_3CH_2\overset{*}{C}H(CH_3)CHO$	$CH_3CH(CH_3)CH_2CHO$	$(CH_3)_3CCHO$
Pentanal	2-Methylbutanal	3-Methylbutanal	2,2-Dimethylpropanal

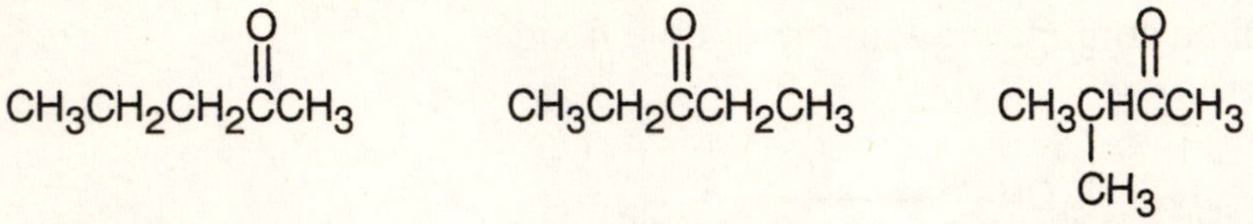

$CH_3CH_2CH_2C(=O)CH_3$	$CH_3CH_2C(=O)CH_2CH_3$	$CH_3CH(CH_3)C(=O)CH_3$
Pentan-2-one	Pentan-3-one	3-Methylbutan-2-one

14.27

(a) 3-Methylcyclohex-3-enone

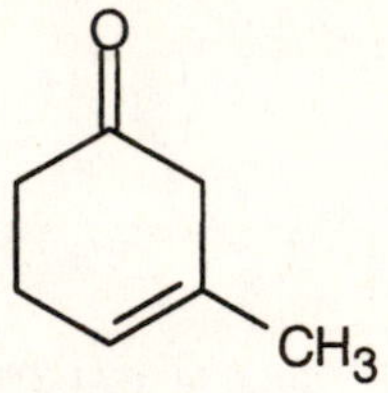

(b) CHO / H–C–OH / CH_2OH

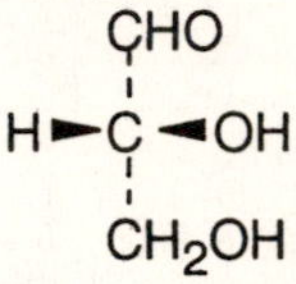

(*R*)-2,3-Dihydroxypropanal (D-Glyceraldehyde)

(c) 5-Isopropyl-2-methyl-cyclohex-2-enone

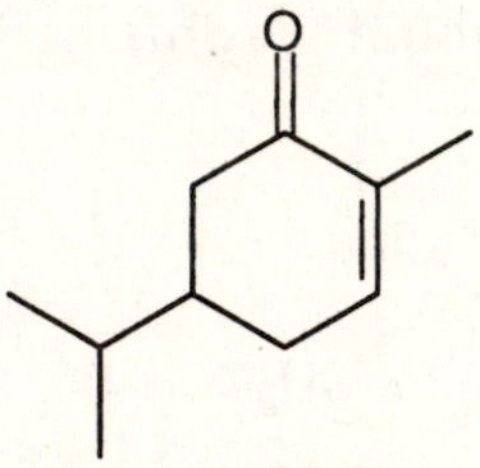

(d) $CH_3CH(CH_3)C(=O)CH_2CH_3$

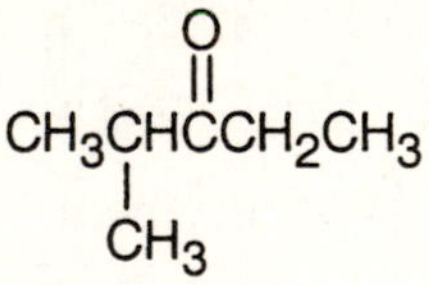

2-Methylpentan-3-one

(e) $CH_3CH(OH)CH_2CHO$

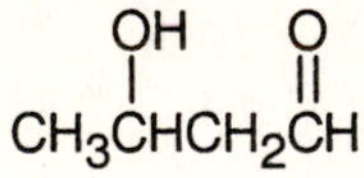

3-Hydroxybutanal

(f) OHC–C_6H_4–CHO

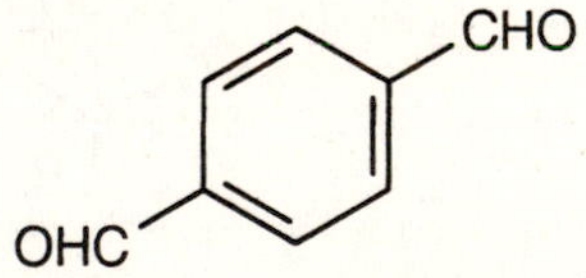

p-Benzenedicarbaldehyde

14.28 (a) The α,ß–unsaturated ketone C_6H_8O contains one ring. Possible structures include:

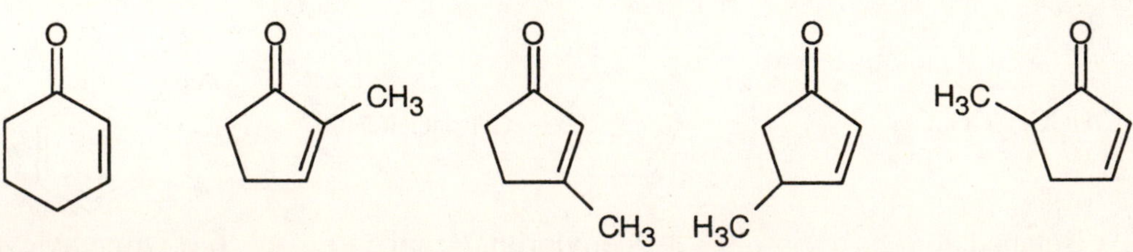

Cyclobutenones and cyclopropenones are also possible.

(b)

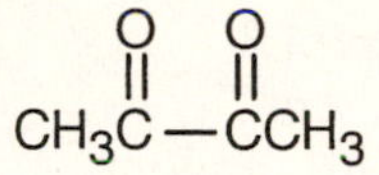

$CH_3\overset{O}{\overset{\|}{C}}-\overset{O}{\overset{\|}{C}}CH_3$ and many other structures.

(c)

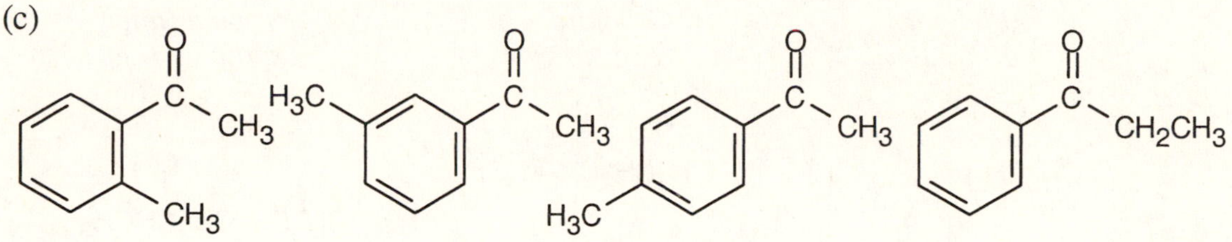

(d)

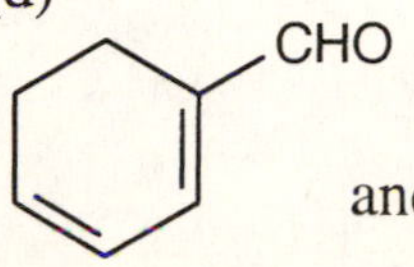

CHO

and many other structures.

14.29 Reactions of phenylacetaldehyde:

(a) $C_6H_5-CH_2CH_2OH$

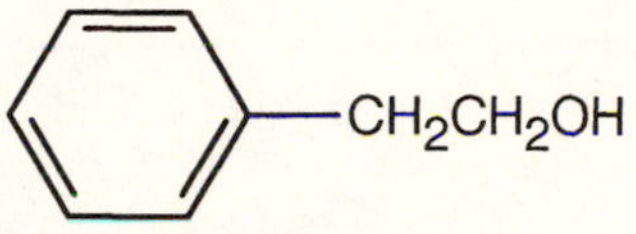

(b) $C_6H_5-CH_2CH(OCH_3)OCH_3$

(c) $C_6H_5-CH_2CH=NCH(CH_3)_2$

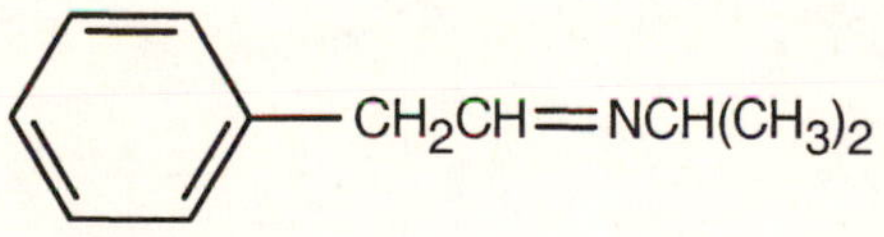

(d) $C_6H_5-CH_2CH(OH)CH_3$

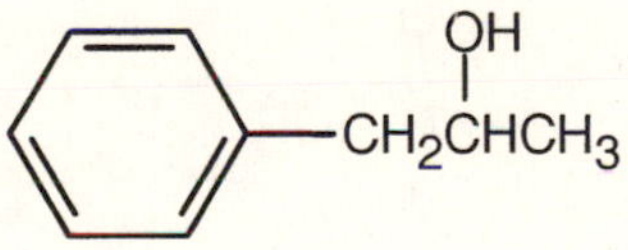

Reactions of acetophenone:

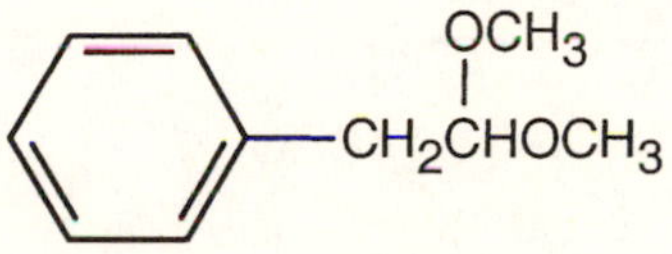

(a) $C_6H_5-C(OH)(H)CH_3$

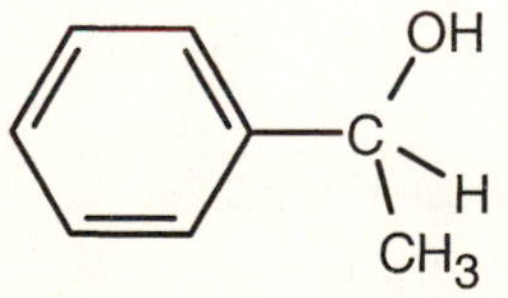

(b) $C_6H_5-C(OCH_3)(OCH_3)CH_3$

(c) $C_6H_5-C(=NCH(CH_3)_2)CH_3$

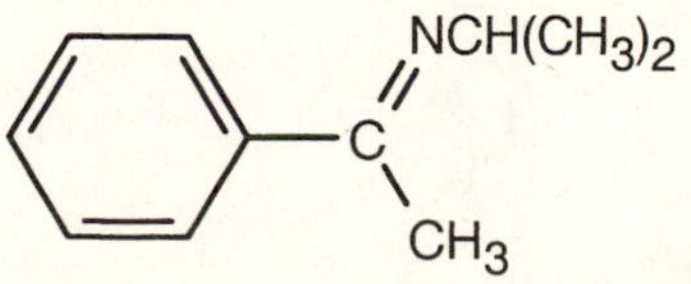

(d) $C_6H_5-C(OH)(CH_3)CH_3$

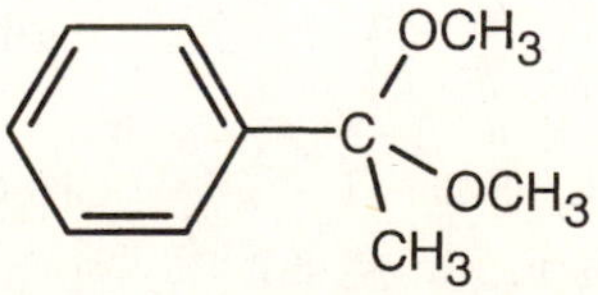

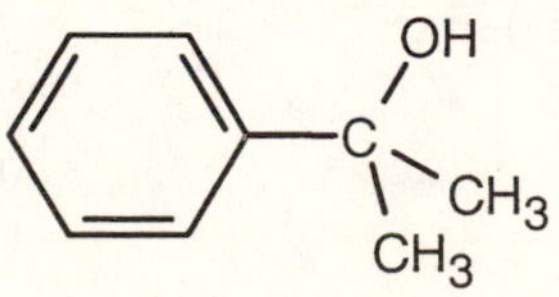

14.30 Remember from Chapter 13:
Primary alcohols are formed from formaldehyde + Grignard reagent.
Secondary alcohols are formed from an aldehyde + Grignard reagent.
Tertiary alcohols are formed from a ketone (or an ester) + Grignard reagent.

	Aldehyde/ Ketone		*Grignard reagent*	*Product (after acidic workup)*
(a)	$CH_3CH(=O)$		$CH_3CH_2CH_2MgBr$	$CH_3CH_2CH_2CH(OH)CH_3$
		or		
	$CH_3CH_2CH_2CH(=O)$		CH_3MgBr	
(b)	CH_2O		$CH_3CH_2CH_2MgBr$	$CH_3CH_2CH_2CH_2OH$
(c)	cyclohexanone (ring C=O)		C_6H_5MgBr	1-phenylcyclohexanol (ring C bearing OH and C_6H_5)
(d)	benzaldehyde (C_6H_5–CHO)		C_6H_5MgBr	$(C_6H_5)_2C(H)OH$
		or		
	$H{-}C(=O){-}OR$		$2\ C_6H_5MgBr$	

14.31 Remember:

$$RCH_2{-}X \;+\; (C_6H_5)_3P{:} \longrightarrow (C_6H_5)_3\overset{+}{P}CH_2R\ X^-$$

alkyl halide + Triphenyl phosphine → phosphonium salt

$$(C_6H_5)_3\overset{+}{P}CH_2R\ X^- \;+\; CH_3CH_2CH_2CH_2^-\ Li^+ \longrightarrow (C_6H_5)_3\overset{+}{P}{-}\overset{-}{C}HR$$

phosphonium salt + Butyllithium → ylide

$$(C_6H_5)_3\overset{+}{P}{-}\overset{-}{C}HR \;+\; O{=}C\langle \longrightarrow R(H)C{=}C\langle$$

ylide + aldehyde or ketone → alkene

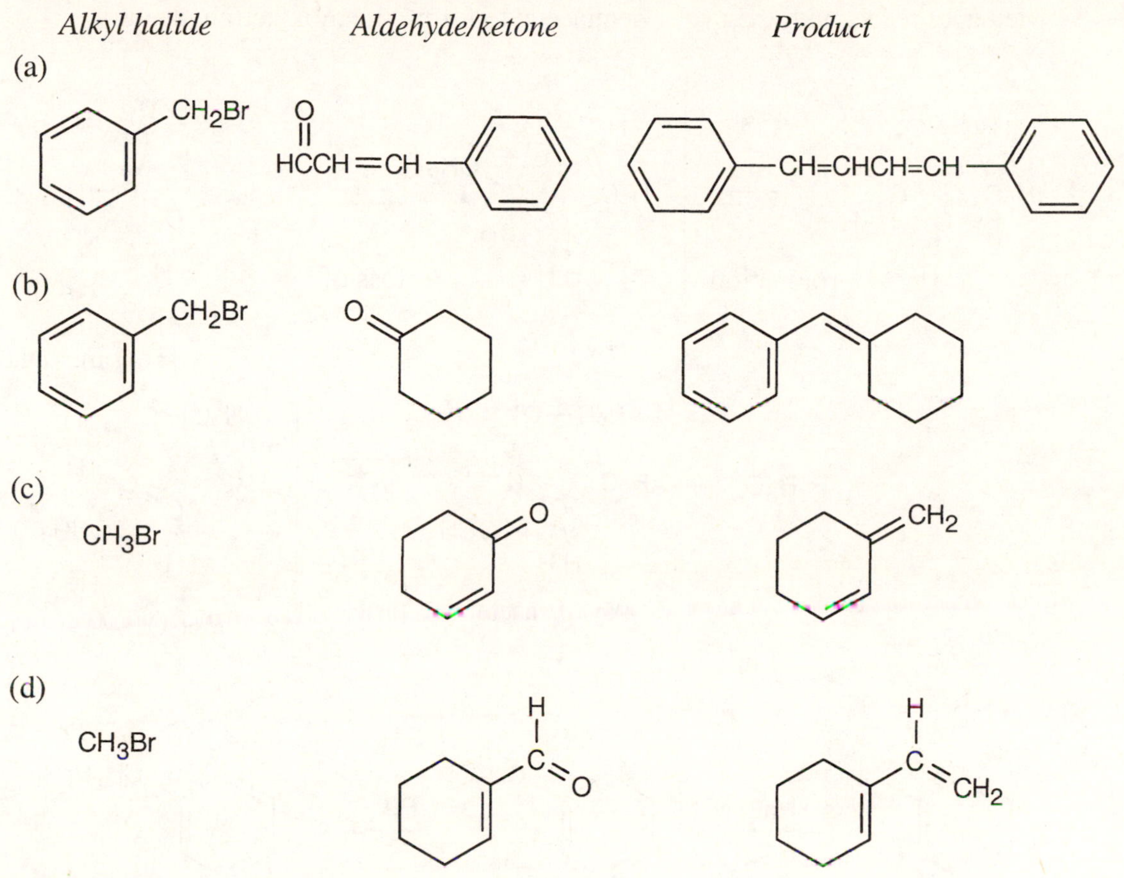

14.32 4-Hydroxybutanal forms a cyclic hemiacetal when the hydroxyl oxygen adds to the aldehyde group.

protonation

addition of –OH

loss of proton

$+ H_3O^+$

hemiacetal

Methanol reacts with the cyclic hemiacetal to form 2-methoxytetrahydrofuran.

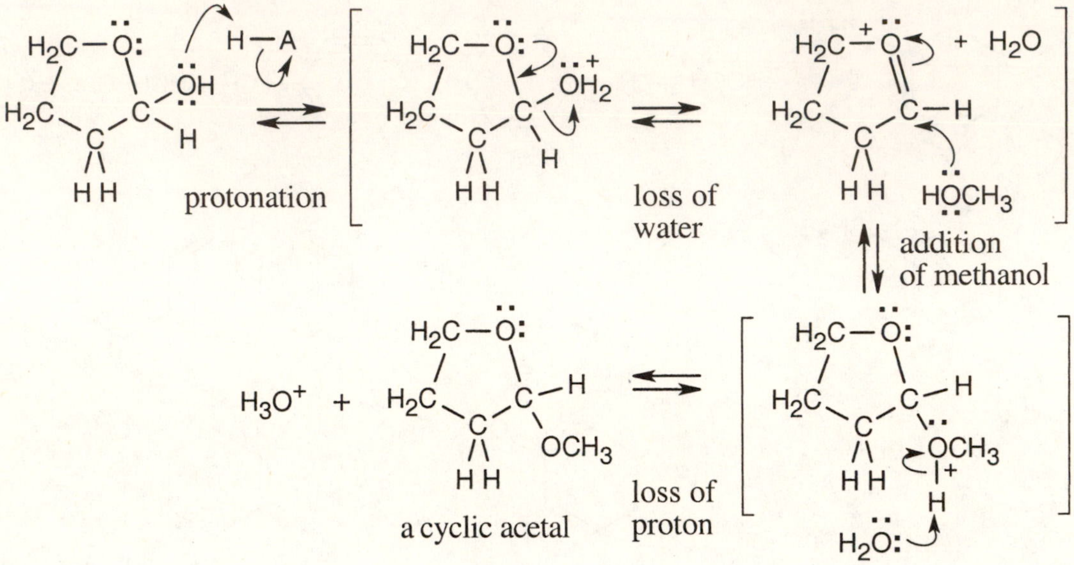

protonation

loss of water

addition of methanol

H_3O^+ +

a cyclic acetal

loss of proton

14.33

(a)

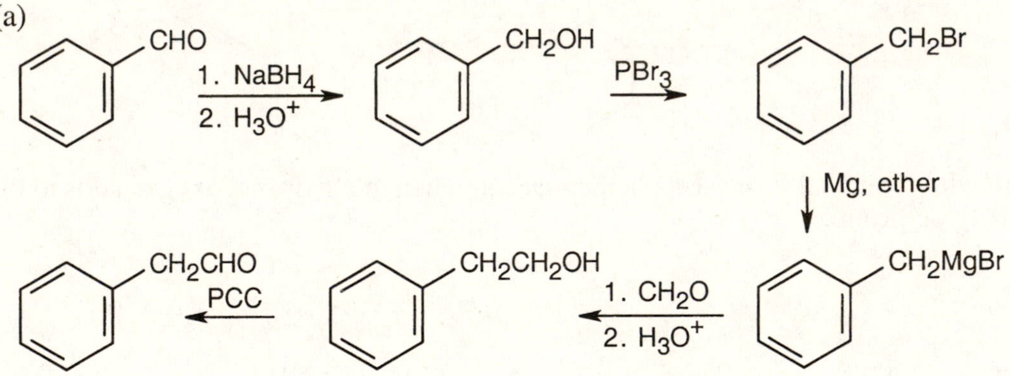

1. $NaBH_4$
2. H_3O^+

PBr_3

Mg, ether

1. CH_2O
2. H_3O^+

PCC

The product resembles the starting material in having an aldehyde group, but a $-CH_2-$ group lies between the aldehyde and the aromatic ring. The product aldehyde results from oxidation of an alcohol that is the product of a Grignard reaction between formaldehyde and benzylmagnesium bromide. The Grignard reagent is formed from benzyl bromide, which results from treatment of benzyl alcohol with PBr_3. Reduction of benzaldehyde yields the alcohol.

(b)

$C_6H_5CHO \xrightarrow[2.\ H_3O^+]{1.\ CH_3MgBr} C_6H_5CH(OH)CH_3 \xrightarrow[H_3O^+]{CrO_3} C_6H_5COCH_3$

$C_6H_5COCH_3$ + pyrrolidine (N–H) $\xrightarrow{H_3O^+}$ $C_6H_5C(NC_4H_8)=CH_2$

When you see a secondary amine and a double bond, you should recognize an enamine. The enamine is formed from an amine and acetophenone. Acetophenone, in turn, results from reaction of benzaldehyde with methylmagnesium bromide, followed by oxidation.

(c)

$C_6H_5CHO \xrightarrow[2.\ H_3O^+]{1.\ C_5H_9MgBr} C_6H_5CH(OH)C_5H_9 \xrightarrow[H_2O]{H_2SO_4} C_6H_5CH=C_5H_8$

A Grignard reaction can form the bond between the aromatic ring and the five-membered ring. Dehydration of the resulting alcohol leads to the desired product.

14.34

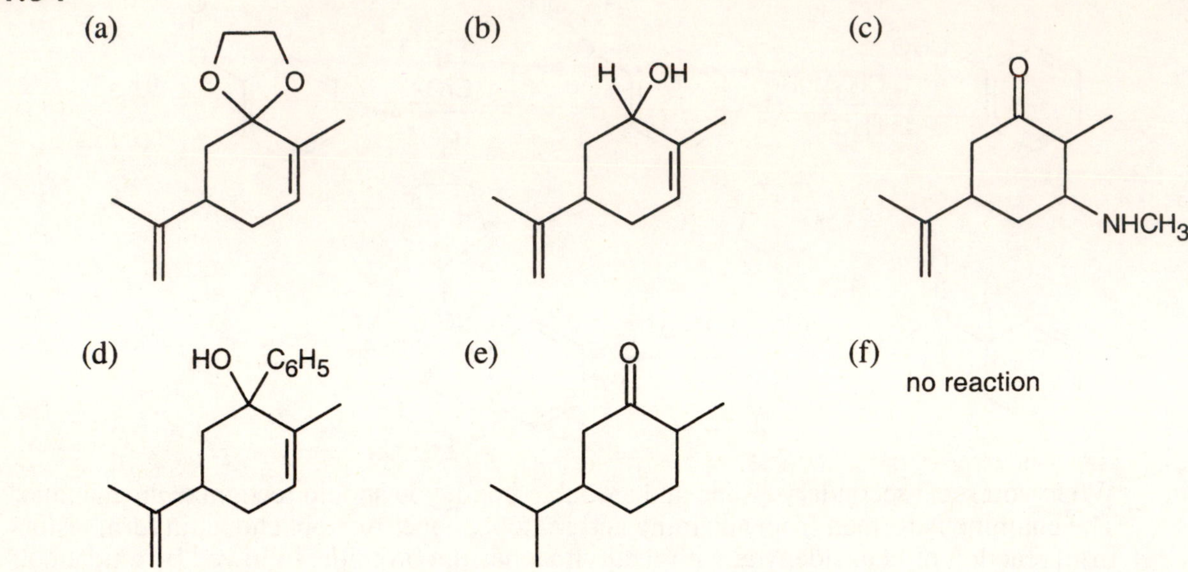

14.35

Conjugate addition of water (1) is followed by a proton shift (2) to give the β-hydroxyacyl CoA product.

14.36

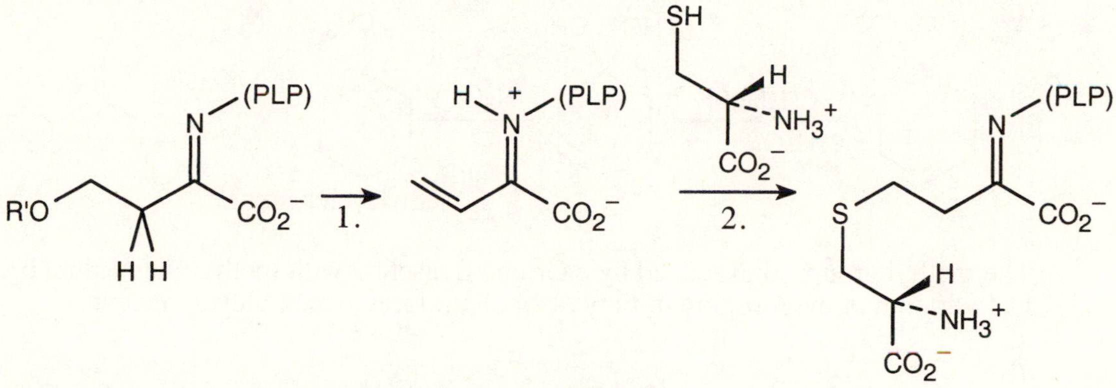

Step 1 is an elimination to form the unsaturated imine, which undergoes conjugate addition with cysteine (Step 2) to yield the observed product.

14.37

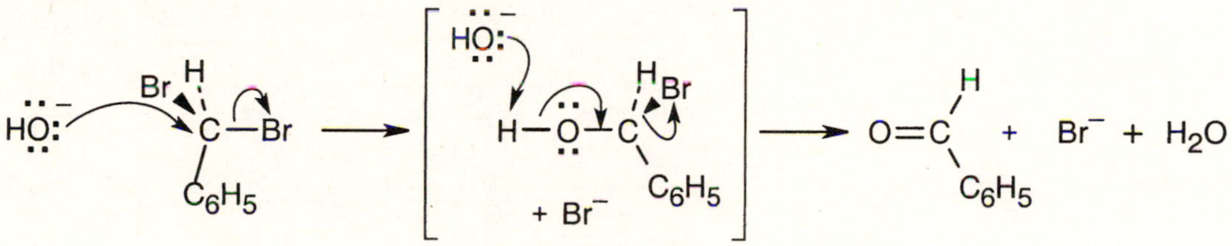

S_N2 substitution of hydroxide ion on $C_6H_5CHBr_2$ yields an unstable bromoalcohol intermediate, which loses Br^- to give benzaldehyde.

14.38

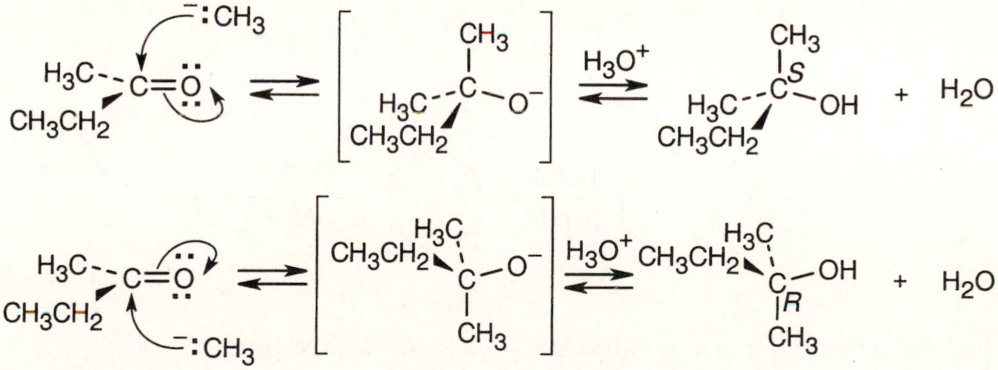

Reaction can occur with equal probability on either side of the planar carbonyl group to yield a racemic product mixture that is optically inactive.

14.39

(a)

$$\text{Cyclohexanone} \xrightarrow[\text{2. } H_3O^+]{\text{1. } CH_3MgBr} \text{1-methylcyclohexanol (HO, CH}_3\text{)} \xrightarrow{H_3O^+} \text{1-Methylcyclohexene}$$

1-Methylcyclohexene

The methyl group is introduced by a Grignard reaction with methylmagnesium bromide. Dehydration of the resulting tertiary alcohol produces 1-methylcyclohexene.

(b)

$$\text{Cyclohexanone} \xrightarrow[\text{2. } H_3O^+]{\text{1. } C_6H_5MgBr} \text{1-phenylcyclohexanol (HO, C}_6\text{H}_5\text{)} \xrightarrow{H_3O^+} \text{1-phenylcyclohexene} \xrightarrow[\text{2. } H_2O_2,\ ^-OH]{\text{1. } BH_3} \textit{trans}\text{-2-phenylcyclohexanol (H, C}_6\text{H}_5\text{, H, OH)}$$

and enantiomer

$$\xrightarrow[H_3O^+]{CrO_3} \text{2-Phenylcyclohexanone}$$

2-Phenylcyclohexanone

Reaction with phenylmagnesium bromide yields a tertiary alcohol that can be dehydrated. The resulting double bond can be treated with BH_3 to give an alcohol that can be oxidized to produce the desired ketone.

(c)

$$\text{Cyclohexanone} \xrightarrow[\text{2. } H_3O^+]{\text{1. } NaBH_4} \text{cyclohexanol (H, OH)} \xrightarrow[\text{pyridine}]{POCl_3} \text{cyclohexene} \xrightarrow[\text{2. } NaHSO_3,\ H_2O]{\text{1. } OsO_4} \textit{cis}\text{-Cyclohexane-1,2-diol (H, OH, OH, H)}$$

cis-Cyclohexane-1,2-diol

Reduction, dehydration and hydroxylation yield the desired product.

(d)

OH (cyclohexanol, from (c)) → 1. PBr_3 2. Mg, ether → MgBr (cyclohexylmagnesium bromide) → 1. Cyclohexanone 2. H_3O^+ → 1-Cyclohexylcyclohexanol

A Grignard reaction forms 1-cyclohexylcyclohexanol.

14.40 The same series of steps used to form an acetal is followed in this mechanism.

:O: H—A → protonation → :OH (+) + $HSCH_2CH_3$ → addition of RSH → HO $\overset{+}{S}CH_2CH_3$ (H, :A⁻) → loss of proton → HO :SCH_2CH_3 + HA hemithioacetal

HO: :SCH_2CH_3 H—A hemithio-acetal → protonation → $H_2\overset{+}{O}$: SCH_2CH_3 → loss of H_2O → $CH_3CH_2\overset{+}{S}$ + H_2O + $HSCH_2CH_3$ → addition of RSH → CH_3CH_2S: $\overset{+}{S}CH_2CH_3$ (H, :A⁻) → loss of proton → CH_3CH_2S: :SCH_2CH_3 HA + thioacetal

14.41

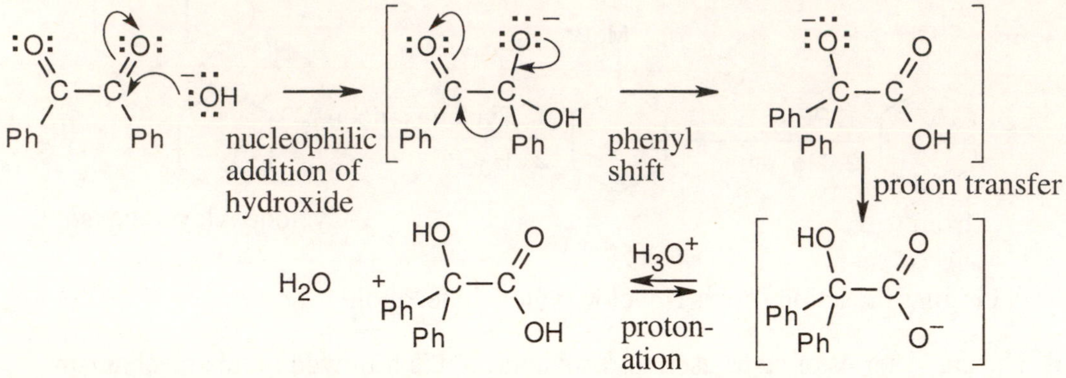

nucleophilic addition of hydroxide

phenyl shift

proton transfer

H_3O^+

protonation

H_2O +

14.42 Even though the product looks unusual, this reaction is made up of steps with which you are familiar.

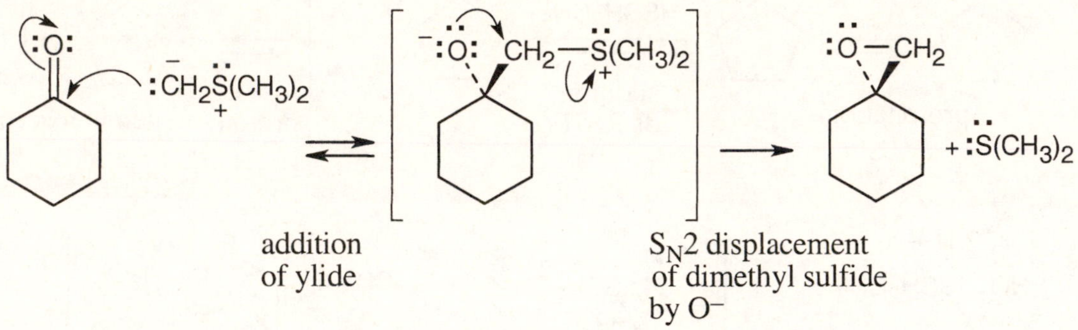

addition of ylide

S_N2 displacement of dimethyl sulfide by O^-

14.43

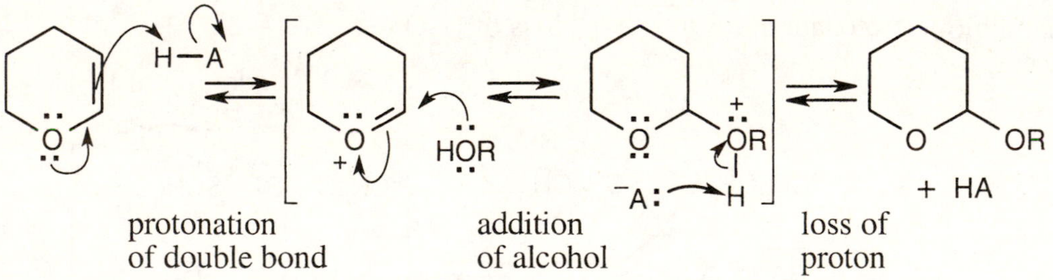

protonation of double bond

addition of alcohol

loss of proton

+ HA

The steps of this reaction are similar to some of the steps in the formation of an acetal from a hemiacetal.

14.44

Benzene → (CH_3CH_2COCl, $AlCl_3$) → $C_6H_5COCH_2CH_3$ → (1. $LiAlH_4$; 2. H_3O^+) → $C_6H_5CH(OH)CH_2CH_3$

Benzene → (1. Br_2, $FeBr_3$; 2. Mg, ether) → C_6H_5MgBr → (1. CH_3CH_2CHO; 2. H_3O^+) → $C_6H_5CH(OH)CH_2CH_3$

1-Phenylpropan-1-ol

$C_6H_5CH(OH)CH_2CH_3$ → (PBr_3) → $C_6H_5CH(Br)CH_2CH_3$ → (Mg, ether) → $C_6H_5CH(MgBr)CH_2CH_3$

$C_6H_5CH(MgBr)CH_2CH_3$ + $(C_6H_5)(p\text{-}(CH_3)_2NCH_2CH_2OC_6H_4)C{=}O$ → (1. ether; 2. H_3O^+) → $HO{-}C(C_6H_5)(p\text{-}C_6H_4OCH_2CH_2N(CH_3)_2){-}C(H)(CH_2CH_3)(C_6H_5)$

→ (H_3O^+) → $(C_6H_5)(p\text{-}(CH_3)_2NCH_2CH_2OC_6H_4)C{=}C(CH_2CH_3)(C_6H_5)$ Tamoxifen

Synthetic analysis: The coupling step is achieved by a Grignard reaction between the illustrated ketone and a Grignard reagent, followed by dehydration. The Grignard reagent is synthesized from 1-phenylpropan-1-ol. which can be prepared from benzene by either of two routes.

14.45

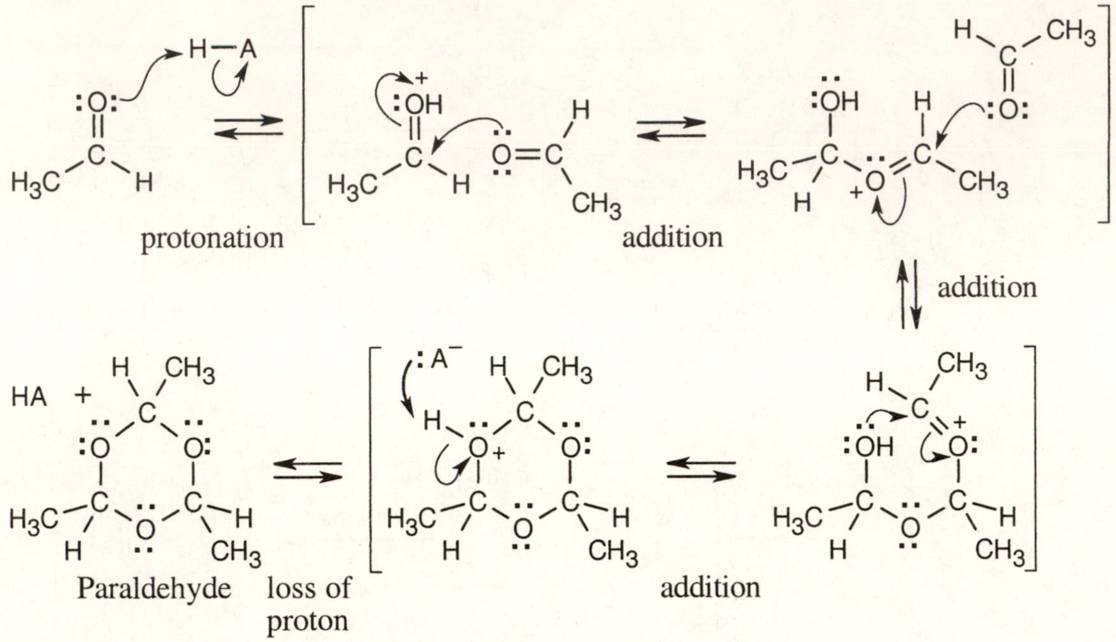

H₃C–CHO ⇌ (protonation) ⇌ (addition) ⇌ (addition) ⇌ (addition) ⇌ (loss of proton) HA + Paraldehyde

protonation

addition

addition

Paraldehyde loss of proton

addition

Protonation makes the carbonyl carbon more electrophilic. Three successive additions of the carbonyl oxygen of acetaldehyde to the electrophilic carbonyl carbon, followed by loss of a proton, give the cyclic product.

14.46

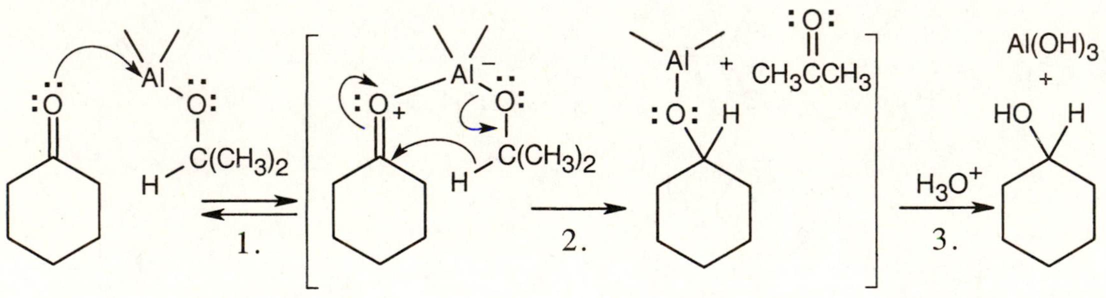

$Al(OH)_3$ + cyclohexanol; CH_3CCH_3; $C(CH_3)_2$; H_3O^+

1. 2. 3.

1. Aluminum, a Lewis acid, complexes with the carbonyl oxygen.
2. Complexation with aluminum makes the carbonyl group more electrophilic and facilitates hydride transfer from isopropoxide.
3. Treatment of the reaction mixture with aqueous acid cleaves the aluminum-oxygen bond and produces cyclohexanol.

Both the Meerwein–Ponndorf–Verley reaction and the Cannizzaro reaction are hydride transfers in which a carbonyl group is reduced by an alkoxide group, which is oxidized. Note that each aluminum triisopropoxide molecule is capable of reducing three ketone molecules.

14.47 (a) Nucleophilic addition of one nitrogen of hydrazine to one of the carbonyl groups, followed by elimination of water, produces an imine.

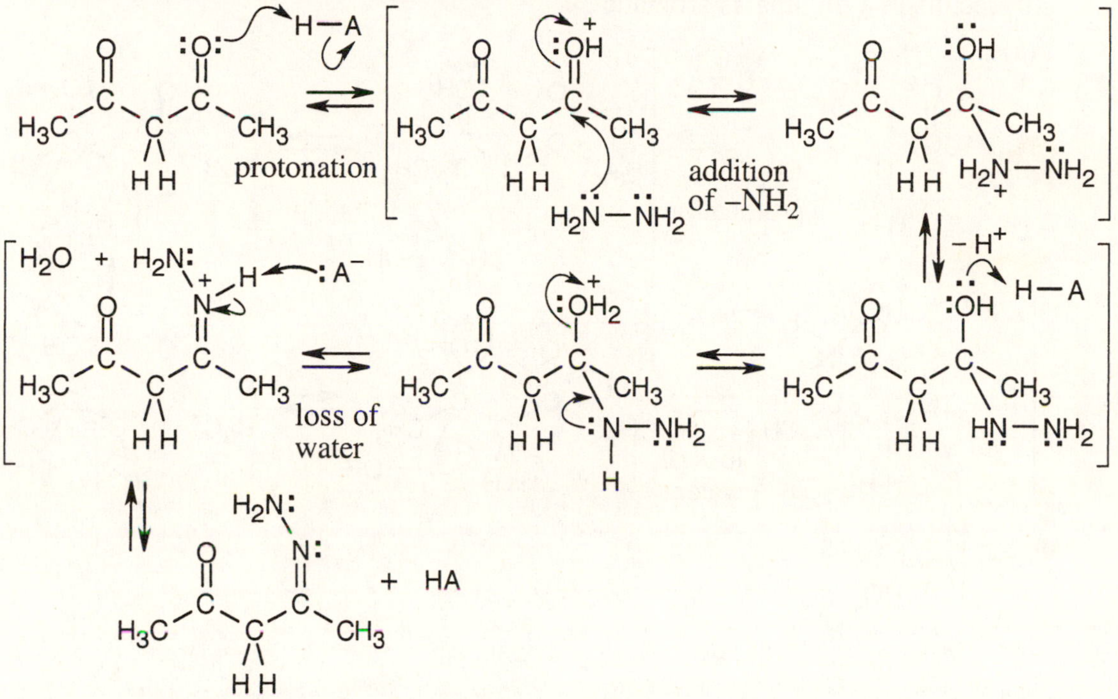

protonation

addition of $-NH_2$

$-H^+$

loss of water

+ HA

(b) In a similar manner, the *second* nitrogen of hydrazine can add to the *second* carbonyl group of 2,4-pentanedione to form the pyrazole.

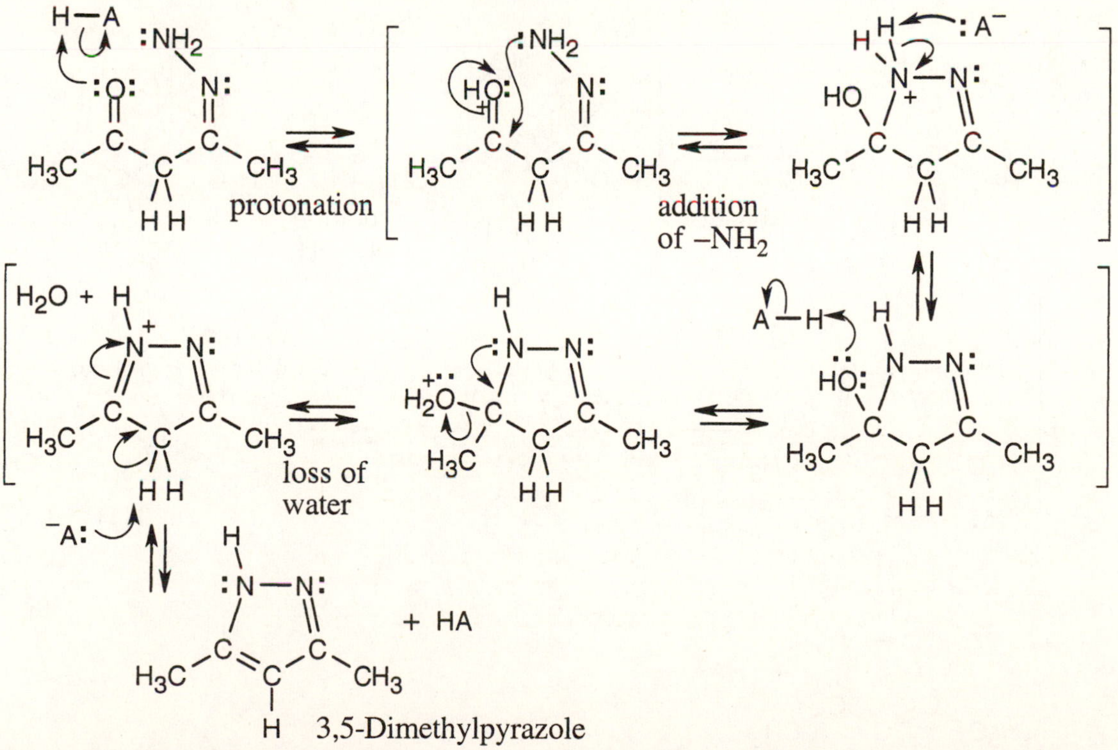

protonation

addition of $-NH_2$

loss of water

+ HA

3,5-Dimethylpyrazole

The driving force behind this reaction is the formation of an aromatic ring. The reactions in this problem are nucleophilic addition of a primary amine, followed by elimination of water to yield an imine or enamine. All of the other steps are protonations and deprotonations.

14.48 The same sequence of steps used in the previous problem leads to the formation of 3,5-dimethylisoxazole when hydroxylamine is the reagent. Loss of a proton in the last step of (b) results in a ring that is aromatic.

(a)

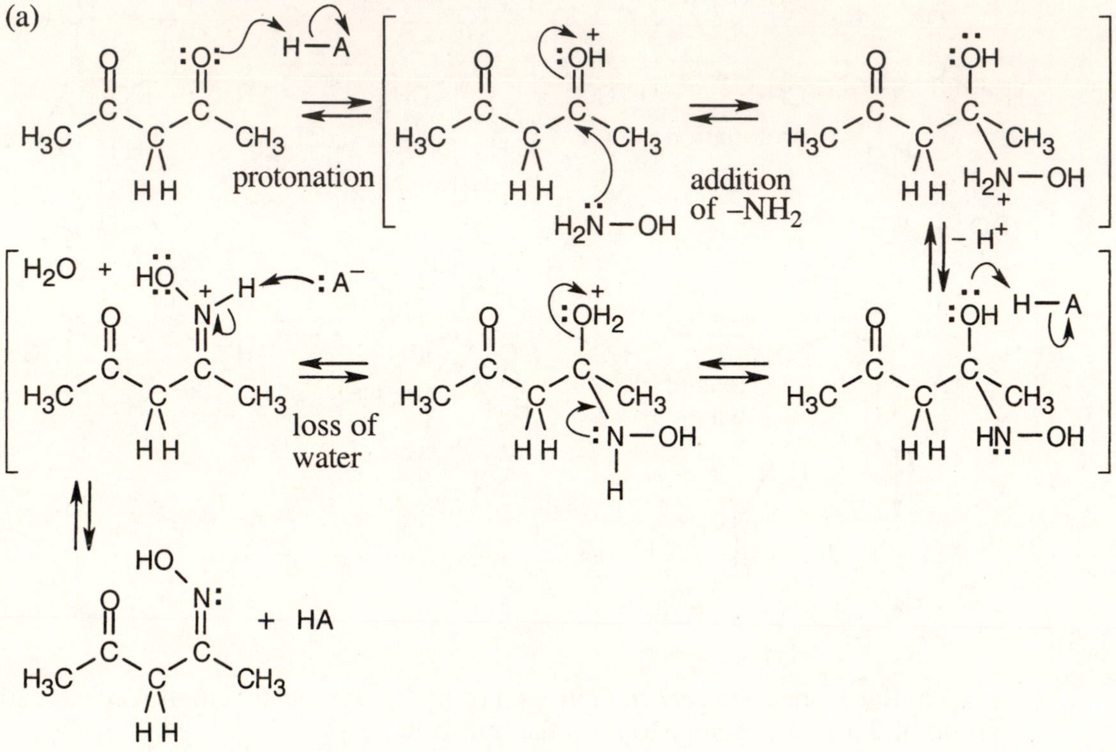

protonation

addition of $-NH_2$

$- H^+$

loss of water

$+ HA$

(b)

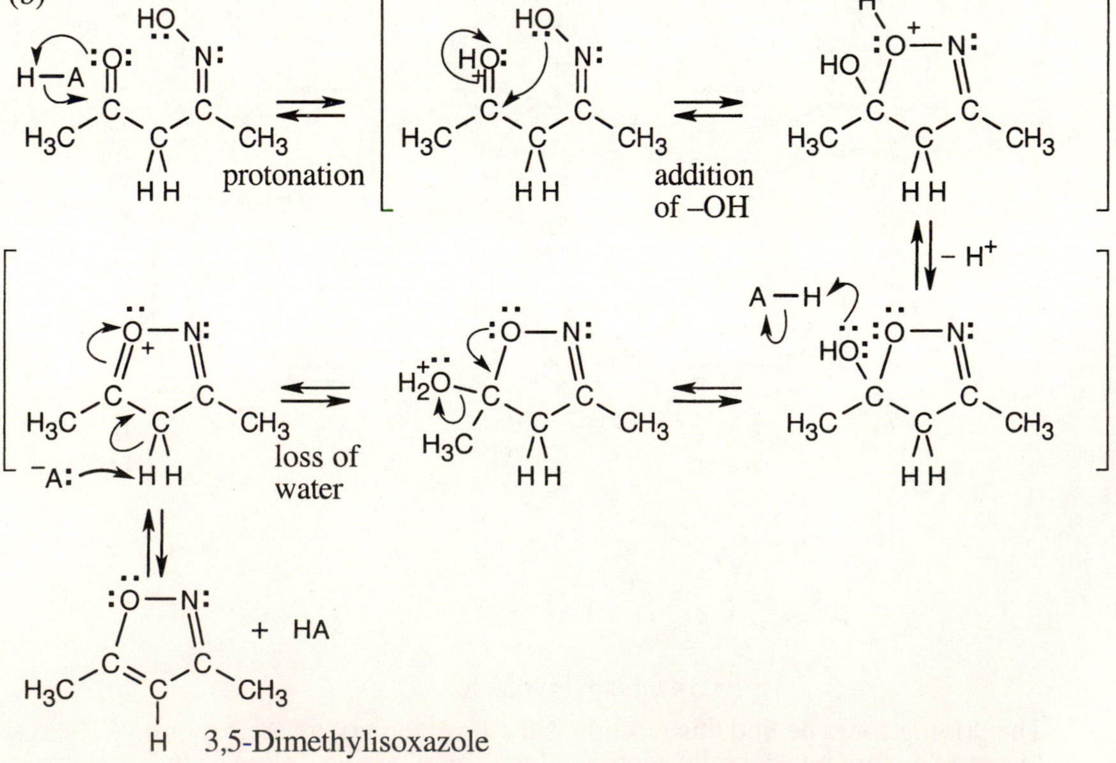

protonation

addition of –OH

$- H^+$

loss of water

$+ HA$

3,5-Dimethylisoxazole

14.49

$$HO{-}O{-}H + {}^{-}:OH \rightleftharpoons {}^{-}:O{-}OH + H_2O$$

Hydrogen peroxide and hydroxide react to form water and hydroperoxide anion.

conjugate addition of hydroperoxide anion

formation of epoxide ring and elimination of ⁻OH

Conjugate addition of hydroperoxide anion is followed by elimination of hydroxide ion, with formation of the epoxide ring.

14.50

rotate 180°

$(Ph)_3P$:

addition of the phosphine nucleophile

rotation of the C–C bond

elimination of triphenylphosphine oxide

$(Ph)_3P{=}O$

The final step is the same as the last step in a Wittig reaction.

14.51

NAD$^+$ H : Base H NH$_2$ + ^-O_2C CO_2^- Glutamate

oxidation of the amine by NAD$^+$

NADH + $\overset{+}{N}H_2$ ^-O_2C CO_2^- $H_2\ddot{O}$: imine

addition of water to form a tetrahedral intermediate

:NH$_2$ ^-O_2C CO_2^- H_2O^+

proton shift

NH_3^+ ^-O_2C CO_2^- HO:

loss of ammonia

H : Base :O$^+$ ^-O_2C CO_2^- + NH$_3$

deprotonation

O ^-O_2C CO_2^- α-Ketoglutarate

14.52 Use Table 14.2 if you need help.

Absorption:	*Due to:*
(a) 1750 cm^{-1}	5-membered ring ketone
1685 cm^{-1}	α,ß-unsaturated ketone
(b) 1720 cm^{-1}	5-membered ring *and* aromatic ketone
(c) 1750 cm^{-1}	5-membered ring ketone
(d) 1705 cm^{-1}, 2720 cm^{-1}, 2820 cm^{-1}	aromatic aldehyde
1715 cm^{-1}	aliphatic ketone

Compounds in parts (b)-(d) also show aromatic ring IR absorptions in the range 1450 cm^{-1} – 1600 cm^{-1} and in the range 690 – 900 cm^{-1}.

14.53

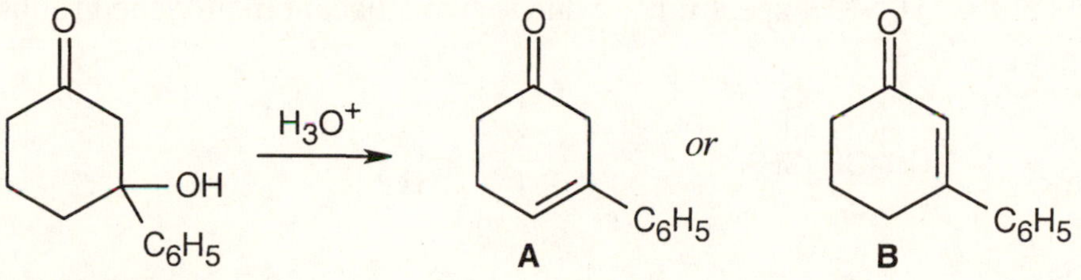

$\xrightarrow{H_3O^+}$ A *or* B

3-Hydroxy-3-phenyl-cyclohexanone

Compound **A** is a cyclic, nonconjugated keto alkene whose carbonyl infrared absorption should occur at 1715 cm^{-1}. Compound **B** is an α,ß-unsaturated, cyclic ketone; additional conjugation with the phenyl ring should lower its IR absorption below 1685 cm^{-1}. Because the actual IR absorption occurs at 1670 cm^{-1}, **B** is the correct structure.

14.54 The molecular weight of Compound A shows that the molecular formula of A is $C_5H_{10}O$ (one degree of unsaturation), and the IR absorption shows that A is an aldehyde. The uncomplicated ^1H NMR is that of 2,2-dimethylpropanal.

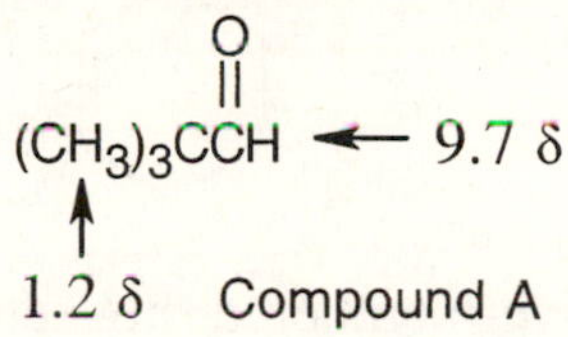

$(CH_3)_3CCH$ (C=O) ← 9.7 δ

1.2 δ Compound A

14.55 The IR of Compound B shows a ketone absorption. The splitting pattern of the ^1H NMR spectrum reveals an isopropyl group and indicates that the compound is a methyl ketone.

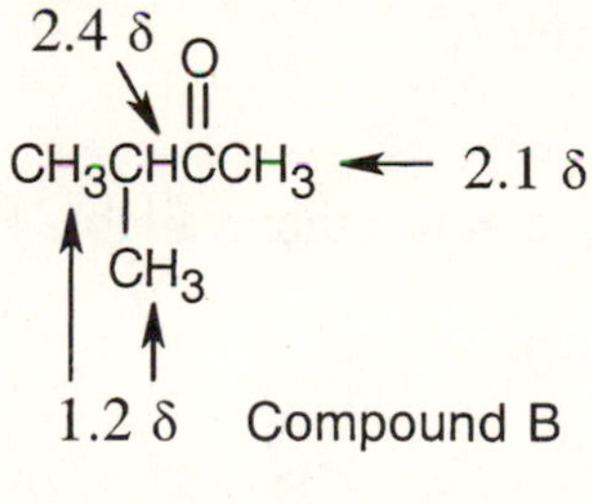

2.4 δ

CH_3CHCCH_3 (C=O, CH_3 branch) ← 2.1 δ

1.2 δ Compound B

14.56 Before looking at the ^1H NMR spectrum, we know that the compound of formula $C_9H_{10}O$ has 5 degrees of unsaturation, and we know from the IR spectrum that the unknown is an aromatic ketone. The splitting pattern in the ^1H NMR spectrum shows an ethyl group next to a ketone, according to chemical shift values.

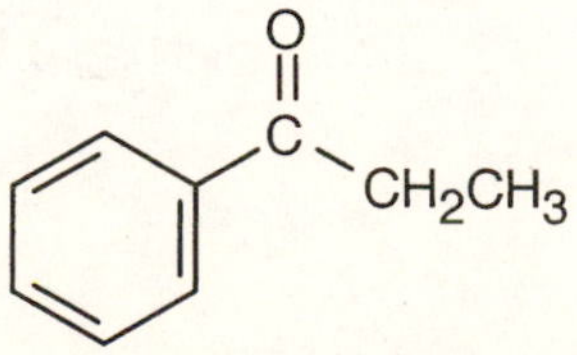

$C_6H_5C(=O)CH_2CH_3$

14.57 The IR absorption is that of an aldehyde that isn't conjugated with the aromatic ring. The two triplets in the ^{1}H NMR spectrum are due to two adjacent methylene groups.

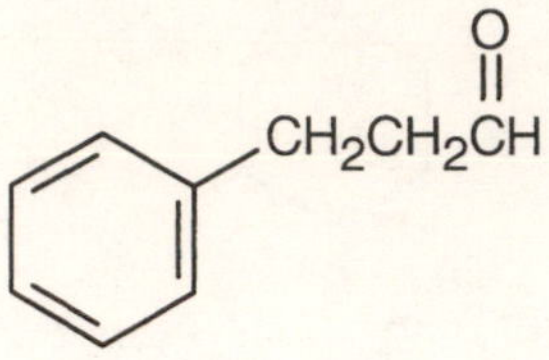

$C_6H_5CH_2CH_2CHO$ (benzene ring–CH_2CH_2CH=O)

14.58 (a) IR shows that the unknown is a ketone, and ^{13}C NMR indicates that the carbonyl group is flanked by a secondary carbon and a tertiary carbon (from ^{13}C NMR).
(b) The unknown is an aldehyde that contains an isopropyl group.
(c) The IR absorption shows that this compound is an α,β-unsaturated ketone, and the molecular formula shows 3 degrees of unsaturation. The ^{13}C NMR spectrum indicates 3 sp^2-hybridized carbons and 3 secondary carbons.

(a) $(CH_3)_2CHC(=O)CH_2CH_3$

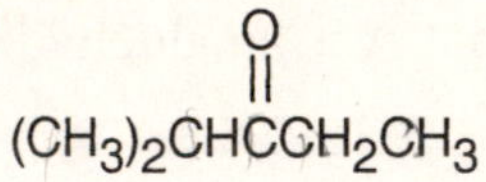

(b) $(CH_3)_2CHCH_2CHO$

(c) 2-cyclohexenone (ring structure with C=O)

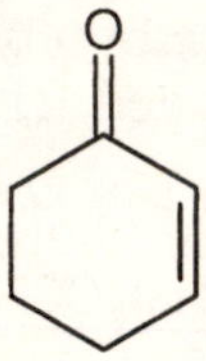

14.59 Compound A has 4 degrees of unsaturation and is a five-membered ring ketone according to the IR spectrum. The ^{13}C NMR spectrum has only three peaks and indicates that A is very symmetrical.

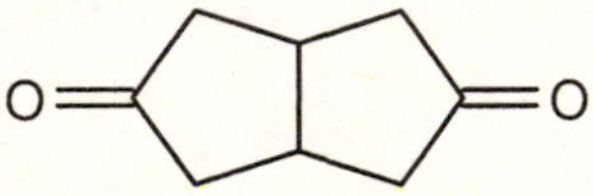

O=(bicyclic diketone)=O Compound A

14.60 As always, calculate the degree of unsaturation first, then use the available IR data to assign the principal functional groups.

(a) $CH_3CH(Cl)C(=O)CH_3$ (a = CH_3 on CHCl, c = CH, b = $COCH_3$)

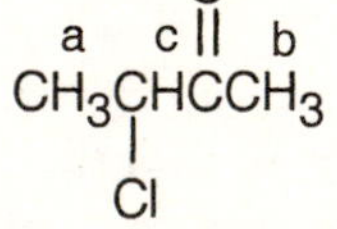

a = 1.7 δ
b = 2.3 δ
c = 4.3 δ

(b) $(CH_3)_3CCH_2C(=O)CH_3$ (a = $(CH_3)_3$, c = CH_2, b = $COCH_3$)

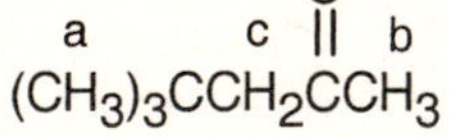

a = 1.0 δ
b = 2.1 δ
c = 2.3 δ

(c) p-ethoxybenzaldehyde: O=C(e H)–C_6H_4(c d)–OCH_2CH_3 (b a)

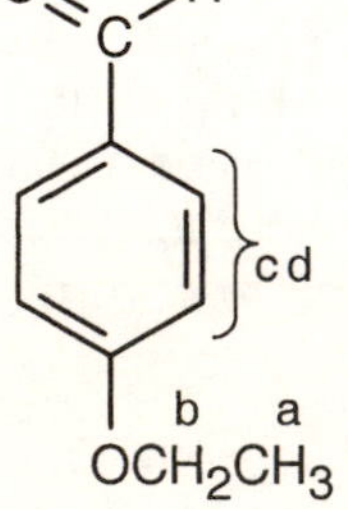

a = 1.5 δ
b = 4.1 δ
c,d = 7.0 δ, 7.8 δ
e = 9.9 δ

14.61

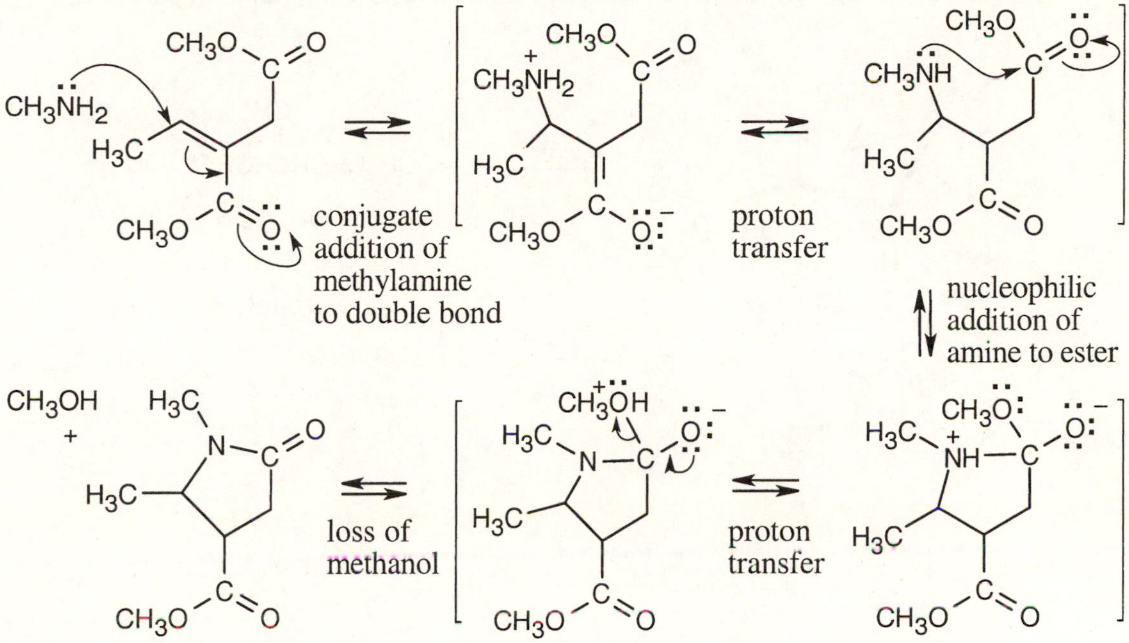

conjugate addition of methylamine to double bond

proton transfer

nucleophilic addition of amine to ester

proton transfer

loss of methanol

In this series of reactions, conjugate addition of an amine to an α,β-unsaturated ester is followed by nucleophilic addition of the amine to a second ester, with loss of methanol.

14.62 Several hydrogens have been omitted to make the drawing less cluttered.

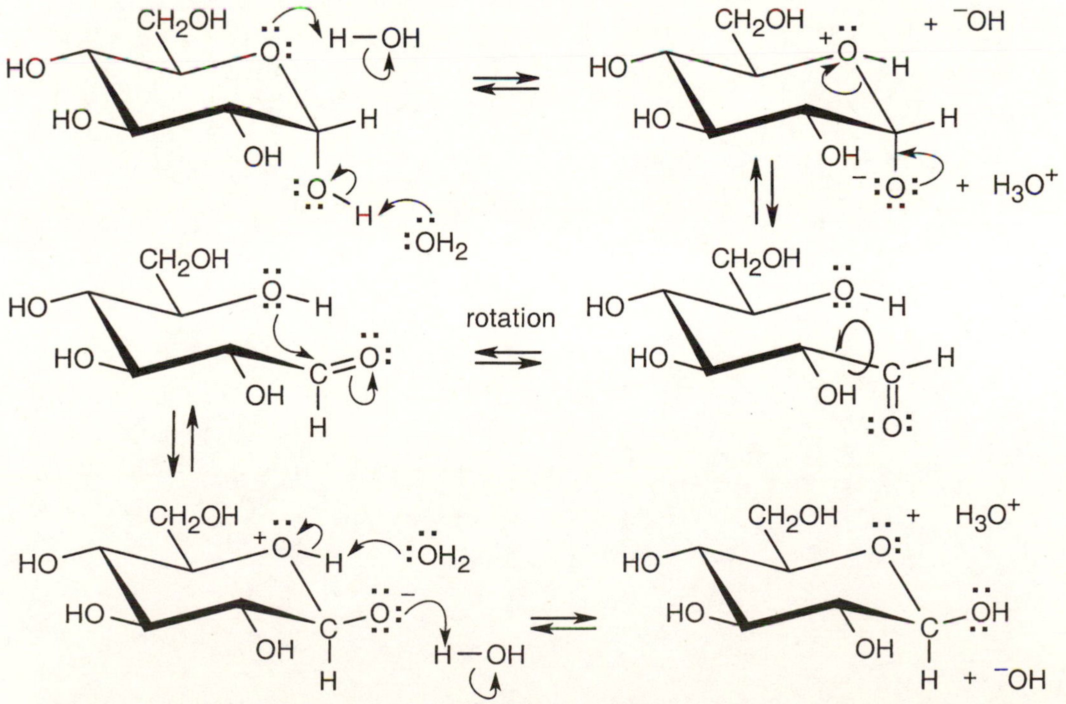

rotation

In this series of equilibrium steps, the hemiacetal ring of α-glucose opens to yield the free aldehyde. Bond rotation is followed by formation of the cyclic hemiacetal of ß-glucose. The reaction is catalyzed by both acid and base.

14.63 The free aldehyde form of glucose (Problem 14.58) is reduced in the same manner described in the text for other aldehydes to produce the polyalcohol sorbitol.

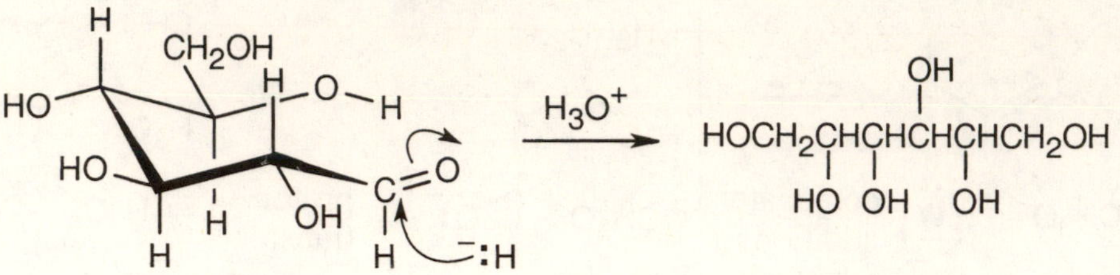

$\xrightarrow{H_3O^+}$ HOCH$_2$CH(OH)CH(OH)CH(OH)CH(OH)CH$_2$OH

Glucose

Sorbitol

Review Unit 6: Alcohols, Ethers, and Related Compounds; Aldehydes and Ketones

Major Topics Covered (with vocabulary):

The –OH group:
alcohol phenol glycol wood alcohol hydrogen bonding alkoxide ion phenoxide ion acidity constant

Alcohols and phenols:
Grignard Reagent lithium aluminum hydride sodium borohydride pyridinium chlorochromate phosphorus oxychloride NAD^+ quinone hydroquinone ubiquinone

Ethers:
Williamson ether synthesis Claisen rearrangement epoxide

Thiols and sulfides:
Thiol sulfide mercapto group alkylthio group disulfide thiolate ion trialkylsulfonium salt sulfoxide sulfone

Aldehydes and ketones:
-carbaldehyde acyl group acetyl group formyl group benzoyl group DIBAH hydrate Tollens reagent

Reactions of aldehydes and ketones:
nucleophilic addition reaction gem diol imine enamine carbinolamine acetal hemiacetal Cannizzaro reaction conjugate addition α,β-unsaturated carbonyl compound

Types of Problems:

After studying these chapters, you should be able to:

- Name and draw structures of alcohols, phenols, ethers, thiols, sulfides, aldehydes and ketones.
- Explain the properties and acidity of alcohols and phenols.
- Prepare all of the types of compounds studied.
- Predict the products of reactions involving alcohols, phenols, ethers, aldehydes and ketones.
- Formulate mechanisms of reactions of these compounds.
- Identify these compounds by spectroscopic techniques.

Points to Remember:

* The great biochemical importance of hydroxyl groups is due to two factors:(1) Hydroxyl groups make biomolecules more soluble because they can hydrogen-bond with water. (2) Hydroxyl groups can be oxidized to aldehydes, ketones and carboxylic acids. The presence of a hydroxyl group in a biological molecule means that all functional groups derived from alcohols can be easily introduced.

* Carbon-carbon bond-forming reactions are always more difficult to learn than functional group transformations because it is often difficult to recognize the components that form a carbon skeleton. The product of a Grignard reaction contains a hydroxyl group bonded to at least one alkyl group (usually two or three). When looking at a product that might have been formed by a Grignard reaction, remember that a tertiary alcohol results from the addition of a Grignard reagent to either a ketone or an ester (the alcohol formed from the ester has two identical –R groups), a secondary alcohol results from addition of a Grignard reagent to an aldehyde, and a primary alcohol results from addition of a Grignard reagent to formaldehyde or to ethylene oxide. Remember that any molecule taking part in a Grignard reaction must not contain functional groups that might also react with the Grignard reagent.

* Ethers are quite unreactive, relative to many other functional groups we study, and are often used as solvents for that reason. Concentrated halogen acids can cleave ethers to alcohols and halides. Remember that the halide is bonded to the less substituted alkyl group when the ethers are primary or secondary alkyl ethers.

* In reactions of aldehydes and ketones, a nucleophile adds to a positively polarized carbonyl carbon to form a tetrahedral intermediate. There are three possible fates for the tetrahedral intermediate: (1) The intermediate can be protonated, as occurs in Grignard reactions, reductions, and cyanohydrin formation. (2) The intermediate can lose water (or ^{-}OH), as happens in imine and enamine formation. (3) The intermediate can lose a leaving group, as occurs in most reactions of carboxylic acid derivatives.

* Many of the reactions in these chapters require acid or base catalysis. An acid catalyst, protonates the carbonyl oxygen, making the carbonyl carbon more reactive toward electrophiles, and/or protonates the tetrahedral intermediate, making loss of a leaving group easier. A base catalyst deprotonates the nucleophile, making it more nucleophilic. The pH optimum for these reactions is a compromise between the two needs.

Self-Test:

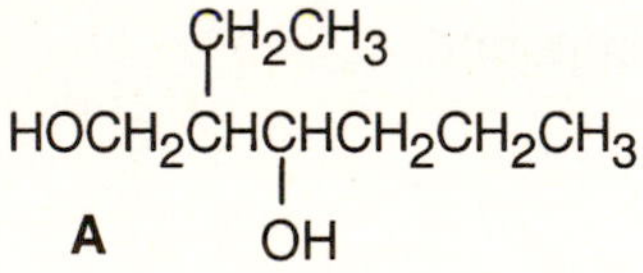

CH_2CH_3

$HOCH_2CHCHCH_2CH_2CH_3$

A OH

an insecticide

Cl — CH_2S — Cl

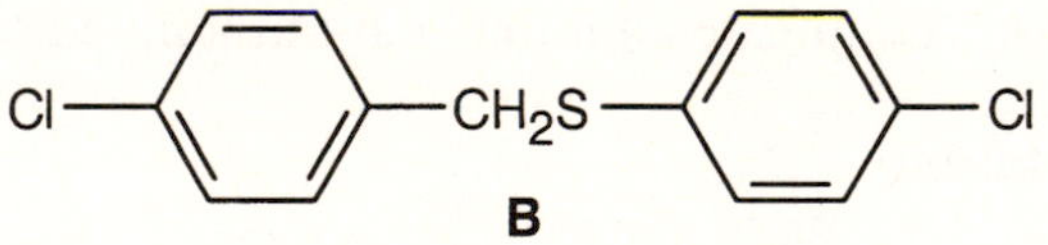

B

Chlorbenside (larvicide)

Provide a IUPAC name for **A**. Would you expect **A** to be water-soluble? Label the hydroxyl groups of **A** as primary, secondary or tertiary. What products are formed when **A** reacts with: (a) CrO_3, H_3O^+ ; (b) PBr_3.

What type of compound is **B**? Name **B**. Synthesize **B** from benzenethiol and benzene. What products are formed when **B** is treated with: (a) CH_3I; (b) H_2O_2, H_2O; (c) product of (b) + CH_3CO_3H.

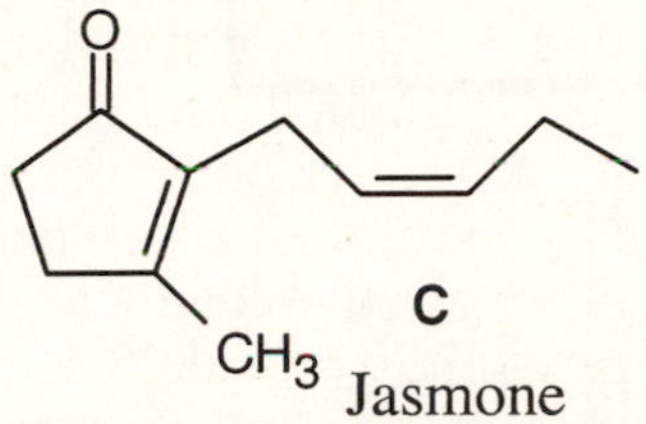

C Jasmone (used in perfumery)

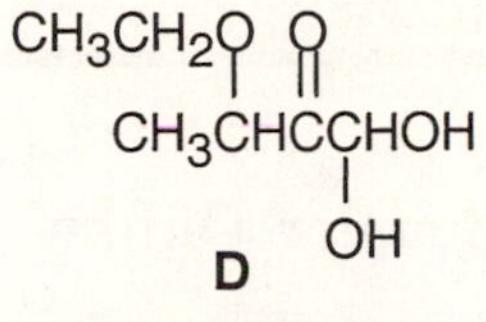

CH_3CH_2O O ‖ $CH_3CHCCHOH$ | OH

D Kethoxal (antiviral)

$COCH_3$, CH_3O, OH, OCH_3

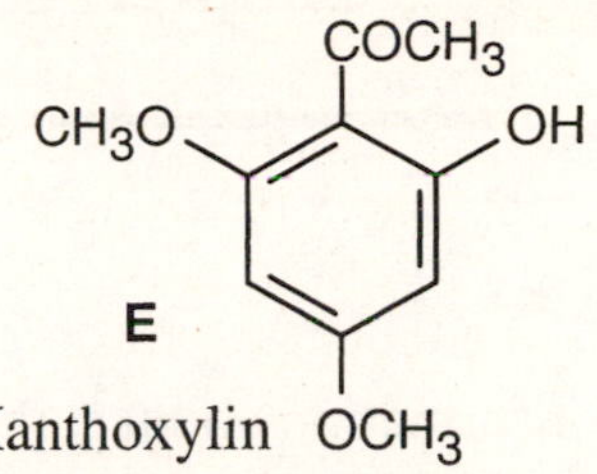

E Xanthoxylin

Predict the products of the reaction of **C** with: (a) $LiAlH_4$, then H_3O^+; (b) C_6H_5MgBr, then H_3O^+; (c) $(CH_3)_2NH$, H_3O^+; (d) CH_3OH, H^+ catalyst (e) 1 equiv. $CH_3CH_2NH_2$, H_3O^+. How would you reduce **C** to yield a saturated ketone? Where would you expect the carbonyl absorption of **C** to occur in its IR spectrum?

Kethoxal (**D**) exists in solution as an equilibrium mixture. With what compound is it in equilibrium. Why does the equilibrium lie on the side of kethoxal?

Name **E**. Describe the IR spectrum and ^1H NMR spectrum of **E**.

Multiple Choice:

1. Hydrogen bonding affects all of the following except:
   (a) boiling point (b) solubility (c) position of –OH absorption in IR spectrum (d) chemical shift of –C–O– carbon in ^{13}C NMR.

2. Which of the following alcohols can't be synthesized by a Grignard reaction?
   (a) Benzyl alcohol (b) Triphenylmethanol (c) 3-Bromo-1-hexanol (d) 1-Hexanol

3. Which alcohol can be formed by three different combinations of carbonyl compound + Grignard reagent?
   (a) 2-Butanol (b) 3-Methyl-3-hexanol (c) Triphenylmethanol (d) 1-Phenylethanol

4. How many diols of the formula $C_4H_{10}O_2$ are chiral?
   (a) 2 (b) 3 (c) 4 (d) 5

5. Which alcohol is the least acidic?
   (a) 2-Propanol (b) Methanol (c) Ethanol (d) 2-Chloroethanol

6. Which of the following compounds can't be reduced to form $C_6H_5CH_2OH$?
   (a) $C_6H_5CO_2H$ (b) C_6H_5CHO (c) $C_6H_5CO_2CH_3$ (d) $C_6H_5OCH_3$

7. The reagent used for dehydration of an alcohol is:
   (a) PCl_3 (b) $POCl_3$ (c) $SOCl_2$ (d) PCC

8. All of the following are products of oxidation of a thiol except:
   (a) a sulfide (b) a disulfide (c) a sulfoxide (d) a sulfone

9. In which of the following nucleophilic addition reactions does the equilibrium lie on the side of the products?
   (a) Acetone + H_2O (b) Acetaldehyde + HBr (c) Propanal + HCN
   (d) 2,2,4,4-Tetramethyl-3-pentanone + HCN

10. Ethers are stable to all of the following reagents except:
   (a) nucleophiles (b) bases (c) strong acids (d) dilute acids

Chapter 15 – Carboxylic Acids and Nitriles

Chapter Outline

I. General information about carboxylic acids and nitriles (Sections 15.1 – 15.4).
  A. Naming carboxylic acids (Section 15.1).
    1. Noncyclic carboxylic acids are named by replacing the *-e* of the corresponding alkane by *-oic acid.*
    2. Compounds that have a carboxylic acid bonded to a ring are named by using the suffix *-carboxylic acid.*
    3. Many carboxylic acids have historical, nonsystematic names.
  B. Naming nitriles.
    1. Simple nitriles are named by adding *-nitrile* to the alkane name.
       The nitrile carbon is C1.
    2. More complex nitriles are named as derivatives of carboxylic acids by replacing *-oic acid* by *-onitrile* or by replacing *-carboxylic acid* by *-carbonitrile.*
       The nitrile carbon is bonded to C1.
  C. Structure and properties of carboxylic acids (Section 15.2).
    1. The carbonyl group of carboxylic acids is sp^2-hybridized and planar.
    2. Carboxylic acids are strongly associated because of hydrogen-bonding, and their boiling points are elevated.
  D. Carboxylic acid acidity (Sections 15.2 – 15.4).
    1. Dissociation of carboxylic acids (Section 15.2).
      a. Carboxylic acids react with bases to form salts that are water-soluble.
      b. Carboxylic acids dissociate slightly in dilute aqueous solution to give H_3O^+ and carboxylate anions.
         The K_a values for carboxylic acids are near 10^{-5}, making them weaker than mineral acids but stronger than alcohols.
      c. The relative strength of carboxylic acids is due to resonance stabilization of the carboxylate anion.
        i. Both carbon-oxygen bonds of carboxylate anions are the same length.
        ii. The bond length is intermediate between single and double bonds.
    2. Biological acids: the Henderson-Hasselbalch equation (Section 15.3)
      a. The pH of biological fluids (7.3) determines the ratio of dissociated to nondissociated forms of carboxylic acids.
      b. This ratio can be calculated by using the Henderson-Hasselbalch equation.

$$\log \frac{[A^-]}{[HA]} = pH - pK_a$$

      c. At physiological pH, carboxylic acids are almost completely dissociated.
    3. Substituent effects on acidity (Section 15.4).
      a. Carboxylic acids differ in acid strength.
        i. Electron-withdrawing groups stabilize carboxylate anions and increase acidity.
        ii. Electron-donating groups decrease acidity.
      b. These inductive effects decrease with increasing distance from the carboxyl group.
      c. Substituent effects in substituted benzoic acids.
        i. Groups that are deactivating in electrophilic aromatic substitution reactions increase the acidity of substituted benzoic acids.
        ii. The acidity of benzoic acids can be used to predict electrophilic reactivity.

II. Carboxylic acids (Sections 15.5 – 15.6).
  A. Preparation of carboxylic acids (Section 15.5).
    1. Methods already studied.
      a. Oxidation of substituted alkylbenzenes.
      b. Oxidation of primary alcohols and aldehydes.
    2. Nitrile hydrolysis.
      a. Nitriles can be hydrolyzed by strong aqueous acids or bases to yield carboxylic acids.
      b. The sequence nitrile formation → nitrile hydrolysis can be used to prepare a carboxylic acid from a halide.
      c. This method is generally limited to compounds that can undergo S_N2 reactions.
    3. Carboxylation of Grignard reagents.
      a. A Grignard reagent can be treated with CO_2 (and protonated) to form a carboxylic acid.
      b. This method is limited to compounds that don't have other functional groups that interfere with Grignard reagent formation.
  B. Reactions of carboxylic acids (Section 15.6).
    1. Carboxylic acids can undergo reactions typical of alcohols and ketones.
    2. Other types of reactions of carboxylic acids:
      a. Alpha substitution.
      b. Decarboxylation.
      c. Nucleophilic acyl substitution.
      d. Reduction.

III. Chemistry of nitriles (Section 15.7).
  A. Preparation of nitriles.
    1. Nitriles can be prepared by S_N2 reaction of ^-CN with a primary alkyl halide.
    2. They can also be prepared by $SOCl_2$ dehydration of primary amides.
  B. Reactions of nitriles.
    1. Nitriles can react with nucleophiles via sp^2-hybridized imine intermediates.
    2. Hydrolysis.
       Aqueous base hydrolyzes nitriles to carboxylates, plus an amine/ammonia.
      a. The reaction involves formation of a hydroxy imine that isomerizes to an amide, which is further hydrolyzed.
      b. Milder conditions allow isolation of the amide.
    3. Reduction.
       $LiAlH_4$ reduces nitriles to primary amines.

IV. Spectroscopy of carboxylic acids and nitriles (Section 15.8).
  A. Infrared spectroscopy.
    1. The O–H absorption occurs at 2500–3300 cm^{-1} and is easy to identify.
    2. The C=O absorption occurs at 1710–1760 cm^{-1}
       The position of this absorption depends on whether the acid is free (1760 cm^{-1}) or associated (1710 cm^{-1}).
    3. Nitriles have an intense absorption at 2250 cm^{-1} that readily identifies them.
  B. NMR spectroscopy.
    1. ^{13}C NMR spectroscopy.
      a. Carboxylic acids absorb between 165–185 δ.
      b. Saturated acids absorb downfield from α,β-unsaturated acids.
      c. Nitrile carbons absorb in the range 115–130 δ
    2. ^1H NMR spectroscopy.
       The carboxylic acid proton absorbs at around 12 δ.

Solutions to Problems

15.1 Carboxylic acids are named by replacing *-e* of the corresponding alkane with *-oic acid.* The carboxylic acid carbon is C1.
When-CO_2H is a substituent of a ring, the suffix *-carboxylic acid* is used; the carboxyl carbon is not numbered in this system.

(a) $CH_3CH(CH_3)CH_2COOH$
3-Methylbutanoic acid

(b) $CH_3CHBrCH_2CH_2COOH$
4-Bromopentanoic acid

(c) $CH_3CH_2CH(CO_2H)CH_2CH_2CH_3$
2-Ethylpentanoic acid

(d) $H_3C(H)C{=}C(H)CH_2CH_2COOH$
(Z)-Hex-4-enoic acid

(e) $CH_3CH(CH_3)CH_2CH(CN)CH_3$
2,4-Dimethylpentanenitrile

(f) cis-1,3-cyclopentane with HO_2C and CO_2H
cis-Cyclopentane-1,3-dicarboxylic acid

15.2

(a) $CH_3CH_2CH_2CH(CH_3)CH(CH_3)CO_2H$
2,3-Dimethylhexanoic acid

(b) $CH_3CH(CH_3)CH_2CH_2CO_2H$
4-Methylpentanoic acid

(c) trans-1,2-cyclobutane with HO_2C, H, H, CO_2H
trans-Cyclobutane-1,2-dicarboxylic acid

(d) benzene ring with CO_2H and OH
o-Hydroxybenzoic acid

(e) CO_2H
(9Z,12Z)-Octadeca-9,12-dienoic acid

(f) $CH_3CH_2CH{=}CHCN$
Pent-2-enenitrile

15.3 Naphthalene is insoluble in water and benzoic acid is only slightly soluble. The *salt* of benzoic acid is very soluble in water, however, and we can take advantage of this solubility in separating naphthalene from benzoic acid.

Dissolve the mixture in an organic solvent, and extract with a dilute aqueous solution of sodium hydroxide or sodium bicarbonate, which will neutralize benzoic acid. Naphthalene will remain in the organic layer, and all of the benzoic acid, now converted to the benzoate salt, will be in the aqueous layer. To recover benzoic acid, remove the aqueous layer, acidify it with dilute mineral acid, and extract with an organic solvent.

15.4

$$Cl_2CHCO_2H + H_2O \xrightleftharpoons{K_a} Cl_2CHCO_2^- + H_3O^+$$

$$K_a = \frac{[Cl_2CHCOO^-][H_3O^+]}{[Cl_2CHCOOH]} = 3.32 \times 10^{-2}$$

	Initial molarity	*Molarity after dissociation*
Cl_2CHCO_2H	0.10 M	0.10 M – y
$Cl_2CHCO_2^-$	0	y
H_3O^+	0	y

$$K_a = \frac{y \cdot y}{0.10 - y} = 3.32 \times 10^{-2}$$

Using the quadratic formula to solve for y, we find that y = 0.0434

$$\text{Percent dissociation} = \frac{0.0434}{0.1000} \times 100\% = 43.4\%$$

15.5

(a)

$$\log \frac{[A^-]}{[HA]} = pH - pK_a = 4.50 - 3.83 = 0.67$$

$$\frac{[A^-]}{[HA]} = \text{antilog}(0.67) = 4.68\text{: } [A^-] = 4.68\,[HA]$$

$$[HA] + [A^-] = 100\%$$

$$[HA] + 4.68\,[HA] = 5.68\,[HA] = 100\%$$

$$[HA] = 100\% \div 5.68 = 18\%$$

$$[A^-] = 100\% - 18\% = 82\%$$

82% of 0.0010 M glycolic acid is dissociated at pH = 4.50

(b)

$$\log \frac{[A^-]}{[HA]} = pH - pK_a = 5.30 - 4.87 = 0.43$$

$$\frac{[A^-]}{[HA]} = \text{antilog}(0.43) = 2.69\text{: } [A^-] = 2.69\,[HA]$$

$$[HA] + [A^-] = 100\%$$

$$[HA] + 2.69\,[HA] = 3.69\,[HA] = 100\%$$

$$[HA] = 100\% \div 3.69 = 27\%$$

$$[A^-] = 100\% - 27\% = 73\%$$

73% of 0.0020 M propanoic acid is dissociated at pH = 5.30.

15.6 You would expect lactic acid to be a stronger acid because the electron-withdrawing inductive effect of the hydroxyl group can stabilize the lactate anion.

15.7

$$HO(O{=})C{-}C({=}O)OH + H_2O \rightleftharpoons H_3O^+ + HO(O{=})C{-}C({=}O)O^- \longleftrightarrow HO(O{=})C{-}C(O^-){=}O$$

The pK_1 of oxalic acid is lower than that of a monocarboxylic acid because the carboxylate anion is stabilized both by resonance and by the electron-withdrawing inductive effect of the nearby second carboxylic acid group.

$$HO(O{=})C{-}C({=}O)O^- + H_2O \rightleftharpoons H_3O^+ + {}^-O(O{=})C{-}C({=}O)O^- \longleftrightarrow O{=}C(O^-){-}C(O^-){=}O$$

The pK_2 of oxalic acid is higher than pK_1 because electrostatic repulsion between the two adjacent negative charges destabilizes the dianion.

15.8 A pK_a of 4.45 indicates that *p*-cyclopropylbenzoic acid is a weaker acid than benzoic acid. This, in turn, indicates that a cyclopropyl group must be electron-donating. Since electron-donating groups increase reactivity in electrophilic substitution reactions, *p*–cyclopropylbenzene should be more reactive than benzene toward electrophilic bromination.

15.9 In part (a), Grignard carboxylation must be used because the starting materials can't undergo S_N2 reactions. In (b), either method can be used.

(a)

$$(CH_3)_3CCl \xrightarrow[3.\ H_3O^+]{1.\ Mg,\ ether;\ 2.\ CO_2,\ ether} (CH_3)_3CCO_2H$$

(b)

$$CH_3CH_2CH_2Br \xrightarrow[3.\ H_3O^+]{1.\ Mg,\ ether;\ 2.\ CO_2,\ ether} \textit{or} \xrightarrow[2.\ H_3O^+]{1.\ NaCN} CH_3CH_2CH_2CO_2H$$

15.10

$$C_6H_5CH_2Br \xrightarrow[3.\ H_3O^+]{1.\ Mg,\ ether;\ 2.\ CO_2,\ ether} \textit{or} \xrightarrow[2.\ H_3O^+]{1.\ NaCN} C_6H_5CH_2CO_2H \xrightarrow[2.\ H_3O^+]{1.\ LiAlH_4} C_6H_5CH_2CH_2OH$$

The alcohol product can be formed by reduction of a carboxylic acid with $LiAlH_4$. The carboxylic acid can be synthesized from benzyl bromide either by Grignard carboxylation or by nitrile hydrolysis.

15.11

$CH_3CH(CH_3)CH_2CH_2C{\equiv}N \xrightarrow[\text{NaOH, } H_2O]{H_3O^+ \text{ or}} CH_3CH(CH_3)CH_2CH_2C(=O)NH_2$

$\xrightarrow[\text{NaOH, } H_2O]{H_3O^+ \text{ or}} CH_3CH(CH_3)CH_2CH_2C(=O)OH$

$CH_3CH(CH_3)CH_2CH_2C(=O)OH \xrightarrow[2.\ H_3O^+]{1.\ LiAlH_4} CH_3CH(CH_3)CH_2CH_2CH_2OH$

15.12

Cyclopentanecarboxylic acid: C=O 1715 cm^{-1}; OH 2500–3300 cm^{-1}

4-Hydroxycyclohexanone: HO 3300–3400 cm^{-1}; C=O 1710 cm^{-1}

The positions of the carbonyl absorptions are too similar to be useful. The –OH absorptions, however, are sufficiently different for distinguishing between the compounds; the broad band of the carboxylic acid hydroxyl group is especially noticeable.

15.13 ^1H NMR:

Cyclopentanecarboxylic acid: OH ← 12 δ

4-Hydroxycyclohexanone: 3.5–4.5 δ → H (on carbon bearing HO); ← 2.0–2.3 δ (hydrogens next to C=O)

The distinctive peak at 12 δ serves to identify the carboxylic acid. For the hydroxy ketone, the absorption of the hydrogen on the oxygen-bearing carbon (3.5–4.5 δ) is significant. The position of absorption of the hydroxyl hydrogen is unpredictable, but addition of D_2O to the sample can be used to identify this peak.

^{13}C NMR:

Cyclopentanecarboxylic acid: C ← 180 δ

4-Hydroxycyclohexanone: 50–60 δ (carbon bearing HO); 210 δ (C=O)

The positions of the carbonyl carbon absorptions can be used to distinguish between these two compounds. The hydroxy ketone also shows an absorption in the range 50–60 δ due to the hydroxyl group carbon.

Visualizing Chemistry

15.14

(a) 3-Bromo-4-methoxybenzoic acid

(b) 3-Methylbut-2-enoic acid

(c) Cyclopenta-1,3-dienecarboxylic acid

(d) (*S*)-3-Cyclopentyl-2-methylpropanoic acid

15.15

(a) (b)

(a) *p*-Bromobenzoic acid is more acidic than benzoic acid because the electron-withdrawing bromine stabilizes the carboxylate anion.
(b) This *p*-substituted aminobenzoic acid is less acidic than benzoic acid because the electron-donating group destabilizes the carboxylate anion.

15.16

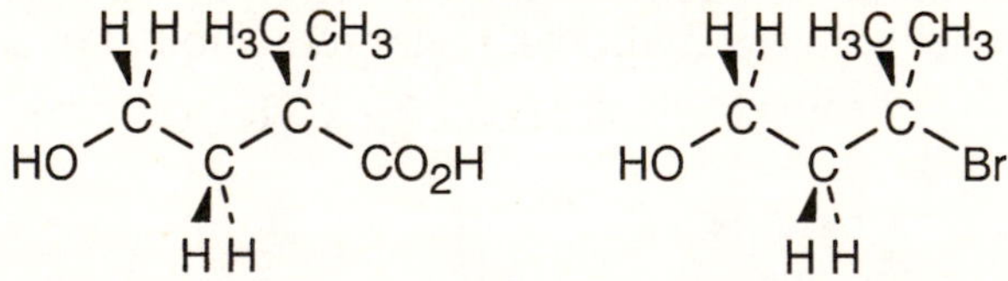

Nitrile hydrolysis can't be used to synthesize the above carboxylic acid because the tertiary halide precursor (shown on the right) doesn't undergo S_N2 substitution with cyanide. Grignard carboxylation also can't be used because the acidic hydroxyl hydrogen interferes with formation of the Grignard reagent. If the hydroxyl group is protected, however, Grignard carboxylation can take place.

15.17 The electrostatic potential maps show that the aromatic ring of anisole is more electron-rich than the aromatic ring of thioanisole, indicating that the methoxyl group is more strongly electron-donating than the methylthio group. Since electron-donating groups decrease acidity, *p*-(methylthio)benzoic acid is likely to be a stronger acid than *p*-methoxybenzoic acid.

Additional Problems

15.18

(a)

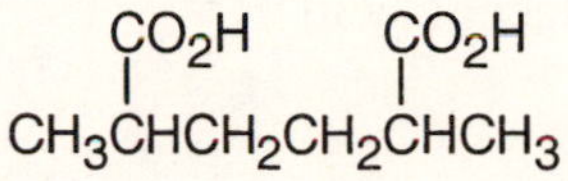

$$\begin{array}{c} \quad CO_2H \qquad\quad CO_2H \\ \quad | \qquad\qquad\quad | \\ CH_3CHCH_2CH_2CHCH_3 \end{array}$$

2,5-Dimethylhexanedioic acid

(b)

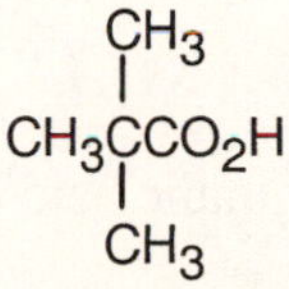

$$\begin{array}{c} CH_3 \\ | \\ CH_3CCO_2H \\ | \\ CH_3 \end{array}$$

2,2-Dimethylpropanoic acid

(c) NC, CO_2H

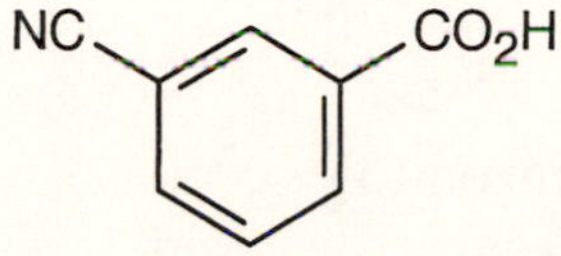

m-Cyanobenzoic acid

(d) CO_2H

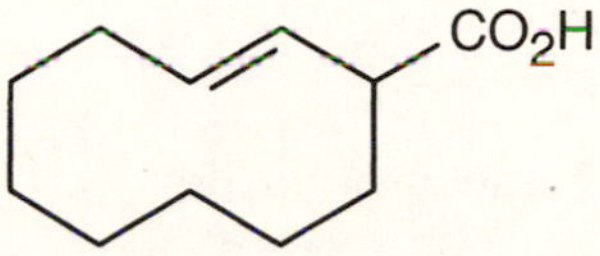

E-Cyclodec-2-enecarboxylic acid

(e)

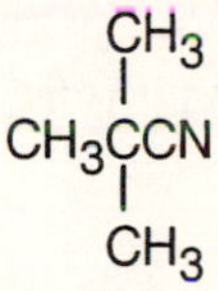

$$\begin{array}{c} CH_3 \\ | \\ CH_3CCN \\ | \\ CH_3 \end{array}$$

2,2-Dimethylpropanenitrile

(f)

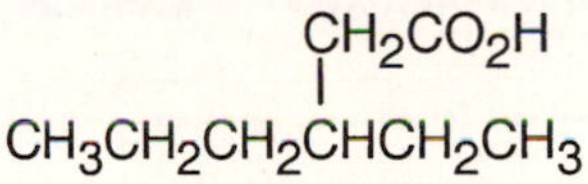

$$\begin{array}{c} CH_2CO_2H \\ | \\ CH_3CH_2CH_2CHCH_2CH_3 \end{array}$$

3-Ethylhexanoic acid

(g)

$$\begin{array}{c} Br \\ | \\ BrCH_2CHCH_2CH_2CO_2H \end{array}$$

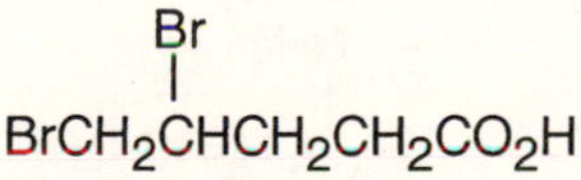

4,5-Dibromopentanoic acid

(h) CN

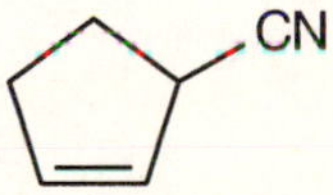

Cyclopent-2-enecarbonitrile

15.19

(a) H, CO_2H, CO_2H, H

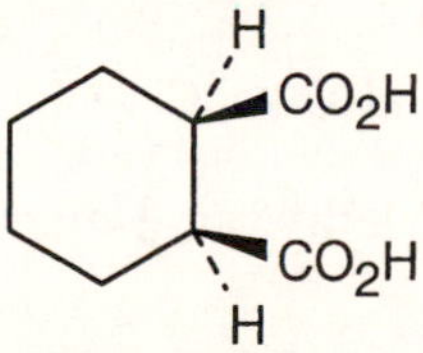

cis-Cyclohexane-1,2-dicarboxylic acid

(b)

$HO_2CCH_2CH_2CH_2CH_2CH_2CO_2H$

Heptanedioic acid

(c)

$CH_3C{\equiv}CCH{=}CHCO_2H$

Hex-2-en-4-ynoic acid

(d)

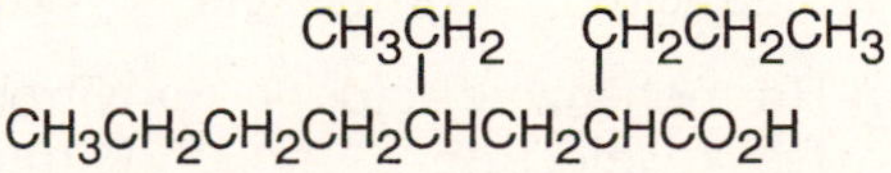

$$\begin{array}{c} \qquad\qquad\qquad CH_3CH_2 \quad CH_2CH_2CH_3 \\ \qquad\qquad\qquad | \qquad\qquad | \\ CH_3CH_2CH_2CH_2CHCH_2CHCO_2H \end{array}$$

4-Ethyl-2-propyloctanoic acid

(e)

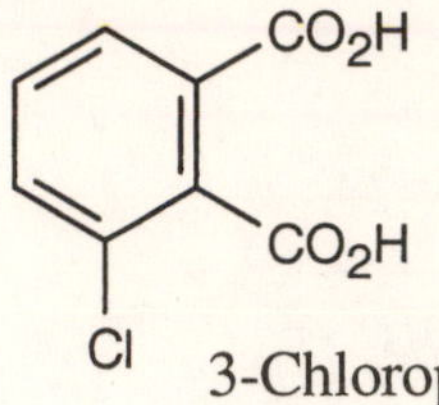

3-Chlorophthalic acid

(f)

$(C_6H_5)_3CCO_2H$
Triphenylacetic acid

(g)

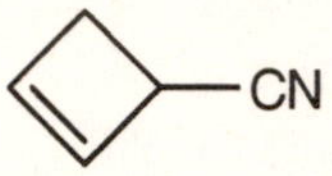

Cyclobut-2-enecarbonitrile

(h)

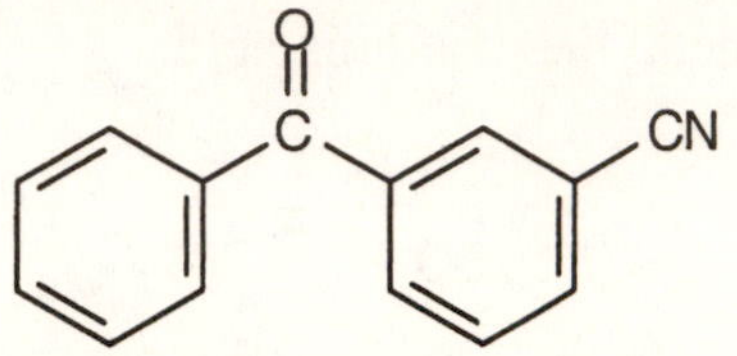

m-Benzoylbenzonitrile

15.20

(a)

$CH_3CH_2CH_2CH_2CH_2CO_2H$
Hexanoic acid

$CH_3CH_2CH_2CH(CH_3)CO_2H$
2-Methylpentanoic acid

$CH_3CH_2CH(CH_3)CH_2CO_2H$
3-Methylpentanoic acid

$CH_3CH(CH_3)CH_2CH_2CO_2H$
4-Methylpentanoic acid

$CH_3CH_2CH(CH_2CH_3)CO_2H$
2-Ethylbutanoic acid

$CH_3CH_2C(CH_3)_2CO_2H$
2,2-Dimethylbutanoic acid

$CH_3CH(CH_3)CH(CH_3)CO_2H$
2,3-Dimethylbutanoic acid

$CH_3C(CH_3)_2CH_2CO_2H$
3,3-Dimethylbutanoic acid

(b)

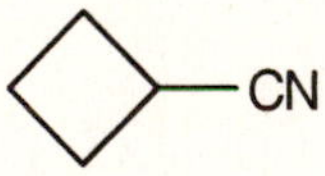

Cyclobutanecarbonitrile

$H_2C{=}CHCH_2CH_2CN$
Pent-4-enenitrile

$H_2C{=}C(CH_3)CH_2CN$
3-Methylbut-3-enenitrile

Other nitriles with the formula C_5H_7N can also be drawn.

15.21

Less acidic ⟶ *More acidic*

(a) CH_3CO_2H (Acetic acid) < HCO_2H (Formic acid) < $HO_2C{-}CO_2H$ (Oxalic acid)

(b) *p*-Bromobenzoic acid (weakly electron-withdrawing substituent) < *p*-Nitrobenzoic acid (strongly electron-withdrawing substituent) < 2,4-Dinitrobenzoic acid (two strongly electron-withdrawing substituents)

(c) $FCH_2CH_2CH_2CO_2H$ < $FCH_2CH_2CO_2H$ < FCH_2CO_2H

In (c), the strongest acid has the most electronegative atom nearest to the carboxylic acid group. The weakest acid has an electronegative atom farthest away from the carboxylic acid group.

15.22 Remember that the conjugate base of a weak acid is a strong base. In other words, the stronger the acid, the weaker the base derived from that acid.

Less basic ⟶ *More basic*

(a) $Mg(OAc)_2$ < $Mg(OH)_2$ < $H_3C^- MgBr^+$

Acetic acid is a much stronger acid than water, which is a much, much stronger acid than methane. The order of base strength is just the reverse.

(b) Sodium *p*-nitrobenzoate < Sodium benzoate < $HC{\equiv}C^- Na^+$

p-Nitrobenzoic acid is stronger than benzoic acid, which is much stronger than acetylene.

(c) $HCO_2^- Li^+$ < $HO^- Li^+$ < $CH_3CH_2O^- Li^+$

$LiOH$ and $LiOCH_2CH_3$ are very similar in basicity.

15.23

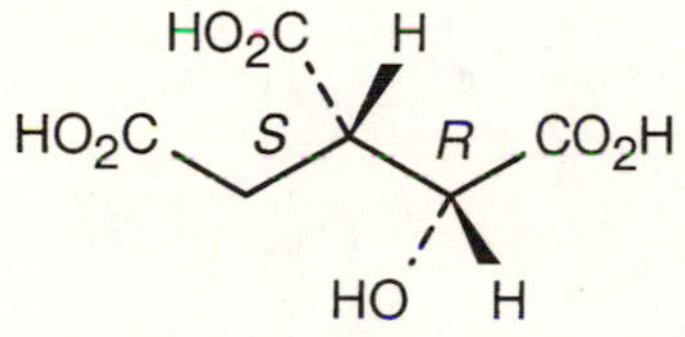

HO2C H
HO2C S R CO2H
HO H

(2*R*,3*S*)-3-Carboxy-2-hydroxypentanedioic acid

15.24

(a)

$CH_3CH_2CH_2CO_2H$ $\xrightarrow[\text{2. } H_3O^+]{\text{1. } LiAlH_4}$ $CH_3CH_2CH_2CH_2OH$

Butan-1-ol

(b)

$CH_3CH_2CH_2CH_2OH$ from (a) $\xrightarrow{PBr_3}$ $CH_3CH_2CH_2CH_2Br$

1-Bromobutane

(c)

$CH_3CH_2CH_2CH_2Br$ from (b) $\xrightarrow[\text{2. } H_3O^+]{\text{1. NaCN}}$ $CH_3CH_2CH_2CH_2CO_2H$

Pentanoic acid

Grignard carboxylation can also be used.

(d)

$CH_3CH_2CH_2CH_2Br$ from (b) $\xrightarrow{K^+ \; ^-OC(CH_3)_3}$ $CH_3CH_2CH{=}CH_2$

But-1-ene

15.25

(a)

$$CH_3CH_2CH_2CH_2OH \xrightarrow[H_3O^+]{CrO_3} CH_3CH_2CH_2CO_2H$$

(b)

$$CH_3CH_2CH_2CH_2Br \xrightarrow{NaOH} CH_3CH_2CH_2CH_2OH \xrightarrow[H_3O^+]{CrO_3} CH_3CH_2CH_2CO_2H$$

(c)

$$CH_3CH_2CH{=}CH_2 \xrightarrow[2.\ H_2O_2,\ ^-OH]{1.\ BH_3,\ THF} CH_3CH_2CH_2CH_2OH \xrightarrow[H_3O^+]{CrO_3} CH_3CH_2CH_2CO_2H$$

(d)

$$CH_3CH_2CH_2Br \xrightarrow[2.\ H_3O^+]{1.\ NaCN} CH_3CH_2CH_2CO_2H$$

or

$$CH_3CH_2CH_2Br \xrightarrow[3.\ H_3O^+]{1.\ Mg,\ ether;\ 2.\ CO_2,\ ether} CH_3CH_2CH_2CO_2H$$

15.26

(a)

$$CH_3CH_2CH_2CN \xrightarrow[H_2O]{NaOH} CH_3CH_2CH_2CO_2H \xrightarrow[2.\ H_3O^+]{1.\ LiAlH_4} CH_3CH_2CH_2CH_2OH$$

(b)

$$\underset{\text{(from a)}}{CH_3CH_2CH_2CH_2OH} \xrightarrow[CH_2Cl_2]{PCC} CH_3CH_2CH_2CHO$$

(c)

$$\underset{\text{(from a)}}{CH_3CH_2CH_2CH_2OH} \xrightarrow{PBr_3} CH_3CH_2CH_2CH_2Br \xrightarrow{NaCN} CH_3CH_2CH_2CH_2CN$$

$$CH_3CH_2CH_2CH_2CN \xrightarrow{H_3O^+\ \text{or}\ NaOH,\ H_2O} CH_3CH_2CH_2CH_2CO_2H$$

15.27

(a)

Benzene $\xrightarrow[AlCl_3]{CH_3Cl}$ toluene (CH_3) $\xrightarrow[H_2O]{KMnO_4}$ benzoic acid (CO_2H) $\xrightarrow[FeCl_3]{Cl_2}$ CO_2H, Cl (meta)

m-Chlorobenzoic acid

(b)

Benzene $\xrightarrow[AlCl_3]{CH_3Cl}$ toluene (CH_3) $\xrightarrow[FeBr_3]{Br_2}$ CH_3, Br (para) $\xrightarrow[H_2O]{KMnO_4}$ CO_2H, Br (para)

p-Bromobenzoic acid

(c)

CO_2H (from a) $\xrightarrow[2.\ H_3O^+]{1.\ LiAlH_4}$ CH_2OH $\xrightarrow{PBr_3}$ CH_2Br $\xrightarrow[3.\ H_3O^+]{1.\ Mg,\ ether;\ 2.\ CO_2,\ ether}$ CH_2CO_2H

or

$\xrightarrow[2.\ H_3O^+]{1.\ NaCN}$

15.28 (a) $K_a = 8.4 \times 10^{-4}$ for lactic acid
$pK_a = -\log(8.4 \times 10^{-4}) = 3.08$

(b) $K_a = 5.6 \times 10^{-6}$ for acrylic acid
$pK_a = -\log(5.6 \times 10^{-6}) = 5.25$

15.29 (a) $pK_a = 3.14$ for citric acid
$K_a = 10^{-3.14} = 7.2 \times 10^{-4}$

(b) $pK_a = 2.98$ for tartaric acid
$K_a = 10^{-2.98} = 1.0 \times 10^{-3}$

15.30

$$\log \frac{[A^-]}{[HA]} = pH - pK_a = 3.00 - 3.42 = -0.42$$

$$\frac{[A^-]}{[HA]} = \text{antilog}\ (-0.42) = 0.38 : [A^-] = 0.38\ [HA]$$

$$[HA] + 0.38\ [HA] = 1.38\ [HA] = 100\%$$

$$[HA] = 100\% \div 1.38 = 72\%$$

$$[A^-] = 100\% - 72\% = 28\%$$

At pH = 3.00, thioglycolic acid is 28% dissociated.

15.31

$$\log \frac{[A^-]}{[HA]} = pH - pK_a = 6.00 - 5.61 = 0.39$$

$$\frac{[A^-]}{[HA]} = \text{antilog}\ (0.39) = 2.45 : [A^-] = 2.45\ [HA]$$

$$[HA] + 2.45\ [HA] = 3.45\ [HA] = 100\%$$

$$[HA] = 100\% \div 3.45 = 29\%$$

$$[A^-] = 100\% - 27\% = 71\%$$

At pH = 6.00, uric acid is 71% dissociated.

Uric acid is acidic because the anion formed by dissociation of any of the three hydrogens is stabilized by resonance. An example of a resonance form:

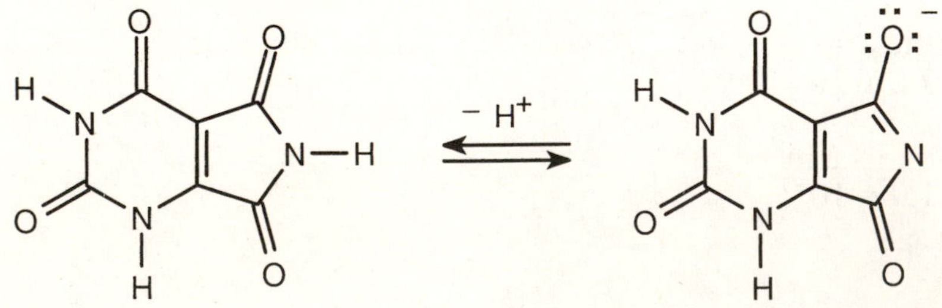

15.32 Inductive effects of functional groups are transmitted through σ bonds. For oxalic acid, the electron-withdrawing inductive effect of one carboxyl group on the ionization of the second group is significant. However, as the length of the carbon chain increases, the effect of one functional group on another decreases. In this example, the influence of the second carboxyl group on the acidity of the first is barely felt by succinic and adipic acids.

15.33

p-Methylbenzoic acid: $H_3C-C_6H_4-CO_2H$

(a) 1. CH_3MgBr; 2. H_3O^+ → $H_3C-C_6H_4-CO_2H$ + CH_4

(b) $KMnO_4$, H_2O → $HO_2C-C_6H_4-CO_2H$

(c) 1. $LiAlH_4$; 2. H_3O^+ → $H_3C-C_6H_4-CH_2OH$

In (a), the acidic proton reacts with the Grignard reagent to form methane.

15.34

(a)

$$CH_3CH_2Br \xrightarrow{\text{Mg, ether}} CH_3CH_2MgBr \xrightarrow[\text{2. } H_3O^+]{\text{1. } ^{13}CO_2\text{, ether}} CH_3CH_2{}^{13}CO_2H$$

(b)

$$CH_3Br \xrightarrow[\text{ether}]{\text{Mg}} CH_3MgBr \xrightarrow[\text{2. } H_3O^+]{\text{1. } ^{13}CO_2\text{, ether}} CH_3{}^{13}CO_2H \xrightarrow[\text{2. } H_3O^+]{\text{1. } LiAlH_4} CH_3{}^{13}CH_2OH$$

$$CH_3{}^{13}CH_2OH \xrightarrow{PBr_3} CH_3{}^{13}CH_2Br \xrightarrow[\text{ether}]{\text{Mg}} CH_3{}^{13}CH_2MgBr \xrightarrow[\text{2. } H_3O^+]{\text{1. } CO_2\text{, ether}} CH_3{}^{13}CH_2CO_2H$$

15.35

Formation of imine:

$$R\text{–}CHO + :NH_2OH \xrightarrow{1.} \left[R\text{–}CH(O^-)\text{–}NH_2OH^+ \xrightarrow{2.} R\text{–}CH(OH)\text{–}NHOH \right] \xrightarrow{3.\ H_3O^+} \left[R\text{–}CH(OH_2^+)\text{–}NHOH \xrightarrow{4.} R\text{–}CH{=}N^+(H)OH \right] \xrightarrow{5.} R\text{–}CH{=}N\text{–}OH + H_3O^+$$

imine

Step 1: Addition of hydroxylamine
Step 2: Proton transfer
Step 3: Protonation of hydroxyl oxygen
Step 4: Loss of water
Step 5: Loss of proton to form imine

Dehydration of imine

$$R\text{–}CH{=}N\text{–}OH + Cl\text{–}S(=O)\text{–}Cl \xrightarrow{1.} \left[Cl^- + R\text{–}CH{=}N\text{–}O\text{–}S(=O)\text{–}Cl \quad :Base \right] \xrightarrow{2.} R\text{–}C{\equiv}N + Cl^- + SO_2$$

Step 1: $SOCl_2$ adds to –OH to make it a better leaving group.
Step 2: Base brings about E2 elimination to form the nitrile.

15.36

$$^-O\text{–}C(=O)\text{–}CH_2\text{–}C(CH_3)(OPO_3^{2-})\text{–}CH_2\text{–}CH_2OP_2O_5^{\,3-} \longrightarrow H_2C{=}C(CH_3)\text{–}CH_2\text{–}CH_2OP_2O_5^{\,3-} + CO_2 + PO_4^{\,3-}$$

15.37

(a)

Methylenecyclohexane (CH_2) $\xrightarrow[\text{2. } H_2O_2,\ ^-OH]{\text{1. } BH_3,\ THF}$ (H, CH_2OH cyclohexane) $\xrightarrow{PBr_3}$ (CH_2Br cyclohexane) $\xrightarrow[\text{2. } H_3O^+]{\text{1. NaCN}}$ (CH_2CO_2H cyclohexane)

Grignard carboxylation can also be used to form the carboxylic acid.

(b)

Methylenecyclohexane (CH_2) $\xrightarrow{HBr}$ (H_3C, Br cyclohexane) $\xrightarrow[\text{ether}]{Mg}$ (H_3C, MgBr cyclohexane) $\xrightarrow[\text{2. } H_3O^+]{\text{1. } CO_2\text{, ether}}$ (H_3C, CO_2H cyclohexane)

Only Grignard carboxylation can be used because ^-CN brings about elimination of the tertiary bromide to form a double bond.

15.38 (a) Grignard carboxylation can't be used to prepare the carboxylic acid because of the acidic phenolic group. Use nitrile hydrolysis.

(b) Either method produces the carboxylic acid. Grignard carboxylation is a better reaction for preparing a carboxylic acid from a secondary bromide. Nitrile hydrolysis produces an optically active carboxylic acid from an optically active bromide.

(c) Neither method of acid synthesis yields the desired product. Any Grignard reagent formed will react with the carbonyl functional group present in the starting material. Reaction with cyanide occurs at the carbonyl functional group, producing a cyanohydrin, as well as at halogen. However, if the ketone is first protected by forming an acetal, either method can be used.

(d) Since the hydroxyl proton interferes with formation of the Grignard reagent, nitrile hydrolysis must be used to form the carboxylic acid.

15.39

$$H_2C{=}CHCH{=}CH_2 \xrightarrow[\text{1,4-addition}]{Br_2,\ CH_2Cl_2} BrCH_2CH{=}CHCH_2Br$$

$$\downarrow \text{2 NaCN, HCN}$$

$$H_2NCH_2CH_2CH{=}CHCH_2CH_2NH_2 \xleftarrow[\text{2. } H_2O]{\text{1. } LiAlH_4} NCCH_2CH{=}CHCH_2CN + 2\ NaBr$$

$$\downarrow H_2\text{, Pd/C}$$

$$H_2NCH_2CH_2CH_2CH_2CH_2CH_2NH_2$$

Hexane-1,6-diamine

15.40 2-Chloro-2-methylpentane is a tertiary alkyl halide and ^-CN is a base. Instead of the desired S_N2 reaction of cyanide with a halide, E2 elimination occurs and yields 2-methylpent-2-ene.

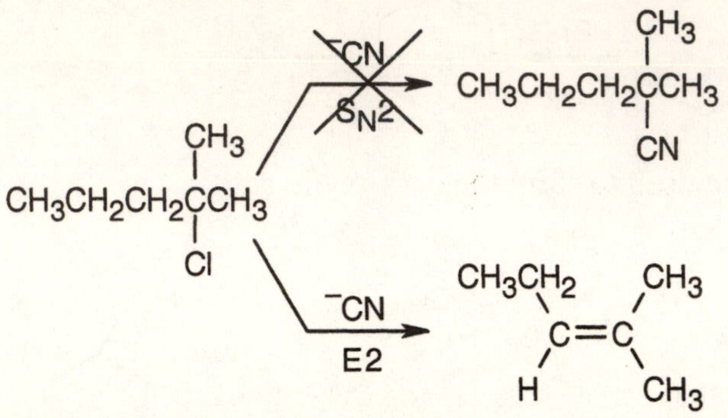

$CH_3CH_2CH_2C(CH_3)(Cl)CH_3$ —(^-CN, S_N2, crossed out)→ $CH_3CH_2CH_2C(CH_3)(CN)CH_3$

$CH_3CH_2CH_2C(CH_3)(Cl)CH_3$ —(^-CN, E2)→ $(CH_3CH_2)(H)C{=}C(CH_3)_2$

15.41

Isobutylbenzene —($ClCOCH_3$, $AlCl_3$)→ 4-isobutylacetophenone —(1. $NaBH_4$; 2. H_3O^+)→ 1-(4-isobutylphenyl)ethanol (OH) —(PBr_3)→ 1-bromo-1-(4-isobutylphenyl)ethane (Br) —(NaCN)→ nitrile (CN) —(1. H_2O, ^-OH; 2. H_3O^+)→ CO_2H

Ibuprofen

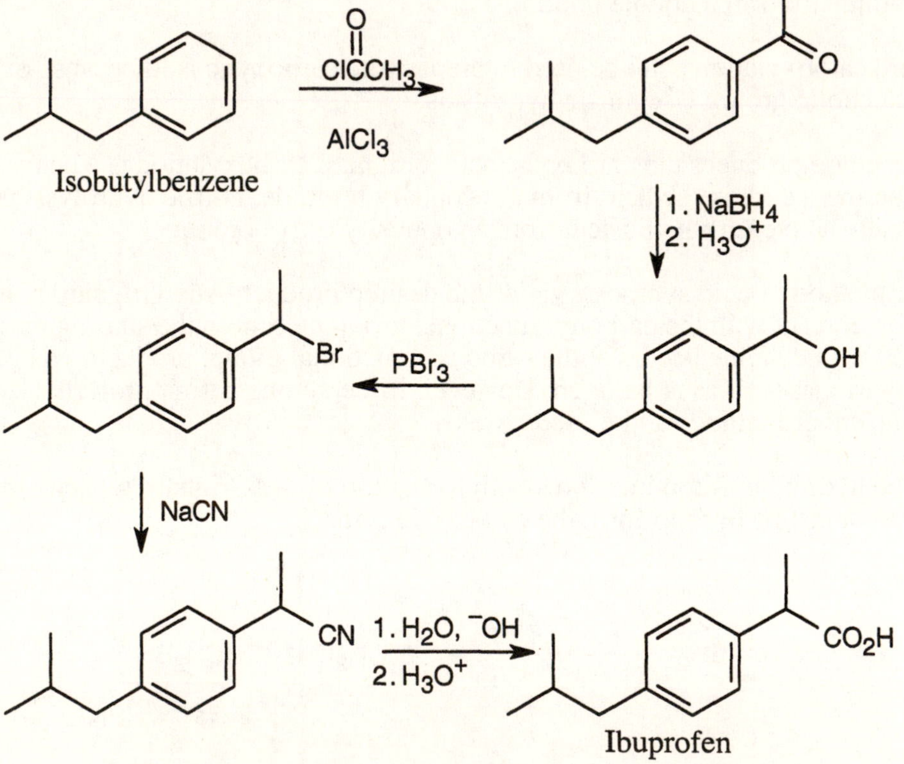

15.42

(a)

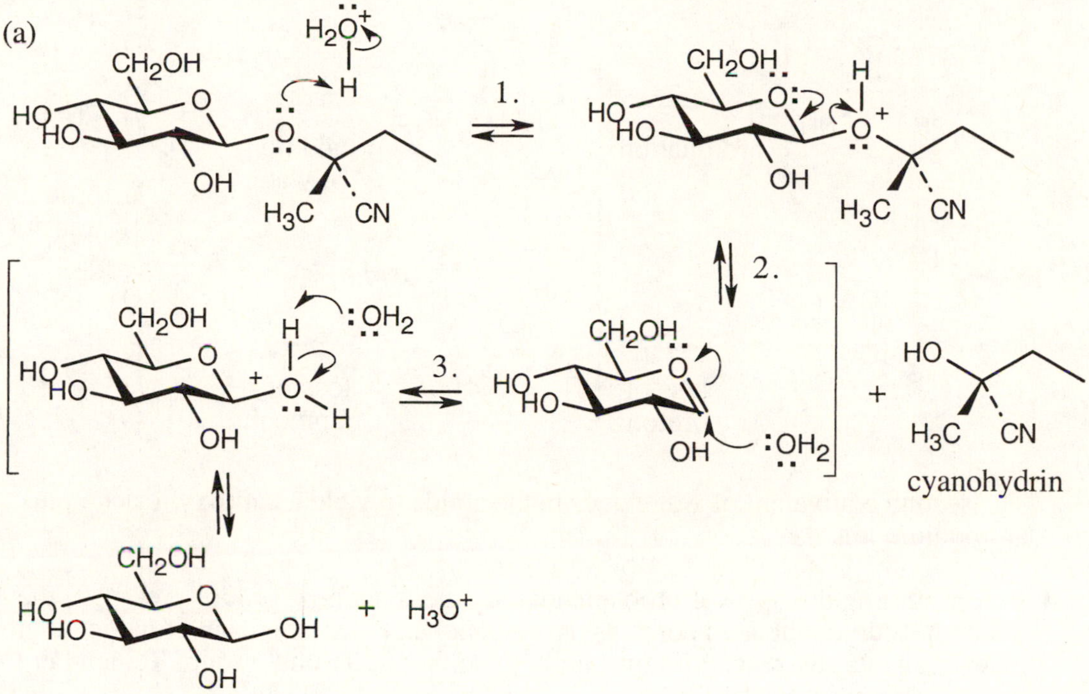

Step 1: Protonation of acetal oxygen.
Step 2: Loss of cyanohydrin.
Step 3: Addition of water, followed by deprotonation.

(b)

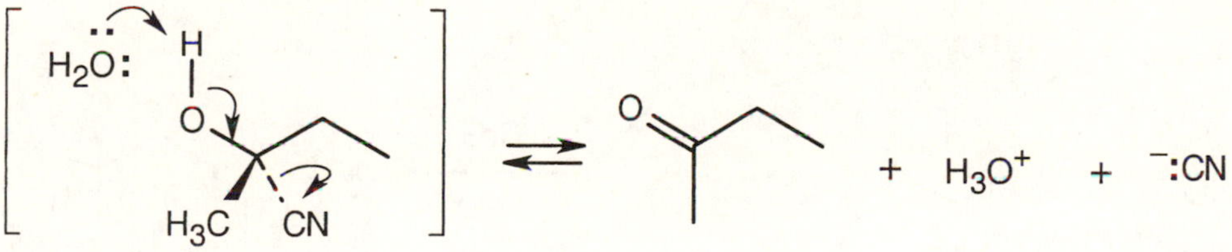

Deprotonation of the cyanohydrin hydroxyl group is followed by loss of $^{-}$CN, forming butan-2-one.

15.43

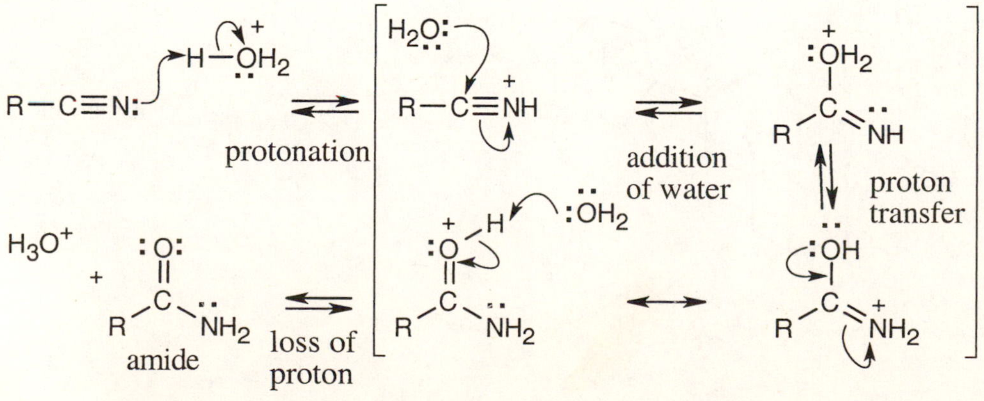

The first equivalent of water adds to a nitrile to produce an amide.

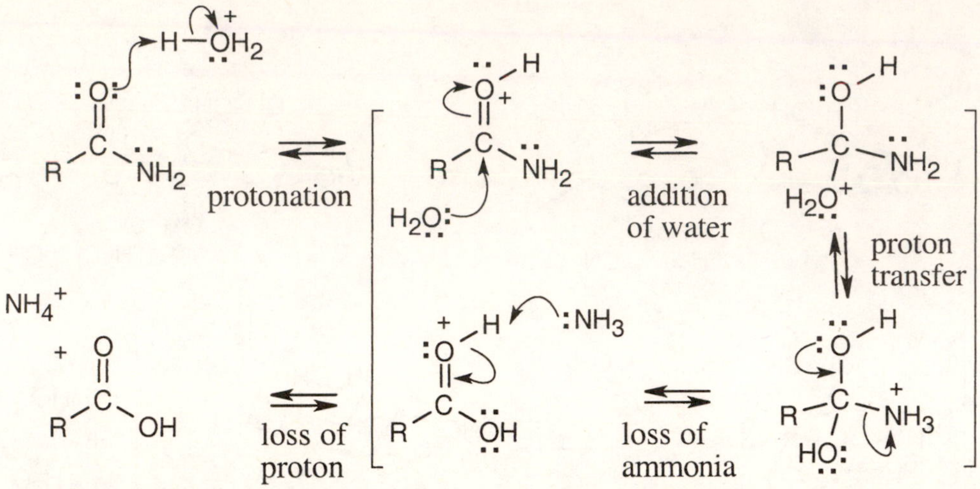

protonation

addition of water

proton transfer

loss of ammonia

loss of proton

The second equivalent of water adds to the amide to yield a carboxylic acid, plus ammonium ion.

15.44 Before starting this type of problem, identify the functional groups present in the starting material. Lithocholic acid contains only alcohol and carboxylic acid functional groups. The given reagents can react with one, both, or neither functional group. Remember to keep track of stereochemistry.

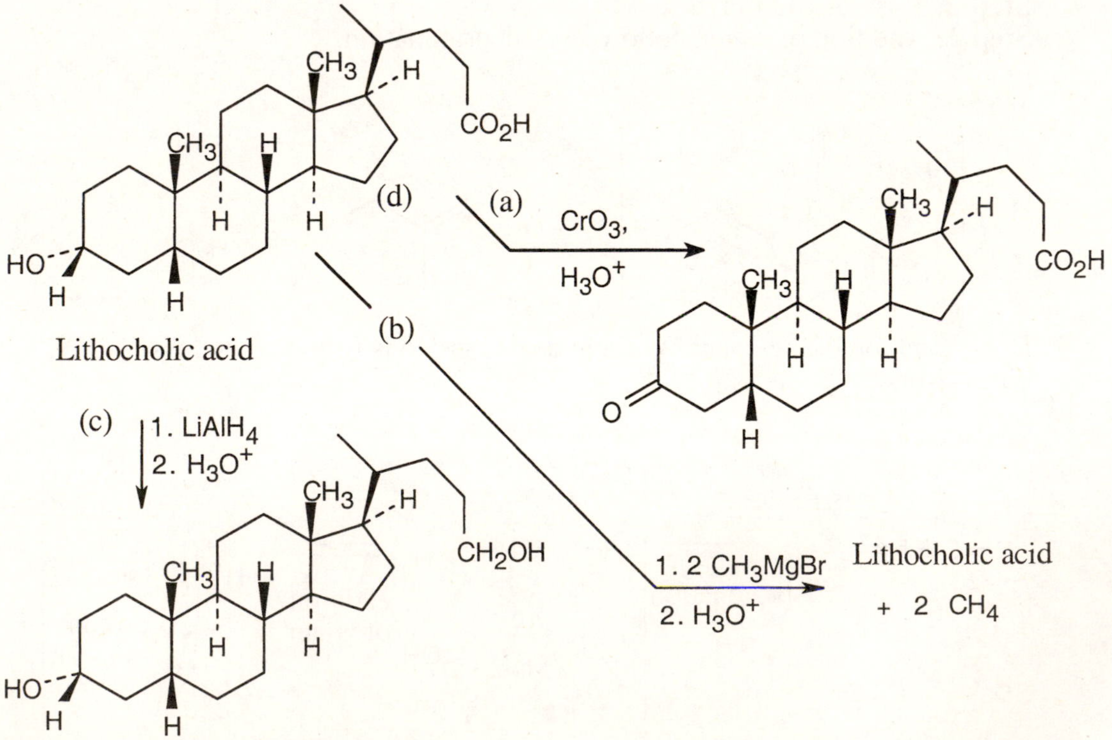

Lithocholic acid

(a) CrO_3, H_3O^+

(b) 1. 2 CH_3MgBr 2. H_3O^+ → Lithocholic acid + 2 CH_4

(c) 1. $LiAlH_4$ 2. H_3O^+

(d)

15.45

Substituent	pK_a	*Acidity*	*E.A.S. reactivity
$—PCl_2$	3.59	Most acidic	Least reactive (most deactivating)
$—OSO_2CH_3$	3.84	↓	↓
—CH=CHCN	4.03		
$—HgCH_3$	4.10		
—H	4.19		
$—Si(CH_3)_3$	4.27	Least acidic	Most reactive (activating)

*Electrophilic aromatic substitution

Recall from Section 15.4 that substituents that increase acidity also decrease reactivity in electrophilic aromatic substitution reactions. Of the above substituents, only $–Si(CH_3)_3$ is an activator.

15.46 As we have seen throughout this book, the influence of substituents on reactions can be due to inductive effects or to resonance effects. For *m*-hydroxybenzoic acid, the negative charge of the carboxylate anion is stabilized by the electron-withdrawing *inductive* effect of –OH, making this isomer more acidic. For *p*-hydroxybenzoic acid, the negative charge of the anion is destabilized by the electron-donating *resonance* effect of –OH that acts over the π electron system of the ring and makes the carboxyl-bearing carbon electron-rich.

HO–C₆H₄–COOH (meta) + H_2O ⇌ HO–C₆H₄–CO_2^- (meta) + H_3O^+

HO–C₆H₄–COOH (para) + H_2O ⇌ HO–C₆H₄–CO_2^- (para) + H_3O^+ ⟷ HO^+=C₆H₄(–)–CO_2^- and other resonance forms

15.47

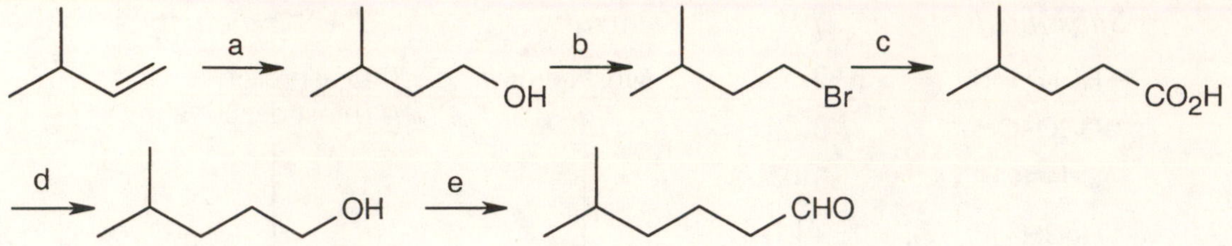

(a) BH_3, THF, then H_2O_2, OH^-; (b) PBr_3; (c) Mg, then CO_2, then H_3O^+ (or ^-CN, then H_3O^+); (d) $LiAlH_4$, then H_3O^+; (e) PCC

15.48

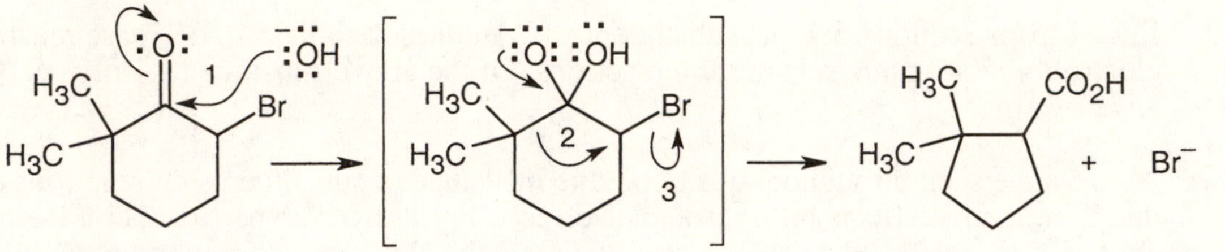

Nucleophilic addition (1), alkyl shift (2) and displacement of bromide (3) lead to the observed product.

15.49 The peak at 1.08 δ is due to a *tert*-butyl group, and the peak at 11.2 δ is due to a carboxylic acid group. The compound is 3,3-dimethylbutanoic acid, $(CH_3)_3CCH_2CO_2H$.

15.50 Either ^{13}C NMR or ^1H NMR can be used to distinguish among these three isomeric carboxylic acids.

Compound	*Number of ^{13}C NMR absorptions*	*Number of ^1H NMR absorptions*	*Splitting of ^1H NMR signals*
$CH_3(CH_2)_3CO_2H$	5	5	1 triplet, peak area 3, 1.0 δ 1 triplet, peak area 2, 2.4 δ 2 multiplets, peak area 4, 1.5 δ 1 singlet, peak area 1, 12.0 δ
$(CH_3)_2CHCH_2CO_2H$	4	4	1 doublet, peak area 6, 1.0 δ 1 doublet, peak area 2, 2.4 δ 1 multiplet, peak area 1, 1.6 δ 1 singlet, peak area 1, 12.0 δ
$(CH_3)_3CCO_2H$	3	2	1 singlet, peak area 9, 1.3 δ 1 singlet, peak area 1, 12.1 δ

15.51 In all of these pairs, different numbers of peaks occur in the spectra of each isomer. (a), (b) Use either ^{1}H NMR or ^{13}C NMR to distinguish between the isomers.

Compound	*Number of ^{13}C NMR absorptions*	*Number of ^{1}H NMR absorptions*
(a)		
p-HO_2C–C_6H_4–CO_2H	3	2
m-HO_2C–C_6H_4–CO_2H	5	4
(b)		
$HO_2CCH_2CH_2CO_2H$	2	2
$CH_3CH(CO_2H)_2$	3	3

(c) Use ^{1}H NMR to distinguish between these two compounds. The carboxylic acid proton of $CH_3CH_2CH_2CO_2H$ absorbs near 12 δ, and the aldehyde proton of $HOCH_2CH_2CH_2CHO$ absorbs near 10 δ and is split into a triplet.

(d) Cyclopentanecarboxylic acid shows four absorptions in both its ^{1}H NMR and ^{13}C NMR spectra. $(CH_3)_2C{=}CHCH_2CO_2H$ shows six absorptions in its ^{13}C NMR and five in its ^{1}H NMR spectrum; one of the ^{1}H NMR signals occurs in the vinylic region (4.5 – 6.5 δ) of the spectrum. The ^{13}C NMR spectrum of the unsaturated acid also shows two absorptions in the C=C bond region (100–150 δ).

15.52 The compound $C_4H_8O_3$ has one degree of unsaturation, consistent with the carboxylic acid absorption seen in the IR spectrum.

$$\overset{a}{CH_3}\overset{b}{CH_2}O\overset{c}{CH_2}CO_2\overset{d}{H}$$

a = 1.3 δ
b = 3.7 δ
c = 4.1 δ
d = 11.1 δ

15.53 A compound with the formula C_4H_7N has two degrees of unsaturation. The IR absorption at 2250 cm^{-1} identifies this compound as a nitrile.

$$\underset{a}{CH_3}\underset{b}{CH_2}\underset{c}{CH_2}C{\equiv}N$$

a = 1.1 δ
b = 1.7 δ
c = 2.3 δ

15.54 Both compounds contain four different kinds of protons (the H_2C= protons are nonequivalent). The carboxylic acid proton absorptions are easy to identify; the other three absorptions in each spectrum are more complex.

It is possible to assign the spectra by studying the methyl group absorptions. The methyl group peak of crotonic acid is split into a doublet by the geminal (CH_3CH=) proton, while the methyl group absorption of methacrylic acid is a singlet. The first spectrum (a) is that of crotonic acid, and the second spectrum (b) is that of methacrylic acid.

15.55 (a) From the formula, we know that the compound has 2 degrees of unsaturation, one of which is due to the carboxylic acid group that absorbs at 183.0 δ. The ^{13}C NMR spectrum also shows that no other sp^2 carbons are present in the sample and indicates that the other degree of unsaturation is due to a ring, which is shown to be a cyclohexane ring by symmetry and by the types of carbons in the structure.

(b) The compound has 5 degrees of unsaturation, and is a methyl-substituted benzoic acid. The symmetry shown by the aromatic absorptions identifies the compound as *p*-methylbenzoic acid.

(a) CO_2H

(b) CO_2H H_3C

15.56 This mechanism divides into two stages. In the first part of the mechanism, an acetal is hydrolyzed to acetone and a dihydroxycarboxylic acid.

H_3C CH_3 H H_2O^+ :O O: CH_3 CH_2CO_2H ⇌ protonation [H_3C CH_3 HO+ O: CH_3 CH_2CO_2H ⇌ ring opening H_2O: H_3C CH_3 OH +O: CH_3 CH_2CO_2H]

⇅ addition of water loss of H+

[CH_3 HO CH_3 OH HO+ CH_3 CH_2CO_2H ⇌ protonation H_2O^+ CH_3 HO H CH_3 OH :O CH_3 CH_2CO_2H ⇌ CH_3 HO CH_3 OH :O CH_3 CH_2CO_2H tetrahedral intermediate]

⇅ loss of alcohol

OH HO CH_3 CH_2CO_2H + [H_2O: H +O=C CH_3 CH_3] ⇌ loss of proton O=C CH_3 CH_3 + H_3O^+

Acetone

In the second stage, the dihydroxycarboxylic acid forms a cyclic ester (a lactone).

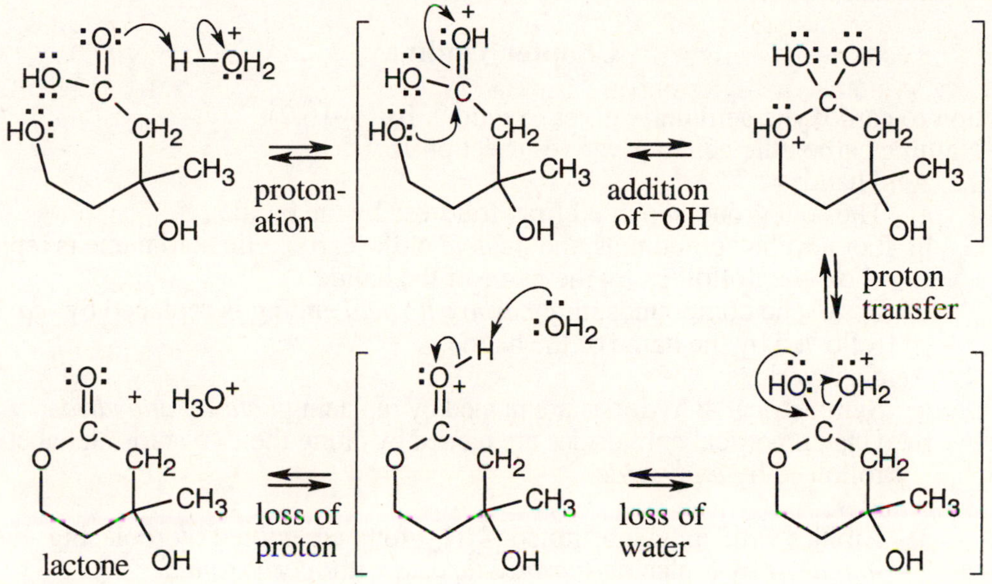

proton-ation

addition of ^{-}OH

proton transfer

loss of water

loss of proton

lactone

15.57

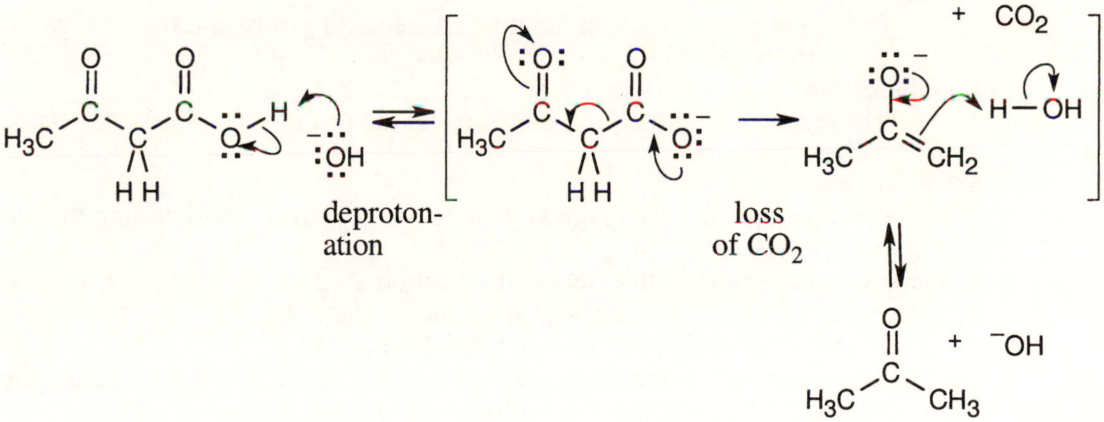

deproton-ation

loss of CO_2

Chapter 16 – Carboxylic Acid Derivatives and Nucleophilic Acyl Substitution Reactions

Chapter Outline

I. Introduction to carboxylic acid derivatives (Sections 16.1 – 16.2).
   A. Naming carboxylic acid derivatives (Section 16.1).
      1. Acid halides.
         a. The acyl group is named first, followed by the halide.
         b. For acyclic compounds, the *-ic acid* of the carboxylic acid name is replaced by *yl* or *-oyl*, followed by the name of the halide.
         c. For cyclic compounds, the *-carboxylic acid* ending is replaced by *-carbonyl,* followed by the name of the halide.
      2. Acid anhydrides.
         a. Symmetrical anhydrides are named by replacing *acid* by *anhydride.*
         b. Unsymmetrical anhydrides are named by citing the two acids alphabetically, followed by *anhydride*.
      3. Amides.
         a. Amides with an unsubstituted $–NH_2$ group are named by replacing *-oic acid* by *-amide* or by replacing *-carboxylic acid* with *-carboxamide.*
         b. If nitrogen is substituted, the nitrogen substituents are named, and an *N-* is put before each.
      4. Esters.
         Esters are named by first identifying the alkyl group and then the carboxylic acid group, replacing *-oic acid* by *-ate.*
      5. Thioesters.
         Thioesters are named like esters, using the prefix *thio-* before the name of the carboxylic acid.
      6. Acyl phosphates.
         Acyl phosphates are named by citing the acyl group and adding the word *phosphate*.
   B. Nucleophilic acyl substitution reactions (Section 16.2).
      1. Mechanism of nucleophilic acyl substitution reactions.
         a. A nucleophile adds to the polar carbonyl group.
         b. The tetrahedral intermediate eliminates one of the two substituents originally bonded to it, resulting in a net substitution reaction.
         c. Reactions of carboxylic acid derivatives take this course because one of the groups bonded to the carbonyl carbon is a good leaving group.
         d. The addition step is usually rate-limiting.
      2. Relative reactivity of carboxylic acid derivatives.
         a. Both steric and electronic factors determine relative reactivity.
            i. Steric hindrance in the acyl group decreases reactivity.
            ii. More polarized acid derivatives are more reactive than less polarized derivatives.
            iii. The effect of substituents on reactivity is similar to their effect on electrophilic aromatic substitution reactions.
         b. It is possible to convert more reactive derivatives into less reactive derivatives.
            i. In order of decreasing reactivity: acid chlorides > acid anhydrides > thioesters > esters > amides.
            ii. Only acyl phosphates, esters,thioesters and amides are found in nature.
      3. Kinds of reactions of carboxylic acid derivatives:
         a. Hydrolysis: reaction with water to yield a carboxylic acid.
         b. Alcoholysis: reaction with an alcohol to yield an ester.

c. Aminolysis: reaction with ammonia or an amine to yield an amide.
d. Reduction.
   i. Reaction with a hydride reducing agent yields an aldehyde or an alcohol.
   ii Amides are reduced to yield amines.

II. Reactions of carboxylic acids and their derivatives (Section 16.3 – 16.8).
   A. Nucleophilic acyl substitution reactions of carboxylic acids (Section 16.3).
      1. Carboxylic acids can be converted to acid chlorides by reaction with $SOCl_2$.
         The reaction proceeds through a chlorosulfite intermediate.
      2. Acid anhydrides are usually formed by heating the corresponding carboxylic acid to remove 1 equivalent of water.
      3. Conversion to esters.
         a. Conversion can be effected by the S_N2 reaction of a carboxylate and an alkyl halide.
         b. Esters can be produced by the acid-catalyzed reaction of a carboxylic acid and an alcohol.
            i. This reaction is known as a Fischer esterification.
            ii. Mineral acid makes the acyl carbon more reactive toward the alcohol.
            iii. All steps are reversible.
            iv. The reaction can be driven to completion by removing water or by using a large excess of alcohol.
            v. Isotopic labelling studies have confirmed the mechanism.
      4. Conversion to amides.
         a. Amides are difficult to form from carboxylic acids because amines convert carboxylic acids to carboxylate salts that no longer have electrophilic carbons.
         b. The reagent DCC can be used; it is used in the laboratory to form peptide bonds.
      5. Reduction to alcohols can be achieved by use of $LiAlH_4$.
   B. Chemistry of carboxylic acid halides (Section 16.4).
      1. Carboxylic acid halides are prepared by reacting carboxylic acids with either $SOCl_2$ or PBr_3.
      2. Acyl halides are very reactive.
         Most reactions occur by nucleophilic acyl substitution mechanisms.
      3. Hydrolysis.
         a. Acyl halides react with water to form carboxylic acids.
         b. The reaction mixture usually contains a base to scavenge the HCl produced.
      4. Alcoholysis.
         a. Acyl halides react with alcohols to form esters.
         b. Base is usually added to scavenge the HCl produced.
         c. Primary alcohols are more reactive than secondary or tertiary alcohols.
            It's often possible to esterify a less hindered alcohol selectively.
      5. Aminolysis.
         a. Acid chlorides react with ammonia and amines to give amides.
         b. Either two equivalents of ammonia/amine must be used, or NaOH must be present, in order to scavenge HCl.
   C. Chemistry of carboxylic acid anhydrides (Section 16.5).
      1. Acid anhydrides can be prepared by reaction of carboxylate anions with acid chlorides.
         Both symmetrical and unsymmetrical anhydrides can be prepared by this route.
      2. Acid anhydrides react more slowly than acid chlorides.
         a. Acid anhydrides undergo most of the same reactions as acid chlorides.
         b. Acetic anhydride is often used to prepare acetate esters.
         c. In reactions of acid anhydrides, one-half of the molecule is "thrown away", making anhydrides inefficient to use.

D. Chemistry of esters (Section 16.6).
  1. Esters can be prepared by:
    a. S_N2 reaction of a carboxylate anion with an alkyl halide.
    b. Fischer esterification.
    c. Reaction of an acid chloride with an alcohol, in the presence of base.
  2. Esters are less reactive than acid halides and anhydrides but undergo the same types of reactions.
  3. Hydrolysis.
    a. Basic hydrolysis (saponification) occurs through a nucleophilic acyl substitution mechanism.
      i. Loss of alkoxide ion yields a carboxylic acid which is deprotonated to give a carboxylate anion.
      ii. Isotope-labelling studies confirm this mechanism.
    b. Acidic hydrolysis can occur by more than one mechanism.
      The usual route is by the reverse of Fischer esterification.
  4. Aminolysis.
    Esters can be converted to amides by heating with ammonia/amines, but it's easier to start with an acid chloride.
  5. Reduction.
    a. $LiAlH_4$ reduces esters to primary alcohols by a route similar to that described for acid chlorides.
    b. If DIBAH at –78°C is used, reduction yields an aldehyde.
  6. Reaction with Grignard reagents.
    Esters react twice with Grignard reagents to produce tertiary alcohols containing two identical substituents.

E. Chemistry of amides (Section 16.7).
  1. Amides are prepared by the reaction of acid chlorides with ammonia/amines.
  2. Hydrolysis.
    a. Hydrolysis (to carboxylic acids) occurs under more severe conditions than needed for hydrolysis of other acid derivatives.
    b. Acid hydrolysis occurs by addition of water to a protonated amide, followed by loss of ammonia or an amine.
    c. Basic hydrolysis occurs by attack of HO^-, followed by loss of $^-NH_2$.
  3. Reduction.
    $LiAlH_4$ reduces amides to amines.

F. Thiol esters and acyl phosphates (Section 16.8).
  1. Nature uses thiol esters and acyl phosphates in nucleophilic acyl substitution reactions because they are intermediate in reactivity between acid anhydrides and esters.
  2. Acetyl CoA is used as an acylating agent.

III. Polyamides and polyesters (Section 16.9).

A. Formation of polyesters and polyamides.
  1. When a diamine and a diacid chloride react, a polyamide is formed.
  2. When a diacid and a diol react, a polyester is formed.
  3. These polymers are called step-growth polymers because each bond is formed independently of the others.

B. Types of polymers.
  1. Nylons are the most common polyamides.
  2. The most common polyester, Dacron, is formed from dimethylterephthalate and ethylene glycol.

IV. Spectroscopy of carboxylic acid derivatives and nitriles (Section 16.10).
   A. Infrared spectroscopy.
      All of these compounds have characteristic carbonyl absorptions that help identify them; these are listed in Table 21.3.
   B. NMR spectroscopy is of limited usefulness in identifying carboxylic acid derivatives.
      1. Hydrogens next to carbonyl groups absorb at around 2.1 δ in a ^{1}H NMR spectrum, but this absorption can't be used to distinguish among carboxylic acid derivatives.
      2. Carbonyl carbons absorb in the range 160–180 δ, but, again, this absorption can't be used to distinguish among carboxylic acid derivatives.

Solutions to Problems

16.1 Table 16.1 lists the suffixes for naming carboxylic acid derivatives. The suffixes used when the functional group is part of a ring are in parentheses.

(a) $CH_3CH(CH_3)CH_2CH_2COCl$

4-Methylpentanoyl chloride

(b) $C_6H_{11}CH_2CONH_2$

Cyclohexylacetamide

(c) $CH_3CH(CH_3)COOCH(CH_3)CH_3$

Isopropyl 2-methylpropanoate

(d) $(C_6H_5CO)_2O$

Benzoic anhydride

(e) $C_5H_9COOCH(CH_3)CH_3$

Isopropyl cyclopentanecarboxylate

(f) $C_5H_9OCOCH(CH_3)CH_3$

Cyclopentyl 2-methylpropanoate

(g) $H_2C{=}CHCH_2CH_2CONHCH_3$

N-Methylpent-4-enamide

(h) H_3C–C(HO)(H)–$COOPO_3^{2-}$

(*R*)-2-Hydroxypropanoyl phosphate

(i) $(H_3C)_2C{=}C(CH_3)COSCH_2CH_3$

Ethyl 2,3-dimethylbut-2-enethioate

16.2

(a)

Phenyl benzoate

(b)

$CH_3CH_2CH_2C(=O)N(CH_3)CH_2CH_3$

N-Ethyl-*N*-methylbutanamide

(c)

$CH_3CH(CH_3)CH_2CH(CH_3)C(=O)Cl$

2,4-Dimethyl-pentanoyl chloride

(d)

Methyl 1-methylcyclo-hexanecarboxylate

(e)

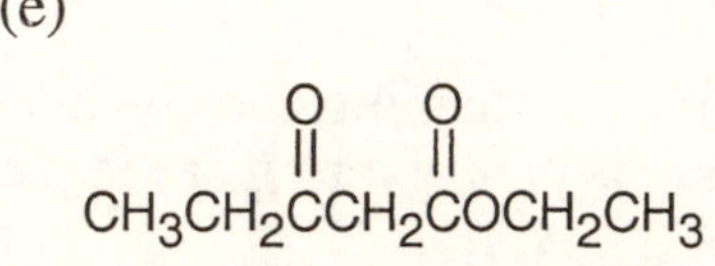

Ethyl 3-oxopenatanoate

(f)

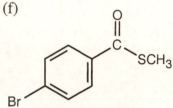

Methyl *p*-bromo-thiobenzoate

(g)

Formic propanoic anhydride

(h)

cis-2-Methylcyclohexane-carbonyl bromide

16.3

addition of methoxide to form a tetrahedral intermediate

elimination of Cl^-

16.4

$$\overset{\delta-}{F_3C}-\overset{\delta+}{C}(=\overset{\delta-}{O})-\overset{\delta-}{OCH_3} \qquad H_3C-\overset{\delta+}{C}(=\overset{\delta-}{O})-\overset{\delta-}{OCH_3}$$

The strongly electron-withdrawing trifluoromethyl group makes the carbonyl carbon more electron-poor and more reactive toward nucleophiles than the methyl acetate carbonyl group.

16.5 Strategy: Identify the nucleophile (boxed) and the leaving group (circled), and replace the leaving group by the nucleophile in the product.

Solution:

(a)

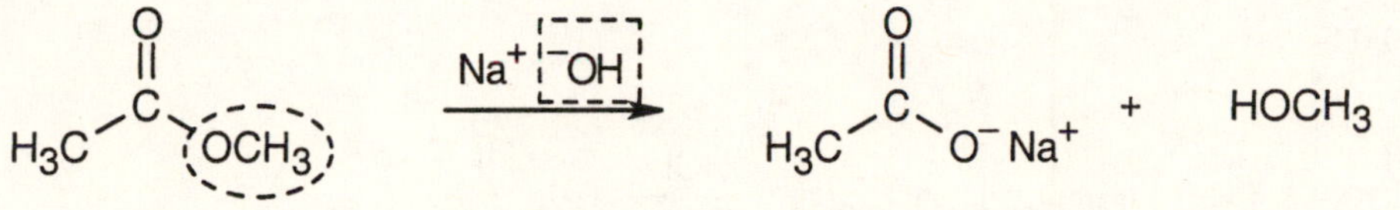

$$H_3C{-}C({=}O){-}OCH_3 \xrightarrow{Na^+\ ^-OH} H_3C{-}C({=}O){-}O^-\ Na^+ + HOCH_3$$

(b)

$$H_3C{-}C({=}O){-}Cl \xrightarrow{2\ NH_3} H_3C{-}C({=}O){-}NH_2 + NH_4^+\ Cl^-$$

(c)

$$H_3C{-}C({=}O){-}O{-}C({=}O){-}CH_3 \xrightarrow{Na^+\ ^-OCH_3} H_3C{-}C({=}O){-}OCH_3 + Na^+\ ^-O{-}C({=}O){-}CH_3$$

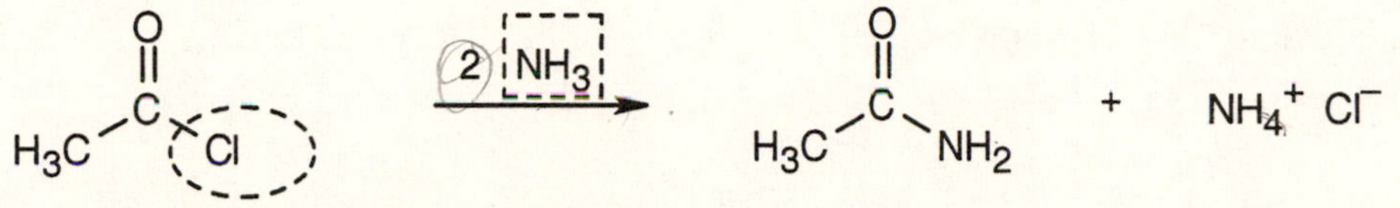

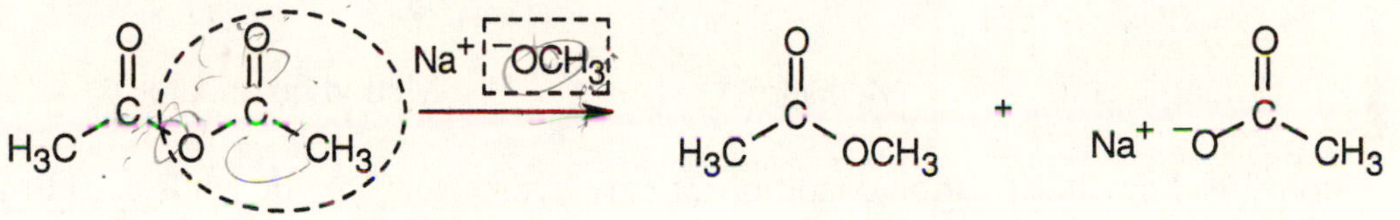

(d)

$$H_3C{-}C({=}O){-}SCH_3 \xrightarrow{CH_3NH_2} H_3C{-}C({=}O){-}NHCH_3 + HSCH_3$$

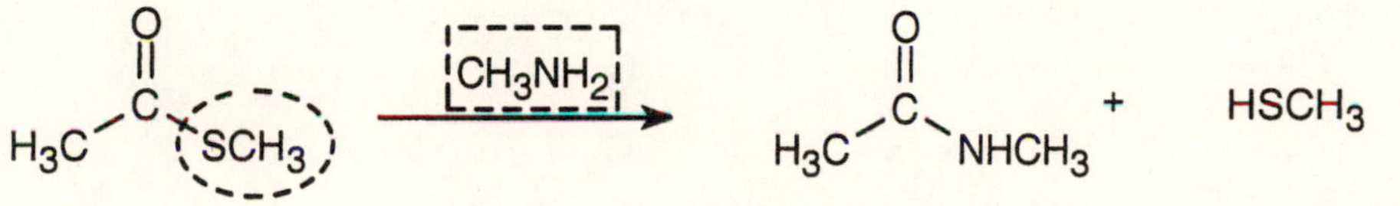

16.6 The structure represents the tetrahedral intermediate in the reaction of methyl cyclopentylacetate with hydroxide, a nucleophile. The products are cyclopentylacetate anion and methanol.

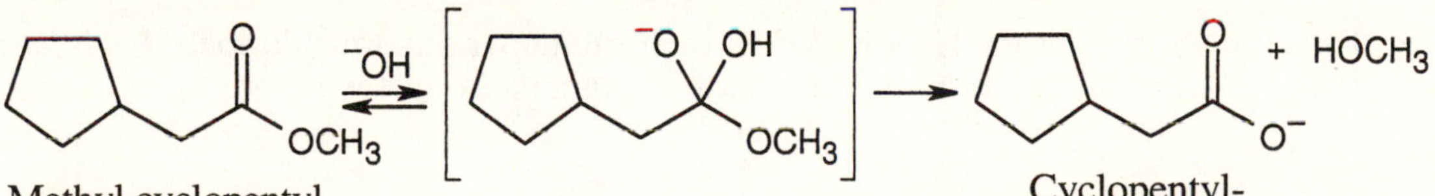

$$\text{Methyl cyclopentylacetate } (C_5H_9CH_2C(=O)OCH_3) \underset{}{\overset{^-OH}{\rightleftharpoons}} \left[C_5H_9CH_2C(O^-)(OH)(OCH_3)\right] \longrightarrow \text{Cyclopentylacetate } (C_5H_9CH_2C(=O)O^-) + HOCH_3$$

Methyl cyclopentyl-acetate

Cyclopentyl-acetate

16.7 **Strategy:** In Fischer esterification, an alcohol undergoes a nucleophilic acyl substitution with a carboxylic acid to yield an ester. The mineral acid catalyst makes the carboxyl group of the acid more nucleophilic. Predicting the products is easier if the two reagents are positioned so that the reacting functional groups point towards each other.

Solution:

(a)

$$H_3C{-}C({=}O){-}OH + HOCH_2CH_2CH_2CH_3 \xrightleftharpoons{HCl} H_3C{-}C({=}O){-}OCH_2CH_2CH_2CH_3 + H_2O$$

Acetic acid, Butan-1-ol, Butyl acetate

(b)

$$CH_3CH_2CH_2{-}C({=}O){-}OH + HOCH_3 \xrightleftharpoons{HCl} CH_3CH_2CH_2{-}C({=}O){-}OCH_3 + H_2O$$

Butanoic acid, Methanol, Methyl butanoate

16.8 Under Fischer esterification conditions, many hydroxycarboxylic acids can form intramolecular esters (lactones).

$$HOCH_2CH_2CH_2CH_2{-}C({=}O){-}OH \xrightleftharpoons{HCl} \text{lactone (six-membered ring: } C({=}O){-}O{-}CH_2{-}CH_2{-}CH_2{-}CH_2\text{)} + H_2O$$

5-Hydroxypentanoic acid, a lactone

16.9 Pyridine neutralizes the HCl byproduct by forming pyridinium chloride. This neutralization removes from the product mixture acid that might cause side reactions.

(a)

$$CH_3CH_2{-}C({=}O){-}Cl + HOCH_3 \xrightarrow{\text{Pyridine}} CH_3CH_2{-}C({=}O){-}OCH_3$$

Propanoyl chloride, Methanol, Methyl propanoate

(b)

$$H_3C{-}C({=}O){-}Cl + HOCH_2CH_3 \xrightarrow{\text{Pyridine}} H_3C{-}C({=}O){-}OCH_2CH_3$$

Acetyl chloride, Ethanol, Ethyl acetate

(c)

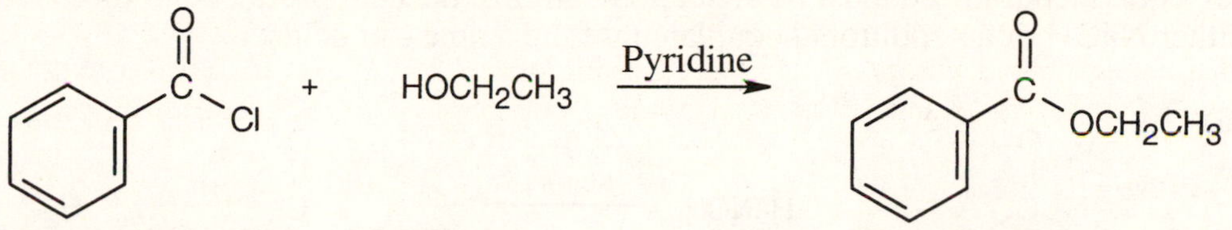
O
C
Cl
+
HOCH2CH3
Pyridine
O
C
OCH2CH3

Benzoyl chloride   Ethanol   Ethyl benzoate

16.10 As explained in the text, only simple, low boiling alcohols are convenient to use in the Fischer esterification reaction. Thus, reaction of cyclohexanol with benzoyl chloride is the preferred method for preparing cyclohexyl benzoate.

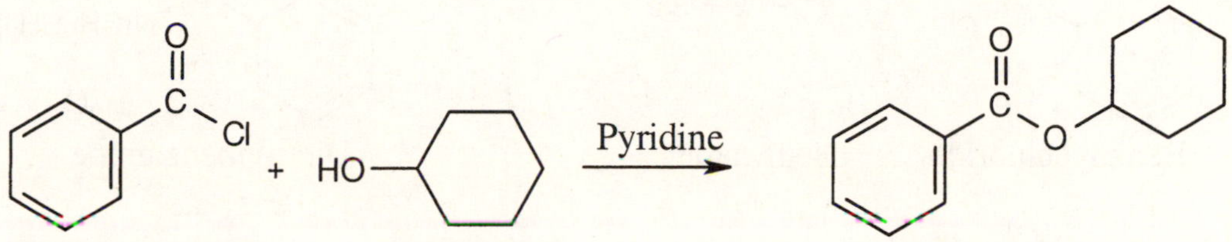
O
C
Cl
+ HO
Pyridine
O
C
O

Benzoyl chloride   Cyclohexanol   Cyclohexyl benzoate

16.11

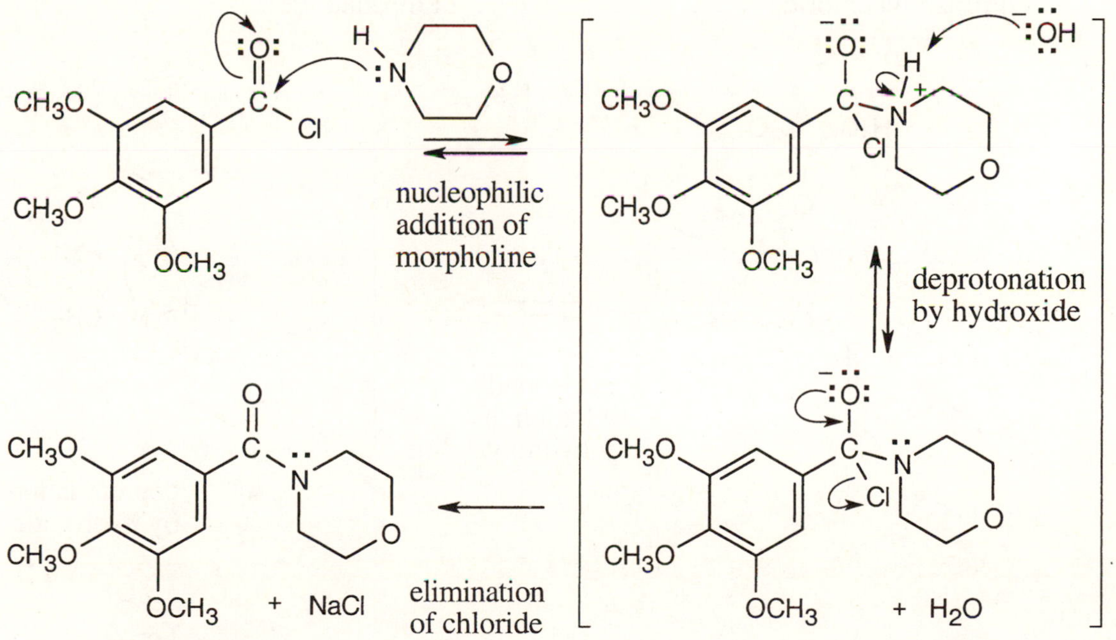
CH3O
CH3O
OCH3
:O:
C
Cl
H
:N
O
nucleophilic
addition of
morpholine
:O:
C
N
H
+
Cl
:OH
deprotonation
by hydroxide
+ H2O
elimination
of chloride
+ NaCl

Trimetozine

16.12 An equivalent of base must be added to neutralize the acid produced in these reactions. Either NaOH or an additional equivalent of the amine can be used.

(a)

$CH_3CH_2C(=O)Cl$ + H_2NCH_3 $\xrightarrow{NaOH}$ $CH_3CH_2C(=O)NHCH_3$ + H_2O + NaCl

Propanoyl chloride — Methylamine — *N*-Methylpropanamide

(b)

Benzoyl chloride + $HN(CH_2CH_3)_2$ $\xrightarrow{NaOH}$ $C_6H_5C(=O)N(CH_2CH_3)_2$ + H_2O + NaCl

Benzoyl chloride — Diethylamine — *N.N*-Diethylbenzamide

(c)

$CH_3CH_2C(=O)Cl$ + 2 NH_3 $\longrightarrow$ $CH_3CH_2C(=O)NH_2$ + NH_4^+ Cl^-

Propanoyl chloride — Propanamide

16.13

Acetic anhydride + p-hydroxyaniline ⇌ (nucleophilic addition of *p*-hydroxyaniline) tetrahedral intermediate + $^-:OH$

⇅ deprotonation by hydroxide

tetrahedral intermediate + H_2O → (loss of acetate) $H_3C-C(=O)-O^-$ + Acetaminophen

nucleophilic addition of *p*-hydroxyaniline

deprotonation by hydroxide

Acetaminophen

loss of acetate

16.14

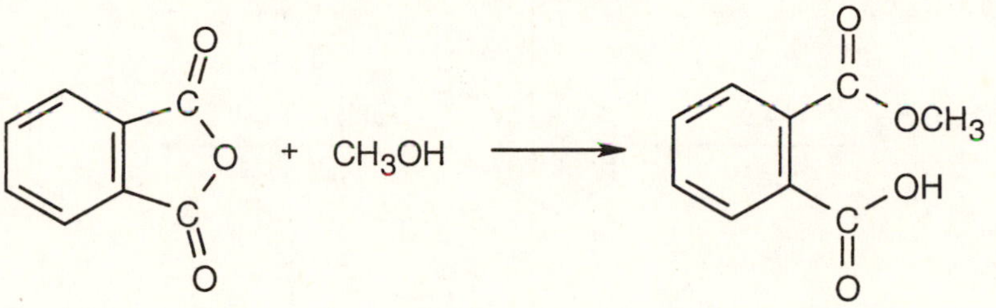

Phthalic anhydride + CH_3OH ⟶ (product with $C(=O)OCH_3$ and $C(=O)OH$ groups)

Phthalic anhydride

The second half of the anhydride becomes a carboxylic acid.

16.15 Acidic hydrolysis of an ester is a reversible reaction because the products are an alcohol and a carboxylic acid. Basic hydrolysis of an ester is irreversible because the products are an alcohol and a carboxylate anion, which has a negative charge and does not react with nucleophiles.

16.16

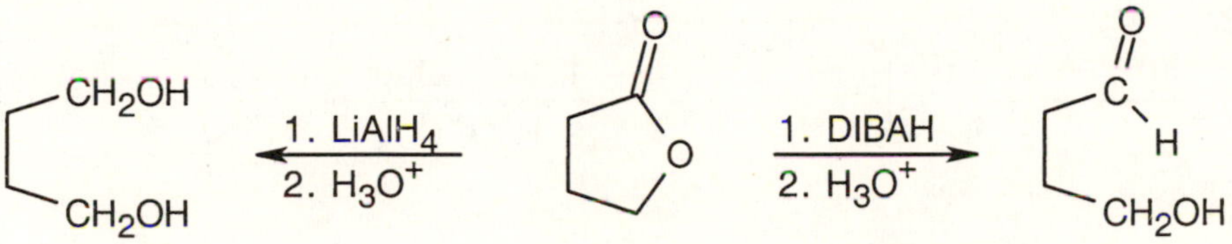

$HOCH_2CH_2CH_2CH_2OH$ ⟵ (1. $LiAlH_4$; 2. H_3O^+) butyrolactone ⟶ (1. DIBAH; 2. H_3O^+) $HOCH_2CH_2CH_2CHO$

The product of reduction of butyrolactone is 4-hydroxybutanal.

16.17 Lithium aluminum hydride reduces an ester to form two alcohols.

(a)

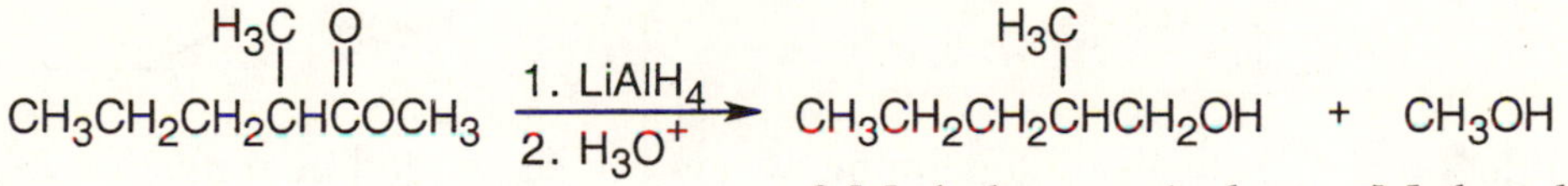

$$CH_3CH_2CH_2CH(CH_3)COCH_3 \xrightarrow[2.\ H_3O^+]{1.\ LiAlH_4} CH_3CH_2CH_2CH(CH_3)CH_2OH + CH_3OH$$

2-Methylpentan-1-ol   Methanol

(b)

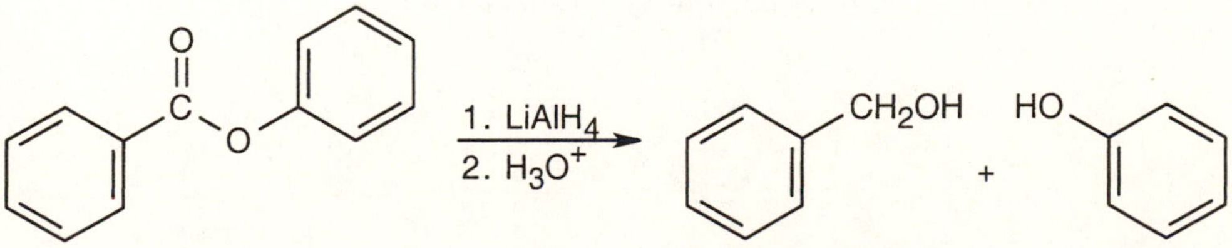

Phenyl benzoate ⟶ (1. $LiAlH_4$; 2. H_3O^+) $C_6H_5CH_2OH$ + HOC_6H_5

Benzyl alcohol   Phenol

16.18

N-Ethylbenzamide —(a) H_3O^+, heat→ benzoic acid (C(=O)OH) —(b) 1. $LiAlH_4$ 2. H_3O^+→ CH_2OH (benzyl alcohol)

N-Ethylbenzamide (C(=O)$NHCH_2CH_3$) —(c) 1. $LiAlH_4$ 2. H_2O→ $CH_2NHCH_2CH_3$

N-Ethylbenzamide

16.19

Br (bromocyclohexane) —Mg, ether→ MgBr —1. CO_2, ether 2. H_3O^+→ C(=O)OH —$SOCl_2$→ C(=O)Cl —$(CH_3)_2NH$, NaOH→ C(=O)$N(CH_3)_2$ —1. $LiAlH_4$ 2. H_2O→ $CH_2N(CH_3)_2$

The product is a *N,N*-disubstituted amine, which can be formed by reduction of an amide. The amide results from treatment of an acid chloride with the appropriate amine. The acid chloride is the product of the reaction of $SOCl_2$ with a carboxylic acid that is formed by carboxylation of the Grignard reagent synthesized from the starting material.

16.20 Even though the entire molecule of coenzyme A is biologically important, we are concerned in this problem only with the –SH group. The remainder of the structure is represented here as "R".

$$H_3C{-}C({=}O){-}O{-}P({=}O)(O^-){-}O{-}\text{Adenosine} \;+\; RS{-}H \;+\; :\text{Base} \xrightarrow{1.} \left[H_3C{-}C(O^-)(SR){-}O{-}P({=}O)(O^-){-}O{-}\text{Adenosine} \right] \xrightarrow{2.} H_3C{-}C({=}O){-}S{-}R \;+\; {}^-O{-}P({=}O)(O^-){-}O{-}\text{Adenosine}$$

Acetyl CoA

Step 1: Nucleophilic addition of the –SR group of CoA (after deprotonation) to acetyl adenylate to form a tetrahedral intermediate.
Step 2: Loss of adenosine monophosphate.

16.21 In each example, if *n* molecules of one component react with *n* molecules of the other component, a polymer with *n* repeating units is formed, and 2*n* small molecules are formed as byproducts; these are shown in each reaction.

(a)

$$BrCH_2CH_2CH_2Br + HOCH_2CH_2CH_2OH \xrightarrow{\text{Base}} \left(CH_2CH_2CH_2OCH_2CH_2CH_2O \right)_n + 2n\ HBr$$

(b)

$$HOCH_2CH_2OH + HO_2C(CH_2)_6CO_2H \xrightarrow[\text{catalyst}]{H_2SO_4} \left(OCH_2CH_2O{-}\overset{O}{\overset{\|}{C}}(CH_2)_6\overset{O}{\overset{\|}{C}} \right)_n + 2n\ H_2O$$

(c)

$$H_2N(CH_2)_6NH_2 + Cl\overset{O}{\overset{\|}{C}}(CH_2)_4\overset{O}{\overset{\|}{C}}Cl \longrightarrow \left(HN(CH_2)_6NH{-}\overset{O}{\overset{\|}{C}}(CH_2)_4\overset{O}{\overset{\|}{C}} \right)_n + 2n\ HCl$$

16.22

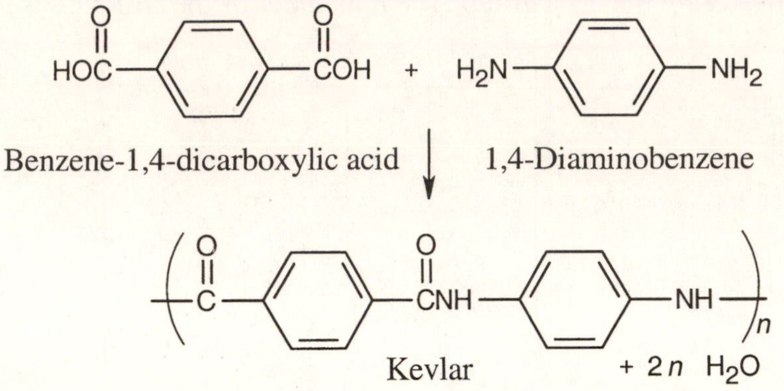

$$HOC(=O)-C_6H_4-C(=O)OH + H_2N-C_6H_4-NH_2$$

Benzene-1,4-dicarboxylic acid ↓ 1,4-Diaminobenzene

$$-\!\!\left(C(=O)-C_6H_4-C(=O)NH-C_6H_4-NH\right)\!\!-_n$$

Kevlar + $2n\ H_2O$

16.23 Use Table 16.2 if you need help.

Absorption	*Functional group present*
(a) 1735 cm^{-1}	Aliphatic ester *or* 6-membered ring lactone
(b) 1810 cm^{-1}	Aliphatic acid chloride
(c) 2500 – 3300 cm^{-1} and 1710 cm^{-1}	Carboxylic acid
(d) 1715 cm^{-1}	Aliphatic ketone *or* 6-membered ring ketone

16.24 (a) 1. IR 1735 cm^{-1} corresponds to an aliphatic ester.

2. $-C(=O)-O-$

3. The remaining five carbons and twelve hydrogens can be arranged in a number of ways to produce a satisfactory structure for this compound. For example:

$$CH_3CH_2C(=O)OCH_2CH_2CH_3 \quad or \quad HC(=O)OCH_2CH_2CH_2CH_2CH_3$$

The structural formula indicates that this compound can't be a lactone.

(b) $CH_3C(=O)N(CH_3)_2$

(c) $CH_3CH{=}CHC(=O)Cl \quad or \quad H_2C{=}C(CH_3)C(=O)Cl$

Visualizing Chemistry

16.25

(a)

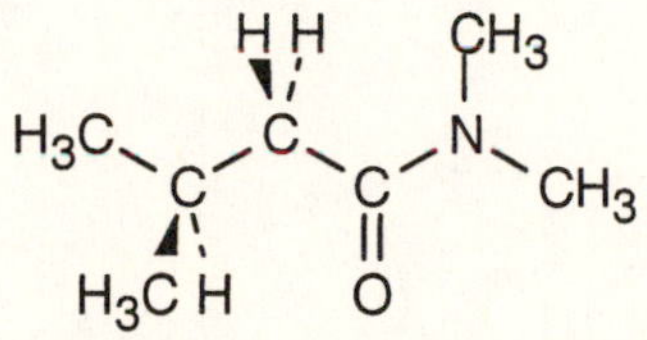

N,N-Dimethyl-3-methylbutanamide

(b)

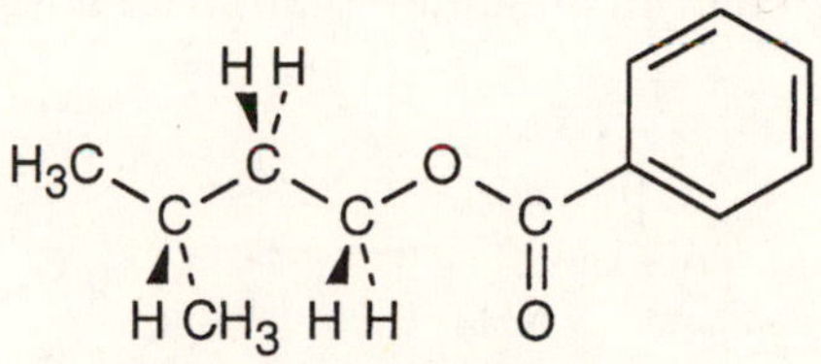

3-Methylbutyl benzoate

16.26

(a)

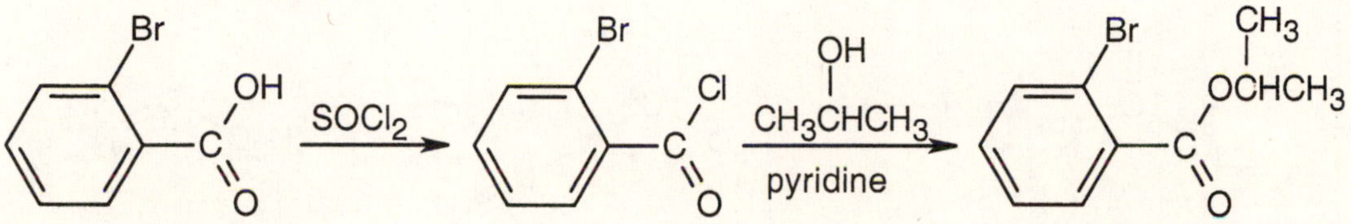

$\xrightarrow{SOCl_2}$ $\xrightarrow[\text{pyridine}]{CH_3CH(OH)CH_3}$

o-Bromobenzoic acid

Propan-2-ol

Isopropyl *o*-bromobenzoate

This compound can also be synthesized by Fischer esterification of *o*-bromobenzoic acid with propan-2-ol and an acid catalyst.

(b)

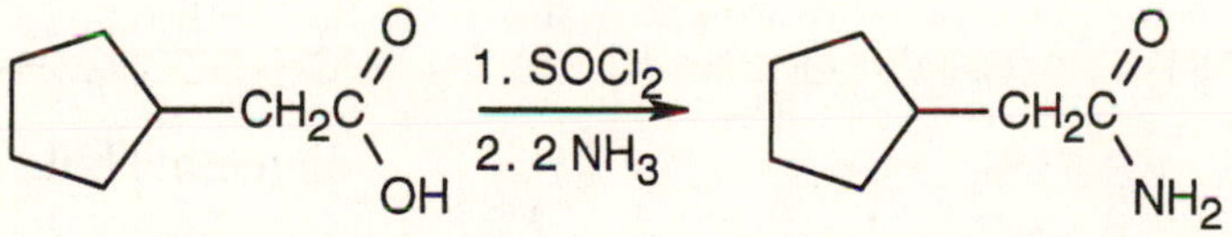

$$C_5H_9{-}CH_2COOH \xrightarrow[\text{2. } 2\ NH_3]{\text{1. } SOCl_2} C_5H_9{-}CH_2CONH_2$$

Cyclopentylacetic acid

Cyclopentylacetamide

16.27

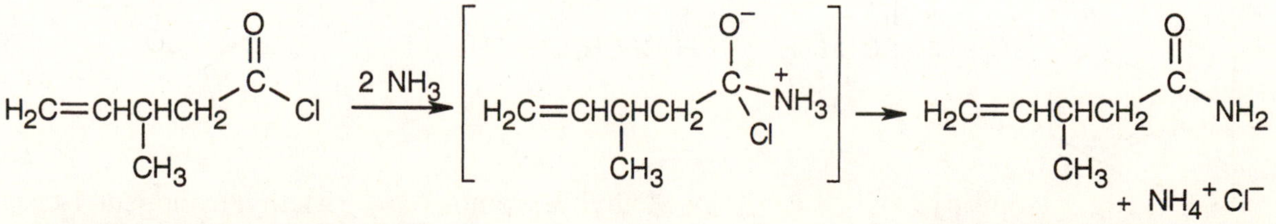

$$H_2C{=}CHCH(CH_3)CH_2COCl \xrightarrow{2\ NH_3} \left[H_2C{=}CHCH(CH_3)CH_2C(O^-)(Cl)\overset{+}{N}H_3\right] \rightarrow H_2C{=}CHCH(CH_3)CH_2CONH_2 + NH_4^+\ Cl^-$$

3-Methylpent-4-enoyl chloride

3-Methylpent-4-enamide

The starting material is 3-methylpent-4-enoyl chloride. Ammonia adds to give the observed tetrahedral intermediate, which eliminates Cl^- to yield the above amide.

16.28 According to the electrostatic potential maps, the carbonyl carbon of acetyl azide is more electron-poor and is more reactive than acetamide in nucleophilic acyl substitution reactions. Resonance donation of nitrogen lone-pair electrons to the carbonyl group is greater in an amide than in an acyl azide.

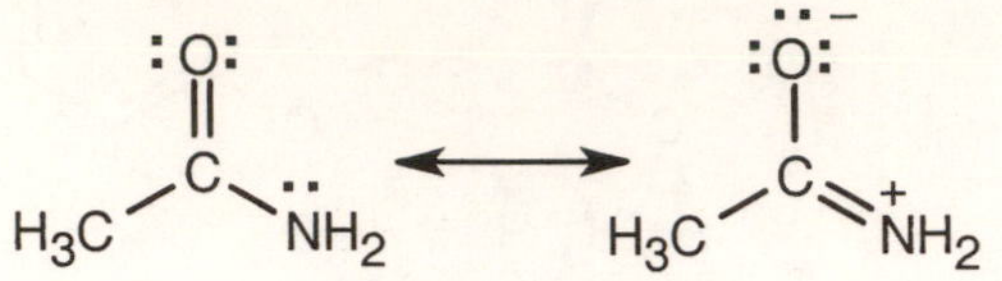

Acetamide

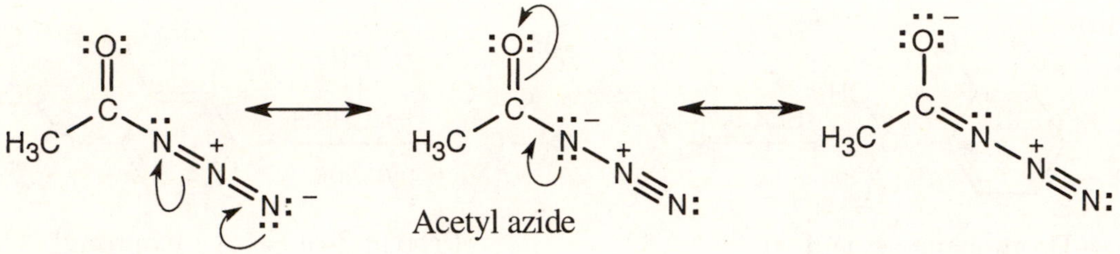

Acetyl azide

Additional Problems

16.29

(a) $CH_3C_6H_4CONH_2$

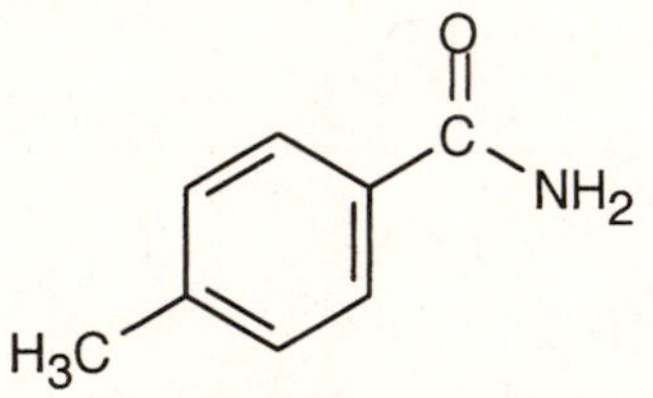

p-Methylbenzamide

(b) $CH_3CH_2CH(CH_2CH_3)CH{=}CHCOCl$

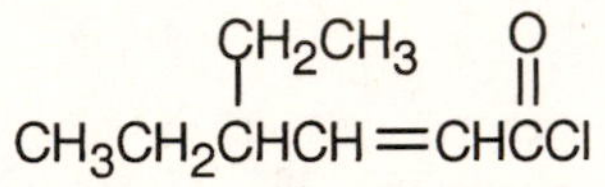

4-Ethylhex-2-enoyl chloride

(c) $CH_3OCOCH_2CH_2COOCH_3$

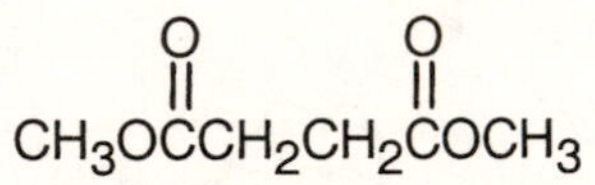

Dimethyl butanedioate
or
Dimethyl succinate

(d) $C_6H_5CH_2CH_2COOCH(CH_3)CH_3$

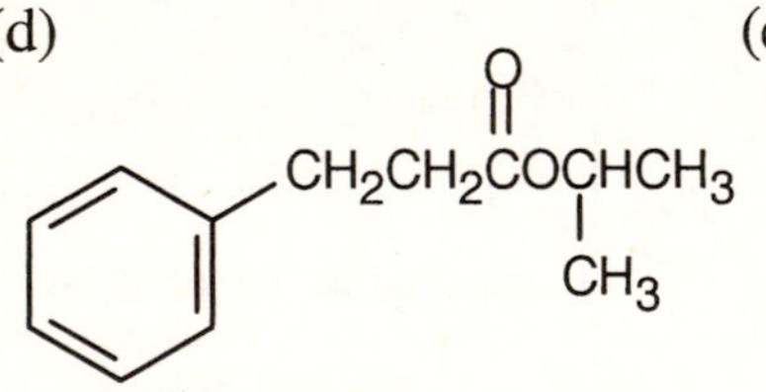

Isopropyl 3-phenylpropanoate

(e) $CH_3CH(Br)CH_2CONHCH_3$

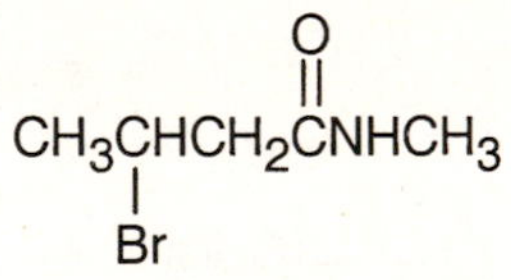

N-Methyl-3-bromo-butanamide

(f) Cyclopentene–$COOCH_3$

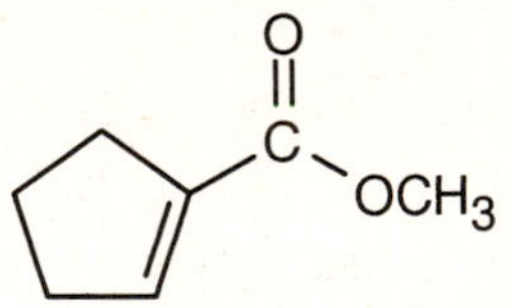

Methyl cyclopent-1-ene-carboxylate

(g) $C_6H_5COOC_6H_5$

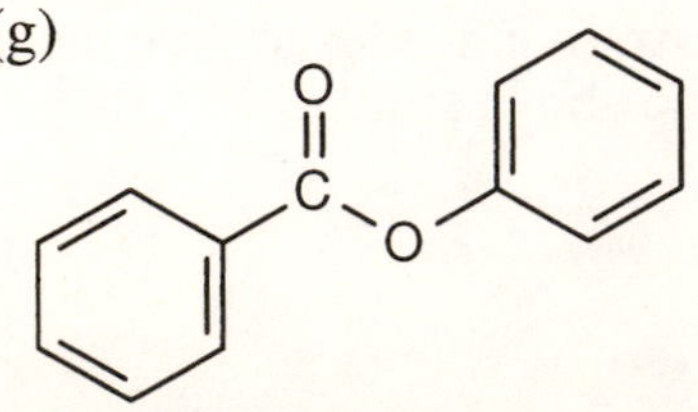

Phenyl benzoate

(h) $C_6H_5COSCH(CH_3)_2$

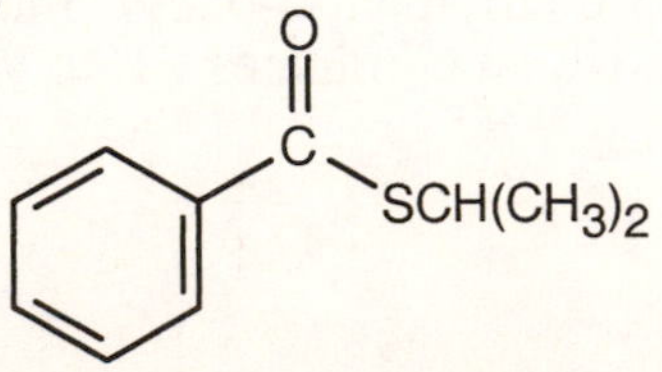

Isopropyl thiobenzoate

16.30

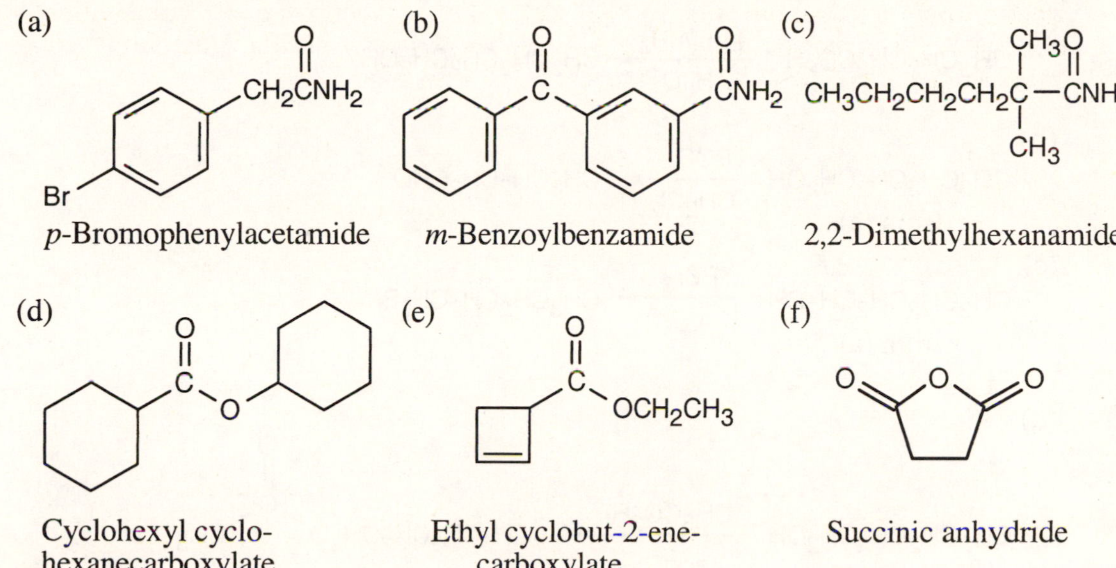

(a) *p*-Bromophenylacetamide (b) *m*-Benzoylbenzamide (c) 2,2-Dimethylhexanamide

(d) Cyclohexyl cyclohexanecarboxylate (e) Ethyl cyclobut-2-enecarboxylate (f) Succinic anhydride

16.31 Many structures can be drawn for each part of this problem.

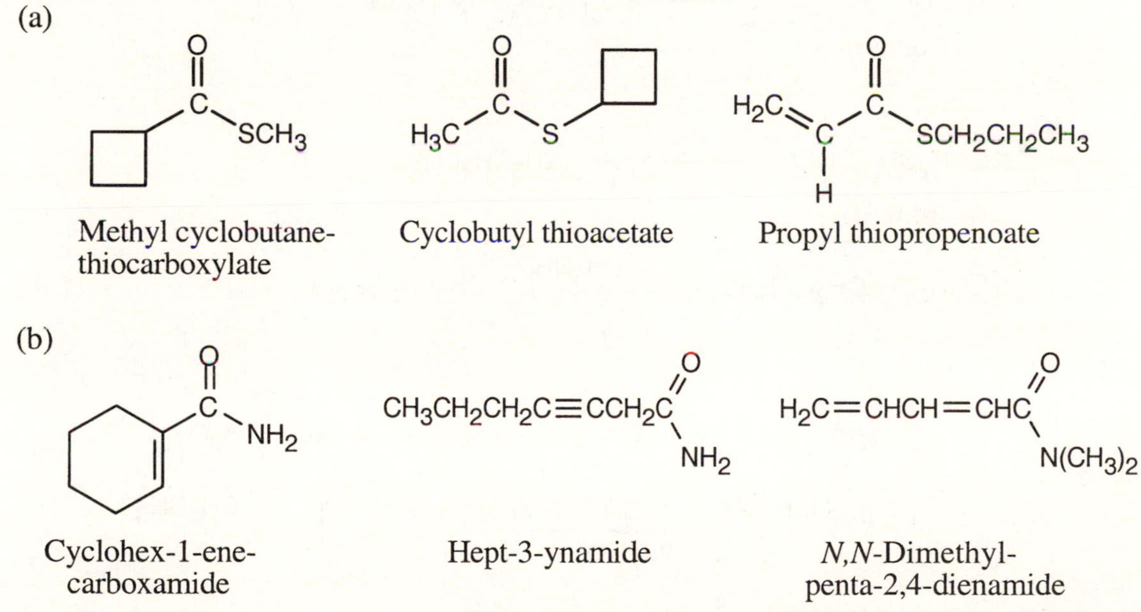

(a) Methyl cyclobutanethiocarboxylate; Cyclobutyl thioacetate; Propyl thiopropenoate

(b) Cyclohex-1-enecarboxamide; Hept-3-ynamide; *N*,*N*-Dimethylpenta-2,4-dienamide

16.32

(a)

$$CH_3CH_2CH_2CO_2H \xrightarrow[2.\ H_3O^+]{1.\ LiAlH_4} CH_3CH_2CH_2CH_2OH$$

(b)

$$\underset{\text{from (a)}}{CH_3CH_2CH_2CH_2OH} \xrightarrow[CH_2Cl_2]{PCC} CH_3CH_2CH_2CHO$$

(c)

$$\underset{\text{from (a)}}{CH_3CH_2CH_2CH_2OH} \xrightarrow{PBr_3} CH_3CH_2CH_2CH_2Br$$

(d)

$$\underset{\text{from (a)}}{CH_3CH_2CH_2CH_2OH} \xrightarrow[\text{acid catalyst}]{CH_3CO_2H} CH_3CH_2CH_2CH_2O{-}\overset{O}{\overset{\|}{C}}{-}CH_3$$

(e)

$$\underset{\text{from (c)}}{CH_3CH_2CH_2CH_2Br} \xrightarrow{NaCN} CH_3CH_2CH_2CH_2CN$$

(f)

$$\underset{\text{from (d)}}{CH_3CH_2CH_2CH_2CN} \xrightarrow{H_3O^+} CH_3CH_2CH_2CH_2CO_2H$$

$$CH_3CH_2CH_2CH_2CO_2H \xrightarrow{SOCl_2} CH_3CH_2CH_2CH_2COCl$$

$$CH_3CH_2CH_2CH_2CONHCH_3 \xleftarrow{2\ CH_3NH_2} CH_3CH_2CH_2CH_2COCl$$

16.33

(a)

$$C_6H_{11}{-}\overset{O}{\overset{\|}{C}}{-}OCH_2CH_3 \xrightarrow[2.\ H_3O^+]{1.\ CH_3CH_2MgBr} C_6H_{11}{-}C(OH)(CH_2CH_3)CH_2CH_3 + CH_3CH_2OH$$

(b)

$$CH_3\overset{CH_3}{\overset{|}{C}}HCH_2CH_2\overset{O}{\overset{\|}{C}}OCH_3 \xrightarrow[2.\ H_3O^+]{1.\ DIBAH} CH_3\overset{CH_3}{\overset{|}{C}}HCH_2CH_2\overset{O}{\overset{\|}{C}}H$$

(c)

$$C_5H_9{-}\overset{O}{\overset{\|}{C}}{-}Cl \xrightarrow{2\ CH_3NH_2} C_5H_9{-}\overset{O}{\overset{\|}{C}}{-}NHCH_3$$

(d)

$$\text{cis-2-methylcyclohexanecarboxylic acid } (CO_2H,\ CH_3) \xrightarrow[H_2SO_4]{CH_3OH} \text{methyl ester } (CO_2CH_3,\ CH_3)$$

(e)

$$H_2C{=}CHCH(CH_3)CH_2\overset{O}{\overset{\|}{C}}OCH_3 \xrightarrow[2.\ H_3O^+]{1.\ LiAlH_4} H_2C{=}CHCH(CH_3)CH_2CH_2OH + CH_3OH$$

(f)

$$\text{cyclohexanol (OH)} \xrightarrow[\text{pyridine}]{(CH_3CO)_2O} \text{cyclohexyl–O–}\overset{O}{\overset{\|}{C}}\text{–}CH_3 + CH_3\overset{O}{\overset{\|}{C}}O^-$$

(g)

$$\text{2-methylbenzamide } (CONH_2,\ CH_3) \xrightarrow[2.\ H_2O]{1.\ LiAlH_4} \text{2-methylbenzylamine } (CH_2NH_2,\ CH_3)$$

(h)

$$\text{(2-bromophenyl)acetic acid } (CH_2CO_2H,\ Br) \xrightarrow{SOCl_2} \text{(2-bromophenyl)acetyl chloride } (CH_2COCl,\ Br)$$

16.34

(a)

$$CH_3\overset{O}{\overset{\|}{C}}OCH_3 \xrightarrow[2.\ H_3O^+]{1.\ LiAlH_4} CH_3CH_2OH + CH_3OH$$

(b)

$$CH_3\overset{O}{\overset{\|}{C}}OCH_3 \xrightarrow[2.\ H_3O^+]{1.\ 2\ CH_3MgBr} CH_3\overset{OH}{\overset{|}{C}}(CH_3)_2 + CH_3OH$$

(c)

$$CH_3\overset{O}{\overset{\|}{C}}OCH_3 \xrightarrow[H_2O]{NaOH} CH_3\overset{O}{\overset{\|}{C}}O^- + CH_3OH$$

(d)

$$CH_3\overset{O}{\overset{\|}{C}}OCH_3 + H_2N\text{–}C_6H_5 \longrightarrow CH_3\overset{O}{\overset{\|}{C}}\text{–}\overset{H}{\overset{|}{N}}\text{–}C_6H_5 + CH_3OH$$

16.35

(a)

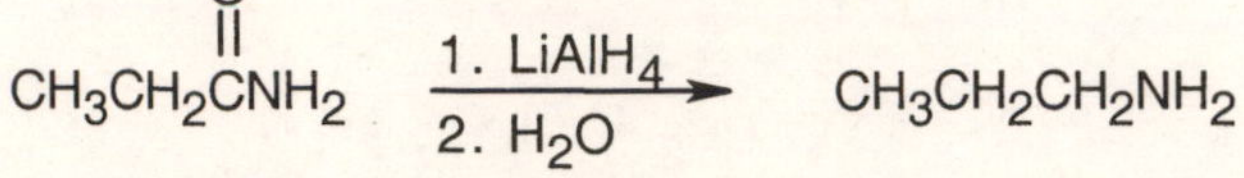

$$CH_3CH_2\overset{O}{\overset{\|}{C}}NH_2 \xrightarrow[2.\ H_2O]{1.\ LiAlH_4} CH_3CH_2CH_2NH_2$$

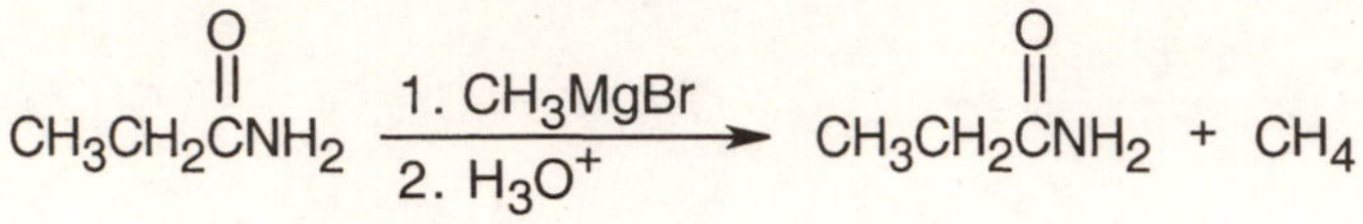

(b)

$$CH_3CH_2\overset{O}{\overset{\|}{C}}NH_2 \xrightarrow[2.\ H_3O^+]{1.\ CH_3MgBr} CH_3CH_2\overset{O}{\overset{\|}{C}}NH_2 + CH_4$$

(c)

$$CH_3CH_2\overset{O}{\overset{\|}{C}}NH_2 \xrightarrow[H_2O]{NaOH} CH_3CH_2\overset{O}{\overset{\|}{C}}O^- + NH_3$$

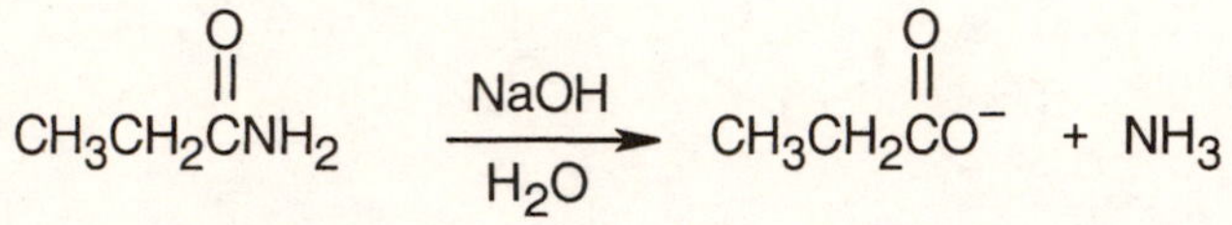

(d) no reaction

16.36

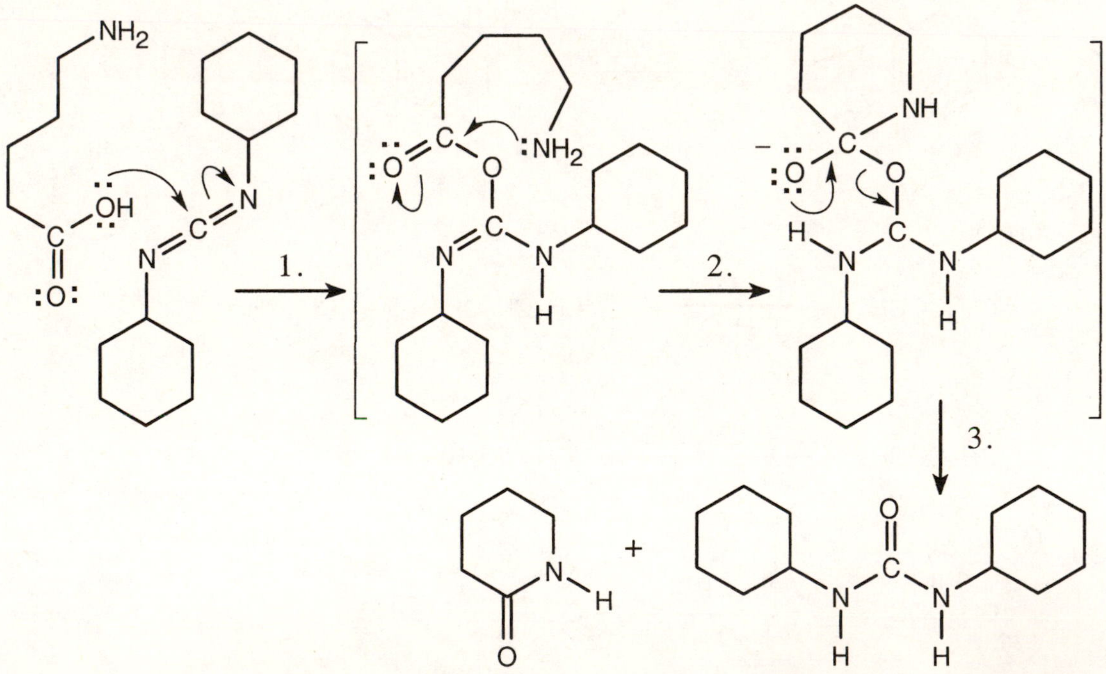

Step 1: The carboxylic acid group adds to DCC to form a reactive intermediate.
Step 2: The amine nitrogen adds to the carbonyl group to yield a tetrahedral intermediate.
Step 3: The intermediate loses dicyclohexylurea to produce the lactam.
Proton transfers occur in steps 1 and 2.

16.37 A negatively charged tetrahedral intermediate is formed when the nucleophile ^{-}OH attacks the carbonyl carbon of an ester. An electron-withdrawing substituent can stabilize this negatively charged tetrahedral intermediate and increase the rate of reaction. (Contrast this effect with substituent effects in electrophilic aromatic substitution, in which positive charge developed in the intermediate is stabilized by electron-*donating* substituents.) Substituents that are deactivating in electrophilic aromatic substitution are activating in ester hydrolysis, as the observed reactivity order shows. The substituents –CN and –CHO are electron-withdrawing; $-NH_2$ is strongly electron-donating.

Most reactive ⟶ *Least reactive*

$$Y = -NO_2 > -C{\equiv}N > -CHO > -Br > -H > -CH_3 > -OCH_3 > -NH_2$$

16.38 The reactivity of esters in saponification reactions is influenced by steric factors. Branching in both the acyl and alkyl portions of an ester hinders reaction of the hydroxide nucleophile. This effect is less pronounced in the alkyl portion of the ester than in the acyl portion because alkyl branching is one atom farther away from the site of reaction, but it is still significant.

Most reactive ⟶ *Least reactive*

$$CH_3COCH_3 > CH_3COCH_2CH_3 > CH_3COCH(CH_3)CH_3 > CH_3COC(CH_3)_2CH_3$$

16.39

2,4,6-Trimethylbenzoic acid

2,4,6–Trimethylbenzoic acid has two methyl groups ortho to the carboxylic acid functional group. These bulky methyl groups block the approach of the alcohol and prevent esterification from occurring under Fischer esterification conditions. Another possible route to the methyl ester:

1. NaOH
2. CH_3I

This route succeeds because reaction occurs farther away from the site of steric hindrance. It is also possible to form the acid chloride of 2,4,6-trimethylbenzoic acid and react it with methanol and pyridine.

16.40

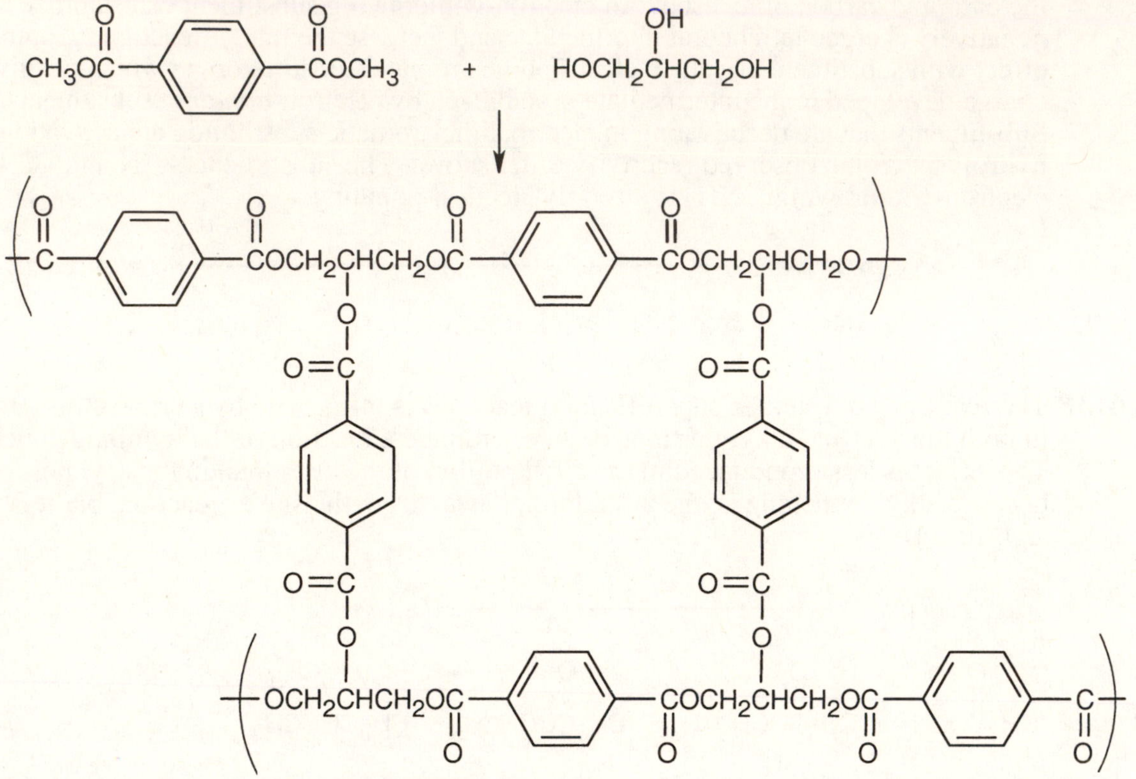

$$CH_3OC(=O)-C_6H_4-C(=O)OCH_3 + HOCH_2CH(OH)CH_2OH \longrightarrow$$

The product of the reaction of dimethyl terephthalate with glycerol has a high degree of cross-linking and is more rigid than Dacron.

16.41

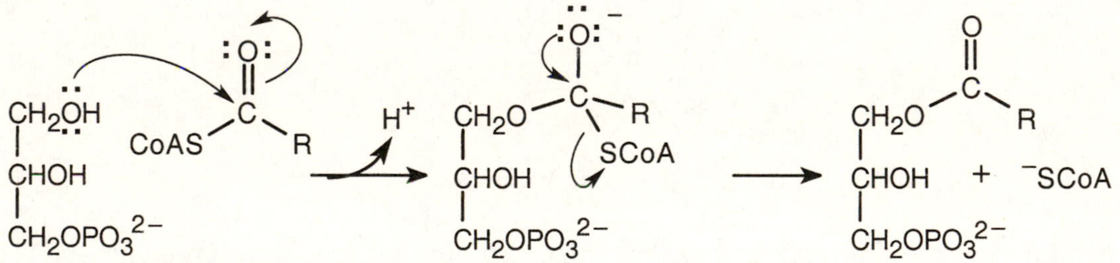

$$HOCH_2CH(OH)CH_2OPO_3^{2-} + CoAS-C(=O)-R \xrightarrow{H^+} RC(O^-)(SCoA)-OCH_2CH(OH)CH_2OPO_3^{2-} \longrightarrow RC(=O)OCH_2CH(OH)CH_2OPO_3^{2-} + {}^-SCoA$$

Addition of –OH to the fatty acyl CoA, followed by loss of –SCoA from the tetrahedral intermediate, produces 1-acylglycerol 3-phosphate.

16.42

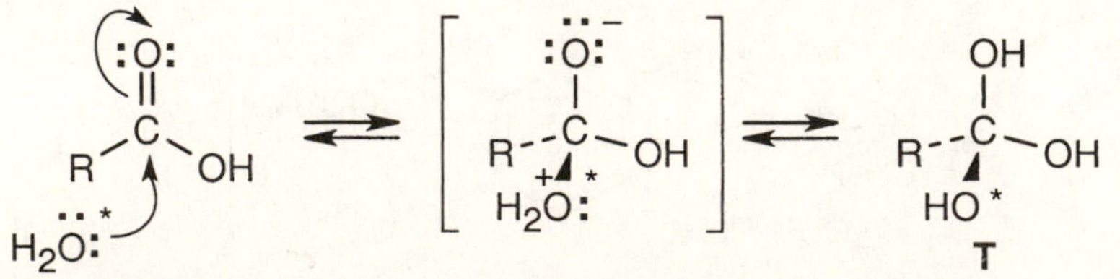

$$RC(=O)OH + H_2O^* \rightleftharpoons \left[R-C(O^-)(OH)-\overset{+}{O}{}^*H_2 \right] \rightleftharpoons R-C(OH)(OH)-{}^*OH \quad \mathbf{T}$$

The tetrahedral intermediate **T** can eliminate any one of the three –OH groups to reform either the original carboxylic acid or labeled carboxylic acid. Further reaction of water with mono-labeled carboxylic acid leads to the doubly labeled product.

16.43

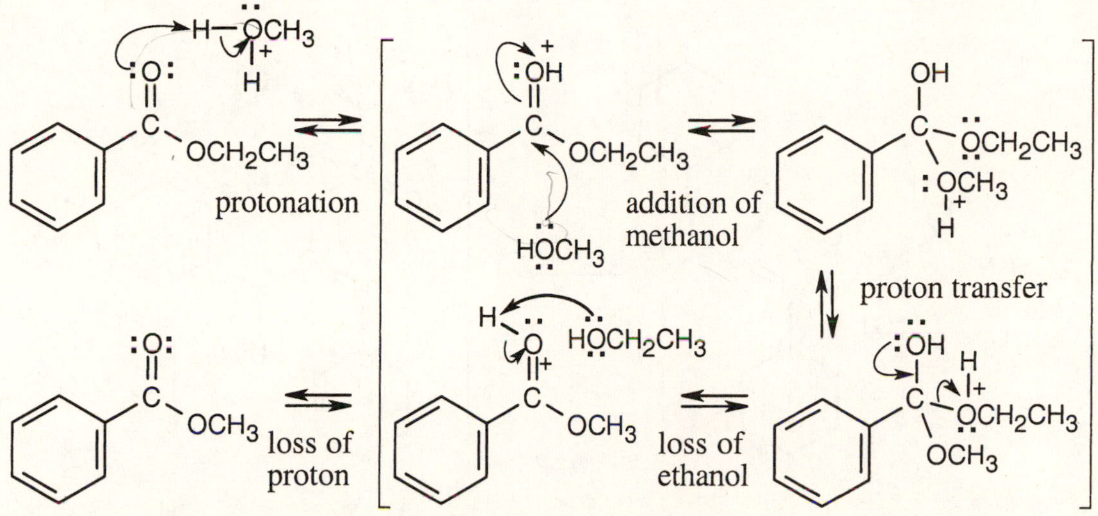

protonation

addition of methanol

proton transfer

loss of ethanol

loss of proton

In acidic methanol, the ethyl ester reacts by a nucleophilic acyl substitution mechanism to yield a methyl ester. The equilibrium favors the methyl ester because of the large excess of methanol present.

16.44

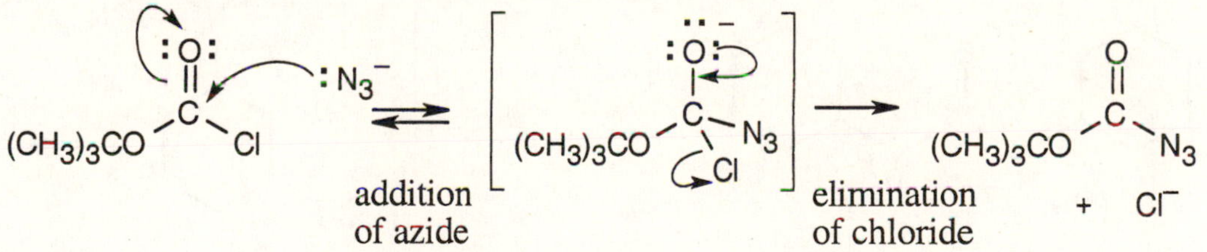

$(CH_3)_3CO-C(=O)-Cl + :N_3^-$ ⇌ (addition of azide) $[(CH_3)_3CO-C(O^-)(N_3)(Cl)]$ → (elimination of chloride) $(CH_3)_3CO-C(=O)-N_3 + Cl^-$

This reaction is a typical nucleophilic acyl substitution reaction, with azide as the nucleophile and chloride as the leaving group.

16.45

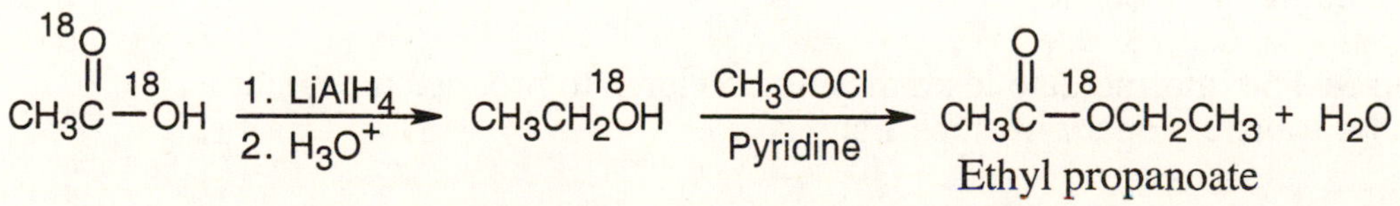

$$CH_3C(={}^{18}O)-{}^{18}OH \xrightarrow[2.\ H_3O^+]{1.\ LiAlH_4} CH_3CH_2{}^{18}OH \xrightarrow[Pyridine]{CH_3COCl} CH_3C(=O)-{}^{18}OCH_2CH_3 + H_2O$$

Ethyl propanoate

Remember that the ^{18}O label appears in both oxygens of the acetic acid starting material.

16.46 Formation of the dipeptide:

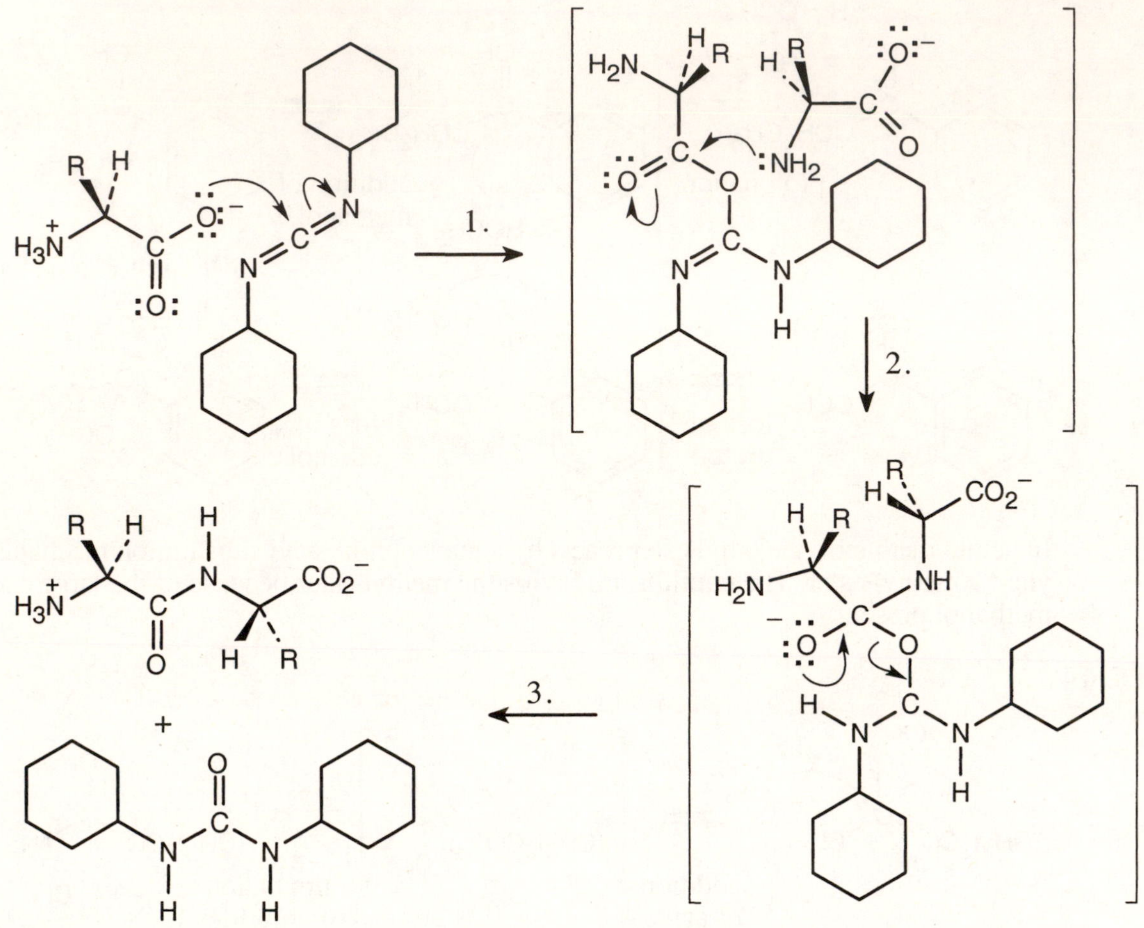

Step 1: The carboxylate group from one amino acid adds to DCC to form a reactive intermediate.

Step 2: The amino group of the second amino acid adds to the carbonyl group to yield a tetrahedral intermediate.

Step 3: The intermediate loses dicyclohexylurea to produce the amide.
Proton transfers occur in steps 1 and 2.

Formation of the 2,5-diketopiperazine:

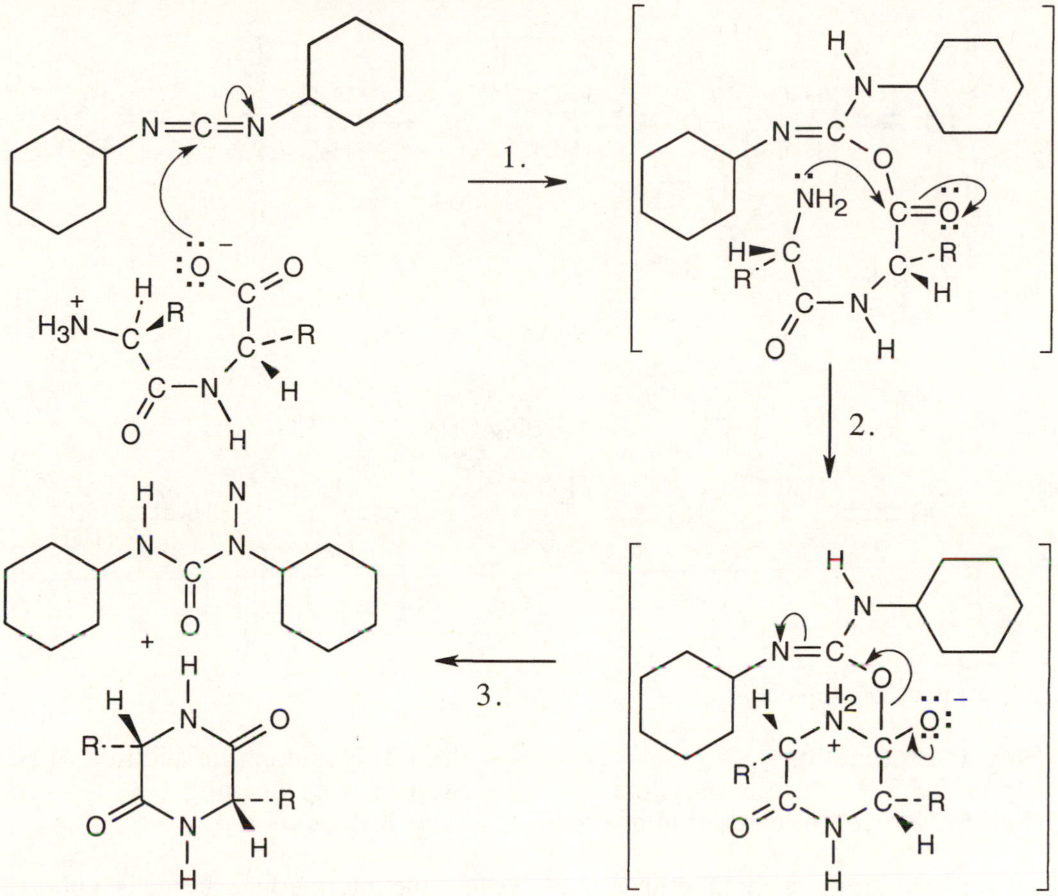

1.

2.

3.

Step 1: Addition of carboxylate to DCC.
Step 2: Intramolecular nucleophilic attack of the amino terminal end of the amide on the acylating agent.
Step 3: Loss of dicyclohexylurea.

16.47

(a)

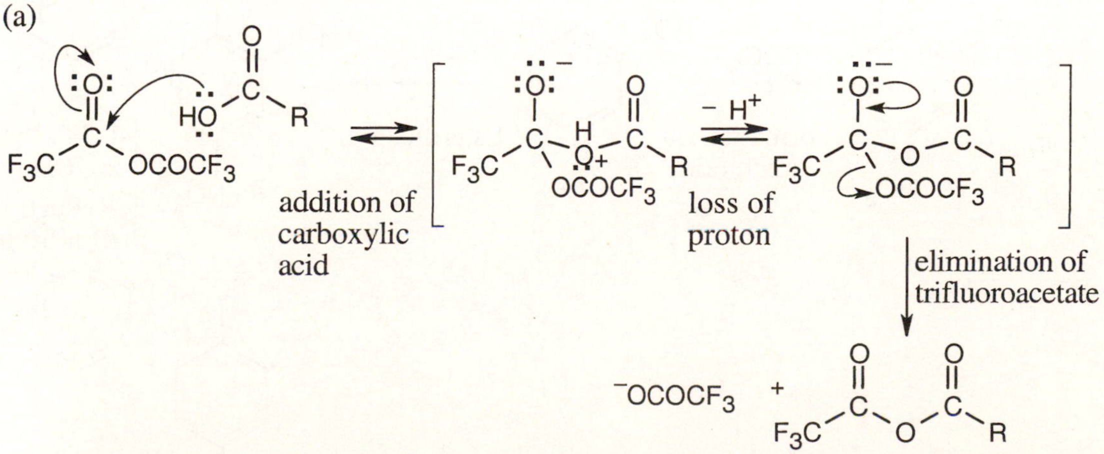

addition of carboxylic acid

$- H^+$

loss of proton

elimination of trifluoroacetate

(b) The electron-withdrawing fluorine atoms polarize the carbonyl group, making it more reactive toward nucleophiles.

(c) Because trifluoroacetate is a better leaving group than other carboxylate anions, the reaction proceeds as indicated.

16.48

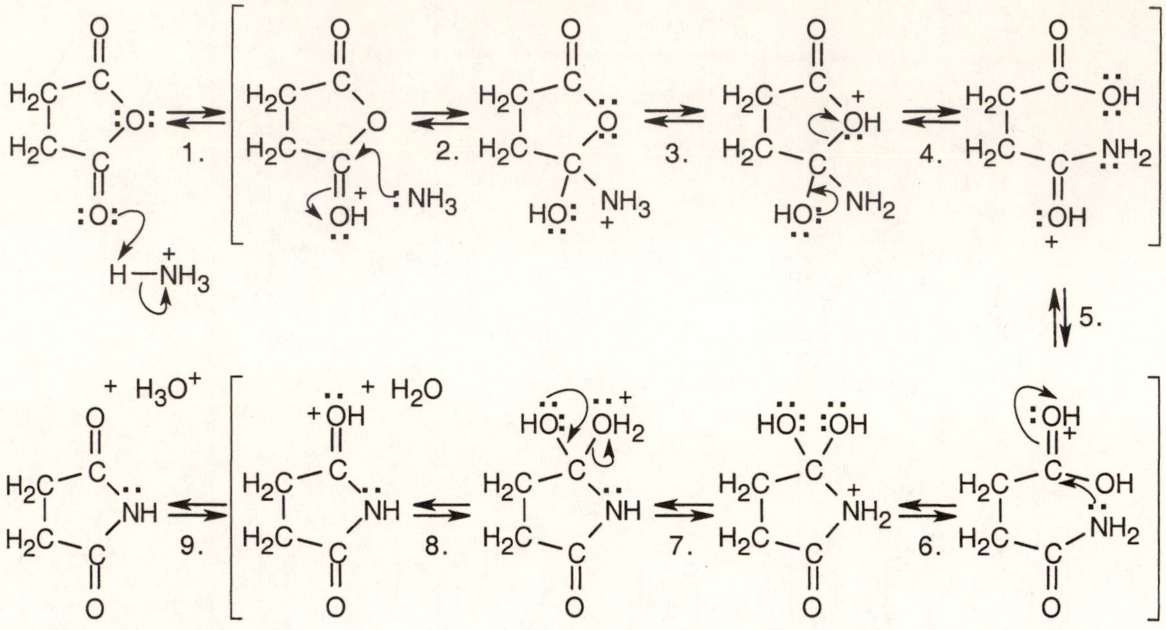

A summary of steps:

Step 1: Protonation
Step 2: Nucleophilic addition of NH_3
Steps 3,5,7,9: Proton transfers
Step 4: Ring opening
Step 6: Nucleophilic addition of $-NH_2$
Step 8: Loss of H_2O

This reaction requires high temperatures because the intermediate amide is a poor nucleophile and the carboxylic acid carbonyl group is unreactive.

16.49

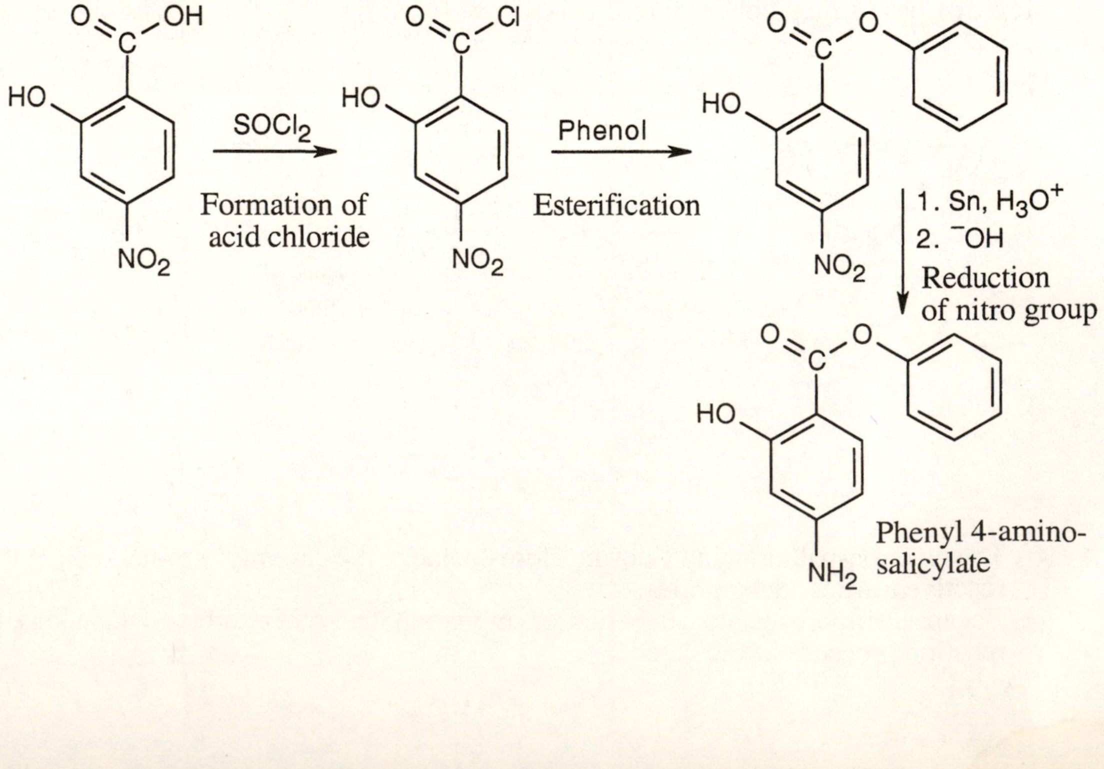

$SOCl_2$ — Formation of acid chloride

Phenol — Esterification

1. Sn, H_3O^+ 2. ^-OH — Reduction of nitro group

Phenyl 4-aminosalicylate

16.50

H_3C–C$_6H_4$–Br (m) →(Mg, ether)→ H_3C–C$_6H_4$–MgBr (m) →(1. CO_2; 2. H_3O^+; Grignard carboxylation)→ H_3C–C$_6H_4$–CO_2H (m)

↓ $SOCl_2$ formation of acid chloride

H_3C–C$_6H_4$–$COCl$ (m) →($HN(CH_2CH_3)_2$, NaOH; formation of amide)→ H_3C–C$_6H_4$–$CON(CH_2CH_3)_2$ (m)

N,N-Diethyl-*m*-toluamide

Synthetic analysis: Grignard carboxylation yields *m*-methylbenzoic acid, which can be converted to an acid chloride and treated with diethylamine.

16.51 (a)

$$H_2\ddot{C}^{-}{-}\overset{+}{N}{\equiv}N{:} \longleftrightarrow H_2C{=}\overset{+}{N}{=}\ddot{N}{:}^{-}$$

The resonance forms show that the carbon of diazomethane is basic, and reaction with an acid can occur to form a methyldiazonium ion.

$$RCO_2H + H_2\ddot{C}^{-}{-}\overset{+}{N}{\equiv}N{:} \longrightarrow RCO_2^{-} + CH_3{-}\overset{+}{N}{\equiv}N{:}$$

(b)

$$RCO_2^{-} + CH_3{-}\overset{+}{N}{\equiv}N{:} \longrightarrow RCOOCH_3 + N_2$$

An S_N2 reaction then occurs in which the carboxylate ion displaces N_2 as the leaving group to form the methyl ester.

16.52

PhC(=O)–C(=O)H + $^{-}$:OH →(nucleophilic addition of hydroxide)→ [PhC(=O)–CH(OH)–O^{-}] →(phenyl shift)→ [PhCH(O^{-})–C(=O)OH]

↓ proton transfer

[PhCH(OH)–C(=O)O^{-}] ⇌(H_3O^+, proton-ation)⇌ H_2O + PhCH(OH)–C(=O)OH

16.53 Both steps involve nucleophilic acyl substitutions.

Formation of acyl phosphate:

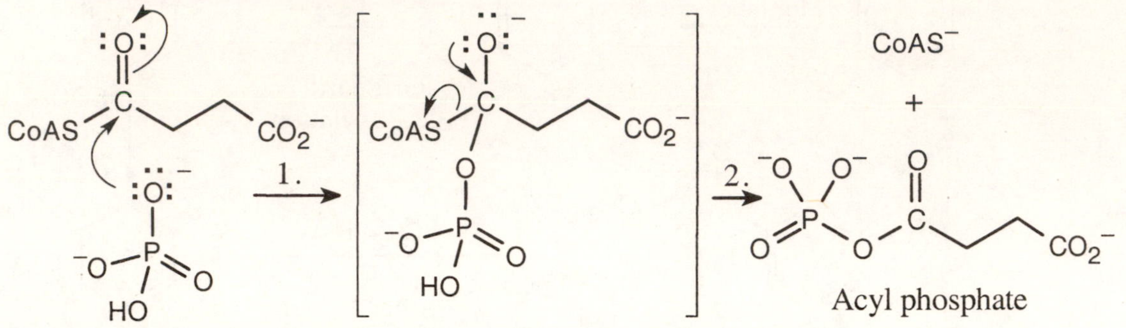

CoAS⁻ + Acyl phosphate

Step 1: Reaction of the phosphate oxygen with the carbonyl carbon of succinyl CoA.
Step 2: Loss of ⁻SCoA from the tetrahedral intermediate, yielding acyl phosphate.

Conversion of acyl phosphate to succinate:

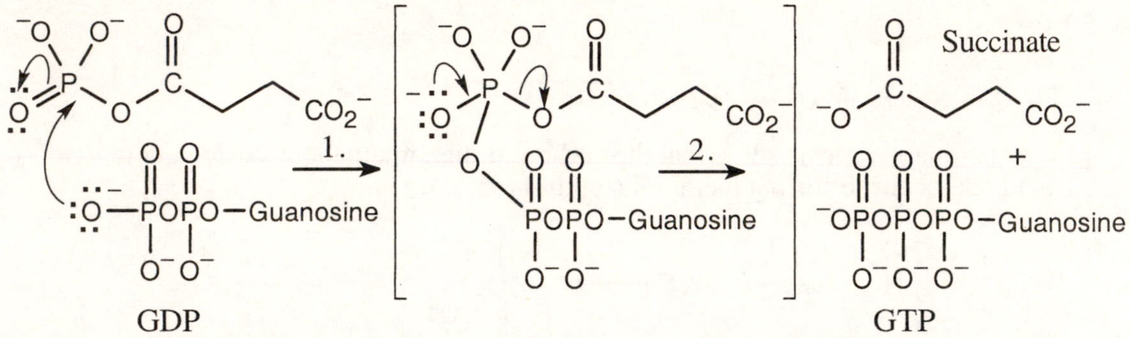

GDP → Succinate + GTP

Step 1: Reaction of the diphosphate oxygen of GDP with the phosphorus of the acyl phosphate to produce an intermediate similar to the intermediates formed in nucleophilic acyl substitutions of carboxylic acid derivatives.
Step 2: Loss of phosphate to form GTP and succinate.

16.54 In all of these reactions, a nucleophile adds to either carbon or phosphorus to form an intermediate that expels a leaving group to give the desired product.

Formation of 1,3-bisphosphoglycerate:

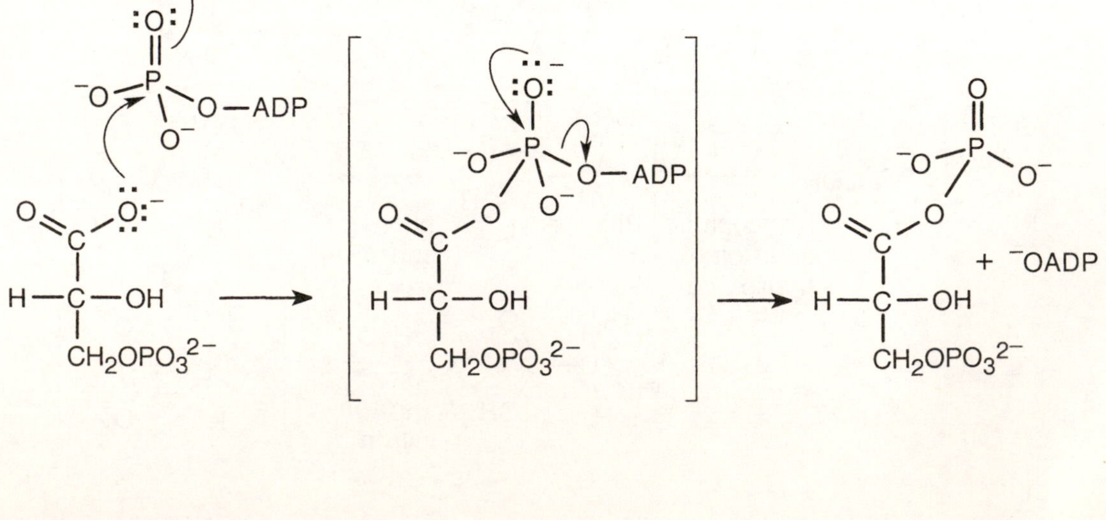

+ ⁻OADP

Route to enzyme-bound thioester:

: Base
Enz
S:
H
O
C
O
P
$^-$O
O$^-$
H—C—OH
$CH_2OPO_3^{2-}$
→
[
Enz
S
$^-$:O:
C
:O:
P
$^-$O
O$^-$
H—C—OH
$CH_2OPO_3^{2-}$
]
→
PO_4^{3-}
+
O=C—S—Enz
H—C—OH
$CH_2OPO_3^{2-}$

Reduction to glyceraldehyde 3-phosphate:

NAD
H
O=C—S—Enz
H—C—OH
$CH_2OPO_3^{2-}$
→
[
H
$^-$:O—C—S—Enz
H—C—OH
$CH_2OPO_3^{2-}$
]
→
O=C—H + HS—Enz
H—C—OH
$CH_2OPO_3^{2-}$
+
NAD^+

16.55 In some of these pairs, IR spectroscopy is sufficient for differentiating between the isomers. For others, either ^{1}H NMR or a combination of ^{1}H NMR and IR data is necessary.

(a)

	$CH_3CH_2C(=O)NHCH_3$ *N*-Methylpropanamide	$CH_3C(=O)N(CH_3)_2$ *N,N*-Dimethylacetamide
IR:	1680 cm^{-1} (*N*-substituted amide)	1650 cm^{-1} (*N,N*-disubstituted amide)
^{1}H NMR:	one methyl group one ethyl group	three methyl groups

(b)

	$HOCH_2CH_2CH_2CH_2C{\equiv}N$ 5-Hydroxypentanenitrile	cyclobutyl–$C(=O)NH_2$ Cyclobutanecarboxamide
IR:	3300–3400 cm^{-1} (hydroxyl) 2250 cm^{-1} (nitrile)	1690 cm^{-1} (amide)

(c)

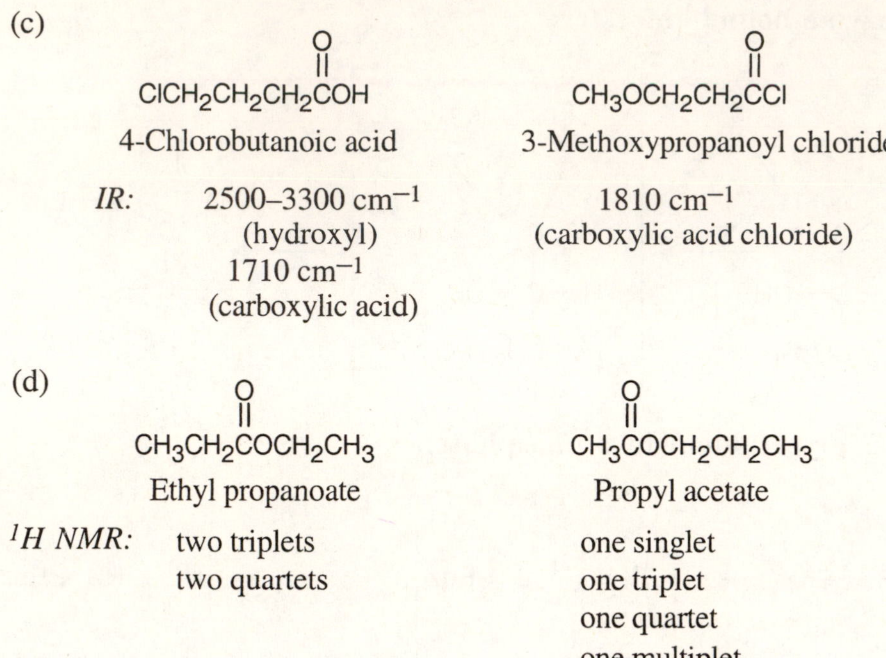

$ClCH_2CH_2CH_2COH$ (with C=O) — 4-Chlorobutanoic acid

$CH_3OCH_2CH_2CCl$ (with C=O) — 3-Methoxypropanoyl chloride

	4-Chlorobutanoic acid	3-Methoxypropanoyl chloride
IR:	2500–3300 cm^{-1} (hydroxyl) 1710 cm^{-1} (carboxylic acid)	1810 cm^{-1} (carboxylic acid chloride)

(d)

$CH_3CH_2COCH_2CH_3$ (with C=O) — Ethyl propanoate

$CH_3COCH_2CH_2CH_3$ (with C=O) — Propyl acetate

	Ethyl propanoate	Propyl acetate
^{1}H NMR:	two triplets two quartets	one singlet one triplet one quartet one multiplet

16.56 The IR spectrum indicates that this compound has a carbonyl group.

$CH_3CH(Cl)C(=O)OCH_3$ (labels: CH_3 a, CH c, OCH_3 b)

a = 1.7 δ

b = 3.8 δ

c = 4.4 δ

16.57

(a) $CH_3CH_2CH_2CCl$ (with C=O; labels a b c)

a = 1.0 δ

b = 1.7 δ

c = 2.8 δ

(b) $N{\equiv}CCH_2COCH_2CH_3$ (with C=O; labels b c a)

a = 1.3 δ

b = 3.6 δ

c = 4.3 δ

(c) $CH_3COCH(CH_3)_2$ (with C=O; labels b c a)

a = 1.3 δ

b = 2.0 δ

c = 5.0 δ

16.58 This a nucleophilic acyl substitution reaction whose mechanism is similar to others we have studied.

$RC(=O)CI_3$ + $^{-}$:OH ⇌ (addition) [$RC(O^{-})(OH)CI_3$ ⟶ (elimination) $RC(=O)OH$ + $^{-}:CI_3$] ⇌ (acid–base reaction) $RC(=O)O^{-}$ + HCI_3

$I_3C:^{-}$ can act as a leaving group because the electron-withdrawing iodine atoms stabilize the carbanion.

Chapter 17 – Carbonyl Alpha-Substitution and Condensation Reactions

Chapter Outline

I. Keto-enol tautomerism (Section 17.1).
   A. Nature of tautomerism.
      1. Carbonyl compounds with hydrogens bonded to their α carbons equilibrate with their corresponding enols.
      2. This rapid equilibration is called tautomerism, and the individual isomers are tautomers.
      3. Unlike resonance forms, tautomers are isomers.
      4. Despite the fact that very little of the enol isomer is present at room temperature, enols are very important because they are reactive.
   B. Mechanism of tautomerism.
      1. In acid-catalyzed enolization, the carbonyl carbon is protonated to form an intermediate that can lose a hydrogen from its α carbon to yield a neutral enol.
      2. In base-catalyzed enol formation, an acid-base reaction occurs between a base and an α hydrogen.
         a. The resultant enolate ion is protonated to yield an enol.
         b. Protonation can occur either on carbon or on oxygen.
         c. Only hydrogens on the α positions of carbonyl compounds are acidic.

II. Enols and enolates (Sections 17.2 - 17.4)
   A. Reactivity of enols (Section 17.2).
      1. The electron-rich double bonds of enols cause them to behave as nucleophiles.
         The electron-donating enol –OH groups make enols more reactive than alkenes.
      2. When an enol reacts with an electrophile, the initial adduct loses –H from oxygen to give a substituted carbonyl compound.
   B Reactions of enols.
      Alpha halogenation of aldehydes and ketones.
         a. Aldehydes and ketones can be halogenated at their α positions by reaction of X_2 in acidic solution.
         b. The reaction proceeds by acid-catalyzed formation of an enol intermediate.
         c. Halogen isn't involved in the rate-limiting step: the rate doesn't depend on the identity of the halogen, but only on [ketone] and $[H^+]$.
         d. α-Bromo ketones are useful in syntheses because they can be dehydrobrominated by base treatment to form α,β-unsaturated ketones.
   C. Enolate ion formation (Section 17.3).
      1. Hydrogens α to a carbonyl group are weakly acidic.
         a. This acidity is due to overlap of a *p* orbital with the carbonyl group *p* orbitals, allowing the carbonyl group to stabilize the negative charge by resonance.
         b. The two resonance forms aren't equivalent; the form with the negative charge on oxygen is of lower energy.
      2. Strong bases are needed for enolate ion formation.
         a. Alkoxide ions are generally too weak to use in enolate formation.
         b. Lithium diisopropylamide (LDA) is used for forming enolates because it is a very strong base, it is soluble in THF, it is hindered and it can be used at low temperatures.
         c. LDA can be used to form the enolates of many different carbonyl compounds.

3. When a hydrogen is flanked by two carbonyl groups, it is much more acidic.
   Both carbonyl groups can stabilize the negative charge.

D. Alkylation reactions of enolate ions (Section 17.4).
   1. General features.
      a. Alkylations are useful because they form a new C–C bond.
      b. Alkylations have the same limitations as S_N2 reactions; the alkyl groups must be methyl, primary, allylic or benzylic.
   2. The malonic ester synthesis.
      a. The malonic ester synthesis is used for preparing a carboxylic acid from a halide while lengthening the chain by two carbon atoms.
      b. Diethyl malonate is useful because its enolate is easily prepared by reaction with sodium ethoxide.
      c. Since diethyl malonate has two acidic hydrogens, two alkylations can take place.
      d. Heating in aqueous HCl causes hydrolysis and decarboxylation of the alkylated malonate.
         Decarboxylations are common only to β-keto acids and malonic acids.
      e. Cycloalkanecarboxylic acids can also be prepared.
   3. The acetoacetic ester synthesis.
      a. The acetoacetic ester synthesis is used for converting an alkyl halide to a methyl ketone, while lengthening the carbon chain by 3 atoms.
      b. As with malonic ester, acetoacetic ester has two acidic hydrogens which are flanked by a ketone and an ester, and two alkylations can take place.
      c. Heating in aqueous HCl hydrolyzes the ester and decarboxylates the acid to yield the ketone.
      d. Most β-keto esters can undergo this type of reaction.
   4. Direct alkylation of ketones, esters, and nitriles.
      a. LDA in a nonprotic solvent can be used to convert the above compounds to their enolates.
      b. Alkylation of an unsymmetrical ketone leads to a mixture of products, but the major product is alkylated at the less hindered position.
   5. Biological alkylations occur occasionally.

III. Carbonyl condensation reactions (Sections 17.5 – 17.12).
   A. Carbonyl condensation reactions take place between two carbonyl components.
      1. One component (the nucleophilic donor) is converted to its enolate and undergoes an α–substitution reaction.
      2. The other component (the electrophilic acceptor) undergoes nucleophilic addition.
   B. The aldol reaction (Sections 17.5. – 17.7).
      1. Characteristics of the aldol reaction (Section 17.5).
         a. The aldol condensation is a base-catalyzed dimerization of two aldehydes or ketones.
         b. The reaction can occur between two components that have α hydrogens.
         c. For simple aldehydes, the equilibrium favors the products, but for other aldehydes and ketones, the equilibrium favors the reactants.
         d. Carbonyl condensation reactions require only a catalytic amount of base.
            Alpha-substitution reactions, on the other hand, use one equivalent of base.
      2. Dehydration of aldol products (Section 17.6).
         a. Aldol products are easily dehydrated to yield α,β-unsaturated aldehydes and ketones.
         b. Dehydration is catalyzed by both acid and base.
         c. Reaction conditions for dehydration are only slightly more severe than for condensation.
         d. Often, dehydration products are isolated directly from condensation reactions.

        e. Conjugated enones are more stable than nonconjugated enones.
        f. Removal of the water byproduct drives the aldol equilibrium towards product formation.
    3. Intramolecular aldol condensations (Section 17.7).
        a. Treatment of certain dicarbonyl compounds with base can lead to cyclic products.
        b. A mixture of cyclic products may result, but the more strain-free ring predominates.

C. The Claisen condensation (Sections 17.8 – 17.12).
    1. Features of the Claisen condensation (Section 17.8).
        a. Treatment of an ester with 1 equivalent of base yields a β-keto ester.
        b. The reaction is reversible and has a mechanism similar to that of the aldol reaction.
        c. A major difference from the aldol condensation is the expulsion of an alkoxide ion from the tetrahedral intermediate of the initial Claisen adduct.
        d. Because the product is acidic, one equivalent of base is needed; proton removal by the base drives the reaction to completion.
        e. Addition of acid yields the final product.
    2. Intramolecular Claisen condensations: the Dieckmann cyclization (Section 17.9).
        a. The Dieckmann cyclization is used to form cyclic β-keto esters.
            i. 1,6-Diesters form 5-membered rings.
            ii. 1,7-Diesters form 6-membered rings.
        b. The mechanism is similar to the Claisen condensation mechanism.
        c. The product β-keto esters can be further alkylated.
           This is a good route to 2-substituted cyclopentanones and cyclohexanones.

D. The Michael reaction (Section 17.10).
    1. The Michael reaction is the conjugate addition of an enolate to an α,β-unsaturated carbonyl compound.
       The highest-yielding reactions occur between stable enolates and unhindered α,β-unsaturated carbonyl compounds.
    2. The mechanism is a conjugate addition of a nucleophilic enolate to the β carbon of a α,β-carbonyl acceptor.
    3. Stable enolates are Michael donors, and α,β-unsaturated compounds are Michael acceptors.

E. The Stork enamine reaction (Section 17.11).
    1. A ketone that has been converted to an enamine can act as a Michael donor in a reaction known as the Stork enamine reaction.
    2. The sequence of reactions in the Stork enamine reaction:
        a. Enamine formation from a ketone.
        b. Michael-type addition to an α,β-unsaturated carbonyl compound.
        c. Enamine hydrolysis back to a ketone.
    3. This sequence is equivalent to the Michael addition of a ketone to an α,β-unsaturated carbonyl compound and provides a 1,5 diketone product.

F. Biological carbonyl condensation reactions (Section 17.12).
    1. Many biomolecules are synthesized by carbonyl condensation reactions.
    2. Aldolase enzymes catalyze the addition of a ketone enolate to an aldehyde.
    3. Acetyl CoA is the major building block for the synthesis of biomolecules.
        a. Acetyl CoA can act as an electrophilic acceptor by being attacked at its carbonyl group.
        b. Acetyl CoA can act as a nucleophilic donor by loss of its acidic α hydrogen.

Solutions to Problems

17.1 Acidic hydrogens in the keto form of each of these compounds are bold.

	Keto Form	*Enol Form*	*Number of Acidic Hydrogens*
(a)	cyclopentanone	1-cyclopentenol	4
(b)	CH_3COSCH_3	$H_2C{=}C(OH)SCH_3$	3
(c)	$CH_3COOCH_2CH_3$	$H_2C{=}C(OH)OCH_2CH_3$	3
(d)	H_3CCH_2CHO	$H_3CCH{=}CH(OH)$	2
(e)	CH_3COOH	$H_2C{=}C(OH)OH$	4
(f)	$C_6H_5CH_2COCH_3$	$C_6H_5CH{=}C(OH)CH_3$ *or* $C_6H_5CH_2C(OH){=}CH_2$	5

In (d) and (f), cis and trans enolates are possible.

17.2

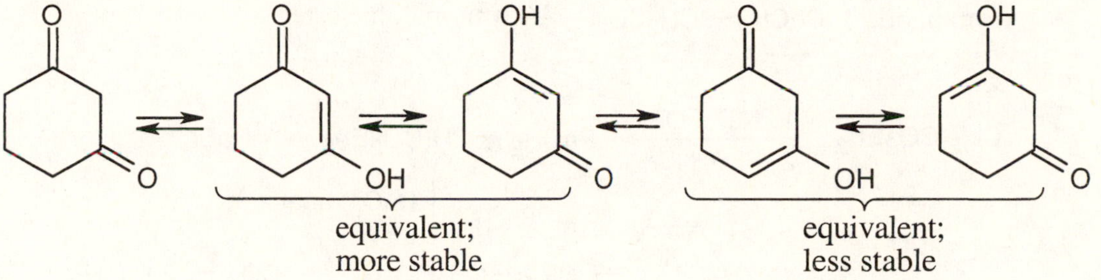

equivalent; more stable

equivalent; less stable

The first two monoenols are more stable because the enol double bond is conjugated with the carbonyl group.

17.3 Hydrogens α to one carbonyl group are weakly acidic. Hydrogens α to two carbonyl groups are much more acidic but are not as acidic as carboxylic acid protons.

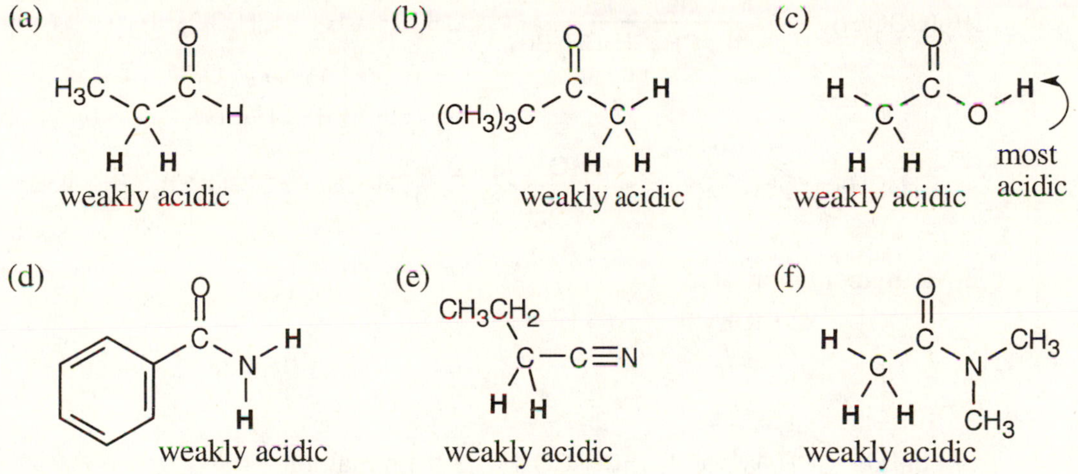

(a) weakly acidic

(b) weakly acidic

(c) weakly acidic; most acidic

(d) weakly acidic

(e) weakly acidic

(f) weakly acidic

17.4 Nitriles are weakly acidic because the nitrile anion can be stabilized by resonance involving the π bond of the nitrile group.

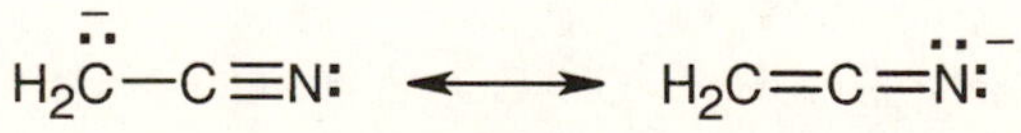

$H_2\ddot{C}^{-}{-}C{\equiv}N: \longleftrightarrow H_2C{=}C{=}\ddot{N}:^{-}$

17.5 The malonic ester synthesis converts a primary or secondary halide into a carboxylic acid with two more carbons (a substituted acetic acid). Identify the component that originates from malonic ester (the acid component). The rest of the molecule comes from the alkyl halide.

(a)

from halide $PhCH_2$ ⁄ CH_2CO_2H from malonic ester

$$CH_2(CO_2Et)_2 \xrightarrow[\text{2. } PhCH_2Br]{\text{1. } Na^+ \ ^-OEt} PhCH_2{-}CH(CO_2Et)_2 + NaBr$$

$$\downarrow H_3O^+, \text{heat}$$

$$C_6H_5{-}CH_2{-}CH_2CO_2H + CO_2 + 2\ EtOH$$

3-Phenylpropanoic acid

(b)

$CH_3CH_2CH_2$ ⁄ $CHCO_2H$ from malonic ester

from halide CH_3 (on the CH carbon)

$$CH_2(COOEt)_2 \xrightarrow[\text{2. } CH_3CH_2CH_2Br]{\text{1. } Na^+ \ ^-OEt} CH_3CH_2CH_2{-}CH(CO_2Et)_2 + NaBr$$

$$\downarrow \text{1. } Na^+ \ ^-OEt;\ \text{2. } CH_3Br$$

$$CH_3CH_2CH_2{-}CH(CH_3)CO_2H \xleftarrow[\text{heat}]{H_3O^+} CH_3CH_2CH_2{-}C(CH_3)(CO_2Et)_2 + NaBr$$

2-Methylpentanoic acid

$+ CO_2 + 2\ EtOH$

(c)

from halide $(CH_3)_2CHCH_2$ ⁄ CH_2CO_2H from malonic ester

$$CH_2(CO_2Et)_2 \xrightarrow[\text{2. } (CH_3)_2CHCH_2Br]{\text{1. } Na^+ \ ^-OEt} (CH_3)_2CHCH_2{-}CH(CO_2Et)_2 + NaBr$$

$$\downarrow H_3O^+, \text{heat}$$

$$(CH_3)_2CHCH_2{-}CH_2CO_2H + CO_2 + 2\ EtOH$$

4-Methylpentanoic acid

17.6

from malonic ester

CO_2H

from halides

$CH_2(CO_2Et)_2$ $\xrightarrow[\text{2. } CH_3CH(CH_3)CH_2Br]{\text{1. } Na^+ \ ^-OEt}$ $CH_3CH(CH_3)CH_2—CH(CO_2Et)_2$ + NaBr

$\downarrow$ 1. $Na^+ \ ^-OEt$ 2. CH_3Br

$CH_3CH(CH_3)CH_2—C(CH_3)(CO_2Et)_2$ + NaBr

$\xrightarrow[\text{heat}]{H_3O^+}$ $CH_3CH(CH_3)CH_2—CH(CH_3)CO_2H$

2,4-Dimethylpentanoic acid

+ CO_2 + 2 EtOH

17.7 As in the malonic ester synthesis, you should identify the structural fragments of the target compound. The acetoacetic ester synthesis converts an alkyl halide into a methyl ketone ("substituted acetone"). The methyl ketone component comes from acetoacetic ester; the other component comes from a halide.

Solution:

(a)

from halide $(CH_3)_2CHCH_2$ + CH_2CCH_3 (C=O) from acetoacetic ester

$CH_3COCH_2CO_2Et$ $\xrightarrow[\text{2. } (CH_3)_2CHCH_2Br]{\text{1. } Na^+ \ ^-OEt}$ $(CH_3)_2CHCH_2—CH(CO_2Et)COCH_3$ + NaBr

$\downarrow$ H_3O^+, heat

$(CH_3)_2CHCH_2—CH_2COCH_3$ + CO_2 + EtOH

5-Methylhexan-2-one

(b)

from halide $C_6H_5CH_2CH_2$ ⁄ $CH_2\overset{O}{\overset{\|}{C}}CH_3$ from acetoacetic ester

$$\underset{CO_2Et}{\underset{|}{CH_2}}\overset{O}{\overset{\|}{C}}CH_3 \xrightarrow[2.\ C_6H_5CH_2CH_2Br]{1.\ Na^+\ ^-OEt} C_6H_5CH_2CH_2-\underset{CO_2Et}{\underset{|}{CH}}\overset{O}{\overset{\|}{C}}CH_3 + NaBr$$

$$\Big\downarrow H_3O^+,\ \text{heat}$$

$$C_6H_5CH_2CH_2-CH_2\overset{O}{\overset{\|}{C}}CH_3 + CO_2 + EtOH$$

5-Phenylpentan-2-one

17.8 The acetoacetic ester synthesis can only be used for certain products:

(1) Three carbons must originate from acetoacetic ester. In other words, compounds of the type $RCOCH_3$ can't be synthesized by the reaction of RX with acetoacetic ester.
(2) Alkyl halides must be primary or methyl.
(3) The acetoacetic ester synthesis can't be used to prepare compounds that are trisubstituted at the α position.

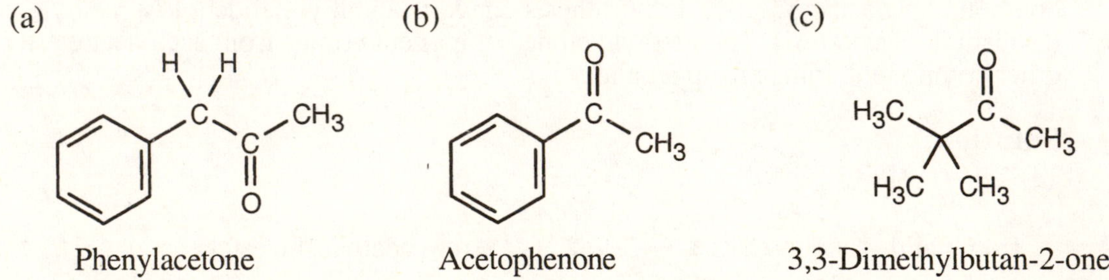

Phenylacetone Acetophenone 3,3-Dimethylbutan-2-one

(a) Phenylacetone can't be produced by an acetoacetic ester synthesis because bromobenzene, the necessary halide, does not enter into S_N2 reactions. [See (2) above.]

(b) Acetophenone can't be produced by an acetoacetic ester synthesis. [See (1) above.]

(c) 3,3-Dimethylbutan-2-one can't be prepared because it is trisubstituted at the α position. [See (3) above.]

17.9

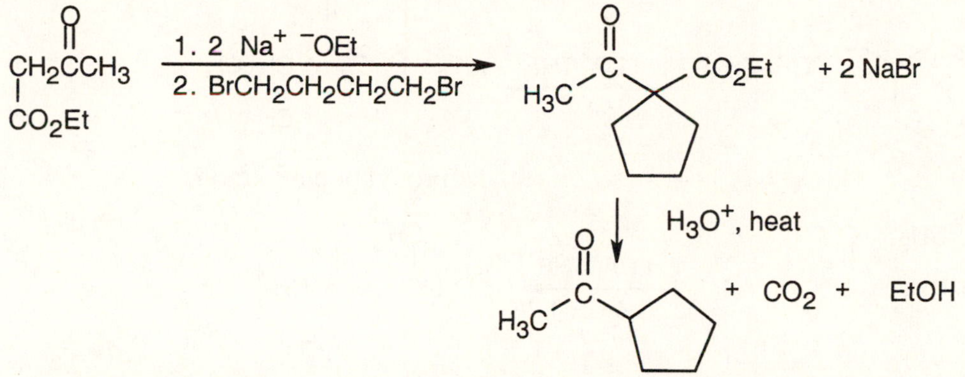

CH_2CCH_3(=O) with CO_2Et $\xrightarrow[\text{2. } BrCH_2CH_2CH_2CH_2Br]{\text{1. 2 } Na^+ \ ^-OEt}$ 1-acetyl-1-(ethoxycarbonyl)cyclopentane (H_3C–C(=O)–, CO_2Et) + 2 NaBr

$\xrightarrow{H_3O^+\text{, heat}}$ H_3C–C(=O)–cyclopentane + CO_2 + EtOH

17.10 Strategy: Direct alkylation is used to introduce substituents α to an ester, ketone or nitrile. Look at the target molecule to identify these substituents. Alkylation is achieved by treating the starting material with LDA, followed by a primary halide.

Solution:

(a)

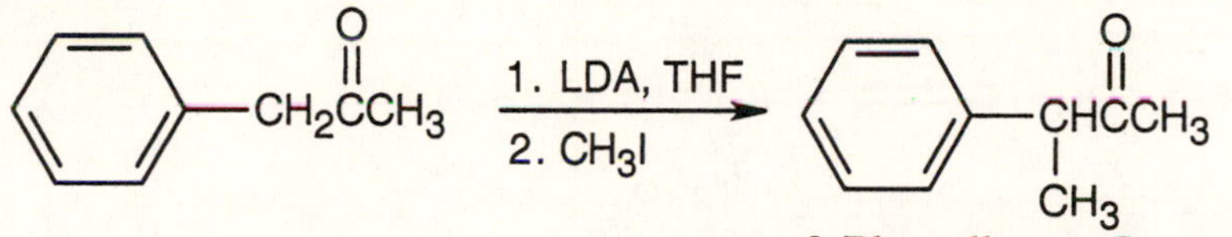

C_6H_5–CH_2CCH_3(=O) $\xrightarrow[\text{2. } CH_3I]{\text{1. LDA, THF}}$ C_6H_5–$CHCCH_3$(=O) with CH_3

3-Phenylbutan-2-one

Alkylation occurs at the carbon next to the phenyl group because the phenyl group can help stabilize the enolate anion intermediate.

(b)

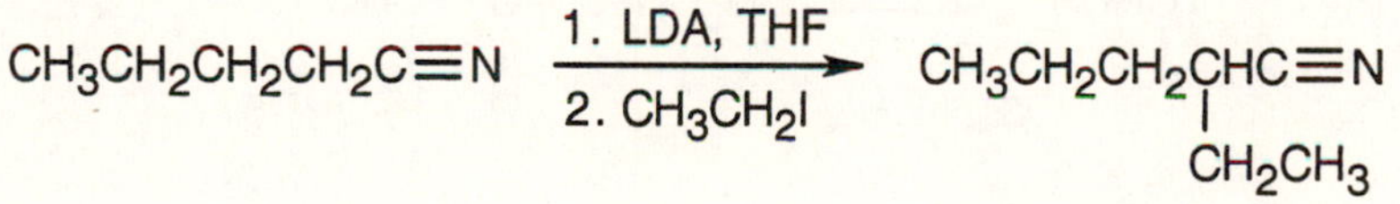

$CH_3CH_2CH_2CH_2C{\equiv}N$ $\xrightarrow[\text{2. } CH_3CH_2I]{\text{1. LDA, THF}}$ $CH_3CH_2CH_2CHC{\equiv}N$ with CH_2CH_3

2-Ethylpentanenitrile

(c)

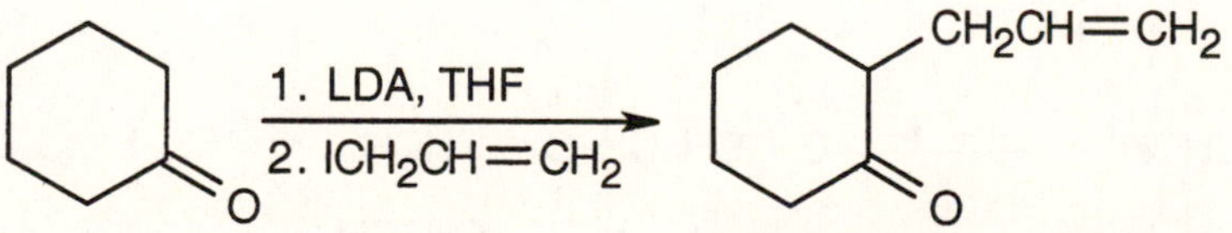

Cyclohexanone $\xrightarrow[\text{2. } ICH_2CH{=}CH_2]{\text{1. LDA, THF}}$ cyclohexanone with $CH_2CH{=}CH_2$

2-Allylcyclohexanone

(d)

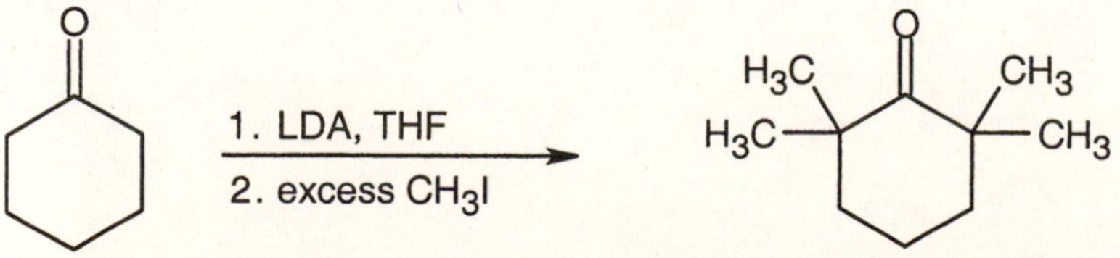

Cyclohexanone $\xrightarrow[\text{2. excess } CH_3I]{\text{1. LDA, THF}}$ cyclohexanone with H_3C, H_3C, CH_3, CH_3

2,2,6,6-Tetramethylcyclohexanone

(e)

$$C_6H_5C(=O)CH_2CH_3 \xrightarrow[\text{2. } CH_3I]{\text{1. LDA, THF}} C_6H_5C(=O)CH(CH_3)CH_3$$

Isopropyl phenyl ketone

(f)

$$CH_3CH(CH_3)CH_2COCH_3 \ (\text{C=O}) \xrightarrow[\text{2. } CH_3CH_2I]{\text{1. LDA, THF}} CH_3CH(CH_3)CH(CH_2CH_3)COCH_3$$

Methyl 2-ethyl-3-methylbutanoate

17.11 When you are first learning the aldol condensation, write all the steps.

(1) Form the enolate of one molecule of the carbonyl compound.

$$CH_3CH_2CH_2CHO + {:}\ddot{O}H^- \rightleftharpoons {}^-{:}CH(CH_2CH_3)CHO + H_2O$$

(2) Have the enolate add to the electrophilic carbonyl group of the second molecule.

$$CH_3CH_2CH_2CHO + {}^-{:}CH(CH_2CH_3)CHO \rightleftharpoons CH_3CH_2CH_2CH(\ddot{O}{:}^-)-CH(CH_2CH_3)CHO$$

(3) Protonate the alkoxide ion.

$$CH_3CH_2CH_2CH(\ddot{O}{:}^-)-CH(CH_2CH_3)CHO + H-OH \rightleftharpoons CH_3CH_2CH_2CH(OH)-CH(CH_2CH_3)CHO + {}^-OH$$

Practice writing out these steps for the other aldol condensations.

(a) See above.

(b)

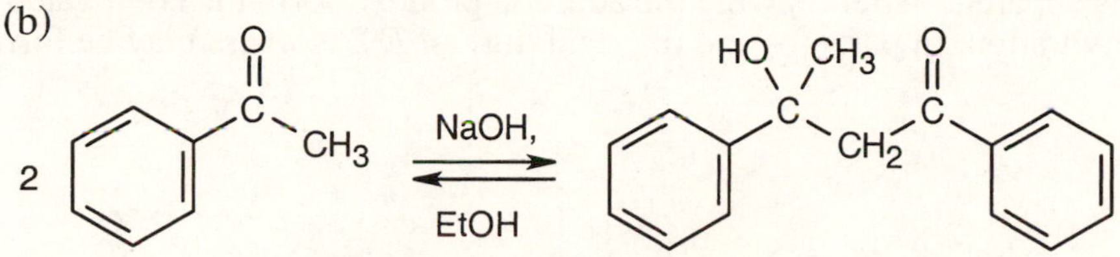

$$2\ C_6H_5COCH_3 \xrightleftharpoons[\text{EtOH}]{\text{NaOH,}} C_6H_5C(OH)(CH_3)CH_2COC_6H_5$$

(c)

$$2\ \text{cyclopentanone} \xrightleftharpoons[\text{EtOH}]{\text{NaOH,}} \text{2-(1-hydroxycyclopentyl)cyclopentanone}$$

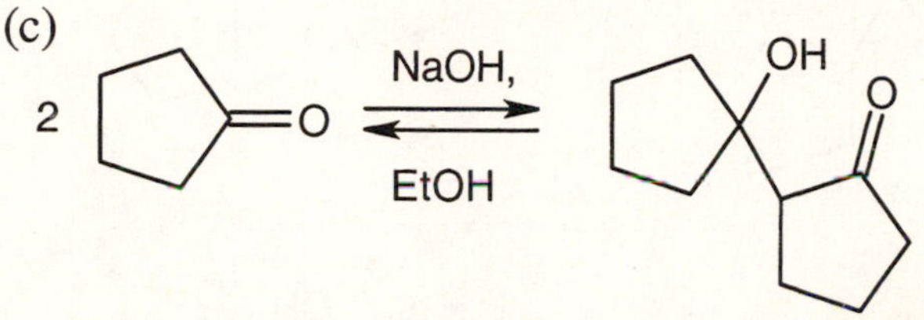

17.12

$$CH_3C(OH)(CH_3)-CH_2CCH_3\ (\text{C=O})$$

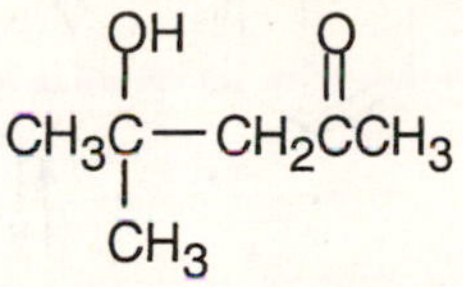

4-Hydroxy-4-methylpentan-2-one

The steps for the reverse aldol are the reverse of those described in Problem 17.11.
(1) Deprotonate the alcohol oxygen.

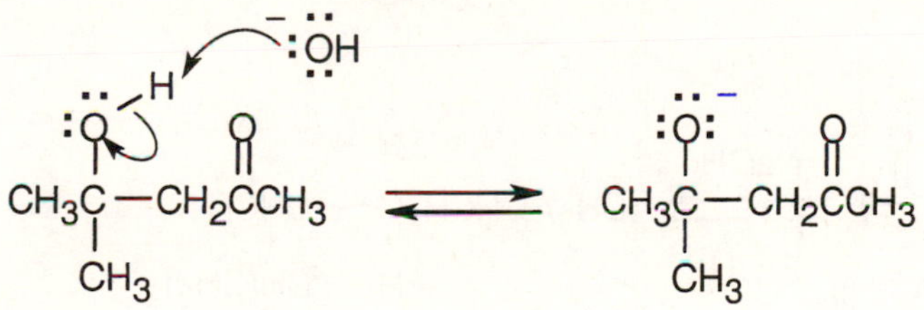

$$CH_3C(OH)(CH_3)-CH_2COCH_3 + {}^-{:}OH \rightleftharpoons CH_3C(O^-)(CH_3)-CH_2COCH_3$$

(2) Eliminate the enolate anion.

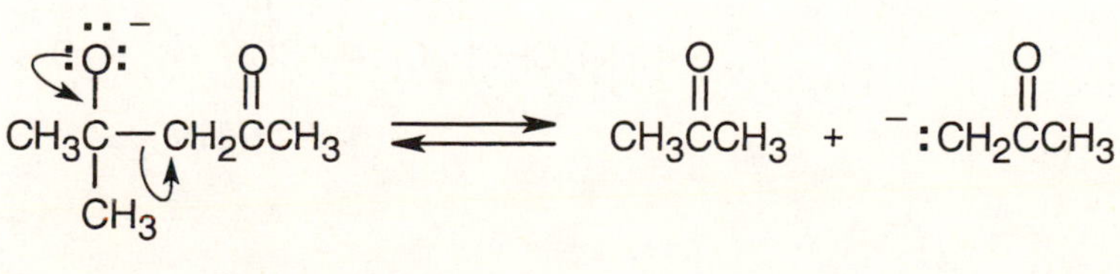

$$CH_3C(O^-)(CH_3)-CH_2COCH_3 \rightleftharpoons CH_3COCH_3 + {}^-{:}CH_2COCH_3$$

(3) Reprotonate the enolate anion.

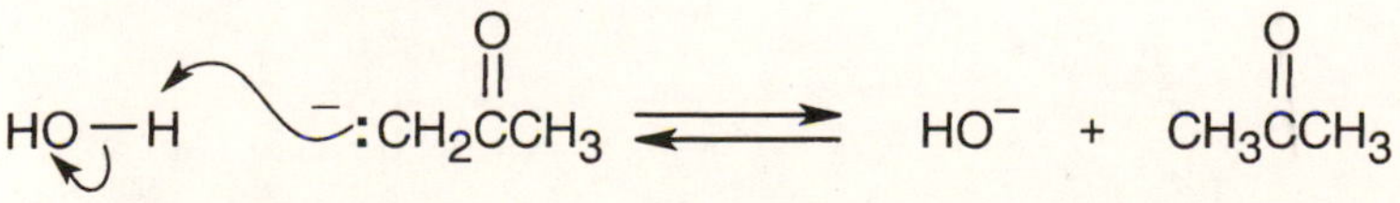

$$HO-H + {}^-{:}CH_2COCH_3 \rightleftharpoons HO^- + CH_3COCH_3$$

17.13 Strategy: As in Problem 17.11, align the two carbonyl compounds so that the location of the new bond is apparent. After drawing the addition product, form the conjugated enone product by dehydration. In parts (b) and (c), a mixture of *E,Z* isomers may be formed.

Solution:

(a)

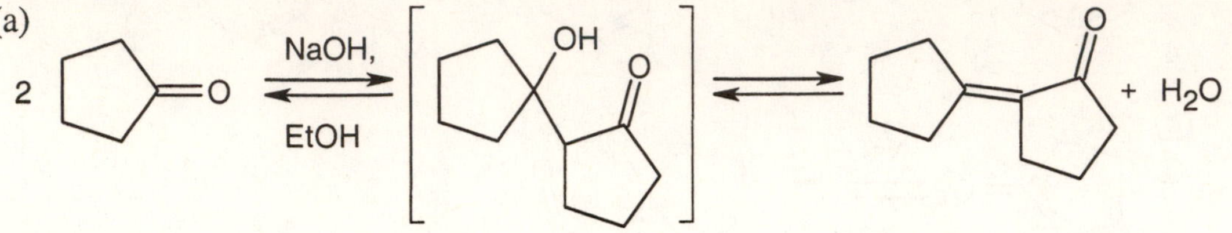

2 cyclopentanone $\xrightleftharpoons[\text{EtOH}]{\text{NaOH,}}$ [aldol addition product: 1-hydroxycyclopentyl attached to C-2 of cyclopentanone] $\rightleftharpoons$ 2-cyclopentylidenecyclopentanone + H_2O

(b)

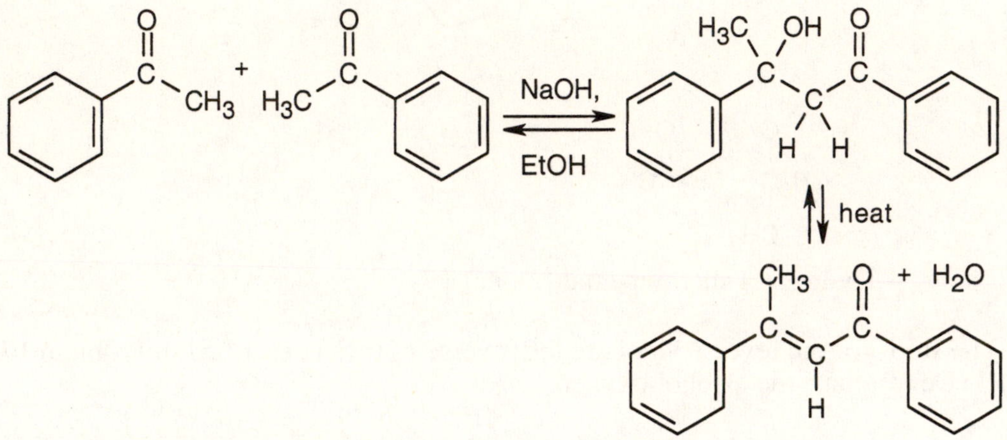

$C_6H_5COCH_3$ + $H_3CCOC_6H_5$ $\xrightleftharpoons[\text{EtOH}]{\text{NaOH,}}$ $C_6H_5C(CH_3)(OH)CH_2COC_6H_5$

$\uparrow\downarrow$ heat

$C_6H_5C(CH_3)=CHCOC_6H_5$ + H_2O

(c)

$$(CH_3)_2CHCH_2\overset{O}{\overset{\|}{C}}H + \underset{CH(CH_3)_2}{\underset{|}{C}}H_2\overset{O}{\overset{\|}{C}}H \xrightleftharpoons[\text{EtOH}]{\text{NaOH,}} (CH_3)_2CHCH_2\underset{H}{\underset{|}{\overset{OH}{\overset{|}{C}}}}-\underset{CH(CH_3)_2}{\underset{|}{C}}H\overset{O}{\overset{\|}{C}}H$$

$\uparrow\downarrow$ heat

$$(CH_3)_2CHCH_2CH=\underset{CH(CH_3)_2}{\underset{|}{C}}\overset{O}{\overset{\|}{C}}H + H_2O$$

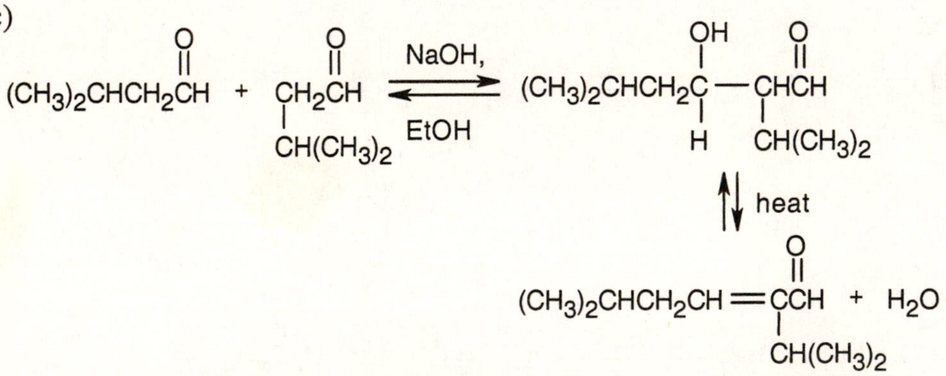

17.14 Including double bond isomers, 4 products can be formed. The major product is formed by reaction of the enolate ion formed by abstraction of a proton at position "a".

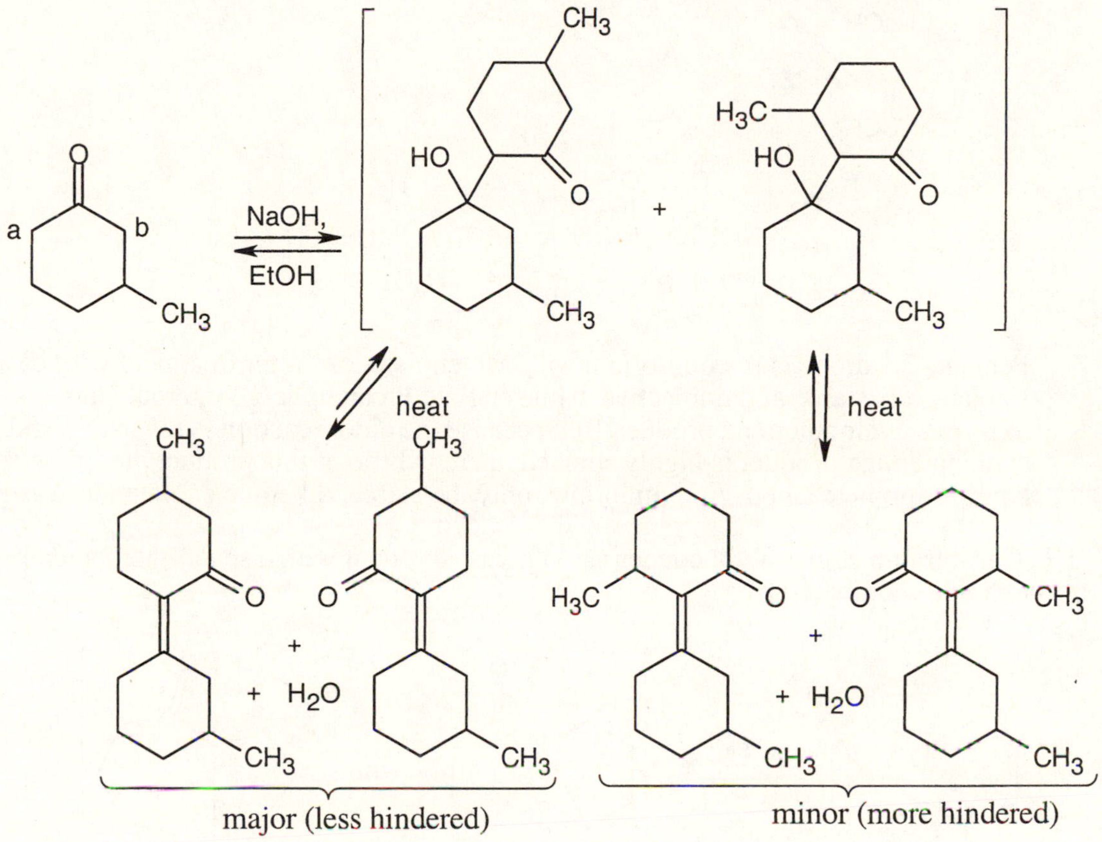

a b CH_3 NaOH, EtOH CH_3 HO O CH_3 + H_3C HO O CH_3

heat heat

CH_3 O + H_2O CH_3 CH_3 O CH_3 H_3C O + H_2O CH_3 O CH_3 + CH_3

major (less hindered) minor (more hindered)

17.15

(a)

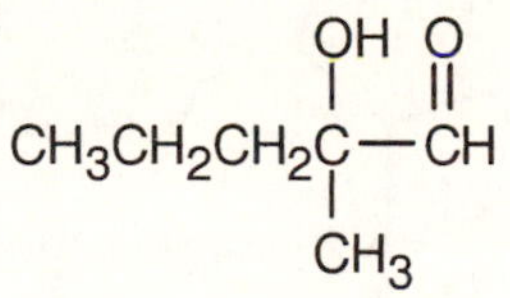

$CH_3CH_2CH_2C(OH)(CH_3)-CHO$ 2-Hydroxy-2-methylpentanal

This is not an aldol product. The hydroxyl group in an aldol product must be ß, not α, to the carbonyl group.

(b)

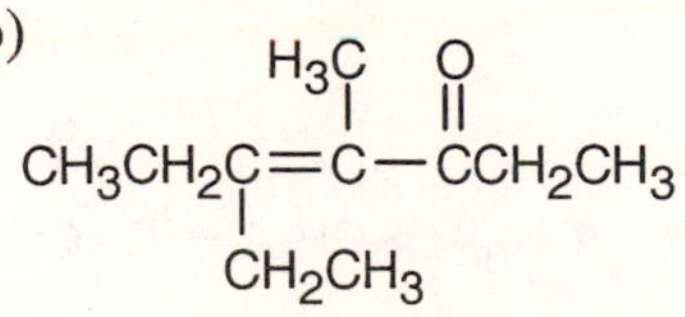

$CH_3CH_2C(CH_2CH_3)=C(CH_3)-CCH_2CH_3$ (C=O) 5-Ethyl-4-methylhept-4-en-3-one

This product results from the aldol self-condensation of pentan-3-one, followed by dehydration.

17.16

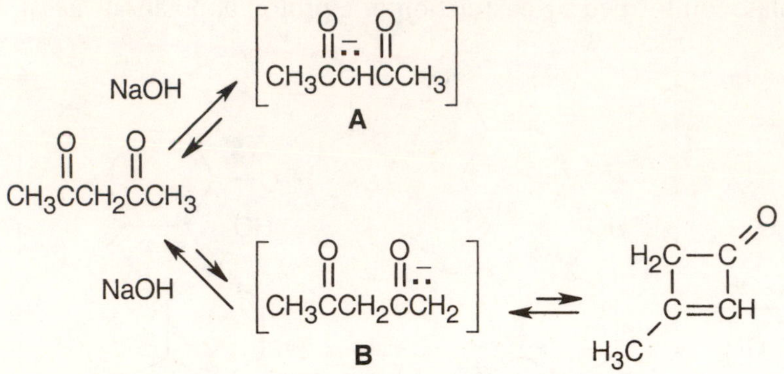

$$CH_3COCH_2COCH_3 \underset{}{\overset{NaOH}{\rightleftharpoons}} \left[CH_3CO\ddot{C}^{-}HCOCH_3\right] \textbf{ A}$$

$$CH_3COCH_2COCH_3 \overset{NaOH}{\rightleftharpoons} \left[CH_3COCH_2CO\ddot{C}^{-}H_2\right] \textbf{ B} \rightleftharpoons \text{(cyclobutenone: } H_2C{-}C{=}O,\ C{=}CH,\ H_3C\text{)}$$

Pentane-2,4-dione is in equilibrium with two enolate ions after treatment with base. Enolate **A** is stable and unreactive, while enolate **B** can undergo internal aldol condensation to form a cyclobutenone product. But, because the aldol reaction is reversible and the cyclobutenone product is highly strained, there is little of this product present when equilibrium is reached. At equilibrium, only the stable, diketone enolate ion **A** is present.

17.17 This intramolecular aldol condensation gives a product with a seven-membered ring fused to a five-membered ring.

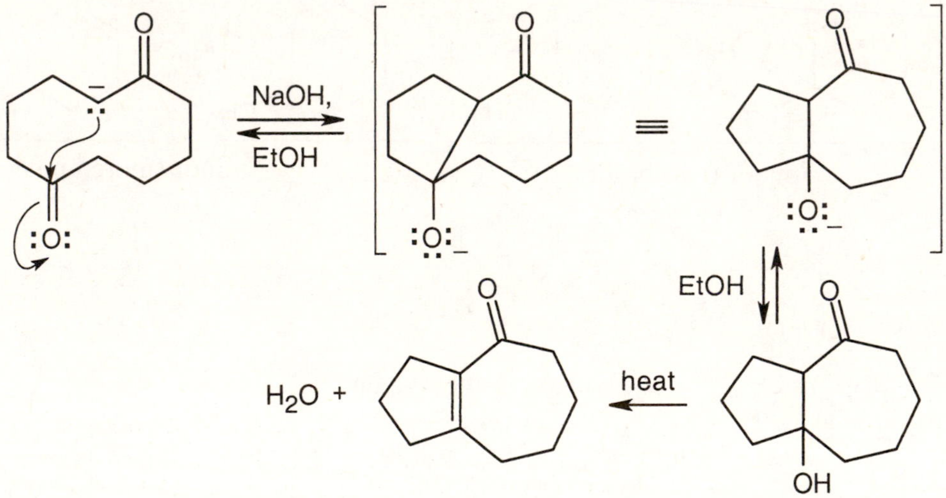

NaOH, EtOH; EtOH; heat; H_2O +

17.18 As in the aldol condensation, drawing the two Claisen components in the correct orientation makes it easier to predict the product.

(a)

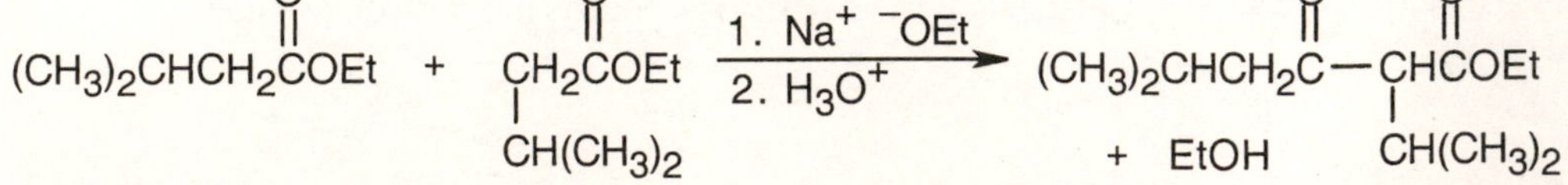

$$(CH_3)_2CHCH_2COEt + CH_2COEt(CH(CH_3)_2) \xrightarrow[2.\ H_3O^+]{1.\ Na^+\ ^-OEt} (CH_3)_2CHCH_2C(=O){-}CH(CH(CH_3)_2)COEt + EtOH$$

(b)

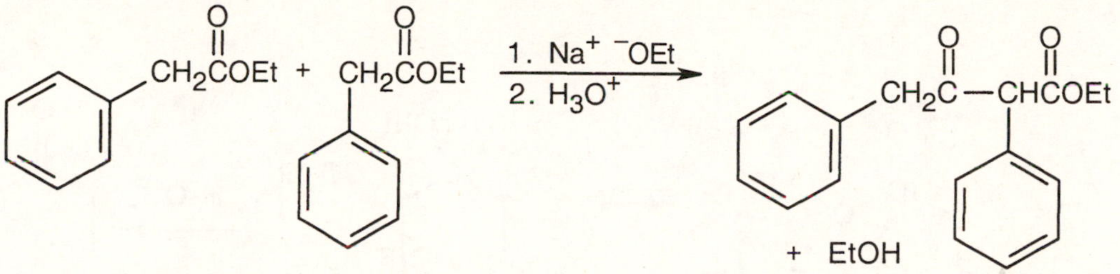

$$C_6H_5CH_2COEt + C_6H_5CH_2COEt \xrightarrow[2.\ H_3O^+]{1.\ Na^+\ ^-OEt} C_6H_5CH_2C(=O)-CH(C_6H_5)COEt + EtOH$$

(c)

$$C_6H_{11}CH_2COEt + C_6H_{11}CH_2COEt \xrightarrow[2.\ H_3O^+]{1.\ Na^+\ ^-OEt} C_6H_{11}CH_2C(=O)-CH(C_6H_{11})COEt + EtOH$$

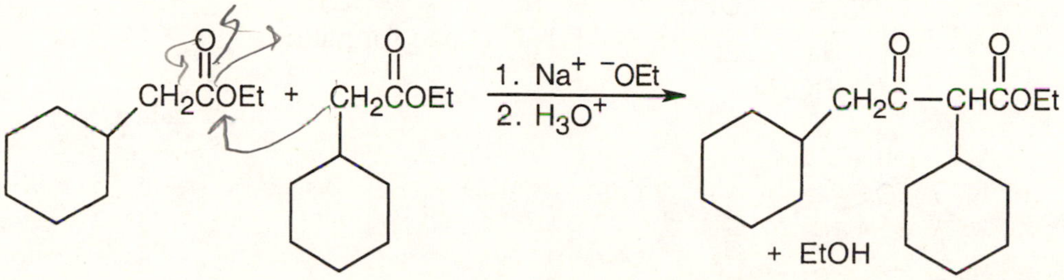

17.19

addition of hydroxide

elimination of acetate

acid-base reaction

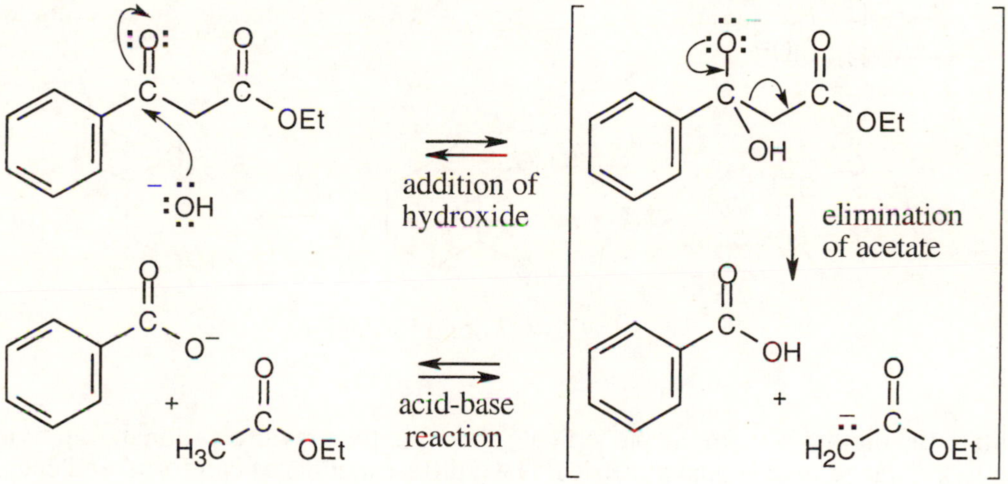

Hydroxide ion can react at two different sites of the ß-keto ester. Abstraction of the acidic α-proton is more favorable but is reversible and does not lead to product. Addition of hydroxide ion to the carbonyl group, followed by irreversible elimination of ethyl acetate anion, accounts for the observed product.

17.20

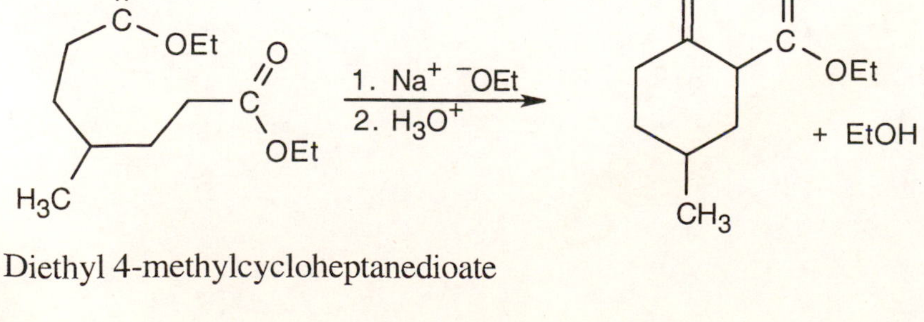

$$\xrightarrow[2.\ H_3O^+]{1.\ Na^+\ ^-OEt} + EtOH$$

Diethyl 4-methylcycloheptanedioate

17.21

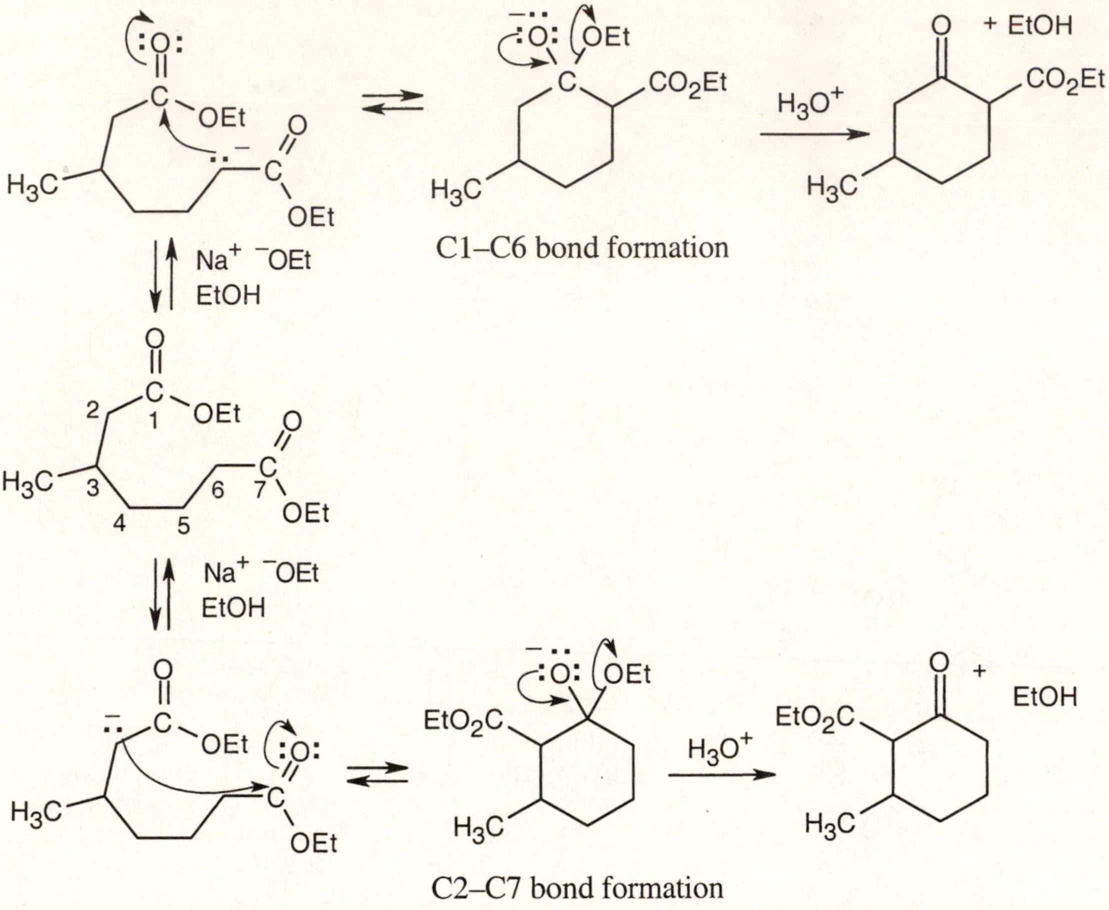

:O:
C
OEt
O
C
OEt
H3C
Na+ −OEt
EtOH
O:
OEt
C
CO2Et
H3C
C1–C6 bond formation
H3O+
O
+ EtOH
CO2Et
H3C
O
C
2
1
OEt
O
C
H3C
3
6
7
OEt
4
5
Na+ −OEt
EtOH
O
C
OEt
:O:
C
H3C
OEt
O:
OEt
EtO2C
H3C
C2–C7 bond formation
H3O+
O
+
EtOH
EtO2C
H3C

Unlike the diethyl 4-methylheptanedioate shown in the previous problem, diethyl 3-methylheptanedioate is unsymmetrical. Two different enolates can form, and each can cyclize to a different product.

17.22 A Michael reaction takes place when a stable enolate ion (Michael donor) adds to the double bond of an α,β-unsaturated carbonyl compound (Michael acceptor). The enolate adds to the double bond of the conjugated system. Predicting Michael products is easier when the donor and acceptor are positioned so that the product is evident.

	Michael Donor	*Michael Acceptor*	*Product*
(a)	$CH_3C(=O)CH_2C(=O)CH_3$	cyclohexenone	$(CH_3CO)_2CH$–cyclohexanone (3-substituted)
(b)	$CH_3C(=O)CH_2C(=O)CH_3$	$H_2C{=}CHC{\equiv}N$	$(CH_3CO)_2CHCH_2CH_2C{\equiv}N$
(c)	$CH_3C(=O)CH_2C(=O)CH_3$	$H_3C(H)C{=}CHC(=O)OEt$	$(CH_3CO)_2CHCH(CH_3)CH_2C(=O)OEt$

17.23

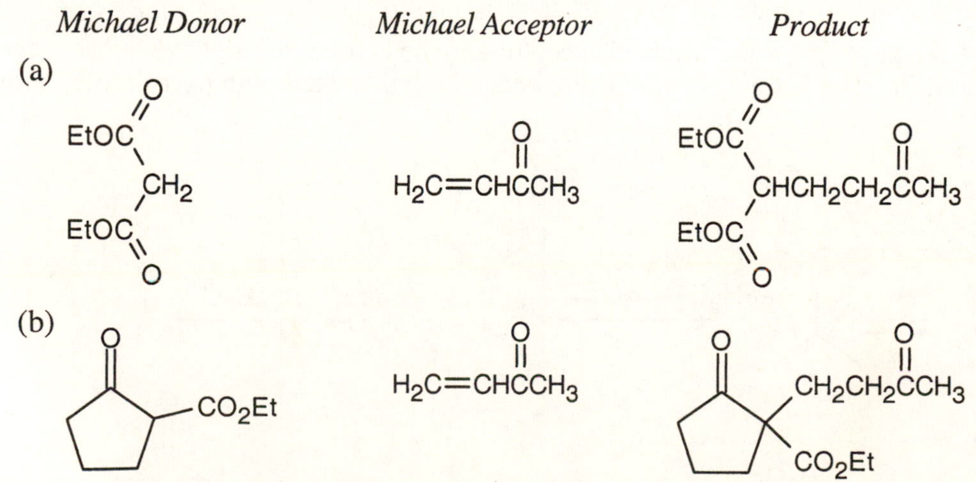

17.24 An enamine is formed from a ketone when it is necessary to synthesize a 1,5-diketone or a 1,5-dicarbonyl compound containing an aldehyde or ketone. The ketone starting material is converted to an enamine in order to increase the reactivity of the ketone and to direct the regiochemistry of addition. The process, as described in Section 17.11, is: (1) conversion of a ketone to its enamine; (2) Michael addition to an α,β-unsaturated carbonyl compound; (3) hydrolysis of the enamine to the starting ketone.

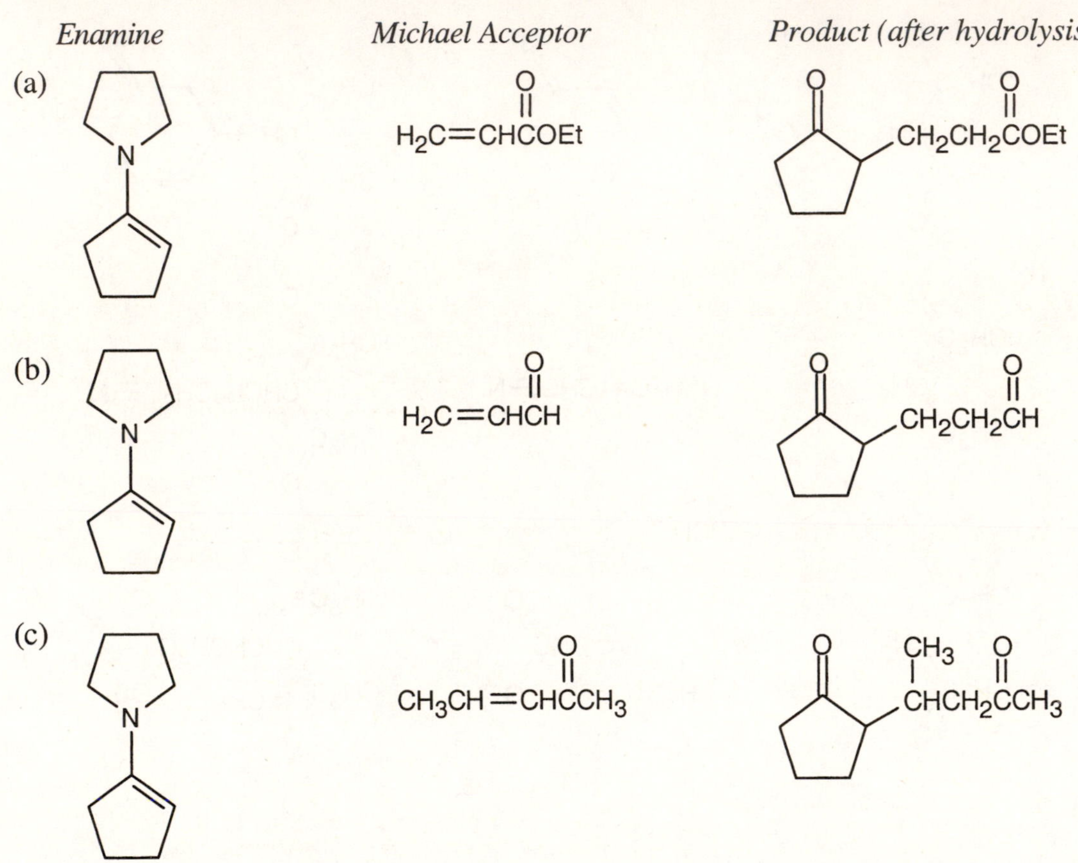

17.25 Analyze the product for the Michael acceptor and the ketone. In (a), the Michael acceptor is propenenitrile. The ketone is cyclopentanone, which is treated with pyrrolidine to form the enamine.

(a)

N H + O ⇌ N + H_2O → $H_2C{=}CHC{\equiv}N$ → [^{+}N $\ddot{C}H_2CHCN$] $\xrightarrow{H_3O^+}$ N H + CH_2CH_2CN

(b)

$$\text{pyrrolidine} + \text{cyclohexanone} \rightleftharpoons \text{enamine} + H_2O \xrightarrow{H_2C{=}CHCOCH_3} \left[\text{iminium}{-}CH_2\ddot{C}HCOCH_3\right] \xrightarrow{H_3O^+} \text{pyrrolidine} + \text{2-}(CH_2CH_2COCH_3)\text{cyclohexanone}$$

Visualizing Chemistry

17.26 (a) Check to see if the target molecule is a methyl ketone or a substituted carboxylic acid. (The target molecule is a methyl ketone, and the reaction is an acetoacetic ester synthesis.) Next, identify the halide or halides that react with acetoacetic ester. (The halide is 1-bromo-3-methylbut-2-ene.) Formulate the reaction, remembering to include a decarboxylation step.

from halide $(CH_3)_2C{=}CHCH_2$ — $CH_2\overset{O}{\overset{\|}{C}}CH_3$ from acetoacetic ester

$$CH_3\overset{O}{\overset{\|}{C}}CH_2CO_2Et \xrightarrow[2.\ (CH_3)_2C=CHCH_2Br]{1.\ Na^+\ ^-OEt} (CH_3)_2C{=}CHCH_2{-}\underset{CO_2Et}{CH}\overset{O}{\overset{\|}{C}}CH_3 + NaBr$$

$$\xrightarrow{H_3O^+,\ heat} CO_2 + EtOH + (CH_3)_2C{=}CHCH_2{-}CH_2\overset{O}{\overset{\|}{C}}CH_3$$

6-Methylhept-5-en-2-one

(b) This product is formed from the reaction of malonic ester with both benzyl bromide and bromomethane.

$C_6H_5CH_2$ — $\underset{CH_3}{CHCOOH}$ from malonic ester; from halides: $C_6H_5CH_2$, CH_3

$$CH_2(CO_2Et)_2 \xrightarrow[2.\ C_6H_5CH_2Br]{1.\ Na^+\ ^-OEt} C_6H_5CH_2{-}CH(CO_2Et)_2 + NaBr$$

$$\xrightarrow[2.\ CH_3Br]{1.\ Na^+\ ^-OEt} C_6H_5CH_2{-}\underset{CH_3}{C}(CO_2Et)_2 + NaBr$$

$$\xrightarrow[heat]{H_3O^+} C_6H_5CH_2{-}\underset{CH_3}{CH}CO_2H$$

2-Methyl-3-phenylpropanoic acid

$+\ CO_2\ +\ 2\ EtOH$

17.27

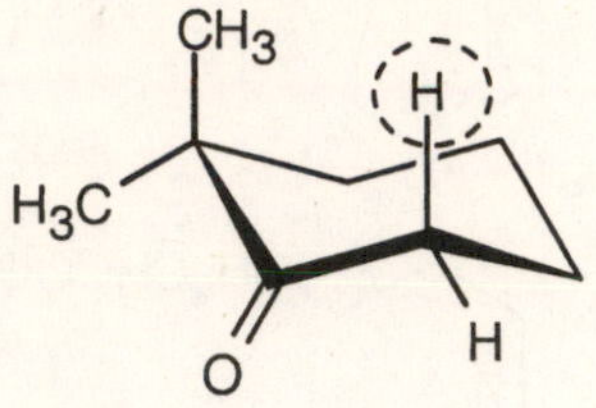

CH_3 H H_3C O H

Enolization can occur on only one side of the carbonyl group because of the two methyl groups on the other side. The circled axial hydrogen is more acidic because the *p* orbital that remains after its removal is aligned for optimum overlap with the π electrons of the carbonyl oxygen.

17.28

(a)

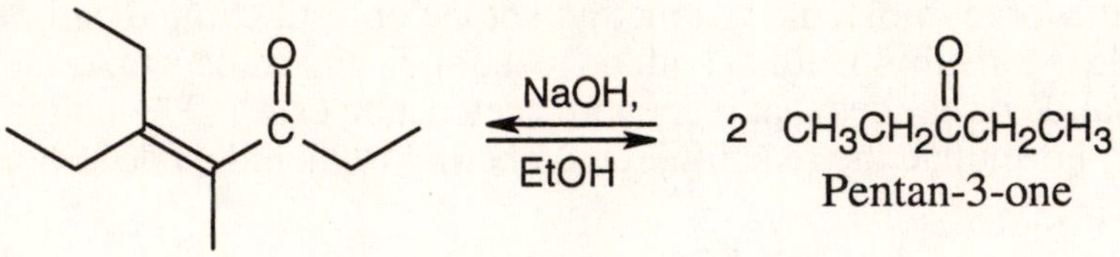

NaOH, EtOH ⇌ 2 $CH_3CH_2CCH_2CH_3$ (C=O)

Pentan-3-one

(b)

NaOH, EtOH ⇌ 2 $HC(=O)CH_2CH(CH_3)_2$

3-Methylbutanal

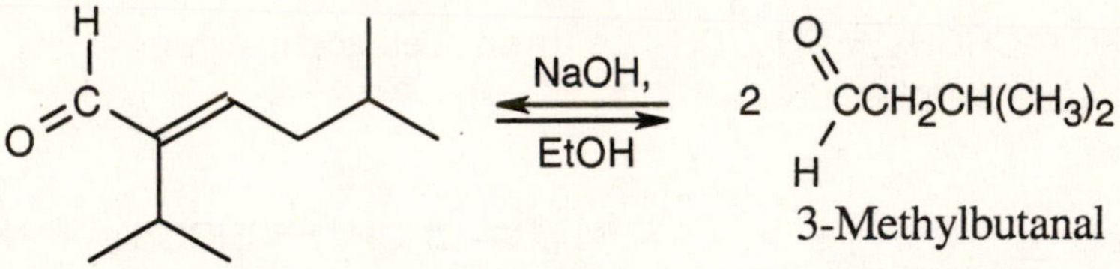

17.29 The enolate of methyl phenylacetate adds to a second molecule of methyl phenylacetate to form the Claisen intermediate that is pictured. Elimination of methoxide (circled) and acidification give the product shown.

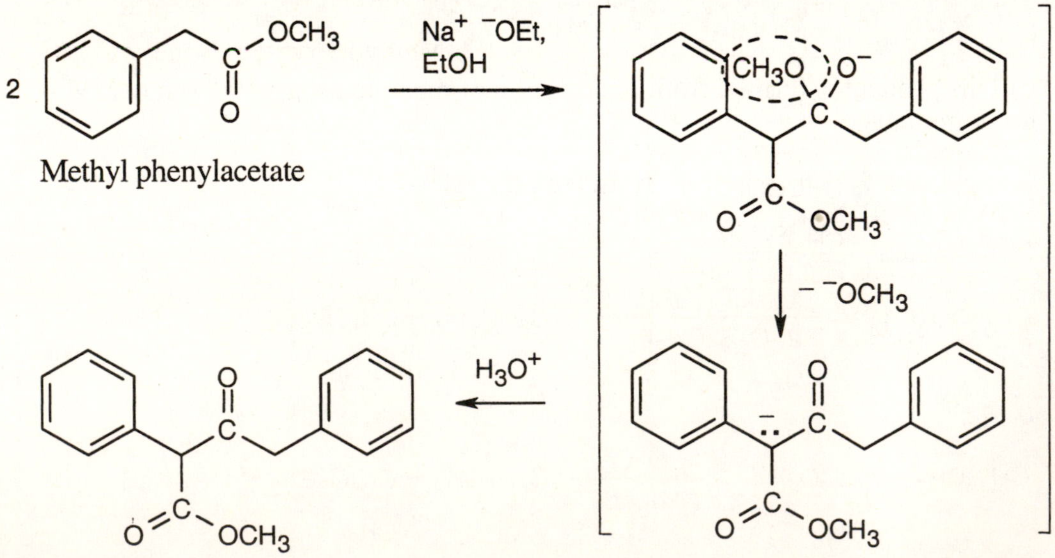

2 Methyl phenylacetate $\xrightarrow[\text{EtOH}]{Na^+ \ ^-OEt,}$ [CH_3O O^- C OCH_3 $\xrightarrow{- \ ^-OCH_3}$ O C OCH_3] $\xrightarrow{H_3O^+}$ O C OCH_3

17.30

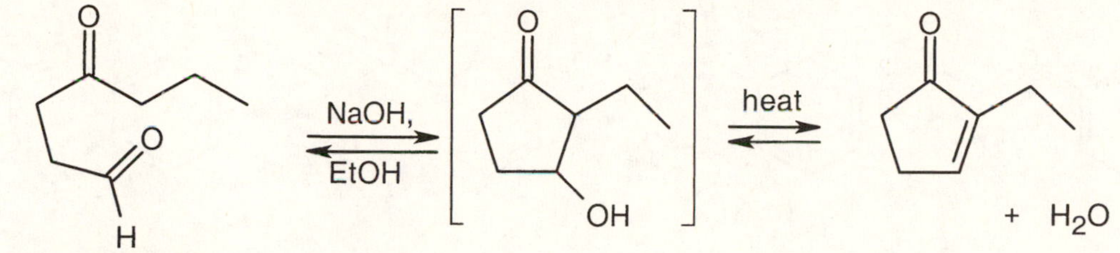

4-Oxoheptanal $\xrightleftharpoons[\text{EtOH}]{\text{NaOH,}}$ [2-ethyl-3-hydroxycyclopentanone] $\xrightleftharpoons{\text{heat}}$ 2-ethylcyclopent-2-enone + H_2O

Additional Problems

17.31 Acidic hydrogens are bold. The most acidic hydrogens are the two between the carbonyl groups in (b) and the hydroxyl hydrogen in (c). The hydrogens in (c) that are bonded to the methyl group are acidic (draw resonance forms to prove it).

(a) $CH_3CH_2\mathbf{C}\mathbf{H}(CH_3)\overset{O}{\overset{\|}{C}}C\mathbf{H}_3$

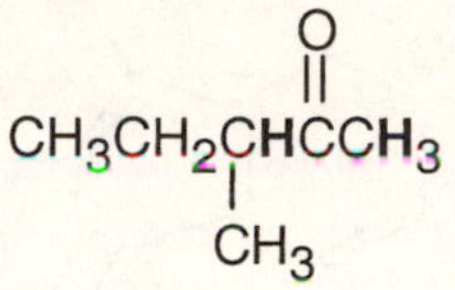

(b) cyclopentane-1,3-dione (H H between the carbonyls; H, H, H, H on ring carbons)

(c) $HOCH_2CH_2\overset{O}{\overset{\|}{C}}C\equiv CCH_3$

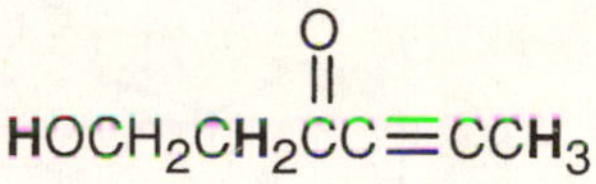

(d) $C_6H_4(CO_2CH_3)(CH_2CN)$

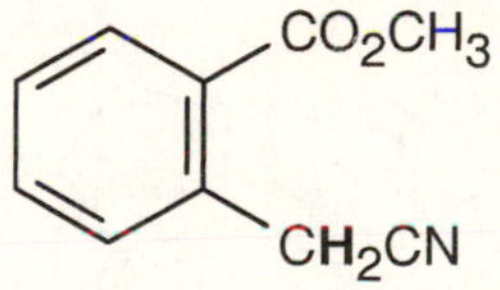

(e) cyclopentyl–C(=O)Cl with H on the ring carbon

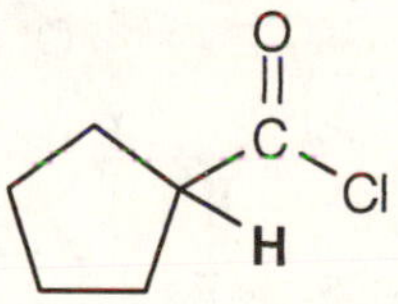

(f) $CH_3CH_2\overset{O}{\overset{\|}{C}}C(CH_3){=}CH_2$

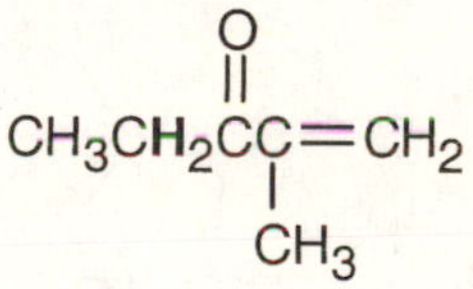

17.32

(a) $H_3C-C(=O)-\ddot{C}^-H-C(=O)-CH_3 \longleftrightarrow H_3C-C(-O^-)=CH-C(=O)-CH_3 \longleftrightarrow H_3C-C(=O)-CH=C(-O^-)-CH_3$

(b) $H_3C-CH=CH-\ddot{C}^-H-C(=O)-CH_3 \longleftrightarrow H_3C-CH=CH-CH=C(-O^-)-CH_3 \longleftrightarrow H_3C-\ddot{C}^-H-CH=CH-C(=O)-CH_3$

(c) $:N\equiv C-\ddot{C}^-H-C(=O)-CH_3 \longleftrightarrow :N\equiv C-CH=C(-O^-)-CH_3 \longleftrightarrow {}^-:\ddot{N}=C=CH-C(=O)-CH_3$

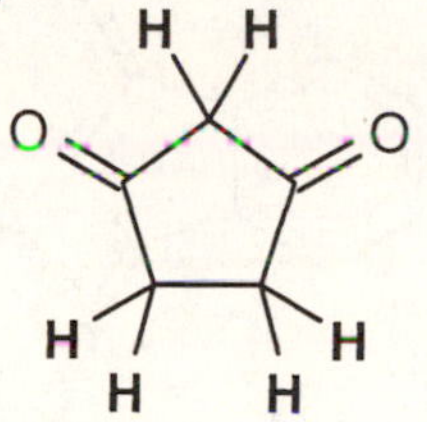

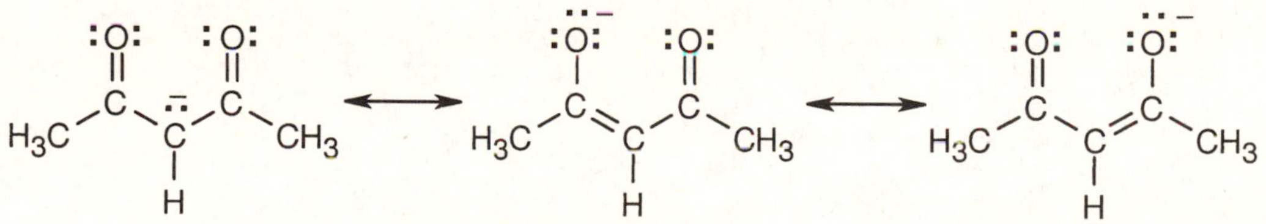

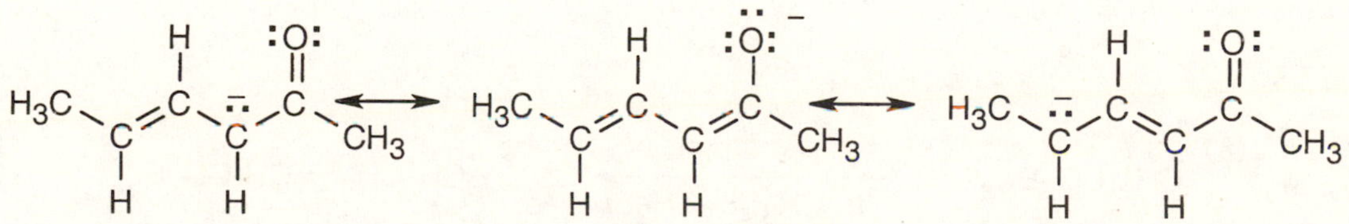

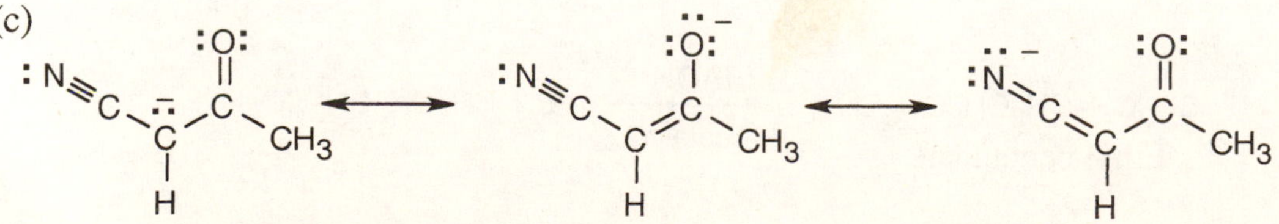

(d)

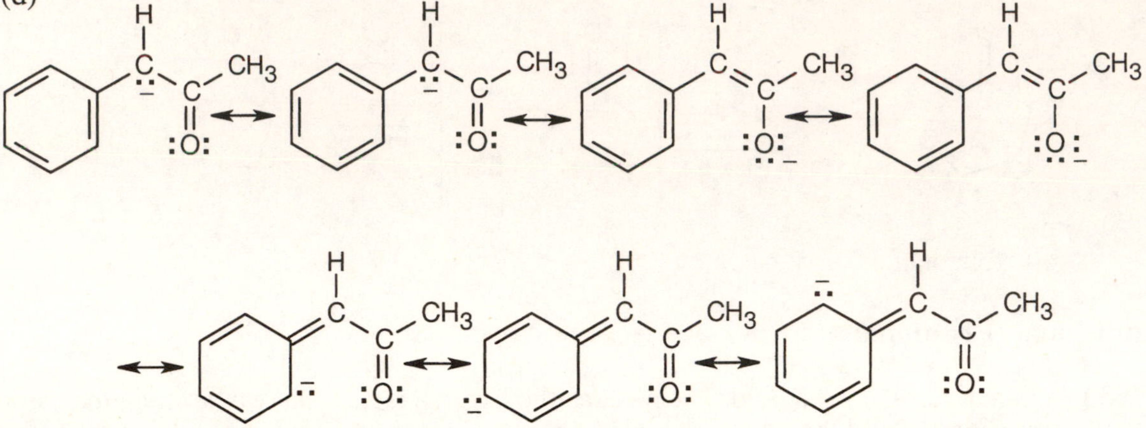

(e)

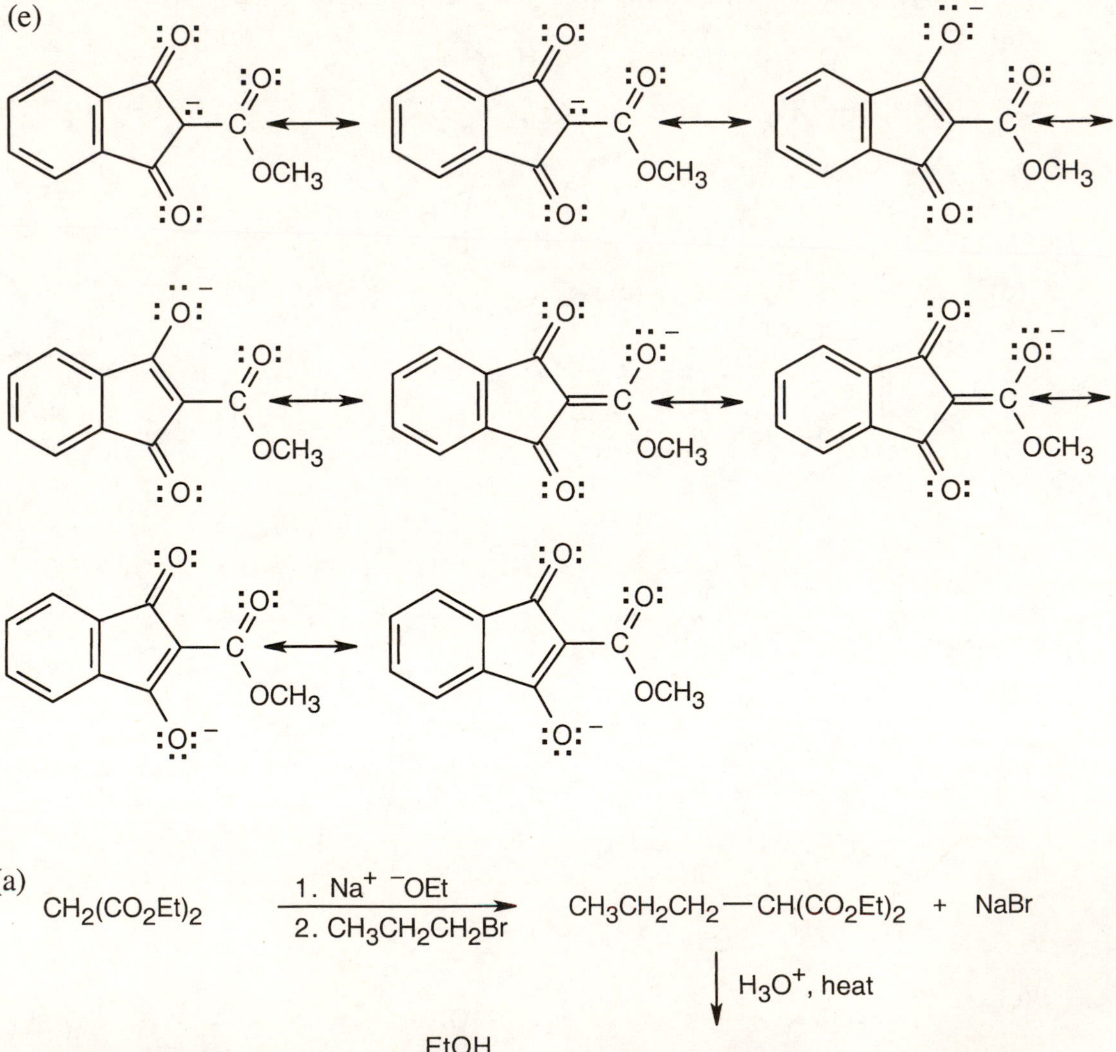

17.33

(a) $CH_2(CO_2Et)_2$ $\xrightarrow[\text{2. } CH_3CH_2CH_2Br]{\text{1. } Na^+ \ ^-OEt}$ $CH_3CH_2CH_2—CH(CO_2Et)_2$ + NaBr

$\downarrow$ H_3O^+, heat

$CH_3CH_2CH_2CH_2CO_2H$ + CO_2 + 2 EtOH

$CH_3CH_2CH_2CH_2CO_2H$ $\xrightarrow[H^+ \text{ catalyst}]{EtOH}$ $CH_3CH_2CH_2CH_2CO_2Et$
Ethyl pentanoate

(b)

$$CH_2(CO_2Et)_2 \xrightarrow[\text{2. } (CH_3)_2CHBr]{\text{1. } Na^+ \ ^-OEt} (CH_3)_2CH{-}CH(CO_2Et)_2 + NaBr$$

$$\downarrow \ H_3O^+, \text{heat}$$

$$(CH_3)_2CHCH_2CO_2Et \xleftarrow[H^+ \text{ catalyst}]{EtOH} (CH_3)_2CHCH_2CO_2H + CO_2 + 2\ EtOH$$

Ethyl 3-methylbutanoate

Some elimination product will also be formed.

(c)

$$CH_2(CO_2Et)_2 \xrightarrow[\text{2. } CH_3CH_2Br]{\text{1. } Na^+ \ ^-OEt} CH_3CH_2CH(CO_2Et)_2 + NaBr \xrightarrow[\text{2. } CH_3Br]{\text{1. } Na^+ \ ^-OEt} CH_3CH_2C(CH_3)(CO_2Et)_2 + NaBr$$

$$\downarrow \ H_3O^+, \text{heat}$$

$$CH_3CH_2CH(CH_3)CO_2Et \xleftarrow[H^+ \text{ catalyst}]{EtOH} CH_3CH_2CH(CH_3)CO_2H + CO_2 + 2\ EtOH$$

Ethyl 2-methylbutanoate

(d) The malonic acid synthesis can't be used to synthesize carboxylic acids such as ethyl 2,2-dimethylpropanoate that are trisubstituted at the alpha position.

17.34

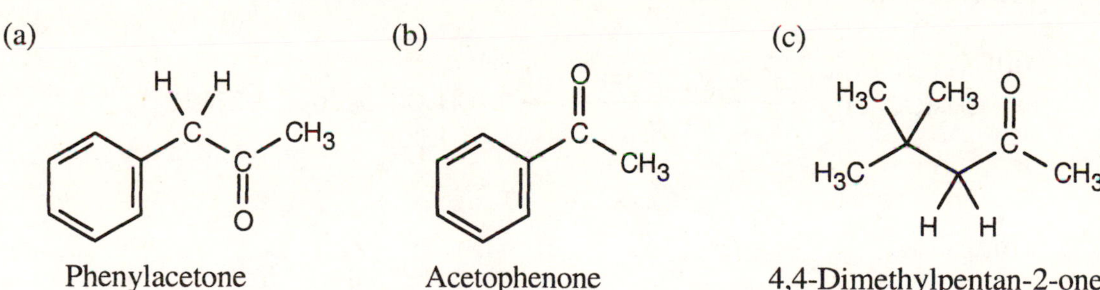

Phenylacetone Acetophenone 4,4-Dimethylpentan-2-one

Look back to Problem 17.8, which describes compounds that can be prepared by an acetoacetic ester synthesis. Neither (a) nor (c) are products of an acetoacetic ester synthesis because the halide component that would be needed for each synthesis doesn't undergo S_N2 reactions. Compound (b) can be prepared by the reaction of acetoacetic ester with 1,5-dibromopentane.

17.35

(a)

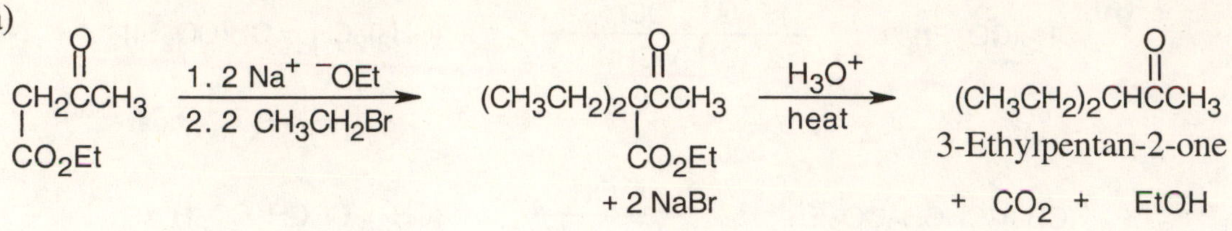

$$CH_2C(=O)CH_3\ (CO_2Et) \xrightarrow[2.\ 2\ CH_3CH_2Br]{1.\ 2\ Na^+\ ^-OEt} (CH_3CH_2)_2C(CO_2Et)C(=O)CH_3 + 2\ NaBr \xrightarrow[heat]{H_3O^+} (CH_3CH_2)_2CHC(=O)CH_3 + CO_2 + EtOH$$

3-Ethylpentan-2-one

(b)

$$CH_2C(=O)CH_3\ (CO_2Et) \xrightarrow[2.\ CH_3CH_2CH_2Br]{1.\ Na^+\ ^-OEt} CH_3CH_2CH_2CH(CO_2Et)C(=O)CH_3 + NaBr$$

$$\xrightarrow[2.\ CH_3Br]{1.\ Na^+\ ^-OEt} CH_3CH_2CH_2C(CH_3)(CO_2Et){-}C(=O)CH_3 + NaBr$$

$$\xrightarrow[heat]{H_3O^+} CH_3CH_2CH_2CH(CH_3){-}C(=O)CH_3 + CO_2 + EtOH$$

3-Methylhexan-2-one

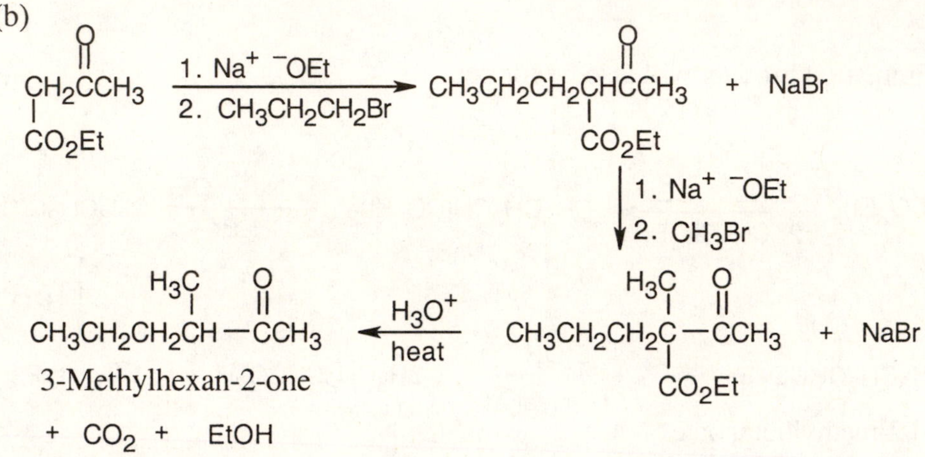

17.36 Use a malonic ester synthesis if the product you want is an α-substituted carboxylic acid or derivative. Use an acetoacetic acid synthesis if the product you want is an α-substituted methyl ketone.

(a)

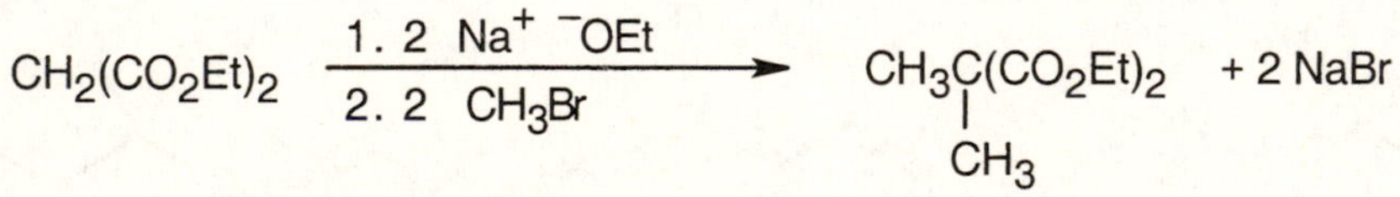

$$CH_2(CO_2Et)_2 \xrightarrow[2.\ 2\ CH_3Br]{1.\ 2\ Na^+\ ^-OEt} CH_3C(CH_3)(CO_2Et)_2 + 2\ NaBr$$

(b)

$$CH_2C(=O)CH_3\ (CO_2Et) \xrightarrow[2.\ BrCH_2(CH_2)_4CH_2Br]{1.\ 2\ Na^+\ ^-OEt} \text{1-(CCH}_3\text{=O)-1-(CO}_2\text{Et)cycloheptane} + NaBr$$

$$\xrightarrow{H_3O^+,\ heat} \text{cycloheptyl–C(=O)CH}_3\ (\text{H}) + CO_2 + EtOH$$

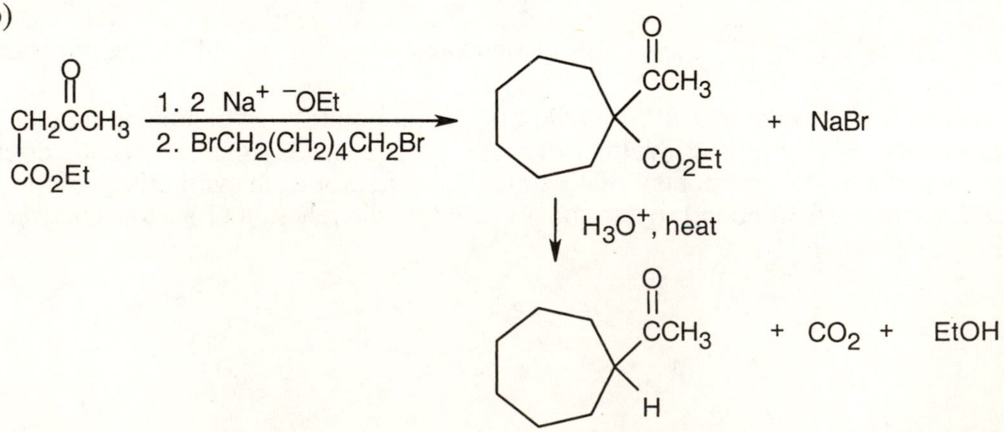

(c)

$CH_2(CO_2Et)_2 \xrightarrow[\text{2. } BrCH_2CH_2CH_2Br]{\text{1. 2 } Na^+ \ ^-OEt}$ 1,1-cyclobutanedicarboxylic acid diethyl ester (CO_2Et, CO_2Et) + 2 NaBr

$\downarrow H_3O^+$, heat

cyclobutanecarboxylic acid (CO_2H, H) + CO_2 + 2 EtOH

(d)

$CH_3COCH_2CO_2Et$ (CH_2CCH_3 with O, CO_2Et) $\xrightarrow[\text{2. } H_2C{=}CHCH_2Br]{\text{1. } Na^+ \ ^-OEt}$ $H_2C{=}CHCH_2CH(CO_2Et)CCH_3$ (with O) + NaBr

$\downarrow H_3O^+$, heat

$H_2C{=}CHCH_2CH_2CCH_3$ (with O) + CO_2 + EtOH

17.37 As always, analyze the product for the carbon–carbon double bond that is formed by dehydration of the initial aldol adduct. Break the bond, and add a carbonyl oxygen to the appropriate carbon to identify the carbonyl reactant(s).

(a)

$CH_3CO(CH_2)_4CHO$ (CCH_3 with O; CHO) $\xrightarrow[\text{2. heat}]{\text{1. NaOH, EtOH}}$ 2-cyclohexenone (O) + H_2O

(b)

$C_6H_5C(O)CH_3$ + $C_6H_5C(CH_3){=}O$ $\xrightarrow[\text{2. heat}]{\text{1. NaOH, EtOH}}$ $C_6H_5C(O)CH{=}C(CH_3)C_6H_5$ (O, C, C, H, CH_3) + H_2O

(c)

$C_6H_4(CH_2CHO)_2$ (CH_2CHO, CH_2CHO) $\xrightarrow[\text{2. heat}]{\text{1. NaOH, EtOH}}$ indene-3-carbaldehyde (CHO) + H_2O

17.38

(a)

$$CH_3CH_2CH_2\overset{O}{\overset{\|}{C}}OEt + \underset{CH_2CH_3}{\underset{|}{CH_2}}\overset{O}{\overset{\|}{C}}OEt \xrightarrow[2.\ H_3O^+]{1.\ NaOEt,\ EtOH} CH_3CH_2CH_2\overset{O}{\overset{\|}{C}}-\underset{CH_2CH_3}{\underset{|}{CH}}\overset{O}{\overset{\|}{C}}OEt + HOEt$$

(b)

cycloheptanone + cycloheptanone $\xrightarrow[2.\ heat]{1.\ NaOH,\ EtOH}$ 2-cycloheptylidenecycloheptanone + H_2O

(c)

3,9-undecanedione $\xrightarrow[2.\ heat]{1.\ NaOH,\ EtOH}$ 3-ethyl-2-methylcyclohex-2-enone + H_2O

(d)

$$C_6H_5CH_2CH_2\overset{O}{\overset{\|}{C}}H + \underset{CH_2C_6H_5}{\underset{|}{CH_2}}\overset{O}{\overset{\|}{C}}H \xrightarrow[2.\ heat]{1.\ NaOH,\ EtOH} C_6H_5CH_2CH_2CH=\underset{CH_2C_6H_5}{\underset{|}{C}}\overset{O}{\overset{\|}{C}}H + H_2O$$

17.39 Enolization at the γ position produces a conjugated enolate anion that is stabilized by delocalization of the negative charge over the π system of five atoms.

$$C_6H_5C(=O)CH=CH\ddot{C}H_2^- \longleftrightarrow C_6H_5C(=O)\ddot{C}H^-CH=CH_2 \longleftrightarrow C_6H_5C(\ddot{O}^-)=CHCH=CH_2$$

17.40

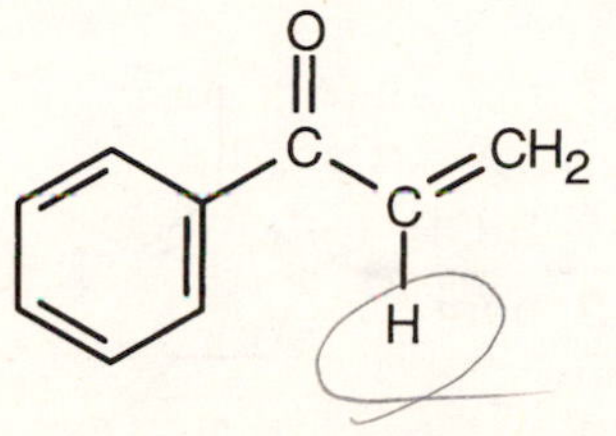

1-Phenylprop-2-enone

The illustrated compound, 1-phenylpropen-2-one, doesn't yield an anion when treated with base because the hydrogen on the α carbon is vinylic and isn't acidic (check Table 17.1 for acidity constants).

17.41 Reaction of (*R*)-2-methylcyclohexanone with aqueous base is shown below. Reaction with aqueous acid proceeds by a related mechanism, through an enol, rather than an enolate ion intermediate.

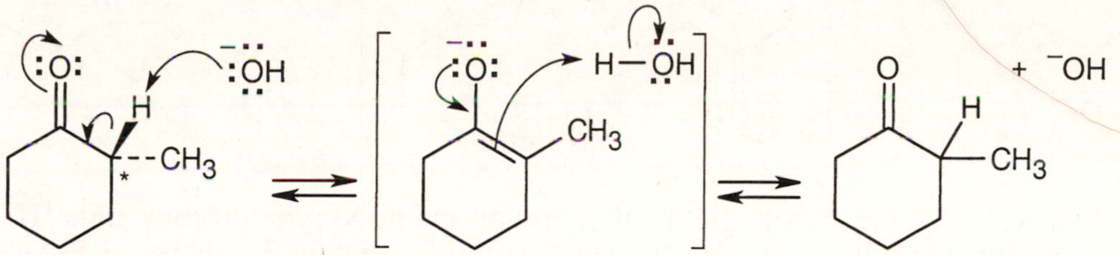

(*R*)-2-Methylcyclohexanone

Carbon 2 loses its chirality when the enolate ion double bond is formed. Protonation occurs with equal probability from either side of the planar sp^2–hybridized carbon 2, resulting in racemic product.

17.42

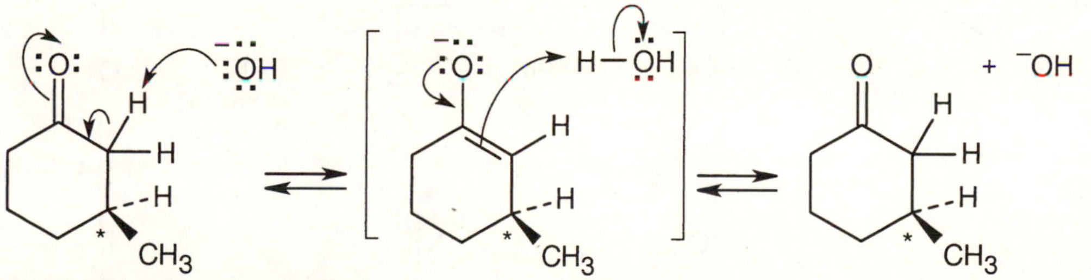

(*S*)-3-Methylcyclohexanone

(*S*)-3-Methylcyclohexanone isn't racemized by base because its chirality center is not involved in the enolization reaction.

17.43

NaOH, EtOH ⇌ [:O: ... −:O:] → 1. EtOH 2. heat → A + H_2O

NaOH, EtOH ⇌ [... −:O:] → 1. EtOH 2. heat → B + H_2O

17.44 Product **A**, which has two singlet methyl groups and no vinylic protons in its ^{1}H NMR, is the major product of the intramolecular cyclization of heptane-2,5-dione.

17.45

:OH — addition of −OH → OH — proton transfer → :O: − — ring opening → :O:

H_2O protonation

−OH deprotonation ⇌ :O: + H_2O ⇌ aldol condensation −:O: → 1. EtOH 2. heat (protonation, dehydration) → A + H_2O

Because all steps in the aldol reaction are reversible, the more stable product is formed at equilibrium.

17.46 The reactive nucleophile in the acid-catalyzed aldol condensation is the *enol* of one of the reactants The electrophile is a reactant with a protonated carbonyl group.

Step 1: Enol formation.

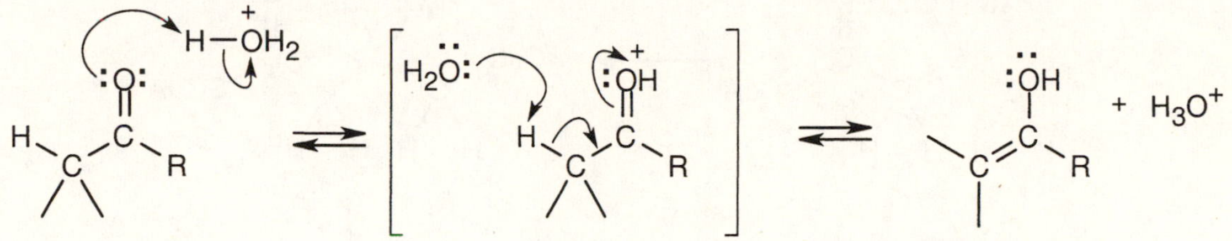

Step 2: Addition of the enol nucleophile to the protonated carbonyl compound.

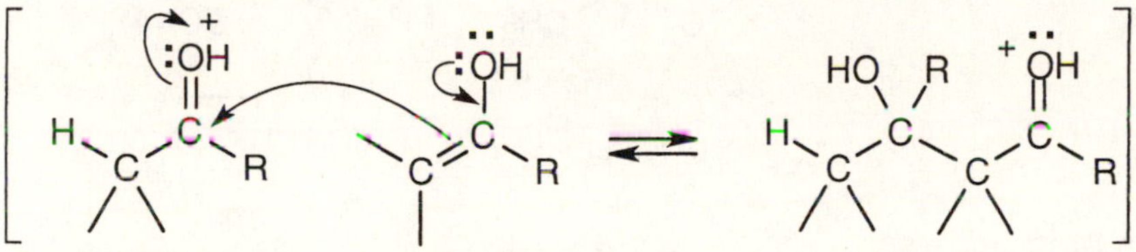

electrophile nucleophile

Step 3: Loss of proton from the carbonyl oxygen.

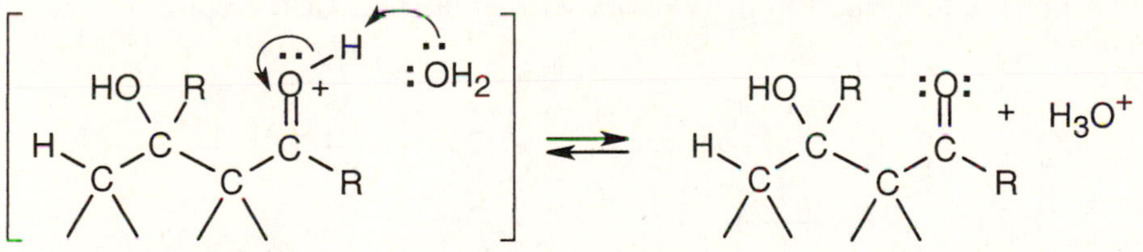

17.47

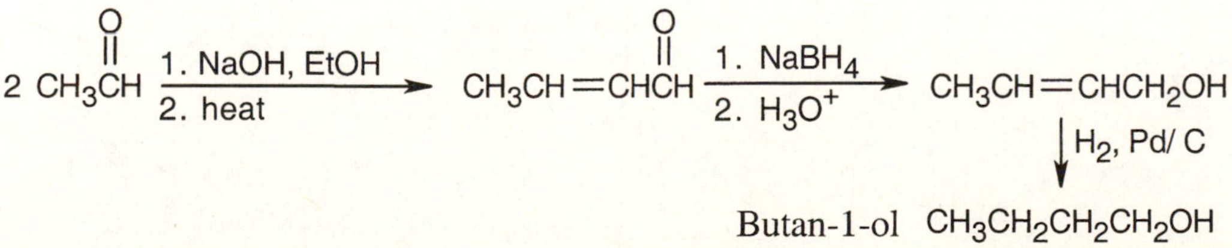

$$2\ CH_3CHO \xrightarrow[2.\ heat]{1.\ NaOH,\ EtOH} CH_3CH{=}CHCHO \xrightarrow[2.\ H_3O^+]{1.\ NaBH_4} CH_3CH{=}CHCH_2OH \xrightarrow{H_2,\ Pd/C} CH_3CH_2CH_2CH_2OH$$

Butan-1-ol $CH_3CH_2CH_2CH_2OH$

17.48 This sequence is a reverse aldol reaction.

: Base

H

H_3C :O: :O:

$^{-}O_2C$ C C C SCoA

H H H H

1.

H_3C :O: :O:

$^{-}O_2C$ C C C SCoA

H H H H

2.

O

$^{-}O_2C$ C C CH_3

H H

Acetoacetate

\+

:O:

H_2C C SCoA

H—A

3.

O

\+

H_3C C SCoA

Acetyl CoA

Step 1: Deprotonation by base.
Step 2: Elimination of acetyl CoA enolate.
Step 3: Protonation of enolate.

17.49 In contrast to the previous problem, this sequence is a reverse Claisen reaction. The first step (not illustrated) is the reaction of HSCoA with a base to form $^{-}$SCoA.

:O: O

H_3C C C C SCoA

H CH_3

CoAS:

1.

:O: :O:

H_3C C C C SCoA

CoAS H CH_3

2.

O

H_3C C SCoA

Acetyl CoA

\+

H—A

:O:

H_3C C C SCoA

H

3.

O

H_3C C C SCoA

H H

Propionyl CoA

Step 1: Addition of $^{-}$SCoA to form a tetrahedral intermediate.
Step 2: Elimination of propionyl CoA anion.
Step 3: Protonation.

17.50

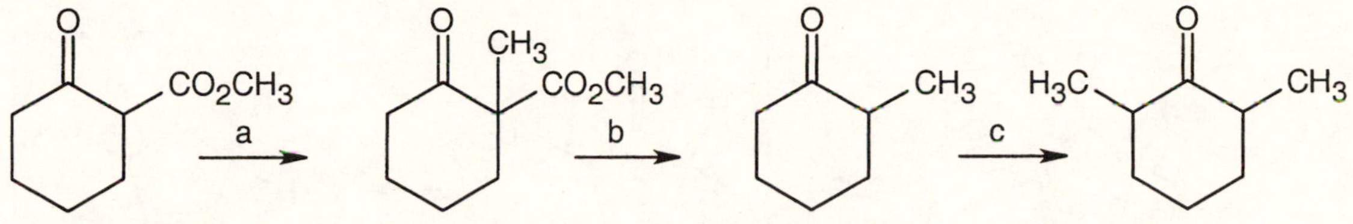

(a) Na^+ ^-OEt, then CH_3I; (b) H_3O^+, heat; (c) LDA, then CH_3I

17.51 Acid-catalyzed equilibrium:

protonation of carbonyl oxygen

abstraction of α proton

enol

protonation at γ position

loss of proton on oxygen

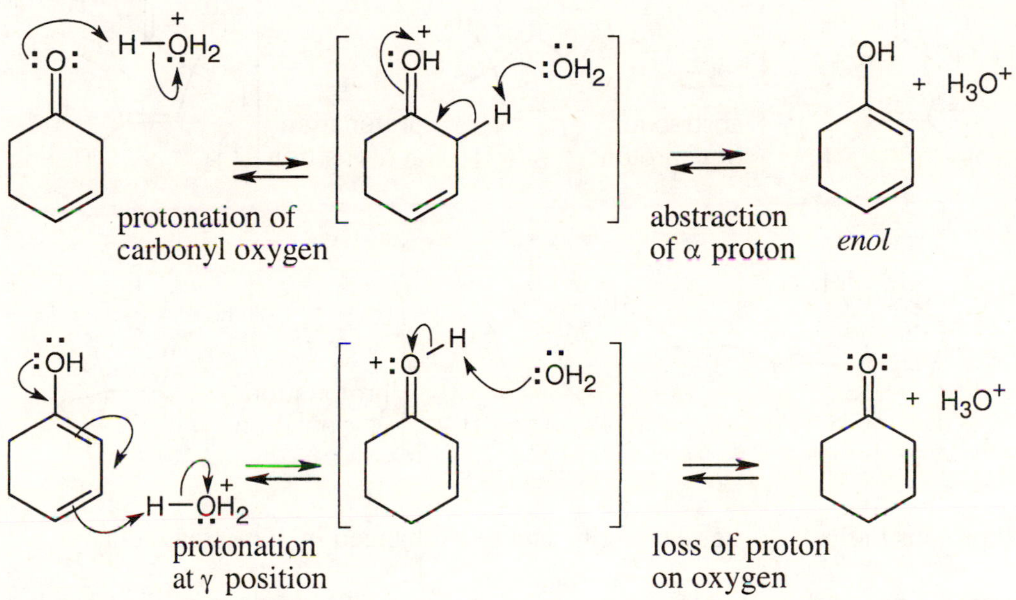

Base-catalyzed equilibrium:

abstraction of α proton

protonation at γ position

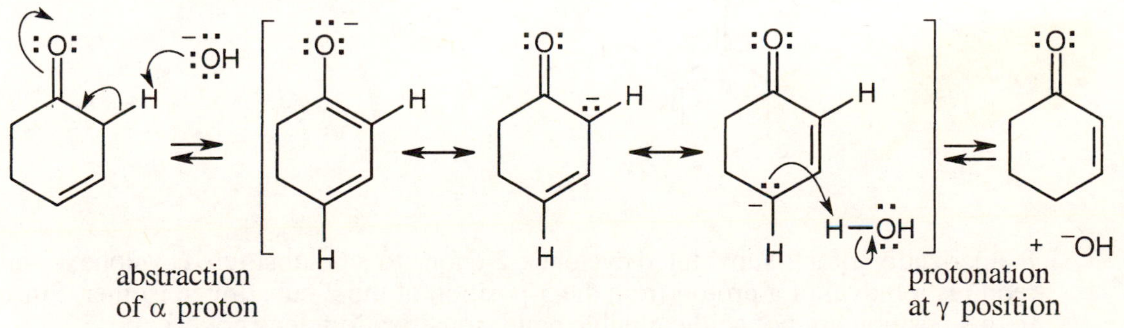

The enolate of cyclohex-3-enone can be protonated at three different positions. Protonation at the γ position yields the α,ß-unsaturated ketone.

17.52

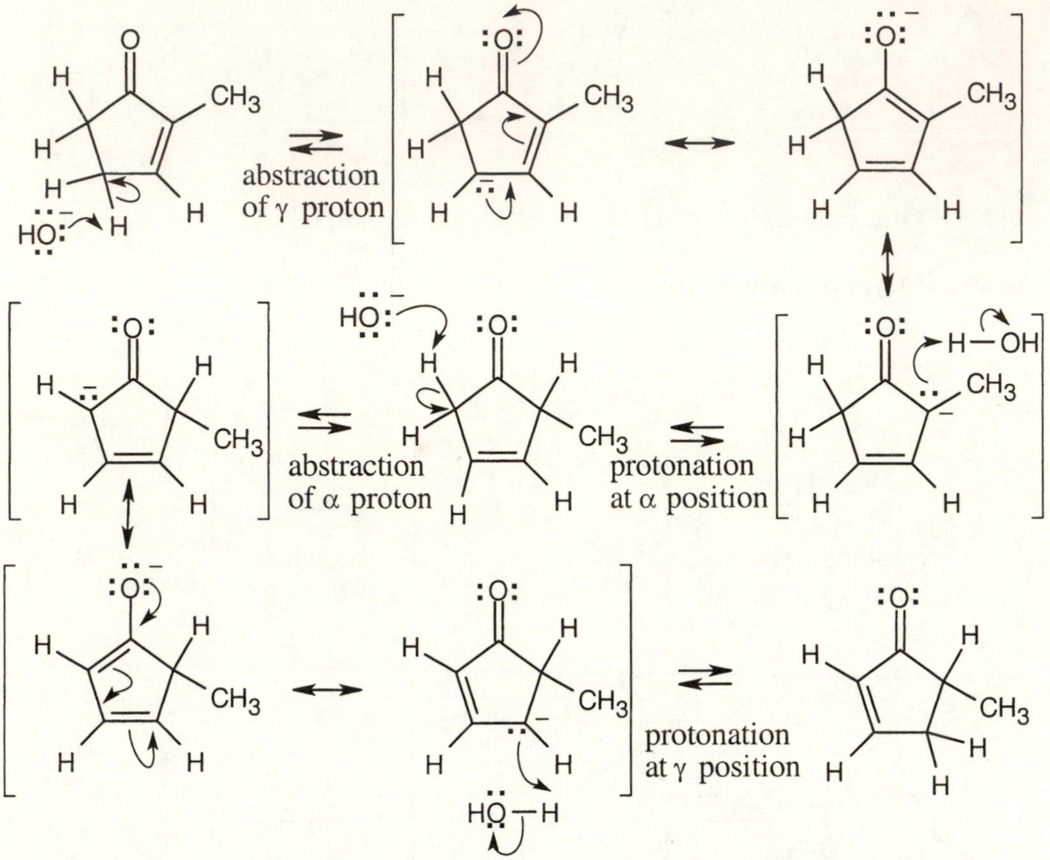

abstraction of γ proton

protonation at α position

abstraction of α proton

protonation at γ position

All protons in the five-membered ring can be exchanged by base treatment.

17.53 Protons α to a carbonyl group or γ to an enone carbonyl group are acidic (Problem 17.51). Thus for 2-methylcyclohex-2-enone, protons at the starred positions are acidic.

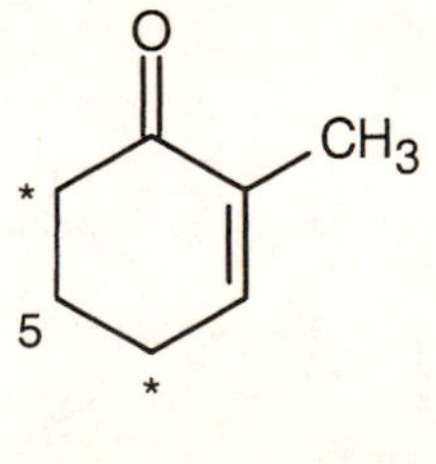

Isomerization of a 2-substituted cyclohex-2-enone to a 6-substituted cyclohex-2-enone requires removal of a proton from the 5-position of the 2-substituted isomer. Since protons in this position are not acidic, double bond isomerization does not occur.

17.54

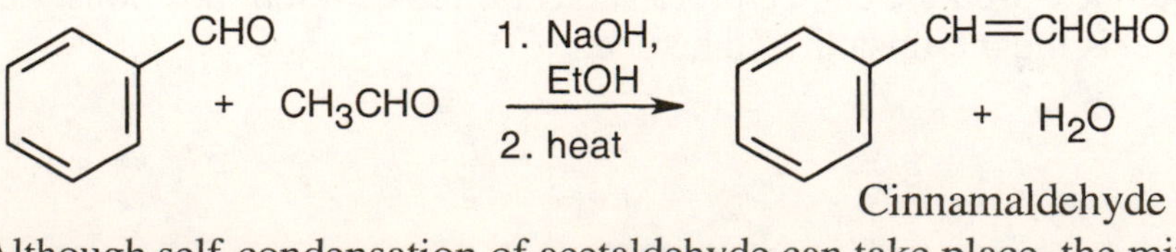

$C_6H_5CHO + CH_3CHO \xrightarrow[\text{2. heat}]{\text{1. NaOH, EtOH}} C_6H_5CH{=}CHCHO + H_2O$

Cinnamaldehyde

Although self-condensation of acetaldehyde can take place, the mixed aldol product predominates.

17.55

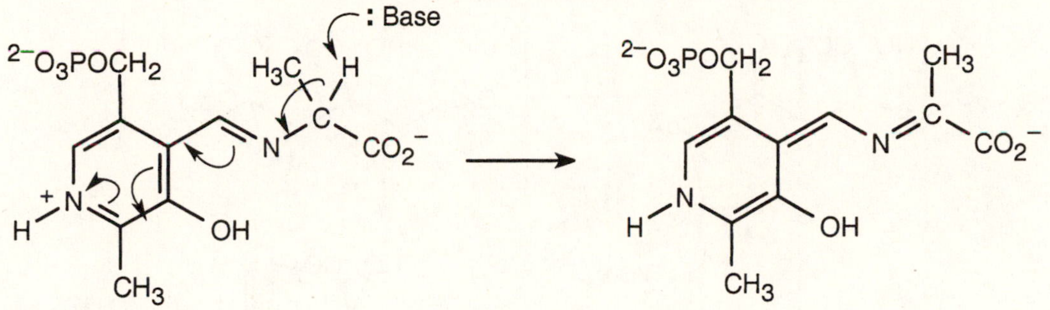

$^{2-}O_3POCH_2$ ... : Base ... H_3C ... H ... N ... C ... CO_2^- ... ^{+}N ... H ... OH ... CH_3 → $^{2-}O_3POCH_2$ ... CH_3 ... N=C ... CO_2^- ... N ... H ... OH ... CH_3

17.56 Decarboxylation, which takes place because of the stability of the resulting anion, is followed by protonation.

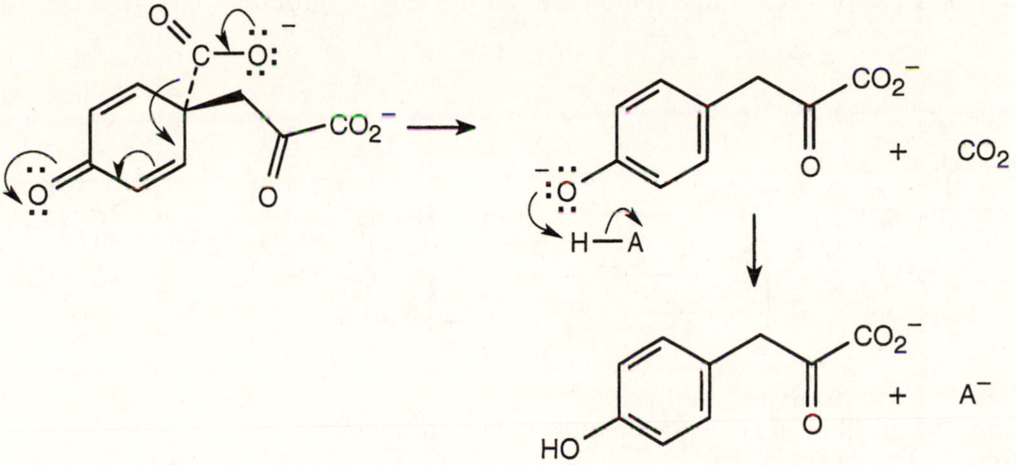

O ... C ... O: ... CO_2^- ... O ... O → CO_2^- ... O ... + CO_2 ... :O ... H—A ↓ CO_2^- ... + A^- ... O ... HO

17.57 Formation of (*S*)-citryl CoA:

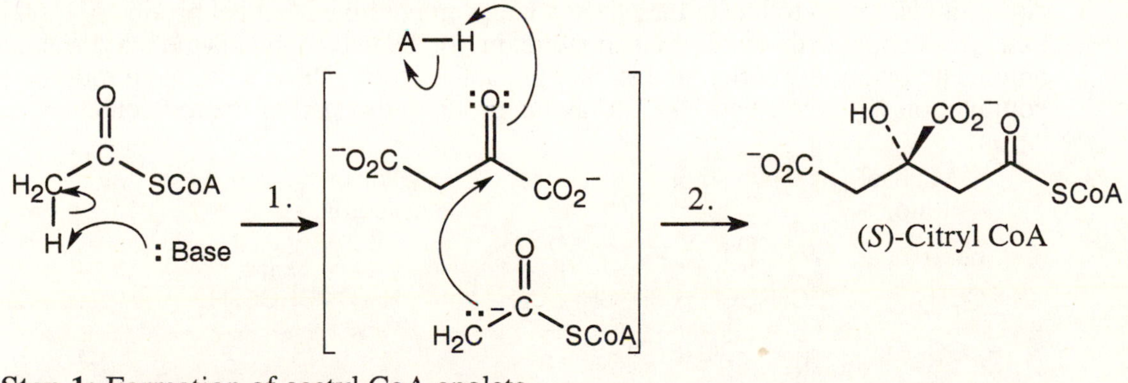

O ... C ... H_2C ... SCoA ... H ... : Base → 1. [A—H ... :O: ... $^{-}O_2C$ ... CO_2^- ... O ... C ... H_2C ... SCoA] → 2. HO ... CO_2^- ... O ... $^{-}O_2C$ ... SCoA

(*S*)-Citryl CoA

Step 1: Formation of acetyl CoA enolate.
Step 2: Aldol-like nucleophilic addition of Acetyl CoA to the carbonyl group of oxaloacetate and protonation.

Loss of CoA to form citrate:

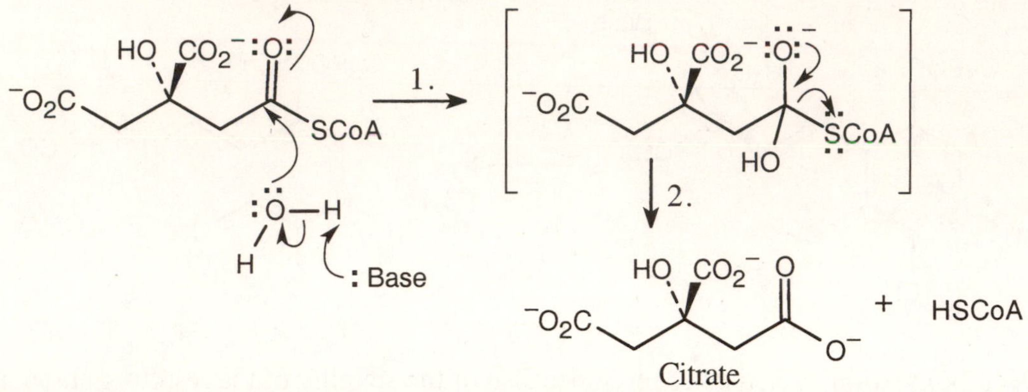

$$^{-}O_2C\text{-}CH_2\text{-}C(OH)(CO_2^-)\text{-}CH_2\text{-}C(=O)SCoA \xrightarrow{1.} [\,^{-}O_2C\text{-}CH_2\text{-}C(OH)(CO_2^-)\text{-}CH_2\text{-}C(O^-)(OH)SCoA\,] \xrightarrow{2.} {}^{-}O_2C\text{-}CH_2\text{-}C(OH)(CO_2^-)\text{-}CH_2\text{-}CO_2^- + HSCoA$$

Citrate

Step 1: Nucleophilic addition of hydroxyl to the carbonyl group of (*S*)-citryl CoA.
Step 2: Loss of $^{-}$SCoA and protonation of the leaving group.

17.58

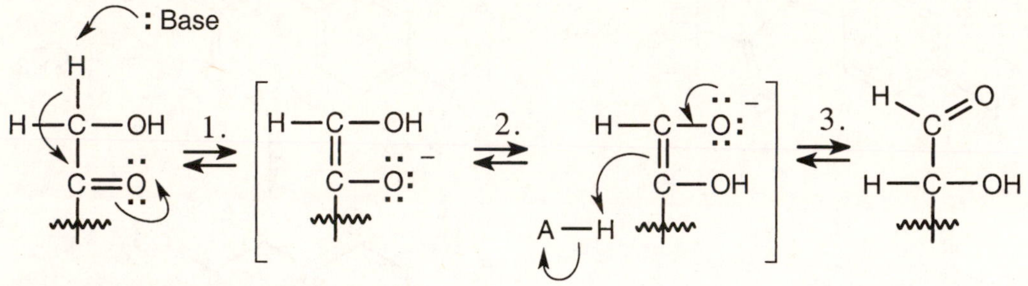

$$H\text{-}CH(OH)\text{-}C{=}O \underset{1.}{\rightleftarrows} [\,H\text{-}C(OH){=}C\text{-}O^- \underset{2.}{\rightleftarrows} H\text{-}C(O^-){=}C\text{-}OH\,] \underset{3.}{\rightleftarrows} H\text{-}C(=O)\text{-}CH\text{-}OH$$

Step 1: Base-catalyzed enolization.
Step 2: Equilibration of two enolates by proton transfer.
Step 3: Protonation.

17.59 Michael reactions occur between stabilized enolate anions and α,ß-unsaturated carbonyl compounds. Learn to locate these components in possible Michael products. Usually, it is easier to recognize the enolate nucleophile; in (a), the nucleophile is the ethyl acetoacetate anion.The rest of the compound is the Michael acceptor. Draw a double bond in conjugation with the electron-withdrawing group in this part of the molecule.

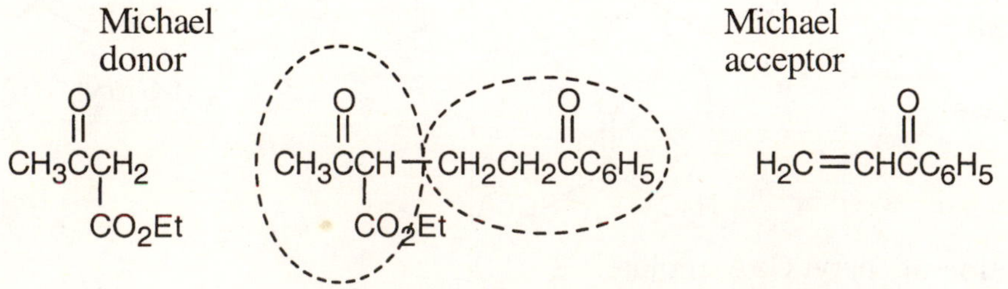

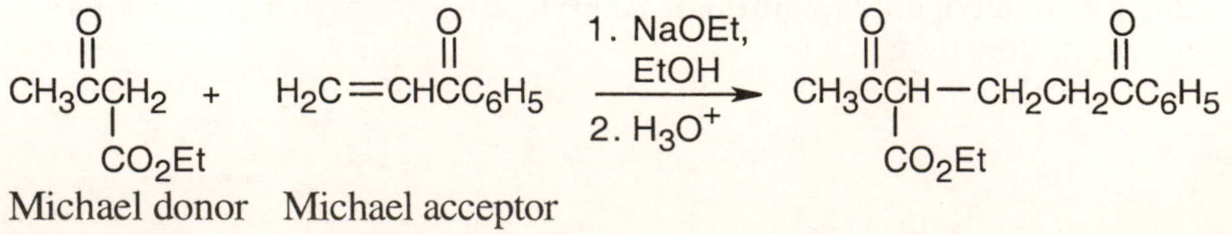

Michael donor: $CH_3C(=O)CH_2CO_2Et$

$CH_3C(=O)CH(CO_2Et)\text{—}CH_2CH_2C(=O)C_6H_5$

Michael acceptor: $H_2C{=}CHC(=O)C_6H_5$

$$\underset{\text{Michael donor}}{CH_3C(=O)CH_2CO_2Et} + \underset{\text{Michael acceptor}}{H_2C{=}CHC(=O)C_6H_5} \xrightarrow[\text{2. } H_3O^+]{\text{1. NaOEt, EtOH}} CH_3C(=O)CH(CO_2Et)\text{—}CH_2CH_2C(=O)C_6H_5$$

(b) When the Michael product has been decarboxylated after the addition reaction, it is more difficult to recognize the original enolate anion.

$$\underset{\text{Michael donor}}{CH_3C(=O)CH_2(CO_2Et)} + \underset{\text{Michael acceptor}}{H_2C{=}CHC(=O)CH_3} \xrightarrow[\text{2. } H_3O^+]{\text{1. NaOEt, EtOH}} CH_3C(=O)CH(CO_2Et){-}CH_2CH_2C(=O)CH_3$$

$$CH_3C(=O)CH(CO_2Et){-}CH_2CH_2C(=O)CH_3 \xrightarrow[\text{heat}]{H_3O^+} CH_3C(=O)CH_2CH_2CH_2C(=O)CH_3 + CO_2 + EtOH$$

(c)

$$\underset{\text{Michael donor}}{EtOC(=O)CH_2(CO_2Et)} + \underset{\text{Michael acceptor}}{H_2C{=}CHC{\equiv}N} \xrightarrow[\text{2. } H_3O^+]{\text{1. NaOEt, EtOH}} EtOC(=O)CH(CO_2Et)CH_2CH_2C{\equiv}N$$

(d)

$$\underset{\text{Michael donor}}{\text{2-(ethoxycarbonyl)cyclohexanone } (CO_2Et)} + \underset{\text{Michael acceptor}}{H_2C{=}CHC(=O)CH_3} \xrightarrow[\text{2. } H_3O^+]{\text{1. NaOEt, EtOH}} \text{2-}(CO_2Et)\text{-2-}(CH_2CH_2C(=O)CH_3)\text{cyclohexanone}$$

17.60

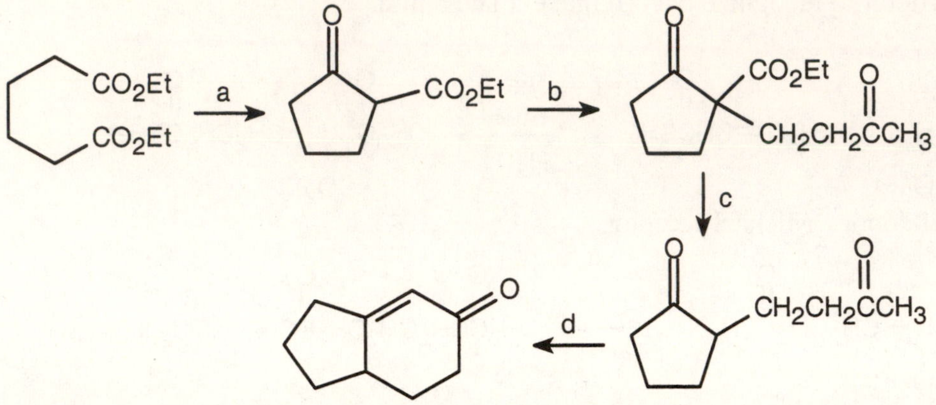

(a) Na^+ ^-OEt (Dieckmann cyclization), then H_3O^+; (b) Na^+ ^-OEt, $H_2C{=}CHCOCH_3$ (Michael reaction), then H_3O^+; (c) H_3O^+, heat (decarboxylation); (d) Na^+ ^-OEt (aldol condensation).

17.61

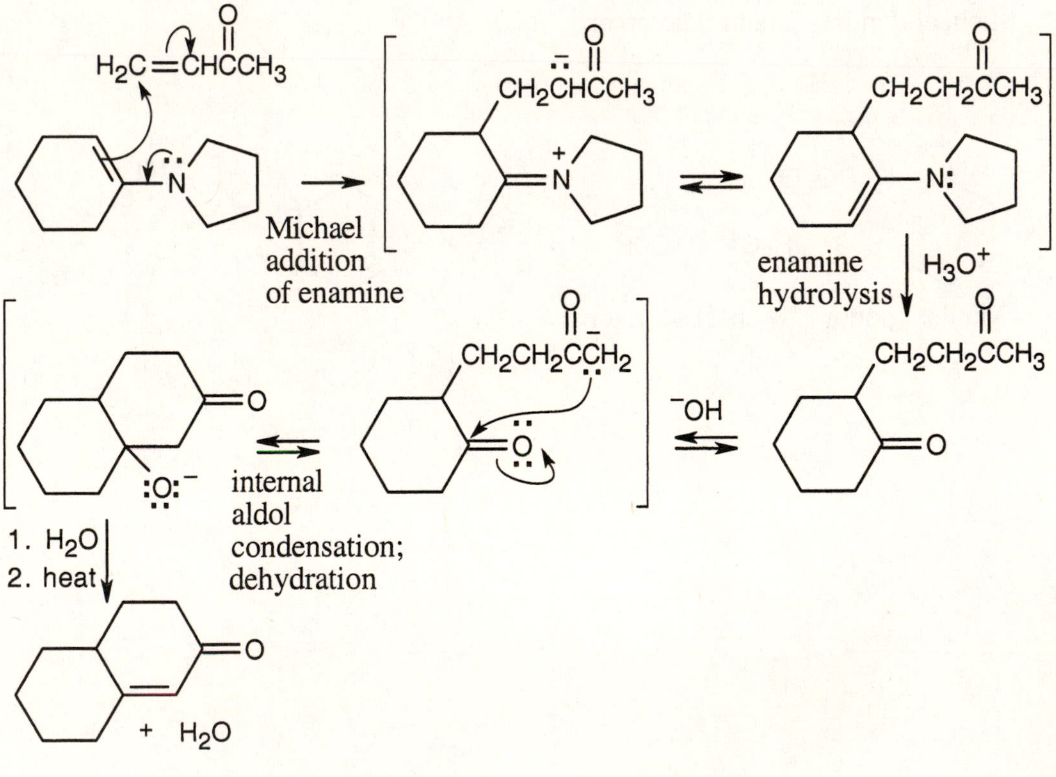

Michael addition of enamine

enamine hydrolysis H_3O^+

^-OH

internal aldol condensation; dehydration

1. H_2O
2. heat

$+ \ H_2O$

17.62

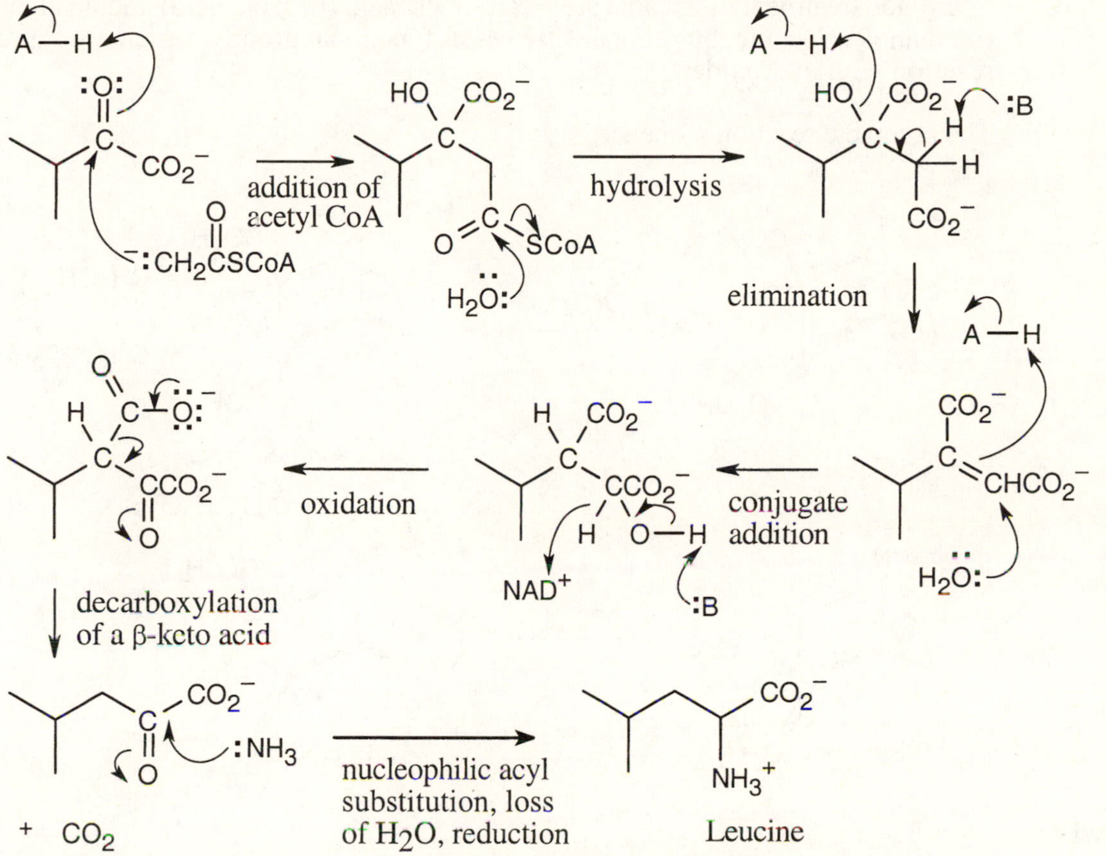

A—H
:O:
C
CO_2^-
addition of acetyl CoA
$:\bar{C}H_2CSCoA$ (C=O)
HO CO_2^-
C
O=C SCoA
$H_2O:$
hydrolysis
A—H
HO CO_2^- :B
C H
H
CO_2^-
elimination
A—H
CO_2^-
C=CHCO$_2^-$
$H_2O:$
conjugate addition
H CO_2^-
C
CCO_2^-
H O—H
NAD^+
:B
oxidation
H C—O:$^-$ (C=O)
C
CCO_2^-
O
decarboxylation of a β-keto acid
CO_2^-
C
O
:NH_3
+ CO_2
nucleophilic acyl substitution, loss of H_2O, reduction
CO_2^-
NH_3^+
Leucine

17.63 Start at the end of the sequence of reactions and work backwards. If necessary, cover up pieces of information you are not using at the moment to keep them from distracting you.

(a) Because the keto acid $C_9H_{13}NO_3$ loses CO_2 on heating, it must be a ß-keto acid. Neglecting stereoisomerism, we can draw the structure of the ß-keto acid as:

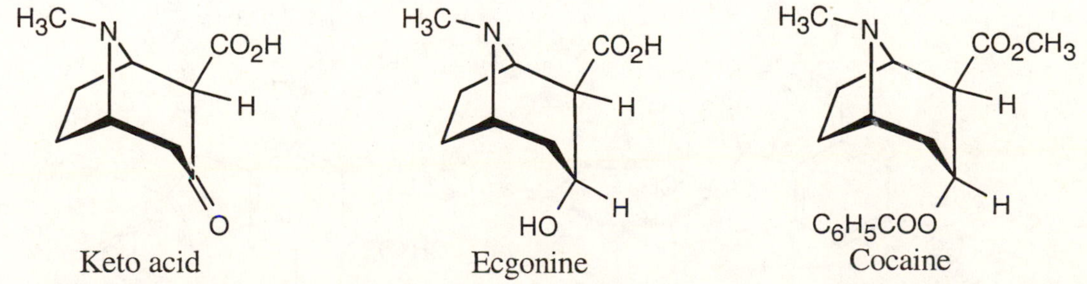

H_3C—N CO_2H H O
Keto acid

H_3C—N CO_2H H HO H
Ecgonine

H_3C—N CO_2CH_3 H C_6H_5COO H
Cocaine

(b) When ecgonine ($C_9H_{15}NO_3$) is treated with CrO_3, the keto acid $C_9H_{13}NO_3$ is produced. Since CrO_3 is used for oxidizing alcohols to carbonyl compounds, ecgonine has the structure shown above. Again, the stereochemistry is unspecified.

(c) Ecgonine contains carboxylic acid and alcohol functional groups. The other products of hydroxide treatment of cocaine are a carboxylic acid (benzoic acid) and an alcohol (methanol). Cocaine thus contains two ester functional groups, which are saponified on reaction with hydroxide.

The complete reaction sequence:

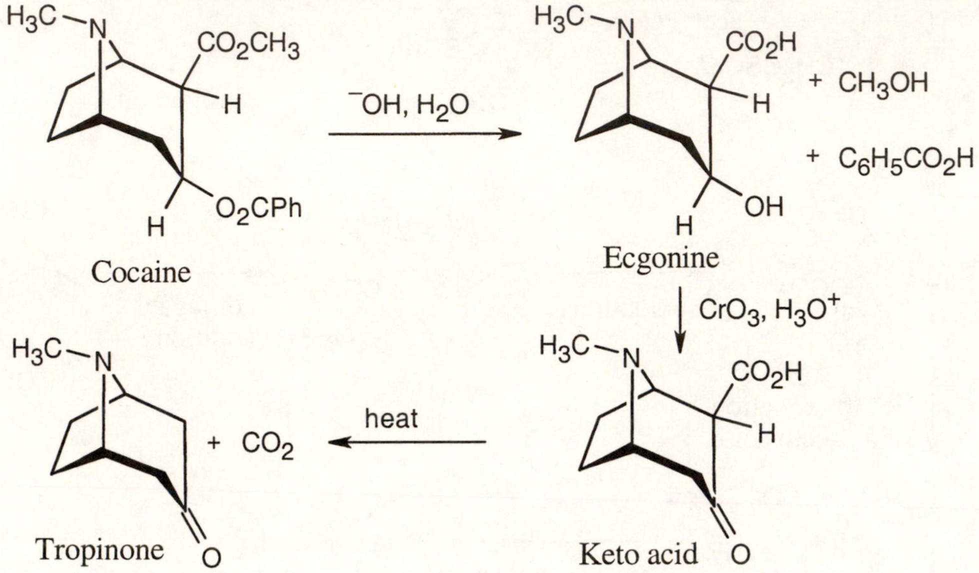

$H_3C{-}N$ CO_2CH_3 H O_2CPh H Cocaine $\xrightarrow{^-OH,\ H_2O}$ $H_3C{-}N$ CO_2H H OH H Ecgonine + CH_3OH + $C_6H_5CO_2H$

Ecgonine $\xrightarrow{CrO_3,\ H_3O^+}$ Keto acid ($H_3C{-}N$, CO_2H, H, O)

Keto acid $\xrightarrow{heat}$ Tropinone ($H_3C{-}N$, O) + CO_2

17.64

OCH3, O, H, H, CH_3O, Cl $\underset{HOC(CH_3)_3}{\overset{K^+\ ^-OC(CH_3)_3}{\rightleftharpoons}}$ carbanion (OCH3, O, H, CH_3O, Cl) + $CH_3OC{\equiv}C{-}\overset{O}{\overset{\|}{C}}{-}CH{=}CHCH_3$

$\downarrow$

OCH3, O, OCH3, $C{=}CH$, $C{=}O$, $HC{=}CH$, H_3C, CH_3O, Cl $\xrightarrow{H_3O^+}$ Griseofulvin (OCH3, O, OCH3, CH3, O, H, CH_3O, Cl)

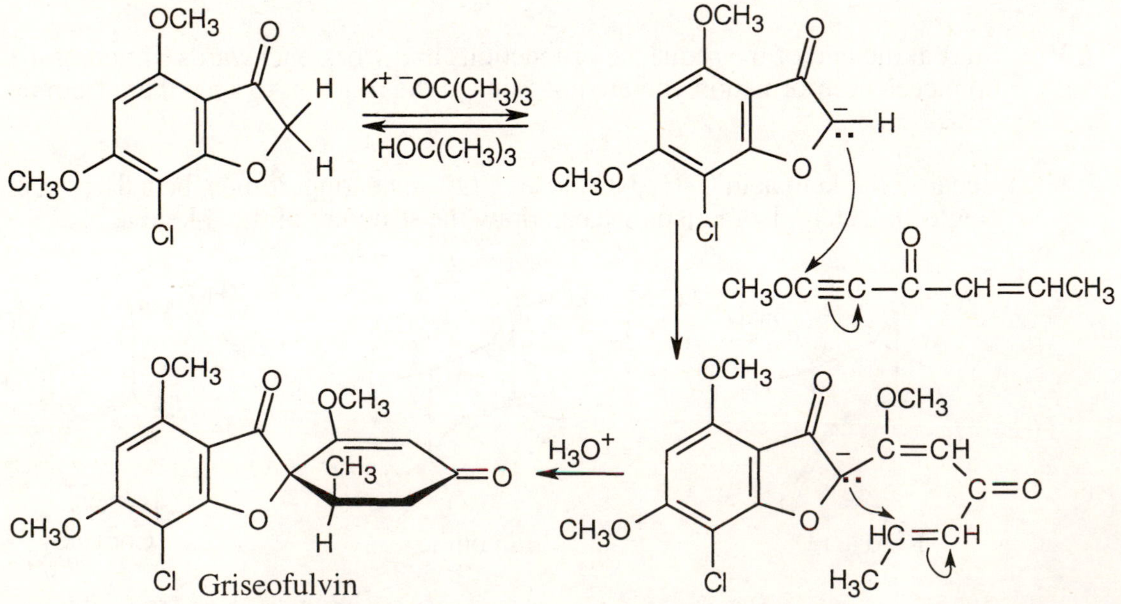

Two Michael reactions are involved in the key step that forms the cyclohexenone ring.

17.65 This problem becomes easier if you draw the starting material so that it resembles the product.

O O ≡ O O $^{-}$OH ⇌ Michael addition

intramolecular aldol reaction

H_2O

HO + $^{-}$OH

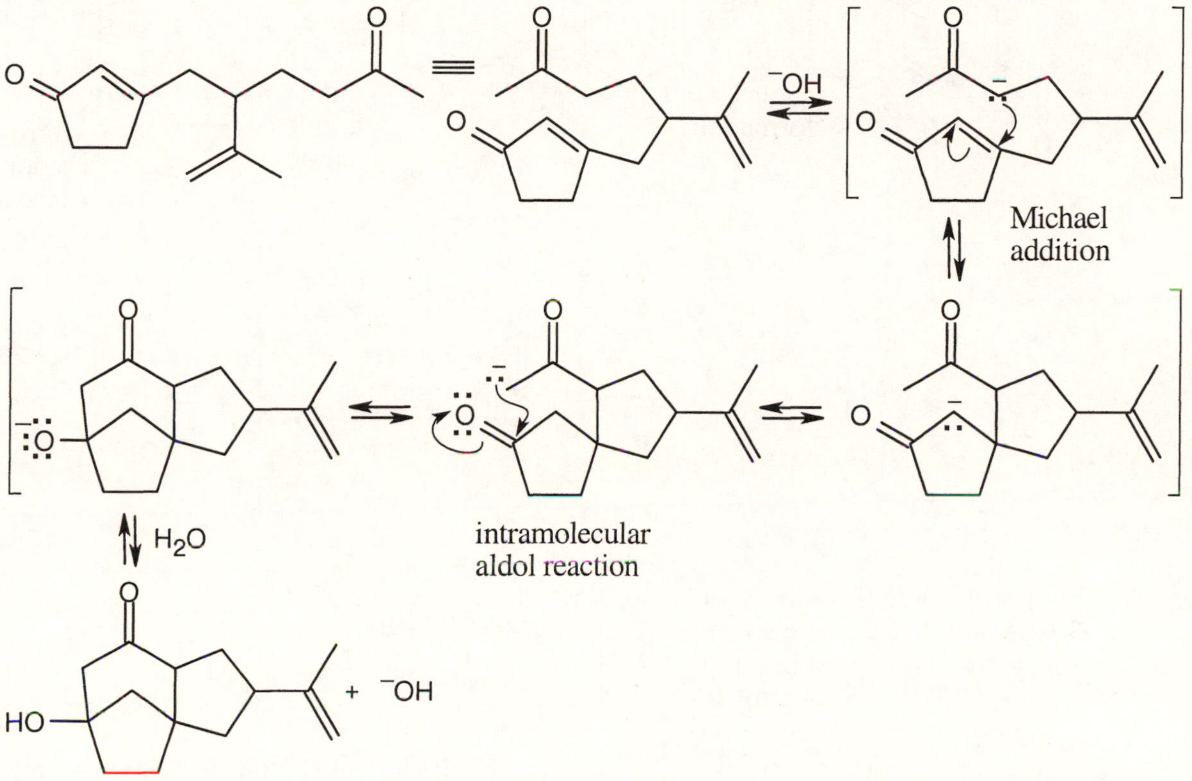

17.66 As in the previous problem, begin by redrawing the starting material so that it resembles the product.

$^{-}$OEt ⇌ enolate formation

Michael addition

EtOH ⇌ protonation + $^{-}$OEt

Michael addition

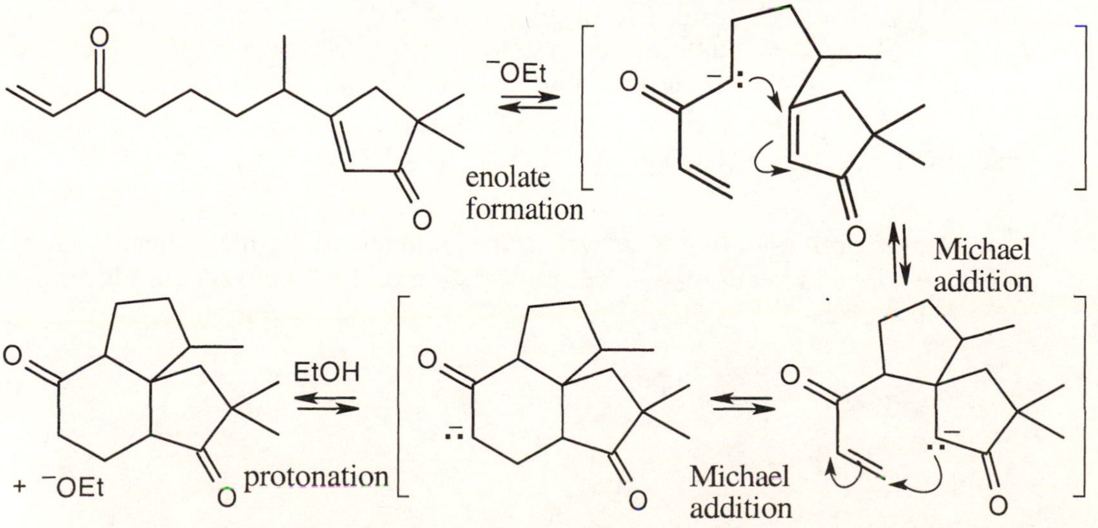

17.67

CO_2Et … $\xrightleftharpoons{^{-}OEt}$ enolate formation … aldol condensation … retro-aldol reaction … $\xrightleftharpoons{EtOH}$ protonation

^{-}OEt +

17.68

$CH_3CNHCCO_2Et$ (with CO_2Et, H) $\xrightarrow{Na^+\ ^{-}OEt}$ formation of enolate [$CH_3CNH\ddot{C}^{-}CO_2Et$ (CO_2Et)]

$\downarrow$ CH_3I alkylation

$CH_3C\text{-}NHC(H_3C)(CO_2Et)\text{-}C(=O)\text{-}OEt$ $\xrightarrow{H_3O^+,\ heat}$ ester hydrolysis, amide hydrolysis, decarboxylation $H_2NCH(CH_3)COOH$

Alanine

$+ \ 2\ EtOH + CO_2 + CH_3CO_2H$

Acid cleaves both ester bonds, as well as the amide bond, by mechanisms that were shown in Figure 16.9 and Section 16.7. Decarboxylation of the β-keto acid produces alanine.

17.69

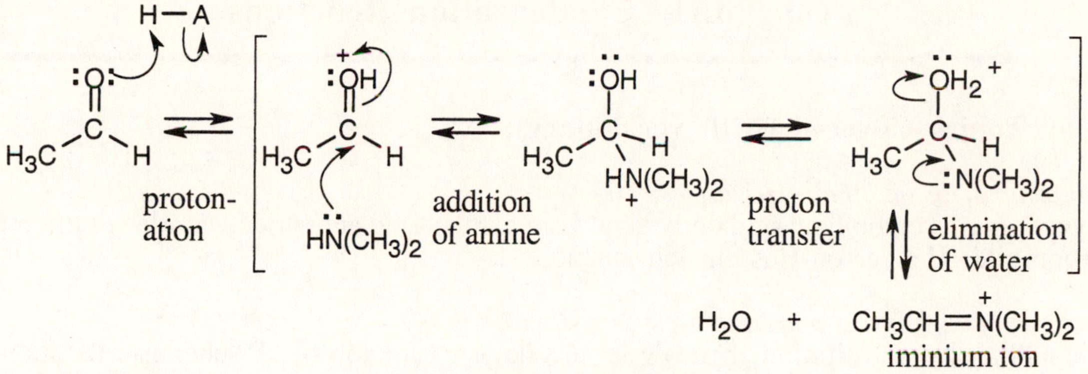

$H-A$

$:O:$ || H_3C-C-H ⇌ [$:\overset{+}{O}H$ || H_3C-C-H ← $\ddot{H}N(CH_3)_2$ ⇌ $:\ddot{O}H$ | H_3C-C-H | $H\overset{+}{N}(CH_3)_2$ ⇌ $\overset{+}{\ddot{O}H_2}$ | H_3C-C-H | $:N(CH_3)_2$]

proton-ation; addition of amine; proton transfer; elimination of water

H_2O + $CH_3CH=\overset{+}{N}(CH_3)_2$
iminium ion

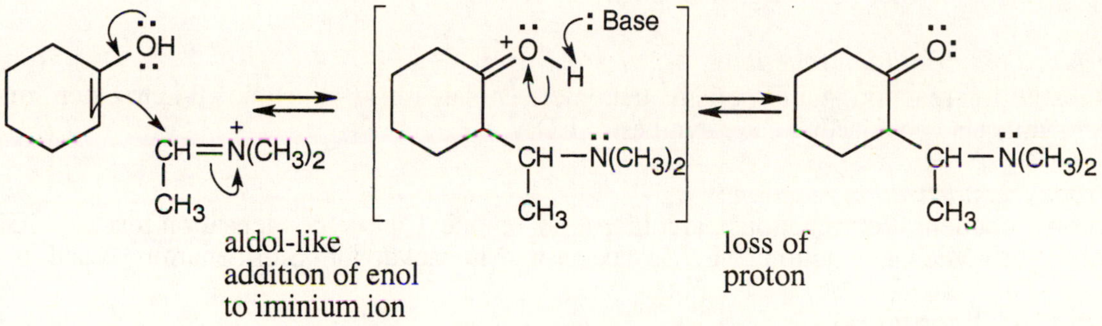

$\ddot{O}H$ + $CH=\overset{+}{N}(CH_3)_2$, CH_3 ⇌ [$:\overset{+}{O}-H$: Base, $CH-\ddot{N}(CH_3)_2$, CH_3] ⇌ $\ddot{O}:$ $CH-\ddot{N}(CH_3)_2$, CH_3

aldol-like addition of enol to iminium ion

loss of proton

17.70

O, H, H, O + H_2NCH_3 + CO_2CH_3, C=O, CO_2CH_3 ⟶ N, CH_3, CO_2CH_3, H, H_3CO_2C, O

[N, CH_3, CO_2CH_3, H, O_2CPh, H
Cocaine]

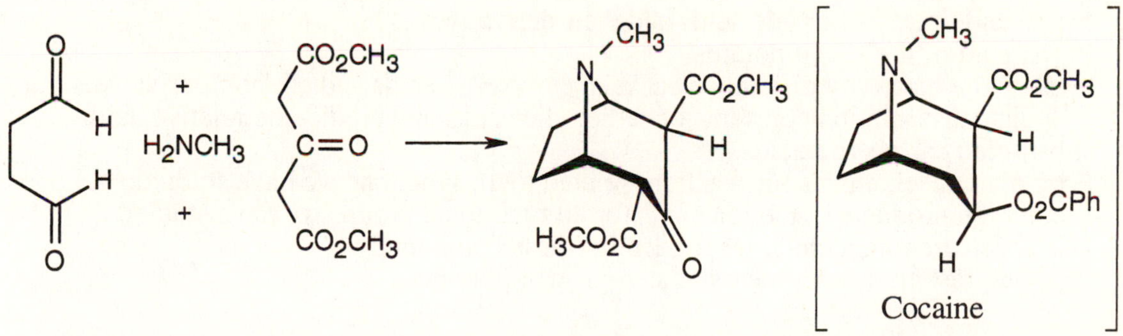

The Mannich reaction occurs between the diester, butanedial, and methylamine.

Review Unit 7: Carboxylic Acids and Their Derivatives; Carbonyl Condensation Reactions

Major Topics Covered (with vocabulary);

Carboxylic acids and their derivatives:
carboxylation carboxylic acid derivative acid halide acid anhydride amide ester nitrile –carbonitrile Henderson-Hasselbalch equation

Reactions of carboxylic acids and their derivatives:
nucleophilic acyl substitution hydrolysis alcoholysis aminolysis Fischer esterification reaction lactone saponification DIBAH lactam thiol ester acyl phosphate polyamide polyester step-growth polymer chain-growth polymer nylon

Carbonyl α-substitution reactions:
α-substitution reaction tautomerism tautomer enolate ion β-diketone β-keto eater malonic ester synthesis acetoacetic eater synthesis LDA

Carbonyl condensation reactions:
carbonyl condensation reactions aldol reaction enone Claisen condensation reaction Dieckmann cyclization Michael reaction Michael acceptor Michael donor Stork enamine reaction

Types of Problems:

After studying these chapters you should be able to:

- Name and draw carboxylic acids and their derivatives.
- Prepare all of these compounds.
- Explain the reactivity difference between carboxylic acids and all their derivatives.
- Calculate dissociation constants of carboxylic acids, and predict the relative acidities of substituted carboxylic acids.
- Formulate mechanisms for reactions related to the reactions we have studied.
- Predict the products of the reactions for all functional groups we have studied.
- Use spectroscopic techniques to identify these compounds.
- Draw representative segments of step-growth polymers.

- Draw keto-enol tautomers of carbonyl compounds, identify acidic hydrogens, and draw the resonance forms of enolates.
- Formulate the mechanisms of acid- and base-catalyzed enolization, of other α-substitution reactions, and of carbonyl condensation reactions.
- Predict the products of α-substitution reactions and carbonyl condensation reactions.
- Use α-substitution reactions and carbonyl condensation reactions in synthesis.

Points to Remember:

* Here are a few reminders for drawing the mechanisms of nucleophilic addition and substitution reactions. (1) When a reaction is acid-catalyzed, none of the intermediates are negatively charged, although, occasionally, a few may be neutral. Check your mechanisms for charge balance. (2) Make sure you have drawn arrows correctly. The point of the arrow shows the new location of the electron pair at the base of the arrow. (3) In a polar reaction, two arrows never point at each other. If you find two arrows pointing at each other, redraw the mechanism.

* Reactions of acyl halides are almost always carried out with an equivalent of base present. The base is used to scavenge the protons produced when a nucleophile adds to an acyl halide. If base were not present, hydrogen ions would protonate the nucleophile and make it unreactive.

* The products of acidic cleavage of an amide are a carboxylic acid and a protonated amine. The products of basic cleavage of an amide are a carboxylate anion and an amine.

* In some of the mechanisms shown in the answers, a series of protonations and deprotonations occur. These steps convert the initial tetrahedral intermediate into an intermediate that more easily loses a leaving group. These deprotonations may be brought about by the solvent, by the conjugate base of the catalyst, by other molecules of the carbonyl compound or may occur intramolecularly. When a "proton transfer" is shown as part of a mechanism, the base that removes the proton has often not been shown. However, it is implied that the proton transfer is assisted by a base: the proton doesn't fly off the intermediate unassisted.

* The most useful spectroscopic information for identifying carbonyl compounds comes from IR spectroscopy and ^{13}C NMR spectroscopy. Carbonyl groups have distinctive identifying absorptions in their infrared spectra. ^{13}C NMR is also useful for identifying aldehydes, ketones, and nitriles, although other groups are harder to distinguish. The ^{1}H NMR absorptions of aldehydes and carboxylic acids are also significant. Look at mass spectra for McLafferty rearrangements and alpha-cleavage reactions of aldehydes and ketones.

* It is unusual to think of a carbonyl compound as an acid, but the protons α to a carbonyl group can be removed by a strong base. Protons α to two carbonyl groups are even more acidic: in some cases, acidity approaches that of phenols. This acidity is the basis for α-substitution reactions of compounds having carbonyl groups. Abstraction by base of an α proton produces a resonance-stabilized enolate anion that can be used in alkylations involving alkyl halides and tosylates.

* Alkylation of an unsymmetrical LDA-generated enolate generally occurs at the less hindered α carbon.

* When you need to synthesize a β-hydroxy ketone or aldehyde or an α,β-unsaturated ketone or aldehyde, use an aldol reaction. When you need to synthesize a β-diketone or β-keto ester, use a Claisen reaction. When you need to synthesize a 1,5-dicarbonyl compound, use a Michael reaction. The Robinson annulation is used to synthesize polycyclic molecules by a combination of a Michael reaction with an aldol condensation.

* In the Claisen condensation, the enolate of the β-dicarbonyl compound is treated with H_3O^+ to yield the neutral product.

Self-Test:

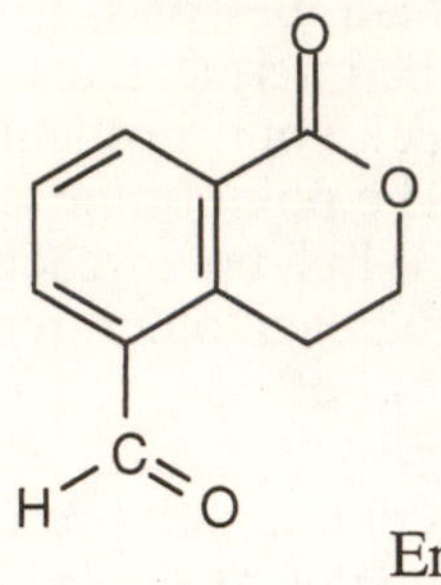

A

Erythrocentaurin
(a bitter tonic)

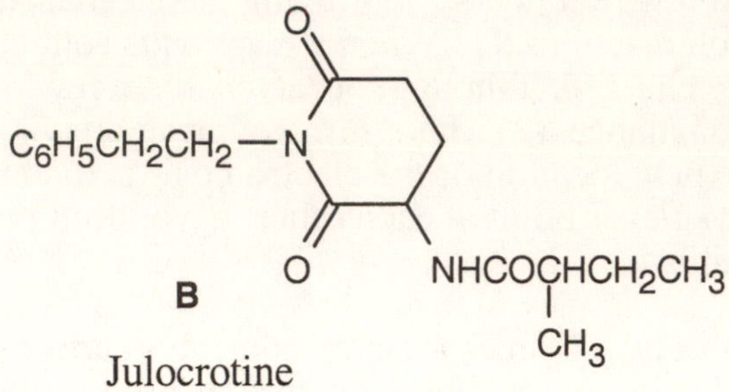

B

Julocrotine

Predict the products of **A** with: (a) $LiAlH_4$, then H_3O^+; (b) C_6H_5MgBr, then H_3O^+; (c) $(CH_3)_2NH$, H_3O^+; (d) CH_3OH, H^+ catalyst. What product(s) would be formed if **A** was treated with Br_2, $FeBr_3$? Where do the carbonyl absorptions occur in the IR spectrum of **A**? Describe the ^{13}C NMR spectrum of **A**.

Identify the carboxylic acid derivatives present in **B**. Show the products of treatment of **B** with (a) ^-OH, H_2O (b) $LiAlH_4$, then H_2O.

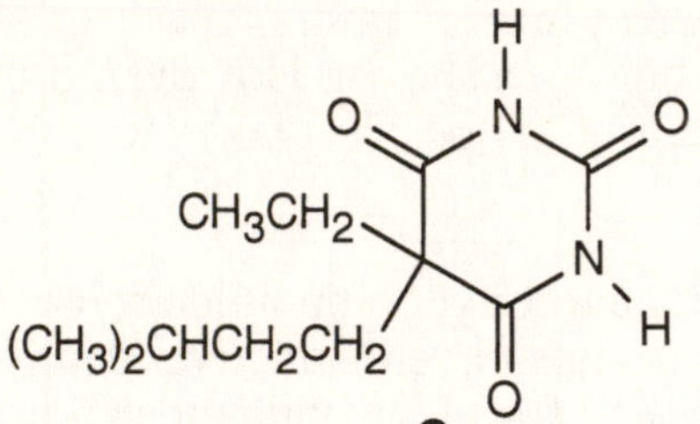

C

Pentymal
(a sedative)

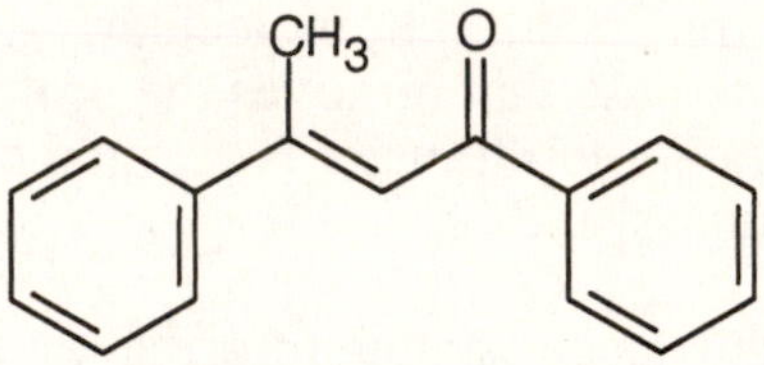

D

Dypnone
(sunscreen)

The six-membered ring in **C** is formed by the cyclization of two difunctional compounds. What are they? What type of reaction occurs to form the ring? The two alkyl groups are introduced into one of the difunctional compounds prior to cyclization. What type of reaction is occurring, and how is it carried out? What type of reaction occurs in the formation of Dypnone (**D**)? Why might **D** be effective as a sunscreen?

Multiple Choice:

1. A nitrile can be converted to all of the following except:
(a) an aldehyde (b) an amide (c) an amine (d) A nitrile can be converted to all of the above compounds.

2. Which of the following *p*- substituted benzoic acids is the least acidic?
(a) $CH_3COC_6H_5CO_2H$ (b) $CH_3OC_6H_5CO_2H$ (c) $BrC_6H_5CO_2H$ (d) $NCC_6H_5CO_2H$

3. Which of the following carboxylic acids can be formed by both Grignard carboxylation and by nitrile hydrolysis?
(a) Phenylacetic acid (b) Benzoic acid (c) Trimethylacetic acid (d) 3-Butynoic acid

4. Acid anhydrides are used mainly for:
(a) synthesizing carboxylic acids (b) forming alcohols (c) introducing acetyl groups
(d) forming aldehydes

5. An infrared absorption at 1650 cm^{-1} indicates the presence of:
(a) aromatic acid chloride (b) *N,N*-disubstituted amide (c) α,β-unsaturated ketone
(d) aromatic ester

6. Which of the following is not likely to be part of a step-growth polymer?
(a) Hexane-1,6-diamine (b) Diphenyl carbonate (c) Ethylene glycol (d) Buta-1,3-diene

7. Which of the following compounds has four acidic hydrogens?
(a) 2-Pentanone (b) 3-Pentanone (c) Acetophenone (d) Phenylacetone

8. If you want to carry out a carbonyl condensation, and you don't want to form α-substitution product, you should:
(a) lower the temperature (b) use one equivalent of base (c) use a catalytic amount of base
(d) use a polar aprotic solvent

9. Which reaction forms a cyclohexenone?
(a) Dieckmann cyclization (b) Michael reaction (c) Claisen condensation
(d) intramolecular aldol condensation

10. All of the following molecules are good Michael donors except:
(a) Ethyl acetoacetate (b) Nitroethylene (c) Malonic ester
(d) Ethyl 2-oxocyclohexanecarboxylate

Chapter 18 – Amines and Heterocycles

Chapter Outline

I. Facts about amines (Section 18.1 –18.5).
   A. Naming amines (Section 18.1).
      1. Amines are classified as primary (RNH_2), secondary (R_2NH), tertiary (R_3N) or quaternary ammonium salts (R_4N^+).
      2. Primary amines are named in two ways:
         a. For simple amines, the suffix *-amine* is added to the name of the alkyl substituent.
         b. For more complicated amines, the $–NH_2$ group is an amino substituent on the parent molecule.
      3. Secondary and tertiary amines:
         a. Symmetrical amines are named by using the prefixes *di-* and *tri-* before the name of the alkyl group.
         b. Unsymmetrical amines are named as *N*-substituted primary amines.
            The largest group is the parent.
      4. The simplest arylamine is aniline.
      5. Heterocyclic amines (nitrogen is part of a ring) have specific parent names.
         The nitrogens receive the lowest possible numbers.
   B. Properties and sources of amines (Section 18.2).
      1. The three amine bonds and the lone pair occupy the corners of a tetrahedron.
      2. An amine with three different substituents is chiral.
         a. The two amine enantiomers interconvert by pyramidal inversion.
         b. This process is rapid at room temperature.
      3. Amines with fewer than 5 carbons are water-soluble and form hydrogen bonds.
      4. Amines have higher boiling points than alkanes of similar molecular weight.
      5. Amines smell really bad.
   C. Amine basicity (Sections 18.3 – 18.5).
      1. The lone pair of electrons makes amines both nucleophilic and basic (Section 18.3).
      2. The basicity constant K_b is the measure of the equilibrium of an amine with water.
         The larger the value of K_b (smaller pK_b), the stronger the base.
      3. More often, K_a is used to describe amine basicity.
         a. K_a is the dissociation constant of the conjugate acid of an amine.
         b. $pK_a + pK_b = 14$ (for aqueous media).
         c. The smaller the value of K_a (larger pK_a), the stronger the base.
      4. Base strength.
         a. Alkylamines have similar basicities.
         b. Arylamines and heterocyclic amines are less basic than alkylamines.
            i. The sp^2 electrons of the pyridine lone pair are less available for bonding.
            ii. The pyrrole lone pair electrons are part of the aromatic ring π system.
         c. Amides are nonbasic.
         d. Some amines are very weak acids.
            LDA is formed from diisopropylamine and acts as a strong base.
      5. Basicity of substituted arylamines (Section 18.4).
         a. Arylamines are less basic than alkylamines for two reasons:
            i. Arylamine lone-pair electrons are delocalized over the aromatic ring and are less available for bonding.
            ii. Arylamines lose resonance stabilization when they are protonated.

b. Electron-donating substituents increase arylamine basicity.
6. Biological amines and the Henderson-Hasselbalch equation (Section 18.5).
a. The Henderson-Hasselbalch equation (Section 15.3) can be used to calculate the percent of protonated vs. unprotonated amines.
b. At physiological pH, most amines exist in the protonated form.

II. Synthesis of amines (Section 18.6).
A. Reduction of amides, nitriles and nitro groups.
1. S_N2 displacement with ^-CN, followed by reduction, turns a primary alkyl halide into an amine with one more carbon atom.
2. Amide reduction converts an amide or nitrile into an amine with the same number of carbons.
3. Arylamines can be prepared by reducing aromatic nitro compounds.
a. Catalytic hydrogenation can be used if no other interfering groups are present.
b. $SnCl_2$ can also be used.
B. S_N2 reactions of alkyl halides.
It is possible to alkylate ammonia or an amine with RX.
Unfortunately, it is difficult to avoid overalkylation.
C. Reductive amination of aldehydes and ketones.
Treatment of an aldehyde or ketone with ammonia or an amine in the presence of a reducing agent yields an amine
a. The reaction proceeds through an imine, which is reduced.
b. $NaBH_4$ or H_2/Ni are the reducing agents most commonly used.

III. Reactions of amines (Section 18.7).
A. Alkylation and acylation.
1. Alkylation of primary and secondary amines is hard to control.
2. Primary and secondary amines can also be acylated.
B. Hofmann elimination.
1. Alkylamines can be converted to alkenes by the Hofmann elimination reaction.
a. The amine is treated with an excess of methyl iodide to form a quaternary ammonium salt.
b. Treatment of the quaternary salt with Ag_2O, followed by heat, gives the alkene.
2. The elimination is an E2 reaction.
3. The less substituted double bond is formed because of the bulk of the leaving group.
4. The reaction was formerly used for structure determination and is rarely used today.
C. Reactions of arylamines.
1. Electrophilic aromatic substitution.
a. Electrophilic aromatic substitutions are usually carried out on *N*-acetylated amines, rather than on unprotected amines.
i. Amino groups are *o,p*-activators, and polysubstitution sometimes occurs.
ii. Friedel-Crafts reactions don't take place with unprotected amines.
b. Aromatic amines are acetylated by treatment with acetic anhydride.
c. The *N*-acetylated amines are *o,p*-directing activators, but are less reactive than unprotected amines.
d. Synthesis of sulfa drugs was achieved by electrophilic aromatic substitution reactions on *N*-protected aromatic compounds.

IV. Heterocyclic amines (Sections 18.8 – 18.9).
A. Pyrrole, imidazole and other 5-membered ring unsaturated heterocycles (Section 18.8).
1. Structures of pyrrole, furan and thiophene.
a. All are aromatic because they have six π electrons in a cyclic conjugated system.
b. Pyrrole is nonbasic because all 5 nitrogen electrons are used in bonding.
c. The carbon atoms in pyrrole are electron-rich and are reactive toward electrophiles.

2. Electrophilic substitution reactions.
   a. All three compounds undergo electrophilic aromatic substitution reactions readily.
   b. Halogenation, nitration, sulfonation and Friedel-Crafts alkylation can take place if reaction conditions are modified.
   c. Reaction occurs at the 2-position because the reaction intermediate from attack is more stable.
3. Imidazole and thiazole.
   A nitrogen in these compounds is basic.

B. Pyridine and pyrimidine.
1. Structure of pyridine.
   a. Pyridine is the nitrogen-containing analog of benzene.
   b. The nitrogen lone pair isn't part of the π electron system .
   c. Pyridine is a stronger base than pyrrole but a weaker base than alkylamines.
2. Electrophilic substitution of pyridine.
   Electrophilic substitutions take place with great difficulty.
   i. The pyridine ring is electron-poor due to the electron-withdrawing inductive effect of nitrogen.
   ii. Acid-base complexation between nitrogen and an electrophile puts a positive charge on the ring.

C. Fused-ring heterocycles (Section 18.9).
1. The reactivity of fused-ring heterocyclic compounds is related to the type of heteroatom and to the size of the ring.
2. Indole has a pyrrole-like nitrogen and undergoes electrophilic aromatic substitutions in the heterocyclic ring.
3. Purines have 4 nitrogens (3 pyridine-like, and one pyrrole-like) in a fused-ring structure.

V. Spectroscopy of amines (Section 18.10).

A. IR spectroscopy.
1. Primary and secondary amines absorb in the region 3300–3500 cm^{-1}.
   a. Primary amines show a pair of bands at 3350 cm^{-1} and 3450 cm^{-1}.
   b. Secondary amines show a single band at 3350 cm^{-1}.
   c. These absorptions are sharper than alcohol absorptions, which also occur in this range.
2. Adding a small amount of HCl causes a broad band in the range 2200–3000 cm^{-1} that is due to the ammonium ion.

B. NMR spectroscopy.
1. ^1H NMR.
   a. Amine protons are hard to identify because they appear as broad signals.
   b. Exchange with D_2O causes the amine signal to disappear and allows identification.
   c. Hydrogens on the carbon next to nitrogen are somewhat deshielded.
2. ^{13}C NMR.
   Carbons next to nitrogen are slightly deshielded.

C. Mass spectrometry.
1. The nitrogen rule: A compound with an odd number of nitrogens has an odd-numbered molecular weight (and molecular ion).
2. Alkylamines undergo α-cleavage and show peaks that correspond to both possible modes of cleavage.

Solutions to Problems

18.1 Facts to remember about naming amines:

(1) Primary amines are named by adding the suffix *-amine* to the name of the alkyl substituent.

(2) The prefix *di-* or *tri-* is added to the names of symmetrical secondary and tertiary amines.

(3) Unsymmetrical secondary and tertiary amines are named as *N*-substituted primary amines. The parent amine has the largest alkyl group.

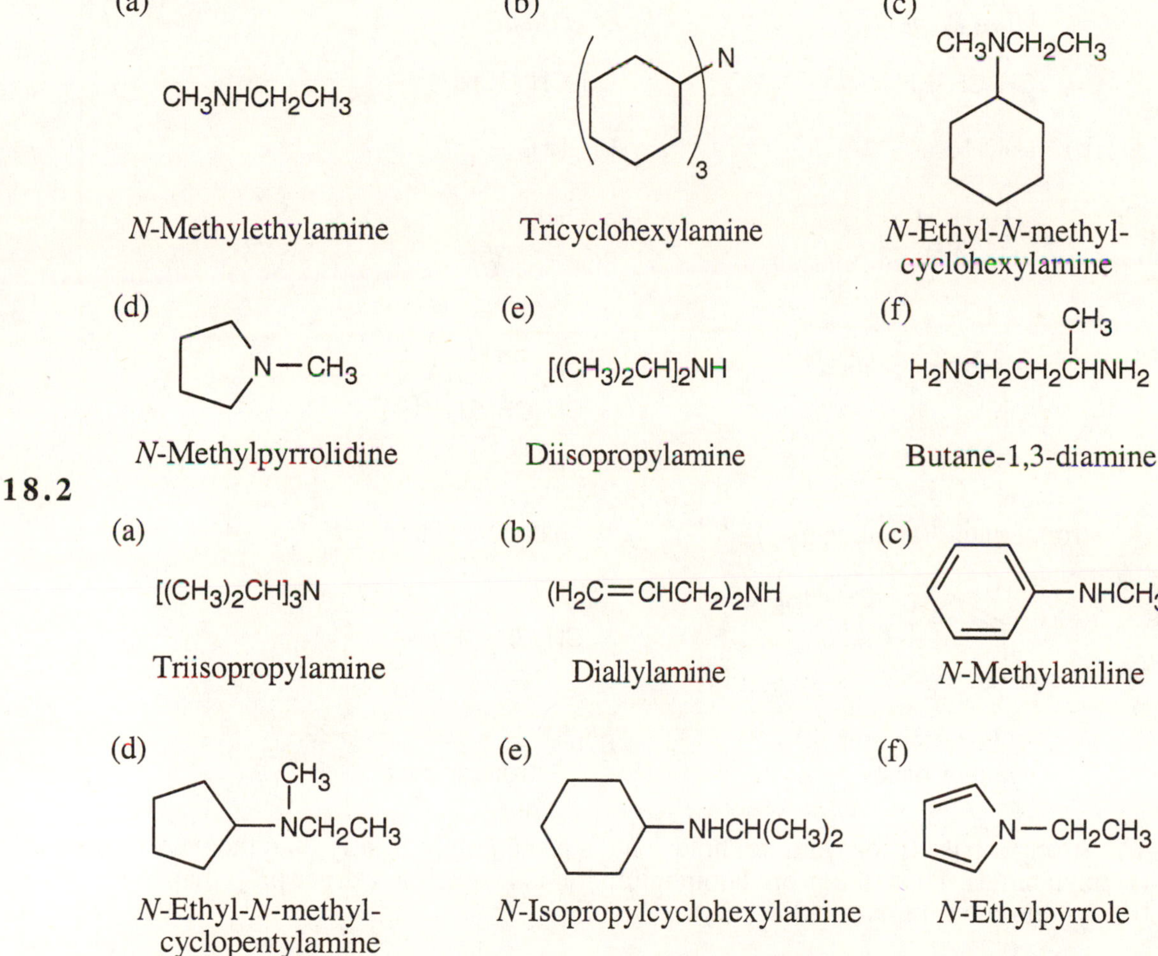

18.3 The numbering of heterocyclic rings is described in Section 18.1.

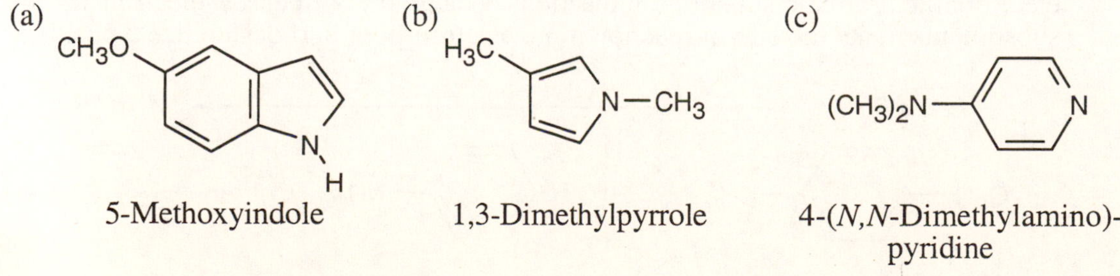

(d)

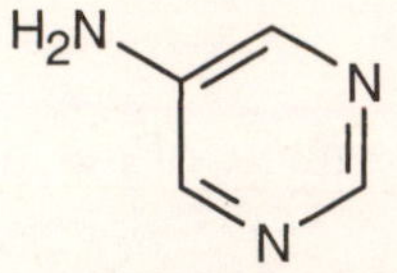

H_2N–(pyrimidine ring)

5-Aminopyrimidine

18.4 Amines are less basic than hydroxide but more basic than amides. The pK_a values of the conjugate acids of the amines in (c) are shown. The larger the pK_a, the stronger the base.

	More Basic	*Less Basic*
(a)	$CH_3CH_2NH_2$	$CH_3CH_2CONH_2$
(b)	NaOH	CH_3NH_2
(c)	CH_3NHCH_3 pK_a = 10.73	pyridine pK_a = 5.25

18.5

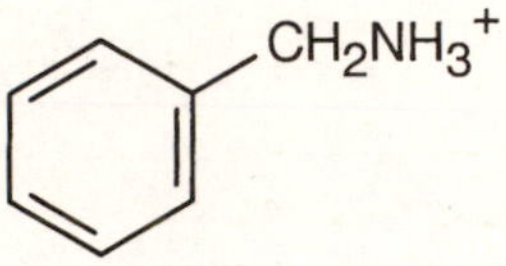

$C_6H_5CH_2NH_3^+$	$CH_3CH_2CH_2NH_3^+$
pK_a = 9.33 stronger acid (smaller pK_a)	pK_a = 10.71 weaker acid (larger pK_a)
$C_6H_5CH_2NH_2$	$CH_3CH_2CH_2NH_2$
pK_b = 14 – 9.33 = 4.67 weaker base	pK_b = 14 – 10.71 = 3.29 stronger base

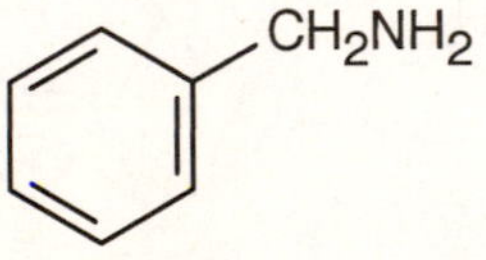

The stronger base (propylamine) holds onto a proton more tightly than the weaker base (benzylamine). Thus, the propylammonium ion is less acidic (larger pK_a) than the benzylammonium ion (smaller pK_a).

To calculate pK_b: $K_a \cdot K_b = 10^{-14}$, pK_a + pK_b = 14 and pK_b = 14 – pK_a.

18.6 The basicity order of substituted arylamines is the same as their reactivity order in electrophilic aromatic substitution reactions because, in both cases, electron-withdrawing substituents make the site of reaction more electron-poor and destabilize a positive charge.

Least Basic ⟶ *Most Basic*

(a) O_2N–C_6H_4–NH_2 < $HC(=O)$–C_6H_4–NH_2 < Br–C_6H_4–NH_2

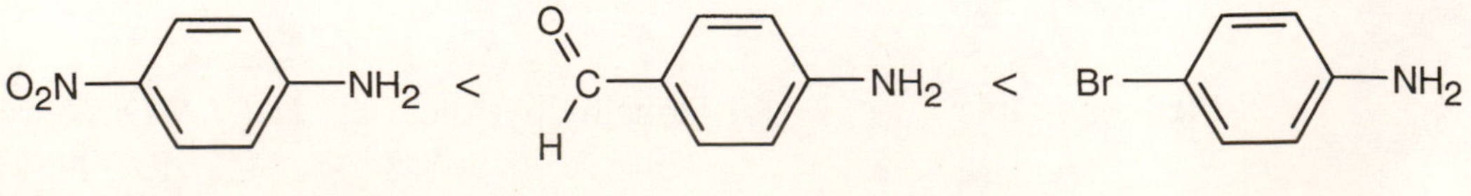

Least Basic ——————————> *Most Basic*

(b)

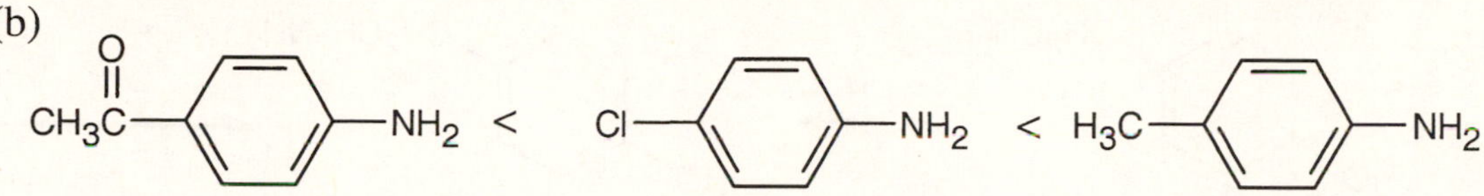

p-CH_3CO–C_6H_4–NH_2 < p-Cl–C_6H_4–NH_2 < p-H_3C–C_6H_4–NH_2

(c)

p-F_3C–C_6H_4–NH_2 < p-H_2FC–C_6H_4–NH_2 < p-H_3C–C_6H_4–NH_2

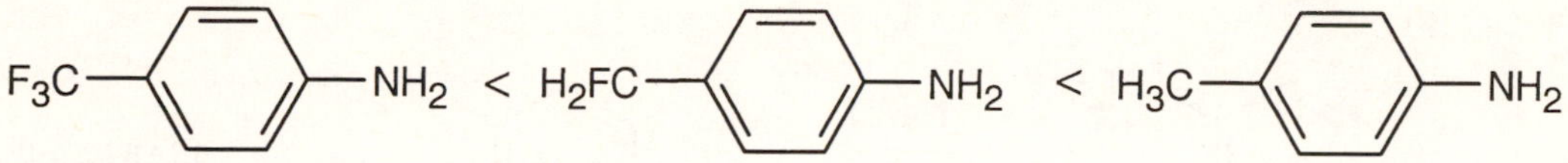

18.7 Use the expressions shown in Section 15.3.

$$\log \frac{[RNH_2]}{[RNH_3^+]} = pH - pK_a = 7.3 - 1.3 = 6.0$$

$$\frac{[RNH_2]}{[RNH_3^+]} = \text{antilog}\ (6.0) = 10^6\text{: } [RNH_2] = 10^6\ [RNH_3^+]$$

At pH = 7.3, virtually 100% of the pyrimidine molecules are in the neutral form.

18.8 Amide reduction can be used to synthesize most amines, but nitrile reduction can be used to synthesize only primary amines. Thus, the compounds in (b) and (d) can be synthesized only by amide reduction.

	Amine	*Nitrile Precursor*	*Amide Precursor*
(a)	$CH_3CH_2CH_2NH_2$	$CH_3CH_2C{\equiv}N$	$CH_3CH_2C(=O)NH_2$
(b)	$(CH_3CH_2CH_2)_2NH$		$CH_3CH_2C(=O)NHCH_2CH_2CH_3$
(c)	C_6H_5–CH_2NH_2	C_6H_5–$C{\equiv}N$	C_6H_5–$C(=O)NH_2$
(d)	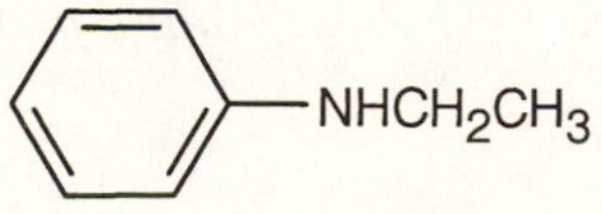 C_6H_5–$NHCH_2CH_3$		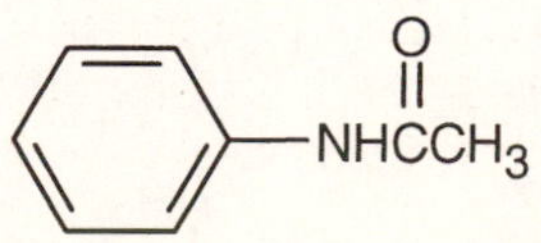 C_6H_5–$NHC(=O)CH_3$

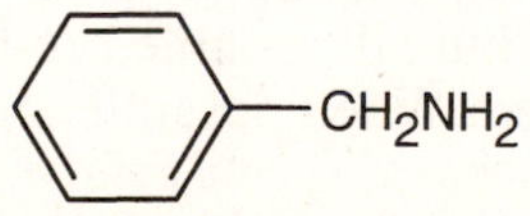

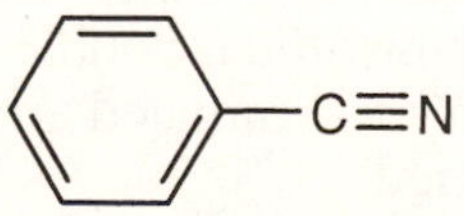

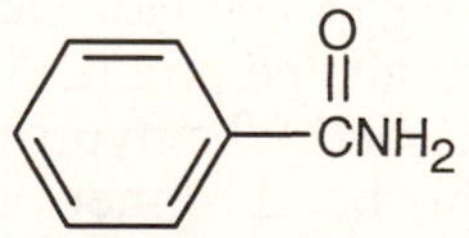

18.9

(3,4-dihydroxyphenyl)CH_2CH_2Br $\xrightarrow{NH_3}$ (3,4-dihydroxyphenyl)$CH_2CH_2NH_2$

(3,4-dihydroxyphenyl)CH_2Br $\xrightarrow{NaCN}$ (3,4-dihydroxyphenyl)CH_2CN $\xrightarrow[2.\ H_2O]{1.\ LiAlH_4,\ ether}$ (3,4-dihydroxyphenyl)$CH_2CH_2NH_2$

(3,4-dihydroxyphenyl)CH_2CO_2H $\xrightarrow[2.\ 2\ NH_3]{1.\ SOCl_2}$ (3,4-dihydroxyphenyl)CH_2CNH_2 (C=O) $\xrightarrow[2.\ H_2O]{1.\ LiAlH_4,\ ether}$ (3,4-dihydroxyphenyl)$CH_2CH_2NH_2$

18.10 Strategy: Look at the target molecule to find the groups bonded to nitrogen. One group comes from the aldehyde/ketone precursor, and the other group comes from the amine precursor. Often, two combinations of amine and aldehyde/ketone are possible.

Solution:

$(CH_3)_2NH$ + $O{=}CH$–(3-methylphenyl) $\xrightarrow{NaBH_4}$ H_3C–$N(CH_3)$–CH_2–(3-methylphenyl)

18.11 Strategy: The Hofmann elimination yields alkenes and amines from larger amines. The major alkene product has the less substituted double bond, but all possible products may be formed. The hydrogens that can be eliminated are starred. When possible, cis and trans double bond isomers are both formed.

Solution:

	Amine	*Alkene Products*	*Amine products*
(a)	$CH_3CH_2CH_2^{*}CHCH_2^{*}CH_2CH_2CH_3$ (NH$_2$ on CH)	$CH_3CH_2CH{=}CHCH_2CH_2CH_2CH_3$	$(CH_3)_3N$
		or	
		$CH_3CH_2CH_2CH{=}CHCH_2CH_2CH_3$	$(CH_3)_3N$

Both hydrogens that might be eliminated are secondary, and both possible products should form in approximately equal amounts.

(b) [structure: cyclohexylamine with starred β carbons] $(CH_3)_3N$ [structure: cyclohexene]

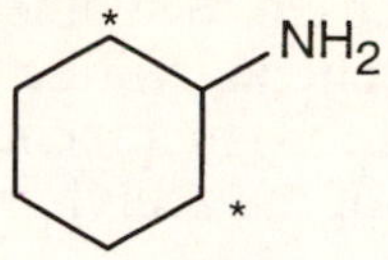

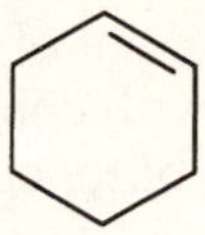

(c) $CH_3CH_2CH_2CH(NH_2)CH_2CH_2CH_3$ → $CH_3CH_2CH{=}CHCH_2CH_2CH_3$ $(CH_3)_3N$

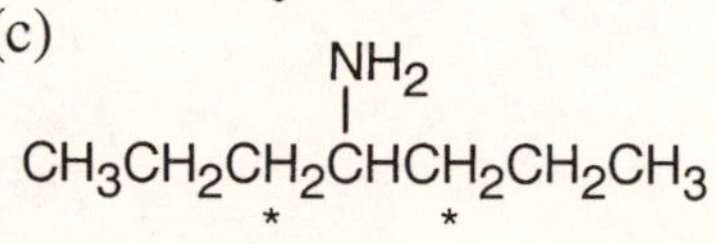

In each of the above reactions, only one product can form.

(d) [structure: N-ethylcyclohexylamine, $NHCH_2CH_3$] $H_2C{=}CH_2$ *(major)* [structure: cyclohexyl-$N(CH_3)_2$]

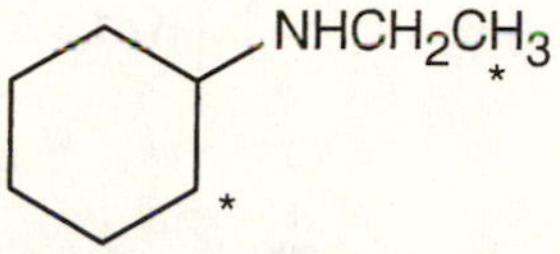

or

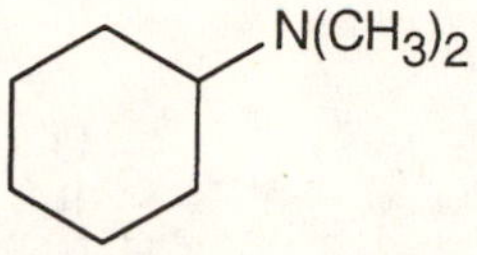

[structure: cyclohexene] *(minor)* $(CH_3)_2NCH_2CH_3$

The first pair of products in (d) results from elimination of a primary hydrogen and are the major products. The second pair of products results from elimination of a secondary hydrogen.

18.12

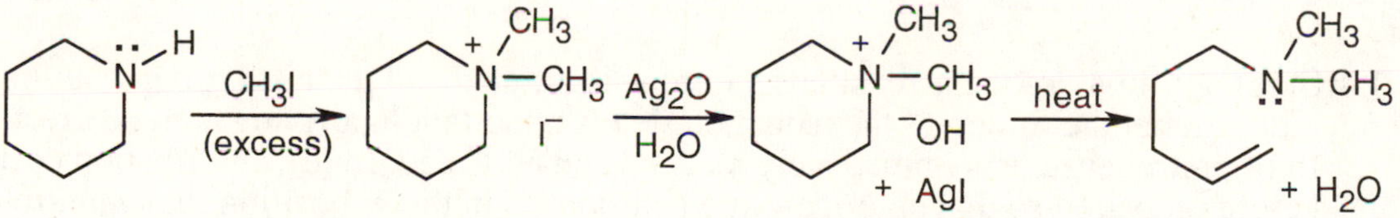

Piperidine $\xrightarrow[\text{(excess)}]{CH_3I}$ N,N-dimethylpiperidinium I^- $\xrightarrow[H_2O]{Ag_2O}$ N,N-dimethylpiperidinium ^-OH + AgI $\xrightarrow{\text{heat}}$ $CH_2{=}CHCH_2CH_2CH_2N(CH_3)_2$ + H_2O

The product contains both the double bond and the tertiary amine in an ring-opened structure. If a second Hofmann elimination is performed, the resulting product is hexa-1,5-diene.

18.13 Strategy: This reaction sequence is similar to the sequence used to synthesize sulfanilamide. Key steps are: (1) treatment of aniline with acetic anhydride to modulate reactivity, (2) reaction of acetanilide with chlorosulfonic acid, (3) treatment of the chlorosulfonate with the heterocyclic base, and (4) removal of the acetyl group.

Solution:

Benzene $\xrightarrow[H_2SO_4]{HNO_3}$ O_2N–C$_6$H$_5$ $\xrightarrow[EtOH]{H_2,\ Pt}$ H_2N–C$_6$H$_5$

$\xrightarrow{1.\ (CH_3CO)_2O}$ CH_3CNH–C$_6$H$_5$ (acetanilide) + CH_3CO_2H

$\xrightarrow{2.\ HOSO_2Cl}$ CH_3CNH–C$_6$H$_4$–SO_2Cl

$\xrightarrow{3.\ H_2N\text{–thiazole}}$ CH_3CNH–C$_6$H$_4$–SO_2-NH–thiazole

$\xrightarrow[H_2O]{NaOH}$ 4. H_2N–C$_6$H$_4$–SO_2-NH–thiazole

Sulfathiazole

18.14 Strategy: In all of these reactions, benzene is nitrated and the nitro group is ultimately reduced, but the timing of the reduction step is important in arriving at the correct product. In (a), nitrobenzene is immediately reduced and alkylated. In (c), chlorination occurs before reduction so that chlorine can be introduced in the *m*-position. In (b), nitrobenzene is reduced and then acetylated, to overcome amine basicity and to control reactivity. The acetyl group is removed in the last step.

Either method of nitro group reduction ($SnCl_2$, H_2) can be used in all parts of this problem; both methods are shown.

Solution:

(a)

Benzene $\xrightarrow[H_2SO_4]{HNO_3}$ C$_6$H$_5$–NO_2 $\xrightarrow[EtOH]{H_2,\ Pt}$ C$_6$H$_5$–NH_2 $\xrightarrow{2\ CH_3Br}$ C$_6$H$_5$–$N(CH_3)_2$

Mono- and trialkylated anilines are also formed.

(b)

NHCOCH$_3$ $\xrightarrow[FeCl_3]{Cl_2}$ NHCOCH$_3$ (Cl) $\xrightarrow[H_2O]{NaOH}$ NH$_2$ (Cl)

(problem 18.13)

(c)

$\xrightarrow[H_2SO_4]{HNO_3}$ NO$_2$ $\xrightarrow[FeCl_3]{Cl_2}$ NO$_2$ (Cl) $\xrightarrow[2.\ NaOH,\ H_2O]{1.\ Sn,\ H_3O^+}$ NH$_2$ (Cl)

18.15

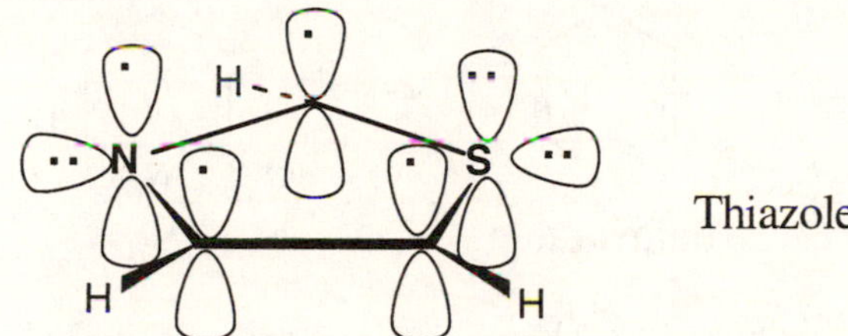

Thiazole

Thiazole contains six π electrons. Each carbon contributes one electron, nitrogen contributes one electron, and sulfur contributes two electrons to the ring π system. Both sulfur and nitrogen have lone electron pairs in sp^2 orbitals that lie in the plane of the ring.

18.16

$$\log \frac{[RNH_2]}{[RNH_3^+]} = pH - pK_a = 7.37 - 6.00 = 1.37$$

$$\frac{[RNH_2]}{[RNH_3^+]} = \text{antilog}\,(1.37) = 23.4\text{:}\;\; [RNH_2] = 23.4\,[RNH_3^+]$$

$$[RNH_3^+] + 23.4\,[RNH_3^+] = 24.4\,[RNH_3^+] = 100\%$$

$$[RNH_3^+] = 100\% \div 24.4 = 4.1\%$$

$$[RNH_2] = 100\% - 4.1 = 96\%$$

4.1% of the imidazole nitrogen of histidine is in the protonated form.

18.17

Attack at C2:

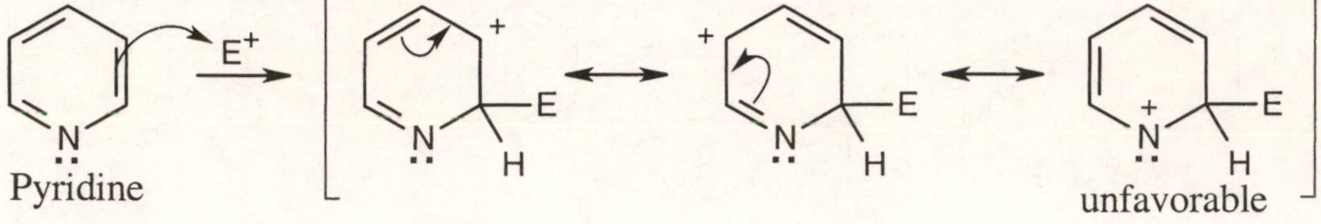

Pyridine

unfavorable

Attack at C3:

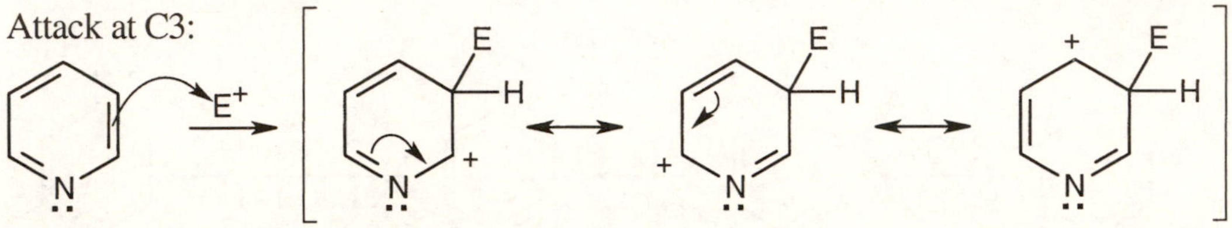

Attack at C4:

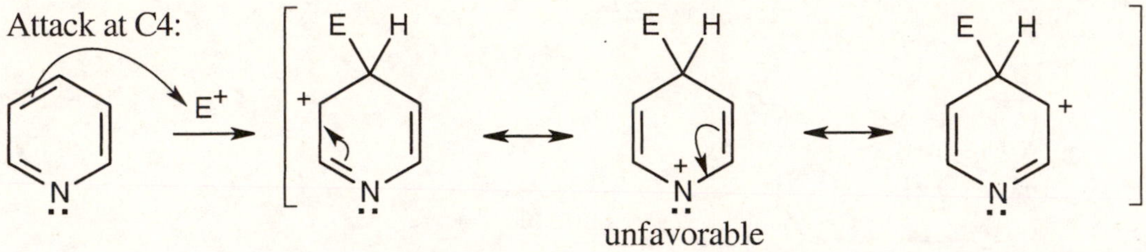

unfavorable

Reaction at C3 is favored over reaction at C2 or C4. The positive charge of the cationic intermediate of reaction at C3 is delocalized over three carbon atoms, rather than over two carbons and the electronegative pyridine nitrogen.

18.18

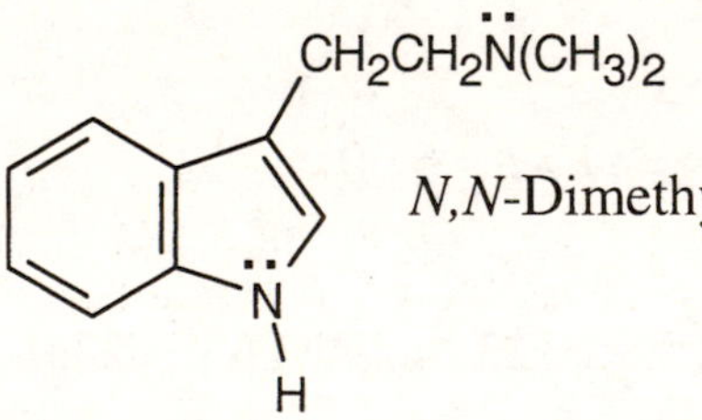

$CH_2CH_2\ddot{N}(CH_3)_2$

N,N-Dimethyltryptamine

The side chain nitrogen atom of *N,N*-dimethyltryptamine is more basic than the ring nitrogen atom because its lone electron pair is more available for donation to a Lewis acid. The aromatic nitrogen electron lone pair is part of the ring π electron system.

18.19

Attack at C2:

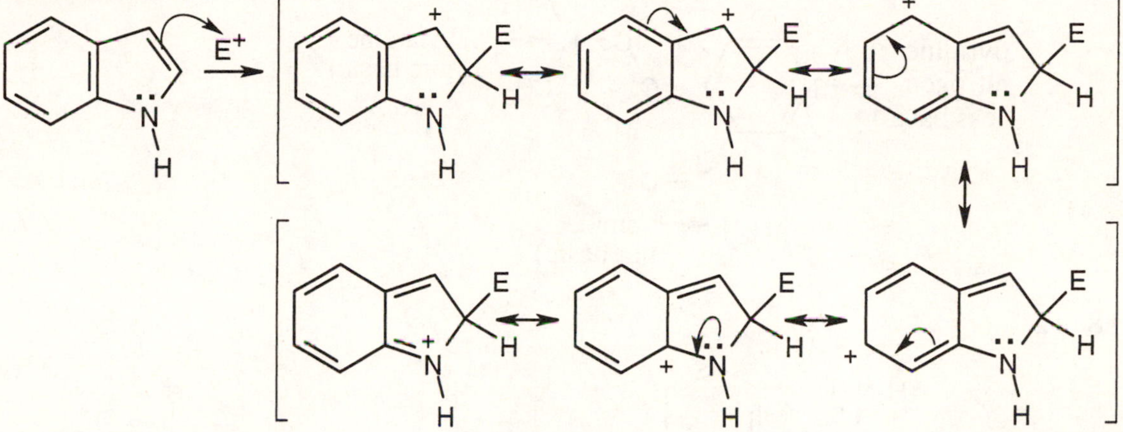

Attack at C3:

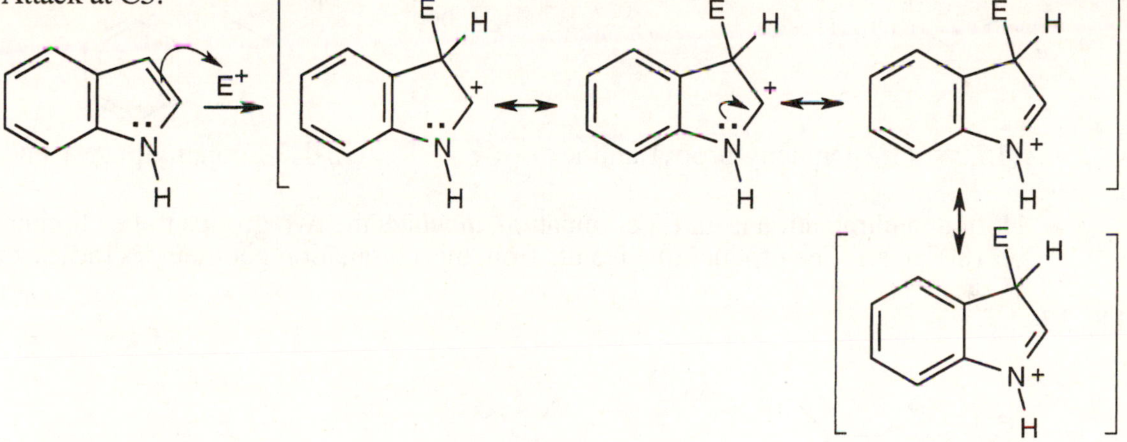

Positive charge can be stabilized by the nitrogen lone-pair electrons in reaction at both C2 and C3. In reaction at C2, however, stabilization by nitrogen destroys the aromaticity of the fused benzene ring. Reaction at C3 is therefore favored, even though the cationic intermediate has fewer resonance forms, because the aromaticity of the six-membered-ring is preserved.

Visualizing Chemistry

18.20

(a)

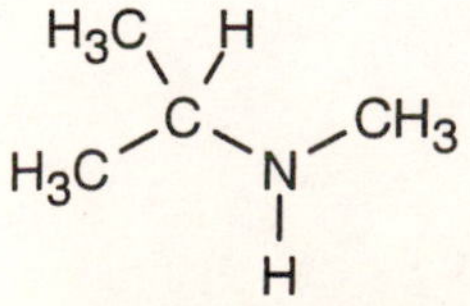

N-Methylisopropylamine
secondary amine

(b)

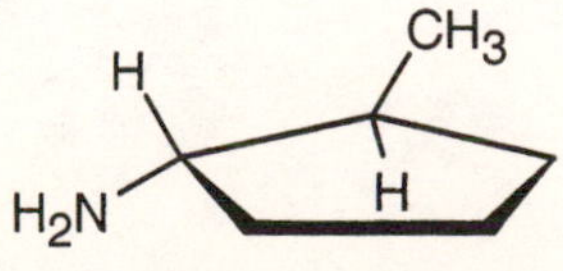

trans-(2-Methylcyclopentyl)amine
primary amine

(c)

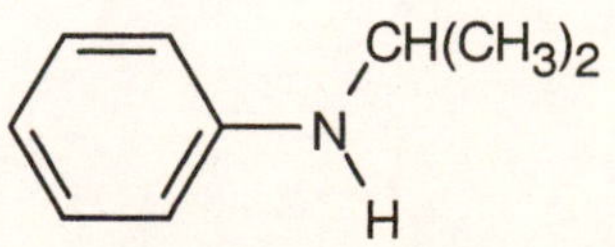

N-Isopropylaniline
secondary amine

18.21

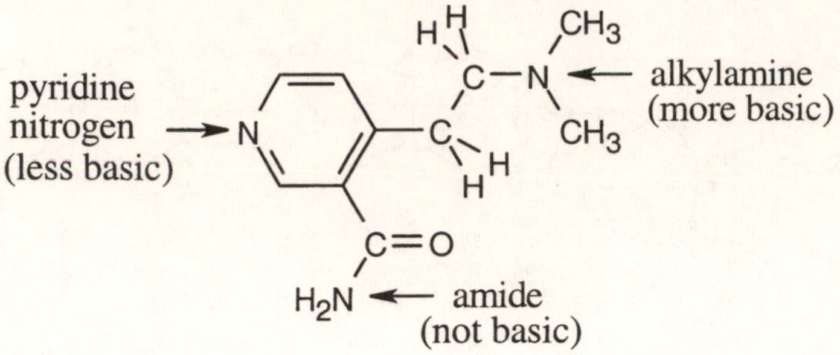

pyridine nitrogen (less basic) → N

alkylamine (more basic)

H_2N ← amide (not basic)

18.22

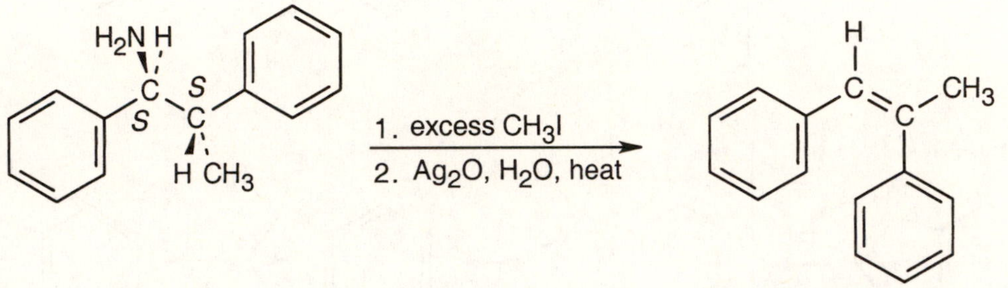

$$\text{(1}S\text{,2}S\text{)-(1,2-Diphenylpropyl)amine} \xrightarrow[\text{2. } Ag_2O,\ H_2O,\ \text{heat}]{\text{1. excess } CH_3I} (Z)\text{-1,2-Diphenylprop-1-ene}$$

(1*S*,2*S*)-(1,2-Diphenylpropyl)amine (*Z*)-1,2-Diphenylprop-1-ene

Hofmann elimination is an E2 elimination, in which the two groups to be eliminated must be 180° apart. The product that results from this elimination geometry is the *Z* isomer.

18.23

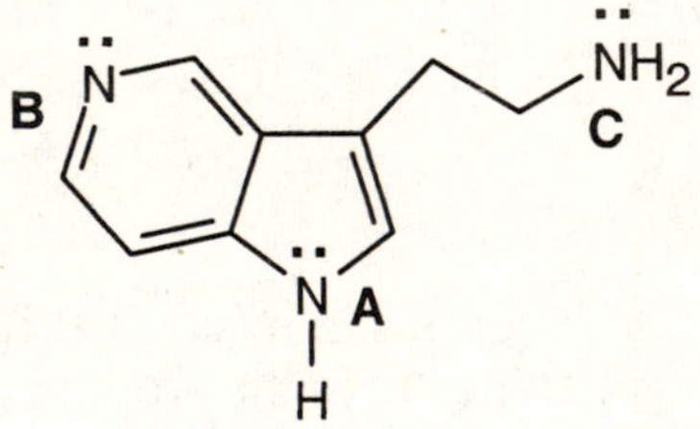

B, C, A

Nitrogen **A** is nonbasic because its electron lone-pair is part of the fused-ring aromatic system and is unavailable for bonding. Nitrogen **B** is less basic than **C** because its electron lone-pair is held closer to the nucleus and is less available for bonding than the lone pair of **C**. Nitrogen **C** is the most basic because it is an alkylamine nitrogen.

Additional Problems

18.24

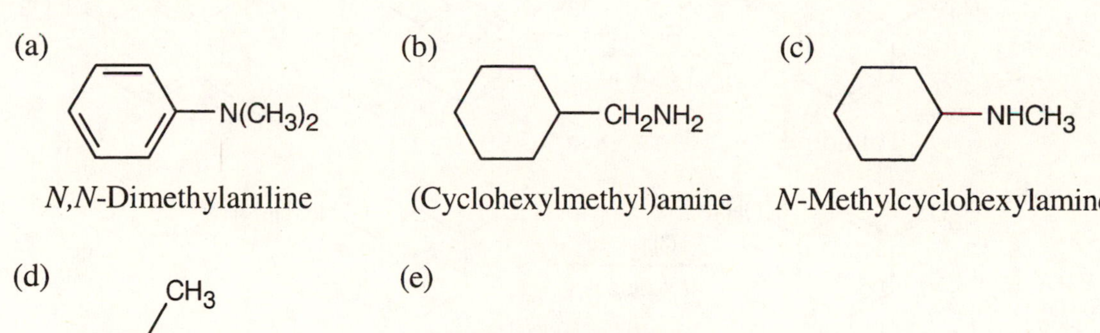

18.25

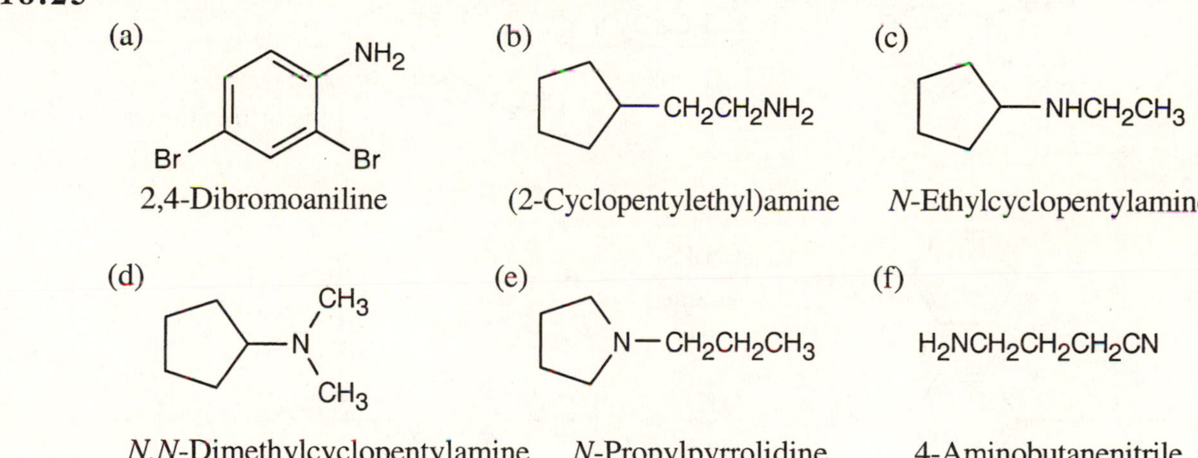

18.26

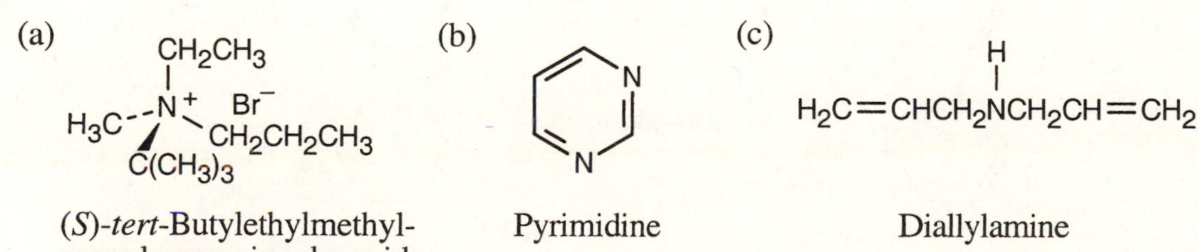

18.27

(a)

m-Toluidine $\xrightarrow[\text{1 mol}]{Br_2}$ 2-bromo-5-methylaniline + 4-bromo-3-methylaniline

m-Toluidine

(b)

H_3C–C_6H_4–NH_2 $\xrightarrow[\text{excess}]{CH_3I}$ H_3C–C_6H_4–$\overset{+}{N}(CH_3)_3I^-$

(c)

H_3C–C_6H_4–NH_2 $\xrightarrow[\text{pyridine}]{CH_3COCl}$ H_3C–C_6H_4–$NHCOCH_3$

(d)

H_3C–C_6H_4–$NHCOCH_3$ $\xrightarrow{HOSO_2Cl}$ H_3C, $NHCOCH_3$, ClO_2S substituted benzene

plus ortho isomer

18.28

(a)

Br–C_6H_4–NH_2 $\xrightarrow[\text{excess}]{CH_3I}$ Br–C_6H_4–$\overset{+}{N}(CH_3)_3I^-$

(b)

Br–C_6H_4–NH_2 $\xrightarrow{HCl}$ Br–C_6H_4–$\overset{+}{N}H_3\ Cl^-$

(c)

Br–C_6H_4–NH_2 $\xrightarrow{CH_3COCl}$ Br–C_6H_4–$NHCOCH_3$

(d)

Br–C_6H_4–NH_2 $\xrightarrow{CH_3MgBr}$ Br–C_6H_4–NH MgBr + CH_4

18.29

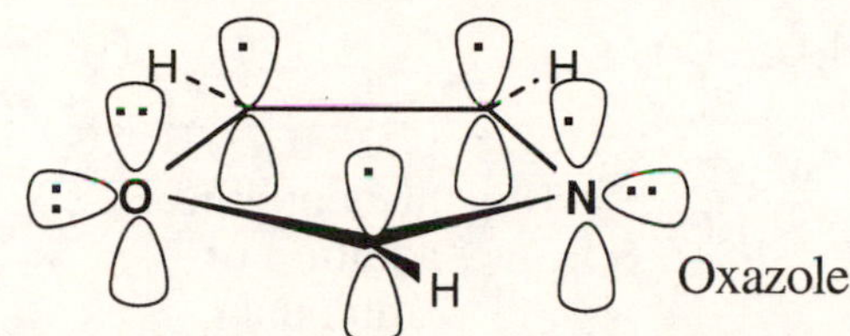

Oxazole is an aromatic 6 π electron heterocycle. Two oxygen electrons and one nitrogen electron are in *p* orbitals that are part of the π electron system of the ring, along with one electron from each carbon. An oxygen lone pair and a nitrogen lone pair are in sp^2 orbitals that lie in the plane of the ring. Since the nitrogen lone pair is available for donation to acids, oxazole is more basic than pyrrole.

18.30

(a)

$$CH_2CH_2CH_2CH_2OH \xrightarrow{PCC} CH_2CH_2CH_2CHO \xrightarrow[NaBH_4]{NH_3} \underset{\text{Butylamine}}{CH_2CH_2CH_2CH_2NH_2}$$

(b)

$$CH_2CH_2CH_2CH_2OH \xrightarrow[H_3O^+]{CrO_3} CH_2CH_2CH_2CO_2H \xrightarrow{SOCl_2} CH_2CH_2CH_2COCl$$

$$CH_2CH_2CH_2COCl \xrightarrow[\text{NaOH}]{CH_2CH_2CH_2CH_2NH_2 \text{ from (a)}} CH_3CH_2CH_2\overset{O}{\overset{\|}{C}}NHCH_2CH_2CH_2CH_3$$

$$CH_3CH_2CH_2\overset{O}{\overset{\|}{C}}NHCH_2CH_2CH_2CH_3 \xrightarrow[2.\ H_2O]{1.\ LiAlH_4} \underset{\text{Dibutylamine}}{CH_3CH_2CH_2CH_2NHCH_2CH_2CH_2CH_3}$$

(c)

$$CH_2CH_2CH_2CH_2OH \xrightarrow{PBr_3} CH_2CH_2CH_2CH_2Br \xrightarrow{NaCN} CH_2CH_2CH_2CH_2CN$$

$$CH_2CH_2CH_2CH_2CN \xrightarrow[2.\ H_2O]{1.\ LiAlH_4} \underset{\text{Pentylamine}}{CH_2CH_2CH_2CH_2CH_2NH_2}$$

(d)

$$\underset{\text{from (c)}}{CH_2CH_2CH_2NH_2} \xrightarrow[CH_3I]{\text{excess}} CH_2CH_2CH_2\overset{+}{N}(CH_3)I^- \xrightarrow[2.\ \text{heat}]{1.\ Ag_2O,\ H_2O} \underset{\text{Propene}}{CH_3CH{=}CH_2} + (CH_3)_3N$$

18.31

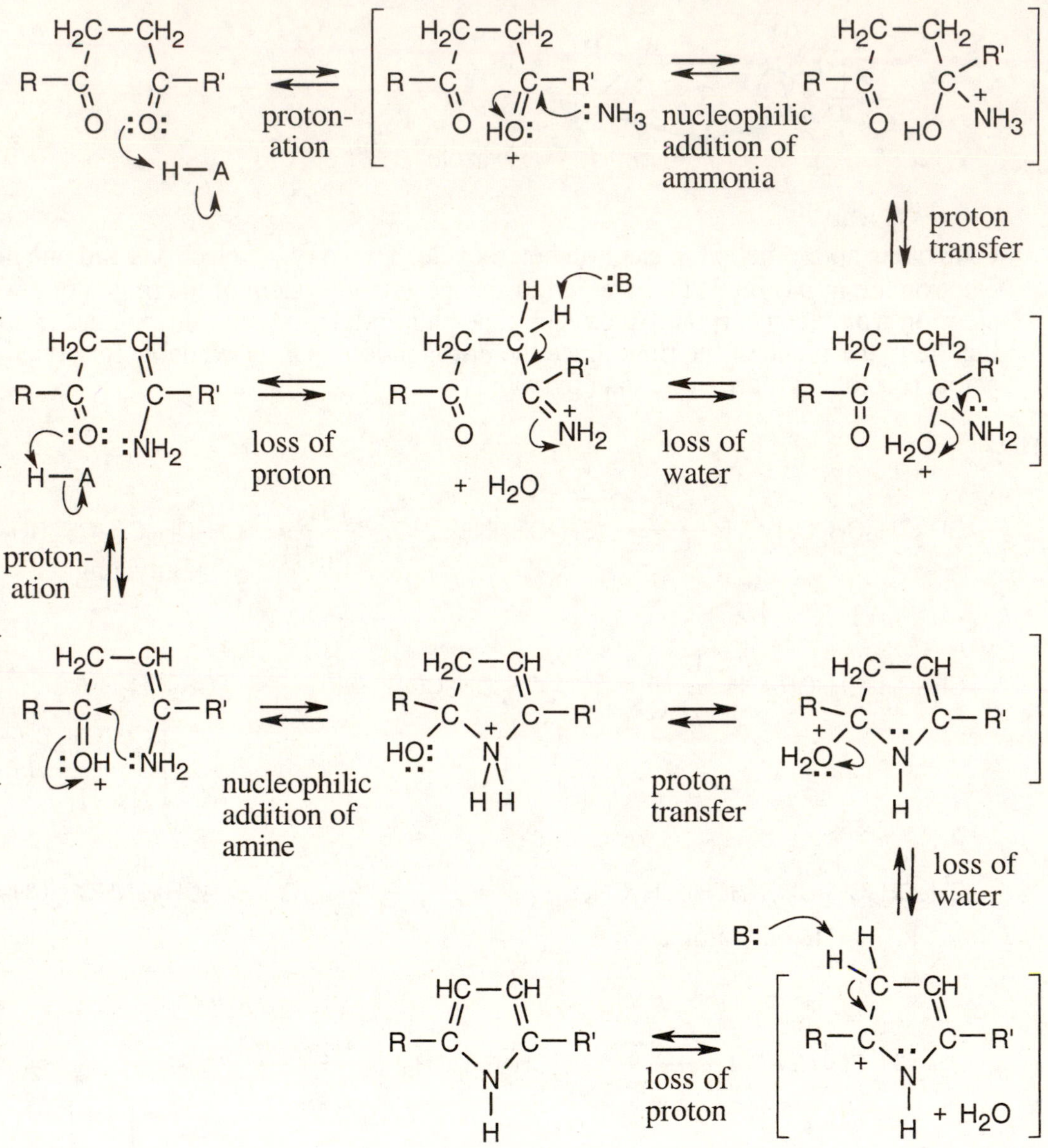

proton-
ation

nucleophilic
addition of
ammonia

proton
transfer

loss of
proton

loss of
water

proton-
ation

nucleophilic
addition of
amine

proton
transfer

loss of
water

loss of
proton

The mechanism consists of the nucleophilic addition of ammonia, first to one of the ketone groups and then to the other, with loss of two equivalents of water.

18.32

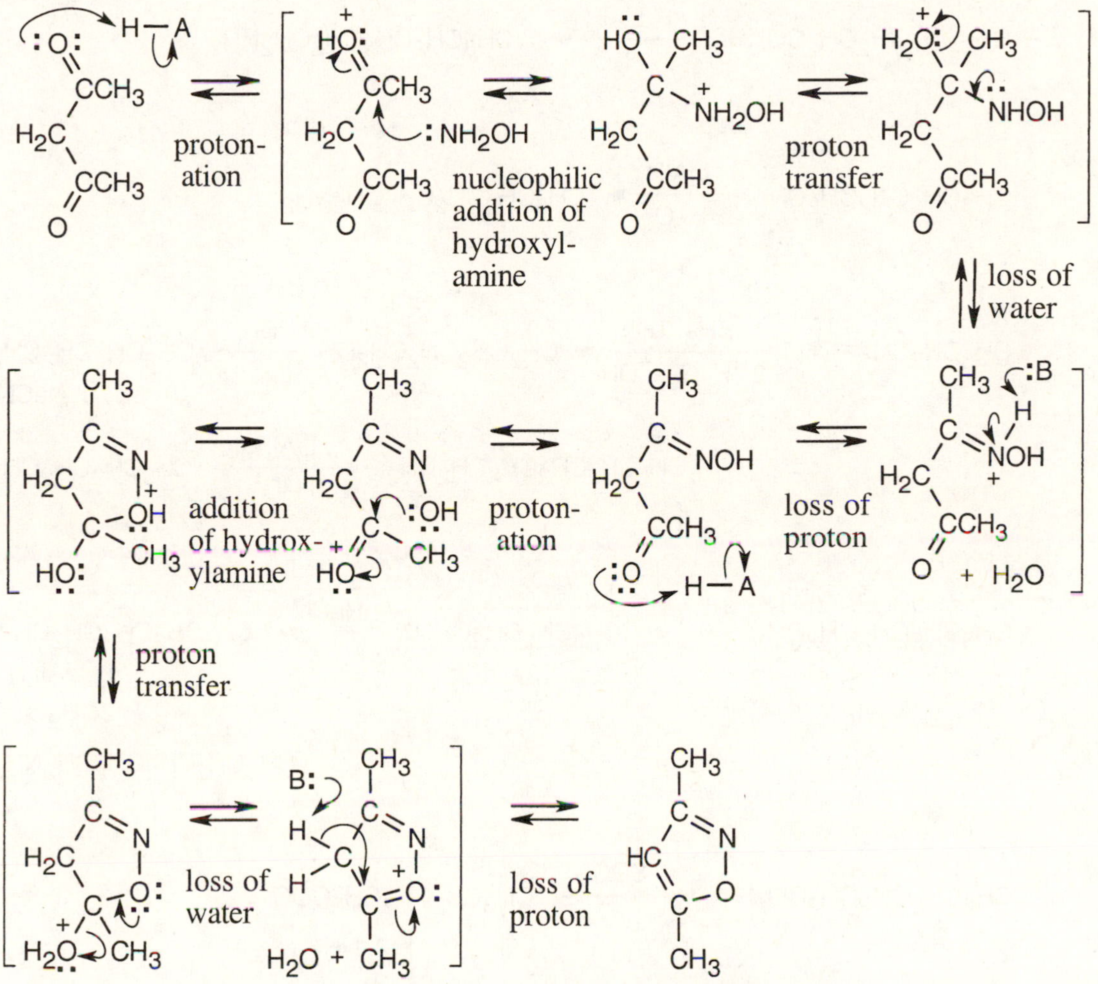

proton-ation

nucleophilic addition of hydroxylamine

proton transfer

loss of water

loss of proton

proton-ation

addition of hydroxylamine

proton transfer

loss of water

loss of proton

3,5-Dimethylisoxazole

This mechanism is virtually identical to the mechanism illustrated in the previous problem and involves two nucleophilic additions to carbonyl groups.

18.33

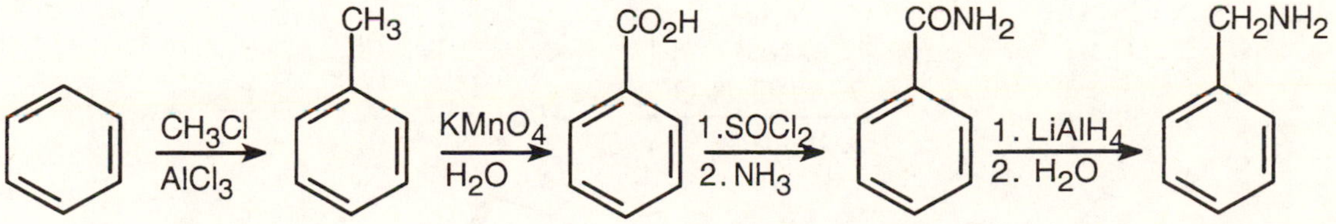

$\xrightarrow[AlCl_3]{CH_3Cl}$ $\xrightarrow[H_2O]{KMnO_4}$ $\xrightarrow[2.\ NH_3]{1.\ SOCl_2}$ $\xrightarrow[2.\ H_2O]{1.\ LiAlH_4}$

CH_3 CO_2H $CONH_2$ CH_2NH_2

18.34

(a)

$$CH_3CH_2CH_2CH_2CONH_2 \xrightarrow[2.\ H_2O]{1.\ LiAlH_4} CH_3CH_2CH_2CH_2CH_2NH_2$$

(b)

$$CH_3CH_2CH_2CH_2CN \xrightarrow[2.\ H_2O]{1.\ LiAlH_4} CH_3CH_2CH_2CH_2CH_2NH_2$$

(c)

$$CH_3CH_2CH{=}CH_2 \xrightarrow[2.\ H_2O_2,\ ^{-}OH]{1.\ BH_3,\ THF} CH_3CH_2CH_2CH_2OH \xrightarrow{PBr_3} CH_3CH_2CH_2CH_2Br$$

$$CH_3CH_2CH_2CH_2Br \xrightarrow{NaCN} CH_3CH_2CH_2CH_2CN$$

$$CH_3CH_2CH_2CH_2CH_2NH_2 \xleftarrow[2.\ H_2O]{1.\ LiAlH_4} CH_3CH_2CH_2CH_2CN$$

(d)

$$CH_3CH_2CH_2CH_2OH \xrightarrow{PBr_3} CH_3CH_2CH_2CH_2Br \xrightarrow{NaCN} CH_3CH_2CH_2CH_2CN$$

$$CH_3CH_2CH_2CH_2CN \xrightarrow[2.\ H_2O]{1.\ LiAlH_4} CH_3CH_2CH_2CH_2CH_2NH_2$$

(e)

$$CH_3CH_2CH_2CH_2CO_2H \xrightarrow{SOCl_2} CH_3CH_2CH_2CH_2COCl$$

$$CH_3CH_2CH_2CH_2COCl \xrightarrow{2\ NH_3} CH_3CH_2CH_2CH_2CONH_2$$

$$CH_3CH_2CH_2CH_2CH_2NH_2 \xleftarrow[2.\ H_2O]{1.\ LiAlH_4} CH_3CH_2CH_2CH_2CONH_2$$

18.35 Hydrogens that can be eliminated are starred. In cases where more than one alkene can form, the alkene with the less substituted double bond is the major product..

Amine $\xrightarrow[3.\ heat]{1.\ excess\ CH_3I\quad 2.\ Ag_2O,\ H_2O}$ *Alkene* + *Amine*

(a) (cyclopentyl)$NHCH_3$ (starred H on both ring carbons adjacent to C–N) → cyclopentene + $N(CH_3)_3$

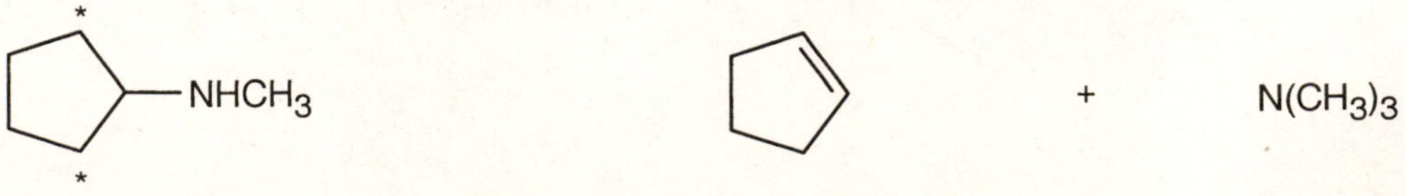

(b)

$$\underset{\text{(}N\text{-phenyl)}}{\mathrm{HN\overset{\overset{*}{CH_3}}{|}CH\underset{*}{}CH_2CH_2CH_2CH_3}}$$

$H_2C{=}CHCH_2CH_2CH_2CH_3$ major

+ $N(CH_3)_2$ (on benzene ring)

$CH_3CH{=}CHCH_2CH_2CH_3$ minor

(c)

$$\mathrm{CH_3\overset{\overset{CH_3}{|}}{C}H\underset{*}{}\underset{\underset{NH_2}{|}}{C}H\underset{*}{}CH_2CH_2CH_3}$$

$$\mathrm{CH_3\overset{\overset{CH_3}{|}}{C}HCH{=}CHCH_2CH_3}$$ major + $N(CH_3)_3$

$$\mathrm{CH_3\overset{\overset{CH_3}{|}}{C}{=}CHCH_2CH_2CH_3}$$ minor

18.36

$$\mathrm{C_6H_5\overset{\overset{O}{||}}{C}CH_3} \xrightarrow{a} \mathrm{C_6H_5\overset{\overset{NH_2}{|}}{C}HCH_3} \xrightarrow{b,\ c} \mathrm{C_6H_5CH{=}CH_2}$$

$$\xrightarrow{d} \mathrm{C_6H_5\overset{O}{CH{-}CH_2}} \xrightarrow{e} \mathrm{C_6H_5\overset{\overset{OH}{|}}{C}HCH_2\underset{\underset{CH_3}{|}}{N}CH_3}$$

(a) NH_3, H_2/Ni; (b) excess CH_3I; (c) Ag_2O, H_2O, heat; (d) RCO_3H (e) $(CH_3)_2NH$
Step (e) is an S_N2 ring opening of the epoxide by nucleophilic substitution of the amine at the primary carbon.

18.37

$$\mathrm{R{-}\overset{\overset{:O:}{||}}{C}{-}\overset{+}{N}H_3}$$

N-Protonation
(no resonance stabilization)

$$\mathrm{R{-}\overset{\overset{:\overset{+}{O}{-}H}{||}}{C}{-}\ddot{N}H_2} \longleftrightarrow \mathrm{R{-}\overset{\overset{:\ddot{O}{-}H}{|}}{C}{=}\overset{+}{N}H_2}$$

O-Protonation
(resonance stabilization)

Protonation occurs on oxygen because an *O*-protonated amide is stabilized by resonance.

18.38

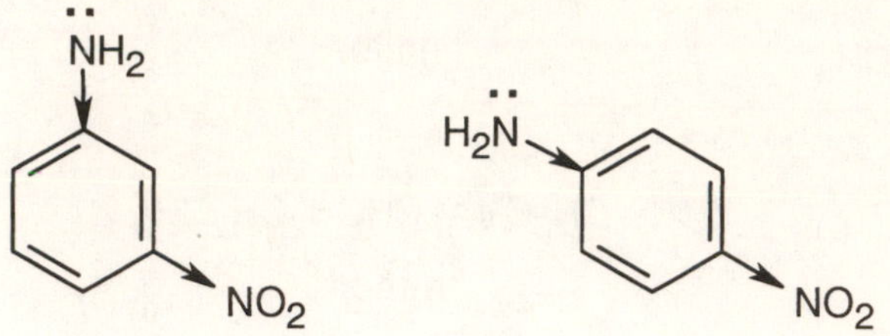

The inductive effect of the electron-withdrawing nitro group makes the amine nitrogens of both *m*-nitroaniline and *p*-nitroaniline less electron-rich and less basic than aniline.

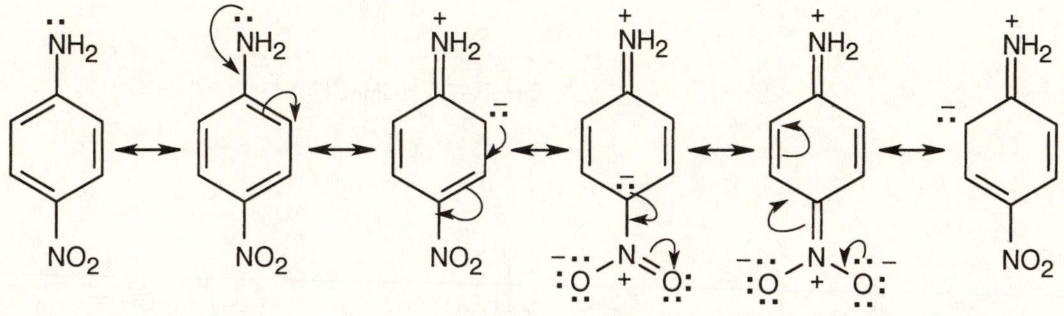

When the nitro group is para to the amino group, conjugation of the amino group with the nitro group can also occur. *p*-Nitroaniline is thus even less basic than *m*–nitroaniline.

18.39

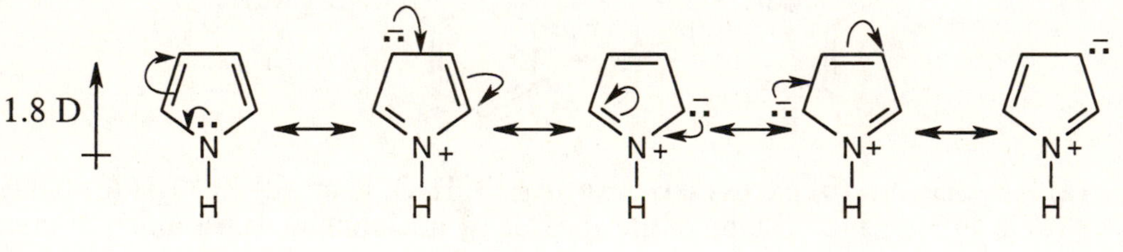

1.8 D

3
The dipole moment of pyrrole points in the direction indicated because resonance structures show that the nitrogen atom is electron-poor.

18.40

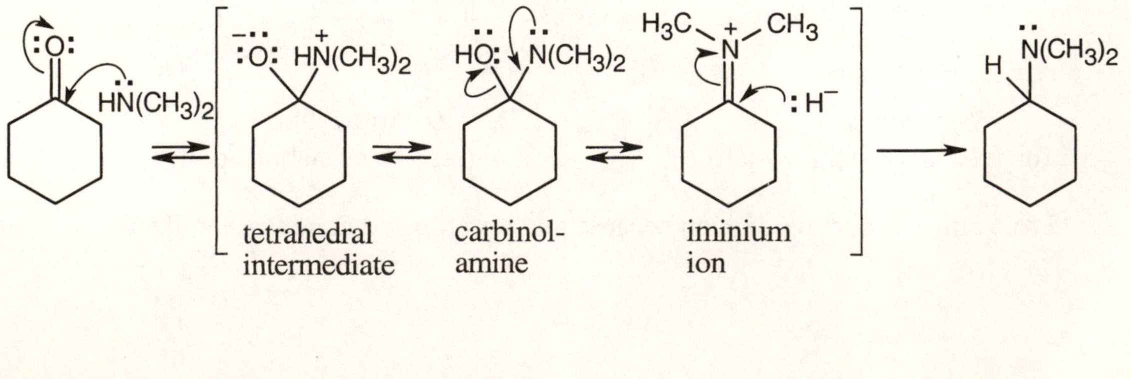

tetrahedral intermediate

carbinol-amine

iminium ion

18.41

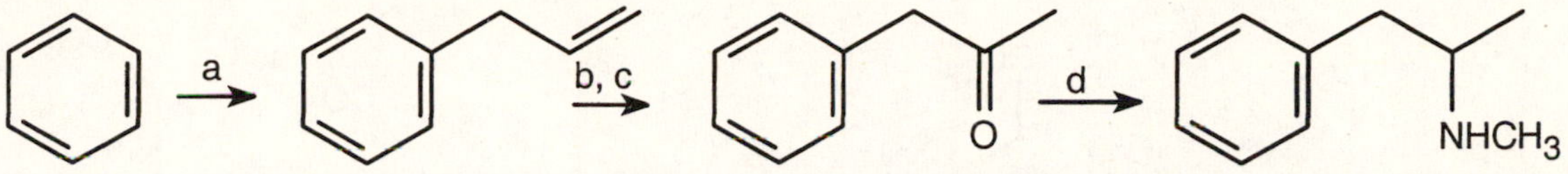

(a) CH_2=$CHCH_2Cl$, $AlCl_3$; (b) $Hg(OAc)_2$, H_2O; Na BH_4; (c) CrO_3, H_3O^+ (d) CH_3NH_2, $NaBH_4$

18.42 Benzaldehyde first reacts with methylamine and $NaBH_4$ in the usual way to give the reductive amination product *N*-methylbenzylamine. This product then reacts further with benzaldehyde in a second reductive amination to give *N*-methyldibenzylamine.

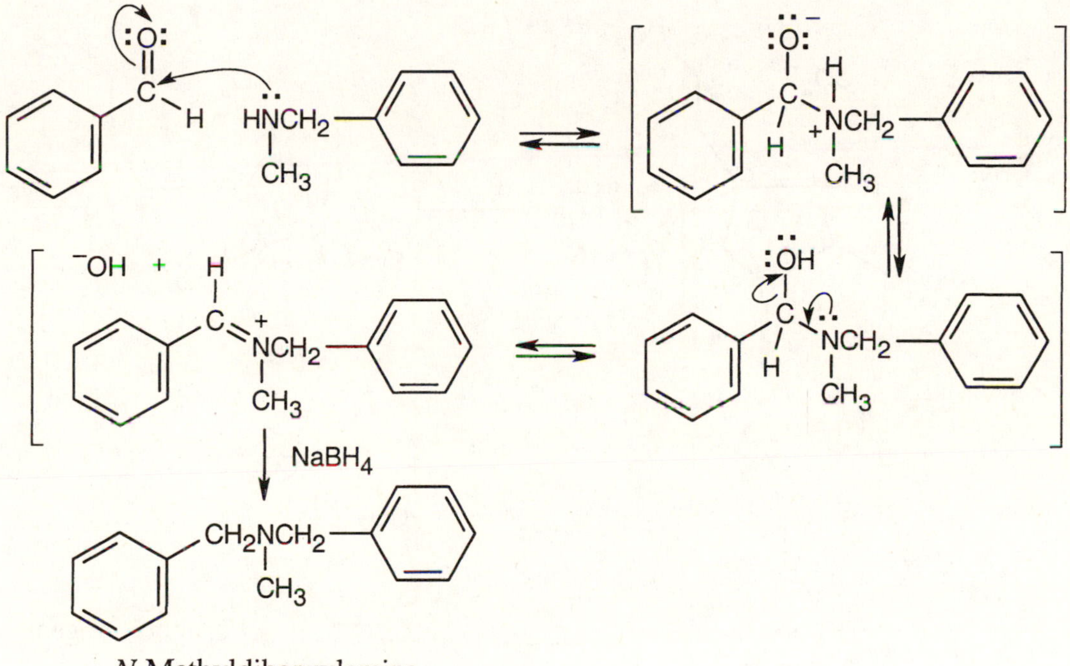

N-Methyldibenzylamine

18.43 The reaction of trimethylamine with ethylene oxide is an S_N2 reaction that opens the epoxide ring.

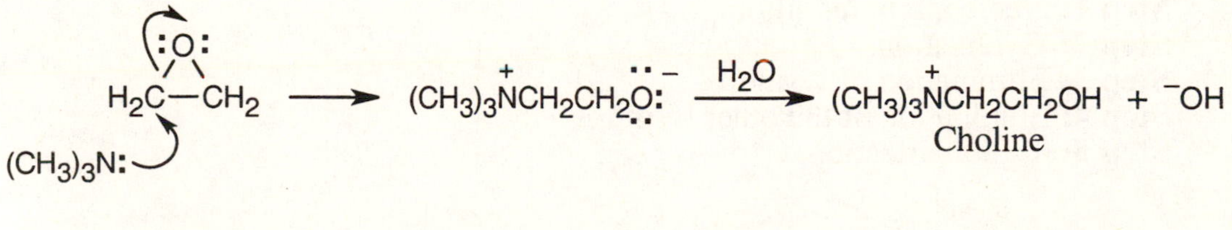

$$(CH_3)_3N: + H_2C\text{—}CH_2 \text{ (epoxide)} \longrightarrow (CH_3)_3\overset{+}{N}CH_2CH_2\ddot{O}:^- \xrightarrow{H_2O} (CH_3)_3\overset{+}{N}CH_2CH_2OH + {}^-OH$$

Choline

18.44

CO_2^- CO_2^- Lys Lys NH_2 H NH_2 H 1. CO_2^- CO_2^- H Lys Lys NH_2 H_2N H 2.

CO_2^- CO_2^- B: H Lys H H NH_2 H 3. CO_2^- CO_2^- H H Lys Lys H_2N H H

4. CO_2^- CO_2^- :B H H NH_2 H 5. CO_2^- CO_2^- $^+NH_3$ H Porphobilinogen

Step 1: Nucleophilic addition.
Step 2: Cyclization.
Step 3: Elimination of lysine.
Step 4: Elimination of the other lysine.
Step 5: Tautomerization.

18.45

$$\text{C}_5\text{H}_9\text{Br} \xrightarrow{\text{Mg}} \text{C}_5\text{H}_9\text{MgBr} \xrightarrow[\text{2. } H_3O^+]{\text{1. } H_2C\overset{O}{-}CHCH_3} \text{C}_5\text{H}_9\text{-}CH_2CH(OH)CH_3 \xrightarrow{\text{PCC}} \text{C}_5\text{H}_9\text{-}CH_2C(=O)CH_3 \xrightarrow[\text{NaBH}_4]{H_2NCH_3} \text{C}_5\text{H}_9\text{-}CH_2CH(CH_3)NHCH_3$$

Cyclopentamine

The last step of the synthesis is a reductive amination of a ketone that is formed by oxidation of the corresponding alcohol. The alcohol results from the Grignard reaction between cyclopentylmagnesium bromide and propylene oxide.

18.46

$$O_2N\text{-}C_6H_4\text{-}COOH \xrightarrow[\text{2. } HOCH_2CH_2N(CH_3)_2,\ \text{Pyridine}]{\text{1. } SOCl_2} O_2N\text{-}C_6H_4\text{-}COOCH_2CH_2N(CH_3)_2 \xrightarrow[\text{EtOH}]{H_2,\ \text{Pt}} H_2N\text{-}C_6H_4\text{-}COOCH_2CH_2N(CH_3)_2 \xrightarrow[\text{NaBH}_4]{CH_3CH_2CH_2CHO} CH_3(CH_2)_3NH\text{-}C_6H_4\text{-}COOCH_2CH_2N(CH_3)_2$$

Tetracaine

18.47

$$\text{Atropine (tropine ester, } OC(=O)CH(C_6H_5)CH_2OH) \xrightarrow[H_2O]{^-OH} \text{Tropine (OH)} + HO_2CCH(C_6H_5)CH_2OH \xrightarrow{H_2SO_4} \text{Tropidene} + H_2O$$

Atropine Tropine Tropidene

Tropic acid

We know the location of the –OH group of tropine because it is stated that tropine is an optically inactive alcohol. This hydroxyl group results from basic hydrolysis of the ester that is composed of tropine and tropic acid.

18.48 The formula $C_9H_{17}N$ indicates two degrees of unsaturation in the product. Both are probably due to rings since the product results from catalytic reduction.

H_2, Pt

CN

O

reduction of nitrile

$:NH_2$

nucleophilic addition

N

OH

H

dehydration

H_2, Pt

N

H

reduction of double bond

N

+ H_2O

18.49 The molecular formula indicates that coniine has one double bond or ring, and the Hofmann elimination product shows that the nitrogen atom is part of a ring.

1. excess CH_3I
2. Ag_2O, H_2O
3. heat

N

H

$(CH_3)_2N$

Coniine

5-(*N*,*N*-Dimethylamino)oct-1-ene

18.50

CO_2Et

CH_2

$CH_3CH_2CH_2C$

O

+

CH_2

CHCN

1. Na^+ ^-OEt, EtOH
2. H_3O^+

Michael addition

H CO_2Et

C

$CH_3CH_2CH_2C$

CH_2

O

CH_2CN

ester hydrolysis, decarboxylation

H_3O^+, heat

H H

C

$CH_3CH_2CH_2C$

CH_2

O

CH_2CN

+ CO_2 + EtOH

H_2/Pt

reduction

H H

C

$CH_3CH_2CH_2C$

CH_2

O

CH_2

H_2N-C

H

H

cyclization

N

H_2, Pt

N

H

Coniine

The mechanism of the last steps is shown in Problem 18.48.

18.51

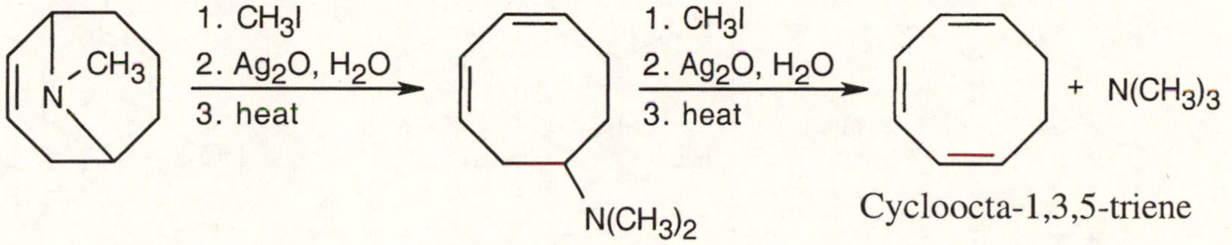

1. CH_3I
2. Ag_2O, H_2O
3. heat

$N(CH_3)_2$

1. CH_3I
2. Ag_2O, H_2O
3. heat

+ $N(CH_3)_3$

Cycloocta-1,3,5-triene

18.52

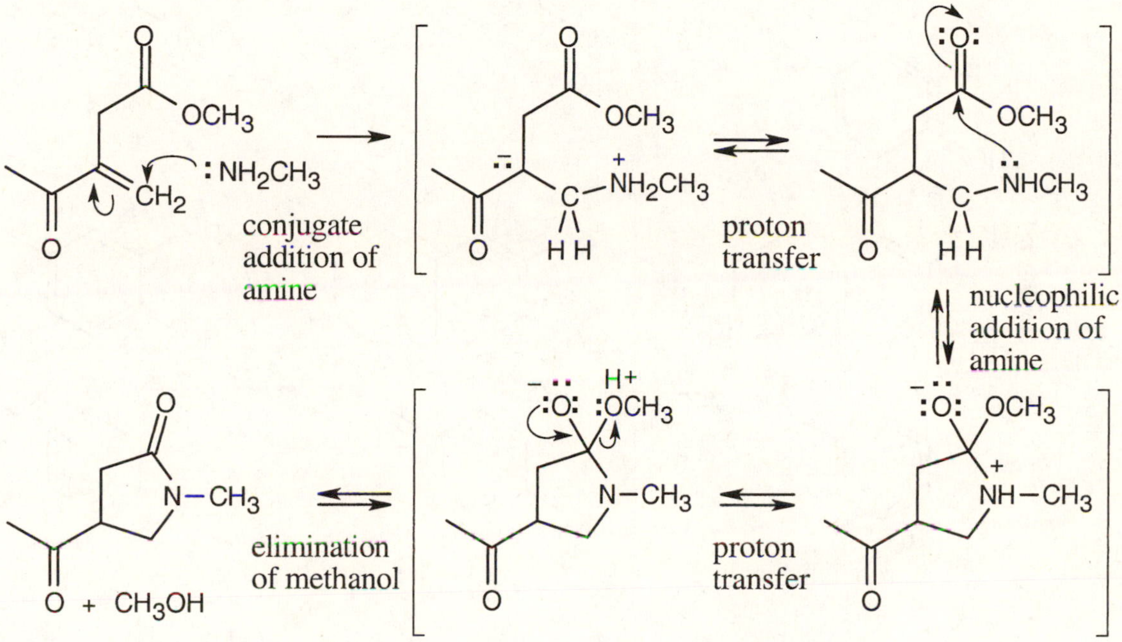

OCH_3 : NH_2CH_3 CH_2

conjugate addition of amine

OCH_3 NH_2CH_3 H H

proton transfer

OCH_3 $NHCH_3$ H H

nucleophilic addition of amine

OCH_3 $NH-CH_3$

proton transfer

H+ OCH_3 $N-CH_3$

elimination of methanol

$N-CH_3$

O + CH_3OH

18.53

Br H_2C CO_2CH_3 NH OH

displacement of Br^- by amine

CO_2CH_3 H N+ OH + Br^-

Et_3N deprotonation

CO_2CH_3 N OH + $Et_3\overset{+}{N}H$

conjugate addition of alcohol

N CO_2CH_3 +OH

N CO_2CH_3 O

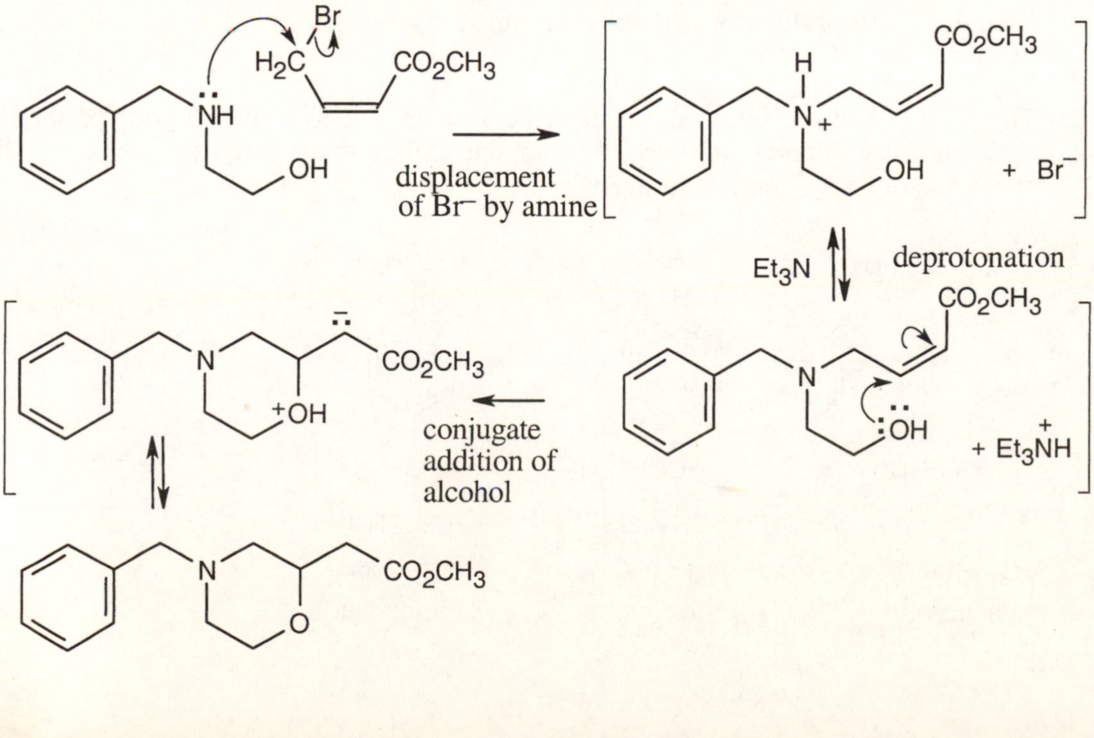

18.54

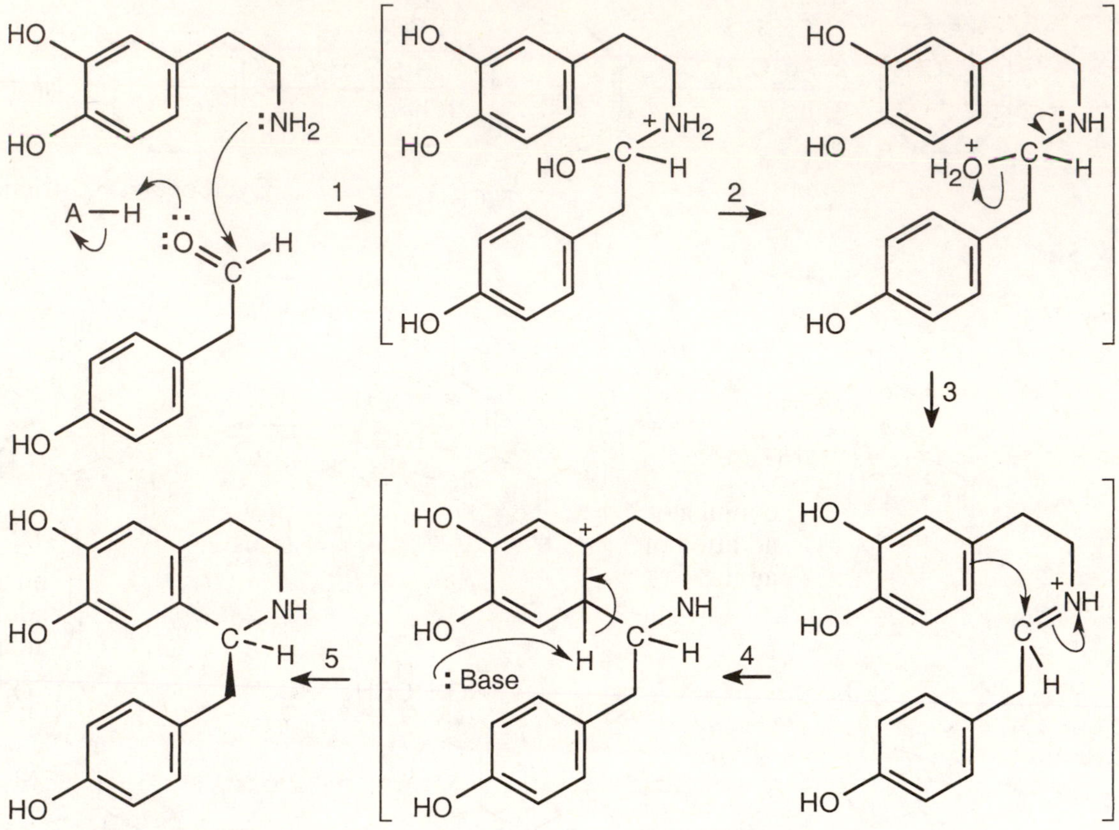

(*S*)-Norcoclaurine

Step 1: Nucleophilic addition of the amine to the protonated aldehyde.
Step 2: Proton transfer.
Step 3: Loss of water.
Step 4: Electrophilic aromatic substitution.
Step 5: Loss of proton.

18.55 The ^{1}H NMR of the amine shows 5 peaks. Two are due to an ethyl group bonded to an electronegative element (oxygen), two are due to 4 aromatic ring hydrogens, and the peak at 3.4 δ is due to 2 amine hydrogens.

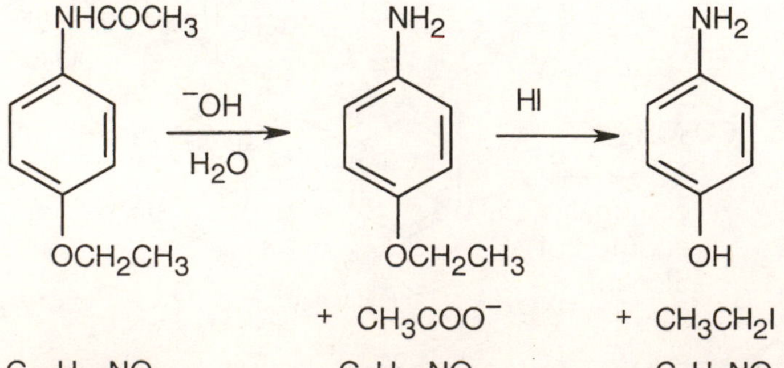

$C_{10}H_{13}NO_2$ Phenacetin $\xrightarrow[H_2O]{^-OH}$ $C_8H_{11}NO$ *p*-Ethoxyaniline + CH_3COO^- $\xrightarrow{HI}$ C_6H_7NO *p*-Aminophenol + CH_3CH_2I

18.56

(a)

$HOCH_2CH_2CH_2NH_2$ (e d a c b)

a = 1.7 δ
b = 2.7 δ
c = 2.9 δ
d = 3.7 δ

(b)

$(CH_3O)_2CHCH_2NH_2$ (c d b a)

a = 1.3 δ
b = 2.8 δ
c = 3.4 δ
d = 4.3 δ

18.57

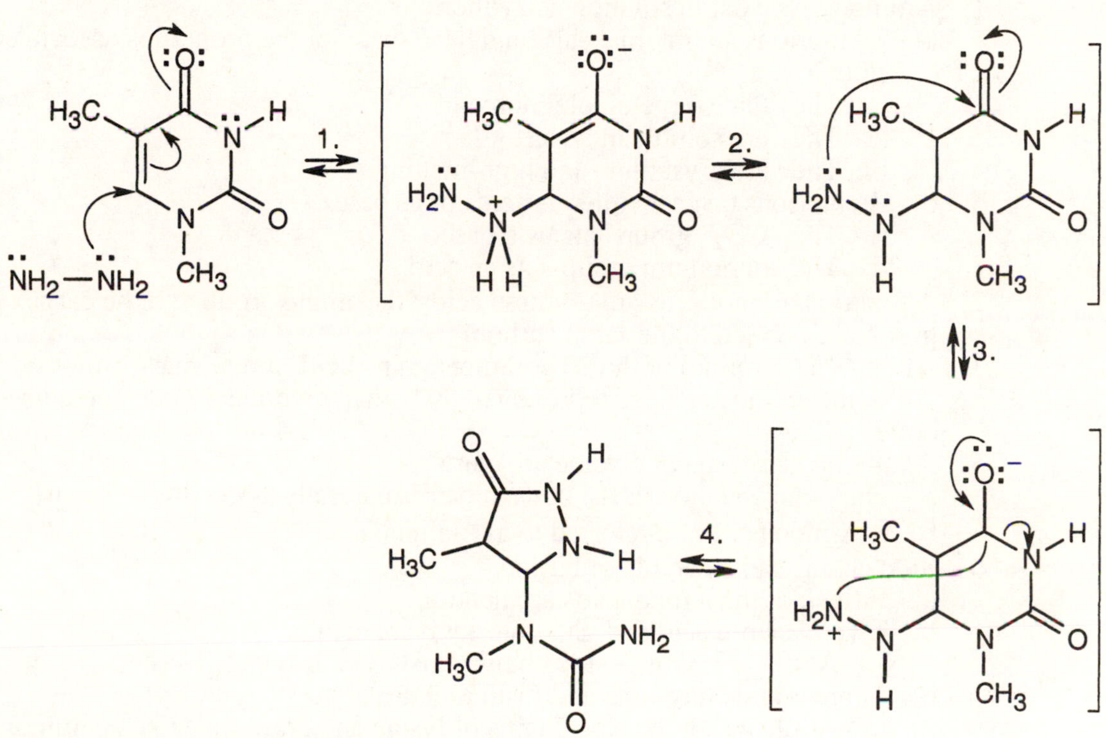

Step 1: Conjugate addition of hydrazine.
Step 2: Proton shift.
Step 3: Nucleophilic acyl substitution.
Step 4: Ring opening, proton shift.

18.58

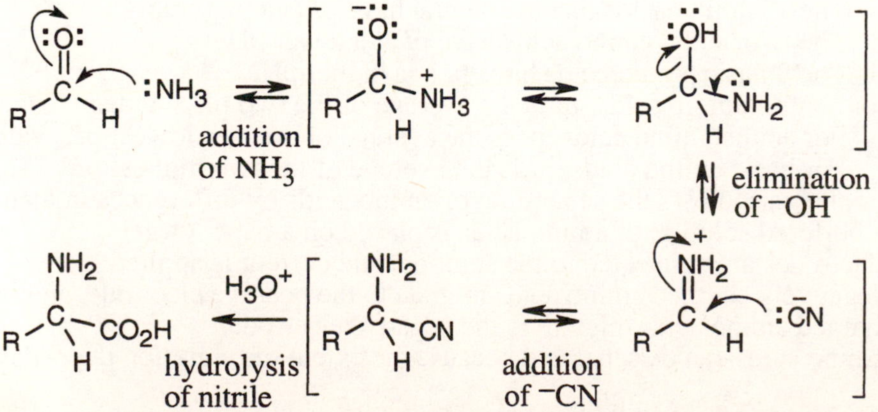

addition of NH_3

elimination of ^{-}OH

addition of ^{-}CN

H_3O^+ hydrolysis of nitrile

The mechanism of acid-catalyzed nitrile hydrolysis is shown in Problem 15.43.

Chapter 19 – Biomolecules: Amino Acids, Peptides, and Proteins

Chapter Outline

I. Amino acids (Sections 19.1 – 19.3).
   A. Structure of amino acids (Section 19.1).
      1. Amino acids exist in solution as zwitterions.
         a. Zwitterions are internal salts and have many of the properties associated with salts.
            i. They have large dipole moments.
            ii. They are soluble in water.
            iii. They are crystalline and high-melting.
         b. Zwitterions can act either as acids or as bases.
            i. The $-CO_2^-$ group acts as a base.
            ii. The ammonium group acts an acid.
      2. All natural amino acids are α-amino acids: the amino group and the carboxylic acid group are bonded to the same carbon.
      3. All but one (proline) of the 20 common amino acids are primary amines.
      4. All of the amino acids are represented by both a three-letter code and a one-letter code.
      5. All amino acids except glycine are chiral.
         a. Only one enantiomer (L) of each pair is naturally-occurring.
         b. α-Amino acids are referred to as L-amino acids.
      6. Side chains can be acidic or basic.
         a. Fifteen of the amino acids are neutral.
         b. Two (aspartic acid and glutamic acid) are acidic.
            At pH = 7.3, their side chains exist as carboxylate ions.
         c. Three (lysine, arginine and histidine) are basic.
            i. At pH = 7.3, the side chains of lysine and arginine exist as ammonium ions.
            ii. Histidine is not quite basic enough to be protonated at pH = 7.3.
            iii. The double-bonded nitrogen in the histidine ring is basic.
         d. Cysteine and tyrosine are weakly acidic.
   B. Isoelectric points (Section 19.2).
      1. The isoelectric point (p*I*) is the pH at which an amino acid exists as a neutral, dipolar zwitterion.
         a. p*I* is related to side chain structure.
            i. The 15 amino acids that are neutral have p*I* near neutrality.
            ii. The two acidic amino acids have p*I* at a lower pH.
            iii. The 3 basic amino acids have p*I* at a higher pH.
         b. For neutral amino acids, p*I* is the average of the two pK_a values.
            i. For acidic amino acids, p*I* is the average of the two lowest pK_a values.
            ii. For basic amino acids, p*I* is the average of the two highest pK_a values.
      2. Electrophoresis allows the separation of amino acids by differences in their p*I*.
         a. A buffered solution of amino acids is placed on a paper or gel.
         b. Electrodes are connected to the solution, and current is applied.
         c. Negatively charged amino acids migrate to the positive electrode, and positively charged amino acids migrate to the negative electrode.
         d. Amino acids can be separated because the extent of migration depends on p*I*.

C. Synthesis of α-amino acids (Section 19.3).
1. The amidomalonate synthesis.
a. An alkyl halide reacts with the anion of diethyl amidomalonate.
b. Hydrolysis of the adduct yields the α-amino acid.
2. Reductive amination.
a. Reductive amination of an α-keto carboxylic acid gives an α-amino acid.
b. This method is related to the biosynthetic pathway for synthesis of amino acids.
3. Both of the methods listed above produce a racemic mixture of amino acids.
4. Enantioselective synthesis of amino acids.
a. Enantioselective hydrogenation of *Z*-enamido acids produces chiral α-amino acids.
b. The most effective catalysts are complexes of rhodium (I), cyclooctadiene and a chiral diphosphine.

II. Peptides (Sections 19.4 – 19.7).
A. Peptide structure (Section 19.4).
1. Peptide bonds.
a. A peptide is an amino acid polymer in which the amine group of one amino acid forms an amide bond with the carboxylic acid group of a second amino acid.
b. This amino acid sequence is known as the backbone of the peptide or protein.
c. Rotation about the amide bond is restricted.
2. The N-terminal amino acid of the polypeptide is always drawn on the left.
3. The C-terminal amino acid of the polypeptide is always drawn on the right.
4. Peptide structure is described by using three-letter codes, or one-letter codes, for the individual amino acids, starting with the N-terminal amino acid.
5. Disulfide bonds.
a. Two cysteines can form a disulfide bond (–S–S–).
b. Disulfide bonds can link two polypeptides or introduce a loop in a polypeptide chain.
B. Structure determination of peptides (Sections 19.5 – 19.6).
1. Amino acid analysis (Section 19.5).
a. Amino acid analysis provides the amount of each amino acid present in a protein or peptide.
b. First, all disulfide bonds are broken and all peptide bonds are hydrolyzed.
c. The mixture is placed on a chromatography column, and the residues are eluted.
d. As each amino acid elutes, it undergoes reaction with ninhydrin, which produces a purple color that is detected and measured spectrophotometrically.
e. Amino acid analysis is reproducible on properly maintained equipment; residues always elute at the same time, and only small sample sizes are needed.
C. Peptide sequencing (Section 19.6).
1. Much peptide sequencing is done by mass spectrometry.
2. The Edman degradation.
a. The Edman degradation removes one amino acid at a time from the $-NH_2$ end of a peptide.
i. The peptide is treated with phenylisothiocyanate, which reacts with the amino-terminal residue.
ii. The PITC derivative is split from the peptide.
iii. The residue undergoes rearrangement to a PTH, which is identified chromatographically.
iv. The shortened chain undergoes another round of Edman degradation.

b. Since the Edman degradation can only be used on peptides containing fewer than 50 amino acids, a protein must be cleaved into smaller fragments.
   i. Partial acid hydrolysis is unselective.
   ii. The enzyme trypsin cleaves proteins at the carboxyl side of arg and lys residues.
   iii. The enzyme chymotrypsin cleaves proteins at the carboxyl side of Phe, Tyr and Trp residues.
c. The complete amino acid sequence of a protein results from determining the individual sequences of peptides and overlapping them.

C. Synthesis of peptides (Section 19.7).
  1. Laboratory synthesis of peptides.
    a. Groups that are not involved in peptide bond formation are protected.
      i. Carboxyl groups are often protected as methyl or benzyl esters.
      ii. Amino groups are protected as Boc derivatives.
    b. The peptide bond is formed by coupling with DCC.
    c. The protecting groups are removed.
      i. Boc groups are removed by brief treatment with trifluoroacetic acid.
      ii. Esters are removed by mild hydrolysis or by hydrogenolysis (benzyl).
  2. Automated peptide synthesis - Merrifield technique.
    a. The carboxyl group of a Boc-protected amino acid is attached to a polystyrene resin.
    b. The resin is washed, and the Boc group is removed.
    c. A second Boc-protected amino acid is coupled to the first, and the resin is washed.
    d. The cycle is repeated as many times as needed.
    e. Finally, treatment with anhydrous HF removes the final Boc group and frees the polypeptide.

III. Proteins (Sections 19.8 – 19.10).
  A. Classification of proteins (Section 19.8).
    1. Proteins can be classified by composition.
      a. Simple proteins yield only amino acids on hydrolysis.
      b. Conjugated proteins yield other non-protein components on hydrolysis.
    2. Proteins can be classified by shape.
      a. Fibrous proteins consist of long, filamentous polypeptide chains.
      b. Globular proteins are compact and roughly spherical.
  B. Protein structure.
    1. Levels of protein structure.
      a. Primary structure refers to the amino acid sequence of a protein.
      b. Secondary structure refers to the organization of segments of the peptide backbone into a regular pattern, such as a helix or sheet.
      c. Tertiary structure describes the overall three-dimensional shape of a protein.
      d. Quaternary structure describes how polypeptide subunits aggregate into a larger structure.
    2. Examples of structural features.
      a. α-Helix
        i. An α-helix is a right-handed coil; each turn of the coil contains 3.6 amino acids.
        ii. The structure is stabilized by hydrogen bonds between amide N–H groups and C=O groups four residues away.

b. β-Pleated sheet.
   i. In a β-pleated sheet, hydrogen bonds occur between residues in adjacent chains.
   ii. In a β-pleated sheet, the peptide chain is extended, rather than coiled.
c. Tertiary structure.
   i. The nonpolar amino acid side chains congregate in the center of a protein to avoid water.
   ii. The polar side chain residues are on the surface, where they can take part in hydrogen bonding and salt bridge formation.
   iii. Other important features of tertiary structure are disulfide bridges, hydrogen bonds between amino acid side chains and salt bridges.

3. Denaturation of proteins.
   a. Modest changes in temperature and pH can disrupt a protein's tertiary structure.
      i. This process is known as denaturation.
      ii. Denaturation doesn't affect protein primary structure.
   b. Denaturation affects both physical and catalytic properties of proteins.
   c. Occasionally, spontaneous renaturation can occur.

C. Enzymes (Sections 19.9 – 19.10).
1. Description of enzymes (Section 19.9).
   a. An enzyme is a substance (usually protein) that catalyzes a biochemical reaction.
   b. An enzyme is specific and usually catalyzes the reaction of only one substrate.
      Some enzymes, such as papain, can operate on a range of substrates.
   c. Function of enzymes.
      i. Enzymes form an enzyme-substrate complex, within which the conversion to product takes place.
      ii. Enzymes accelerate the rate of reaction by lowering the energy of the transition state.
   d. Enzymes are grouped into 6 classes according to the reactions they catalyze.
      i. Oxidoreductases catalyze oxidations and reductions.
      ii. Transferases catalyze the transfer of a group from one substrate to another.
      iii. Hydrolases catalyze hydrolysis reactions.
      iv. Isomerases catalyze isomerizations.
      v. Ligases catalyze bond formation between two molecules.
      vi. Lyases catalyze the loss of a small molecule from a substrate.
   e. The name of an enzyme has two parts, ending with *-ase*.
      i. The first part identifies the substrate.
      ii. The second part identifies the enzyme's class.
   f. Most enzymes are globular proteins, and many consist of a protein portion (apoenzyme) and a cofactor.
      i. Cofactors may be small organic molecules (coenzymes) or inorganic ions.
      ii. Many coenzymes are vitamins.
2. How enzymes work – citrate synthase (Section 19.10).
   a. Citrate synthase catalyzes the aldol-like addition of acetyl CoA to oxaloacetate to produce citrate.
   b. Functional groups in a cleft of the enzyme bind oxaloacetate.
   c. Functional groups in a second cleft bind acetyl CoA.
      The two reactants are now in close proximity.
   d Two enzyme amino acid residues generate the enol of acetyl CoA.
   e. The enol undergoes nucleophilic addition to the ketone carbonyl group of oxaloacetate.
   f. Two enzyme amino acid residues deprotonate the enol and protonate the carbonyl oxygen.
   g. Water hydrolyzes the thiol ester, releasing citrate and CoA.

Solutions to Problems

19.1 Amino acids with aromatic rings: Phe, Tyr, Trp, His.

Amino acids containing sulfur: Cys, Met.

Amino acids that are alcohols: Ser, Thr. (Tyr is a phenol.)

Amino acids having hydrocarbon side chains: Ala, Ile, Leu, Val, Phe.

19.2

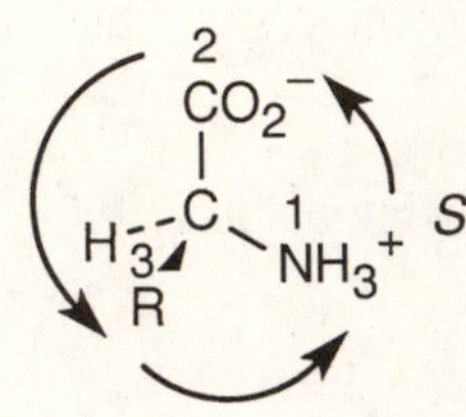

For most L-amino acids:

Group	*Priority*
$-NH_2$	1
$-COOH$	2
$-R$	3
$-H$	4

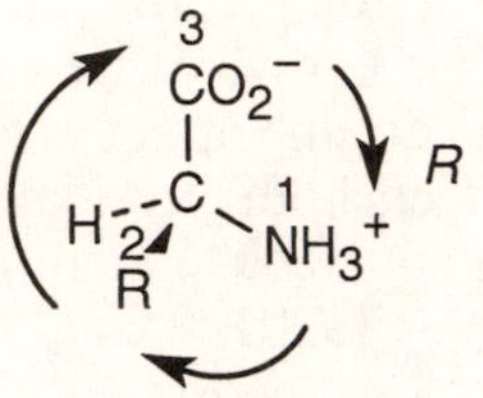

For cysteine:

Group	*Priority*
$-NH_2$	1
$-CH_2SH$	2
$-COOH$	3
$-H$	4

19.3

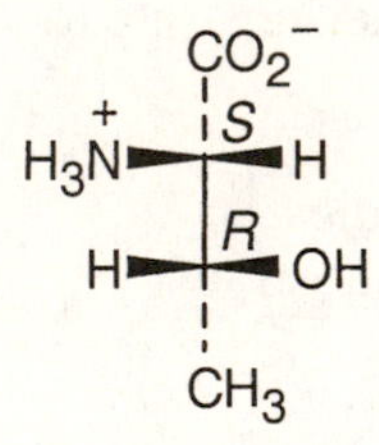

L-Threonine

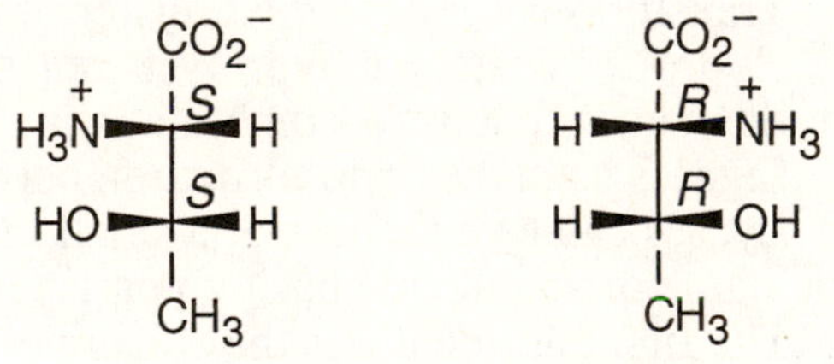

Diastereomers of L-Threonine

19.4 On the low pH (acidic) side of pI, a protein has a net positive charge, and on the high pH (basic) side of pI, a protein has a net negative charge. Thus hemoglobin (pI = 6.8) has a net positive charge at pH = 5.3 and a net negative charge at pH = 7.3.

19.5

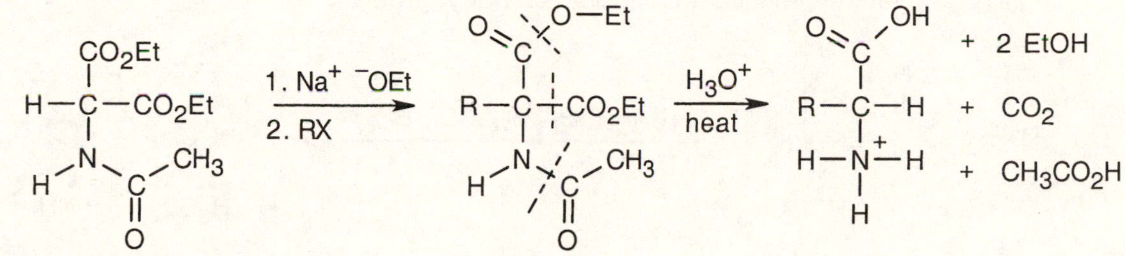

$$\mathrm{H{-}C(CO_2Et)_2{-}NHC(=O)CH_3} \xrightarrow[\text{2. RX}]{\text{1. Na}^+\ ^-\text{OEt}} \mathrm{R{-}C(CO_2Et)_2{-}NHC(=O)CH_3} \xrightarrow[\text{heat}]{\mathrm{H_3O^+}} \mathrm{R{-}CH(NH_3^+){-}C(=O)OH} + 2\ \mathrm{EtOH} + \mathrm{CO_2} + \mathrm{CH_3CO_2H}$$

In the amidomalonate synthesis, shown above, an alkyl halide RX is converted to $RCH(NH_3^+)CO_2H$. Choose an alkyl halide that completes the structure of the desired amino acid.

	Amino Acid	*Halide*
(a) 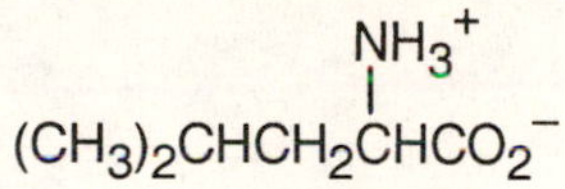	$(CH_3)_2CHCH_2CH(NH_3^+)CO_2^-$ Leucine	$(CH_3)_2CHCH_2Br$
(b)	4-imidazolyl–$CH_2CH(NH_3^+)CO_2^-$ Histidine	4-imidazolyl–CH_2Br
(c)	3-indolyl–$CH_2CH(NH_3^+)CO_2^-$ Tryptophan	3-indolyl–CH_2Br
(d) 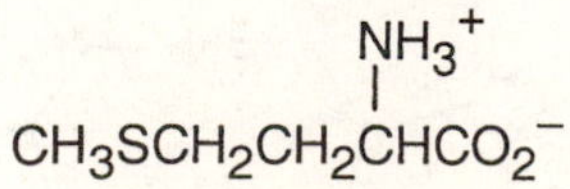	$CH_3SCH_2CH_2CH(NH_3^+)CO_2^-$ Methionine	$CH_3SCH_2CH_2Br$

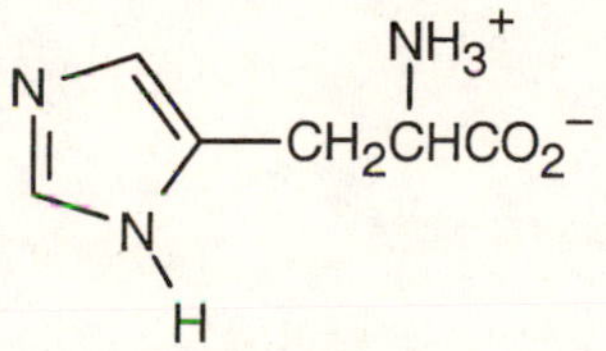

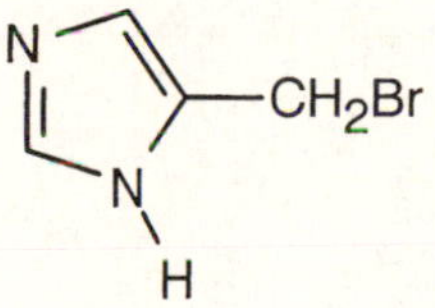

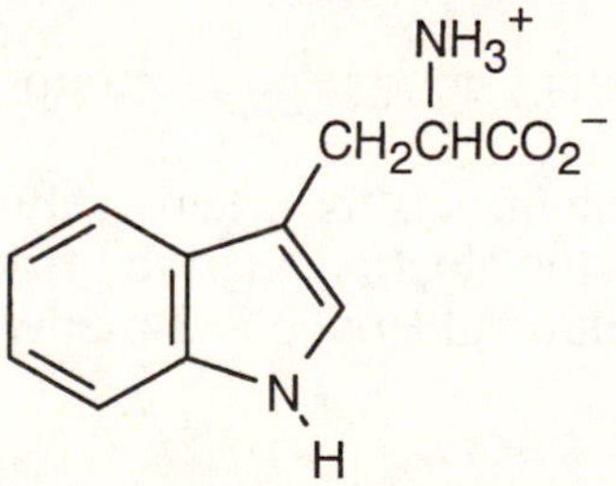

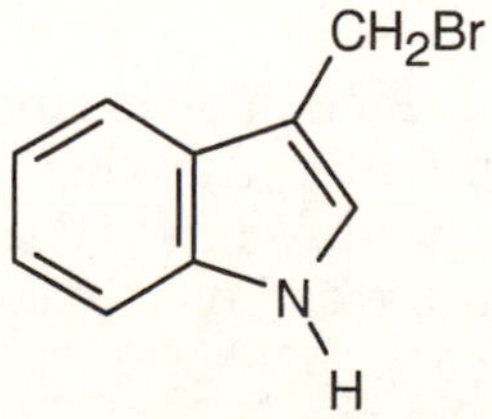

19.6 The precursor to an amino acid prepared by enantioselective hydrogenation has a *Z* double bond conjugated with a carboxylic acid carbonyl group.

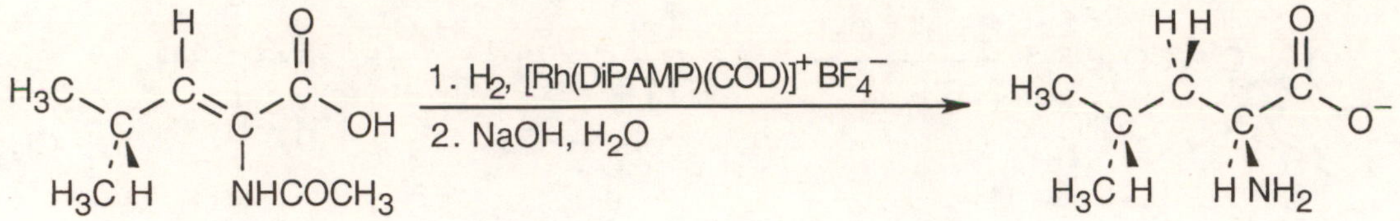

1. H_2, $[Rh(DiPAMP)(COD)]^+BF_4^-$
2. NaOH, H_2O

19.7 Val–Tyr–Gly (VYG) Tyr–Gly–Val (YGV) Gly–Val–Tyr (GVY)
Val–Gly–Tyr (VGY) Tyr–Val–Gly (YVG) Gly–Tyr–Val (GYV)

19.8

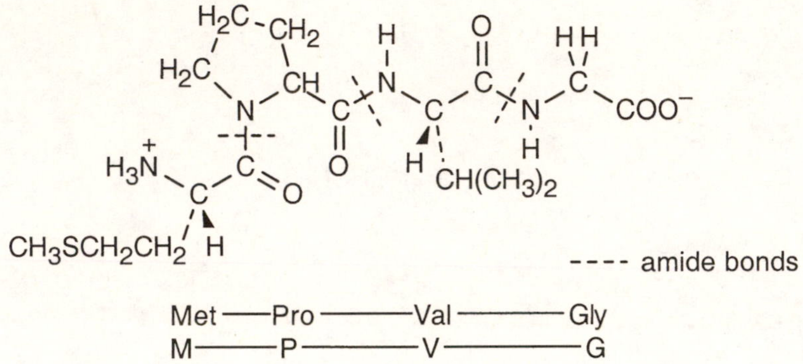

---- amide bonds

Met——Pro——Val——Gly
M——P——V——G

19.9

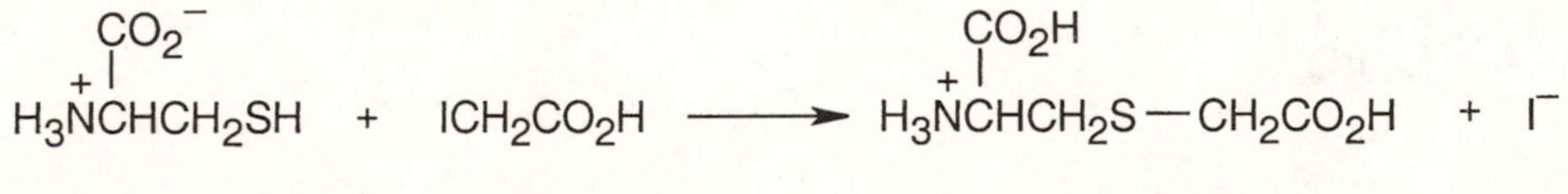

$$H_3\overset{+}{N}CH(CO_2^-)CH_2SH + ICH_2CO_2H \longrightarrow H_3\overset{+}{N}CH(CO_2H)CH_2S{-}CH_2CO_2H + I^-$$

The cysteine sulfur is a good nucleophile, and iodide is a good leaving group.

19.10 One product of the reaction of an amino acid with ninhydrin is the extensively conjugated purple ninhydrin product. The other major product is the aldehyde derived from the side chain of the amino acid. When valine reacts, the resulting aldehyde is 2-methylpropanal. The other products are carbon dioxide and water.

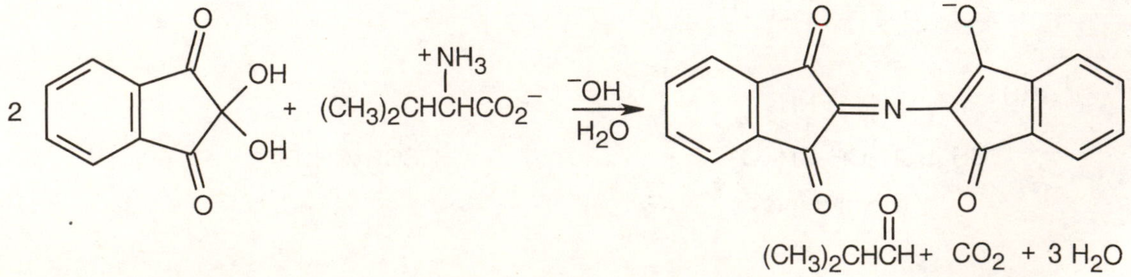

$$2\ \text{ninhydrin} + (CH_3)_2CHCH(\overset{+}{N}H_3)CO_2^- \xrightarrow[H_2O]{^-OH} \text{purple ninhydrin product} + (CH_3)_2CHCHO + CO_2 + 3\ H_2O$$

19.11 Trypsin cleaves peptide bonds at the carboxyl side of lysine and arginine. Chymotrypsin cleaves peptide bonds at the carboxyl side of phenylalanine, tyrosine and tryptophan.

Asp–Arg–Val–Tyr–Ile–His–Pro–Phe —Trypsin→ Asp–Arg + Val–Tyr–Ile–His–Pro–Phe

Asp–Arg–Val–Tyr–Ile–His–Pro–Phe —Chymotrypsin→ Asp–Arg–Val–Tyr + Ile–His–Pro–Phe

19.12 The part of the PTH derivative that lies to the right of the indicated dotted lines comes from the N-terminal residue. Complete the structure to identify the amino acid, which in this problem is methionine.

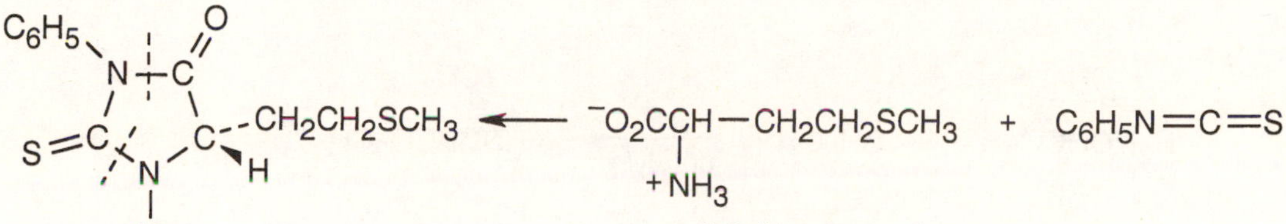

PTH derivative (C_6H_5, $S{=}C$, N–H, $C{=}O$; $CH_2CH_2SCH_3$, H) ← $^{-}O_2CCH(^{+}NH_3){-}CH_2CH_2SCH_3$ + $C_6H_5N{=}C{=}S$

Methionine

19.13 The N-terminal residue of angiotensin II is aspartic acid. Replace the –R group of the PTH derivative in Figure 19.3 with $–CH_2CO_2H$ to arrive at the correct structure.

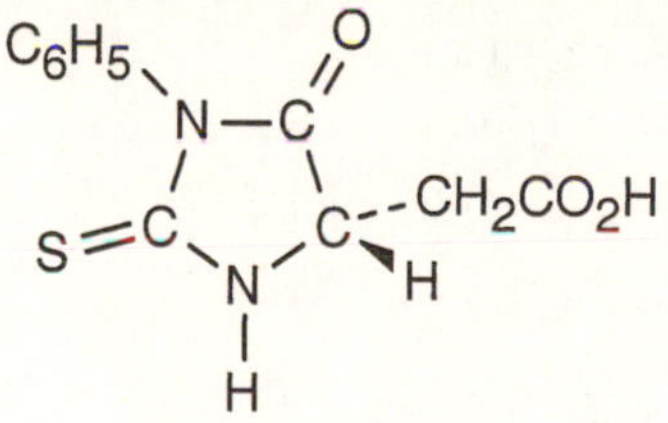

PTH derivative (C_6H_5, $S{=}C$, N–H, $C{=}O$; CH_2CO_2H, H)

19.14

$(CH_3)_3CO{-}C(=O){-}O{-}C(=O){-}OC(CH_3)_3$ + $H_2NCHCO_2^-$ (R) ⇌(1.) [$(CH_3)_3CO{-}C(O^-)({-}^{+}NH(H){-}CHCO_2^-(R)){-}O{-}C(=O){-}OC(CH_3)_3$; Et_3N: ⇌(2.) $(CH_3)_3CO{-}C(O^-)(HNCHCO_2H(R)){-}O{-}C(=O){-}OC(CH_3)_3$ + Et_3NH^+] →(3.) $(CH_3)_3CO{-}C(=O){-}NHCHCO_2^-$ (R) + CO_2 + $HOC(CH_3)_3$

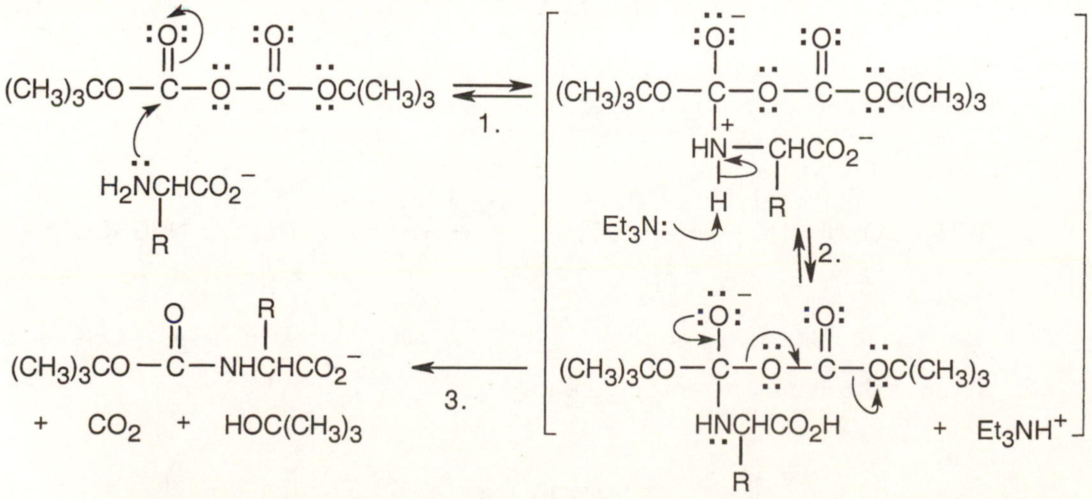

Step 1: Nucleophilic addition of amino acid nitrogen.
Step 2: Deprotonation.
Step 3: Loss of CO_2 and *tert*-butoxide.

19.15

$$\text{Leu} = \overset{+}{H_3N}\underset{}{\overset{CH_2CH(CH_3)_2}{\overset{|}{C}}}HCO_2^- = \overset{+}{H_3N}\overset{R}{\overset{|}{C}}HCO_2^- \qquad R = CH_2CH(CH_3)_2$$

1. Protect the amino group of leucine.

$$(CH_3)_3CO\overset{O}{\overset{\|}{C}}O\overset{O}{\overset{\|}{C}}OC(CH_3)_3 + \underset{\text{Leu}}{H_2N\overset{R}{\overset{|}{C}}HCO_2^-} \xrightarrow{Et_3N} (CH_3)_3CO\overset{O}{\overset{\|}{C}}NH\overset{R}{\overset{|}{C}}HCO_2^- + CO_2 + HOC(CH_3)_3$$

2. Protect the carboxylic acid group of alanine.

$$\underset{\text{Ala}}{H_2N\overset{CH_3}{\overset{|}{C}}HCO_2H} + CH_3OH \xrightarrow[\text{catalyst}]{H^+} H_2N\overset{CH_3}{\overset{|}{C}}HCO_2CH_3$$

3. Couple the protected amino acids with DCC.

$$(CH_3)_3CO\overset{O}{\overset{\|}{C}}NH\underset{R}{\underset{|}{C}}HCO_2^- + H_2N\underset{CH_3}{\underset{|}{C}}H\overset{O}{\overset{\|}{C}}OCH_3 + C_6H_{11}{-}N{=}C{=}N{-}C_6H_{11}$$

$$\downarrow$$

$$(CH_3)_3CO\overset{O}{\overset{\|}{C}}NH\underset{R}{\underset{|}{C}}H\overset{O}{\overset{\|}{C}}{-}NH\underset{CH_3}{\underset{|}{C}}H\overset{O}{\overset{\|}{C}}OCH_3 + C_6H_{11}{-}NH{-}\overset{O}{\overset{\|}{C}}{-}NH{-}C_6H_{11}$$

4. Remove the leucine protecting group.

$$(CH_3)_3CO\overset{O}{\overset{\|}{C}}NH\underset{R}{\underset{|}{C}}H\overset{O}{\overset{\|}{C}}{-}NH\underset{CH_3}{\underset{|}{C}}H\overset{O}{\overset{\|}{C}}OCH_3 \xrightarrow{CF_3CO_2H} \overset{+}{H_3N}\underset{R}{\underset{|}{C}}H\overset{O}{\overset{\|}{C}}{-}NH\underset{CH_3}{\underset{|}{C}}H\overset{O}{\overset{\|}{C}}OCH_3 + (CH_3)_2C{=}CH_2 + CO_2$$

5. Remove the alanine protecting group.

$$\overset{+}{H_3N}\underset{R}{\underset{|}{C}}H\overset{O}{\overset{\|}{C}}{-}NH\underset{CH_3}{\underset{|}{C}}H\overset{O}{\overset{\|}{C}}OCH_3 \xrightarrow[\text{2. } H_3O^+]{\text{1. NaOH, } H_2O} \overset{+}{H_3N}\underset{(CH_3)_2CHCH_2}{\underset{|}{C}}H\overset{O}{\overset{\|}{C}}{-}NH\underset{CH_3}{\underset{|}{C}}H\overset{O}{\overset{\|}{C}}OH + CH_3OH$$

Leu—Ala

19.16 (a) Pyruvate decarboxylase is a lyase.
(b) Chymotrypsin is a hydrolase.
(c) Alcohol dehydrogenase is an oxidoreductase.

Visualizing Chemistry

19.17

(a) Isoleucine

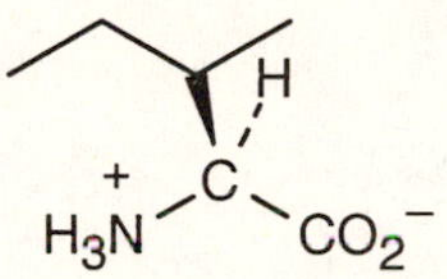

(b) Histidine

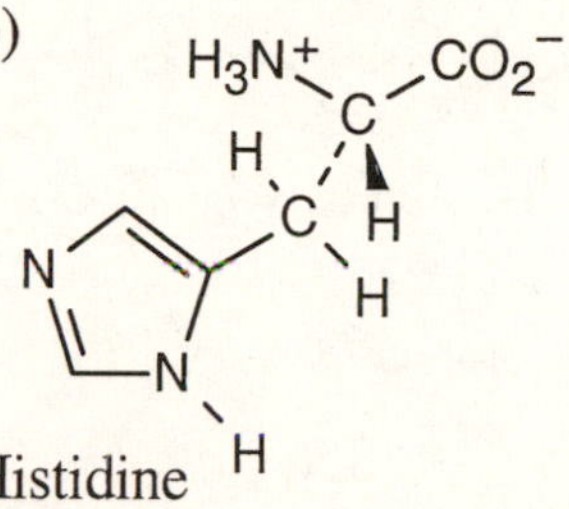

(c) Glutamine

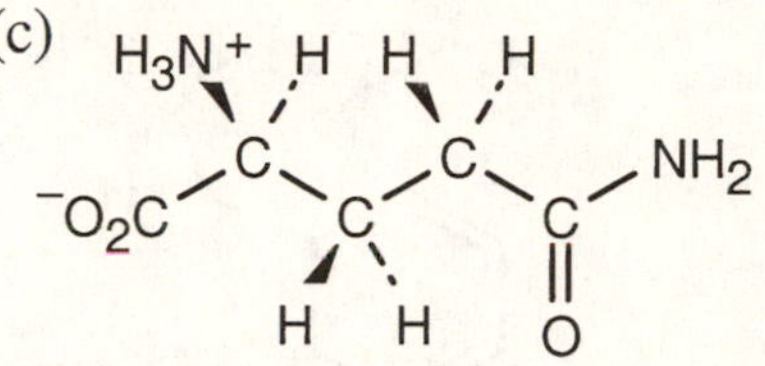

19.18

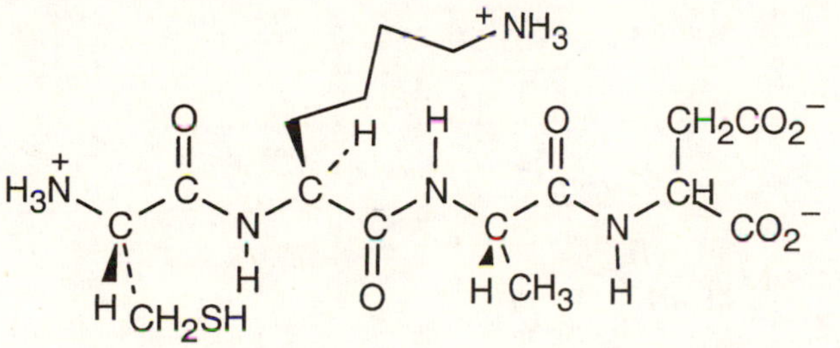

Cys——Lys——Ala——Asp

C——K——A——D

19.19

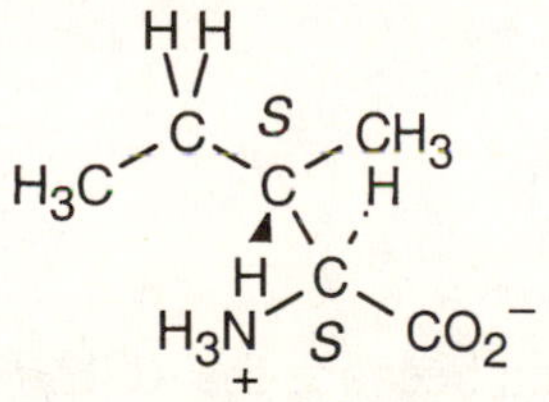

19.20 It's possible to identify this representation of valine as the D enantiomer by noting the configuration at the chirality center. The configuration is *R*, and thus the structure is D-valine.

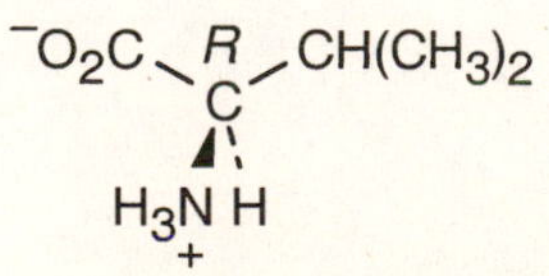

D-Valine
(*R*)-Valine

19.21 After identifying the amino acid residues, notice that the tetrapeptide has been drawn with the amino terminal residue on the right. To name the sequence correctly, the amino terminal residue must be cited first. Thus, the tetrapeptide should be named Ser–Leu–Phe–Ala.

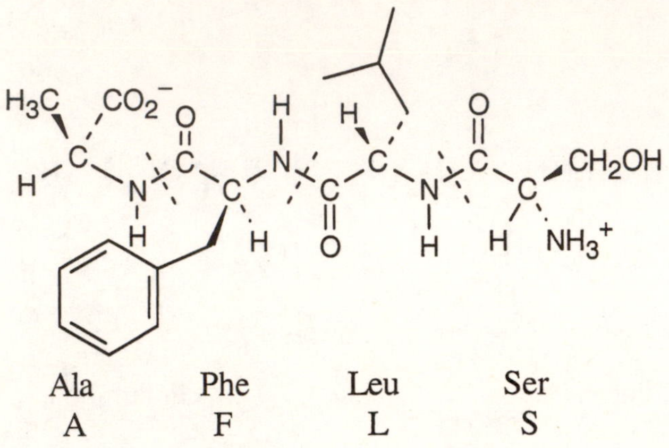

Ala A Phe F Leu L Ser S

Additional Problems

19.22

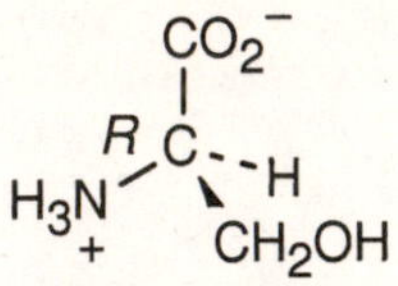

(*R*)-Serine 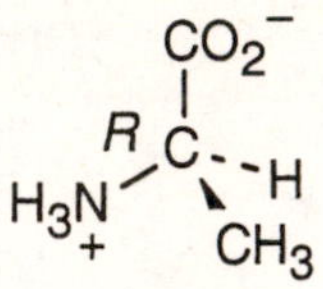 (*R*)-Alanine

Both (*R*)-serine and (*R*)-alanine are D-amino acids.

19.23

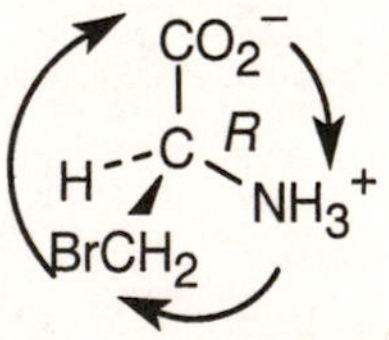

L-Bromoalanine
(*R*)-Bromoalanine

This L "amino acid" also has an *R* configuration because the $–CH_2Br$ "side chain" is higher in priority than the $–CO_2H$ group.

19.24

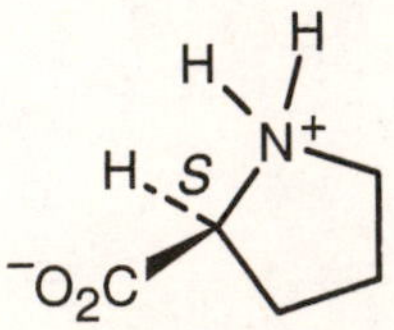

(*S*)-Proline

19.25

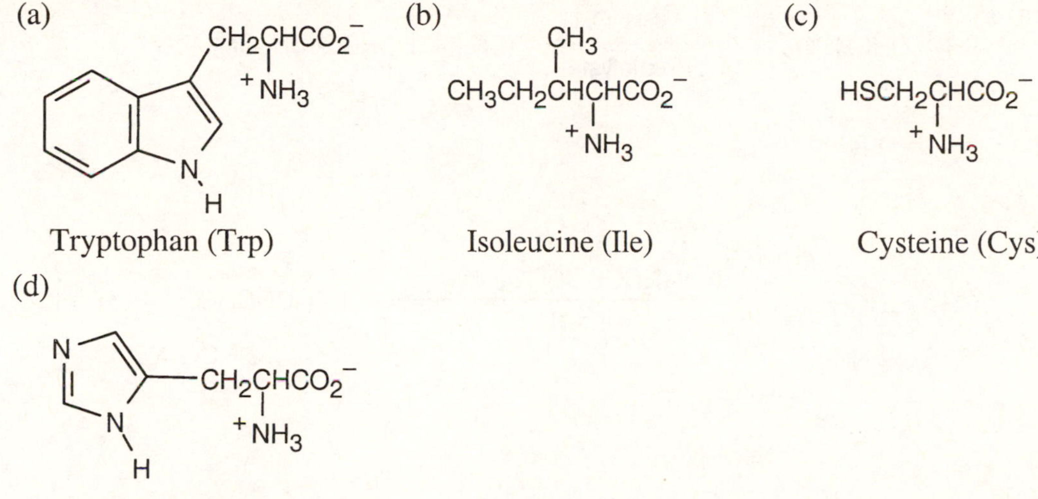

Tryptophan (Trp) Isoleucine (Ile) Cysteine (Cys)

Histidine (His)

19.26

H_2A^+ (protonated) $\underset{K_{a1}}{\overset{-\,H^+}{\rightleftharpoons}}$ HA (neutral) $\underset{K_{a2}}{\overset{-\,H^+}{\rightleftharpoons}}$ A^- (deprotonated)

protonated neutral deprotonated

At pH = 2.50:

$$\log\frac{[HA]}{[H_2A^+]} = pH - pK_{a1} = 2.50 - 1.99 = 0.51; \quad \frac{[HA]}{[H_2A^+]} = 3.24$$

At pH = 2.50, approximately three times as many proline molecules exist in the neutral form as exist in the protonated form.

At pH = 9.70:

$$\log\frac{[A^-]}{[HA]} = pH - pK_{a2} = 9.70 - 10.60 = -0.90; \quad \frac{[A^-]}{[HA]} = 0.126$$

At pH = 9.70, the ratio of deprotonated proline to neutral proline is approximately 1:8.

19.27 (a)

Val–Leu–Ser	V-L-S	Ser–Val–Leu	S-V-L
Val–Ser–Leu	V-S-L	Leu–Val–Ser	L-V-S
Ser–Leu–Val	S-L-V	Leu–Ser–Val	L-S-V

(b)

Ser–Leu–Leu–Pro	S-L-L-P	Leu–Leu–Ser–Pro	L-L-S-P
Ser–Leu–Pro–Leu	S-L-P-L	Leu–Leu–Pro–Ser	L-L-P-S
Ser–Pro–Leu–Leu	S-P-P-L	Leu–Ser–Leu–Pro	L-S-L-P
Pro–Leu–Leu–Ser	P-L-L-S	Leu–Ser–Pro–Leu	L-S-P-L
Pro–Leu–Ser–Leu	P-L-S-L	Leu–Pro–Leu–Ser	L-P-L-S
Pro–Ser–Leu–Leu	P-S-L-L	Leu–Pro–Ser–Leu	L-P-S-L

19.28

(a)

$$(CH_3)_2CHCHCO_2^- \ (\text{with } {}^+NH_3) \xrightarrow[H^+ \text{ catalyst}]{CH_3CH_2OH} (CH_3)_2CHCHCO_2CH_2CH_3 \ (\text{with } {}^+NH_3)$$

L-Valine

(b)

$$(CH_3)_2CHCHCO_2^- \ (\text{with } {}^+NH_3) \xrightarrow[Et_3N]{(CH_3)_3COC(=O)OC(=O)OC(CH_3)_3} (CH_3)_2CHCHCO_2^- \ (\text{with } NHC(=O)OC(CH_3)_3) + HOC(CH_3)_3 + CO_2$$

(c)

$$(CH_3)_2CHCHCO_2^- \ (\text{with } {}^+NH_3) \xrightarrow{KOH,\ H_2O} (CH_3)_2CHCHCO_2^-K^+ \ (\text{with } NH_2) + H_2O$$

(d)

$$(CH_3)_2CHCHCO_2^- \ (\text{with } {}^+NH_3) \xrightarrow[2.\ H_2O]{1.\ CH_3COCl,\ \text{pyridine}} (CH_3)_2CHCHCO_2^- \ (\text{with } NHC(=O)CH_3)$$

19.29 The diethylamidomalonate anion is formed by treating diethylamidomalonate with sodium ethoxide.

(a)

$$\left[{}^-{:}C(CO_2Et)_2(NHC(=O)CH_3) \right] \xrightarrow{(CH_3)_2CHCH_2Br} (CH_3)_2CHCH_2-C(CO_2Et)_2-NHC(=O)CH_3$$

$$\xrightarrow{H_3O^+} CH_3CO_2H + CO_2 + 2\ EtOH + (CH_3)_2CHCH_2CHCO_2H \ (\text{with } {}^+NH_3)$$

Leucine

(b) $[^{-}:C(CO_2Et)_2NHCOCH_3]$ + 3-(bromomethyl)indole (CH_2Br) → indol-3-yl-$CH_2-C(CO_2Et)_2-NHCOCH_3$

$\xrightarrow{H_3O^+}$ CH_3CO_2H + CO_2 + 2 EtOH + indol-3-yl-$CH_2CH(^{+}NH_3)CO_2H$

Tryptophan

19.30

(a)

$$CH_3SCH_2CH_2\overset{O}{\overset{\|}{C}}CO_2H \xrightarrow[NaBH_4]{NH_3} CH_3SCH_2CH_2\overset{^{+}NH_3}{\overset{|}{C}}HCO_2^-$$

Methionine

(b)

$$CH_3CH_2\overset{H_3C}{\overset{|}{C}}H\overset{O}{\overset{\|}{C}}CO_2H \xrightarrow[NaBH_4]{NH_3} CH_3CH_2\overset{H_3C}{\overset{|}{C}}H\overset{^{+}NH_3}{\overset{|}{C}}HCO_2^-$$

Isoleucine

19.31

(a) N-acetyl-2,3-dihydropyrrole-2-carboxylic acid ($C-OH$, $NCOCH_3$) $\xrightarrow[\text{2. NaOH, } H_2O]{\text{1. } H_2\text{, [Rh(DiPAMP)(COD)]}^+ BF_4^-}$ proline carboxylate (H, $C(=O)O^-$, NH)

(b) $(CH_3)(H_3C)C=C(NHCOCH_3)C(=O)OH$ $\xrightarrow[\text{2. NaOH, } H_2O]{\text{1. } H_2\text{, [Rh(DiPAMP)(COD)]}^+ BF_4^-}$ $H_3C-CH(CH_3)-C(H)(NH_2)-C(=O)O^-$

19.32

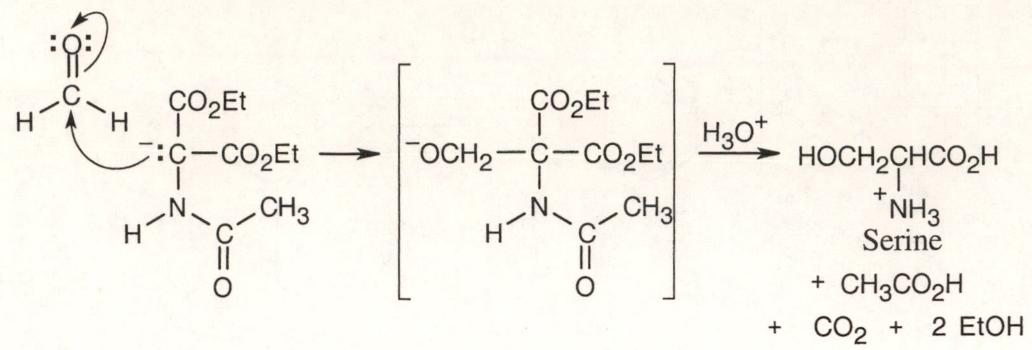

$$H_2C{=}O + {}^-\!:C(CO_2Et)_2NHCOCH_3 \longrightarrow \left[{}^-OCH_2{-}C(CO_2Et)_2NHCOCH_3\right] \xrightarrow{H_3O^+} HOCH_2CH(\overset{+}{N}H_3)CO_2H$$

Serine

$+ CH_3CO_2H$

$+ CO_2 + 2\ EtOH$

19.33

(a)

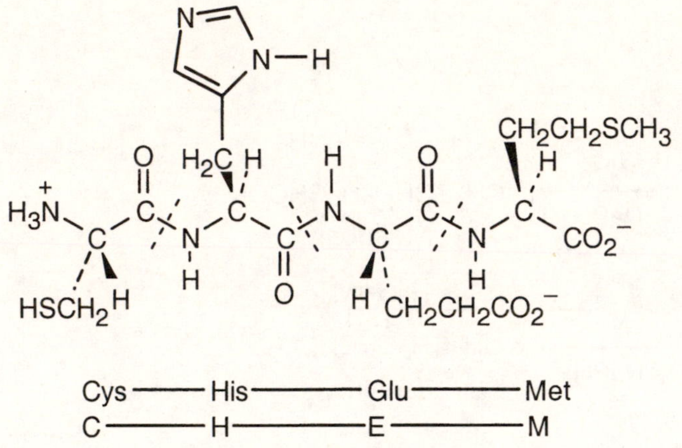

Cys——His——Glu——Met

C——H——E——M

(b)

Glu——Ala——Ser——Tyr

E——A——S——Y

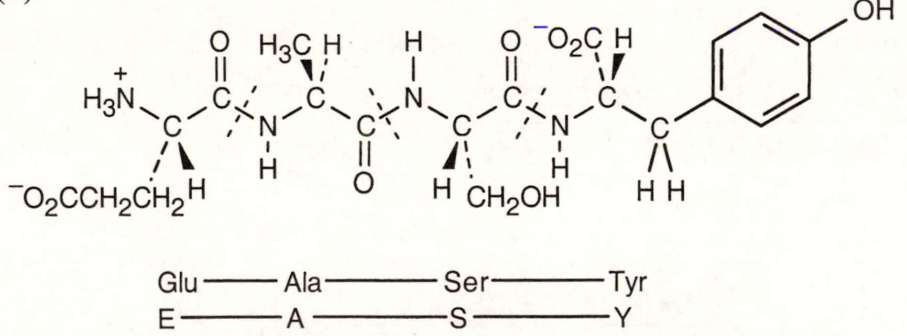

19.34

The tripeptide is cyclic.

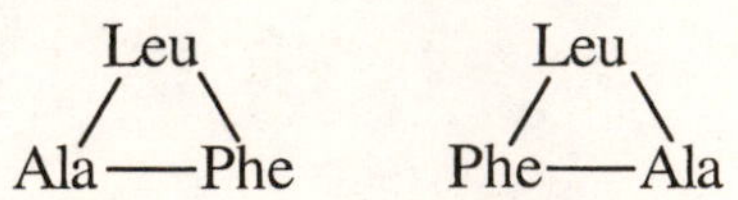

Leu / Ala—Phe (cyclic)   Leu / Phe—Ala (cyclic)

19.35

Step 1: Valine is protected as its Boc derivative.

$$(CH_3)_3COC(=O)OC(=O)OC(CH_3)_3 + \text{Val} \xrightarrow{Et_3N} (CH_3)_3COC(=O)\text{—Val—OH}$$

(Boc—Val—OH)

Step 2: Boc–Val bonds to the polymer in an S_N2 reaction.

$$\text{Boc—Val—OH} + ClCH_2\text{—(Polymer)} \xrightarrow{\text{Base}} \text{Boc—Val—}OCH_2\text{—(Polymer)}$$

Step 3: The polymer is first washed, then is treated with CF_3CO_2H to cleave the Boc group.

$$\text{Boc—Val—}OCH_2\text{—(Polymer)} \xrightarrow[\text{2. } CF_3CO_2H]{\text{1. wash}} \text{Val—}OCH_2\text{—(Polymer)}$$

Step 4: A Boc-protected Ala is coupled to the polymer-bound valine by reaction with DCC. The polymer is washed.

$$\text{Boc—Ala} + \text{Val—}OCH_2\text{—(Polymer)} \xrightarrow[\text{2. wash}]{\text{1. DCC}} \text{Boc—Ala—Val—}OCH_2\text{—(Polymer)}$$

Step 5: The polymer is treated with CF_3CO_2H to remove Boc.

$$\text{Boc—Ala—Val—}OCH_2\text{—(Polymer)} \xrightarrow{CF_3CO_2H} \text{Ala—Val—}OCH_2\text{—(Polymer)}$$

Step 6: A Boc-protected Phe is coupled to the polymer by reaction with DCC. The polymer is washed.

$$\text{Boc—Phe} + \text{Ala—Val—}OCH_2\text{—(Polymer)} \xrightarrow[\text{2. wash}]{\text{1. DCC}} \text{Boc—Phe—Ala—Val—}OCH_2\text{—(Polymer)}$$

Step 7: Treatment with anhydrous HF removes the Boc group and cleaves the ester bond between the peptide and the polymer.

$$\text{Boc—Phe—Ala—Val—}OCH_2\text{—(Polymer)} \xrightarrow{HF}$$

$$\text{Phe—Ala—Val} + (CH_3)_2C{=}CH_2 + CO_2 + HOCH_2\text{—(Polymer)}$$

19.36

Peptide $\xrightarrow{\text{PITC}}$ *Phenylthiohydantoin* + *Shortened Peptide*

(a)

Ile–Leu–Pro–Phe
I—L—P—F

C_6H_5 N—C(=O) ; CH_3 | C--$CHCH_2CH_3$; S=C ; N ; H ; H

Leu–Pro–Phe
L—P—F

(b)

Asp–Thr–Ser–Gly–Ala
D—T—S—G—A

C_6H_5 N—C(=O) ; C--CH_2CO_2H ; S=C ; N ; H ; H

Thr–Ser–Gly–Ala
T—S—G—A

19.37 Aldehydes and ketones can undergo nucleophilic addition reactions. In particular, aldehydes and ketones can react with amines to form imines and enamines, reactions that might compete with formation of amide bonds between amino acids. Because of this reactivity, aldehydes and ketones are unlikely to be found in amino acid side chains.

19.38 A proline residue in a polypeptide chain interrupts α-helix formation because the amide nitrogen of proline has no hydrogen that can contribute to the hydrogen-bonded structure of an α-helix.

19.39

Phe-⌇-Leu—Met—Lys-┆-Tyr-⌇-Asp—Gly—Gly—Arg-┆-Val—Ile—Pro—Tyr

Cleaved by trypsin = ----
Cleaved by chymotrypsin = ⌇⌇⌇

19.40 (a) Hydrolases catalyze the cleavage of bonds by addition of water (hydrolysis).
(b) Lyases catalyze the elimination of a small molecule (H_2O, CO_2) from a molecule.
(c) Transferases catalyze the transfer of a functional group between substrates.

19.41 Amino acids with polar side chains are likely to be found on the outside of a globular protein, where they can form hydrogen bonds with water and with each other. Amino acids with nonpolar side chains are found on the inside of a globular protein, where they can avoid water. Thus, aspartic acid (b) and lysine (d) are found on the outside of a globular protein, and valine (a) and phenylalanine (c) are likely to be found on the inside.

19.42

Formation of cation:

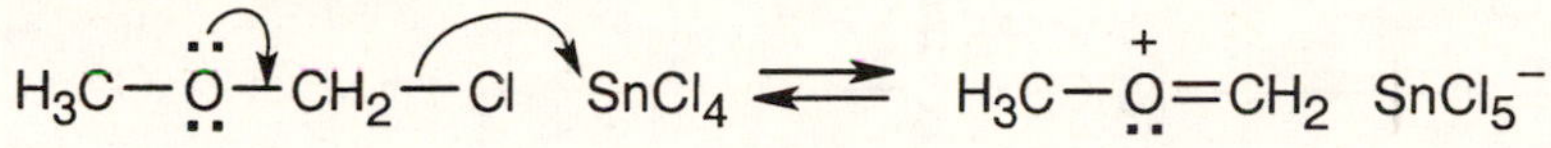

$$H_3C-\ddot{O}-CH_2-Cl \quad SnCl_4 \rightleftharpoons H_3C-\overset{+}{O}=CH_2 \quad SnCl_5^-$$

Electrophilic aromatic substitution:

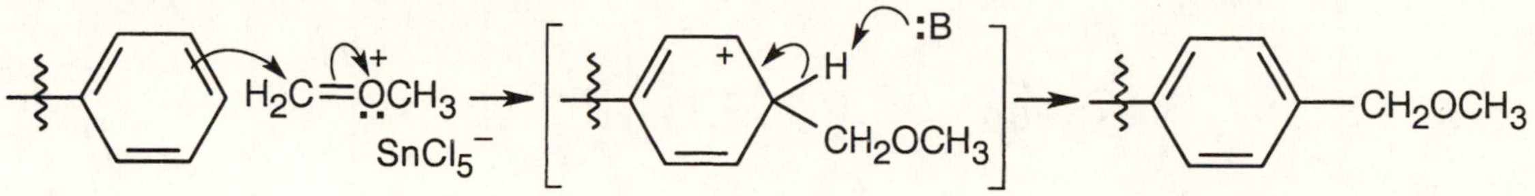

$$H_2C=\overset{+}{O}CH_3 \quad SnCl_5^- \rightarrow [\ldots CH_2OCH_3] \rightarrow -CH_2OCH_3$$

Protonation of the ether oxygen, followed by displacement of methanol by Cl⁻.

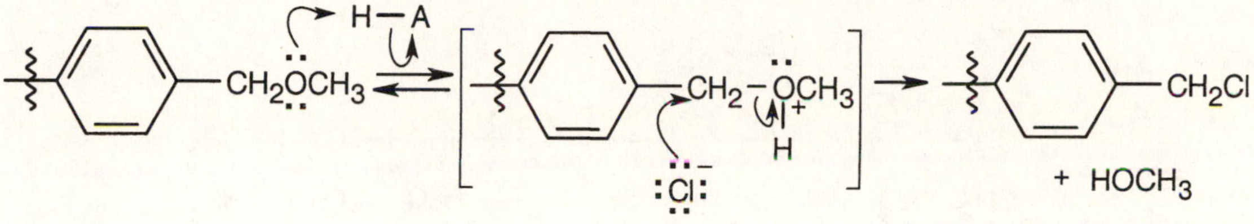

$$-CH_2OCH_3 \xrightleftharpoons{H-A} [-CH_2-\overset{+}{O}(H)CH_3 \quad :Cl:^-] \rightarrow -CH_2Cl + HOCH_3$$

19.43

$$\text{Fmoc-NHCHRCO-} \xrightarrow{1.} \text{dibenzofulvene} + {}^-O_2C\text{-NHCHRCO-} + H_2N^+\text{(piperidinium)}$$

$$CO_2 + H_2NCHRCO- + H-N\text{(piperidine)} \xleftarrow{2.}$$

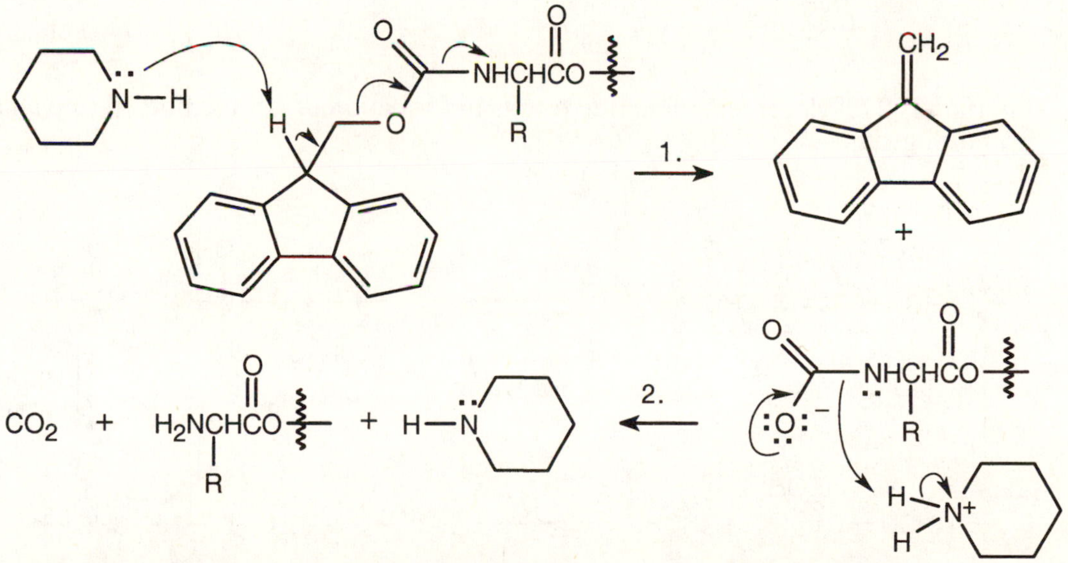

Step 1: Piperidine brings about elimination of the carboxylated peptide.
Step 2: Loss of CO_2.

The Fmoc group is acidic because the resulting anion is similar to the cyclopentadienyl anion, which is resonance-stabilized and is aromatic.

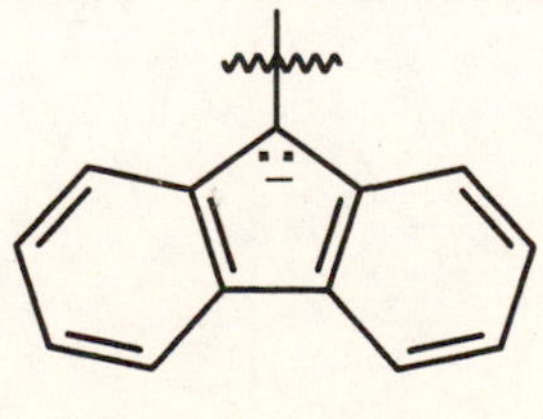

19.44

(a)

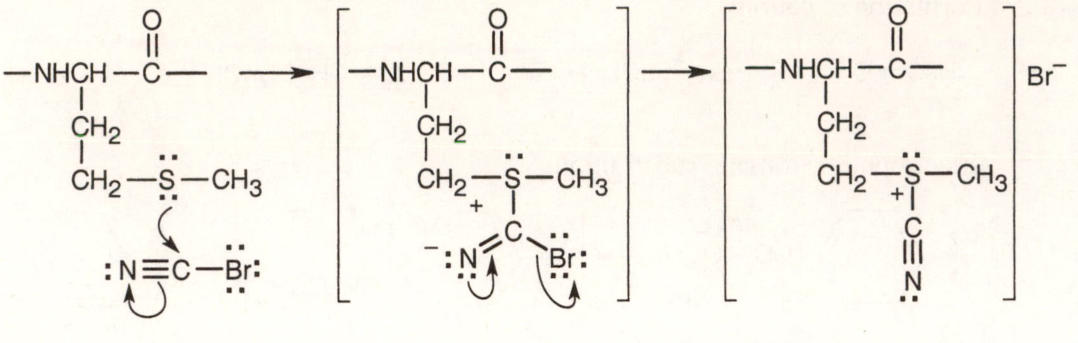

The first step is a substitution similar to the nucleophilic acyl substitution reactions that we studied in Chapter 16.

(b)

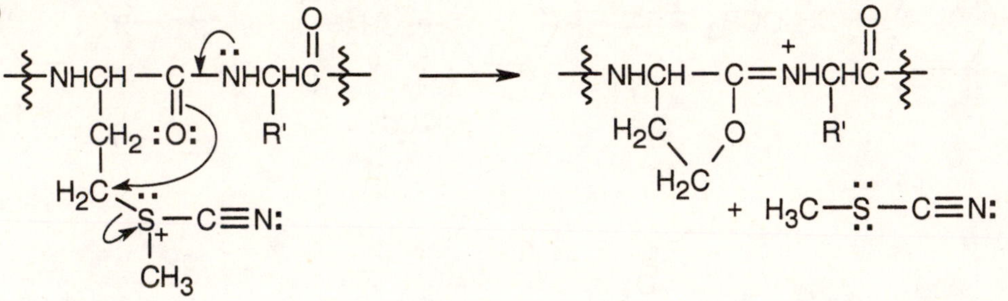

Internal S_N2 displacement of sulfide results in formation of a 5-membered ring and an iminium group.

(c)

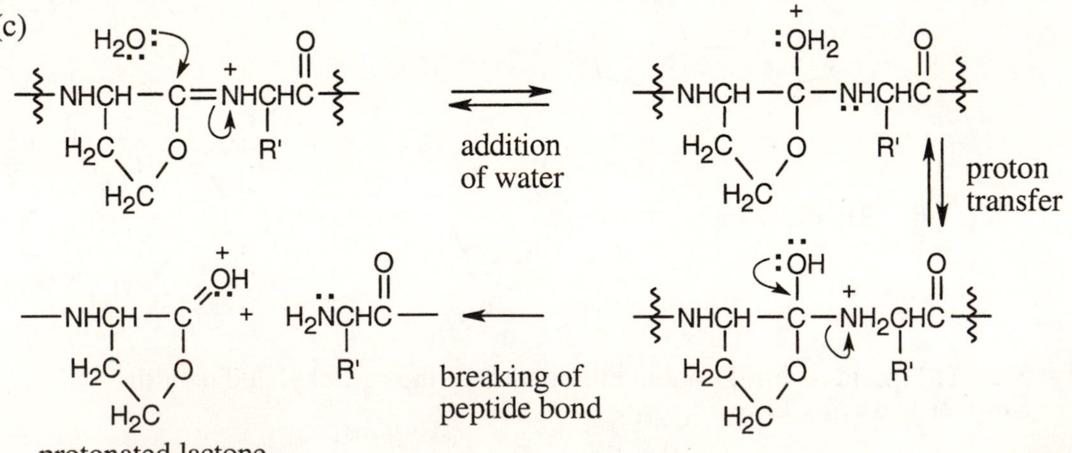

addition of water

proton transfer

breaking of peptide bond

protonated lactone

In this sequence of steps, water adds to the imine double bond, and the peptide bond is cleaved.

(d)

$$\text{-NHCH}(\text{H}_2\text{C})\text{-C}(=\overset{+}{\text{O}}\text{H})\text{-O-CH}_2 \;(\text{lactone}) + \text{H}_2\ddot{\text{O}}: \underset{\text{of water}}{\overset{\text{addition}}{\rightleftharpoons}} \text{-NHCH-C}(\text{H}_2\overset{+}{\text{O}})(\text{OH})\text{-O-} \underset{\text{transfer}}{\overset{\text{proton}}{\rightleftharpoons}} \text{-NHCH-C}(\text{HO:})(\text{OH})(\overset{+}{\text{O}}\text{H}) \underset{\text{opening}}{\overset{\text{ring}}{\rightleftharpoons}} \text{-NHCH}(\text{H}_2\text{C-CH}_2\text{OH})\text{-C}(=\overset{+}{\text{O}}\text{H})\text{OH} + \text{B:}$$

$$\rightleftharpoons \underset{\text{from (c)}}{\text{H}_3\overset{+}{\text{N}}\text{CH(R)C(=O)-}} + \text{-NHCH}(\text{H}_2\text{C-CH}_2\text{OH})\text{-C(=O)OH}$$

Water opens the lactone ring to give the product shown.

19.45

$$\text{H}_2\ddot{\text{N}}\text{-C}(=\ddot{\text{N}}\text{H})\text{-}\ddot{\text{N}}\text{HR} \overset{\text{H}^+}{\rightleftharpoons} \left[\text{H}_2\ddot{\text{N}}\text{-C}(=\overset{+}{\text{N}}\text{H}_2)\text{-}\ddot{\text{N}}\text{HR} \longleftrightarrow \text{H}_2\overset{+}{\text{N}}\text{=C}(\ddot{\text{N}}\text{H}_2)\text{-}\ddot{\text{N}}\text{HR} \longleftrightarrow \text{H}_2\ddot{\text{N}}\text{-C}(\ddot{\text{N}}\text{H}_2)\text{=}\overset{+}{\text{N}}\text{HR} \longleftrightarrow \text{H}_2\ddot{\text{N}}\text{-}\overset{+}{\text{C}}(\ddot{\text{N}}\text{H}_2)\text{-}\ddot{\text{N}}\text{HR} \right]$$

The protonated guanidino group can be stabilized by resonance.

19.46 100 g of cytochrome *c* contains 0.43 g iron, or 0.0077 mol Fe:

$$0.43 \text{ g Fe} \times \frac{1 \text{ mol Fe}}{55.8 \text{ g Fe}} = 0.0077 \text{ mol Fe}$$

Assuming that each mole of protein contains 1 mol Fe, then mol Fe = mol protein.

$$\frac{100 \text{ g Cytochrome } c}{0.0077 \text{ mol Fe}} = \frac{13{,}000 \text{ g Cytochrome } c}{1 \text{ mol Fe}}$$

Cytochrome C has a minimum molecular weight of 13,000 g/ mol.

19.47

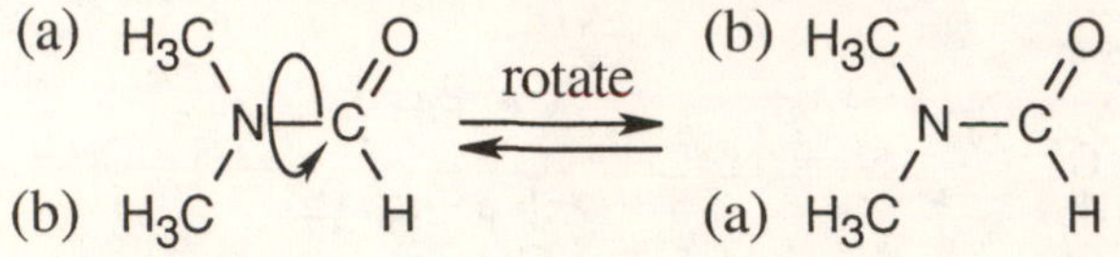

(a) H_3C / (b) H_3C N–C(=O)H ⇌ (rotate) (b) H_3C / (a) H_3C N–C(=O)H

The presence of two methyl absorptions in the ^{1}H NMR at room temperature shows that the two methyl groups of *N,N*–dimethylformamide are non-equivalent and that there is a barrier to rotation around the CO–N bond. This barrier is due to the partial double-bond character of the CO–N bond, as indicated by the two resonance forms. Rotation to interconvert the two methyl groups is slow at room temperature, but heating to 180° supplies enough energy to allow rapid rotation and to cause the two NMR absorptions to coalesce.

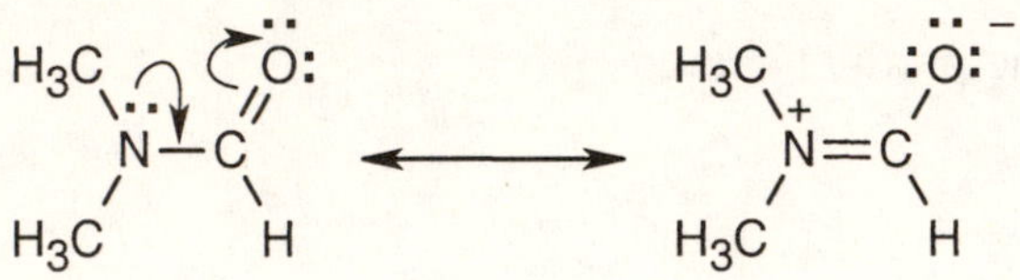

$(H_3C)_2\ddot{N}–CH=O$ ⟷ $(H_3C)_2N^+=CH–O^-$

19.48

(a)

Step 1: ninhydrin (OH, OH) ⇌ – H_2O; Step 2: + H_2NCHCO_2H (R) ⇌ $NH_2^+CHCO_2^-$ (R), OH; Step 3 ⇌ $^+OH_2$, NHCHCO$_2^-$ (R); Step 4 ⇌ =$\overset{+}{N}$HCHCO$_2^-$ (R) + H_2O

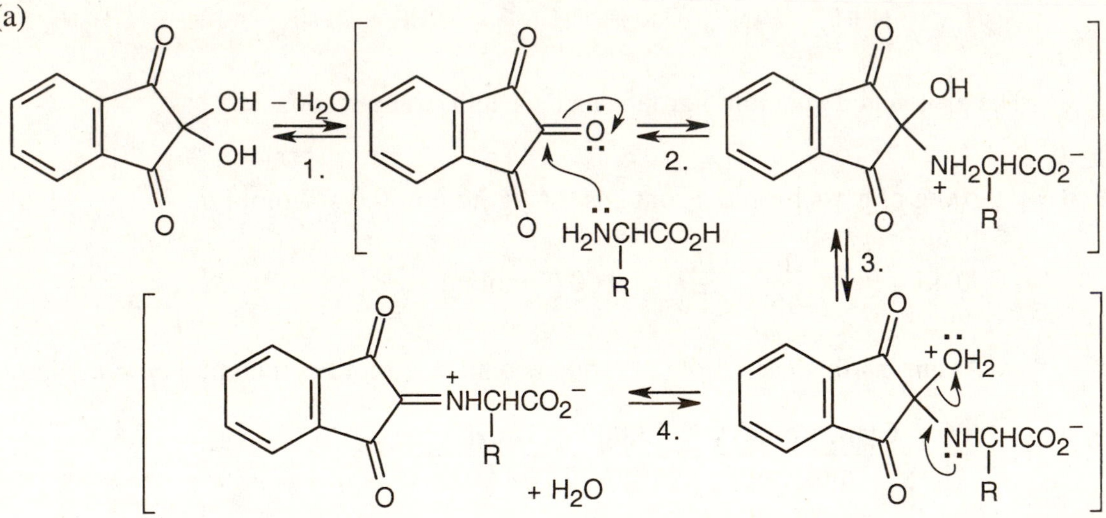

The steps involved in imine formation: (1) dehydration, (2) nucleophilic addition of the amino group of the amino acid, (3) proton transfer, and (4) loss of water.

(b)

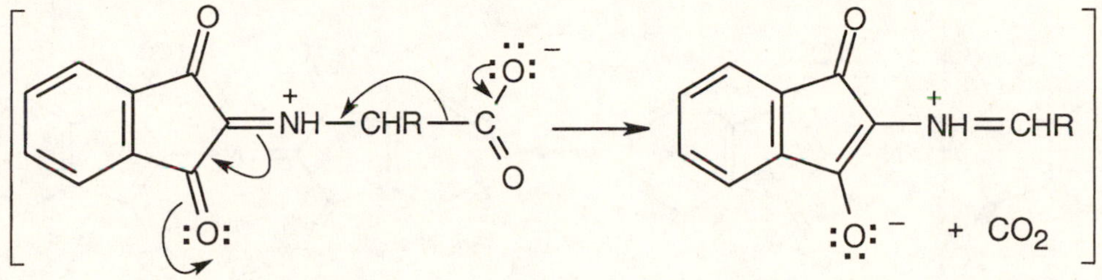

Decarboxylation produces a new imine.

(c)

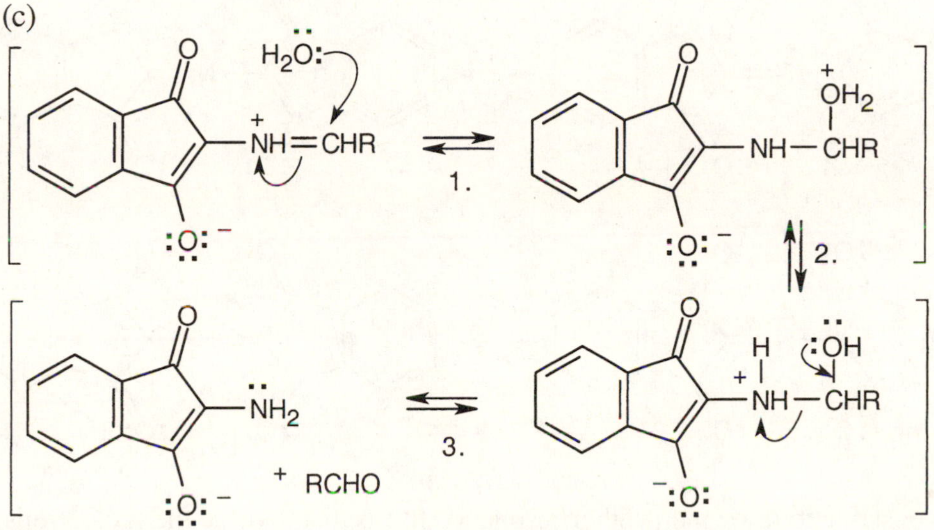

Hydrolysis of this imine occurs by addition of water (1), proton transfer (2), and bond cleavage (3) to yield an aldehyde and an amine.

(d)

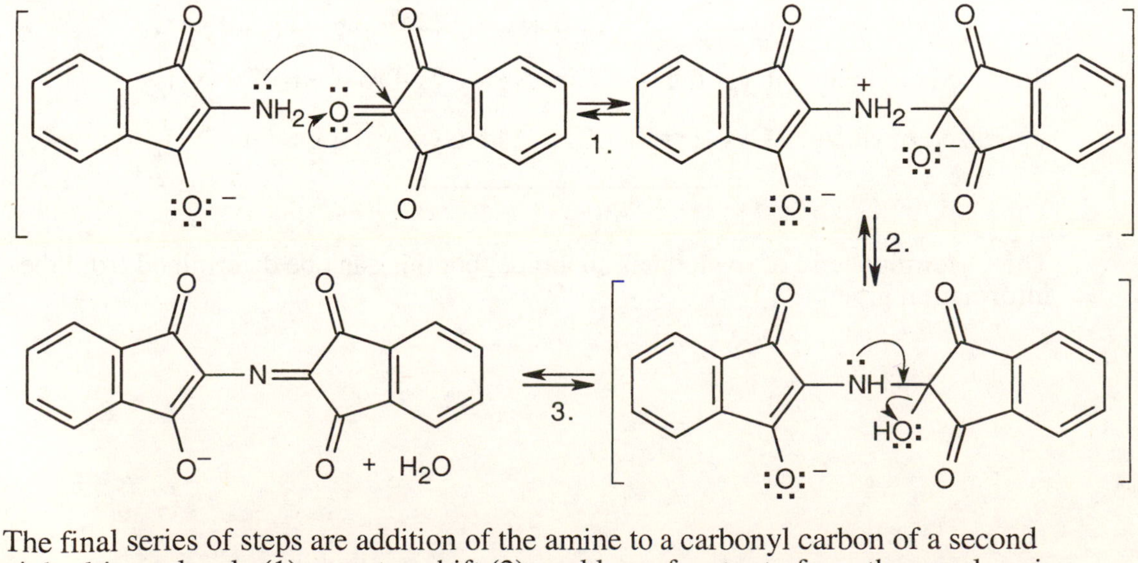

The final series of steps are addition of the amine to a carbonyl carbon of a second ninhydrin molecule (1), a proton shift (2), and loss of water to form the purple anion. Notice that the amino nitrogen is all that remains of the original amino acid.

19.49

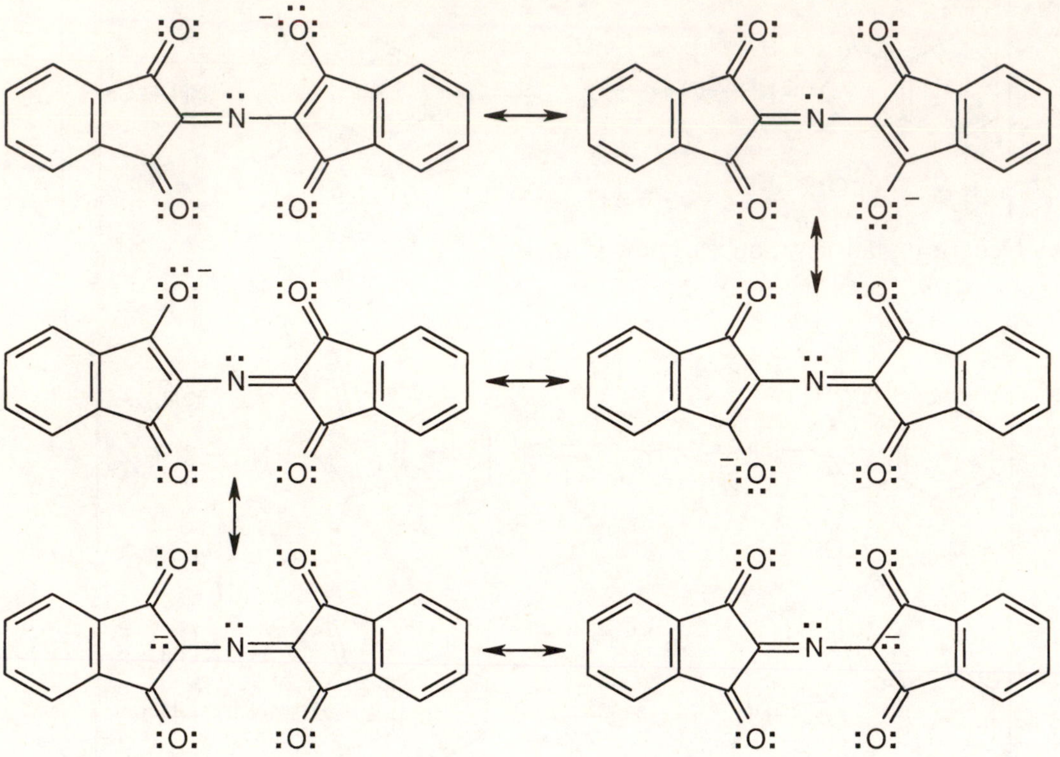

It is also possible to draw many other resonance forms that involve the π electrons of the aromatic 6-membered rings.

19.50 Ser–Ile–Arg–Val–Val–Pro–Tyr–Leu–Arg
S—I—R—V—V—P—Y—L—R

19.51

C—Y—I—Q—N—C—P—L—G—NH_2

Reduced oxytocin: Cys–Tyr–Ile–Gln–Asn–Cys–Pro–Leu–Gly–NH_2

Oxidized oxytocin: Cys–Tyr–Ile–Gln–Asn–Cys–Pro–Leu–Gly–NH_2
(S—S disulfide bridge between the two Cys residues)

The C–terminal end of oxytocin is an amide, but this can't be determined from the information given.

19.52

(a)

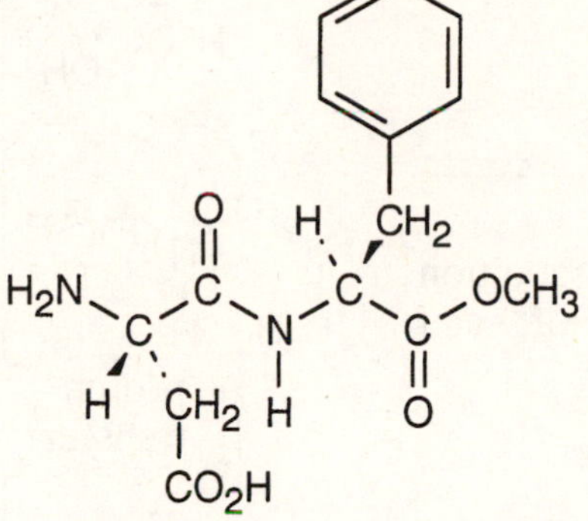

Aspartame (nonzwitterionic form)

(b)

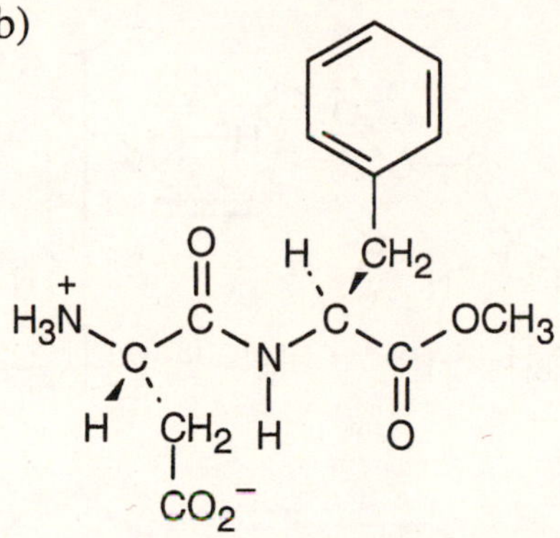

Aspartame at p*I* (pH = 5.9)

(c)

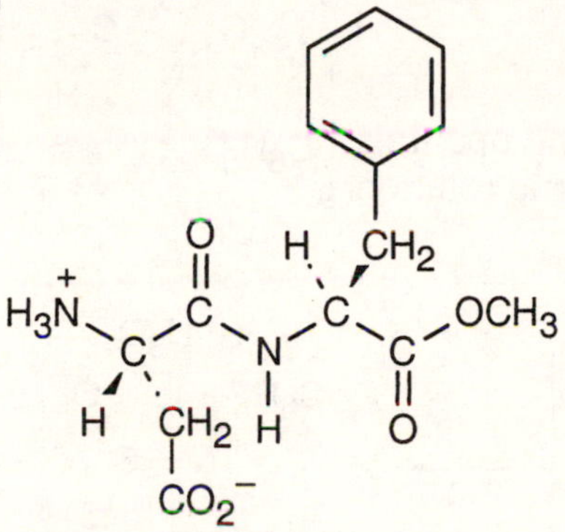

Aspartame at pH = 7.3

19.53

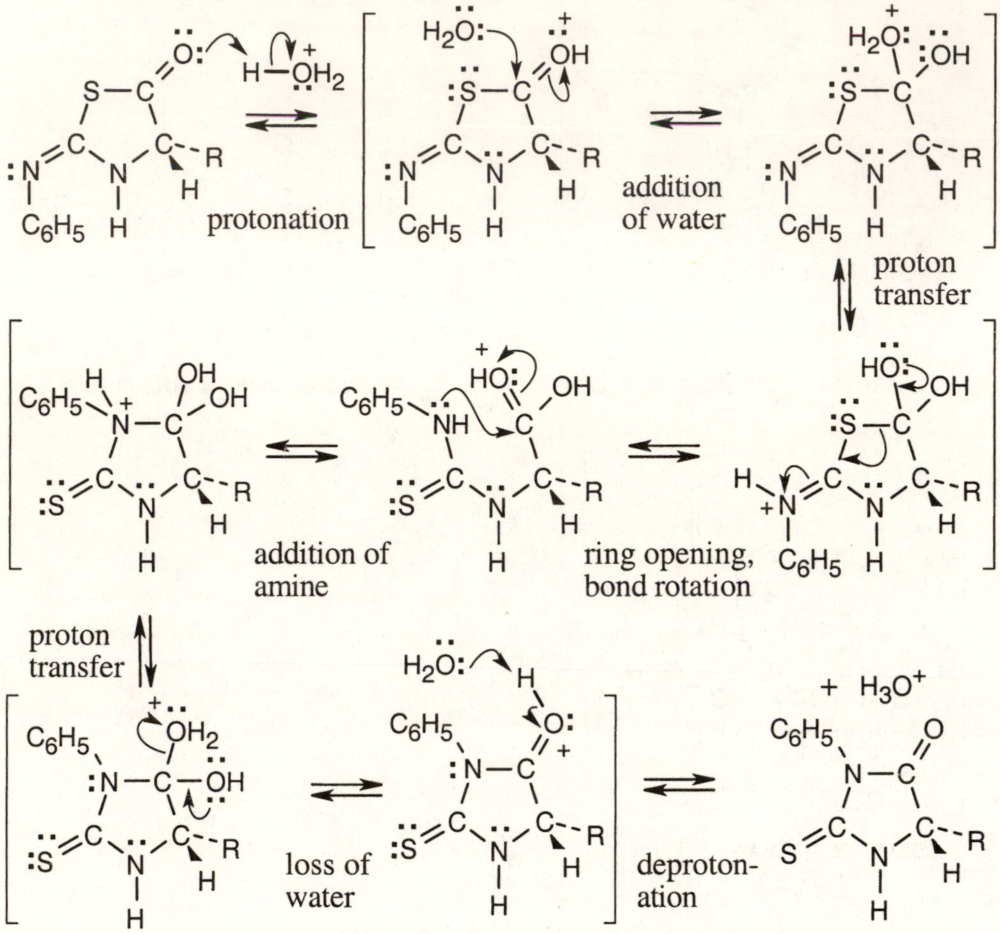

protonation

addition of water

proton transfer

ring opening, bond rotation

addition of amine

proton transfer

loss of water

deprotonation

$+ H_3O^+$

19.54

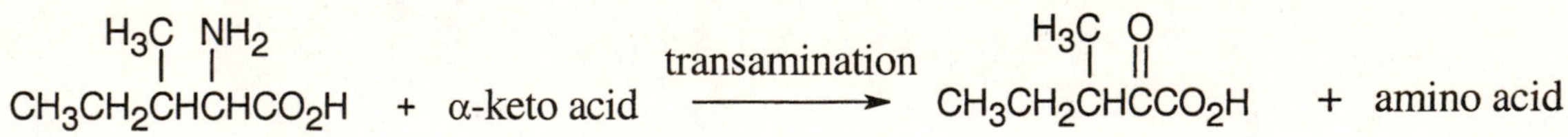

$$CH_3CH_2CH(CH_3)CH(NH_2)CO_2H + \alpha\text{-keto acid} \xrightarrow{\text{transamination}} CH_3CH_2CH(CH_3)C(=O)CO_2H + \text{amino acid}$$

Chapter 20 – Amino Acid Metabolism

Chapter Outline

I. Overview of metabolism and biochemical energy (Section 20.1).
   A. Metabolism.
      1. The reactions that take place in the cells of organisms are collectively called metabolism.
         a. The reactions that produce smaller molecules from larger molecules are called catabolism.
         b. The reactions that build larger molecules from smaller molecules are called anabolism.
      2. Catabolism can be divided into four stages:
         a. In digestion, bonds in food are hydrolyzed to yield sugars, fatty acids, and amino acids.
         b. These small molecules are degraded to acetyl CoA.
         c. In the citric acid cycle, acetyl CoA is catabolized to CO_2, and energy is produced.
         d. Energy from the citric acid cycle enters the electron transport chain, where ATP is synthesized.
   B. Biochemical energy.
      1. ATP, a phosphoric acid anhydride, is the storehouse for biochemical energy.
      2. The breaking of a P–O bond of ATP can be coupled with an energetically unfavorable reaction, so that the overall energy change is favorable.
      3. The resulting phosphates are much more reactive than the original compounds.

II. Catabolism of amino acids (Sections 20.2 –20.4).
   A. The pathway to amino acid catabolism:
      1. The amino group is removed as ammonia.
      2. Ammonia is removed as urea.
      3. What remains is converted to a compound that enters the citric acid cycle.
   B. Deamination (Section 20.2)
      1. Transamination is the first step.
         a. The $-NH_2$ group of an amino acid adds to enzyme-bound PLP to form an amino acid imine.
         b. Deprotonation of the imine causes rearrangement to an α-keto acid imine.
         c. Reprotonation of the PLP carbon leads to tautomerization.
         d. The second imine is hydrolyzed to give an α-keto acid and an amino derivative of pyridoxal phosphate.
         e. The pyridoxal derivative transfers its amino group to α-ketoglutarate, to regenerate PLP-enzyme and form glutamate.
      3. Deamination.
         The glutamate from transamination undergoes oxidative deamination to yield ammonium ion and α-ketoglutarate.
   C. The urea cycle (Section 20.3).
      1. Urea reacts with bicarbonate to form carbamoyl phosphate.
         Two ATPs are also involved.
      2. Carbamoyl phosphate reacts with the terminal amino group of ornithine to yield citrulline.
      3. Citrulline undergoes condensation with aspartate to form argininosuccinate.
      4. Elimination from argininosuccinate gives fumarate and arginine.
      5. Hydrolysis of arginine regenerates ornithine and produces urea.

D. The carbon chains (Section 20.4).
   1. General features.
      a. Carbon chains are converted to one of 7 citric acid cycle intermediates.
      b. Amino acids that are directly converted to citric acid cycle intermediates are glucogenic.
      c. Amino acids that are converted to acetoacetate or acetyl CoA are ketogenic.
      d. Some amino acids are both glucogenic and ketogenic.
   2. Catabolism to pyruvate (6 amino acids).
      a. Alanine is transaminated.
      b. Serine undergoes PLP-catalyzed dehydration.
      c. Cysteine is first oxidized to a sulfinate, then it is transaminated, and finally loses SO_2 to give pyruvate.
   3. Asparagine and aspartate are converted to either oxaloacetate or to fumarate.
   4. Histidine is converted to glutamate by a four-step reaction sequence.
      a. Elimination of ammonia gives *trans*-urocanate.
      b. Addition of water, followed by tautomerization, yields imidazolone 5-propionate.
      c. Hydrolysis gives *N*-formaminoglutamate.
      d. The coenzyme tetrahydrofolate brings about loss of the formamino group to produce glutamate.

III. Biosynthesis of amino acids (Section 20.5)
   A. General features.
      1. Humans are able to synthesize only 11 of 20 amino acids.
      2. The other 9 (essential amino acids) must be provided in the diet.
   B. Alanine, aspartate and glutamate are synthesized from either pyruvate or from citric acid cycle intermediates.
   C. Asparagine and glutamine are synthesized from aspartate and glutamate.
   D. Arginine and proline are synthesized from glutamate.
   E. Threonine is synthesized in plants from aspartate by a five-step sequence.
      1. Aspartate is phosphorylated to yield aspartyl β-phosphate.
      2. Aspartyl β-phosphate is reduced to aspartate semialdehyde.
      3. Further reduction gives homoserine, which is phosphorylated to produce phosphohomoserine.
      4. The final step involves elimination of phosphate, addition of water and rearrangement to form threonine.

Solutions to Problems

20.1

$$^{-}O-C(=O)-OH \xrightarrow{ATP \;\to\; ADP} {}^{2-}O_3PO-C(=O)-OH$$

20.2

$^{2-}O_3PO$ … $\ddot{N}H_2$ … CO_2^- … CO_2^- … CH_3 → $^{2-}O_3PO$ … CO_2^- … + H_2O … CO_2^- … CH_3

PMP   α-Ketoglutarate   α-Keto acid imine tautomer

Nucleophilic acyl substitution, followed by loss of water, forms the imine tautomer.

$H_2\ddot{N}$–Enz

$^{2-}O_3PO$ … H H … CO_2^- … CO_2^- … CH_3 → + $H_3\overset{+}{N}$–Enz … $^{2-}O_3PO$ … H … CO_2^- … CO_2^- … CH_3

α-Keto acid imine tautomer   α-Keto acid imine

A lysine residue deprotonates the carbon next to the ring, leading to tautomerization.

$^{2-}O_3PO$ … H … CO_2^- … H–$\overset{+}{N}H_2$–Enz … CO_2^- … CH_3 → + $H_2\ddot{N}$–Enz … $^{2-}O_3PO$ … H … H CO_2^- … CO_2^- … CH_3

α-Keto acid imine   PLP glutamate imine

Enzymatic protonation of the keto acid imine yields PLP glutamate imine.

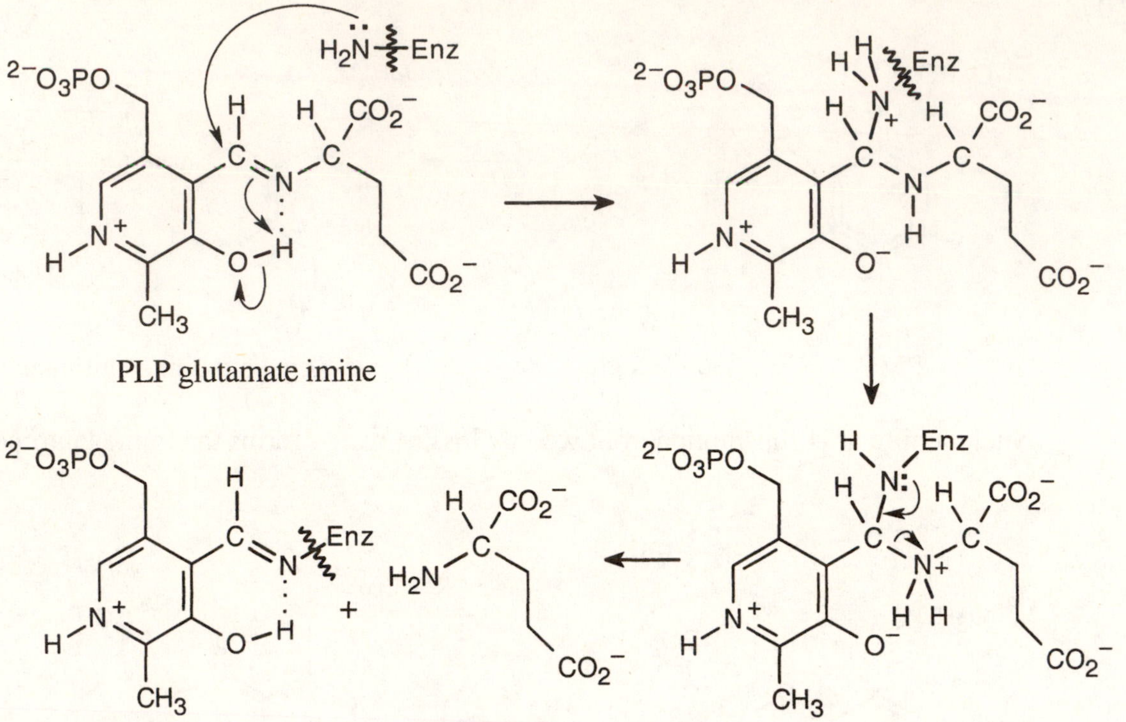

PLP glutamate imine

PLP imine

Glutamate

Addition of the enzyme, followed by loss of glutamate, regenerates PLP imine.

20.3

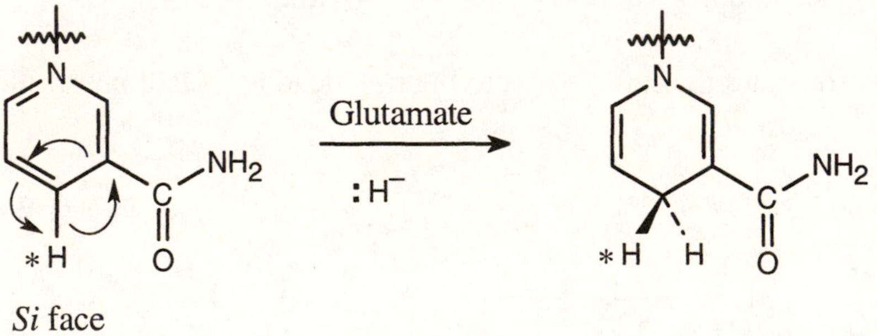

Glutamate

$:H^-$

Si face

The face that lies in the front of the plane of the page is the *Si* face.Since the added hydride ion is transferred to the rear of NAD^+, it is transferred to the *Re* face.

20.4

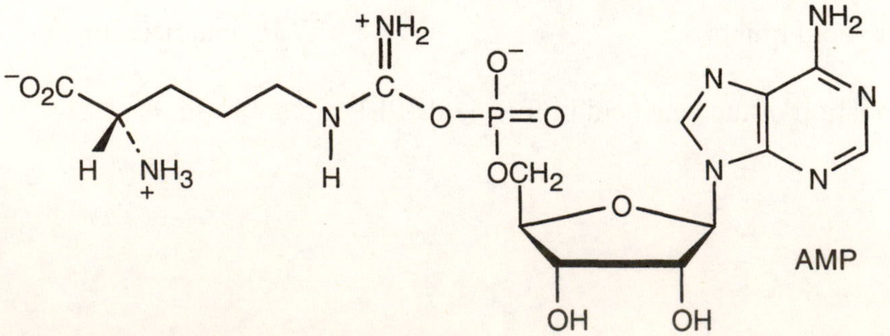

AMP

20.5

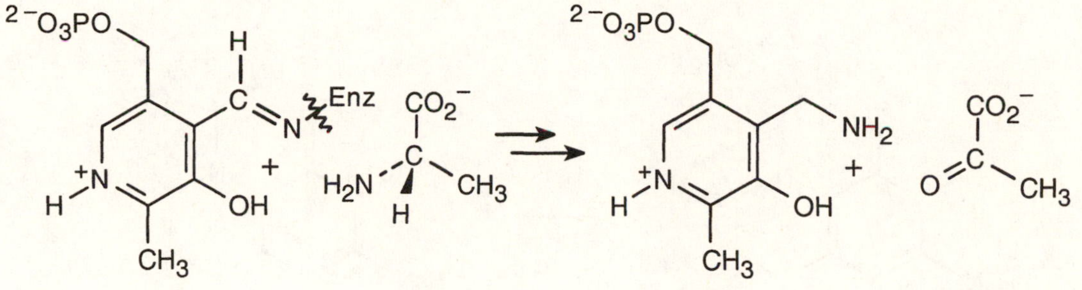

PLP–Enz Alanine PMP Pyruvate

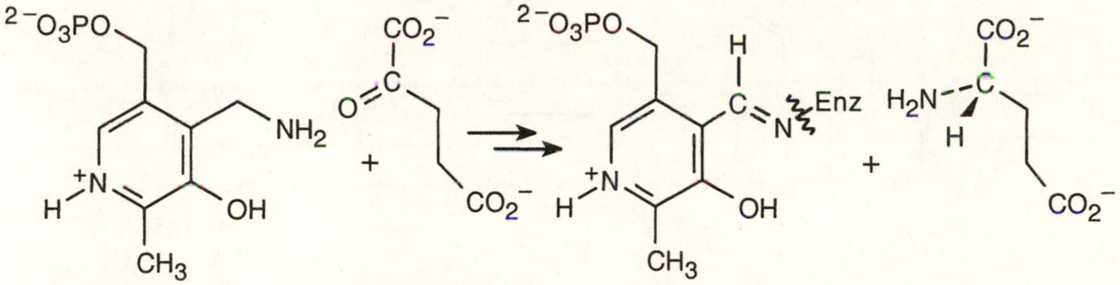

PMP α-Keto-glutarate PLP–Enz Glutamate

The detailed mechanisms of these transformations appear in Figure 20.2 and in Problem 20.2.

20.6 This is a typical nucleophilic acyl substitution reaction, with water displacing ammonia.

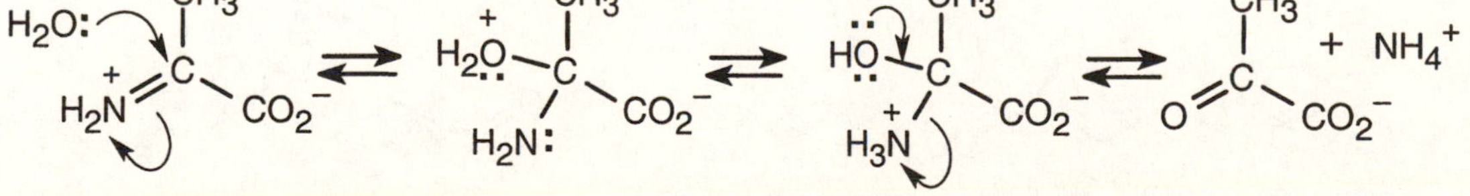

20.7 This mechanism is almost the same as the mechanism in the previous problem.

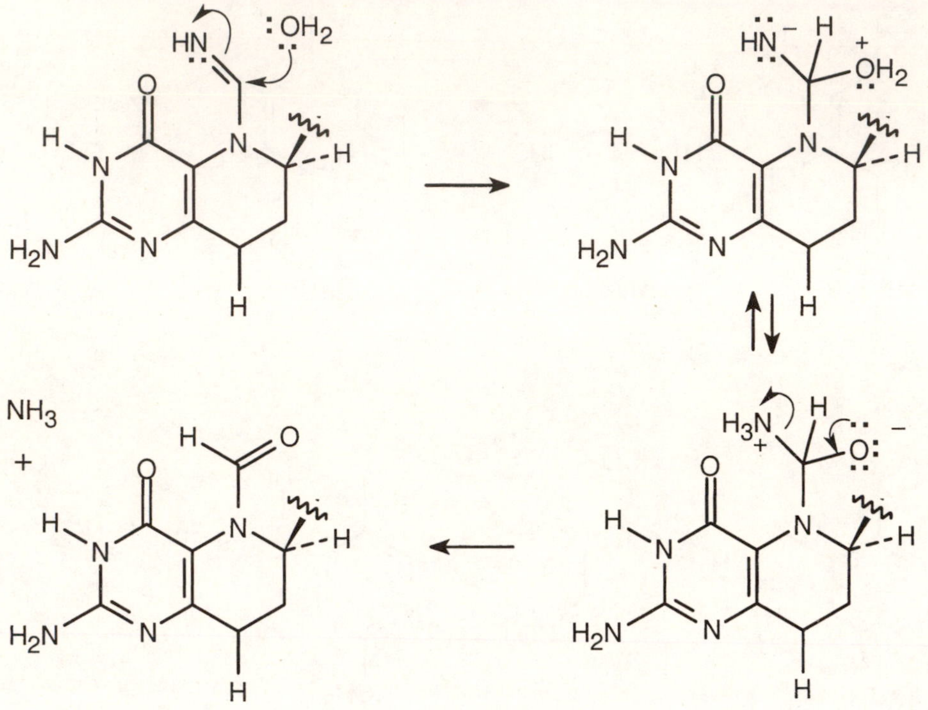

This reaction is nucleophilic acyl substitution.

20.8

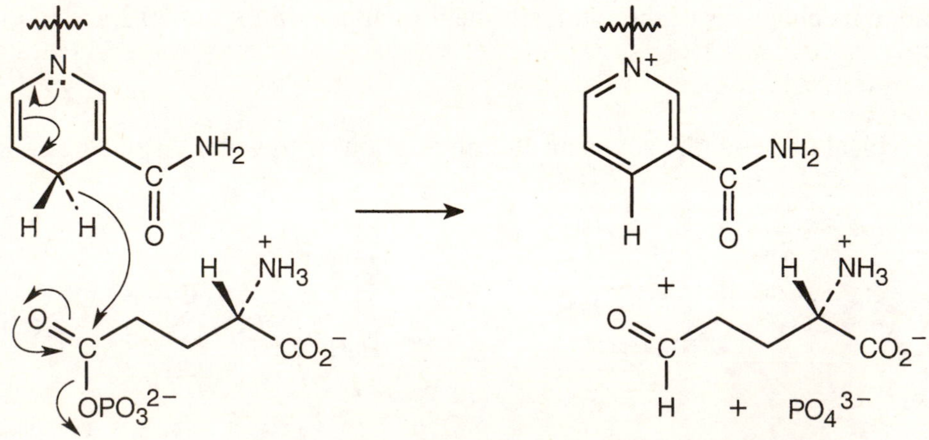

20.9

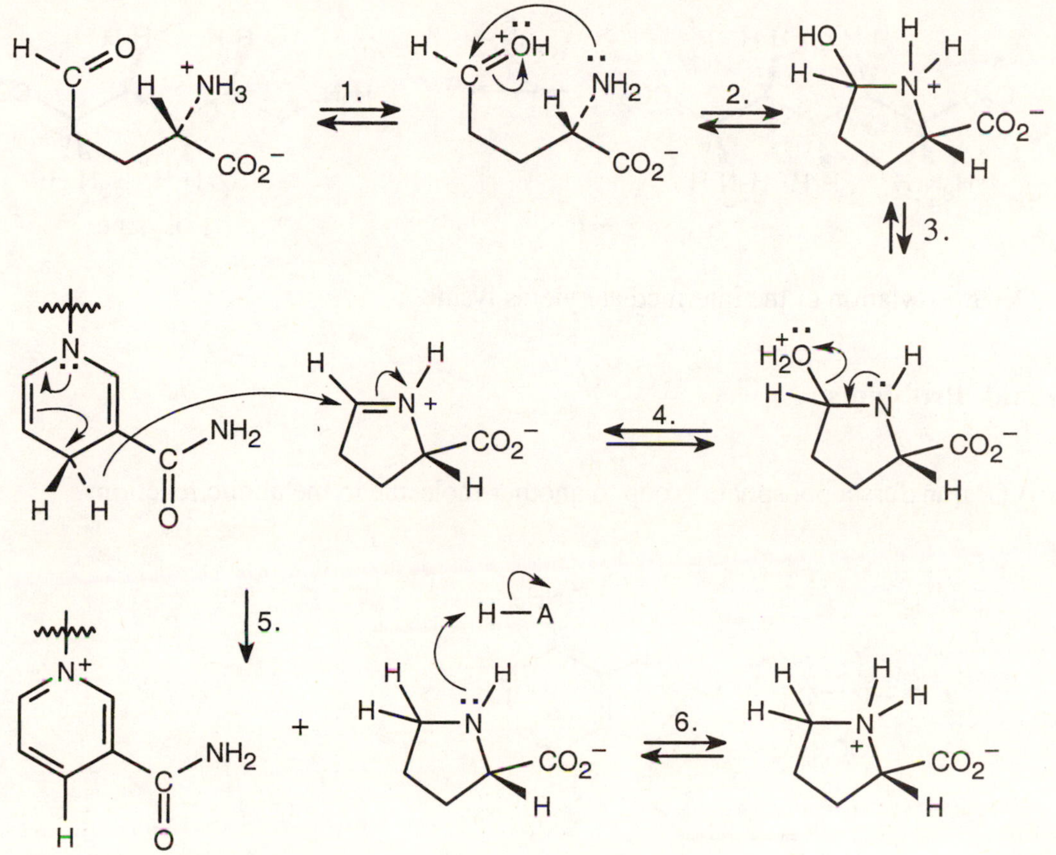

Step 1: Proton shift.
Steps 2-4: Nucleophilic acyl substitution (addition of amine, proton shift, loss of H_2O).
Step 5: NADH reduction.
Step 6: Protonation.

Visualizing Chemistry

20.10

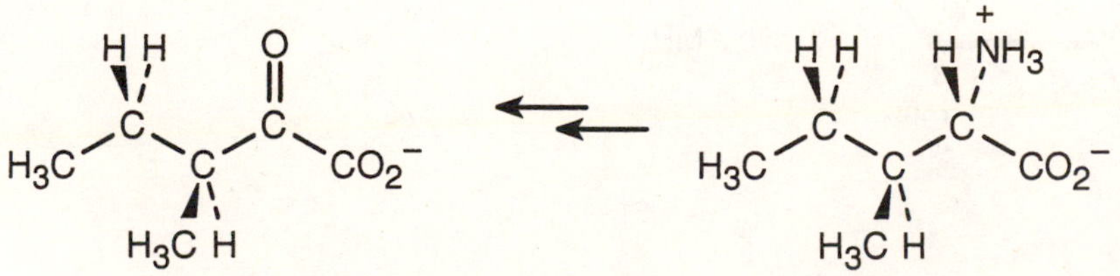

Isoleucine

20.11

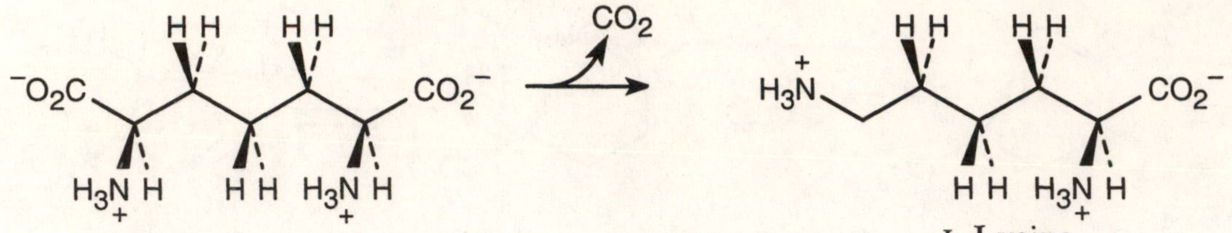

$^{-}O_2C$–CH($\overset{+}{N}H_3$)–CH₂–CH₂–CH($\overset{+}{N}H_3$)–CO_2^- $\xrightarrow{-CO_2}$ $H_3\overset{+}{N}$–CH₂–CH₂–CH₂–CH₂–CH($\overset{+}{N}H_3$)–CO_2^-

L-Lysine

Decarboxylation of the intermediate yields lysine.

Additional Problems

20.12 ATP transfers a phosphate group to another molecule in metabolic reactions.

20.13

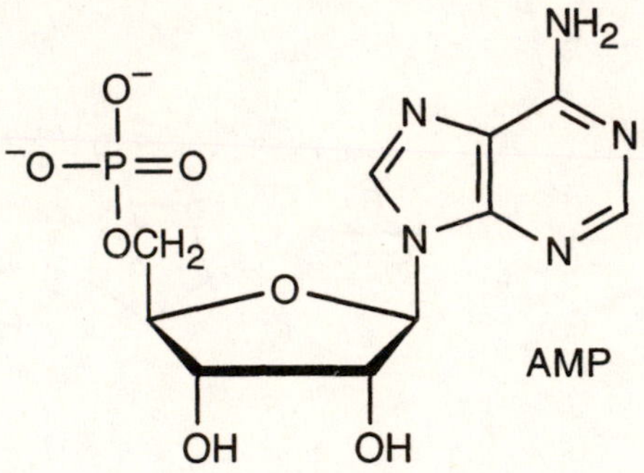

AMP

20.14

cyclic AMP

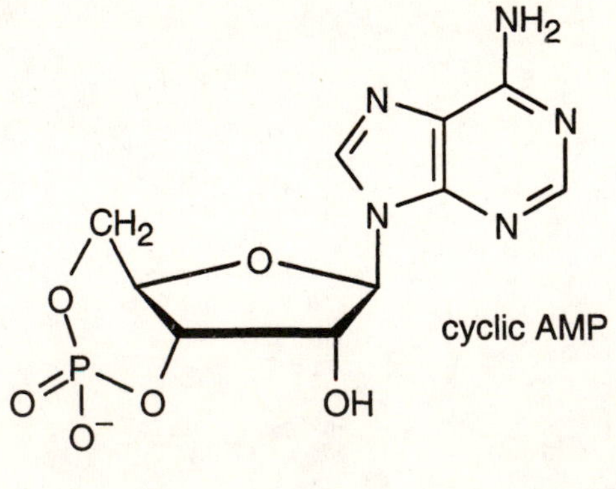

20.15

(1)

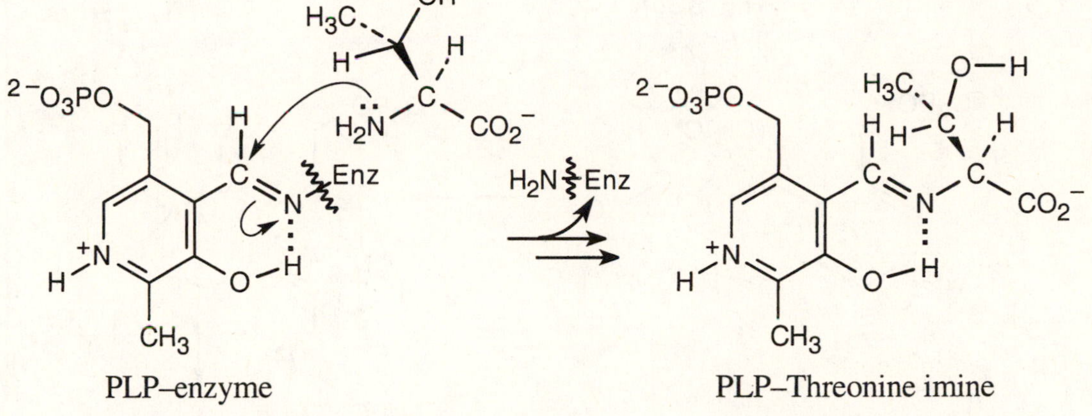

PLP–enzyme

H_2N–Enz

PLP–Threonine imine

(2)

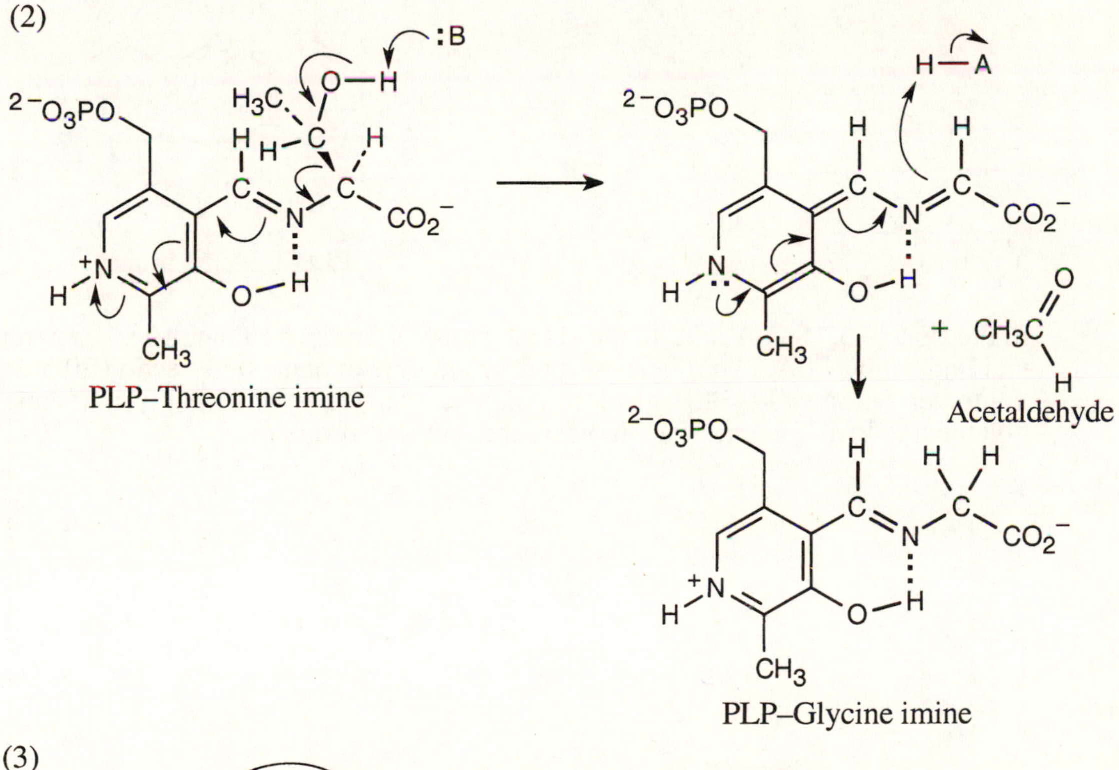

PLP–Threonine imine

Acetaldehyde

PLP–Glycine imine

(3)

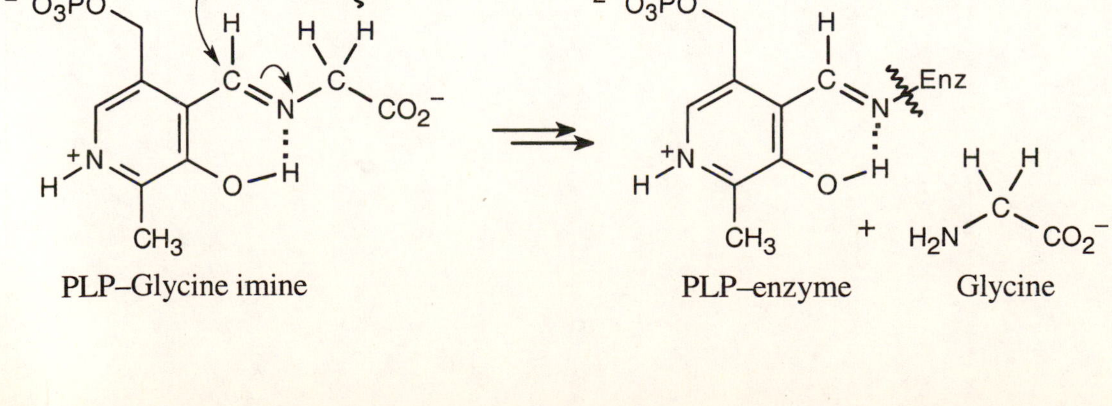

PLP–Glycine imine

PLP–enzyme

Glycine

20.16

$H—SCoA$

B:

$^{2-}O_3PO$

CH_3C (=O)

CO_2^-

^{+}N

CH_3

$H_2\overset{+}{N}$–Enz

CH_3CSCoA (=O)

Acetyl CoA

H_2N–Enz

PLP–Glycine imine

In the first step, $^{-}$SCoA adds to the ketone group to bring about a retro-Claisen reaction, yielding acetyl CoA. (The "heart-shaped" arrows are explained in Section 20.5.) Tautomerization yields PLP–glycine imine. The conversion of PLP–glycine imine to glycine and PLP–enzyme was shown in the previous problem.

20.17

(1)

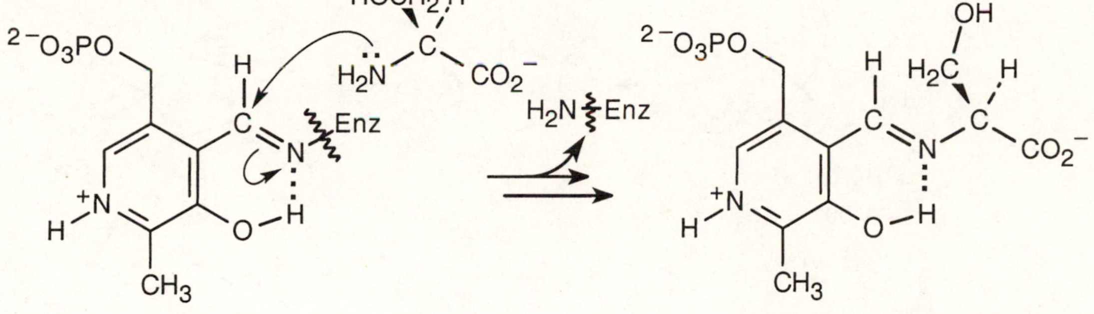

PLP–enzyme

PLP–Serine imine

(2)

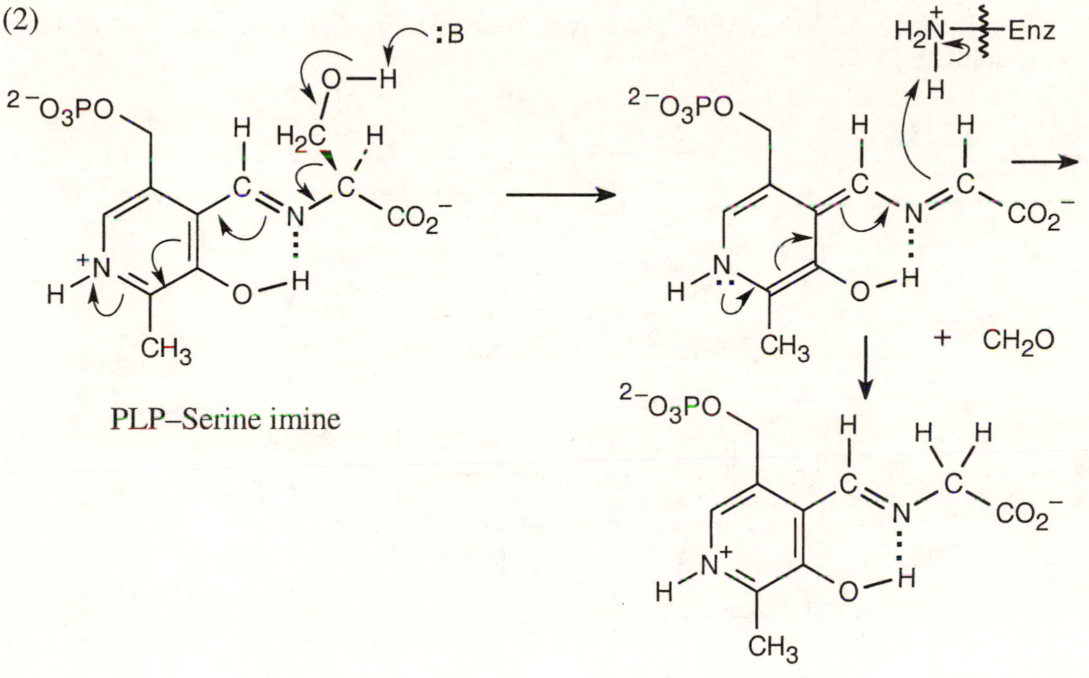

PLP–Serine imine

$+ \ CH_2O$

PLP–Glycine imine

20.18

(a)

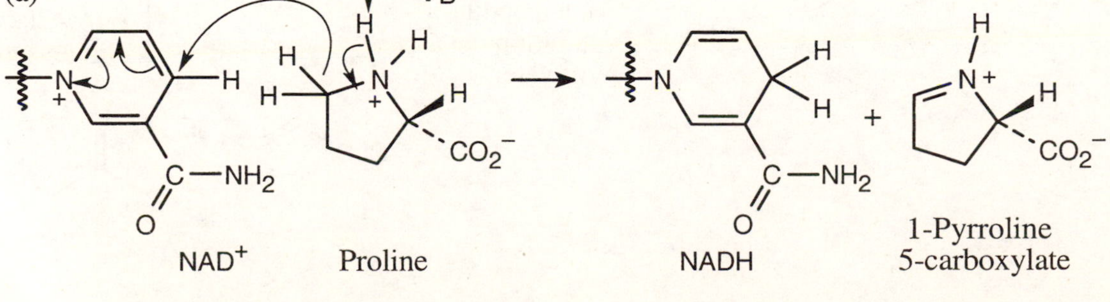

NAD^+

Proline

NADH

1-Pyrroline
5-carboxylate

(b)

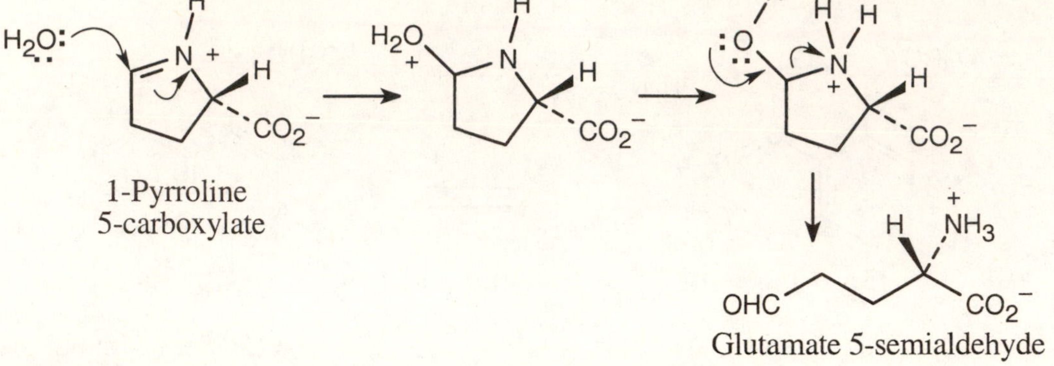

1-Pyrroline 5-carboxylate

Glutamate 5-semialdehyde

(c) Either NAD^+ or $NADP^+$ is likely to be used to oxidize glutamate 5-semialdehyde to glutamate.

20.19

(a)

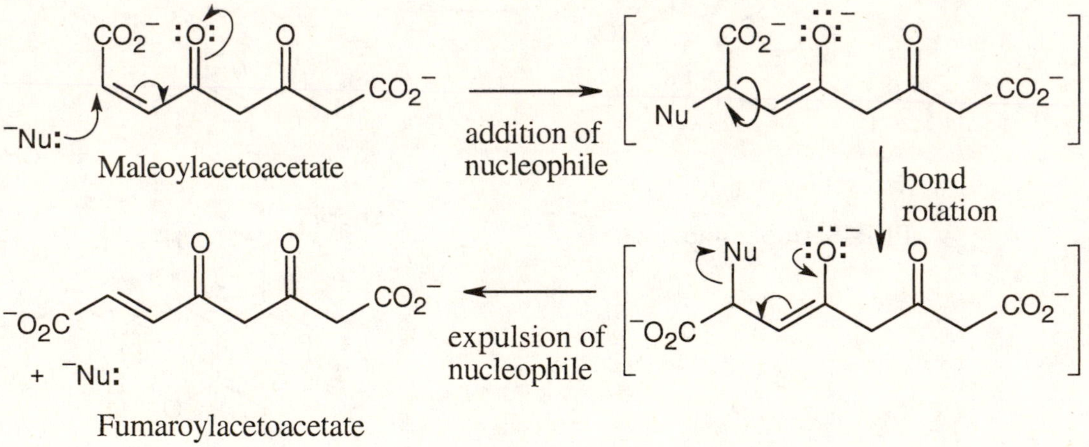

Maleoylacetoacetate

addition of nucleophile

bond rotation

expulsion of nucleophile

Fumaroylacetoacetate

(b)

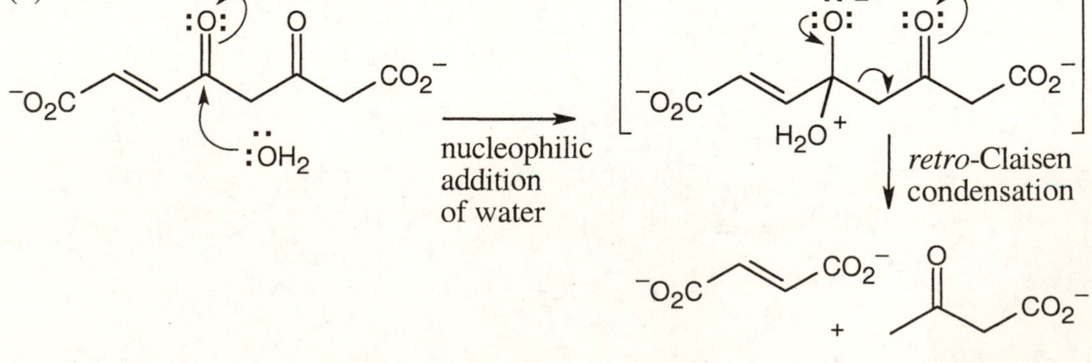

nucleophilic addition of water

retro-Claisen condensation

20.20

(a)

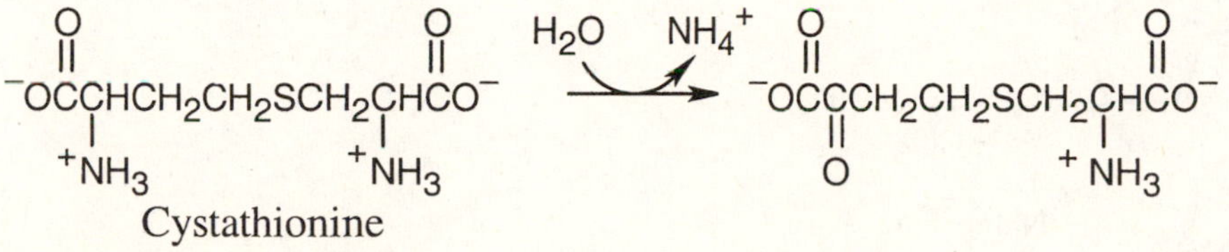
Cystathionine

(b)

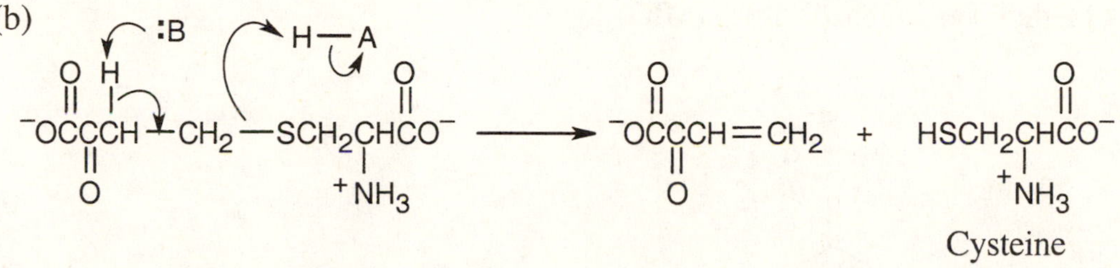
Cysteine

(c)

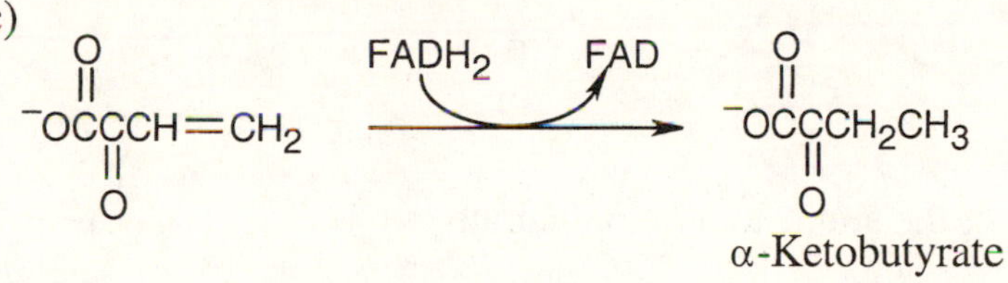
α-Ketobutyrate

The product of double-bond reduction is α-ketobutyrate. $FADH_2$ is the necessary enzyme cofactor.

20.21

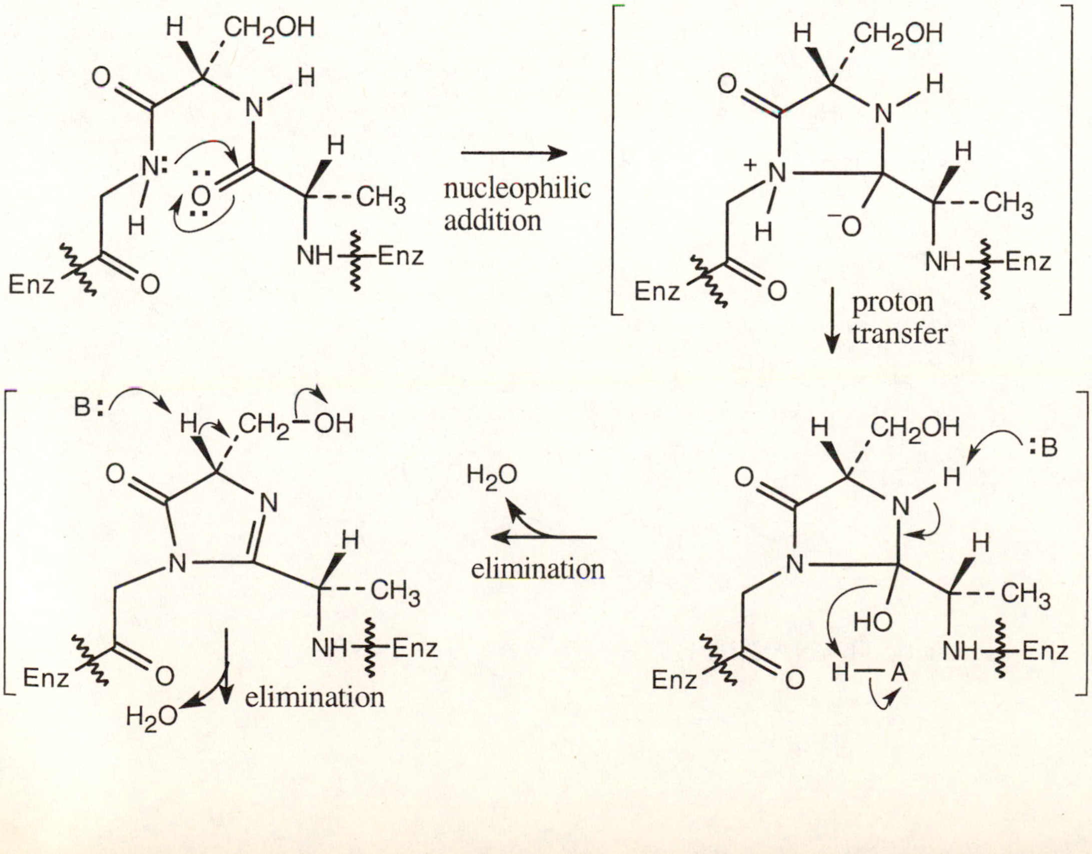
nucleophilic addition
proton transfer
elimination
elimination

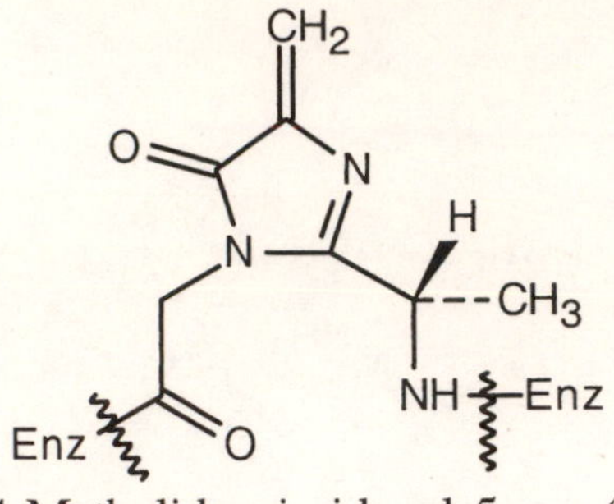

4-Methylideneimidazol-5-one (MIO)

20.22

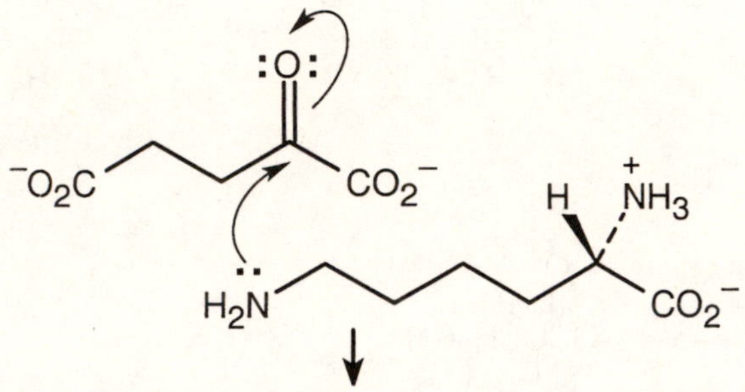

Nucleophilic addition of the amine to α-ketoglutarate.

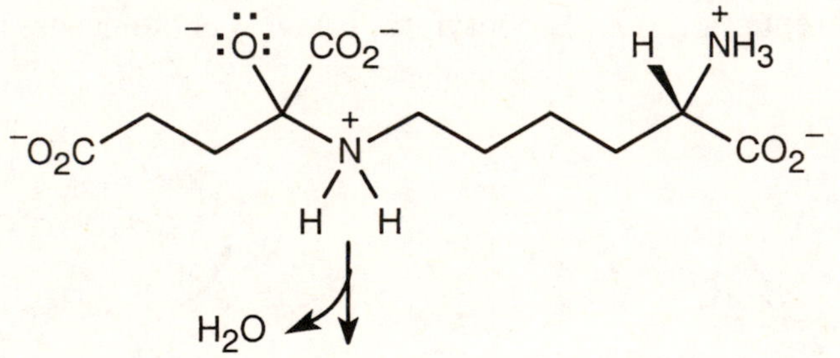

Loss of water.

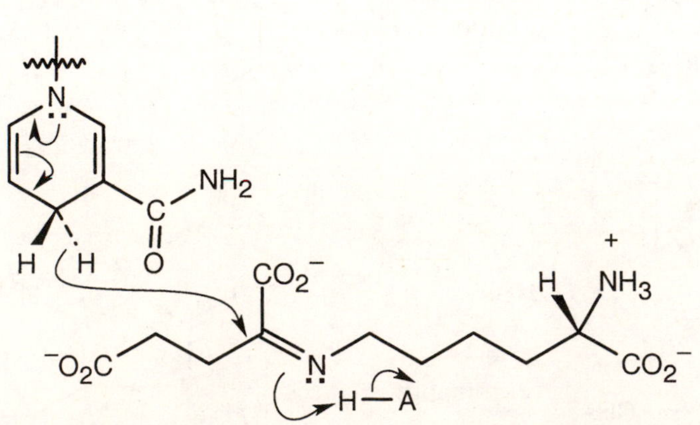

Reduction by NADPH/H^+

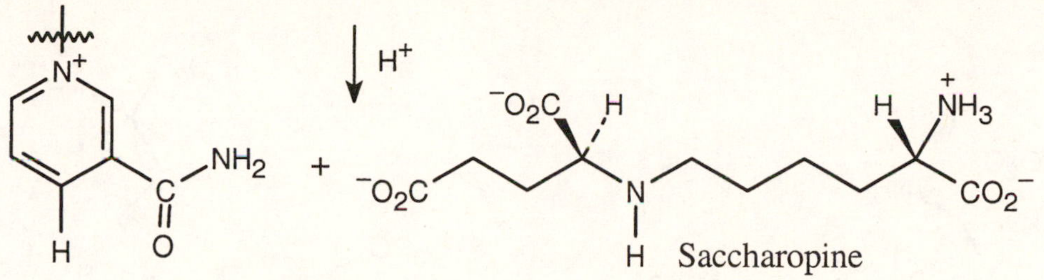

$\downarrow H^+$

Saccharopine

20.23

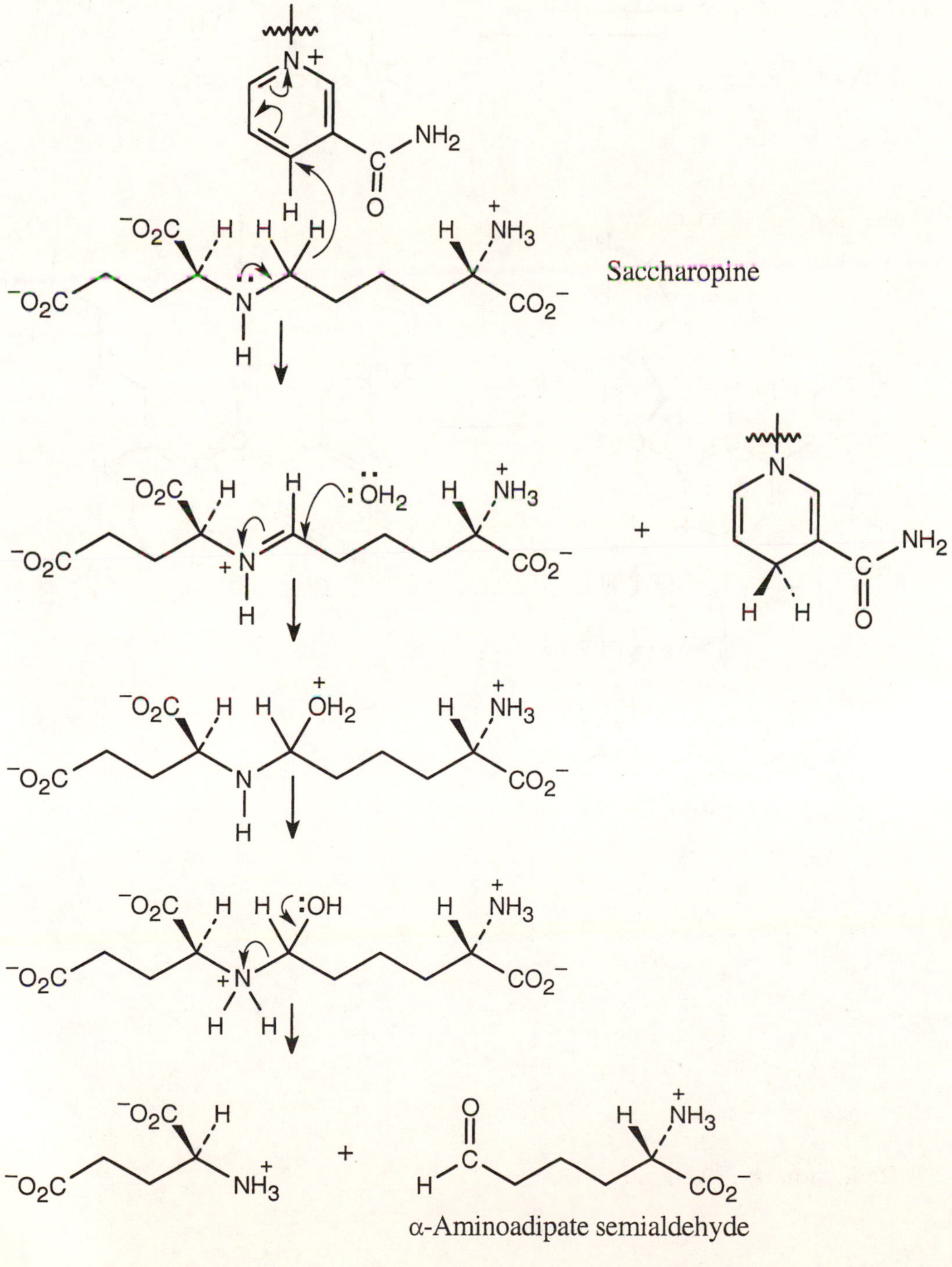

Saccharopine

α-Aminoadipate semialdehyde

20.24

$^{-}O_2C$ $\overset{+}{N}H_3$ H

$\overset{+}{N}H_3$ H $^{-}O_2C$

Enz–PLP Enz–NH_2

$^{2-}O_3PO$ H H $^{-}O_2C$ $\overset{+}{N}H_3$ H C N C C O O H O H $\overset{+}{N}$ H CH_3

CO_2

CO_2^- $\overset{+}{N}H_3$ H $^{2-}O_3PO$ H C N C H N H O H CH_3 H–A

$^{-}O_2C$ $\overset{+}{N}H_3$ H $^{-}O_3PO$ H C N H C H H $\overset{+}{N}$ H O H CH_3

Enz–NH_2 Enz–PLP

H $\overset{+}{N}H_3$ $H_3\overset{+}{N}$ CO_2^-

Lysine

20.25

1
2 hydrolysis
H_2O HSCoA
A—H
:O:
C
CO_2^-
:CH_2CSCoA
^-O CO_2^-
O=C SCoA
HO CO_2^- :B
H
H
CO_2^-
elimination 3
CO_2^-
H—A
C=CHCO$_2^-$
$H_2\ddot{O}$:
4 conjugate addition
H CO_2^-
C CO_2^-
C
H OH
NADH/H$^+$ NAD$^+$
5 oxidation
O
H C—$\ddot{O}$:$^-$
C CO_2^-
C
O H—A
6 decarboxylation of a β-keto acid
CO_2^-
C
O
+ CO_2

Review Unit 8: Amines, Amino Acids, and Amino Acid Metabolism

Major Topics Covered (with vocabulary):

Amines:
primary, secondary, tertiary amine quaternary ammonium salt arylamine heterocyclic amine pyramidal inversion K_b reductive amination Hofmann elimination reaction nitrogen rule pyrrole thiophene furan pyridine fused-ring heterocycle pyrimidine purine

Amino acids:
amino acid zwitterion amphoteric α-amino acid side chain isoelectric point (p*I*) electrophoresis Henderson-Hasselbalch equation amidomalonate synthesis reductive amination enantioselective synthesis

Peptides:
residue backbone *N*-terminal amino acid C-terminal amino acid disulfide link amino acid analysis Edman degradation phenylthiohydantoin trypsin chymotrypsin peptide synthesis protection BOC derivative DCC Merrifield solid-phase technique

Proteins:
fibrous protein globular protein primary structure secondary structure tertiary structure quaternary structure α-helix β-pleated sheet salt bridge enzyme cofactor turnover number coenzyme vitamin isomerase hydrolase ligase lyase oxidoreductase transferase denaturation

Metabolic pathways:
metabolism anabolism catabolism digestion phosphoric acid anhydride ATP NAD^+ $NADH/H^+$ citric acid cycle electron transport chain

Amino acid metabolism:
deamination transamination pyridoxal phosphate (PLP) urea cycle glucogenic amino acids ketogenic amino acids

Types of Problems:

After studying these chapters, you should be able to:

- Name and draw amines, and classify amines as primary, secondary, tertiary , quaternary, arylamines, or heterocyclic amines.
- Predict the basicity of alkylamines, arylamines and heterocyclic amines.
- Synthesize alkylamines and arylamines by several routes.
- Predict the products of reactions involving alkylamines and arylamines.
- Use diazonium salts in reactions involving arylamines, including diazo coupling reactions.
- Propose mechanisms for reactions involving alkylamines and arylamines.
- Identify amines by spectroscopic techniques.
- Draw orbital pictures of heterocycles and explain their acid-base properties.

- Identify the common amino acids and draw them with correct stereochemistry in dipolar form.
- Explain the acid-base behavior of amino acids.
- Synthesize amino acids.
- Draw the structure of simple peptides.
- Deduce the structure of peptides and proteins.
- Outline the synthesis of peptides.
- Explain the classification of proteins and the levels of structure of proteins.
- Draw structures of reaction products of amino acids and peptides.

- Explain the basic concepts of metabolism, and understand the energy relationships of biochemical reactions.
- Outline the mechanisms of transamination, oxidative deamination and the urea cycle.
- Explain the catabolic and biosynthetic pathways for amino acid side chains.

Points to Remember:

* In many of the mechanisms in this group of chapters, the steps involving proton transfer are not explicitly shown. The proton transfers occur between the proton and the conjugate base with the most favorable p*K* of those present in the solution. These steps have been omitted at times to simplify the mechanisms.

* For an amine, the larger the value of pK_a of its ammonium ion, the stronger the base. The smaller the value of pK_b of the amine, the stronger the base.

* At physiological pH, the side chains of the amino acids aspartic acid and glutamic acid exist as anions, and the side chains of the amino acids lysine and arginine exist as cations. The imidazole ring of histidine exists as a mixture of protonated and neutral forms.

* Since the amide backbone of a protein is neutral and uncharged, the isoelectric point of a protein or peptide is determined by the relative numbers of acidic and basic amino acid residues present in the peptide.

* Note the difference between transamination and oxidative deamination. Transamination is a reaction in which an amino group of an α-amino acid is transferred to α-ketoglutarate, yielding an α-keto acid and glutamate. In oxidative deamination, glutamate loses its amino group in an NAD^+-dependent reaction that regenerates α-ketoglutarate and produces NH_4^+.

* The catabolism of amino acids involves three parts: Removal of ammonia; excretion of ammonia; catabolism of the side chain.

Self-Test:

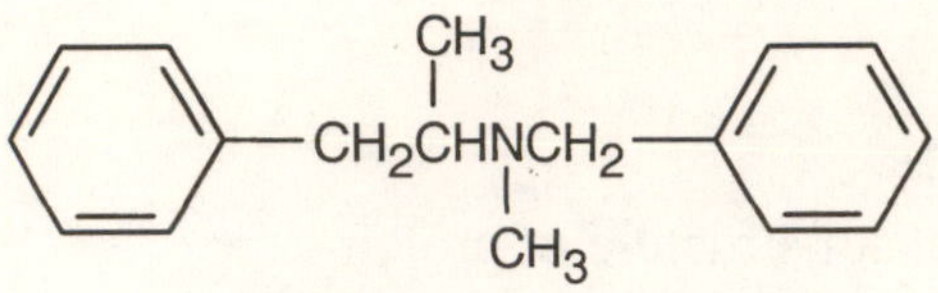

A
Benzphetamine
(an appetite suppressant)

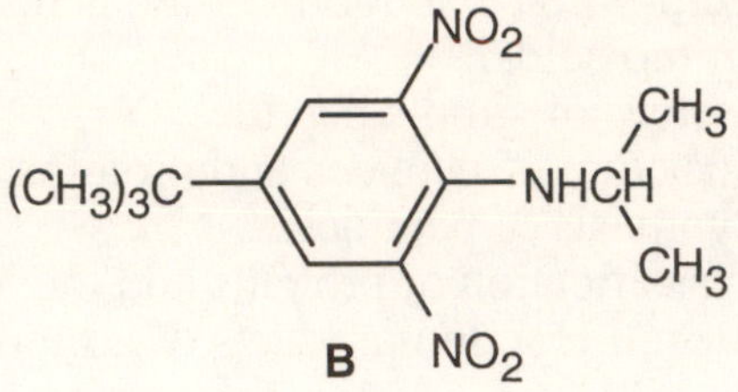

B
Butralin
(an herbicide)

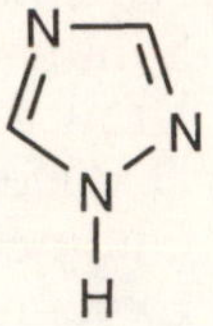

C
1,2,4-Triazole

What type of amine is **A**? Do you expect it to be more or less basic than ammonia? Than aniline? What product do you expect from Hofmann elimination of **A**? What significant absorptions might be seen in the IR spectrum of **A**? What information can be obtained from the mass spectrum? Plan a synthesis of **B** from benzene.

One of the nitrogens of 1,2,4-triazole (**C**) is less basic than the others. Which one is it? What product do you expect when **C** is treated with Br_2, $FeBr_3$?

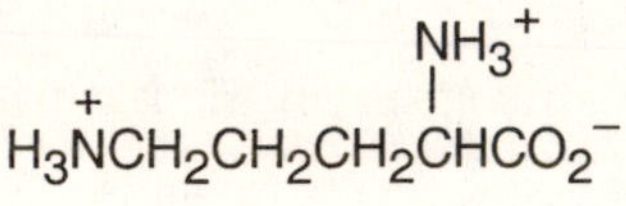

$H_3\overset{+}{N}CH_2CH_2CH_2CH(NH_3^+)CO_2^-$

D
Ornithine

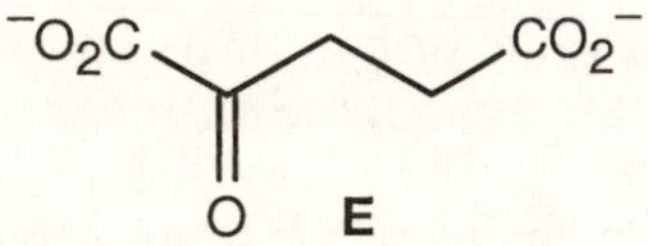

E

Ornithine (**D**) is a nonstandard amino acid that occurs in metabolic processes. Which amino acid does it most closely resemble? Estimate pK_a values and pI for ornithine, and draw the major form present at pH = 2, pH = 6, and pH = 11. If ornithine were a component of proteins, how would it affect the tertiary structure of a protein?

Compound **E** is an important intermediate in amino acid metabolism. Identify it and explain its function. **E** is also an end product of catabolism of carbon chains in amino acid metabolism. Name 3 amino acids whose side chains are catabolized to **E.**

Tyr–Gly–Gly–Phe–Leu–Arg–Arg–Ile–Arg–Pro–Lys–Leu–Lys–Trp–Asp–Asn–Gln
Porcine Dynorphin (**F**)

Dynorphin (**F**)is a neuropeptide. Indicate the *N*-terminal end and the C-terminal end. Show the products of cleavage with: (a) carboxypeptidase; (b) trypsin; (c) chymotrypsin. Show the *N*-phenylthiohydantoin that results from treatment of **F** with phenyl isothiocyanate. Do you expect **F** to be an acidic, a neutral or a basic peptide?

Multiple Choice:

1. The ammonium ion of which of the following amines has the smallest value of pK_a?
(a) Methylamine (b) Trimethylamine (c) Aniline (d) *p*-Bromoaniline

2. Which of the following heterocyclic amines is not aromatic?
(a) pyrrole (b) pyrimidine (c) pyridine (d) piperidine

3. In order for the percent protonated amine to be 99% and the percent unprotonated amine to be 1%, the value of (pK_a – pH) should be approximately:
(a) 10^2 (b) 2 (c) –2 (d) It depends on the amine

4. To find the p*I* of an acidic amino acid:
(a) find the average of the two lowest pK_a values (b) find the average of the two highest pK_a values (c) find the average of all pK_a values (d) use the value of the pK_a of the side chain.

5. Which of the following techniques can synthesize a single enantiomer of an amino acid?
(a) Hell-Volhard-Zelinskii reaction (b) reductive amination (c) amidomalonate synthesis (d) hydrogenation of a *Z* enamido acid

6. The purple product that results from the reaction of ninhydrin with an amino acid contains which group of the amino acid?
(a) the amino group (b) the amino nitrogen (c) the carboxylic acid group (d) the side chain

7. Which of the following reagents is not used in peptide synthesis?
(a) Phenylthiohydantoin (b) Di-*tert*-butyl dicarbonate (c) Benzyl alcohol (d) Dicyclohexylcarbodiimide

8. Which structural element is not present in myoglobin?
(a) salt bridges (b) regions of α-helix (c) hydrophobic regions (d) quaternary structure

9. In which part of the urea cycle is ATP transformed to AMP?
(a) ornithine —> citrulline (b) citrulline —> arginosuccinate (c) arginosuccinate —> arginine (d) arginine —> ornithine

10. Which of the following amino acids is strictly ketogenic?
(a) phenylalanine (b) tyrosine (c) leucine (d) isoleucine

Chapter 21 – Biomolecules: Carbohydrates

Chapter Outline

I. Classification of carbohydrates(Section 21.1).
   A. Simple *vs*. complex:
      1. Simple carbohydrates can't be hydrolyzed to smaller units.
      2. Complex carbohydrates are made up of two or more simple sugars linked together.
         a. A disaccharide is composed of two monosaccharides.
         b. A polysaccharide is composed of three or more monosaccharides.
   B. Aldoses *vs*. ketoses:
      1. A monosaccharide with an aldehyde carbonyl group is an aldose.
      2. A monosaccharide with a ketone carbonyl group is a ketose.
   C. *Tri-*, *tetr-* , *pent-*, etc. indicate the number of carbons in the monosaccharide.

II. Monosaccharides (Sections 21.2 – 21.7).
   A. Configurations of monosaccharides (Section 21.2 – 21.4).
      1. Fischer projections (Section 21.2).
         a. Each chirality center of a monosaccharide is represented by a pair of crossed lines.
            i. The horizontal line represents bonds coming out of the page.
            ii. The vertical line represents bonds going into the page.
         b. Allowed manipulations of Fischer projections:
            i. A Fischer projection can be rotated on the page by 180°, but not by 90° or 270°.
            ii. Holding one group steady, the other three groups can be rotated clockwise or counterclockwise.
         c. Rules for assigning *R,S* configurations.
            i. Assign priorities to the substituents in the usual way.
            ii. Perform one of the two allowed motions to place the lowest priority group at the top of the Fischer projection.
            iii. Determine the direction of rotation of the arrow that travels from group 1 to group 2 to group 3, and assign *R* or *S* configuration.
         d. Carbohydrates with more than one chirality center are shown by stacking the centers on top of each other.
            The carbonyl carbon is placed at or near the top of the Fischer projection.
      2. D,L sugars (Section 21.3).
         a. (*R*)-Glyceraldehyde is also known as D-glyceraldehyde.
         b. In D sugars, the –OH group farthest from the carbonyl group points to the right.
            Most naturally-occurring sugars are D sugars.
         c. In L sugars, the –OH group farthest from the carbonyl group points to the left.
         d. D,L designations refer only to the configuration farthest from the carbonyl carbon and are unrelated to the direction of rotation of plane-polarized light.
      3. Configurations of the aldoses (Section 21.4).
         a. There are 4 aldotetroses – D and L erythrose and threose.
         b. There are 4 D,L pairs of aldopentoses: ribose, arabinose, xylose and lyxose.
         c. There are 8 D,L pairs of aldohexoses : allose, altrose, glucose, mannose, gulose, idose, galactose, and talose.
         d. A scheme for drawing and memorizing the D-aldohexoses:
            i. Draw all –OH groups at C5 pointing to the right.
            ii. Draw the first four –OH groups at C4 pointing to the right and the second four pointing to the left.

iii. Alternate –OH groups at C3: two right, two left, two right, two left.
iv. Alternate –OH groups at C2: right, left, etc.
v. Use the mnemonic " All altruists gladly make gum in gallon tanks" to assign names.

B. Cyclic structures of monosaccharides (Section 21.5).
1. Hemiacetal formation.
a. Monosaccharides are in equilibrium with their internal hemiacetals.
i. Glucose exists primarily as a six-membered pyranose ring, formed by the –OH group at C5 and the aldehyde.
ii. Fructose exists primarily as a five-membered furanose ring.
b. Structure of pyranose rings.
i. Pyranose rings have a chair-like geometry.
ii. The hemiacetal oxygen is at the right rear for D-sugars.
iii. An –OH group on the right in a Fischer projection is on the bottom face in a pyranose ring, and an –OH group on the left is on the top face.
iv. For D sugars, the $-CH_2OH$ group is on the top.
2. Mutarotation.
a. When a monosaccharide cyclizes, a new chirality center is generated.
i. The two diastereomers are anomers.
ii. The form with the anomeric –OH group trans to the $-CH_2OH$ group is the α anomer (minor anomer).
iii. The form with the anomeric –OH group cis to the $-CH_2OH$ group is the β anomer (major anomer).
b. When a solution of either pure anomer is dissolved in water, the optical rotation of the solution reaches a constant value.
i. This process is called mutarotation.
ii. Mutarotation is due to the reversible opening and recyclizing of the hemiacetal ring and is catalyzed by both acid and base.

C. Reactions of monosaccharides (Section 21.6).
1. Ester and ether formation.
a. Esterification occurs by treatment with an acid anhydride or acid chloride.
b. Ethers are formed by treatment with methyl iodide and Ag_2O.
c. Ester and ether derivatives are crystalline and easy to purify.
2. Glycoside formation.
a. Treatment of a hemiacetal with an alcohol and an acid catalyst yields an acetal.
i. Acetals aren't in equilibrium with an open-chain form.
ii. Aqueous acid reconverts the acetal to a monosaccharide.
b. These acetals, called glycosides, occur in nature.
3. Phosphorylation.
a. Many carbohydrates are linked by their anomeric center to other biological molecules.
b. A monosaccharide is phosphorylated by ATP to yield a glycosyl phosphate.
c. The glycosyl phosphate reacts with a second nucleotide diphosphate in order to activate the anomeric –OH for a nucleophilic substitution reaction.
4. Reduction of monosaccharides.
Reaction of a monosaccharide with $NaBH_4$ yields an alditol.
5. Oxidation of monosaccharides.
a. Several mild reagents can oxidize the carbonyl group to a carboxylic acid (aldonic acid).
i. Tollens reagent, Fehling's reagent and Benedict's reagent all serve as tests for reducing sugars.
ii. All aldoses and some ketoses are reducing sugars, but glycosides are nonreducing.

iii. In the laboratory, aqueous Br_2 is used to oxidize aldoses (not ketoses).
b. The more powerful oxidizing agent, dilute HNO_3, oxidizes aldoses to dicarboxylic acids (aldaric acids).
c. The $-CH_2OH$ of an aldose can be oxidized enzymatically.
D. Eight essential monosaccharides (Section 21.7).
1. Glucose, galactose, mannose and xylose are monosaccharides.
2. Fucose is a deoxy sugar.
3. *N*-Acetylglucosamine and *N*-acetylgalactosamine are amino sugars.
4. *N*-Acetylneuraminic acid is the parent compound of the sialic acids.
III. Other carbohydrates (Sections 21.8 – 21.10).
A. Disaccharides (Section 21.8).
1. Cellobiose and maltose.
a. Cellobiose and maltose contain a 1,4'-glycosidic acetal bond between two glucose monosaccharide units.
The prime (') shows that the glycosidic bond is between two different sugars.
b. Maltose consists of two glucopyranose units joined by a 1,4'-α-glycosidic bond.
c. Cellobiose consists of two glucopyranose units joined by a 1,4'-β-glycosidic bond.
d. Both maltose and cellobiose are reducing sugars and exhibit mutarotation.
e. Humans can't digest cellobiose but can digest maltose.
2. Lactose.
a. Lactose consists of a unit of galactose joined by a β-glycosidic bond between C1 and C4 of a glucose unit.
b. Lactose is a reducing sugar found in milk.
3. Sucrose.
a. Sucrose is a disaccharide that yields glucose and fructose on hydrolysis.
a. Sucrose is called "invert sugar" because the sign of rotation changes when sucrose is hydrolyzed.
b. Sucrose is one of the most abundant pure organic chemicals in the world.
b. The two monosaccharides are joined by a glycosidic link between C1 of glucose and C2 of fructose.
c. Sucrose isn't a reducing sugar and doesn't exhibit mutarotation.
B. Polysaccharides and their synthesis (Section 21.9).
1. Polysaccharides have a reducing end and undergo mutarotation, but aren't considered to be reducing sugars because of their size.
2. Important polysaccharides.
a. Cellulose.
i. Cellulose consists of thousands of D-glucose units linked by 1,4'-β-glycosidic bonds.
ii. In nature, cellulose is used as structural material.
b. Starch.
i. Starch consists of thousands of D-glucose units linked by 1,4'-α-glycosidic bonds.
ii. Starch can be separated into amylose (water-soluble) and amylopectin (water-insoluble) fractions.
Amylopectin contains 1,6'-α-glycosidic branches.
iii. Starch is digested in the mouth by glycosidase enzymes, which only cleave α-glycosidic bonds.
c. Glycogen.
i. Glycogen is an energy-storage polysaccharide.
ii. Glycogen contains both 1,4'- and 1,6'-links.

3. An outline of the glycan assembly method of polysaccharide synthesis.
   a. A glycal (a monosaccharide with a C1–C2 double bond) is protected at C6 by formation of a silyl ether and at C3-C4 by formation of a cyclic carbonate ester.
   b. The protected glycal is epoxidized.
   c. Treatment of the glycal epoxide (in the presence of $ZnCl_2$) with a second glycal having a free C6 hydroxyl group forms a disaccharide.
   d. The process can be repeated.

C. Cell surface carbohydrates and carbohydrate vaccines (Section 21.10).
   1. Polysaccharides are involved in cell-surface recognition.
      a. Polysaccharide markers on the surface of red blood cells are responsible for blood-group incompatibility.
      b. Red blood cells have two types of markers (antigenic determinants) – A and B.
      c. Unusual carbohydrates are components of these markers.
   2. Possible anticancer vaccines have been synthesized from antibodies to cell-surface polysaccharides found on the surface of cancer cells.

Solutions to Problems

21.1

(a) Threose *an aldotetrose*

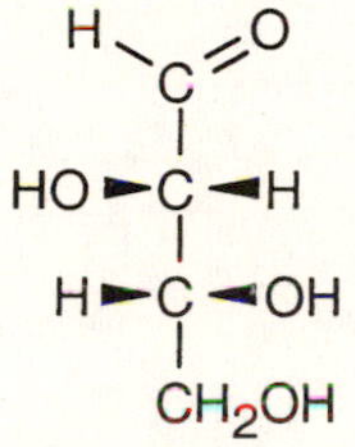

(b) Ribulose *a ketopentose*

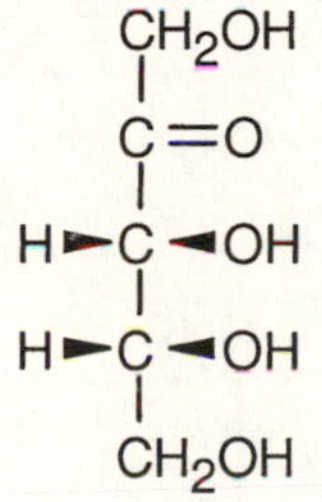

(c) Tagatose *a ketohexose*

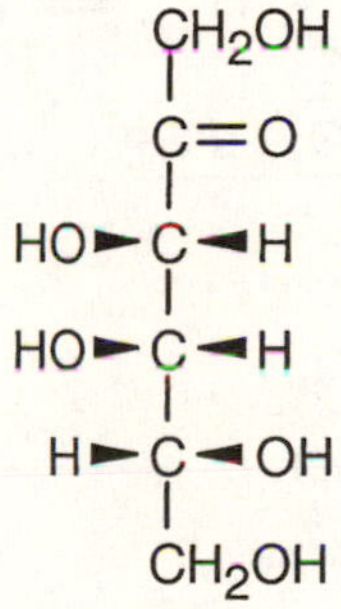

(d) 2-Deoxyribose *an aldopentose*

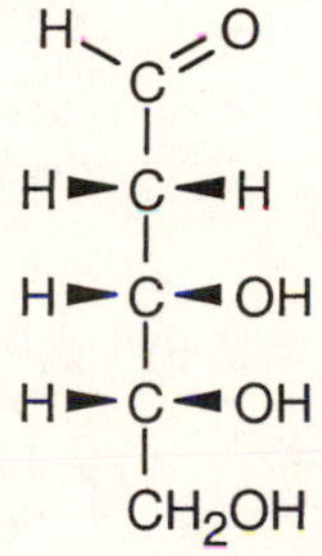

21.2 Horizontal bonds of Fischer projections point out of the page, and vertical bonds point into the page.

(a)

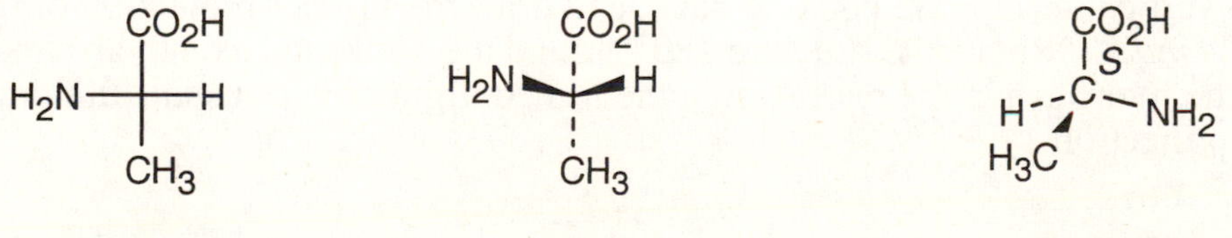

(b)

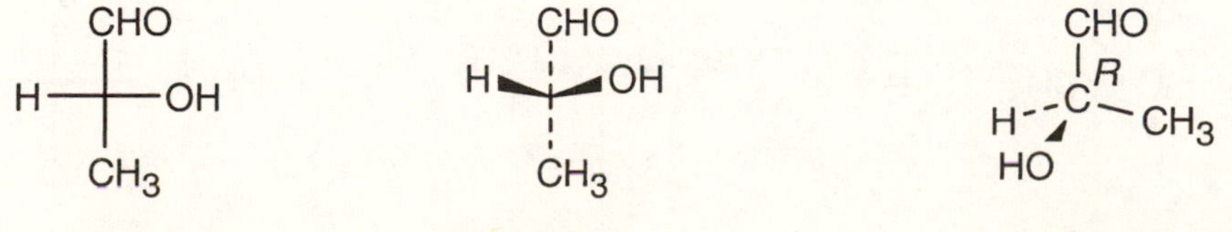

(c)

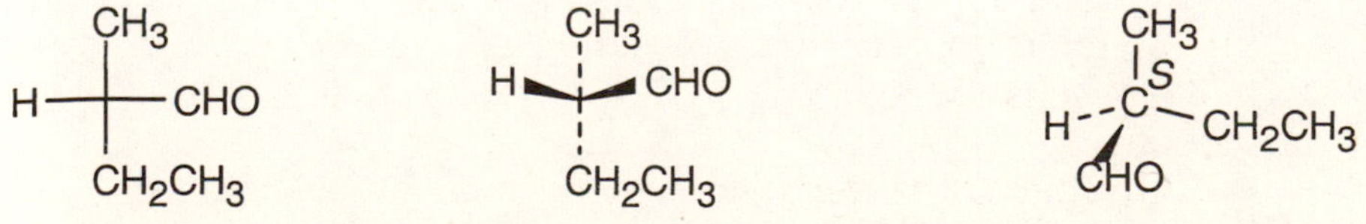

21.3 **Strategy:** To decide if two Fischer projections are identical, use the two allowable rotations to superimpose two groups of each projection. If the remaining groups are also superimposed after rotation, the projections represent the same enantiomer.

Solution:
Since –H is in the same position in both A and B, keep it steady, and rotate the other three groups. If, after rotation, all groups are superimposed, the two projections are identical. If only two groups are superimposed, the projections are enantiomers. Thus, A is identical to B.

A: CHO (top), HO (left), H (right, boxed, steady), CH_2OH (bottom) ≡ B: OH (top), $HOCH_2$ (left), H (right), CHO (bottom)

B: OH (top), $HOCH_2$ (left), H (right), CHO (bottom, boxed, steady) ≡ C: H (top), HO (left), CH_2OH (right), CHO (bottom)

D: CH_2OH (top), H (left), CHO (right), OH (bottom); 180° → OH (top), OHC (left), H (right), CH_2OH (bottom) ≢ B: OH (top), $HOCH_2$ (left), H (right), CHO (bottom)

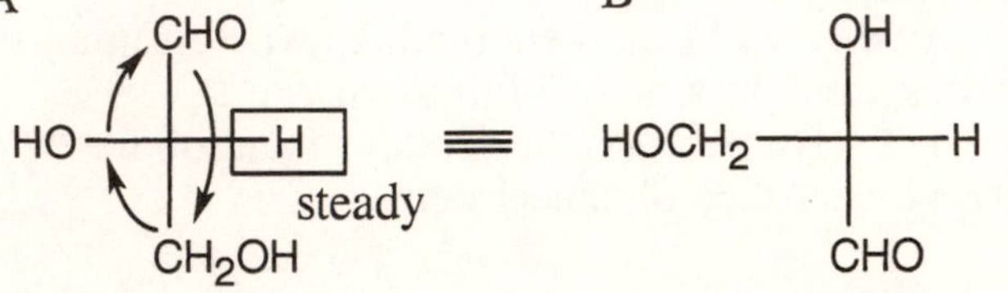

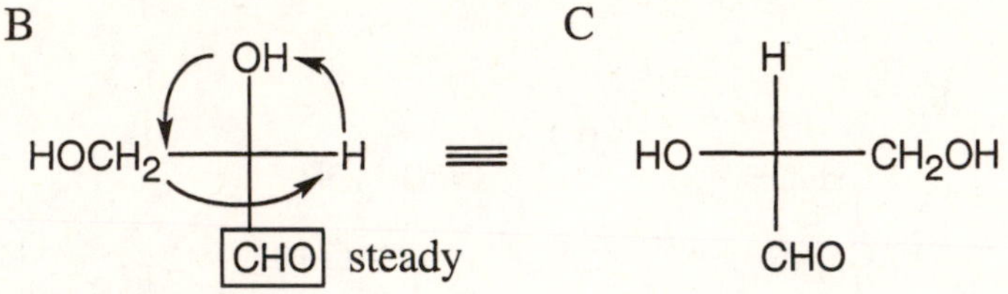

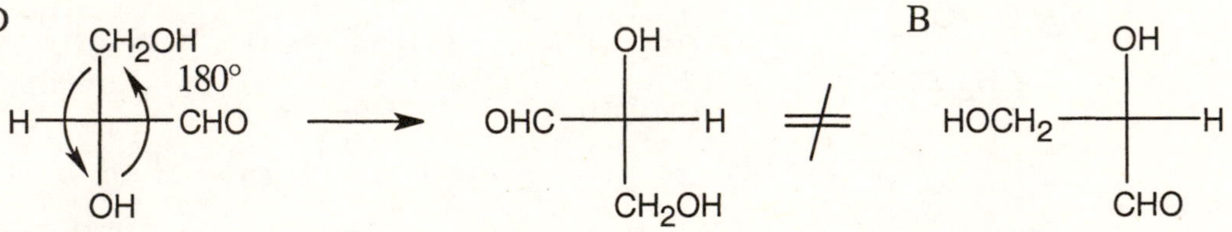

Projections A, B and C are identical, and D is their enantiomer.

21.4 The easiest way to solve this problem is to build a model, assign *R* or *S* configuration to the chirality center, manipulate the model so that two horizontal groups are pointing out and two vertical groups are pointing back, and draw the Fischer projection of the model. *Without a model*: Rotate the structure 180° around the horizontal axis to arrive at a drawing having the hydrogen at the rear. Assign the *R,S* configuration as usual, and draw the Fischer projection

$HOCH_2$--C(Cl)(H)--CH_3 = $HOCH_2$, CH_3, H, Cl about C = Fischer projection: H (top), $HOCH_2$ (left), CH_3 (right), Cl (bottom); *R*

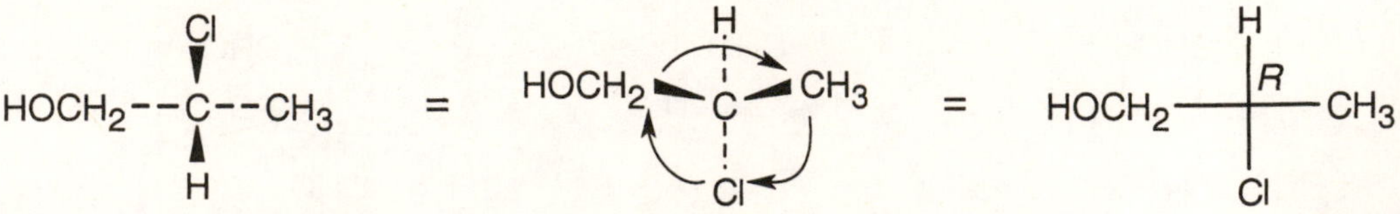

21.5 Draw the skeleton of the Fischer projection and add the –CHO and –CH_2OH groups to the top and bottom, respectively. Look at each carbon from the direction in which the –H and –OH point out of the page, and draw what you see on the Fischer projection.

View C3 from this side; –OH is on the right.

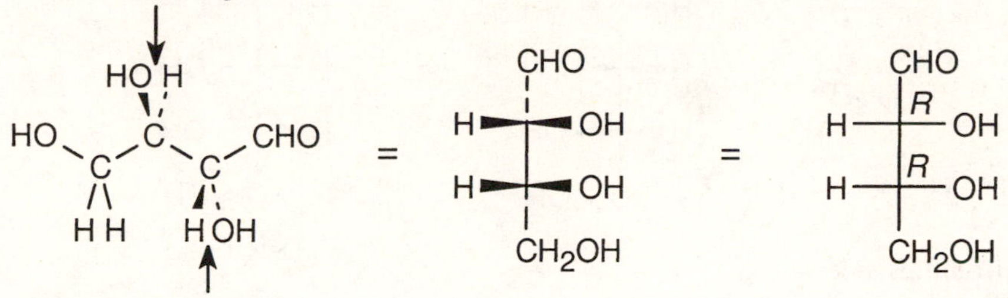

View C2 from this side; –OH is on the right.

21.6 The hydroxyl group bonded to the chiral carbon farthest from the carbonyl group points to the right in a D sugar, and points to the left in an L sugar.

(a) L-Erythrose (b) D-Xylose (c) D-Xylulose

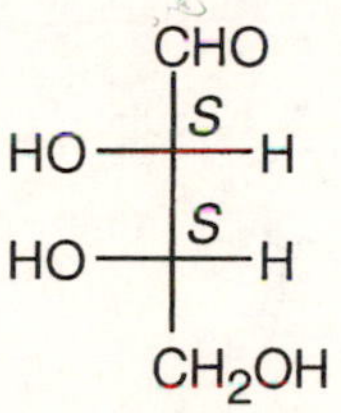

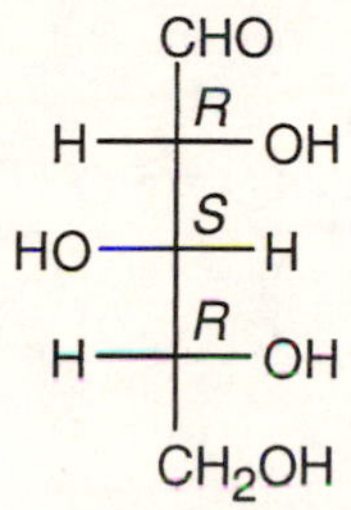

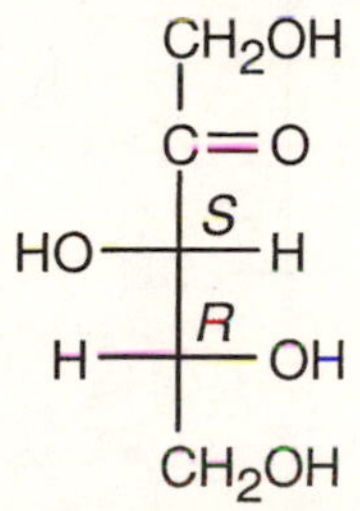

21.7

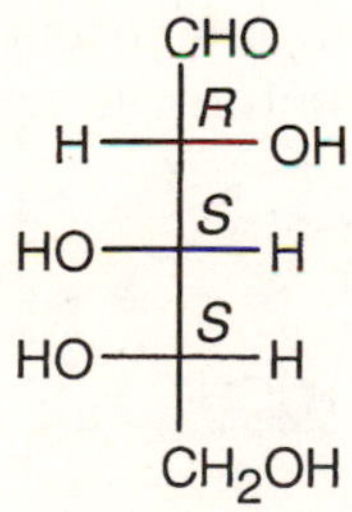

L-(+)-Arabinose

21.8

(a) L-Xylose (b) L-Galactose (c) L-Allose

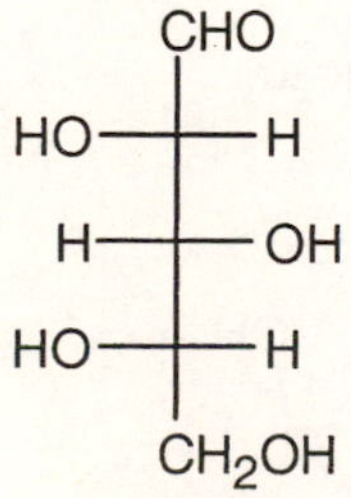

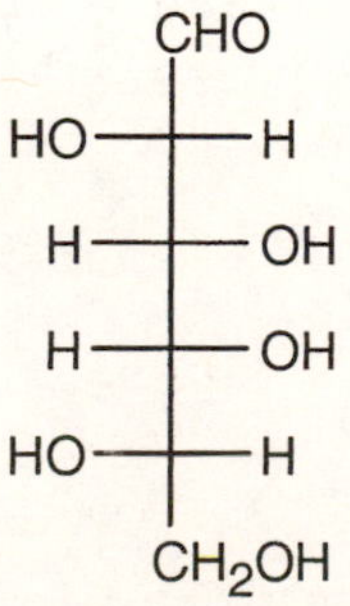

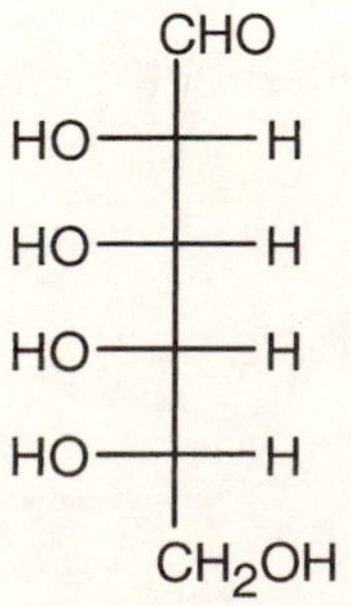

21.9 An aldoheptose has 5 chirality centers. Thus, there are $2^5 = 32$ aldoheptoses — 16 D aldoheptoses and 16 L aldoheptoses.

21.10 See Problem 21.5 for the method of solution.

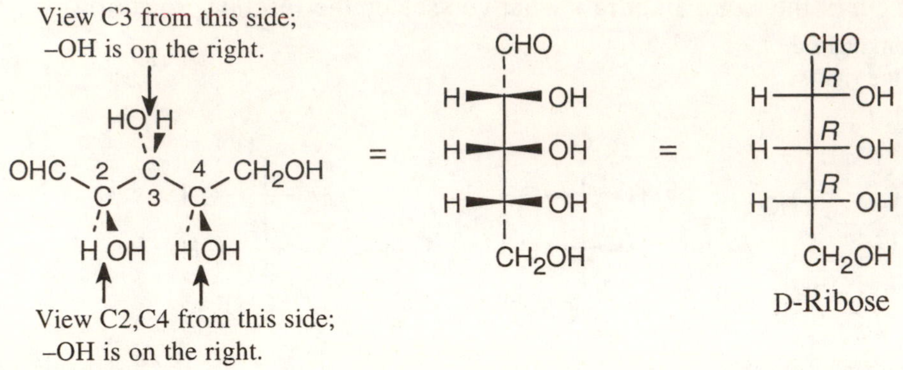

View C3 from this side; –OH is on the right.

HO H

OHC 2 C 4 CH2OH C 3 C

H OH H OH

View C2,C4 from this side; –OH is on the right.

CHO / H–OH / H–OH / H–OH / CH2OH = CHO / H–R–OH / H–R–OH / H–R–OH / CH2OH

D-Ribose

21.11 The steps for drawing a furanose are similar to the steps for drawing a pyranose. Ring formation occurs between the –OH group at C4 and the carbonyl carbon.

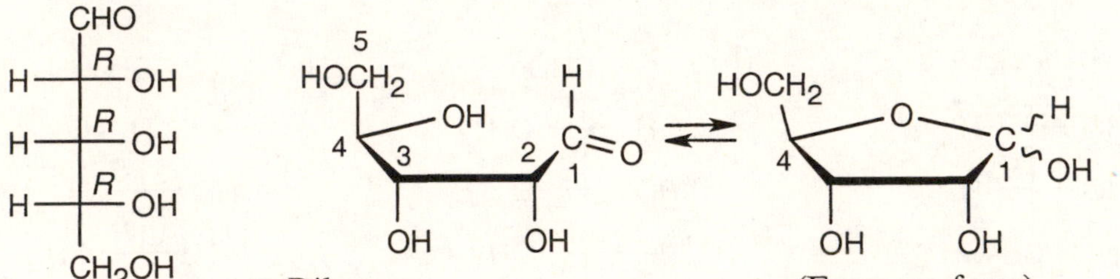

CHO / H–R–OH / H–R–OH / H–R–OH / CH2OH

D-Ribose

(Furanose form)

21.12 The furanose of fructose results from ring formation between the –OH group at C5 and the ketone at C2. In the α anomer, the anomeric –OH group is trans to the C6 –CH_2OH group, and in the β anomer the two groups are cis. In the pyranose form, cyclization occurs between the –OH group at C6 and the ketone. The more stable chair conformations are shown.

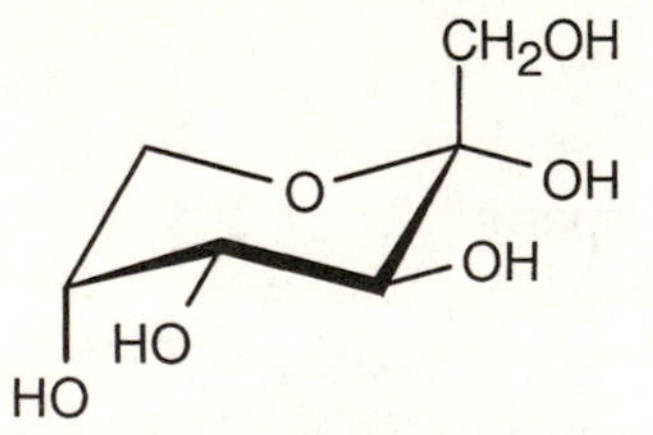

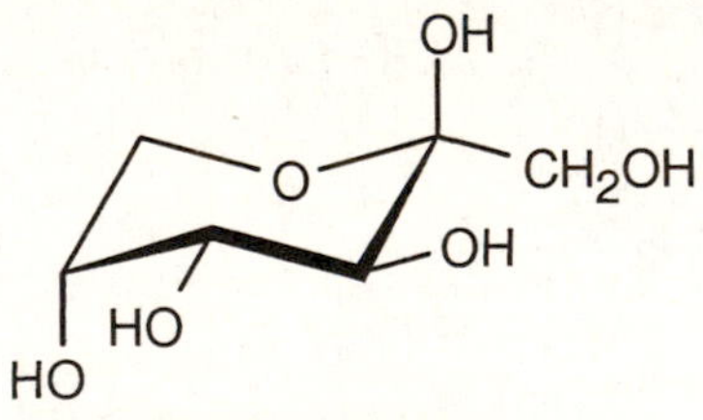

α-D-Fructopyranose

β-D-Fructopyranose

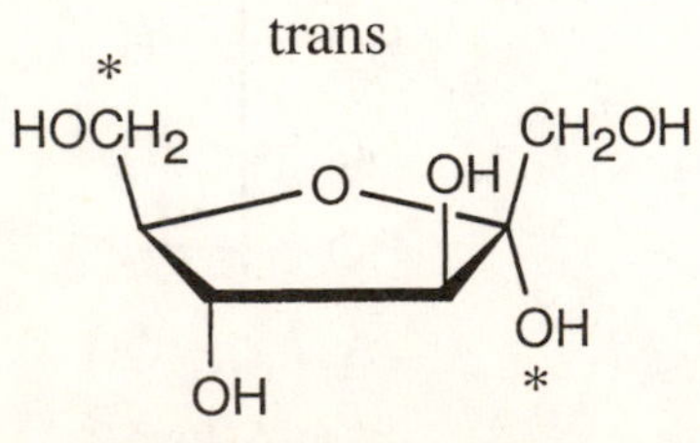

trans

α-D-Fructofuranose

cis

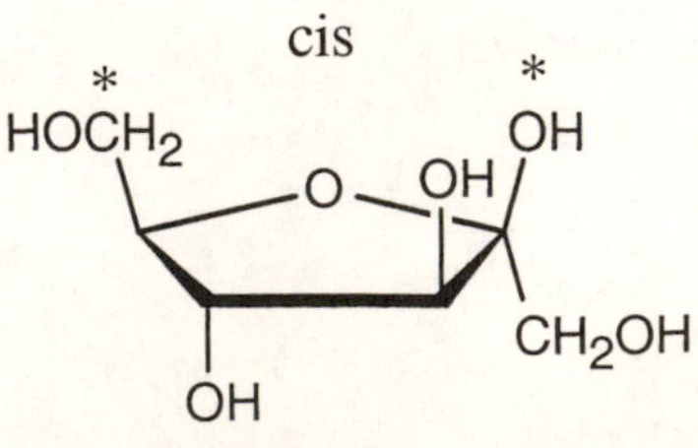

β-D-Fructofuranose

21.13 Strategy: There are two ways to draw these anomers: (1) Draw the Fischer projection, lay it on its side, form the pyranose ring, and convert it to a chair, remembering that the anomeric –OH group is cis to the C6 group. (2) Draw β-D-glucopyranose, and exchange the hydroxyl groups that differ between glucose and the other two hexoses.

Solution:

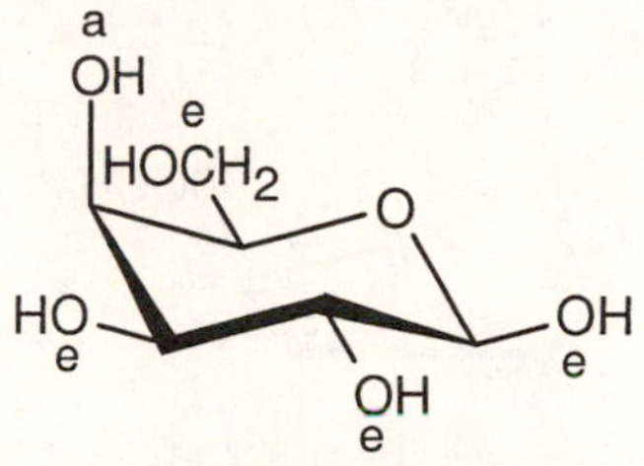

β-D-Galactopyranose 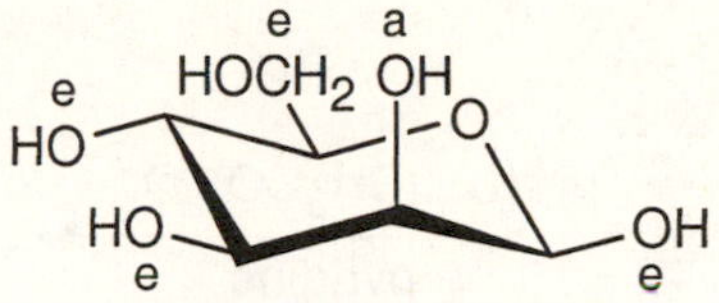 β-D-Mannopyranose

β-D-Galactopyranose and β-D-mannopyranose each have one hydroxyl group in the axial position and are therefore of similar stability.

21.14 The previous problem shows a drawing of β-D-galactopyranose. In this problem, invert the configuration at each chirality center of the D enantiomer and perform a ring-flip to arrive at the structure of the L enantiomer.

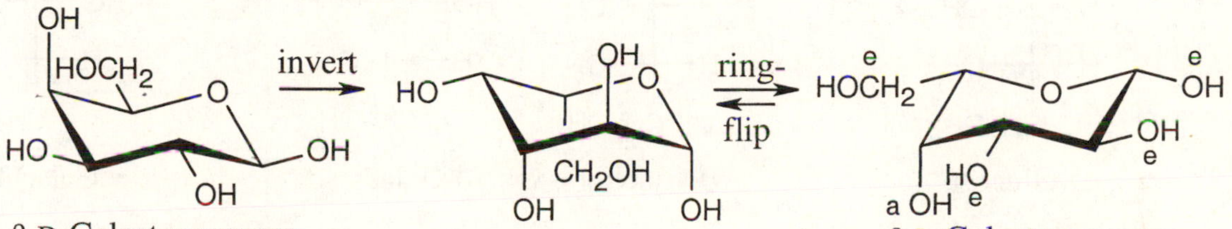

β-D-Galactopyranose invert ring-flip β-L-Galactopyranose

All substituents, except for the –OH at C4, are equatorial in the more stable conformation of β-L-galactopyranose.

21.15 The monosaccharide is the pyranose form of a D-aldohexose. It is an α-anomer because the anomeric hydroxyl group is trans to the group at C6. Comparing the model with α-D-glucopyranose, one can see that all groups have the same axial/equatorial relationship, except for the hydroxyl group at C3, which is axial in the model and equatorial in α-D-glucopyranose. The monosaccharide is α-D-allopyranose.

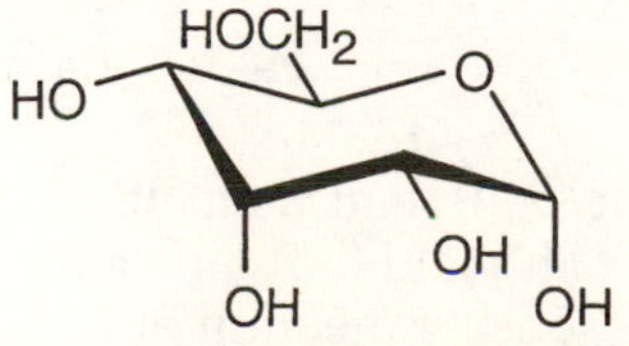

α-D-Allopyranose 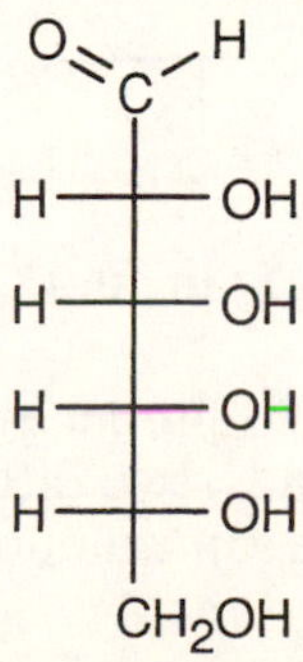 D-Allose

21.16

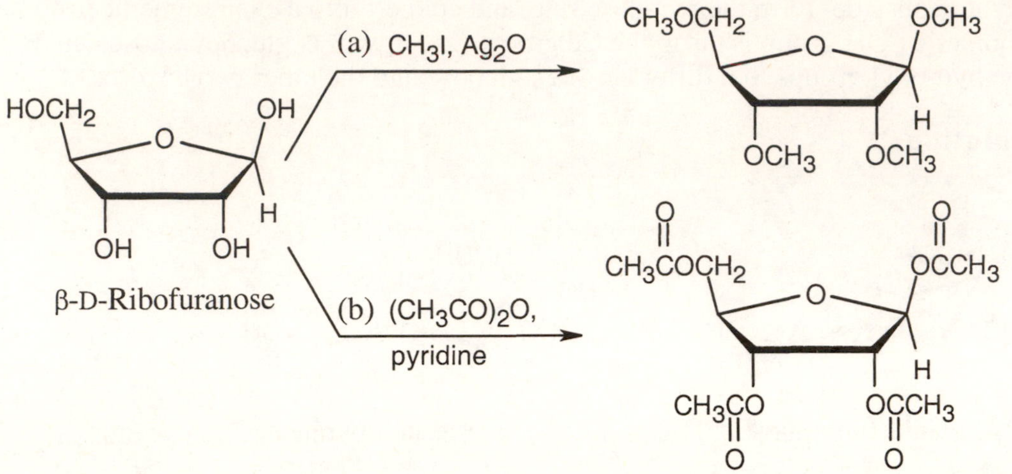

β-D-Ribofuranose

(a) CH_3I, Ag_2O

(b) $(CH_3CO)_2O$, pyridine

21.17

D-Glucose → (1. $NaBH_4$ 2. H_2O) → D-Glucitol

D-Galactose → (1. $NaBH_4$ 2. H_2O) → Galactitol

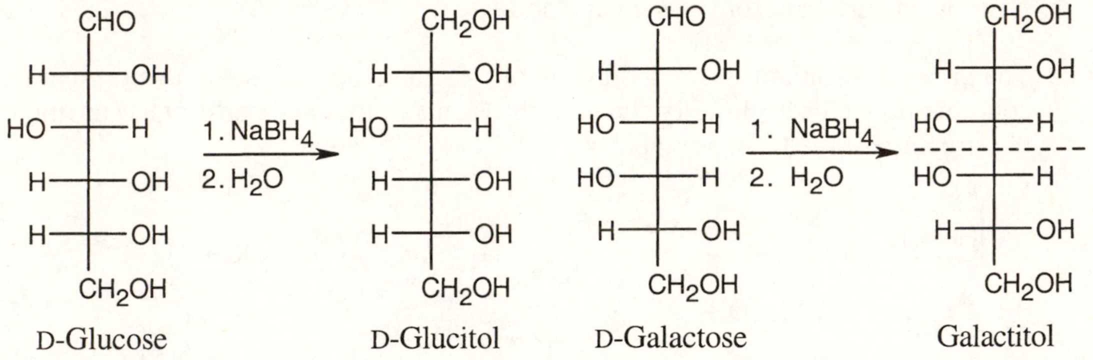

Reaction of D-galactose with $NaBH_4$ yields an alditol that has a plane of symmetry and is a meso compound. D-Glucitol has no plane of symmetry.

21.18

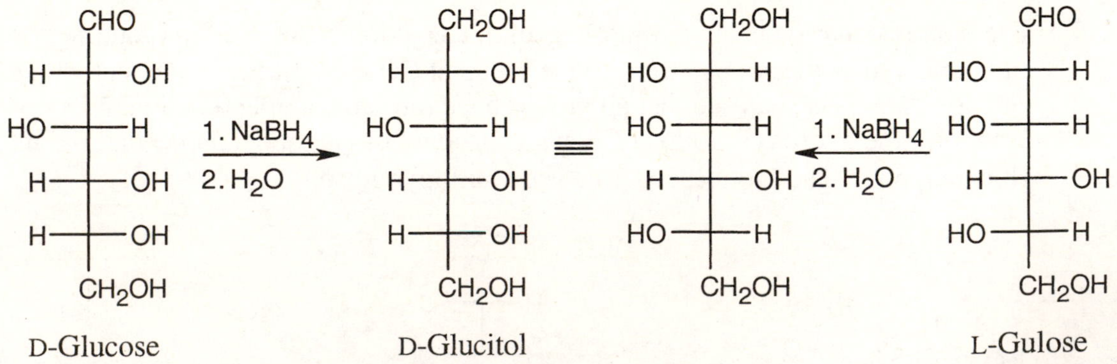

D-Glucose → (1. $NaBH_4$ 2. H_2O) → D-Glucitol ≡ ← (1. $NaBH_4$ 2. H_2O) L-Gulose

Reaction of an aldose with $NaBH_4$ produces a polyol (alditol). Because an alditol has the same functional group at both ends, two different aldoses can yield the same alditol. Here, L-gulose and D-glucose form the same alditol (rotate the Fischer projection of L-gulitol 180° to see the identity).

21.19

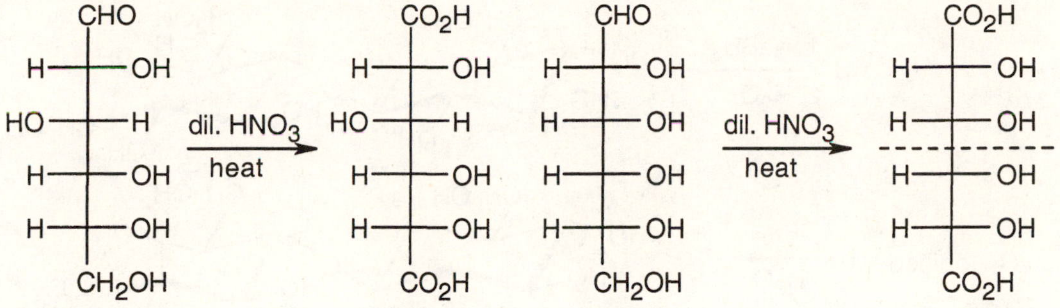

D-Glucose (CHO; H–OH; HO–H; H–OH; H–OH; CH_2OH) $\xrightarrow[\text{heat}]{\text{dil. } HNO_3}$ Glucaric acid (CO_2H; H–OH; HO–H; H–OH; H–OH; CO_2H)

D-Allose (CHO; H–OH; H–OH; H–OH; H–OH; CH_2OH) $\xrightarrow[\text{heat}]{\text{dil. } HNO_3}$ Allaric acid (CO_2H; H–OH; H–OH; H–OH; H–OH; CO_2H)

Allaric acid has a plane of symmetry and is an optically inactive meso compound. Glucaric acid has no symmetry plane.

21.20 D-Allose and D-galactose yield meso aldaric acids. All other D-hexoses produce optically active aldaric acids on oxidation.

21.21

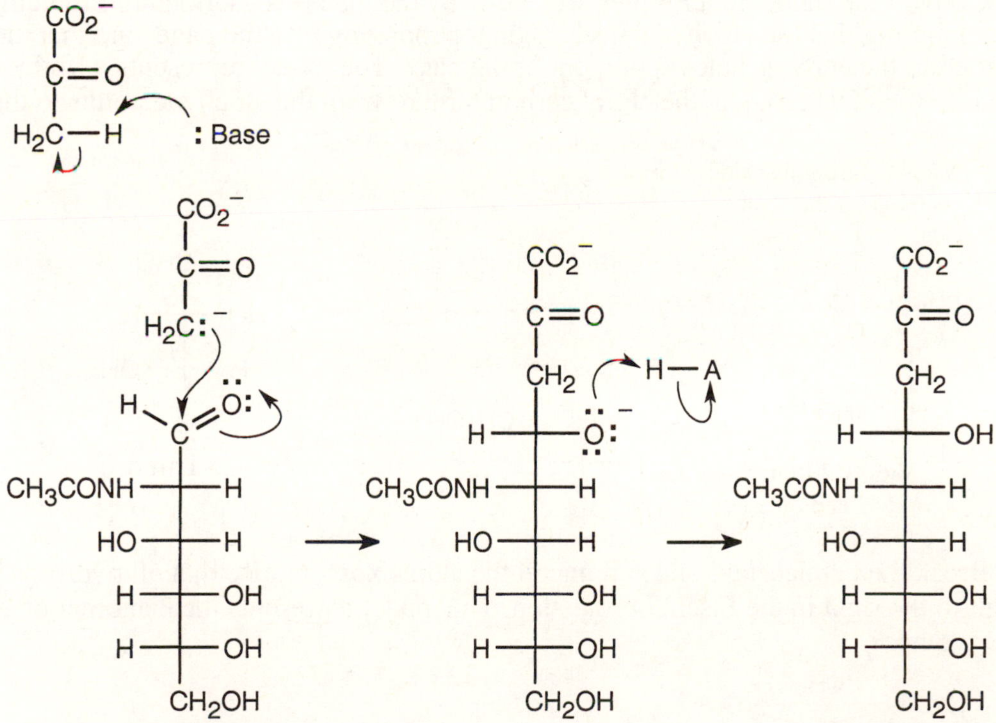

CO_2^-; C=O; H_2C—H   : Base

CO_2^-; C=O; $H_2C:^-$ + H–C=O (aldehyde of N-acetylmannosamine: CH₃CONH–H; HO–H; H–OH; H–OH; CH_2OH) $\longrightarrow$ CO_2^-; C=O; CH_2; H–O^-; CH₃CONH–H; HO–H; H–OH; H–OH; CH_2OH (+ H—A) $\longrightarrow$ CO_2^-; C=O; CH_2; H–OH; CH₃CONH–H; HO–H; H–OH; H–OH; CH_2OH

N-Acetylmannosamine   *N*-Acetyl-D-neuraminic acid

21.22

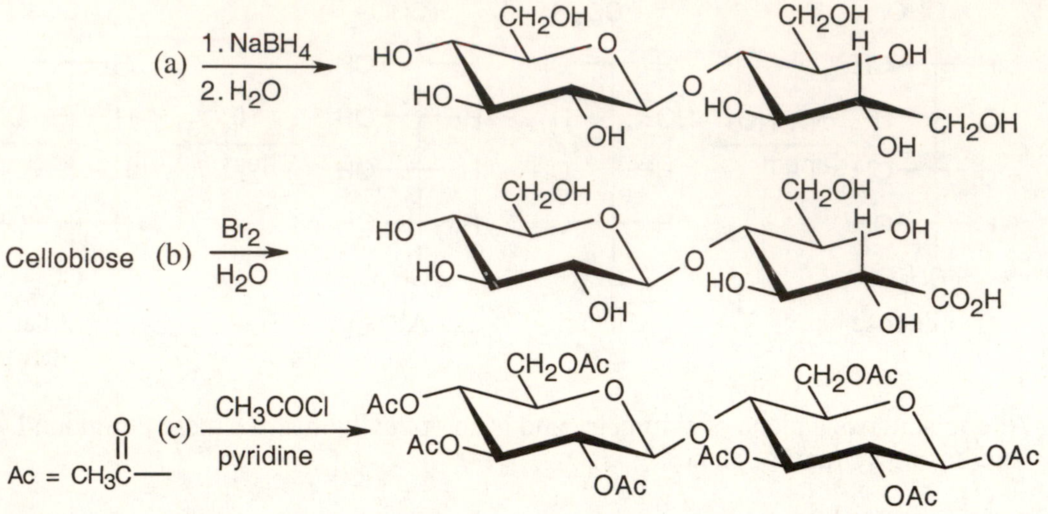

Cellobiose (a) 1. $NaBH_4$, 2. H_2O →
(b) Br_2, H_2O →
(c) CH_3COCl, pyridine →

Ac = $CH_3C(=O)$—

Visualizing Chemistry

21.23 (a) Convert the model to a Fischer projection by the method described in Problem 21.5, remembering that the aldehyde group is on top, pointing into the page, and that the groups bonded to the carbons below point out of the page. The model represents a D-aldose because the –OH group at the chiral carbon farthest from the aldehyde points to the right.

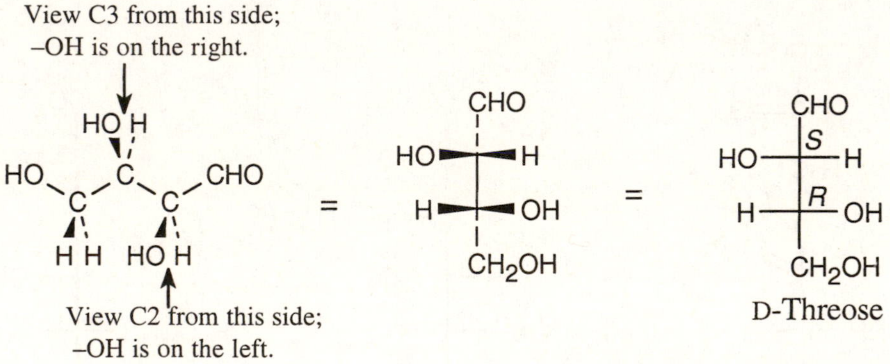

View C3 from this side; –OH is on the right.

View C2 from this side; –OH is on the left.

D-Threose

(b) Break the hemiacetal bond and uncoil the aldohexose. Notice that all hydroxyl groups point to the right in the Fischer projection. The model represents the β anomer of D-allopyranose.

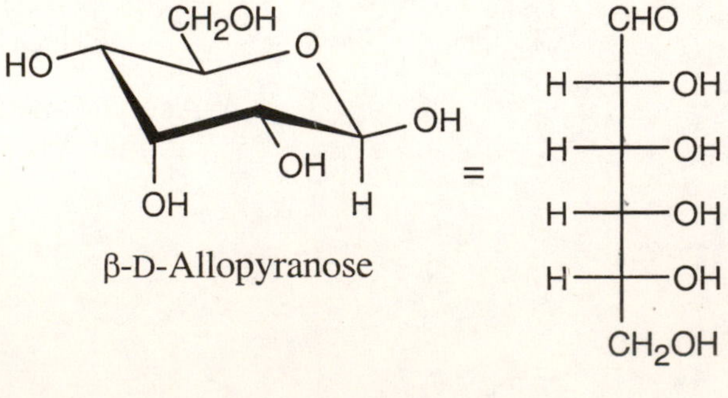

β-D-Allopyranose

21.24 The hints in the previous problem also apply here. Molecular models are also helpful.

(a)

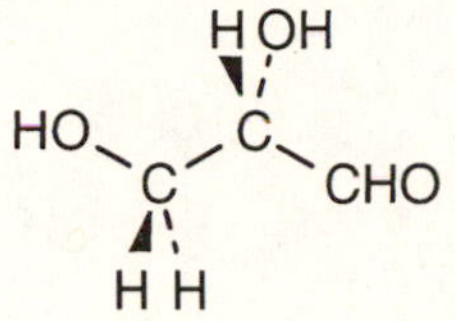

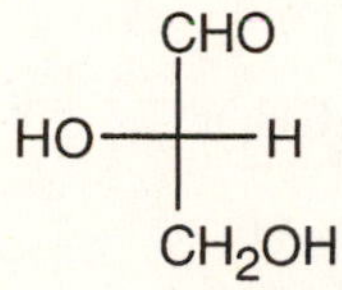

L-Glyceraldehyde

(b)

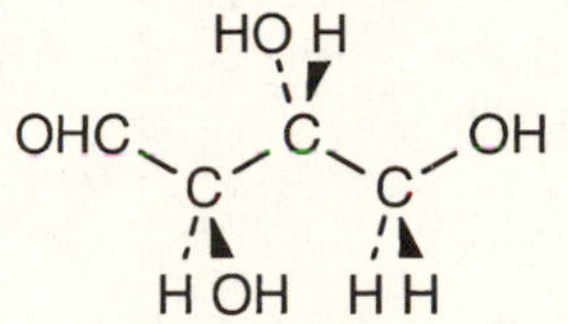

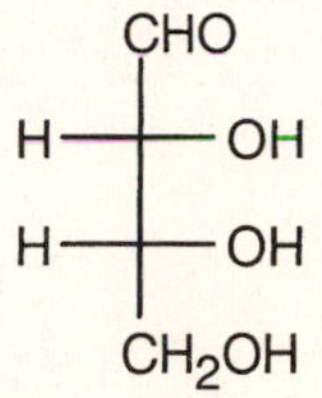

D-Erythrose

21.25 The structure represents an α anomer because the anomeric –OH group and the $-CH_2OH$ group are trans. The compound is α-L-mannopyranose because it differs from L-glucose only at C2.

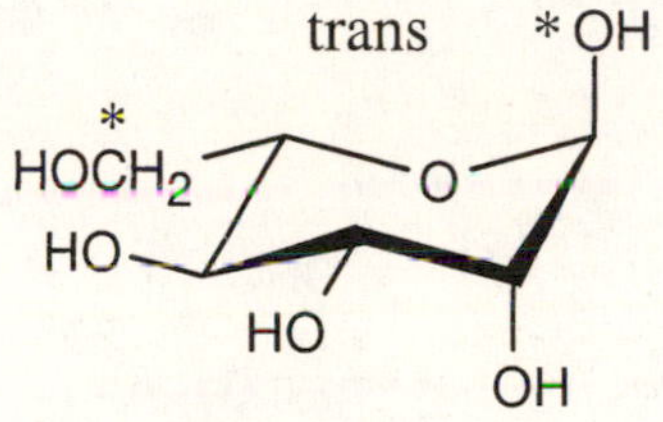

α-L-Mannopyranose

21.26

(a)

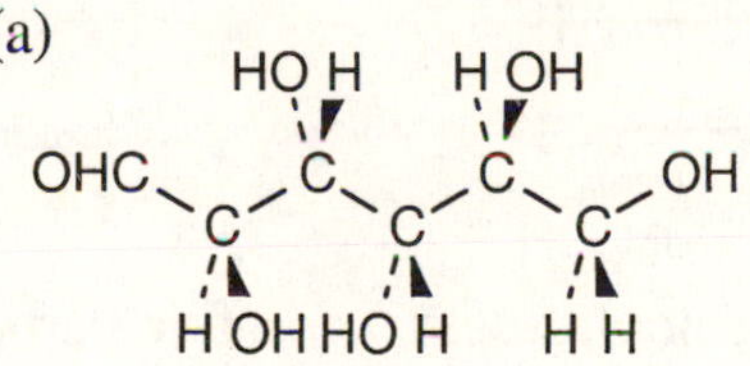

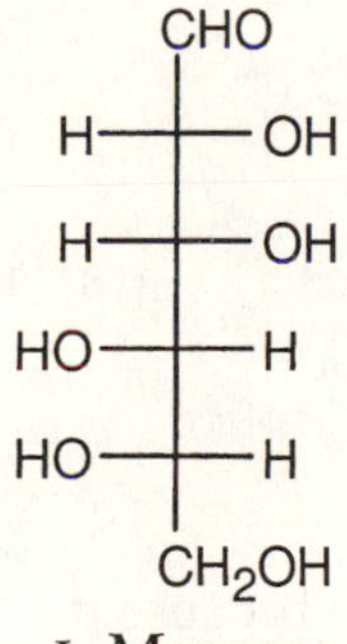

L-Mannose	D-Mannose (enantiomer)	D-Glucose (diastereomer)
CHO	CHO	CHO
H—OH	HO—H	H—OH
H—OH	HO—H	HO—H
HO—H	H—OH	H—OH
HO—H	H—OH	H—OH
CH_2OH	CH_2OH	CH_2OH

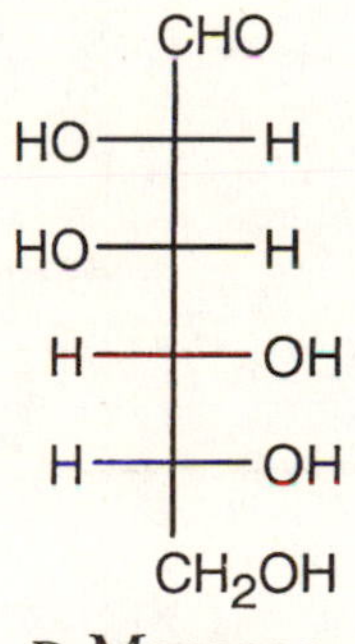

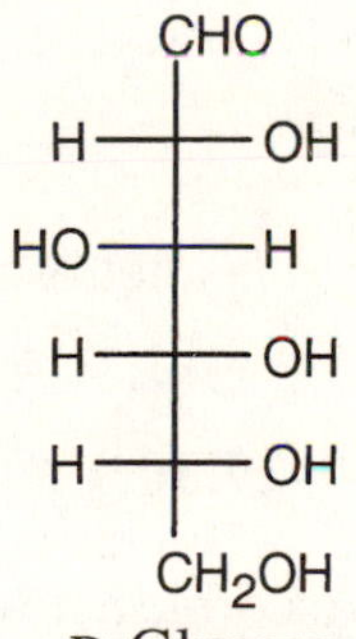

(b) The model represents an L-aldohexose because the hydroxyl group on the chiral carbon farthest from the aldehyde group points to the left.

(c) This is tricky! The furanose ring of an aldohexose is formed by connecting the –OH group at C4 to the aldehyde carbon. The best way to draw the anomer is to lie L-mannose on its side and form the ring. All substituents point down in the furanose, and the anomeric –OH and the $-CH(OH)CH_2OH$ group are cis.

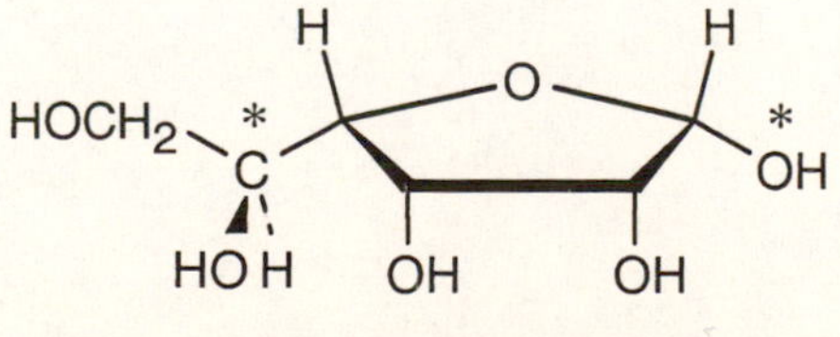

β-L-Mannofuranose

Additional Problems

21.27

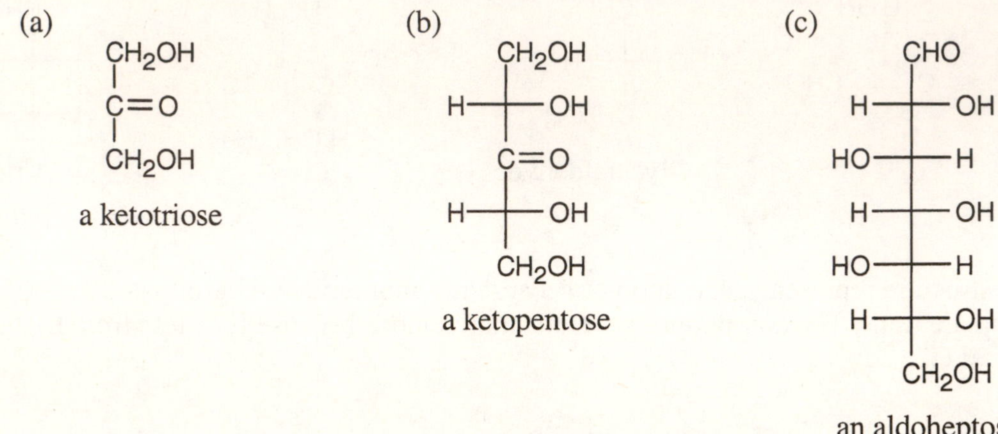

(a) a ketotriose (b) a ketopentose (c) an aldoheptose

21.28

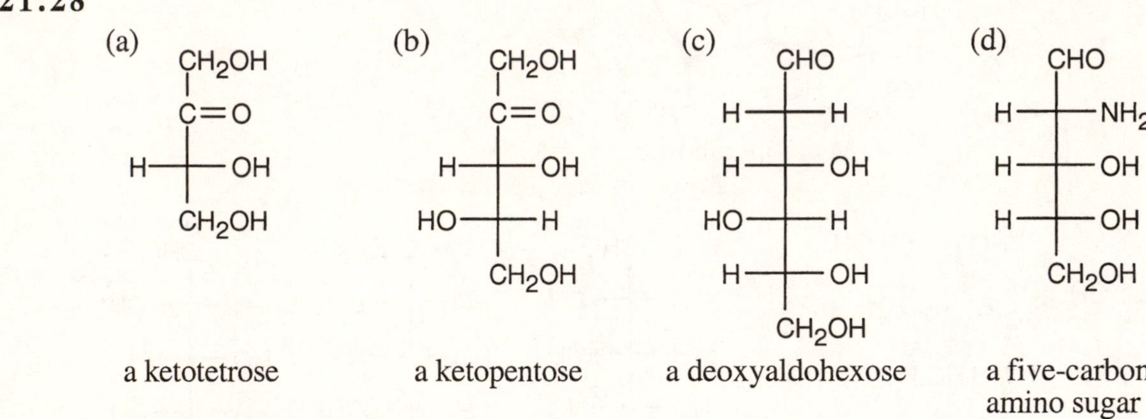

(a) a ketotetrose (b) a ketopentose (c) a deoxyaldohexose (d) a five-carbon amino sugar

21.29 – 29.30

Ascorbic acid has an L configuration because the hydroxyl group at the lowest chirality center points to the left.

L-Ascorbic acid

21.31

(a)

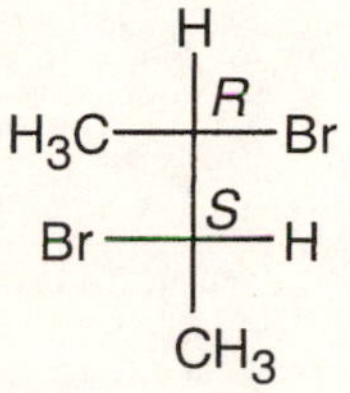

(b)

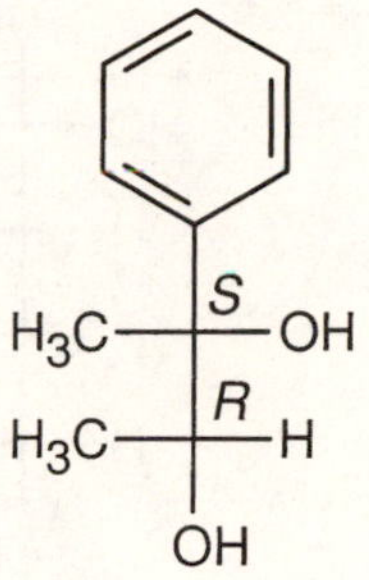

(c)

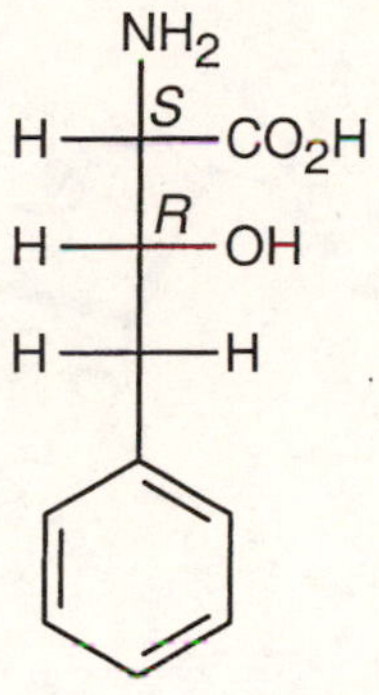

21.32

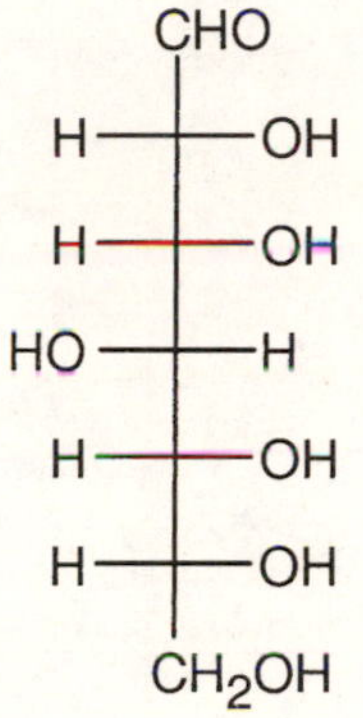

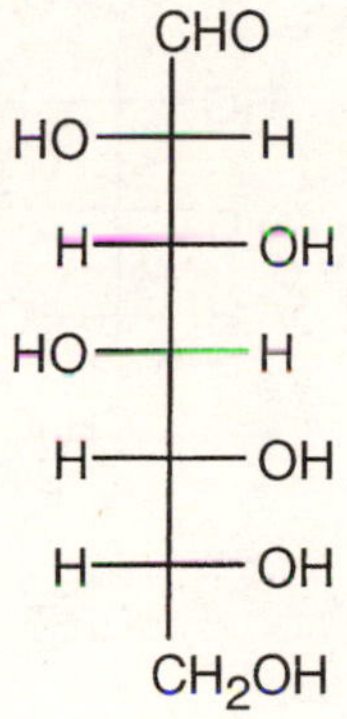

21.33

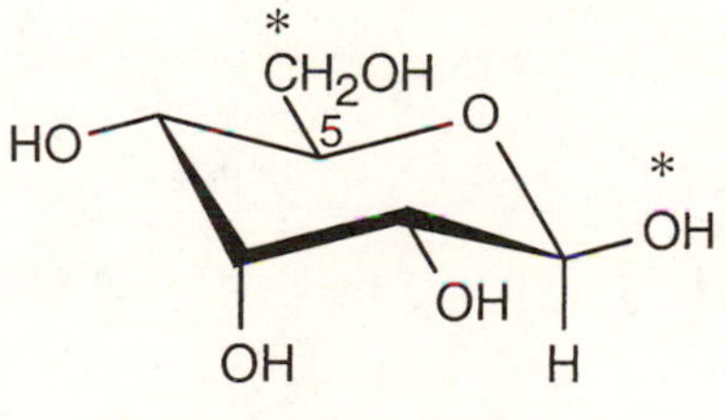

β-D-Allopyranose

This structure is a pyranose (6-membered ring) and is a β anomer (the C1 hydroxyl group and the $–CH_2OH$ groups are cis). It is a D sugar because the -O- at C5 is on the right in the uncoiled form.

21.34

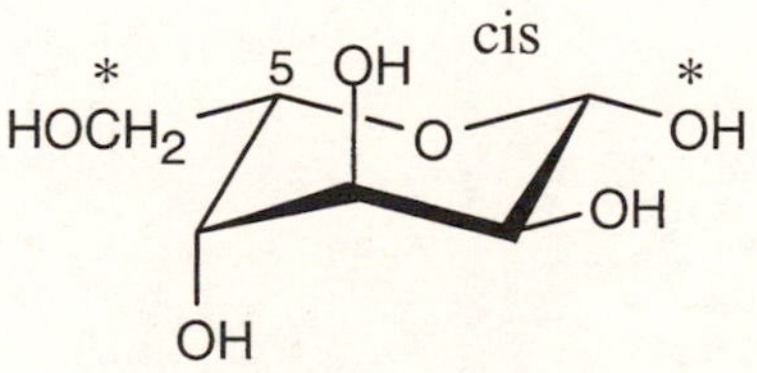

β-L-Gulopyranose

This sugar is a β-pyranose. It is an L sugar because the -O- at C5 points to the left in the uncoiled form. It's also possible to recognize this as an L sugar by the fact that the configuration at C5 is *S*.

21.35

(a)

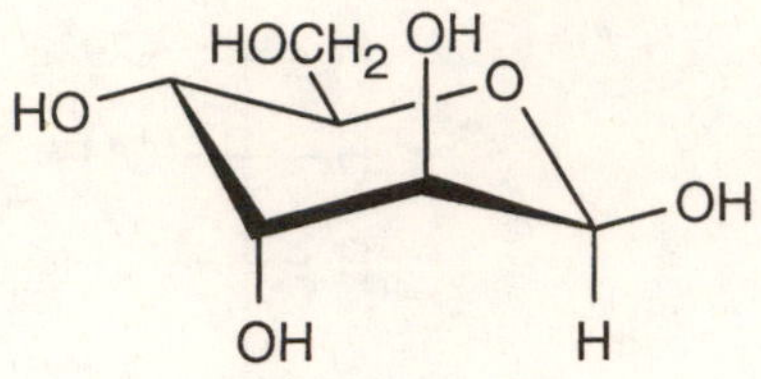

HO, HOCH$_2$, OH, O, OH, OH, H

β-D-Altropyranose

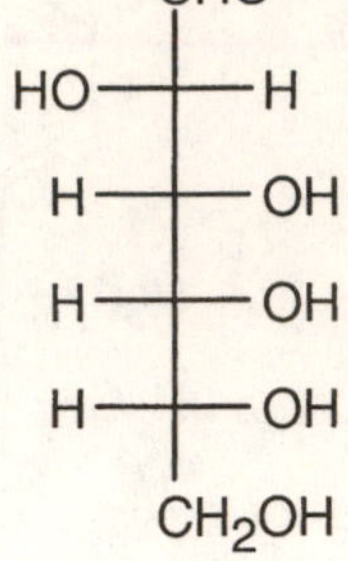

CHO
HO—H
H—OH
H—OH
H—OH
CH_2OH

(b)

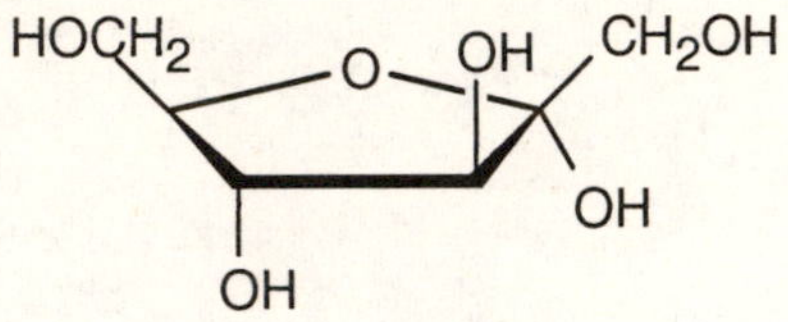

HOCH$_2$, O, OH, CH_2OH, OH, OH

α-D-Fructofuranose

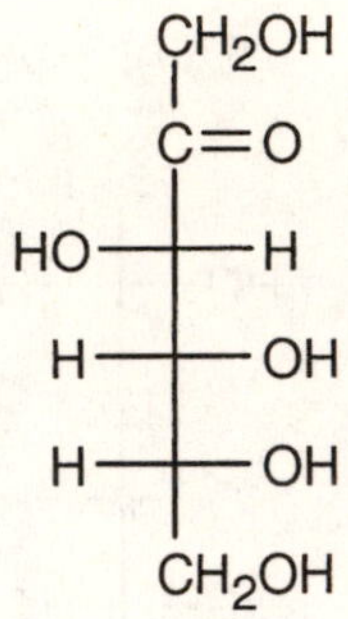

CH_2OH
C=O
HO—H
H—OH
H—OH
CH_2OH

(c)

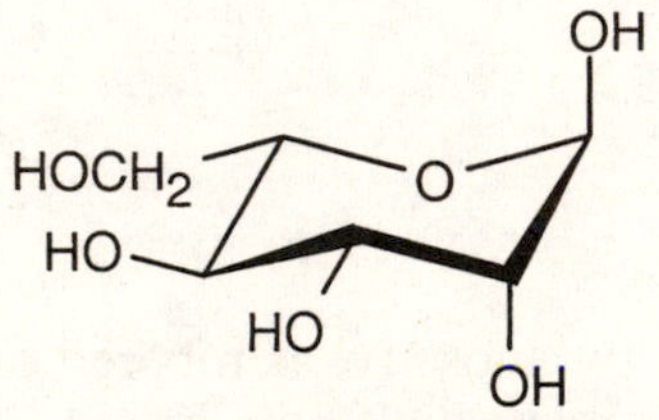

OH, HOCH$_2$, O, HO, HO, OH

α-L-Mannopyranose

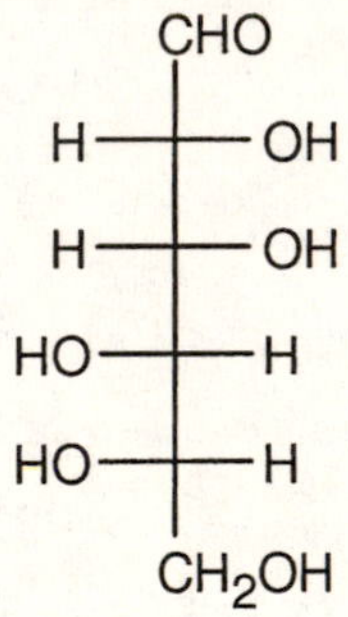

CHO
H—OH
H—OH
HO—H
HO—H
CH_2OH

21.36

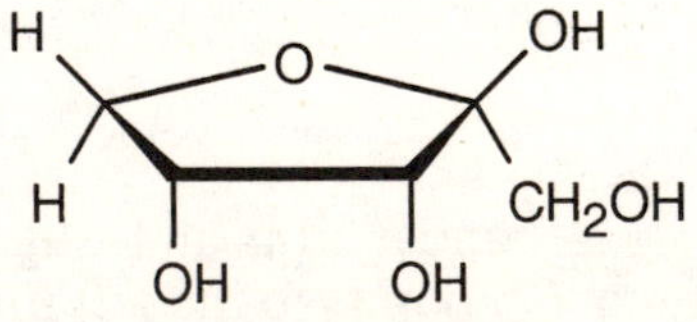

H, O, OH, H, CH_2OH, OH, OH

β-D-Ribulofuranose

21.37–27.38

(a) 1. $NaBH_4$ 2. H_2O; (b) dil. HNO_3; (c) Br_2, H_2O; (d) CH_3CH_2OH, HCl; (e) CH_3I, Ag_2O; (f) $(CH_3CO)_2O$, pyridine

β-D-Talopyranose

and α anomer

21.39 D-Ribose and L-xylose are diastereomers and differ in all physical properties (or if they have identical physical properties in one category, it is a coincidence).

21.40

D-Galactose

α-D-Galactopyranose
$[\alpha]_D = +150.7°$

β-D-Galactopyranose
$[\alpha]_D = +52.8°$

Let x be the percent of D-galactose present as the α anomer and y be the percent of D-galactose present as the ß anomer.

$150.7°x + 52.8°y = 80.2° \qquad x + y = 1; \quad y = 1 - x$
$150.7°x + 52.8°(1-x) = 80.2°$
$97.9°x = 27.4°$
$x = 0.280$
$y = 0.720$

28.0% of D-galactose is present as the α anomer, and 72.0% is present as the β anomer.

21.41–21.43 Four D-2-ketohexoses are possible.

D-Psicose	D-Fructose	D-Sorbose	D-Tagatose
CH_2OH	CH_2OH	CH_2OH	CH_2OH
C=O	C=O	C=O	C=O
H—OH	HO—H	H—OH	HO—H
H—OH	H—OH	HO—H	HO—H
H—OH	H—OH	H—OH	H—OH
CH_2OH	CH_2OH	CH_2OH	CH_2OH

D-Psicose and D-Fructose → 1. $NaBH_4$ 2. H_2O → Allitol + Altritol

D-Sorbose and D-Tagatose → 1. $NaBH_4$ 2. H_2O → Gulitol + Iditol

Allitol	Altritol	Gulitol	Iditol
CH_2OH	CH_2OH	CH_2OH	CH_2OH
H—OH	HO—H	H—OH	HO—H
H—OH	H—OH	H—OH	H—OH
H—OH	H—OH	HO—H	HO—H
H—OH	H—OH	H—OH	H—OH
CH_2OH	CH_2OH	CH_2OH	CH_2OH

21.44 The two lactones are formed between a carboxylic acid and a hydroxyl group 4 carbons away. When the lactones are reduced with sodium amalgam, the resulting hexoses have an aldehyde at one end and a hydroxyl group at the other end.

dil. HNO_3 → + Na(Hg) Na(Hg)

rotate 180° ≡

D-Glucose L-Gulose

21.45

1. $NaBH_4$ 2. H_2O → rotate 180° ≡ ← 1. $NaBH_4$ 2. H_2O

D-Talose D-Altrose

21.46

dil. HNO_3 → rotate 180° ≡ ← dil. HNO_3

D-Allose L-Allose

CHO / H—OH / HO—H / HO—H / H—OH / CH_2OH (D-Galactose) $\xrightarrow{\text{dil. } HNO_3}$ CO_2H / H—OH / HO—H / HO—H / H—OH / CO_2H rotate 180° ≡ CO_2H / HO—H / H—OH / H—OH / HO—H / CO_2H $\xleftarrow{\text{dil. } HNO_3}$ CHO / HO—H / H—OH / H—OH / HO—H / CH_2OH (L-Galactose)

D-Galactose

L-Galactose

21.47

CHO / HO—H / HO—H / H—OH / CH_2OH (D-Lyxose) $\xrightarrow{\text{dil. } HNO_3}$ CO_2H / HO—H / HO—H / H—OH / CO_2H rotate 180° ≡ CO_2H / HO—H / H—OH / H—OH / CO_2H $\xleftarrow{\text{dil. } HNO_3}$ CHO / HO—H / H—OH / H—OH / CH_2OH (D-Arabinose)

D-Lyxose

D-Arabinose

21.48 (a) D-Galactose gives the same aldaric acid as L-galactose.

CHO / HO—H / H—OH / H—OH / HO—H / CH_2OH (L-Galactose) $\xrightarrow{\text{dil. } HNO_3}$ CO_2H / H—OH / HO—H / HO—H / H—OH / CO_2H rotate 180° ≡ CO_2H / HO—H / H—OH / H—OH / HO—H / CO_2H $\xleftarrow{\text{dil. } HNO_3}$ CHO / H—OH / HO—H / HO—H / H—OH / CH_2OH (D-Galactose)

L-Galactose

D-Galactose

(b) The other aldohexose is a D-sugar.

(c)

OH, CH_2OH, O, HO, OH, OH

β-D-Galactopyranose

21.49 The hard part of this problem is determining where the glycosidic bond occurs on the second glucopyranose ring. Treatment with iodomethane, followed by hydrolysis, yields a tetra-*O*-methyl glucopyranose and a tri-*O*-methyl glucopyranose. The oxygen in the tri-*O*-methylated ring that is not part of the hemiacetal group and is not methylated is the oxygen that forms the acetal bond. In this problem, the C6 oxygen forms the glycosidic link.

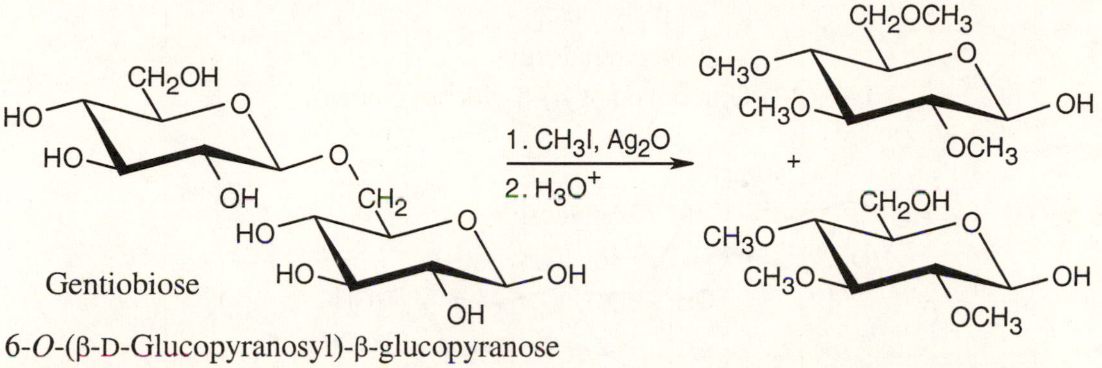

Gentiobiose

6-*O*-(β-D-Glucopyranosyl)-β-glucopyranose

1. CH_3I, Ag_2O
2. H_3O^+

21.50 Amygdalin has the same carbohydrate skeleton as gentiobiose. Draw the cyanohydrin of benzaldehyde, and form a bond between the hemiacetal oxygen and the carbonyl carbon of benzaldehyde, with elimination of water.

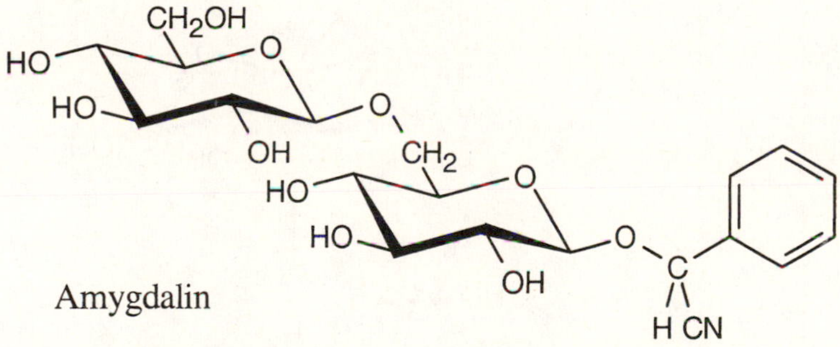

Amygdalin

21.51 Since trehalose is a nonreducing sugar, the two glucose units must be connected through an oxygen atom at the anomeric carbon of each glucose. There are three possible structures for trehalose: The two glucopyranose rings can be connected (α,α), (β,β), or (α,β).

21.52 Since trehalose is not cleaved by ß-glycosidases, it must have an α,α glycosidic linkage.

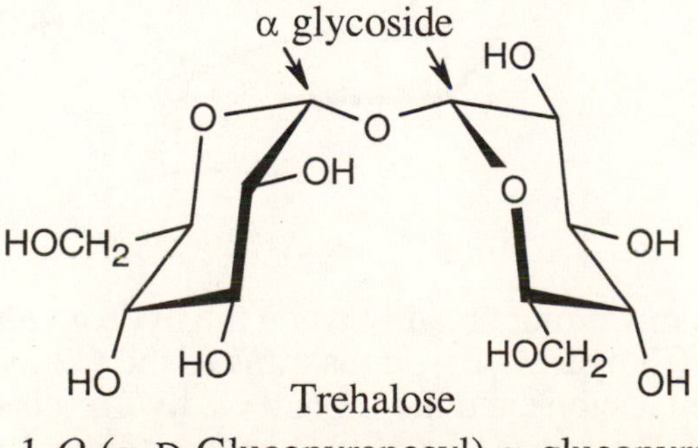

α glycoside

Trehalose

1-*O*-(α-D-Glucopyranosyl)-α-glucopyranose

21.53

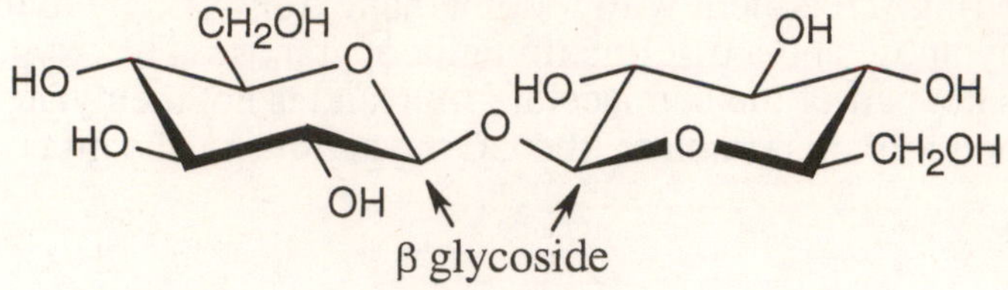

CH$_2$OH HO HO OH O O OH HO O OH CH$_2$OH

β glycoside

Neotrehalose

1-*O*-(β-D-Glucopyranosyl)-β-glucopyranose

CH$_2$OH HO HO OH O OH HO O OH CH$_2$OH

α glycoside

β glycoside

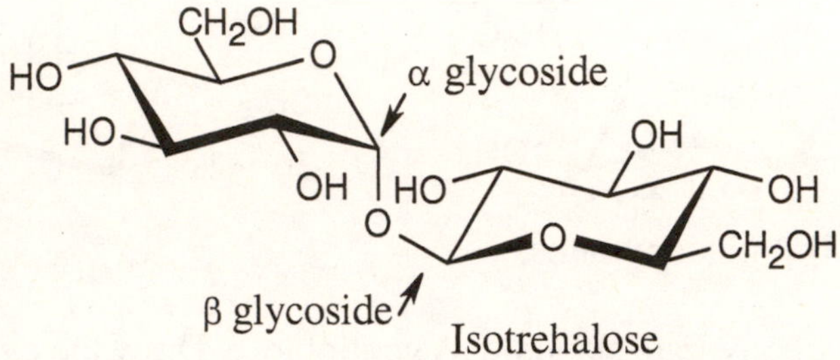

Isotrehalose

1-*O*-(α-D-Glucopyranosyl)-β-glucopyranose

21.54

Glucopyranose is in equilibrium with glucofuranose

Reaction with two equivalents of acetone occurs by the mechanism we learned for acetal formation (Sec 14.8)

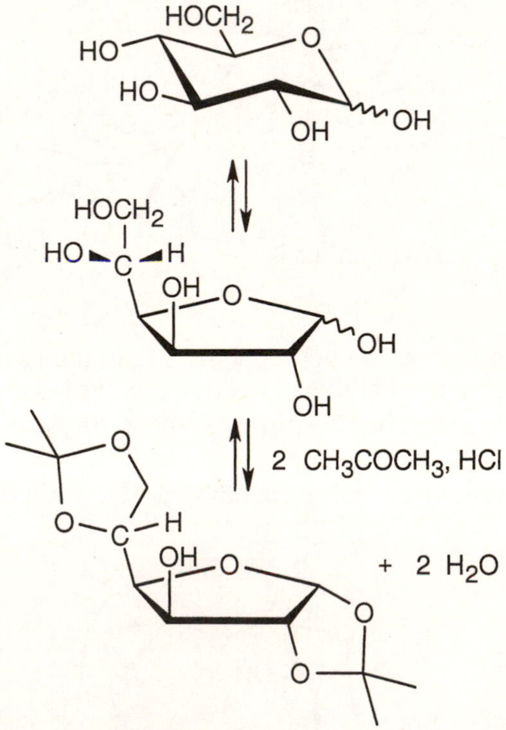

2 CH_3COCH_3, HCl

+ 2 H_2O

A five-membered acetal ring forms much more readily when the hydroxyl groups are cis to one another. In glucofuranose, the C3 hydroxyl group is trans to the C2 hydroxyl group, and acetal formation occurs between acetone and the C1 and C2 hydroxyls of glucofuranose. Since the C1 hydroxyl group is part of the acetone acetal, the furanose is no longer in equilibrium with the free aldehyde, and the diacetone derivative is not a reducing sugar.

21.55

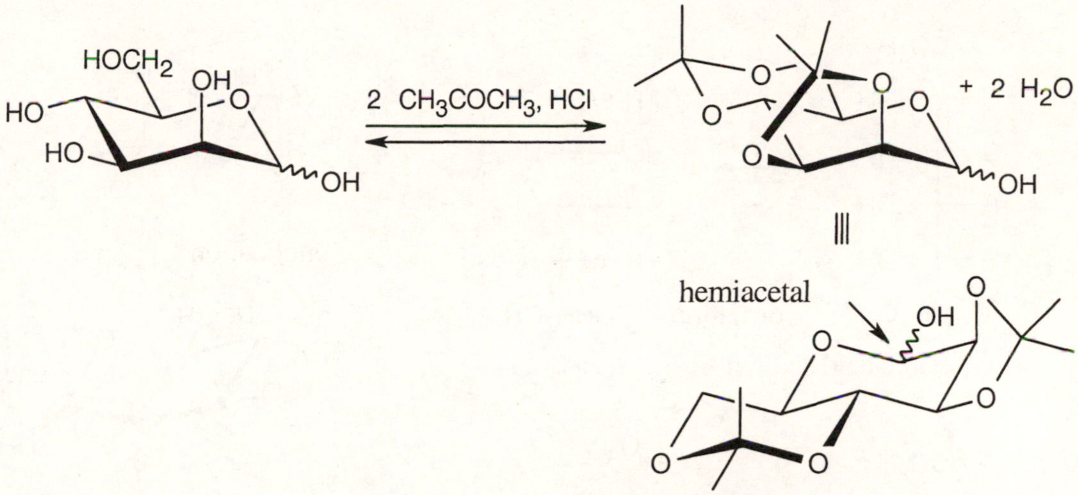

HOCH2, OH, HO, HO, O, OH — 2 CH_3COCH_3, HCl ⇌ + 2 H_2O

hemiacetal

2,3:4,6-Diacetone mannopyranoside

Acetone forms an acetal with the hydroxyl groups at C2 and C3 of D-mannopyranoside because the hydroxyl groups at these positions are cis to one another. The pyranoside ring is still a hemiacetal that is in equilibrium with free aldehyde, which is reducing toward Tollens' reagent.

21.56 Dilute base abstracts a proton α to the carbonyl carbon, forming an enolate. The enolate double bond can be protonated from either side, giving either mannose or glucose as the product.

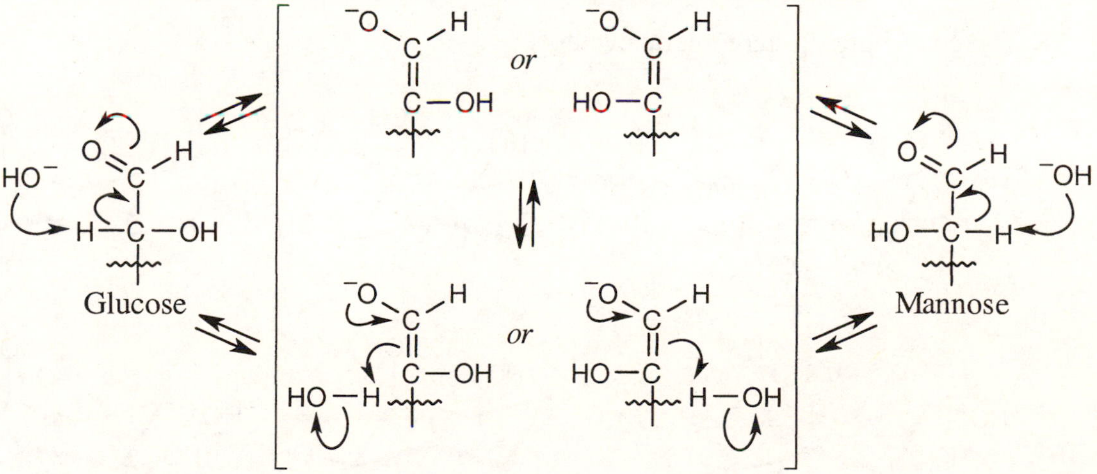

Glucose ⇌ [enolate *or* enolate] ⇌ Mannose

21.57

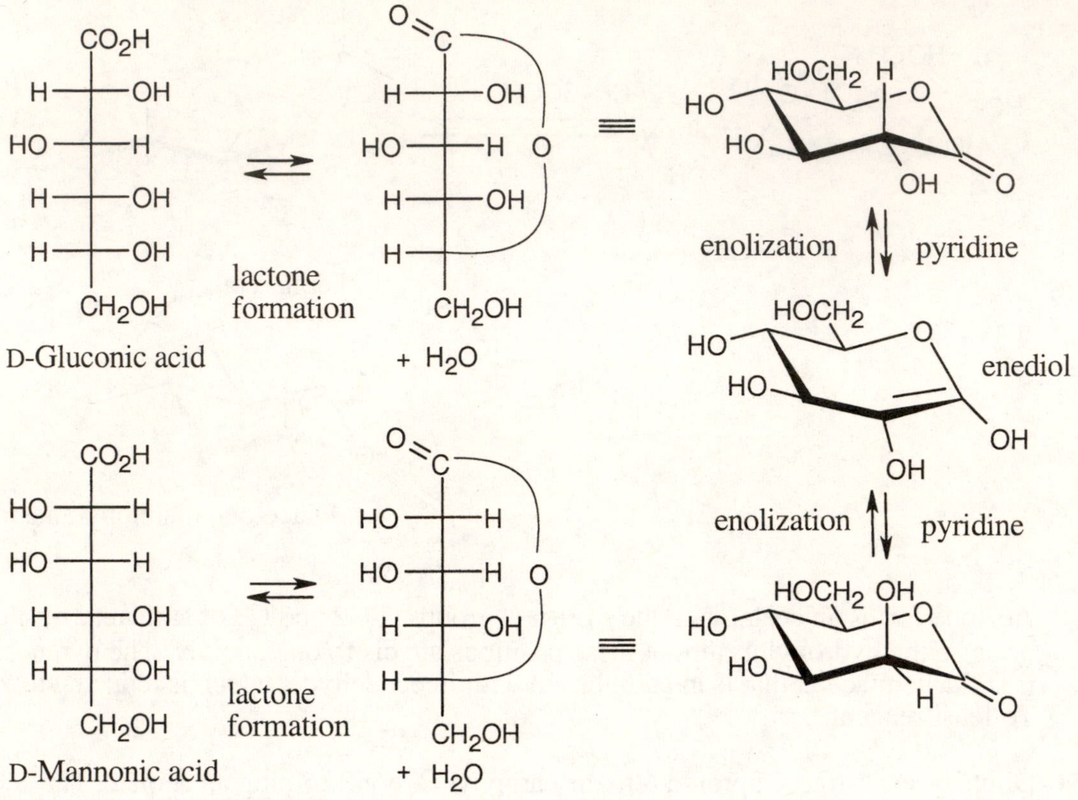

CO2H / H—OH / HO—H / H—OH / H—OH / CH2OH

D-Gluconic acid ⇌ (lactone formation) lactone + H_2O ≡ lactone ⇌ (enolization, pyridine) enediol ⇌ (enolization, pyridine) lactone

D-Mannonic acid ⇌ (lactone formation) lactone + H_2O

Isomerization at C2 occurs because the enediol can be reprotonated on either side of the double bond.

21.58 There are eight diastereomeric cyclitols.

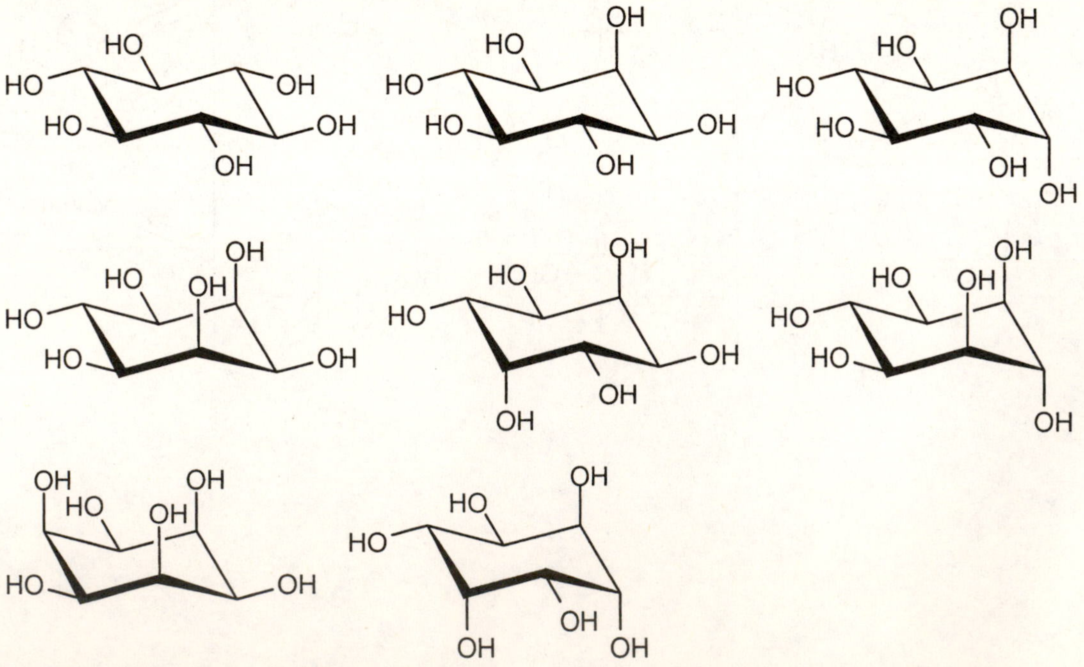

21.59 The products of Kiliani–Fischer reaction of D-ribose have the same configuration at C3, C4 and C5 as D-ribose.

D-Ribose (CHO; H–OH; H–OH; H–OH; CH_2OH) $\xrightarrow{\text{Kiliani–Fischer}}$ D-Allose (CHO; H–OH; H–OH; H–OH; H–OH; CH_2OH) + D-Altrose (CHO; HO–H; H–OH; H–OH; H–OH; CH_2OH)

The aldopentose L-xylose has the same configuration at C3, C4 and C5 as L-idose and L-gulose.

L-Xylose (CHO; HO–H; H–OH; HO–H; CH_2OH) $\xrightarrow{\text{Kiliani–Fischer}}$ L-Idose (CHO; H–OH; HO–H; H–OH; HO–H; CH_2OH) + L-Gulose (CHO; HO–H; HO–H; H–OH; HO–H; CH_2OH)

21.60 The aldopentoses have the same configurations at C3 and C4 as D-threose.

D-Xylose (CHO; H–OH; HO–H; H–OH; CH_2OH) *or* D-Lyxose (CHO; HO–H; HO–H; H–OH; CH_2OH) $\xrightarrow{\text{Wohl degradation}}$ D-Threose (CHO; HO–H; H–OH; CH_2OH)

21.61

	CHO			CO_2H	
	H — — OH			H — — OH	
D-Ribose	H — — OH	$\xrightarrow[\text{heat}]{\text{dil. } HNO_3}$		H — — OH	
	H — — OH			H — — OH	
	CH_2OH			CO_2H	
	A			**B**	

A $\xrightarrow{\text{Kiliani–Fischer}}$ **C** + **D**

CO_2H		CHO		CHO		CO_2H
HO — — H		HO — — H		H — — OH		H — — OH
H — — OH	$\xleftarrow[\text{heat}]{\text{dil. } HNO_3}$	H — — OH	+	H — — OH	$\xrightarrow[\text{heat}]{\text{dil. } HNO_3}$	H — — OH
H — — OH		H — — OH		H — — OH		H — — OH
H — — OH		H — — OH		H — — OH		H — — OH
CO_2H		CH_2OH		CH_2OH		CO_2H
E		**C**		**D**		**F**
		D-Altrose		D-Allose		

Because **A** is oxidized to an optically inactive aldaric acid, the possible structures are D-ribose and D-xylose. Chain extension of D-xylose, however, produces two hexoses that, when oxidized, yield optically active aldaric acids.

21.62

(a)

H–C=O		CH_2OH		H–C=O		H–C=NNHPh
H — — OH		C=O		HO — — H		C=NNHPh
HO — — H	*or*	HO — — H	*or*	HO — — H	$\xrightarrow{2\ H_2NNHPh}$	HO — — H
H — — OH		H — — OH		H — — OH		H — — OH
H — — OH		H — — OH		H — — OH		H — — OH
CH_2OH		CH_2OH		CH_2OH		CH_2OH
D-Glucose		D-Fructose		D-Mannose		An osazone

(b)

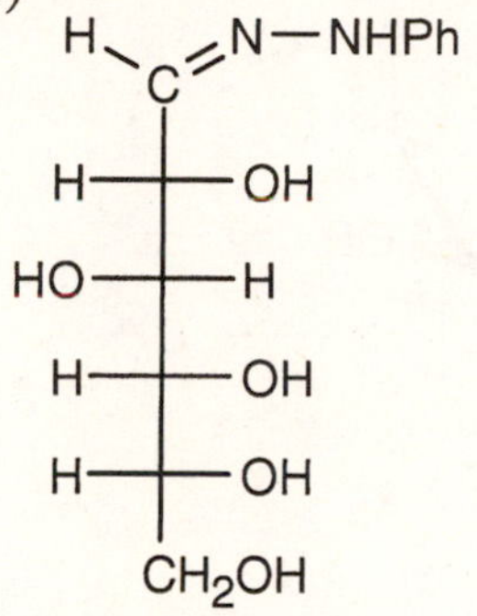

H–C=N–NHPh; H–C–OH; HO–C–H; H–C–OH; H–C–OH; CH_2OH

phenylhydrazone

(c)

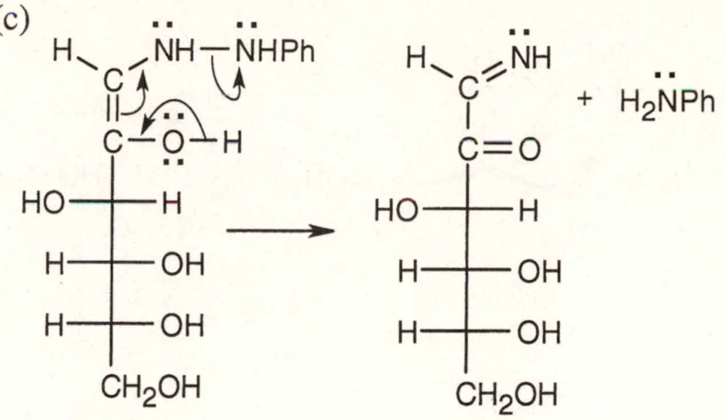

H–C(=C–OH)–NH–NHPh; HO–C–H; H–C–OH; H–C–OH; CH_2OH → H–C=NH; C=O; HO–C–H; H–C–OH; H–C–OH; CH_2OH + H_2NPh

enol keto imine

(d)

PhNH–NH_2 + H–C=NH; C=O; HO–C–H; H–C–OH; H–C–OH; CH_2OH ⇌ PhNHN(H)–C(H)–NH_2; C=O; HO–C–H; H–C–OH; H–C–OH; CH_2OH ⇌ $PhNHN^-$–C(H)–NH_3^+; C=O; HO–C–H; H–C–OH; H–C–OH; CH_2OH ⇌ PhNHN=C(H) + NH_3; C=O; HO–C–H; H–C–OH; H–C–OH; CH_2OH

PhNHN=C(H); PhNH–NH_2 + C=O; HO–C–H; H–C–OH; H–C–OH; CH_2OH ⇌ PhNHN=C(H); $PhNH\overset{+}{N}H_2$–C–O^-; HO–C–H; H–C–OH; H–C–OH; CH_2OH ⇌ PhNHN=C(H); PhNHNH–C–OH; HO–C–H; H–C–OH; H–C–OH; CH_2OH ⇌ PhNHN=C(H) + H_2O; PhNHN=C; HO–C–H; H–C–OH; H–C–OH; CH_2OH

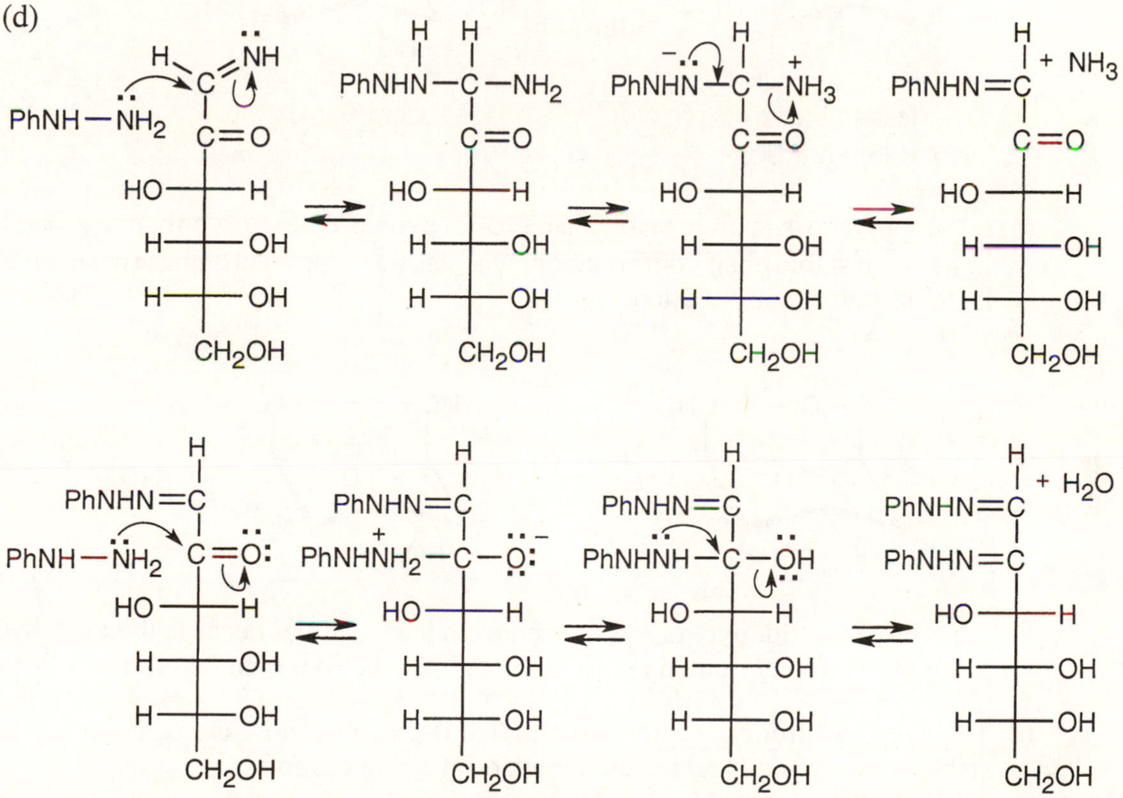

In these last steps, two nucleophilic addition reactions take place to yield imine products. The mechanism has been worked out in greater detail in Section 14.7, but the essential steps are additions of phenylhydrazine, first to the imine, then to the ketone. Proton transfers are followed by eliminations, first of ammonia, then of H_2O.

21.63 (a)

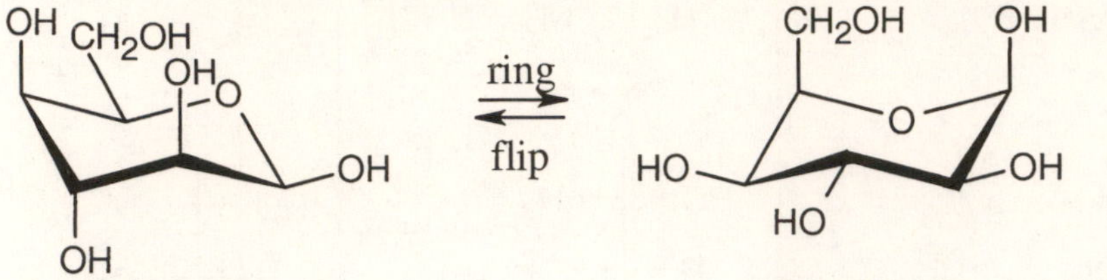

OH CH2OH OH O OH ⇌ (ring flip) CH2OH OH O HO OH HO

less stable β-D-Idopyranose — more stable

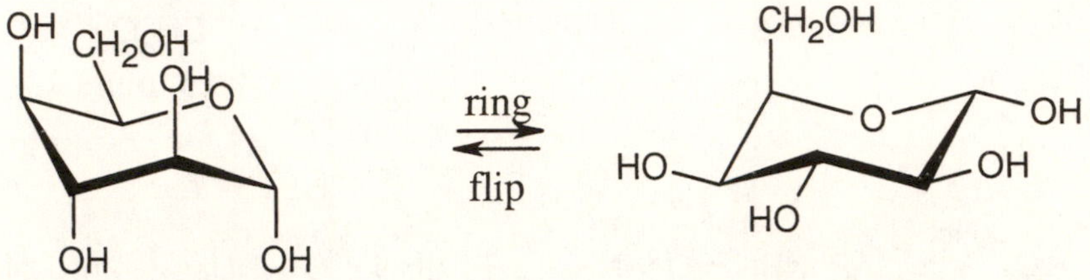

OH CH2OH OH O OH OH ⇌ (ring flip) CH2OH O OH HO OH HO

less stable α-D-Idopyranose — more stable

(b) α-D-Idopyranose is more stable than β-D-idopyranose because only one group is axial in its more stable chair conformation, whereas β-D-idopyranose has two axial groups in its more stable conformation.

(c)

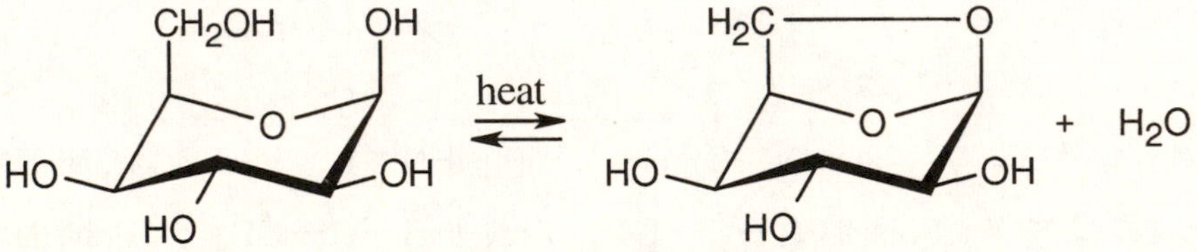

CH2OH OH O HO OH HO $\xrightarrow{\text{heat}}$ H2C—O O HO OH HO + H_2O

1,6-Anhydro-D-idopyranose is formed from the β anomer because the axial hydroxyl groups on carbons 1 and 6 are close enough for the five-membered ring to form.

(d) The hydroxyl groups at carbons 1 and 6 of D-glucopyranose are equatorial in the most stable conformation and are too far apart for a ring to form.

21.64

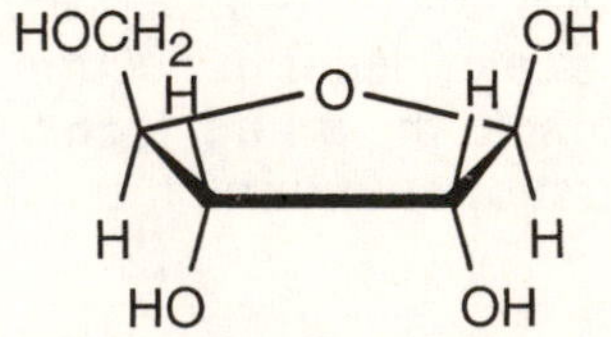

HOCH2 OH H O H H H HO OH

D-Ribofuranose is the sugar present in acetyl CoA.

21.65 Cleavage of fructose 1,6-bisphosphate occurs by a retro-aldol reaction.

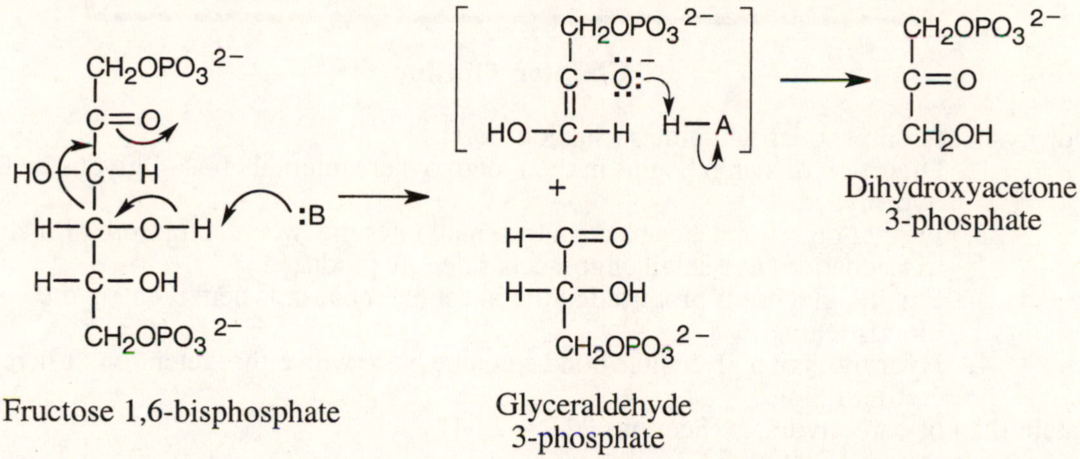

$CH_2OPO_3^{2-}$
C=O
HO—C—H
H—C—O—H :B
H—C—OH
$CH_2OPO_3^{2-}$

Fructose 1,6-bisphosphate

→

[$CH_2OPO_3^{2-}$ / C—Ö:⁻ / ‖ / HO—C—H H—A]

+

H—C=O
H—C—OH
$CH_2OPO_3^{2-}$

Glyceraldehyde 3-phosphate

→

$CH_2OPO_3^{2-}$
C=O
CH_2OH

Dihydroxyacetone 3-phosphate

Chapter 22 – Carbohydrate Metabolism

Chapter Outline

I. Hydrolysis of complex carbohydrates (Section 22.1).
1. Digestion of starch begins in the mouth, where internal (1—›4')glycoside links are hydrolyzed.
2. Further digestion takes place in the small intestine, where a mixture of maltose, trisaccharides and small oligosaccharides are produced
3. Finally, glucose is produced in the intestinal mucosa, where it enters the bloodstream.
4. Hydrolysis of a glycosidic bond can take place with either retention or inversion of configuration.

II. Catabolism of carbohydrates (Sections 22.2 – 22.4).
A. Glycolysis (Section 22.2).
1. Glycolysis is a 10-step series of reactions that converts glucose to pyruvate.
2. Steps 1–3: Phosphorylation and isomerization.
   a. Glucose is phosphorylated at the 6-position by reaction with ATP.
      The enzyme hexokinase is involved.
   b. Glucose 6-P is isomerized to fructose 6-P by glucose-6-P isomerase.
   c. Fructose 6-P is phosphorylated to yield fructose 1,6-bisphosphate.
      ATP and phosphofructokinase are involved.
3. Steps 4–5: Cleavage and isomerization.
   a. Fructose 1,6-bisphosphate is cleaved to glyceraldehyde 3-phosphate and dihydroxyacetone phosphate.
      The reaction is a reverse aldol reaction catalyzed by aldolase.
   b. Dihydroxyacetone phosphate is isomerized to glyceraldehyde 3-phosphate.
   c. The net result is production of two glyceraldehyde 3-phosphates, both of which pass through the rest of the pathway.
4. Steps 6–8: Oxidation and phosphorylation.
   a. Glyceraldehyde 3-phosphate is both oxidized and phosphorylated to give 1,3-bisphosphoglycerate.
      Oxidation occurs via a hemithioacetal to yield a product that forms the mixed anhydride.
   b. The mixed anhydride reacts with ADP to form ATP and 3-phosphoglycerate
      The enzyme phosphoglycerate kinase is involved.
   c. 3-Phosphoglycerate is isomerized to 2-phosphoglycerate by phosphoglycerate mutase.
5. Steps 9–10: Dehydration and dephosphorylation.
   i. 2-Phosphoglycerate is dehydrated by enolase to give PEP.
   ii. Pyruvate kinase catalyzes the transfer of a phosphate group to ADP, with formation of pyruvate.

B. The conversion of pyruvate to acetyl CoA (Section 22.3).
1. The conversion pyruvate → acetyl CoA is catalyzed by an enzyme complex called pyruvate dehydrogenase complex.
2. Step 1: Addition of thiamine.
   A nucleophilic ylide group on thiamine diphosphate adds to the carbonyl group of pyruvate to yield a tetrahedral intermediate.
3. Step 2: Decarboxylation.
4. Step 3: Reaction with lipoamide.
   The enamine product of decarboxylation reacts with lipoamide, displacing sulfur and opening the lipoamide ring.

5. Step 4: Elimination of thiamine diphosphate.
6. Step 5: Acyl transfer.
    i. Acetyl dihydrolipoamide reacts with coenzyme A to give acetyl CoA.
    ii. The resulting dihydrolipoamide is reoxidized to lipoamide by FAD.
    iii. $FADH_2$ is reoxidized to FAD by NAD^+.
7. Other fates of pyruvate.
    i. In the absence of oxygen, pyruvate is reduced to lactate.
    ii. In bacteria, pyruvate is fermented to ethanol.

C. The citric acid cycle (Section 22.4).

1. Characteristics of the citric acid cycle.
    a. The citric acid cycle is a closed loop.
    b. The intermediates are constantly regenerated.
    c. The cycle operates as long as NAD^+ and FAD are available, which means that oxygen must also be available.
2. Steps 1–2: Addition to oxaloacetate.
    a. Acetyl CoA adds to oxaloacetate to form citryl CoA, which is hydrolyzed to citrate, in a reaction catalyzed by citrate synthase.
    b. Citrate is isomerized to isocitrate by aconitase.
        The reaction is an E2 dehydration, followed by conjugate addition of water.
3. Steps 3–4: Oxidative decarboxylations.
    a. Isocitrate is oxidized by isocitrate dehydrogenase to give a ketone that loses CO_2 to give α-ketoglutarate.
    b. α-Ketoglutarate is transformed to succinyl CoA in a reaction catalyzed by a multienzyme dehydrogenase complex.
4. Steps 5–6: Hydrolysis and dehydrogenation of succinyl CoA.
    a. Succinyl CoA is converted to an acyl phosphate, which transfers a phosphate group to GDP in a reaction catalyzed by succinyl CoA synthase.
    b. Succinate is dehydrogenated by FAD and succinate dehydrogenase to give fumarate.
5. Steps 7–8: Regeneration of oxaloacetate.
    a. Fumarase catalyzes the addition of water to fumarate to produce L-malate.
    b. L-malate is oxidized by NAD^+ and malate dehydrogenase to complete the cycle.
6. Some conclusions about biological chemistry.
    a. The mechanisms of biochemical reactions are almost identical to the mechanisms of laboratory reactions.
    b. Most metabolic pathways are linear.
        i. Linear pathways make sense when a multifunctional molecule is undergoing transformation.
        ii. Cyclic pathways may be more energetically feasible when a molecule is small.

III. Biosynthesis of glucose: Gluconeogenesis.(Sections 22.5).

1. Step 1: Carboxylation.
    Pyruvate is carboxylated in a reaction that uses biotin and ATP.
2. Step 2: Decarboxylation and phosphorylation.
    Concurrent decarboxylation and phosphorylation produce phosphoenolpyruvate.
3. Steps 3–4: Hydration and isomerization.
    a. Conjugate addition of water gives 2-phosphoglycerate.
    b. Isomerization produces 3-phosphoglycerate.
4. Steps 5–7: Phosphorylation, reduction and tautomerization.
    a. Reaction of 3-phosphoglycerate with ATP yields an acyl phosphate.
    b. The acyl phosphate is reduced by $NADPH/H^+$ to an aldehyde.
    c. The aldehyde tautomerizes to dihydroxyacetone phosphate.

5. Step 8: Aldol condensation.
   a. Dihydroxyacetone phosphate and glyceraldehyde 3-phosphate condense to form fructose 1,6-bisphosphate.
   b. This condensation involves the imine of dihydroxyacetone phosphate, which forms an enamine that takes part in the condensation.
6. Steps 9-10: Hydrolysis and isomerization.
   a. Fructose 1,6-bisphosphate is hydrolyzed to fructose 6-phosphate.
   b. Fructose 6-phosphate isomerizes to glucose 6-phosphate.
7. Step 11: Hydrolysis to glucose.

Solutions to Problems

22.1 ATP is produced in step 7 (3-phosphoglyceroyl phosphate —> 3-phosphoglycerate) and in step 10 (phosphoenolpyruvate —> pyruvate).

22.2 **Step 1** is a nucleophilic acyl substitution at phosphorus (*phosphate transfer*) by the –OH group at C6 of glucose, with ADP as the leaving group.

Step 2 is an *isomerization*, in which the pyranose ring of glucose 6-phosphate opens, keto-enol tautomerism causes isomerization to fructose 6-phosphate, and a furanose ring is formed.

Step 3 is a substitution, similar to the one in step 1, involving the –OH group at C1 of fructose 6-phosphate (*phosphate transfer*).

Step 4 is a *retro-aldol reaction* that cleaves fructose 1,6-bisphosphate to glyceraldehyde 3-phosphate and dihydroxyacetone phosphate.

Step 5 is an *isomerization* of dihydroxyacetone phosphate to glyceraldehyde 3-phosphate that occurs by keto-enol tautomerization.

Step 6 begins with a nucleophilic addition reaction to the aldehyde group of glyceraldehyde 3-phosphate by a thiol group of an enzyme to form a hemithioacetal, which is *oxidized* by NAD^+ to an acyl thioester. *Nucleophilic acyl substitution* by phosphate yields the product 1,3-bisphosphoglycerate.

Step 7 is a nucleophilic acyl substitution reaction at phosphorus, in which ADP reacts with 1,3-diphosphoglycerate, yielding ATP and 3-phosphoglycerate (*phosphate transfer*).

Step 8 is an *isomerization* of 3-phosphoglycerate to 2-phosphoglycerate.

Step 9 is an *E1cB elimination* of H_2O to form phosphoenolpyruvate.

Step 10 is a substitution reaction at phosphorus that forms ATP and enolpyruvate, which tautomerizes to pyruvate (*phosphate transfer*).

22.3 By performing the substitution test described in Section 9.13, you can see that the *pro-R* hydrogen is removed in Step 5 of glycolysis.

22.4 As shown below, the *Si* face of NAD^+ faces out from the plane of the paper. Hydride adds to this face, and the added hydride (circled) lies in front of the plane of the paper.

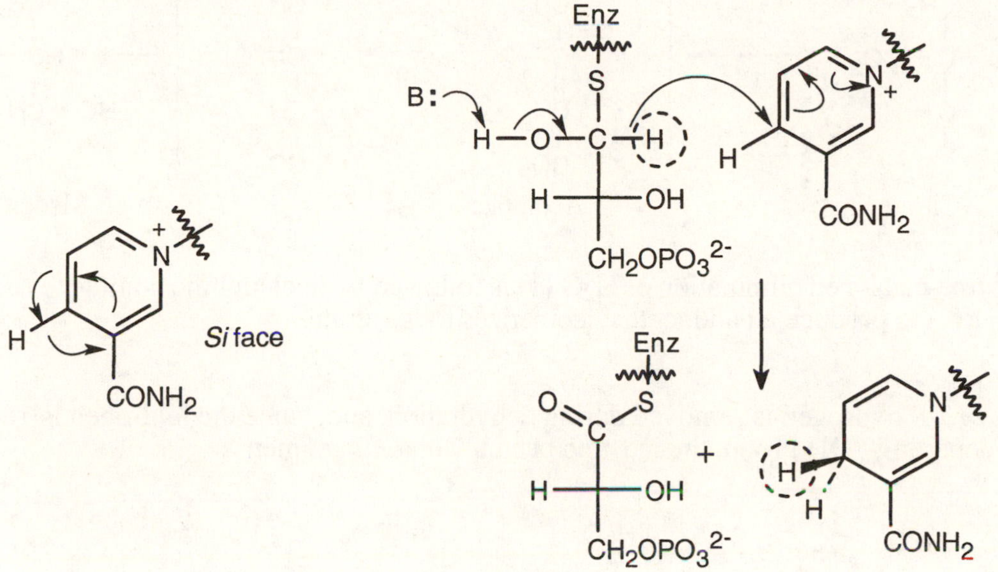

Enz
S
B:
H—O—C—H
H—OH
$CH_2OPO_3^{2-}$
H
$CONH_2$
H
Si face
$CONH_2$
Enz
O
S
C
+
H—OH
$CH_2OPO_3^{2-}$
H
H
$CONH_2$

22.5

1 CHO, 2 H—C—OH, 3 HO—C—H, 4 H—C—OH, 5 H—C—OH, 6 CH_2OH
Glucose

→→

1 $CH_2OPO_3^{2-}$, 2 C=O, 3 HO—C—H, 4 H—C—OH, 5 H—C—OH, 6 $CH_2OPO_3^{2-}$
Fructose 1,6-bisphosphate

Aldolase →

1 $CH_2OPO_3^{2-}$, 2 C=O, 3 CH_2OH
Dihydroxy-acetone phosphate

Triose phosphate isomerase →

1 $CH_2OPO_3^{2-}$, 2 CHOH, 3 CHO

+ 4 CHO, 5 H—C—OH, 6 $CH_2OPO_3^{2-}$
Glyceraldehyde 3-phosphate

3,4 CHO, 2,5 CHOH, 1,6 $CH_2OPO_3^{2-}$
Glyceraldehyde 3-phosphate

→→

3,4 CO_2^-, 2,5 C=O, 1,6 CH_3
Pyruvate

Pyruvate dehydrogenase complex →

SCoA, 2,5 C=O, 1,6 CH_3 + 2 CO_2
Acetyl CoA

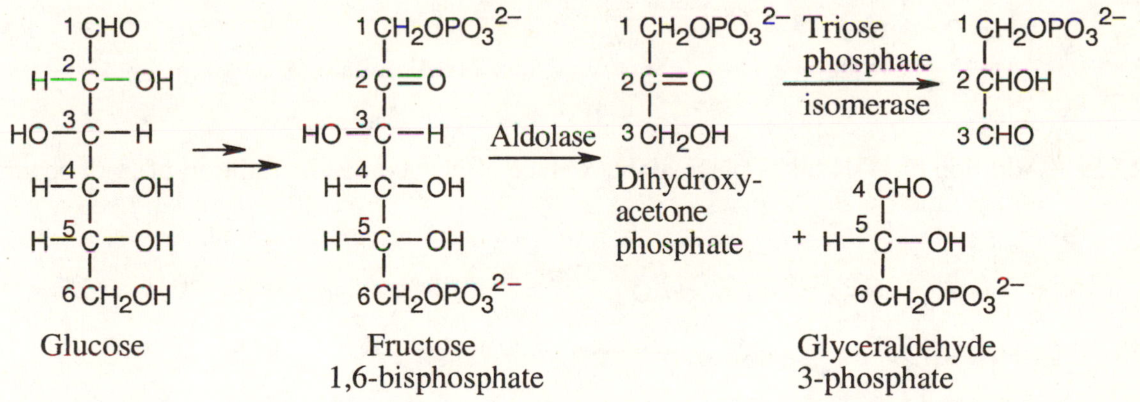

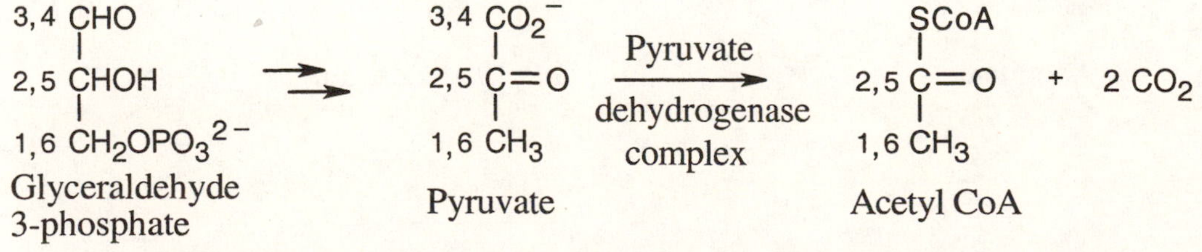

Carbons 1 and 6 of glucose end up as $-CH_3$ groups of acetyl CoA, and carbons 3 and 4 of glucose end up as CO_2.

22.6 Citrate and isocitrate are tricarboxylic acids.

22.7

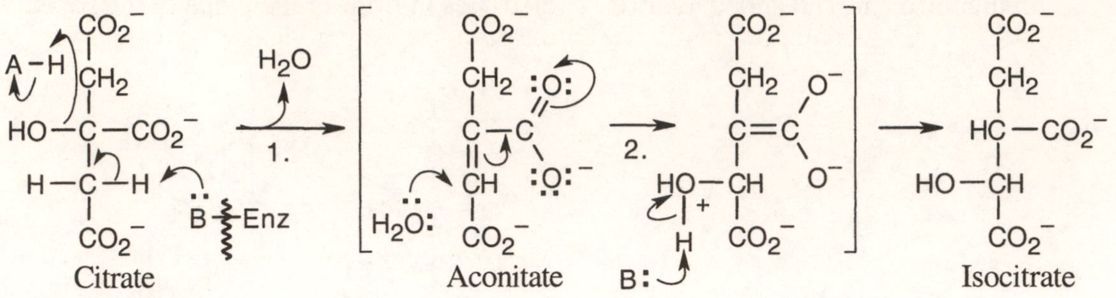

Citrate → (1. H_2O) → [Aconitate → 2.] → Isocitrate

Enzyme-catalyzed elimination of H_2O (1) is followed by nucleophilic conjugate addition of water (2) to produce an adduct that isomerizes to isocitrate.

22.8 The *pro-R* hydrogen is removed during dehydration, and, since the hydrogen is 180° from the departing –OH group, the reaction occurs with *anti* geometry.

Citrate (*pro-R*, *pro-S*) → (H_2O) → *cis*-Aconitate

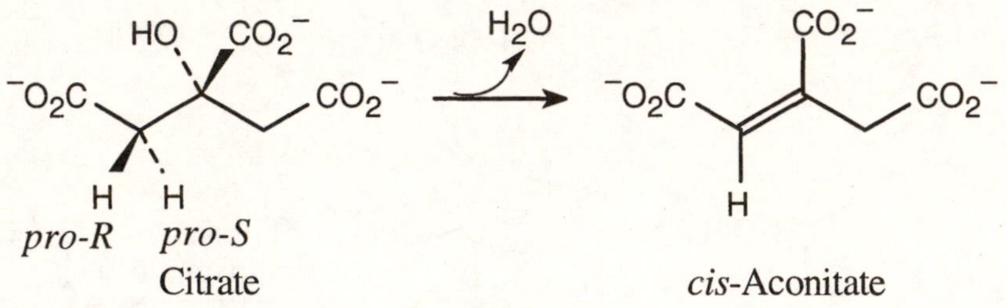

22.9 Addition of –OH takes place at the *Re* face of *cis*-aconitate. Addition of H_2O occurs with *anti* geometry.

cis-Aconitate → (H_2O) → (2*R*,3*S*)-Isocitrate

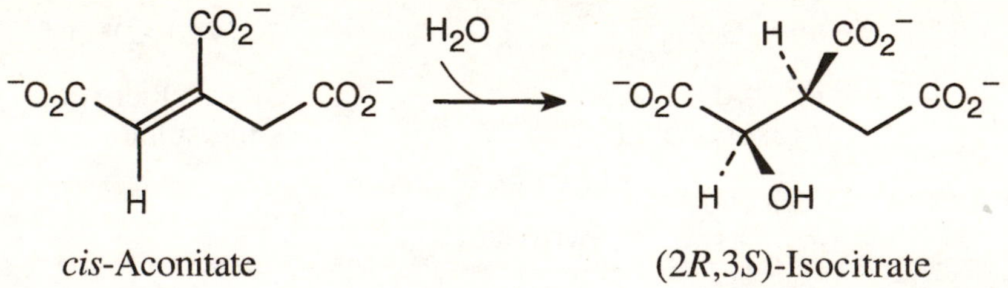

22.10 In Step 1, 1,3-bisphosphoglycerate reacts with a cysteine residue of the enzyme in a nucleophilic acyl substitution reaction, with loss of phosphate. In Step 2, reduction by NADH in a second nucleophilic acyl substitution reaction yields glyceraldehyde 3-phosphate.

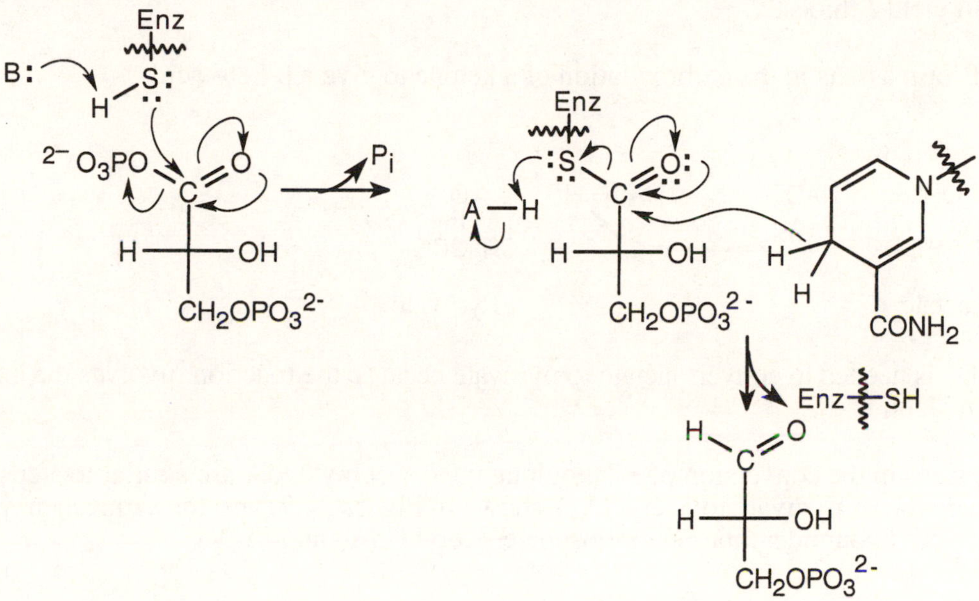

Visualizing Chemistry

22.11 The intermediate is (*S*)-malate.

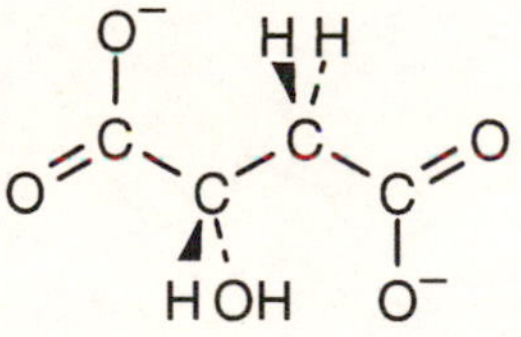

(*S*)-Malate

22.12 The intermediate is derived from D-erythrose.

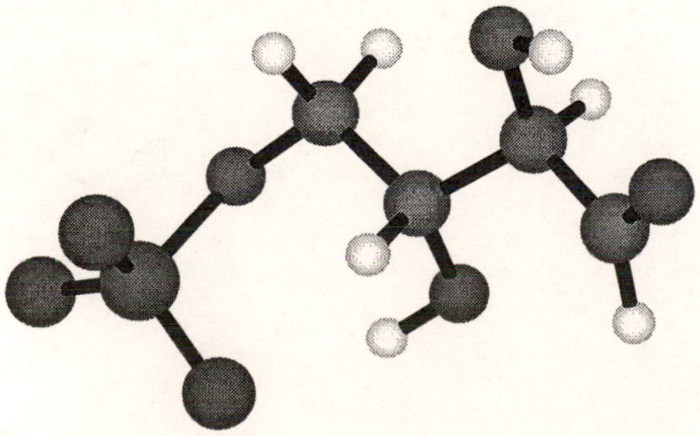

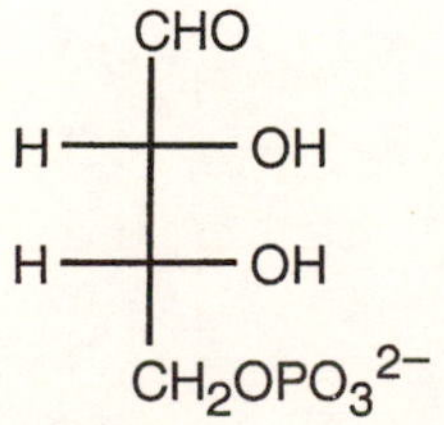

D-Erythrose 4-phosphate

Additional Problems

22.13 (a) ATP is used to phosphorylate an alcohol.

(b) Thiamin diphosphate is associated with the oxidative decarboxylation of an α-keto acid to yield a thioester.

(c) Biotin assists in the carboxylation of a ketone to give a β-keto acid.

22.14

$$\underset{\text{Lactate}}{CH_3CH(OH)CO_2^-} \xrightarrow[\text{NAD}^+ \;\rightarrow\; \text{NADH/H}^+]{} \underset{\text{Pyruvate}}{CH_3C(=O)CO_2^-}$$

NAD^+ is needed to convert lactate to pyruvate because the reaction involves the oxidation of an alcohol.

22.15 The steps in the conversion of α-ketoglutarate to succinyl CoA are similar to steps in the conversion of pyruvate to acetyl CoA shown in Figure 22.7, and the same coenzymes are involved: lipoamide, thiamin diphosphate, acetyl CoA and NAD^+.

H—A
^-O
C
O
N—R
S
R'
CH_3
CO_2^-
1
OH
CO_2^-
S
N—R
+
R'
CH_3
2

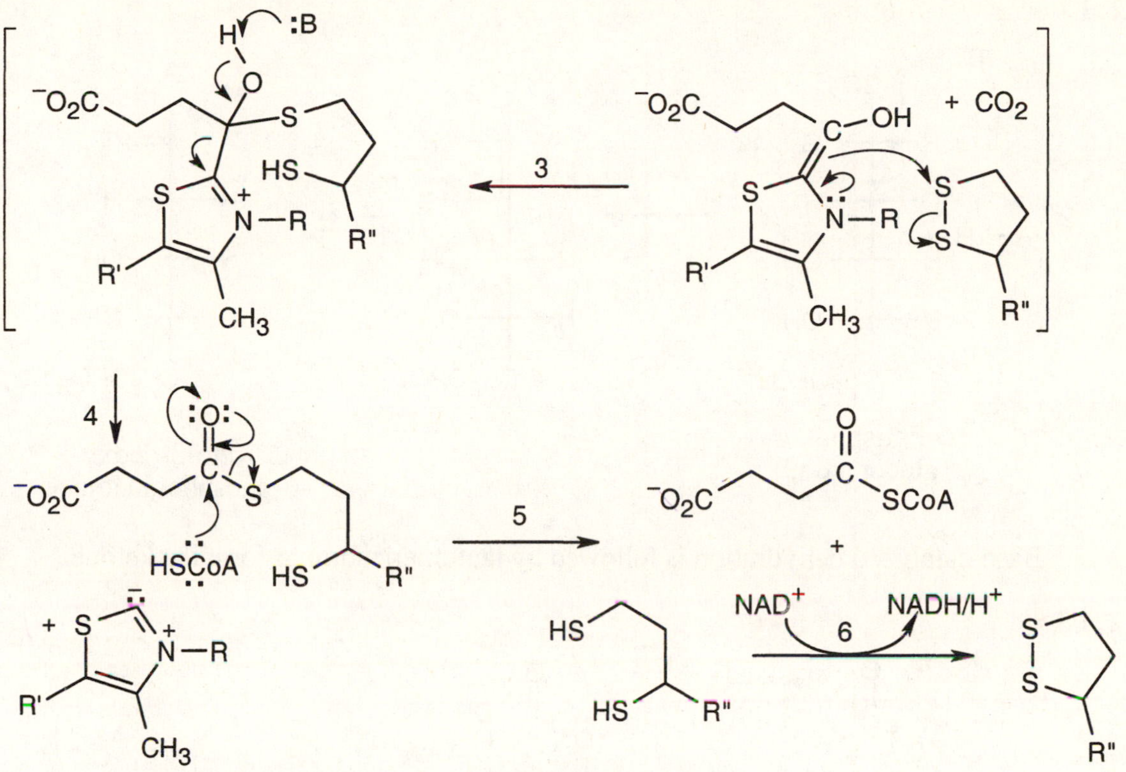

An outline of the mechanism: (1) nucleophilic addition of thiamine diphosphate; (2) decarboxylation; (3) addition of double bond to lipoamide, with ring opening; (4) elimination of thiamine diphosphate; (5) nucleophilic addition of acetyl CoA to succinyl lipoamide and elimination of succinyl dihydrolipoamide to give succinyl CoA; (6) reoxidation of dihydrolipoamide to lipoamide.

22.16

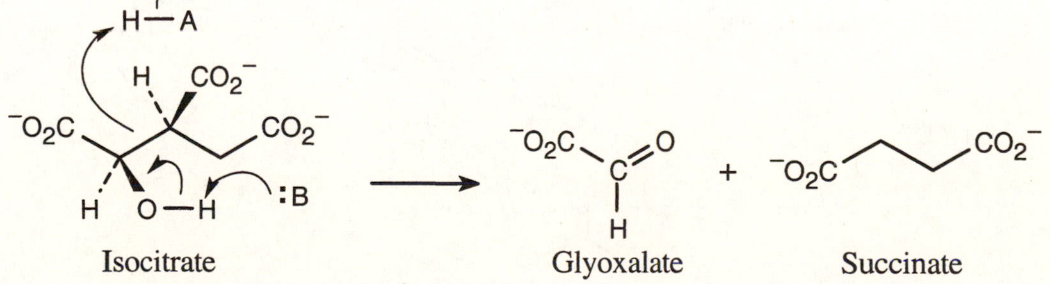

Isocitrate → Glyoxalate + Succinate

The reaction is a retro aldol reaction.

22.17

B:
CO_2^-
H—OH
HO—H
A—H
H—OH
H—OH
$CH_2OPO_3^{2-}$

6-Phospho-
gluconate

H_2O

CO_2^-
—OH
—H
H—OH
H—OH
$CH_2OPO_3^{2-}$

CO_2^-
=O
H—H
H—OH
H—OH
$CH_2OPO_3^{2-}$

2-Keto-3-deoxy-
6-phosphogluconate

Base-catalyzed dehydration is followed by tautomerization to form the ketone.

22.18

H—A
O
^-O C C=O:
CH_3
S
N—R
R'
CH_3

O
:O: C OH
C—CH_3
S
N—R
R'
CH_3

CO_2
(a)

HO
A—H
C—CH_3
S
N—R
R'
CH_3

HETPP

B:
H—O
H—C—CH_3
S
N—R
R'
CH_3

(b)

H—A
O
H_3C—C
H
H
N
H
$CONH_2$
+
S
N—R
R'
CH_3

(c)

H OH
C
H_3C H
+
NAD^+

22.19

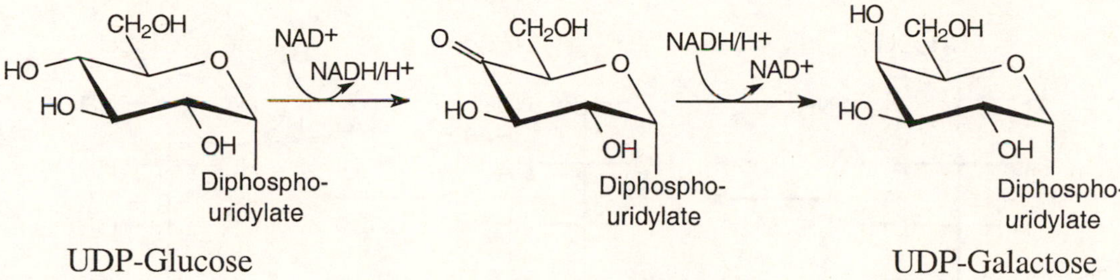

UDP-Glucose → (NAD+ / NADH/H+) → → (NADH/H+ / NAD+) → UDP-Galactose

Oxidation of the C4 hydroxyl group by NAD^+ forms a ketone plus NADH, and reduction of the ketone by NADH yields UDP-galactose. The result is an epimerization at carbon 4 of the pyranose ring. The mechanism of NAD^+/NADH oxidations has been illustrated in many problems.

22.20

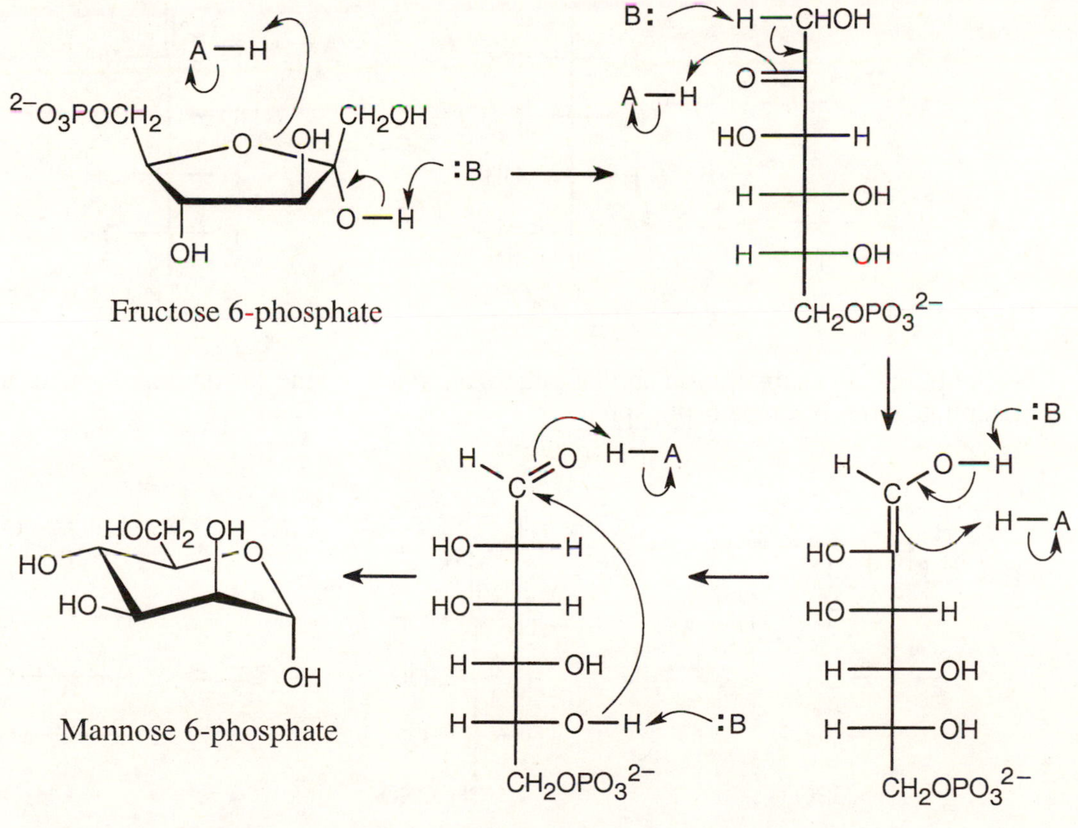

Fructose 6-phosphate

Mannose 6-phosphate

All of these reactions are acid/base catalyzed enolizations or hemiacetal openings/formations.

22.21

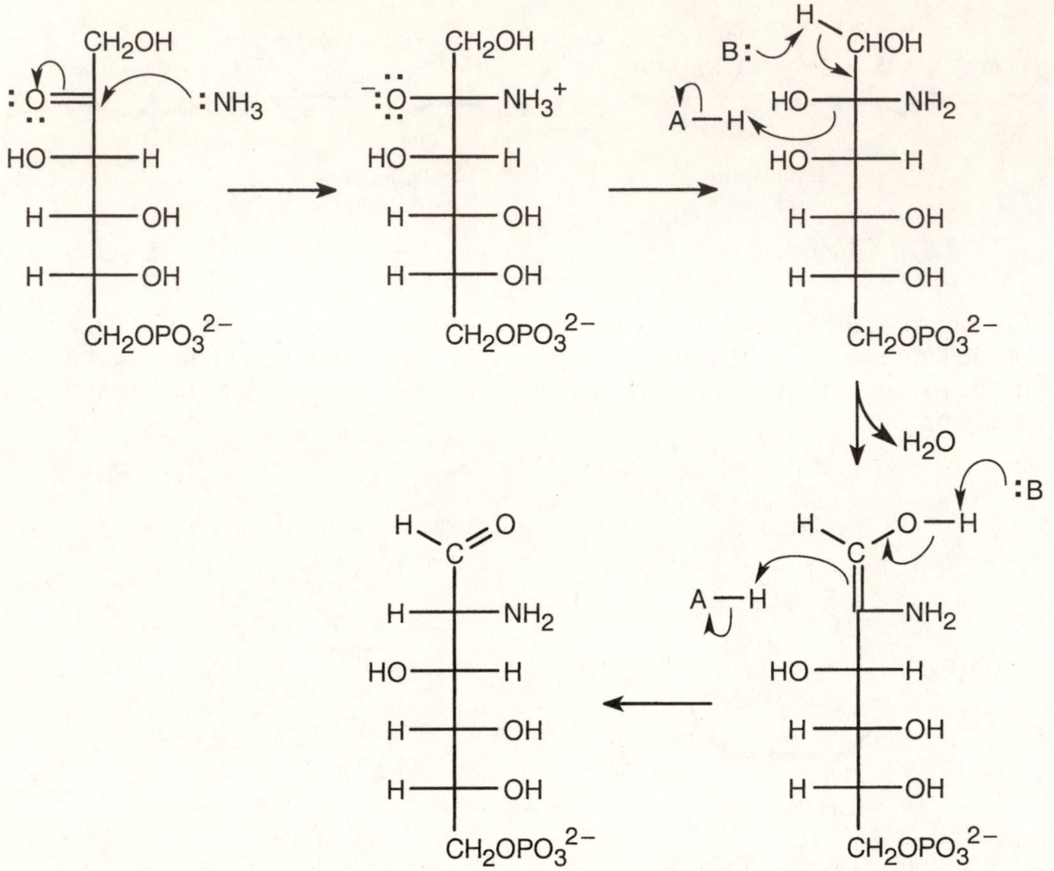

Nucleophilic acyl substitution and tautomerization lead to the formation of glucosamine 6-phosphate from fructose 6-phosphate.

22.22

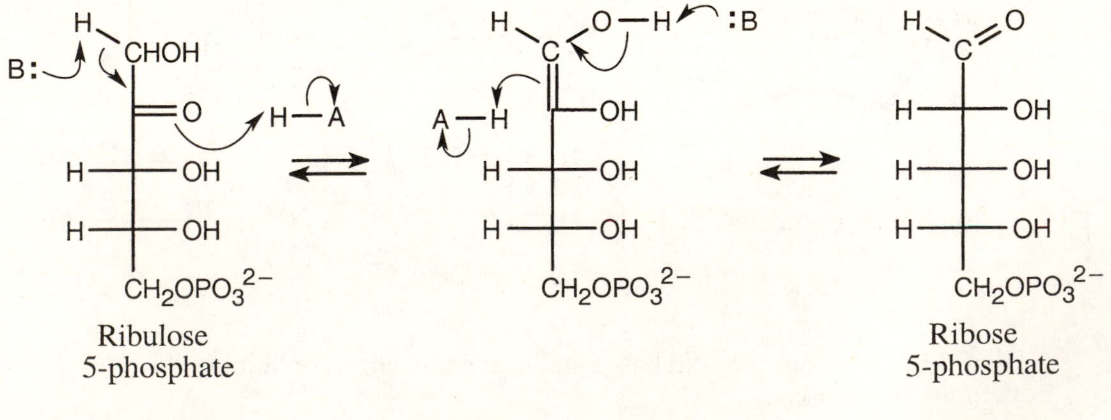

Ribulose 5-phosphate

Ribose 5-phosphate

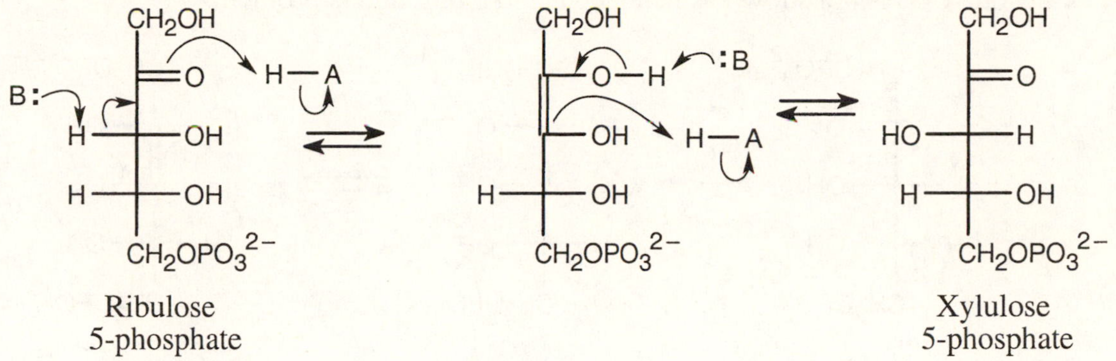

CH2OH, B:, H, O, H—A, OH, O—H, :B, HO

Ribulose 5-phosphate

Xylulose 5-phosphate

22.23

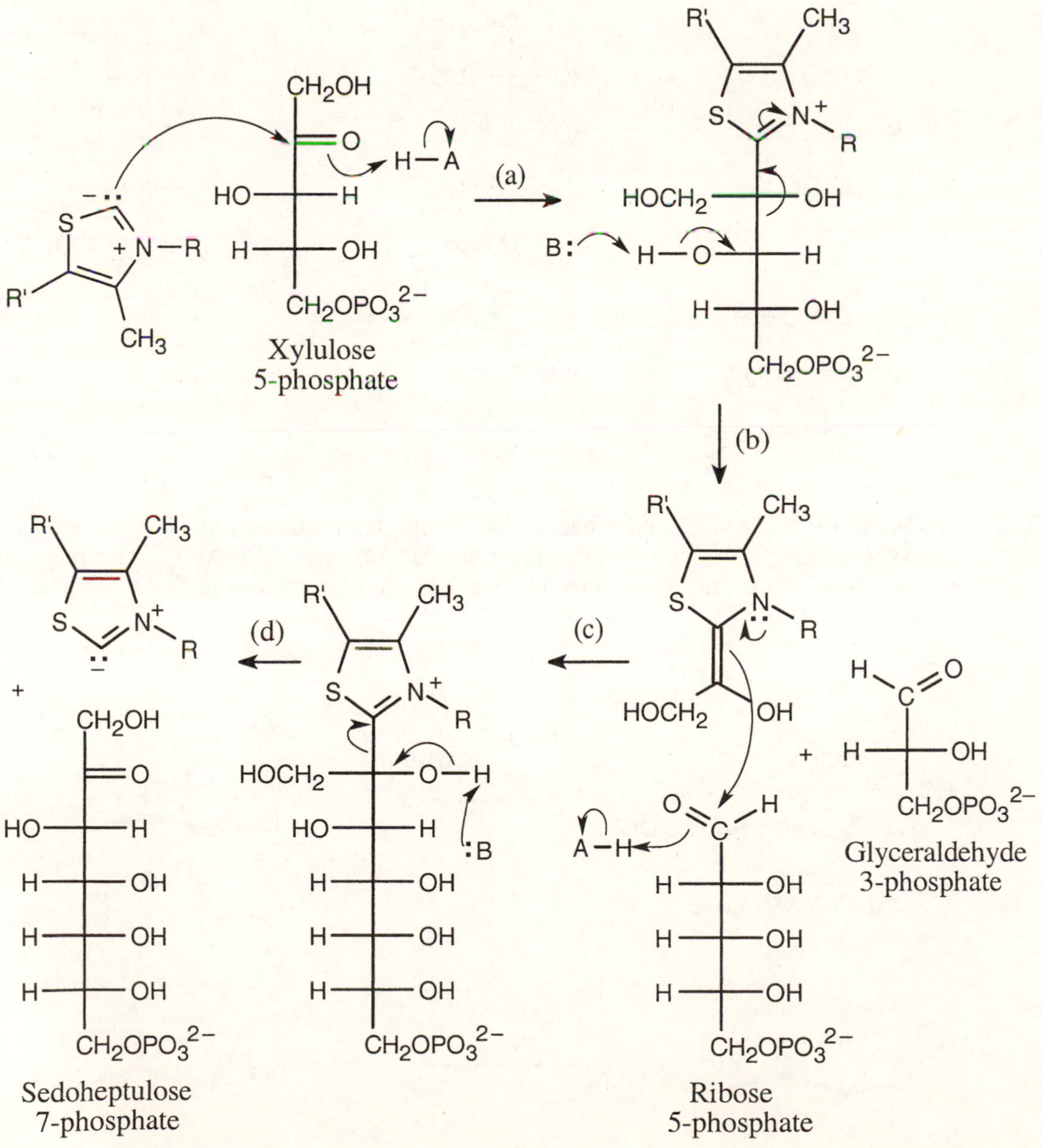

R', CH3, S, N+, R, CH2OH, O, H—A, (a), HO, H, OH, HOCH2, B:, $CH_2OPO_3^{2-}$

Xylulose 5-phosphate

(b)

(c)

(d)

+

Glyceraldehyde 3-phosphate

Sedoheptulose 7-phosphate

Ribose 5-phosphate

22.24 Note: CO_2 is not shown as being bound to biotin in this problem.

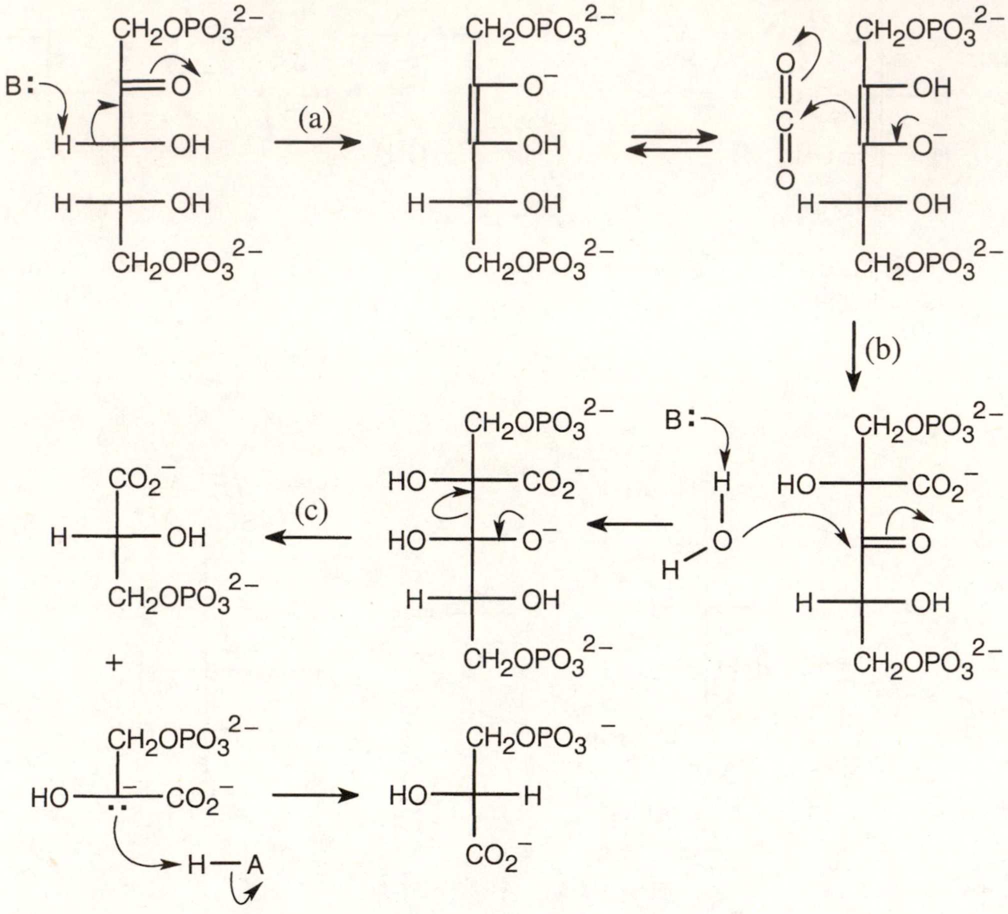

(a)

(b)

(c)

22.25 (a) Oxidation by $NADP^+$, elimination, and conjugate reduction by NADPH give the observed product. Notice that there is no net consumption of $NADP^+$. The mechanism of $NADP^+$ oxidations and reductions has been shown many times in this book and also appears in part (c).

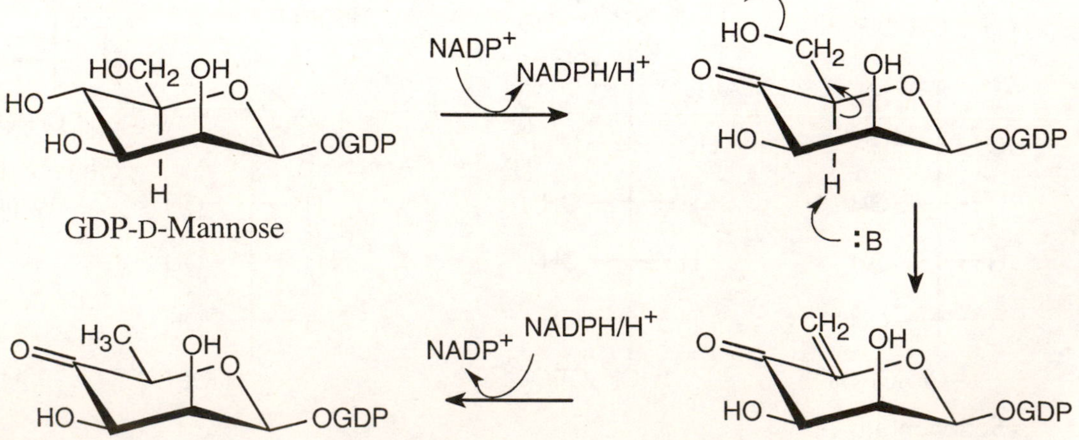

GDP-D-Mannose

(b) Two epimerizations, both α to the carbonyl group, cause a change in stereochemistry.

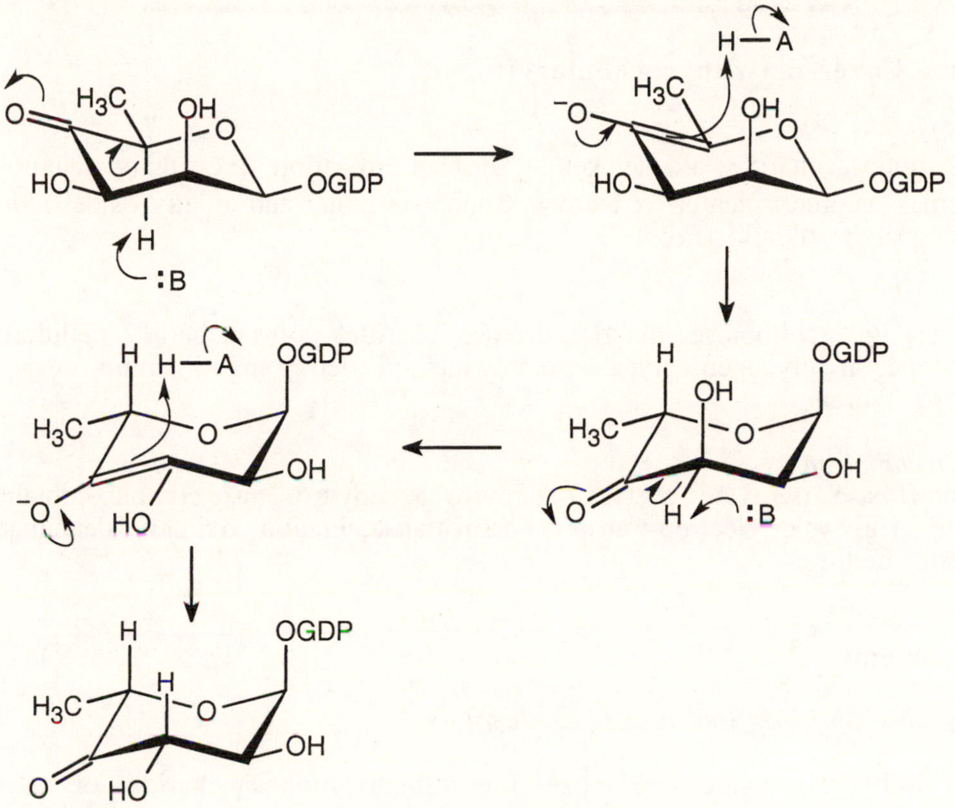

(c) Reduction by NADPH forms GDP-L-fucose.

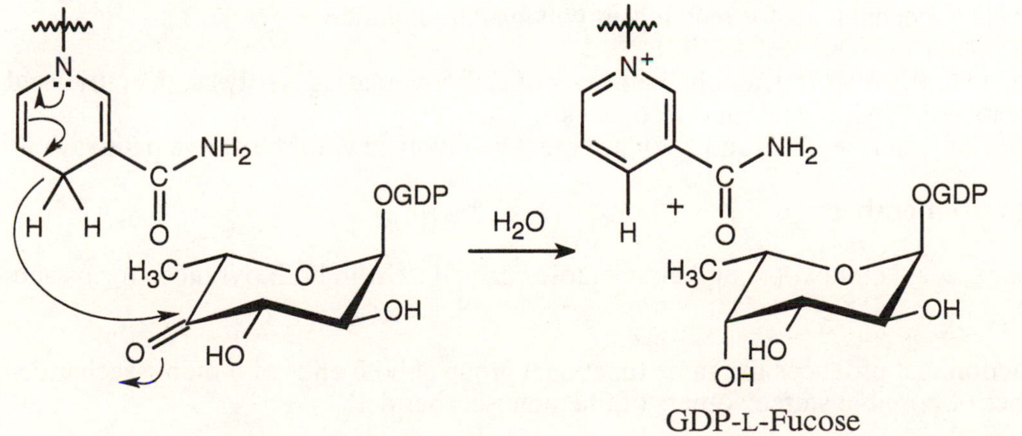

GDP-L-Fucose

Review Unit 9: Carbohydrates and Carbohydrate Metabolism

Major Topics Covered (with vocabulary):

Monosaccharides:
carbohydrate monosaccharide aldose ketose Fischer projection D,L sugars pyranose furanose anomer anomeric center α anomer β anomer mutarotation glycoside aldonic acid alditol reducing sugar aldaric acid

Other sugars:
disaccharide 1,4' link cellobiose maltose lactose sucrose polysaccharide cellulose amylose amylopectin glycogen glycal assembly method deoxy sugar amino sugar cell-surface carbohydrate

Carbohydrate metabolism:
glycolysis Schiff base pyruvate acetyl CoA pyruvate dehydrogenase complex thiamine lipoamide citric acid cycle electron-transport chain transamination oxidative deamination gluconeogenesis biotin

Types of Problems:

After studying these chapters, you should be able to:

- Classify carbohydrates as aldoses, ketoses, D or L sugars, monosaccharides, or polysaccharides.
- Draw monosaccharides as Fischer projections or chair conformations.
- Predict the products of reactions of monosaccharides and disaccharides.
- Deduce the structures of monosaccharides and disaccharides.
- Formulate mechanisms for reactions involving carbohydrates.

- Understand the major metabolic pathways of carbohydrates: glycolysis, the citric acid cycle, pyruvate conversion, and gluconeogenesis.
- Identify the intermediates and enzyme cofactors involved with the above pathways.

Points to Remember:

* Aldohexoses, ketohexoses and aldopentoses can all exist in both pyranose forms and furanose forms.

* A reaction that produces the same functional group at both ends of a monosaccharide halves the number of possible stereoisomers of the monosaccharide.

* The reaction conditions that form a glycoside are different from those that form a polyether, even though both reactions, technically, form –OR bonds.

* Look at the steps of glycolysis, and then look at the steps of gluconeogenesis. Several steps in one pathway are the exact reverse of steps in the other pathway because the energy required for these steps is small. Other, high-energy transformations must occur by steps that are not the exact reverse and that require different enzymes. Gluconeogenesis is a metabolic pathway that takes place mainly during fasting and strenuous exercise because dietary sources of carbohydrates are usually available.

* The conversion pyruvate → acetyl CoA is catalyzed by pyruvate dehydrogenase complex. The conversion acetyl CoA → carbohydrates doesn't occur in animals because they can obtain carbohydrates from food and don't usually need to synthesize carbohydrates. Only plants can, at times, use acetyl CoA to synthesize carbohydrates.

Self-Test:

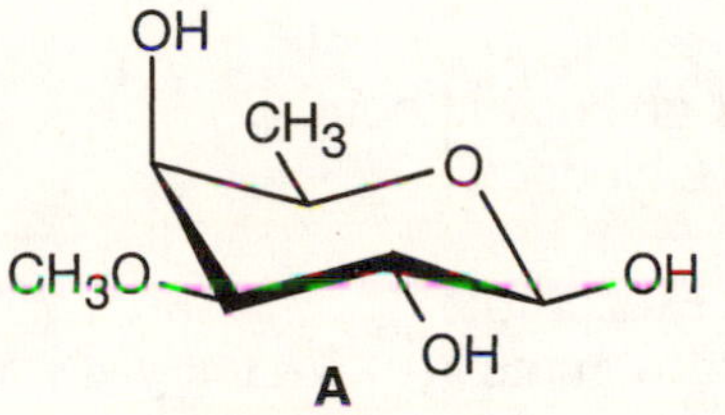

OH, CH_3, O, CH_3O, OH, OH

A

Digitalin
(hydrolysis product of digitoxigenin, a heart medication)

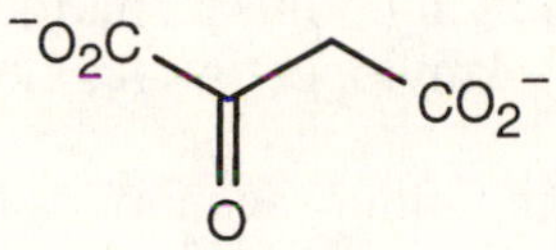

$^{-}O_2C$, CO_2^{-}, O

C

Digitalin (**A**) is related to which D-aldohexose? Provide a name for **A**, including the configuration at the anomeric carbon. Predict the products of the reaction of **A** with: (a) CH_3OH, H^+ catalyst; (b) CH_3I, Ag_2O.

Vicianose (**B**) is a disaccharide associated with a natural product found in seeds. Treatment of **B** with CH_3I and Ag_2O, followed by hydrolysis, gives 2,3,4-tri-*O*-methyl-D-glucose and 2,3,4-tri-*O*-methyl-D-arabinose. What is the structure of **B**? Is **B** a reducing sugar?

Name **C**. **C** is an intermediate in two carbohydrate metabolic pathways. Name them. **C** appears in one of these pathways but not in its reverse. Why might this be so? **C** is the product of transamination of what amino acid?

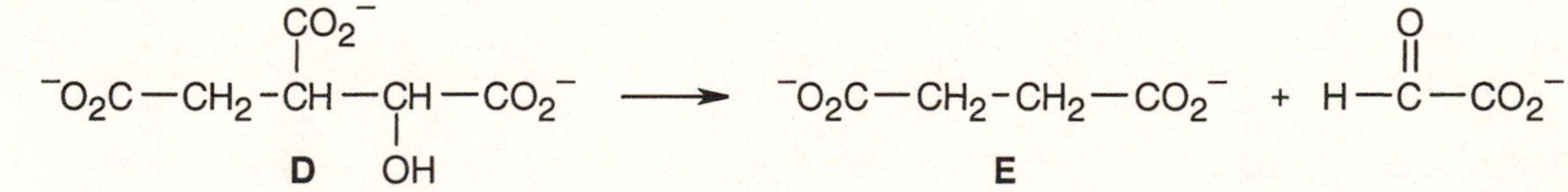

$$^{-}O_2C-CH_2-CH(CO_2^{-})-CH(OH)-CO_2^{-} \longrightarrow {}^{-}O_2C-CH_2-CH_2-CO_2^{-} + H-C(=O)-CO_2^{-}$$

D (reactant), **E** (product)

The above reaction is part of a metabolic pathway that occurs in plants. Identify **D** and **E**. What type of reaction is taking place? Do think that NAD^+, FAD, or ATP are needed for this reaction to occur?

Multiple choice:

1. The enantiomer of α-D-glucopyranose is:
(a) β-D-Glucopyranose (b) α-L-Glucopyranose (c) β-L-Glucopyranose (d) none of these

2. All of the following reagents convert an aldose to an aldonic acid except:
(a) dilute HNO_3 (b) Fehling's reagent (c) Benedict's reagent (d) aqueous Br_2

3. Which two aldoses yield D-lyxose after chain-shortening?
(a) D-Glucose and D-Mannose (b) D-Erythrose and D-Threose (c) D-Galactose and D-Altrose
(d) D-Galactose and D-Talose

4. All of the following disaccharides are reducing sugars except:
(a) Cellobiose (b) Sucrose (c) Maltose (d) Lactose

5. Which of the following polysaccharides contains β-glycosidic bonds?
(a) Amylose (b) Amylopectin (c) Cellulose (d) Glycogen

6. All of the following monosaccharides are essential except for:
(a) *N*-Acetyl-D-glucosamine (b) *N*-Acetyl-D-galactosamine (c) *N*-Acetyl-D-neuraminic acid
(d) *N*-Acetyl-D-mannosamine

7. How many molecules of ATP are produced directly as a result of glycolysis?
(a) 0 (b) 2 (c) 4 (d) ATP is consumed, not produced.

8. Which of the following enzyme cofactors is not involved in the conversion of pyruvate to acetyl CoA?
(a) Thiamine pyrophosphate (b) Pyridoxal phosphate (c) Lipoamide (d) NAD^+

9. Which of the following steps of the citric acid cycle doesn't produce reduced coenzymes?
(a) Isocitrate → α-Ketoglutarate (b) α-Ketoglutarate → Succinyl CoA
(c) Fumarate → Malate (d) Succinate → Fumarate

10. The amino acid aspartate can be metabolized as what citric acid cycle intermediate after transamination?
(a) Oxaloacetate (b) Malate (c) α-Ketoglutarate (d) Succinate

Chapter 23 – Biomolecules: Lipids and Their Metabolism

Chapter Outline

I. Esters (Sections 23.1 – 23.3).
  A. Waxes, fats and oils (Section 23.1).
    1. Waxes are esters of long-chain fatty acids with long-chain alcohols.
    2. Fats and oils are triacylglycerols.
      a. Hydrolysis of a fat yields glycerol and three fatty acids.
      b. The fatty acids need not be the same.
    3. Fatty acids.
      a. Fatty acids are even-numbered,unbranched long-chain (C_{12}–C_{20}) carboxylic acids.
      b. The most abundant saturated fatty acids are palmitic (C_{16}) and stearic (C_{18}) acids.
      c. The most abundant unsaturated fatty acids are oleic and linoleic acids (both C_{18}).
        Linolenic and arachidonic acids are polyunsaturated fatty acids.
      d. Unsaturated fatty acids are lower-melting than saturated fatty acids because the double bonds keep molecules from packing closely.
      e. The C=C bonds can be hydrogenated to produce higher-melting fats.
        Occasionally, cis-trans bond isomerization takes place.
  B. Soap (Section 23.2).
    1. Soap is a mixture of the sodium and potassium salts of fatty acids produced by hydrolysis of animal fat.
    2. Soap acts as a cleanser because the two ends of a soap molecule are different.
      a. The hydrophilic carboxylate end dissolves in water.
      b. The hydrophobic hydrocarbon tails solubilize greasy dirt.
      c. In water, the hydrocarbon tails aggregate into micelles, where greasy dirt can accumulate.
    3. Soaps can form scum when they encounter Mg^{2+} and Ca^{2+} salts.
      This problem is circumvented by detergents, which don't form insoluble metal salts.
  3. Phospholipids (Section 23.3).
    1. Glycerophospholipids.
      a. Glycerophospholipids consist of glycerol, two fatty acids (at C1 and C2 of glycerol), and a phosphate group bonded to an amino alcohol at C3 of glycerol.
      b. Glycerophospholipids comprise the major lipids in cell membranes.
        The phospholipid molecules are organized into a lipid bilayer, which has polar groups on the inside and outside, and nonpolar tails in the middle.
    2. Sphingomyelins.
      a. Sphingomyelins have sphingosine as their backbone.
      b. They are abundant in brain and nerve tissue.

II. Lipid metabolism (Sections 23.4 – 23.6).
  A. Catabolism of triacylglycerols (Sections 23.4 – 23.5).
    1. Triacylglycerols are first hydrolyzed in the stomach and small intestine to yield glycerol plus fatty acids (Section 23.4).
      a. The reaction is catalyzed by a lipase.
        Aspartic acid, serine and histidine bring about reaction.
      b. Glycerol is phosphorylated and oxidized and enters glycolysis.
        i. The mechanism of oxidation involves a hydride transfer to NAD^+.
        ii. The addition to NAD^+ is stereospecific.

c. Fatty acids are degraded by β oxidation, a 4-step spiral that results in the cleavage of an *n*-carbon fatty acid into *n*/2 molecules of acetyl CoA.
c. Before entering β oxidation, a fatty acid is first converted to its fatty-acyl CoA.
2. Steps of β oxidation (Section 23.5).
   a. Introduction of a double bond conjugated with the carbonyl group.
      i. The reaction is catalyzed by acyl CoA dehydrogenase.
      ii. The enzyme cofactor FAD is also involved.
      iii. The mechanism involves abstraction of the *pro-R* α and β hydrogens, resulting in formation of a trans double bond.
   b. Conjugate addition of water to form an alcohol.
      The reaction is catalyzed by enoyl CoA hydratase, and results in a *syn* addition of water.
   c. Alcohol oxidation.
      i. The reaction is catalyzed by L-3-hydroxyacyl CoA dehydrogenase.
      ii. The cofactor NAD^+ is reduced to NADH/H^+ at the same time.
      iii. The mechanism of the reaction resembles a Cannizzaro reaction, followed by conjugate addition of hydride to NAD^+.
   d. Cleavage of acetyl CoA from the chain.
      i. The reaction, which is catalyzed by β-keto thiolase, is a retro-Claisen reaction.
      ii. Nucleophilic addition of coenzyme A to the keto group is followed by loss of acetyl CoA enolate, leaving behind a chain-shortened fatty-acyl CoA.
3. An *n*-carbon fatty acid yields *n*/2 molecules of acetyl CoA after (*n*/2 – 1) passages of β oxidation.
   a. Since most fatty acids have an even number of carbons, no carbons are left over after β oxidation.
   b. Those with an odd number of carbons require further steps for degradation.

B. Biosynthesis of fatty acids (Section 23.6).
1. All common fatty acids have an even number of carbons because they are synthesized from acetyl CoA.
2. Steps 1–2: Acyl transfers convert acetyl CoA to more reactive species.
   a. Acetyl CoA is converted to acetyl ACP.
   b. The acetyl group of acetyl ACP is transferred to a synthase enzyme.
3. Steps 3–4: Carboxylation and acyl transfer.
   a. Acetyl CoA reacts with bicarbonate to yield malonyl CoA and ADP.
      The coenzyme biotin, a CO_2 carrier, transfers CO_2 in a nucleophilic acyl substitution reaction.
   b. Malonyl CoA is converted to malonyl ACP.
   c. At this point, both acetyl groups and malonyl groups are bound to the synthase enzyme.
4. Step 5: Condensation.
   a. A Claisen condensation forms acetoacetyl CoA from acetyl synthase and malonyl ACP.
   b. The reaction proceeds through an intermediate β-keto acid that loses CO_2 to give acetoacetyl CoA.
5. Steps 6–8: Reduction and dehydrogenation.
   a. The ketone group of acetoacetyl CoA is reduced by NADPH.
   b. The β-hydroxy thiol ester is dehydrated.
   c. The resulting double bond is hydrogenated by NADPH to yield butyryl ACP.
6. The steps are repeated with butyryl synthase and malonyl ACP to give a six-carbon unit.

7. Fatty acids up to palmitic acid are synthesized by this route.
   Elongation of palmitic acid and larger acids occurs with acetyl CoA units as the two-carbon donor.

III. Prostaglandins and other eicosanoids (Section 23.7).
   A. Prostaglandins are C_{20} lipids that contain a C_5 ring and two side chains.
   B. Prostaglandins are present in small amounts in all body tissues and fluids.
   C. Prostaglandins have many effects: they lower blood pressure, affect blood platelet aggregation, affect kidney function and stimulate uterine contractions.
   D. Eicosanoids are named by their ring system, substitution pattern and number of double bonds.
   E. Eicosanoids are biosynthesized from arachidonic acid, which is synthesized from linoleic acid.
      1. The transformation from arachidonic acid is catalyzed by the cyclooxygenase (COX) enzyme.
      2. One form of the COX enzyme catalyzes the usual functions, and a second form produces additional prostaglandin as a result of inflammation.

IV. Terpenoids (Section 23.8).
   A. Facts about terpenoids.
      1. Terpenoids occur as essential oils in lipid extractions of plants.
      2. Terpenoids are small organic molecules with diverse structures.
      3. All terpenoids are structurally related.
         Terpenoids arise from the 5-carbon precursor isopentenyl diphosphate.
      4. Terpenoids are classified by the number of 5-carbon multiples they contain.
         a. Monoterpenes are synthesized from two units.
         b. Sesquiterpenes are synthesized from three units.
         c. Larger terpenes occur in both animals and plants.
   B. Biosynthesis of terpenoids.
      1. Nature uses the isoprene equivalent isopentenyl diphosphate (IPP) to synthesize terpenes.
         IPP is biosynthesized by two routes that depend on the organism and the structure of the terpenoid.
            i. The mevalonate pathway produces sesquiterpenes and triterpenes in most animals and plants.
            ii. The 1-deoxyxylulose 5-phosphate pathway gives monoterpenes, diterpenes, and tetraterpenes.
      2. The mevalonate pathway.
         a. Acetyl CoA undergoes Claisen condensation to form acetoacetyl CoA.
         b. Another acetyl CoA undergoes an aldol-like addition to acetoacetyl CoA to give (3*S*)-3-hydroxy-3-methylglutaryl CoA (HMG-CoA).
         c. HMG CoA is reduced by NADPH, yielding (*R*)-mevalonate.
         d. Phosphorylation and decarboxylation convert (*R*)-mevalonate to IPP.
      3. Conversion of IPP to terpenoids.
         a. IPP is isomerized to dimethylallyl diphosphate (DMAPP) by a carbocation pathway.
         b. The C=C bond of IPP displaces the PPO^- group of dimethallyl diphosphate, to form geranyl diphosphate, the precursor to all monoterpenes.
         c. Geranyl diphosphate reacts with IPP to yield farnesyl diphosphate, the precursor to sesquiterpenes.
         d. GPP is isomerized and cyclizes on the way to yielding many monoterpenes.

V. Steroids (Section 23.9 – 23.10).
  A. Facts about steroids (Section 23.9).
    1. Steroids are found in the lipid extracts of plants and animals.
    2. Steroids have a tetracyclic ring structure.
      The rings adopt chair conformations but are unable to undergo ring-flips.
    3 Steroids function as hormones in humans.
  B. Stereochemistry of steroids.
    1. Two cyclohexane rings can be joined either cis or trans.
      a. In a trans-fused ring, the groups at the ring junction are trans.
      b. In cis-fused rings, the groups at the ring junction are cis.
      c. Cis ring fusions usually occur between rings A and B.
    2. In both kinds of ring fusions, the angular methyl groups usually protrude above the rings.
    3. Steroids with A–B trans systems are more common.
    4. Substituents can be either axial or equatorial.
      Equatorial substituents are more favorable.
  C. Types of steroid hormones.
    1. Sex hormones.
      a. Androgens (testosterone, androsterone) are male sex hormones.
      b. Estrogens (estrone, estradiol) and progestins are female sex hormones.
    2. Adrenocortical hormones.
      a. Mineralocorticoids (aldosterone) regulate cellular Na^+ and K^+ balance.
      b. Glucocorticoids (hydrocortisone) regulate glucose metabolism and control inflammation.
    3. Synthetic steroids.
      Oral contraceptives and anabolic steroids are examples of synthetic steroids.
  D. Steroid biosynthesis (Section 23.10).
    1. All steroids are biosynthesized from squalene.
    2. Squalene is first epoxidized to form 2,3-oxidosqualene.
    3. Nine additional steps are needed to form lanosterol.
      a. The first several steps are cyclization reactions.
      b. The last steps are hydride and methyl shifts.
    4. Other enzymes convert lanosterol to cholesterol.

Solutions to Problems

23.1

$$CH_3(CH_2)_{18}C(=O)O(CH_2)_{31}CH_3$$

Carnauba wax

from C_{20} acid

from C_{32} alcohol

23.2

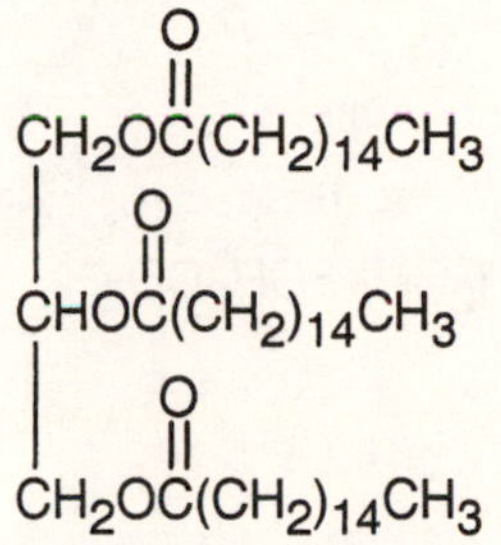

$$\begin{array}{l} CH_2OC(=O)(CH_2)_{14}CH_3 \\ | \\ CHOC(=O)(CH_2)_{14}CH_3 \\ | \\ CH_2OC(=O)(CH_2)_{14}CH_3 \end{array}$$

Glyceryl tripalmitate

$$\begin{array}{ll} CH_2OC(=O)(CH_2)_7CH{=}CH(CH_2)_7CH_3 & \text{(cis)} \\ | & \\ CHOC(=O)(CH_2)_7CH{=}CH(CH_2)_7CH_3 & \text{(cis)} \\ | & \\ CH_2OC(=O)(CH_2)_7CH{=}CH(CH_2)_7CH_3 & \text{(cis)} \end{array}$$

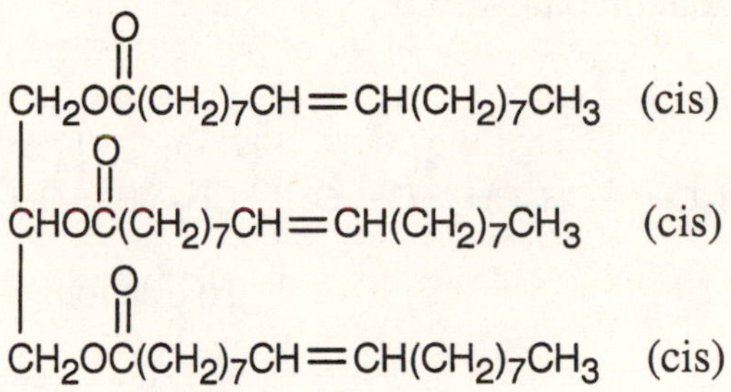

Glyceryl trioleate

Glyceryl tripalmitate is higher melting because it is saturated.

23.3

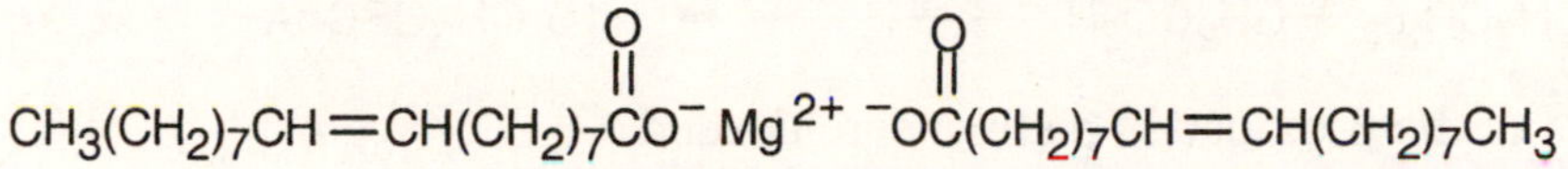

$$CH_3(CH_2)_7CH{=}CH(CH_2)_7C(=O)O^-\ Mg^{2+}\ {}^-OC(=O)(CH_2)_7CH{=}CH(CH_2)_7CH_3$$

Magnesium oleate

The double bonds are cis.

23.4

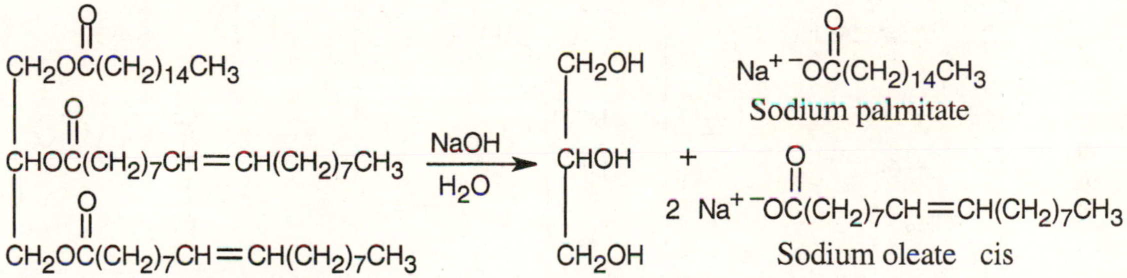

$$\begin{array}{l} CH_2OC(=O)(CH_2)_{14}CH_3 \\ | \\ CHOC(=O)(CH_2)_7CH{=}CH(CH_2)_7CH_3 \\ | \\ CH_2OC(=O)(CH_2)_7CH{=}CH(CH_2)_7CH_3 \end{array} \xrightarrow[H_2O]{NaOH} \begin{array}{l} CH_2OH \\ | \\ CHOH \\ | \\ CH_2OH \end{array} + \begin{array}{l} Na^+\ {}^-OC(=O)(CH_2)_{14}CH_3 \\ \text{Sodium palmitate} \\ 2\ Na^+\ {}^-OC(=O)(CH_2)_7CH{=}CH(CH_2)_7CH_3 \\ \text{Sodium oleate cis} \end{array}$$

Glyceryl dioleate monopalmitate (cis double bonds) — Glycerol

23.5

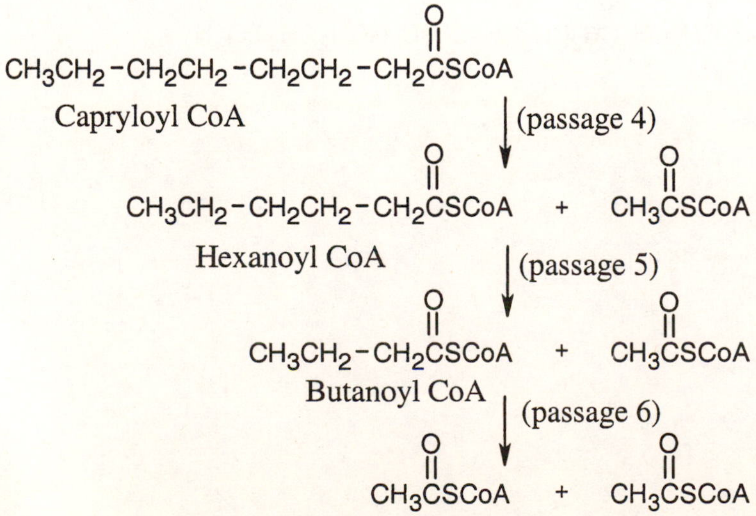

$$CH_3CH_2{-}CH_2CH_2{-}CH_2CH_2{-}CH_2C(=O)SCoA$$

Capryloyl CoA

↓ (passage 4)

$$CH_3CH_2{-}CH_2CH_2{-}CH_2C(=O)SCoA + CH_3C(=O)SCoA$$

Hexanoyl CoA

↓ (passage 5)

$$CH_3CH_2{-}CH_2C(=O)SCoA + CH_3C(=O)SCoA$$

Butanoyl CoA

↓ (passage 6)

$$CH_3C(=O)SCoA + CH_3C(=O)SCoA$$

23.6 A fatty acid with *n* carbons yields *n*/2 acetyl CoA molecules after (*n*/2 – 1) passages of the β-oxidation pathway.

(a)

$CH_3CH_2-CH_2CH_2-CH_2CH_2-CH_2CH_2-CH_2CH_2-CH_2CH_2-CH_2CH_2-CH_2CO_2H$

$\xrightarrow{\beta \text{ oxidation}}$ $8\ CH_3\overset{O}{\overset{\|}{C}}SCoA$

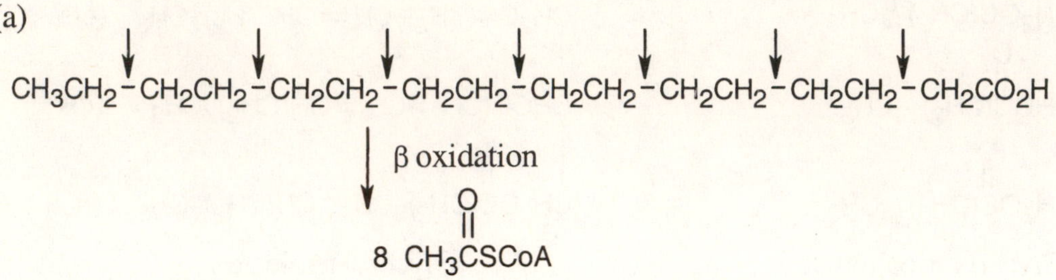

Seven passages of the β-oxidation pathway are needed.

(b)

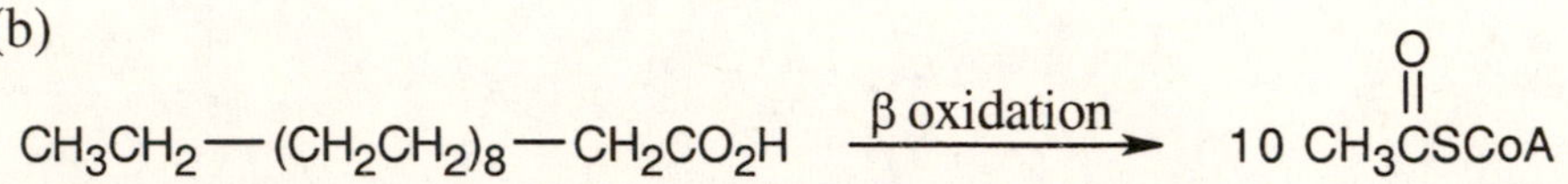

$CH_3CH_2-(CH_2CH_2)_8-CH_2CO_2H \xrightarrow{\beta \text{ oxidation}} 10\ CH_3\overset{O}{\overset{\|}{C}}SCoA$

Nine passages of the β-oxidation pathway are needed.

23.7 β-Hydroxybutyryl ACP resembles the β-hydroxy ketones that were described in Chapter 17 and that dehydrate readily by an E1cB mechanism.

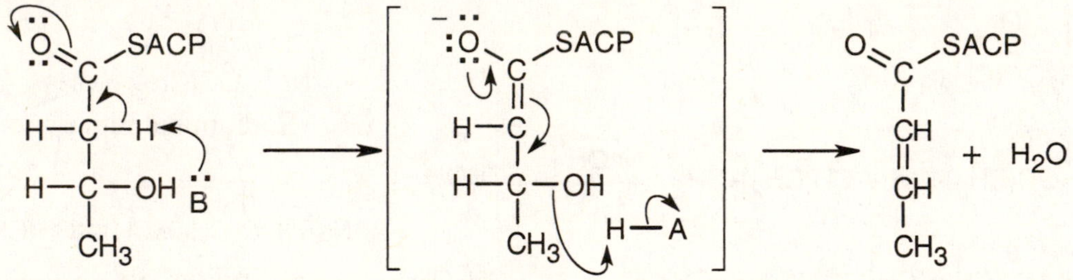

$O{=}C(SACP)-CH_2-CH(OH)-CH_3 \xrightarrow{:B} [\,^{-}O-C(SACP){=}CH-CH(OH)-CH_3\,] \xrightarrow{H-A} O{=}C(SACP)-CH{=}CH-CH_3 + H_2O$

23.8 A fatty acid synthesized from $^{13}CH_3CO_2H$ has an alternating labeled and unlabeled carbon chain. The carboxylic acid carbon is unlabeled.

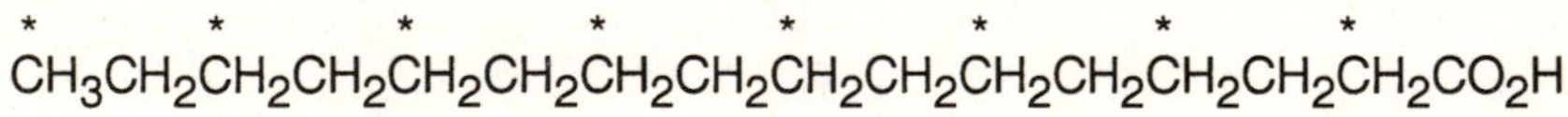

$\overset{*}{C}H_3CH_2\overset{*}{C}H_2CH_2\overset{*}{C}H_2CH_2\overset{*}{C}H_2CH_2\overset{*}{C}H_2CH_2\overset{*}{C}H_2CH_2\overset{*}{C}H_2CH_2\overset{*}{C}H_2CO_2H$

23.9 The face in front of the plane of the page is the *re* face. Since addition occurs from behind the plane of the page, it occurs at the *si* face.

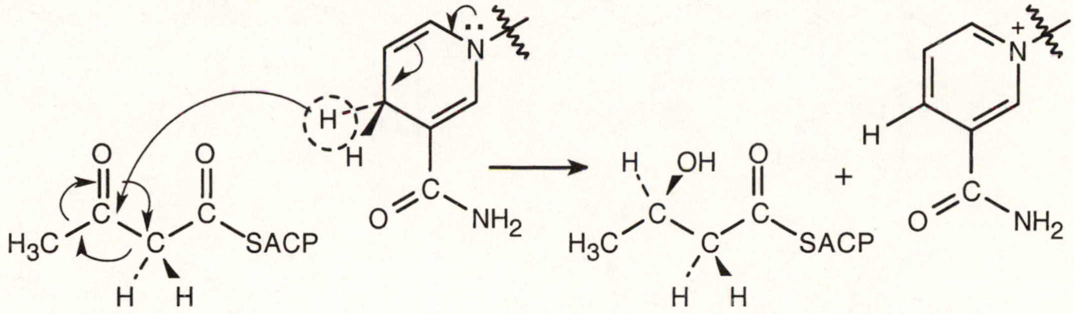

H_3C SACP H H O O NH$_2$ N H H → H OH O H_3C SACP H H + N H O NH$_2$

23.10

O H COOH R R R S H OH H H OH

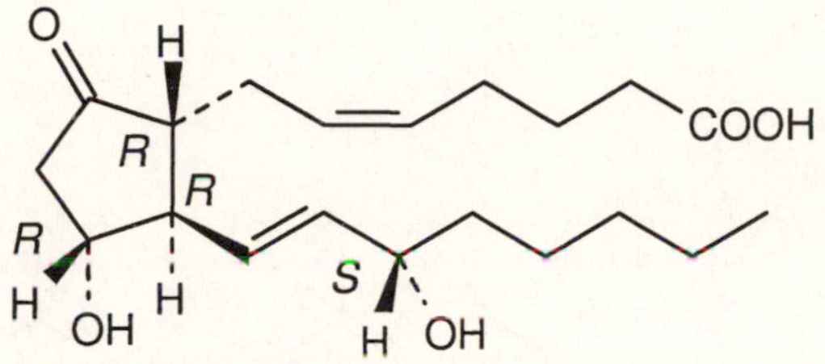

Prostaglandin E_2

23.11 The *pro-S* hydrogen (green) ends up cis to the methyl group, and the *pro-R* hydrogen (red) ends up trans.

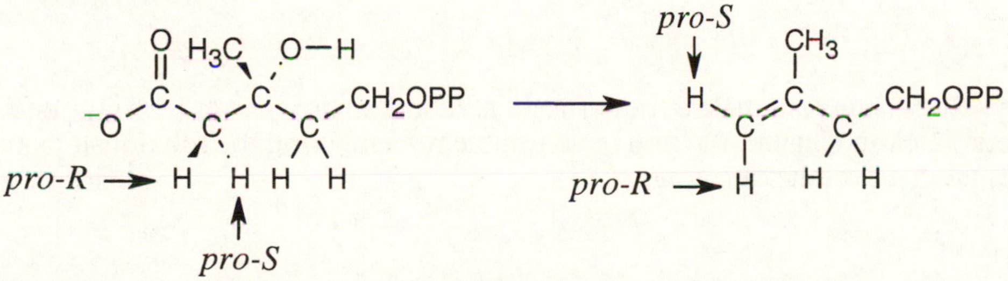

O H_3C O–H $^-$O CH_2OPP *pro-R* → H H H H *pro-S* → *pro-S* H CH_3 CH_2OPP *pro-R* → H H H

23.12 Strategy: As described in Worked Example 23.1, draw the diphosphate precursor so that it resembles the product. Often, the precursor is linalyl diphosphate, which results from isomerization of geranyl diphosphate (the mechanism is shown in Figure 23.15). In (a), it's not easy to see the relationship, but once you've arrived at the product, rotate the structure.

Solution:

(a)

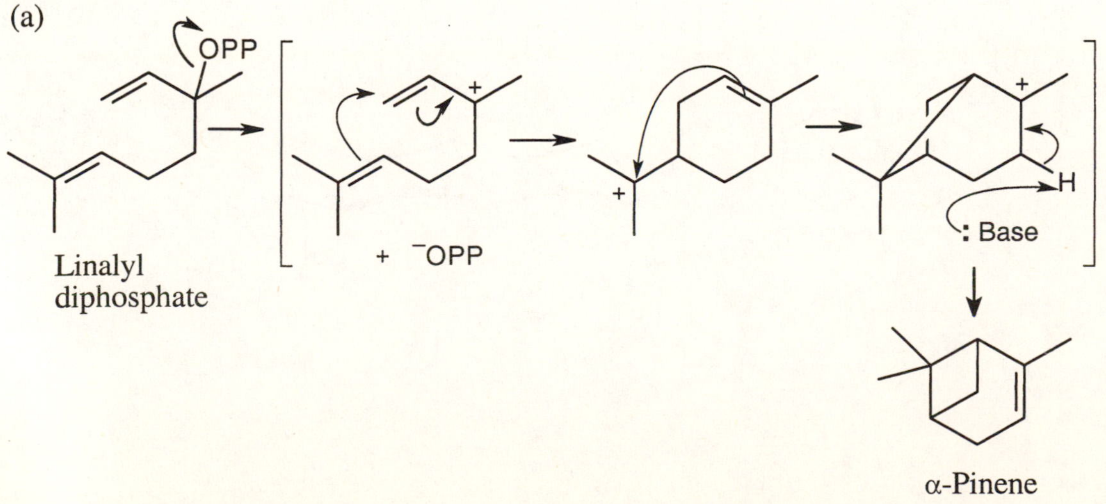

OPP → + + $^-$OPP → + → + H : Base

Linalyl diphosphate

α-Pinene

(b)

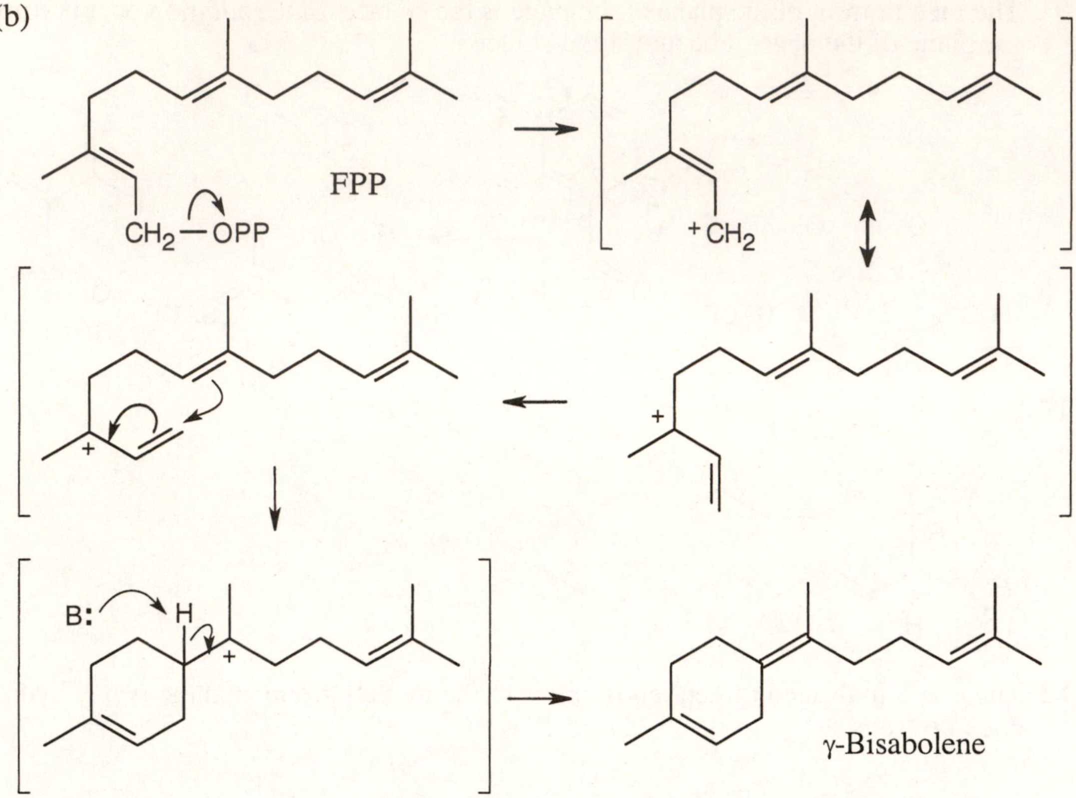

FPP

CH_2—OPP

$^+CH_2$

B:

H

γ-Bisabolene

23.13 Strategy: Both ring systems are trans-fused, and both hydrogens at the ring junctions are axial. Refer back to Chapter 4 if you have trouble remembering the relationships of substituents on a cyclohexane ring.

Solution:

(a)

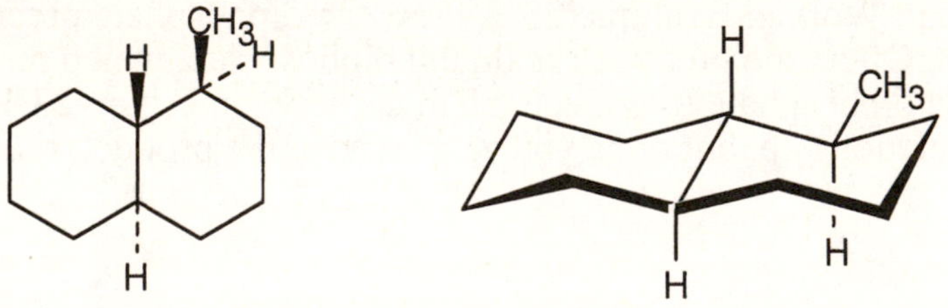

CH_3 H H H

H CH_3 equatorial

H H

(b)

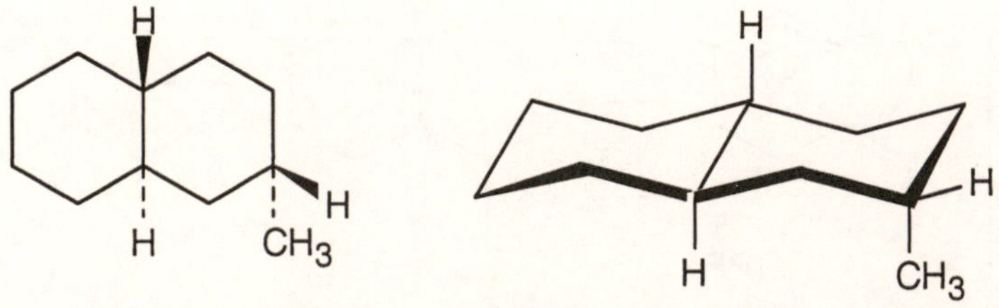

H H CH_3

H H H CH_3 axial

23.14 Draw the three-dimensional structure and note the relationship of the hydroxyl group to groups whose orientation is known.

OH ← equatorial

Lithocholic acid

23.15

Lanosterol

1. Two methyl groups at C4.
2. One methyl group at C14.
3. C5–C6 single bond.
4. C8–C9 double bond.
5. Double bond in side chain

Cholesterol

1. Two hydrogens at C4.
2. One hydrogen at C14.
3. C5–C6 double bond
4. C8–C9 single bond.
5. Saturated side chain.

Visualizing Chemistry

23.16

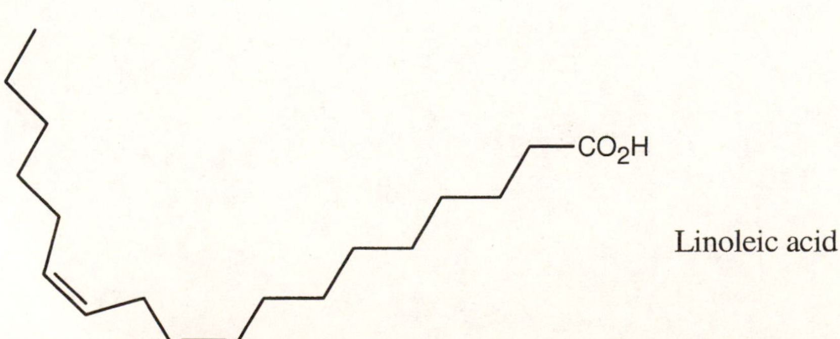

An unsaturated fat such as linoleic acid is more likely to be found in peanut oil.

23.17

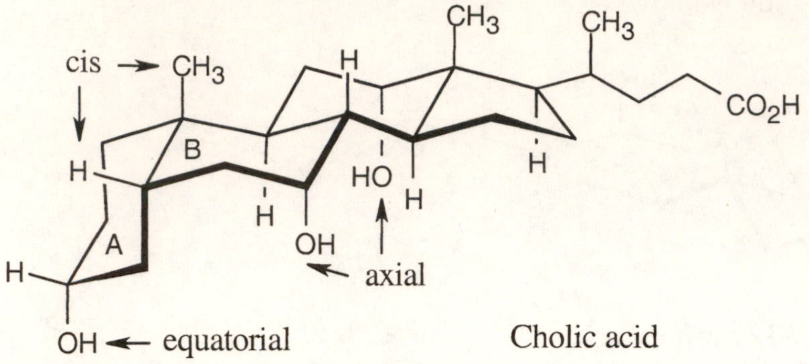

cis → CH_3
H
CH_3
CH_3
CO_2H
B
H
HO
H
H
H
A
OH
H
axial
OH ← equatorial

Cholic acid

Cholic acid is an A–B cis steroid because the groups at the fusion of ring A and ring B have a cis relationship.

23.18

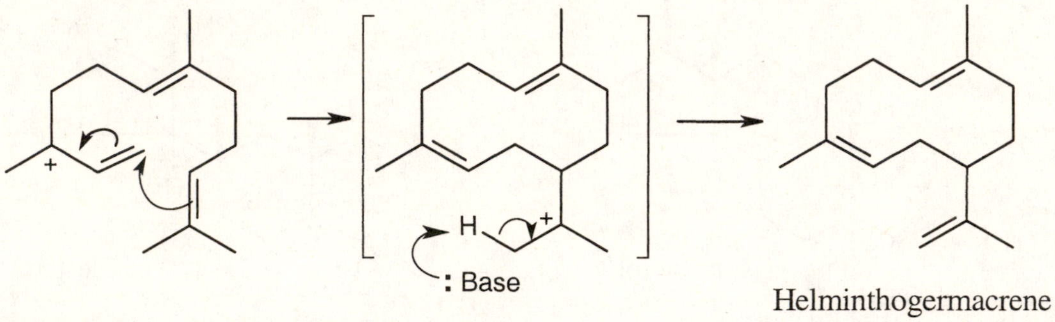

+
H
+
: Base

Helminthogermacrene

Draw farnesyl diphosphate in the configuration that resembles the product, then draw its allylic isomer. (The mechanism for the formation of the isomer is shown in Problem 23.12. Also, to save space, the cation formed by dissociation of –OPP is shown as the starting material.) In this reaction, a cyclization, followed by loss of a proton to form the double bond, gives helminthogermacrene.

Additional Problems

23.19

$$\begin{array}{l} CH_2OC(O)(CH_2)_{16}CH_3 \\ |\\ {}^{*}CHOC(O)(CH_2)_{16}CH_3 \\ | \\ CH_2OC(O)(CH_2)_7CH{=}CH(CH_2)_7CH_3 \ \text{(cis)} \end{array} \quad or \quad \begin{array}{l} CH_2OC(O)(CH_2)_{16}CH_3 \\ | \\ CHOC(O)(CH_2)_7CH{=}CH(CH_2)_7CH_3 \ \text{(cis)} \\ | \\ CH_2OC(O)(CH_2)_{16}CH_3 \end{array}$$

optically active — optically inactive

1. ^{-}OH, H_2O
2. H_3O^+

$$\begin{array}{l} CH_2OH \\ | \\ CHOH \\ | \\ CH_2OH \end{array} + \underset{\text{Oleic acid}}{HOC(O)(CH_2)_7CH{=}CH(CH_2)_7CH_3 \ \text{(cis)}} + 2\ \underset{\text{Stearic acid}}{HOC(O)(CH_2)_{16}CH_3}$$

Four different groups are bonded to the central glycerol carbon atom in the optically active fat.

23.20

$$CH_3(CH_2)_{14}C(=O)OCH_2(CH_2)_{14}CH_3$$

Cetyl palmitate

23.21 Fats and plasmalogens are esters of a glycerol molecule that has carboxylic acid ester groups at C2 and C3. The third group bonded to glycerol, however, differs with the type of lipid; fats have a carboxylic acid ester at C1, and plasmalogens have a vinyl ether in that position.

23.22

$$\begin{array}{l} CH_2OH + HOC(O)R'' \\ | \\ CHOH \ + HOC(O)R' \\ | \\ CH_2OH + HC(O)CH_2R \end{array} \xleftarrow{H_3O^+} \begin{array}{l} CH_2OC(O)R'' \\ | \\ CHOC(O)R' \\ | \\ CH_2OCH{=}CHR \end{array} \xrightarrow[H_2O]{NaOH} \begin{array}{l} CH_2OH + Na^+ \ {}^{-}OC(O)R'' \\ | \\ CHOH \ + Na^+ \ {}^{-}OC(O)R' \\ | \\ CH_2OCH{=}CHR \end{array}$$

Basic hydrolysis cleaves the carboxylic acid ester bonds but doesn't affect the ether bond. Acidic hydrolysis cleaves all three groups bonded to glycerol and produces an aldehyde from the vinyl ether group.

23.23

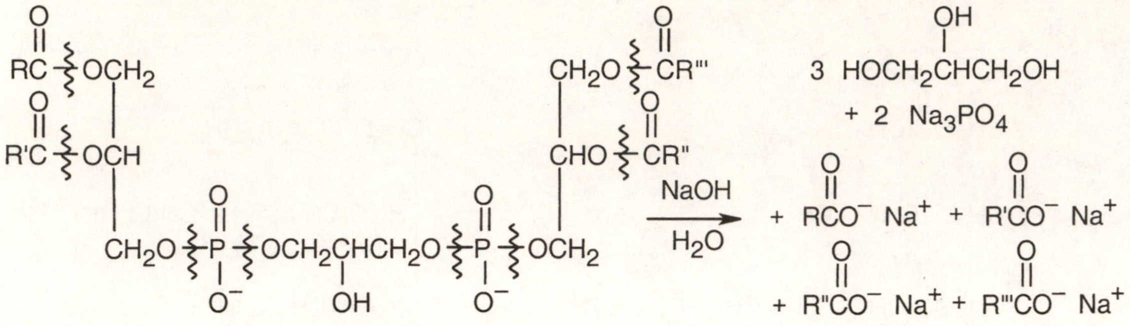

$$\text{a cardiolipin: } RC(=O)\text{-}OCH_2\text{-}CH(OC(=O)R')\text{-}CH_2O\text{-}P(=O)(O^-)\text{-}OCH_2CH(OH)CH_2O\text{-}P(=O)(O^-)\text{-}OCH_2\text{-}CH(OC(=O)R'')\text{-}CH_2OC(=O)R''' \xrightarrow[H_2O]{NaOH} 3\ HOCH_2CH(OH)CH_2OH + 2\ Na_3PO_4 + RCO_2^-\ Na^+ + R'CO_2^-\ Na^+ + R''CO_2^-\ Na^+ + R'''CO_2^-\ Na^+$$

a cardiolipin

Saponification of a cardiolipin yields 4 different carboxylates, 3 equivalents of glycerol and two equivalents of phosphate.

23.24

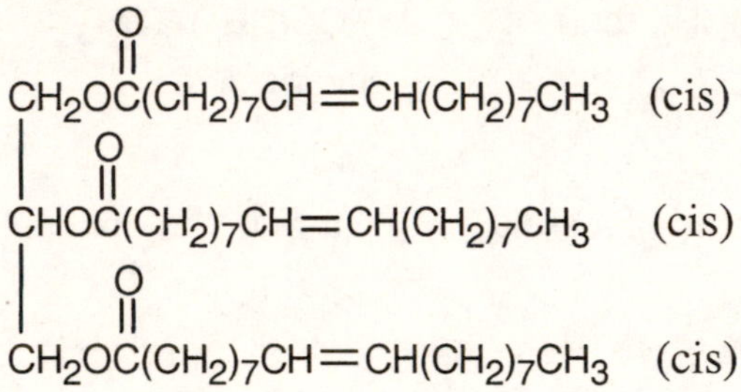

$$CH_2OC(=O)(CH_2)_7CH{=}CH(CH_2)_7CH_3 \text{ (cis)}$$
$$CHOC(=O)(CH_2)_7CH{=}CH(CH_2)_7CH_3 \text{ (cis)}$$
$$CH_2OC(=O)(CH_2)_7CH{=}CH(CH_2)_7CH_3 \text{ (cis)}$$

Glyceryl trioleate

(a)

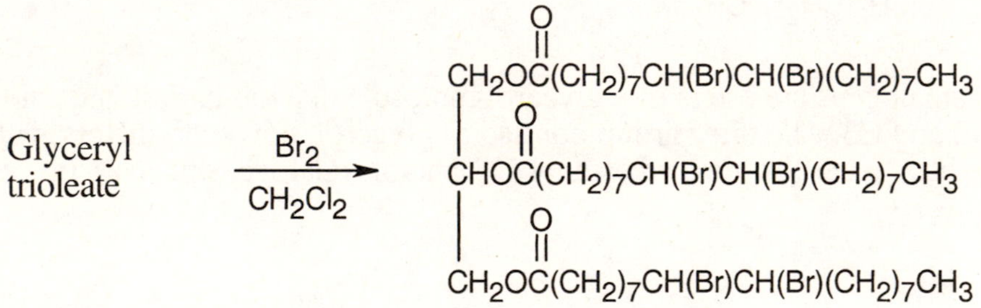

$$\text{Glyceryl trioleate} \xrightarrow[CH_2Cl_2]{Br_2} \begin{array}{l} CH_2OC(=O)(CH_2)_7CH(Br)CH(Br)(CH_2)_7CH_3 \\ CHOC(=O)(CH_2)_7CH(Br)CH(Br)(CH_2)_7CH_3 \\ CH_2OC(=O)(CH_2)_7CH(Br)CH(Br)(CH_2)_7CH_3 \end{array}$$

(b)

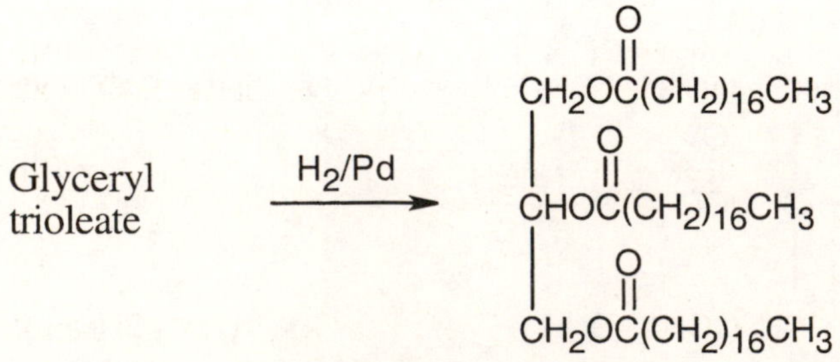

$$\text{Glyceryl trioleate} \xrightarrow{H_2/Pd} \begin{array}{l} CH_2OC(=O)(CH_2)_{16}CH_3 \\ CHOC(=O)(CH_2)_{16}CH_3 \\ CH_2OC(=O)(CH_2)_{16}CH_3 \end{array}$$

(c)

$$\text{Glyceryl trioleate} \xrightarrow[H_2O]{NaOH} \begin{array}{l} CH_2OH \\ CHOH \\ CH_2OH \end{array} + 3\ Na^+\ {}^-O_2C(CH_2)_7CH{=}CH(CH_2)_7CH_3$$

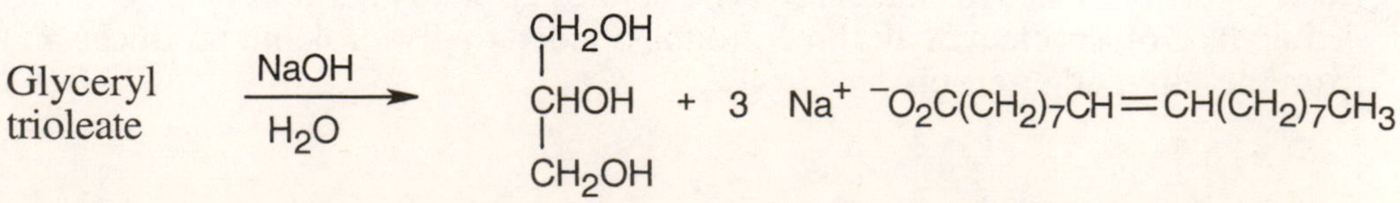

(d)

$$\text{Glyceryl trioleate} \xrightarrow[\text{2. } H_3O^+]{\text{1. } LiAlH_4} \begin{array}{c} CH_2OH \\ | \\ CHOH \\ | \\ CH_2OH \end{array} + 3\ HOCH_2(CH_2)_7CH{=}CH(CH_2)_7CH_3$$

(e)

$$\text{Glyceryl trioleate} \xrightarrow[\text{2. } H_3O^+]{\text{1. } CH_3MgBr} \begin{array}{c} CH_2OH \\ | \\ CHOH \\ | \\ CH_2OH \end{array} + 3\ HOC(CH_3)_2(CH_2)_7CH{=}CH(CH_2)_7CH_3$$

23.25

$$CH_3(CH_2)_7CH{=}CH(CH_2)_7CO_2H \quad \text{(cis)}$$

Oleic acid

(a)

$$\text{Oleic acid} \xrightarrow[HCl]{CH_3OH} CH_3(CH_2)_7CH{=}CH(CH_2)_7CO_2CH_3$$

Methyl oleate

(b)

$$\text{Methyl oleate from (a)} \xrightarrow{H_2/Pd} CH_3(CH_2)_{16}CO_2CH_3$$

Methyl stearate

(c)

$$2\ CH_3(CH_2)_{16}CO_2CH_3 \text{ (from (b))} \xrightarrow[\text{2. } H_3O^+]{\text{1. } Na^+\ {}^-OCH_3} CH_3(CH_2)_{16}\overset{O}{\overset{\|}{C}}CH(CO_2CH_3)(CH_2)_{15}CH_3 + HOCH_3$$

$$\xrightarrow{H_3O^+,\ \text{heat}} CH_3(CH_2)_{16}\overset{O}{\overset{\|}{C}}CH_2(CH_2)_{15}CH_3 + CO_2 + HOCH_3$$

Pentatriacontan-18-one

This synthesis uses a Claisen condensation, followed by a β-keto ester decarboxylation.

23.26

CO_2H

Eicosa-5,8,11,14,17-pentaenoic acid

23.27

(a)

$$CH_3CH_2CH_2CH_2CH_2\overset{O}{\overset{\|}{C}}SCoA \xrightarrow[\text{Acyl-CoA dehydrogenase}]{FAD \;\; FADH_2} CH_3CH_2CH_2CH{=}CH\overset{O}{\overset{\|}{C}}SCoA$$

(b)

$$CH_3CH_2CH_2CH{=}CH\overset{O}{\overset{\|}{C}}SCoA + H_2O \xrightarrow[\text{hydratase}]{\text{Enoyl-CoA}} CH_3CH_2CH_2\overset{OH}{\overset{|}{C}}HCH_2\overset{O}{\overset{\|}{C}}SCoA$$

(c)

$$CH_3CH_2CH_2\overset{OH}{\overset{|}{C}}HCH_2\overset{O}{\overset{\|}{C}}SCoA \xrightarrow[\beta\text{-Hydroxyacyl-CoA dehydrogenase}]{NAD^+ \;\; NADH/H^+} CH_3CH_2CH_2\overset{O}{\overset{\|}{C}}CH_2\overset{O}{\overset{\|}{C}}SCoA$$

23.28

sn-Glycerol 1-phosphate: C1 $CH_2OPO_3^{2-}$; C2 (R) HO–C–H; C3 CH_2OH

sn-Glycerol 2,3-diacetate: C1 CH_2OH; C2 (R) $CH_3\overset{O}{\overset{\|}{C}}O$–C–H; C3 $CH_2O\overset{O}{\overset{\|}{C}}CH_3$

sn-Glycerol 1-phosphate *sn*-Glycerol 2,3-diacetate

23.29–23.31

Remember that a compound with *n* chirality centers can have a maximum of 2^n stereoisomers. Not all the possible stereoisomers of these compounds are found in nature or can be synthesized. Some stereoisomers have highly strained ring fusions; others contain 1,3-diaxial interactions.

(a)

H3C, CH3, H3C, H3C, OH, OPP

Guaiol
(8 possible stereoisomers)

(b)

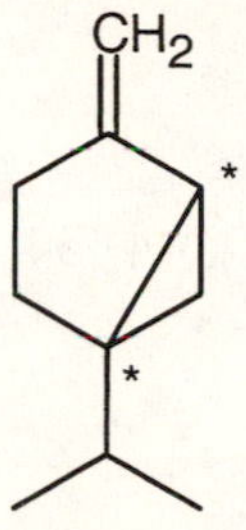

Sabinene
(4 possible stereoisomers)

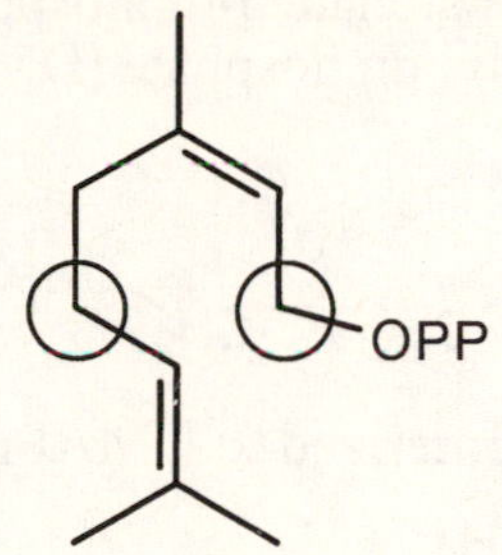

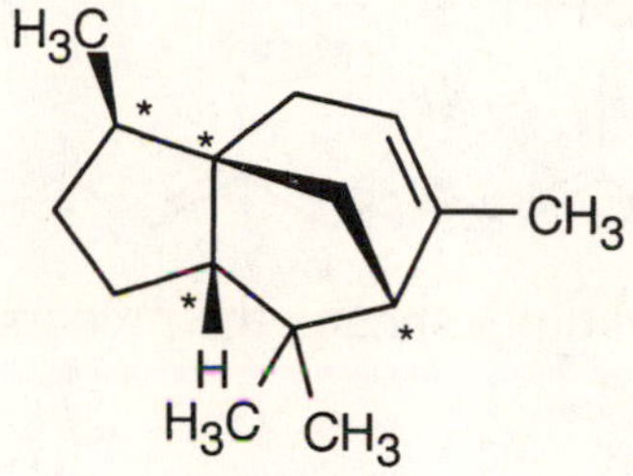

Cedrene
(16 possible stereoisomers)

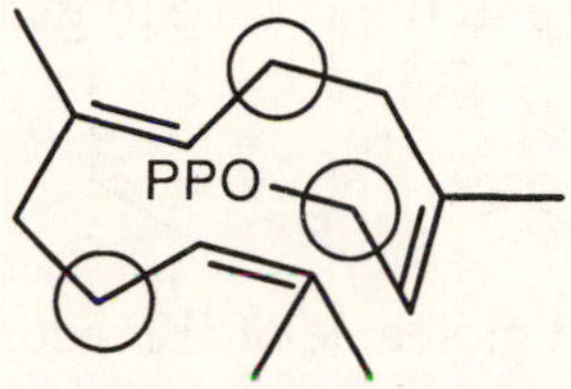

If carbon 1 of each diphosphate were isotopically labeled, the labels would appear at the circled positions of the terpenoids.

23.32

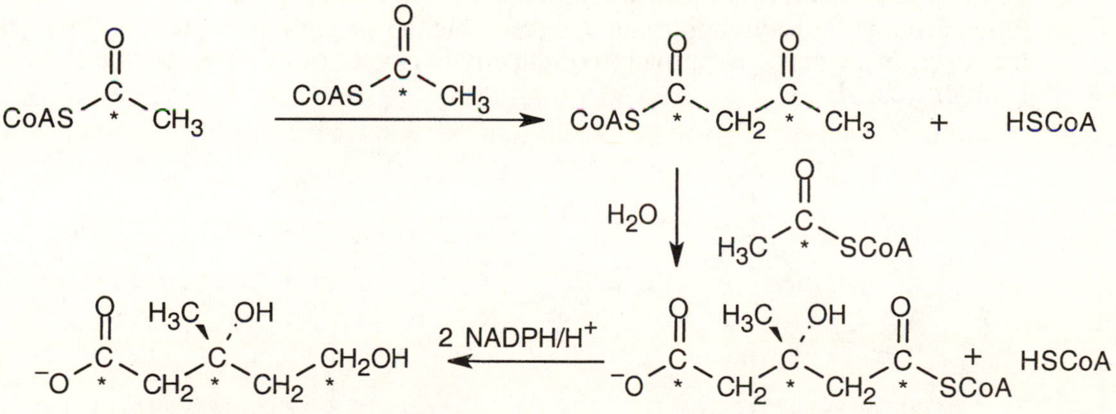

$CoAS{-}\overset{*}{C}(=O){-}CH_3 \xrightarrow{CoAS{-}\overset{*}{C}(=O){-}CH_3} CoAS{-}\overset{*}{C}(=O){-}CH_2{-}\overset{*}{C}(=O){-}CH_3 + HSCoA$

$\xrightarrow[H_2O]{H_3C{-}\overset{*}{C}(=O){-}SCoA}$ $^{-}O{-}\overset{*}{C}(=O){-}CH_2{-}\overset{*}{C}(H_3C)(OH){-}CH_2{-}\overset{*}{C}(=O){-}SCoA + HSCoA$

$\xrightarrow{2\ NADPH/H^+}$ $^{-}O{-}\overset{*}{C}(=O){-}CH_2{-}\overset{*}{C}(H_3C)(OH){-}CH_2{-}\overset{*}{C}H_2OH$

(*R*)-Mevalonate + $^{-}SCoA$

23.33 First, mevalonate 5-diphosphate is converted to isopentenyl diphosphate (IPP) and dimethallyl diphosphate (DMAPP).

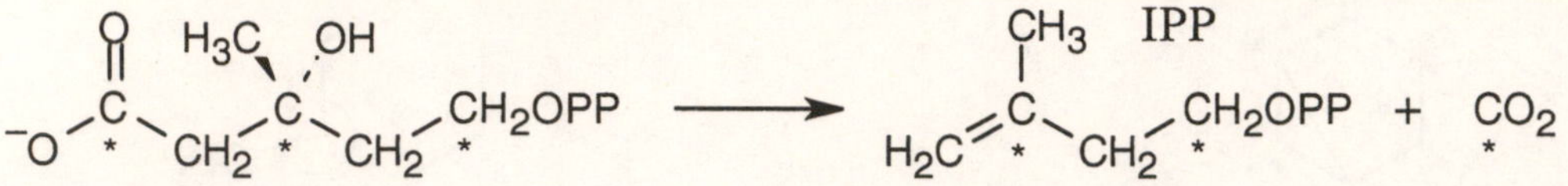

IPP is isomerized to DMAPP.

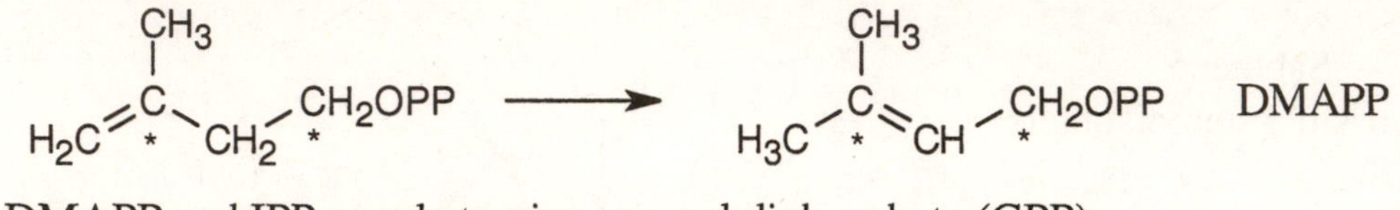

DMAPP and IPP couple to give geranyl diphosphate (GPP).

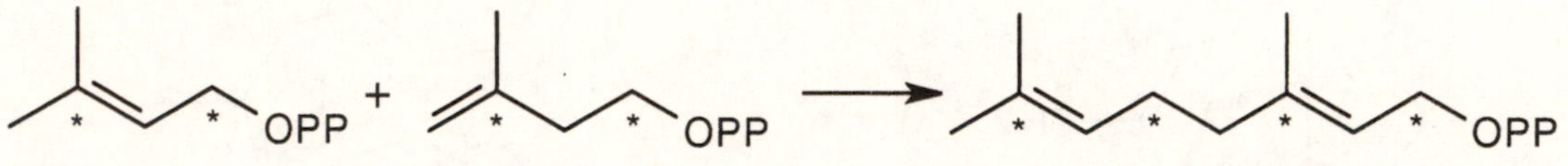

A second molecule of IPP adds to GPP to give farnesyl diphosphate, the precursor to α-cadinol.

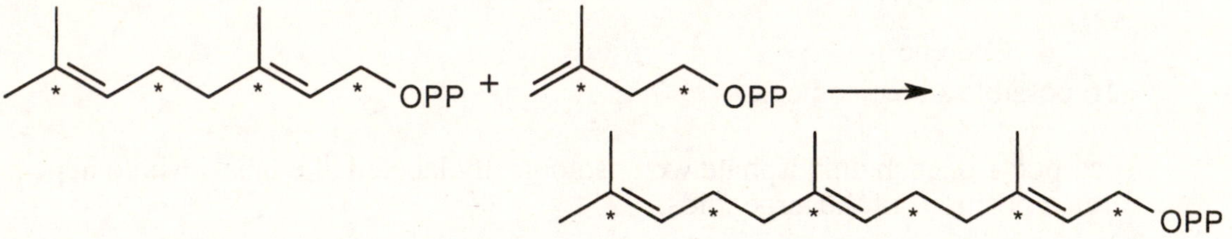

Notice that the ^{14}C labels are located at two different positions: (1) at the carbon to which –OPP was bonded; (2) at the carbon bonded to the methyl group.
Now, arrange farnesyl diphosphate to resemble the skeleton of α-cadinol. The first step in the reaction sequence is formation of the allylic isomer of FPP, which was shown in Problem 23.11.

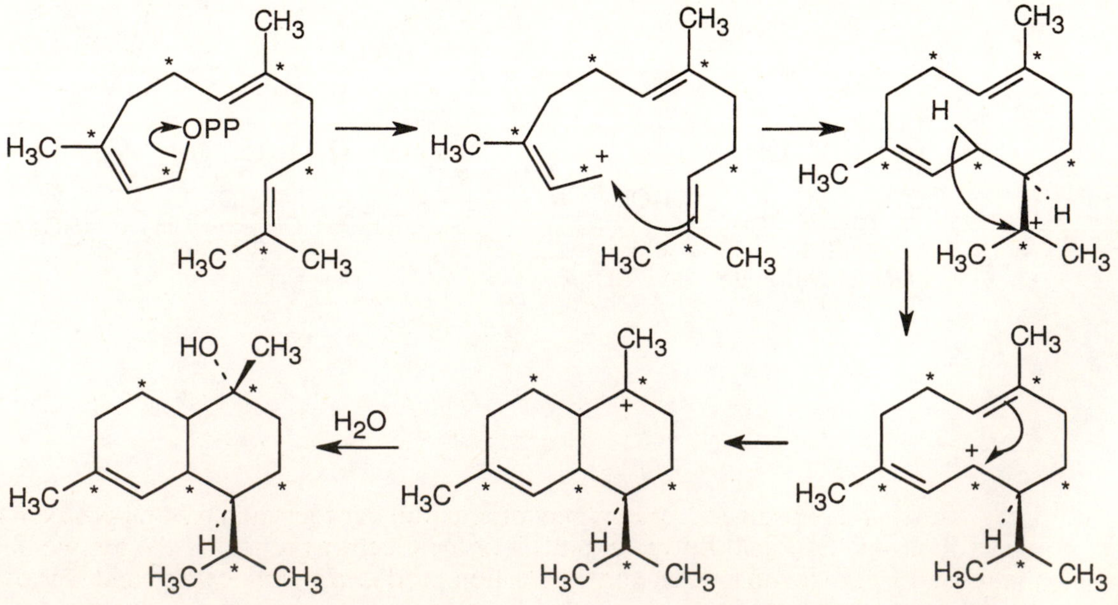

23.34 Farnesyl diphosphate (from the previous problem) dimerizes to form squalene.

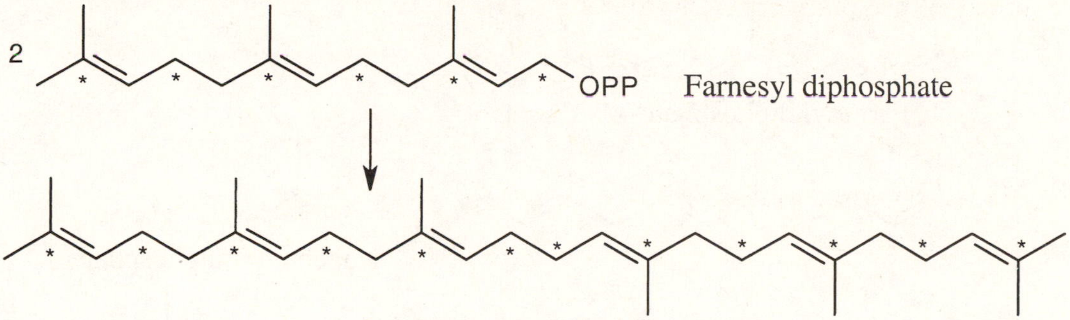

2 OPP Farnesyl diphosphate

23.35 Squalene is converted to lanosterol by the series of steps pictured in Figure 23.19.

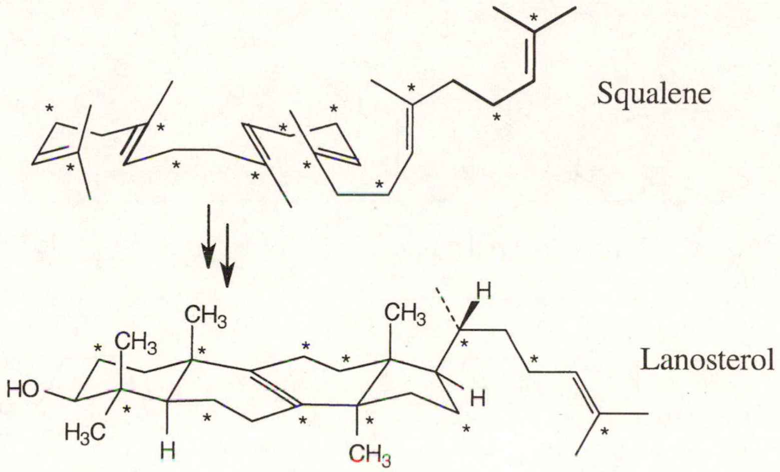

Squalene

CH3 CH3 CH3 H HO H3C H CH3 H

Lanosterol

23.36

CH3 H3C H3C PPO H3C

CH3 H3C H3C + H3C + PPO$^-$

H3C H3C + H3C H3C

H3C H3C H3C H3C H + B:

H3C H3C H3C H2C

Caryophyllene

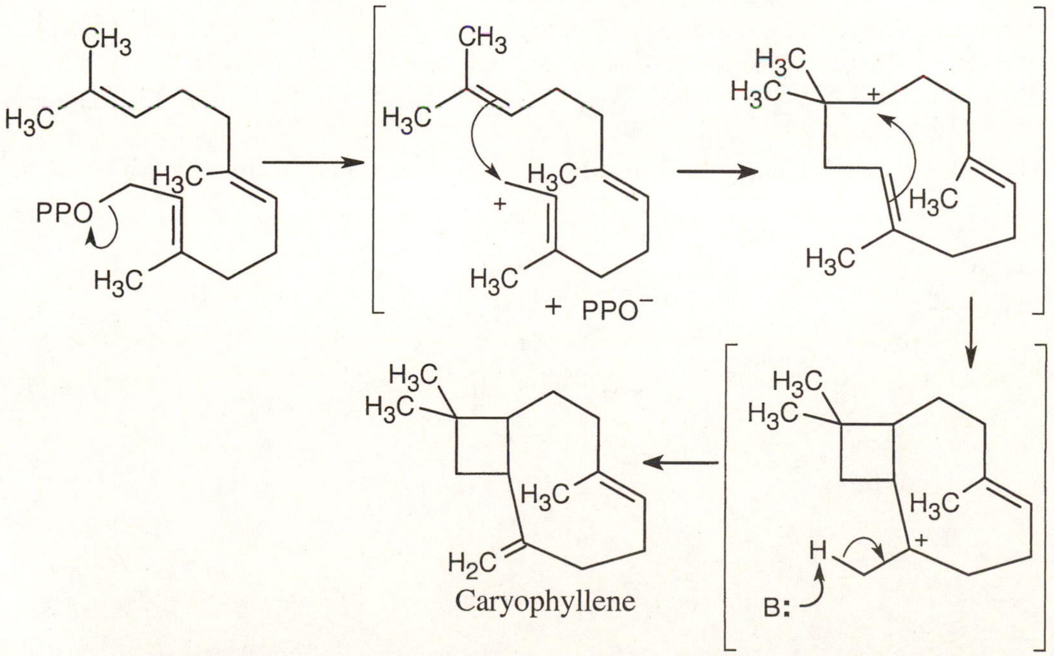

Draw farnesyl diphosphate in the correct orientation in order to make this problem much easier. Loss of $^-$OPP is followed by attack of the electrons of one of the double bonds on the resulting carbocation. A second cyclization produces a carbocation, and loss of a proton yields caryophyllene.

23.37

Farnesyl diphosphate

OPP

PP_i

OPP

Isopentenyl diphosphate

OPP

:B

OPP

The precursor to flexibilene is formed from the reaction of farnesyl diphosphate and isopentenyl diphosphate.

OPP

B:

+ $^{-}$OPP

Flexibilene

The precursor cyclizes by the now-familiar mechanism to produce flexibilene.

23.38

$H_2\ddot{O}^+$—H

ψ–Ionone

:$\ddot{O}H_2$

β–Ionone

+ H_3O^+

Acid protonates a double bond, and the electrons of a second double bond attack the carbocation. Deprotonation yields β-ionone.

23.39

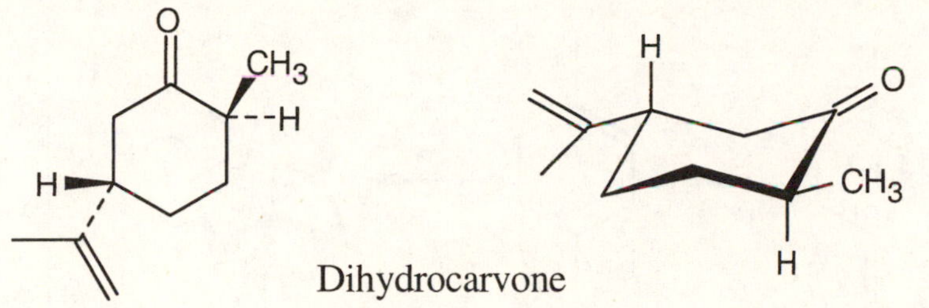

Dihydrocarvone

The two hydrocarbon substituents are equatorial in the most stable chair conformation.

23.40

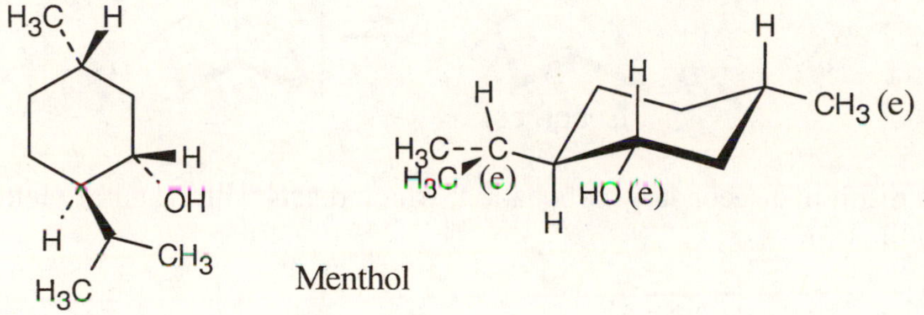

Menthol

All ring substituents are equatorial in the most stable conformation of menthol.

23.41

(a) 1 equiv. $(CH_3CO)_2O$

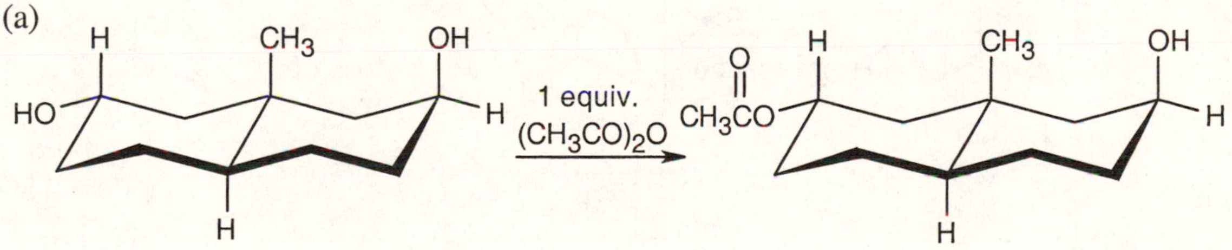

(b) 1 equiv. $(CH_3CO)_2O$

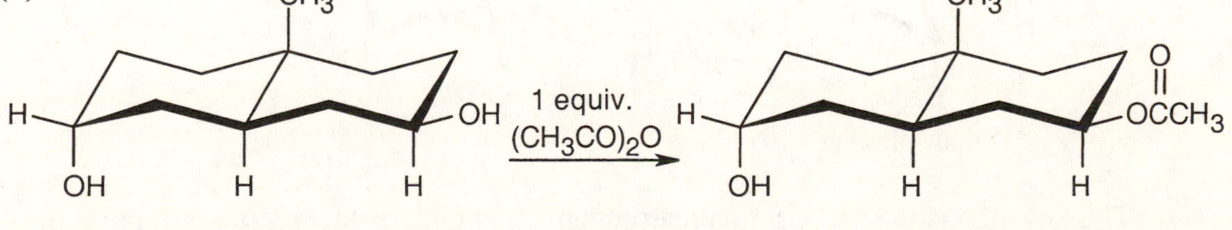

As always, use the stereochemistry of the groups at the ring junction to label the other substituents.

23.42

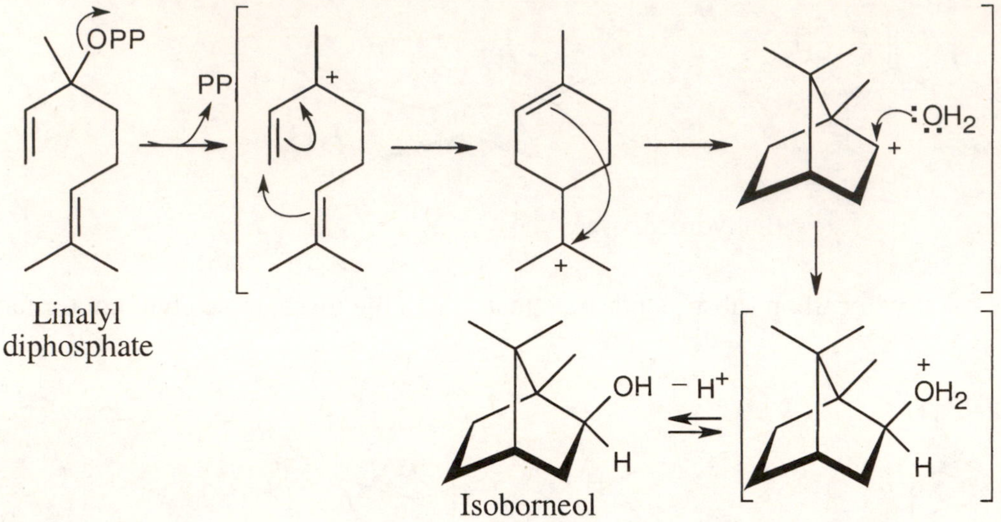
OPP
PP
+
:OH2
+
+
Linalyl diphosphate
OH
- H+
+
OH2
H
H
Isoborneol

The cyclizations result in a secondary carbocation, which reacts with water to yield the secondary alcohol.

23.43

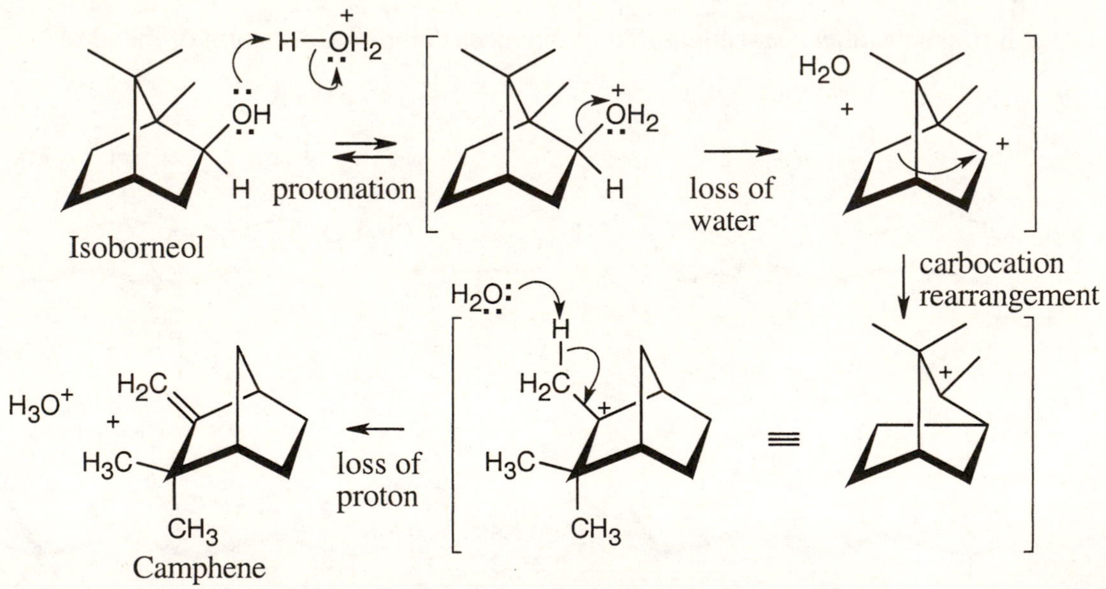
H—OH2
OH
H protonation
Isoborneol
+
OH2
H
loss of water
H2O
+
+
carbocation rearrangement
H2O:
H
H2C
+
H3C
CH3
H3O+
+
H2C
H3C
CH3
Camphene
loss of proton

The key step is the carbocation rearrangement, which occurs by the migration of one of the ring bonds.

23.44

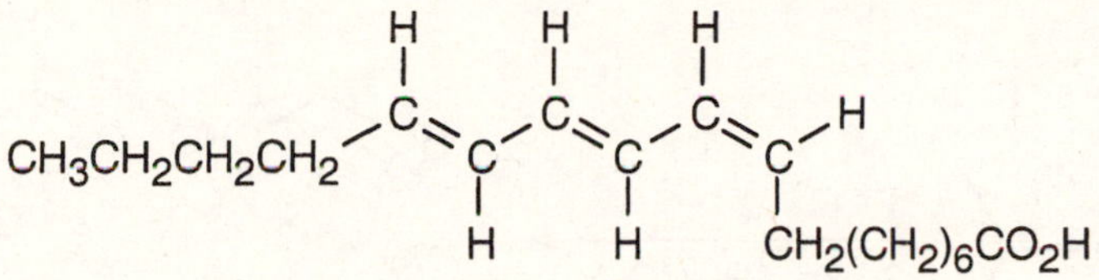

$$CH_3CH_2CH_2CH_2-CH=CH-CH=CH-CH=CH-CH_2(CH_2)_6CO_2H$$

(9*Z*,11*E*,13*E*)-Octadeca-9,11,13-trienoic acid
(Eleostearic acid)

1. O_3
2. Zn, CH_3CO_2H

$$CH_3CH_2CH_2CH_2CHO + OHC-CHO + OHC-CHO + OHC(CH_2)_7CO_2H$$

The stereochemistry of the double bonds can't be determined from the information given.

23.45 This mechanism also appears in Problem 23.37

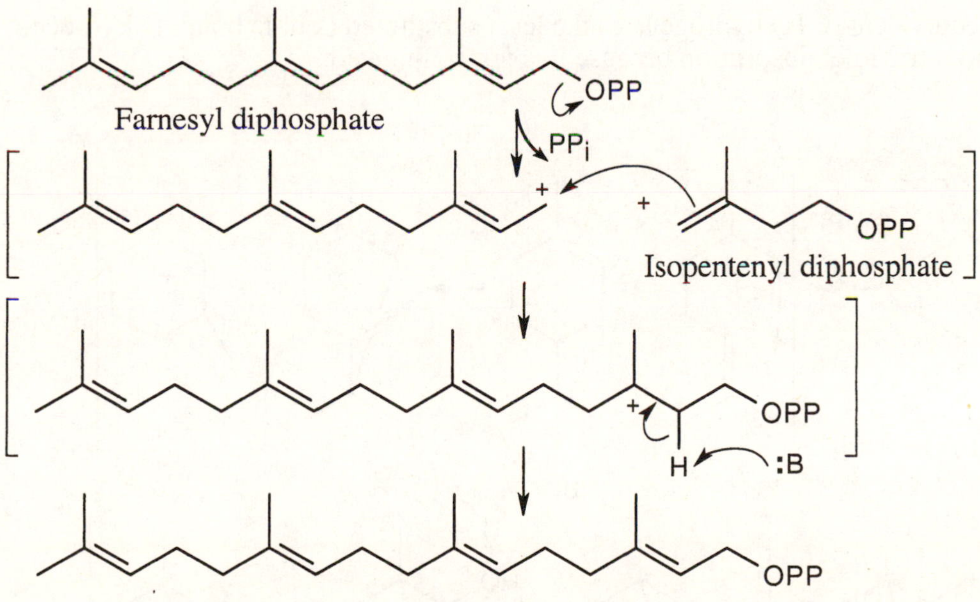

OPP

Farnesyl diphosphate

PP_i

OPP

Isopentenyl diphosphate

OPP

H :B

OPP

Geranylgeranyl diphosphate (GGPP)

23.46

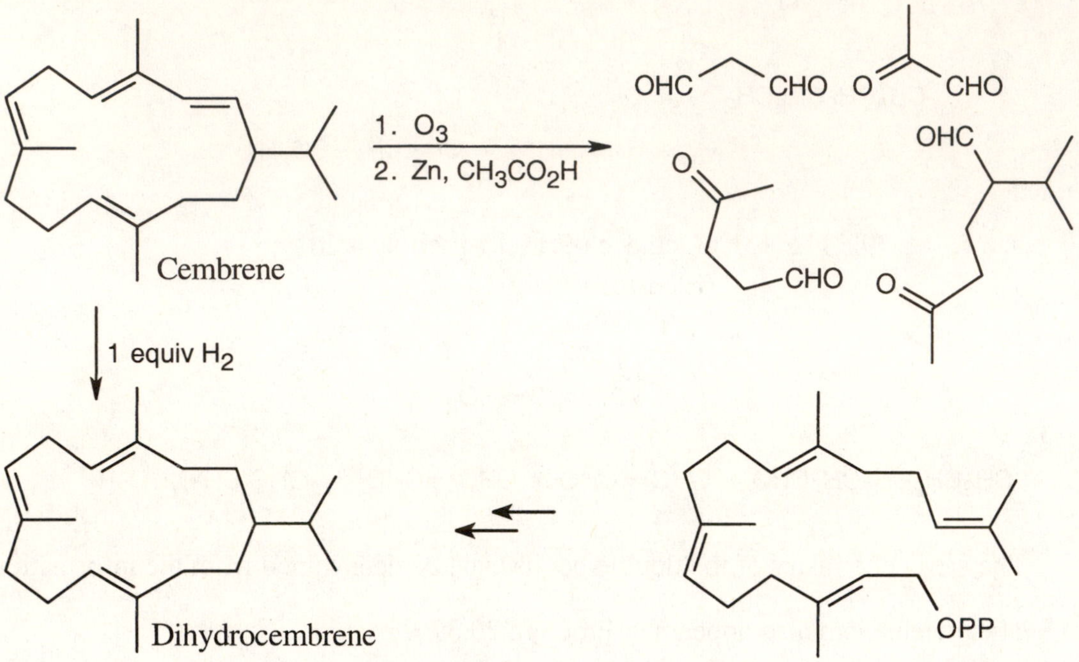

Cembrene

1. O_3
2. Zn, CH_3CO_2H

OHC CHO O CHO OHC O CHO O

1 equiv H_2

Dihydrocembrene

OPP

One equivalent of H_2 hydrogenates the least substituted double bond. Dihydrocembrene has no ultraviolet absorption because it is not conjugated.

23.47

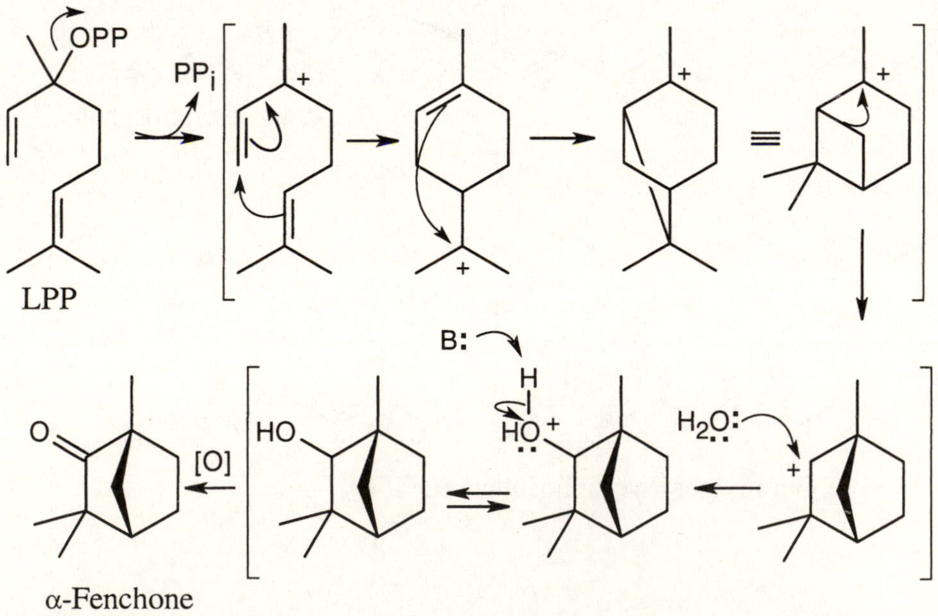

OPP PP$_i$ + + + +

LPP

B: H HO+ H_2O: HO O [O] +

α-Fenchone

The mechanism follows the usual path: cyclization of linalyl diphosphate, followed by attack of the π electrons of the second double bond, produce an intermediate carbocation. A carbocation rearrangement occurs, and the resulting carbocation reacts with water to form an alcohol that is oxidized to give α-fenchone.

23.48

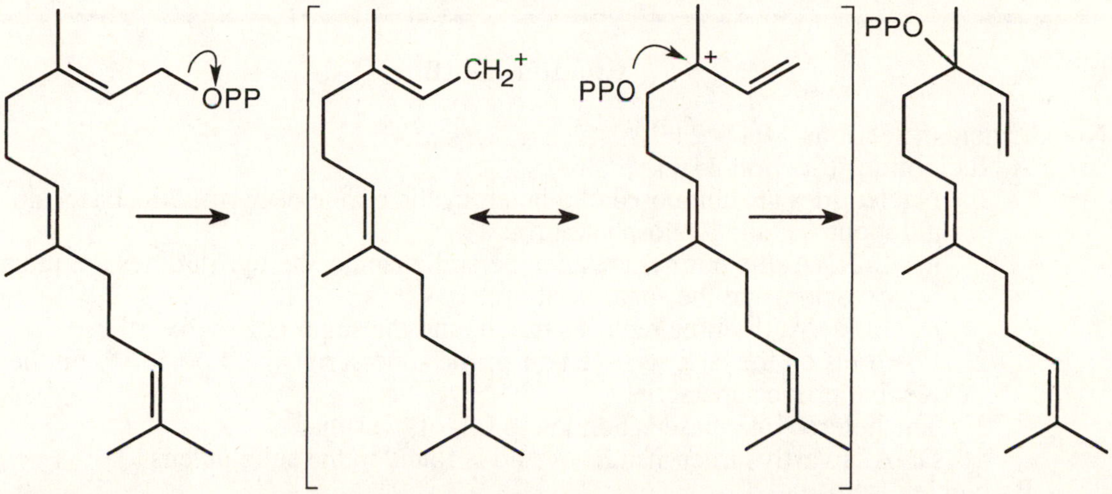
OPP
CH_2^+
PPO
+
PPO

In this series of steps, dissociation of diphosphate ion allows bond isomerization to take place, making it possible for ring formation to occur. This mechanism is very similar to the mechanism shown in Figure 23.15.

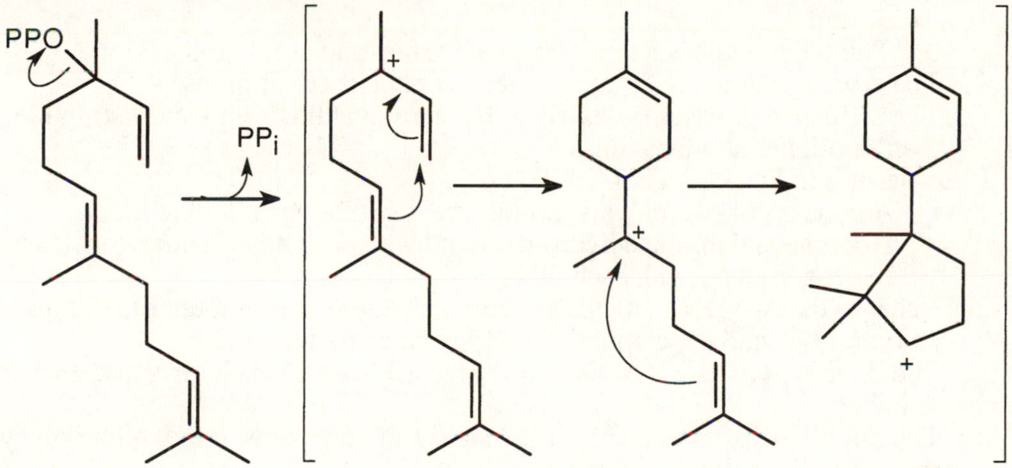
PPO
PP_i
+
+
+

Two cyclizations produce the trichodiene ring skeleton and a secondary carbocation.

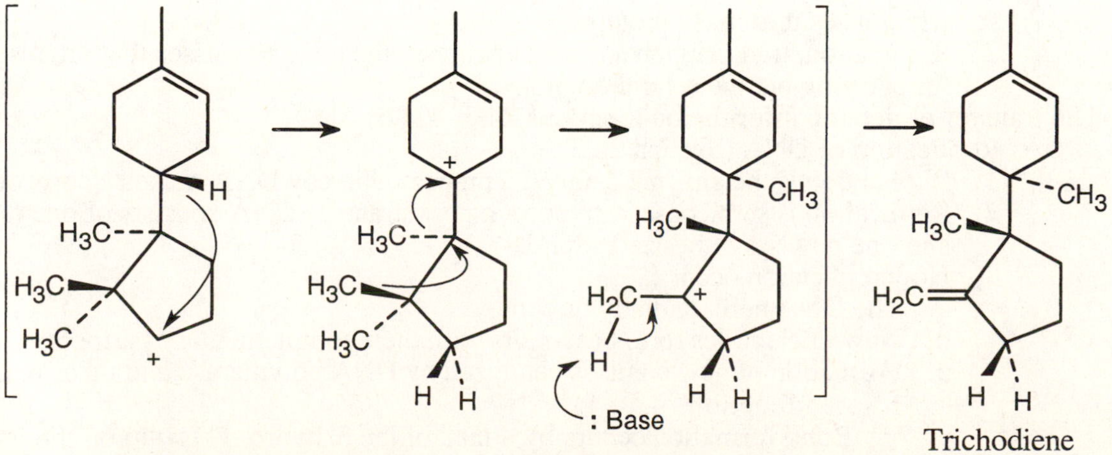
H
H_3C
H_3C
H_3C
+
H_3C
H_3C
H_3C
+
H H
CH_3
H_3C
H_2C
+
H
: Base
H H
CH_3
H_3C
H_2C
H H
Trichodiene

A hydride shift, two methyl shifts, and loss of $-H^+$ yield trichodiene.

Chapter 24 – Biomolecules: Nucleic Acids and Their Metabolism

Chapter Outline

I. Nucleic acids (Sections 24.1 – 24.2).
  A. Nucleotides (Section 24.1).
    1. Nucleotides are composed of a heterocyclic purine or pyrimidine base, an aldopentose, and a phosphate group.
      a. In RNA, the purines are adenine and guanine, the pyrimidines are uracil and cytosine, and the sugar is ribose.
      b. In DNA, thymine replaces uracil, and the sugar is 2'-deoxyribose.
    2. Positions on the base receive non-prime superscripts, and positions on the sugar receive prime superscripts.
    3. The heterocyclic base is bonded to C1' of the sugar.
    4. DNA is vastly larger than RNA and is found in the cell nucleus.
  B. Nucleic acids.
    1. Nucleic acids are composed of nucleotides connected by a phosphate ester bond between the 5' ester of one nucleotide and the 3' hydroxyl group of another.
      a. One end of the nucleic acid polymer has a free hydroxyl group and is called the 3' end.
      b. The other end has a free phosphate group and is called the 5' end.
    2. The structure of a nucleic acid depends on the order of bases.
    3. The sequence of bases is described by starting at the 5' end and listing the bases by their one-letter abbreviations.
  C. Base-pairing in DNA (Section 24.2).
    1. DNA consists of two polynucleotide strands coiled in a double helix.
       Adenine and thymine hydrogen-bond with each other, and cytosine and guanine hydrogen-bond with each other.
    2. Because the two DNA strands are complementary, the amount of A equals the amount of T, and the amount of C equals the amount of G.
    3. The double helix is 2.0 Å wide, there are 10 bases in each turn, and each turn is 3.4 Å in height.
    4. The double helix has a major groove and a minor groove into which polycyclic aromatic molecules can intercalate.
  D. The "central dogma" of molecular genetics.
    1. The function of DNA is to store genetic information and to pass it on to RNA, which uses it to make proteins.
    2. Replication, transcription and translation are the three processes that are responsible for carrying out the central dogma.

II. The transfer of genetic information (Sections 24.3 – 24.5).
  A. Replication of DNA (Section 24.3).
    1. Replication is the enzyme-catalyzed process whereby DNA makes a copy of itself.
    2. Replication is semiconservative: each new strand of DNA consists of one old strand and one newly synthesized strand.
    3. How replication occurs:
      a. The DNA helix partially unwinds.
      b. New nucleotides form base-pairs with their complementary partners.
      c. Formation of new bonds is catalyzed by DNA polymerase and takes place in the 5' $\rightarrow$ 3' direction.
         Bond formation occurs by attack of the 3' hydroxyl group on the 5' triphosphate, with loss of a diphosphate leaving group.

- d. Both new chains are synthesized in the 5' $\rightarrow$ 3' direction.
  - i. One chain is synthesized continuously.
  - ii. The other strand is synthesized in small pieces, which are later joined by DNA ligase enzymes.

C. Transcription (Section 24.4).

1. There are 3 types of RNA:
   - a. Messenger RNA (mRNA) carries genetic information to ribosomes where protein synthesis takes place.
   - b. Ribosomal RNA (rRNA), complexed with protein, comprises the physical makeup of the ribosomes.
   - c. Transfer RNA (tRNA) brings amino acids to the ribosomes, where they are joined to make proteins.
2. mRNA is synthesized in the nucleus by transcription of DNA.
   - a. The DNA partially unwinds, forming a "bubble".
   - b. Ribonucleotides form base pairs with their complementary DNA bases.
   - c. Bond formation occurs in the 5' $\rightarrow$ 3' direction.
   - d. Only one strand of DNA (the template strand) is transcribed.
   - e. Thus, the synthesized mRNA is a copy of the coding strand (with U replacing T).
3. DNA contains "promoter sites", which indicate where mRNA synthesis is to begin, and base sequences that indicate where mRNA synthesis stops.
4. Synthesis of mRNA is not necessarily continuous.
   - a. Often, synthesis begins in a region of DNA called an exon and is interrupted by a seemingly nonsensical region of DNA called an intron.
   - b. In the final mRNA, the nonsense sections have been removed and the remaining pieces have been spliced together.

D. Translation (Section 24.5).

1. Translation is the process in which proteins are synthesized at the ribosomes by using mRNA as a template.
2. The message delivered by mRNA is contained in "codons" – 3-base groupings that are specific for an amino acid.
   - a. Amino acids are coded by 61 of the possible 64 codons.
   - b. The other 3 codons are "stop" codons.
3. Each tRNA is responsible for bringing an amino acid to the growing protein chain.
   - a. A tRNA has a cloverleaf-shaped secondary structure and consists of 70–100 ribonucleotides.
   - b. Each tRNA contains an anticodon complementary to the mRNA codon.
4. The protein chain is synthesized by enzyme-catalyzed peptide bond formation.
5. A 3-base "stop" codon on mRNA signals when synthesis is complete.

III. DNA technology (Sections 24.6 – 24.8).

A. DNA sequencing (Section 24.6).

1. Before sequencing, the DNA chain is cleaved at specific sites by restriction endonucleases.
   - a. The restriction endonuclease recognizes both a sequence on the coding strand and its complement on the template strand.
   - b. The DNA strand is cleaved by several different restriction endonucleases, to produce fragments that overlap those from a different cleavage.
2. Maxam–Gilbert DNA sequencing.

   This method uses chemical techniques.

3. Sanger dideoxy DNA sequencing.
   a. The following mixture is assembled:
      i. The restriction fragment to be sequenced.
      ii. A primer (a small piece of DNA whose sequence is complementary to that on the 3' end of the fragment).
      iii. The 4 DNA nucleotide triphosphates.
      iv. Small amounts of the four dideoxynucleotide triphosphates, each of which is labeled with a different fluorescent dye.
   b. DNA polymerase is added to the mixture, and a strand begins to grow from the end of the primer.
   c. Whenever a dideoxynucleotide is incorporated, chain growth stops.
   d. When reaction is complete, the fragments are separated by gel electrophoresis.
   e. Because fragments of all possible lengths are represented, the sequence can be read by noting the color of fluorescence of each fragment.

B. DNA synthesis (Section 24.7).
1. DNA synthesis is based on principles similar to those for protein synthesis.
2. The following steps are needed:
   a. The nucleosides are protected and bound to a silica support.
      i. Adenine and cytosine bases are protected by benzoyl groups.
      ii. Guanine is protected by an isobutyryl group.
      iii. Thymine isn't protected.
      iv. The 5' –OH group is protected as a DMT ether.
   b. The DMT group is removed.
   c. The polymer-bound nucleoside is coupled with a protected nucleoside containing a phosphoramidite group.
      i. One of the phosphoramidite oxygens is protected as a β-cyano ether.
      ii. Tetrazole catalyzes the coupling.
   d. After coupling, the phosphite is oxidized to a phosphate with I_2.
   e. Steps b – d are repeated until the desired chain is synthesized.
   f. All protecting groups are removed and the bond to the support is cleaved by treatment with aqueous ammonia.

C. The polymerase chain reaction (Section 24.8).
1. The polymerase chain reaction (PCR) can produce vast quantities of a DNA fragment.
2. The key to PCR is *Taq* DNA polymerase, a heat-stable enzyme.
3. Steps in PCR:
   a. The following mixture is heated to 95°C (a temperature at which DNA becomes single-stranded);
      i. *Taq* polymerase.
      ii. Mg^{2+} ion.
      iii. The 4 deoxynucleotide triphosphates.
      iv. A large excess of two oligonucleotide primers, each of which is complementary to the ends of the fragment to be synthesized.
   b. The temperature is lowered to 37°C–50°C, causing the primers to hydrogen-bond to the single-stranded DNA.
   c. After raising the temperature to 72°C, *Taq* catalyzes the addition of further nucleotides, yielding two copies of the original DNA.
   d. The process is repeated until the desired quantity of DNA is produced.

IV. Nucleotide biochemistry (Sections 24.9 – 24.10).
  A. Catabolism of nucleotides (Section 24.9).
    1. General features.
      a. Catabolism of nucleotides is complicated.
      b. Nucleic acids are cleaved to nucleotides in the intestines.
      c. First, nucleotides are dephosphorylated, and the resulting nucleosides are cleaved to produce sugars and bases.
    2. Catabolism of purines.
      a. Guanosine.
        i. Guanosine is phosphorylated to give guanine and ribose 1-phosphate.
        ii. Hydrolysis of guanine yields xanthine.
        iii. Oxidation of xanthine gives uric acid, which is excreted.
      b. Other purine nucleosides are catabolized by a slightly different pathway.
    3. Catabolism of pyrimidines: uridine.
      a. Catabolism includes phosphorylation, reduction, hydrolysis, decarboxylation,transamination, and oxidation.
      b. Catabolism of other pyrimidines follows a similar path.
  B. Biosynthesis of nucleotides (Section 24.10).
    1. Pyrimidine biosynthesis: uridine monophosphate.
      a. Carbamoyl phosphate and aspartate combine to form carbamoyl aspartate.
      b. Carbamoyl aspartate cyclizes, yielding dihydroorotate.
      c. Dihydroorotate is dehydrogenated to form orotate.
      d. Reaction with PRPP gives orotidine monophosphate.
      e. Decarboxylation yields uridine monophosphate.
    2. Purine biosynthesis.
      a. Purine nucleotides are synthesized by attachment of an $-NH_2$ group to a phosphoribose, followed by buildup of the purine ring system.
      b. A complicated series of steps produces inosine monophosphate (IMP).
      c. IMP is converted to AMP by a two-step process:
        i. Reaction with aspartate to yield adenylosuccinate.
        ii. Elimination of fumarate to give AMP.
      d. GMP is synthesized from IMP by oxidation/hydrolysis, followed by amination.

Solutions to Problems

24.1

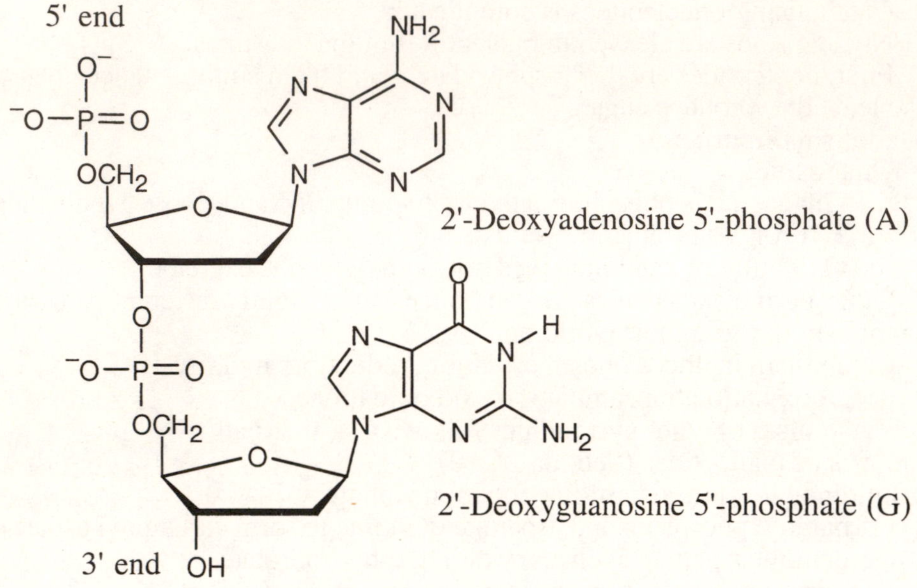

5' end

2'-Deoxyadenosine 5'-phosphate (A)

2'-Deoxyguanosine 5'-phosphate (G)

3' end

24.2

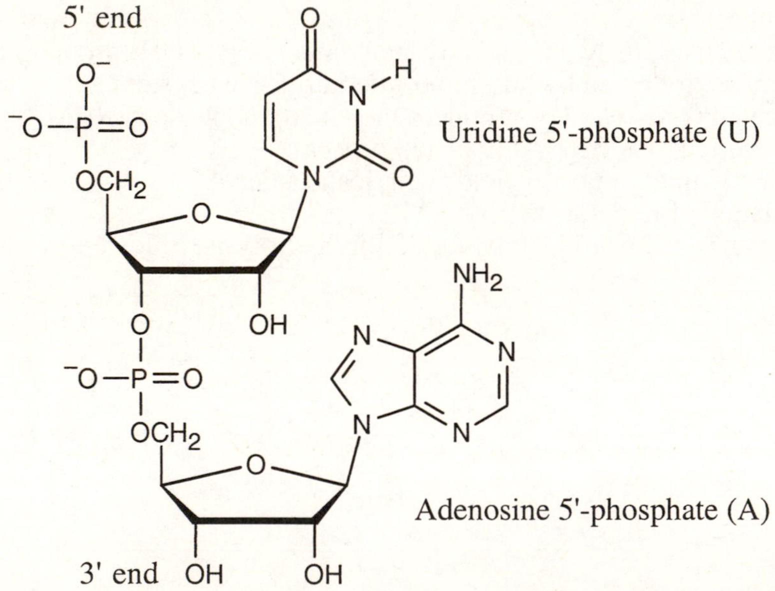

5' end

Uridine 5'-phosphate (U)

Adenosine 5'-phosphate (A)

3' end

24.3 DNA (5' end) GGCTAATCCGT (3' end) is complementary to
DNA (3' end) CCGATTAGGCA (5' end)

Remember that the complementary strand has the 3' end on the left and the 5' end on the right.

24.4

Uracil — Adenine

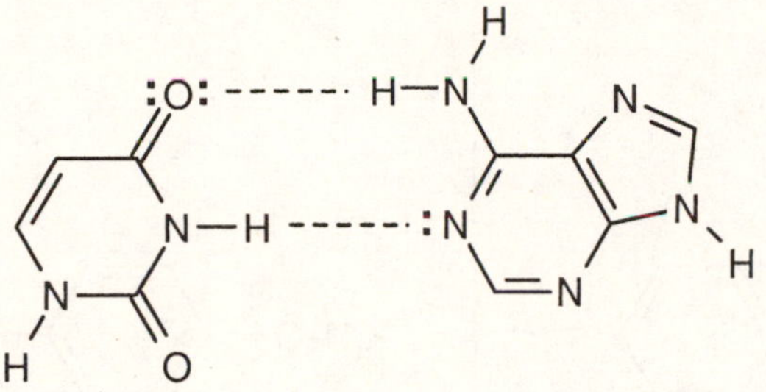

24.5 DNA (5' end) GATTACCGTA (3' end) is complementary to
RNA (3' end) CUAAUGGCAU (5' end)

24.6 RNA (5' end) UUCGCAGAGU (3' end)
DNA (3' end) AAGCGTCTCA (5' end) template strand

24.7 Several different codons can code for the same amino acid. The corresponding anticodon follows the slash mark after each codon. The mRNA codons are written with the 5' end on the left and the 3' end on the right, and the tRNA anticodons have the 3' end on the left and the 5' end on the right.

Amino acid:	Ala	Phe	Leu	Tyr
Codon sequence/ tRNA anticodon:	GCU/CGA GCC/CGG GCA/CGU GCG/CGC	UUU/AAA UUC/AAG	UUA/AAU UUG/AAC CUU/GAA CUC/GAG CUA/GAU CUG/GAC	UAU/AUA UAC/AUG

24.8–24.9

The mRNA base sequence:	(5' end)	CUU–AUG–GCU–UGG–CCC–UAA (3' end)
The amino acid sequence:		Leu—Met—Ala—Trp—Pro–(stop)
The DNA sequence: (template strand)	(3' end)	GAA–TAC–CGA–ACC–GGG–ATT (5' end)

24.10

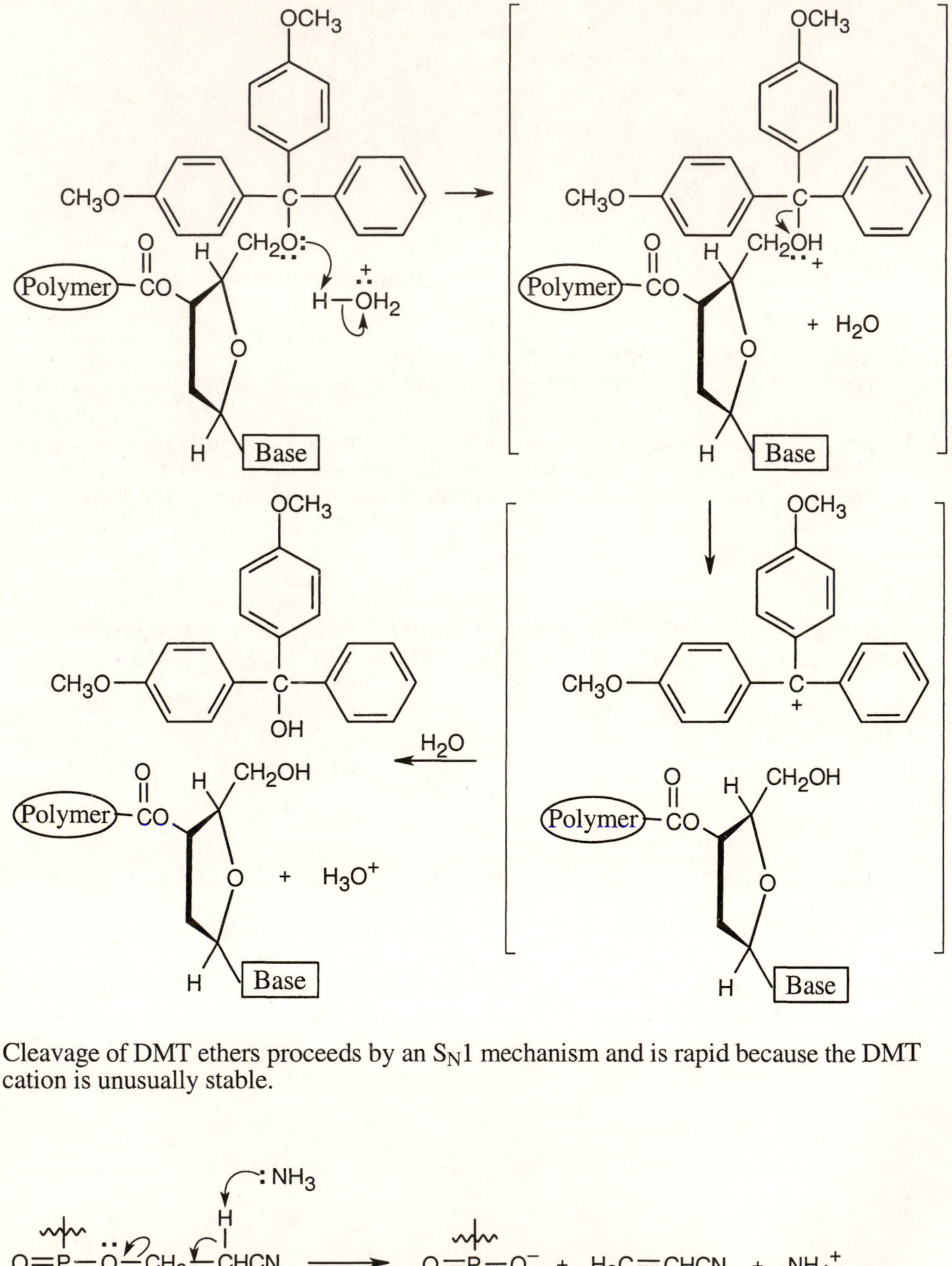

Cleavage of DMT ethers proceeds by an S_N1 mechanism and is rapid because the DMT cation is unusually stable.

24.11

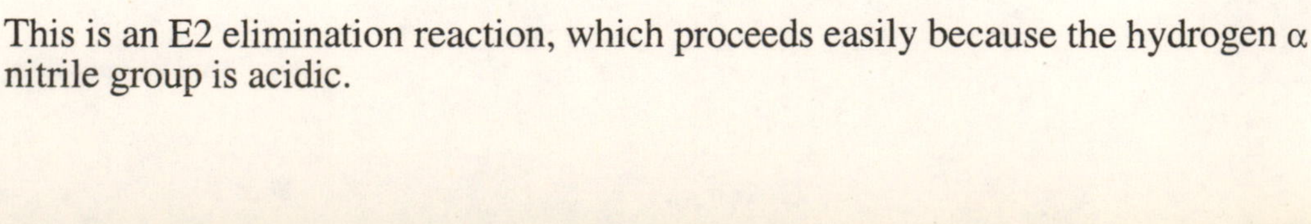

This is an E2 elimination reaction, which proceeds easily because the hydrogen α to the nitrile group is acidic.

24.12 The first series of steps are very much like the mechanism shown for transamination of amino acids described in Section 20.2. The steps include: (1) nucleophilic addition of β-alanine to enzyme-bound PLP; (2) deprotonation of the carbon α to the amino group; (3) reprotonation of the C–C double bond and tautomerization; (4) hydrolysis.

$^{2-}O_3PO$ H CO_2^- H_2N Enz

PLP–enzyme

H_2N-Enz

1.

$^{2-}O_3PO$ H CO_2^- H H H_2N Enz

PLP–imine

2.

$H_2\overset{+}{N}$ Enz H

$^{2-}O_3PO$ H CO_2^- H

Aldehyde imine

3.

$^{2-}O_3PO$ H H CO_2^- $:OH_2$ H

Aldehyde imine tautomer

4.

$^{2-}O_3PO$ H H NH_2 + CO_2^- H

PMP

Malonic semialdehyde

The mechanism of the transamination of α-ketoglutarate to form glutamate and PLP is shown in Problem 20.2.

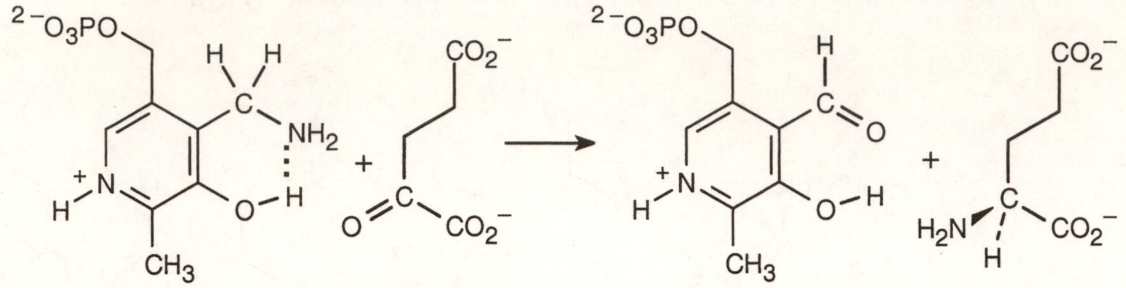

PMP α-Ketoglutarate PLP Glutamate

24.13 The same series of steps illustrated in Fig. 24.13 also take place in the catabolism of thymine. These include: (1) reduction; (2) hydrolysis; (3) hydrolysis, then decarboxylation; (4) transamination; (5) oxidation.

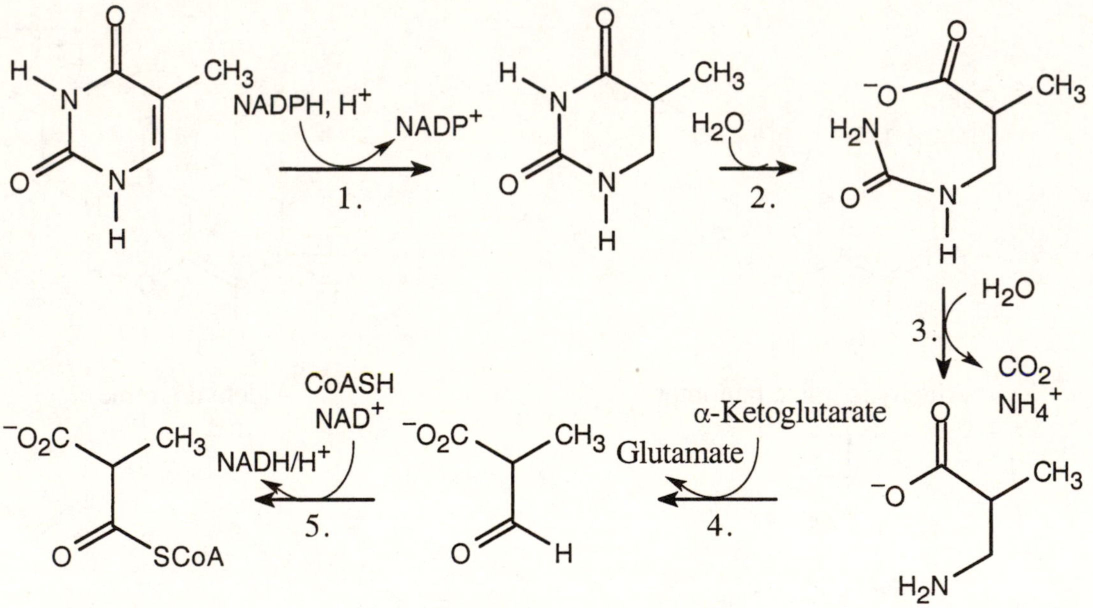

NADPH, H^+ → $NADP^+$ (1.)

H_2O (2.)

H_2O → CO_2, NH_4^+ (3.)

α-Ketoglutarate → Glutamate (4.)

CoASH, NAD^+ → NADH/H^+ (5.)

24.14 This sequence is similar, but not identical, to the conversion of UTP to CTP. The steps of the mechanism: (1) phosphorylation of inosine monophosphate by reaction with GTP; (2) acid-catalyzed addition of aspartate; (3) loss of phosphate.

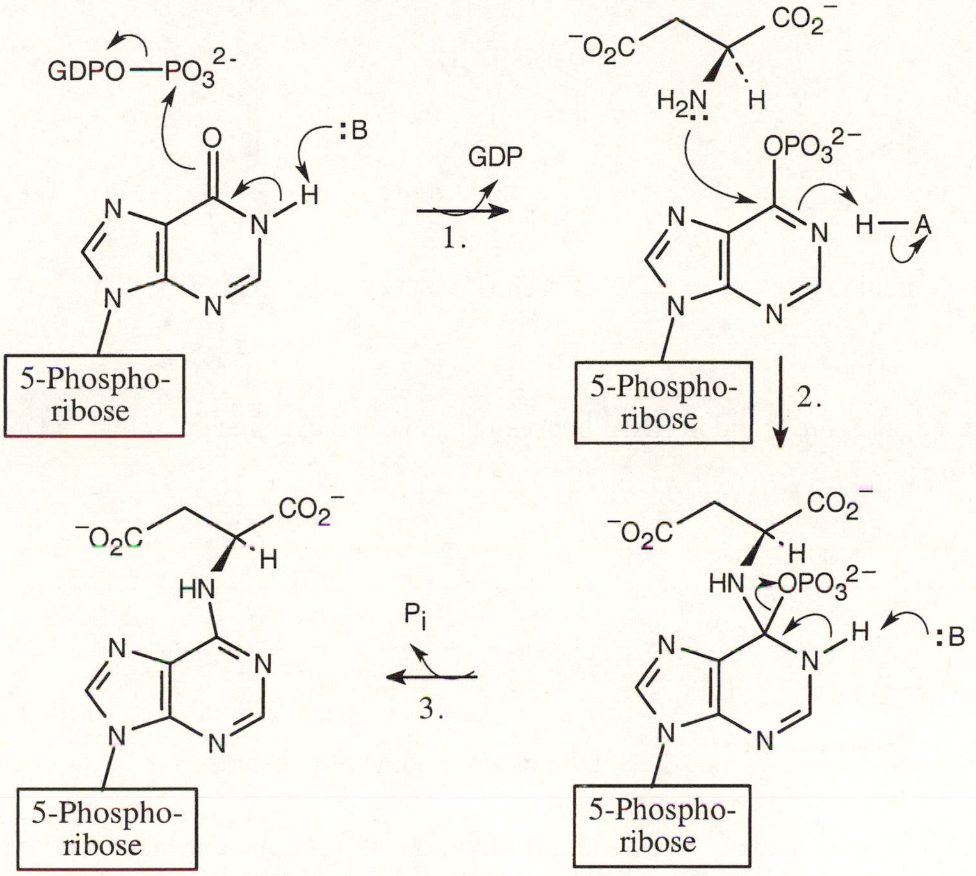

GDPO—PO_3^{2-} :B GDP 1. ^-O_2C CO_2^- H_2N H OPO_3^{2-} H—A 2. 3. P_i HN :B

5-Phospho-ribose

24.15 This reaction is an E1cB elimination.

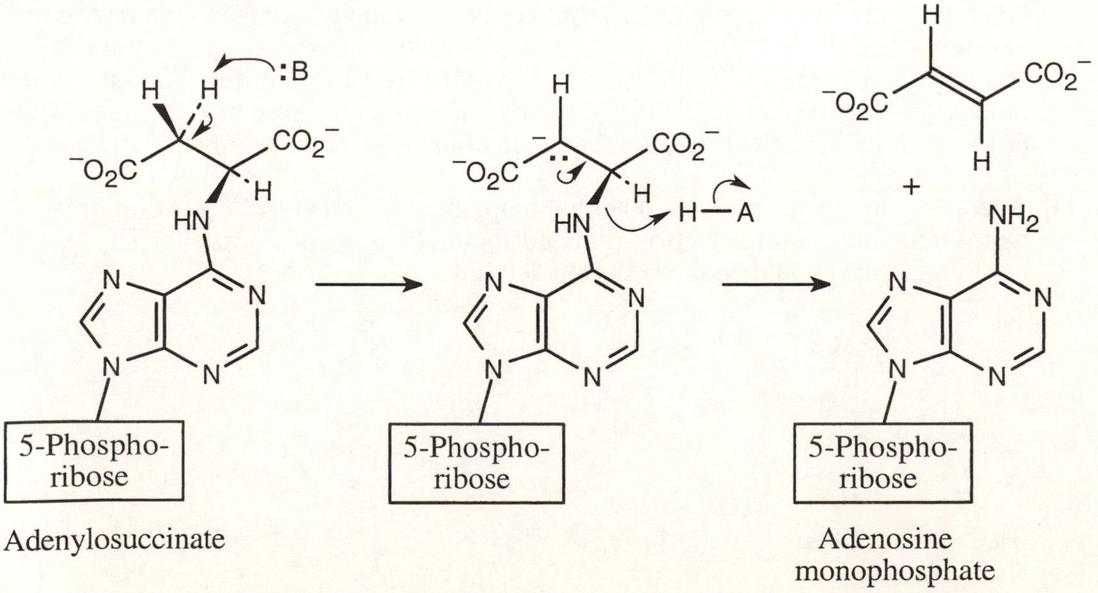

:B H H ^-O_2C CO_2^- HN H—A + NH_2

5-Phospho-ribose

Adenylosuccinate

Adenosine monophosphate

Visualizing Chemistry

24.16

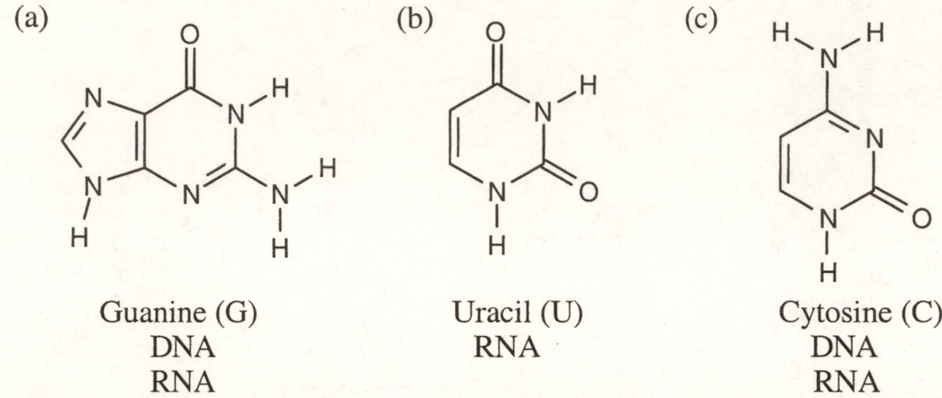

Guanine (G) DNA RNA — Uracil (U) RNA — Cytosine (C) DNA RNA

All three bases are found in RNA, but only guanine and cytosine are found in DNA.

24.17

2'3'-Dideoxythymidine 5'-phosphate

The triphosphate made from 2'3'-dideoxythymidine 5' phosphate is labeled with a fluorescent dye and used in the Sanger method of DNA sequencing. Along with the restriction fragment to be sequenced, a DNA primer, and a mixture of the four dNTPs, small quantities of the four labeled dideoxyribonucleotide triphosphates are mixed together. DNA polymerase is added, and a strand of DNA complementary to the restriction fragment is synthesized. Whenever a dideoxyribonucleotide is incorporated into the DNA chain, chain growth stops. The fragments are separated by electrophoresis, and each terminal dideoxynucleotide can be identified by the color of its fluorescence. By identifying these terminal dideoxynucleotides, the sequence of the restriction fragment can be read.

24.18 According to the electrostatic potential map, the nitrogen at the 7 position of 9-methylguanine is more electron-rich and should be more nucleophilic. Thus 9-methylguanine should be the better nucleophile.

9-Methylguanine

9-Methyladenine

Additional Problems

24.19 The DNA that codes for natriuretic peptide (32 amino acids) consists of 99 bases; 3 bases code for each of the 32 amino acids in the chain (96 bases), and a 3-base "stop" codon is also needed.

24.20 *Position 9:*

Horse amino acid = Gly				Human amino acid = Ser					
mRNA codons (5' —> 3'):									
GGU	GGC	GGA	GGG	UCU	UCC	UCA	UCG	AGU	AGC
DNA bases (template strand 3' —> 5'):									
CCA (underlined)	CCG (underlined)	CCT	CCC	AGA	AGG	AGT	AGC	TCA (underlined)	TCG (underlined)

The underlined horse DNA base triplets differ from their human counterparts (also underlined) by only one base.

Position 30:

Horse amino acid = Ala				Human amino acid = Thr			
mRNA codons (5' —> 3'):							
GCU	GCC	GCA	GCG	ACU	ACC	ACA	ACG
DNA bases (template strand 3' —> 5'):							
CGA	CGG	CGT	CGC	TGA	TGG	TGT	TGC

Each of the above groups of DNA bases from horse insulin has a counterpart in human insulin that differs from it by only one base. It is possible that horse insulin DNA differs from human insulin DNA by only two bases out of 159!

24.21 The percent of A always equals the percent of T, since A and T are complementary. The percent G equals the percent C for the same reason. Thus, sea urchin DNA contains about 32% each of A and T, and about 18% each of G and C.

24.22 Even though the stretch of DNA shown contains UAA in sequence, protein synthesis doesn't stop. The codons are read as 3-base individual units from start to end, and, in this mRNA sequence, the unit UAA is read as part of two codons, not as a single codon.

24.23 Restriction endonucleases cleave DNA base sequences that are palindromes, meaning that the sequence reads the same as the complement when both are read in the (5') to (3') direction. Thus, the sequence in (c), CTCGAG is recognized. The sequence in (a), GAATTC, is also a palindrome and is recognized by a restriction endonuclease. The sequence in (b) is not recognized.

24.24–24.26

		(a)	(b)	(c)	(d)
mRNA codon :	(5'–3')	AAU	GAG	UCC	CAU
Amino acid:		Asn	Glu	Ser	His
DNA sequence:	(3'–5')	TTA	CTC	AGG	GTA
tRNA anticodon:	(3'–5')	UUA	CUC	AGG	GUA

The DNA sequence of the template strand is shown.

24.27–24.28 UAC is a codon for tyrosine. It was transcribed from ATG of the template strand of a DNA chain.

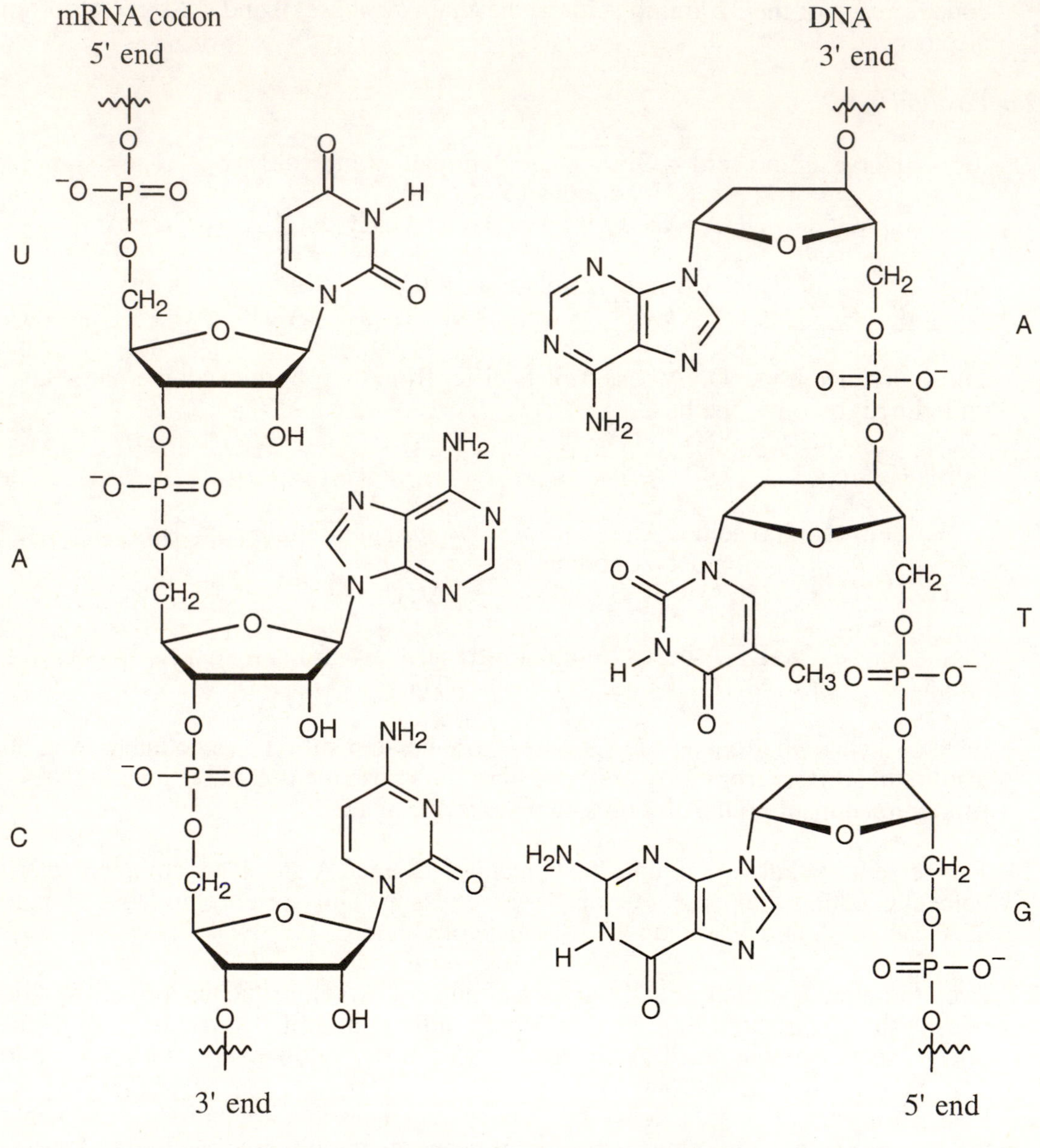

mRNA codon
5' end

U
A
C

3' end

DNA
3' end

A
T
G

5' end

24.29 Tyr——Gly————Gly————Phe————Met (stop) is coded by

Tyr	Gly	Gly	Phe	Met	(stop)
UAC	GGU	GGU	UUU	AUG	UAA
UAU	GGC	GGC	UUC		UAG
	GGA	GGA			UGA
	GGG	GGG			

A total of 2 x 4 x 4 x 2 x 1 x 3 = 194 different mRNA sequences can code for metenkephalin!

24.30 Angiotensin II:

	Asp	Arg	Val	Tyr	Ile	His	Pro	Phe	(stop)
mRNA sequence: (5'–3')	GAU	CGU	GUU	UAU	AUU	CAU	CCU	UUU	UAA
	GAC	CGC	GUC	UAC	AUC	CAC	CCC	UUC	UAG
		CGA	GUA		AUA		CCA		UGA
		CGG	GUG				CCG		
		AGA							
		AGG							

As in the previous problem, many mRNA sequences (13,824) can code for angiotensin II.

24.31 DNA coding strand (5'–3'): CTT— CGA—CCA— GAC—AGC—TTT
mRNA (5'–3'): CUU—CGA—CCA—GAC—AGC—UUU
Amino acid sequence: Leu——Arg——Pro——Asp——Ser——Phe
The mRNA sequence is the complement of the DNA template strand, which is the complement of the DNA coding strand. Thus, the mRNA sequence is a copy of the DNA coding strand, with T replaced by U.

24.32 mRNA sequence (5'–3'): CUA—GAC—CGU—UCC—AAG—UGA
Amino Acid: Leu——Asp——Arg——Ser——Lys (stop)

24.33

	Original Sequence	*Miscopied Sequence*
DNA coding strand (5'–3'):	-CAA-CCG-GAT-	-CGA-CCG-GAT-
mRNA sequence (5'–3'):	-CAA-CCG-GAU-	-CGA-CCG-GAU-
Amino acid sequence:	-Gln—Pro—Asp-	-Arg—Pro—Asp-

If this gene sequence were miscopied in the indicated way, a glutamine in the original protein would be replaced by an arginine in the mutated protein.

24.34 1. First, protect the nucleotides.

(a) Bases are protected by amide formation.

Adenine
NH_2
C_6H_5COCl
pyridine
$NHCOC_6H_5$

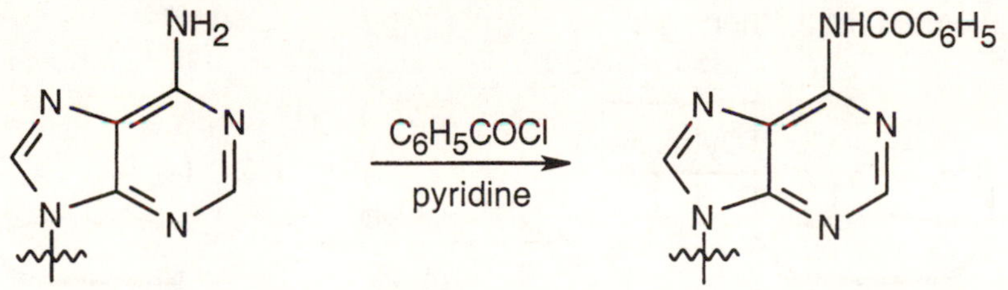

Cytosine
NH_2
C_6H_5COCl
pyridine
$NHCOC_6H_5$

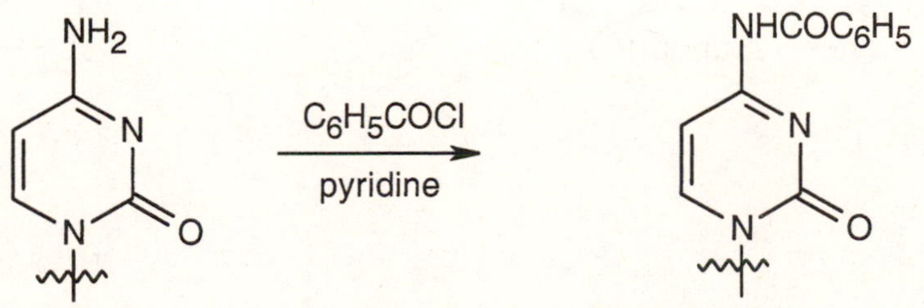

Guanine
NH_2
$(CH_3)_2CHCOCl$
pyridine
$NHCOCH(CH_3)_2$

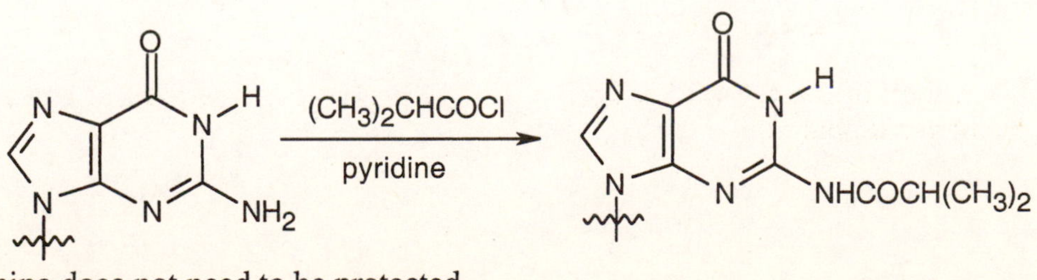

Thymine does not need to be protected.

(b) The 5' hydroxyl group is protected as its *p*-dimethoxytrityl (DMT) ether.

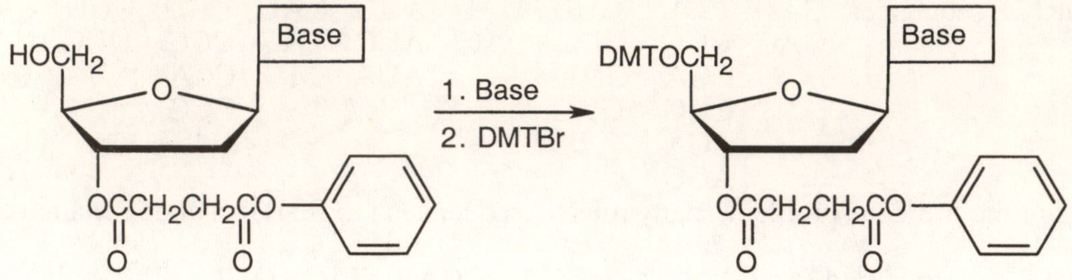

Base
$HOCH_2$
O
$OCCH_2CH_2CO$
O O
1. Base
2. DMTBr
Base
$DMTOCH_2$
O
$OCCH_2CH_2CO$
O O

2. Attach a protected 2-deoxycytidine nucleoside to the polymer support.

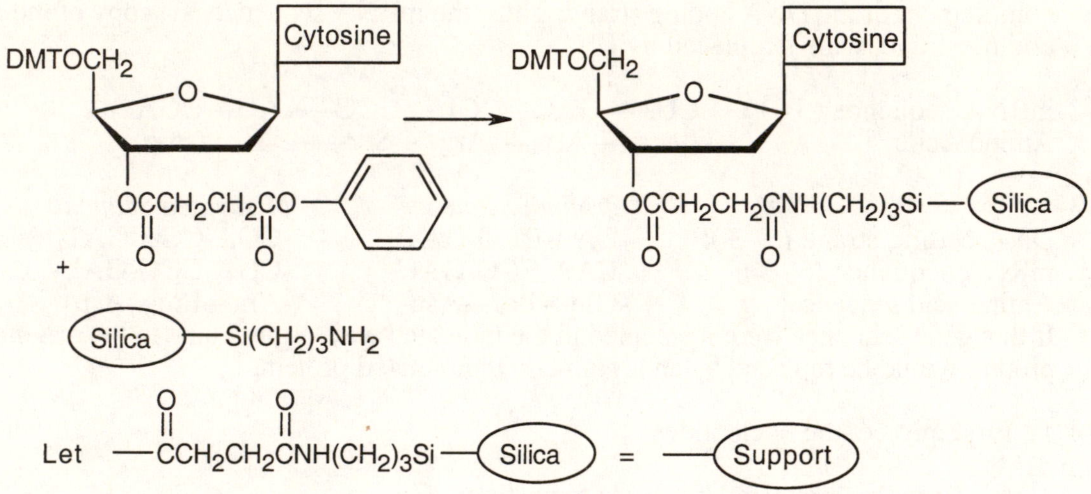

Cytosine
$DMTOCH_2$
O
$OCCH_2CH_2CO$
O O
+
Silica $-Si(CH_2)_3NH_2$

Cytosine
$DMTOCH_2$
O
$OCCH_2CH_2CNH(CH_2)_3Si$ — Silica
O O

Let $-CCH_2CH_2CNH(CH_2)_3Si$ — Silica = — Support

3. Cleave the DMT ether.

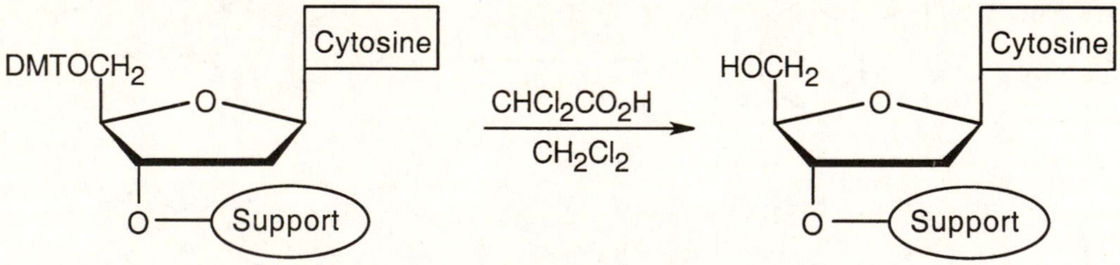

Cytosine
$DMTOCH_2$
O
O — Support
$CHCl_2CO_2H$
CH_2Cl_2
Cytosine
$HOCH_2$
O
O — Support

4. Couple protected 2'-deoxythymidine to the polymer-2'-deoxycytidine. (The nucleosides have a phosphoramidite group at the 3' position.)

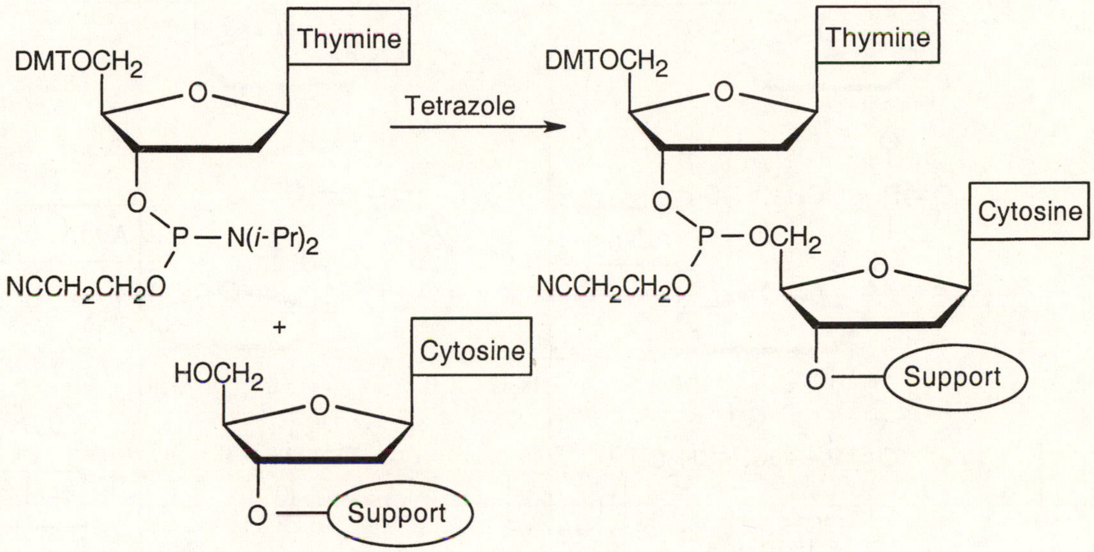

5. Oxidize the phosphite product to a phosphate triester, using iodine.

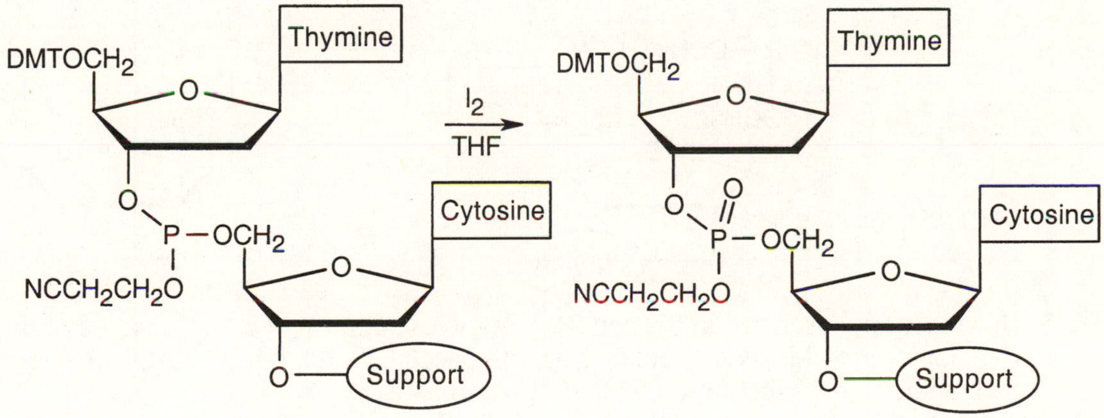

6. Repeat steps 3–5 with protected 2'-deoxyadenosine and protected 2'-deoxyguanosine.

7. Cleave all protecting groups with aqueous ammonia to yield the desired sequence.

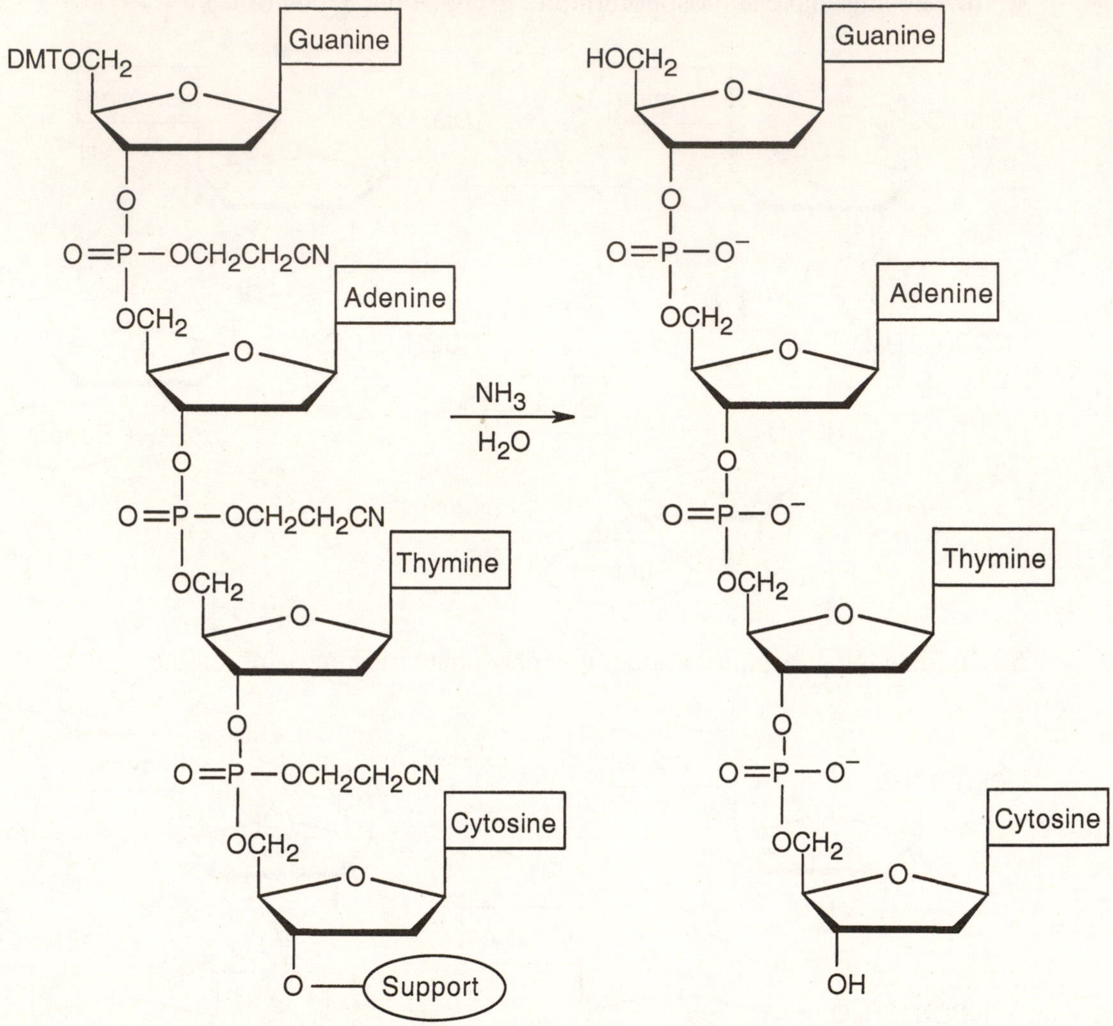

DMTOCH2 Guanine
O=P—OCH2CH2CN Adenine
O=P—OCH2CH2CN Thymine
O=P—OCH2CH2CN Cytosine
O—Support

$\xrightarrow[H_2O]{NH_3}$

HOCH2 Guanine
O=P—O⁻ Adenine
O=P—O⁻ Thymine
O=P—O⁻ Cytosine
OH

24.35 This reaction, described in Section 24.4, involves addition of a thiol residue of the enzyme to malonic semialdehyde, yielding a hemithioacetal. Oxidation by NAD^+, followed by nucleophilic acyl substitution by CoA, gives malonyl CoA.

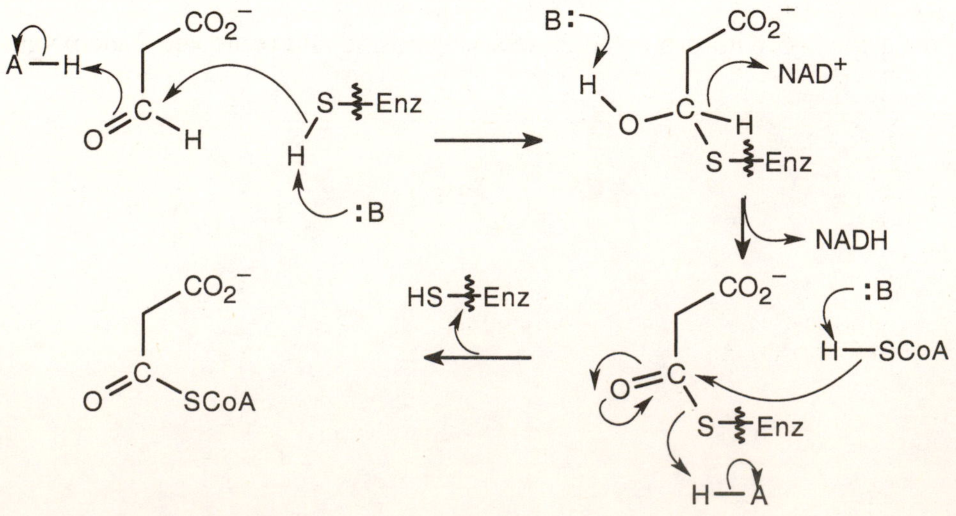

A—H, CO_2^-, S—Enz, :B → B:, CO_2^-, NAD^+, S—Enz → NADH → CO_2^-, :B, H—SCoA, S—Enz, H—A → HS—Enz → CO_2^-, SCoA

24.36 The steps: (1) Nucleophilic substitution at ATP, with PP_i as a leaving group; (2) addition of the amide group of glutamine to the enolate bond; (3) loss of AMP; (4) hydrolysis.

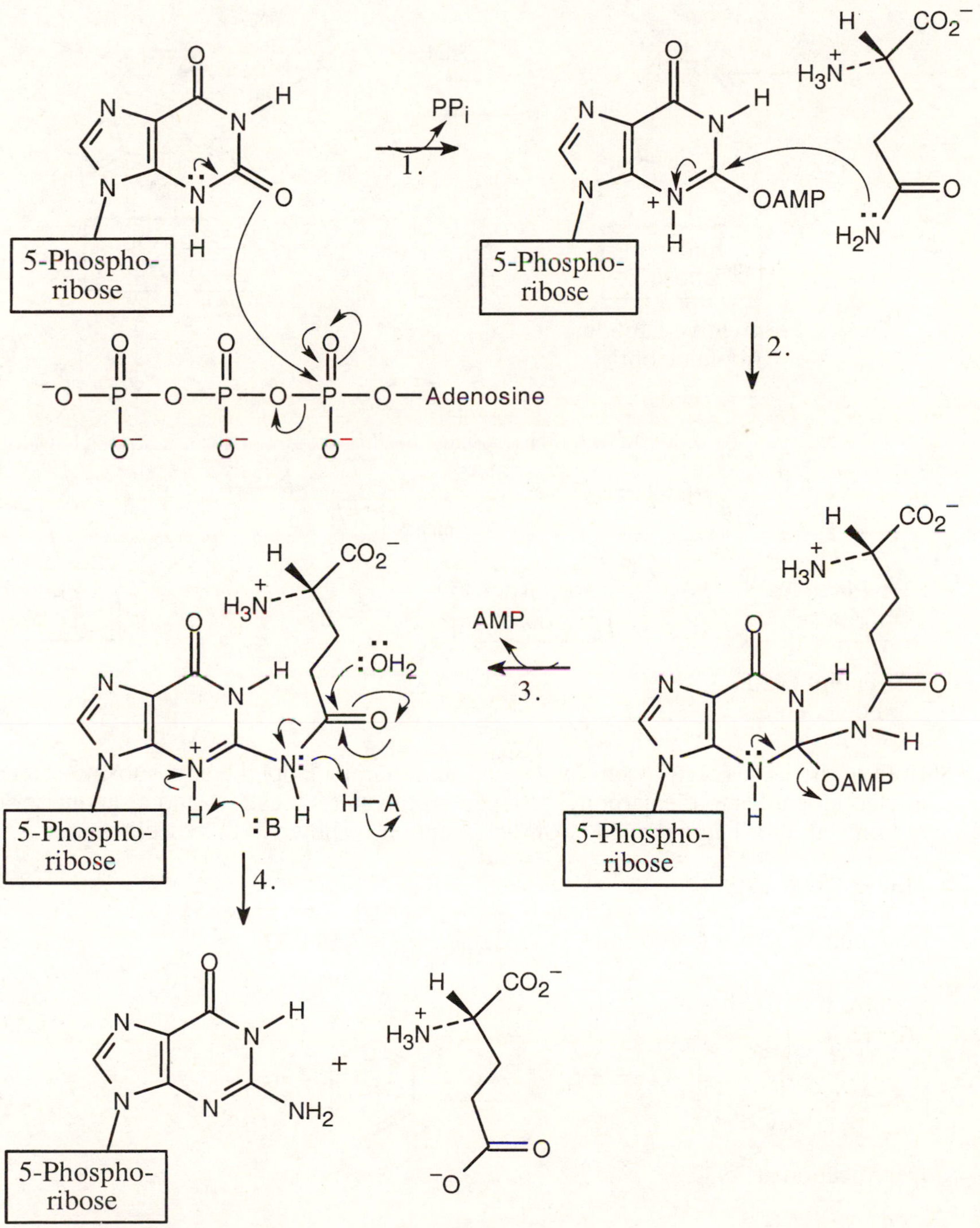

Guanosine monophosphate

Glutamate

24.37 The steps: (1) phosphorylation; (2) cyclization; (3) loss of phosphate; (4) tautomerization.

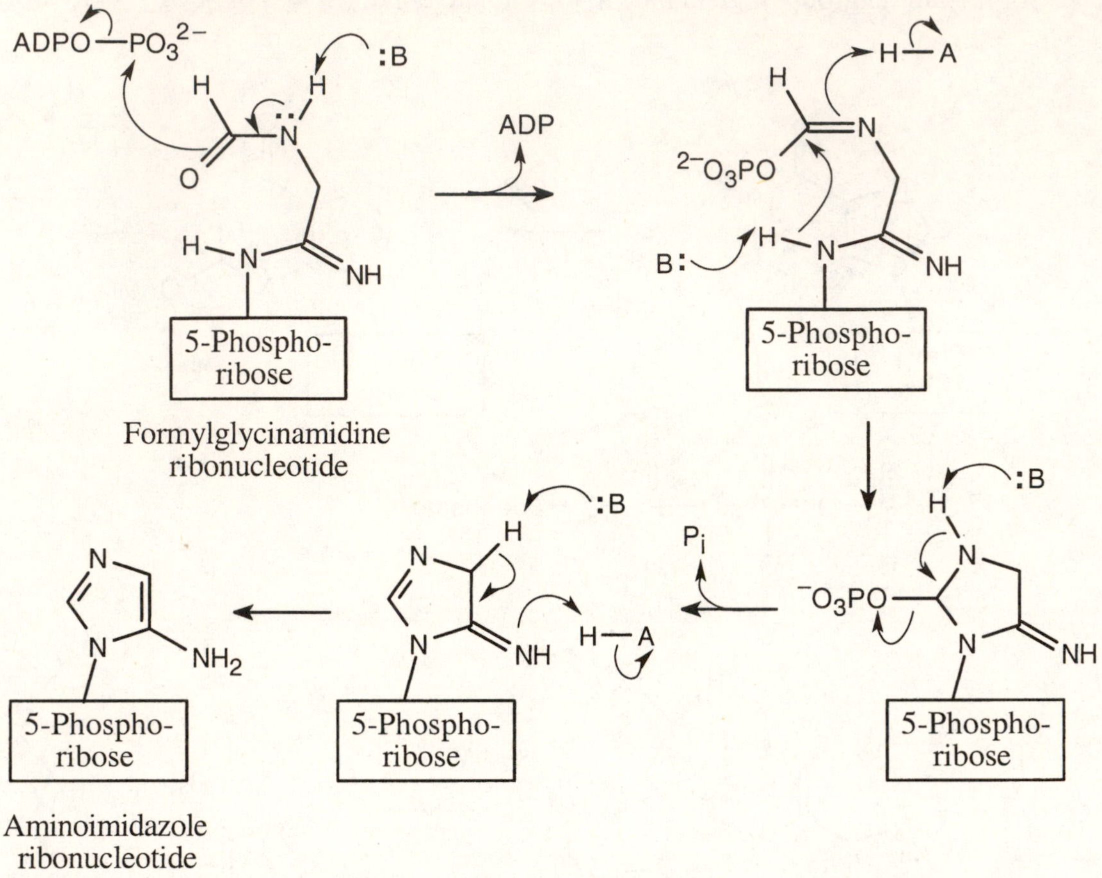
ADPO—PO3 2−
:B
ADP
H—A
B:
NH
5-Phospho-ribose
Formylglycinamidine ribonucleotide
Pi
NH2
Aminoimidazole ribonucleotide

24.38 Both of these cleavages occur by the now-familiar nucleophilic acyl substitution route. A nucleophile adds to the carbonyl group, a proton shifts location, and a second group is eliminated. the "heart-shaped arrow" convention is shown in this mechanism

Deprotection at 1:

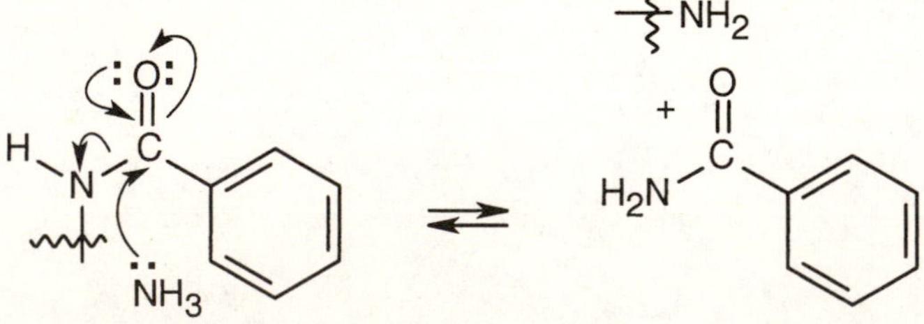
NH3
NH2
H2N

Deprotection at 2:

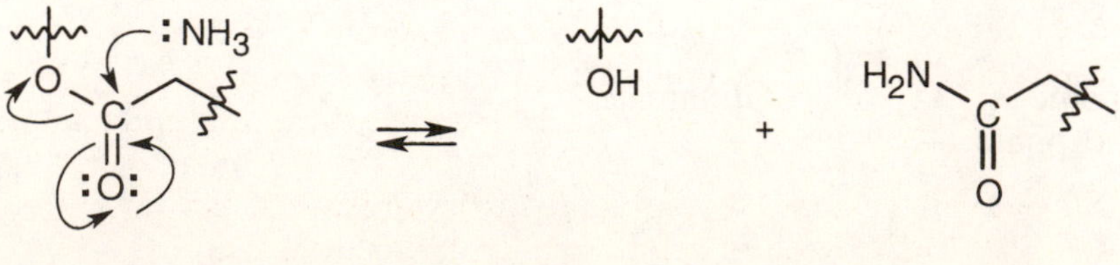
:NH3
OH
H2N

Review Unit 10: Lipids and Nucleic Acids

Major Topics Covered (with vocabulary):

Lipids:
wax fat oil triacylglycerol fatty acid polyunsaturated fatty acid soap saponification micelle glycerophospholipid sphingomyelin lipid bilayer sphingosine β-oxidation pathway prostaglandin eicosanoid terpenoid monoterpene sesquiterpene isopentenyl diphosphate steroid hormone sex hormone adrenocortical hormone androgen estrogen mineralocorticoid glucocorticoid squalene lanosterol

Nucleic acids and nucleotides:
nucleoside nucleotide deoxyribonucleic acid (DNA) ribonucleic acid (RNA) adenine guanine thymine cytosine 3' end 5' end base pairing double helix complementary pairing major groove minor groove intercalation

Nucleic acids and heredity:
replication semiconservative DNA polymerase replication fork DNA ligase transcription mRNA rRNA tRNA coding strand template strand promoter sites exon intron translation codon anticodon

DNA technology:
DNA sequencing restriction endonuclease restriction fragment palindrome Sanger dideoxy method DNA synthesis DMT ether phosphoramidite phosphite polymerase chain reaction (PCR)

Biochemistry of nucleotides:
nuclease xanthine molybdenum cofactor uric acid

Types of Problems:

After studying these chapters you should be able to:

- Draw the structures of fats, oils, steroids and other lipids.
- Determine the structure of a fat.
- Predict the products of reactions of fats and steroids.
- Locate the 5-carbon units in terpenoids.
- Understand the mechanism of terpenoid and steroid biosynthesis.
- Draw the structures and conformations of steroids and other fused-ring systems.

- Draw purines, pyrimidines, nucleosides, nucleotides, and representative segments of DNA and their complements.
- List the base sequence that codes for a given amino acid or peptide.
- Deduce an amino acid sequence from a given mRNA sequence (and *vice versa*).
- Draw the anticodon sequence of tRNA, given the mRNA sequence.
- Outline the process of DNA sequencing, and deduce a DNA sequence from an electrophoresis pattern.
- Outline the method of DNA synthesis, and formulate the mechanisms of synthetic steps.
- Be able to outline the pathways of nucleotide catabolism and biosynthesis.

Points to Remember:

* When trying to locate the 5-carbon units in a terpenoid, look for an isopropyl group first; at least one should be apparent. After finding it, count 5 carbons, and locate the second unit. If there are two possibilities for the second unit, choose the one that has the double bond in the correct location.

* In general, the reactions of steroids that are presented in this book are familiar and uncomplicated. Keeping track of the stereochemistry of the tetracyclic ring system is somewhat more complicated.

* In situations where base-pairing occurs, such as replication, transcription or translation, a polynucleotide chain (written with the 5' end on the left and the 3' end on the right) pairs with a second chain (written with the 3' end on the left and the 5' end on the right). Base pairing is complementary, and the two chains are always read in opposite directions.

Self-test:

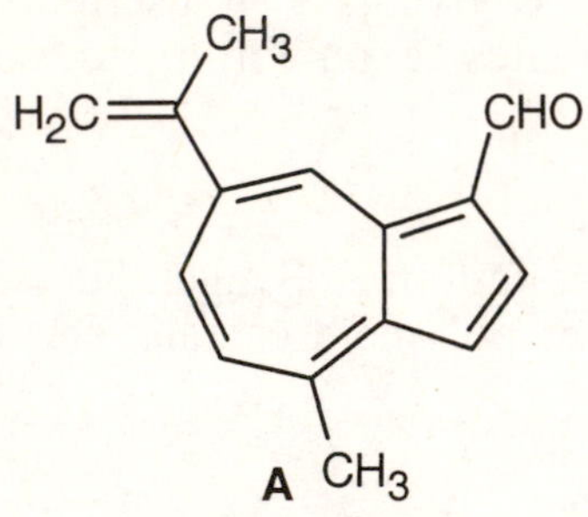

A
Lactaroviolin
(an antibiotic)

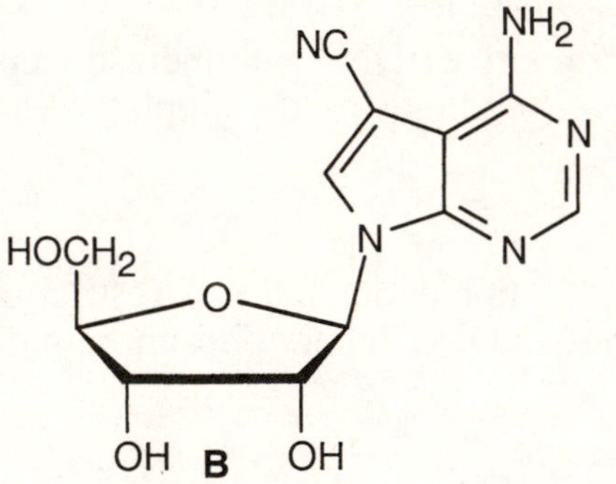

B
Toyocamycin
(an antibiotic)

What type of terpenoid is **A**? Show the location of the 5-carbon units.

Toyocamycin (**B**) is related to which nucleoside? What are the differences between **C** and the nucleoside?

3' 5'
–ACG–CCT–TAG–GGC–TTA–GGA–
C

C represents a segment of the template strand of a molecule of DNA. Draw: (a) the coding strand; (b) the mRNA that is synthesized from **C** during transcription; (c) the tRNA anticodons that are complementary to the mRNA codons; (d) the amino acids (use one-letter codes) that form the peptide that **C** codes for.

Multiple choice:

1. Which type of molecule is most likely to be found in a lipid bilayer?
(a) triacylglycerol (b) prostaglandin (c) sphingomyelin (d) triterpene

2. Which of the following terpenes might have been formed by a tail-to-tail coupling?
(a) monoterpene (b) sesquiterpene (c) diterpene (d) triterpene

3. Prostaglandins and related compounds have all of the following structural features in common except:
(a) cis double bonds (b) a carboxylic acid group (c) a C_{20} chain (d) hydroxyl groups

4. A sphingomyelin contains all of the following components except:
(a) amide (b) carboxylic acid ester (c) phosphate group (d) amino alcohol

5. Which nucleic acid has nonstandard bases, in addition to the usual bases?
(a) DNA (b) mRNA (c) rRNA (d) tRNA

6. Which base doesn't need a protecting group in DNA synthesis?
(a) Thymine (b) Cytosine (c) Adenine (d) Guanine

7. Which amino acid has only one codon?
(a) Tyrosine (b) Arginine (c) Lysine (d) Tryptophan

8. Androstenedione, implicated in steroid misuse among athletes, is a:
(a) bile steroid (b) male sex hormone (c) adrenocortical hormone (d) synthetic steroid

9. A person suffering from gout, an inability to properly excrete uric acid, is advised to avoid food containing:
(a) prostaglandins (b) cholesterol (c) purines (d) terpenes

10. Which of the following cofactors is not involved in fatty acid biosynthesis?
(a) biotin (b) NAD^+ (c) acetyl ACP (d) ATP

Functional-Group Synthesis

The following table summarizes the synthetic methods by which important functional groups can be prepared. The functional groups are listed alphabetically, followed by reference to the appropriate text section and a brief description of each synthetic method.

Acetals, $R_2C(OR')_2$

(Sec. 14.8)	from ketones and aldehydes by acid-catalyzed reaction with alcohols

Acid anhydrides, RCO_2COR'

(Sec. 16.3)	from dicarboxylic acids by heating
(Sec. 16.4)	from acid chlorides by reaction with carboxylate salts

Acid chlorides, RCOCl

(Sec. 16.3)	from carboxylic acids by reaction with $SOCl_2$

Alcohols, ROH

(Sec. 7.4)	from alkenes by oxymercuration/demercuration
(Sec. 7.4)	from alkenes by hydroboration/oxidation
(Sec. 7.7)	from epoxides by acid-catalyzed ring opening with either H_2O or HX
(Sec. 7.7)	from alkenes by hydroxylation with OsO_4
(Sec. 10.5)	from alkyl halides and tosylates by S_N2 reaction with hydroxide ion
(Secs. 13.3, 14.6)	from ketones and aldehydes by reduction with $NaBH_4$ or $LiAlH_4$
(Secs. 13.3, 14.6)	from ketones and aldehydes by addition of Grignard reagents
(Secs. 13.3, 16.3)	from carboxylic acids by reduction with either $LiAlH_4$
(Secs. 13.3, 16.6)	from esters by reduction with $LiAlH_4$
(Secs. 13.6, 16.6)	from esters by reaction with Grignard reagents
(Sec. 13.9)	from ethers by acid-induced cleavage

Aldehydes, RCHO

(Secs. 13.5, 14.2)	from primary alcohols by oxidation
(Secs. 14.2, 16.6)	from esters by reduction with DIBAH [$HAl(i\text{-}Bu)_2$]

Alkanes, RH

(Sec. 7.5)	from alkenes by catalytic hydrogenation
(Sec. 10.3)	from alkyl halides by protonolysis of Grignard reagents

Alkenes, $R_2C{=}CR_2$

(Secs. 7.1, 10.11)	from alkyl halides by treatment with strong base (E2 reaction)
(Secs. 7.1, 13.4)	from alcohols by dehydration
(Sec. 7.12)	from alkynes by catalytic hydrogenation using the Lindlar catalyst
(Sec. 14.9)	from ketones and aldehydes by treatment with alkylidenetriphenylphosphoranes (Wittig reaction)
(Sec. 17.2)	from α-bromo ketones by heating with pyridine
(Sec. 18.7)	from amines by methylation and Hofmann elimination

Amides, $RCONH_2$

(Sec. 15.7)	from nitriles by partial hydrolysis with either acid or base
(Sec. 16.3)	from a carboxylic acid and an amine by treatment with dicyclohexylcarbodiimide (DCC)
(Sec. 16.4)	from acid chlorides by treatment with an amine or ammonia
(Sec. 16.5)	from acid anhydrides by treatment with an amine or ammonia

Amines, RNH_2

(Sec. 14.11)	from conjugated enones by addition of primary or secondary amines
(Secs. 16.7, 18.6)	from amides by reduction with $LiAlH_4$
(Secs. 15.7, 18.6)	from nitriles by reduction with $LiAlH_4$
(Sec. 18.6)	from primary alkyl halides by treatment with ammonia
(Sec. 18.6)	from ketones and aldehydes by reductive amination with an amine and $NaBH_4$

Amino Acids, $RCH(NH_2)CO_2H$

(Sec. 19.3)	from α-keto acids by reductive amination
(Sec. 19.3)	from primary alkyl halides by alkylation with diethyl acetamidomalonate
(Sec. 19.3)	from *(Z)*-amido acids by enantioselective hydrogenation

Arenes, Ar–R

(Sec. 8.7)	from arenes by Friedel–Crafts alkylation with an alkyl halide
(Sec. 8.9)	from aryl alkyl ketones by catalytic reduction of the keto group

Arylamines, $Ar–NH_2$

(Secs. 8.6, 18.6)	from nitroarenes by reduction with either Fe, Sn, or H_2/Pd.

Arenesulfonic acids $Ar–SO_3H$

(Sec. 8.6)	from arenes by electrophilic aromatic substitution with SO_3/H_2SO_4

Carboxylic acids, RCO_2H

(Sec. 8.9)	from arenes by side-chain oxidation with $Na_2Cr_2O_7$ or $KMnO_4$
(Sec. 14.3)	from aldehydes by oxidation
(Sec. 15.5)	from alkyl halides by conversion to Grignard reagents followed by reaction with CO_2
(Sec. 15.7)	from nitriles by acid or base hydrolysis
(Sec. 16.4)	from acid chlorides by reaction with aqueous base
(Sec. 16.5)	from acid anhydrides by reaction with aqueous base
(Sec. 16.6)	from esters by hydrolysis with aqueous base
(Sec. 16.7)	from amides by hydrolysis with aqueous base

Disulfides, RS–SR'

(Sec. 13.6)	from thiols by oxidation with bromine

Enamines, $RCH=CRNR_2$

(Sec. 14.7)	from ketones or aldehydes by reaction with secondary amines

Epoxides, $R_2C\overset{O}{—}CR_2$

(Sec. 7.6)	from alkenes by treatment with a peroxyacid
(Sec. 7.6)	from halohydrins by treatment with base

Esters, RCO_2R'

(Sec. 16.3)	from carboxylic acid salts by S_N2 reaction with primary alkyl halides
(Sec. 16.3)	from carboxylic acids by acid-catalyzed reaction with an alcohol (Fischer esterification)
(Sec. 16.4)	from acid chlorides by base-induced reaction with an alcohol
(Sec. 16.5)	from acid anhydrides by base-induced reaction with an alcohol
(Sec. 17.4)	from alkyl halides by alkylation with diethyl malonate
(Sec. 17.4)	from esters by alkylation of their enolate ions

Ethers, R–O–R'

(Sec. 7.6)	from alkenes by epoxidation with peroxyacids
(Sec. 13.8)	from primary alkyl halides by S_N2 reaction with alkoxide ions (Williamson ether synthesis)

Grignard reagents, RMgX

(Sec. 10.3)	from organohalides by treatment with magnesium

Halides, alkyl, $R_3C–X$

(Sec. 6.6)	from alkenes by electrophilic addition of HX
(Sec. 7.2)	from alkenes by addition of halogen
(Sec. 7.3)	from alkenes by electrophilic addition of hypohalous acid (HOX) to yield halohydrins
(Sec. 7.12)	from alkynes by addition of halogen
(Sec. 7.12)	from alkynes by addition of HX
(Sec. 10.2)	from alcohols by reaction with HX
(Sec. 10.2)	from alcohols by reaction with $SOCl_2$
(Sec. 10.2)	from alcohols by reaction with PBr_3
(Sec. 17.2)	from ketones by α-halogenation with bromine

Halides, aryl, Ar–X

(Sec. 8.6)	from arenes by electrophilic aromatic substitution with halogen

Halohydrins, $R_2CXC(OH)R_2$

(Sec. 7.4)	from alkenes by electrophilic addition of hypohalous acid (HOX)

Imines. $R_2C{=}NR'$

(Sec. 14.7)	from ketones or aldehydes by reaction with primary amines

Ketones, $R_2C=O$

(Sec. 8.7)	from arenes by Friedel–Crafts acylation reaction with an acid chloride
(Secs. 13.5, 14.2)	from secondary alcohols by oxidation
(Sec. 17.4)	from primary alkyl halides by alkylation with ethyl acetoacetate
(Sec. 17.4)	from ketones by alkylation of their enolate ions with primary alkyl halides

Nitriles, $R-C\equiv N$

(Secs. 11.6, 15.7)	from primary alkyl halides by S_N2 reaction with cyanide ion
(Sec. 15.7)	from primary amides by dehydration with $SOCl_2$
(Sec. 17.4)	from nitriles by alkylation of their α-anions with primary alkyl halides

Nitroarenes, $Ar-NO_2$

(Sec. 8.6)	from arenes by electrophilic aromatic substitution with nitric/sulfuric acids

Quinones, O=⟨cyclohexa-2,5-diene-1,4-dione⟩=O

(Sec. 13.5)	from phenols by oxidation with Fremy's salt $[(KSO_3)_2NO]$

Sulfides, R–S–R'

(Sec. 13.10)	from thiols by S_N2 reaction of thiolate ions with primary alkyl halides

Sulfones, $R-SO_2-R'$

(Sec. 13.10)	from sulfides or sulfoxides by oxidation with peroxyacids

Sulfoxides, R–SO–R'

(Sec. 13.10)	from sulfides by oxidation with H_2O_2

Thiols, R–SH

(Sec. 13.6)	from primary alkyl halides by S_N2 reaction with hydrosulfide anion
(Sec. 13.6)	from primary alkyl halides by S_N2 reaction with thiourea, followed by hydrolysis

Functional-Group Reactions

The following table summarizes the reactions of important functional groups. The functional groups are listed alphabetically, followed by a reference to the appropriate text section.

Acetal

1. Hydrolysis to yield a ketone or aldehyde plus alcohol (Sec. 14.8)

Acid anhydride

1. Hydrolysis to yield a carboxylic acid (Sec. 16.5)
2. Alcoholysis to yield an ester (Sec. 16.5)
3. Aminolysis to yield an amide (Sec. 16.5)

Acid chloride

1. Friedel–Crafts reaction with an aromatic compound to yield an aryl ketone (Sec. 8.7)
2. Hydrolysis to yield a carboxylic acid (Sec. 16.4)
3. Alcoholysis to yield an ester (Sec. 16.4)
4. Aminolysis to yield an amide (Sec. 16.4)

Alcohol

1. Conversion to an alkyl halide (Secs. 10.2, 13.4)
   a. Reaction of a tertiary alcohol with HX
   b. Reaction of a primary or secondary alcohol with $SOCl_2$
   c. Reaction of a primary or secondary alcohol with PBr_3
2. Acidity (Sec. 13.2)
3. Dehydration to yield an alkene (Sec. 13.4)
4. Oxidation (Sec. 13.5)
   a. Reaction of a primary alcohol to yield an aldehyde or acid
   b. Reaction of a secondary alcohol to yield a ketone
5. Reaction with a primary alkyl halide to yield an ether (Sec. 13.8)
6. Reaction with a carboxylic acid to yield an ester (Sec. 16.3)
7. Reaction with an acid chloride to yield an ester (Sec. 16.4)
8. Reaction with an acid anhydride to yield an ester (Sec. 16.5)

Aldehyde

1. Oxidation to yield a carboxylic acid (Sec. 14.3)
2. Nucleophilic addition reactions
   a. Reduction to yield a primary alcohol (Secs. 13.3, 14.6)
   b. Reaction with a Grignard reagent to yield a secondary alcohol (Secs. 13.3, 14.6)
   c. Grignard reaction of formaldehyde to yield a primary alcohol (Sec. 14.6)
   d. Reaction with an alcohol to yield an acetal (Sec. 14.8)
   e. Wittig reaction to yield an alkene (Sec. 14.9)
   f. Reaction with an amine to yield an imine or enamine (Sec. 14.7)
3. Aldol reaction to yield a β-hydroxy aldehyde (Sec. 17.5)

Alkane

1. Radical halogenation to yield an alkyl halide (Sec. 5.3)

Alkene

1. Electrophilic addition of HX to yield an alkyl halide (Sec. 6.6)
2. Electrophilic addition of halogen to yield a 1,2-dihalide (Sec. 7.2)
3. Oxymercuration/demercuration to yield an alcohol (Sec. 7.4)
   Markovnikov regiochemistry is observed.
4. Hydroboration/oxidation to yield an alcohol (Sec. 7.4)
   Non-Markovnikov regiochemistry is observed
5. Hydrogenation to yield an alkane (Sec. 7.5)
6. Reaction with a peroxyacid to yield an epoxide (Sec. 7.6)
7. Hydroxylation to yield a 1,2-diol (Sec. 7.7)

Alkyne

1. Electrophilic addition of HX to yield a vinylic halide (Sec. 7.12)
2. Electrophilic addition of halogen to yield a dihalide (Sec. 7.12)
3. Hydrogenation over Lindlar catalyst to yield a cis alkene (Sec. 7.12)

Amide

1. Hydrolysis to yield a carboxylic acid (Sec. 16.7)
2. Reduction with $LiAlH_4$ to yield an amine (Sec. 16.7)
3. Dehydration to yield a nitrile (Sec. 16.7)

Amine

1. Nucleophilic acyl substitution reactions
   a. Reaction with an acid chloride to yield an amide (Sec. 16.4)
   b. Reaction with an acid anhydride to yield an amide (Sec. 16.5)
2. Basicity (Sec. 18.3)
3. S_N2 alkylation of an alkyl halide to yield an amine (Sec. 18.6)
4. Hofmann elimination to yield an alkene (Sec. 18.7)

Arene

1. Oxidation of an alkylbenzene side chain to yield a benzoic acid (Sec. 8.9)
2. Reduction of an aryl alkyl ketone to yield an arene (Sec. 8.9)
3. Electrophilic aromatic substitution
   a. Halogenation (Sec. 8.6)
   b. Nitration (Sec. 8.6)
   c. Sulfonation (Sec. 8.6)
   d. Friedel–Crafts alkylation (Sec. 8.7)
   e. Friedel–Crafts acylation (Sec. 8.7)

Carboxylic acid

1. Acidity (Secs. 15.3–15.4)
2. Reduction with $LiAlH_4$ to yield a primary alcohol (Secs. 13.3, 16.3)
3. Nucleophilic acyl substitution reactions (Sec. 16.3)
   a. Conversion to an acid chloride (Sec. 16.3)
   b. Conversion to an acid anhydride (Sec. 16.3)
   c. Conversion to an ester (Sec. 16.3)
      (1) Fischer esterification (Sec. 16.3)
      (2) S_N2 reaction with an alkyl halide (Sec. 16.3)

Diene

1. Conjugate addition of HX and X_2 (Sec. 7.11)

Epoxide

1. Acid-catalyzed ring opening with aqueous acid to yield a 1,2-diol (Sec. 7.7)

Ester

1. Hydrolysis to yield a carboxylic acid (Sec. 16.6)
2. Reduction to yield a primary alcohol (Secs. 13.3, 16.6)
3. Partial reduction with DIBAH to yield an aldehyde (Sec. 16.6)
4. Grignard reaction to yield a tertiary alcohol (Secs. 13.3, 16.6)
5. Claisen condensation to yield a β-keto ester (Sec. 17.8)

Ether

1. Acid-induced cleavage to yield an alcohol and an alkyl halide (Sec. 13.9)

Grignard reagent

1. Reduction by treatment with acid to yield an alkane (Sec. 10.3)
2. Nucleophilic addition to a carbonyl compound to yield an alcohol (Secs. 13.3, 16.6)
3. Reaction with carbon dioxide to yield a carboxylic acid (Sec. 15.5)

Halide, alkyl

1. Reaction with magnesium to form a Grignard reagent (Sec. 10.3)
2. Reduction to yield an alkane (Sec. 10.3)
3. Nucleophilic substitution (S_N1 or S_N2) (Secs. 10.5–10.8)
4. Dehydrohalogenation to yield an alkene (E1, E1cB, or E2) (Secs. 10.10–10.12)

Halohydrin

1. Conversion to an epoxide (Sec. 7.6)

Ketone

1. Nucleophilic addition reactions
   a. Reduction to yield a secondary alcohol (Secs. 13.3, 16.6)
   b. Reaction with a Grignard reagent to yield a tertiary alcohol (Secs. 13.3, 16.6)
   c. Reaction with an alcohol to yield an acetal (Sec. 14.8)
   d. Wittig reaction to yield an alkene (Sec. 14.9)
   e. Reaction with an amine to yield an imine or enamine (Sec. 14.7)
2. Aldol reaction to yield a β-hydroxy ketone (Sec. 17.5)
3. Alpha bromination (Sec. 17.2)

Nitrile

1. Hydrolysis to yield a carboxylic acid (Secs. 15.7, 18.6)
2. Reduction to yield a primary amine (Secs. 15.7, 18.6)

Nitroarene

1. Reduction to yield an arylamine (Secs. 8.6, 18.6)

Phenol

1. Acidity (Sec. 13.2)
2. Reaction with an acid chloride to yield an ester (Sec. 16.4)
3. Reaction with an alkyl halide to yield an ether (Sec. 13.8)
4. Oxidation to yield a quinone (Sec. 13.5)

Quinone

1. Reduction to yield a hydroquinone (Sec. 13.5)

Sulfide

1 Reaction with an alkyl halide to yield a sulfonium salt (Sec. 13.10)
2. Oxidation to yield a sulfoxide (Sec. 13.10)
3. Oxidation to yield a sulfone (Sec. 13.10)

Thiol

1. Reaction with an alkyl halide to yield a sulfide (Sec. 13.10)
2. Oxidation to yield a disulfide (Sec. 13.6)

Reagents in Organic Chemistry

The following table summarizes the uses of some important reagents in organic chemistry. The reagents are listed alphabetically, followed by a brief description of the uses of each and references to the appropriate text sections.

Acetic anhydride, $(CH_3CO)_2O$
- Reacts with alcohols to yield acetate esters (Section 16.5)
- Reacts with amines to yield acetamides (Section 16.5).

Aluminum chloride, $AlCl_3$
- Acts as a Lewis acid catalyst in Friedel–Crafts alkylation and acylation reactions of aromatic compounds (Section 8.7).

Ammonia, NH_3
- Reacts with acid chlorides and acid anhydrides to yield amides (Sections 16.4 and 16.5).

Borane, BH_3
- Adds to alkenes, giving alkylboranes that can be oxidized with alkaline H_2O_2 to yield alcohols (Section 7.4).

Bromine, Br_2
- Adds to alkenes, yielding 1,2-dibromides (Section 7.2).
- Adds to alkynes, yielding either 1,2-dibromoalkenes or 1,1,2,2-tetrabromoalkanes (Section 7.12).
- Reacts with arenes in the presence of $FeBr_3$ catalyst to yield bromoarenes (Section 8.6).
- Oxidizes thiols to disulfides (Section 13.6)
- Reacts with ketones in acetic acid solvent to yield α-bromo ketones (Section 17.2).
- Oxidizes aldoses to aldonic acids (Section 21.6).

Di-*tert*-butoxy dicarbonate, ($\underline{t}$-BuOCO)$_2$O
- Reacts with amino acids to give Boc protected amino acids suitable for use in peptide synthesis (Section 19.7).

Butyllithium, $CH_3CH_2CH_2CH_2Li$
- Reacts with alkyltriphenylphosphonium salts to yield alkylidenephosphoranes (Wittig reagents (Section 14.9).
- A strong base; reacts with dialkylamines to yield lithium dialkylamide bases such as LDA [lithium diisopropylamide] (Section 17.3).

Carbon dioxide, CO_2
- Reacts with Grignard reagents to yield carboxylic acids (Section 15.5).

Chlorine, Cl_2
- Reacts with alkanes in the presence of light to yield chloroalkanes by a radical chain reaction pathway (Section 5.3).
- Adds to alkenes to yield 1,2-dichlorides (Section 7.2).
- Reacts with arenes in the presence of $FeCl_3$ catalyst to yield chloroarenes (Section 8.6).

***m*-Chloroperoxybenzoic acid, *m*-$ClC_6H_4CO_3H$**
- Reacts with alkenes to yield epoxides (Section 7.6).

Chromium trioxide, CrO_3
- Oxidizes alcohols in aqueous acid to yield carbonyl-containing products. Primary alcohols yield carboxylic acids, and secondary alcohols yield ketones (Section 13.5).

Dichloroacetic acid, Cl_2CHCO_2H
- Cleaves DMT protecting groups in DNA synthesis (Section 24.7).

Dicyclohexylcarbodiimide (DCC), $C_6H_{11}-N=C=N-C_6H_{11}$
- Couples an amine with a carboxylic acid to yield an amide (Section 16.3).

Diethyl acetamidomalonate, $CH_3CONHCH(CO_2Et)_2$
- Reacts with alkyl halides in a common method of α-amino acid synthesis (Section 19.3).

Diisobutylaluminum hydride (DIBAH), $(i\text{-Bu})_2AlH$
- Reduces esters to yield aldehydes (Section 16.6).

Ethylene glycol, $HOCH_2CH_2OH$
- Reacts with ketones or aldehydes in the presence of an acid catalyst to yield acetals (Section 14.8).

Ferric bromide, $FeBr_3$
- Acts as a catalyst for the reaction of arenes with Br_2 to yield bromoarenes (Section 8.6).

Ferric chloride, $FeCl_3$
- Acts as a catalyst for the reaction of arenes with Cl_2 to yield chloroarenes (Section 8.6).

Grignard reagent, RMgX
- Reacts with acids to yield alkanes (Section 10.3).
- Adds to ketones, aldehydes, esters to yield alcohols (Sections 13.3 and 14.6).

Hydrogen bromide, HBr
- Adds to alkenes with Markovnikov regiochemistry to yield alkyl bromides (Section 6.6).
- Adds to alkynes to yield either bromoalkenes or 1,1-dibromoalkanes (Section 7.12).
- Reacts with alcohols to yield alkyl bromides (Sections 10.2 and 10.8).
- Cleaves ethers to yield alcohols and alkyl bromides (Section 13.9).

Hydrogen chloride, HCl
- Adds to alkenes with Markovnikov regiochemistry to yield alkyl chlorides (Section 6.6).
- Adds to alkynes to yield either chloroalkenes or 1,1-dichloroalkanes (Section 7.12).
- Reacts with alcohols to yield alkyl chlorides (Sections 10.2 and 10.8).

Hydrogen iodide, HI
- Cleaves ethers to yield alcohols and alkyl iodides (Section 13.9).

Hydrogen peroxide, H_2O_2
- Oxidizes organoboranes to yield alcohols (Section 7.4).
- Oxidizes sulfides to sulfoxides (Section 13.10).

Iodine, I_2
- Reacts with arenes in the presence of CuCl or H_2O_2 to yield iodoarenes (Section 8.6).
- Oxidizes thiols to disulfides (Section 13.6)

Iodomethane, CH_3I
- Reacts with alkoxide anions to yield methyl ethers (Section 13.8).
- Reacts with carboxylate anions to yield methyl esters (Section 16.3).
- Reacts with enolate ions to yield α-methylated carbonyl compounds (Section 17.4).
- Reacts with amines to yield methylated amines (Section 18.7).

Iron, Fe
- Reacts with nitroarenes in the presence of aqueous acid to yield anilines (Section 8.6).

Lindlar catalyst
- Acts as a catalyst for the partial hydrogenation of alkynes to yield cis alkenes (Section 7.12).

Lithium aluminum hydride, $LiAlH_4$
- Reduces ketones, aldehydes, esters, and carboxylic acids to yield alcohols (Sections 13.3, 14.6, and 16.6).
- Reduces nitriles to yield amines (Section 15.7).
- Reduces amides to yield amines (Section 16.7).

Lithium diisopropylamide (LDA), LiN(*i*-Pr)$_2$
- Reacts with carbonyl compounds (aldehydes, ketones, esters) to yield enolate ions (Sections 17.3 and 17.4).

Magnesium, Mg
- Reacts with organohalides to yield Grignard reagents (Section 10.3).

Mercuric acetate, $Hg(O_2CCH_3)_2$
- Adds to alkenes in the presence of water, giving α-hydroxy organomercury compounds that can be reduced with $NaBH_4$ to yield alcohols (Section 7.3).

Nitric acid, HNO_3
- Reacts with arenes in the presence of sulfuric acid to yield nitroarenes (Section 8.6).
- Oxidizes aldoses to yield aldaric acids (Section 21.6).

Osmium tetroxide, OsO_4
- Adds to alkenes to yield 1,2-diols (Section 7.7).

Palladium on carbon, Pd/C
- Acts as a hydrogenation catalyst for reducing carbon–carbon multiple bonds. Alkenes and alkynes are reduced to alkanes (Sections 7.5 and 7.12).
- Acts as a hydrogenation catalyst for reducing aryl ketones to alkylbenzenes (Section 8.9).
- Acts as a hydrogenation catalyst for reducing nitroarenes to yield anilines (Section 24.6).

Peroxyacetic acid, CH_3CO_3H
- Oxidizes sulfoxides to yield sulfones (Section 13.10)

Phenylisothiocyanate, $C_6H_5{-}N{=}C{=}S$
- Used in the Edman degradation of peptides to identify N-terminal amino acids (Section 19.6).

Phosphorus oxychloride, $POCl_3$
- Dehydrates secondary and tertiary and alcohols to yield alkenes (Section 13.4).

Phosphorus tribromide, PBr_3
- Reacts with alcohols to yield alkyl bromides (Section 10.2).

Platinum oxide (Adam's catalyst), PtO_2
- Acts as a hydrogenation catalyst in the reduction of alkenes to yield alkanes (Sections 7.5)

Potassium nitrosodisulfonate (Fremy's salt), $K(SO_3)_2NO$
- Oxidizes phenols to yield quinones (Section 13.5).

Potassium permanganate, $KMnO_4$
- oxidizes aromatic side chains to yield benzoic acids (Section 8.9).

Potassium *tert*-butoxide, KO-*t*-Bu
- Reacts with alkyl halides to yield alkenes (Sections 10.10 and 10.11).

Pyridine, C_5H_5N
- Acts as a basic catalyst for the reaction of alcohols with acid chlorides to yield esters (Section 16.4).
- Acts as a basic catalyst for the reaction of alcohols with anhydrides to yield esters (Section 16.5).
- Reacts with α-bromo ketones to yield α,β-unsaturated ketones (Section 17.2).

Pyridinium chlorochromate (PCC), $C_5H_6NCrO_3Cl$
- Oxidizes primary alcohols to yield aldehydes and secondary alcohols to yield ketones (Section 13.5).

Pyrrolidine, C_4H_8N
- Reacts with ketones to yield enamines for use in the Stork enamine reaction (Sections 14.7 and 17.11).

Silver oxide, Ag_2O
- Catalyzes the reaction of monosaccharides with alkyl halides to yield ethers (Section 13.8).
- Reacts with tetraalkylammonium salts to yield alkenes (Hofmann elimination; Section 18.7).

Sodium bisulfite, $NaHSO_3$
- Reduces osmate esters, prepared by treatment of an alkene with osmium tetroxide, to yield 1,2-diols (Section 7.7).

Sodium borohydride, $NaBH_4$
- Reduces organomercury compounds, prepared by oxymercuration of alkenes, to convert the C–Hg bond to C–H (Section 7.4).
- Reduces ketones and aldehydes to yield alcohols (Sections 13.3 and 14.6).
- Reduces quinones to yield hydroquinones (Section 13.5).
- Reacts with ketones and aldehydes in the presence of ammonia to yield amines by reductive amination (Section 18.6).

Sodium dichromate, $Na_2Cr_2O_7$
- Oxidizes alkylbenzenes to yield benzoic acids (Section 8.9).
- Oxidizes primary alcohols to yield carboxylic acids and secondary alcohols to yield ketones (Section 13.5).

Sodium hydride, NaH
- Reacts with alcohols to yield alkoxide anions (Section 13.2).

Stannous chloride, $SnCl_2$
- Reduces quinones to yield hydroquinones (Section 13.5).

Sulfur trioxide, SO_3
- Reacts with arenes in sulfuric acid solution to yield arenesulfonic acids (Section 8.6).

Sulfuric acid, H_2SO_4
- Reacts with alcohols and water to yield alkenes (Section 5.5).
- Catalyzes the reaction of nitric acid with aromatic rings to yield nitroarenes (Section 8.6).
- Catalyzes the reaction of SO_3 with aromatic rings to yield arenesulfonic acids (Section 8.6).

Tetrazole
- Acts as a coupling reagent for use in DNA synthesis (Section 24.7).

Thionyl chloride, $SOCl_2$
- Reacts with primary and secondary alcohols to yield alkyl chlorides (Section 10.2).
- Reacts with carboxylic acids to yield acid chlorides (Section 16.3).

Thiourea, H_2NCSNH_2
- Reacts with primary alkyl halides to yield thiols (Section 13.6).

Trifluoroacetic acid, CF_3CO_2H
- Acts as a catalyst for cleaving *tert*-butyl ethers, yielding alcohols and 2-methylpropene (Section 13.9).
- Acts as a catalyst for cleaving the *t*-Boc protecting group from amino acids in peptide synthesis (Section 19.7).

Triphenylphosphine, $(C_6H_5)_3P$
- Reacts with primary alkyl halides to yield the alkyltriphenylphosphonium salts used in Wittig reactions (Section 14.9).

Zinc, Zn
- Reduces disulfides to yield thiols (Section 13.6).

Named Reactions in Organic Chemistry

Acetoacetic ester synthesis (Section 17.4): a multistep reaction sequence for converting a primary alkyl halide into a methyl ketone having three more carbon atoms in the chain.

$$RCH_2X + CH_3\text{–}\overset{O}{\overset{\|}{C}}\text{–}\overset{-}{\ddot{C}}H\text{–}\overset{O}{\overset{\|}{C}}\text{–}OCH_3 \xrightarrow[2.\ H_3O^+,\ heat]{1.\ Heat} RCH_2\text{–}CH_2\overset{O}{\overset{\|}{C}}CH_3 + CO_2 + CH_3OH$$

Adams' catalyst (Section 7.5): PtO_2, a catalyst used for the hydrogenation of carbon–carbon double bonds.

Aldol condensation reaction (Section 17.5): the nucleophilic addition of an enol or enolate ion to a ketone or aldehyde, yielding a β-hydroxy ketone.

$$2\ R\text{–}\overset{O}{\overset{\|}{C}}\text{–}\overset{|}{\underset{|}{C}}\text{–}H \xrightarrow{NaOH} R\text{–}\overset{O}{\overset{\|}{C}}\text{–}\overset{|}{\underset{|}{C}}\text{–}\overset{OH}{\overset{|}{\underset{\underset{R}{|}}{C}}}\text{–}\overset{|}{\underset{|}{C}}\text{–}H$$

Amidomalonate amino acid synthesis (Section 19.3): a multistep reaction sequence, similar to the malonic ester synthesis, for converting a primary alkyl halide into an amino acid.

$$RCH_2X + {}^{-}\!:\!\underset{\underset{NHAc}{|}}{C}(CO_2Et)_2 \xrightarrow[2.\ H_3O^+,\ heat]{1.\ mix} RCH_2\text{–}\underset{\underset{NH_2}{|}}{C}H\overset{O}{\overset{\|}{C}}OH + CO_2 + 2\ EtOH$$

Cannizzaro reaction (Section 14.10): the disproportionation reaction that occurs when a nonenolizable aldehyde is treated with base.

$$2\ R_3\overset{O}{\overset{\|}{C}}H \xrightarrow[2.\ H_3O^+]{1.\ HO^-} R_3\overset{O}{\overset{\|}{C}}OH + R_3CCH_2OH$$

Claisen condensation reaction (Section 17.8): a nucleophilic acyl substitution reaction that occurs when an ester enolate ion attacks the carbonyl group of a second ester molecule. The product is a β-keto ester.

$$2\ R\text{–}CH_2\text{–}\overset{O}{\overset{\|}{C}}\text{–}OCH_3 \xrightarrow[2.\ H_3O^+]{1.\ HO^-} R\text{–}CH_2\text{–}\overset{O}{\overset{\|}{C}}\text{–}\underset{\underset{R}{|}}{C}H\text{–}\overset{O}{\overset{\|}{C}}\text{–}OCH_3 + CH_3OH$$

Dieckmann reaction (Section 17.9): the intramolecular Claisen condensation reaction of a 1,6- or 1,7-diester, yielding a cyclic β-keto ester.

CO_2Et CO_2Et — 1. NaOEt; 2. H_3O^+ → O, CO_2Et + EtOH

Edman degradation (Section 19.6): a method for cleaving the N-terminal amino acid from a peptide by treatment of the peptide with *N*-phenylisothiocyanate.

Ph–N=C=S + H_2N–CH(R)–C(=O)–NH–⌇ → Ph–N (S, N–H, R, H, O) + H_2N–⌇

Fischer esterification reaction (Section 16.3): the acid-catalyzed reaction between a carboxylic acid and an alcohol, yielding the ester.

$$\text{R–C(=O)–OH} + \text{R'–OH} \xrightarrow{\text{H}^+,\ \text{heat}} \text{R–C(=O)–OR'} + \text{H}_2\text{O}$$

Friedel–Crafts reaction (Section 8.7): the alkylation or acylation of an aromatic ring by treatment with an alkyl- or acyl chloride in the presence of a Lewis-acid catalyst.

R–Cl, $AlCl_3$ → R — Alkylation

R–C(=O)–Cl, $AlCl_3$ → C(=O)–R — Acylation

Glycal assembly method (Section 21.9): a method of polysaccharide synthesis in which a glycal is converted into its epoxide, which is then opened by reaction with an alcohol.

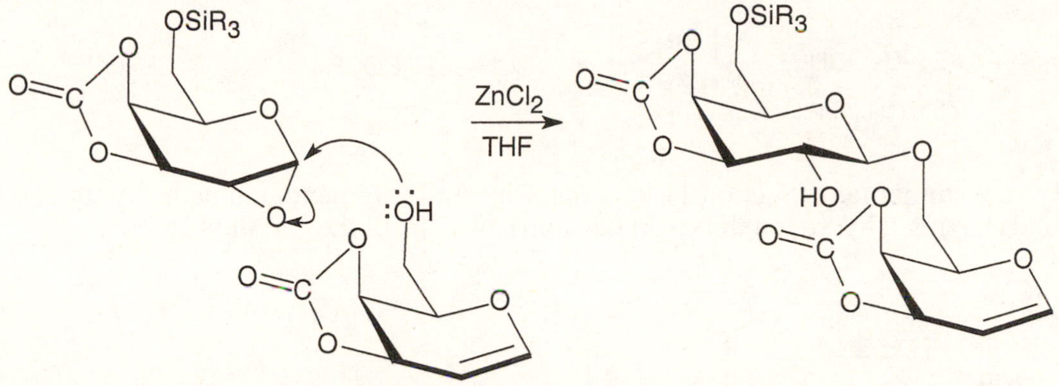

Grignard reaction (Section 10.3, 14.6): the nucleophilic addition reaction of an alkylmagnesium halide to a ketone, aldehyde, or ester carbonyl group.

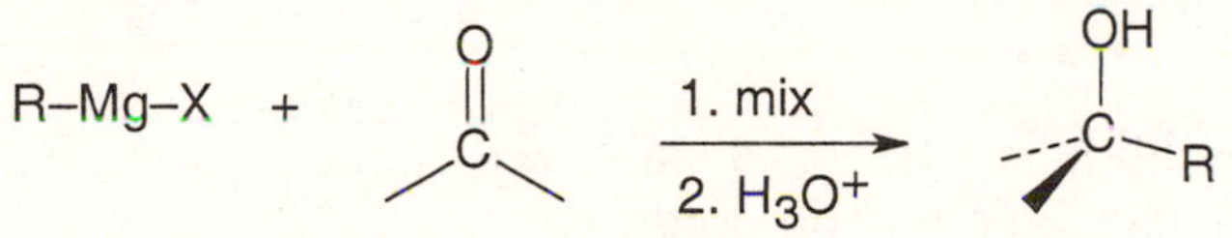

Grignard reagent (Section 10.3): an organomagnesium halide, RMgX, prepared by reaction between an organohalide and magnesium metal. Grignard reagents add to carbonyl compounds to yield alcohols.

Hofmann elimination (Section 18.7): a method for effecting the elimination reaction of an amine to yield an alkene. The amine is first treated with excess iodomethane, and the resultant quaternary ammonium salt is heated with silver oxide.

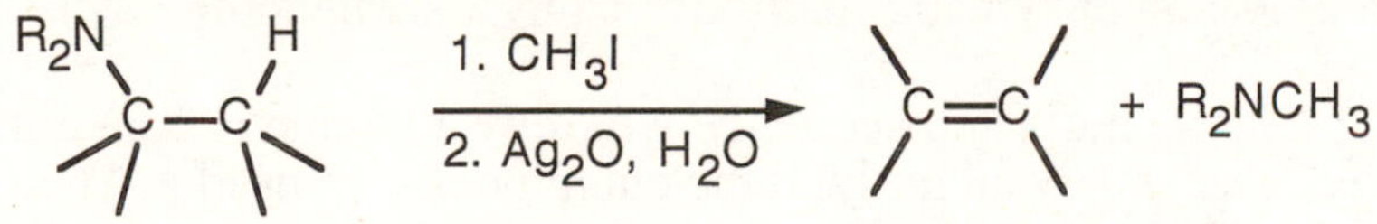

Knowles amino acid synthesis (Section 19.3): an enantioselective method of amino acid synthesis by hydrogenation of a Z enamido acid with a chiral hydrogenation catalyst.

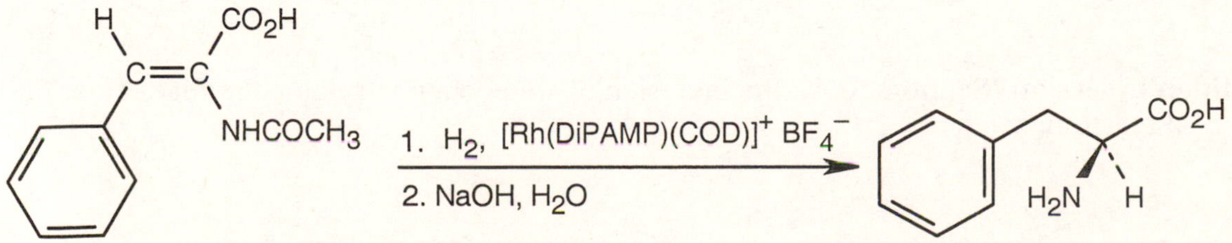

Malonic ester synthesis (Section 17.4): a multistep sequence for converting an alkyl halide into a carboxylic acid with the addition of two carbon atoms to the chain.

$$\text{R–CH}_2\text{–X} + {}^{-}\text{:}\underset{\displaystyle \text{CO}_2\text{CH}_3}{\text{CH}}\text{–}\overset{\displaystyle \text{O}}{\overset{\|}{\text{C}}}\text{–OCH}_3 \xrightarrow[\text{2. H}_3\text{O}^+\text{, heat}]{\text{1. heat}} \text{RCH}_2\text{–CH}_2\overset{\displaystyle \text{O}}{\overset{\|}{\text{C}}}\text{OCH}_3 + \text{CO}_2 + \text{CH}_3\text{OH}$$

McLafferty rearrangement (Section 11.8): a mass spectral fragmentation pathway for carbonyl compounds having a hydrogen three carbon atoms away from the carbonyl carbon.

$$\text{(ketone with H three carbons from C=O)} \xrightarrow{e^-} \text{(alkene)} + \text{(enol radical cation, } \text{H–O}^{+\cdot}\text{)}$$

Merrifield solid-phase peptide synthesis (Section 19.7): a rapid and efficient means of peptide synthesis in which the growing peptide chain is attached to an insoluble polymer support.

Michael reaction (Section 17.10): the 1,4-addition reaction of a stabilized enolate anion such as that from a 1,3-diketone to an α,β-unsaturated carbonyl compound.

$$(\text{–C(=O)})_2\text{CH}_2 + \text{H}_2\text{C=CH}\overset{\displaystyle \text{O}}{\overset{\|}{\text{C}}}\text{R} \xrightarrow{\text{Base}} (\text{–C(=O)})_2\text{CH–CH}_2\text{CH}_2\overset{\displaystyle \text{O}}{\overset{\|}{\text{C}}}\text{R}$$

Sanger dideoxy method (Section 24.6): an enzymatic method for DNA sequencing.

Stork enamine reaction (Section 17.11): a multistep sequence whereby a ketone is converted into an enamine by treatment with a secondary amine, and the enamine is then used in Michael reactions.

$$\text{(ketone)} + \text{HNR}_2 \longrightarrow \text{(enamine, NR}_2\text{)} \xrightarrow[\text{2. H}_3\text{O}^+]{\text{1. H}_2\text{C=CHCOR'}} \text{(1,5-dicarbonyl, R')}$$

Walden inversion (Section 10.4): the inversion of stereochemistry at a chirality center during an S_N2 reaction.

$$\text{Nu:}^- + \text{C–X} \longrightarrow \text{Nu–C} + \text{:X}^-$$

Williamson ether synthesis (Section 13.8): a method for preparing an ether by treatment of a primary alkyl halide with an alkoxide ion.

$$R{-}O^- \ Na^+ + R'CH_2Br \longrightarrow R{-}O{-}CH_2R' + NaBr$$

Wittig reaction (Section 14.9): a general method of alkene synthesis by treatment of a ketone or aldehyde with an alkylidenetriphenylphosphorane.

$$RC(=O)R' + \rangle C{=}PPh_3 \longrightarrow (R)(R')C{=}C\langle + Ph_3P{=}O$$

Abbreviations

Å	symbol for Angstrom unit (10^{-8} cm = 10^{-10} m)
A	adenine
Ac–	acetyl group, $CH_3\overset{O}{\overset{\parallel}{C}}—$
ACP	acyl carrier protein
ADP	adenosine 5'-diphosphate
AMP	adenosine 5'-monophosphate
cAMP	3',5'-cyclic adenosine monophosphate
Ar–	aryl group
ATP	adenosine 5'-triphosphate
at. no.	atomic number
at. wt.	atomic weight
$[\alpha]_D$	specific rotation
Boc	*tert*-butoxycarbonyl group, $(CH_3)_3CO\overset{O}{\overset{\parallel}{C}}—$
bp	boiling point
n-Bu	*n*-butyl group, $CH_3CH_2CH_2CH_2$–
sec-Bu	*sec*-butyl group, $CH_3CH_2CH(CH_3)$–
t-Bu	*tert*-butyl group, $(CH_3)_3C$–
C	cytosine
c	speed of light
cm	centimeter
cm^{-1}	wavenumber, or reciprocal centimeter
CoA	coenzyme A
CoQ	coenzyme Q
CDP	cytidine 5'-diphosphate
CMP	cytidine 5'-monophosphate
COX	cyclooxygenase
CTP	cytidine 5'-triphosphate
D	stereochemical designation of carbohydrates and amino acids
d	deoxy
DCC	dicyclohexylcarbodiimide, C_6H_{11}–N=C=N–C_6H_{11}
δ	chemical shift in ppm downfield from TMS

Δ	symbol for heat; also symbol for change
ΔH	heat of reaction, or enthalpy of reaction
DHAP	dihydroxyacetone phosphate
DHF	dihydrofolate
dm	decimeter (0.1 m)
DMAPP	dimethylallyl diphosphate
DMF	dimethylformamide, $(CH_3)_2NCHO$
DMSO	dimethyl sulfoxide, $(CH_3)_2SO$
DNA	deoxyribonucleic acid
DXP	deoxyxylulose phosphate
(E)	entgegen, stereochemical designation of double bond geometry
E_{act}	activation energy
E1	unimolecular elimination reaction
E1cB	unimolecular elimination reaction through the conjugate base
E2	bimolecular elimination reaction
Et	ethyl group, CH_3CH_2–
FAD	flavin adenine dinucleotide
$FADH_2$	reduced flavin adenine dinucleotide
FMN	flavin mononucleotide
$FMNH_2$	reduced flavin mononucleotide
FPP	farnesyl diphosphate
G	guanine
g	gram
GABA	γ-aminobutyric acid
GAP	glyceraldehyde 3-phosphate
GDP	guanosine 5'-diphosphate
GMP	guanosine 5'-monophosphate
GTP	guanosine 5'-triphosphate
HETPP	hydroxyethylthiamin diphosphate
HMG	3-hydroxy-3-methylglutaryl
hν	symbol for light
Hz	Hertz, or cycles per second (s^{-1})
i-	iso
IPP	isopentenyl diphosphate
IR	infrared

J	Joule
J	symbol for coupling constant
K	Kelvin temperature
K_a	acid dissociation constant
kJ	kilojoule
L	stereochemical designation of carbohydrates and amino acids
LT	leukotriene
Me	methyl group, CH_3–
mg	milligram (0.001 g)
MHz	megahertz (10^6 s^{-1})
mL	milliliter (0.001 L)
mm	millimeter (0.001 m)
mp	melting point
μg	microgram (10^{-6} g)
MW	molecular weight
n-	normal, straight-chain alkane or alkyl group
NAD^+	nicotinamide adenine dinucleotide
NADH	reduced nicotinamide adenine dinucleotide
$NADP^+$	nicotinamide adenine dinucleotide phosphate
NADPH	reduced nicotinamide adenine dinucleotide phosphate
ng	nanogram (10^{-9} gram)
nm	nanometer (10^{-9} meter)
NMR	nuclear magnetic resonance
P_i	phosphate ion, PO_4^{3-} (or hydrogen phosphate ion, $HOPO_3^{2-}$)
–OAc	acetate group, $-O\overset{\displaystyle O}{\overset{\|\|}{C}}CH_3$
PCC	pyridinium chlorochromate
PEP	phosphoenolpyruvate
PG	prostaglandin
Ph	phenyl group, $-C_6H_5$
pH	measure of acidity of aqueous solution
pK_a	measure of acid strength (= –log K_a)
PLP	pyridoxal phosphate
PMP	pyridoxamine phosphate
pm	picometer (10^{-12} m)
PP_i	diphosphate ion

ppm	parts per million
n-Pr	*n*-propyl group, $CH_3CH_2CH_2$–
i-Pr	isopropyl group, $(CH_3)_2CH$–
pro-R	designation of a prochirality center
pro-S	designation of a prochirality center
R–	symbol for a generalized alkyl group
(R)	*rectus*, designation of chirality center
Re face	a face of a planar, sp^2-hybridized carbon atom
RNA	ribonucleic acid
mRNA	messenger ribonucleic acid
rRNA	ribosomal ribonucleic acid
tRNA	transfer ribonucleic acid
(S)	*sinister*, designation of chirality center
SAH	*S*-adenosylhomocysteine
SAM	*S*-adenosylmethionine
sec-	secondary
Si face	a face of a planar, sp^2-hybridized carbon atom
S_N1	unimolecular substitution reaction
S_N2	bimolecular substitution reaction
T	thymine
tert-	tertiary
THF	tetrahydrofuran
TMP	thymidine 5'-monophosphate
TMS	tetramethylsilane nmr standard, $(CH_3)_4Si$
Tos	tosylate group, $-S(=O)_2-C_6H_4-CH_3$
TPP	thiamin diphosphate
TX	thromboxane
U	uracil
UDP	uridine 5'-diphosphate
UMP	uridine 5'-monophosphate
UTP	uridine 5'-triphosphate
UV	ultraviolet
X–	halogen group (–F, –Cl, –Br, –I)

(Z)	zusammen, stereochemical designation of double bond geometry
R–⫶	an organic part-structure
$>$	greater than
$<$	less than
$\longrightarrow$	chemical reaction in direction indicated
$\rightleftarrows$	reversible chemical reaction
$\longleftrightarrow$	resonance symbol
⤻	direction of electron flow in polar reaction
⤻	direction of electron flow in radical reaction
▬	single bond coming out of the plane of the paper
- - - - -	single bond receding into the plane of the paper
……	partial bond
$\delta+$, $\delta-$	partial charge
‡	denoting the transition state

Infrared Absorption Frequencies

Functional Group		Frequency (cm^{-1})	Text Section
Alcohol	–O–H	3300–3600 (s)	13.11
	$\rangle$C–O–	1050 (s)	
Aldehyde	–CO–H	2720, 2820 (m)	14.12
aliphatic	$\rangle$C=O	1725 (s)	
aromatic		1705 (s)	
Alkane			11.8
	$\rangle$C–H	2850–2960 (s)	
	$\rangle$C–C$\langle$	800–1300 (m)	
Alkene			11.8
	=C(H)	3020–3100 (s)	
	$\rangle$C=C$\langle$	1650–1670 (m)	
	$RCH{=}CH_2$	910, 990 (m)	
	$R_2C{=}CH_2$	890 (m)	
Alkyne	≡C–H	3300 (s)	11.8
	–C≡C–	2100–2260 (m)	
Alkyl bromide			11.8
	$\rangle$C–Br	500–600 (s)	
Alkyl chloride			11.8
	$\rangle$C–Cl	600–800 (s)	

Amine, *primary*	$-N\begin{matrix}H\\H\end{matrix}$	3400, 3500 (s)	18.10
secondary	$\rangle N-H$	3350 (s)	
Aromatic ring	Ar–H	3030 (m)	11.8
monosubstituted	Ar–R	690–710 (s)	
		730–770 (s)	
o-disubstituted		735–770 (s)	
m-disubstituted		690–710 (s)	
		810–850 (s)	
p-disubstituted		810–840 (s)	
Carboxylic acid	–O–H	2500–3300 (broad)	15.8
associated	$\rangle C=O$	1710 (s)	
free		1760 (s)	
Acid anhydride	$\rangle C=O$	1760, 1820 (s)	16.10
Acid chloride			16.10
aliphatic	$\rangle C=O$	1810 (s)	
aromatic		1770 (s)	
Amide			16.10
aliphatic	$\rangle C=O$	1810 (s)	
aromatic		1770 (s)	
N-substituted		1680 (s)	
N,N-disubstituted		1650 (s)	
Ester			16.10
aliphatic	$\rangle C=O$	1735 (s)	
aromatic		1720 (s)	

Ether			13.11
	—O—C— (with two further bonds on C)	1050–1150 (s)	
Ketone			14.12
aliphatic	>C=O	1715 (s)	
aromatic		1690 (s)	
6-memb. ring		1715 (s)	
5-memb. ring		1750 (s)	
Nitrile			15.8
aliphatic	–C≡N	2250 (m)	
aromatic		2230 (m)	
Phenol	–O–H	3500 (s)	13.11

(s) = strong; (m) = medium intensity

Proton NMR Chemical Shifts

Type of Proton		Chemical Shift (δ)	Text Section
Alkyl, primary	R–CH_3	0.7–1.3	12.9
Alkyl, secondary	R–CH_2–R	1.2–1.4	12.9
Alkyl tertiary	R_3C–**H**	1.4–1.7	12.9
Allylic	–C=C–C–**H**	1.6–1.9	12.9
α to carbonyl	–C(=O)–C–**H**	2.0–2.3	14.12
Benzylic	Ar–C–**H**	2.3–3.0	12.9
Acetylenic	R–C≡C–**H**	2.5–2.7	12.9
Alkyl chloride	Cl–C–**H**	3.0–4.0	12.9
Alkyl bromide	Br–C–**H**	2.5–4.0	12.9
Alkyl iodide	I–C–**H**	2.0–4.0	12.9
Amine	N–C–**H**	2.2–2.6	18.10
Alcohol	HO–C–**H**	3.5–4.5	13.11
Ether	RO–C–**H**	3.5–4.5	13.11
Vinylic	–C=C–**H**	5.0–6.0	12.9
Aldehyde	R–C(=O)–**H**	6.5–8.0	12.9
Aromatic	Ar–**H**	9.7–10.0	14.12
Carboxylic acid	R–C(=O)–O–**H**	11.0–12.0	15.8
Phenol	R–O–**H**	3.5–4.5	13.11
Alcohol	Ar–O–**H**	2.5–6.0	13.11

Nobel Prize Winners in Chemistry

1901 **Jacobus H. van't Hoff** (The Netherlands):
"for the discovery of laws of chemical dynamics and of osmotic pressure"

1902 **Emil Fischer** (Germany):
"for syntheses in the groups of sugars and purines"

1903 **Svante A. Arrhenius** (Sweden):
"for his theory of electrolytic dissociation"

1904 **Sir William Ramsey** (Britain):
"for the discovery of gases in different elements in the air and for the determination of their place in the periodic system"

1905 **Adolf von Baeyer** (Germany):
"for his researches on organic dyestuffs and hydroaromatic compounds"

1906 **Henri Moissan** (France):
"for his research on the isolation of the element fluorine and for placing at the service of science the electric furnace that bears his name"

1907 **Eduard Buchner** (Germany):
"for his biochemical researches and his discovery of cell-less formation"

1908 **Ernest Rutherford** (Britain):
"for his investigation into the disintegration of the elements and the chemistry of radioactive substances"

1909 **Wilhelm Ostwald** (Germany):
"for his work on catalysis and on the conditions of chemical equilibrium and velocities of chemical reactions"

1910 **Otto Wallach** (Germany):
"for his services to organic chemistry and the chemical industry by his pioneer work in the field of alicyclic substances"

1911 **Marie Curie** (France):
"for her services to the advancement of chemistry by the discovery of the elements radium and polonium"

1912 **Victor Grignard** (France):
"for the discovery of the so-called Grignard reagent, which has greatly helped in the development of organic chemistry"

Paul Sabatier (France):
"for his method of hydrogenating organic compounds in the presence of finely divided metals"

1913 **Alfred Werner** (Switzerland):
"for his work on the linkage of atoms in molecules by which he has thrown new light on earlier investigations and opened up new fields of research especially in inorganic chemistry"

1914 **Theodore W. Richards** (U.S.):
"for his accurate determinations of the atomic weights of a great number of chemical elements"

1915 **Richard M. Willstätter** (Germany):
"for his research on plant pigments, principally on chlorophyll"

1916 No award

1917 No award

1918 **Fritz Haber** (Germany):
"for the synthesis of ammonia from its elements, nitrogen and hydrogen"

1919 No award

1920 **Walther H. Nernst** (Germany):
"for his thermochemical work"

1921 **Frederick Soddy** (Britain):
"for his contributions to the chemistry of radioactive substances and his investigations into the origin and nature of isotopes"

1922 **Francis W. Aston** (Britain):
"for his discovery, by means of his mass spectrograph, of the isotopes of a large number of nonradioactive elements, as well as for his discovery of the whole-number rule"

1923 **Fritz Pregl** (Austria):
"for his invention of the method of microanalysis of organic substances"

1924 No award

1925 **Richard A. Zsigmondy** (Germany):
for his demonstration of the heterogeneous nature of colloid solutions, and for the methods he used, which have since become fundamental in modern colloid chemistry"

1926 **Theodor Svedberg** (Sweden):
"for his work on disperse systems"

1927 **Heinrich O. Wieland** (Germany):
"for his research on bile acids and related substances"

1928 **Adolf O. R. Windaus** (Germany):
"for his studies on the constitution of the sterols and their connection with the vitamins"

1929 **Arthur Harden** (Britain):
Hans von Euler-Chelpin (Sweden):
"for their investigation on the fermentation of sugar and of fermentative enzymes"

1930 **Hans Fischer** (Germany):
"for his researches into the constitution of hemin and chlorophyll, and especially for his synthesis of hemin"

1931 **Frederich Bergius** (Germany):
Carl Bosch (Germany):
"for their contributions to the invention and development of chemical high-pressure methods"

1932 **Irving Langmuir** (U.S.):
"for his discoveries and investigations in surface chemistry"

1933 No award

1934 **Harold C. Urey** (U.S.):
"for his discovery of heavy hydrogen"

1935 **Frederic Joliot** (France):
Irene Joliot-Curie (France):
"for their synthesis of new radioactive elements"

1936 **Peter J. W. Debye** (Netherlands/U.S.):
"for his contributions our knowledge of molecular structure through his investigations on dipole moments and on the diffraction of X rays and electrons in gases"

1937 **Walter N. Haworth** (Britain):
"for his researches into the constitution of carbohydrates and vitamin C"

Paul Karrer (Switzerland):
"for his researches into the constitution of carotenoids, flavins, and vitamins A and B"

1938 **Richard Kuhn** (Germany):
"for his work on carotenoids and vitamins"

1939 **Adolf F. J. Butenandt** (Germany):
"for his work on sex hormones"

Leopold Ruzicka (Switzerland):
"for his work on polymethylenes and higher terpenes"

1940 No award

1941 No award

1942 No award

1943 **Georg de Hevesy** (Hungary):
"for his work on the use of isotopes as tracer elements in researches on chemical processes"

1944 **Otto Hahn** (Germany):
"for his discovery of the fission of heavy nuclei"

1945 **Artturi I. Virtanen** (Finland):
"for his researches and inventions in agricultural and nutritive chemistry, especially for his fodder preservation method"

1946 **James B. Sumner** (U.S.):
"for his discovery that enzymes can be crystallized"

John H. Northrop (U.S.):
Wendell M. Stanley (U.S.):
for their preparation of enzymes and virus proteins in a pure form"

1947 **Sir Robert Robinson** (Britain):
"for his investigations on plant products of biological importance, particularly the alkaloids"

1948 **Arne W. K. Tiselius** (Sweden):
"for his researches on electrophoresis and adsorption analysis, especially for his discoveries concerning the complex nature of the serum proteins"

1949 **William F. Giauque** (U.S.):
"for his contributions in the field of chemical thermodynamics, particularly concerning the behavior of substances at extremely low temperatures"

1950 **Kurt Alder** (Germany):
Otto P. H. Diels (Germany):
"for their discovery and development of the diene synthesis"

1951 **Edwin M. McMillan** (U.S.):
Glenn T. Seaborg (U.S.):
"for their discoveries in the chemistry of the transuranium elements"

1952 **Archer J. P. Martin** (Britain):
Richard L. M. Synge (Britain):
"for their development of partition chromatography"

1953 **Hermann Staudinger** (Germany):
"for his discoveries in the field of macromolecular chemistry"

1954 **Linus C. Pauling** (U.S.):
"for his research into the nature of the chemical bond and its application to the elucidation of the structure of complex substances"

1955 **Vincent du Vigneaud** (U.S.):
"for his work on biochemically important sulfur compounds, especially for the first synthesis of a polypeptide hormone"

1956 **Sir Cyril N. Hinshelwood** (Britain):
Nikolai N. Semenov (U.S.S.R.):
"for their research in clarifying the mechanisms of chemical reactions in gases"

1957 **Sir Alexander R. Todd** (Britain):
"for his work on nucleotides and nucleotide coenzymes"

1958 **Frederick Sanger** (Britain):
"for his work on the structure of proteins, particularly insulin"

1959 **Jaroslav Heyrovsky** (Czechoslovakia):
"for his discovery and development of the polarographic method of analysis"

1960 **Willard F. Libby** (U.S.):
"for his method to use carbon-14 for age determination in archaeology, geology, geophysics, and other branches of science"

1961 **Melvin Calvin** (U.S.):
"for his research on the carbon dioxide assimilation in plants"

1962 **John C. Kendrew** (Britain):
Max F. Perutz (Britain):
"for their studies of the structures of globular proteins"

1963 **Giulio Natta** (Italy):
Karl Ziegler (Germany):
"for their work in the controlled polymerization of hydrocarbons through the use of organometallic catalysts"

1964 **Dorothy C. Hodgkin** (Britain):
"for her determinations by X-ray techniques of the structures of important biochemical substances, particularly vitamin B_{12} and penicillin"

1965 **Robert B. Woodward** (U.S.):
"for his outstanding achievements in the 'art' of organic synthesis"

1966 **Robert S. Mulliken** (U.S.):
"for his fundamental work concerning chemical bonds and the electronic structure of molecules by the molecular orbital method"

1967 **Manfred Eigen** (Germany):
Ronald G. W. Norrish (Britain):
George Porter (Britain):
"for their studies of extremely fast chemical reactions, effected by disturbing the equilibrium with very short pulses of energy"

1968 **Lars Onsager** (U.S.):
"for his discovery of the reciprocal relations bearing his name, which are fundamental for the thermodynamics of irreversible processes"

1969 **Sir Derek H. R. Barton** (Britain):
Odd Hassel (Norway):
"for their contributions to the development of the concept of conformation and its application in chemistry"

1970 **Luis F. Leloir** (Argentina):
"for his discovery of sugar nucleotides and their role in the biosynthesis of carbohydrates"

1971 **Gerhard Herzberg** (Canada):
"for his contributions to the knowledge of electronic structure and geometry of molecules, particularly free radicals"

1972 **Christian B. Anfinsen** (U.S.):
"for his work on ribonuclease, especially concerning the connection between the amino acid sequence and the biologically active conformation"

Stanford Moore (U.S.):
William H. Stein (U.S.):
"for their contribution to the understanding of the connection between chemical structure and catalytic activity of the active center of the ribonuclease molecule"

1973 **Ernst Otto Fischer** (Germany):
Geoffrey Wilkinson (Britain):
"for their pioneering work, performed independently, on the chemistry of the organometallic sandwich compounds"

1974 **Paul J. Flory** (U.S.):
"for his fundamental achievements, both theoretical and experimental, in the physical chemistry of macromolecules"

1975 **John Cornforth** (Australia/Britain):
"for his work on the stereochemistry of enzyme-catalyzed reactions"

Vladimir Prelog (Yugoslavia/Switzerland):
"for his work on the stereochemistry of organic molecules and reactions"

1976 **William N. Lipscomb** (U.S.):
"for his studies on the structures of boranes illuminating problems of chemical bonding"

1977 **Ilya Pregogine** (Belgium):
"for his contributions to nonequilibrium thermodynamics, particularly the theory of dissipative structures"

1978 **Peter Mitchell** (Britain):
"for his contribution to the understanding of biological energy transfer through the formulation of the chemiosmotic theory"

1979 **Herbert C. Brown** (U.S.):
"for his application of boron compounds to synthetic organic chemistry"

Georg Wittig (Germany):
"for developing phosphorus reagents, presently bearing his name"

1980 **Paul Berg** (U.S.):
"for his fundamental studies of the biochemistry of nucleic acids, with particular regard to recombinant DNA"

Walter Gilbert (U.S.)
Frederick Sanger (Britain):
"for their contributions concerning the determination of base sequences in nucleic acids"

1981 **Kenichi Fukui** (Japan)
Roald Hoffmann (U.S.):
for their theories, developed independently, concerning the course of chemical reactions"

1982 **Aaron Klug** (Britain):
"for his development of crystallographic electron microscopy and his structural elucidation of biologically important nucleic acid – protein complexes"

1983 **Henry Taube** (U.S.):
"for his work on the mechanisms of electron transfer reactions, especially in metal complexes"

1984 **R. Bruce Merrifield** (U.S.):
"for his development of methodology for chemical synthesis on a solid matrix"

1985 **Herbert A. Hauptman** (U.S.):
Jerome Karle (U.S.):
"for their outstanding achievements in the development of direct methods for the determination of crystal structures"

1986 **John C. Polanyi** (Canada):
"for his pioneering work in the use of infrared chemiluminescence in studying the dynamics of chemical reactions"

Dudley R. Herschbach (U.S.):
Yuan T. Lee (U.S.):
"for their contributions concerning the dynamics of chemical elementary processes"

1987 **Donald J. Cram** (U.S.):
Jean-Marie Lehn (France):
Charles J. Pedersen (U.S.):
"for their development and use of molecules with structure-specific interactions of high selectivity"

1988 **Johann Deisenhofer** (Germany):
Robert Huber (Germany):
Hartmut Michel (Germany):
"for their determination of the structure of the photosynthetic reaction center of bacteria"

1989 **Sidney Altman** (U.S.):
Thomas R. Cech (U.S.):
"for their discovery of catalytic properties of RNA"

1990 **Elias J. Corey** (U.S.):
"for his development of the theory and methodology of organic synthesis"

1991 **Richard R. Ernst** (Switzerland):
"for his contributions to the development of the methodology of high resolution NMR spectroscopy"

1992 **Rudolph A. Marcus** (U.S.):
"for his contributions to the theory of electron-transfer reactions in chemical systems"

1993 **Kary B. Mullis** (U.S.):
"for his development of the polymerase chain reaction"

Michael Smith (Canada):
"for his fundamental contributions to the establishment of oligonucleotide-based site-directed mutagenesis and its development for protein studies"

1994 **George A Olah** (U.S.):
"for pioneering research on carbocations and their role in the chemical reactions of hydrocarbons"

1995 **F. Sherwood Rowland** (U.S.)
Mario Molina (U.S.)
Paul Crutzen (Germany)
"for their work in atmospheric chemistry, particularly concerning the formation and decomposition of ozone"

1996 **Robert F. Curl, Jr.** (U.S.)
Harold W. Kroto (U.K.)
Richard E. Smalley (U.S.)
"for their discovery of carbon atoms bound in the form of a ball (fullerenes)."

1997 **Paul D. Boyer** (U.S.)
John E. Walker (U.K.)
"for having elucidated the mechanism by which ATP synthase catalyzes the synthesis of adenosine triphosphate, the energy currency of living cells"

Jens C. Skou (Denmark)
"for his discovery of the ion-transporting enzyme Na^+–K^+ ATPase, the first molecular pump"

1998 **Walter Kohn** (U.S.)
John A. Pople (U.S.)
"to Walter Kohn for his development of the density-functional theory and to John Pople for his development of computational methods in quantum chemistry"

1999 **Ahmed H. Zewail** (Egypt, U.S.)
"for his studies of the transition states of chemical reactions using femtosecond spectroscopy."

2000 **Alan J. Heeger** (U.S.)
Alan G. MacDiarmid (U.S.)
Hideki Shirakawa (Japan)
"for opening and developing the important new field of electrically conductive polymers"

2001 **William S. Knowles** (U.S.)
Ryoji Noyori (Japan)
K. Barry Sharpless (U.S.)
"for their work on chirally catalysed hydrogenation and oxidation reactions"

2002 **John B. Fenn** (U.S.)
Koichi Tanaka (Japan)
"for their development of soft desorption ionisation methods for mass spectrometric analyses of biological macromolecules"

Kurt Wüthrich (Switzerland)
"for his development of nuclear magnetic resonance spectroscopy for determining the three-dimensional structure of biological macromolecules in solution"

2003 **Peter Agre** (U.S.)
"for the discovery of water channels in cell membranes"
Roderick MacKinnon (U.S.)
"for structural and mechanistic studies of ion channels in cell membranes"

2004 **Aaron Ciechanover** (Israel)
Avram Hershko (Israel)
Irwin Rose (U.S.)
"for the discovery of ubiquitin-mediated protein degradation"

2005 **Yves Chauvin** (France)
Robert H. Grubbs (U.S.)
Schrock, Richard R. (U.S.)
"for the development of the metathesis method in organic synthesis"

Answers to Multiple-Choice Questions in Review Units 1–10

Review Unit 1:	1. d	2. b	3. a	4. c	5. d	6. b	7. a	8. d	9. a	10. c	
Review Unit 2:	1. c	2. a	3. d	4. c	5. d	6. b	7. a	8. b	9. c	10. b	
Review Unit 3:	1. c	2. b	3. a	4. b	5. a	6. d	7. d	8. c	9. d	10. b	
Review Unit 4:	1. d	2. b	3. c	4. c	5. b	6. a	7. d	8. c	9. a	10. d	
Review Unit 5:	1. b	2. b	3. c	4. a	5. d	6. d	7. c	8. a	9. b	10. c	11. c
Review Unit 6:	1. d	2. c	3. b	4. b	5. a	6. d	7. b	8. a	9. c	10. c	
Review Unit 7:	1. a	2. b	3. a	4. c	5. b	6. d	7. b	8. c	9. d	10. b	
Review Unit 8:	1. c	2. d	3. b	4. a	5. d	6. b	7. a	8. d	9. b	10. c	
Review Unit 9:	1. b	2. a	3. d	4. b	5. c	6. d	7. b	8. b	9. c	10. a	
Review Unit 10:	1. c	2. d	3. a	4. b	5. d	6. a	7. d	8. b	9. c	10. b	